Digital Circuits and Design

Digital Circuits and Design

D. P. Kothari
Director Research, GPGI, Nagpur
Former Director-In-Charge, Indian Institute of Technology Delhi
Former Vice Chancellor, VIT, Vellore and Former Principal, VNIT, Nagpur

J. S. Dhillon
Professor
Department of Electrical and Instrumentation Engineering
Sant Longowal Institute of Engineering and Technology
Longowal, Punjab, India

PEARSON
Delhi • Chennai

Senior Editor—Acquisitions: Anita Yadav
Editor—Production: M. Balakrishnan

ISBN 978-93-325-4353-9

First Impression

Published by Pearson India Education Services Pvt. Ltd,
CIN: U72200TN2005PTC057128, formerly known as TutorVista Global Pvt. Ltd,
licensee of Pearson Education in South Asia.

Head office: A-8 (A), 7th Floor, Knowledge Boulevard, Sector 62, Noida 201 309,
Uttar Pradesh, India.
Registered office: Module G4, Ground Floor, Elnet Software City, TS-140, Blocks 2 & 9,
Rajiv Gandhi Salai, Taramani, Chennai 600 113, Tamil Nadu, India.
Fax: 080-30461003, Phone: 080-30461060
www.pearson.co.in, Email: companysecretary.india@pearson.com

Compositor: SRS Technologies, Puducherry
Printed in India by Rahul Print O Pack

To my wife Shobha Kothari and my daughters
Seema Mundhada and Shikha Bhangadia

—D. P. Kothari

To my wife Amrit Kaur and my children
Gunpreet Singh and Gurpreet Kaur

—J. S. Dhillon

Brief Contents

Contents

Preface

The vast practicability and utility of digital principles and systems may be seen from the wide variety of applications in the field of industrial machinery, computers, microprocessors, household appliances, digital televisions, medical equipment, digital communication, Internet, e-banking, e-business, e-governance among others. The areas of applications of digital electronics have been increasing steadily, resulting in an unparalleled interest in the subject of digital electronics. In fact, digital systems have conquered every field of life through a digital revolution.

One of the important reasons for the unprecedented growth of digital electronics is the advent of integrated circuits (ICs). Developments in the IC technology have made it possible to fabricate complex digital circuits, such as microprocessors, memories, complex-programmable logic devices (CPLDs) and field-programmable gate arrays (FPGAs). The emergence of various programmable logic devices has resulted in significant changes in the design methodologies of digital systems. Therefore, it is essential to have a strong foundation of the basic digital techniques for making effective use of automation in digital design.

This book is a comprehensive study of the principles and techniques of modern digital systems. It teaches the principles of digital systems and covers thoroughly both traditional and modern methods of applying digital design and development techniques. The book is intended for the students of three- and four-year diploma/degree programmes in technology, engineering, and computer science. Although a background in basic electronics is helpful, most of the material requires no training of electronics.

Chapter 1 introduces the fundamental concepts of digital electronics, advantages of digital systems, and the basic digital circuits. Various number systems and commonly used codes in digital systems and computers have been discussed in Chapter 2. Error-detecting and error-correcting codes have also been discussed in the chapter. Basic digital logics, universal gates that are building blocks of combinational circuits have been presented in Chapter 3.

Chapter 4 deals with the conventional methods of combinational circuits' design such as algebraic method, K-map simplification and Quine–McCluskey method. Hazards in combinational digital circuits and design of hazard-free circuit have been illustrated.

Chapter 5 discusses the logic circuit design for arithmetic operations such as adders, subtractors, and comparators. Code conversion is also presented along with circuit design for multiplication and division. Arithmetic logic unit design concept has been detailed in this chapter.

Chapter 6 presents data-processing devices and their applications for the design of digital systems. Basic concepts of decoders, encoders, multiplexers and demultiplexers are discussed in this chapter.

Chapter 7 introduces the basic building block of a sequential circuit—the flip-flops. All types of flip-flops with their characteristic tables, excitation tables, state diagrams and triggering methods have been illustrated. Flip-flop conversion has also been detailed.

Sequential circuit's analysis and design have been presented in Chapter 8. Analysis and design of synchronous sequential as well as asynchronous sequential circuits has been discussed. State reductions techniques are detailed along with state assignment procedures. Hazards in sequential circuits have also been dealt with in this chapter. Registers like parallel-in-parallel-out (PIPO), serial-in-serial-out (SISO), parallel-in-serial-out (PISO), serial-in-parallel-out (SIPO) and universal registers are discussed in Chapter 9. Chapter 9 discusses registers, shift registers, ring counter, Johnson counter and sequence generator. Chapter 10 discusses asynchronous and synchronous counters design. BCD counters and hybrid counters are presented in this chapter.

Chapter 11 deals with semiconductor memories which have assumed an important role in present-day digital systems. This chapter has thoroughly been revised to include various semiconductor memory devices which are being used currently. Programming techniques used for programmable ROMs and erasing techniques used for erasable programmable ROMs have also been discussed. The chapter presents various programmable logic devices (PLDs), such as programmable logic array (PLA), programmable array logic (PAL), electrically erasable PLD (EEPLD), CPLDs and FPGA devices. Static random access memory (SRAM), dynamic random access memory (DRAM), first-in-first-out (FIFO), last-in-first-out (LIFO) and associative memories are also presented in this chapter.

Chapter 12 discusses the analog-to-digital (A/D) and digital-to-analog (D/A) converters those form an important part of many digital systems. The commonly used techniques for A/D and D/A conversions have been discussed.

The VHDL, a hardware description language, has been introduced in Chapter 13. Combinational logic circuits: truth table, arithmetic circuits, decoders, multiplexers, priority encoder, digital comparators, BCD-to-seven segment decoder, and Sequential logic circuits: Latches, flip-flops, shift registers, registers and counters have been described using VHDL.

Based on semiconductor devices, various digital circuits, referred to as logic families, have been discussed in Chapter 14. CMOS logic has now almost replaced the earlier most commonly used TTL logic. However, because of its higher speed of operation and driving capabilities, TTL logic is still preferred in many designs. Even a number of CMOS devices are available which are TTL compatible. This chapter presents the interfacing problems between ICs of the same logic family and between those of different logic families to obtain maximum benefits in the design of digital systems.

Chapter 15 deals with timing circuits and their applications which are essential to a digital system. Astable, monostable and bistable multivibrators have been presented in this chapter. Astable and monostable multivibrators are discussed using operational amplifiers and timer 555.

The authors would like to acknowledge all the graduate students who gave us the insight to compile the teaching notes in the book form. We are indebted to our colleagues and authorities of Sant Longowal Institute of Engineering and Technology, Longowal, Vindhya Institute of Technology and Science, Indore, Raisoni Group of Institutions, Nagpur, VIT University, Vellore, JB Institute of Engineering and Technology, Hyderabad, MVSR Engineering College, Hyderabad, Wainganga College of Engineering and Management, Nagpur and Gaikwad-Patil Group of Institutions, Nagpur, for their encouragement and fruitful suggestions. We would like to thank Ms Anita Yadav and Mr M. Balakrishnan of Pearson Education for extending their support to publish the book. Finally we would like to thank our respective families for their love and support. We hope this book will challenge the readers to delve into an insightful understanding of principles of Digital Electronics. We will welcome constructive criticism and appraisal by the readers.

D. P. Kothari
J. S. Dhillon

About the Authors

D. P. Kothari is presently Director Research of Gaikwad Patil Group of Institutions, Nagpur. He obtained his B.E. (Electrical) in 1967, M.E. (Power Systems) in 1969 and Ph.D. in 1975 from Birla Institute of Technology and Sciences (BITS), Pilani, Rajasthan. From 1969 to 1977, he was involved in teaching and development of several courses at BITS Pilani. Prior to assuming charge as Director Research of GPGI, Nagpur, Dr Kothari served as Vice Chancellor, VIT, Vellore, Director-In-Charge and Deputy Director (Administration) as well as Head in the Centre of Energy Studies at the Indian Institute of Technology Delhi, Delhi and as Principal, Visvesvaraya Regional College of Engineering, Nagpur. He was also a visiting professor at the Royal Melbourne Institute of Technology, Melbourne, Australia, during 1982–83 and 1989, for two years. He was also a NSF Fellow at Purdue University, Indiana, USA, in 1992.

Dr Kothari, who is a recipient of the most Active Researcher Award, has published and presented 780 research papers in various national as well as international journals, conferences, guided 42 Ph.D. scholars and 65 M.Tech. students, and has authored 38 books in various allied areas. He has delivered several keynote addresses and invited lectures at both national and international conferences. He has also delivered 42 video lectures on YouTube with maximum of 40,000 hits!

Dr Kothari is a Fellow of the National Academy of Engineering (FNAE), Fellow of Indian National Academy of Science (FNASc), Fellow of Institution of Engineers (FIE), Fellow IEEE and Hon. Fellow ISTE.

His many awards include the National Khosla Award for Lifetime Achievements in Engineering (2005) from IIT, Roorkee. The University Grants Commission (UGC), Government of India, has bestowed the UGC National Swami Pranavandana Saraswati Award (2005) in the field of education for his outstanding scholarly contributions.

He is also the recipient of the Lifetime Achievement Award (2009) conferred by the World Management Congress, New Delhi, for his contribution to the areas of educational planning and administration. Recently he has received Excellent Academic Award at IIT Guwahati by NPSC–2014.

J. S. Dhillon is presently working as a Professor in Electrical and Instrumentation Engineering Department at Sant Longowal Institute of Engineering and Technology, Longowal, where he also served as Head of Electrical and Instrumentation Engineering Department (October 2002–October 2005), Dean (Academics) (March 2010–June 2012) and Head of Computer Science and Engineering Department (July 2012–October 2013). Earlier, he served as an Assistant Professor (December 1992–July 2002) Giani Zail Singh College of Engineering & Technology, Bathinda, Lecturer (July 1987–November 1992), Thapar Institute of Engineering and Technology, Patiala.

He received his B.E. (Electrical) (1983) from Guru Nanak Dev Engineering College, Ludhiana (GNDEC), M.E. (Systems) (1987), Punjab Agricultural University, Ludhiana (PAU), Ph.D. (1996) Thapar University, Patiala. His research activities include Microprocessor/Microcontroller applications, Power System Optimization, Neural Networks, Fuzzy Theory and Soft Computing Applications. Professor Dhillon has published and presented 118 research papers in various national and international journals/proceedings/conferences. He has co-authored two books and supervised seven Ph.D. and 26 M.E. scholars. He is a Life Member at the Institute of Engineers (India), ISTE, SSI.

1

Introduction

CHAPTER OBJECTIVES

The main goal of this chapter is to impart knowledge about the signals. Readers will be able to discuss the following aspects:

- Brief history of electronics, integrated circuits (ICs) and computers
- Analog signal, waveforms and definitions, instantaneous, average and RMS quantities
- Digital systems and signals
- Logic levels and pulse waveforms
- Advantages, limitations and future of digital electronics
- Purpose of memory
- Basic components of a computer

1.1 | HISTORY OF DIGITAL ELECTRONICS SYSTEMS

The term 'electronics' has been coined from electron—the key particle of an atom. The rate of flow of electrons through a conductor provides electric current (current). The electric current can be produced with the help of batteries and generators either in the form of direct current (DC) or alternating current (AC) at 50 Hz (cycles/s). Various applications can be achieved by using current such as illumination, heating, driving the motors. The generation, control, transmission and utilization of both DC and AC at 50-Hz magnitudes ranging from a few microamperes to kiloamperes come under the purview of electrical engineering.

The flow of electrons can be regulated and controlled in electronic devices under the stimuli of external influences which may be called signal, message, data or any other appropriate name suitable for the application. The controlled flow of electrons by various methods and forms is known as electronics. Electronics proves its strength in different walks of life like household goods, transports, health, communication, entertainment, multimedia, Internet. Common electronic gadgets such as radio and television receivers, audio and video tape recorder, frequency synthesizer, calculators, cameras, musical doorbells, and so on are also in use. Perhaps one of the major achievements of electronics is computer which has revolutionized the world. In medical diagnostics and treatments as well as in laboratory practice and industrial operation, reliability and precision are the two important factors. In medical diagnostics and surgery, all scanning devices and techniques apply electronics. In ultrasound, a pulse is sent through the body and as it passes through the tissues, its attenuation is measured and this helps to produce an image of the tissues or to detect abnormalities in a human body. In tomography, cross-sectional images (obtained by X-rays) are combined to produce a three-dimensional picture of the interior of the body. The electrocardiograph (ECG) is used for the diagnosis of heart and circulatory ailments. The electroencephalograph (EEG) records the activity of brain.

The development of communication facilities is another important gift of electronics in the twentieth century. People around the world are brought closer and communication among them is made easier through wireless communications, emails and Internet. Aircraft receives the information on the weather and terminal traffic with constant use of radio communication. Satellites also have brought revolution in the field of communications. Photographs and data captured through satellites provide valuable information about the earth surface and of outer space. All the space voyages to moon or mars utilized the application of electronics to send signals if they want to make contacts with others. Radar had changed the war profile in the Second World War and since then electronics is being used in warfare which coined a term 'electronic warfare'.

Electronics is often applied in industrial applications in counting, sorting, illumination control, welding control, liquid-gaseous flow control, automatically regulating temperature and humidity and in early warning system. Electronics has made precision measurements possible not only for electrical quantities but also for other non-electrical quantities such as colour, weight, light intensity, odour, time measurements and many others. Electronic circuits together with digital displays have made measuring gadgets much more reliable, accurate and easy to handle. In fact, in any field where there is a need for precision and reliability, electronics is an automatic choice.

The subject matter of electronics engineering may be classified according to components, techniques (circuits) and systems. Components are basic building blocks that are combined using the proper technique to yield a system. The basic components are divided into two major groups: passive and active. Passive components are resistors, capacitors and inductors. Active devices are diodes (two terminals), transistors (three terminals) and vacuum tubes (two or more terminals). Due to the phenomenal growth of semiconductor technology, semiconductor devices have replaced the vacuum tubes in most operations. The semiconductor devices have less weight, occupying small space and are more reliable. So, it is possible to accommodate thousands of transistors in a very small space, giving rise to integrated circuit (IC) chip. These chips are small systems which are able to perform specific function in small space. Thus, small size, reliable and high-speed computers are fabricated using IC chips. Rightly, therefore, the major thrust in the course will centre on semiconductor devices, the circuit-building blocks and electronic systems made using these blocks.

1.1.1 | Evolution of Electronics

- In 1890, Heinrich Rudolf Hertz performed the first experiment on generation of electromagnetic waves.
- In 1894, Sir J. C. Bose first showed the propagation of radio waves. At the same time, Marconi also postulated the theory of radio-wave propagation.
- In 1895, H. A. Lorentz postulated the existence of discrete charges, called electrons.
- In 1897, J. J. Thomson experimentally verified the existence of an electron.
- In 1897, Karl Ferdinand Braun made the first electron tube.
- In 1904, John A. Fleming, a British scientist, invented the diode, called valve.

- In 1906, Lee de Forest, an American electrical engineer, put a third electrode (called grid) into the Fleming valve and invented triode tube, which he called an audion. The audion was the first amplifier.
- In 1909, the charge on electron was measured by the American physicist Robert A. Millikan.
- In 1912, first application of radio and birth of Institute of Radio Engineers (IRE) in USA.
- In 1918, the super heterodyne receiver was invented by US army major Edwin Armstrong.
- In 1925, black-and-white television (TV) was introduced in public by John Logie Baird. Physicist Julius Edgar Lilienfeld filed a patent for a field-effect transistor (FET) in Canada in 1925, which was intended to be a solid-state replacement for the triode.
- In 1933, FM (frequency modulation) is invented by Edwin Howard Armstrong.
- In 1940, the term 'RADAR' was coined in 1940 by the US Navy as an acronym for 'radio detection and ranging'.
- In 1950, colour television (TV) was introduced by Radio Corporation of America (RCA) laboratories.
- In 1963, birth of Institution of Electrical and Electronics Engineers (IEEE) took place.

1.1.2 | Evolution of Transistors

- In 1947, Walter Brattain and John Bardeen invented point-contact transistor. William Shockley discovered junction (December) transistor. In 1956, all three of them were honoured and awarded the Nobel Prize in Physics for their commendable work and that was the first Nobel Prize in engineering devices.
- In 1950, first junction transistor was invented.
- In 1951, transistor was produced commercially (first germanium and then silicon).
- In 1956, PNPN triggering transistor known as thyristor or silicon-controlled rectifier (SCR) developed by the Bell laboratories.
- In 1958, Julian C. Kilby (Texas Instruments USA) gave the idea of monolithic.
- In 1960, the MOSFET (metal–oxide–semiconductor field-effect transistor) was first proposed by Dawon Kahng.
- In 1961, Fairchild and Texas Instruments commercially produced ICs.

1.1.3 | Evolution of ICs

- In 1951, discrete transistor was invented.
- In 1958, first IC was developed by Jack Kilby at Texas Instruments and Robert Norton Noyce at Fairchild Semiconductor.
- In 1960, small-scale integration (SSI) (<100 components per chip) was proposed.
- In 1966, medium-scale integration (MSI) (>100 but <1000) was proposed.
- In 1969, large-scale integration (LSI) (>1000 but <10,000) was proposed.
- In 1971, the Intel 4004 was a 4-bit central processing unit (CPU) was released by Intel Corporation. It was the first complete CPU on one chip, and also the first commercially available microprocessor.
- In 1975, very large-scale integration (VLSI) (>10,000) was proposed.

- In 1976, the MCS-48 microcontroller (μC) series, Intel's first microcontroller, was released. Its first members were 8048, 8035 and 8748.
- In 1980, 106 components per chip, typical VLSI chip size in 1978, 3×5 mm^2 area, 0.1 mm thick. Total 30,000 components, that is, 2000 components/mm^2 were proposed.
- In 1998, ultra large-scale integration (ULSI), (>10,000 components/mm^2) was proposed. 10^8 components per chip. Entire computer on single chip (6×6 mm^2) area.

1.1.4 | Evolution of Computers

- In 1617, John Napier, a Scottish mathematician, developed Napier Bones. The bones are set of 11 rods with numbers marked on them in such a way that by simply placing the rods side by side products and quotients of large numbers can be obtained.
- In 1632, William Oughtred, an English mathematician, developed Oughtred slide rule that consists of two movable rulers placed side by side. By sliding the rulers, one can quickly multiply and divide numbers.
- In 1642, Blaise Pascal, the great French mathematician, invented a Pascal's Calculator that may be considered as the first adding machine. The driven registered numbers by rotating a cogwheel gear by one to 10 steps with a carryover ratchet to operate the next higher digit wheel when the given cogwheel exceeded 10 units. The odometer is an example of a device.
- In 1801, Joseph Marie Jacquard, a French weaver, developed the first punched card machine called Jacquard's Loom. The pattern woven by the loom that was determined by the placement of holes in a control card: only those threads whose guiding hook encountered a hole in the card could enter the pattern.
- In 1833, Charles Babbage, also known as the father of computers, followed the difference engine with a much deeper and more general conception. It is based on principle that for certain formulas the difference between values is constant.
- In 1890, Dr Herman Hollerith, a statistician used 3×5 inch punched cards to record the data and constructed a box to sort the data and manually fed, electromagnetic counting machine to tabulate the data.
- In 1944, Howard G. Aiken, a professor of mathematics at Harvard University, invented a computer capable of automatically performing a long sequence of arithmetic and logical operations. It was an electromechanical (relay device. Mark I was built by IBM Corporation).
- *1946–1954, First-generation computer:* Electronic computer was made using vacuum tubes or valves. Fixed point arithmetic was used. Magnetic tape/drum was used as secondary memory. No subroutine linkage was developed. Punched cards and paper tapes were invented to feed programs and data and to get results. Examples are ENIAC (electronic numerical integrator and calculator), EDSAC (electronic-delayed storage automatic computer), EDVAC (electronic discrete variable automatic computer), UNIVAC (UNIVersal Automatic Computer), IBM (International Business Machines Corporation) 704, IBM 709, etc.
- *1955–1964, Second-generation computer:* After the discovery of transistor, vacuum tubes were replaced by transistors, and computers used magnetic core storage. Computers were widely used in industry and commercial organizations. Computers were used for the preparation of pay roll, inventory control, marketing, production

planning, general ledger system research, scientific and engineering analysis and design, etc. A few examples of second-generation computers are IBM 7070, IBM 7090, IBM 7080, IBM 7094, IBM 1620.

- *1965–1974, Third-generation computer:* Third-generation computers used integrated solid-state circuitries, when IC technology was developed then thousands of transistors were fabricated on single IC. So a number of transistors were replaced by single IC. Hence the size of computer was reduced by large amount in third-generation computer. Microprogramming parallel processing, multiprogramming multiuser system, primary and secondary memory, cache memory, etc. were introduced. Few examples of third-generation computers are PDP 11, CDC 7600, IBM 370 series.
- *1975–1995, Fourth-generation computer:* Circuits for different arithmetic and logical operations were fabricated on a single IC called as microprocessor. VLSI technology is used in fourth-generation computers. The 32-bit and 64-bit RISC (reduced instruction set computing) and CISC (complex instruction set computing) microprocessors are used in this generation. Era of minicomputers and microcomputers started during this period.
- *1995–Till date, Fifth-generation computer:* Computers consist of first level cache, second level cache, memory management unit (MMU), multiple pipe lines, floating point unit (FPU), MMX (MultiMedia eXtension) pipeline, and other important chips like advanced interrupt controller and cache controller, etc. It can execute several instructions per clock cycle, so their speed is very high. Computers use data flow architecture due to which processors can fetch 20 to 30 instructions in advanced in sequential manner and put them in an instruction pool. Intelligent programming is used. Extensive parallel processing computations such as pipe lining, multiprocessor system are used. Computers use ULSI, so their size is very small.

1.2 | SIGNAL AND SYSTEMS

A signal is a set of information or data, for example, a telephone or a television signal, monthly sales of a corporation or daily closing prices of a stock market. The signals are functions of the independent variable 'time'. An electrical charge is distributed over a body; the signal is charge density which is a function of space rather than time. Signals can broadly be classified as (i) continuous-time and discrete-time signals, (ii) analog and digital signals and (iii) periodic and aperiodic signals. A signal that is specified for every value of time, t (Figure 1.3) is a continuous-time signal. A signal that is specified only at discrete values of time, t (Figure 1.4) is a discrete-time signal. Telephone and video camera outputs are continuous-time signals, whereas the quarterly gross national product (GNP), monthly sales of a Corporation, and stock market daily averages are discrete-time signals. A function $y(t)$ is periodic with time period, T if $y(t) = y(t+T)$ for all values of t. T is a least positive number. A signal that is not periodic is known as aperiodic signal. A system is an entity that processes a set of signals as input to generate another set of signals as outputs. A system may be made up of physical components, as in electrical, mechanical, hydraulic or pneumatic.

1.3 | ANALOG SIGNALS AND SYSTEMS

Electricity is a coveted form of energy, since it can be generated centrally in bulk and transmitted economically over long distances. Electricity can be easily and efficiently used for various applications in both industries and domestic purpose. Electrical energy is produced in generators, transformed to an appropriate voltage level by the transformers and then dispatched (transmitted) through transmission lines for final distribution to the loads (consumers). Most of the world's electrical energy is generated and distributed by means of voltages and currents which vary sinusoidally with time.

1.3.1 | Direct Signals

A signal whose magnitude remains fixed with time is called a direct signal (Figure 1.1a). The systems or networks in which magnitude of current remains constant are known as direct current systems.

1.3.2 | Alternating Signal

The signal whose magnitude changes with time and direction reverse periodically is known as alternating signal (Figures 1.1b–1.1d). The systems or networks in which the magnitude of current varies in a repetitive manner are known as alternating current (AC) system. Sinusoidal signals (excitations) are important for number of reasons.

- Sinusoidal signals of adjustable frequency are used for testing electronic equipments.
- Most periodic functions can be represented as sum of sinusoids (sinusoidal waves).
- Mathematically, sine and cosine functions can be represented by the exponential functions using Euler's theorem.

$$e^{j\omega t} = \cos \omega t + j \sin \omega t \tag{1.1}$$

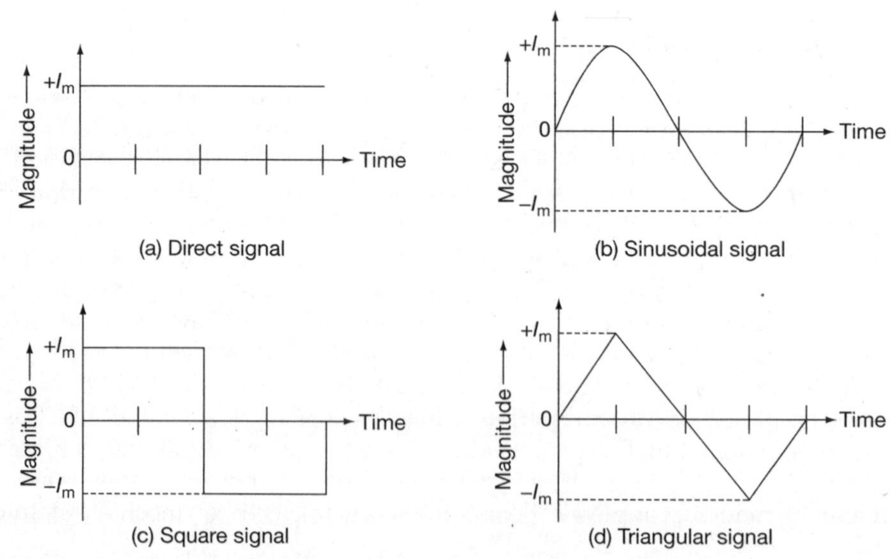

FIGURE 1.1 | Time-varying signals

- Exponential function remains exponential after performing differentiation and integration. Sinusoidal function can be represented as a complex number, so it is easy to solve the AC circuits in steady state.
- The time required for completion of one cycle (period) changes, when the frequency of a sine wave changes. So, it is useful to express points on the sine wave in terms of an angular measurement in degrees or radians. The angular or phase measurement of a sine wave is independent of frequency.

A sine wave of voltage is produced by rotating electromechanical machines (AC generators). As the rotor of the AC generator goes through a full 360° of rotation, the resulting voltage output is on full cycle of a sine wave. Thus, the angular measurement of a sine wave can be related to the angular rotation of a generator. Angular velocity is defined as the ratio of angular displacement to the time taken to undergo the displacement.

$$\text{Angular velocity}, \omega = \frac{\text{Angular displacement}, \theta}{\text{Time taken}, t}$$

$$\theta = \omega t \tag{1.2}$$

Time period, T is defined as the time taken to complete one revolution, that is, to cover 360° or 2π. Angular velocity is rewritten as

$$\omega = \frac{2\pi}{T} \tag{1.3}$$

Frequency is defined as the reciprocal of time. That is $f = \dfrac{1}{T}$

$$\omega = 2\pi f \tag{1.4}$$

$$1° = \left(\frac{\pi}{180}\right) \text{radians and } \pi = \frac{\text{Circumference of a circle}}{\text{Diameter of a circle}}$$

EXAMPLE 1.1 The maximum value of sinusoidal alternating current having frequency 50 Hz is 25 A. Find the equation for instantaneous value of alternating current. Determine current at 0.04 s. Find the time taken at which current is 12.5 A.

SOLUTION Maximum current $I_m = 25$ A

Angular speed, $\omega = 2\pi f$

Frequency, $f = 50$ Hz

$$\omega = 2\pi f = 2 \times 3.14159 \times 50 = 314.159 \text{ rad/s}$$

Time, $t = 0.004$ s

Instantaneous current $i = I_m \sin \omega t = 25 \sin(314.159 \times 0.04) = 25 \sin(1.256)6 = 23.776$ A

(*Note:* Angle is in radian)

Given $i = 12.5$ A

$$i = I_m \sin \omega t \Rightarrow 12.5 = 25\sin(314.159t) \Rightarrow \sin(314.159t) = \frac{12.5}{25}$$

$$(314.159t) = \sin^{-1} 0.5 \Rightarrow t = \frac{1}{314.159}\sin^{-1} 0.5 = 1.667\times10^{-3} = 1.667 \text{ ms}$$

1.3.3 | Sinusoidal Signal

An alternating signal is one in which the signal (voltage or current) magnitude varies with respect to time in a repetitive manner. But, in direct signal system, the signal (voltage or current) magnitude does not vary with respect to time. The sine wave represents the most common type of alternating signal (voltage or current). It is also called a sinusoidal waveform or sinusoidal (Figure 1.2). Other types of electrical signals known as non-sinusoidal signals are made up of many individual sine waves called harmonics.

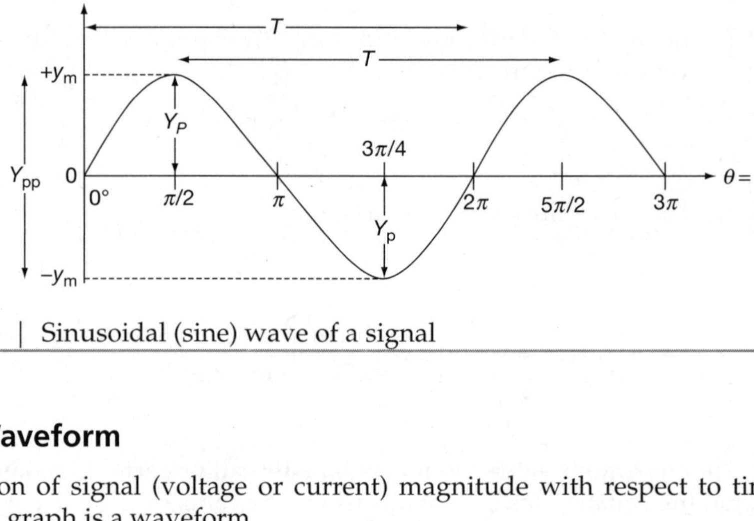

FIGURE 1.2 | Sinusoidal (sine) wave of a signal

1.3.4 | Waveform

The variation of signal (voltage or current) magnitude with respect to time or rotation shown on a graph is a waveform.

1.3.5 | Cycle

Cycle is the repetition of a variable quantity, recurring at equal intervals.

1.3.6 | Time Period

The time period is the time taken to complete one complete cycle. The period can be measured from one zero crossing to the corresponding zero crossing in the next cycle. The period can also be measured from the positive or negative peak (maximum or minimum point) of magnitude in one cycle to the positive or negative peak in the next cycle, respectively.

1.3.7 | Frequency

The frequency is the number of cycles that the signal completes in 1 s.

$$f = \frac{1}{T} \tag{1.5}$$

where f is the frequency in hertz (Hz) and T is the time period of one cycle in seconds.

Power frequencies vary from 16 to 80 Hz. The unit of frequency is hertz (Hz) in memory of Heinrich Rudolf Hertz who demonstrated experimentally the existence of electromagnetic radiation predicted by Maxwell in 1865.

1.3.8 | Peak Value

The peak value of a sine wave is the amount of signal (voltage or current) at the positive or negative maximum points referenced to the zero crossing. Y_p is its peak value.

1.3.9 | Peak-to-Peak Value

The peak-to-peak value is the signal (Y_{pp}) (voltage or current) from positive peak to the negative peak. It is always twice the peak value ($Y_{pp} = 2Y_p$).

1.3.10 | Instantaneous Value

Instantaneous value of a sine wave varies from one instant to the next. The instantaneous value of the signal as a sine wave is given by

$$y = y_m \sin(\omega t + \theta) \tag{1.6}$$

where y is the instantaneous value of the signal, y_m is the maximum magnitude of the signalm ω is the angular velocity and θ is the phase shift.

Similarly, the instantaneous value of the current as a sine wave can be given by

$$i = I_m \sin \omega t \tag{1.7}$$

where i is the instantaneous value of voltage, I_m is the maximum magnitude of the voltage and ω is the angular velocity.

1.3.11 | Periodic Functions

A function $y(t)$ is periodic with time period, T if $y(t) = y(t+T)$ for all values of t. T is a least positive number.

A function $y(t)$ is said to have half-wave symmetry if $y(t) = -y\left(t + \frac{T}{2}\right)$ for all values of t. Time period, T is a least positive number.

The general periodic function $y(t)$, with time period T, has an average value Y_{av} is given by

$$\text{Average value of wave } Y_{av} = \frac{1}{T}\int_0^T y(t)\mathrm{d}t \tag{1.8}$$

Average value of signal represented by sine wave is $Y_{av} = \frac{2}{\pi} y_m$

The general periodic function $y(t)$, with time period T, has root mean square (effective) value Y_{RMS} given by

$$\text{Root mean square (RMS) value } Y_{RMS} = \sqrt{\frac{1}{T}\int_0^T y(t)^2 \, dt} \tag{1.9}$$

Root mean square value of signal represented by sine wave is $Y_{RMS} = \frac{\sqrt{2} y_m}{2}$

The form factor of a wave is given by

$$\text{Form factor} = \frac{\text{RMS value}}{\text{Average value}} \tag{1.10}$$

The peak factor of a wave is given by

$$\text{Peak factor} = \frac{\text{Peak value}}{\text{RMS value}} \tag{1.11}$$

Peak factor is also known as crest factor. The peak value is the maximum value.

EXAMPLE 1.2 Find the periodicity of the sinusoidal function.

SOLUTION Let a sinusoidal function $y(t) = \sin \omega t$

$$y(t) = \sin\left(\frac{2\pi}{T} t\right) \qquad \because \omega = \left(\frac{2\pi}{T}\right)$$

Replace t by $(t + T)$ in the above equation

$$y(t+T) = \sin\left(\frac{2\pi}{T}(t+T)\right)$$

$$y(t+T) = \sin\left(\frac{2\pi}{T} t + \frac{2\pi}{T} T\right)$$

$$y(t+T) = \sin\left(\frac{2\pi}{T} t + 2\pi\right)$$

$$y(t+T) = \sin\left(\frac{2\pi}{T} t\right) = y(t) \qquad [\because \sin(2\pi + \theta) = \sin\theta]$$

Hence, the function is periodic over time period T. Time period T is 2π.

1.4 | DIGITAL SYSTEM AND SIGNALS

Electronics circuits and systems are of two kinds—Analog systems and Digital systems. Analog system is one in which voltage and current vary continuously through specific range (e.g. –10 to 10 V). It has infinite values in the range (Figure 1.3). Digital circuit is one in which voltage level takes finite number of distinct values (Figure 1.4).

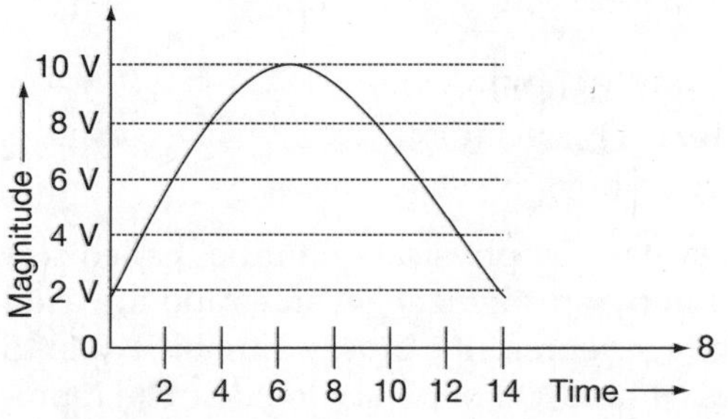

FIGURE 1.3 | Continuous analog signal

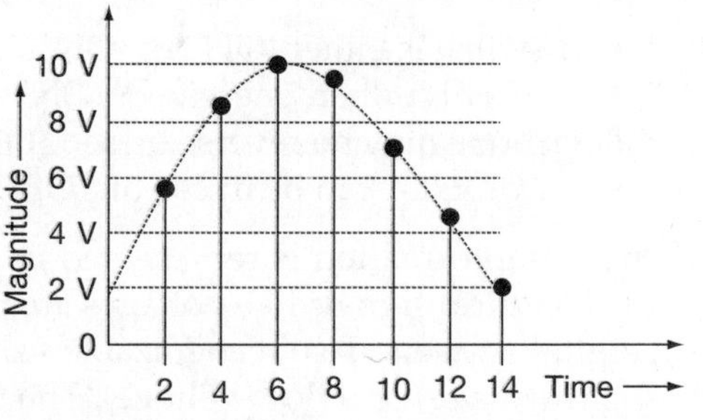

FIGURE 1.4 | Digital signal

In recent years, there has been a rapid increase in the use of digital systems like digital computers, calculators, watches, etc. The trend towards digital rather than analog systems is due to the availability of low-cost and high-speed digital computers. Limitation of digital system is compatibility with real-time applications. The real world is analog. Analog information cannot be processed directly with digital systems. First, it is converted into digital form and then processed using digital techniques. Finally digital information is reconverted into analog information to interpret properly (Figure 1.5). The conversion techniques from analog to digital (A/D) and digital to analog (D/A) require appreciations. The analog and digital techniques can be employed in the same system and such a system is called hybrid system.

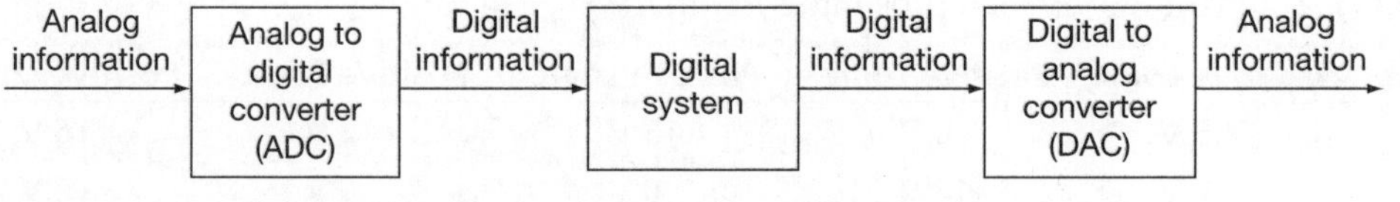

FIGURE 1.5 | Block diagram of real-time applications

1.5 | LOGIC LEVELS AND PULSE WAVEFORMS

Digital systems use digital signals. The decimal digits 0, 1, 2, 3, 4, 5, 6, 7, 8, 9 provide 10 discrete values. The signal in electronic digital systems has only two discrete values and are said to be binary. The digital-system designer uses binary signals because of high reliability and stability of the two discrete values. Its primary advantage is that it requires only two digits 1 and 0. The ideal two-stated device should have the following characteristics: (a) be turned on or off by binary 1 and 0 logic levels, (b) change from one state to the other instantaneously, and (c) have infinite resistance when off and zero resistance when on. Many electrical devices have two stable states that can be easily distinguished (Table 1.1).

TABLE 1.1 | Binary signal

Binary	Switch	Relay	Diode	Bipolar Transistor
0	OFF	De-energized	Reverse biased	Cut-off state
1	ON	Energized	Forward biased	Saturated state

- A switch is either 'OFF' or 'ON'.
- A relay is either 'energized' (ON) or 'de-energized' (OFF)
- A diode may be reverse biased (OFF) or forward biased (ON)
- A transistor can be in cut-off (OFF) or in saturation (ON) state

Binary information is represented in a digital system by physical quantities called signals. Electrical signals like voltages are presented in two recognizable values and are used in digital systems. Two recognizable values of voltage represent a binary variable equal to 1 (high voltage) or 0 (low voltage). The two states ('1' and '0') of binary logic can be represented in many ways in terms of voltage levels as given in Table 1.2.

So, the input terminals of digital systems accept binary signal within allowable tolerance. Therefore, two-state devices are used to represent the two binary digits 1 and 0 by two different voltage levels, called HIGH and LOW voltage levels, respectively. High-voltage level represents 1 logic. Low-voltage level represents 0 logic. Such a system is called the positive logic system. On the other hand, if the HIGH voltage level represents 0 logic and the LOW voltage level represents 1, then the system is called the negative logic system. The binary digits 0 and 1 are represented by the logic voltage levels 0 and +5 V, respectively. Hence, in a positive logic system, 1 is represented by +5 V (HIGH) and 0 is represented by 0 V (LOW). In a negative logic system, 0 is represented by +5 V (HIGH) and 1 is represented by 0 V (LOW). Both positive and negative logics are used in digital systems (Table 1.2), but the positive logic is more common. One can be changed into another by applying duality.

TABLE 1.2 | Voltage levels represented in binary

Digit	Negative Logic	Positive Logic	Positive Logic	Positive Logic	Positive Logic
1	–4.5 V	0 V	+5 V	+12 V	+6 V
0	0 V	–12 V	0 V	0 V	–6 V

A particular digital system employs a signal of 5 V to represent a binary 1 and 0.5 V for binary 0 as shown in Figure 1.6. In reality, because of circuit variations, 0 and 1 would be represented by voltage ranges instead of particular voltage levels, usually, any voltage between 0 and 0.8 V represents the logic 0 and any voltage between 2 and 5 V represents the logic 1. Normally, all input and output signals fall within one of these ranges except during transition from one level to another. The range between 0.8 and 2 V is called the indeterminate range. As the signal falls between 0.8 and 2 V, the response is not predictable. Each binary value has an acceptable deviation from nominal (fixed) value. The intermediate region has considerable separation. To go from region I to region II, there is need to cross over the intermediate region.

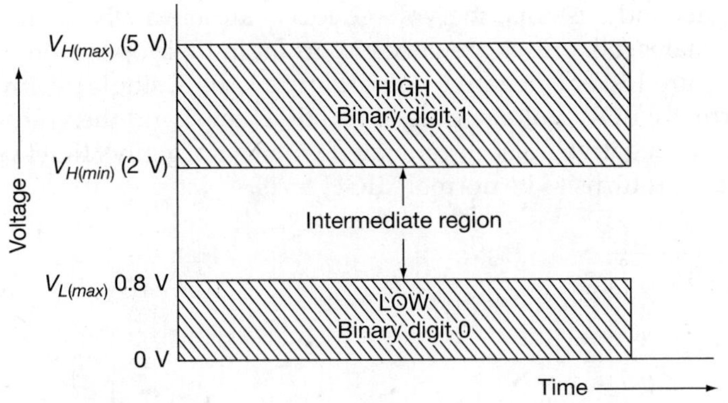

FIGURE 1.6 | Signal representing binary digits

Digital circuits are designed to give response to input voltages that are within the specified range. So, the accurate values of voltages are not important. The circuit responds to all input voltages in the allowed range, i.e. a voltage of 0 V gives the same response as a voltage of 0.2 V or 0.5 V or 0.8 V. Similarly, a voltage of 2 V gives the same response as a voltage of 2.5 V or 3.5 V or 4.5 V or 5 V.

A binary digital signal voltage may be either unipolar or bipolar. A unipolar signal has only one polarity voltage, either positive or negative, and it switches between this voltage and (nominally) zero voltage. A bipolar signal switches between a positive voltage and a negative voltage. Figure 1.7 shows an example of unipolar binary digital voltages, and Figure 1.8 shows an example of a bipolar binary digital voltage.

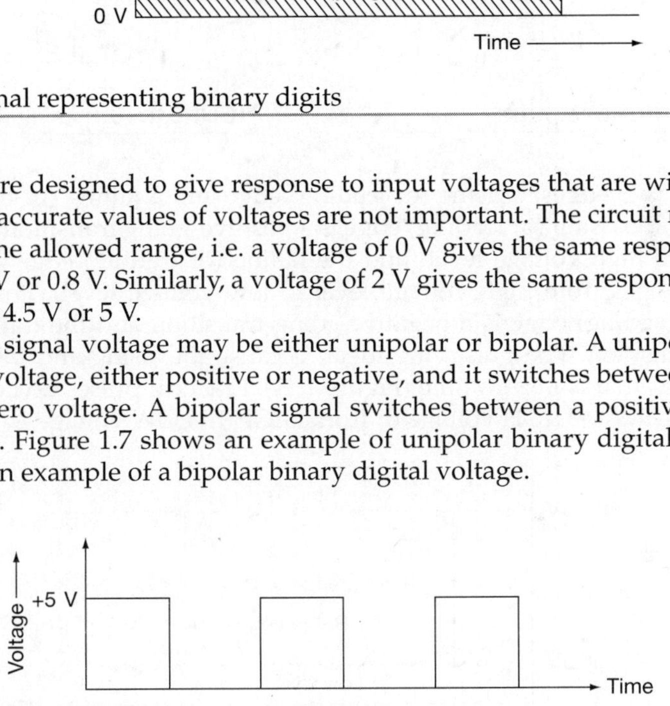

FIGURE 1.7 | Unipolar signal

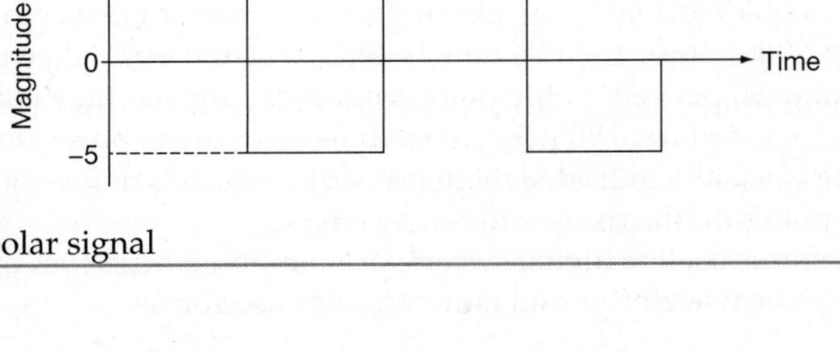

FIGURE 1.8 | Bipolar signal

In digital circuits and systems, the voltage levels are normally changing between the HIGH and LOW states. So, pulses are very important in their operation. A pulse may be a positive pulse (Figure 1.9) or a negative pulse (Figure 1.10). A single positive pulse is generated when a normally LOW voltage goes to its HIGH level and then returns to its normal LOW level. A single negative pulse is generated when a normally HIGH voltage goes to its LOW level and then returns to its normal HIGH level.

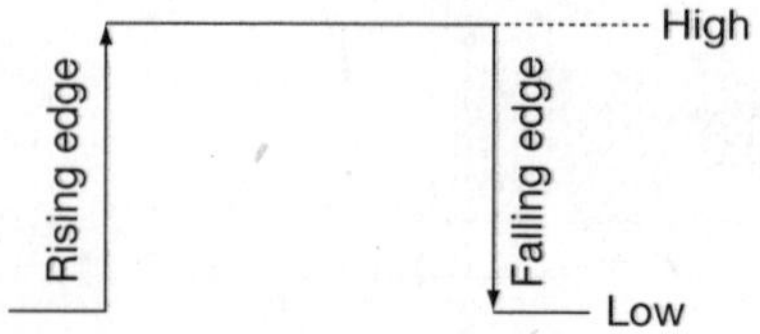

FIGURE 1.9 | Positive pulse

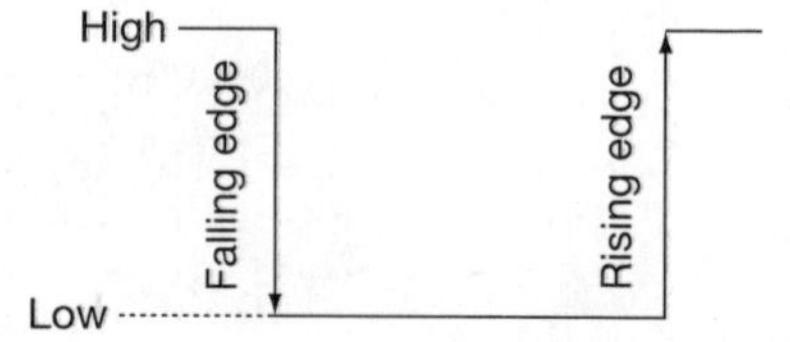

FIGURE 1.10 | Negative pulse

A pulse has two edges: a rising (or leading) edge and a falling (or trailing) edge. For a positive pulse, the rising (or leading) edge is a positive-going transition (changes from low voltage level to high voltage level) and the falling (or trailing) edge is a negative-going transition (changes from high voltage level to low voltage level). For a negative pulse, the rising (or leading) edge is a negative-going transition and the trailing edge is a positive-going transition. The pulses are ideal, because the rising and falling edges change instantaneously, that is in zero time (Figure 1.11). Practical pulses do not change instantaneously from LOW to HIGH voltage or from HIGH to LOW voltage.

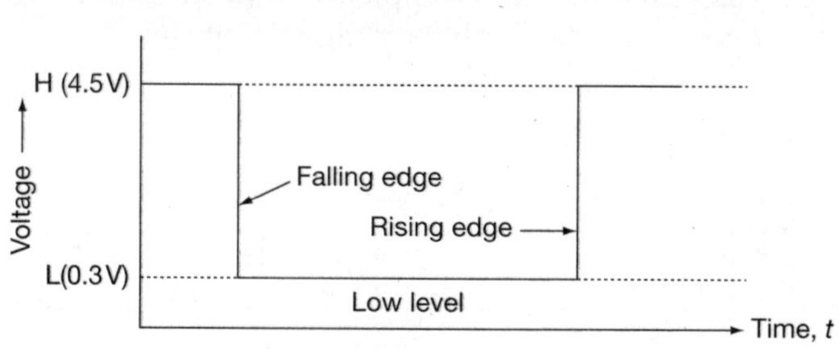

FIGURE 1.11 | Ideal pulse and its transition

A non-ideal pulse has finite rise and fall times (Figure 1.12). The time taken by the pulse to rise from LOW to HIGH is called the rise time. The time taken by the pulse to go from HIGH to LOW is called the fall time. Due to the nonlinearities that commonly occur at the bottom and top of the pulse, the rise time is defined as the time taken by the pulse to rise from 10 per cent to 90 per cent of the pulse amplitude. The fall time is defined as the time taken by the pulse to fall from 90 per cent to 10 per cent of the pulse amplitude. The duration of the pulse is usually indicated by pulse width, which is defined as the time between the 50 per cent points on the rising and falling edges.

Most of waveforms used in digital systems are composed of a series of pulses and can be classified as periodic waveforms and non-periodic waveforms.

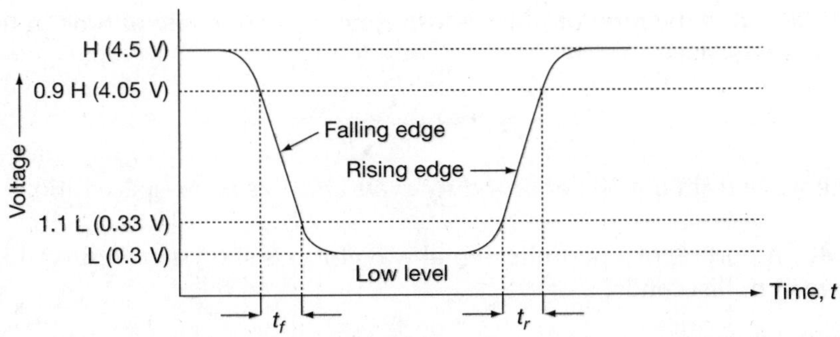

FIGURE 1.12 | Non-ideal pulse and its transition

Periodic waveform is one which repeats itself at regular intervals of time called the period (Figure 1.13). A *non-periodic waveform* does not repeat itself at regular intervals between the pulses (Figure 1.14).

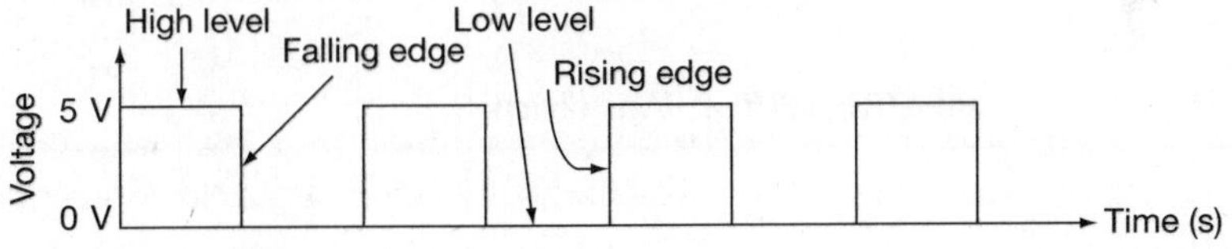

FIGURE 1.13 | Pulse train (square wave)

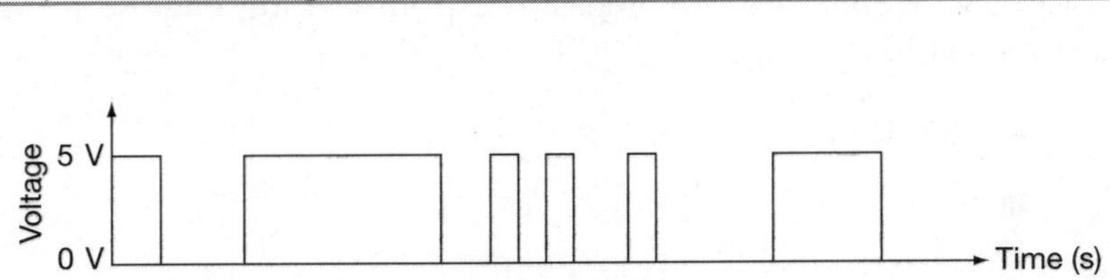

FIGURE 1.14 | Pulse train (non-periodic)

Frequency: The reciprocal of the time period (T) is called the frequency (f) of the periodic waveform.

$$f = \frac{1}{T} \tag{1.12}$$

where f is the frequency in hertz (Hz) and T is the time period of one cycle in seconds.

Duty cycle: An important characteristic of the periodic pulse waveform is its duty cycle. Duty cycle is defined as the ratio of the pulse width time (t_W) (i.e. active time) to the time period (T) of the pulse form. It is expressed in percentage.

$$\text{Duty cycle} = \frac{t_W}{T} \times 100\% \tag{1.13a}$$

Mark-space ratio: It is the ratio of pulse width time (t_W) to the rest of time in time period ($T - t_W$). It is expressed as

$$\text{Mark-space ratio} = \frac{t_W}{T - t_W} \tag{1.13b}$$

For square wave form has 50 per cent duty cycle and 1 as mark-space ratio.

EXAMPLE 1.3 A portion of a periodic digital waveform is shown in Figure 1.15. The measurements are in milliseconds.

Determine the following:
(a) Time period
(b) Frequency
(c) Duty cycle

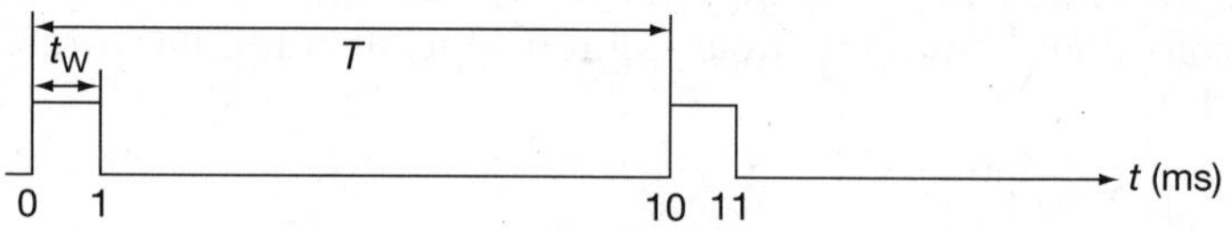

FIGURE 1.15 | Waveform

SOLUTION

(a) The time period is measured from the edge of one pulse to the corresponding edge of the next pulse. In this case, T is measured from leading edge to leading edge, as indicated. T equals 10 ms.

(b) $f = \dfrac{1}{T} = \dfrac{1}{10\text{ ms}} = 100\text{ Hz}$

(c) $\text{Duty cycle} = \left(\dfrac{t_W}{T}\right)100\% = \left(\dfrac{1\text{ ms}}{10\text{ ms}}\right)100\% = 10\%$

1.6 | DIGITAL WAVEFORM AND BINARY INFORMATION

Binary information that is handled by digital systems appears as waveforms that represent sequence of bits. When the waveform is HIGH, a binary 1 is present; when the waveform is LOW, a binary 0 is present. Each bit in a sequence occupies a defined time interval called a bit time.

The clock in digital systems, in all waveforms is synchronized with a basic timing waveform called the clock. The clock is a periodic waveform in which each interval between pulses (the period) equals the time for a bit.

An example of a clock waveform is shown in Figure 1.16. Notice that, in this case, each change in level of waveform A occurs at the leading edge of the clock waveform. In other cases, level changes occur at the trailing edge of the clock. During each bit time of the clock, waveform A is either HIGH or LOW. These HIGHs and LOWs represent a sequence of bits

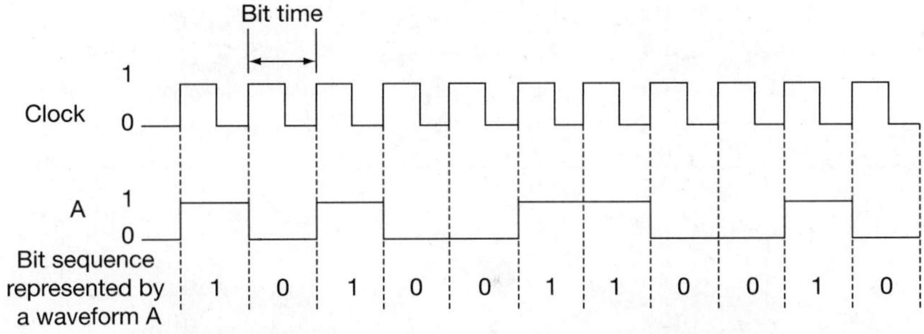

FIGURE 1.16 | Waveform representing bit sequence by a waveform

as indicated. A group of several bits can be used as a piece of binary information, such as a number or a letter. The clock waveform itself does not carry information.

A timing diagram is a graph of digital waveforms showing the actual time relationship of two or more waveforms and how each waveform changes in relation to the others. By looking at a timing diagram, you can determine the states (HIGH or LOW) of all the waveforms at any specific point in time and the exact time that a waveform changes state relative to the other waveforms. The diagram (Figure 1.17) is an example of a timing diagram made up of four waveforms. From this timing diagram you can see, for example, that the three waveforms A, B and C are High only during bit time 7 (shaded area) and they all change back to LOW at the end of bit time 7.

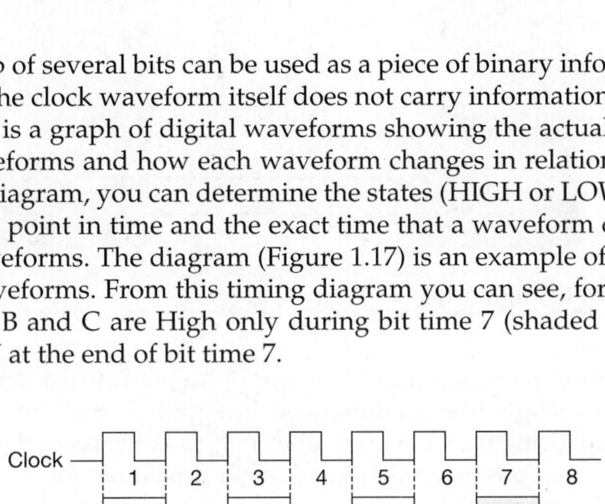

FIGURE 1.17 | Timing diagram

1.6.1 | Data Transfer

Data refers to groups of bits that convey some type of information. Binary data, which are represented by digital waveforms, must be transferred from one circuit to another within a digital system or from one system to another in order to accomplish a given purpose. For example, numbers stored in binary form in the memory of a computer must be transferred to the computer's CPU in order to be added. The sum of the addition must then be transferred to a monitor for display and/or transferred back to the memory. In computer systems, as illustrated in Figures 1.18 and 1.19, binary data are transferred in two ways serial and parallel.

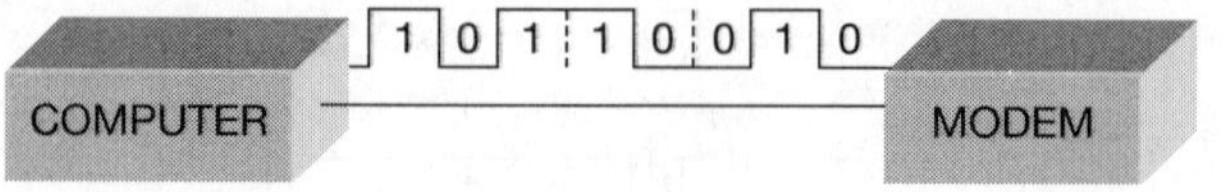

FIGURE 1.18 | Serial data transfer

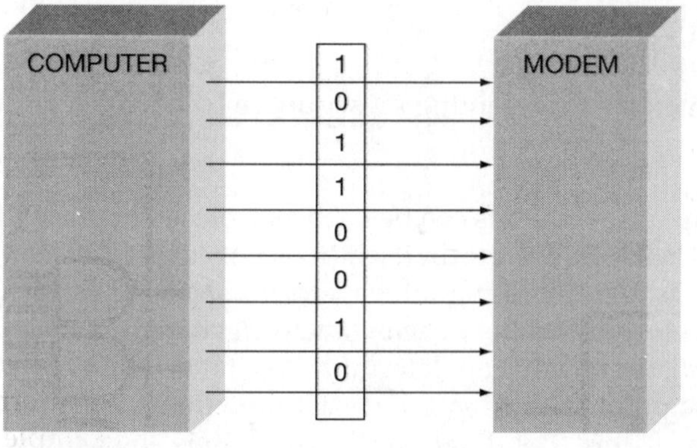

FIGURE 1.19 | Parallel data transfer

When bits are transferred in serial form from one point to another, they are sent one bit at a time along a single line, as illustrated in Figure 1.18 for the case of a computer-to-modem transfer. During the time interval from t_0 to t_1, the second bit is transferred, and so on. The transfer eight bits in series, it takes eight time intervals.

When the bits are transferred in parallel form, all the bits are in group are sent out on separate line at the same time. There is one line for each bit, as shown in Figure 1.19 for the example of eight bits being transferred from a computer to a printer. To transfer eight bits in parallel, it takes one time interval compared to eight time intervals for the serial transfer.

To summarize, an advantage of serial transfer of binary data is that a minimum of only one line is required. In parallel transfer, a number of lines equal to the number of bits to be transferred at one time are required. A disadvantage of serial transfer is that it takes longer to transfer a given number of bits than with parallel transfer at the same clock frequency. For example, if one bit can be transferred in 1 μs, then it takes 8 μs to serially transfer eight bits but only 1 μs to parallel transfer eight bits. A disadvantage of parallel transfer is that it takes more lines than serial transfer.

1.7 | ADVANTAGES OF DIGITAL TECHNOLOGY

Recent advances in electronics, digital techniques are replacing analog methods. The advantages of digital technology are enumerated as follows:

1. *Easy design of digital systems:* Switching circuits are used in digital systems, where exact values of voltage or current are not important, only the range in which values of voltage or current fall is necessary.

2. *Easy information storage:* Special devices and circuits can latch onto digital information to hold the information for as long as necessary. Mass storage techniques are capable to store billions of bits of information in a small physical space. In contrast, analog storage capabilities are limited.
3. *Greater accuracy and precision:* Once a signal is digitized, the contained information does not deteriorate and it is processed. In analog systems, the voltage and current signals have tendency to be disturbed by the effects of temperature, humidity and component tolerance variations in the circuits that process the signal.
4. *Easily programmable:* Digital systems are designed whose operation is controlled by a set of stored instructions in some sequence called a program. Analog systems can also be programmed, but the variety of the available operations is limited.
5. *Less affected by noise:* Spurious fluctuations in voltage are not critical in digital systems because the exact value of a voltage is not important as long as the noise is not large enough to change state from a HIGH level to a LOW level or vice versa.
6. *IC Technology:* No doubt analog circuitry has also benefited from the development of IC technology. But relative complexity and use of analog devices prevent their economic integration. Analog systems do not achieve the same high degree of integration as that of digital systems.

1.8 | LIMITATIONS OF DIGITAL TECHNOLOGY

Digital technology has very small number of drawbacks. The two major problems are observed in digital technology. First problem is that the real world is analog. Second problem is that a digitized signal takes time in processing. Physical quantities are analog in nature. Analog quantities are often the inputs and outputs that are being monitored, operated on, and controlled by a system. Some examples are temperature, pressure, position, velocity, liquid level, flow rate, and so on. These quantities are expressed digitally, for example, the temperature is 61°C (or more precise 60.8°C); it is a digital approximation to a quantity. To take advantage of digital techniques when dealing with analog inputs and outputs, four steps must be followed (Figure 1.5):

- Convert the physical variable to an electrical analog signal.
- Convert the electrical analog signal into digital form.
- Process the digital information.
- Convert the digital outputs back to analog signal.

There are many kinds of devices that convert various physical variables into electrical analog signals. Such devices are known as transducers. On a car, there are sensors for fluid level, temperature, velocity, acceleration, pressure and flow rate, etc.

Compact discs (CDs) have replaced cassette tapes because they provide a much better means for recording and playing back music. In this, sounds from instruments and human voices produce an analog voltage signal in a microphone. This analog signal is converted to a digital format using an analog to digital conversion (ADC) process. The digital information is stored on the CD's surface. During playback, the CD player takes the digital information from the CD surface and converts it into an analog signal that is then amplified and fed to a speaker, where it can be listenable by the human ear (Figure 1.20).

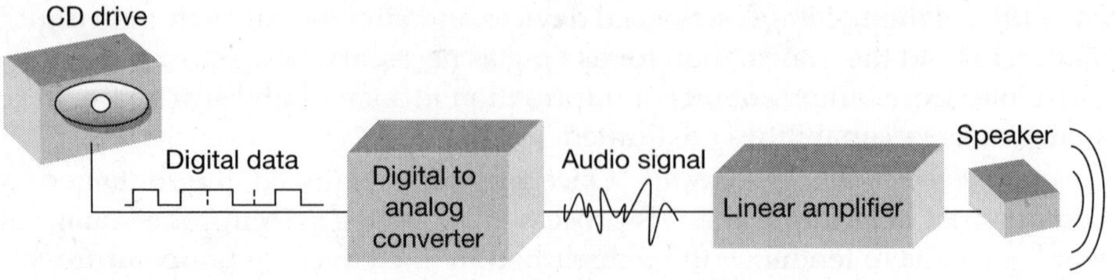

FIGURE 1.20 | Basic block diagram of a CD player

Another drawback to digital systems is that processing these digitized signals takes time. Conversion between the digital and analog forms of information adds complexity and expense to a system. The more precise the numbers need to be, the longer it takes to process them. In many applications, these factors are outweighed by the numerous advantages of using digital techniques, and so the conversion between analog and digital quantities has become quite commonplace in the current technology.

1.9 | ADVANCES IN DIGITAL TECHNOLOGY

The advances in digital technology over the past three decades have been nothing short of extra special, and more is coming. Day-to-day items are changing from analog format to digital in these days. An indoor/outdoor wireless digital thermometer can easily be purchased in affordable price. Cars become digitally controlled vehicles. Digital audio has provided CD and MP3 player. Digital video invented digital versatile disc (DVD). Digital home video and still cameras; digital recording with systems like TiVo, digital cellular phones; and digital imaging in X-ray, magnetic resonance imaging (MRI), and ultrasound systems in hospitals are some of the applications that are available in the market due to digital revolution. Telephone and television systems will go digital. The growth rate in the digital realm continues to be staggering. Automobiles are going to be equipped with a system such as global positioning system (GPS) installed on car dashboard into a hub for wireless communication, information and navigation. Human beings may be using voice commands to send or retrieve e-mail, call for a traffic report, check on the car's maintenance needs, or just switch radio stations. Cars can report their exact location in case of emergency or mechanical breakdown. Telephones will able to receive, sort and respond to incoming calls like a well-trained secretary. The digital television revolution will provide not only higher definition of the picture but also more flexibility in programming. We are able to select the programs that we are interested to view and load them into our television's memory, allowing to pause or replay scenes as per convenience, very much like viewing a DVD today. As virtual reality continues to improve, we will be able to interact with the subject matter we are studying. This may not sound exciting when studying electronics, but imagine studying history from the standpoint of being a participant, or learning proper techniques for everything from athletics to survey through simulations based on your actual performance.

Digital technology will continue its high-speed incursion into current areas of our lives as well as break new ground in ways we may never have considered. These applications are based on the principles presented in this text. The software tools to develop complex systems are constantly being upgraded and are available to anyone over the web. We will study the technical underpinnings necessary to communicate with any of these tools, and prepare you for fascinating and rewarding career.

1.10 | DIGITAL INFORMATION STORAGE

As an input signal is applied to devices or circuits, the output changes in response to the input, and when the input signal is removed the output returns to its original state. These circuits do not have the property of memory because their outputs revert back to normal.

Digital circuits do have memory. When an input is applied to digital circuit, the output will change its state. On removing the input, the circuit will remain in the new state. This property of retaining its response to a momentary input is called memory.

Memory devices and circuits play an important role in digital systems because they provide a means for storing binary numbers either temporarily or permanently, with the ability to change the stored information at any time. The various memory elements include magnetic and optical types and those that utilize electronic latching circuits. Block diagram of memory is shown in Figure 1.21.

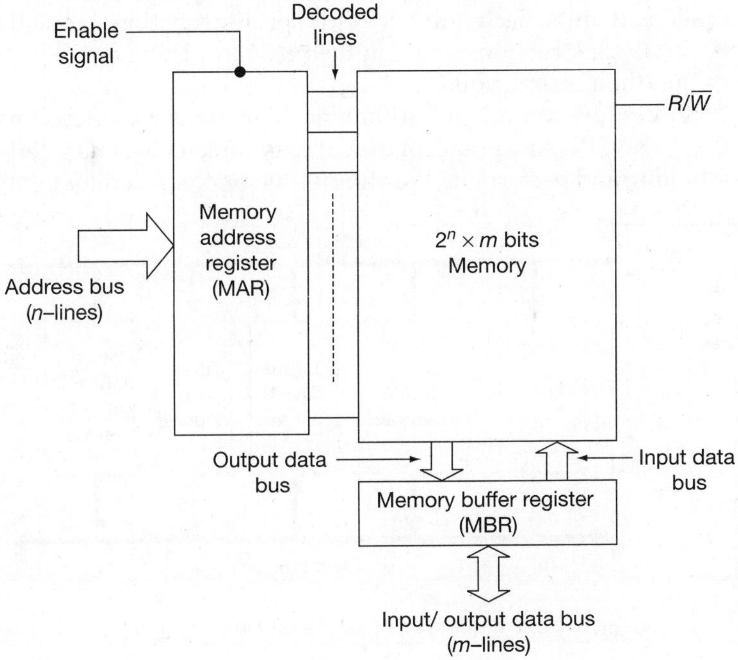

FIGURE 1.21 | RAM memory organization

1.11 | DIGITAL COMPUTING SYSTEMS

Digital techniques have found their way into innumerable areas of technology, but the area of automatic digital computers is by far the most notable and most extensive. Although digital computers affect some part of all of our lives, it is doubtful that many of us know exactly what a computer does. In simplest terms, a computer is a system of hardware that performs arithmetic operations, manipulates data (usually in binary form) and makes decisions.

For the most part, human beings can do whatever computers can do, but computers can do it with much greater speed and accuracy, in spite of the fact that computers perform their entire calculations and operations one step at a time. For example, a human being can take a list of 10 numbers and find their sum all in one operation by listing the numbers one over the other and adding them column by column. A computer, on the other hand, can add numbers only two at a time, so that adding this same list of numbers will take nine actual addition steps. Of course, the fact that the computer requires only a few nanoseconds per step makes up for this apparent inefficiency.

A computer is faster and more accurate than people are, but unlike most of us, it must be given a complete set of instructions so that it should be instructed exactly what to do at each step of its operation. This set of instructions, called a program, is prepared by one or more persons for each job the computer is to do. Programs are placed in the computer's memory unit in binary-coded form with each instruction having a unique code. The computer takes these instruction codes from memory one at a time and performs the operation called for by the code. There are several types of computer systems, but each can be broken down into the functional units. Each unit performs specific functions, and all units function together to carry out the instructions given in the program. There are five major parts of a digital computer and their instructions.

As Figure 1.22 shows, the control and arithmetic/logic units are often considered as one unit, called the CPU. The CPU contains all of the circuitry for fetching and interpreting instructions and for controlling and performing the various operations called for by the instructions.

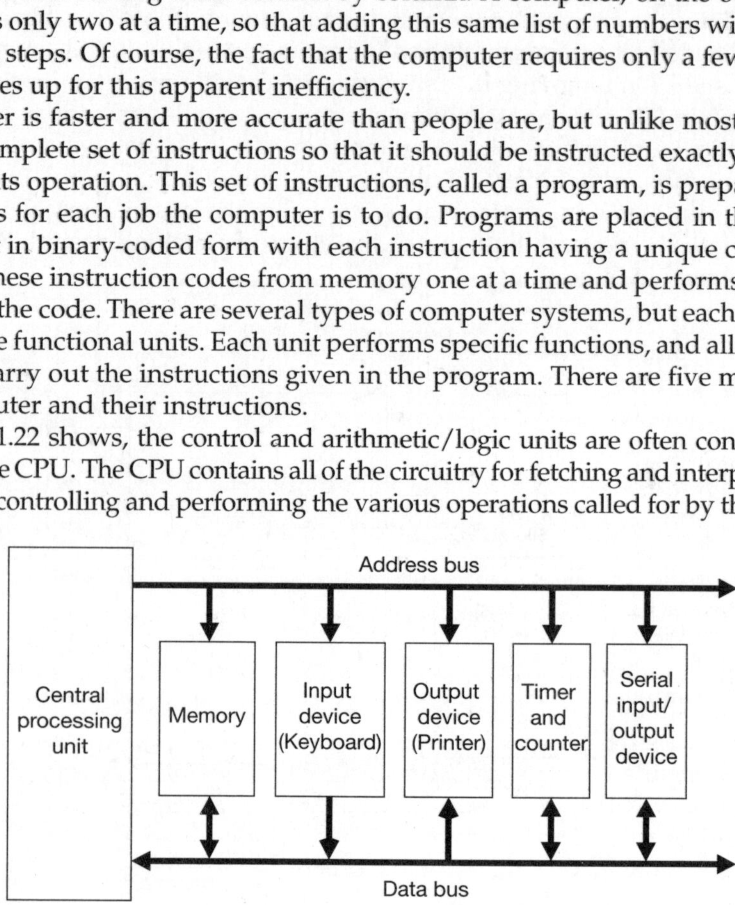

FIGURE 1.22 | General-purpose microprocessor system

The major functions of each unit are:

1. *Input unit:* Through this unit, a complete set of instructions and data is fed into the computer system and into the memory unit, to be stored until needed. The information typically enters the input unit from a keyboard or a disc.
2. *Memory unit:* The memory stores the instructions and data received from the input unit. It stores the results of arithmetic operations received from the arithmetic unit. It also supplies information to the output unit.

3. *Control unit:* This unit takes instructions from the memory unit one at a time and interprets them. Then it sends appropriate signals to all the other units to cause the specific instruction to be executed.
4. *Arithmetic and logic unit:* All arithmetic calculations and logical decisions are performed in this unit, which can then send results to the memory unit to be stored.
5. *Output unit:* This unit takes data from memory unit and prints out, displays or otherwise presents the information to the operator.

1.11.1 | Advances in Computing Systems

All computers are made up of the basic units described above, but they can differ as to physical size, operating speed, memory capacity and computational power, as well as other characteristics. Computer systems are configured in many and various ways today, with many common characteristics and distinguishing differences. Large computer systems that are permanently installed in multiple cabinets are used by corporations and universities for information technology support. Desktop personal computers are used in our homes and offices to run useful application programs that enhance our lives and provide communication with other computers. Portable computers are found in PDAs (Personal Digital Assistants) and specialized computers are found in video game systems. The most prevalent form of computers can be found performing dedicated routine tasks in appliances and systems all around us. Today, all but the largest of these systems utilize technology that has evolved from the invention of the microprocessor. The microprocessor is essentially a CPU in an IC that can be connected to the other blocks of a computer system. Computers that use a microprocessor as their CPU are usually referred to as microcomputers. The general-purpose microcomputers (e.g., PCs, PDAs) perform a variety of tasks in a wide range of applications depending on the software (programs) they are running. Contrast these with the dedicated computers that are doing things such as operating your car's engine, controlling your car's antilock braking system, or running your microwave oven. These computers cannot be programmed by the user, but simply perform their intended control task: they are referred to as microcontrollers (Figure 1.23). Since these microcontrollers are an integral part of a bigger system and serve a dedicated purpose, they also are called embedded controllers. Microcontrollers generally have all the elements of a complete computer (CPU, memory and input/output ports), all contained in a single IC. You can find them embedded in your kitchen appliances, entertainment equipment, photocopiers, automatic teller machines (ATMs), automated manufacturing equipment, medical instrumentation and much, much more.

So you see, even people who do not own a PC or use one at work or college are using microcomputers every day because so many modern consumer electronic devices, appliances, office equipment and much more are built around embedded microcontrollers. If you work, play or go to college in this digital age, there is no way that you could avoid a computer; definitely you will use a microcomputer somewhere.

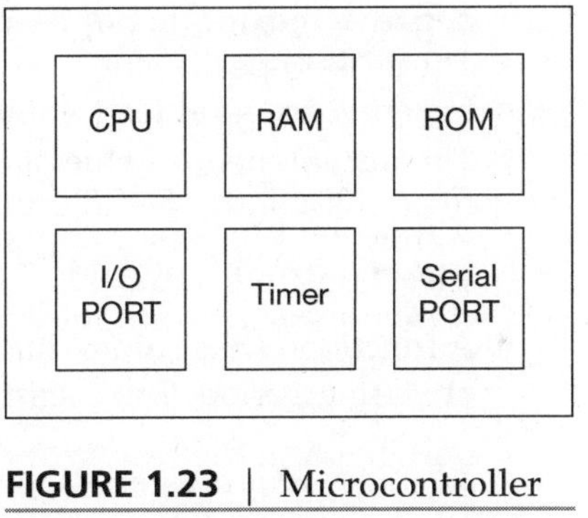

FIGURE 1.23 | Microcontroller

SUMMARY

- The rate of flow of electrons through a conductor provides electric current (current).
- The electric direct current (DC) can be produced with the help of batteries or generators.
- The electric alternating current (AC) can be produced with the help of generators or alternators.
- Electrical energy is generated and distributed by means of voltages and currents which vary sinusoidally with time.
- The generation, control, transmission and utilization of both direct current (DC) and alternating current (AC) at 50-Hz magnitudes at large value of A or kA come under the purview of electrical engineering.
- The controlled flow of electrons by various methods and forms is known as electronics.
- Direct current (DC) is a current whose magnitude remains constant with time.
- Direct current systems are the systems or networks in which magnitude of current remains constant with time.

Analog signal and system

- Alternating current (AC) is electric current whose magnitude changes with time and direction reverses periodically.
- AC system is the systems or networks in which the magnitude of current varies in a repetitive manner.
- $\theta = \omega t$
- Time period, T is defined as time taken to complete one revolution, i.e. to cover 360° or 2π.
- Frequency is defined as the reciprocal of time, i.e. $f = 1/T$.
- $\omega = \dfrac{2\pi}{T}$
- $\omega = 2\pi f$
- Waveform is the variation of voltage or current magnitude with respect to time or rotation shown on a graph.
- Cycle is the repetition of a variable quantity, recurring at equal intervals.
- The peak value of a sine wave is the amount of voltage or current at the positive or negative maximum points referenced to the zero crossing.
- The peak-to-peak value is the voltage or current from positive peak to the negative peak.
- Instantaneous value of a sine wave varies from one instant to the next.
- The instantaneous value of the signal as a sine wave is given by $y = y_{\mathrm{m}} \sin(\omega t + \theta)$.
- A function $y(t)$ is periodic with time period, T if $y(t) = y(t+T)$ for all values of t. T is a least positive number.
- A function $y(t)$ is said to have half-wave symmetry if $y(t) = -y\left(t + \dfrac{T}{2}\right)$ for all values of t. Time period, T is a least positive number.
- Average value of a general periodic function $y(t)$, with time period T is $Y_{\mathrm{av}} = \dfrac{1}{T}\int_0^T y(t)\mathrm{d}t$.

- Average value of voltage represented by sine wave $V_{av} = \frac{2}{\pi} V_m$.
- Average value of current represented by sine wave $I_{av} = \frac{2}{\pi} I_m$.
- RMS (effective) value of a general periodic function $y(t)$, with time period T is $Y_{RMS} = \sqrt{\frac{1}{T}\int_0^T y(t)^2 dt}$.
- RMS value of voltage represented by sine wave $V = \frac{\sqrt{2}V_m}{2}$
- RMS value of current represented by sine wave $I = \frac{\sqrt{2}I_m}{2}$
- Form factor $= \frac{\text{RMS value}}{\text{Average value}}$
- Peak factor $= \frac{\text{Peak value}}{\text{RMS value}}$

Digital signals and systems

- Digital circuit is one in which voltage level takes finite number of distinct values.
- Digital systems use digital signals.
- The decimal digits 0, 1, 2, 3, 4, 5, 6, 7, 8, 9 provide 10 discrete values.
- The signal in electronic digital systems has only two discrete values and these values are said to be binary.
- Pulse is a signal in which voltage levels are normally changing between the HIGH and LOW states.
- The pulses are ideal, because the rising and falling edges change instantaneously, that is in zero time.
- Non-ideal pulses do not change instantaneously from LOW to HIGH voltage or from HIGH to LOW voltage.
- A positive pulse is generated when a normally LOW voltage goes to its HIGH level and then returns to its normal LOW level.
- A positive pulse is generated when the rising (or leading) edge is a positive-going transition and the falling (or trailing) edge is a negative-going transition.
- A single negative pulse is generated when a normally HIGH voltage goes to its LOW level and then returns to its normal HIGH level.
- A negative pulse is generated when the rising (or leading) edge is a negative-going transition and the trailing edge is a positive-going transition.
- *Periodic waveform* is one which repeats itself at regular intervals of time called the period.
- *Non-periodic waveform* does not repeat itself at regular intervals between the pulses.
- Frequency is reciprocal of the time period (T) of the periodic waveform.

$$f = \frac{1}{T}$$

- Duty cycle is defined as the ratio of the time of pulse width (t_W) (i.e. active time) to the time period (T) of the pulse form. Duty cycle $= \frac{t_W}{T} \times 100\%$.
- Mark-space ratio is the ratio of pulse width time (t_W) to the rest of time of time period ($T - t_w$). Mark-space ratio $= \frac{t_W}{T - t_W}$.
- A timing diagram is a graph of digital waveforms showing the actual time relationship of two or more waveforms and how each waveform changes in relation to the others.
- A unipolar signal has only one polarity voltage, either positive or negative, and it switches between this voltage and (nominally) zero voltage.
- A bipolar signal switches between a positive voltage and a negative voltage.

MULTIPLE CHOICE QUESTIONS

1.1 A quantity having discrete voltage values is
(a) a digital quantity
(b) an analog quantity
(c) a binary quantity
(d) a natural quantity

1.2 A quantity having continuous voltage values is
(a) a digital quantity
(b) an analog quantity
(c) a binary quantity
(d) a natural quantity

1.3 The term bit means
(a) a small amount of data
(b) a 1 or a 0
(c) binary digit
(d) none of above

1.4 The time interval on the leading edge of a pulse between 10 per cent and 90 per cent of the amplitude is the
(a) rise time
(b) fall time
(c) pulse width
(d) period

1.5 The time interval on the trailing edge of a pulse between 90 per cent and 10 per cent of the amplitude is the
(a) rise time (b) fall time
(c) pulse width (d) period

1.6 A pulse in a certain waveform occurs every 100 ms. The frequency is
(a) 1 Hz (b) 10 Hz
(c) 100 Hz (d) 1 kHz

1.7 The frequency of waveform is 50 Hz. The time period of wave is
(a) 2 ms (b) 20 ms
(c) 200 ms (d) 2000 ms

1.8 In a certain digital waveform, the period is twice the pulse width. The duty cycle is
(a) 50 per cent
(b) 100 per cent
(c) 200 per cent
(d) none of above

1.9 In a certain digital waveform, if the duty cycle is 50 per cent, the mark-space ratio is
(a) 0 (b) 1
(c) 2 (d) 4

1.10 In a bipolar voltage signal, the voltage varies in
(a) Positive half
(b) Negative half
(c) positive as well as negative half
(d) positive or negative half

Answers

1.1 (a) 1.2 (b) 1.3 (c) 1.4 (a) 1.5 (b) 1.6 (b) 1.7 (b)
1.8 (a) 1.9 (b) 1.10 (c)

QUESTIONS

1.1 For each of the following, state whether it is an analog or a digital quantity:

(i) The pages in a book. (D)
(ii) The temperature at night. (A)
(iii) The waveband switch on a radio receiver. (D)
(iv) The ink left in a fountain pen. (A)
(v) A hand calculator. (D)
(vi) The height of a child. (A)
(vii) The indicator on a garage petrol pump. (D)
(viii) The speed of a car. (A)
(ix) The number of seconds in a minute. (D)
(x) Binary numbers. (D)
(xi) A digital watch. (D)
(xii) An electric light switch. (D)
(xiii) A dimmer switch. (A)
(xiv) Altitude of an aircraft. (A)
(xv) Pressure in a bicycle tyre. (A)
(xvi) Timer setting on microwave oven. (D)
(xvii) Current through a speaker. (A)
(xviii) Width of a piece of lumber. (D)
(xix) Time displayed on a quartz watch. (D)
(xx) Altitude above sea level measured on a staircase. (A)

1.2 A digital signal voltage varies between the values +9 V and +0.2 V. Is this
(a) a unipolar or a bipolar signal?
(b) If the +9 V represents logic 1, is this an example of positive or negative logic?

1.3 Draw a bipolar binary digital voltage that varies between ±12 V with a clock frequency of 10 MHz. Determine the periodic time of the waveform.

1.4 Discuss the reasons why a digital computer employs digital electronic circuitry and not analog circuits. Give a reason why binary digital circuitry is employed and not decimal digital.

1.5 Draw a positive unipolar digital waveform which has 8 bits alternately 0 and 1 if logic 1 is at 4.5 V and logic 0 is at 0.2 V and the bit width is 1 μs.

1.6 Draw a clock waveform that has a periodic time of 240 ns. What is its frequency?

1.7 A repetitive waveform is at +5 V for 1 μs and at 0 V for 2 μs. Calculate the following:
(a) Its duty cycle.
(b) Its mark-space ratio.

1.8 Draw and label a digital signal that has a duty cycle of 60 per cent, a pulse width of 200 ns, a rise time of 20 ns, and a fall time of 30 ns.

1.9 A digital transmission system transmits bits are (a) 4800 bits/s (bps), (b) 140 Mbps and (c) 565 Mbps. Calculate the bit width (or bit time) in each case.

1.10 A pulse waveform with a frequency of 10 kHz is applied to the input of a counter. During 100 ms, how many pulses are counted?

2

Number System

CHAPTER OBJECTIVES

The main goal of this chapter is to impart knowledge about the number system. Readers will be able to discuss:

- Conversion of number system from decimal to base-r and base-r to decimal.
- Complements of the base-r number system
- Arithmetic operation of base-r number system like addition and subtraction.
- Conversion of binary number system to decimal number system and vice versa.
- Arithmetic operation of binary number system like addition, subtraction, multiplication and division
- Conversion of octal number system to decimal and binary number system and vice versa.
- Arithmetic operation of octal number system like addition and subtraction
- Conversion of hexadecimal number system to decimal, binary and octal number system and vice versa.
- Arithmetic operation of hexadecimal number system like addition and subtraction
- Weighted, non-weighted, self-complementary and cyclic codes
- BCD code and its arithmetic operations
- Excess-3 code and its arithmetic operations
- Error-detecting and correcting codes like check sum and Hamming codes
- Multi-precision numbers

The decimal number system is ingrained in our culture. It is difficult to switch over to any other number system because of its simplicity. People are very comfortable with the decimal number system, but digital systems force them to use the binary system. Although the binary number system has many practical advantages and is widely used in digital computers. In many cases it is very convenient to work with decimal numbers, especially when communication between man and machine is extensive. Since most of the numerical data generated by man are in decimal numbers. To process the data on machines there is need to convert decimal data to binary. To simplify the communication process between man and machine, several systems of numeric codes have been devised to represent decimal numbers as a series of binary coded decimal (BCD) codes.

2.1 | DECIMAL NUMBER SYSTEM

The decimal number system contains 10 unique symbols; 0, 1, 2, 3, 4, 5, 6, 7, 8 and 9. Each symbol in the number system is called a digit. The number may be integer (123), fraction (0.25) or mixed (123.25).

Let a decimal number: 123.25. It can be written as:

$$123.25 = 1 \times 10^2 + 2 \times 10^1 + 3 \times 10^0 + 2 \times 10^{-1} + 5 \times 10^{-2}$$

The decimal number has two parts:

- integer portion (123)
- fractional portion (.25)

Both portions are combined with radix point or decimal point.

Further, first digit of integer portion is 3 and weight assigned to it is 10^0 (10^{1-1}). Second digit of integer portion is 2 and weight assigned to it is 10^1 (10^{2-1}). Third digit of integer portion is 1 and weight assigned to it is 10^2 (10^{3-1}). Similarly, first digit of fractional portion is 2 and weight assigned to it is 10^{-1}. Second digit of fractional portion is 5 and weight assigned to it is 10^{-2}.

The number system in which the weight of each digit depends on the relative position within the number is called positional number system.

$$(N)_{10} = \underbrace{d_{n-1}d_{n-2}\ldots d_i \ldots d_1 d_0}_{\text{Integer portion}} \;.\; \underbrace{d_{-1}d_{-2}\ldots d_{-j}\ldots d_{-m}}_{\text{Fractional portion}}$$

$$(N)_{10} = d_{n-1}r^{n-1} + d_{n-2}r^{n-2} + \cdots + d_1 r^1 + d_0 r^0 + d_{-1}r^{-1} + d_{-2}r^{-2} + \cdots + d_{-m}r^{-m}$$

$$(N)_{10} = \sum_{i=1}^{n} d_{i-1}r^{i-1} + \sum_{j=1}^{m} d_{-j}r^{-j} \tag{2.1}$$

The general form of number system can be written as:

where r is radix or base of the number system.

n is number of digits in integer portion or having assigned weight $\geq r^0$.

m is number of digits in fractional portion or having assigned weight $< r^0$.

d_{n-1} is the most significant digit (MSD).

d_{-m} is the least significant digit (LSD).

Observations:

- Base-r (or radix-r) number system depends on the value of r.
  - $r = 2$; Binary-number system
  - $r = 8$; Octal-number system
  - $r = 10$; Decimal-number system
  - $r = 16$; Hexadecimal-number system.
- Number of unique digits in base-r (or radix-r) number system is r.
  - Binary number system has two unique digits.
  - Octal number system has eight unique digits.
  - Decimal number system has 10 unique digits.
  - Hexadecimal number system has 16 unique digits.

- Base-r (or radix-r) number system has 0 to $(r - 1)$ distinct digits
  - Binary number system has 0 and 1 distinct digits.
  - Octal number system has 0, 1, 2, 3, 4, 5, 6 and 7 distinct digits.
  - Decimal number system has 0, 1, 2, 3, 4, 5, 6, 7, 8 and 9 distinct digits.
  - Hexadecimal number system has 0, 1, 2, 3, 4, 5, 6, 7, 8, 9, A, B, C, D, E and F distinct digits.
- Assigned weight to n^{th} digit of integer portion of base-r (or radix-r) number is r^{n-1}
- Assigned weight to m^{th} digit of fractional portion of base-r (or radix-r) number is r^{-m}
- Position of digits is counted from radix point (or decimal point).
- For integer portion, the position of digit is counted from right to left with respect to decimal point.
- For fractional portion, the position of digit is counted from left to right with respect to decimal point.
- MSD is the digit to which highest weight is assigned in the number.
- LSD is the digit to which lowest weight is assigned in the number
- Range of based-r numbers is 0 to $(r^n - r^{-m})_{10}$, where $(n + m)$ digits are combined out of which n digits represent in positive integer portion and m digits represent fractional portion.

  e.g. $(d_1 d_0)_{10} \Rightarrow$ Range is $0-(10^2 - 10^0) \Rightarrow 0 - 99$

  $(d_1 d_0 . d_{-1} d_{-2})_{10} \Rightarrow$ Range is $0-(10^2 - 10^{-2}) \Rightarrow 0 - 99.99$

Consider a decimal number $(9876.2045)_{10}$

Weight to digit ⇒	10^3	10^2	10^1	10^0		10^{-1}	10^{-2}	10^{-3}	10^{-4}
Digit position ⇒	4th	3rd	2nd	1st		1st	2nd	3rd	4th
	Left ←			Right		Left		→	Right
	9	8	7	6	.	2	0	4	5
	Integer portion					Fractional portion			
	MSD				Decimal Point				LSD

9 is MSD because weight assigned to it has highest value, i.e. 10^3. 5 is LSD because weight assigned to it has lowest value, i.e. 10^{-4}.

$$(9876.2045)_{10} = 9 \times 10^3 + 8 \times 10^2 + 7 \times 10^1 + 6 \times 10^0 + 2 \times 10^{-1} + 0 \times 10^{-2} + 4 \times 10^{-3} + 5 \times 10^{-4}$$

$$(9876.2045)_{10} = 9000 + 800 + 70 + 6 + 0.2 + 0.00 + 0.004 + 0.0005$$

Consider another decimal number $(7645)_{10}$ which has no fractional portion.

$$(7645)_{10} = 7 \times 10^3 + 6 \times 10^2 + 4 \times 10^1 + 5 \times 10^0$$

7 is MSD because weight assigned to it has highest value, i.e. 10^3. 5 is LSD because weight assigned to it has lowest value, i.e. 10^0

Weight to digit ⇒	10^3	10^2	10^1	10^0
Digit position ⇒	4th	3rd	2nd	1st
	Left ←			Right
	7	6	4	5
	Integer portion			

Consider a number $(0.0325)_{10}$ which has only fractional portion.

$$(0.0325)_{10} = 0\times10^{-1} + 3\times10^{-2} + 2\times10^{-3} + 5\times10^{-4}$$

3 is MSD because weight assigned to it has highest value, i.e. 10^{-2}. 5 is LSD because weight assigned to it has lowest value, i.e. 10^{-4}.

Table 2.1 gives the information about the various number systems and set of distinct digits in that number system. Table 2.2 shows the minimum and maximum values can be represented when number of digits is combined to represent integer number in base-r number system. As the base-r is larger (like hexadecimal system) then maximum number obtained by combining n-digit representing positive integer represents biggest number as decimal number. So, it is prevalent to represent the numbers in hexadecimal (hex) number system on paper as it requires lesser space to represent bigger number but electronics circuits always deal with binary number system to process the data.

TABLE 2.1 | Characteristics of various number systems to represent numbers

Number System	Base or Radix	Range of Distinct Digits in Number System	Set of Distinct Digits in Number System	Example	Place Value of Integer Having *n*-Digits	Place Value of Fraction Having *m*-Digits
Binary	2	0 to (2-1)	0,1	1011.101	2^0 to $2^{+(n-1)}$	2^{-1} to 2^{-m}
Base 3	3	0 to (3-1)	0, 1, 2	2101.201	3^0 to $3^{+(n-1)}$	3^{-1} to 3^{-m}
Base 4	4	0 to (4-1)	0, 1, 2, 3	3123.032	4^0 to $4^{+(n-1)}$	4^{-1} to 4^{-m}
Base 5	5	0 to (5-1)	0, 1, 2, 3, 4	4123.031	5^0 to $8^{+(n-1)}$	5^{-1} to 8^{-m}
Base 6	6	0 to (6-1)	0, 1, 2, 3, 4, 5	5432.012	6^0 to $6^{+(n-1)}$	6^{-1} to 6^{-m}
Base 7	7	0 to (7-1)	0, 1, 2, 3, 4, 5, 6	6543.012	7^0 to $7^{+(n-1)}$	7^{-1} to 7^{-m}
Octal	8	0 to (8-1)	0, 1, 2, 3, 4, 5, 6, 7	7653.012	8^0 to $8^{+(n-1)}$	8^{-1} to 8^{-m}
Decimal	10	0 to (10-1)	0, 1, 2, 3, 4, 5, 6, 7, 8, 9	9874.123	10^0 to $10^{+(n-1)}$	10^{-1} to 10^{-m}
Hexadecimal	16	0 to (16-1)	0, 1, 2, 3, 4, 5, 6, 7, 8, 9 A, B, C, D, E, F	3FA9.6BC	16^0 to $16^{+(n-1)}$	16^{-1} to 16^{-m}

2.1.1 | Conversion of Base-*r* Number to Decimal Number

A base-r (radix-r) number can be converted to decimal number by the positional weights method. Each digit of the number is multiplied by its position weight, $(d_n \times r^{n-1})$ and the product terms are added to find the decimal number. Mathematically, it can be represented by:

$$(N)_{10} = \sum_{i=1}^{n} d_i r^{i-1} + \sum_{j=1}^{m} d_{-j} r^{-j} \tag{2.2}$$

where d_i is i^{th} digit of integer portion in base-r number.
i is the position of base-r digit in integer portion.

TABLE 2.2 | Minimum and maximum values of base-*r*

	2-Digits Integer of Base-*r*		4-Digits Integer of Base-*r*	
	Minimum Value	Maximum Value	Minimum Value	Maximum Value
Base-2	$(00)_2 \Leftrightarrow (0)_{10}$	$(11)_2 \Leftrightarrow (2^2 - 1)_{10}$	$(0000)_2 \Leftrightarrow (0)_{10}$	$(1111)_2 \Leftrightarrow (2^4 - 1)_{10}$
Base-3	$(00)_3 \Leftrightarrow (0)_{10}$	$(22)_3 \Leftrightarrow (3^2 - 1)_{10}$	$(0000)_3 \Leftrightarrow (0)_{10}$	$(2222)_3 \Leftrightarrow (3^4 - 1)_{10}$
Base-4	$(00)_4 \Leftrightarrow (0)_{10}$	$(33)_4 \Leftrightarrow (4^2 - 1)_{10}$	$(0000)_4 \Leftrightarrow (0)_{10}$	$(3333)_4 \Leftrightarrow (4^4 - 1)_{10}$
Base-5	$(00)_5 \Leftrightarrow (0)_{10}$	$(44)_5 \Leftrightarrow (5^2 - 1)_{10}$	$(0000)_5 \Leftrightarrow (0)_{10}$	$(4444)_5 \Leftrightarrow (5^4 - 1)_{10}$
Base-6	$(00)_6 \Leftrightarrow (0)_{10}$	$(55)_6 \Leftrightarrow (6^2 - 1)_{10}$	$(0000)_6 \Leftrightarrow (0)_{10}$	$(5555)_6 \Leftrightarrow (6^4 - 1)_{10}$
Base-7	$(00)_7 \Leftrightarrow (0)_{10}$	$(66)_7 \Leftrightarrow (7^2 - 1)_{10}$	$(0000)_7 \Leftrightarrow (0)_{10}$	$(6666)_7 \Leftrightarrow (7^4 - 1)_{10}$
Base-8	$(00)_8 \Leftrightarrow (0)_{10}$	$(77)_8 \Leftrightarrow (8^2 - 1)_{10}$	$(0000)_8 \Leftrightarrow (0)_{10}$	$(7777)_8 \Leftrightarrow (8^4 - 1)_{10}$
Base-10	$(00)_{10} \Leftrightarrow (0)_{10}$	$(99)_{10} \Leftrightarrow (10^2 - 1)_{10}$	$(0000)_{10} \Leftrightarrow (0)_{10}$	$(9999)_{10} \Leftrightarrow (10^4 - 1)_{10}$
Base-16	$(00)_{16} \Leftrightarrow (0)_{10}$	$(FF)_{16} \Leftrightarrow (16^2 - 1)_{10}$	$(0000)_{16} \Leftrightarrow (0)_{10}$	$(FFFF)_{16} \Leftrightarrow (16^4 - 1)_{10}$

r^{i-1} is the weight assigned to the position of ith base-*r* digit.
d_{-j} is j^{th} digit of fractional portion in base-*r* number (if any)
j is the position of base-*r* digit in fractional portion.
r^{-j} is the weight assigned to the position of jth base-*r* digit.

- Position of base-*r* digits is always counted from decimal point.
- For integer portion, the position of digit is counted from right to left starting from decimal point. On moving one digit from right to left, weight assigned to digit increases by multiple of *r* that is the base value of the number system.
- For fractional portion, the position of digit is counted from left to right for fractional portion starting from decimal point. On moving one digit from left to right, weight assigned to digit decrease by of r^{-1}, where *r* is the base value of the number system.

EXAMPLE 2.1 Convert $(11011.\,101)_3$ to decimal.

SOLUTION Base-3 number system has three distinct digits 0, 1 and 2.

	81	27	9	3	1		1/3	1/9	1/27
Weight to digits ⇒	3^4	3^3	3^2	3^1	3^0		3^{-1}	3^{-2}	3^{-3}
Position of digits ⇒	5th	4th	3rd	2nd	1st		1st	2nd	3rd
Digits ⇒	1	1	0	1	1	.	1	0	1
	d_5	d_4	d_3	d_2	d_1		d_{-1}	d_{-2}	d_{-3}

$$(N)_{10} = \underbrace{\sum_{i=1}^{5} d_i 3^{i-1}}_{\text{Integer portion}} + \underbrace{\sum_{j=1}^{3} d_{-j} 3^{-j}}_{\text{Fractional portion}}$$

$$(N)_{10} = 1 \times 3^0 + 1 \times 3^1 + 0 \times 3^2 + 1 \times 3^3 + 1 \times 3^4 + 1 \times 3^{-1} + 0 \times 3^{-2} + 1 \times 3^{-3}$$

$$(N)_{10} = (1+3+27+81)+\left(\frac{1}{3}+\frac{1}{27}\right)$$

$$(N)_{10} = (112.37037)_{10} \quad \textbf{Ans.}$$

EXAMPLE 2.2 Convert $(1310214)_5$ to decimal.

SOLUTION Base-5 number has no fraction part. Decimal point is considered after the right most digit, i.e. 4. Base-5 number system has 5 distinct digits 0, 1, 2, 3 and 4.

Weight to digits ⇒	5^6	5^5	5^4	5^3	5^2	5^1	5^0
Position of digits ⇒	7th	6th	5th	4th	3rd	2nd	1st
Digits ⇒	1	3	1	0	2	1	4
	d_7	d_6	d_5	d_4	d_3	d_2	d_1

$$(N)_{10} = \sum_{i=1}^{11} d_i 5^{i-1}$$

$$(N)_{10} = 4\times 5^0 + 1\times 5^1 + 2\times 5^2 + 0\times 5^3 + 1\times 5^4 + 3\times 5^5 + 1\times 5^6$$

$$(N)_{10} = (4+5+50+625+9375+15625)$$

$$(N)_{10} = (25684)_{10} \quad \textbf{Ans.}$$

EXAMPLE 2.3 Convert $(6666)_7$ to decimal.

SOLUTION Base-7 number system has 7 distinct digits 0, 1, 2, 3, 4, 5 and 6. There is no fractional part. Decimal point is considered after the rightmost digit.

Weight to digits ⇒	7^3	7^2	7^1	7^0
Position of digits ⇒	4th	3rd	2nd	1st
Digits ⇒	6	6	6	6
	d_{-2}	d_{-3}	d_{-4}	d_{-5}

$$(N)_{10} = \sum_{i=1}^{4} d_i 7^{i-1}$$

$$(N)_{10} = 6\times 7^0 + 6\times 7^1 + 6\times 7^2 + 6\times 7^3$$

$$(N)_{10} = 6+42+294+2058$$

$$(N)_{10} = (2400)_{10} \quad \textbf{Ans.}$$

2.1.2 | Conversion from Decimal Number to Base-*r* Number

Decimal number has two parts. One is integer part and another is fractional part. Both the parts of decimal number are converted separately in to base-*r* number. So, for any decimal number, conversion into base-*r* is carried out separately for integer portion and fractional portion of decimal number and then both conversions are combined.

Conversion of integer term of decimal number to base-r number The method of conversion from decimal positive integer to base-*r* (radix-*r*) number is to divide successively by *base* and resulting remainders are the digits for base-*r* number and quotient are used for successive operation. First remainder will act as LSD of base-*r* number. The process is repeated till quotient becomes zero. The arithmetic process is given below in Algorithm 2.1.

Algorithm 2.1 | Conversion of decimal positive integer into base-*r* integer

Step 1: Consider decimal positive integer number as dividend.
Step 2: Divide the dividend by 'base-*r*' value. Find the quotient and the remainder. Consider 1st obtained remainder as LSD of base-*r* number system.
Step 3: Consider quotient obtained in step 2 as dividend for successive operation.
Step 4: If dividend becomes zero then STOP and consider last obtained remainder as MSD, otherwise go to step 2 and repeat.

EXAMPLE 2.4 Convert decimal integer $(41)_{10}$ into base-3 number.

SOLUTION Successive division is performed by taking divisor as 3.

Dividend	÷	Divisor	=	Quotient	: Remainder	Base-3 number digits
41	÷	3	=	13	: $2 \rightarrow d_1$ (LSD)	$d_1 = 2$
13	÷	3	=	4	: $1 \rightarrow d_2$	$d_2 = 1$
4	÷	3	=	1	: $1 \rightarrow d_3$	$d_3 = 1$
1	÷	3	=	0	: $1 \rightarrow d_4$	$d_4 = 1$

Base-3 number = $(d_4d_3d_2d_1)_3$

$(41)_{10} = (1112)_3$ **Ans.**

Conversion of factional term of decimal number to base-r number The conversion of a decimal fraction to base-*r* number is to multiply successively by base-*r* value *r* and use integers part of that number as the digits for base-*r* number and remained fraction part is taken as multiplicand for successive multiplication. The process is repeated till the fraction term becomes zero. The arithmetic process is given below in Algorithm 2.2.

Algorithm 2.2 | Conversion of decimal fraction to fraction of base-*r* number

Step 1: Consider fractional portion of decimal number as multiplicand.
Step 2: Multiply the multiplicand by *r*, i.e. base value. Separate integer and fractional portion. Consider the first integer as first fractional digit.
Step 3: Consider fractional portion obtained in step 2 as multiplicand for successive operation.
Step 4: If multiplicand is zero (or desired accuracy) then STOP, otherwise go to step 2 and repeat.

EXAMPLE 2.5 Convert $(0.125)_{10}$ decimal number to 4-base number.

SOLUTION Successive multiplication is performed by 4.

Multiplicand	× 4	=	Integer	+	Fraction	Base-4 digit
0.125	× 4	=	0	+	0.50	$d_{-1} \to 0$ (MSD)
0.50	× 4	=	2	+	0.00	$d_{-2} \to 2$

Base-4 number = $(0.d_{-1}d_{-2})_4$
$(0.125)_{10} = (0.02)_4$

EXAMPLE 2.6 Convert the following decimal number

(a) 128 to base-5 number (b) 19.56 to base 7 number.

SOLUTION

(a) Given decimal number is $(128)_{10}$. Division is performed by 5.

Dividend	÷	Divisor	=	Quotient	: Remainder	Base-5 digits
128	÷	5	=	25	: $3 \to d_1$ (LSD)	$d_1 \to 3$
25	÷	5	=	5	: $0 \to d_2$	$d_2 \to 0$
5	÷	5	=	1	: $0 \to d_3$	$d_3 \to 0$
1	÷	5	=	0	: $1 \to d_4$	$d_4 \to 1$

Base-5 number = $(d_4d_3d_2d_1)_5$
$(128)_{10} = (1003)_5$ **Ans.**

(b) Integer portion of decimal number 19.56 is 19. Integer is converted into base-7 number by performing successive division by 7.

Dividend	÷	Divisor	=	Quotient	: Remainder	Base-7 digits
19	÷	7	=	2	: $5 \to d_1$ (LSB)	$d_1 \to 5$
2	÷	7	=	0	: $2 \to d_2$	$d_2 \to 2$

Base-7 number = $(d_2d_1)_7$
$(19)_{10} = (25)_7$

Fractional portion of decimal number 19.56 is 0.56. Fraction is converted into base-7 number by performing successive multiplication by 7.

Multiplicand	× 7	=	Integer	+	Fraction	Base-7 digit
0.56	× 7	=	3	+	.92	$d_{-1} \to 3$
0.92	× 7	=	6	+	.44	$d_{-2} \to 6$
0.44	× 7	=	3	+	.088	$d_{-3} \to 3$

Base-7 number = $(0.d_{-1}d_{-2}d_{-3})_7$
$(0.56)_{10} = (0.363)_7$

Combining integer $(19)_{10} = (25)_7$ and fraction portion $(0.56)_{10} = (0.363)_7$ to get the number $(19.56)_{10} = (25.363)_7$

2.1.3 | Base-*r* Arithmetic

Base-*r* addition and subtraction operations are performed in a manner similar to decimal arithmetic. After arithmetic operation, decimal number is obtained. The decimal number needs to be converted into desired base-*r* by subtracting the base value *r* from each decimal digit. It results the base-*r* digit along with carry. Decimal digit and its equivalent in base-*r* is given in Table 2.3. Decimal 5 needs conversion if base is less than or equivalent to 5 whereas 5 is also valid digit for base-6, base-7, base-8 and base-9 number system. Digit 5 needs conversion for base-2, base-3, base-4 and base-5.

$$(5)_{10} \Leftrightarrow (101)_2 \Leftrightarrow (12)_3 \Leftrightarrow (11)_4 \Leftrightarrow (10)_5$$

For subtraction when minuend is lesser than subtrahend then there is need of borrow. Borrow is taken in base-*r*.

TABLE 2.3 | Decimal digit equivalent base-*r* digits

Decimal	Base-2	Base-3	Base-4	Base-5	Base-6	Base-7	Base-8	Base-9
0	0	0	0	0	0	0	0	0
1	1	1	1	1	1	1	1	1
2	10	2	2	2	2	2	2	2
3	11	10	3	3	3	3	3	3
4	100	11	10	4	4	4	4	4
5	101	12	11	10	5	5	5	5
6	110	20	12	11	10	6	6	6
7	111	21	13	12	11	10	7	7
8	1000	22	20	13	12	11	10	8
9	1001	100	21	14	13	12	11	10

2.1.3.1 | Base-*r* Addition

Addition is performed in the similar procedure as is done in decimal arithmetic. Resulting sum should not be more than the largest digit of Base-*r*. If the digit is larger than largest digit of Base-*r* distinct digits then subtract *r* (base value) to get the required digit with carry over as 1. Single digit addition of base-3 is given below.

	2nd	1st	2nd	1st	2nd	1st	2nd	1st	2nd	1st	2nd	1st
Carry over	0		0									
Augend		0		0		0		1		1		2
Addend	+	0	+	1	+	2	+	1	+	2	+	2
Sum	0	0	0	1	0	2	0	2	1	0	1	1

For example, when digits 1 and 2 of Base-3 are added, will result digit 3 which is larger than 2 (largest digit among set of base-3 distinct digits), so subtract 3 (Base) to get 0 with carry over as 1. Single-digit addition for base-4, base-5, base-6, base-7, base-8 and base-10 are given below:

	Base-4		Base-5		Base-6		Base-7		Base-8		Base-10	
	2nd	1st	2nd	1st	2nd	1st	2nd	1st	2nd	1st	2nd	1st
Carry over	1		1		1		1		1		1	
Augend		3		3		5		6		7		9
Addend	+	3	+	4	+	2	+	6		6		9
Sum	1	2	1	2	1	1	1	5	1	5	1	8

- Generally, addition in Base-*r* number system is performed in two steps. (i) Add carry to augend bit to get partial sum and (ii) add addend to partial sum to get the final sum.
- Carry-over in previous step must be taken care. Start operation from LSD (least significant digit). Consider 0 as previous carry for first digit (base-*r*).
- Numbers of digits of augend and addend must be same. If not, complete by filling zero(s) in such a manner so that weightage should remain same.

In case, numbers of digits are combined to represent a number. Carry over are performed in the same manner as in decimal arithmetic. Decimal and Base-3, Base-4 and Base-6 addition is elaborated with an example.

	Decimal number		Base-3 number				Base-4 number				Base-6 number		
Digits	2nd	1st	4th	3rd	2nd	1st	4th	3rd	2nd	1st	3rd	2nd	1st
Carry	1	0	1	1	0			1	1	0	1	0	0
Augend	1	5		1	2	0		0	3	3		2	3
Addend	2	6		2	2	2		1	2	2		4	2
Sum	4	1	1	1	1	0		2	2	1	1	0	5

Addition of first digit column generates carry for second digit of higher weightage. Addition of second digit column generates carry for third digit and so on.

2.1.3.2 | Base-*r* Subtraction

The base-*r* subtraction is performed in a manner similar to that in decimal subtraction. Borrow taken in decimal system adds 10 (base value) whereas in base-*r* the borrow taken adds *r* (base value). Single-bit subtraction for base-3 is performed and is given below.

	Case-I		Case-II		Case-III		Case-IV		Case-V		Case-VI	
Digit	2nd	1st	2nd	1st	2nd	1st	2nd	1st	2nd	1st	2nd	1st
Borrow							–1	10	–1	10	–1	10
Minuend		0		1		2		0		0		1
Subtrahend	–	0	–	1	–	2	–	1	–	2	–	2
Difference	0	0	0	0	0	0	1	2	1	1	1	2

- In first three possible cases there is no need of borrow from next bit.
- In case-IV, there is need of borrow from next bit.
- Borrow is taken from the next higher place valued digit as $(10)_3$ or $(3)_{10}$ (base value 3).
- Taken borrow is added to minuend and sum is obtained $(10)_3 + (0)_3 = (10)_3$ or $(3)_{10} + (0)_{10} = (3)_{10}$
- Finally, subtrahend is subtracted from partial sum to get the difference,
- i.e. $(10)_3 - (1)_3 = (2)_3$ or $(3)_{10} - (1)_{10} = (2)_{10}$
- Borrow taken by previous digit column is subtracted from minuend of present digit column.

> - When minuend is larger than or equal to subtrahend then there will be no borrow.
> - When minuend is smaller than subtrahend then there is need of borrow.
> - Borrow is taken from the next higher place valued digit as *r* (base value).
> - Taken borrow is added to minuend and sum is obtained.
> - Subtrahend is subtracted from partial sum to get the difference.

Decimal, Base-3, base-5 and base-7 subtraction is elaborated with an example in which $(17)_{10}$ is subtracted from $(26)_{10}$.

	Decimal number		Base-3			Base-5			Base-7	
Digits	2nd	1st	3rd	2nd	1st	3rd	2nd	1st	2nd	1st
							10			
Borrow	–1	10				–1	–1	10		
Minuend	2	6	2	2	2	1	0	1	3	5
Subtrahend	1	7	1	2	2	0	3	2	2	3
Difference	0	9	1	0	0	0	1	4	1	2

2.1.4 | Complement Form

Complements are used in digital computers:

- for simplifying the subtraction operation and
- to represent signed numbers.

There are two types of complements for each base-*r* system.

1. $(r - 1)$'s complement or radix-minus-one complement
2. *r*'s complement or true complement.

When the value of base is substituted as 10, then 9's and 10's complement form are obtained.

2.1.4.1 | (*r* – 1)'s Complement (Radix-1 Complement)

Let *N* is number in base-*r*.

$(r - 1)$'s complement of base-*r* number *N* is defined as $(r^n)_r - (r^{-m})_r - (N)_r$. (2.3)

where N is a positive number with base r.
n is number of digits in integer part.
m is number of digits in fractional part.

EXAMPLE 2.7 Find 9's complement of the following decimal number.

(a) 3475 (b) 783.45

SOLUTION

(a) Decimal number, $N = 3475$ and $r - 1 = 9$, as base r is 10
Number of digits in integer part, $n = 4$
Number of digits in fractional part, $m = 0$

$$\begin{aligned} \text{9's complement of } 3475 &= (10^4)_{10} - (10^{-0})_{10} - (3475)_{10} \\ &= (10000 - 1)_{10} - (3475)_{10} \\ &= (9999)_{10} - (3475)_{10} \\ &= (6524)_{10} \quad \textbf{Ans.} \end{aligned}$$

> ***Note:*** In case, no base is mentioned it means number is decimal and base is 10.

(b) Decimal number, $N = 783.45$ and $r = 10$, so base is 10
Number of digits in integer part, $n = 3$
Number of digits in fractional part, $m = 2$

$$\begin{aligned} \text{9's complement of } 783.45 &= 10^3 - 10^{-2} - 783.45 \\ &= 1000 - 0.01 - 783.45 \\ &= 999.99 - 783.45 = 216.54 \quad \textbf{Ans.} \end{aligned}$$

> ***Rule:*** $(r - 1)$'s complement of a base-r number is obtained by subtracting each digit of base-r number from $(r - 1)$ or the largest digit from set of base-r distinct digits.

For instance, 9's complement of 783.45 can be obtained as follow:

$$(N)_{10} = 783.45$$

Nine's complement of $(N)_{10} = 216.54$

$(\because\ 9 - 7 = 2, 9 - 8 = 1, 9 - 3 = 6, 9 - 4 = 5 \text{ and } 9 - 5 = 4)$

Two's complement of base-3 number $(1002.221)_3$ is $(1220.001)_3$

$(\because\ 2 - 1 = 1, 2 - 0 = 2, 2 - 0 = 2, 2 - 2 = 0, 2 - 2 = 0, 2 - 2 = 0, 2 - 1 = 1)$

Three's complement of base-4 number $(1230.023)_4$ is $(2103.310)_4$

$(\because\ 3 - 1 = 2, 3 - 2 = 1, 3 - 3 = 0, 3 - 0 = 3, 3 - 0 = 3, 3 - 2 = 1, 3 - 3 = 0)$

Four's complement of base-5 number $(2130.423)_5$ is $(2314.021)_5$

$(\because\ 4 - 2 = 2, 4 - 1 = 3, 4 - 3 = 1, 4 - 0 = 4, 4 - 4 = 0, 4 - 2 = 2, 4 - 3 = 1)$

Five's complement of base-6 number $(2345.201)_6$ is $(3210.354)_6$

$(\because\ 5 - 2 = 3, 5 - 3 = 2, 5 - 4 = 1, 5 - 5 = 0, 5 - 2 = 3, 5 - 0 = 5, 5 - 1 = 4)$

Six's complement of base-7 number $(2345.231)_7$ is $(4321.435)_7$

$(\because\ 6 - 2 = 4, 6 - 3 = 3, 6 - 4 = 2, 6 - 5 = 1, 6 - 2 = 4, 6 - 3 = 3, 6 - 1 = 5)$

EXAMPLE 2.8 Find radix-1 complement of the following given base number.

(a) $(3425)_7$ (b) $(423.41)_5$

SOLUTION

(a) Base-7 number, $N = (3425)_7$ and $r - 1 = 6$, as base-r is 7
Number of digits in integer part, $n = 4$
Number of digits in fractional part, $m = 0$
Six's complement of $(3475)_7 = (7^4)_7 - (7^{-0})_7 - (3425)_7$
$= (10000 - 1)_7 - (3425)_7$
$= (6666)_7 - (3425)_7$
$= (3241)_7$ **Ans.**

(b) Base-5 number, $N = (423.41)_5$ and base, $r = 5$, so base-1 is 4
Number of digits in integer part, $n = 3$
Number of digits in fractional part, $m = 2$
Four's complement of $(423.41)_5 = \left(5^3 - 5^{-2}\right)_5 - (423.41)_5$
$= (1000.00 - 0.01)_5 - (423.41)_5$
$= (444.44 - 423.41)_5$
$= (021.03)_5$ **Ans.**

2.1.4.2 | *r's* Complement (True Complement)

Given a positive number N in base r:

$$r\text{'s complement of } N = \begin{cases} (r^n)_r - (N)_r & \text{for } N \neq 0 \\ 0 & \text{for } N = 0 \end{cases} \tag{2.4}$$

where r is base
N is positive number in base r.
n is number of digits in integer part.

EXAMPLE 2.9 Find 10's complement of the following decimal number.

(a) 3475 (b) 783.45 (c) 0.2345

SOLUTION

(a) Base, $r = 10$, Number of integer digits, $n = 4$, Number, $N = 3475$.
Zero's complement of $3475 = 10^4 - 3475$
$= 10000 - 3475$
$= (9999 + 1) - 3475$
$= (9999 - 3475) + 1$
$= 6524 + 1$
$= 6525$ **Ans.**

> ***Note:*** In case, no base is mentioned it means number is decimal and base is 10.

(b) Base, $r = 10$, Number of integer digits, $n = 3$, Number, $N = 783.45$
Ten's complement of $783.45 = 10^3 - 783.45$

$= 1000.00 - 783.45$
$= (999.99 - 0.01) - 783.45$
$= (999.99 - 783.45) + 0.01$ => (Note)
$= 216.54 + 0.01$
$= 216.55$ **Ans.**

> ***Rule:*** *r*'s complement of a base-*r* number is obtained by adding 1 (one) to least significant digit (LSD) to its (*r* – 1)'s complement

(c) Base, $r = 10$, Number of integer digits, $n = 0$,
Number, $N = -0.2345$
Ten's complement of $0.2345 = 10^0 - 0.2345$
$= 1.0000 - 0.2345$
$= (0.9999 + 0.0001) - 0.2345$
$= (0.9999 - 0.2345) + 0.0001$ => (Note)
$= 0.7654 + 0.0001$
$= 0.7655$ **Ans.**

EXAMPLE 2.10 Find true complement of the following given base number.

(a) $(3425)_6$ (b) $(211.11)_3$ (c) $(0.234)_5$

SOLUTION

(a) Base, $r = 6$, Number of integer digits, $n = 4$, Base-6 number, $N = (3425)_6$.
Six's complement of $(3425)_6 = (6^4)_6 - (3425)_6$
$= (10000)_6 - (3425)_6$
$= (5555 + 1)_6 - (3425)_6$
$= (5555 - 3425)_6 + (1)_6$
$= (2130 + 1)_6$
$= (2131)_6$ **Ans.**

> ***Rule:*** *r*'s complement of a base-*r* number is obtained by adding 1 (one) to least significant digit (LSD) to its (*r* – 1)'s complement.

(b) Base, $r = 3$, Number of integer digits, $n = 3$, Number, $N = (211.11)_3$
Three's complement of $(211.11)_3 = (3^3)_3 - (211.11)_3$
$= (1000.00)_3 - (211.11)_3$
$= (222.22 - 0.01)_3 - (211.11)_3$
$= (222.22 - 211.11)_3 + (0.01)_3$ => (Note)
$= (011.11)_3 + (0.01)_3$
$= (011.12)_3$ **Ans.**

(c) Base, $r = 5$, Number of integer digits, $n = 0$, Number, $N = (0.234)_5$
Five's complement of $(0.234)_5 = (5^0)_5 - (0.234)_5$
$= (1.000)_5 - (0.234)_5$
$= (0.444 + 0.001)_5 - (0.234)_5$
$= (0.444 - 0.234)_5 + (0.001)_5$ => (Note)
$= (0.210 + 0.001)_5$
$= (0.211)_5$ **Ans.**

EXAMPLE 2.11 Find 10's complement of the following decimal numbers.

(a) 4609 (b) 1506.047

SOLUTION

(a) Number, $(N)_{10} = 4609$
Nine's complement of $(4609)_{10} = 5390$ $(9 - 4 = 5; 9 - 6 = 3; 9 - 0 = 9; 9 - 9 = 0)$
Ten's complement of $(4609)_{10} = 5390 + 1$
$= 5391$ **Ans.**

(b) Number, $(N)_{10} = 1506.049$
Nine's complement of $(1506.049)_{10} = 8493.950$ $(9 - 1 = 8; 9 - 5 = 4; 9 - 0 = 9; 9 - 6 = 3; 9 - 0 = 9; 9 - 4 = 5; 9 - 9 = 0)$
Ten's complement of $(1506.049)_{10} = 8493.950 + 0.001$
$= 8493.951$ **Ans.**

EXAMPLE 2.12 Find true complement of the following given base numbers.

(a) $(4302)_5$ (b) $(1506.042)_7$

SOLUTION

(a) Four's complement of $(4302)_5 = (0142)_5$ $(4 - 4 = 0; 4 - 3 = 1; 4 - 0 = 4; 4 - 2 = 2)$
Five's complement of $(4302)_5 = (0142)_5 + (0001)_5$
$= (0143)_5$ **Ans.**

(b) Six's complement of $(1506.042)_7 = (5160.624)_7$ $(6 - 1 = 5; 6 - 5 = 1, 6 - 0 = 6; 6 - 6 = 0; 6 - 6 = 0; 6 - 2 = 4; 6 - 4 = 2)$
Seven's complement of $(1506.042)_7 = (5160.624)_7 + (0.001)$
$= (5160.625)_7$ **Ans.**

2.1.5 | Base-*r* Subtraction using Complement

Complements are used to perform subtraction and to represent signed numbers. Positive numbers are known as unsigned numbers. Signed numbers can be positive or negative numbers. Moreover, subtraction is performed by doing addition using complement.

2.1.5.1 | Subtraction with (*r* – 1)'s Complement

To perform subtraction of $(M - N)_r$, Algorithm 2.3 is given below. Let $(M)_r$ and $(N)_r$ both are positive binary numbers.

Consider each base-*r* number is of *n*-digits, the value in $(n + 1)^{th}$ digit is considered as end carry.

Algorithm 2.3 | Subtraction using (*r* – 1)'s complement method

Step 1: Take (*r* – 1)'s complement of the subtrahend, *N*. Subtract each digit from (*r* – 1).
Step 2: Add minuend $(M)_r$ and (*r* – 1)'s complement of the subtrahend, $(N)_r$ obtained in Step 1.

(*Continued*)

Algorithm 2.3 | (Continued)

Step 3: Check end carry obtained after Step 2.
3.1 If an end carry occurs, then add '1' to the least significant digit of the number obtained after Step 2 and place positive sign (+ve).
3.2 If an end carry does not occur then take $(r - 1)$'s complement of the number obtained in Step 2 and place negative sign (–ve).

EXAMPLE 2.13 Perform subtraction of given base numbers using $(r - 1)$'s complement.

(a) $(110 - 10)_5$ (b) $(111 - 0.111)_7$

SOLUTION

> **Note:** Number of digits of Minuend *M* and subtrahend *N* must be same. Otherwise, complete by filling zero(s). Weightage of digit should remain same.

(a) Make three digits of base-5 numbers, $(M)_5$ and $(N)_5$ each.

$(M)_5 = 1\ 1\ 0$
$(N)_5 = 0\ 1\ 0$

Take 4's complement of base-5 number, *N* and add it to base-5 number, *M*.
Take 4's complement $(010)_5 = (434)_5$

$(M)_5 =$		1	1	0
4's complement of $(N)_5 =$		4	3	4
$(M)_5 +$ (4's complement of $(N)_5$) =	1	0	4	4

End carry (digit in 4th digit after addition) is one, so add 1 to LSD of the sum, $(011)_5$ and place positive sign.

Carry over		1	1	0
$(M)_5 +$ (4's complement of $(N)_5$) =		0	4	4
+		0	0	1
$(M)_5 - (N)_5 =$	+	1	0	0 **Ans.**

(b) Make three integer digits and three fractional digit of base-7 numbers, $(M)_7$ and $(N)_7$ each.

$(M)_7 = 1\ 1\ 1\ .\ 0\ 0\ 0$
$(N)_7 = 0\ 0\ 0\ .\ 1\ 1\ 1$

Take 6's complement of, $(N)_7$. and add it to $(M)_7$.
Six's complement of $(000.111)_7 = (666.555)_7$

Carry over		1	1					
$(M)_7 =$		1	1	1	.	0	0	0
6's complement of $(N)_7 =$		6	6	6	.	5	5	5
$(M)_7 +$ (6's complement of $(N)_7$) =	1	1	1	0	.	5	5	5

End carry (digit in seventh digit after addition) is one. So add 1 to LSD of the sum, $(110.555)_7$ and place positive sign

$(M)_7 + (6\text{'s complement of } (N)_7) =$		1	1	0	.	5	5	5	
	+	0	0	0	.	0	0	1	
$(M)_7 - (N)_7 =$	+	1	1	0	.	5	5	6	**Ans.**

2.1.5.2 | Subtraction with *r*'s Complement

To perform the subtraction of two positive base-*r* numbers, $(M - N)_r$ using *r*'s complement is given below as Algorithm 2.4.

Algorithm 2.4 | Subtraction using *r*'s complement method

Step 1: Take *r*'s complement of subtrahend, $(N)_r$. Subtract each digit from $(r - 1)$ and add 1 to LSD.
Step 2: Add the minuend, $(M)_r$ and *r*'s complement of subtrahend, $(N)_r$ obtained in Step 1.
Step 3: Check the result obtained in Step 2 for an end carry
3.1 If an end carry occurs then discard it and place positive sign (+ve)
3.2 If an end carry does not occur then take *r*'s complement of the number obtained in Step 2 and place negative sign (–ve)

EXAMPLE 2.14 Perform subtraction of given base number using *r*'s complement.

(a) $(110 - 10)_6$ (b) $(111 - 0.111)_9$

SOLUTION

(a) Make three digits of $(M)_6$ and $(N)_6$ each.

$(M)_6 =$ 1 1 0
$(N)_6 =$ 0 1 0

Take 6's complement of $(N)_6$ and add it to $(M)_6$.
Five's complement of $(010)_6 = (545)_6$
Six's complement of $(010)_6 = (550)_6$

Carry over	1	1		
$(M)_6 =$		1	1	0
6's complement of $(N)_6 =$		5	5	0
$(M)_6 + (6\text{'s complement of } (N)_6 =$	1	1	0	0

End carry (digit in 4th digit after addition) is one, so discard and place positive sign before the sum.

$(M)_6 - (N)_6 = +$ 1 0 0 **Ans.**

(b) Make the three integer digits and three fractional digit of $(M)_9$ and $(N)_9$ each.

$$(M)_9 = 1\ 1\ 1\ \cdot\ 0\ 0\ 0$$
$$(N)_9 = 0\ 0\ 0\ \cdot\ 1\ 1\ 1$$

Take 9's complement of $(N)_9$. and add it to $(M)_9$.
Eight's complement of $(000.111)_9 = (888.777)_9$
Nine's complement of $(000.111)_9 = (888.778)_9$

Carry over	1	1	1				
$(M)_9 =$		1	1	1 ·	0	0	0
Nine's complement of $(N)_9 =$		8	8	8 ·	7	7	8
$(M)_9 +$ (9's complement of $(N)_9$) $=$	1	1	1	0 ·	7	7	8

End carry (digit in seventh digit after addition) is one, so discard it and place positive sign

$$(M)_9 - (N)_9 = +\ 1\ 1\ 0\ \cdot\ 7\ 7\ 8$$ **Ans.**

2.2 | BINARY NUMBER SYSTEM

The binary number system is the language of the digital computer. Its primary advantage is that it requires only two digits: 1 and 0. The binary system is a positional weighted system.

- The base or radix of binary system is 2.
- The symbols or digits in binary number system are 0 and 1.

Number of binary digits can be combined to get a number. For example $(1111)_2$: four binary digits are combined to get this number. The binary number can also be represented as 1111 B. B denotes for binary number system. More examples of binary number are 101101B, $(1001.101)_2$, 00110.1101 B.

Binary system is highly advantageous for computer operations. But it is difficult to recognize the binary data in simpler way by the people using the computer. The decimal number $(1623)_{10}$ is one that a person can read once and both remember and conceive of as a size of number. So, in decimal system it is simple to compare. The binary number of value equal to decimal number 1623 D is 11001010111 B, while 1298 D in decimal is equal to 10100010010 B in binary. It is difficult to compare both numbers in binary. So, the result of a binary computation must be converted to decimal before being read out to the operator.

'bit' is a binary digit. Combined 4 bits form a nibble and 8 bits form a byte. Combined 16 bits form 16-bit word. Similarly, combined 32-bit form 32-bit word.

1 bit is 1 binary digit.
1 nibble is 4-bit binary data.
1 byte is 8-bit binary data
1 word is 16, 32 or 64-bit binary data.

On combining 10 bits of binary number, $1\,\text{k (kilo)} = 2^{10} \Leftrightarrow 1.024 \times 10^3$, 10-bit binary words are obtained.

On combining 20 bits of binary number, $1\,\text{M(Mega)} = 2^{20} \Leftrightarrow 1.048\,576 \times 10^6$, 20-bit binary words are obtained.

On combining 30 bits of binary number, $1\,\text{G (Giga)} = 2^{30} \Leftrightarrow 1.073741824 \times 10^9$, 30-bit binary words are obtained.

On combining 40 bits of binary number, $1\,\text{T (Tera)} = 2^{40} \Leftrightarrow 1.099511627776 \times 10^{12}$, 40-bit binary words are obtained.

2.2.1 | Binary to Decimal Conversion

A binary number can be converted to decimal by the positional weights method. Each digit of the number is multiplied by its position weight, $\left(d_n \times 2^{n-1}\right)$ and the product terms are added to find the decimal number. Mathematically, it is represented by:

$$(N)_{10} = \sum_{i=1}^{n} d_i 2^{i-1} + \sum_{j=1}^{m} d_{-j} 2^{-j} \tag{2.5}$$

where d_i is i^{th} digit of integer portion in binary number.
i is the position of binary digit in integer portion.
2^{i-1} is the weight assigned to the position of ith binary digit.
d_{-j} is j^{th} digit of fractional portion in binary number (if any)
j is the position of binary digit in fractional portion.
2^{-j} is the weight assigned to the position of jth binary digit in fractional portion.

- Position of binary digits is always counted from decimal point.
- The position of digit is counted from right to left, for integer portion starting from decimal point.
- The position of digit is counted from left to right for fractional portion starting from decimal point.

EXAMPLE 2.15 Convert $(11011.\,101)_2$ to decimal.

SOLUTION Positional weight method is applied

	16	8	4	2	1		0.5	0.25	0.125
Weight to digits ⇒	2^4	2^3	2^2	2^1	2^0		2^{-1}	2^{-2}	2^{-3}
Position of digits ⇒	5th	4th	3rd	2nd	1st		1st	2nd	3rd
Digits ⇒	1	1	0	1	1	.	1	0	1
	d_5	d_4	d_3	d_2	d_1		d_{-1}	d_{-2}	d_{-3}

$$(N)_{10} = \underbrace{\sum_{i=1}^{5} d_i 2^{i-1}}_{\text{Integer portion}} + \underbrace{\sum_{j=1}^{3} d_{-j} 2^{-j}}_{\text{Fractional portion}}$$

$$(N)_{10} = 1 \times 2^0 + 1 \times 2^1 + 0 \times 2^2 + 1 \times 2^3 + 1 \times 2^4 + 1 \times 2^{-1} + 0 \times 2^{-2} + 1 \times 2^{-3}$$

$$(N)_{10} = (1 + 2 + 8 + 16) + \left(\frac{1}{2} + \frac{1}{8}\right)$$

$(N)_{10} = (27.625)_{10}$ **Ans.**

EXAMPLE 2.16 Convert 11001010111B to decimal.

SOLUTION The binary number has no fraction position. Decimal point is after the LSB, i.e. right-most end.

	1024	512	256	128	64	32	16	8	4	2	1
Weight to digits ⇒	2^{10}	2^9	2^8	2^7	2^6	2^5	2^4	2^3	2^2	2^1	2^0
Position of digits ⇒	11th	10th	9th	8th	7th	6th	5th	4th	3rd	2nd	1st
Digits ⇒	1	1	0	0	1	0	1	0	1	1	1
	d_{11}	d_{10}	d_9	d_8	d_7	d_6	d_5	d_4	d_3	d_2	d_1

$$(N)_{10} = \sum_{i=1}^{11} d_i 2^{i-1}$$

$$(N)_{10} = 1\times2^0 + 1\times2^1 + 1\times2^2 + 0\times2^3 + 1\times2^4 + 0\times2^5 + 1\times2^6 + 0\times2^7 + 0\times2^8 + 1\times2^9 + 1\times2^{10}$$

$$(N)_{10} = (1+2+4+16+64+512+1024)$$

$$(N)_{10} = (1623)_{10} \quad \textbf{Ans.}$$

EXAMPLE 2.17 Convert 0.10011 B to decimal.

SOLUTION

	1		0.5	0.25	0.125	0.0625	0.03125
Weight to digits ⇒	2^0		2^{-1}	2^{-2}	2^{-3}	2^{-4}	2^{-5}
Position of digits ⇒	1st		1st	2nd	3rd	4th	5th
Digits ⇒	0	.	1	0	0	1	1
	d_1		d_{-1}	d_{-2}	d_{-3}	d_{-4}	d_{-5}

$$(N)_{10} = \sum_{i=1}^{1} d_i 2^{i-1} + \sum_{j=1}^{5} d_{-j} 2^{-j}$$

$$(N)_{10} = 0\times2^0 + 1\times2^{-1} + 0\times2^{-2} + 0\times2^{-3} + 1\times2^{-4} + 0\times2^{-5}$$

$$(N)_{10} = \left(\frac{1}{2} + \frac{1}{16} + \frac{1}{32}\right)$$

$$(N)_{10} = (0.5 + 0.0625 + 0.03125)$$

$$(N)_{10} = (0.59375)_{10} \quad \textbf{Ans.}$$

2.2.2 | Decimal to Binary Conversion

Binary number has two parts. One is integer part and another is fractional part. Both the parts of decimal number are converted separately in to binary number. So, any decimal number, conversion into binary (base-2) is carried out separately for integer portion and fractional portion of decimal number and then both conversions are combined to get the number.

Conversion of integer term of decimal number to binary number The method of conversion from decimal to binary is to divide successively by 2 and resulting remainders are the digits for binary number and quotient are used for successive operation. First division gives remainder as LSB of binary number. The process is repeated till quotient becomes zero. The arithmetic process is elaborated in Algorithm 2.1.

For example, integer $(41)_{10}$ is equivalent to $(101001)_2$ binary number and is detailed below:

Dividend	÷	Divisor	=	Quotient	: Remainder	Binary digits
41	÷	2	=	20	: $1 \rightarrow d_1$ (LSB)	$d_1 = 1$
20	÷	2	=	10	: $0 \rightarrow d_2$	$d_2 = 0$
10	÷	2	=	5	: $0 \rightarrow d_3$	$d_3 = 0$
5	÷	2	=	2	: $1 \rightarrow d_4$	$d_4 = 1$
2	÷	2	=	1	: $0 \rightarrow d_5$	$d_5 = 0$
1	÷	2	=	0	: $1 \rightarrow d_6$ (MSB)	$d_6 = 1$

$$\text{Binary number} = (d_6d_5d_4d_3d_2d_1)_2$$
$$(41)_{10} = (101001)_2$$

Conversion of fractional term of decimal number to binary number The conversion of a decimal fraction to binary is to multiply successively by 2 and use integer part of the number as the digits for binary number and remained fraction part is taken for successive operation. The process is repeated till the fraction becomes zero. The arithmetic process is given in Algorithm 2.2. Conversion of decimal number $(0.25)_{10}$ into binary number is detailed below.

Multiplicand	×	2	=	Integer	+	Fraction	Binary digit
0.25	×	2	=	0	+	.50	$d_{-1} \rightarrow 0$
0.50	×	2	=	1	+	.00	$d_{-2} \rightarrow 1$

$$\text{Binary number} = (0.d_{-1}d_{-2})_2$$
$$(0.25)_{10} = (0.01)_2$$

EXAMPLE 2.18 Convert the following decimal number to binary:

(a) 128 (b) 19.56

SOLUTION

(a) Decimal number is $(128)_{10}$. Successive division is performed by dividing the number by 2.

Dividend	÷	Divisor	=	Quotient	: Remainder	Binary digits
128	÷	2	=	64	: $0 \rightarrow d_1$ (LSB)	$d_1 \rightarrow 0$
64	÷	2	=	32	: $0 \rightarrow d_2$	$d_2 \rightarrow 0$
32	÷	2	=	16	: $0 \rightarrow d_3$	$d_3 \rightarrow 0$
16	÷	2	=	8	: $0 \rightarrow d_4$	$d_4 \rightarrow 0$

8	÷	2	=	4	$: 0 \rightarrow d_5$	$d_5 \rightarrow 0$
4	÷	2	=	2	$: 0 \rightarrow d_6$	$d_6 \rightarrow 0$
2	÷	2	=	1	$: 0 \rightarrow d_7$	$d_7 \rightarrow 0$
1	÷	2	=	0	$: 1 \rightarrow d_8$ (MSB)	$d_8 \rightarrow 1$

Binary number = $(d_8d_7d_6d_5d_4d_3d_2d_1)_2$

$(128)_{10} = (10000000)_2$ **Ans.**

(b) Integer portion of decimal number 19.56 = 19

Dividend	÷	Divisor	=	Quotient	: Remainder	Binary digits
19	÷	2	=	9	$: 1 \rightarrow d_1$ (LSB)	$d_1 \rightarrow 1$
9	÷	2	=	4	$: 1 \rightarrow d_2$	$d_2 \rightarrow 1$
4	÷	2	=	2	$: 0 \rightarrow d_3$	$d_3 \rightarrow 0$
2	÷	2	=	1	$: 0 \rightarrow d_4$	$d_4 \rightarrow 0$
1	÷	2	=	**0**	$: 1 \rightarrow d_5$	$d_5 \rightarrow 1$

Binary number = $(d_5d_4d_3d_2d_1)_2$

$(19)_{10} = (10011)_2$

Fractional portion of decimal number 19.56 = 0.56

Multiplicand	×	2	=	Integer	+	Fraction	Binary digit
0.56	×	2	=	1	+	.12 (MSB)	$d_{-1} \rightarrow 1$
0.12	×	2	=	0	+	.24	$d_{-2} \rightarrow 0$
0.24	×	2	=	0	+	.48	$d_{-3} \rightarrow 0$
0.48	×	2	=	0	+	.96	$d_{-4} \rightarrow 0$
0.96	×	2	=	1	+	.92	$d_{-5} \rightarrow 1$
0.92	×	2	=	1	+	.84	$d_{-6} \rightarrow 1$

Binary number = $(0.d_{-1}d_{-2}d_{-3}d_{-4}d_{-5}d_{-6})_2$

$(0.56)_{10} = (0.100011)_2$

Note: Fractional part will never become zero, so 0.56 fraction cannot be expressed exactly in binary equivalent. In such a case, the process of multiplication can be terminated after certain significant digits giving desired accuracy or the successive fraction number starts repeating.

Combining integer and fractional part of number 15.56, to get binary number $(19.56)_{10} = (10011.100011)_2$ **Ans.**

2.3 | BINARY ARITHMETIC

Binary addition and subtraction operation are commonly used in all type of microprocessors, microcontrollers, computers etc. These operations are performed in a manner similar to decimal arithmetic.

2.3.1 | Binary Addition

Generally, addition in binary is performed in two steps; (i) Add carry to augend bit to get partial sum (ii) add addend to partial sum to get the final sum and carry-over in previous step must be taken care. Start operation from LSB (least significant bit). Consider 0 as previous carry for first bit (binary digit). The number of bits for augend and addend must be same otherwise if not, complete by filling zero(s) in such a manner so Single digit addition is given below.

Augend	0	0	1	1
Addend	+ 0	+ 1	+ 0	+ 1
Sum	0 0	0 1	0 1	1 0
	Carry ↑	Carry ↑	Carry ↑	Carry ↑

In case, numbers of bits are combined to represent a number. Carry over are performed in the same manner as in decimal arithmetic. Decimal and binary addition is elaborated with an example that adds $(15)_{10}$ and $(26)_{10}$.

Decimal number

Digits	2nd	1st
Previous Carry	1	
Augend	1	5
Addend	2	6
Sum	4	1

Binary number

Binary digits	6th	5th	4th	3rd	2nd	1st
Previous carry		1	1	1	0	
Augend		0	1	1	1	1
Addend		1	1	0	1	0
Sum	1	0	1	0	0	1

Explanation:

In first bit column: $0 + 1 = 1$ with no carry over to second bit column.

In second bit column:

(i) Carry-over from first bit is added to augend of second bit, i.e. $0 + 1 = 1$ with no carry over to next bit column.

(ii) Addend of second bit is further added to partial sum obtained in step i, i.e. $1 + 1 = 10$. One carry-over is for third bit.

In third bit column:

(i) Carry-over from second bit is added to augend of third bit, i.e. $1 + 1 = 10$. One carry-over is for next bit column.

(ii) Addend of third bit is further added to partial sum obtained in step i, i.e. $0 + 0 = 0$. No carry-over to next bit but one carry over has already been obtained in step (i) to fourth bit column.

In fourth bit column:

(i) Carry-over from third bit is added to augend of fourth bit, i.e. $1 + 1 = 10$. One carry-over is for next bit column.

(ii) Addend of fourth bit is further added to partial sum obtained in step i, i.e. $0 + 1 = 1$ with no carry-over to next bit but one carry-over has already been obtained in step i, to fifth bit column.

In fifth bit column:

(i) Carry-over from fourth bit is added to augend of fifth bit, i.e. 1 + 0 = 1 with no carry-over to next bit column.
(ii) Addend of fifth bit is further added to partial sum obtained in step i, i.e. 1 + 1 = 0 with one carry-over to next bit, i.e. sixth bit column.

Bitwise operation is explained below in tabular form.

	6th	5th	5th	4th	4th	3rd	3rd	2nd	2nd	1st
First step										
Carry from previous bit		1		1		1		0		0
Augend	+	0	+	1	+	1	+	1	+	1
Partial Sum	0	1	1	0	1	0	0	1	0	1
Second step										
Partial Sum	0	1	1	0	1	0	0	1	0	1
Addend	+	1	+	1	+	0	+	1	+	0
Sum	1	0	1	1	1	0	1	0	0	1

Addition of first bit column generates carry for second bit. Addition of second bit column generates carry for third bit and so on.

EXAMPLE 2.19 Add the binary numbers 1101.101 and 111.011.

SOLUTION Start operation of addition from LSB (least significant bit). LSB is considered as first bit column.

Binary digit (bit) column number	8th	7th	6th	5th	4th		3rd	2nd	1st
Carry from previous column (bit)		1	1	1	1		1	1	
Augend		1	1	0	1	.	1	0	1
Addend	+	0	1	1	1	.	0	1	1
Sum	1	0	1	0	1	.	0	0	0

2.3.2 | Binary Subtraction

The binary subtraction is performed in a manner similar to that in decimal subtraction. Borrow taken in decimal system adds 10 (base value) whereas in binary the borrow taken adds 2 (base value). Single bit subtraction is performed and is given below.

Borrow							–1	10
Minuend		0		1		1		0
Subtrahend	–	0	–	1	–	0	–	1
Difference	0	0	0	0	0	1	1	1
	↑ Borrow		↑ Borrow		↑ Borrow		↑ Borrow	

- In first three possible cases, there is no need of borrow from next bit.
- In the fourth case, there is a need of borrow from next bit.
- Borrow is taken from the next digit as 2 (base value).
- Taken borrow is added to minuend and sum is obtained (2 + 0 = 2).
- Finally, subtrahend is subtracted from partial sum to get the difference, i.e. (2 − 1 = 1).
- Borrow taken by previous digit column is subtracted from minuend of present digit column.

To memorize

0 – 0 = 0 with no borrow
1 – 1 = 0 with no borrow
1 – 0 = 1 with no borrow
0 – 1 = 1 with one borrow from next higher place valued bit.

Decimal and binary subtraction is elaborated with an example.

	Decimal number	
Digits	2nd	1st
Borrow	–1	10
Minuend	1	2
Subtrahend	0	6
Difference	0	6

	Binary number				
Binary digits	5th	4th	3rd	2nd	1st
Borrow		–1	–1		
Minuend		1	1	0	0
Subtrahend		0	1	1	0
Difference		0	1	1	0

Explanation:
In first digit column: 0 – 0 = 0
In second digit column: 0 – 1 = 1 with one borrow from third bit.
In third digit column:

(i) Borrow taken by second digit is subtracted from minuend of third bit, i.e. 1 – 1 = 0 with no borrow.
(ii) Subtrahend of third bit is further subtracted from difference obtained in step (i), i.e. 0 – 1 = 1 with one borrow from fourth bit.

In fourth digit column:

(i) Borrow taken by third digit is subtracted from minuend of fourth bit, i.e. 1 – 1 = 0 with no borrow.
(ii) Subtrahend of fourth bit is further subtracted from difference obtained in step (i), i.e. 0 – 0 = 0 with no borrow.

EXAMPLE 2.20 Subtract $(111.111)_2$ from $(1010.11)_2$.

SOLUTION

Binary digit (bit) column number	8th	7th	6th	5th	4th		3rd	2nd	1st
Borrow from next column (bit)		–1	–1	–1	–1		–1	–1	
Minuend		1	0	1	0	.	1	1	0
Subtrahend	–	0	1	1	1	.	1	1	1
Difference		0	0	1	0	.	1	1	1

Explanation:
In first digit column: 0 – 1 = 1 with one borrow from second bit.
In second digit column:

(i) Borrow taken for first digit is subtracted from minuend of second bit, i.e. 1 – 1 = 0 with no borrow.
(ii) Subtrahend of second bit is further subtracted from difference obtained in step (i), i.e. 0 – 1 = 1 with one borrow from third bit.

In third digit column:

(i) Borrow taken for second digit is subtracted from minuend of third bit, i.e. 1 – 1 = 0 with no borrow.
(ii) Subtrahend of third bit is further subtracted from difference obtained in step (i), i.e. 0 – 1 = 1 with one borrow from fourth bit.

In fourth digit column:

(i) Borrow taken by third digit is subtracted from minuend of fourth bit, i.e. 0 – 1 = 1 with one borrow from fifth bit.
(ii) Subtrahend of fourth bit is further subtracted from difference obtained in step (i), i.e. 1 – 1 = 0 with no borrow but already has been taken from fifth bit in step 1.

2.4 | SIGNED NUMBERS

Unsigned numbers are those numbers which can represent only positive numbers. In all the number system both positive and negative numbers are possible. Signed numbers are those numbers which represents both positive and negative number. To represents signed numbers there are different methods

1. Sign magnitude representation
2. One's complement representation
3. Two's complement representation

In computation, design of computers prefers complement representation to present signed numbers. Complement method is used to perform subtraction where difference is obtained by performing addition and sign is also taken care simultaneously.

2.4.1 | Sign Magnitude Representation

In sign magnitude representation (Figure 2.1) an extra bit is reserved at the MSB to represent sign of that number. This additional bit is known as sign bit. Binary 1 is by convention used to represent minus sign and binary 0 is used for plus sign.

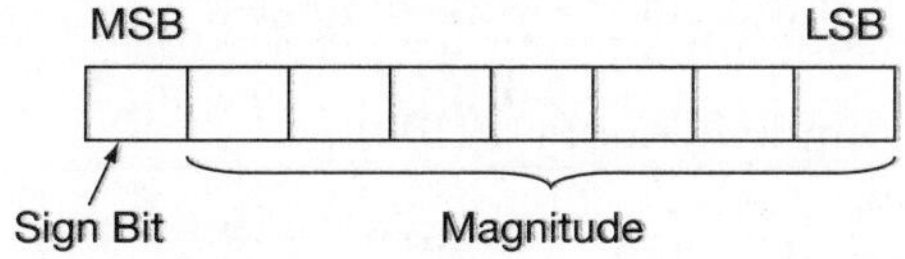

FIGURE 2.1 | Sign magnitude representation

A n-bit signed binary number the most significant bit (MSB) represents the sign of the number and remaining $(n - 1)$ bits express the magnitude of the number in binary. For example, consider an 8-bit binary number in sign magnitude representation.

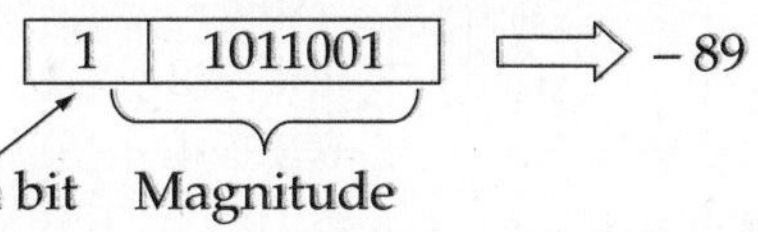

In general, the sign magnitude representation can accommodate numbers in the range

$$-\left(2^{n-1}-1\right) \text{ to} +\left(2^{n-1}-1\right)$$

where n is the number of bits combined to represent a binary number.

Table 2.4 presents signed numbers in all the three forms. Two's complement present only one value for positive and negative zero hence is preferred than other methods.

TABLE 2.4 | Signed magnitude representation

Decimal Signed Integer	Signed Magnitude Representation	One's Complement Representation	Two's Complement Representation
+0	0 000	0 000	0 000
−0	1 000	1 111	0 000
+3	0 011	0 011	0 011
−3	1 011	1 100	1 101
+13	0 1101	0 1101	0 1101
−13	1 1101	1 0010	1 0011
+127	0 111 1111	0 111 1111	0 111 1111
−127	1 111 1111	1 000 0000	1 000 0001

2.4.2 | One's Complement (Radix-minus-one Complement)

One's complement of binary number N is defined as $((2^n)_2 - (2^{-m})_2 - (N)_2)$. (2.6)

where N is a positive binary number with base 2.

n is number of bits in integer part of binary number N.

m is number of bits in fractional part of binary number N.

In general, the sign magnitude representation can accommodate numbers in the range

$$-\left(2^{n-1}-1\right) \text{ to} +\left(2^{n-1}-1\right)$$

where n is the number of bits combined to represent a binary number. Most significant binary digit (MSB) is considered as sign bit. 0 bit represents positive binary number and 1 bit represents for negative numbers.

EXAMPLE 2.21 Find 1's complement of $(101100)_2$.

SOLUTION Binary number $N = 101100$

Number of integer binary digits, $n = 6$

Number of fractional digits, $m = 0$ (no fractional part)

One's complement of $(101100)_2 = (2^6 - 2^{-0})_2 - (101100)_2$

$= (1000000 - 1)_2 - (101100)_2$

$= (111111 - 101100)_2$

$= 010011$ **Ans.**

	−1	−1	−1	−1	−1	−1	
$2^6 =$	1	0	0	0	0	0	0
$-2^{-0} =$ −	0	0	0	0	0	0	1
$2^6 - 1 =$	0	1	1	1	1	1	1

$2^6 - 1 =$		1	1	1	1	1	1
$-N$	−	1	0	1	1	0	0
$(2^6 - 1) - N =$		0	1	0	0	1	1

EXAMPLE 2.22 Find 1's complement of $(0.0110)_2$.

SOLUTION Binary number, $N = 0.0110$

Number of integer binary digits, $n = 0$ (no integer part)

Number of fractional digits $m = 4$

One's complement of $(0.0110)_2 = (2^0 - 2^{-4}) - 101100$

$= (1 - 0.0001) - 0.0110$

$= 0.1111 - 0.\ 0110$

$= 0.1001$ **Ans.**

		−1	−1	−1	−1	−1
$2^0 =$		1	.	0	0	0	0
-2^{-4}	−	0	.	0	0	0	1
$1 - 2^{-4} =$		0	.	1	1	1	1
$1 - 2^{-4} =$		0	.	1	1	1	1
$-N$	−	0	.	0	1	1	0
$(2^0 - 2^{-4}) - N =$		0	.	1	0	0	1

EXAMPLE 2.23 Find 1's complement of $(1010.011)_2$.

SOLUTION Binary Number, $N = 1010.011$

Number of integer binary digits, $n = 4$

Number of fractional digits, $m = 3$

One's complement of $(1010.011)_2 = (2^4 - 2^{-3}) - 1010.011$

$= (10000 - 0.001) - 1010.011$

$= 1111.111 - 1010.011$

$= 0101.100$ **Ans.**

> ***Alternate method to find 1's complement***
> - One's complement of a binary number is obtained by subtracting each bit of binary number from 1 (radix – 1). (1 – 0 = 1 and 1 – 1 = 0)
> - Simple way to get 1's complement of a binary number is to change each bit of binary number from '1' to '0' and '0' to '1'.

The 1's complement of 10101.0101 can be obtained as:

$$
\begin{array}{rcccccccccc}
N = & 1 & 0 & 1 & 0 & 1 & . & 0 & 1 & 0 & 1 \\
& \downarrow & \downarrow & \downarrow & \downarrow & \downarrow & & \downarrow & \downarrow & \downarrow & \downarrow \\
\text{One's complement of } N = & 0 & 1 & 0 & 1 & 0 & . & 1 & 0 & 1 & 0
\end{array}
$$

2.4.3 | Two's Complement (True Complement)

Given a positive binary number N in base 2:

$$\text{Two's complement of binary number } N = \begin{cases} 2^n - N & \text{for } N \neq 0 \\ 0 & \text{for } N = 0 \end{cases} \tag{2.7}$$

where N is positive binary number in base 2.

n is number of bits in integer part of binary number N.

In general, the sign magnitude representation can accommodate numbers in the range

$$-(2^{n-1}) \text{ to } +(2^{n-1} - 1)$$

where n is the number of bits combined to represent a binary number. Most significant binary digit (MSB) is considered as sign bit. 0 bit represents positive binary number and 1 bit represents for negative numbers. Zero has one representation in 2's complement representation.

Two's complement of 1000 0000 is 1000 0000 which is not possible in 1's complement.

One's complement of 1000 0000 is 0111 1111 which is not negative number hence is invalid.

EXAMPLE 2.24 Find 2's complement of $(101100.0110)_2$.

SOLUTION Binary number, $N = 101100.0110$

Number of integer digits in binary number, $n = 6$

$$\begin{aligned}\text{Two's complement of } (101100.0110)_2 &= (2^6)_2 - (101100.0110)_2 \\ &= 1000000.0000 - 101100.0110 \\ &= (111111.1111 + 0.0001) - 101100.0110 \\ &= (111111.1111 - 101100.0110) + 0.0001 \\ &= 010011.1001 + 0.0001 \\ &= 010011.1010 \quad \textbf{Ans.}\end{aligned}$$

> ***Note:*** $(010011.1001)_2$ is 1's complement of $(101100.0110)_2$

> ***Alternate Method-I to find 2's Complement***
> - Find 1's complement of binary number by changing each binary digit '0' to '1' and '1' to '0'
> - Add 1 to the LSB (least significant bit). LSB may be integer part or in fractional part

EXAMPLE 2.25 Find 2's complement of 101100.

SOLUTION

$N =$	1	0	1	1	0	0	
One's complement of $N =$	0	1	0	0	1	1	
+	0	0	0	0	0	1	
Two's complement of $N =$	0	1	0	1	0	0	**Ans.**

EXAMPLE 2.26 Find 2's complement of $(0.0110)_2$.

SOLUTION

$N =$	0	.	0	1	1	0	
One's complement of $N =$	1	.	1	0	0	1	
+	0	.	0	0	0	1	
Two's complement of $N =$	1	.	1	0	1	0	**Ans.**

> ***Alternate Method-II to find 2's Complement***
> Considering from LSB, write each digit as such up to first appearance of digit '1'. After that, convert '0' to '1' and '1' to '0' and write.

For example, 2's complement of 101100 can be obtained as follows.

Binary number, $N =$	1	0	1	1	0	0	
	↓	↓	↓				
Two's complement of binary number, $N =$	0	1	0	1	0	0	**Ans.**

EXAMPLE 2.27 Find 2's complement of 100100.0101.

SOLUTION

$$\begin{array}{rcccccccccccl}
\text{Binary number, } N = & 1 & 0 & 0 & 1 & 0 & 0 & . & 0 & 1 & 0 & 1 & \\
 & \downarrow & \downarrow & \downarrow & \downarrow & \downarrow & \downarrow & & \downarrow & \downarrow & \downarrow & & \\
\text{Two's complement of binary number, } N = & 0 & 1 & 1 & 0 & 1 & 1 & . & 1 & 0 & 1 & 1 & \textbf{Ans.}
\end{array}$$

Advantage of 2's complement method is that there is unique zero. The 2's complement of zero is zero.

$$\begin{array}{rcccccccc}
\text{Let binary number, } N = & 0 & 0 & 0 & 0 & 0 & 0 & 0 & 0 \\
 & \downarrow & \downarrow & \downarrow & \downarrow & \downarrow & \downarrow & \downarrow & \downarrow \\
\text{One's complement of binary number, } N = & 1 & 1 & 1 & 1 & 1 & 1 & 1 & 1
\end{array}$$

$$\begin{array}{rcccccccc}
\text{Let binary number, } N = & 0 & 0 & 0 & 0 & 0 & 0 & 0 & 0 \\
\text{Two's complement of binary number, } N = & 0 & 0 & 0 & 0 & 0 & 0 & 0 & 0
\end{array}$$

2.5 | BINARY SUBTRACTION USING COMPLEMENT

Complements are used to perform subtraction and to represent signed numbers. Subtraction using complement needs addition.

2.5.1 | Subtraction with 1's Complement

The subtraction of binary numbers M and N can be performed as described in Algorithm 2.3. Binary subtraction using 1's complement method basically performs addition of minuend and 1's complement of subtrahend to find subtraction. If carry appears that is added to LSB of sum otherwise take 1's complement to find the subtraction.

EXAMPLE 2.28 Perform subtraction of binary numbers using 1's complement.

(a) 1010 – 1011 (b) 111 – 0.111

SOLUTION

(a) Make four digits of M and N each.

$$\begin{array}{rcccc}
\text{Binary number, } M = & 1 & 0 & 1 & 0 \\
\text{Binary number, } N = & 0 & 1 & 0 & 0
\end{array}$$

Take 1's complement of N and add it to M.
One's complement of $(0100)_2 = (1011)_2$

$$\begin{array}{rccccc}
\text{Binary number, } M = & & 1 & 0 & 1 & 0 \\
\text{1's complement of binary number, } N = & & 0 & 1 & 0 & 0 \\
\hline
M + (\text{1's complement of binary number, } N) = & 0 & 1 & 1 & 1 & 0 \\
\hline
\end{array}$$

End carry (digit in fifth digit after addition) is zero, so take 1's complement of the sum, 1110 and place negative sign

Binary number M + (1's complement of binary number N) =	1	1	1	0	
$(M)_2 - (N)_2 =$ −	0	0	0	1	**Ans.**

> ***Note:*** Number of digits of Minuend M and subtrahend N must be same. Otherwise, complete by filling zero(s). Weightage of digit should remain same.

(b) Make three-integer digits and three fractional digit of M and N each.

Binary number M = 1 1 1 . 0 0 0

Binary Number N = 0 0 0 . 1 1 1

Take 1's complement of N and it to M.
One's complement of $(000.111)_2 = (111.000)_2$

Binary number, M =		1	1	1	.	0	0	0
One's complement of binary number, N =		1	1	1	.	0	0	0
M + (1's complement of binary number, N) =	1	1	1	0	.	0	0	0

End carry (digit in seventh digit after addition) is one. So add 1 to LSB of the sum, $(110.000)_2$ and place positive sign

Binary number M + (1's complement of binary number, N) =		1	1	0	.	0	0	0
	+	0	0	0	.	0	0	1
$(M)_2 - (N)_2 =$	+	1	1	0	.	0	0	1 **Ans.**

2.5.2 | Binary Subtraction with 2's Complement

The subtraction of two positive binary numbers, (M ~ N) is performed as given in Algorithm 2.4. Let binary number M is minuend and Binary number N is subtrahend. Binary subtraction using 2's complement method basically performs addition of minuend and 2's complement of subtrahend to find subtraction. If carry appears that is discarded otherwise take 2's complement to find the subtraction. This method is better as compared to 1's complement.

EXAMPLE 2.29 Perform subtraction of binary number using 2's complement.

(a) 110 – 10 (b) 111 – 0.111

SOLUTION

(a) Make three bits binary number, M and N each.

Binary number, M = 1 1 0

Binary number, N = 0 1 0

Take 2's complement of binary number, N and add it to binary number, M.
One's complement of $(010)_2 = (101)_2$
Two's complement of $(010)_2 = (110)_2$

Binary number, M =		1	1	0
Two's complement of Binary number, N =		1	1	0
M + (2's complement of Binary number, N) =	1	1	0	0

End carry (digit in fourth digit after addition) is one, so discard and place positive sign before the sum.

$(M)_2 - (N)_2 \quad + \ 1\ 0\ 0$ **Ans.**

(b) Make the three integer binary digits and three fractional binary digit of M and N each.

Binary number, M = 1 1 1 . 0 0 0
Binary number, N = 0 0 0 . 1 1 1

Take 2's complement of N and it to M.
One's complement of $(000.111)_2 = (111.000)_2$
Two's complement of $(111.000)_2 = (111.001)_2$

Binary number, M =		1	1	1	.	0	0	0
Two's complement of Binary number, N =		1	1	1	.	0	0	1
M + (2's complement of Binary number, N) =	1	1	1	0	.	0	0	1

End carry (digit in seventh digit after addition) is one, so discard it and place positive sign

$(111.000)_2 - (000.111)_2 = \ + \ 1\ 1\ 0\ .\ 0\ 0\ 1$ **Ans.**

2.6 | BINARY MULTIPLICATION

Binary multiplication is performed in the same manner as is done in decimal system. Multiplication table of binary digits is given by:

$$0 \times 0 = 0$$
$$1 \times 1 = 1$$
$$0 \times 1 = 0$$
$$1 \times 0 = 0$$

Consider multiplication of 1101 by 110.

Multiplicand				1	1	0	1
Multiplier				×	1	1	0
				0	0	0	0
			1	1	0	1	
		1	1	0	1		
Product	1	0	0	1	1	1	0

Ans.

In multiplication if multiplier bit is 1, then add the multiplicand to partial sum otherwise no need to add and give one shift to left.

EXAMPLE 2.30 Multiply $(101.101)_2$ by $(1.01)_2$.

SOLUTION

Multiplicand			1	0	1	1	0	1	
Multiplier				×		1	0	1	
			1	0	1	1	0	1	
		0	0	0	0	0	0		
	1	0	1	1	0	1			
Product	1	1	1	0	0	0	0	1	**Ans.**

Place decimal point after (3 + 2) 5 digits.

$(101.101)_2 \times (1.01)_2 = (111.00001)_2$ **Ans.**

Computer can perform addition of two operands at a time. The above procedure of multiplication is sequenced for successive addition of multiplicand by recognizing the bit of multiplier and giving the left shift to partial sum and multipliers. Let multiplier is of 4 bits (1 Nibble) and set counter to 4. Multiplicand should be of 8 bits (1 byte) for 4-bit (1 Nibble) multiplier. For 8-bit (1 Byte) multiplier, set counter to 8 and multiplicand should be of 16 bits (2 bytes). Number of bits of multiplicand is twice the number of bits of multiplier to represent them. Basically multiplication requires, left shift to multiplier and partial sum is required which is explained below.

Shift to left by one bit Shift to left is performed by shifting lower bit to next higher bit place value. In other word data is multiplied by 2. For 4-bit data the left shift can be performed by shifting the bits one by one to their next higher place valued bit. The process is started from most significant bit (MSB) having highest weight or place value. Sequence of steps are explained below

(i) $Carry \leftarrow b_3;$ (ii) $b_3 \leftarrow b_2;$ (iii) $b_2 \leftarrow b_1$ (iv) $b_1 \leftarrow b_0;$ (v) $b_0 \leftarrow 0$

	2^4	2^3	2^2	2^1	2^0	Place value
Steps	Carry	b_3	b_2	b_1	b_0	Bits
		0	1	0	1	Data
Carry $\leftarrow b_3$;	**0**					b_3 is shifted to carry
$b_3 \leftarrow b_2$;	0	**1**				b_2 is shifted to b_3
$b_2 \leftarrow b_1$;	0	1	**0**			b_1 is shifted to b_2
$b_1 \leftarrow b_0$;	0	1	0	**1**		b_0 is shifted to b_1
$b_0 \leftarrow 0$;	0	1	0	1	**0**	0 is shifted to b_0
	0	1	0	1	0	data after 1 bit left shift

For 8-bit data the left shift can be performed by shifting the bits to their next higher place value starting from most significant bit (MSB). Sequence steps are explained below

(i) $Carry \leftarrow b_7;$ (ii) $b_7 \leftarrow b_6;$ (iii) $b_6 \leftarrow b_5;$ (iv) $b_5 \leftarrow b_4;$ (v) $b_4 \leftarrow b_3;$

(vi) $b_3 \leftarrow b_2;$ (vii) $b_2 \leftarrow b_1$ (viii) $b_1 \leftarrow b_0;$ (ix) $b_0 \leftarrow 0$

Carry	b_7	b_6	b_5	b_4	b_3	b_2	b_1	b_0	Bits
	0	1	0	1	1	0	1	1	Data
0	1	0	1	1	0	1	1	0	Data after 1 bit left shift

Flow chart is given in Figure 2.2 to explain the steps. Stepwise computation procedure is given in Table 2.5 to explain the product of (0101) and (0011) to get (0000 1111)

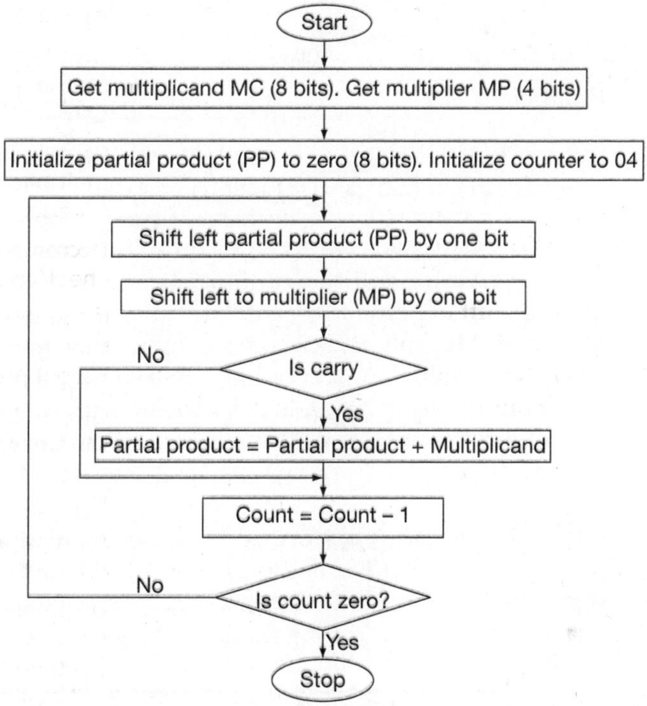

FIGURE 2.2 | Flow chart for binary multiplication

TABLE 2.5 | Multiplication of two numbers

Partial Product		Multiplicand		Carry	Multiplier	Counter	Comments
0000	0000	0000	0101	–	0011	4	
				0	0110		Give left shift to Multiplier and check Carry
0000	0000					3	Carry is zero, decrement counter by 1. Check counter is zero, if no then give left shift to Partial product
				0	1100		Give left shift to Multiplier, and check Carry

(Continued)

TABLE 2.5 | (Continued)

Partial Product	Multiplicand	Carry	Multiplier	Counter	Comments
0000 0000				2	Carry is zero, decrement counter by 1. Check counter is zero, if no then give left shift to Partial Product
		1	1001		Give left shift to Multiplier and check Carry
0000 0000 0000 0101					Carry is One, add multiplicand to Partial Product
0000 0101				1	Decrement counter by 1. Check counter is zero.
0000 1010					If counter is not zero then give left shift to partial product
		1	0011		Give left shift to Multiplier and check Carry
0000 1010 0000 0101					Carry is One, add Multiplicand to Partial Product.
0000 1111				0	Decrement counter by 1. Check counter is zero if yes then stop.

2.7 | BINARY DIVISION

Binary division is performed in the same manner as is done in decimal system. Division table of binary digits is given by:

$0 \div 0 = 0$

$1 \div 1 = 1$

$0 \div 1 = 0$

$1 \div 0$ is not possible being infinite value.

```
                     1 0 1   ←———— Quotient
Divisor ——→ 1 0 1 ) 1 1 0 0 1  ←—— Dividend
                    1 0 1
                    -----
                      0 1 0
                      0 0 0
                      -----
                        1 0 1
                        1 0 1
                        -----
                          ×   ←——— Remainder
```

$(11001)_2 \times (101)_2 = (101)_2 \times (101)_2 + (000)_2$ **Ans.**

EXAMPLE 2.31 Divide $(11101)_2$ by $(1100)_2$.

SOLUTION

	1	0	.	0	1	1			
1 1 0 0	1	1	1	0	1	.	0	0	0
	1	1	0	0					
		0	1	0	1				
		0	0	0	0				
			1	0	1	0			
			0	0	0	0			
			1	0	1	0	0		
				1	1	0	0		
				1	0	0	0	0	
					1	1	0	0	
					0	1	0	0	

$(11101)_2 \times (1100)_2 = (10.011)_2 \times (1100)_2 + (0100)_2$ **Ans.**

Computer can perform subtraction of two operands at a time. The above procedure of division is sequenced for successive subtraction of divisor from dividend and giving the left shift to dividend. If subtraction is performed then 1 is added to quotient otherwise zero. Let divisor is of 4 bits (1 Nibble) and set counter to 4 for successive operations. Dividend should be of 8 bits (1 byte) for 4 bits (1 Nibble) divisor. For 8 bits (1 Byte) divisor, set counter to 8 and dividend should be of 16 bits (2 bytes). Number of bits of dividend is twice the number of bits of divisor to represent them. Basically division requires left shift to dividend, subtract divisor from dividend. If possible then quotient is incremented and this process repeated for the number of divisors. Flow chart is given in Figure 2.3 to explain the steps for division. Stepwise computation procedure is given in Table 2.6 to explain the division of (0000 1111) and (0011) to get (0101) as quotient and (0000) as remainder.

2.8 | OCTAL NUMBER SYSTEM

The octal number has a base or radix of 8. Eight different symbols are used to represent the octal number system. These eight digits are 0, 1, 2, 3, 4, 5, 6 and 7. Octal symbols or digits are represented by unique combination of 3-bit binary numbers. Since $2^3 = 8$, so range of data obtained from the combination of 3 bits is from 0 to $7(2^3 - 1)$.

The number 238 is not a valid octal number. '8' digit is not a symbol of octal numbers. It is a base of octal number system. 238 is not a valid binary number. However, 238 is a valid decimal number.

$(6237)_8$ is a valid octal number

Weight to digit ⇒	8^3	8^2	8^1	8^0
Digit position ⇒	4th	3rd	2nd	1st
	Left ←			Right
	6	2	3	7
	Integer portion			

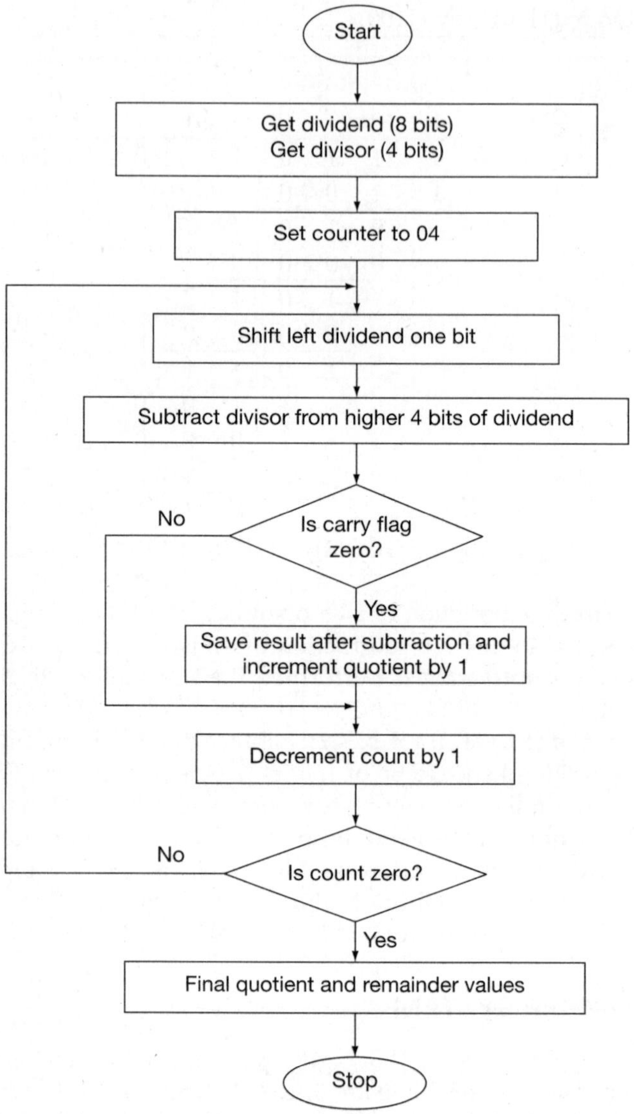

FIGURE 2.3 | Flow chart for binary division

TABLE 2.6 | Divide two binary unsigned (positive) numbers

Carry	Dividend		Divisor	Counter	Comments
	DH	DL			
	0000	1111	0011	4	
	0001	1110			Give left shift to dividend
	0011				Subtract divisor from higher nibble (4 bits) of dividend (DH).

(*Continued*)

TABLE 2.6 | (Continued)

Carry	Dividend DH	Dividend DL	Divisor	Counter	Comments
1				3	Check carry, if it is one then divisor is more than DH so it is not divisible and decrement counter. Check counter
	0011	1100			If counter is not zero, then give left shift to dividend.
	0011				Subtract divisor from higher nibble (4 bits) of dividend (DH).
0	0000	1101		2	Check carry, if it is zero, perform subtraction, increment to lower nibble of dividend (DL) and decrement to Counter. Check counter
	0001	1010			Counter is not zero then give left shift to dividend
	0011				Subtract divisor from higher nibble (4 bits) of dividend (DH).
1				1	Check carry, if it is one then divisor is more than DH so it is not divisible and decrement counter. Check counter
	0011	0100			Counter is not zero then give left shift to dividend
	0011				Subtract divisor from higher nibble (4 bits) of dividend (DH).
0	0000	0101		0	Check carry if it is zero perform subtract and increment to lower nibble of dividend (DL) and decrement to Counter. Check counter
	Remainder	Quotient			Counter is zero then STOP

Number 6 is MSD because weight assigned to it has highest value, i.e. 8^3.
Number 7 is LSD because weight assigned to it has lowest value, i.e. 8^0.
Consider a octal number $(4376.2045)_8$

Weight to digit ⇒	8^3	8^2	8^1	8^0		8^{-1}	8^{-2}	8^{-3}	8^{-4}
Digit position ⇒	4th	3rd	2nd	1st		1st	2nd	3rd	4th
	Left ←			→ Right	·	Left →			→ Right
	4	3	7	6		2	0	4	5
	Integer portion					Fractional portion			
	MSD				Decimal Point				LSD

Number 4 is MSD because weight assigned to it has highest value, i.e. 8^3.
Number 5 is LSD because weight assigned to it has lowest value, i.e. 8^{-4}.

2.8.1 | Octal to Binary Conversion

Three binary digits are combined to get the values from 0 to 7 in decimal. 0 to 7 are the symbols of octal system. The octal digits, 0 to 7 can be written in combination of three binary digits (3 bits) as given in Table 2.7. Hence, octal number is converted into equivalent binary number by replacing octal digits to its 3 bits equivalent binary number.

TABLE 2.7 | Decimal binary numbers equivalent to octal

Decimal	Octal	Binary		
		2^2	2^1	2^0
0	0	0	0	0
1	1	0	0	1
2	2	0	1	0
3	3	0	1	1
4	4	1	0	0
5	5	1	0	1
6	6	1	1	0
7	7	1	1	1

EXAMPLE 2.32 Convert $(74765)_8$ into binary.

SOLUTION Represent each octal digit equivalent to its 3 bits binary number.

Octal digit	↔	Equivalent 3-bit binary number
4	↔	100
5	↔	101
6	↔	110
7	↔	111

Octal Number:	7	4	7	6	5
Three-bit Binary Number:	111	100	111	110	101

So, $(74765)_8 = (111\ 100\ 111\ 110\ 101)_2$ **Ans.**

EXAMPLE 2.33 Convert $(576.234)_8$ into binary.

SOLUTION

Octal Number:	5	7	6	.	2	3	4
Three-bit Binary Number:	101	111	110	.	010	011	100

So, $(576.234)_8 = (101\ 111\ 110.\ 010\ 011\ 100)_2$ **Ans.**

2.8.2 | Binary to Octal Conversion

The conversion from binary to octal is done by partitioning the binary number into groups of three binary digits each and replacing each three-bit binary group to its equivalent octal digit. Partitioning is started from decimal point and proceeds to left for integers and to right for fractions. Complete combinations of 3 bits by adding zeros in such a manner that number should not change.

EXAMPLE 2.34 Convert $(111100111110001)_2$ into octal.

SOLUTION

Three-bit binary number:	111	100	111	110	001
Octal number:	7	4	7	6	1

So $(111\ 100\ 111\ 110\ 001)_2 = (74761)_8$ **Ans.**

EXAMPLE 2.35 Convert $(10011)_2$ into octal.

SOLUTION

Three-bit binary number:	010	011
Octal number:	2	3

So, $(10\ 011)_2 = (23)_8$ **Ans.**

EXAMPLE 2.36 Convert $(1011.1111)_2$ into octal.

SOLUTION

Three-bit binary number:	001	011	.	111	100
Octal Number:	1	3	.	7	4

So, $(1011.1111)_2 = (13.74)_8$ **Ans.**

EXAMPLE 2.37 Convert $(435.251)_8$ to decimal.

SOLUTION

	64	8	1		0.125	0.015625
Weight to digits ⇒	8^2	8^1	8^0		8^{-1}	8^{-2}
Position of digits ⇒	3rd	2nd	1st		1st	2nd
Digits ⇒	4	3	5	.	2	5
	d_3	d_2	d_1		d_{-1}	d_{-2}

$$(N)_{10} = \underbrace{\sum_{i=1}^{3} d_i 8^{i-1}}_{\text{Integer portion}} + \underbrace{\sum_{j=1}^{2} d_{-j} 8^{-j}}_{\text{Fractional portion}}$$

$$(N)_{10} = 5\times 8^0 + 3\times 8^1 + 4\times 8^2 + 2\times 8^{-1} + 5\times 8^{-2}$$

$$(N)_{10} = (5+24+256)+(0.25+0.078125)$$

$(N)_{10} = (285.328125)_{10}$ **Ans.**

EXAMPLE 2.38 Convert 129.56 into octal.

SOLUTION Decimal number = $(129.56)_{10}$. Integer portion of decimal number 129.56 = 129

Dividend	÷ Divisor	=	Quotient	: Remainder	Octal digits
129	÷ 8	=	16	: $1 \rightarrow d_1$ (LSD)	$d_1 \rightarrow 1$
16	÷ 8	=	2	: $0 \rightarrow d_2$	$d_2 \rightarrow 0$
2	÷ 8	=	0	: $2 \rightarrow d_3$(MSD)	$d_3 \rightarrow 2$

Octal number = $(d_3d_2d_1)_8$

$(129)_{10} = (201)_8$

Fractional portion of decimal number 129.56 = 0.56

Multiplicand	×	8	=	Integer	+	Fraction	Octal digit
0.56	×	8	=	4	+	.48	$d_{-1} \rightarrow 4$
0.48	×	8	=	3	+	.84	$d_{-2} \rightarrow 3$
0.84	×	8	=	6	+	.72	$d_{-3} \rightarrow 6$
0.72	×	8	=	5	+	.76	$d_{-4} \rightarrow 5$

Octal number = $(0.d_{-1}d_{-2}d_{-3}d_{-4})_8$

$(0.56)_{10} = (0.4365)_8$

Combining integer and fractional part of number 129.56, to get

$(129.56)_{10} = (201.4365)_8$ **Ans.**

2.8.3 | Octal Arithmetic

Octal (base-8) addition and subtraction operations are performed in a manner similar to decimal arithmetic.

In case, a number of digits are combined to represent a number. Carry over are performed in the same manner as in decimal arithmetic. Addition of first digit column generates carry for second digit of higher weightage. Addition of second digit column generates carry for third digit and so on. Decimal and octal addition to add $(15)_{10}$ and $(26)_{10}$ is elaborated with an example which is given below.

	Decimal number		Octal number		
Digits	2nd	1st	3rd	2nd	1st
Carry	1	0	0	1	0
Augend	1	5		1	7
Addend	2	6		3	2
Sum	4	1	0	5	1

Procedure to find 7's complement

Seven's complement of an octal number is obtained by subtracting each octal digit from 7 (radix-1).

Simple way to get 7's complement of an octal number is to subtract each digit of octal number from '7'.

e.g. $7-0=7$, $7-1=6$, $7-2=5$, $7-3=4$, $7-4=3$, $7-5=2$, $7-6=1$ and $7-7=0$

The 7's complement of 64125.0135 can be obtained as:

$N=$	6	4	1	2	5 .	0	1	3	5
	↓	↓	↓	↓	↓	↓	↓	↓	↓
	(7 – 6)	(7 – 4)	(7 – 1)	(7 – 2)	(7 – 5)	(7 – 0)	(7 – 1)	(7 – 3)	(7 – 5)
7's complement of $N=$	1	3	6	5	2 .	7	6	4	2

> ***Procedure to find 8's Complement***
>
> Find 7's complement of octal number by subtracting each octal digit from largest digit of octal number i.e. '7'
>
> Add 1 to the LSD (least significant digit) LSD may be integer part or in fractional part.

EXAMPLE 2.39 Find 8's complement of $(0.0441)_2$.

SOLUTION

$$
\begin{array}{rl}
(N)_8 = & 0\ .\ 0\ 4\ 4\ 1 \\
\text{Seven's complement of } (N)_8 = & 7\ .\ 6\ 3\ 3\ 6 \\
+ & 0\ .\ 0\ 0\ 0\ 1 \\
\text{Eight's complement of } (N)_8 = & 7\ .\ 6\ 3\ 3\ 7 \quad \textbf{Ans.}
\end{array}
$$

EXAMPLE 2.40 Subtract $(2011)_8$ from $(1516)_8$ using 7's complement.

SOLUTION

Take 7's complement of $(N)_8$ and add it to $(M)_8$.
Seven's complement of $(2011)_8 = (5766)_8$

$$
\begin{array}{rccccc}
\text{Carry} & & 1 & 1 & 1 & \\
(M)_8 = & & 1 & 5 & 1 & 6 \\
\text{7's complement of } (N)_8 = & & 5 & 7 & 6 & 6 \\
\hline
(M)_8 + (\text{7's complement of } (N)_8) = & 0 & 7 & 5 & 0 & 4 \\
\hline
\end{array}
$$

End carry (digit in fifth digit after addition) is zero, so take 7's complement of $(7504)_8$ and place negative sign

$$
\begin{array}{rccccc}
(M)_8 + (\text{7's complement of } (N)_8) = & & 7 & 5 & 0 & 4 \\
(M)_8 - (N)_8 = & - & 0 & 2 & 7 & 3 \quad \textbf{Ans.}
\end{array}
$$

EXAMPLE 2.41 Subtract $(0.456)_8$ from $(131)_8$ using 8's complement.

SOLUTION

Make the three integer digits and three fractional digit of $(M)_8$ and $(N)_8$ each.

$$(M)_8 = 1\ 3\ 1\ .\ 0\ 0\ 0$$
$$(N)_8 = 0\ 0\ 0\ .\ 4\ 5\ 6$$

Take 8's complement of $(N)_8$ and add it to $(M)_8$
Seven's complement of $(000.456)_8 = (777.321)_8$
Eight's complement of $(000.456)_8 = (777.322)_8$

Carry over	1	1	1					
$(M)_8 =$		1	3	1	.	0	0	0
Eight's complement of $(N)_8 =$		7	7	7	.	3	2	2
$(M)_8 +$ (8's complement of $(N)_8$) $=$	1	1	3	0	.	3	2	2

End carry (digit in seventh digit after addition) is one, so discard it and place positive sign

$$(M)_8 - (N)_8 = +\ 1\ 3\ 0\ .\ 3\ 2\ 2 \quad \textbf{Ans.}$$

2.9 | HEXADECIMAL NUMBER SYSTEM

A typical microcomputer memory location might hold the binary number 10011110. This long string of 0's and 1's is hard to remember and difficult to enter on a keyboard. Most microcomputer systems use hexadecimal notation to simply remembering and entering binary numbers.

Hexadecimal number system has a base or radix of 16. Sixteen different symbols are used to represent the hexadecimal (hex) numbers. These sixteen digits are 0, 1, 2, 3, 4, 5, 6, 7, 8, 9, *A*, *B*, *C*, *D*, *E* and *F* (Table 2.8). Hexadecimal symbols or digits are represented by unique combination of 4-bit binary numbers, because 2^4 is 16.

Range of data obtained from the combination of 4 bits is from 0 to $15(2^4 - 1)$.

TABLE 2.8 | Decimal hexadecimal numbers equivalent binary

Decimal	Hexadecimal	Binary Numbers			
		2^3	2^2	2^1	2^0
0	0	0	0	0	0
1	1	0	0	0	1
2	2	0	0	1	0
3	3	0	0	1	1
4	4	0	1	0	0
5	5	0	1	0	1
6	6	0	1	1	0
7	7	0	1	1	1

(Continued)

TABLE 2.8 | (Continued)

Decimal	Hexadecimal	Binary Numbers			
		2^3	2^2	2^1	2^0
8	8	1	0	0	0
9	9	1	0	0	1
10	*A*	1	0	1	0
11	*B*	1	0	1	1
12	*C*	1	1	0	0
13	*D*	1	1	0	1
14	*E*	1	1	1	0
15	*F*	1	1	1	1

$(237A)_{16}$ or 237AH is a valid hexadecimal (hex) number. H stands for hexadecimal code.

Weight to digit ⇒ 16^3 16^2 16^1 16^0
Digit position ⇒ 4th 3rd 2nd 1st
Left ← Right
2 3 7 *A*
Integer portion

2 is MSD because weight assigned to it has highest value, i.e. 16^3.
A is LSD because weight assigned to it has lowest value, i.e. 16^0.
Consider a hex (hexadecimal) EFAB.2CD5H

Weight to digit ⇒ 16^3 16^2 16^1 16^0 16^{-1} 16^{-2} 16^{-3} 16^{-4}
Digit position ⇒ 4th 3rd 2nd 1st 1st 2nd 3rd 4th
Left ← Right Left → Right
E *F* *A* *B* . 2 *C* *D* 5
Integer portion Fractional portion
MSD Decimal point LSD

E is MSD because weight assigned to it has highest value, i.e. 16^3.
5 is LSD because weight assigned to it has lowest value, i.e. 16^{-4}.

2.9.1 | Hexadecimal to Binary Conversion

Four binary digits (bits) are combined to get the values from 0 to 15 in decimal. Further, 0 to 15 are the symbols (digits) of hexadecimal system. The hexadecimal digits, 0 to 9, *A*, *B*, *C*, *D*, *E* and *F* can be written in a combination of 4-binary digits (4 bits) equivalent to binary number. Table 2.8 gives the equivalent values in binary. In short, hexadecimal numbers are known as hex numbers. Hence, hexadecimal (hex) number is converted into equivalent binary number by replacing each hexadecimal (hex) digit to its 4 bits equivalent binary number.

EXAMPLE 2.42 Convert $(94E65)_{16}$ into binary.

SOLUTION Represent each octal digit equivalent to its 4-bit binary number.

Octal digit	↔	Equivalent 4-bit binary number
4	↔	0100
5	↔	0101
6	↔	0110
9	↔	1001
E	↔	1110

Hexadecimal number:	9	4	E	6	5
Four-bit binary number:	1001	0100	1110	0110	0101

So, $(94E65)_{16} = (1001\ 0100\ 1110\ 0110\ 0101)_2$ **Ans.**

EXAMPLE 2.43 Convert 5AB6.234H into binary.

SOLUTION

Hexadecimal number:	5	A	B	6	.	2	3	4
Four-bit binary number:	0101	1010	1011	0110	.	0010	0011	0100

So, 5AB6.234H = 0101 1010 1011 0110. 0010 0011 0100B **Ans.**

2.9.2 | Binary to Hexadecimal Conversion

The conversion from binary to hexadecimal is done by partitioning the binary number into groups of four binary digits (4 bits) each and replacing each 4-bit binary group to its equivalent hexadecimal (hex) digit. Partitioning is started from decimal point and proceeds to left for integers and to right for fractions. Complete combinations of 4 bits by adding zeros in such a manner that number should not change its value.

EXAMPLE 2.44 Convert $(111100111110001)_2$ into hexadecimal.

SOLUTION

Four-bit group of binary number:	0111	1001	1111	0001
Hex number:	7	9	F	1

So, $(111\ 100\ 111\ 110\ 001)_2 = (79F1)_{16}$ **Ans.**

EXAMPLE 2.45 Convert 10011B into hexadecimal.

SOLUTION

Four-bit group of binary number:	0001	0011
Hex number:	1	3

So 10 011B = 13H **Ans.**

EXAMPLE 2.46 Convert 1011.1111B into hexadecimal.

SOLUTION

Four-bit group of binary number:	1011 (←)	.	1111 (→)
Hex number:	B	.	F

So, 1011.1111B = B.FH **Ans.**

EXAMPLE 2.47 Convert $(435.251)_{16}$ to decimal.

SOLUTION

Weight to digits ⇒	16^2	16^1	16^0		16^{-1}	16^{-2}
Position of digits ⇒	3rd	2nd	1st		1st	2nd
Digits ⇒	4	3	5	.	2	5
	d_3	d_2	d_1		d_{-1}	d_{-2}

$$(N)_{10} = \underbrace{\sum_{i=1}^{3} d_i 16^{i-1}}_{\text{Integer portion}} + \underbrace{\sum_{j=1}^{2} d_{-j} 16^{-j}}_{\text{Fractional portion}}$$

$$(N)_{10} = 5\times16^0 + 3\times16^1 + 4\times16^2 + 2\times16^{-1} + 5\times16^{-2}$$

$$(N)_{10} = (5+48+1024)+(0.125+0.01953125)$$

$$(N)_{10} = (1077.14453125)_{10}$$ **Ans.**

EXAMPLE 2.48 Convert 129.56 into hex.

SOLUTION Decimal number = $(129.56)_{10}$. Integer portion of decimal number 129.56 = 129.

Dividend	÷ 16	= Quotient	: Remainder	Hex digits
129	÷ 16	= 8	: 1 → d_1 (LSD)	$d_1 \rightarrow 1$
8	÷ 16	= 0	: 8 → d_2(MSD)	$d_2 \rightarrow 8$

Octal number = $(d_2 d_1)$

$(129)_{10} = (81)_{16}$ or 81H

Fractional portion of decimal number 129.56 = 0.56

Multiplicand	× 16	=	Integer	+	Fraction	Hex digit
0.56	× 16	=	8	+	.96	$d_{-1} \rightarrow 8$
0.96	× 16	=	15	+	.36	$d_{-2} \rightarrow F$
0.36	× 16	=	5	+	.76	$d_{-3} \rightarrow 5$
0.76	× 16	=	12	+	.16	$d_{-4} \rightarrow C$
0.16	× 16	=	2	+	.56	$d_{-5} \rightarrow 2$
0.56	× 16	=	8	+	.96	$d_{-6} \rightarrow 8$

Octal number = $(0.d_{-1}d_{-2}d_{-3}d_{-4}\,d_{-5}\,d_{-6})_{16}$

$(0.56)_{10} = (0.8F5C28)_{16}$ or 0.8F5C28H

> ***Note:*** Fractional part will never become zero, so 0.56 fraction cannot be expressed exactly in binary equivalent. In such a case, the process of multiplication can be terminated after certain significant digits giving desired accuracy or the successive fraction number starts repeating.

Combining integer and fractional part of number 129.56, to get

$(129.56)_{10} = (81.8F5C28)_{16}$ or 81.8F5C28H **Ans.**

2.9.3 | Hexadecimal to Octal Conversion

To convert the hexadecimal (hex) number to octal number, the easiest method is to first convert the given hexadecimal number to binary number and then binary number is converted into the octal number.

EXAMPLE 2.49 Convert 5AB6.23H into octal number.

SOLUTION

Given hexadecimal number:	5	A	B	6		.	2		3	
Convert each hex number into 4-bit binary number:	0101	1010	1011	0110		.	0010		0011	
Binary number:	0101101010110110					.	00100011			
Make group of 3 bits of binary number:	101	101	010	110	110	.	001	000	110	
	5	5	2	6	6	.	1	0	6	

So, 5AB6.23H = 01011010 1011 0110. 0010 0011B = 55266.106O **Ans.**

2.9.4 | Octal to Hexadecimal Conversion

To convert the octal number to hexadecimal number, the easiest method is to first convert the given octal number to binary number and then binary number is converted into the hexadecimal (hex) number.

EXAMPLE 2.50 Convert 675.306O into hex number.

SOLUTION

Given octal number:	6	7	5	.	3	0	6
Convert each octal number into 3-bit binary number:	110	111	101	.	011	000	110
Binary number:	110111101			.	011000110		
Make group of 4 bits of binary number:	0001	1011	1101	.	0110	0011	0000
	1	B	D	.	6	3	0

So, 675.306O = 110111101.01100011 B = 1BD.63H **Ans.**

2.9.5 | Hexadecimal Arithmetic

Hexadecimal also known as hex (base-16) addition and subtraction operations are performed in a manner similar to decimal arithmetic.

2.9.5.1 | Hexadecimal Addition

Addition is performed as is done in decimal arithmetic. Resulting sum should not be more than the largest digit of hexadecimal (Base-16). If the digit is not a valid hex digit, then subtract '10H' (base value) to get the required digit with carry over as 1. Single hexadecimal (hex) digit addition of base-16 is given below. When digits $(1)_H$ and $(F)_H$ are added, will result digit $(16)_{10}$ which is larger than F (largest digit of hex distinct digit set), so subtract $(16)_{10}$ (Base) to get 0 with carry over as 1.

	2nd	1st	2nd	1st	2nd	1st	2nd	1st	2nd	1st	2nd	1st	2nd	1st	2nd	1st
Carry over	1		1		1		1		1		1		1		1	
Augend		1		2		3		4		4		B		A		F
Addend	+	F	+	E	+	D	+	C		D		8		B		F
Sum	1	0	1	0	1	0	1	0	1	1	1	3	1	5	1	E

In case, numbers of digits are combined to represent a number. Carry over are performed in the same manner as in decimal arithmetic. Addition of first digit column generates carry for second digit of higher weightage. Addition of second digit column generates carry for third digit and so on. Decimal and hexadecimal addition is elaborated with an example that adds $(15)_{10}$ and $(26)_{10}$.

	Decimal number		Hex number		
Digits	2nd	1st	3rd	2nd	1st
Carry	1	0	0	1	0
Augend	1	5		0	F
Addend	2	6		1	A
Sum	4	1	0	2	9

2.9.5.2 | Hexadecimal Subtraction

The hexadecimal (hex, base-16) subtraction is performed in a manner similar to that in decimal subtraction. Borrow taken in decimal system adds $(10)_{10}$ (base value) whereas in hex the borrow taken adds $(10)_H$ (base value). Single hex digit subtraction is performed and is given below for base 16.

Digit	2nd	1st	2nd	1st	2nd	1st	2nd	1st	2nd	1st	2nd	1st
Borrow	−1	10	−1	10	−1	10	−1	10	−1	10	−1	10
Minuend		1		2		3		4		5		6
Subtrahend	−	F	−	E	−	D	−	C	−	D	−	7
Difference	1	2	1	4	1	6	1	8	1	7	1	F

- Number 1 is less than F, borrow is taken from the next higher place valued digit as $(10)_H$ (hexadecimal base value).
- Taken borrow is added to minuend and sum is obtained, i.e. $(10 + 1 = 11)_H$
- Finally, subtrahend is subtracted from partial sum to get the difference, i.e. $((11)_H - (F)_H = (2)_H)$. Note that 2 is a valid octal digit as well as of decimal. Note F is valid hex digit and is equivalent to $(15)_{10}$ decimal system.
- Borrow taken by previous digit column is subtracted from minuend of present digit column.

Decimal and hex subtraction is given below that subtracts $(26)_{10}$ from $(41)_{10}$

	Decimal number		hex number		
Digit	2nd	1st	3rd	2nd	1st
Borrow	–1	10		–1	10
Minuend	4	1		2	9
Subtrahend	2	6	–	1	A
Difference	1	5		0	F

Method to find F's complement

F's complement of a hex number is obtained by subtracting each hex digit from *F* (radix-1).

For example, $F - 0 = F$, $F - 1 = E$, $F - 2 = D$, $F - 3 = C$, $F - 4 = B$, $F - 5 = A$, $F - 6 = 9$ and $F - F = 0$

The F's complement of (64125.0135)H can be obtained as:

$N =$	6	4	1	2	5	.	0	1	3	5
	↓	↓	↓	↓	↓		↓	↓	↓	↓
	$(F - 6)$	$(F - 4)$	$(F - 1)$	$(F - 2)$	$(F - 5)$		$(F - 0)$	$(F - 1)$	$(F - 3)$	$(F - 5)$
F's complement of $N =$	9	B	E	D	A	.	F	E	C	A

EXAMPLE 2.51 Find 16's complement of $(0.0441)_H$.

SOLUTION

$$
\begin{array}{rl}
(N)_H = & 0\ .\ 0\ 4\ 4\ 1 \\
F\text{'s complement of } (N)_H = & F\ .\ F\ B\ B\ E \\
+ & 0\ .\ 0\ 0\ 0\ 1 \\
\hline
\text{Sixteen's complement of } (N)_H = & F\ .\ F\ B\ B\ F \quad \textbf{Ans.}
\end{array}
$$

EXAMPLE 2.52 Perform subtraction of hex numbers: $(1516 - 2011)_H$ using F's complement.

SOLUTION

Take F's complement of $(N)_H$ and add it to $(M)_H$.
F's complement of 2011H = DFEEH

$$\begin{array}{rccccc} & & 1 & 1 & 1 & \\ (M)_H = & & 1 & 5 & 1 & 6 \\ F\text{'s complement of } (N)_H = & & D & F & E & E \\ \hline (M)_H + (F\text{'s complement of } (N)_H) = & 0 & F & 5 & 0 & 4 \\ \hline \end{array}$$

End carry (digit in fifth digit after addition) is zero, so take F's complement of $(F504)_H$ and place negative sign

$$\begin{array}{rccccc} (M)_H + (F\text{'s complement of } (N)_H) = & & F & 5 & 0 & 4 \\ (M)_H - (N)_H = & - & 0 & A & F & B \end{array} \quad \textbf{Ans.}$$

EXAMPLE 2.53 Subtract 0.456_H from 131_H using 16's complement.

SOLUTION

Make the three integer digits and three fractional digit of $(M)_H$ and $(N)_H$ each.

$$\begin{array}{rccccccc} (M)_H = & 1 & 3 & 1 & . & 0 & 0 & 0 \\ (N)_H = & 0 & 0 & 0 & . & 4 & 5 & 6 \end{array}$$

Take 16's complement of $(N)_H$. and add it to $(M)_H$.
F's complement of 000.456H = FFF.BA9H
Sixteen's complement of 000.456H = FFF.BAAH

$$\begin{array}{rcccccccc} \text{Carry over} & 1 & 1 & 1 & & & & & \\ (M)_H = & & 1 & 3 & 1 & . & 0 & 0 & 0 \\ \text{Sixteen's complement of } (N)_H = & & F & F & F & . & B & A & A \\ (M)_H + (16\text{'s complement of } (N)_H) = & 1 & 1 & 3 & 0 & . & B & A & A \end{array}$$

End carry (digit in 7th digit after addition) is one, so discard it and place positive sign

$$(M)_H - (N)_H = +\ 1\ 3\ 0\ .\ 3\ 2\ 2 \quad \textbf{Ans.}$$

2.10 | BINARY CODES

Computers, microprocessors, microcontrollers and other digital circuits process data in the binary format. Different codes are used to represent the data. Data may be numeric, alphanumeric or special characters. To interpret the data, knowledge of binary code is necessary.

Binary	1000001
Decimal	65
Octal	101
Hex	41
BCD	41
ASCII	'A'

The binary code can be interpreted in given codes such as decimal, octal, hex, BCD (binary-coded decimal) and ASCII (American Standard Code for Information Interchange).

Binary codes can be classified as numeric codes and alphanumeric codes. Alphanumeric codes represent alphanumeric information, i.e. letters of the alphabet and decimal numbers as a sequence of 0s and 1s. Numeric codes represent numeric information, i.e. only numbers as a series of 0s and 1s. Numeric codes used to represent decimal digits are called BCD codes.

BCD code is one, in which the digits of a decimal number are encoded—one at a time—into groups of four binary digits. These codes combine the features of decimal and binary numbers. There are a large number of BCD codes. In order to represent decimal digits 0, 1, 2,..., 9, it is necessary to use a sequence of at least four binary digits. Such a sequence of binary digits which represents a decimal digit is called a code word.

2.10.1 | Weighted and Non-weighted Code

Decimal digits are represented by group of 4 bits. The BCD codes may be weighted or non-weighted. The weighted codes may be either positively weighted or negatively weighted. Classification of binary codes is given in Figure 2.4.

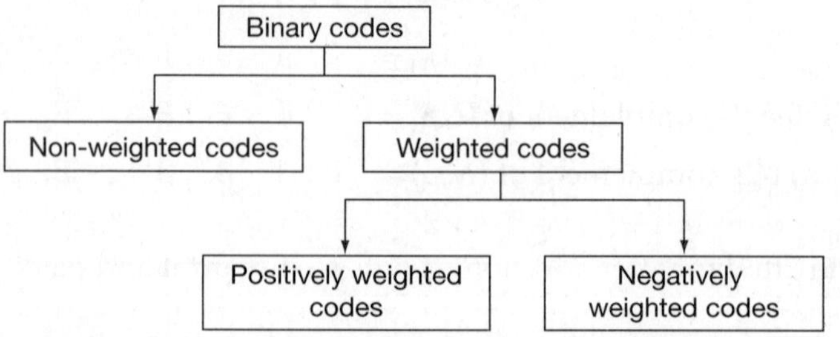

FIGURE 2.4 | Classification of binary codes

2.10.1.1 | Weighted Codes

The weighted codes are those which obey the position-weighting principle. Each position of the number represents a specific weight. For each group of four bits, the sum of the weights of those positions where the binary digit is 1 is equal to the decimal digit. These are several weighted codes.

$$(d)_{10} = w_4 \times b_4 + w_3 \times b_3 + w_2 \times b_2 + w_1 \times b_1 \qquad (2.14)$$

where

w_1, w_2, w_3 and w_4 are the weights assigned to first, second, third and fourth bits, respectively.

b_1, b_2, b_3, and b_4 are the first, second, third and fourth bits which can be either 0 or 1.

Positively weighted code Positively weighted codes are those in which all the weights assigned to the binary digits are positive. There are only 17 positively weighted codes. In every positively weighted code, the first weight must be 1, the second weight must be either 1 or 2, and the sum of all the weights must be equal to or greater than 9. The codes 8421, 7421, 6311, 5311, 5211, 4311, 4221, 3321, 2421 and 5421 are some of the positively weighted codes available which are given in Table 2.9 to represent 0 to 9 decimal numbers.

Let 2421 binary code. This code assigns $w_1 = 1$ to first bit, $w_2 = 2$ to second bit, $w_3 = 4$ to third bit, and $w_4 = 2$ to fourth bit, respectively.

$$(1011)\, of\ binary\ code\ 2421 = 2\times1 + 4\times0 + 2\times1 + 1\times1 \Leftrightarrow (5)_{10}$$

TABLE 2.9 | Positively weighted binary codes

Decimal	8421	7421	6311	5311	5211	4311	4221	3321	2421	5421
0	0000	0000	0000	0000	0000	0000	0000	0000	0000	0000
1	0001	0001	0001	0001	0001	0001	0001	0001	0001	0001
2	0010	0010	0011	0011	0011	0011	0010	0010	0010	0010
3	0011	0011	0100	0100	0101	0100	0011	0011	0011	0011
4	0100	0100	0101	0101	0111	0101	1000	0101	0100	0100
5	0101	0101	0111	1000	1000	1010	0111	1010	1011	1000
6	0110	0110	1000	1001	1010	1011	1100	1100	1100	1001
7	0111	1000	1001	1011	1100	1100	1101	1101	1101	1010
8	1000	1001	1011	1100	1110	1110	1110	1110	1110	1011
9	1001	1010	1100	1101	1111	1111	1111	1111	1111	1100

Negatively weighted code Negatively weighted codes, some of the weights assigned to the binary digits must be negative. The codes 84-2-1, 74-2-1, 731-2, 642-3, 631-1 and 443-2 are some of the negatively weighted codes. Table 2.10 shows some negatively weighted binary codes which represent 0 to 9 decimal digits.

Let 84-2-1 binary code. This code assigns $w_1 = -1$ to first bit, $w_2 = -2$ to second bit, $w_3 = 4$ to third bit, and $w_4 = 8$ to fourth bit, respectively.

$$(1111)\, of\ binary\ code\ 84-2-1 = 8\times1 + 4\times1 + (-2)\times1 + (-1)\times1 \Leftrightarrow (9)_{10}$$

TABLE 2.10 | Negatively weighted binary codes

Decimal	Negatively Weighted Codes						Non-weighted Codes	
	84-2-1	74-2-1	731-2	642-3	631-1	443-2	Excess-3	Gray
0	0000	0000	0000	0000	0000	0000	0011	0000
1	0111	0111	0010	0101	0010	0011	0100	0001
2	0110	0110	0111	0010	0101	1001	0101	0011
3	0101	0101	0100	1001	0100	0010	0110	0010
4	0100	0100	0110	0100	0110	1000	0111	0110
5	1011	1010	1001	1011	1001	0111	1000	0111
6	1010	1001	1011	0110	1011	1101	1001	0101
7	1001	1000	1000	1101	1010	0110	1010	0100
8	1000	1111	1101	1010	1101	1100	1101	1100
9	1111	1110	1111	1111	1111	1111	1100	1101

2.10.1.2 | Non-weighted Codes

Non weighted codes are codes do not obey the position-weighting principle. Excess-3 (XS-3) code and Gray code are non-weighted codes. Table 2.10 shows some non-weighted binary codes which represent 0 to 9 decimal digits.

2.10.2 | Sequential Codes

A sequential code is one in which each succeeding code word is binary number which is greater than its preceding code word. Consecutive number is one greater than the previous number represented by binary codes. Such a code facilitates mathematical manipulation of data. The 8421 and XS-3 codes are sequential. The codes 5211, 2421 and 642-3 are not sequential (Table 2.11).

TABLE 2.11 | Sequential and non-sequential binary codes

Decimal	Sequential Binary Codes		Non-sequential Binary Codes		
	8421	XS-3	5211	2421	642-3
0	0000	0011	0000	0000	0000
1	0001	0100	0001	0001	0101
2	0010	0101	0011	0010	0010
3	0011	0110	0101	0011	1001
4	0100	0111	0111	0100	0100
5	0101	1000	1000	1011	1011
6	0110	1001	1010	1100	0110
7	0111	1010	1100	1101	1101
8	1000	1101	1110	1110	1010
9	1001	1100	1111	1111	1111

2.10.2.1 | Self-complementary Codes

A code is said to be self-complementary, if the code word of the 9's complement of decimal digit can be obtained by interchanging all the 0s and 1s of 4-bit binary code representation of that digit. 9's complement of 9 is 0. Similarly, 9's complement of 8 is 1, 7 is 2, 6 is 3 and 5 is 4. Therefore, in a self-complementing code 2421, the code for 9 (1111) is 1's complement of the code for 0 (0000), the code for 8 (1110) is 1's complement of the code for 1 (0001), the code for 7 (1101) is 1's complement of the code for 2 (0010), the code for 6 (1100) is 1's complement of the code for 3 (0011), and the code for 5 (1011) is 1's complement of the code for 4 (0100).

The 2421, 5211, 84-2-1, 731-2, 642-3, 83-2-1 and XS-3 are self-complementing codes. The 8421, 5421 and 74-2-1 codes are not self-complementing. Tables 2.12 and 2.13 show the self-complementary codes.

The self-complementing property is desirable in a code when the 9's complement must be found such as in 9's complement subtraction. For a code to be self-complementing, the sum of all its weights must be 9. This is because whatever may be the weights, 0 is to be represented by 0000 and since in a self-complementary code, the code for 9 is the complement of the code for 0, 9 has to be represented by 1111. There are four (2421, 5211, 3321, 4311) positively weighted self-complementing codes. There are 13 negatively-weighted self-complementing codes.

TABLE 2.12 | Self-complementary and non-self-complementary Binary codes

Decimal	Self-complementary Binary Codes								Non-self-complementary Binary Codes	
	4311	4221	3321	2421	5211	84-2-1	731-2	642-3	8421	5421
0	0000	0000	0000	0000	0000	0000	0000	0000	0000	0000
1	0001	0001	0001	0001	0001	0111	0010	0101	0001	0001
2	0011	0010	0010	0010	0011	0110	0111	0010	0010	0010
3	0100	0011	0011	0011	0101	0101	0100	1001	0011	0011
4	0101	1000	0101	0100	0111	0100	0110	0100	0100	0100
5	1010	0111	1010	1011	1000	1011	1001	1011	0101	1000
6	1011	1100	1100	1100	1010	1010	1011	0110	0110	1001
7	1100	1101	1101	1101	1100	1001	1000	1101	0111	1010
8	1110	1110	1110	1110	1110	1000	1101	1010	1000	1011
9	1111	1111	1111	1111	1111	1111	1111	1111	1001	1100

TABLE 2.13 | Self-complementary and non-self-complementary binary codes

Decimal	Self-complementary Binary Codes					Non-self-complementary Binary Codes		
	Positively Weighted		Negatively Weighted		Non-weighted	Positively Weighted		Negatively Weighted
	2421	5211	84-2-1	731-2	XS-3	8421	5421	74-2-1
0	0000	0000	0000	0000	0011	0000	0000	0000

(Continued)

TABLE 2.13 | (Continued)

Decimal	Self-complementary Binary Codes					Non-self-complementary Binary Codes		
	Positively Weighted		Negatively Weighted		Non-weighted	Positively Weighted		Negatively Weighted
1	0001	0001	0111	0010	0100	0001	0001	0111
2	0010	0011	0110	0111	0101	0010	0010	0110
3	0011	0101	0101	0100	0110	0011	0011	0101
4	0100	0111	0100	0110	0111	0100	0100	0100
5	1011	1000	1011	1001	1000	0101	1000	1010
6	1100	1010	1010	1011	1001	0110	1001	1001
7	1101	1100	1001	1000	1010	0111	1010	1000
8	1110	1110	1000	1101	1101	1000	1011	1111
9	1111	1111	1111	1111	1100	1001	1100	1110

2.10.2.2 | Cyclic Codes

Cyclic codes are those in which each successive code word differs from the preceding one in only one bit position. They are also called unit distance codes. The Gray code is a cyclic code. It is often used for translating an analog quantity such as shaft position into a digital form. Table 2.10 shows gray code.

2.11 | BCD CODE

Binary to decimal conversion is quite difficult. In calculators, games and digital instruments, where there are frequent user inputs and outputs in decimal, a special code is used to represent decimal numbers. This code is called BCD code. BCD code is also known as 8421 code. It is a weighted code. For conversion, the decimal digit is translated directly to its 4-bit BCD equivalent or vice versa (Table 2.14).

One disadvantage of using BCD code is encountered when 9's complement of the number is computed.

TABLE 2.14 | BCD code

Decimal Digit	BCD Code			
	8s	4s	2s	1s
0	0	0	0	0
1	0	0	0	1
2	0	0	1	0
3	0	0	1	1
4	0	1	0	0
5	0	1	0	1
6	0	1	1	0

(*Continued*)

TABLE 2.14 | (Continued)

Decimal Digit	BCD Code			
	8s	4s	2s	1s
7	0	1	1	1
8	1	0	0	0
9	1	0	0	1

Decimal to BCD conversion Each decimal digit (0 to 9) is represented in its equivalent 4-bit BCD code given in Table 2.14.

Decimal number:	3	6	9	1	.	7
BCD:	0011	0110	1001	0001	.	0111

BCD to Decimal conversion Each 4-bit BCD code is represented in its equivalent decimal digit (0 to 9) given in Table 2.14. Start grouping from decimal point for integer part as well as for fractional part.

BCD:	0101	0110	1001	1000	.	0100
Decimal number:	5	6	9	8	.	4

The binary data can be considered as BCD data by making group of 4 bits starting from decimal point for integer part as well as for fractional part. The following binary numbers represented in 4-bit combination are not valid BCD digits however these are valid hexadecimal digits.

1010, 1011, 1100, 1101, 1110 and 1111

EXAMPLE 2.54 Convert 100111000010.1101 into BCD.

SOLUTION

Binary number:	100111000010			.	1101
Make group of 4-:	(1001)	(1100)	(0010)	.	(1101)
Decimal number:	9	Not valid	2	.	Not valid

So binary number 100111000010.1101 cannot be represented in BCD. **Ans.**

EXAMPLE 2.55 Convert 10010110.01100010 into BCD.

SOLUTION

Binary number:	10010110		.	01100010	
Make group of 4-:	(1001)	(0110)	.	(0110)	(0010)
Decimal number:	9	6	.	6	2

So, 10010110.01100010B = 96.62 BCD **Ans.**

2.11.1 | BCD Addition

BCD addition procedure is explained below:

- Add augend and addend digit pair one by one and observe carry from every addition of digit-pairs.
- There is carry over after adding each digit pair or sum of each digit pair is more than 9 then add 0110 $(6)_{10}$ else add $(0)_{10}$ to each digit.
- Propagate carry to next digit if there occurs.

Six is added to skip 6 hexadecimal digits *A*, *B*, *C*, *D*, *E* and *F* represented by 4-bit 1010, 1011, 1100, 1101, 1110 and 1111, respectively.

EXAMPLE 2.56 Perform BCD addition of the following

(a) 99 + 87 (b) 768.5 + 645.9

SOLUTION

(a)

		Decimal		BCD code		
Digits		2nd	1st	3rd	2nd	1st
Carry		1		1	1	
Augend		9	9		1001	1001
Addend		8	7		1000	0111
Sum	1	8	6	1	0010	0000
Carry						
					0010	0000
					0110	0110
Result				1	1000	0110
BCD				1	8	6

Explanation:

- First digit is less than (1001) but carry is 1 so add (0110)
- Second digit is less than (1001) but carry is 1 so add (0110)
- Carry is propagated after conversion

(b)

		Decimal				BCD numbers				
Digits		4th	3rd	2nd	1st	5th	4th	3rd	2nd	1st
Carry	1	1	1	1		0	0	0	0	
Augend		7	6	8	.5		0111	0110	1000	.0101
Addend		6	4	5	.9		0110	0100	0101	.1001
Sum	1	4	1	4	.4	0	1101	1010	1101	.1110

Carry	1	1	1	1	
Augend	0	1101	1010	1101	.1110
Addend		0110	0110	0110	.0110
Sum	1	0100	0001	0100	0100
	1	4	1	4	.4

Explanation:

- First digit is more than 1001, so add (0110)
- Second digit is more than 1001, so add $(0110)_{BCD}$
- Third digit is more than 1001, so add $(0110)_{BCD}$
- Fourth digit is more than 1001, but carry is 0 so add $(0110)_{BCD}$

2.11.2 | BCD Subtraction

BCD subtraction is performed preferably by 9's complement or 10's complement method. Best is 10's complement method because it gives only zero. Conventional method requires conversion from binary digit value to BCD digit value which is explained below

- Subtract subtrahend from minuend BCD digit wise. Binary subtraction is performed.
- If borrow is taken for the BCD digit pair to subtract, then subtract $(0110)_{BCD}$

EXAMPLE 2.57 Perform subtraction of BCD numbers using conventional method.

(a) (48 – 39) (b) (126 – 59)

SOLUTION

(a)

	Decimal		BCD	
Digits	2nd	1st	2nd	1st
Borrow	–1	10	–1	
Minuend	4	8	0100	1000
Subtrahend	3	9	0011	1001
Difference	0	9	0000	1111
				–0110
			0000	1001

Explanation:

- Minuend of first digit is less than subtrahend digit so needs borrow, so subtract $(0110)_{BCD}$ from the digit
- Minuend of second digit is larger than subtrahend, so there is no need of borrow

(b)

Digits	Decimal 3rd	2nd	1st	BCD 3rd	2nd	1st
	−1	10				
Borrow		−1	10	−1	−1	
Minuend	1	2	6	0001	0010	0110
Subtrahend	0	5	9	0000	0101	1001
Difference	0	6	7	0000	1100	1101
					−0110	−0110
				0000	0110	0111

Explanation:

- Minuend of first digit is less than subtrahend digit, so needs borrow; so subtract $(0110)_{BCD}$ from the digit
- Minuend of second digit is less than subtrahend digit, so needs borrow, hence so subtract $(0110)_{BCD}$ from the digit
- Minuend of third digit is larger than subtrahend, so there is no need of borrow

2.11.3 | BCD Subtraction using 9's Complement

To perform subtraction of BCD numbers *M* and *N*, procedure is outlined in algorithm 2.5 given below. Let (*M*) and (*N*) both are positive BCD numbers represented in 4-bit binary code.

Algorithm 2.5 | Subtraction of BCD number using 9's complement method

Step 1: Take 9's complement of the subtrahend, $(N)_{BCD}$. Subtract each digit from $(1001)_2$.

Step 2: Add minuend (*M*) and 9's complement of the subtrahend. (*N*) obtained in Step 1. Addition is performed by adding digits one by one.

Step 3: There is carry over after adding each digit pair or sum of each digit pair is more than 9 then add 0110 else add (0000) to each digit. Propagate carry to next digit if there occurs.

Step 4: Cheek end carry obtained after Step 3.

4.1 If an end carry occurs, then add '1' to the least significant digit of the number obtained after Step 3 and place positive sign (+ve).

4.2 If an end carry does not occur then take 9's complement of the number obtained in Step 3 and place negative sign (−ve).

In BCD number is of *n*-digits, the value in $(n + 1)^{th}$ digit is considered as end carry. However, BCD digit is represented by 4-bit binary code.

EXAMPLE 2.58 Perform subtraction of BCD numbers using 9's complement.

(a) $(64 - 25)_{BCD}$ (b) $(504.4 - 686.5)_{BCD}$

SOLUTION

(a) Take 9's complement of $(N)_{BCD} = 25$ and add it to $(M)_{BCD} = 64$
Nine's complement of 25 = 74
BCD code of 9's complement of 25 = 0111 0100

				⇔			
	0				0	0	
(M) =		6	4	⇔		0110	0100
Nine's complement of (N) =		7	4			0111	0100
(M) + (9's complement of (N)) =	1	3	8		0	1101	1000

		1101	1000
Convert binary into BCD		0110	0000
	1	0011	1000

End carry (digit in third digit after addition) is one, so add 1 to the LSD place positive sign

(M) + (9's complement of (N)) =		3	8				0011	1000
Add 1 to LSD =	+		1			+		1
(M) − (N) =	+	3	9	**Ans.**	⇔		0011	1001 **Ans.**
							3	9

(b) Take 9's complement of $(N)_{BCD} = 686.5$ and add it to $(M)_{BCD} = 504.4$
Nine's complement of 686.5 = 313.4
BCD code of 9's complement of 25 = 0011 0001 0011.0100

Carry	0					0				
(M) =	5	0	4	.4	⇔	0101	0000	0100	.0100
Nine's complement of (N) =	3	1	3	.4	⇔	0011	0001	0011	.0100
(M) + (Nine's complement of (N)) =	8	1	7	.8	⇔	1000	0001	0111	.1000

End carry (digit in fourth digit after addition) is one, so take 9's complement of the result and place negative sign.

(M) + (9's complement of (N)) =		8	1	7	.8			1000	0001	0111	.1000
So $(M)_8 - (N)_8$ =	−	1	8	2	1	**Ans.**	−	0001	1000	0010	.0001 **Ans.**

2.11.4 | BCD Subtraction using 10's Complement

To perform subtraction of BCD numbers *M* and *N*, procedure is outlined in algorithm 2.6. Let (*M*) and (*N*) both are positive BCD numbers represented by 4-bit binary code.

Algorithm 2.6 | Subtraction of BCD numbers using 10's complement method

Step 1: Take 10's complement of the subtrahend, $(N)_{BCD}$. Subtract each digit from $(1001)_2$ and add 1 to least significant digit (LSD).
Step 2: Add minuend (M) and 10's complement of the subtrahend. (N) obtained in Step 1. Addition is performed by adding digit pairs one by one. Observe carry after every addition of digit pairs.
Step 3: There is carry over after adding each digit pair or sum of each digit pair is more than 9 then add $(0110)_2$ else add $(0000)_2$ to each digit. Propagate carry to next digit if there occurs.
Step 4: Cheek end carry obtained after Step 3.
4.1 If an end carry occurs, then discard it and place positive sign (+ve).
4.2 If an end carry does not occur then take 10's complement of the number obtained in Step 3 and place negative sign (–ve).

In case, BCD number is of n-digits, the value in $(n + 1)^{th}$ digit is considered as end carry.

Example 2.59 Perform subtraction of BCD numbers using 10's complement.

(a) $(64 - 25)_{BCD}$ (b) $(504.4 - 686.5)_{BCD}$

Solution

(a) Take 10's complement of $(N)_{BCD} = 25$ and add it to $(M)_{BCD} = 64$
10's complement of 25 = 75
BCD code of 10's complement of 25 = 0111 0101

		0		0	0	
(M) =		6	4		0110	0100
Ten's complement of (N) =		7	5		0111	0101
(M) + (10's complement of (N)) =	1	3	9	0	1101	1001
					1101	1001
					0110	0000
				1	0011	1001

End carry (digit in third digit after addition) is one, so discard carry and place positive sign

$(M) - (N) = +\ 3\ 9$ **Ans.** 0011 1001 **Ans.**
3 9

(b) Take 10's complement of $(N)_{BCD} = 686.5$ and add it to $(M)_{BCD} = 504.4$
Ten's complement of 686.5 = 313.5
BCD code of 10's complement of 686.5 = 0011 0001 0011. 0101

	Decimal	BCD
Carry	0	0
$(M) =$	5 0 4 .4	0101 0000 0100 .0100
Ten's complement of $(N) =$	3 1 3 .5	0011 0001 0011 .0101
$(M) +$ (10's complement of $(N)) =$	8 1 7 .9	1000 0001 0111 .1001

End carry (digit in fourth digit after addition) is one, so take 10's complement of the result and place negative sign.

	Decimal	BCD
$(M) +$ (10's complement of $(N)) =$	8 1 7 .9	1000 0001 0111 .1001
So $(M)_8 - (N)_8 =$	– 1 8 2 .1 **Ans.**	– 0001 1000 0010 .0001 **Ans.**

2.12 | EXCESS-3 CODE

Excess-3 (XS-3) code is a kind of BCD code. Each decimal digit is coded into 4-bit binary code. The code for each digit is obtained by adding decimal 3 to the natural BCD code of the digit.

For example,

Decimal number = 3
BCD code = 0011
BCD code + 3 = (0011 + 0011)
Excess-3 (XS-3) code = 0110

For example,

Decimal number = 9
BCD code = 1001
BCD code + 3 = (1001 + 0011)
Excess-3 (XS-3) code = 1100

Excess-3 code is not a weighted code. Excess-3 is self-complementary code. It means 1's complementary of excess-3 code of a digit yields 9's complement of the digit itself (Table 2.15).

TABLE 2.15 | Self-complementary code

Decimal Digit	Excess-3 (XS-3) Code	1's Complement of Excess-3 Digit	Excess-3 (XS-3) Code	Decimal Digit	Remarks
0	0011	1100	1100	9	9's complement of 0 is 9
1	0100	1011	1011	8	9's complement of 1 is 8
2	0101	1010	1010	7	9's complement of 2 is 7
3	0110	1001	1001	6	9's complement of 3 is 6
4	0111	1000	1000	5	9's complement of 4 is 5

(Continued)

TABLE 2.15 | (Continued)

Decimal Digit	Excess-3 (XS-3) Code	1's Complement of Excess-3 Digit	Excess-3 (XS-3) Code	Decimal Digit	Remarks
5	1000	0111	0111	4	9's complement of 5 is 4
6	1001	0110	0110	3	9's complement of 6 is 3
7	1010	0101	0101	2	9's complement of 7 is 2
8	1011	0100	0100	1	9's complement of 8 is 1
9	1100	0011	0011	0	9's complement of 9 is 0

The following codes in binary cannot be represented in Excess-3 Code.

0000, 0001, 0010, 1101, 1110, 1111

EXAMPLE 2.60 Convert 3691.7 BCD number into excess-3 code.

SOLUTION

BCD number:	3	6	9	1	.	7
Excess-3 number:	0110	1001	1100	0100	.	1010

So, 3691.7 BCD1 = 0110100111000100.1010 Ecess-3 **Ans.**

EXAMPLE 2.61 Convert 10010110.0110 excess-3 code into BCD code.

SOLUTION

Excess-3 number:	10010110		.	0110
Make group of 4 bits:	(1001)	(0110)	.	(0110)
BCD number:	6	3	.	3

So, 10010110.0110 BCD = 63.3 Excess-3 code **Ans.**

2.12.1 | Xcess-3 (XS-3) Addition

- Addition is performed by adding augend and addend digit pair one by one. Carry occurred after addition of every digit pair is observed. Binary addition is performed and as a result binary result will occur.
- If there is carry over after the addition of one digit pair (4 bits), then add $(0011)_2$ to the resulting sum.
- If there is no carry over after the addition of one digit pair (4 bits), then subtract $(0011)_2$ from the resulting sum.

EXAMPLE 2.62 Perform Excess-3 addition of the following

(a) 45 + 26 (b) 99 + 87

SOLUTION

(a)

	Decimal		XS-3 code		
Digits	2nd	1st	3rd	2nd	1st
Carry	1	0	0	1	
Augend	4	5		0111	1000
Addend	2	6		0101	1001
Sum	7	1	0	1101	0001
Carry					
				1101	0001
				−0011	+0011
				1010	0100
Digit				7	1

Explanation:

- 1 as carry after addition of first digit pair, so add (0011) to sum (0001)
- 0 as carry after addition of second digit pair, so subtract (0011) from sum (1101)

(b)

	Decimal			XS-3 code		
Digits		2nd	1st	3rd	2nd	1st
Carry		1		1	1	
Augend		9	9		1100	1100
Addend		8	7		1011	1010
Sum	1	8	6	1	1000	0110
Carry						
					1000	0110
					+0011	+0011
				1	1011	1001
				1	8	6

Explanation:

- 1 as carry after addition of first digit pair, so add (0011) to sum (0110)
- 1 as carry after addition of second digit pair, so add (0011) to sum (1000)

2.12.2 | Excess-3 (XS-3) Subtraction

XS-3 subtraction is performed preferably by 9's complement or 10's complement method. 10's complement method is better because it gives only 0 after complement.

Conventional method requires conversion from binary digit value to XS-3 digit value which is explained below

- Subtract subtrahend from minuend digit wise. Then requirement of borrow is considered. Binary subtraction is performed and result is also in binary.
- If borrow is needed, then subtract (0011) from the resulting difference of digits.
- If no borrow is needed, then add (0011) to the resulting difference of digits.

EXAMPLE 2.63 | Subtract 39 from 48 using Excess-3 code.

SOLUTION

	Decimal			BCD	
Digits	2nd	1st		2nd	1st
Borrow	–1	10	0	–1	
Minuend	4	8		0111	1011
Subtrahend	3	9		0110	1100
Difference	0	9		0000	1111
				0000	1111
				0011	–0011
				0011	1100

Explanation:

- Minuend of first digit is less than subtrahend digit, so there is need of borrow hence subtract (0011) from the digit
- Minuend of second digit is larger than subtrahend, so there is no need of borrow hence add 0011

EXAMPLE 2.64 | Subtract 59 from 126 in Excess-3 code.

SOLUTION

	Decimal				BCD		
Digits	3rd	2nd	1st		3rd	2nd	1st
	–1	10					
Borrow		–1	10	0	–1	–1	
Minuend	1	2	6		0100	0101	1001
Subtrahend	0	5	9		0011	1000	1100
Difference	0	6	7		0000	1100	1101

0000	1100	1101
+0011	−0011	−0011
0011	1001	1010
0	6	7

Explanation:

- Minuend of first digit is less than subtrahend digit, so there is need of borrow hence subtract (0011) from the digit
- Minuend of second digit is less than subtrahend digit, so there is need of borrow hence subtract (0011) from the digit
- Minuend of third digit is larger than subtrahend, so there is no need of borrow hence add 0011

2.12.3 | Excess-3 (XS-3) Subtraction using 9's Complement

To perform subtraction of M and N in excess-3 code, procedure is given in algorithm 2.7. Let (M) and (N) both are positive decimal numbers represented in XS-3 code. Consider each number is of n-digits, the value in $(n + 1)^{th}$ digit is considered as end carry.

Algorithm 2.7 | Subtraction using 9's complement method in Excess-3 code

Step 1: Take 9's complement of the subtrahend, (*N*). Take 1's complement of each XS-3 code of number.
Step 2: Add minuend (*M*) and 9's complement of the subtrahend. (*N*) obtained in Step 1. Addition is performed by adding single digit at a time.
Step 3: Carry over after addition of each digit pair is observed
  3.1 If there is carry over after the addition of one digit pair, then add $(0011)_2$ to the resulting sum.
  3.2 If there is no carry over after the addition of one digit pair, then subtract $(0011)_2$ from the resulting sum.
Step 4: Cheek end carry obtained after Step 3.
  4.1 If an end carry occurs, then add '1' to the least significant digit of the number obtained after Step 3 and place positive sign (+ve).
  4.2 If an end carry does not occur, then take 9's complement or 1's complement of XS-3 code of the number obtained in Step 3 and place negative sign (−ve).

Example 2.65 Perform subtraction of XS-3 numbers using 9's complement.
(a) (64 – 25) (b) (504.4 – 686.5)

Solution

(a) Take 9's complement of (N) = 25 and add it to (M) = 64
Nine's complement of 25 = 74
XS-3 code of 9's complement of 25 = 1010 0111

XS-3 code of 25 = 0101 1000
One's complement of XS-3 code of 25 = 1010 0111

		0		1	0	
$(M) =$		6	4		1001	0111
Nine's complement of $(N) =$		7	4		1010	0111
$(M) +$ (9's complement of $(N)) =$	1	3	8	1	0011	1110
					0011	1110
					+0011	−0011
				1	0110	1011

End carry (digit in third digit after addition) is one, so add 1 to the LSD place positive sign

$M +$ (9's complement of $N) =$		3	8			0110	1011
Add 1 to LSD =	+		1		+		1
$M - N =$	+	3	9	**Ans.**		0110	1100 **Ans.**
						3	9

(b) Take 9's complement of $N = 686.5$ and add it to $M = 504.4$
Nine's complement of 686.5 = 313.4
XS-3 code of 9's complement of 686.5 = 0110 0100 0110. 0111
XS-3 code of 686.5 = 1001 1011 1001.1000
One's complement of XS-3 code of 686.5 = 0110 0100 0110. 0111

Carry	0				0 0	0	0	
$M =$	5	0	4	.4	1000	0011	0111	.0111
Nine's complement of $N =$	3	1	3	.4	0110	0100	0110	.0111
$M +$ (9's complement of $N) =$	8	1	7	.8	1110	0111	1101	.1110
					1110	0111	1101	.1110
					−0011	−0011	−0011	.0011
					1011	0100	1010	.1011

End carry (digit in fourth digit after addition) is one, so take 9's complement of the result and place negative sign.

$M +$ (9's complement of $N) =$		8	1	7	.8		1011	0100	1010	.1011
So $M - N =$	−	1	8	2	.1 **Ans.**	−	0100	1011	0101	.0100 **Ans.**
							1	8	2	.1

2.12.4 | Excess-3 (XS-3) Subtraction using 10's Complement

To perform subtraction of (M – N), algorithm 2.8 is given below. Let (M) and (N) both are positive decimal numbers represented in XS-3 code. Consider each number is of n-digits, the value in $(n + 1)^{th}$ digit is considered as end carry.

Algorithm 2.8 | Subtraction using 10's complement method in XS-3 code

Step 1: Take10's complement of the subtrahend (N). Take 2's complement of XS-3 code of the number.
Step 2: Add minuend (M) and 10's complement of the subtrahend. (N) obtained in Step 1. Addition is performed by adding single digit at a time.
Step 3: Carry over after addition of each digit pair is observed
3.1 If there is carry over after the addition of one digit pair, then add $(0011)_2$ to the resulting sum.
3.2 If there is no carry over after the addition of one digit pair then subtract $(0011)_2$ from the resulting sum.
Step 4: Cheek end carry obtained after Step 3.
4.1 If an end carry occurs, then discard it and place positive sign (+ve).
4.2 If an end carry does not occur, then take 10's complement or 2's complement of XS-3 code of the number obtained in Step 3 and place negative sign (–ve).

EXAMPLE 2.66 Perform subtraction of numbers using 10's complement using XS-3 code.

(a) (64 – 25) (b) $(504.4 - 686.5)_{BCD}$

SOLUTION

(a) Take 10's complement of (N) = 25 and add it to (M) = 64
Ten's complement of 25 = 75
XS-3 code of 10's complement of 25 = 1010 1000
XS-3 code of 25 = 0101 1000
Two's complement of XS-3 code of 25 = 1010 1000

Carry		0			1	0	
(M) =		6	4			1001	0111
10's complement of (N) =		7	5		+	1010	1000
(M) + (10's complement of (N)) =	1	3	9		1	0011	1111
						0011	1111
						+0011	–0011
					1	0110	1100

End carry (digit in third digit after addition) is one, so discard carry and place positive sign

(M) – (N) = + 3 9 **Ans.** 0110 1100 **Ans.**
3 9

(b) Take 10's complement of $(N) = 686.5$ and add it to $(M) = 504.4$
Ten's complement of $686.5 = 313.5$
XS-3 code of 10's complement of $686.5 = 0110\ 0100\ 0110.1000$

Carry	0					0	0	0	0	
$(M) =$	5	0	4	.4		1000	0011	0111	.0111	
Ten's complement of $(N) =$	3	1	3	.5		0110	0100	0110	.1000	
(M) + (10's complement of (N)) =	8	1	7	.9		1110	0111	1101	.1111	
					0					
						1110	0111	1101	.1111	
						−0011	−0011	−0011	−.0011	
						1011	0100	1010	.1100	

End carry (digit in fourth digit after addition) is one, so take 10's complement of the result and place negative sign.

(M) + (10's complement of (N)) =		8	1	7	.9		1011	0100	1010	.1100
So, $(M)_8 - (N)_8 =$	−	1	8	2	.1 **Ans.**	−	0100	1011	0101	.0100 **Ans.**
							1	8	2	.1

2.13 | GRAY CODE

The gray code is a non-weighted code and non-sequential code. It is not a BCD code. It is a cyclic code because successive code words in gray code differ in one bit position only.

- Gray code is a unit distance code
- Gray code is used extensively for shaft encoders.
- Gray code is not suitable for arithmetic operations.
- Gray code is a reflective code. N-LSBs for 2^n through $(2^{n-1} - 1)$ are the mirror image of 0 through 2^n.

A 2-bit gray code can be obtained by reflecting (2-1) bit code about an axis at the end of (2-1) bit code and putting MSB as 0 above axis and 1 below axis (Table 2.16).

TABLE 2.16 | Two-bit Gray Code

Decimal	Gray Code		Remarks
	G_1	G_0	
0	0	0	
1	0	1	Axis at end of 1-bit code
2	1	1	
3	1	0	

A 3-bit gray code can be obtained by reflecting (3-1) bit code about an axis at the end of (3-1) bit code and putting MSB as 0 above axis and 1 below axis (Table 2.17).

TABLE 2.17 | 3-bit Gray code

Decimal	Gray Code			Remarks
	G_2	G_1	G_0	
0	0	0	0	
1	0	0	1	Axis at end of 1-bit code
2	0	1	1	
3	0	1	0	Axis at end of 2-bit code
4	1	1	0	
5	1	1	1	Axis at end of 1-bit code
6	1	0	1	
7	1	0	0	

A n-bit gray code can be obtained by reflecting (n – 1) bit code about an axis at the end of (n – 1) bit code and putting MSB as 0 above axis and 1 below axis. Table 2.18 shows 4-bit gray code.

Table 2.18 | Four-bit Gray code

Decimal	Gray Code				Remarks
	G_3	G_2	G_1	G_0	
0	0	0	0	0	
1	0	0	0	1	Axis at end of 1-bit code
2	0	0	1	1	
3	0	0	1	0	Axis at end of 2-bit code
4	0	1	1	0	
5	0	1	1	1	Axis at end of 1-bit code
6	0	1	0	1	
7	0	1	0	0	Axis at end of 3-bit code
8	1	1	0	0	
9	1	1	0	1	Axis at end of 1-bit code
10	1	1	1	1	
11	1	1	1	0	Axis at end of 2-bit code
12	1	0	1	0	
13	1	0	1	1	Axis at end of 1-bit code
14	1	0	0	1	
15	1	0	0	0	

2.13.1 | Binary to Gray Code Conversion

Consider a n-bit binary number which is represented by $B_n, B_{n-1}, B_{n-2},\ldots, B_2, B_1$, where B_n is the MSB and B_1 is LSB. Consider also n-bit gray number which is represented by $G_n, G_{n-1}, G_{n-2},\ldots, G_2, G_1$, where G_n is MSB and G_1 is the LSB.

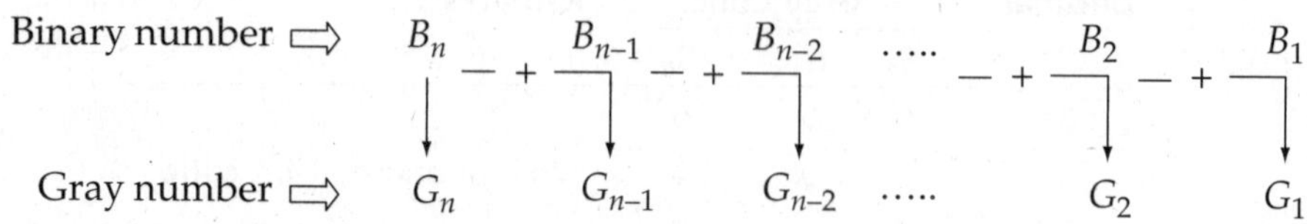

The conversion procedure is outlined below:

1. Write MSB of binary number (B_n) as the MSB of Gray number (G_n):

$$B_n \rightarrow G_n$$

2. Perform addition bit-wise, ignoring carry if any

$$B_i + B_{i-1} \rightarrow G_{i-1};\ i = n, n-1, n-2, \ldots, 3, 2$$

EXAMPLE 2.67 Convert binary 1001B into Gray code.

SOLUTION

	B_4	B_3	B_2	B_1
Given binary number is	1	0	0	1

Following the conversion procedure, binary bits are obtained

$$\begin{array}{llll} G_4 = B_4 = 1 & G_3 = B_4 + B_3 & G_2 = B_3 + B_2 & G_1 = B_2 + B_1 \\ & G_3 = 1 + 0 & G_2 = 0 + 0 & G_1 = 0 + 1 \\ & G_3 = 1 & G_2 = 0 & G_1 = 1 \end{array}$$

	G_4	G_3	G_2	G_1
Hence, gray number becomes:	1	1	0	1

2.13.2 | Gray to Binary Code Conversion

Consider n-bit binary number which is represented by $B_n, B_{n-1}, B_{n-2},\ldots, B_2, B_1$, where B_n is the MSB and B_1 is LSB. Also consider n-bit gray number which is represented by $G_n, G_{n-1}, G_{n-2},\ldots, G_2, G_1$, where G_n is the MSB and B_1 is the LSB.

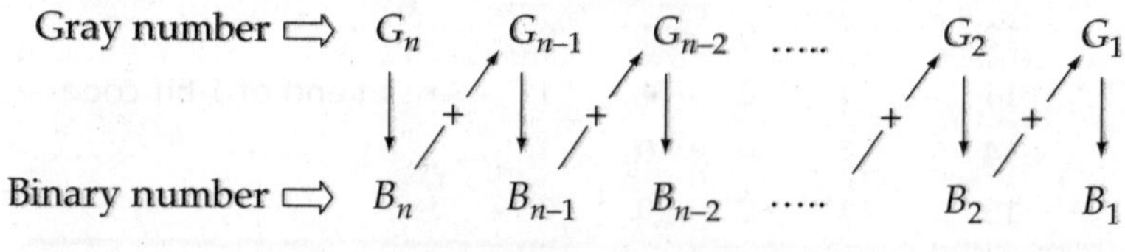

The conversion procedure is outlined below:

1. Write MSB of gray number (G_n) as the MSB of binary number (B_n):

$$G_n \rightarrow B_n$$

2. Perform addition bit wise, ignoring carry if any

$$B_i + G_{i-1} \rightarrow B_{i-1};\ i = n, n-1, n-2, \ldots, 3, 2$$

EXAMPLE 2.68 Convert the gray code to 1001 into binary code.

SOLUTION

$$\begin{array}{lcccc} & G_4 & G_3 & G_2 & G_1 \\ \text{Given gray number is} & 1 & 1 & 0 & 1 \end{array}$$

Following the conversion procedure, binary bits are obtained.

$$\begin{array}{llll} B_4 = G_4 & B_3 = B_4 + G_3 & B_2 = B_3 + G_2 & B_1 = B_2 + G_1 \\ B_4 = 1 & B_3 = 1 + 1 & B_2 = 0 + 0 & B_1 = 0 + 1 \\ & B_3 = 0 & B_2 = 0 & B_1 = 1 \end{array}$$

$$\begin{array}{lcccc} & B_4 & B_3 & B_2 & B_1 \\ \text{Hence, binary number becomes:} & 1 & 0 & 0 & 1 \end{array}$$

Table 2.19 gives the representation of the decimal numbers through 15 in the binary, octal, hexadecimal, BCD codes, excess-3 code (XS-3) and gray codes. From decimal number 8 to 15, two octal digits are required to represent. From decimal 10 to 15 BCD and XS-3 code requires two digits represented by 4-bit code each.

TABLE 2.19 | Decimal numbers 1 through 15 in various binary codes

Decimal	Binary	Octal	Hexadecimal	BCD	XS-3	Gray
0	0	0	0	0000	0011	0000
1	1	1	1	0001	0100	0001
2	10	2	2	0010	0101	0011
3	11	3	3	0011	0110	0010
4	100	4	4	0100	0111	0110
5	101	5	5	0101	1000	0111
6	110	6	6	0110	1001	0101
7	111	7	7	0111	1010	0100
8	1000	10	8	1000	1011	1100
9	1001	11	9	1001	1100	1101
10	1010	12	A	0001 0000	0100 0011	1111
11	1011	13	B	0001 0001	0100 0100	1110

(Continued)

TABLE 2.19 | (Continued)

Decimal	Binary	Octal	Hexadecimal	BCD	XS-3	Gray
12	1100	14	*C*	0001 0010	0100 0101	1010
13	1101	15	*D*	0001 0011	0100 0110	1011
14	1110	16	*E*	0001 0100	0100 0111	1001
15	1111	17	*F*	0001 0101	0100 1000	1000

2.14 | ALPHANUMERIC CODE

Codes that contain both alphabetic characters and numbers are needed when computers communicate with input/output devices (I/O devices). Such codes are called alphanumeric codes.

2.14.1 | American Standard Code for Information Interchange (ASCII) Code

American Standard Code for Information Interchange (ASCII) code is widely used alphanumeric code.

- It is pronounced as 'as-k-ee' code.
- ASCII code is 7-bit code.
- It gives code for $2^7 = 128$ characters.
- Table 2.20 shows ASCII code and list of abbreviations are given in Table 2.21.

TABLE 2.20 | ASCII code

MSB LSB	000	001	010	011	100	101	110	111
0000	NUL	DEL	Space	0	@	P		
0001	SOH	DC1	!	1	A	Q	a	Q
0010	STX	DC2	"	2	B	R	b	R
0011	ETX	DC3	#	3	C	S	c	S
0100	EOT	DC4	$	4	D	T	d	T
0101	END	NAK	%	5	E	U	e	U
0110	ACK	SYN	&	6	F	V	f	V
0111	BEL	ETB	'	7	G	W	g	W
1000	BS	CAN	(	8	H	X	h	X
1001	HT	EM	)	9	I	Y	i	Y
1010	LF	SUB	*	:	J	Z	j	Z
1011	VT	ESC	+	;	K	[	k	{
1100	FF	FS	'	<	L	\	l	|
1101	CR	GS	-	=	M	]	m	}
1110	SO	RS	.	>	N	^	n	~
1111	S1	US	/	?	O	_	o	DLE

TABLE 2.21 | Abbreviations used in Table 2.20

ACK	Acknowledge	EM	End of medium	NAK	Negative acknowledge
BEL	Bell	ENQ	Enquiry	NUL	Null
BS	Backspace	EOT	End of transmission	RS	Record separator
CAN	Cancel	ESC	Escape	SI	Shift in
CR	Carriage return	ETB	End of transmission block	SO	Shift out
DC1	Direct control 1	EXT	End of Text	SOH	Start of heading
DC2	Direct control 2	FF	Form feed	STX	Start text
DC3	Direct control 3	FS	Form separator	SUB	Substitute
DC4	Direct control 4	GS	Group separator	SYN	Synchronous idle
DEL	Delete idle	HT	Horizontal tab	US	Unit Separator
DLE	Data link escape	LF	Line feed	VT	Vertical tab

2.14.2 | Extended Binary-coded Decimal Interchange Code (EBCDIC)

Extended Binary Coded Decimal Interchange Code (EBCDIC) is an alphanumeric code.

- It is pronounced as **'eb-si-dic'** code.
- EBCDIC code is 8-bit code.
- It gives code for $2^8 = 256$ characters.
- Table 2.22 shows EBCDIC code for characters.

TABLE 2.22 | EBCDIC

Least Significant Hexadecimal Digit (LSD)	Most Significant Hexadecimal Digit (MSD)															
	0	1	2	3	4	5	6	7	8	9	A	B	C	D	E	F
0	NUL	DLE	DS		SP	&							[	]	\	0
1	SOH	DC1	SOS				/		a	j	~		A	J		1
2	STX	DC2	FS	SYN					b	k	s		B	K	S	2
3	ETX	DC3							c	l	t		C	L	T	3
4	PF	RES	BYP	PN					d	m	u		D	M	U	4
5	HT	NL	LF	RS					e	n	v		E	N	V	5
6	LC	BS	EOB	YC					f	o	w		F	O	W	6
7	DEL	IL	PRE	EOT					g	p	x		G	P	X	7
8		CAN							h	q	y		H	Q	Y	8
9		EM							i	r	z		I	R	Z	9
A	SMM	CC	SM		φ	!	|	:								
B	VT				.	$	,	#								
C	FF	IFS		DC4	<	*	%	@								
D	CR	IGS	ENQ	NAK	(	)	-	,								
E	SO	IRS	ALK		+	;	>	=								
F	SI	IUS	BEL	SUB	|		?	.								

2.14.3 | Unicode Characters

Basically, computers deal with numbers. Machines store letters, numeric and special characters by assigning a number for each one. Before invent of unicode, there were many different encoding systems for assigning these numbers. No single encoding could contain enough characters: For instance for a language like English no single encoding was adequate for all the letters, punctuation, and technical symbols in common use. These encoding systems are in conflict with one another. That is, two encodings can use the same number for two different characters, or use different numbers for the same character. Computers particularly servers need to support many different encodings because data is passed between different encodings or platforms. Hence data always runs the risk of corruption.

Unicode provides a unique number for every character, irrespective of the platform and the program written in any programming language. The Unicode Standard has been adopted by such industry leaders as Apple, Hewlett–Packard (HP), International Business Machines Corporation (IBM), Microsoft, Oracle, SAP (Systems, Applications and Products in Data Processing), Sun, Sybase, Unisys etc. Unicode is required by modern standards such as Extensible Markup Language (XML), Java, JavaScript, LDAP (Lightweight Directory Access Protocol), CORBA 3.0 (Common Object Request Broker Architecture), Wireless Markup Language (WML), etc., and is the official way to implement ISO/IEC 10646. It is supported in many operating systems, all modern browsers, and many other products. The emergence of the Unicode Standard, and the availability of tools supporting it, is among the most significant recent global software technology trends. Incorporating Unicode into client–server or multi-tiered applications and websites offers significant cost savings over the use of legacy character sets. Unicode enables a single software product or a single website to be targeted across multiple platforms, languages and countries without re-engineering. It allows data to be transported through many different systems without corruption.

Unicode defines a code space in the range 0hex to 10FFFFhex. Normally a Unicode code point is referred to by writing 'U+' followed by its hexadecimal number. For code points in the Basic Multilingual Plane (BMP), four digits are used (e.g. U+0058 for the character Latin capital letter X); for code points outside the BMP, five or six digits are used, as required (e.g. U+E0001 for the character language tag and U+10FFFD for the character private use character-10FFFD). Older versions of the standard have used similar notations, but with slightly different rules.

The Unicode Standard, the latest version of Unicode, consists of a repertoire of more than 109,000 characters covering 93 scripts, a set of code charts for visual reference, an encoding methodology and set of standard character encodings, an enumeration of character properties such as upper and lower case, a set of reference data computer files, and a number of related items, such as character properties, rules for normalization, decomposition, collation, rendering, and bidirectional display order (for the correct display of text containing both right-to-left scripts, such as Arabic and Hebrew, and left-to-right scripts). Latest Unicode is Unicode 6.0.

2.15 | ERROR DETECTION CODES

Binary data is transmitted from one place to another. The transmission of data (Figure 2.5) may be serial or parallel. Binary is also processed. During transmission or processing, binary datum is susceptible to noise that can alter or distort its contents.

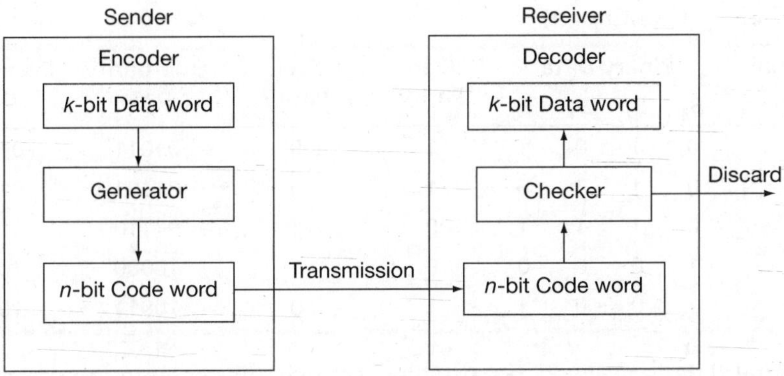

FIGURE 2.5 | Data transmission

The 1′s may be changed to 0′s and 0′s to 1′s. Noise causes error in data. Because digital systems must be accurate to the digit, errors can pose a serious problem in processing the data. Many schemes exist to detect the occurrence of a single-bit error in a binary word, so that whenever such an error occurs the concerned binary word can be corrected and can be retransmitted. Error detecting codes are parity, check sums, and block parity. Five-bit BCD codes, Ring counter code. Hamming code is 7-bit error-correcting code.

2.15.1 | Parity

The simple technique to detect errors is parity bit technique. Parity bit technique adds an extra bit, known as the parity bit, to each word which is being transmitted. There are two types of parity—odd parity and even parity.

Odd parity The parity bit is set to a 0 or a 1 at the sender such that the total number of 1 bit in the word including the parity bit is an odd number.

Even parity The parity bit is set to a 0 or a 1 at the sender such that the total number of 1 bit in the word including the parity bit is an even number.

Odd parity is preferred than even parity because even parity does not detect the situation where all 0s are created by short-circuit or some other fault condition. Table 2.23 shows the parity bits to be added to transmit decimal digits 0 through 9 in binary.

TABLE 2.23 | Parity bit data

Decimal	Binary Data				Odd Parity	Even Parity	Odd-parity Data	Even-parity Data
	B_3	B_2	B_1	B_0				
0	0	0	0	0	1	0	00001	00000
1	0	0	0	1	0	1	00010	00011
2	0	0	1	0	0	1	00100	00101
3	0	0	1	1	1	0	00111	00110
4	0	1	0	0	0	1	01000	01001

(Continued)

TABLE 2.23 | (Continued)

Decimal	Binary Data				Odd Parity	Even Parity	Odd-parity Data	Even-parity Data
	B_3	B_2	B_1	B_0				
5	0	1	0	1	1	0	01011	01010
6	0	1	1	0	1	0	01101	01100
7	0	1	1	1	0	1	01110	01111
8	1	0	0	0	0	1	10000	10001
9	1	0	0	1	1	0	10011	10010

As the digital data is received, a parity checking circuit generates an error signal if the total number of 1s is even in an odd-parity system or odd in an even-parity system. This parity check detects a single-bit error. It cannot detect two or more errors within the data word.

If the code have the property by which the occurrence of any single bit error transform a valid code word into an invalid one, it is said to be an error-detecting (single bit) code. In general, to obtain an n-bit error-detecting code, not more than half of the possible 2^n combinations of digits are used. To obtain an error-detecting code for 8 decimal digits, at least 4 binary digits are to be used. The code words are chosen in such a manner that in order to change one valid code word into another valid code word, at least two digits must be complemented. A code is an error-detecting code, if and only if its minimum distance is two or more. The distance between two words is defined as the number of digits that must change in a word so that the other word results. For example, the distance between 00110 and 10100 is 2, and the distance between 011100 and 100000 is 4.

2.15.2 | Block Parity

When several binary words are transmitted or stored in succession, the resulting collection of bits can be regarded as a block of data, having rows and columns. Parity bits can then be assigned to both rows and columns. This scheme is able to correct any single error occurring in a data word and to detect any two errors in a word. The parity row is often called a parity word. Such a block parity technique, also called word parity, is widely used for data stored on magnetic tapes.

For example, six 8-bit words in succession can be found in to a 6 × 8 block for transmission. Parity bits are added so that odd parity is maintained both row-wise and column-wise and the block is transmitted as 7 × 9 blocks as shown in Table 2.24. At the receiving end, parity is checked both row-wise and column-wise and supposes errors are detected as shown in Table 2.25. These single-bit errors detected can be corrected by complementing the error bit. In Table 2.25, parity errors in the third row and fourth column mean that the fourth bit in the third row is in error. It can be corrected by complementing it so 0 is complemented to 1.

TABLE 2.24 | Block parity

Row \ Column	8	7	6	5	4	3	2	1	Parity Bit
1	0	1	0	1	1	0	1	1	0
2	1	0	0	1	0	1	0	1	1

(Continued)

TABLE 2.24 | (Continued)

Row \ Column	8	7	6	5	4	3	2	1	Parity Bit
3	0	1	1	0	1	1	1	0	0
4	1	1	0	1	0	0	1	1	0
5	1	0	0	0	1	1	0	1	1
6	0	1	1	1	0	1	1	1	1
Parity row	0	1	1	1	0	1	1	0	0

TABLE 2.25 | Block parity with error

Row \ Column	8	7	6	5	4	3	2	1	Parity Column	Remarks
1	0	1	0	1	1	0	1	1	0	
2	1	0	0	1	0	1	0	1	1	
3	0	1	1	0	[0]	1	1	0	0	Parity error in third row
4	1	1	0	1	0	0	1	1	0	
5	1	0	0	0	1	1	0	1	1	
6	0	1	1	1	0	1	1	1	1	
Parity row	0	1	1	1	0	1	1	0	0	
					Parity error in fourth column					

Limitation: In Table 2.26, there are two errors detected at third and sixth column. But fourth row gives correct parity. So it becomes difficult to find which bit is to be corrected either third bit of fourth row or sixth bit of sixth row. Hence, two errors as shown in Table 2.25 can only be detected but not corrected. To correct the bit(s), column-wise error bits and row wise error bits should be same.

TABLE 2.26 | Limitation of block parity with error

Row \ Column	8	7	6	5	4	3	2	1	
1	0	1	0	1	1	0	1	1	0
2	1	0	0	1	0	1	0	1	1
3	0	1	1	0	1	1	1	0	0
4	1	1	1	1	0	1	1	1	0
5	1	0	0	0	1	1	0	1	1
6	0	1	1	1	0	1	1	1	1
Parity row	0	1	1	1	0	1	1	0	0
				Parity error in third and sixth columns					

2.15.3 | Five-bit Codes

Some 5-bit BCD codes that have parity contained within each code for ease of error detection are shown in Table 2.27. These are 63210, 2-out-5, shift-counter and 51111.

The **63210** is a weighted code (except for the decimal digit 0). It has the useful error-detecting property that there are exactly two 1's in each code group. This code is used for storing data on magnetic tapes.

The **2-out-of-5** code is a non-weighted code. It has exactly two 1's in each code group of 5 bits. This code is used in the telephone and communication industries. At the receiving end, the receiver can check the number of 1s in each character received.

The **shift-counter code**, also called the Johnson code or ring tail code, has the bit patterns produced by a 5-bit Johnson counter. The 51111 code is similar to Johnson code but is weighted.

TABLE 2.27 | Five-bit codes

Decimal	63210	2-out-of-5	Shift-counter or Ring Tail	51111
0	00001	00011	00000	00000
1	00011	00101	00001	00001
2	00101	00110	00011	00011
3	01001	01001	00111	00111
4	01010	01010	01111	01111
5	01100	01100	11111	10000
6	10001	10001	11110	11000
7	10010	10010	11100	11100
8	10100	10100	11000	11110
9	11000	11000	10000	11111

2.15.4 | The Biquinary Code

The biquinary code shown in Table 2.28 is a weighted 7-bit BCD code. It is a parity data code. Seven-bit code group consists of a 2-bit subgroup and a 5-bit subgroup, and each of these subgroups contains a single 1. Further, each code group has exactly two 1s and each subgroup has exactly one 1. Thus, it has the error checking feature. The weights of the bit positions are 50 43210. Since there are two positions with weight 0, it is possible to encode decimal 0 with a group containing 1's, unlike other weighted codes. The biquinary code is used in the Abacus.

TABLE 2.28 | Biquinary code

Decimal Digit	Biquinary Code						
	5	0	4	3	2	1	0
0	0	1	0	0	0	0	1
1	0	1	0	0	0	1	0
2	0	1	0	0	1	0	0
3	0	1	0	1	0	0	0

(Continued)

TABLE 2.28 | (Continued)

Decimal Digit	Biquinary Code						
	5	0	4	3	2	1	0
4	0	1	1	0	0	0	0
5	1	0	0	0	0	0	1
6	1	0	0	0	0	1	0
7	1	0	0	0	1	0	0
8	1	0	0	1	0	0	0
9	1	0	1	0	0	0	0

2.15.5 | The Ring Counter Code

A 10-bit ring counter produces a sequence of 10-bit groups. Each group of 10 bits has a single 1. The ring-counter code shown in Table 2.29 is the code obtained by assigning a decimal digit to each of those ten patterns. It is a weighted code (9 8 7 6 5 4 3 2 1 0) because each bit position has a weight equal to one of the 10 decimal digits. A ring counter code is not efficient. Binary 10 bits in combination gives 1024 (2^{10}) numbers however ring-counter code gives only 10 numbers. However, it has excellent error-detecting properties and is easier to implement.

TABLE 2.29 | Ring counter code

Decimal Digit	Ring Counter Code									
	9	8	7	6	5	4	3	2	1	0
0	0	0	0	0	0	0	0	0	0	1
1	0	0	0	0	0	0	0	0	1	0
2	0	0	0	0	0	0	0	1	0	0
3	0	0	0	0	0	0	1	0	0	0
4	0	0	0	0	0	1	0	0	0	0
5	0	0	0	0	1	0	0	0	0	0
6	0	0	0	1	0	0	0	0	0	0
7	0	0	1	0	0	0	0	0	0	0
8	0	1	0	0	0	0	0	0	0	0
9	1	0	0	0	0	0	0	0	0	0

2.15.6 | Check Sums

Simple parity cannot detect two errors within the same word. One way overcoming this difficulty is to use a sort of two-dimensional parity. As each word is transmitted, it is added to the sum of the previously transmitted words, and the sum is retained at the transmitter end. At the end of transmission, the sum (called the checksum) up to that time is sent to the receiver. The receiver can check its sum with the transmitted sum. If the two sums are different, the receiver can ask for retransmission of the entire data. This is the type of transmission used in teleprocessing systems.

The checksum is used in the Internet by several protocols although not at the data link layer. Like linear and cyclic codes, the checksum is based on the concept of redundancy. Several protocols still use the checksum for error detection.

Let data is a list of five 4-bit numbers that is required to send to a destination. In addition to sending these numbers, the sum of the numbers will be sent. For example, if the set of numbers is (7, 11, 12, 0, 6). The sender sends set of number is (7, 11, 12, 0, 6, −36), where −36 is the sum of the original numbers with negative sign. The negative (complement) of the sum is called the checksum. The receiver adds all the numbers received including the checksum. If the result is 0, it assumes no error; otherwise, there is an error.

One's complement The example discussed above has one major drawback. All of the data can be written as a 4-bit word (they are less than 15) except for the checksum. One solution is to use one's complement arithmetic. In this arithmetic, n-bits can be represented as unsigned numbers between 0 and $(2^n - 1)$. If the number is represented by more than n-bits, the extra leftmost bits need to be added to the n rightmost bits (wrapping). In one's complement arithmetic, a negative number can be represented by inverting all bits (changing a 0 to a 1 and a 1 to a 0). This is the same as subtracting the number from $(2^n - 1)$.

Checksum procedure

Sender end:

- Add all the n-bit numbers.
- Wrap the $(m - n)$ extra bits by adding them to the remaining n-bits. m is more than n. It is called as n-bit wrapped sum.
- Take 1's complement of n-bit wrapped sum.
- It is treated as checksum to send

Receivers end:

- Add all the n-bit numbers.
- Wrap the $(m - n)$ extra bits more than n by adding them to the remaining n-bits. m is more than n. It is called as n-bit wrapped sum.
- Take 1's complement of n-bit wrapped sum
- Result is zero then data is correct.
- Receiver drops checksum and receives correct data.

Figure 2.6 shows the process at the sender and at the receiver end. The sender initializes the checksum to 0 and adds all data items and the checksum. The result is 36. However, 36 cannot be expressed in 4 bits. The extra two (6-4) bits are wrapped and added with the sum to create the wrapped sum value 6. The wrapped sum is then complemented, resulting in the checksum value 9 (15 – 6 = 9). The sender now sends six data items to the receiver including the checksum 9.

The receiver follows the same procedure as the sender. It adds all the data items (including the checksum); the result is 45. The sum is wrapped and becomes 15. The wrapped sum is complemented and becomes 0. Since the value of the checksum is 0, this means that the data is not corrupted. The receiver drops the checksum and keeps the other data items. If the checksum is not zero, the entire packet is dropped.

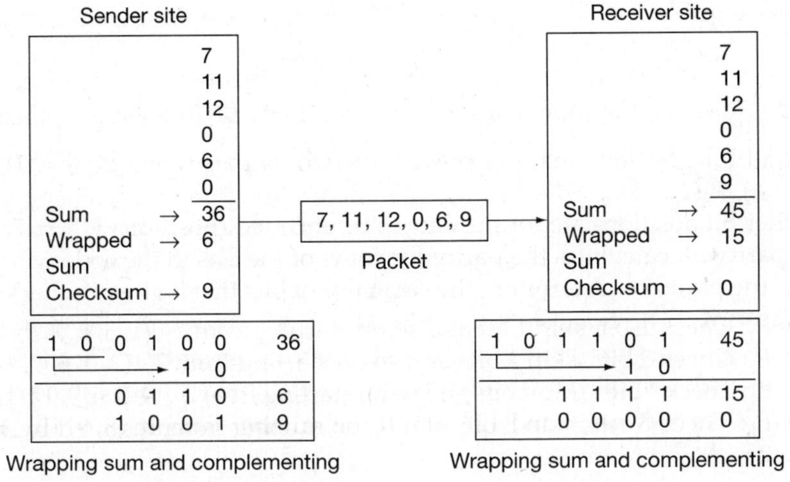

FIGURE 2.6 | Checksum-based transmission

2.15.7 | Error-correcting Code

Error-correcting codes are those codes which detect the error as well as correct it. The hamming code detects and corrects errors. Other error-correcting codes available are block code, convolution code, cyclic code, etc.

Hamming code detects an error and indicates which bit is in error. To generate Hamming code for a group of M-bit information, K number of parity bits is added to form an $(M + K)$ bit code. The location of each of $(M + K)$ bits within code word is assigned a decimal value, starting by assigning a 1 to the MSB and $(M + K)$ to the LSB. K parity checks are performed on selected bits of each code word. The result of each parity check is recorded as 1 or 0 depending whether the error has or has not been detected. The value of K should satisfy the inequality.

$$2^K \geq M + K + 1 \tag{2.15}$$

This code is used for detecting and correcting a single bit error on a message of any length. The parity bits $P_1, P_2, P_3, P_4, \ldots$ are placed at each 2^n bit position, where $n = 0, 1, 2, \ldots, K - 1$. Suppose original message is 4 bits longer. The value of K must be chosen to satisfy Eq. (2.15) where $M = 4$.

$$2^K \geq 4 + K + 1$$
$$2^K \geq 5 + K$$

Therefore, the minimum value of K for which above inequality is satisfied is 3, so number of parity bits, $K = 3$,

The parity bits P_1, P_2 and P_3 are placed in locations 2^0, 2^1 and 2^2, i.e., 1, 2 and 4, respectively. Hence, the final code is

	M_4	M_3	M_2	2^2	$(M_1)2^0$	2^1	2^0
Bit number	7	6	5	4	3	2	1
Code word	D_7	D_6	D_5	P_4	D_3	P_2	P_1

In the above expressions, M_4, M_3, M_2, and M_1 represent the message bits. M_1 is LSB and M_4 is MSB.

Sender End To create the code word to be transmitted, the following is considered.

1. Mark all bit positions that are powers of two as parity bits $P_1(2^0 = 1)$, $P_2(2^1 = 2)$, $P_4(2^2 = 4)$, $P_8(2^3 = 8)$, …
2. All other bit positions are for the data to be encoded (positions 3, 5, 6, 7, 9, 10, …)
3. Each parity bit calculates the parity for some of the bits in the code word. The position of the parity bit determines the sequence of bits that it already checks and skips.
   Parity P_1: Check 1-bit, skip 1 bit and so on starting from 2^0. (1, 3, 5, 7, 9, 11, …)
   Parity P_2: Check 2 bits, skip 2 bits and so on starting from 2^1. (2, 3, 6, 7, 10, 11, …)
   Parity P_4: Check 4 bits, skip 4 bits and so on starting from 2^2. (4, 5, 6, 7, 12, 13, 14, 15, …)
   Parity P_8: Check 8 bits, skip 8 bits and so on starting from 2^3. (8, 9, 10, 11, 12, 13, 14, 15, 24, 25…)
4. For even parity or, set a parity bit to 1 if the total number of ones in the positions it checks is odd. Set a parity bit to 0 if the total number of ones in the position, it checks is even.
5. For odd parity, set a parity bit to 0 if the total number of ones in the positions it checks is odd. Set a parity bit to 1 if the total number of ones in the position it checks is even.

Receiver End

1. Each parity bit calculates the parity for some of the bits in the code word. The position of the parity bit determines the sequence of bits that it already checks and skips.
   Parity P_1: Check 1 bit, skip 1 bit and so on starting from 2^0. (1, 3, 5, 7, 9, 11, …)
   Parity P_2: Check 2 bits, skip 2 bits and so on starting from 2^1. (2, 3, 6, 7, 10, 11, …Parity P_4: Check 4 bits, skip 4 bits and so on starting from 2^2. (4, 5, 6, 7, 12, 13, 14, 15, …)
   Parity P_8: Check 8 bits, skip 8 bits and so on starting from 2^3. (8, 9, 10, 11, 12, 13, 14, 15, 24, 25…)
2. For even parity or, set a parity bit to 1 if the total number of ones in the positions it checks is odd. Set a parity bit to 0 if the total number of ones in the position, it checks is even.
3. For odd parity, set a parity bit to 0 if the total number of ones in the positions it checks is odd. Set a parity bit to 1 if the total number of ones in the position it checks is even.
4. Check 3-bit parity code ($C_4C_2C_1$) and detect error in the bit by converting binary 3-bit data to decimal. Alter the decoded bit. The explanation is given below.

	M_4	M_3	M_2	2^2	M_1	2^1	2^0	
Bit number	7	6	5	4	3	2	1	
Code word	D_7	D_6	D_5	P_4	D_3	P_2	P_1	No error
Error code	111	110	101	100	011	010	001	000

5. Remove the parity bits P_1 $(2^0 = 1)$, $P_2(2^1 = 2)$, $P_4(2^2 = 4)$, $P_8(2^3 = 8)$.
6. All other bit positions are data positions (positions 3, 5, 6, 7) which were transmitted.

EXAMPLE 2.69 What is the size of code word if data size is 5 bits?

SOLUTION Size of data, $M = 5$. No. of parity bits, $K = ?$

$$2^K \geq 5 + K + 1$$

$$2^K \geq 6 + K$$

Let $K = 1$; then $2^1 < 6 + 1$

$K = 2$; then $2^2 < 6 + 2$

$K = 3$; then $2^3 < 6 + 3$

$K = 4$; then $2^4 \geq 6 + 4$

So, number of parity bits, $K = 4$. Size of code word is $M + K = 9$ $(5 + 4)$ **Ans.**

EXAMPLE 2.70 Construct code word to transmit 1010 data with odd parity.

SOLUTION No. data of data, $M = 4$. No. of parity bits, $K = 3$ ($\because 2^3 \geq 4 + 3$). So, size of code word = 7.

So	M_4	M_3	M_2	2^2	M_1	2^1	2^0
Bit number	7	6	5	4	3	2	1
Data and parity bits	1	0	1	P_4	0	P_2	P_1
Code word	1	0	1	1	0	0	1

Because

To find Parity bit P_1, bits are to be checked	7	5	3	1
Number of 1's becomes 3 (odd), if $P_1 = 1$	1	1	0	1
To find Parity bit P_2, bits are to be checked	7	6	3	2
Number of 1's becomes odd, if $P_2 = 0$	1	0	0	0
To find Parity bit P_4, bits are to be checked	7	6	5	4
Number of 1's becomes odd, if $P_4 = 1$	1	0	1	1

Code word to be transmitted is (1011001)B **Ans.**

EXAMPLE 2.71 Construct code word to transmit 10111010 data with even parity.

SOLUTION No. of data bits, $M = 8$. No. of parity bits, $K = 4$ ($\because 2^4 \geq 8 + 4$). So, size of code word = 12.

So	M_8	M_7	M_6	M_5	2^3	M_4	M_3	M_2	2^2	M_1	2^1	2^0
Bit number	12	11	10	9	8	7	6	5	4	3	2	1
Data and parity bits	1	0	1	1	P_8	1	0	1	P_4	0	P_2	P_1
Code word	1	0	1	1	1	1	0	1	1	0	0	1

Because

To find parity bit P_1, bits are to be checked	11	9	7	5	3	1
Number of 1's becomes 4 (even), if $P_1 = 1$	0	1	1	1	0	1
To find parity bit P_2, bits are to be checked	11	10	7	6	3	2
Number of 1's becomes 2 (even), if $P_2 = 0$	0	1	1	0	0	0
To find parity bit P_4, bits are to be checked		12	7	6	5	4
Number of 1's becomes 4 (even), if $P_4 = 1$		1	1	0	1	1
To find parity bit P_8, bits are to be checked		12	11	10	9	8
Number of 1's becomes 4 (even), if $P_8 = 1$		1	0	1	1	1

Code word to be transmitted is (101111011001)B **Ans.**

EXAMPLE 2.72 A 7-bit hamming code is received as 1100111. Find error if any. Parity checks are created by odd parity.

SOLUTION

Positions of Bits	M_4	M_3	M_2	2^2	M_1	2^1	2^0	
Bit number	7	6	5	4	3	2	1	
Data received	1	1	0	0	1	1	1	
$C_4C_2C_1$	111	110	101	100	011	010	001	000

Because

To find parity bit P_1, bits are to be checked	7	5	3	1
Number of 1's becomes 3 (odd), if $C_1 = 0$	1	0	1	1
To find parity bit P_2, bits are to be checked	7	6	3	2
Number of 1's becomes 4 (even), if $C_2 = 1$	1	1	1	1
To find parity bit P_4, bits are to be checked	7	6	5	4
Number of 1's becomes 2 (even) so $C_4 = 1$	1	1	0	0

$C_4C_2C_1 = 110$, so error is in sixth bit. The correct transmitted data is 1001. **Ans.**

EXAMPLE 2.73 A 7-bit hamming code is received as 1001110. Find error if any. Parity checks are created by odd parity.

SOLUTION

Positions of Bits	M_4	M_3	M_2	2^2	M_1	2^1	2^0	
Bit number	7	6	5	4	3	2	1	
Data received	1	0	0	1	1	1	0	
$C_4C_2C_1$	111	110	101	100	011	010	001	000

Because

To find parity bit P_1, bits are to be checked	7	5	3	1
Number of 1's becomes 2 (even), if $C_1 = 1$	1	0	1	0
To find parity bit P_2, bits are to be checked	7	6	3	2
Number of 1's becomes 3 (odd), if $C_2 = 0$	1	0	1	1
To find parity bit P_4, bits are to be checked	7	6	5	4
Number of 1's becomes 2 (even) so $C_4 = 1$	1	0	0	1

$C_4C_2C_1 = 101$, so error is in fifth bit. The correct transmitted data is 1011. **Ans.**

2.16 | MULTI-PRECISION NUMBERS

A form of arithmetic two or more words may be used to represent each number. A double-precision number uses twice as many bits as a single-precision value, so it can represent quantities much more exactly. The precision of a value describes the number of digits that are used to express that value. The term double precision is something of a misnomer because the precision is not really double. The word double derives from the fact that a double-precision number uses twice as many bits as a regular number. For example, if a single-precision number requires 16 bits, its double-precision counterpart will be 32 bits long. Similarly, if a single-precision number requires 32 bits, its double-precision counterpart will be 64 bits long. The extra bits increase not only the precision but also the range of magnitudes that can be represented. The exact amount by which the precision and range of magnitudes are increased depends on what format the program is using to represent floating-point values. If a single-precision number requires 16 bits, its triple-precision will be 48 bits long and so on.

In computer science, double-precision arithmetic indicates that calculations are performed on numbers whose digits of precision are limited only by the available memory of the host computing system. This contrasts with the faster fixed-precision arithmetic found in most arithmetic logic units (ALU) hardware, which typically offers between 16 and 64 bits of precision. Multiple-precision arithmetic is also called bignum arithmetic, arbitrary precision arithmetic, and sometimes infinite-precision arithmetic.

Several modern programming languages have built-in support for bignums, and others have libraries available for multiple-precision integer and floating-point math. Rather than store values as a fixed number of binary bits related to the size of the processor register, these implementations typically use variable-length arrays of digits.

2.16.1 | Floating-point Numbers

To represent very large integer numbers, many bits are required. There is also a problem when number with both integer and fractional parts, such as 32.6581 need to be represented. The floating-point number system, based on scientific notation, is capable of representing very large and very small numbers without an increase in the number of bits and also for representing numbers that have both integer and fractional components as shown in Figure 2.7.

A floating-point number (also known as a real number) consists of three parts: (i) a sign, (ii) the mantissa and (iii) exponent.

Mantissa is the part of a floating-point number that represents the magnitude of the number and is between 0 and 1.

The exponent is the part of a floating-point number that represents the number of places that the decimal point (or binary point) is to be moved.

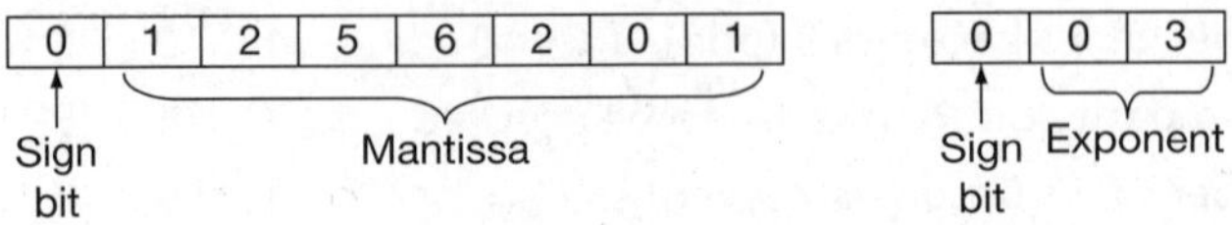

FIGURE 2.7 | Floating-point representation

A decimal example will be helpful in understanding the basic concept of floating-point numbers. Let's consider a decimal number which, in integer form, is 421,605,700. The mantissa is .4216057 and the exponent is 9. When the integer is expressed as a floating-point number, it is normalized by moving the decimal point to the left of all the digits so that the mantissa is a fractional number and the exponent is the power of 10. The floating-point number is written as 0.4216057×10^9.

To store a floating-point numbers there is need of two registers. The first register store mantissa and second stores the exponent (the position of radix or base point). For example, + 125.6201 real number is represented in floating-point representation as 0.1256201×10^3.

To store positive as well as negative exponents, sign bit is removed and exponent is called as biased exponent. Hence, total range 00 to 99 is divided into two parts. The number 50 is taken as origin and is known as bias, i.e. 0. The positive exponents are represented in the register in the range of numbers from 50 to 99. The subtraction of 50 gives the positive values from 00 to 49. Negative exponents are represented in the register in the range of 00 to 49. The subtraction of 50 gives the negative values in the range of –50 to –l. It is shown in Figure 2.8.

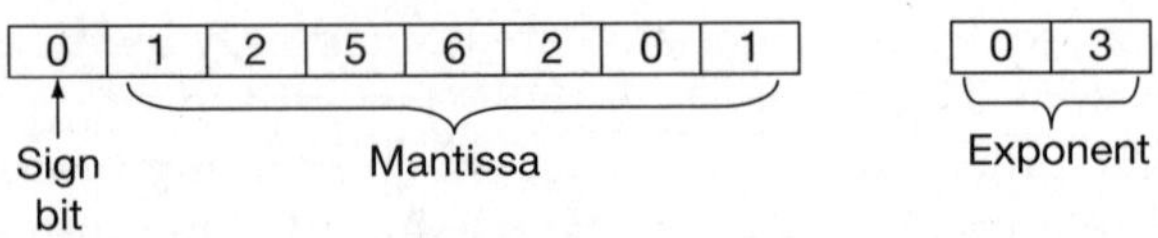

FIGURE 2.8 | Floating-point representation

To illustrate consider, floating-point number 0.4216057×10^9.
Mantissa component = 0.4216057
Actual exponent = +9
Biased exponent = 9 + 50 = 57.
Consider another floating-point number .1256201
Mantissa component = $.1256201 \times 10^{-3}$.
Actual exponent = –3
Biased exponent = – 3 + 50 = 47.
Floating-point number is always represented in the following form: $M \times r^E$
where M is mantissa portion, r is the radix or base and E is exponent or co-efficient
The number can be represented in the normalized form as:
$0.00125 \times 10^{-7} \Leftrightarrow 0.125 \times 10^{-9}$ and $0.000236 \times 10^7 \Leftrightarrow 0.236 \times 10^4$

2.16.2 | Binary Floating-point Numbers

Binary floating-point numbers can be represented by defining three parts of the binary number: number, (i) the sign bit (*S*) is the left-most bit, (ii) the exponent (*E*) includes the next eight bits, and (iii) the mantissa or fractional part (*M*) includes the remaining 23 bits as shown in Figure 2.9.

0	110 1101 0100000000000000	10001011
Sign bit (1bit)	Mantissa ((1+23) bits)	Exponent (8 bits)

FIGURE 2.9 | Floating-point representation

In the mantissa or fractional part, the binary point is understood to be to the left of the 23 bits. Effectively, there are 24 bits in the mantissa because in any binary number the left most (most significant) bit is always a 1. Therefore, this 1 does not occupy an actual bit position.

A biased exponent is represented by 8 bits, which is obtained by adding 127 to the actual exponent. The bias allows very large or very small numbers without using a separate sign bit for the exponents. The biased exponent allows a range of actual exponent values from −126 to + 128.

To express a binary number in floating-point format, consider a binary number 1011 0110 101 as an example. First, it can be expressed as 1 plus a fractional binary number by moving the binary point 10 places to the left and then multiplying by the appropriate power of two.

$$101\ 1011\ 0101 = 1.0110\ 1101\ 01 \times 2^{10}$$

Presuming that this is a positive number, the sign bit (*S*) is 0. The exponent, 10, is expressed as a biased exponent by adding it to 137 (10 + 127 = 137). The biased exponent (E) is expressed as the binary number 10001001. The mantissa is the fractional part (*M*) of the binary number, .0110 1101 01. Because there is always a 1 to the left of the binary point in the power-of-two expression, it is not included in the mantissa. The complete floating-point number is 1 bit (sign bit), 8 bits (biased exponent) and 23 bits (mantissa).

0	011 0110 1010 0000 0000 0000	10001001
Sign 1 bit	Mantissa (1 + 23) bits	Exponent (8 bits)

To evaluate a binary number that is in floating-point format is considered here. The general approach to determine the value of a floating-point number is expressed by following formula:

$$\text{Number} = (-1)^S \left(1 + \sum_{i=1}^{23} b_{-i} 2^{-i}\right)(2^{E-127})$$

To illustrate, consider the following floating-point binary number:

1	1000 1010 0001 0000 0000 000	10001111
Sign 1 bit	Mantissa (1 + 23) bits	Exponent (8 bits)

Sign bit = 1, Exponent = 10001111 = 143

$$\text{Number} = (-1)^1 (1.100\ 0101\ 0000\ 1000\ 0000\ 0000)(2^{143-127})$$

$$\text{Number} = (-1)(1.1000\ 1010\ 0001\ 0000\ 0000\ 000) \times 2^{16}$$

$$\text{Number} = -(11000101000010000)$$

This floating-point binary number is equivalent to −100,880 in decimal. Since the exponent can be any number between −126 and +128, extremely large and small numbers can be expressed. A 32-bit floating-point number can replace a binary integer number having 129 bits. Because the exponent determines the position of the binary point, numbers containing both integer and fractional parts can be represented.

There are two exceptions to the format for floating-point numbers: The number 0.0 is represented by all 0's, and ∞ is represented by all 1's in the exponent and all 0's in the mantissa.

2.16.3 | IEEE Standard for Floating-point Representation

For binary floating-point numbers, the format is defined by ANSI/IEEE Standard 754-1985 in three forms: single-precision, double-precision, and extended-precision. These all have the same basic formats except for the number of bits. Single-precision floating-point numbers have 32 bits, double-precision numbers have 64 bits, and extended-precision numbers have 80 bits.

Single-Precision Floating-Point Standard The IEEE (Institute of Electrical and Electronics Engineers) single-precision floating-point standard representation requires a 32-bit word, which may be represented as numbered from 0 to 31, left to right. The first bit is the sign bit, S, the next eight bits are the exponent bits, *E*, and the final 23 bits are the fraction *F*:

S	EEEE EEEE		FFFF FFFF FFFF FFFF FFFF FFF
0	1	8	9 31

The value of binary number represented by the word may be determined as follows:

- If exponent is 255 and fraction part is nonzero, then it is not a number (NaN).
- If exponent is 255 and *F* is zero and S is 1, then $V = -\infty$.
- If exponent is 255 and *F* is zero and S is 0, then $V = \infty$.
- If $0 < \text{exponent } i < 255$ then $\text{number} = (-1)^S \left(1 + \sum_{i=1}^{23} b_{-i} 2^{-i}\right)(2^{E-127})$.

  where '1.fraction' is intended to represent the binary number created by prefixing Fraction with an implicit leading 1 and a binary point.
- If exponent is 0 and fraction is nonzero, then $\text{Number} = (-1)^S (0.\text{Fraction})(2^{-126})$
- These are 'un-normalized' values.
- If exponent is 0 and fraction is zero and sign bit is 1, then number = −0.
- If exponent is 0 and fraction is zero and sign bit is 0, then number = 0.

Double-Precision Floating-Point Standard The IEEE double-precision floating-point standard representation requires a 64 bit word, which may be represented as numbered from 0 to 63, left to right. The first bit is the sign bit, S, the next eleven bits are the exponent bits, *E*, and the final 52 bits are the fraction *F*'.

S	EEEE EEEE EEE	FFFF FFFF FFFF FFFF FFFF FFFF FFFF FFFF FFFF FFFF FFFF FFFF FFF
0	1 11	12 63

The value of binary number represented by the word may be determined as follows:

- If exponent is 2047 and fraction part is nonzero, then it is not a number (NaN).
- If exponent is 2047 and F is zero and S is 1, then $V = -\infty$.
- If exponent is 2047 and F is zero and S is 0, then $V = \infty$.
- If $0 <$ exponent < 2047, then number $= (-1)^S \left(1 + .\sum_{i=1}^{52} b_{-i} 2^{-i}\right)(2^{E-1023})$.

  where '1.fraction' is intended to represent the binary number created by prefixing Fraction with an implicit leading 1 and a binary point.
- If exponent is 0 and fraction is nonzero, then number $= (-1)^S$ (0.Fraction)(2^{-1022}). These are 'un-normalized' values.
- If exponent is 0 and fraction is zero and Sign bit is 1, then number = –0.
- If exponent is 0 and fraction is zero and Sign bit is 0, then number = 0.

2.16.4 | Arithmetic Operations with Floating-point Numbers

Addition, subtraction, multiplication and division can be performed with floating-point notations. These operations are explained below.

Addition: Addition with floating-point number can be done in four steps:

(i) Make the exponent part of augend and addend equal.
(ii) Select the smaller value exponent either of augend or addend and give shift to that mantissa toward right and increment the corresponding exponent by one.
(iii) Mantissa of augend and addend are added.
(iv) Normalize the result if required.

EXAMPLE 2.74 Add 245630 and 344 using floating-number notation.

SOLUTION Augend 245630 is represented in floating-point number $= 0.245630 \times 10^6$

Addend 344 is represented in floating-point number $= 0.344 \times 10^3$

Exponent of addend is smaller than the exponent of augend. So given shift to mantissa and add one to exponent.

Addend 344 is represented in floating-point number $= 0.0344 \times 10^4$

Addend 344 is represented in floating-point number $= 0.00344 \times 10^5$

Addend 344 is represented in floating-point number $= 0.000344 \times 10^6$

$$245630 + 344 = 0.245630 \times 10^6 + 0.000344 \times 10^6$$

$$= 0.245974 \times 10^6 \Rightarrow 245974$$ **Ans.**

Subtraction: Subtraction with floating-point number can be done in four steps:

(i) Make the exponent part of minuend and subtrahend equal.
(ii) Select the smaller value exponent either of minuend or subtrahend and give shift to that mantissa toward right and increment the corresponding exponent by one.
(iii) Mantissa of subtrahend is subtracted from minuend.
(iv) Normalize the result if required.

EXAMPLE 2.75 Subtract 0.2456×10^{-6} from 0.344×10^{-4} using floating-number notation.

SOLUTION Let $M = 0.2456\times10^{-6}$

$N = 0.344\times10^{-4}$

Exponent of M is smaller than the exponent of N, so given shift to mantissa and add one to exponent.

$M = 0.2456\times10^{-6} \Leftrightarrow 0.02456\times10^{-5} \Leftrightarrow 0.002456\times10^{-4}$

$M - N = 0.002456\times10^{-4} - 0.344000\times10^{-4}$

$= -0.341544\times10^{-4}$ **Ans.**

Multiplication: Floating-point multiplication can be obtained by the following three steps:

(i) multiply the multiplicand mantissa and multiplier mantissa portions,
(ii) add multiplicand and multiplier exponents and
(iii) normalize the result.

EXAMPLE 2.76 Multiply 0.25×10^{-6} and 0.344×10^{-3} using floating-number notation.

SOLUTION Let $M = 0.25\times10^{-6}$

$N = 0.34\times10^{-3}$

$M\times N = (0.25\times0.34)\times10^{-6+(-3)} \Leftrightarrow 0.085\times10^{-9} \Leftrightarrow 0.85\times10^{-10}$ **Ans.**

Division: Floating-point division can be obtained by the following steps

(i) divide the dividend mantissa by divisor mantissa portions,
(ii) subtract divisor exponent from dividend exponent and
(iii) normalize the result obtained.

EXAMPLE 2.77 Divide 0.25×10^{-3} by 0.344×10^{-5} using floating-number notation.

SOLUTION Let $M = 0.25\times10^{-3}$

$N = 0.34\times10^{-5}$

$M\div N = (0.25\div0.34)\times10^{-3-(-5)} \Leftrightarrow 0.7353\times10^{+2}$ **Ans.**

EXAMPLE 2.78 Given that $32 = (200)_R$, find the value of R.

SOLUTION 32 is a decimal number.

$32 = 2\times R^2 + 0\times R^1 + 0\times R^0 \Leftrightarrow 32 = 2\times R^2 \Leftrightarrow R^2 = 16 \Leftrightarrow R = +4$

Hence, $R = 4$ **Ans.**

EXAMPLE 2.79 Given that $(2X)_8 = (36)_X$, find the value of X.

SOLUTION $2\times8 + X = 3\times X + 6$

$16 + X = 3X + 6$

$2X = 10$

Hence, $X = 5$ **Ans.**

EXAMPLE 2.80 Given that $(222)_X = (152)_7$, find the value of X.

SOLUTION $2X^2 + 2X + 2 = 7^2 + 5\times 7 + 2$

$2X^2 + 2X + 2 = 86$

$2X^2 + 2X - 84 = 0$

$$X^2 + X - 42 = 0 \Rightarrow R = \frac{-1 \pm \sqrt{(1)^2 - 4\times 1\times(-42)}}{2\times 1} = \frac{-1\pm 13}{2} = 6 \text{ or } -7$$

Hence, $X = 6$ **Ans.**

EXAMPLE 2.81 Given that $(292)_{10} = 565$, in some number system, find the base of that system.

SOLUTION Let 565 has base-R.

$292 = 5\times R^2 + 6\times R^1 + 5\times R^0$

$5R^2 + 6R - 287 = 0$

$R = \frac{-6 \pm \sqrt{(6)^2 - 4\times 5\times(-287)}}{2\times 5} = \frac{-6\pm 76}{10} = 7$ (Ignore negative value as base cannot be negative.)

Hence, $R = 7$ **Ans.**

EXAMPLE 2.82 Determine the missing number of the series: 10010, 200, 102, /, 30, 24, 22, 20.

SOLUTION Let 10010 is in binary. So, $1\times 2^4 + 0\times 2^3 + 0\times R^2 + 1\times 2^1 + 0\times 2^0 = 16 + 2 = (18)_{10}$.

Number 200 is of Base-3, so $2\times 3^2 + 0\times 3^1 + 0\times 3^0 = 18 + 0 + 0 = (18)_{10}$

Number 102 is of Base-4, so $1\times 4^2 + 0\times 4^1 + 2\times 4^0 = 16 + 0 + 2 = (18)_{10}$

Number 30 is of Base-6, so $3\times 6^1 + 0\times 6^0 = 18 + 0 = (18)_{10}$

Number 24 is of Base-7, so $2\times 7^1 + 4\times 7^0 = 14 + 4 = (18)_{10}$

Number 22 is of Base-8, so $2\times 8^1 + 2\times 8^0 = 16 + 2 = (18)_{10}$

Number 20 is of Base-9, so $2\times 9^1 + 0\times 9^0 = 18 + 0 = (18)_{10}$

So, find the equivalent of $(18)_{10}$ in base 5.

Dividend	÷	Divisor	=	Quotient	: Remainder, Base-5
18	÷	5	=	3	: $3 \rightarrow d_1$ (LSD)
3	÷	5	=	0	: $31 \rightarrow d_2$

Base-5 number = $(d_2 d_1)_3$

$(18)_{10} = (33)_5$ **Ans.**

EXAMPLE 2.83 Find the possible base in arithmetic operation $3234 - 1531 = 1303$.

SOLUTION Let base be R.

$$(3R^3 + 2R^2 + 3R + 4) - (R^3 + 5R^2 + 3R + 1) = (R^3 + 3R^2 + 3)$$
$$(R^3 - 3R^2 + 3) = (R^3 + 3R^2 + 3)$$
$$R^3 = 6R^2 \Leftrightarrow R = 6 \quad \textbf{Ans.}$$

EXAMPLE 2.84 Find the possible base in arithmetic operation $1234 + 4531 = 6065$.

SOLUTION Let base be R.

$$(R^3 + 2R^2 + 3R + 4) + (4R^3 + 5R^2 + 3R + 1) = (6R^3 + 6R + 5)$$
$$(5R^3 + 7R^2 + 6R + 5) = (6R^3 + 6R + 5)$$
$$7R^2 = R^3 \Leftrightarrow R = 7 \quad \textbf{Ans.}$$

EXAMPLE 2.85 Find the possible base in arithmetic operation $33 + 53 + 44 + 24 + 12 = 210$.

SOLUTION Let base be R.

$$(3R + 3) + (5R + 3) + (4R + 4) + (2R + 4) + (R + 2) = (2R^2 + R)$$
$$(15R + 16) = (2R^2 + R) \Rightarrow 2R^2 - 14R - 16 = 0$$
$$R^2 - 7R - 8 = 0 \Rightarrow R = \frac{-(-7) \pm \sqrt{(7)^2 - 4 \times 1 \times (-8)}}{2 \times 1} = \frac{7 \pm 9}{2} = 8 \text{ or } -1$$
$$7R^2 = R^3 \Leftrightarrow R = 7 \quad \textbf{Ans.}$$

EXAMPLE 2.86 Find the possible base in arithmetic operation $51/3 = 13$.

SOLUTION Let base be R.

$$\frac{5R + 1}{3} = R + 3 \Rightarrow (5R + 1) = 3(R + 3)$$
$$R = 8 \quad \textbf{Ans.}$$

EXAMPLE 2.87 Find the possible base in arithmetic operation $44/4 = 11$.

SOLUTION Let base be R.

$$\frac{4R + 4}{4} = R + 1 \Rightarrow (4R + 4) = 4(R + 1)$$
$$4R + 4 = 4R + 4$$

This equation is valid for $R \geq 5$ because 4 is not a valid digit of base 2, 3 and 4. **Ans.**

EXAMPLE 2.88 Find the possible base-*R* arithmetic operation $\sqrt{(51)_R} = (6)_{10}$.

SOLUTION

$$\sqrt{5R+1} = 6 \Rightarrow (5R+1) = 36 \Rightarrow 5R = 35$$
$$R = 7 \quad \textbf{Ans.}$$

SUMMARY

- Base-*r* (or radix-*r*) number system depends on the value of *r*.
  - Binary-number system has Base-2
  - Octal-number system has Base-8
  - Decimal-number system has Base-10
  - Hexadecimal-number system has Base-16
- Base-*r* (or radix-*r*) number system has *r* number of unique digits.
- Base-*r* number system has 0 to $(r - 1)$ distinct digits.
- Base or radix point separates the integer and fraction parts.
- Assigned weight to n^{th} digit of integer portion of base-*r* number is r^{n-1}.
- Assigned weight to m^{th} digit of fractional portion of base-*r* number is r^{-m}.
- Position of digits is counted from radix point.
- For integer portion, the position of digit is counted from right to left with respect to decimal point.
- For fractional portion, the position of digit is counted from left to right with respect to decimal point.
- Most significant digit (MSD) is the digit to which highest weight is assigned in the number or the extreme left digit in any number is the MSD.
- Least significant digit (LSD) is the digit to which lowest weight is assigned in the number or the extreme right digit is the LSD.
- Range of based-*r* numbers is 0 to $(r^n - r^{-m})_{10}$, where $(n + m)$ digits are combined out of which *n* digits represent in positive integer portion and *m* digits represent fractional portion.
- A base-*r* (radix-*r*) number can be converted to decimal number by the positional weights method.
- $(N)_{10} = \sum_{i=1}^{n} d_i r^{i-1} + \sum_{j=1}^{m} d_{-j} r^{-j}$
- So, for any decimal number, conversion into base-*r* is carried out separately for integer portion and fractional portion of decimal number and then both conversions are combined.
- The method of conversion from decimal positive integer to base-*r* (radix-*r*) number is to divide successively by *base* and resulting remainders are the digits for base-*r* number and quotient are used for successive operation. First remainder will act as LSD of base-*r* number. The process is repeated till quotient becomes zero.
- The conversion of a decimal fraction to base-*r* number is to multiply successively by base-*r* value *r* and use integers part of that number as the digits for base-*r* number and remained fraction part is taken as multiplicand for successive multiplication. The process is repeated till the fraction term becomes zero.

- Radix-minus-one or $(r - 1)$'s complement of base-r number N is defined as $(r^n)_r - (r^{-m})_r - (N)_r$.
- $(r - 1)$'s complement of a base-r number is obtained by subtracting each digit of base-r number from $(r - 1)$ or the largest digit from set of base-r distinct digits.
- Simple way to get 1's complement of a binary number is to change each bit of binary number from '1' to '0' and '0' to '1'.
- True or r's complement of $N = \begin{cases} (r^n)_r - (N)_r & \text{for } N \neq 0 \\ 0 & \text{for } N = 0 \end{cases}$
- r's complement of a base-r number is obtained by adding 1 (one) to LSD to its $(r - \mathbf{1})$'s complement.
- Considering from LSB, write each digit as such up to first appearance of digit '1'. After that, convert '0' to '1' and '1' to '0' and write.
- Binary codes may be numeric or alphanumeric.
- 1 bit is 1 binary digit.
- 1 Nibble is 4-bit binary data.
- 1 Byte is 8-bit binary data.
- 1 word is 16-, 32- or 64-bit binary data.
- On combining 10 bits of binary number, $1\,\text{k} = 2^{10}$ 10-bit binary words are obtained.
- On combining 20 bits of binary number, $1\,\text{M} = 2^{20}$, 20-bit binary words are obtained.
- On combining 30 bits of binary number, $1\,\text{G} = 2^{30}$, 30-bit binary words are obtained.
- On combining 40 bits of binary number, $1\,\text{T} = 2^{40}$ 40-bit binary words are obtained.
- Double precision or triple precision or multi-precision are used to represent large or very small numbers.
- Double-precision notation requires two storage locations (registers) to represent each word of binary number.
- In the floating-point notation, both very large and very small binary numbers can be represented conveniently.
- The main advantage of octal and hexadecimal systems is their ease of conversion to and from binary.
- An octal number is one-third the length of binary number and a hex number is one-fourth the length of the corresponding binary number.
- Replace each octal digit by its three-bit binary equivalent, to convert an octal number into binary.
- To convert a binary number, into Octal, starting from the decimal point make groups of 3 bits on both sides of the decimal point and replace each three-bit group by an octal digit.
- To convert a binary number into hex, starting from the decimal point, make groups of 4 bits on both sides of decimal point and replace 4-bit group by a hex digit.
- Replace each hex digit by its 4-bit binary equivalent, to convert a hex number into binary.
- The hex system is particularly useful for human communication with computers.
- Alphanumeric codes represent the letters of the alphabet and decimal number as a sequence of 0's and l's.
- A sequence of binary digits that represents a decimal digit is called a code word.
- Numeric codes used to represent decimal numbers are called BCD codes.

- The BCD codes may be weighted or non-weighted. The weighted codes obey the positive weighting principle.
- In every positively weighted code, one of the weights must be l, the second must be either 1 or 2 and the sum of all the weights must be equal to or greater than 9.
- Positively weighted codes are those codes in which all the weights assigned to the binary digits are positive only.
- Negatively weighted codes are codes in which at least one of the weights assigned to the binary digits is negative.
- There are 17 positively weighted codes, out of which only four are self-complementing and there are 13 negatively weighted self-complementing codes.
- In sequential code, each succeeding code word is one binary number greater than its preceding code word.
- In self-complementing code, the code word for the 9's complement of *N* is the 1's complement of code word for *N*.
- For a code to be self-complementing, the sum of positional weights must be 9.
- In a cyclic code, each successive code word differs in only one digit position from its preceding code word.
- The 8421, BCD is less efficient than pure binary, but its main advantage is its ease of conversion to and from decimal.
- Each one of the 4-bit BCD codes has 6 illegal states in 4-bit binary code.
- Excess-3 (XS-3) BCD code is a non-weighted code, but is used for mathematical operations because it is sequential code.
- In Excess-3 (XS-3) code, each binary code word is the corresponding 8421 code word plus 0011.
- Where special additive requirements exist, Excess-3 (XS-3) code can be used.
- Where mechanical systems are used, binary coded Gray codes are helpful.
- The gray code is a unit distance code and its main advantage is its ease of conversion from binary and vice versa.
- The Gray code is not BCD; it is not sequential; it is non-weighted; but it is cyclic and reflective.
- Gray code is not a BCD code; it is not sequential but cyclic and reflective.
- Gray code cannot be used for mathematical operations. It is used in instrumentation and acquisition systems where linear or angular displacement is measured. It is used shaft encoders, I/O devices, A/D converters and other peripheral devices.
- The simplest technique for detecting a single error is that of adding an extra bit, known as parity bit, to each word being transmitted.
- For even parity, the parity bit should be set to a 0 or an 1, so that the total number of 1s in the word to be transmitted including the parity bit is even.
- For odd parity, the parity bit should be set to a 0 or an 1, so that the total number of, 1s in word to be transmitted including the parity bit is odd.
- Odd parity is used more often than even parity, because even parity does not detect a situation where all 0s are created by a short circuit or any other fault condition.
- A code is said to be error detecting, if it possesses the property such that the occurrence of a single error transforms a valid code word into an invalid one.
- Check sums provide a sort of two-dimensional parity.
- Block parity can correct any single error in a data word and detect any two errors a data word.

- A code is said to be error correcting, if the correct code word can always be deduced from an erroneous word.
- The minimum distance of a code is the smallest number of bits by which any code words differ.
- The minimum distance of both BCD and XS-3 codes is l.
- The number of digits that must change in a word to get the other word is the distance between those two words.
- For single-bit error detection, the minimum distance between code words must be two or more, and for single-bit error correction it must be three or more.
- Hamming codes can, not only detect errors but also correct them.
- In the Hamming code, to each group of m data bits, k parity bits located at positions 2^{k-1} are added to form an $(m + k)$-bit code word.
- The ASCII code and the EBCDIC code are the most popular, modem alphanumeric codes.
- The EBCDIC code does not use a straight binary sequence for representing but uses BCD as the basis of binary assignment.

MULTIPLE CHOICE QUESTIONS

2.1 Which one of the following statement is true for 2-out-of 5 codes?
(a) It has even parity
(b) It has exactly two 1's in each code group
(c) Both (a) and (b)
(d) None of the above.

2.2 By taking 1's complement again of the 1's complement of a binary one gets.
(a) Two's complement
(b) The original number
(c) The sign magnitude form of the number
(d) Both (b) and (c).

2.3 The 7's complement of the number 45 is
(a) 43 (b) 32
(c) 33 (d) 42

2.4 The number of bits in the binary representation of the decimal number 19 is
(a) 3 (b) 4
(c) 5 (d) 6

2.5 In sign magnitude numbers,
(a) MSB stands for the sign
(b) MSB is always 0
(c) MSB is always 1
(d) MSB is part of the magnitude

2.6 The value of the binary 11111 is
(a) 2^5
(b) $2^5 - 1$
(c) $2^5 + 1$
(d) $2^4 - 1$

2.7 The Gray code for number 0110 is
(a) 0110 (b) 0101
(c) 1001 (d) 1100

2.8 Number125 in BCD code is
(a) 0001 0010 0101
(b) 0100 0101 1000
(c) 1110 1101 1010
(d) 1110 1101 1011

2.9 The 2's complement representation in 8 bit format is 11010000. The equivalent hexadecimal representation is
(a) −30 (b) +30
(c) −50 (d) +50

2.10 Two's complement can be obtained from 1's complement by
(a) Adding 1
(b) Subtracting 1
(c) Ignoring the MSB
(d) None of above.

2.11 Decimal equivalent of $(2CD.2F)_{16}$ is
(a) $(717)_{10}$ (b) $(0.1836)_{10}$
(c) $(717.1836)_{10}$ (d) $(717.18)_{10}$

2.12 If $(2X)_8 = (34)_X$, then find the value of X.
(a) 6
(b) 12
(c) 4
(d) None of the above

2.13 Which addition is correct?
(a) 1011 + 0101 = 1111
(b) $(23)_8 + (76)_8 = (121)_8$
(c) BC25H + 1B2EH = D653H
(d) $(2011)_3 + (2011)_3 = (4022)_3$

2.14 The sum of weights in a self-complementing code is
(a) −9 (b) 9
(c) 10 (d) 16

2.15 In 2's complement subtraction, if end carry occurs
(a) Add it with the result
(b) Take 2's complement of result
(c) Discard it
(d) None of the above

Answers

2.1 (c)	2.2 (b)	2.3 (a)	2.4 (c)	2.5 (d)	2.6 (b)	2.7 (b)
2.8 (a)	2.9 (a)	2.10 (a)	2.11 (c)	2.12 (a)	2.13 (b)	2.14 (b)
2.15 (c)						

QUESTIONS

2.1 Briefly define each of the following:
(a) Radix or base
(b) Number
(c) Two's complement
(d) Block parity

2.2 What is meant by a positional weighted system?

2.3 What is meant by 'base' or 'radix' of a number system?

2.4 How do you convert a decimal number to an equivalent number in any other base system?

2.5 How do you convert a number in any base system to an equivalent decimal number?

2.6 Explain: (a) bit; (b) byte; (c) nibble; and (d) word length

2.7 What are the advantages of (a) the binary number system and (b) the hexadecimal number system?

2.8 Discuss the computer method of multiplication and division.

2.9 What do you mean by double precision numbers and floating-point numbers?

2.10 How can negative numbers be represented?

2.11 How 1's complement and 2'scomplent of binary number can be obtained? Compare both.

2.12 How binary subtraction is performed using the 1's and 2's complement method?

2.13 What is end around carry? When does it come into picture?

2.14 How are binary and octal numbers related?

2.15 What are the advantages of octal and hexadecimal systems? Where are they used?

2.16 How hexadecimal number is converted to octal and vice versa?

2.17 What are the weights of the MSB of (a) a 10-bit binary number and (b) an 8-bit hex number?

2.18 Why can't the parity method detect even number of error in transmitted data?

2.19 What is the advantage of using Gray code? Enumerate the steps required for the conversion of binary number into its Gray code equivalent.

2.20 What are the various methods used to represent signed numbers, in digital systems?

2.21 Write a short note on:
(a) Hamming code
(b) Sign magnitude representation
(c) Block parity
(d) Floating-point representation

2.22 Why octal number system is used in digital systems?

2.23 Explain the procedure to convert: (a) Binary to Gray code (b) Gray code to Binary.

2.24 Define number. Give a general expression for a number to represent in base-*r* system?

2.25 Distinguish between error-correcting codes and error detecting codes.

2.26 How can the parity bit detect a single bit error in data unit?

2.27 What is the difference between an even parity and odd parity?

2.28 List the steps involved in creating a Hamming code.

2.29 Explain briefly: (a) true complement (b) radix-1 complement.

2.30 What do you understand by alphanumeric code? Discuss its area of applications.

PROBLEMS

2.1 Convert the following decimal numbers to binary:
(a) 136 (b) 0.723 (c) 1247.625

2.2 Convert the following numbers to decimal:
(a) $(101101.\ 10\ 10)_2$
(b) $(721)_8$
(c) $(12FA)_{16}$

2.3 Convert the following numbers to binary:
(a) $(FE\ 12)_{16}$ (b) $(713)_8$ (c) 723

2.4 Obtain 1's and 2's complement of following binary number:
(a) 10110101 (b) 10101.0 1 (c) 1100101 (d) 1110101

2.5 Find 8's complement of $(421)_8$.

2.6 Perform the subtraction with following binary numbers using (i) 1's complement (ii) 2's complement:
(a) 10101 – 0101 (b) 0110 – 1010

2.7 What is the advantage of using Gray code? Enumerate the steps required for the conversion of binary number into its Gray code equivalent. Convert 1001011B to Gray code.

2.8 Convert the following pairs of decimal numbers in excess-3 numbers and perform their addition in excess-3 code:
(a) 38, 37 (b) 129, 131

2.9 Convert the following pairs of decimal numbers in BCD numbers and perform their addition in BCD code.
(a) 48, 87 (b) 129, 171

2.10 What are the self-complementary codes? Is the following codes are self-complementary codes?
(a) 2, 4, 2, 1 (b) 8 , 4, −2 , −1 (c) BCD code

2.11 Find decimal equivalent of binary number (1101 1100), if binary number is
(a) Sign magnitude number
(b) Straight Binary number
(c) Two's complement representation
(d) Excess-3 number

2.12 Find the value of X if $(35)_X + (24)_X = (103)_X$

2.13 Construct the even parity seven bit Hamming code for a word 1001.

2.14 A 7-bit Hamming code is received as 0010010. Is there any error? If yes, locate the position of the error bit using parity checks assuming code was created using an even parity.

2.15 Convert the following hexadecimal number into octal.
(a) F51 (b) 2B3C (c) 21 (d) 9AC

2.16 Subtract the following decimal numbers by the 9's and 10's complement methods.
(a) 274 − 86 (b) 93 − 615 (c) 574.6 − 275.7 (d) 376.3 − 765.6

2.17 Convert the following binary numbers to decimal.
(a) 1011 (b) 1101 101 (c) 1101.11 (d) 1101110.011

2.18 Convert the following decimal numbers to binary:
(a) 37 (b) 128 (c) 197.56 (d) 205.15

2.19 Add the following binary numbers.
(a) 11011 + 1101
(b) 1011 + 1101 + 1001 + 1111
(c) 101111.101 + 1101111.01

2.20 Subtract the following binary numbers.
(a) 1011 − 101 (b) 10110 − 1011 (c) 1100.10 − 111.01

2.21 Multiply the following binary numbers.
(a) 1101 × 101 (b) 110001 × 10

2.22 Divide the following binary numbers.
(a) 1010 by 11 (b) 11110 by 101

2.23 Find the 16 bit 2's complement for the following decimal numbers.
(a) −37 (b) −173 (c) −65.5 (d) −197.5

2.24 Find the 16 bit, l's complement form of the following decimal numbers.
(a) −97 (b) −224 (c) −205.75 (d) −29.375

2.25 Subtract the following decimal number using 1's complement and 2's complement method.
(a) 49 – 16 (b) 27 – 75 (c) 125.3 – 46.4

2.26 Represent the following decimal numbers in 10-bit, (i) sign magnitude form, (ii) sign l's complement and (iii) sign 2's complement form:
(a) +14 (b) +27 (c) +45 (d) −17 (e) −47 (f) −76

2.27 Convert the following octal numbers to hexadecimal number:
(a) 256 (b) 2035 (c) 1762.46

2.28 Convert the following hexadecimal number to octal numbers:
(a) 4AB (b) 42FD (c) 4F7.A8 (d) BC70.0E

2.29 Convert the following octal numbers to decimal:
(a) 463 (b) 2056 (c) 3057.65 (d) 6534.04

2.30 Convert the following decimal numbers to octal:
(a) 287 (b) 3965 (c) 430.7 (d) 7485.65

2.31 The following bit stream is encoded using block parity and even parity. Correct the error; if any
11000011 11110011 10110010 00001010 10101010
00101011 10100111 01001011 111100001

2.32 Detect errors if any in the following message assuming it is in ASCII code with odd parity:
00001011100000110 100111000101010

2.33 If a word has 32 bits, how many bits would you allocate for mantissa and exponent? Explain your choice.

2.34 Why can't the parity method detect even number of error in transmitted data?

2.35 Is it possible to construct 4-3-2-1 weighted code? Justify your answer.

2.36 Show how the following floating-point arithmetic's are performed.
(a) $(0.7743 \times 10^{-2}) \times (0.1554 \times 10^{5})$
(b) $(0.7743 \times 10^{-2}) + (0.1554 \times 10^{1})$
(c) $(0.7743 \times 10^{-2}) / (0.1554 \times 10^{-5})$
(d) $(0.3344 \times 10^{-2}) - (0.8877 \times 10^{-1})$

2.37 Is it possible to construct following code? Justify your answers.
(a) 7421 (b) 7, 4, −2, −1 (c) 5421 (d) 6311
(e) 9876543210

2.38 The following words are to be sent using odd parity. What is the value of the parity bit in each case, respectively?
(a) 1010100 (b) 1110101 (c) 1100011 (d) 1111110

2.39 Add the following number in the given base without converting to decimal.
(a) $(345)_6$ and $(54.4)_6$ (b) $(652)_7$ and $(532)_7$

2.40 Use 9's complement and 10's complement to perform subtraction with given decimal numbers.
(a) 7265 – 5763 (b) 1324 – 4279

3

Digital Logic

CHAPTER OBJECTIVES

The main goal of this chapter is to impart knowledge about the digital logics. Readers will be able to discuss the following aspects in this chapter:

- Basic gates
- Universal logic gates
- Positive and negative logic
- Inhibit circuits
- Integrated circuits-7400 series
- ANSI | IEEE Standard logic signals
- Pulsed operation of logic gates

3.1 | BASIC GATES

In 1854, a mathematician named George Boole described the way logical decisions are made based on true or false circumstances. For instance, an animal is either alive or dead, a light is either on or off, a door is locked or unlocked; So, Boolean logic is the method to take logical decisions based on true or false circumstances. Boolean algebra uses symbols to represent a logical expression that has one of two possible values: true or false. The logical expression should be door is closed, push button is not pressed, or water level is low. Writing such expressions is very tedious, so symbols such as *X*, *Y* and *Z* can be substituted.

Boolean algebra differs from ordinary algebra because Boolean constants and variables are allowed to have only binary values either 0 or 1. Boolean variables are often used to represent the voltage level present on a wire or at the input/output terminals of a circuit. In a certain digital system, the Boolean value of 0 is generally assigned to any voltage in the range from 0 to 0.8 V, whereas, the Boolean value of 1 is assigned to any voltage in the range 2 to 5 V. So, Boolean 0 and 1 represent the state of a voltage variable, which is called its logic level. A voltage in a digital circuit is said to be at the logic 0 level or the logic 1 level, depending on its actual numerical value. In digital logic, several other terms are used synonymously with 0/LOW/OFF and 1/HIGH/ON.

Basically, Boolean algebra expresses the relationship between a logic circuit's inputs and outputs. Logical expressions describe the relationship between a logic circuit's output (the decision) and its inputs (the circumstances). Basic logic gates are the fundamental building blocks from which all of the logic circuits and digital systems are constructed. Logic gates produce one output level when some combinations of input levels are given. A different output level is produced when other combinations of input levels are fed. There are three basic types of gates-AND, OR and NOT. These basic operations are called logic operations. Digital circuits called logic gates can be constructed from diodes, transistors and resistors connected so that the circuit output is the result of a basic logic; operation (OR, AND, NOT) performed on the inputs. The interconnection of gates to perform a variety of logical operations is called logic design.

Boolean algebra describes and analyses these basic logic gates, then later to analyse and design combinations of logic gates connected as logic circuits. A truth table describes the dependence of a logic circuit's output on the logic level presents at the circuit's inputs. The number of input combinations will equal to 2^N for an N-input truth table. The list of all possible combinations follows the binary counting sequence.

3.1.1 | OR Gate

OR gate has two or more inputs and only one output. The logic symbol of two inputs OR gate is given in Figure 3.1. Truth table of 2-inputs OR-gate is given in Table 3.1. For 2-inputs, there are four (2^2) possible combinations. OR gate gives output as 1, if any one input of OR gate is a 1, otherwise output is 0.

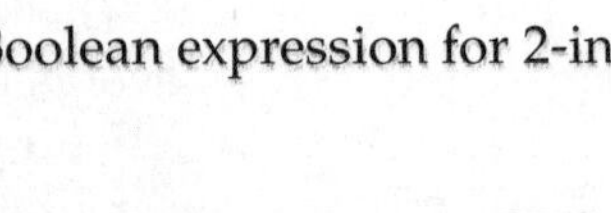

FIGURE 3.1 | Symbol of 2-input OR logic gate

TABLE 3.1 | Truth table of 2-input OR gate

Inputs		Output
A	*B*	*X*
0	0	0
0	1	1
1	0	1
1	1	1

Boolean expression for 2-input OR gate

$$X = A + B \tag{3.1}$$

where + sign represents OR logic.

Truth table is better explained with the help of electrical circuit shown in Figures 3.2 and 3.3.

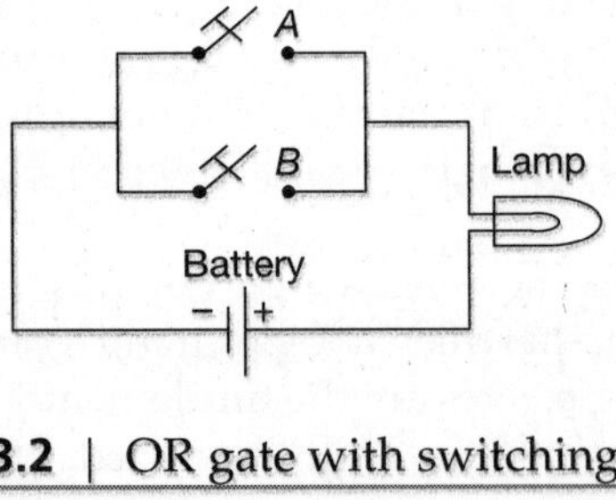

FIGURE 3.2 | OR gate with switching circuit

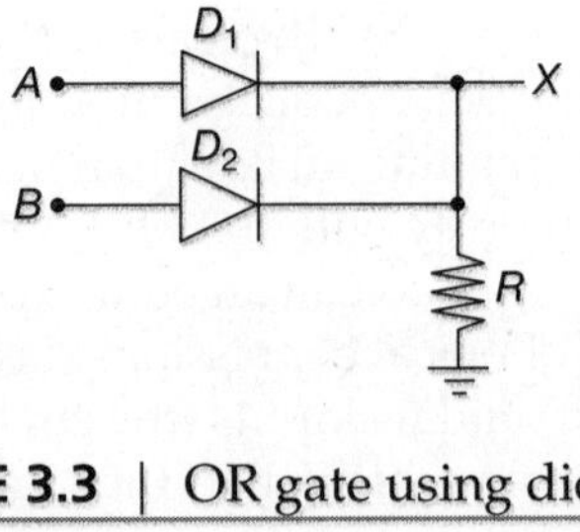

FIGURE 3.3 | OR gate using diodes

Electric Circuit for OR gate: Circuit is given in Figure 3.2. Switches act as inputs. Output of circuit is operation of Lamp.

- When both the switches A and B are open (OFF) (i.e. $A = 0, B = 0$), the lamp does not glow (X = 0).
- When switch A is closed (ON) and switch B is open (OFF) (i.e. $A = 1$, B = 0), the lamp glows (X = 1).
- If switch A is open (OFF) and switch B is closed (ON), (i.e. $A = 0$, B = 1), the lamp glows (X = 1).
- If both the switches A and B are closed (ON), (i.e. $A = 1$. $B = 1$), the lamp glows ($X = 1$).
- It is concluded, if any one switch is ON, the lamp glows.

OR Gate Using Diode: Circuit is given in Figure 3.3.

- When A is at 0 V and B is at 0 V (i.e. $A = 0, B = 0$), both the diodes D_1 and D_2 are OFF. No current flows through R. Hence output voltage X is 0V (X = 0).
- When A is at 0 V and B is at 5 V (i.e. $A = 0$, $B = 1$), diode D_1 is OFF and D_2 is ON. Current flows through R. Hence output voltage X is 5 V (X = 1).
- When A is at 5 V and B is at 0 V (i.e. $A = 1$, $B = 0$), diode D_1 is ON and D_2 is OFF. Current flows through R. Output voltage, X is 5V. (i.e. X = 1).
- When A is at 5 V and B is at 5 V (i.e. $A = 1$, B = 1), diodes D_1 and D_2 are ON. X = 5 − 0.7 = 4.3 V. (X = 1).
- It is concluded, if any one diode is ON, then X is at 5 V.

Three Input OR-gate is shown in Figure 3.4. Truth table of 3-inputs OR gate is given in Table 3.2. For 3-inputs, there are eight (2^3) possible combinations.

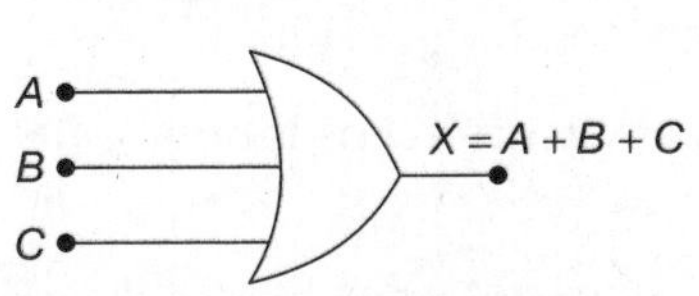

FIGURE 3.4 | Symbol of 3-input OR logic gate

TABLE 3.2 | Truth table of 3-input OR-gate

Inputs			Output
A	*B*	*C*	*X*
0	0	0	0
0	0	1	1
0	1	0	1
0	1	1	1
1	0	0	1
1	0	1	1
1	1	0	1
1	1	1	1

Boolean expression for 3-input OR gate

$$X = A + B + C \tag{3.2}$$

3.1.2 | AND Gate

AND gate has two or more inputs and only one output. The logic symbol of two inputs AND gate is given in Figure 3.5. Truth table of 2-inputs AND gate is given in Table 3.3.

AND gate gives output as 1, if all the inputs of AND gate are 1, otherwise output is 0. For 2-inputs, there are four (2^2) possible combinations.

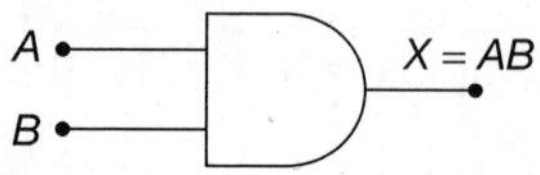

FIGURE 3.5 | Symbol of 2-inputs AND logic gate

TABLE 3.3 | Truth table of 2-input AND gate

Inputs		Output
A	***B***	***X***
0	0	0
0	1	0
1	0	0
1	1	1

Boolean expression for 2-input OR gate

$$X = AB \tag{3.3}$$

Truth table is better explained with the help of electrical circuit shown in Figures 3.6 and 3.7.

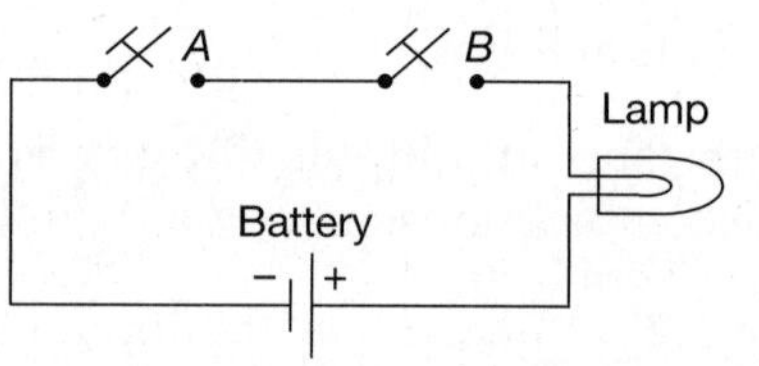

FIGURE 3.6 | AND gate with switching circuit

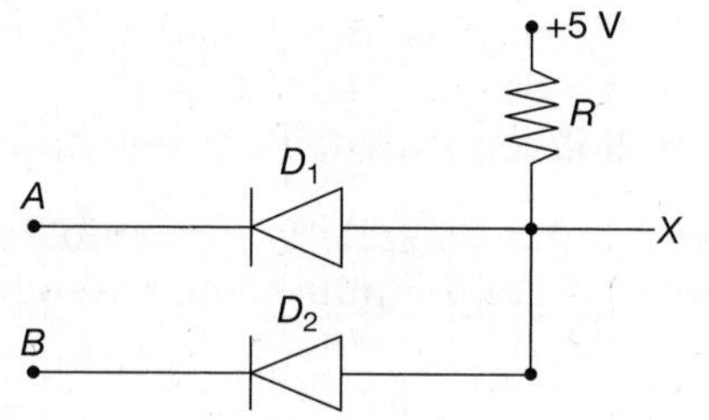

FIGURE 3.7 | AND gate using diodes

Electric Circuit for AND gate: Circuit is given in Figure 3.6. Switches act as inputs. Output of circuit is operation of bulb.

- Switches A and B are open (i.e. $A = 0$, $B = 0$), the lamp does not glow ($X = 0$).
- Switch A is open and switch B is closed (i.e. $A = 0$, $B = 1$) the lamp does not glow ($X = 0$).
- Switch A is closed and switch B is open (i.e. $A = 1$, $B = 0$) the lamp does not glow ($X = 0$).
- Switches A and B are closed (i.e. $A = 1$. $B = 1$). The lamp glows ($X = 1$).
- It is concluded that if both switches are closed the lamp glows.

Diode Circuit for AND gate: Circuit is given in Figure 3.7.

- When A and B are at 0 V, Diodes D_1 and D_2 are ON. Output X is at 0 V (approx.).
- When A is at 0 V and B is at 5 V, Diode D_1 is ON and D_2 is OFF. Output X is at 0 V (approx.).
- When B is at 5 V and A is at 0 V, Diode D_1 is OFF and D_2 is ON so output, X is at 0 V (approx.).

- When A and B are at 5 V, D_1 and D_2 are OFF. No current flows through R. So X is at 5 V.
- It is concluded that if both diodes are off then X is at 5 V.

Three Input AND-gate is shown in Figure 3.8. Truth table of 3-input AND-gate is given in Table 3.4. For 3-inputs, there are eight (2^3) possible combinations.

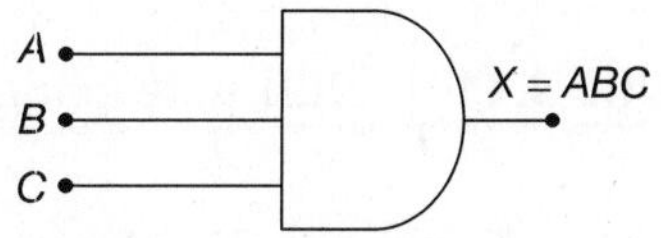

FIGURE 3.8 | Symbol of 3-input AND logic gate

TABLE 3.4 | Truth table of 3-input AND-gate

Inputs			Output
A	***B***	***C***	***X***
0	0	0	0
0	0	1	0
0	1	0	0
0	1	1	0
1	0	0	0
1	0	1	0
1	1	0	0
1	1	1	1

Boolean expression for 3-input OR gate

$$X = ABC \tag{3.4}$$

3.1.3 | NOT Gate

NOT gate is known as inverter. It has only one input and one output. The logic symbol is given in Figure 3.9. Truth table is given in Table 3.5.

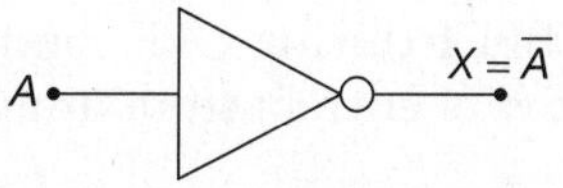

FIGURE 3.9 | Symbol of NOT logic gate

TABLE 3.5 | Truth table of NOT gate

Input	Output
A	***X***
0	1
1	0

Boolean expression for NOT gate is

$$X = \bar{A} \tag{3.5}$$

Truth table is better explained with the help of electrical circuit shown in Figures 3.10–3.12.

Electric Circuit for NOT gate: Circuit is given in Figure 3.10. Switch acts as input. Output of circuit is operation of bulb.

- Switch A is open (i.e. $A = 0$), the lamp glows ($X = 1$).
- Switch A is closed (i.e. $A = 1$) the lamp does not glow ($X = 0$).

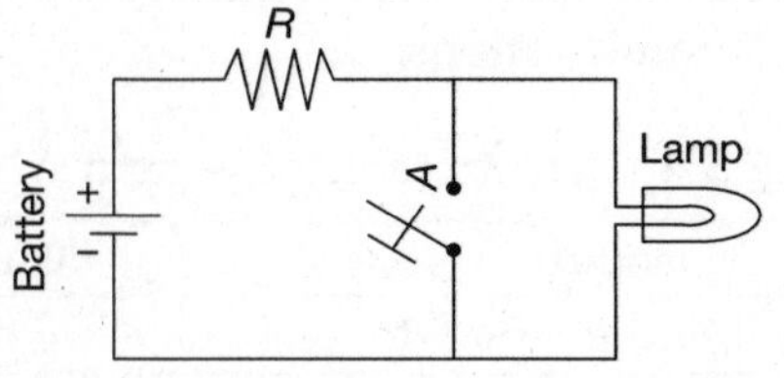

FIGURE 3.10 | NOT gate with switching circuit

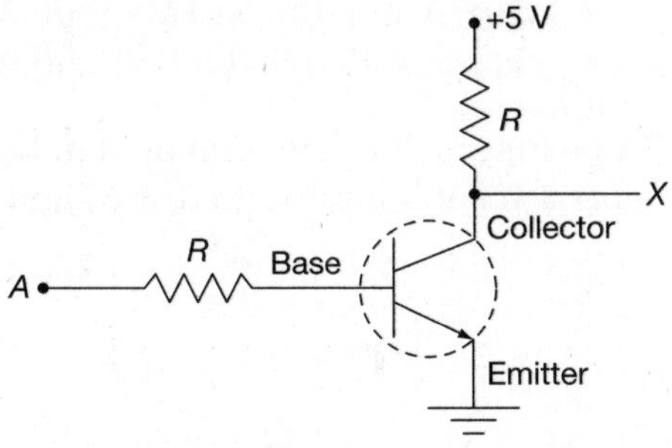

FIGURE 3.11 | NOT gate using BJT

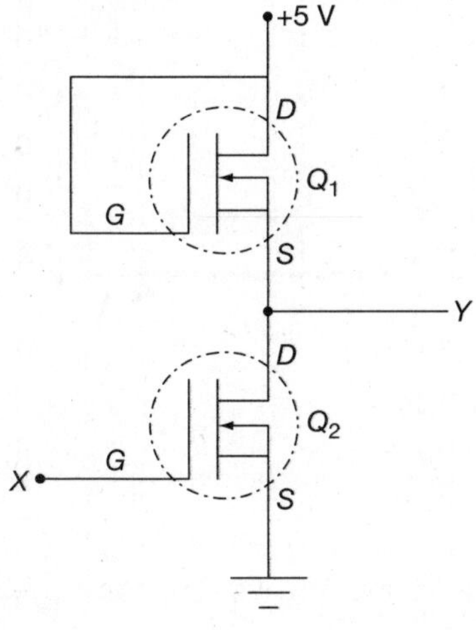

FIGURE 3.12 | Inverter using NMOSFET

Bipolar Junction Transistor Circuit for NOT gate: Circuit is given in Figure 3.11.

- When input A is at 0 V (i.e. $A = 0$). Bipolar junction transistor (BJT) remains OFF because transistor is not biased. No current will flow through *R*. So there will be no voltage drop across *R*. Hence output *X* is at +5 V (approx.) (i.e. $X = 1$)
- When A is at 5 V (i.e. $A = 1$). Bipolar junction transistor (BJT) is ON because transistor is biased. Current will flow through R. So there will be voltage drop across *R*. Hence output *X* is at +0 V (approx.) (i.e. X = 0)

NMOSFET Circuit for NOT gate: Circuit is given in Figure 3.12, in which n-type metal oxide semiconductor field effect transistor (NMOSFET), Q_1 remains ON and offers resistance of 1 kΩ. So, MOSFET Q_1 is acting as resistor. MOSFET during OFF condition offers 10^6 Ω resistance and acts as OFF switch providing no flow of current from drain to source.

- When input A is at 0 V (i.e. $A = 0$). MOSFET remains OFF because transistor is not biased. No current will flow through Q_2. So there will be no voltage drop across Q_1. Hence, output *X* is at +5 V (approx.) (i.e. $X = 1$)
- When A is at 5 V (i.e. A = 1). MOSFET is ON because transistor is biased. Current will flow through Q_2. So, there will be voltage drop across Q_1. Hence, output *X* is at +0 V (approx.) (i.e. $X = 0$)

3.1.4 | NAND Gate

NAND gate is combination of AND gate and NOT gate as shown in Figure 3.13. The logic symbol of NAND gate is given in Figure 3.14. Truth table of NAND gate is given in Table 3.6. NAND gate has two or more inputs and only one output: For 2-inputs, there are four (2^2) possible combinations.

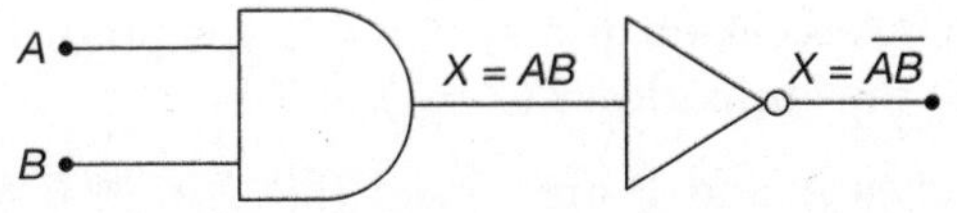

FIGURE 3.13 | NAND logic from AND logic gate and NOT logic gate

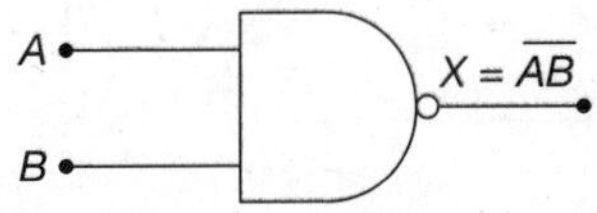

FIGURE 3.14 | Symbol of 2-input NAND logic gate

Boolean expression for 2-inputs NAND gate

$$X = \overline{AB} \tag{3.6}$$

TABLE 3.6 | Truth table of NAND gate

Input		Output
A	**B**	**X**
0	0	1
0	1	1
1	0	1
1	1	0

Truth table is better explained with the help of electrical circuit using switches, BJT and n-type MOS field effect transistor (MOSFET), which are shown in Figures 3.15, 3.16 and 3.17, respectively.

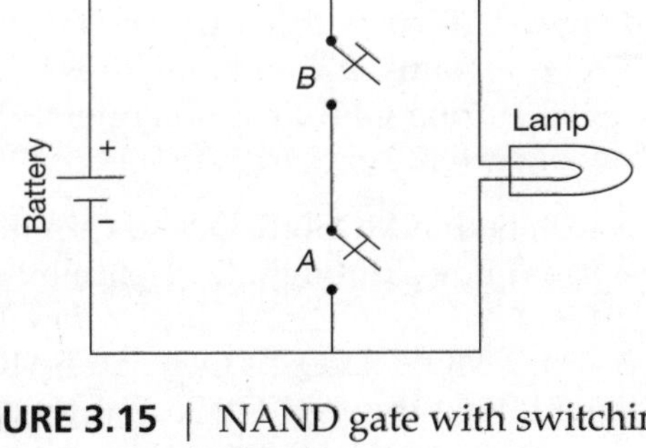

FIGURE 3.15 | NAND gate with switching circuit

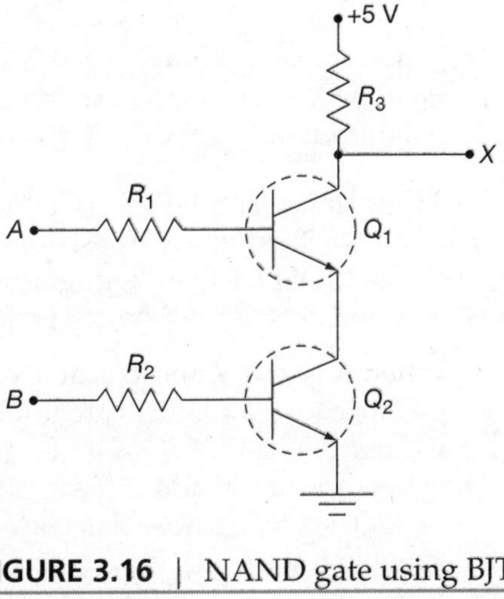

FIGURE 3.16 | NAND gate using BJT

Electric Circuit for NAND gate: Circuit is given in Figure 3.15. Switches act as inputs. Output of circuit is operation of bulb.

- Switches A and B are open (i.e. $A = 0$, $B = 0$), the lamp glows ($X = 1$).
- Switch A is open and switch B is closed (i.e. $A = 0$, $B = 1$) the lamp glows ($X = 1$).

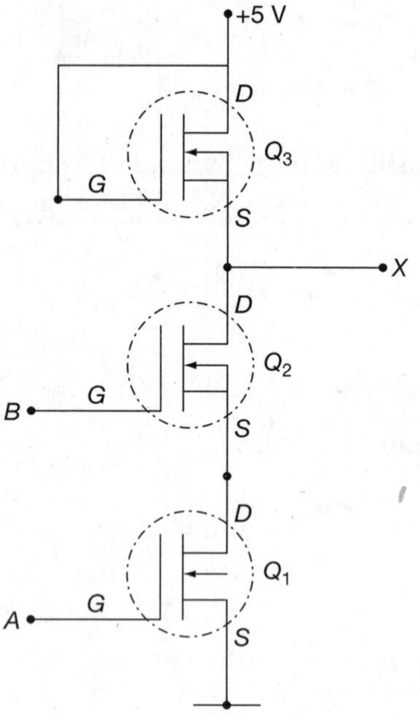

FIGURE 3.17 | NAND gate using NMOSFET

- Switch A is closed and switch B is open (i.e. $A = 1$, $B = 0$) the lamp glows ($X = 1$).
- Switches A and B are closed (i.e. $A = 1$, $B = 1$). The lamp does not glow ($X = 0$). Because current follows least resistance path through closed switches A and B.

NAND Gate Using BJT: Circuit is given in Fig. 3.16.

- When A is at 0 V and B is at 0 V (i.e. $A = 0$, $B = 0$), both the bipolar junction transistors (BJT), Q_1 and Q_2 remain OFF because transistors are not biased. No current flows through resistor, R_3. Hence, output voltage X is 5 V ($X = 1$).
- When A is at 0 V and B is at 5 V (i.e. $A = 0$, $B = 1$), bipolar junction transistor Q_1, is OFF because transistor is not biased. Bipolar junction transistor Q_2 is ON because transistor is biased. Hence, no current flows through resistor, R_3. So, output voltage X is 5 V (X = 1).
- When A is at 5 V and B is at 0 V (i.e. $A = 1$, $B = 0$), bipolar junction transistor Q_1, is ON because transistor is not biased. Bipolar junction transistor Q_2 is OFF because transistor is not biased. Hence, no current flows through resistor, R_3. So, output voltage X is 5 V (X = 1).
- When A is at 5 V and B is at 5 V (i.e. $A = 1$, $B = 1$), both bipolar junction transistors, Q_1 and Q_2 are ON because transistors are biased. Current flows through resistor, R_3. So, output voltage X is 0 V (X = 0).

NAND Gate Using NMOSFET: Circuit is given in Figure 3.17, in which n-type metal oxide semiconductor field effect transistors (NMOSFET), Q_3 remains ON and offers resistance of 1 kΩ. So, MOSFET Q_3 is acting as resistor. MOSFET during OFF condition offers 10^6 Ω resistance and acts as OFF switch providing no flow of current from drain to source.

- When A is at 0 V and B is at 0 V (i.e. $A = 0$, $B = 0$), both NMOSFET Q_1 and Q_2 remain OFF because MOSFETs are not biased. No current flows through, Q_3. Hence output voltage X is 5 V ($X = 1$).
- When A is at 0 V and B is at 5 V (i.e. $A = 0$, $B = 1$), NMOSFET Q_1, is OFF and NMOSFET, Q_2 is ON because it is biased. Hence, current flows through, Q_3. So, output voltage X is 5 V ($X = 1$).
- When A is at 5 V and B is at 0 V (i.e. $A = 1$, $B = 0$), NMOSFET, Q_1, is ON because it is biased and NMOSFET, Q_2 is OFF. Hence, current flows through Q_3. So, output voltage X is 5 V ($X = 0$).
- When A is at 5 V and B is at 5 V (i.e. $A = 1$, $B = 1$), both NMOSFETs Q_1 and Q_2 are ON because these are biased. Current flows through Q_3, Q_1 and Q_2. So, output voltage X is 0 V ($X = 0$).

Three inputs NAND gate is shown in Figure 3.18. Truth table for 3-inputs NAND gate is given in Table 3.7. For 3-inputs, there are eight (2^3) possible combinations.

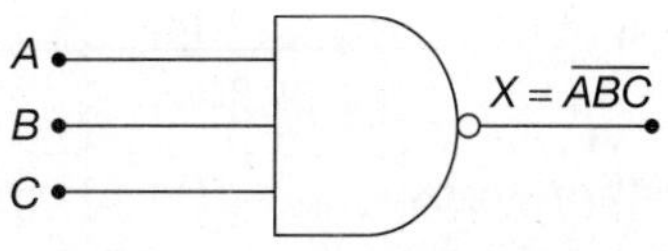

FIGURE 3.18 | Symbol of 3-input NAND logic gate

TABLE 3.7 | Truth table of 3-input NAND-gate

Inputs			Output
A	***B***	***C***	***X***
0	0	0	1
0	0	1	1
0	1	0	1
0	1	1	1
1	0	0	1
1	0	1	1
1	1	0	1
1	1	1	0

Boolean expression for 3-inputs NAND gate

$$X = \overline{ABC} \tag{3.7}$$

NAND gate is known as universal gate because three basic gates OR, AND and NOT gates can be constructed using NAND gate only.

3.1.5 | NOR Gate

NOR gate is combination of OR gate and NOT gate as shown in Figure 3.19. The logic symbol of NOR gate is given in Figure 3.20. Truth table of NOR gate is given in Table 3.8. NOR gate has two or more inputs and only one output. For 2-inputs, there are four (2^2) possible combinations.

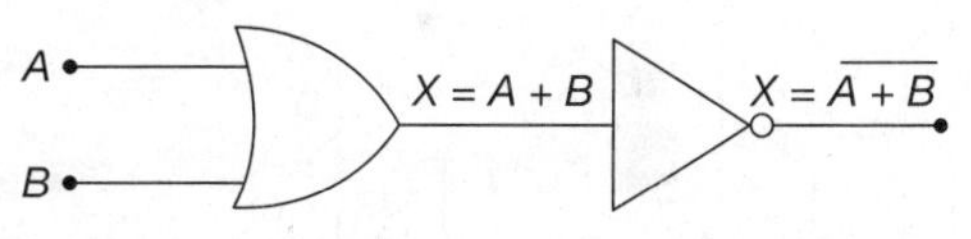

FIGURE 3.19 | NOR logic from OR logic gate and NOT logic gate

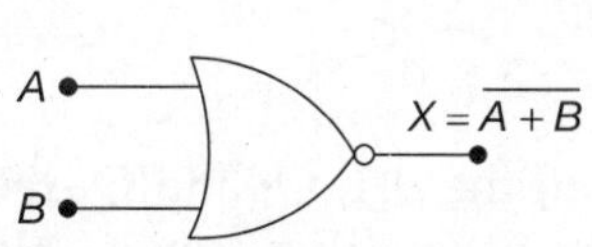

FIGURE 3.20 | Symbol of 2-input NOR logic gate

TABLE 3.8 | Truth table of 2-inputs NOR gate

Input		Output
A	***B***	***X***
0	0	1
0	1	0
1	0	0
1	1	0

Boolean expression for 2-inputs NOR gate

$$X = \overline{A + B} \tag{3.8}$$

Truth table is better explained with the help of electrical circuit using switches, BJT and n-type MOSFET, which are shown in Figures 3.21, 3.22 and 3.23, respectively.

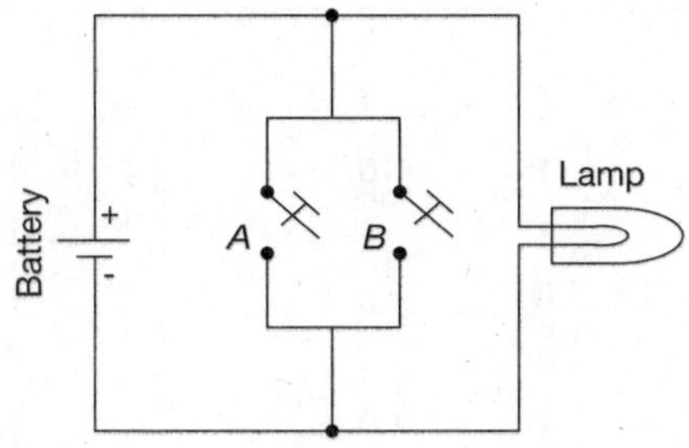

FIGURE 3.21 | NOR gate with switching circuit

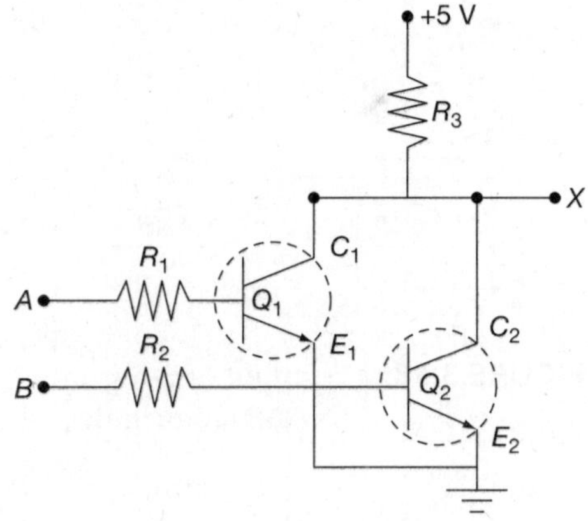

FIGURE 3.22 | NOR gate using BJT

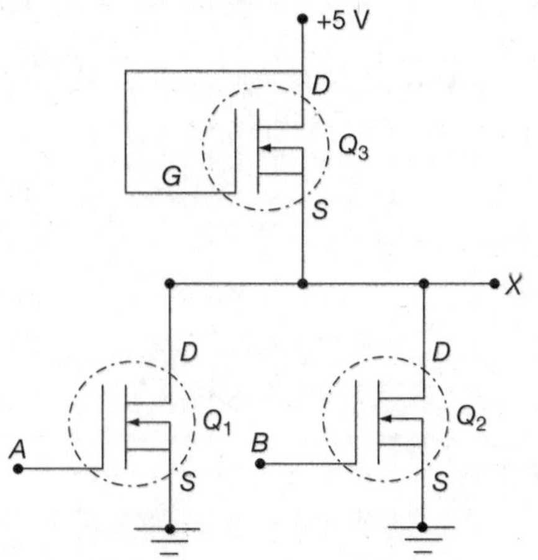

FIGURE 3.23 | NOR gate using NMOSFET

Electric Circuit for NOR gate: Circuit is given in Figure 3.21. Switches act as inputs. Output of circuit is operation of bulb.

- Switches A and B are open (i.e. $A = 0$, $B = 0$), the lamp glows ($X = 1$).
- Switch A is open and switch B is closed (i.e. $A = 0$, $B = 1$) the lamp does not glow ($X = 0$). Because current follows least resistance path through closed switch B.
- Switch A is closed and switch B is open (i.e. $A = 1$, $B = 0$) and the lamp glows ($X = 0$). Because current follows least resistance path through closed switch A.
- Switches A and B are closed (i.e. $A = 1$. $B = 1$). The lamp does not glow ($X = 0$). Because current follows least resistance path through closed switch A and B.

NOR Gate Using BJT: Circuit is given in Figure 3.22.

- When A is at 0 V and B is at 0 V (i.e. $A = 0$, $B = 0$), both the BJT, Q_1 and Q_2 remain OFF because transistors are not biased. No current flows through resistor, R_3. Hence, output voltage X is 5 V ($X = 1$).
- When A is at 0 V and B is at 5 V (i.e. $A = 0$, $B = 1$), BJT, Q_1, is OFF because transistor is not biased. BJT, Q_2 is ON because transistor is biased. Hence, current flows through resistor, R_3 following path through transistor Q_2. So, output voltage X is 5 V ($X = 1$).
- When A is at 5 V and B is at 0 V (i.e. $A = 1$, $B = 0$), BJT, Q_1, is ON because transistor is not biased. BJT, Q_2 is OFF because transistor is not biased. Hence, current flows through resistor, R_3 following path through transistor Q_1. So, output voltage X is 5 V ($X = 1$).

- When A is at 5 V and B is at 5 V (i.e. $A = 1, B = 1$), both BJTs, Q_1 and Q_2 are ON because transistors are biased. Current flows through resistor, R_3 following path through transistors Q_1 and Q_2. So, output voltage X is 0 V ($X = 0$).

NOR Gate Using NMOSFET: Circuit is given in Figure 3.23, in which n-type metal oxide semiconductor field effect transistors (NMOSFET), Q_3 remains ON and offers resistance of 1 kΩ.

- When A is at 0 V and B is at 0 V (i.e. $A = 0, B = 0$), both NMOSFET Q_1 and Q_2 remain OFF because MOSFETs are not biased. No current flows through resistor. Hence, output voltage X is 5 V (X = 1).
- When A is at 0 V and B is at 5 V (i.e. $A = 0, B = 1$), NMOSFET Q_1, is OFF and NMOSFET, Q_2 is ON because it is biased. Hence, current flows through Q_3 and Q_2. So, output voltage X is 0 V ($X = 0$).
- When A is at 5 V and B is at 0 V (i.e. $A = 1, B = 0$), NMOSFET, Q_1, is ON because it is biased and NMOSFET, Q_2 is OFF. Hence, current flows through Q_3 and Q_1. So, output voltage X is 0 V ($X = 0$).
- When A is at 5 V and B is at 5 V (i.e. $A = 1, B = 1$), both NMOSFETs, Q_1 and Q_2 are ON because these are biased. Current flows through, Q_3. So, output voltage X is 0 V ($X = 0$).

Three inputs NOR gate is shown in Figure 3.24. Truth table for 3-inputs NOR gate is given in Table 3.9. For 3-inputs, there are eight (2^3) possible combinations.

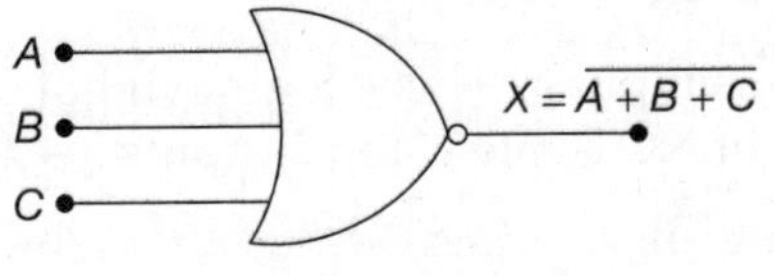

FIGURE 3.24 | Symbol of 3-input NOR logic gate

TABLE 3.9 | Truth table of 3-input NOR-gate

Inputs			Output
A	*B*	*C*	*X*
0	0	0	1
0	0	1	0
0	1	0	0
0	1	1	0
1	0	0	0
1	0	1	0
1	1	0	0
1	1	1	0

Boolean expression for 3-inputs NOR gate

$$X = \overline{A + B + C} \tag{3.9}$$

NOR gate is known as universal gate because three basic gates OR, AND and NOT gates can be constructed using NOR-gates only.

3.1.6 | EXCLUSIVE-OR Gate

An X-OR (Exclusive-OR) gate is a two inputs and one output logic circuit. When inputs are different, output is logic 1. When inputs are same, then output is logic 0. X-OR gate is called an anti-coincidence or inequality detector. The logic symbol of X-OR gate is given in

Figure 3.25. Truth table of X-OR gate is given in Table 3.10. For 2-inputs, there are four (2^2) possible combinations. Boolean expressions for 2-inputs X-OR gate are given as follows:

$$X = A\bar{B} + \bar{A}B \tag{3.10}$$

$$X = A \oplus B \tag{3.11}$$

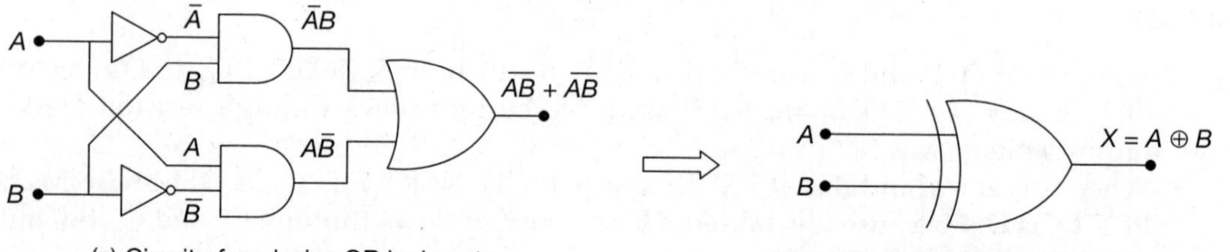

(a) Circuit of exclusive OR logic gate. (b) Symbol of exclusive-OR logic gate.

FIGURE 3.25 | Exclusive-OR logic gate

TABLE 3.10 | Truth table of 2-inputs X-OR gate

Input		Output
A	*B*	*X*
0	0	0
0	1	1
1	0	1
1	1	0

Inverter is obtained from 2-inputs X-OR gate by fixing one input as 1. The logic symbol of X-OR gate as inverter is given in Figure 3.26. Truth table of NOT gate using 2-inputs X-OR gate is given in Table 3.11.

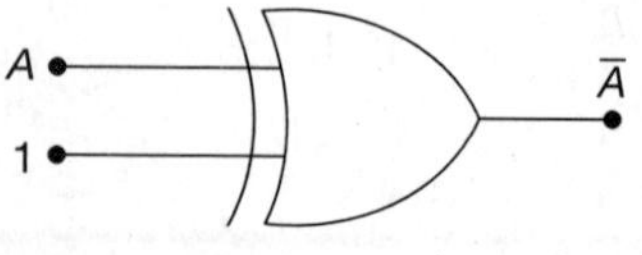

FIGURE 3.26 | X-OR logic gate as inverter

TABLE 3.11 | Truth table of NOT gate using 2-inputs X-OR gate

Inputs		Output
A	1	$X = A \oplus 1$
0	1	1
1	1	0

Three inputs X-OR gate is shown in Figure 3.27. Truth tables for 3-inputs X-OR gate is given in Table 3.12. For 3-inputs, there are eight (2^3) possible combinations.

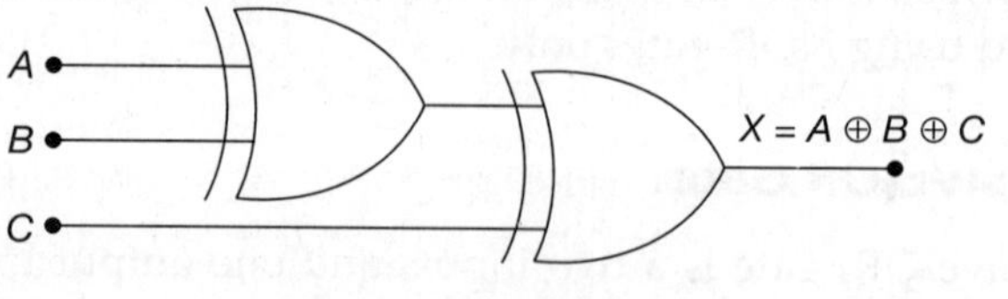

FIGURE 3.27 | Symbol of 3-input X-OR logic gate

TABLE 3.12 | Truth table of 3-input X-OR-gate

Inputs			Output
A	*B*	*C*	*X*
0	0	0	0
0	0	1	1
0	1	0	1
0	1	1	0
1	0	0	1
1	0	1	0
1	1	0	0
1	1	1	1

Let $A = 0, B = 1$ and $C = 0$

So $X = (A \oplus B) \oplus C$

$X = (0 \oplus 1) \oplus 0$

$X = 1 \oplus 0$

$X = 1$

Let $A = 0, B = 1$ and $C = 1$

So $X = (A \oplus B) \oplus C$

$X = (0 \oplus 1) \oplus 1$

$X = 1 \oplus 1$

$X = 0$

3.1.7 | EXCLUSIVE-NOR Gate

An X-NOR (Exclusive-NOR) gate is a two inputs and one output logic circuit. When inputs are different, output is logic 0. When inputs are same, then output is logic 1. X-NOR gate is called coincidence or equality detector. The logic symbol of 2-inputs X-NOR gate is shown in Figures 3.28 and 3.29. Truth table of 2-inputs X-NOR gate is given in Table 3.13. For 2-inputs, there are four (2^2) possible combinations. Boolean expressions for 2-inputs X-NOR gate are given as follows:

$$X = AB + \overline{A}\overline{B} \tag{3.12}$$

$$X = \overline{A \oplus B} \tag{3.13}$$

$$X = A \odot B \tag{3.14}$$

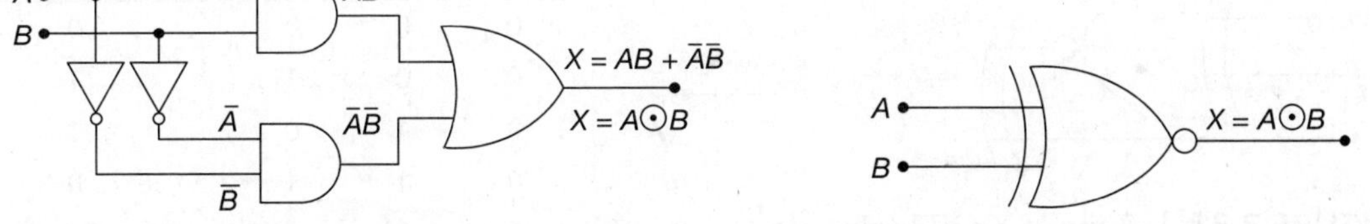

(a) Circuit of exclusive-NOR logic gate circuit

(b) Symbol of exclusive-NOR logic gate

FIGURE 3.28 | Exclusive-NOR logic gates

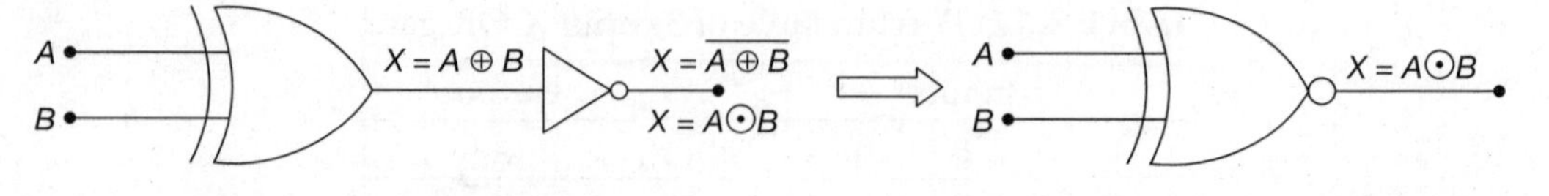

(a) Circuit of exclusive-NOR logic gate (b) Symbol of exclusive-NOR logic gate

FIGURE 3.29 | Exclusive-NOR logic gates

TABLE 3.13 | Truth table of 2-inputs X-NOR gate

Input		Output
A	***B***	***X***
0	0	1
0	1	0
1	0	0
1	1	1

Inverter is obtained from 2-inputs X-NOR gate by fixing one input as 0. The logic symbol of X-NOR gate as inverter is given in Figure 3.30. Truth table of NOT gate using 2-inputs X-NOR gate is given in Table 3.14.

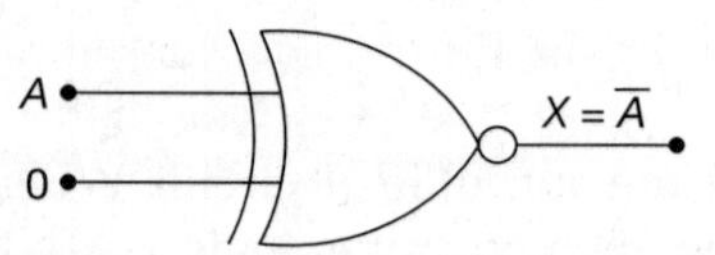

FIGURE 3.30 | X-NOR logic gate as inverter

TABLE 3.14 | Truth table of NOT gate using 2-inputs X-NOR gate

Inputs		Output
A	**0**	$X = A \odot 0$
0	0	1
1	0	0

Three inputs X-OR gate are shown in Figure 3.31. Truth table for 3-inputs X-NOR gate is given in Table 3.15. For 3-inputs, there are eight (2^3) possible combinations.

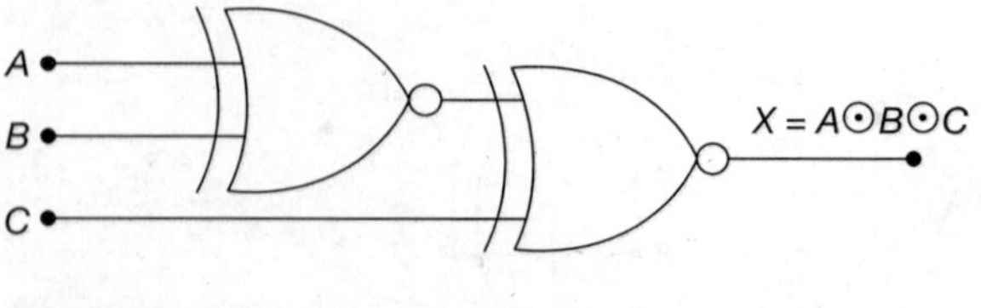

FIGURE 3.31 | 3-input X-OR logic gate

TABLE 3.15 | Truth table of 3-input X-NOR-gate

Inputs			Output
A	***B***	***C***	***X***
0	0	0	0
0	0	1	1
0	1	0	1
0	1	1	0
1	0	0	1
1	0	1	0
1	1	0	0
1	1	1	1

Let. $A = 0, B = 0$ and $C = 0$

So $X = (A \odot B) \odot C$

$X = (0 \odot 0) \odot 0$

$X = 1 \odot 0$

$X = 0$

Let $A = 0, B = 1$ and $C = 1$

So $X = (A \odot B) \odot C$

$X = (0 \odot 1) \odot 0$

$X = 0 \odot 0$

$X = 1$

Universal gates NAND and NOR gates are known as universal gates because three basic gates OR, AND and NOT gates can be constructed using either NAND or NOR-gates only.

NOT gate using NAND gate NOT gate operation is obtained from 2-inputs NAND gate having same inputs. Circuit is given in Figure 3.32. There is need of one NAND to construct NOT gate. Truth Table 3.16 gives the justification. Single variable A has two (2^1) possibilities. Output X is same as that obtained from NOT gate shown in Table 3.5.

Output of AND operation of two same inputs does not change, e.g. $AA = A$.

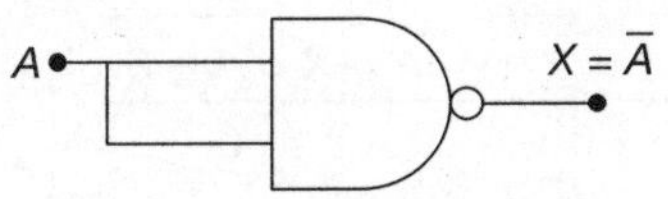

FIGURE 3.32 | NOT logic gate from NAND logic gate

TABLE 3.16 | Truth table of NOT gate from NAND gate

Input	Intermediate Output		Output
A	$A = AA$	$\overline{AA}$	X
0	0	1	1
1	1	0	0

AND gate using NAND gate AND gate operation is obtained from 2-inputs NAND gate and NOT gate. Circuit is given in Figure 3.33. There is need of two NAND gates to construct AND gate. Truth table 3.17 gives the justification. Two variables A and B have four (2^2) possibilities. Output X is same as that obtained from AND gate shown in Table 3.3.

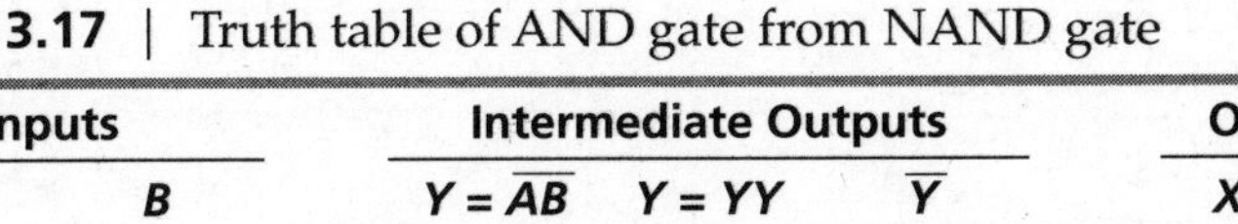

FIGURE 3.33 | AND logic gate from NAND logic gate

TABLE 3.17 | Truth table of AND gate from NAND gate

Inputs		Intermediate Outputs			Output
A	B	$Y = \overline{AB}$	$Y = YY$	$\overline{Y}$	$X = AB$
0	0	1	1	0	0
0	1	1	1	0	0
1	0	1	1	0	0
1	1	0	0	1	1

OR gate using NAND gate OR gate operation is obtained by feeding two inverted inputs to 2-inputs NAND gate. Inverted inputs are obtained from NOT gates using NAND gates. Circuit is given in Figure 3.34. There is need of three NAND to construct OR-gate. Truth Table 3.18 gives the justification. Two variables A and B have four (2^2) possibilities. Output X is same as that obtained from OR gate shown in Table 3.1.

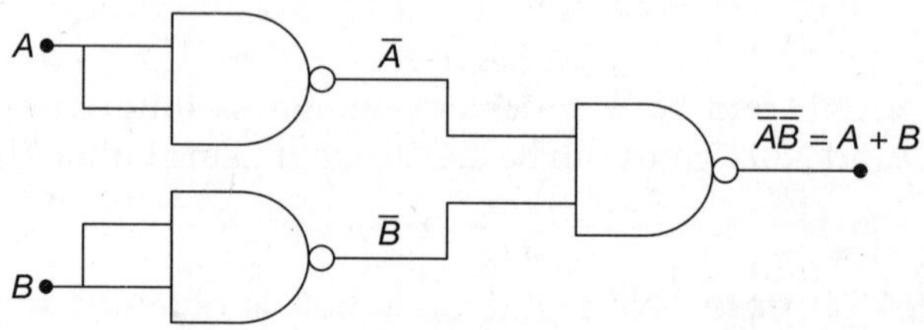

FIGURE 3.34 | OR logic gate from NAND logic gate

TABLE 3.18 | Truth table of OR gate from NAND gate

Inputs		**Intermediate Outputs**				**Output**
A	B	$\overline{A} = \overline{AA}$	$\overline{B} = \overline{BB}$	$\overline{A}\,\overline{B}$	$\overline{\overline{A}\,\overline{B}}$	$X = A + B$
0	0	1	1	1	0	0
0	1	1	0	0	1	1
1	0	0	1	0	1	1
1	1	0	0	0	1	1

NOT gate using NOR gate NOT gate operation is obtained from 2-inputs NOR gate having same inputs. Circuit is given in Figure 3.35. One NOR gate constructs NOT gate. Truth Table 3.19 gives the justification. Single variable A has two (2^1) possibilities. Output X is same as that obtained from NOT gate shown in Table 3.5.

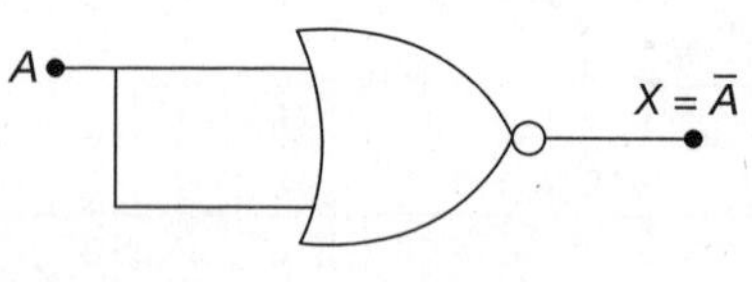

FIGURE 3.35 | NOT logic gate from NOR logic gate

TABLE 3.19 | Truth table of NOT gate from NOR gate

Input	**Intermediate Output**		**Output**
A	$A = A + A$	$\overline{A + A}$	X
0	0	1	1
1	1	0	0

Output of OR operation of two same inputs does not change, e.g. $A + A = A$.

OR gate using NOR gate OR gate operation is obtained from 2-inputs NOR gate and NOT gate. Circuit is given in Figure 3.36. Two NOR gates construct OR gate. Truth Table 3.20 gives the justification. Two variables A and B have four (2^2) possibilities. Output X is same as that obtained from OR gate shown in Table 3.1.

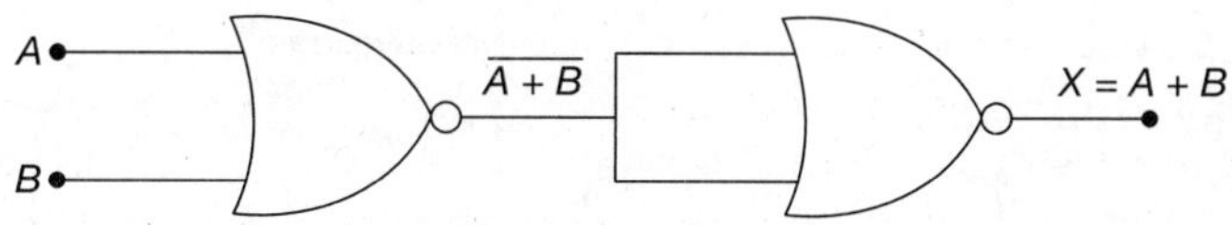

FIGURE 3.36 | OR logic gate from NOR logic gates

TABLE 3.20 | Truth table of OR gate from NOR gate

Inputs		Intermediate Outputs			Output
A	B	$Y = \overline{A+B}$	YY	$\overline{YY}$	$X = A + B$
0	0	1	1	0	0
0	1	0	0	1	1
1	0	0	0	1	1
1	1	0	0	1	1

AND gate using NOR gate OR gate operation is obtained by feeding two inverted inputs to 2-inputs NOR gate. Inverted inputs are obtained from NOT gates using NOR gates. Circuit is given in Figure 3.37. Three NOR gates construct AND-gate. Truth table 3.21 gives the justification. Two variables A and B have four (2^2) possibilities. Output X is same as that obtained from AND gate shown in Table 3.3.

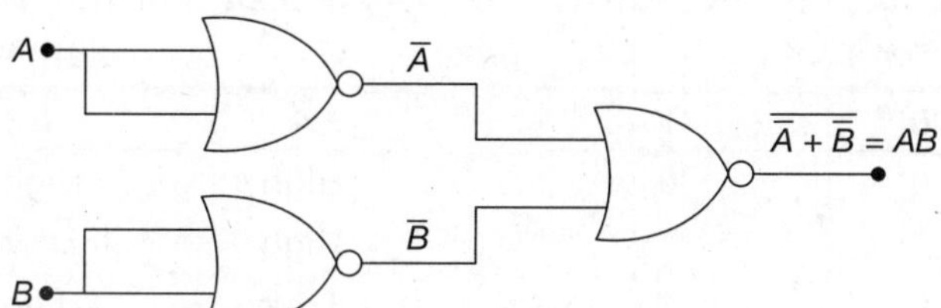

FIGURE 3.37 | AND logic gate from NOR logic gates

TABLE 3.21 | Truth table of AND gate from NOR gate

Inputs		Intermediate Outputs				Output
A	B	$\bar{A} = \overline{A+A}$	$\bar{B} = \overline{B+B}$	$\bar{A}+\bar{B}$	$\overline{\bar{A}+\bar{B}}$	$X = AB$
0	0	1	1	1	0	0
0	1	1	0	1	0	0
1	0	0	1	1	0	0
1	1	0	0	0	1	1

3.2 | POSITIVE LOGIC AND NEGATIVE LOGIC

In positive logic, 0 defines low-voltage state and 1 defines high-voltage state as shown in Figure 3.38. In contrast, negative logic is totally opposite to the positive logic. In negative logic, 0 defines the high-voltage state and 1 defines the low-voltage state as shown in Figure 3.39.

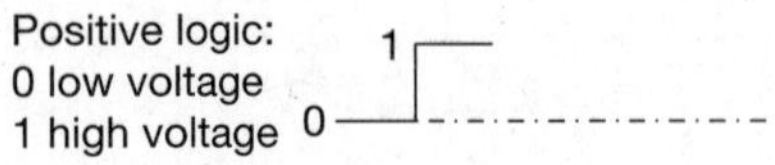

FIGURE 3.38 | Positive logic

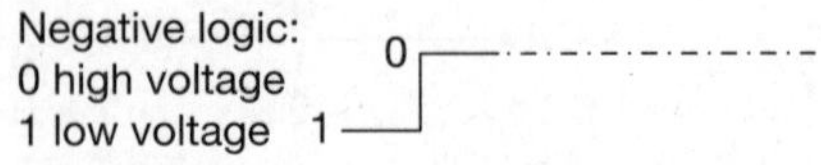

FIGURE 3.39 | Negative logic

An OR gate in a positive logic as shown in Tables 3.22 and 3.23, becomes an AND gate in a negative logic as depicted in Tables 3.24 and 3.25.

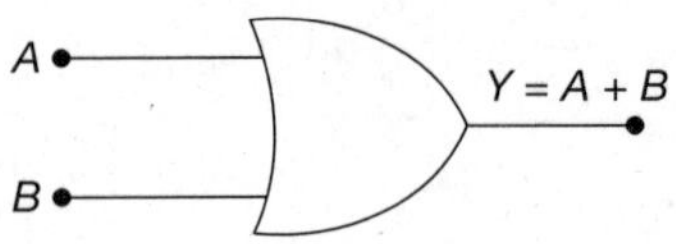

FIGURE 3.40 | Symbol of 2-input OR logic gate

TABLE 3.22 | Truth table of OR gate in positive logic

A	*B*	*Y*
Low	Low	Low
Low	High	High
High	Low	High
High	High	High

TABLE 3.23 | Truth table of OR gate in positive logic

A	*B*	*Y*
0	0	0
0	1	1
1	0	1
1	1	1

TABLE 3.24 | Truth table of OR gate in negative logic

A	*B*	*Y*
High	High	High
High	Low	High
Low	High	High
Low	Low	Low

TABLE 3.25 | Truth table of AND gate in positive logic

A	*B*	*Y*
1	1	1
1	0	0
0	1	0
0	0	0

Similarly, an AND in a positive logic becomes an OR gate in a negative logic. NOR gate in a positive logic becomes NAND gate in a negative logic. On the other hand, NAND gate in positive logic becomes NOR gate in negative logic. NOT gate act as inverter both in positive logic and negative logic.

Change of positive logic to negative logic is termed as duality. To obtain dual term change 0 to 1, 1 to 0, AND to OR and OR to AND, NAND to NOR, NOR to NAND (Table 3.26). Dual of NOT gate is the same NOT gate.

TABLE 3.26 | Dual of binary expression

Operator	Dual
AND	OR
OR	AND
NAND	NOR
NOR	NAND
0	1
1	0

Expression		Dual	Expression		Dual
$\overline{0}=1$	⇒	$\overline{1}=0$	$A\cdot B=B\cdot A$	⇒	$A+B=B+A$
$0\cdot 1=0$	⇒	$1+0=1$	$A\cdot(B\cdot C)=(A\cdot B)\cdot C$	⇒	$A+(B+C)=(A+B)+C$
$0\cdot 0=0$	⇒	$1+1=1$	$A\cdot(B+C)=A\cdot B+A\cdot C$	⇒	$A+(BC)=(A+B)(A+C)$
$\overline{0+1}=0$	⇒	$\overline{1\cdot 0}=1$	$A\cdot(B+C)=A\cdot B+A\cdot C$	⇒	$A+(BC)=(A+B)\cdot(A+C)$
$\overline{0\cdot 0}=1$	⇒	$\overline{1+1}=0$	$A\cdot(A+B)=A$	⇒	$A+A\cdot B=A$
$A\cdot 0=0$	⇒	$A+1=1$	$\overline{AB}=\bar{A}+\bar{B}$	⇒	$\overline{A+B}=\bar{A}\cdot\bar{B}$
$A\cdot A=A$	⇒	$A+A=A$	$(\bar{A}+B)(A+C)=AB+\bar{A}C$	⇒	$(\bar{A}B)+(AC)=(A+B)(\bar{A}+C)$
$A\cdot\bar{A}=0$	⇒	$A+\bar{A}=1$	$\overline{\overline{AB}+\bar{A}+AB}=0$	⇒	$\overline{\overline{(A+B)}\bar{A}(A+B)}=0$

3.3 | INHIBIT CIRCUITS

AND, OR, NAND, NOR, exclusive OR and exclusive NOR gates can be used to control the flow of an input logic signal through to the input to the output. This is shown in Figures 3.41 and 3.42 where a logic signal A is applied to one input of each of the above mentioned logic gates. The other input of each gate acts as the control input B. The logic level at this control input determines whether the input signal is enabled to reach the output or inhibited from reaching the output. Table 3.27 demonstrates the inhibit or enable controlled signal for AND logic gate.

Table 3.28 demonstrates the inhibit or enable controlled signal for OR logic gate.

It can be noticed that when AND and NOR gates are inhibited, they produce a constant LOW output. Conversely, the NAND and OR gates produce a constant HIGH output in the inhibited condition.

There will be many situations in digital circuit design, where the passage of a logic signal is either enabled or inhibited depending on the conditions present at one or more control inputs.

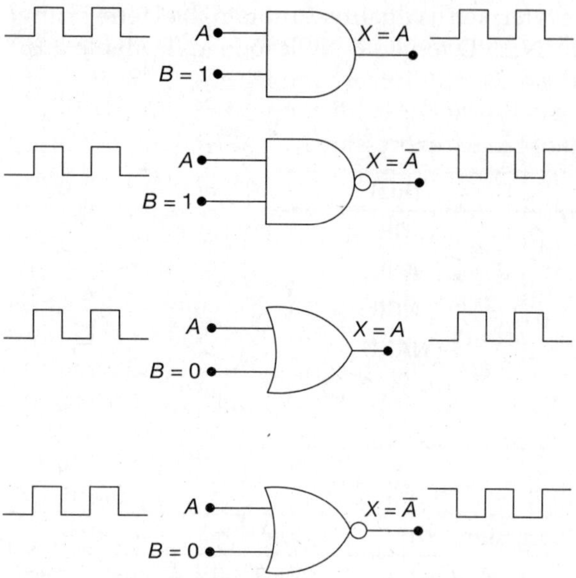

FIGURE 3.41 | Enable signal

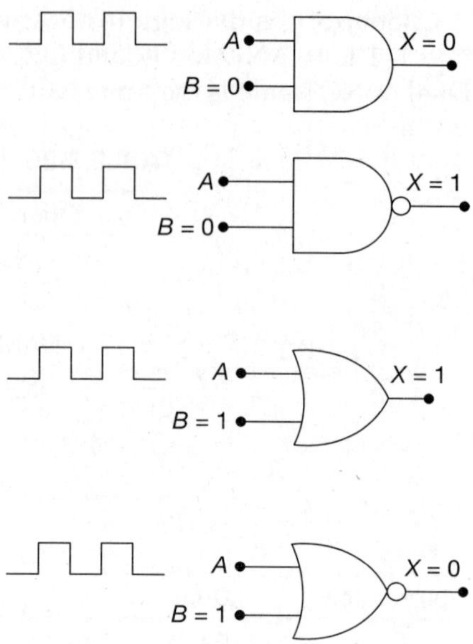

FIGURE 3.42 | Inhibit signal

TABLE 3.27 | Inhibit or transfer signal for AND gate

Action	Input		Output	Remarks
	Data A	Control B	$X = A \cdot B$	
Inhibit	0	0	0	Output locked at 0
	1	0	0	
Enable	0	1	0	Data transferred unaltered as A
	1	1	1	

TABLE 3.28 | Inhibit or transfer signal for OR gate

Action	Input		Output	Remarks
	Data A	Control B	$X = A + B$	
Enable	0	0	0	Data transferred unaltered as A
	1	0	1	
Inhibit	0	1	1	Output locked at 1
	1	1	1	

3.4 | 7400-SERIES INTEGRATED CIRCUITS

There are a number of logic families. The logic gates discussed so far and many other logic circuits are available in each one of those logic families. Each logic family has its own merits and demerits. There are transistor-transistor logic (TTL), emitter-coupled logic (ECL), integrated injection logic (IIL), metal oxide semiconductor logic (MOS), complementary metal-oxide semiconductor (CMOS). Three families of digital logic circuits CMOS (complementary metal-oxide semiconductor), bipolar, and BiCMOS are in use. They differ in the types or circuit components used internally to implement the logic function. CMOS is implemented with a type of field-effect transistor (FET), bipolar also known as TTL, logic uses bipolar junction transistors and BiCMOS employs a combination of both. The logic families differ in operational parameters such as switching speed (propagation time), power consumption, and noise immunity but not in the basic logic operation.

TTL is the most popular of the logic families. This popular series of chips contains a variety of gates whose inputs and outputs are accessible through connections to external pins. The chips are designed by 7400-series numbers, each of which refers to a particular set or combination of logic gates implemented by a chip. For example, IC-7400 contains four two-input NAND gates, IC-7410 contains three NAND gates having three-inputs. The voltage level used to represent 0 and 1 logic levels in 7400 series logic are 0 V and +5 V, respectively. The 0 V level means ground. It does not mean open circuit or no connection. In fact, if the input to a TTL gate is left open, the gate behaves as if logic 1 were connected to the input. All 7400 series chips require +5 volts DC power connection designated V_{cc}, and ground connections. Figure 3.43 shows some of the most commonly standard sequence. Few available IC gates are enlisted in Table 3.29. Dual, triple, quad or hex represents that 2, 3, 4, or 6 number of gates are available in the IC. 2, 3, 4, 5, 8 and 13 -input means each gate which is available in IC has 2, 3, 4, 5, 8 and 13 number of inputs, respectively. Pin configurations of Hex inverter (7404), Triple 3-input NOR gate (74287), Quad 2-input EX-OR gate (7486), Quad 2-input exclusive NOR gate (74810), Quad 2-input AND gate (7408) and Quad 2-input NAND gate (7400) are shown in Figures 3.43(a), 3.43(b), 3.43(c), 3.43(d), 3.43(e) and 3.43(f), respectively. The groove or dot on the chip identifies the top of a chip and the numbers begin from upper left-hand pin, run down across and end up in the upper right-hand pin. Each pin is the terminal for an input, or output of a gate or power supply connection. ICs are available in different package styles, and the pin assignments vary with the style selected.

TABLE 3.29 | Few available IC gates

IC Number	Description	IC Number	Description
7400	Quad 2-input NAND gate	7427	Triple 3-input NOR gate
7402	Quad 2-input NOR gate	7430	8-input NAND gate
7404	Hex inverter	7432	Quad 2-input OR gate
7408	Quad 2-input AND gate	7486	Quad 2-input EX-OR gate
7410	Triple 3-input NAND gate	74133	13-input NAND gate
7411	Triple 3-input AND gate	74135	Quad 2-input EX-OR/NOR gate
7420	Dual 4-input NAND gate	74287	Triple 3-input NOR gate
7421	Dual 4-input AND gate	74810	Quad 2-input EX-NOR gate

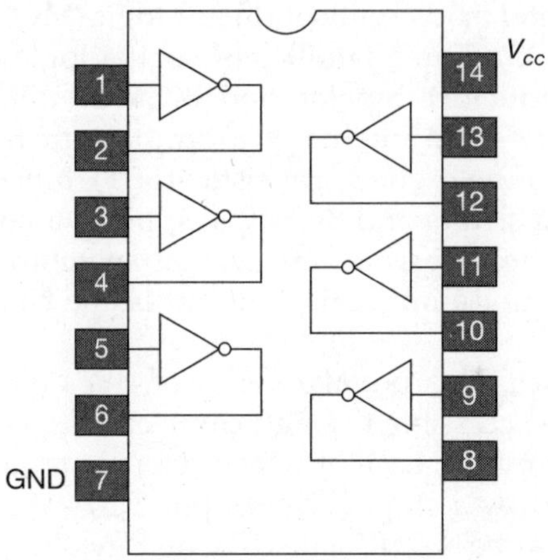

(a) 7404 hex inverter

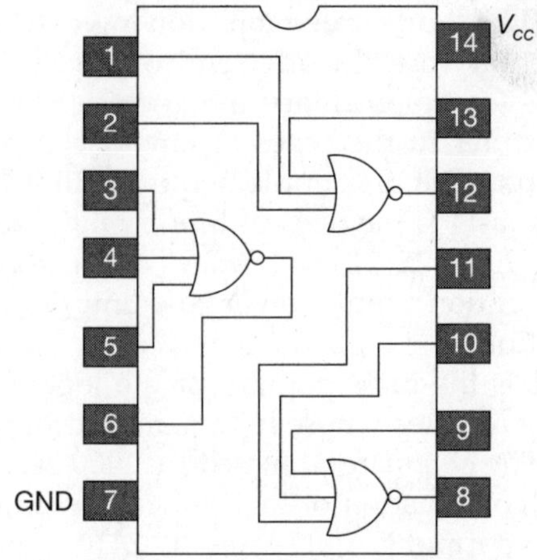

(b) 74287 triple three-input NOR gate

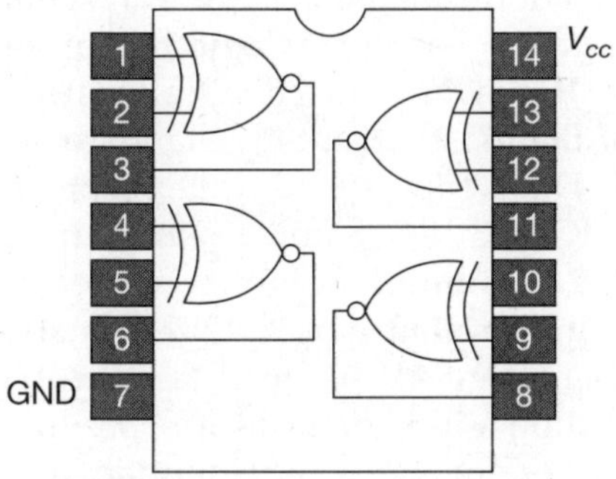

(c) 7486 quad exclusive OR gate

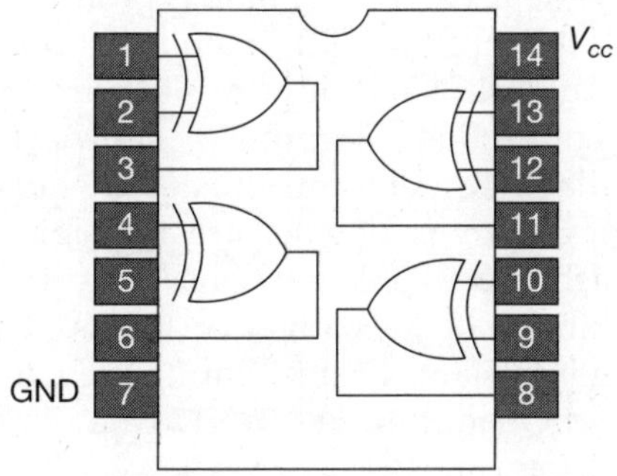

(d) 74810 quad exclusive NOR gate

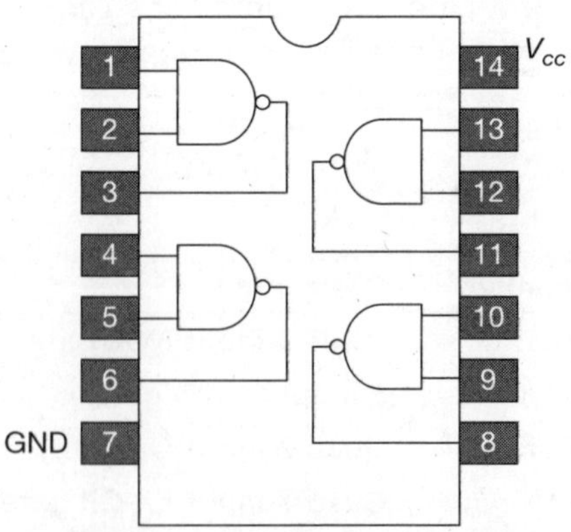

(e) 7400 quad two-input AND gate

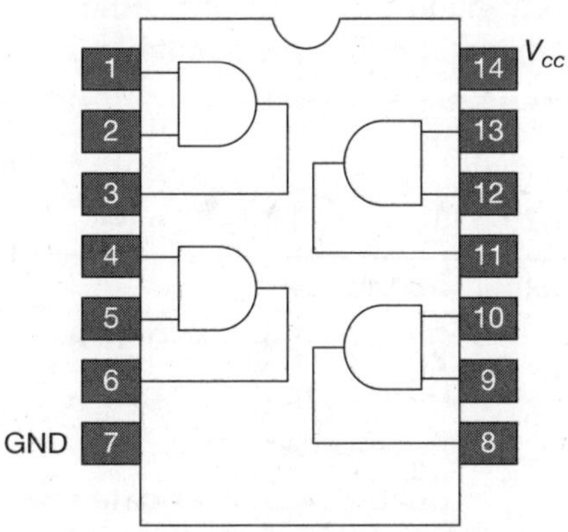

(f) 7408 quad two-input NAND gate

FIGURE 3.43 | Pin configuration of IC gates

Table 3.30 gives the different types of TTL series IC of Inverter. Various symbols used to represent types of TTL series and CMOS series are given in Table 3.31. Advanced CMOS numbering uses 5-digit device number beginning with digits 11 which is different from TTL, 74C, 74HC/HCT device numbering. For example, hex inverter 74AC11**004** is similar of 74HC**04.**

TABLE 3.30 | Types of TTL series IC

TTL Series	Prefix	Example of IC	Function
Standard TTL	74	7404	Hex Inverter
Schottky TTL	74S	74S04	Hex Inverter
Low power Schottky TTL	74LS	74LS04	Hex Inverter
Advanced Schottky TTL	74AS	74AS04	Hex Inverter
Advanced Low power Schottky TTL	74ALS	74ALS04	Hex Inverter

TABLE 3.31 | Symbols used for types of TTL and CMOS series IC numbers

Letters	Description	Letters	Description
ALS	Advanced low-power Schottky	ALVC	Advanced low-voltage CMOS
AS	Advanced Schottky	AUC	Advanced ultra low-voltage CMOS
F	Fast	AUP	Advanced ultra low-power CMOS
H	High speed	AVC	Advanced very low-voltage CMOS
L	Low power	CD4000	Standard CMOS
LS	Low-power Schottky	FCT	Fast CMOS technology
S	Schottky	HC	High-speed CMOS
AC	Advanced CMOS	HCT	High-speed CMOS with TTL compatible inputs
ACT	Advanced CMOS with TTL compatible inputs	LV-A	Low-voltage CMOS
AHC	Advanced high-speed CMOS	LV-AT	Low-voltage CMOS with TTL compatible inputs
AHCT	Advanced high-speed CMOS with TTL compatible inputs	LVC	Low-voltage CMOS

All the logic gates are specified small-scale-integration (SSI). SSI chips are used for interfaces. Modern digital systems most of functions performed by medium-scale-integration (MSI), large-scale integration devices (LSI), very-large-scale integration (VLSI), ultra-large-scale integration (ULSI) and giga-scale integration (GSI) (Table 3.32).

TABLE 3.32 | Integration complexity

Complexity	Gates Per Chip
Small-scale-integration (SSI)	Less than 12
Medium-scale-integration (MSI)	12 to 99
Large-scale integration devices (LSI)	100–9999
Very-Large-scale integration (VLSI)	10,000–99,999
Ultra-Large-scale integration (ULSI)	1,00,000–9,99,999
Giga-scale integration (GSI)	>10,00,000

3.5 | ANSI/IEEE STANDARD LOGIC SYMBOLS

The symbols used to represent logic gates have been in use for many years and will, no doubt, continue to be used. However, a new standard issued jointly by the American National Standards Institute (ANSI) and the Institute for Electrical and Electronics Engineers (IEEE) is now being used by many industries and manufacturers of electronic devices. It is called the ANSI/IEEE standard. Instead of using distinctive symbols for various types of gates, the new standard depicts all gates in rectangular outlines.

Characters called qualifying symbols are placed inside the rectangular outlines to indicate the type of logic operations performed. Figure 3.44 represents the OR logic gate symbol and Figure 3.45 gives the ANSI/IEEE OR logic gate.

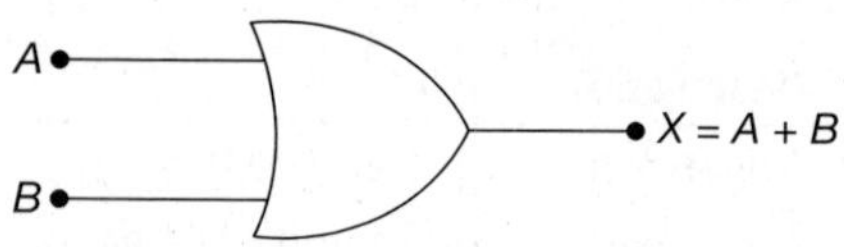

FIGURE 3.44 | Symbol of OR logic gate

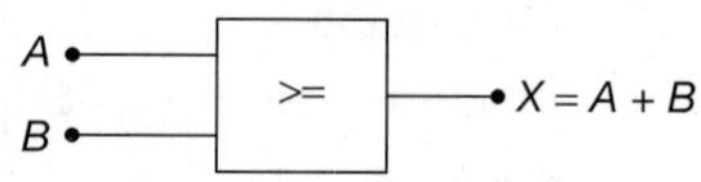

FIGURE 3.45 | Symbol of ANSI/IEEE OR logic gate

Figure 3.46 represents the AND logic gate symbol and Figure 3.47 gives the ANSI/IEEE AND logic gate.

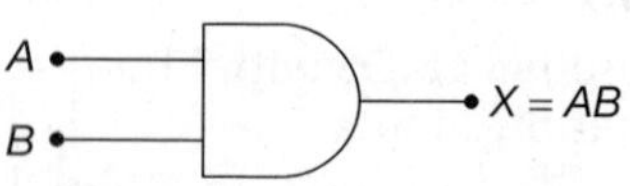

FIGURE 3.46 | Symbol of AND logic gate

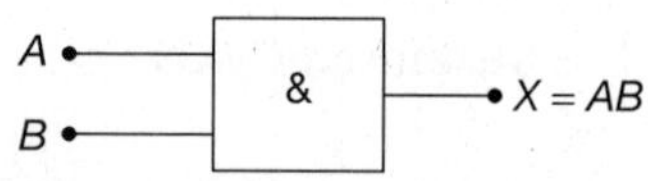

FIGURE 3.47 | Symbol of ANSI/IEEE AND logic gate

Figure 3.48 represents the NOT (Inverter) logic gate symbol and Figure 3.49 gives the ANSI/IEEE NOT logic gate.

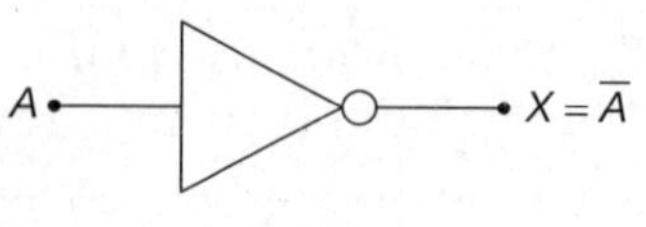

FIGURE 3.48 | Symbol of NOT logic gate

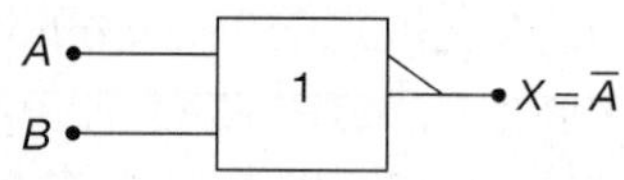

FIGURE 3.49 | Symbol of ANSI/IEEE NOT logic gate

Figure 3.50 represents the NAND logic gate symbol and Figure 3.51 gives the ANSI/IEEE NAND logic gate.

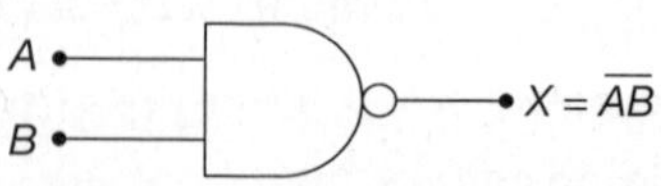

FIGURE 3.50 | Symbol of NAND logic gate

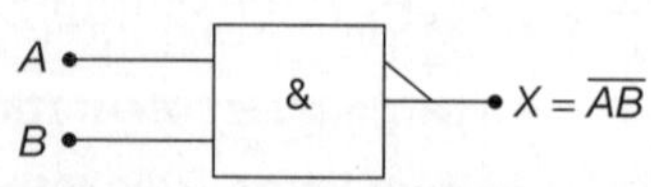

FIGURE 3.51 | Symbol of ANSI/IEEE NAND logic gate

Figure 3.52 represents the NOR logic gate symbol and Figure 3.53 gives the ANSI/IEEE NOR logic gate.

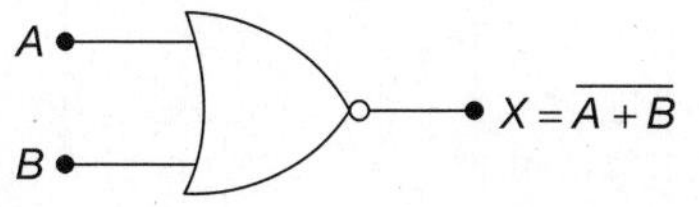

FIGURE 3.52 | Symbol of NOR logic gate

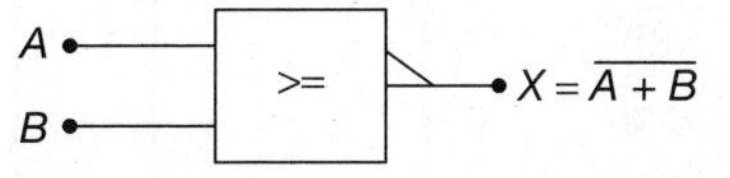

FIGURE 3.53 | Symbol of ANSI/IEEE NOR logic gate

Figure 3.54 represents the exclusive OR logic gate symbol and Figure 3.55 gives the ANSI/IEEE exclusive OR logic gate.

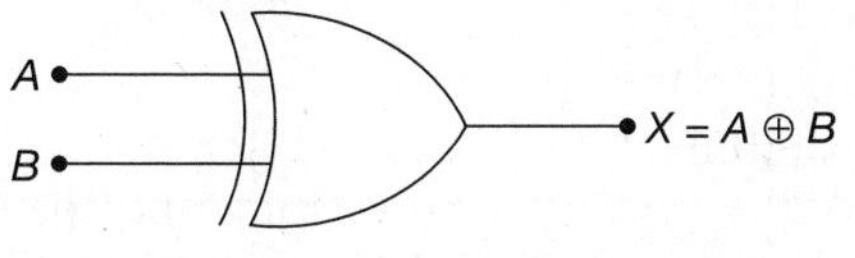

FIGURE 3.54 | Symbol of exclusive-OR logic gate

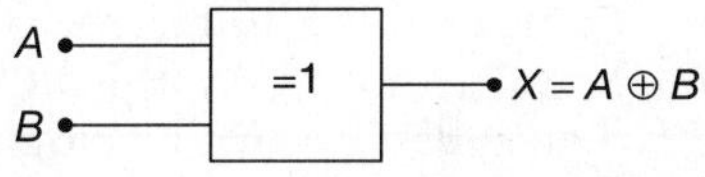

FIGURE 3.55 | Symbol of ANSI/IEEE XOR logic gate

Figure 3.56 represents the exclusive-NOR logic gate symbol and Figure 3.57 gives the ANSI/IEEE exclusive-NOR logic gate.

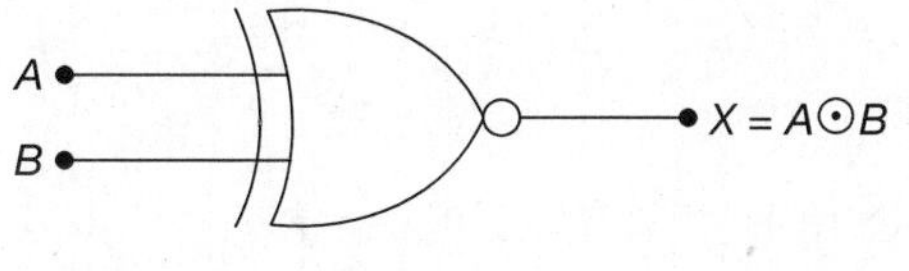

FIGURE 3.56 | Symbol of exclusive-NOR logic gate

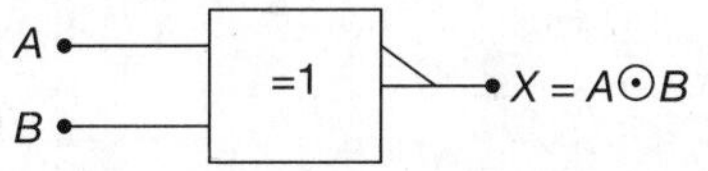

FIGURE 3.57 | Symbol ANSI/IEEE exclusive NOR logic gate

3.6 | PULSED OPERATION OF LOGIC GATES

Generally, in many cases, the inputs to a gate are not stationary (Constant) levels. But these are voltages those change frequently between two logic levels; 1 (high) and 0 (low). Such voltages are classified as pulse waveforms. Irrespective of the nature of inputs either constant levels or pulsed levels, all the logic gates obey the truth table operation. The timing diagrams of various gates for different inputs are given below as examples.

EXAMPLE 3.1 Draw the output waveform of NOT gate. The NOT gate input A is varying according to the timing diagrams shown in Figure 3.58(a).

SOLUTION In Figure 3.58(a), waveform A starts out as LOW. When input is HIGH, the output of NOT gate is LOW and when input is LOW, the output of NOT gate is HIGH. So, NOT gate output in the start is HIGH as shown in Figure 3.58(a).

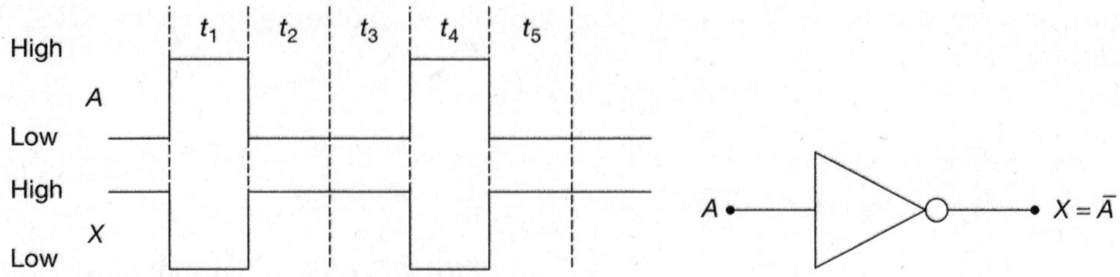

(a) Input and output pulses (b) NOT logic gate

FIGURE 3.58 | Pulsed output of NOT logic gate

At time t_1, input A is HIGH, the NOT gate output X is LOW.
At time t_2, input A is LOW, the NOT gate output X is HIGH.
At time t_3: input A is LOW, the NOT gate output X is HIGH.
At time t_4: input A is HIGH, the NOT gate output X is LOW.
At time t_5: input A is LOW, the NOT gate output X is HIGH.

EXAMPLE 3.2 Draw the output waveform of AND gate. The AND gate inputs A and B are varying according to the timing diagrams shown in Figure 3.59(a).

SOLUTION In Figure 3.59(a), waveforms A and B start out as LOW. If any input of AND gate is LOW, the AND gate output is LOW. Output of AND gate is HIGH only when both inputs are HIGH as shown in Figure 3.59(a).

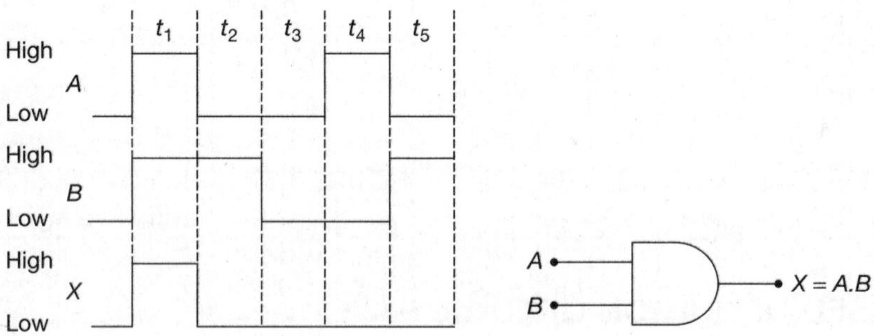

(a) Input and output pulses (b) AND logic gate

FIGURE 3.59 | Pulsed output of AND logic gate

At time t_1, both inputs are HIGH so AND gate output X is HIGH.
At time t_2, input A is LOW and Input B is HIGH, so AND gate output X is LOW.
At time t_3, both inputs are LOW so AND gate output X is also LOW.
At time t_4, input A is HIGH and Input B is LOW, so AND gate output X is LOW.
At time t_5, input A is LOW and input B is HIGH. AND gate output X is LOW.

EXAMPLE 3.3 Draw the output waveform of OR gate. The OR gate inputs A and B are varying according to the timing diagrams shown in Figure 3.60(a).

SOLUTION In Figure 3.60(a), waveforms A and B start out as LOW. When any one input is HIGH, the output of OR gate is HIGH as shown in Figure 3.60(a).

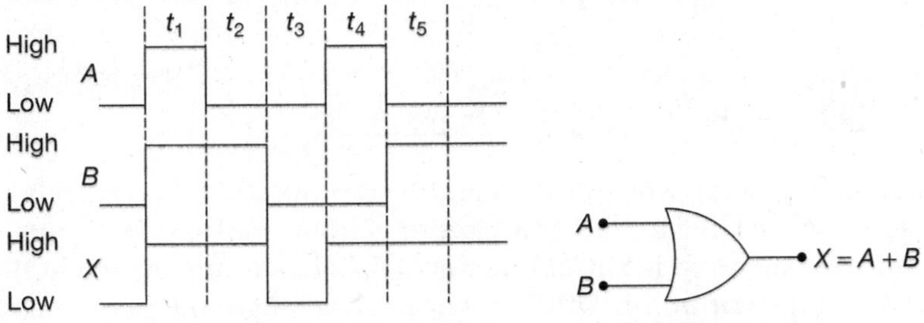

(a) Input and output pulses

(b) OR logic gate

FIGURE 3.60 | Pulsed output of OR logic gate

At time t_1, both inputs are HIGH so OR gate output X is HIGH.
At time t_2, input A is LOW and input B is HIGH, so OR gate output X is HIGH.
At time t_3, both inputs of OR gate are LOW, so OR gate output X is also LOW.
At time t_4, input A is HIGH and input B is LOW, so OR gate output X is HIGH.
At time t_5, input A is LOW and input B is HIGH. so OR gate output X is HIGH.

EXAMPLE 3.4 Draw the output waveform of NAND gate. The NAND gate inputs A and B are varying according to the timing diagrams shown in Figure 3.61(a).

SOLUTION In Figure 3.61(a), waveforms A and B start out as LOW. If any input of NAND gate is LOW, the NAND gate output is HIGH. Output of NAND gate is LOW only when both inputs are HIGH. So, output of NAND gate is HIGH in the start as shown in Figure 3.61(a).

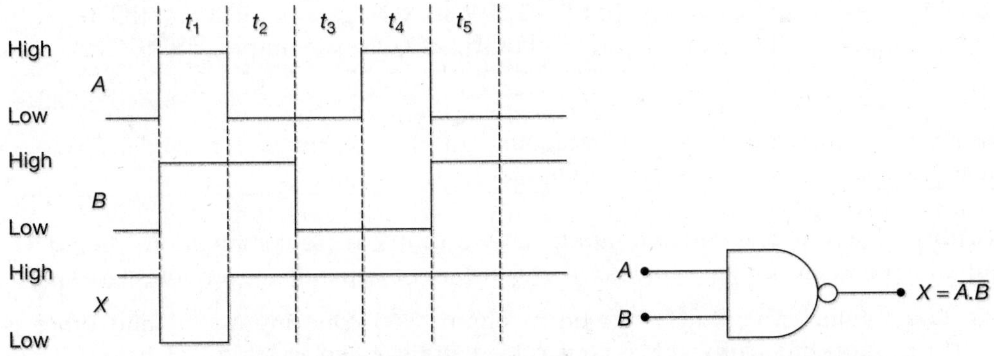

(a) Input and output pulses

(b) NAND logic gate

FIGURE 3.61 | Pulsed output of NAND logic gate

At time t_1 both inputs are HIGH so NAND gate output X is LOW.
At time t_2, input A is LOW and Input B is HIGH, so NAND gate output X is HIGH.
At time t_3, both inputs are LOW so NAND gate output X is also HIGH.
At time t_4, input A is HIGH and Input B is LOW, so NAND gate output X is HIGH.
At time t_5, input A is LOW and input B is HIGH. NAND gate output X is HIGH.

EXAMPLE 3.5 Draw the output waveform of XOR gate. The XOR gate inputs A and B are varying according to the timing diagrams shown in Figure 3.62(a).

SOLUTION In Figure 3.62(a), waveforms A and B start out as LOW. When both the inputs are same either LOW or HIGH, XOR gate output is LOW. When both inputs of XOR gate are different (means one input is HIGH other is LOW or vice versa), then XOR gate output is HIGH. So, XOR gate output X is LOW in the start as shown in Figure 3.62(a).

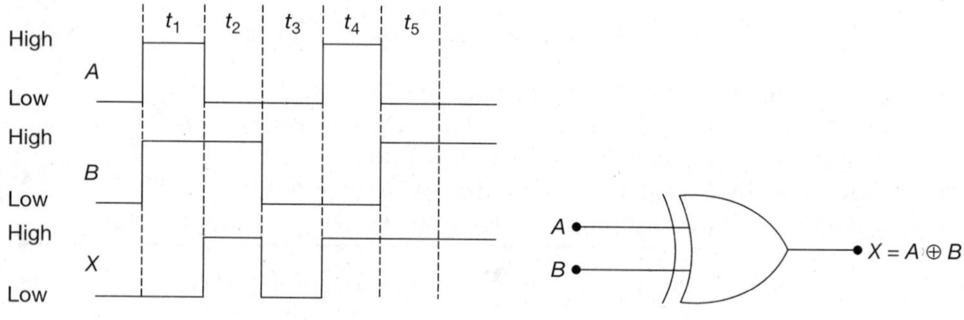

(a) Input and output pulses

(b) X-OR logic gate

FIGURE 3.62 | Pulsed output of X-OR logic gate

At time t_1, both inputs are HIGH so XOR gate output X is LOW.
At time t_2, input A is LOW and input B is HIGH, so XOR gate output X is HIGH.
At time t_3, both inputs of XOR gate are LOW, so XOR gate output X is also LOW.
At time t_4, input A is HIGH and input B is LOW, so XOR gate output X is HIGH.
At time t_5, input A is LOW and input B is HIGH. so XOR gate output X is HIGH.

EXAMPLE 3.6 Show that (i) $A \cdot 0 = 0$ (inhibition), (ii) $A \cdot 1 = A$ (transfer or enable), (iii) $A \cdot A = A$ (transfer or enable) and (iv) $A \cdot \overline{A} = 0$ (inhibition).

SOLUTION There is one variable input, A. Two (2^1) possible combinations (0 and 1) are required to evaluate. Other signal is either constant or depends on variable signal A.

(i) Logic gate with inputs is given in Figure 3.63. One input is A and other is 0. Corresponding truth table is constructed and is given as Table 3.33, below. It can be observed that the output is 0 when the input A is either 0 or 1.

Hence, $A \cdot 0 = 0$ and AND gate inhibits the output signal to 0.

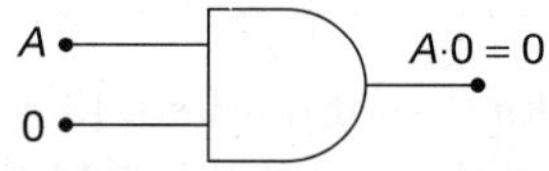

FIGURE 3.63 | AND logic gate

TABLE 3.33 | Truth table of AND gate

Inputs		Output
A	**0**	**$A \cdot 0 = 0$**
0	0	0
1	0	0

(ii) Logic gate with inputs is given in Figure 3.64. One input is A and other is 1. Corresponding truth table is constructed and is given as Table 3.34, below. It can be observed that the output is 0 when the input A is 0 and the output is 1 when the input A is 1. So output is similar to A.

Hence $A \cdot 1 = A$ and AND gate transfers (enables) the signal A to the output.

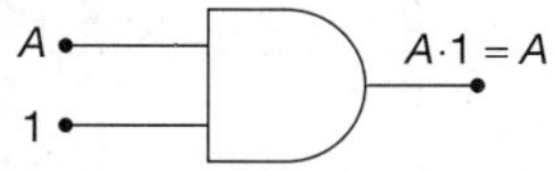

FIGURE 3.64 | AND logic gate

TABLE 3.34 | Truth table of AND gate

Inputs		Output
A	**1**	**$A \cdot 1 = A$**
0	1	0
1	1	1

(iii) Logic gate with inputs is given in Figure 3.65. Both inputs are same as A. Corresponding truth table is constructed and is given as Table 3.35 below. It can be observed that the output is 0 when the input A is 0 and the output is 1 when the input A is 1. So, output is similar to A.

Hence, $A \cdot A = A$ and AND gate transfers (enables) the signal A to the output.

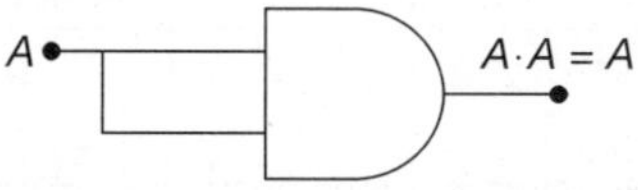

FIGURE 3.65 | AND logic gate

TABLE 3.35 | Truth table of AND gate

Inputs		Output
A	**A**	**$A \cdot A = A$**
0	0	0
1	1	1

(iv) Logic gate with inputs is given in Figure 3.66. One input is A and other is complement of A. When A is 0 its complement is 1. If A is 1 its complement is 0. Corresponding truth table is constructed and is given as Table 3.36, below. It can be observed that the output is 0 in both the possible values of A (either 0 or 1) and its complement (1 and 0).

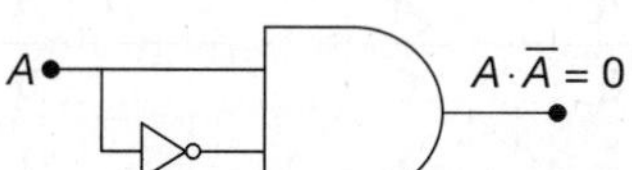

FIGURE 3.66 | AND logic gate

TABLE 3.36 | Truth table of AND gate

Inputs		Output
A	**$\bar{A}$**	**$A \cdot \bar{A} = 1$**
0	1	0
1	0	0

Hence, $A \cdot \bar{A} = 0$ and AND gate inhibits the output signal to 0.

EXAMPLE 3.7 Show that (i) $A + 0 = A$ (transfer), (ii) $A + 1 = 1$ (inhibition), (iii) $A + A = A$ (transfer or enable) and (iv) $A + \bar{A} = 1$ (inhibition).

SOLUTION There is one variable input, A. Two (2^1) possible combinations (0 and 1) are required to evaluate. Other signal is either constant or depends on variable signal A.

(i) Logic gate with inputs is given in Figure 3.67. Input A is either 0 or 1. Corresponding truth table is constructed and is given as Table 3.37. It can be observed that the output is 0 when the input A is 0 and the output is 1 when the input A is 1. So, output is similar to A input.
Hence, $A+0=A$ and OR gate transfers (enables) the signal A to the output.

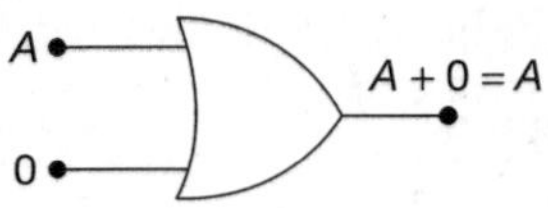

FIGURE 3.67 | 2-input OR logic gate

TABLE 3.37 | Truth table of OR gate

Inputs		Output
A	0	$A + 0 = A$
0	0	0
1	0	1

(ii) Logic gate with inputs is given in Figure 3.68. Input A is either 0 or 1. Corresponding truth table is constructed and is given as Table 3.38 below. It can be observed that the output is 0 when the value of A is either 1 or 0.
Hence, $A+1=1$ and OR inhibits the output signal to 1.

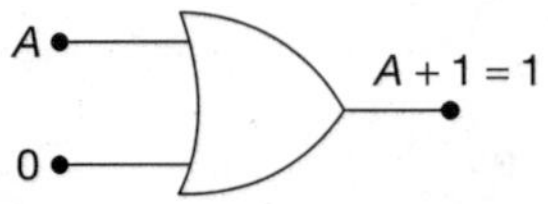

FIGURE 3.68 | 2-input OR logic gate

TABLE 3.38 | Truth table of OR gate

Inputs		Output
A	1	$A + 1 = 1$
0	1	1
1	1	1

(iii) Logic gate with inputs is given in Figure 3.69. Input A is either 0 or 1. Corresponding truth table is constructed and is given as Table 3.39, below. It can be observed that the output is 0 when the input A is 0 and the output is 1 when the input A is 1. So, output is similar to A.
Hence, $A+A=A$ and OR gate transfers (enables) the signal A to the output.

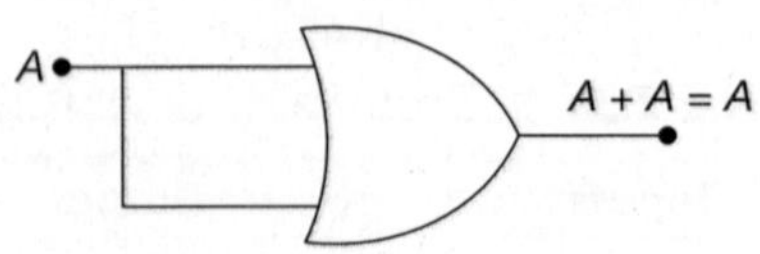

FIGURE 3.69 | OR gate

TABLE 3.39 | Truth table of OR gate

Inputs		Output
A	A	$A + A = A$
0	0	0
1	1	1

(iv) Logic gate with inputs is given in Figure 3.70. One input is A and other is complement of A. When A is 0 its complement is 1. If A is 1 its complement is 0. Corresponding truth table is constructed and is given as Table 3.40 below. It can be observed that the output is 1 in both the possible values of A (either 0 or 1) and its complement (1 and 0).

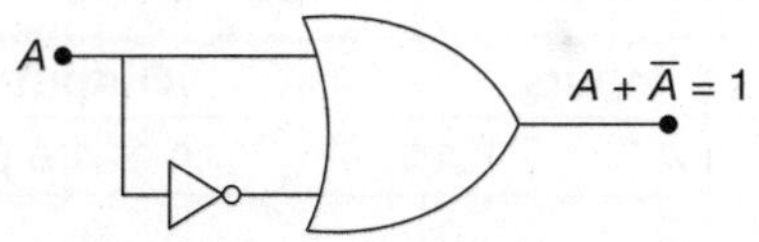

FIGURE 3.70 | OR gate

TABLE 3.40 | Truth table of OR gate

Inputs		Output
A	$\bar{A}$	$A + \bar{A} = 1$
0	1	1
1	0	1

Hence, $A + \bar{A} = 1$ and OR gate inhibits the output signal to 1.

EXAMPLE 3.8 Show that (i) $A \oplus 0 = A$ (Complement), (ii) $A \oplus 1 = \bar{A}$ (Transfer), (iii) $A \oplus A = 0$ (inhibit to low) and (iv) $A \oplus \bar{A} = 1$ (inhibit to high).

SOLUTION There is one variable input, A. Two (2^1) possible combinations (0 and 1) are required to evaluate. Other signal is either constant or depends on variable signal A.

(i) Logic gate with inputs is given in Figure 3.71. Input A is either 0 or 1. Corresponding truth table is constructed and is given as Table 3.41. It can be observed that the output is 0 when the input A is 0 and the output is 1 when the input A is 1. So output is similar to A.

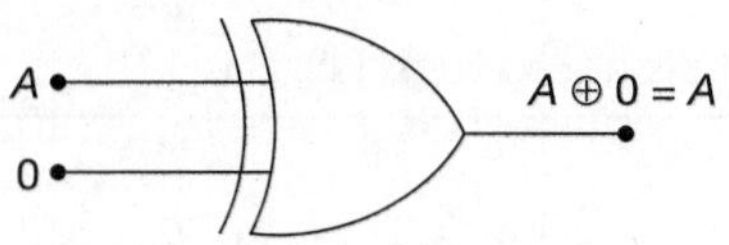

FIGURE 3.71 | 2-input XOR logic gate

TABLE 3.41 | Truth table of XOR gate

Inputs		Output
A	0	$A \oplus 0 = A$
0	0	0
1	0	1

Hence, $A \oplus 0 = A$ and XOR gate transfers (enables) the signal A to the output.

(ii) Logic gate with inputs is given in Figure 3.72. Input A is either 0 or 1. Corresponding truth table is constructed and is given as Table 3.42. It can be observed that the output is 1 when the input A is 0 and the output is 0 when the input A is 1. So, output is complement of input.

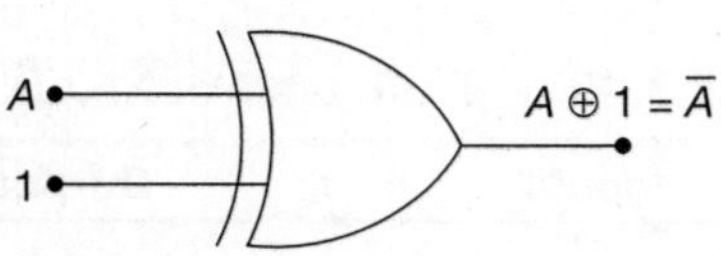

FIGURE 3.72 | 2-input XOR logic gate

TABLE 3.42 | Truth table of XOR gate

Inputs		Output
A	1	$A \oplus 1 = \bar{A}$
0	1	1
1	1	0

Hence, $A \oplus 1 = \bar{A}$ and XOR gate inverts the signal A to its complement.

(iii) Logic gate with inputs is given in Figure 3.73. Input A is either 0 or 1. Corresponding truth table is constructed and is given as Table 3.43 below. It can be observed that the output is 0 when the value of A is either 1 or 0.

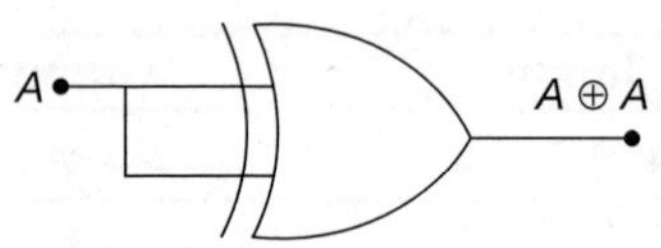

FIGURE 3.73 | 2-input XOR logic gate

TABLE 3.43 | Truth table of XOR gate

Inputs		Output
A	A	$A \oplus A = 0$
0	0	0
1	1	0

Hence, $A \oplus A = 0$ and XOR inhibits the output signal to low (0).

(iv) Logic gate with inputs is given in Figure 3.74. One input is A and other is complement of A. When A is 0 its complement is 1. If A is 1 its complement is 0. Corresponding truth table is constructed and is given as Table 3.44 below. It can be observed that the output is 1 in both the possible values of A (either 0 or 1) and its complement (1 and 0).

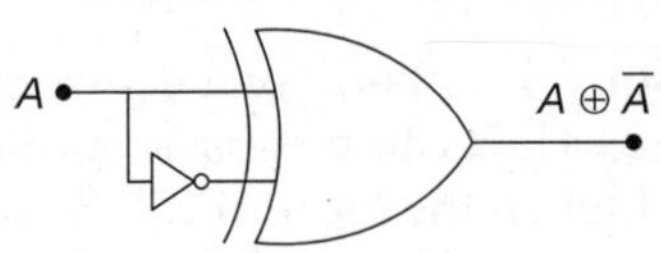

FIGURE 3.74 | 2-input XOR logic gate

TABLE 3.44 | Truth table of XOR gate

Inputs		Output
A	$\bar{A}$	$A \oplus \bar{A} = 1$
0	1	1
1	0	1

Hence, $A \oplus \bar{A} = 1$ and XOR gate inhibits output signal to high (1).

EXAMPLE 3.9 Show that (i) $A \odot 0 = \bar{A}$ (Complement), (ii) $A \odot 1 = A$ (Transfer), (iii) $A \odot A = 1$ (inhibit to low) and (iv) $A \odot \bar{A} = 0$ (inhibit to high).

SOLUTION There is one variable input, A. Two (2^1) possible combinations (0 and 1) are required to evaluate. Other signal is either constant or depends on variable signal A.

(i) Logic gate with inputs is given in Figure 3.75. Input A is either 0 or 1. Corresponding truth table is constructed and is given as Table 3.45. It can be observed that the output is 1 when the input A is 0 and the output is 0 when the input A is 1. So, output is complement of input.

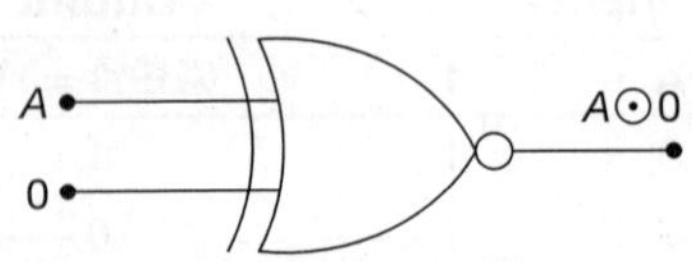

FIGURE 3.75 | 2-input XNOR Logic gate

TABLE 3.45 | Truth table of XNOR gate

Inputs		Output
A	0	$A \odot 0 = \bar{A}$
0	0	1
1	0	0

Hence, $A \odot 0 = \bar{A}$ and XOR gate inverts the signal A to its complement.

(ii) Logic gate with inputs is given in Figure 3.76. Input A is either 0 or 1. Corresponding truth table is constructed and is given as Table 3.46. It can be observed that the output is 0 when the input A is 0 and the output is 1 when the input A is 1. So output is similar to A.

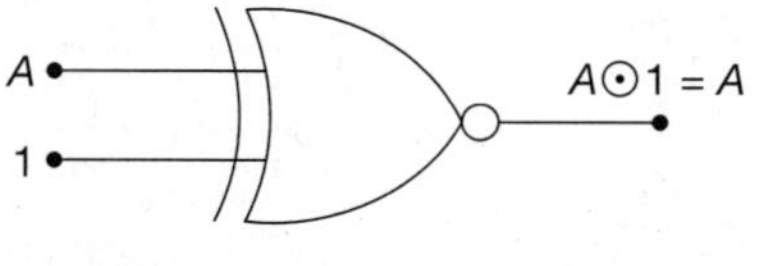

FIGURE 3.76 | 2-input XNOR logic gate

TABLE 3.46 | Truth table of XNOR gate

Inputs		Output
A	1	$A \odot 1 = A$
0	1	0
1	1	1

Hence, $A \odot 1 = A$ and XOR gate transfers (enables) the signal A to the output.

(iii) Logic gate with inputs is given in Figure 3.77. Input A is either 0 or 1. Corresponding truth table is constructed and is given as Table 3.47 below. It can be observed that the output is 0 when the value of A is either 1 or 0.

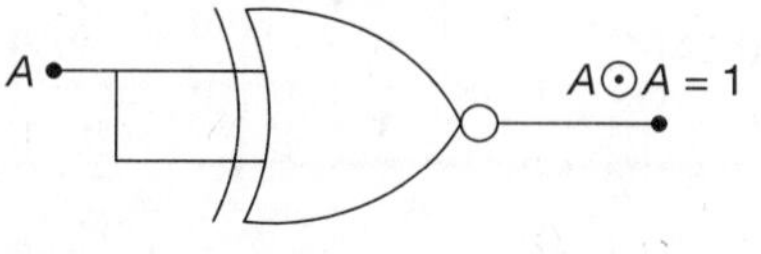

FIGURE 3.77 | 2-input XNOR logic gate

TABLE 3.47 | Truth table of XNOR gate

Inputs		Output
A	A	$A \odot A = 1$
0	0	1
1	1	1

Hence, $A \odot A = 1$ and XOR inhibits the output signal to high (1).

(iv) Logic gate with inputs is given in Figure 3.78. One input is A and other is complement of A. When A is 0 its complement is 1. If A is 1 its complement is 0. Corresponding truth table is constructed and is given as Table 3.48 below. It can be observed that the output is 1 in both the possible values of A (either 0 or 1) and its complement (1 and 0).

Hence, $A \odot \bar{A} = 0$ and XOR gate inhibits output signal to low (0).

TABLE 3.48 | Truth table of XNOR gate

Inputs		Output
A	$\bar{A}$	$A \odot \bar{A} = 0$
0	1	0
1	0	0

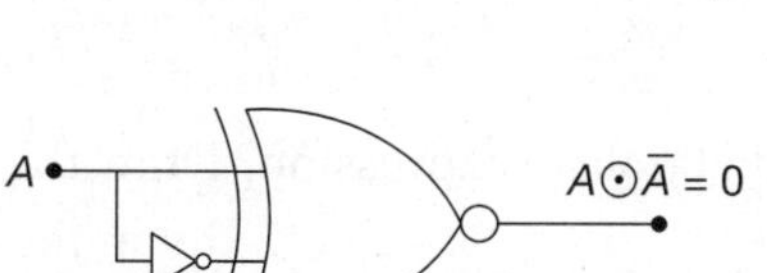

FIGURE 3.78 | 2-input XNOR logic gate

EXAMPLE 3.10 Show that $AB + (\overline{A+B})$ is equivalent to $A \odot B$.

SOLUTION There are two inputs A and B. Four (2^2) possible combinations are required to evaluate. Truth table gives the output corresponding to inputs those are 00, 01, 10 and 11. Inputs give binary counting sequence of 2-digit combinations. Outputs given in truth Tables 3.49 and 3.50 are same. Circuit diagram is given in Figure 3.79.

TABLE 3.49 | Truth table of XNOR gate

Inputs		Outputs
A	*B*	$A \odot B$
0	0	1
0	1	0
1	0	0
1	1	1

TABLE 3.50 | Truth table of Boolean logic

Input		Intermediate Outputs			Outputs
A	*B*	$A + B$	AB	$\overline{A + B}$	$AB + (\overline{A + B})$
0	0	0	0	1	1
0	1	1	0	0	0
1	0	1	0	0	0
1	1	1	1	0	1

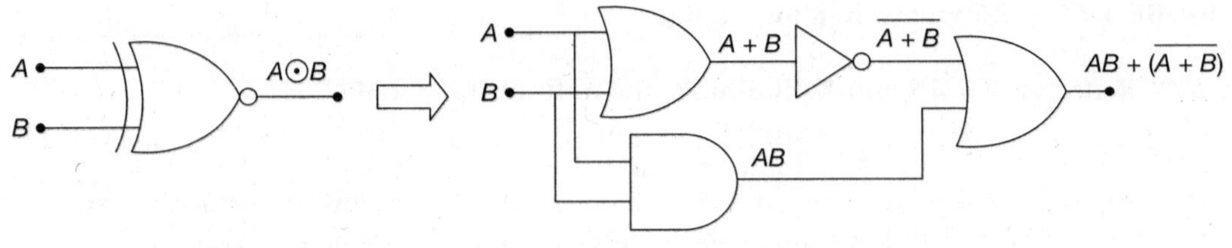

FIGURE 3.79 | Equivalent circuit diagrams

EXAMPLE 3.11 What will be the output *Y* of the given Boolean expression? Draw the logic circuit diagram $Y = \overline{\left(\left(\overline{AB}\right)\left(\overline{AB}\right)\right) + \left(\overline{A+B}\right)}$.

SOLUTION

Logic circuit diagram: To draw the circuit diagram let *X* is the output of a NAND gate and *W* is output of NOR gate.

$$X = \overline{AB} \text{ and } W = \overline{A+B}$$

Z is the output of NAND gate and *Y* is the output of NOR gate

$$Z = \overline{XX} \text{ and } Y = \overline{W+Z}$$

Logic circuit diagram is given the Figure 3.80.

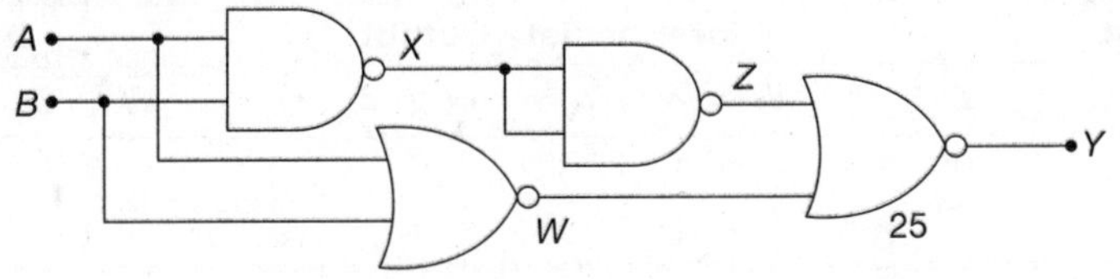

FIGURE 3.80 | Logic circuit diagram

Truth Table: There are two inputs, *A* and *B*. Four (2^2) possible combinations are required to evaluate. Truth Table 3.51 gives the output corresponding to inputs those are 00, 01, 10 and 11.

TABLE 3.51 | Truth table of Example 3.12

Input		Intermediate Output			Output
A	B	$X = \overline{AB}$	$W = \overline{A+B}$	$Z = \overline{XX}$	$Y = \overline{W+Z}$
0	0	1	1	0	0
0	1	1	0	0	1
1	0	1	0	0	1
1	1	0	0	1	0

EXAMPLE 3.12 Prove $\overline{(X \odot Y \odot Z)} = X \odot Y \oplus Z$? Draw the logic circuit diagram.

SOLUTION

Truth Table: There are three inputs, *X*, *Y* and *Z*. Eight (2^3) possible combinations are required to evaluate. Truth table gives the output corresponding to inputs those are 000, 001, 010, 011, 100, 101, 110 and 111.

Exclusive-NOR gate: If inputs are same output is '1' and if inputs are different output is '0'.

Exclusive-OR gate: If inputs are same output is '0' and if inputs are different output is '1'.

Accordingly truth table is prepared and is given in Table 3.52.

$$\overline{(X \odot Y \odot Z)} = X \odot Y \oplus Z, \text{ where } W = X \odot Y$$

$$\overline{(W \odot Z)} = W \oplus Z$$

TABLE 3.52 | Truth table of Example 3.13

Input			Intermediate Output		Outputs	
X	Y	Z	$W = X \odot Y$	$W \odot Z$	$\overline{W \odot Z}$	$W \oplus Z$
0	0	0	1	0	1	1
0	0	1	1	1	0	0
0	1	0	0	1	0	0
0	1	1	0	0	1	1
1	0	0	0	1	0	0
1	0	1	0	0	1	1

(Continued)

TABLE 3.52 | (Continued)

Input			Intermediate Output		Outputs	
X	Y	Z	$W = X \odot Y$	$W \odot Z$	$\overline{W \odot Z}$	$W \oplus Z$
1	1	0	1	0	1	1
1	1	1	1	1	0	0

Outputs $\overline{(W \odot Z)}$ and $W \oplus Z$ are same.

Diagram is given in Figure 3.81.

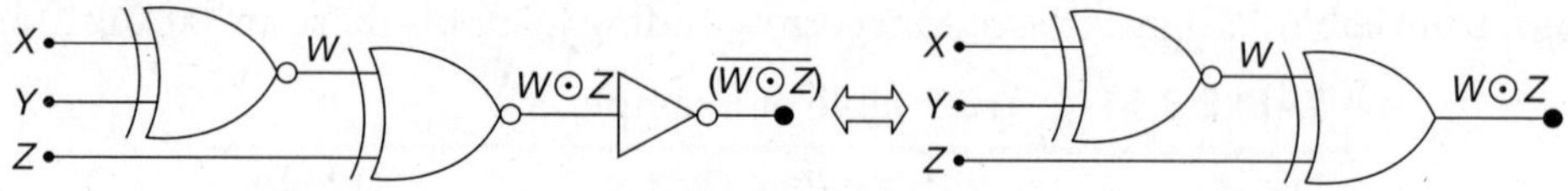

FIGURE 3.81 | Logic circuit diagram of Example 3.13

EXAMPLE 3.13 Find dual of the given Boolean expression: $Y=\overline{\left(\overline{(\overline{AB})(\overline{AB})}\right)+\left(\overline{A+B}\right)}$.

SOLUTION NOT gate will remain same.

Change AND gate to OR gate so AB is changed to $A + B$.
Change OR gate to AND gate so $A + B$ is changed to AB.

Expression can be rewritten as dual of $Y=\overline{\left(\overline{(\overline{A+B})+(\overline{A+B})}\right)\left(\overline{AB}\right)}$

EXAMPLE 3.14 What is the Boolean function of the circuit shown in Figure 3.82?

SOLUTION

Series circuit follows AND logic. Parallel circuit follows OR logic.
B and C switches are in parallel can be represented by OR logic, i.e. $B + C$.

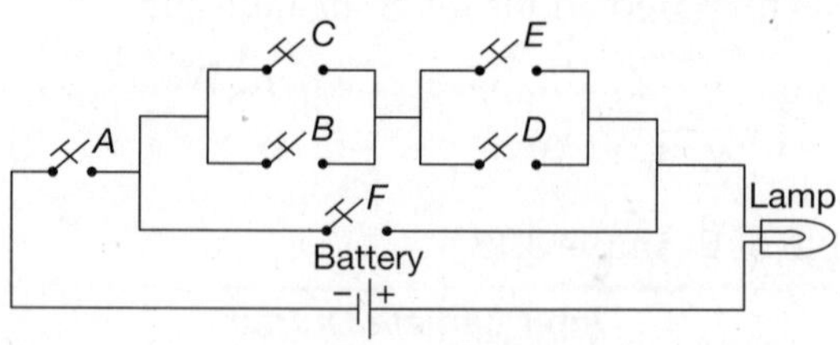

FIGURE 3.82 | Switching circuit of example 3.15

D and E switches are in parallel can be represented by OR logic, i.e. $D+E$. Combination of B and C switches is in series with combination of D and E switches can be represented by AND logic, i.e. $(B + C)\cdot(D + E)$. Combination of B, C, D and E switches is in parallel with F can be represented by OR logic, i.e. $(B+C)\cdot(D+E)+F$. Switch A is in series with combination of B, C, D, E and F switches hence the Boolean function for the given circuit is $A\cdot((B+C)\cdot(D+E)+F)$.

SUMMARY

- Logic gates are the fundamental building blocks of digital systems.
- AND, OR and NOT are the basic types of gates. The interconnection of gates to perform a variety of logical expression is called logic design.
- A truth table lists all possible combinations of inputs and corresponding outputs.
- In a positive logic system, the higher of the two voltage levels represents logic 1 and the lower of the two voltages represents logic 0.
- In a negative logic system, the higher of the two voltage levels represents logic 0 and the lower of the two voltages represents logic 1.
- An inverter performs the NOT operation.
- A NOT gate complements its input. It produces a 1 when the input is a 0. It produces a 0 when the input is a 1.
- An AND gate is an all, or nothing gate, it produces a 1, when all its inputs are 1. In all other cases, its output is 0.
- An OR gate is any or all gate; it produces a 1 if any one of its inputs is 1. It gives 0 when all its inputs are 0.
- A NAND gate is effectively an AND gate followed by NOT gate. NAND gate produces a 1, when all its inputs are 0. In all other cases, its output is 0.
- A NOR gate is effectively an OR gate followed by an inverter. It produces a I when all of its inputs are a 0. It produces a 0 when any of its inputs is a l. It therefore also acts as a negative AND gate.
- A NOR gate can be used as an inverter by tying all its input terminals together and feeding the signal to be inverted to the common terminal, or by tying all but one input terminal to logic 0 and feeding the signal to be inverted to the remaining terminal.
- NAND and NOR gates are called universal gates. Any circuit of any complexity can be realized by using only NAND gates or only NOR gates.
- AND, OR, NAND, and NOR gates can have any number of inputs.
- An X-OR gate is an anti-coincidence gate. It is an inequality detector. It produces a 1, only when its two inputs are not equal, i.e. when one is a 0 and the other is a 1.
- An X-OR gate can be used as an inverter by connecting one input terminal to logic 1 and feeding the signal to be inverted to the other terminal. It is a controlled inverter.
- An X-NOR gate is a coincidence gate it is an equality detector It produces a 1, only when its two inputs are equal, i.e. when both inputs are a 0 or a 1.
- Three or more input X-OR or X-NOR gates do not exist.
- An X-NOR gate can be used as an inverter by connecting one input terminal to logic 0 and feeding the signal to be inverted to the other terminal. It is a controlled inverter.
- When three or more variables are to be X-ORed or X-NORed, a number of two-input X-OR or X-NOR gates may be used.
- All logic gates have only one output.
- AND, OR, NAND, and NOR gates can be used either to enable or inhibit the passage of an input signal.
- When AND and OR gates are enabled, the output follows the input exactly.
- When NAND and NOR gates are enabled, the output is the exact inverse of the input signal.
- AND and NOR gates produce a constant LOW output when they are in the inhibited condition.

- NAND and OR gates produce a constant HIGH output when they are in the inhibited condition.
- All logic gates obey their truth table operations regardless of whether their inputs are constant levels or pulsed levels.

MULTIPLE CHOICE QUESTIONS

3.1 When the input to an inverter is HIGH (l), the output is
(a) HIGH or 1
(b) LOW or 1
(c) HIGH or 0
(d) LOW or 1

3.2 An inverter performs an operation known as
(a) complementation
(b) assertion
(c) inversion
(d) both answers (a) and (c)

3.3. The output of an AND gate with inputs *A*, *B*, and *C* is a 1 (HIGH) when
(a) $A = 1, B = 1$ and $C = 1$
(b) $A = 0, B = 1$ and $C = 1$
(c) $A = 1, B = 0$ and $C = 0$
(d) $A = 0, B = 0$ and $C = 0$

3.4 The output of an OR gate with inputs *A*, *B*, and *C* is a 1 (HIGH) when
(a) $A = 1, B = 1$ and $C = 1$
(b) $A = 1, B = 0$ and $C = 0$
(c) $A = 1, B = 1$ and $C = 0$
(d) All of above

3.5 A pulse is applied to each of 2-input NAND gate. One pulse goes HIGH at $t = 0$ and goes back low at $t = 1$ ms. The other pulse goes HIGH at $t = 0.8$ ms and goes LOW at $t = 3$ ms. The pulse can be described as follows:
(a) It goes LOW at $t = 0$ and back HIGH at $t = 3$ ms.
(b) It goes LOW at $t = 0.8$ ms and back HIGH at $t = 3$ ms.
(c) It goes LOW at $t = 0.8$ ms and back HIGH at $t = 1$ ms.
(d) It goes LOW at $t = 0.8$ ms and back LOW at $t = 1$ ms.

3.6 A pulse is applied to each of 2-input NOR gate. One pulse goes HIGH at $t = 0$ and goes back low at $t = 1$ ms. The other pulse goes HIGH at $t = 0.8$ ms and goes LOW at $t = 3$ ms. The pulse can be described as follows:
(a) It goes LOW at $t = 0$ and back HIGH at $t = 3$ ms.
(b) It goes LOW at $t = 0.8$ ms and back HIGH at $t = 3$ ms.
(c) It goes LOW at $t = 0.8$ ms and back HIGH at $t = 1$ ms.
(d) It goes HIGH at $t = 0.8$ ms and back LOW at $t = 1$ ms.

3.7 A pulse is applied to each of 2-input exclusive OR gate. One pulse goes HIGH at $t = 0$ and goes back low at $t = 1$ ms. The other pulse goes HIGH at $t = 0.8$ ms and goes LOW at 3 ms. The pulse can be described as follows:
(a) It goes HIGH at $t = 0$ and back LOW at $t = 3$ ms.
(b) It goes HIGH at $t = 0$ and back LOW at $t = 0.8$ ms.
(c) It goes HIGH at $t = 1$ and back LOW at $t = 3$ ms.
(d) Both (b) and (c)

3.8 A positive-going pulse is applied to an inverter. The time interval from the leading edge of the input to the leading edge of output is 7 ns. The parameter is:
(a) speed-power product
(b) propagation delay, t_{PHL}
(c) propagation delay, t_{PLH}
(d) pulse width

3.9 The number of terminals of input terminals which a NOT gate can have is

(a) 1 (b) 2
(c) 3 (d) Any number

3.10 The output of an AND logic gate, if one of its input terminal is connected to logic '0' is
(a) 0 (b) 1
(c) X (d) complement of X

3.11 A positive-OR gate is equivalent to
(a) Negative-OR gate
(b) Positive AND gate
(c) Negative AND gate
(d) Negative NOR gate

3.12 An exclusive-OR gate will have output as 1 positive-OR gate is equivalent to
(a) when all the inputs are 0
(b) when all the inputs are 1
(c) when even number of inputs are 1
(d) when odd number of inputs are 1

3.13 The output of a logic gate is '1' when all its inputs are at logic '0'. Then the gate is either
(a) a NAND or an exclusive-OR gate
(b) a NOR or an exclusive-NOR gate
(c) a OR or an exclusive-NOR gate
(d) a AND or an exclusive-OR gate

3.14 NOR gate inhibits to LOW when control line is
(a) either 0 or complement of data
(b) either 1 or complement of data
(c) either 0 or data
(d) either 1 or data

3.15 Complement of data can be obtained from
(a) NOR gate by controlling the input by either setting Low or having same to data.
(b) NAND gate by controlling the input by either setting HIGH or having same to data.
(c) EX-NOR gate by controlling the input by either setting HIGH or having same to data.
(d) all of above

Answers

3.1 (d)	3.2 (d)	3.3 (a)	3.4 (d)	3.5 (c)	3.6 (a)	3.7 (d)
3.8 (b)	3.9 (a)	3.10 (a)	3.11 (c)	3.12 (d)	3.13 (b)	3.14 (b)
3.15 (d)						

QUESTIONS

3.1 What are logic gates? Name the basic gates. What is logic design?

3.2 What are discrete logic gates?

3.3 What is the purpose of a truth table?

3.4 What are (a) level logic, (b) positive logic system, and (c) negative logic system?

3.5 What is dual of logic?

3.6 What is the logic levels used in TTL logic system?

3.7 Define the function performed by the following gates: (a) AND (b) OR (c) NOT-gates.

3.8 What is the only input combination that will produce a HIGH at the output of a six-inputs-AND gate?

3.9 One of the inputs to an AND gate is permanently set to LOW, what will be the shape of the output waveform when the remaining inputs are applied?

3.10 What is the only input combination that will produce a LOW at the output of a five-input OR gate?

3.11 If one of the inputs to an OR gate is permanently kept HIGH what would be the shape of the output waveform when the remaining inputs are applied?

3.12 Which logic gate is called (a) any or all gate (b) all or nothing gate (c) inverter*

3.13 How is logical addition different from ordinary addition?

3.14 How is logical multiplication different from ordinary multiplication?

3.15 What is the maximum number of inputs a NOT gate can have?

3.16 What are universal gates? Why these gates are called as universal gates?

3.17 What are (a) bubbled AND gate (b) bubbled OR gate?

3.18 What is the only set of input conditions that will produce a HIGH output from a three input NOR gate?

3.19 What is the only set of input conditions that will produce a LOW output from a three input NAND gate?

3.20 What type of gate is equivalent to a NAND gate followed by an inverter?

3.21 What type of gate is equivalent to a NOR gate followed by an inverter?

3.22 Which gates can be used as inverters in addition to the NOT gate and how?

3.23 Give the IC numbers of the following TTL gates: AND, OR, NOT, EX-OR, EX-NOR and NAND.

3.24 Draw logic diagrams to realize the following expressions: (a) $A \oplus B \oplus C \oplus D$ (b) $A \odot B \odot C$

3.25 What is the purpose of pulsed operation of a logic gate?

PROBLEMS

3.1 Prove the following using truth tables:

(a) $\overline{X \oplus Y} = X \odot Y$

(b) $\overline{X \oplus Y \oplus Z} = X \odot Y \oplus Z$

3.2 Show that a positive-logic NAND gate is a negative-logic NOR gate and vice versa.

3.3 What will be the output y of the given Boolean expression? Draw the logic circuit diagram $Y = \overline{\left(\left(\overline{A+B}\right)+\left(\overline{A+B}\right)\right)\left(\overline{AB}\right)}$.

3.4 Draw the logic circuit diagram and construct truth table for the following Boolean functions:

(a) $AB + A(B + \bar{C})$ (b) $\bar{A}\bar{B} + \bar{A}B + AB)$

(c) $(A + C)(\bar{A} + B)$ (d) $AB\left(\overline{ABC} + \bar{A}B\bar{C}\right)$

3.5 What will be the pulse train at the output of the logic circuit diagram drawn from the following Boolean function? Pulse train of A and B of same frequency at the start pulse for A is HIGH and for B is LOW.

$(A + B) \oplus \bar{B}$

4

Combinational Logic Design

CHAPTER OBJECTIVES

The main goal of this chapter is to impart knowledge about the combinational logic circuit design using digital logics. Readers will be able to discuss the following aspects in this chapter:

- Boolean laws and theorems
- Algebraic simplifications using Boolean laws and theorems
- Standard representations of logic functions
- Algebraic simplifications using Karnaugh Map method following product-of-sum (POS) and sum-of-product (SOP)
- Algebraic simplifications using Karnaugh Map method considering don't care conditions
- Algebraic simplifications using Quine–McCluskey method
- Variable elimination method
- Design of digital logic circuit using universal gates
- Hazard in logic circuits

4.1 | COMBINATIONAL CIRCUITS

Boolean algebra uses symbols and logical operators (AND, OR and NOT) to represent a logical expression that has one of two possible binary values: 0 (false) or 1(true). A binary variable is like a variable in regular algebra, except it can only have the values of zero (0) or one (1). Binary variables are designated by names, symbols, letters, bars or their combinations. Boolean algebra expresses the relationship between a logic circuit's inputs and outputs. So, logical expressions describe the relationship between a logic circuit's outputs and its inputs. Three logic functions (AND, OR, and NOT or complement) provide the foundation for all digital system's analysis and design.

A function is a term used in mathematics and logic to denote a relationship of binary input and output. A tabular representation of the combinations that a group of input binary variables can assume to represent output is called a truth table. An output variable X is said to be a function of a set of input variables, A, B, and C. Mathematically it is represented as $X = f(A,B,C)$. X is binary output of logical function. A, B and C are the binary input variables. Function can be AND, OR, NOT, NAND, NOR, X-OR, and X-NOR logical functions. Binary variables can have either zero (0) or one (1) value.

Basic logic gates are the fundamental building blocks from which all of the logic circuits and digital systems are constructed. Logic gates produce one output level when some combinations of input levels are given. The interconnection of gates to perform a variety of logical operations is called logic design.

Constant: A fixed or non-changeable quantity is termed as constants. 0 and 1 are treated as binary logic constants.

Variable: A changeable quantity is termed as variable. In binary logic a variable can have 0 or l value. A symbol, letter, number, or short name (mnemonic) is assigned to distinguish one variable from another.

Complement: Complement means to complete the binary set. In binary algebra, the complement is the opposite. For example, the complement of *X* (primed) is written $\bar{X}$ or *X'* (un-primed). The complement of 1 is 0 and complement of 0 is 1. The terms complement and inverse can be used interchangeably. Complement of function is obtained by NOT gate.

Literal: A literal is a Boolean variable (primed) or its complement (un-primed). For instance, let *X* be a binary variable, then both *X* and complement of *X*, would be literals.

Product term: A product term is a literal or the logical AND operation of multiple literals. Logical AND operation gives the output that is equivalent to binary product of single binary digits. So, it is known as logical product. For instance, let *X*, *Y* and *Z* be binary variables. Then a representative product term could be *X*, *XY*, or *X'YZ*.

Sum term: A sum term is a literal or the logical OR operation of multiple literals. Logical OR operation is termed as sum because of the sign '+'; however it has no analogy with the binary single digit sum. It is also known as logical sum. Let *X*, *Y* and *Z* are binary variables. Then a representative sum term could be $X, X+\bar{Y}$, or $\bar{X}+\bar{Y}+Z$.

Theorem: A postulate or group of postulates that have been 'proven'. If a single postulate or a set of postulates is assumed to be true, then it may be used to 'prove' a theorem. Perfect induction, proof by trying all of the possible combinations, may be used to prove a postulate.

Perfect induction: A method of proof by examination of all possible input variable combinations, therefore showing the output variable value in every possible condition. It is an exhaustive proof of a postulate or theorem.

Truth table: A tabular shows the 'truth' relationship between a set of input and output variables. For all possible input variable combinations, the output is examined. Truth is defined as a 1 (for positive logic) or 0 (for negative logic) of binary variable.

Logic diagram: A drawing made up of logic symbols, distinctive shape or IEEE symbol, showing input and output connections between various logic functions. A logic diagram and the logic expression(s) have the direct relationship.

Logical/Boolean Expression: Two logical variables are combined with logical operator to form logical expression.

4.2 | BOOLEAN LAWS AND THEOREMS

There are some basic postulates or theorems followed to analyses or design the digital circuit are given below for reference.

1. Law of Intersection: $A \cdot 1 = A$
   $A \cdot 0 = 0$
2. Law of Union: $A + 1 = 1$
   $A + 0 = A$
3. Law of Identity: $A \cdot 1 = A$
   $A + 1 = 1$
4. Law of Null: $A \cdot 0 = 0$
   $A + 0 = A$
5. Law of Tautology or Idempotence: $A \cdot A = A$
   $A + A = A$
6. Law of Complement or Negation: $A \cdot \bar{A} = 0$
   $A + \bar{A} = 1$
7. Law of Double Negation or Involution: $\bar{\bar{A}} = A$
8. Law of Commutation: $A + B = B + A$
   $A \cdot B = B \cdot A$
9. Law of Association: $(A \cdot B) \cdot C = A \cdot (B \cdot C)$
   $(A + B) + C = A + (B + C)$
10. Law of Distribution: $A \cdot B + A \cdot C = A \cdot (B + C)$
    $(A + B)(A + C) = A + BC$
11. Law of Absorption or Redundancy: $A \cdot (A + B) = A$
    $A + AB = A$
    $AB + \bar{B} = A + \bar{B}$
    $A\bar{B} + B = A + B$
12. De Morgan's Theorem-I: $\overline{A + B} = \bar{A} \cdot \bar{B}$
    De Morgan's Theorem-II: $\overline{A \cdot B} = \bar{A} + \bar{B}$
13. Consensus Theorem: $AB + \bar{A}C + BC = AB + \bar{A}C$
    $(A + B)(\bar{A} + C)(B + C) = (A + B)(\bar{A} + C)$
14. Transposition Theorem: $AB + \bar{A}C = (A + C)(\bar{A} + B)$

Boolean algebra is also known as switching algebra. Some fundamental laws (relations) of Boolean algebra have been presented with logic gates justified by Truth tables. These are explained in the ensuing section.

4.2.1 | Law of Intersection

Intersection is AND operation of two inputs when one of them is either 0 or 1. Mathematical logic expressions are given in Eqs. (4.1) and (4.2). Logic diagram of Eq. (4.1) is given in Figure 4.1 and its truth table is given in Table 4.1. Logic diagram of Eq. (4.2) is given in Figure 4.2 and its truth table is given in Table 4.2. For one binary variable, A there are two possible values considered in truth tables.

$$A \cdot 1 = A \tag{4.1}$$

$$A \cdot 0 = 0 \tag{4.2}$$

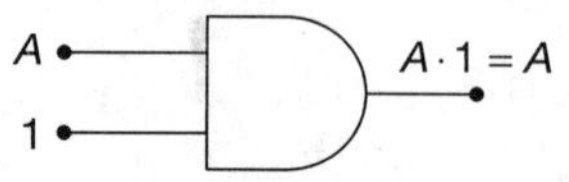

FIGURE 4.1 | Intersection law

TABLE 4.1 | Law of intersection or identity

Inputs		Output
A	1	$A \cdot 1 = A$
0	1	0
1	1	1

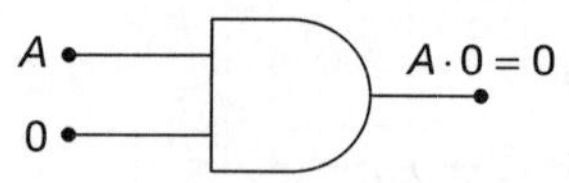

FIGURE 4.2 | Intersection law

TABLE 4.2 | Law of intersection or null

Inputs		Output
A	0	$A \cdot 0 = 0$
0	0	0
1	0	0

4.2.2 | Law of Union

Union is OR operation of two inputs when one of them is either 0 or 1. Mathematical logic expressions are given in Eqs. (4.3) and (4.4). Logic diagram of Eq. (4.3) is given in Figure 4.3 and its truth table is given in Table 4.3. Logic diagram of Eq. (4.4) is given in Figure 4.4 and its truth table is given in Table 4.4. For one binary variable, A there are two possible values considered in truth tables.

$$A + 1 = 1 \tag{4.3}$$

$$A + 0 = A \tag{4.4}$$

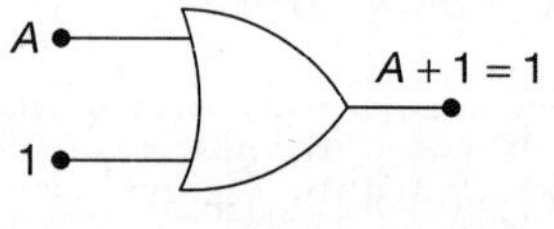

FIGURE 4.3 | Union law

TABLE 4.3 | Law of union or identity

Inputs		Output
A	1	$A + 1 = 1$
0	1	1
1	1	1

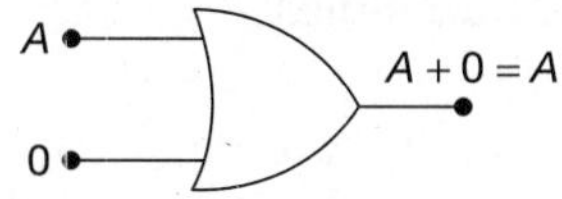

FIGURE 4.4 | Union law

TABLE 4.4 | Law of union or null

Inputs		Output
A	0	$A + 0 = A$
0	0	0
1	0	1

4.2.3 | Law of Identity

Either OR or AND operation having two inputs when one of them is 1 (identity) states Identity law. Mathematical logic expressions are given in Eqs. (4.5) and (4.6). Logic diagram of Eq. (4.5) is given in Figure 4.5 and its truth table is given in Table 4.1. Logic diagram of Eq. (4.6) is given in Figure 4.6 and its truth table is given in Table 4.3. For one binary variable A, there are two possible values considered in truth tables.

$$A \cdot 1 = A \tag{4.5}$$

$$A + 1 = 1 \tag{4.6}$$

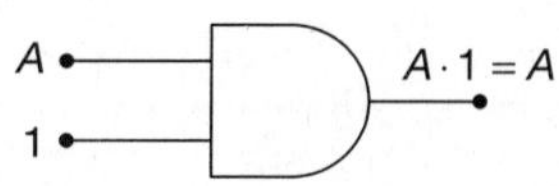

FIGURE 4.5 | Identity law

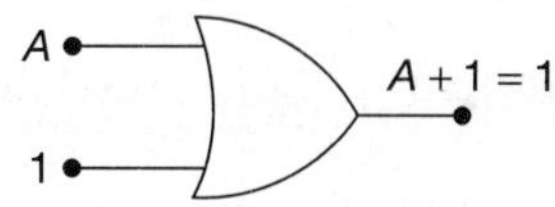

FIGURE 4.6 | Identity law

4.2.4 | Law of Null

Either OR or AND operation of two inputs when one of them is 0 (null) states Null law. Mathematical logic expressions are given in Eqs. (4.7) and (4.8). Logic diagram of Eq. (4.7) is given in Figure 4.7 and its truth table is given in Table 4.2. Logic diagram of Eq. (4.8) is given in Figure 4.8 and its truth table is given in Table 4.4. For one binary variable A, there are two possible values considered in truth tables.

$$A \cdot 0 = 0 \tag{4.7}$$

$$A + 0 = A \tag{4.8}$$

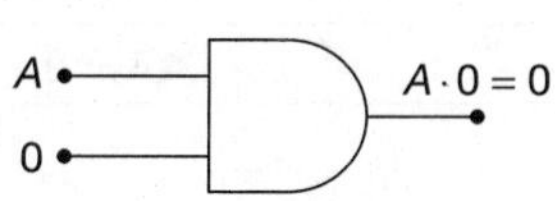

FIGURE 4.7 | Null law

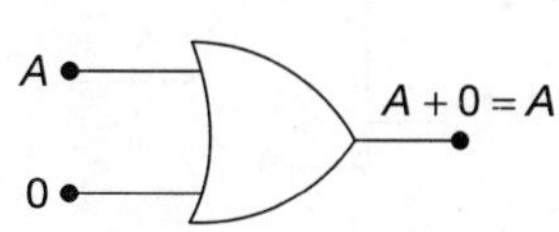

FIGURE 4.8 | Null law

4.2.5 | Law of Tautology or Idempotence

Either OR or AND operation of two inputs when both inputs are same states tautology or idempotence. Mathematical logic expressions are given in Eqs. (4.9) and (4.10). Logic diagram of Eq. (4.9) is given in Figure 4.9 and its truth table is given in Table 4.5. Logic diagram of Eq. (4.10) is given in Figure 4.10 and its truth table is given in Table 4.6. For one binary variable *A*, there are two possible values considered in truth tables.

$$A \cdot A = A \tag{4.9}$$

$$A + A = A \tag{4.10}$$

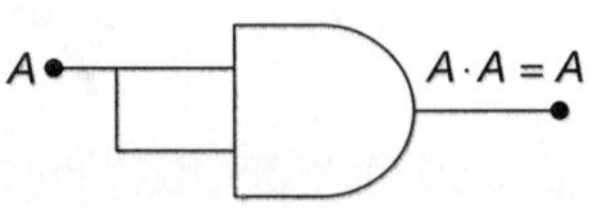

FIGURE 4.9 | Tautology law

TABLE 4.5 | Law of tautology or idempotence

Inputs		Output
A	A	$A \cdot A = A$
0	0	0
1	1	1

FIGURE 4.10 | Tautology law

TABLE 4.6 | Law of tautology or idempotence

Inputs		Output
A	A	$A + A = A$
0	0	0
1	1	1

4.2.6 | Law of Complement or Negation

Either AND or OR operation of two inputs when one input is complement of second input states complement or negation law. Mathematical logic expressions are given in Eq. (4.11) and Eq. (4.12). Logic diagram of Eq. (4.11) is given in Figure 4.11 and its truth table is given in Table 4.7. Logic diagram of Eq. (4.12) is given in Figure 4.12 and its truth table is given in Table 4.8. For one binary variable *A*, there are two possible values considered in truth tables.

$$A \cdot \bar{A} = 0 \tag{4.11}$$

$$A + \bar{A} = 1 \tag{4.12}$$

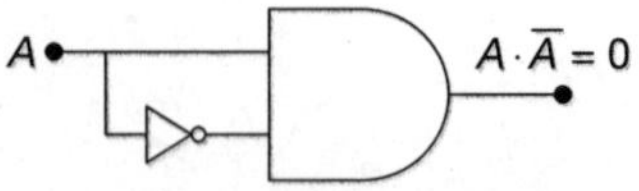

FIGURE 4.11 | Complement law

TABLE 4.7 | Law of complement or negation

Inputs		Output
A	$\bar{A}$	$A\bar{A} = 0$
0	1	0
1	0	0

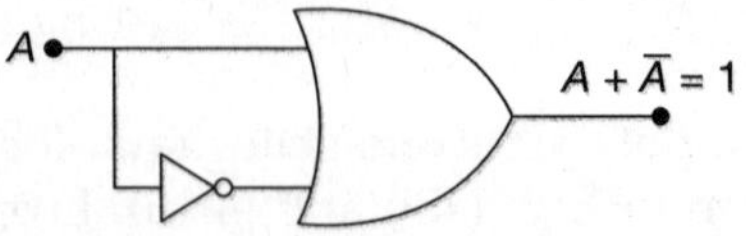

FIGURE 4.12 | Complement law

TABLE 4.8 | Law of complement or negation

Inputs		Output
A	$\bar{A}$	$A + \bar{A} = 1$
0	1	1
1	0	1

4.2.7 | Law of Double Negation or Involution

Double negation or involution states that two times negation nullify the complement. In other words, after complement the input two times, the output becomes equal to input. Logic diagram of Eq. (4.13) is given in Figure 4.13 and its truth table is given in Table 4.9.

$$\bar{\bar{A}} = A \tag{4.13}$$

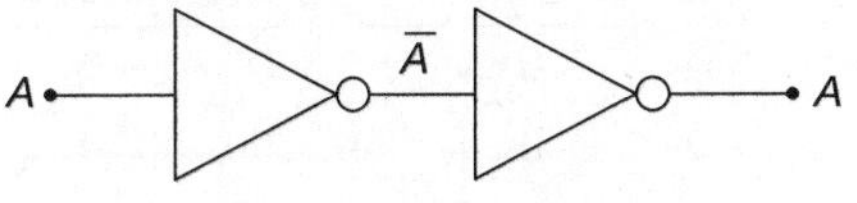

FIGURE 4.13 | Double negation

TABLE 4.9 | Law of double negation or involution

Inputs	Intermediate Output	Output
A	$\bar{A}$	$\bar{\bar{A}} = A$
0	1	0
1	0	1

4.2.8 | Law of Commutation

It applies on two inputs. Law of commutation is stated below by Eqs. (4.14) and (4.15).

$$A + B = B + A \tag{4.14}$$

$$A \cdot B = B \cdot A \tag{4.15}$$

AND or OR logic operations of two variables can be done in any order, the result remains same. Logic diagram of Eq. (4.14) is given in Figure 4.14 and its truth table is given in Table 4.10. Logic diagram of Eq. (4.15) is given in Figure 4.15 and its truth table is given in Table 4.10.

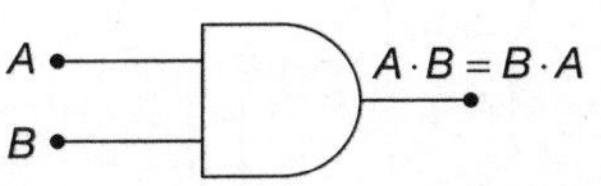

FIGURE 4.14 | Commutation law

TABLE 4.10 | Law of commutation

Inputs		Outputs			
A	B	$A+B$	$B+A$	AB	BA
0	0	0	0	0	0
0	1	1	1	0	0
1	0	1	1	0	0
1	1	1	1	1	1

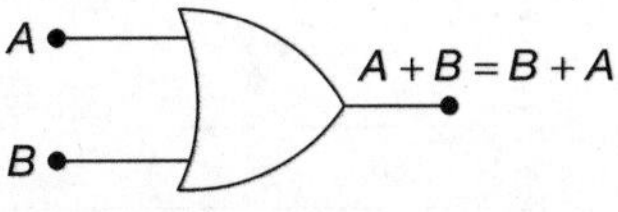

FIGURE 4.15 | Commutation law

4.2.9 | Law of Association

It applies on three inputs. Any two inputs can be combined to get the result. Law of association is stated below:

$$(A \cdot B) \cdot C = A \cdot (B \cdot C) \tag{4.16}$$

$$(A + B) + C = A + (B + C) \tag{4.17}$$

Logic diagram of Eq. (4.16) is given in Figure 4.16 and its truth table is given in Table 4.11. Logic diagram of Eq. (4.17) is given in Figure 4.17 and its truth table is given in Table 4.12. For three variables, eight possible combinations are considered in truth tables.

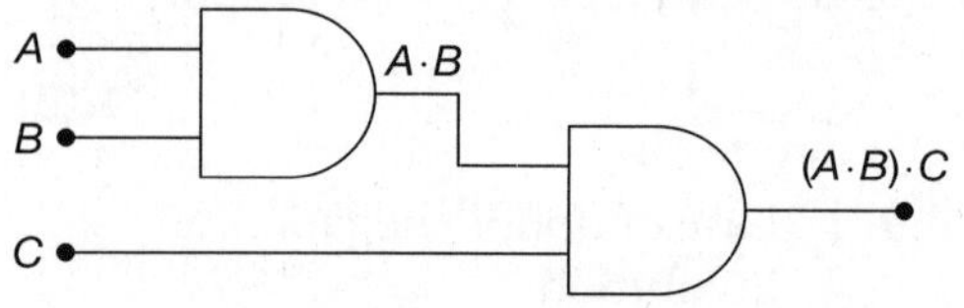

(a) Logic diagram for $(A \cdot B) \cdot C$

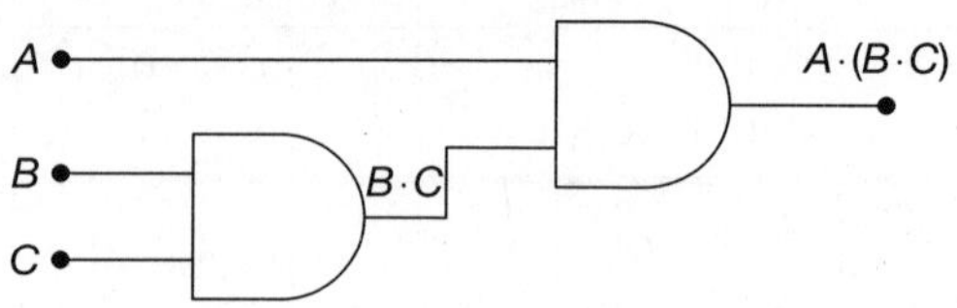

(b) Logic diagram for $A \cdot (B \cdot C)$

FIGURE 4.16 | Association law

TABLE 4.11 | Law of association

Inputs			Intermediate Outputs		Outputs	
A	***B***	***C***	***AB***	***BC***	***(AB)C***	***A(BC)***
0	0	0	0	0	0	0
0	0	1	0	0	0	0
0	1	0	0	0	0	0
0	1	1	0	1	0	0
1	0	0	0	0	0	0
1	0	1	0	0	0	0
1	1	0	1	0	0	0
1	1	1	1	1	1	1

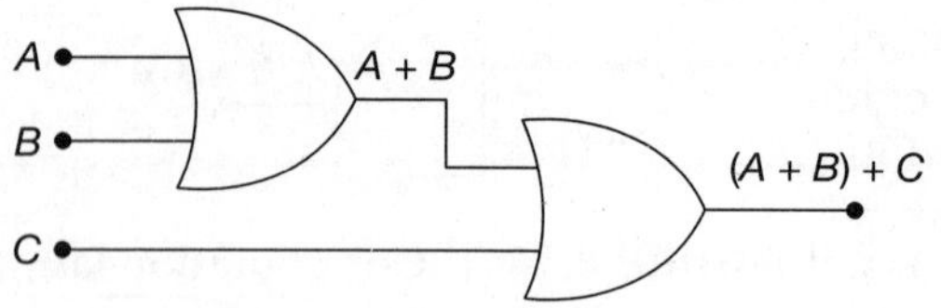

(a) Logic diagram for $(A + B) + C$

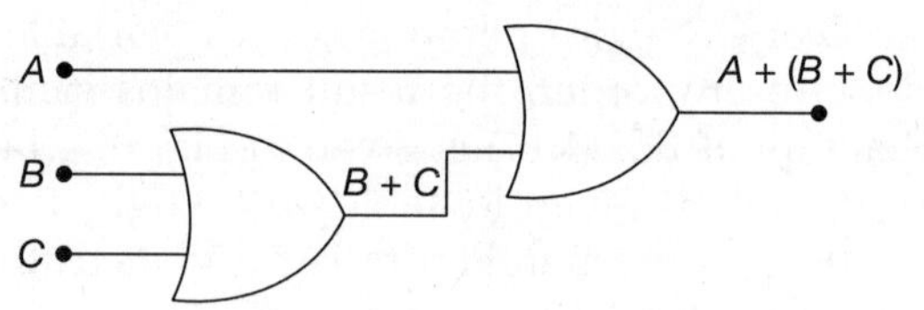

(b) Logic diagram for $A + (B + C)$

FIGURE 4.17 | Association law

TABLE 4.12 | Law of association

Inputs			Intermediate Outputs		Outputs	
A	***B***	***C***	***A + B***	***B + C***	***(A + B) + C***	***A + (B + C)***
0	0	0	0	0	0	0
0	0	1	0	1	1	1
0	1	0	1	1	1	1
0	1	1	1	1	1	1
1	0	0	1	0	1	1
1	0	1	1	1	1	1
1	1	0	1	1	1	1
1	1	1	1	1	1	1

4.2.10 | Law of Distribution

Three inputs are required. Law of distribution is stated below:

$$A \cdot B + A \cdot C = A \cdot (B + C) \tag{4.18}$$

$$(A+B)(A+C) = A + BC \tag{4.19}$$

Logic diagram of Eq. (4.18) is given in Figure 4.18 and its truth table is given in Table 4.13. Logic diagram of Eq. (4.19) is given in Figure 4.19 and its truth table is given in Table 4.14.

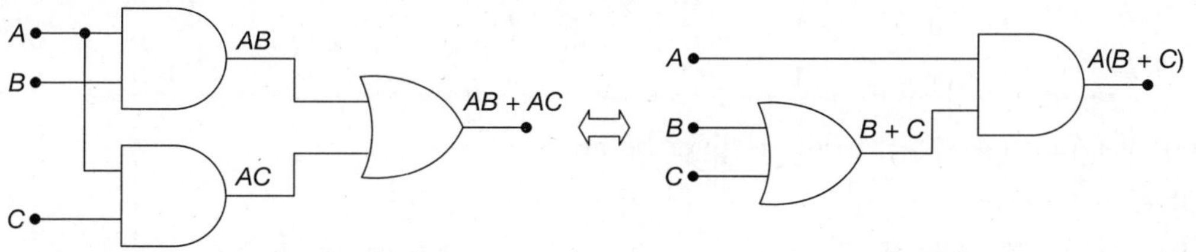

FIGURE 4.18 | Distribution law

TABLE 4.13 | Law of distribution

Inputs			Intermediate Outputs			Outputs	
A	*B*	*C*	*AB*	*AC*	*B* + *C*	*AB* + *AC*	*A*(*B* + *C*)
0	0	0	0	0	0	0	0
0	0	1	0	0	1	0	0
0	1	0	0	0	1	0	0
0	1	1	0	0	1	0	0
1	0	0	0	0	0	0	0
1	0	1	0	1	1	1	1
1	1	0	1	0	1	1	1
1	1	1	1	1	1	1	1

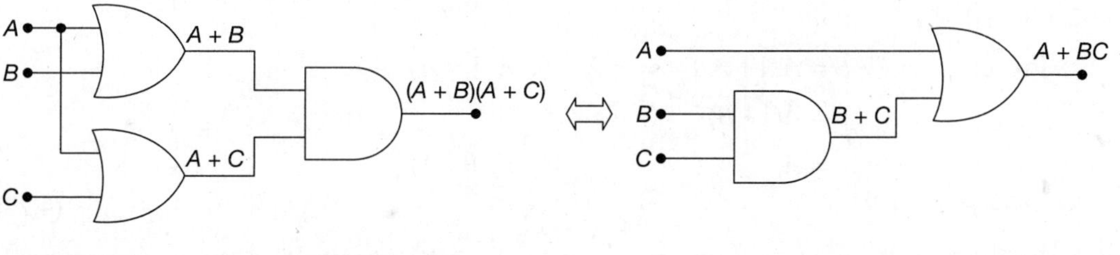

FIGURE 4.19 | Distribution law

TABLE 4.14 | Law of distribution

Inputs			Intermediate Outputs			Outputs	
A	***B***	***C***	***A* + *B***	***A* + *C***	***BC***	**(*A* + *B*)(*A* + *C*)**	***A* + *BC***
0	0	0	0	0	0	0	0
0	0	1	0	1	0	0	0
0	1	0	1	0	0	0	0
0	1	1	1	1	1	1	1
1	0	0	1	1	0	1	1
1	0	1	1	1	0	1	1
1	1	0	1	1	0	1	1
1	1	1	1	1	1	1	1

Proof of $(A+B)(A+C)=A+BC$ is given below

$\text{LHS}=(A+B)(A+C)$

$\text{LHS}=AA+AC+AB+BC$ $\quad(\because A\cdot A=A)$

$\text{LHS}=A+AC+AB+BC$ $\quad(\because A\cdot 1=A)$

$\text{LHS}=A(1+C)+AB+BC$ $\quad\because AB+AC=A(B+C)$

$\text{LHS}=A+AB+BC$ $\quad(\because C+1=1)$

$\text{LHS}=A(1+B)+BC$ $\quad\because AB+AC=A(B+C)$

$\text{LHS}=A+BC=\text{RHS}$ $\quad(\because B+1=1)$

4.2.11 | Law of Absorption

1. $A\cdot(A+B)=A$ (4.20)

   Proof: $A\cdot(A+B)=A\cdot A+A\cdot B$

   $=A+AB$ $\quad(\because AA=A)$

   $=A\cdot 1+A\cdot B$ $\quad(\because A\cdot 1=A)$

   $=A(1+B)$ $\quad(\because A\cdot 1=A)$

   $=A$ $\quad(\because 1+B=1)$

2. $A+AB=A$ (4.21)

   Proof: $A+AB=A\cdot 1+A.B$ $\quad(\because A\cdot 1=A)$

   $=A(1+B)$

   $=A$ $\quad(\because 1+B=1)$

3. $AB+\overline{B}=A+\overline{B}$ (4.22)

   Proof: $AB+\overline{B}=(A+\overline{B})(B+\overline{B})$ $\quad(\because A+BC=(A+B)(A+C))$

   $=A+\overline{B}$ $\quad(\because B+\overline{B}=1 \text{ and } A\cdot 1=A)$

4. $A\overline{B}+B=A+B$ (4.23)

Proof: $A\overline{B}+B=(A+B)(\overline{B}+B)$ $(\because A+BC=(A+B)(A+C))$

$=A+B$ $(\because B+\overline{B}=1 \text{ and } A\cdot 1=A)$

4.2.12 | Consensus Theorem

1. $AB+\overline{A}C+BC=AB+\overline{A}C$ (4.24)

Proof: $AB+\overline{A}C+BC=AB+\overline{A}C+BC(\overline{A}+A)$ $(\because A+\overline{A}=1)$

$=AB+\overline{A}C+\overline{A}BC+ABC$

$=(AB+ABC)+(\overline{A}C+\overline{A}BC)$

$=AB(1+C)+\overline{A}C(1+B)$ $(\because 1+B=1,\ 1+C=1)$

$=AB+\overline{A}C$

2. $(A+B)(\overline{A}+C)(B+C)=(A+B)(\overline{A}+C)$ (4.25)

Proof: $\text{LHS}=(A+B)(\overline{A}+C)(B+C)$

$=(A\overline{A}+AC+B\overline{A}+BC)(B+C)$

$=(AC+\overline{A}B+BC)(B+C)$ $(\because A\overline{A}=0,\ A+0=A,\ AB=BA)$

$=(ACB+\overline{A}BB+BCB+ACC+\overline{A}BC+BCC)$

$=(ACB+\overline{A}B+BC+AC+\overline{A}BC+BC)$

$(\because BB=B,\ CC=C,\ ACB=ABC)$

$=(AC+ABC+\overline{A}B+\overline{A}BC+BC)$ $(\because BC+BC=BC)$

$=AC(1+B)+\overline{A}B(1+C)+BC$

$=AC+\overline{A}B+BC$ $(\because 1+C=1,\ 1+B=1)$

$\text{RHS}=(A+B)(\overline{A}+C)$

$=A\overline{A}+AC+B\overline{A}+BC$

$=(AC+\overline{A}B+BC)$ $(\because A\overline{A}=0,\ AB=BA)$

Hence, LHS = RHS.

4.2.13 | Transposition Theorem

$AB+\overline{A}C=(A+C)(\overline{A}+B)$ (4.26)

Proof: $(A+C)(\overline{A}+B)=A\overline{A}+AB+C\overline{A}+BC$

$=(\overline{A}C+AB+BC)$ $(\because A\overline{A}=0,\ \overline{A}C=C\overline{A})$

$$= AB + \overline{A}C + BC(\overline{A} + A) \qquad (\because A + \overline{A} = 1)$$

$$= AB + \overline{A}C + \overline{A}BC + ABC$$

$$= (AB + ABC) + (\overline{A}C + \overline{A}BC)$$

$$= AB(1 + C) + \overline{A}C(1 + B) \qquad (\because 1 + B = 1,\ 1 + C = 1)$$

$$= AB + \overline{A}C$$

4.2.14 | De Morgan's Theorem-I

NOR logic is equivalent to AND logic with inverted inputs. De Morgan's first law is stated below:

$$\overline{A+B} = \overline{A} \cdot \overline{B} \tag{4.27}$$

Logic diagram of Eq. (4.27) is given in Figure 4.20 and its truth table is given in Table 4.15.

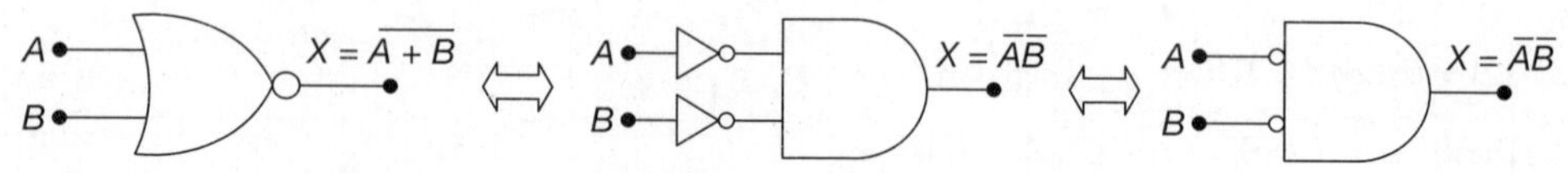

FIGURE 4.20 | De Morgan's theorem-I

TABLE 4.15 | De Morgan's first law

Inputs		Intermediate Outputs			Outputs	
A	B	$A+B$	$\overline{A}$	$\overline{B}$	$\overline{A+B}$	$\overline{A}\,\overline{B}$
0	0	0	1	1	1	1
0	1	1	1	0	0	0
1	0	1	0	1	0	0
1	1	1	0	0	0	0

4.2.15 | De Morgan's Theorem-II

NAND logic is equivalent to OR logic with inverted inputs. De Morgan's second law is stated below:

$$\overline{A \cdot B} = \overline{A} + \overline{B} \tag{4.28}$$

Logic diagram of Eq. (4.19) is given in Figure 4.21 and its truth table is given in Table 4.16.

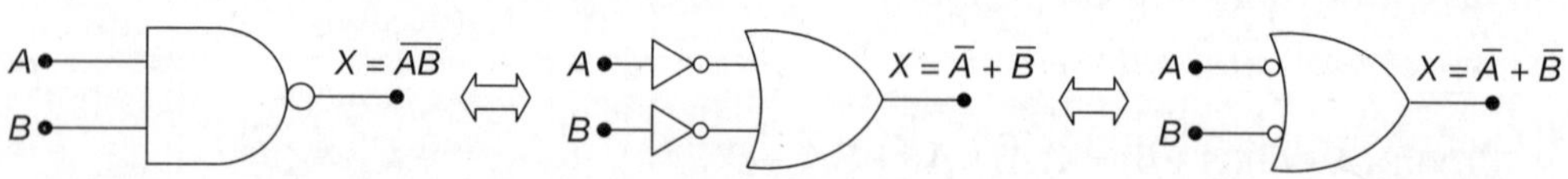

FIGURE 4.21 | De Morgan's theorem-II

TABLE 4.16 | De Morgan's second law

Inputs		Intermediate Outputs			Outputs	
A	B	AB	$\bar{A}$	$\bar{B}$	$\overline{AB}$	$\bar{A}+\bar{B}$
0	0	0	1	1	1	1
0	1	0	1	0	1	1
1	0	0	0	1	1	1
1	1	1	0	0	0	0

e.g.

$$\overline{(AB)(CD)(EFG)} = \left(\overline{AB}\right)+\left(\overline{CD}\right)+\left(\overline{EFG}\right)$$
$$= \left(\bar{A}+\bar{B}\right)+\left(\bar{C}+\bar{D}\right)+\left(\bar{E}+\bar{F}+\bar{G}\right)$$

Rule to De Morganize the expression:

1. Identify the different terms.
2. Find dual of the expression. (Dual of OR operator is AND operator or vice versa).
3. Complement each term.
4. Complement the expression.
5. Repeat the process considering term(s) as expressions till all the terms are De Morganized.

EXAMPLE 4.1 De Morganize the expression $\overline{A+B+C} = \bar{A}\cdot\bar{B}\cdot\bar{C}$.

SOLUTION

- Three terms A, B and C.
- Find dual of expression: $\overline{A\cdot B\cdot C}$
- Complement each term: $\overline{\bar{A}\cdot\bar{B}\cdot\bar{C}}$
- Take complement of the expression: $\overline{\overline{\bar{A}\cdot\bar{B}\cdot\bar{C}}} \Rightarrow \bar{A}\cdot\bar{B}\cdot\bar{C}$

  So $\overline{A+B+C} = \bar{A}\cdot\bar{B}\cdot\bar{C}$ **Ans.**

EXAMPLE 4.2 De Morganize the expression $\overline{ABC} = \bar{A}+\bar{B}+\bar{C}$.

SOLUTION

- Three terms A, B and C.
- Find dual of expression: $\overline{A+B+C}$
- Complement each term: $\overline{\bar{A}+\bar{B}+\bar{C}}$
- Take complement of the expression: $\overline{\overline{\bar{A}+\bar{B}+\bar{C}}} \Rightarrow \bar{A}+\bar{B}+\bar{C}$

  So, $\overline{ABC} = \bar{A}+\bar{B}+\bar{C}$ **Ans.**

EXAMPLE 4.3 De Morganize the expression $\overline{AB+CD+EFG}$.

SOLUTION

- Expression has three terms, i.e. AB, CD and EFG
- Find dual of expression: $\overline{(AB)\cdot(CD)\cdot(EFG)}$
- Complement each term: $\overline{\overline{(AB)}\cdot\overline{(CD)}\cdot\overline{(EFG)}}$
- Take complement of the expression: $\overline{\overline{\left(\overline{AB}\right)\cdot\left(\overline{CD}\right)\cdot\left(\overline{EFG}\right)}} \Rightarrow \left(\overline{AB}\right)\cdot\left(\overline{CD}\right)\cdot\left(\overline{EFG}\right)$
- Further three terms AB, CD and EFG are identified to apply De Morgan theorem. AB term has A and B literals; CD term has C and D literals; EF G has E, F and G literals.
- Find dual of each term: $\Rightarrow (\overline{A+B})(\overline{C+D})(\overline{E+F+G})$
- Complement each term: $\Rightarrow \left(\overline{\overline{A}+\overline{B}}\right)\left(\overline{\overline{C}+\overline{D}}\right)\left(\overline{\overline{E}+\overline{F}+\overline{G}}\right)$
- Take complement of each expression: $\Rightarrow \left(\overline{\overline{\overline{A}+\overline{B}}}\right)\left(\overline{\overline{\overline{C}+\overline{D}}}\right)\left(\overline{\overline{\overline{E}+\overline{F}+\overline{G}}}\right)$

$\overline{AB+CD+EFG} = \left(\overline{A}+\overline{B}\right)\left(\overline{C}+\overline{D}\right)\left(\overline{E}+\overline{F}+\overline{G}\right)$ **Ans.**

EXAMPLE 4.4 Using Boolean algebra, write an equivalent expression for $\overline{Y}\cdot Z\cdot 0$.

SOLUTION $\overline{Y}\cdot Z\cdot 0 = \overline{Y}\cdot(Z\cdot 0)$

$= \overline{Y}\cdot 0 \qquad (\because Z\cdot 0 = 0)$

$= 0 \qquad (\because \overline{Y}\cdot 0 = 0)$

> term + 0 = term
> term + 1 = 1
> (term) · 0 = 0
> (term) · 1 = term

Logical expression is having AND operation with 0, results 0. So, $\overline{Y}\cdot Z\cdot 0 = 0$ **Ans.**

EXAMPLE 4.5 Using Boolean algebra, write an equivalent expression for $(X+\overline{Y}+Z)\cdot 1$.

SOLUTION $\left(X+\overline{Y}+Z\right)\cdot 1 = \left(X+\overline{Y}+Z\right)$ **Ans.** $\qquad (\because A\cdot 1 = A)$

Logical expression $\left(X+\overline{Y}+Z\right)$ is having AND operation with 1, results same expression $\left(X+\overline{Y}+Z\right)$.

EXAMPLE 4.6 Using Boolean algebra, write an equivalent expression for $\overline{Y}\cdot Z+1$.

SOLUTION $\overline{Y}\cdot Z+1 = (\overline{Y}+1)(Z+1) \ (\because Z+1=1)$

$= (1)(1)$

$= 1$ **Ans.**

EXAMPLE 4.7 Using Boolean algebra, write an equivalent expression for $X+\overline{Y}+\overline{Z}+1$.

SOLUTION

$$
\begin{aligned}
X+\overline{Y}+\overline{Z}+1 &= X+\overline{Y}+(\overline{Z}+1) \\
&= X+(\overline{Y}+1) && (\because \overline{Z}+1=1) \\
&= X+1 && (\because \overline{Y}+1=1) \\
&= 1 \quad \textbf{Ans.} && (\because X+1=1)
\end{aligned}
$$

> (term) + (term) = term
> (term) + ($\overline{\text{term}}$) = 1
> (term) · (term) = term
> (term) · ($\overline{\text{term}}$) = 0

EXAMPLE 4.8 Using Boolean algebra, write an equivalent expression for $X\overline{Y}Z+X\overline{Y}Z$.

SOLUTION

$$
\begin{aligned}
X\overline{Y}Z+X\overline{Y}Z &= X\overline{Y}Z(1+1) \\
&= X\overline{Y}Z \quad \textbf{Ans.} && (\because 1+1=1)
\end{aligned}
$$

EXAMPLE 4.9 Using Boolean algebra, write an equivalent expression for $(X\overline{Y}Z)(X\overline{Y}Z)$.

SOLUTION

$$
\begin{aligned}
\left(X\overline{Y}Z\right)\left(X\overline{Y}Z\right) &= (XX)\left(\overline{Y}\overline{Y}\right)(ZZ) \\
&= X\overline{Y}Z \quad \textbf{Ans.} && (\because A\cdot A=A)
\end{aligned}
$$

EXAMPLE 4.10 Using Boolean algebra, write an equivalent expression for $XZ+\overline{XZ}$.

SOLUTION

$$
\begin{aligned}
XZ+\overline{XZ} &= XZ+\overline{X}+\overline{Z} && (\because \overline{XZ}=\overline{X}+\overline{Z}) \\
&= \left(X+\overline{X}\right)\left(Z+\overline{X}\right)+\overline{Z} && \left(\because XZ+\overline{X}=\left(X+\overline{X}\right)\left(Z+\overline{X}\right)\right) \\
&= \left(Z+\overline{X}\right)+\overline{Z} = \left(Z+\overline{Z}\right)+\overline{X} && (\because X+\overline{X}=1) \\
&= 1+\overline{X} && (\because Z+\overline{Z}=1) \\
&= 1 \quad \textbf{Ans.} && (\because 1+\overline{X}=1)
\end{aligned}
$$

> ***Note:***
> $XZ+X\overline{Z}\neq 1$

EXAMPLE 4.11 Apply De Morgan's Theorem to $x+y$.

SOLUTION

- Expression has x and y terms.
- Find dual of expression: xy
- Complement each term: $\overline{x}\,\overline{y}$
- Complement the expression: $\overline{\overline{x}\,\overline{y}}$

$x+y=\overline{\overline{x}\,\overline{y}}$ **Ans.**

> expression + 0 = expression
> expression + 1 = 1
> (expression) · 0 = 0
> (expression) · 1 = expression

EXAMPLE 4.12 Apply De Morgan's Theorem to $xy + xz$.

SOLUTION

(expression) + (expression) = expression
(expression) + $\overline{\text{(expression)}}$ = 1
(expression) · (expression) = expression
(expression) · $\overline{\text{(expression)}}$ = 0

- Expression has xy and xz terms.
- Find dual of expression: $(xy)(xz)$
- Complement each term: $\left(\overline{xy}\right)\left(\overline{xz}\right)$
- Complement the expression: $\overline{\left(\overline{xy}\right)\left(\overline{xz}\right)}$
- Further two terms xy and xz are identified to apply De Morgan's theorem. xy term has x and y literals; xz term has x and z literals.
- Find dual of each term: $\Rightarrow \overline{(\overline{x+y})(\overline{x+z})}$
- Complement each term: $\Rightarrow \overline{\left(\overline{\bar{x}+\bar{y}}\right)\left(\overline{\bar{x}+\bar{z}}\right)}$
- Complement of the term those are De Morganized: $\Rightarrow \overline{\left(\overline{\overline{\bar{x}+\bar{y}}}\right)\left(\overline{\overline{\bar{x}+\bar{z}}}\right)}$

  On simplification $\Rightarrow \overline{\left(\bar{x}+\bar{y}\right)\left(\bar{x}+\bar{z}\right)}$

  $\Rightarrow xy + xz = \overline{\left(\bar{x}+\bar{y}\right)\left(\bar{x}+\bar{z}\right)}$ **Ans.**

EXAMPLE 4.13 Apply De Morgan's Theorem to $W + Q$.

SOLUTION $W+Q \Rightarrow WQ \Rightarrow \bar{W}\bar{Q} \Rightarrow \overline{\bar{W}\bar{Q}}$

Hence $W+Q = \overline{\bar{W}\bar{Q}}$ **Ans.**

EXAMPLE 4.14 Apply De Morgan's Theorem to $(A+B+C)D$.

SOLUTION $(A+B+C)D \Rightarrow (A+B+C)+D \Rightarrow \left(\overline{A+B+C}\right)+\bar{D} \Rightarrow \overline{\left(\overline{A+B+C}\right)+\bar{D}}$

$$\overline{\left(\overline{A+B+C}\right)+\bar{\bar{D}}} \Rightarrow \overline{\left(\overline{ABC}\right)+\bar{\bar{D}}} \Rightarrow \overline{\left(\overline{\bar{A}\bar{B}\bar{C}}\right)+\bar{D}} \Rightarrow \overline{\left(\overline{\overline{\bar{A}\bar{B}\bar{C}}}\right)+\bar{D}} \Rightarrow \overline{\left(\bar{A}\bar{B}\bar{C}\right)+\bar{D}}$$

Hence, $(A+B+C)D = \overline{\left(\bar{A}\bar{B}\bar{C}\right)+\bar{D}}$ **Ans.**

EXAMPLE 4.15 Apply De Morgan's Theorem to $\left[\overline{\left(\overline{A+B}\right)+\bar{C}}\right]$.

SOLUTION $\left[\overline{\left(\overline{A+B}\right)+\bar{C}}\right] \Rightarrow \left[\overline{\left(\overline{A+B}\right)\bar{\bar{C}}}\right] \Rightarrow \left[\overline{\left(\overline{\overline{A+B}}\right)\bar{\bar{\bar{C}}}}\right] \Rightarrow \left[\overline{\overline{(A+B)C}}\right] \Rightarrow (A+B)C \Rightarrow \left(\overline{\bar{A}\bar{B}}\right)C$

So, $\left[\overline{\left(\overline{A+B}\right)+\bar{C}}\right] = (A+B)C$ **Ans.**

EXAMPLE 4.16 Simplify the expression $\bar{a}\bar{b}\bar{c} + a\bar{b}\bar{c} + a\bar{b}c$ using Boolean algebra.

SOLUTION

$$\bar{a}\bar{b}\bar{c} + a\bar{b}\bar{c} + a\bar{b}c = \bar{b}\bar{c}\left(\bar{a} + a\right) + a\bar{b}c$$

$$= \bar{b}\bar{c} + a\bar{b}c \qquad (\because a + \bar{a} = 1)$$

$$= \bar{b}\left(\bar{c} + ac\right)$$

$$= \bar{b}\left(\bar{c} + a\right)\left(\bar{c} + c\right) \qquad \left(\because \bar{c} + ac = \left(\bar{c} + c\right)\left(\bar{c} + a\right) = \left(\bar{c} + a\right)\right)$$

$$= \left(\bar{b}\bar{c} + a\bar{b}\right) \quad \textbf{Ans.}$$

EXAMPLE 4.17 Simplify the expression $\left(x\left(\overline{xy}\right)\right)\left(y\left(\overline{xy}\right)\right)$ using Boolean algebra.

SOLUTION

$$\left(x\left(\overline{xy}\right)\right)\left(y\left(\overline{xy}\right)\right) = \left(x\left(\bar{x} + \bar{y}\right)\right)\left(y\left(\bar{x} + \bar{y}\right)\right) \qquad \left(\because \overline{xy} = \bar{x} + \bar{y}\right)$$

$$= \left(x\bar{x} + x\bar{y}\right)\left(\bar{x}y + y\bar{y}\right) \qquad \left(\because x\bar{x} = 0 \text{ and } y\bar{y} = 0\right)$$

$$= \left(0 + x\bar{y}\right)\left(\bar{x}y + 0\right)$$

$$= \left(x\bar{y}\right)\left(\bar{x}y\right)$$

$$= \left(x\bar{x}\right)\left(\bar{y}y\right) \qquad \left(\because x\bar{x} = 0 \text{ and } y\bar{y} = 0\right)$$

$$= 0$$

EXAMPLE 4.18 Simplify the expression $f = xyz + \bar{x}z$ using Boolean algebra.

SOLUTION

$$f = xyz + \bar{x}z$$

$$f = \left(xy + \bar{x}\right)z$$

$$f = \left(\bar{x} + xy\right)z \qquad \left(\because \left(\bar{x} + xy\right) = \bar{x} + y\right)$$

$$f = \left(\bar{x}z + yz\right) \quad \textbf{Ans.}$$

EXAMPLE 4.19 Simplify the expression $f = abc + a\bar{b}c + \bar{a}$ using Boolean algebra.

SOLUTION

$$f = abc + a\bar{b}c + \bar{a}$$

$$f = ac(b + \bar{b}) + \bar{a} \qquad \left(\because b + \bar{b} = 1\right)$$

$$f = ac + \bar{a} \qquad \left(\because \left(\bar{a} + ac\right) = \bar{a} + c\right)$$

$$f = c + \bar{a} \quad \textbf{Ans.}$$

EXAMPLE 4.20 Simplify the expression $f = \overline{rst}\left(\overline{r+s+t}\right)$ using Boolean algebra.

SOLUTION

$f = \overline{rst}\left(\overline{r+s+t}\right)$

$f = \left(\overline{r}+\overline{s}+\overline{t}\right)\left(\overline{r}\,\overline{s}\,\overline{t}\right)$ $\quad \left(\because \overline{rst} = \overline{r}+\overline{s}+\overline{t}\right)$ and $\left(\overline{r+s+t} = \overline{r}\,\overline{s}\,\overline{t}\right)$

$f = \overline{r}\,\overline{r}\,\overline{s}\,\overline{t} + \overline{r}\,\overline{s}\,\overline{s}\,\overline{t} + \overline{r}\,\overline{s}\,\overline{t}\,\overline{t}$ $\quad \left(\because \overline{r}\,\overline{r} = \overline{r}\right)$ $\left(\overline{s}\,\overline{s} = \overline{s}\right)$ and $\left(\overline{t}\,\overline{t} = \overline{t}\right)$

$f = \overline{r}\,\overline{s}\,\overline{t} + \overline{r}\,\overline{s}\,\overline{t} + \overline{r}\,\overline{s}\,\overline{t}$ $\quad (\because a + a = a)$

$f = \overline{r}\,\overline{s}\,\overline{t}$ **Ans.**

EXAMPLE 4.21 Simplify the expression $f = (b+\overline{c})(\overline{b}+c) + \overline{\overline{a}+b+c}$ using Boolean algebra.

SOLUTION

$f = (b+\overline{c})(\overline{b}+c) + \overline{\overline{a}+b+c}$ $\quad \left(\because \overline{a+b+c} = \overline{a}+\overline{b}+\overline{c}\right)$

$f = (b\overline{b} + bc + \overline{b}\overline{c} + c\overline{c}) + \overline{\overline{a}}\overline{b}\overline{c}$

$f = (0 + bc + \overline{b}\overline{c} + 0) + a\overline{b}\overline{c}$ $\quad \left(\because b\overline{b} = 0 \text{ and } c\overline{c} = 0\right)$

$f = bc + \overline{b}\overline{c} + a\overline{b}\overline{c}$

$f = bc + (1+a)\overline{b}\overline{c}$ $\quad (\because 1 + a = 1 \text{ and } 1 \cdot a = a)$

$f = bc + \overline{b}\overline{c}$ **Ans.**

4.3 | SUM-OF-PRODUCT AND PRODUCT-OF-SUM FORM

- Boolean expression can be expressed in a *standard* or *canonical* or *expanded sum* (OR) *of products (AND)-SOP form*
- *Sum-of-products (SOP):* A SOP is the logical OR operation of multiple product terms. Each product term is the logical AND operation of binary literals. For example, $XY + X\overline{Y} + YZ$ is a SOP expression.
- A standard SOP form is one in which a number of product terms, each one of which contains all the variables of the function either in complemented (un-primed, $\overline{x}$) or non-complemented (primed or normal, x) form, are summed together.
- Boolean expression can be expressed in a *standard* or *canonical* or *expanded product* (AND) *of sums (OR)-POS form.*
- *Product-of-sum (POS):* A POS is the logical AND operation of multiple ORrd terms. Each sum term is the logical OR operation of binary literals. For example, $(X+\overline{Y})(X+Y+Z)(\overline{X}+Y+\overline{Z})$ is a POS expression.
- A standard POS form is one in which a number of sum terms, each one of which contains all the variables of the function either in complemented (un-primed, $\overline{x}$) or un-complemented (primed or normal, x) form, are multiplied together.
- Complement of standard SOP form gives standard POS form of that function and vice versa.
- Each of the product terms in the standard SOP form is called a *minterm.*
- Each of the sum terms in the standard POS form is called a *maxterm.*

- The minterms and maxterms are usually represented as binary words in terms of 1s and 0s, instead of actual variables.
- *Minterm:* A minterm is a special case product (AND) term. A minterm is a product term that contains all of the input variables (each literal no more than once) that makes up a Boolean expression.
- For minterms, the binary words are formed by representing
  - each non-complemented variable by a 1
  - each complemented variable by a 0, and
  - the decimal equivalent of this binary word is expressed as a subscript of lower-case m_j, i.e. m_0, m_1, m_2, m_3,... etc.
  - For n variables there are 2^n possible minterms. Minterms (decimal value) varies from 0 to $(2^n - 1)$.
    - For two variables, there are $2^2 = 4$ possible minterms; these are m_0, m_1, m_2 and m_3 minterms (see Table 4.19).
    - For three variables, there are $2^3 = 8$ possible minterms; these are m_0, m_1, m_2, m_3, m_4, m_5, m_6 and m_7 minterms (see Table 4.20).
    - For four variables, there are $2^4 = 16$ possible minterms; these are m_0, m_1, m_2, m_3, m_4, m_5, m_6, m_7, m_8, m_9, m_{10}, m_{11}, m_{12}, m_{13}, m_{14} and m_{15} minterms (see Table 4.21).
- *Maxterm:* A maxterm is a special case sum (OR) term. A maxterm is a sum term that contains all of the input variables (each literal no more than once) that make up a Boolean expression.
- For maxterms, the binary words are formed by representing
  - each non-complemented variable by a 0
  - each complemented variable by a 1, and
  - the decimal equivalent of this binary word is expressed as a subscript of the upper-case letter M_j, i.e. M_0, M_1, M_2, ... etc.
  - For n variables, there are 2^n possible maxterms. Maxterms (decimal value) vary from 0 to $(2^n - 1)$.
    - For two variables, there are $2^2 = 4$ possible maxterms; these are M_0, M_1, M_2 and M_3 maxterms (see Table 4.19).
    - For three variables, there are $2^3 = 8$ possible maxterms, and these are M_0, M_1, M_2, M_3, M_4, M_5, M_6 and M_7 are maxterms (see Table 4.20).
    - For four variables, there are $2^4 = 16$ possible maxterms; these are M_0, M_1, M_2, M_3, M_4, M_5, M_6, M_7, M_8, M_9, M_{10}, M_{11}, M_{12}, M_{13}, M_{14} and M_{15} maxterms (see Table 4.21).
- Boolean function which is not in the standard form, can be converted to standard form by *expanding* the function.
- A standard SOP form can always be converted to a standard POS form, by treating the missing minterms of the SOP form as the maxterms of the POS form.
- Similarly, a standard POS form can always be converted to a standard SOP form, by treating the missing maxterms of the POS form as the minterms of the corresponding SOP form.
- *Canonical SOP:* A canonical sum of products is a complete set of minterms that defines when an output variable is a logical l. Each minterm corresponds to the row in the truth table where the output function is l; that is, the SOP for the output, f is

$$f = \overline{X}\overline{Y}\overline{Z} + \overline{X}YZ + XY\overline{Z} + XYZ$$

- A SOP expression contains only OR (sum) operations at the 'outermost' level

$$f = \bar{X}\bar{Y}\bar{Z} + \bar{X}YZ + XY\bar{Z}$$

Each term that is summed must be a product of literals

- The advantage is that any SOP expression can be implemented using a two-level circuit (Figure 4.22)
  - o literals and their complements at the '0th' level
  - o AND gates at the first level
  - o a single OR gate at the second level
- *Canonical POS:* A canonical POS is a complete set of maxterms that defines when an output is a logical 0. Each maxterm corresponds to a row in the truth table where the output is a 0; that is, the POS for the output, *f* is

$$f = (\bar{X} + \bar{Y} + Z)(\bar{X} + Y + \bar{Z})(X + \bar{Y} + \bar{Z})(X + \bar{Y} + Z)$$

- A POS expression contains–only AND (product) operations at the 'outermost' level

$$f = (\bar{X} + \bar{Y} + \bar{Z})(\bar{X} + Y + Z)(X + Y + \bar{Z})$$

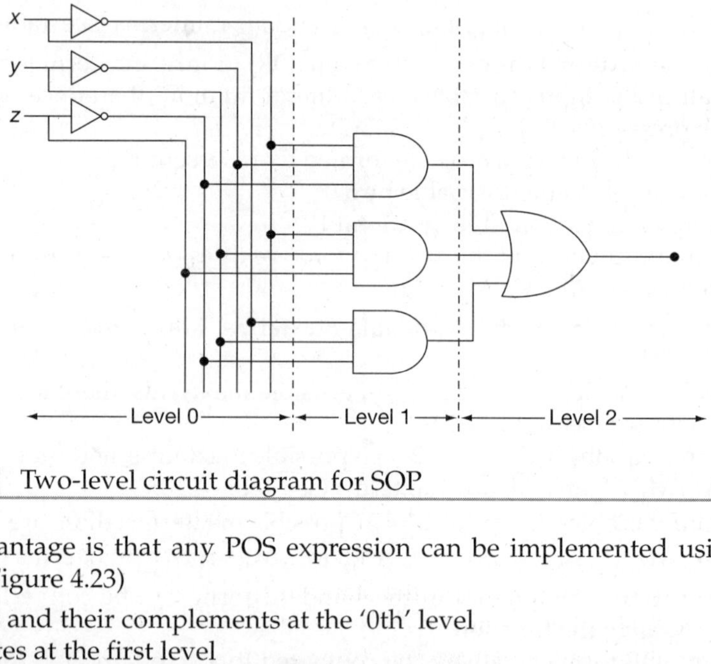

FIGURE 4.22 | Two-level circuit diagram for SOP

- The advantage is that any POS expression can be implemented using a two-level circuit (Figure 4.23)
  - o literals and their complements at the '0th' level
  - o OR gates at the first level
  - o a single AND gate at the second level

Expansion of a Boolean expression in standard SOP form or minterms:

1. Write down all the terms.
2. If one or more variables are missing in any product term, expand that term by performing AND operation with $(X + \bar{X})$ of each one. X is the missing variable and its complement, $\bar{X}$. It will not change the any term as $(X \cdot 1 = X)$ because $(X + \bar{X} = 1)$.
3. Expand the terms and drop out the redundant terms as $(X + X = X)$.

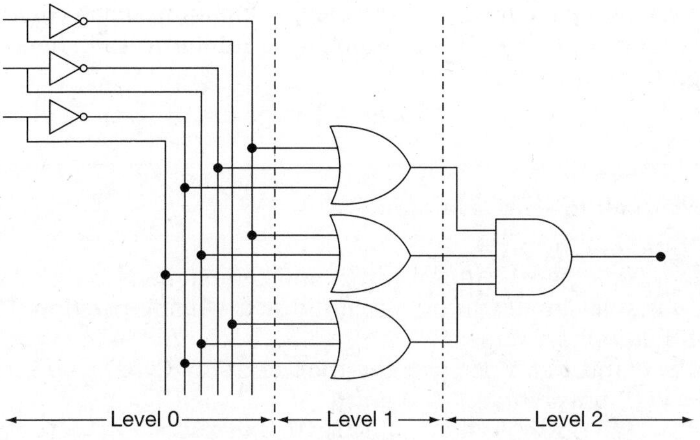

FIGURE 4.23 | Two-level circuit diagram for POS

Expansion of a Boolean expression in standard POS form or maxterms:

1. Write down all the terms.
2. If one or more variables are missing in any sum term, expand that term by performing OR operation with $(X\bar{X})$ of each one. X is the missing variable and its complement, $\bar{X}$. It will not change the any term as $(X + 0 = X)$ because $(X\bar{X} = 0)$.
3. Apply law of distribution following the relations
   a. $A + B\bar{B} = (A + B)(A + \bar{B})$
   b. $A + B + C\bar{C} = (A + B + C)(A + B + \bar{C})$
   c. $A + B + C + D\bar{D} = (A + B + C + D)(A + B + C + \bar{D})$
4. Drop out the redundant terms as $(XX = X)$.

EXAMPLE 4.22 Expand $\bar{A} + \bar{B}$ to minterms and maxterms.

SOLUTION

- The given expression has two-variables, A and B.
- First term has A variable and variable B is missing; so, do AND operation with $(B + \bar{B})$.
- Second term has B variable and variable A is missing; so, do AND operation with $(A + \bar{A})$.

Therefore,

$$\begin{aligned}\bar{A} + \bar{B} &= \bar{A}(B + \bar{B}) + (A + \bar{A})\bar{B} \\ &= \bar{A}B + \bar{A}\bar{B} + A\bar{B} + \bar{A}\bar{B} \\ &= \bar{A}B + \bar{A}\bar{B} + A\bar{B} \\ &= 01 + 00 + 10 \\ &= m_1 + m_0 + m_2 \\ &= \Sigma m(0, 1, 2)\end{aligned}$$

Two (2) binary variables have four (2^2) possible combinations. Each combination has decimal value in the range of 0 to 3 ($2^2 - 1$). So missing minterm is m_3, hence maxterm is M_3. Change sum sign to product sign.

$$\bar{A}+\bar{B}=\prod M(3)$$

Alternative Method: To write standard SOP

1. Consider terms one by one.
2. Replace un-complemented literal with 1 and complemented literal with 0.
3. Complete variables in a term by placing *X* at the blank position. Position of the variable should remain same.
4. Replace *X* by 0 and 1 and perform the OR operation. Other values remain same.
   - X has two (2^1) possibilities, i.e. 0 and 1
   - XX has four (2^2) possibilities, i.e. 00, 01, 10 and 11
   - XXX has eight (2^3) possibilities, i.e. 000, 001, 010, 011, 100, 101, 110, and 111
5. Replace 1 with un-complemented literal and 0 with complemented literal.

   or

   write minterms as per the m and decimal value of term in subscript
6. Drop out the redundant terms.

EXAMPLE 4.23 Expand $\bar{A}+\bar{B}$ to minterms and maxterms.

SOLUTION There are two binary variables $\bar{A}$ and $\bar{B}$. $f(A,B)=\bar{A}+\bar{B}$

- Consider first term $\bar{A}$.
- Replace $\bar{A}$ with 0.
- Complete the terms by placing $X \Rightarrow 0X$.
- Replace X by 0 and 1 by performing OR operation $\Rightarrow 00+01$.
- Write minterms $\Rightarrow m_0+m_1=\sum m(0,1)$ or $\Rightarrow \bar{A}\bar{B}+\bar{A}B$.
- Consider second term $\bar{B}$.
- Replace $\bar{B}$ with 0.
- Complete the terms by placing $X \Rightarrow X0$ (Note: position of variable remains fixed).
- Replace X by 0 or 1 by performing OR operation $\Rightarrow 00+10$
- Write minterms $\Rightarrow m_0+m_2=\sum m(0,2)$ or $\Rightarrow \bar{A}\bar{B}+A\bar{B}$

  $\bar{A}+\bar{B}=\bar{A}\bar{B}+\bar{A}B+A\bar{B}=\sum m(0,1,2)$ **Ans.**

EXAMPLE 4.24 Expand $A+\bar{B}\bar{C}$ to minterms and maxterms.

SOLUTION The expression is in SOP.

- The given expression has three variables, *A*, *B* and *C*.
- First term has *A* variable and variables *B* and *C* are missing; so, do AND operation with and $(B+\bar{B})$ and $(C+\bar{C})$.

- Second term has B and C variables and variable A is missing; so, multiply it by $(A+\bar{A})$.

Therefore,

$$
\begin{aligned}
A+\bar{B}\bar{C} &= A(B+\bar{B})(C+\bar{C})+(A+\bar{A})\bar{B}\bar{C} \\
&= (AB+A\bar{B})(C+\bar{C})+A\bar{B}\bar{C}+\bar{A}\bar{B}\bar{C} \\
&= ABC+AB\bar{C}+A\bar{B}C+A\bar{B}\bar{C}+A\bar{B}\bar{C}+\bar{A}\bar{B}\bar{C} \\
&= ABC+AB\bar{C}+A\bar{B}C+A\bar{B}\bar{C}+\bar{A}\bar{B}\bar{C} \\
&= (111+110+101+100+000) \\
&= m_7+m_6+m_5+m_4+m_0 \\
& \sum m(0,4,5,6,7)
\end{aligned}
$$

Three (3) binary variables have eight (2^3) possible combinations. Each combination has decimal value in the range of 0 to 7 (2^3-1). So missing minterms are m_1, m_2 and m_3, hence maxterms are M_1, M_2 and M_3.

$$A+\bar{B}\bar{C} = \prod M(1,2,3)$$

Alternative Method:

$$
\begin{aligned}
&f(A,B,C) = A+\bar{B}\bar{C} \\
&f(A,B,C) = 1XX+X00 \\
&f(A,B,C) = (100+101+110+111)+(000+100) \\
&f(A,B,C) = (m_4+m_5+m_6+m_7+m_0+m_4) = \sum m(0,4,5,6,7) \quad \textbf{Ans.}
\end{aligned}
$$

Or

$$f(A,B,C) = (A\bar{B}\bar{C}+A\bar{B}C+AB\bar{C}+ABC)+(\bar{A}\bar{B}\bar{C}+A\bar{B}\bar{C})$$

$$f(A,B,C) = \bar{A}\bar{B}\bar{C}+A\bar{B}\bar{C}+A\bar{B}C+AB\bar{C}+ABC \quad \textbf{Ans.}$$

EXAMPLE 4.25 Expand $(\bar{A}+B)(B+\bar{C})$ to maxterms.

SOLUTION The expression is in POS

Replace complemented variable with 1 and un-complemented variable with 0.

$$
\begin{aligned}
(\bar{A}+B)(B+\bar{C}) &= (\bar{A}+B+C\bar{C})(B+\bar{C}+A\bar{A}) \\
\bar{A}+B+C\bar{C} &= (\bar{A}+B+C)(\bar{A}+B+\bar{C}) \\
B+\bar{C}+A\bar{A} &= (B+\bar{C}+A)(B+\bar{C}+\bar{A}) = (A+B+\bar{C})(\bar{A}+B+\bar{C}) \\
(\bar{A}+B)(B+\bar{C}) &= (\bar{A}+B+C)(\bar{A}+B+\bar{C})(A+B+\bar{C})(\bar{A}+B+\bar{C})
\end{aligned}
$$

$$(\because (\bar{A}+B+\bar{C})(\bar{A}+B+\bar{C}) = (\bar{A}+B+\bar{C}))$$

$$(\bar{A}+B)(B+\bar{C}) = (\bar{A}+B+C)(\bar{A}+B+\bar{C})(A+B+\bar{C}) \quad \textbf{Ans.}$$

$$(\bar{A}+B)(B+\bar{C}) = (1+0+0)(1+0+1)(0+0+1)$$

$$(\bar{A}+B)(B+\bar{C}) = (M_1)(M_4)(M_5) = \prod M(1,4,5)$$

EXAMPLE 4.26 Write the function from the Truth Table 4.17.

TABLE 4.17 | Truth table

Input			Output
x	*y*	*z*	*f*
0	0	0	0
0	0	1	0
0	1	0	1
0	1	1	1
1	0	0	0
1	0	1	0
1	1	0	1
1	1	1	1

SOLUTION Table 4.18 gives the minterms and maxterms also.

TABLE 4.18 | Fundamental/Standard product for three inputs

Decimal Value	Inputs			Output	Fundamental Product	Minterm	Fundamental Sum	Maxterm
	x	*y*	*z*	*f*				
0	0	0	0	0	$\bar{x}\bar{y}\bar{z}$	m_0	$x+y+z$	M_0
1	0	0	1	0	$\bar{x}\bar{y}z$	m_1	$x+y+\bar{z}$	M_1
2	0	1	0	1	$\bar{x}y\bar{z}$	m_2	$x+\bar{y}+z$	M_2
3	0	1	1	1	$\bar{x}yz$	m_3	$x+\bar{y}+\bar{z}$	M_3
4	1	0	0	0	$x\bar{y}\bar{z}$	m_4	$\bar{x}+y+z$	M_4
5	1	0	1	0	$x\bar{y}z$	m_5	$\bar{x}+y+\bar{z}$	M_5
6	1	1	0	1	$xy\bar{z}$	m_6	$\bar{x}+\bar{y}+z$	M_6
7	1	1	1	1	xyz	m_7	$\bar{x}+\bar{y}+\bar{z}$	M_7

A canonical SOP is a complete set of minterms that defines when an output variable is a logical 1.

$$f_1 = \bar{x}y\bar{z} + \bar{x}yz + xy\bar{z} + xyz$$

$$f_1 = m_2 + m_3 + m_6 + m_7 = \sum m(2,3,6,7) \tag{4.29}$$

A canonical POS is a complete set of maxterms that defines when an output is a logical 0.

$$f_2 = (x+y+z)(x+y+\bar{z})(\bar{x}+y+z)(\bar{x}+y+\bar{z}) \tag{4.30}$$

$$f_2 = M_0 + M_1 + M_4 + M_5 = \prod M(0,1,4,5)$$

Obtain complement of function by performing OR operation of minterms when an output of those minterms is logical 0. Minterms m_0, m_1, m_4 and m_5 represent logic 0 (Table 4.18). So, complement of function is given as:

$$\overline{f_3} = \overline{x}\,\overline{y}\,\overline{z} + \overline{x}\,\overline{y}z + x\overline{y}\,\overline{z} + x\overline{y}z$$

> A complement of canonical SOP is a complete set of minterms that defines when an output variable is a logical 1.

Complement the function

$$\overline{\overline{f_3}} = \overline{\overline{x}\,\overline{y}\,\overline{z} + \overline{x}\,\overline{y}z + x\overline{y}\,\overline{z} + x\overline{y}z}$$

Applying De Morgan's Theorem-I

$$f_3 = \left(\overline{\overline{x}\,\overline{y}\,\overline{z}}\right)\left(\overline{\overline{x}\,\overline{y}z}\right)\left(\overline{x\overline{y}\,\overline{z}}\right)\left(\overline{x\overline{y}z}\right)$$

$$f_3 = (x+y+z)(x+y+\overline{z})(\overline{x}+y+z)(\overline{x}+y+\overline{z}) \tag{4.31}$$

Equations (4.30) and (4.31) are same, so $f_2 = f_3$.

Obtain complement of function by performing AND operation of maxterms when an output of those maxterms is logical 1. Maxterms M_2, M_3, M_6 and M_7 represent logic 1 (Table 4.18) . So, complement of function is given as:

$$\overline{f_4} = (x+\overline{y}+z)(x+\overline{y}+\overline{z})(\overline{x}+\overline{y}+z)(\overline{x}+\overline{y}+\overline{z})$$

Complement the function

$$\overline{\overline{f_4}} = \overline{(x+\overline{y}+z)(x+\overline{y}+\overline{z})(\overline{x}+\overline{y}+z)(\overline{x}+\overline{y}+\overline{z})}$$

> A complement of canonical POS is a complete set of maxterms that defines when an output variable is a logical 0.

Applying De Morgan's theorem-II

$$f_4 = \left(\overline{x+\overline{y}+z}\right)+\left(\overline{x+\overline{y}+\overline{z}}\right)+\left(\overline{\overline{x}+\overline{y}+z}\right)+\left(\overline{\overline{x}+\overline{y}+\overline{z}}\right)$$

$$f_4 = \overline{x}y\overline{z} + \overline{x}yz + xy\overline{z} + xyz \tag{4.32}$$

Equations (4.29) and (4.32) are same, so $f_1 = f_4$.

Conversion between SOP and POS canonical forms:

1. Write complement of function by performing OR operation of minterms (SOP) when an output variable corresponding to minterm is a logical 0 (instead of logical 1).
2. Complement the function.
3. Apply DeMorgan theorem-II

Conversion between POS and SOP canonical forms:

1. Write complement of function by performing AND operation of maxterms when an output variable corresponding to maxterm is a logical 1 (instead of logical 0).
2. Complement the function.
3. Apply DeMorgan theorem-I

4.4 | KARNAUGH MAP (K-MAP)

- It originated from the 'map method' proposed by Edward Veitch also called the 'Veitch Diagram' and then modified by Maurice Karnaugh 1953.
- Developed by Karnaugh in 1953 that he presented in his paper entitled 'The map method for synthesis of combinational logic circuit'.
- Boolean expressions can be simplified algebraically. The effectiveness of algebraic simplification depends on the knowledge and ability to apply Boolean algebraic rules, laws and theorems.
- The Karnaugh map (K-map) is a systematic method of simplifying Boolean expressions.
- It is a pictorial form of a truth table and could handle up to six variables.
- It is an array of cells/squares in which each cell represents a binary value of the input variables.
- The Karnaugh map reduces extensive calculations by taking advantage of humans' pattern-recognition capability, also permitting the rapid identification and elimination of potential race conditions.
- The K-map is a chart or a graph, composed of an arrangement of adjacent cells, each representing a particular combination of variables in SOP form.
- K-map can easily be applied for problems involving up to six variables.
- An n variable function can have 2^n possible combinations of product terms in SOP form or 2^n possible combinations of POS form.
- Since the K-map is a graphical representation of Boolean expressions,
  - a two-variable K-map has $2^2 = 4$ cells or squares,
  - a three-variable K-map has $2^3 = 8$ cells or squares,
  - a four-variable K-map has $2^4 = 16$ cells or squares,
  - a four-variable K-map has $2^5 = 32$ cells or squares, and so on.

4.4.1 | K-Map Set-Up

- The number of cell/square in a K-map is equal to 2^n, where n is the number of input variables.
- The map is drawn to show the relationship between squares and input variables. Variables are assigned to row and column. Binary marking are placed in each row and column using reflected (Gray) code sequence.
- A Karnaugh map comprises a box for every line in the truth table. The binary value for each box is the binary value of the input terms in the corresponding table row and column.
- Unlike a truth table, in which the input values typically follow a standard binary sequence (00, 01, 10, 11), the Karnaugh map's input values must be ordered such that the values for adjacent columns vary by only a single bit, for example, 00, 01, 11, and 10. This ordering is known as a Gray code.

4.4.1.1 | Two-Variable K-map using Minterms

A two-variable K-map is shown in Figure 4.24 and corresponding minterms are shown in Table 4.19. Simplified form of two-variable K-map set up is given in Figure 4.25. Rows represent variables having MSB positions.

- K-map has two columns. First column represents 0 ($\overline{y}$) and second column represents 1 (y).
- K-map has two rows. First row represents 0 ($\overline{x}$) and second row represents 1 (x).
- Find the value of each cell or square as row value and column value (like matrix element)
    - First row and first column represents m_0 or 00 or ($\overline{x}\,\overline{y}$)
    - First row and second column represents m_1 or 01 or ($\overline{x}y$)
    - Second row and first column represents m_2 or 10 or ($x\overline{y}$)
    - Second row and second column represents m_3 or 11 or (xy)

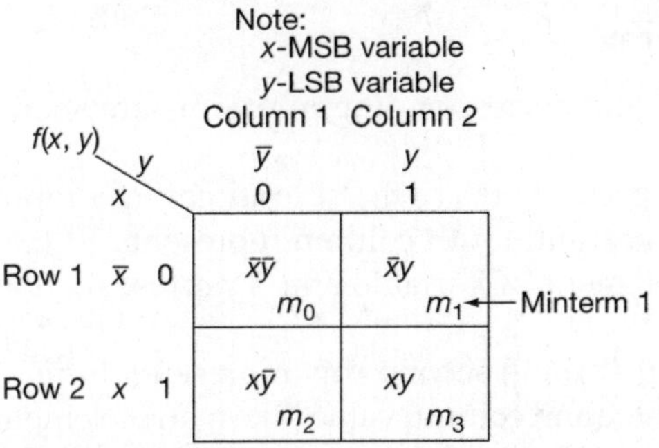

FIGURE 4.24 | Two-variable K-map

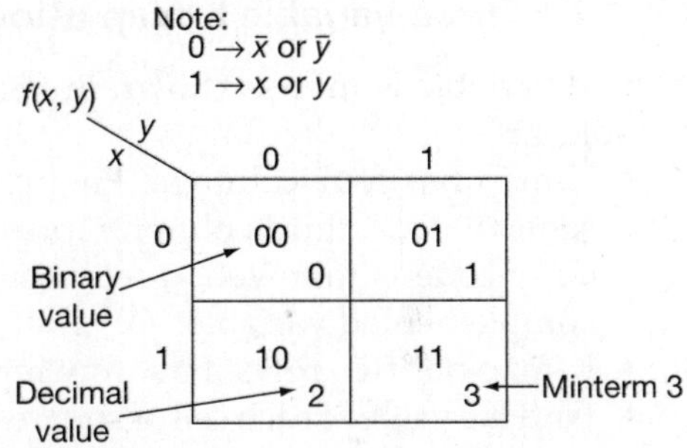

FIGURE 4.25 | Two-variable K-map

TABLE 4.19 | Fundamental/Standard product for two inputs

Decimal Value	Inputs x	y	Fundamental Product	Minterm	Fundamental Sum	Maxterm
0	0	0	$\overline{x}\,\overline{y}$	m_0	$x+y$	M_0
1	0	1	$\overline{x}y$	m_1	$x+\overline{y}$	M_1
2	1	0	$x\overline{y}$	m_2	$\overline{x}+y$	M_2
3	1	1	xy	m_3	$\overline{x}+\overline{y}$	M_3

4.4.1.2 | Two-Variable K-map using Maxterms

A two-variable K-map is shown in Figure 4.26 and corresponding maxterms are shown in Table 4.19. Simplified form of two-variable K-map set up is given in Figure 4.27.

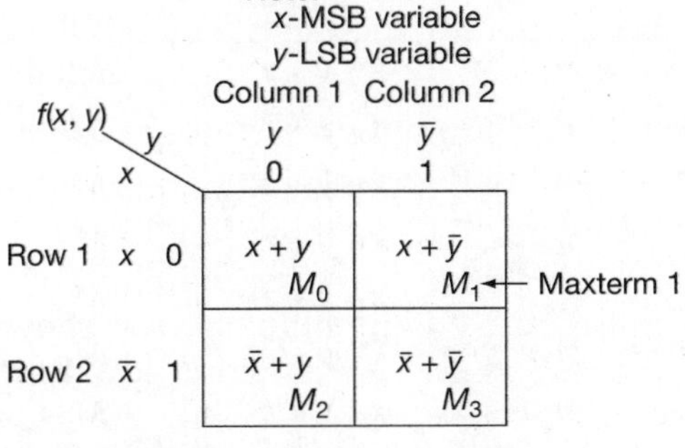

FIGURE 4.26 | Two-variable K-map

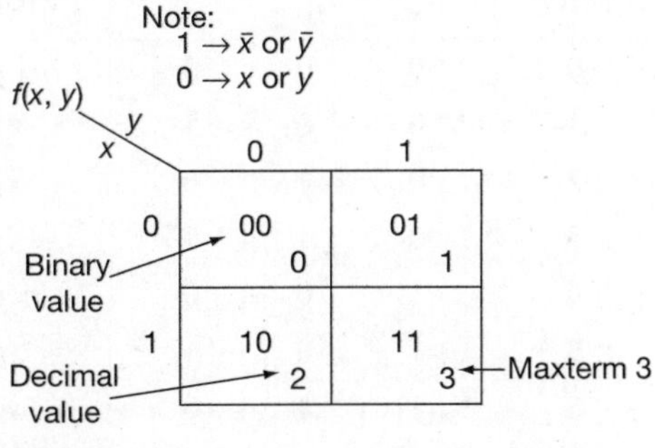

FIGURE 4.27 | Two-variable K-map

- K-map has two columns. First column represents 0 (y) and second column represents 1 ($\overline{y}$).
- K-map has two rows. First row is represents by 0 (x) and second row represents 1 ($\overline{x}$).
- Find the value of each cell or square as row value and column value (like matrix element)
  - First row and first column represents M_0 or (00) or $(x + y)$
  - First row and second column represents M_1 or (01) or $(x + \overline{y})$
  - Second row and first column represents M_2 or (10) or $(\overline{x} + y)$
  - Second row and second column represents M_3 or (11) or $(\overline{x} + \overline{y})$

4.4.1.3 | Three-Variable K-map using Minterms

A three-variable K-map is shown in Figure 4.28 and corresponding minterms are shown in Table 4.20.

- K-map has four columns. First column represents 00 $(\overline{y}\,\overline{z})$, second column represents 01 $(\overline{y}z)$, third column represents 11 (yz), and fouth column represents 10 $(y\overline{z})$. Gray code is followed. 0 represents complement of variable and 1 represents non-complemented variable.
- K-map has two rows. First row represents 0 $(\overline{x})$ and second row represents 1 (x).
- Find the value of each cell or square as row value and column value (like matrix element)
  - First row and first column represents m_0 or 000 or $(\overline{x}\,\overline{y}\,\overline{z})$
  - Second row and first column represents m_4 or 100 or $(x\overline{y}\,\overline{z})$

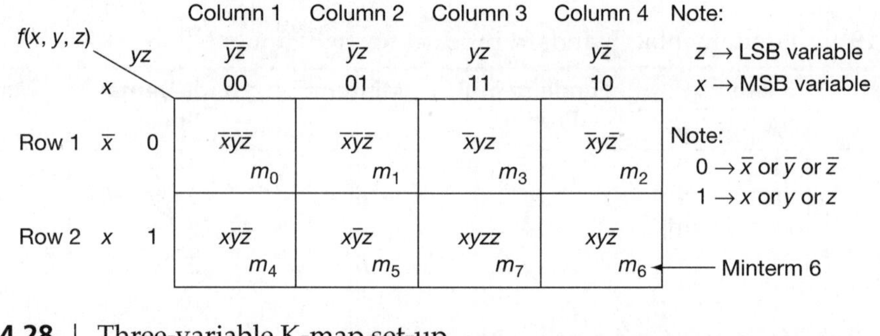

FIGURE 4.28 | Three-variable K-map set-up

TABLE 4.20 | Fundamental/Standard product for three inputs

Decimal Value	Inputs x	y	z	Fundamental Product	Minterm	Fundamental Sum	Maxterm
0	0	0	0	$\overline{x}\,\overline{y}\,\overline{z}$	m_0	$x+y+z$	M_0
1	0	0	1	$\overline{x}\,\overline{y}z$	m_1	$x+y+\overline{z}$	M_1
2	0	1	0	$\overline{x}y\overline{z}$	m_2	$x+\overline{y}+z$	M_2
3	0	1	1	$\overline{x}yz$	m_3	$x+\overline{y}+\overline{z}$	M_3
4	1	0	0	$x\overline{y}\,\overline{z}$	m_4	$\overline{x}+y+z$	M_4
5	1	0	1	$x\overline{y}z$	m_5	$\overline{x}+y+\overline{z}$	M_5
6	1	1	0	$xy\overline{z}$	m_6	$x+y+z$	M_6
7	1	1	1	xyz	m_7	$\overline{x}+\overline{y}+\overline{z}$	M_7

4.4.1.4 | Three-Variable K-map using Maxterms

A three-variable K-map is shown in Figure 4.29 and corresponding fundamental terms maxterms are shown in Table 4.20.

- K-map has four columns. First column represents (00) $(y+z)$, second column represents (01) $(y+\overline{z})$, third column represents (11) $(\overline{y}+\overline{z})$, and fourth column represents (10) $(\overline{y}+z)$. Gray code is followed. 1 represents complement of variable and 0 represents non-complemented variable.
- K-map has two rows. First row represents 0 (x) and second row represents 1 $(\overline{x})$.
- Find the value of each cell or square as row value and column value (like matrix element)
    - First row and second column represents M_1 or (001) or $(x+y+\overline{z})$
    - Second row and second column represents M_5 or (101) or $(\overline{x}+y+\overline{z})$

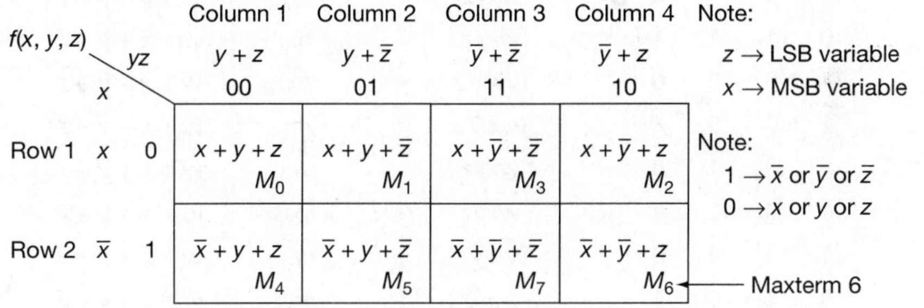

FIGURE 4.29 | Four-variable K-map set-up

4.4.1.5 | Four-Variable K-map using Minterms

A four-variable K-map is shown in Figure 4.30 and corresponding fundamental terms minterms are shown in Table 4.21.

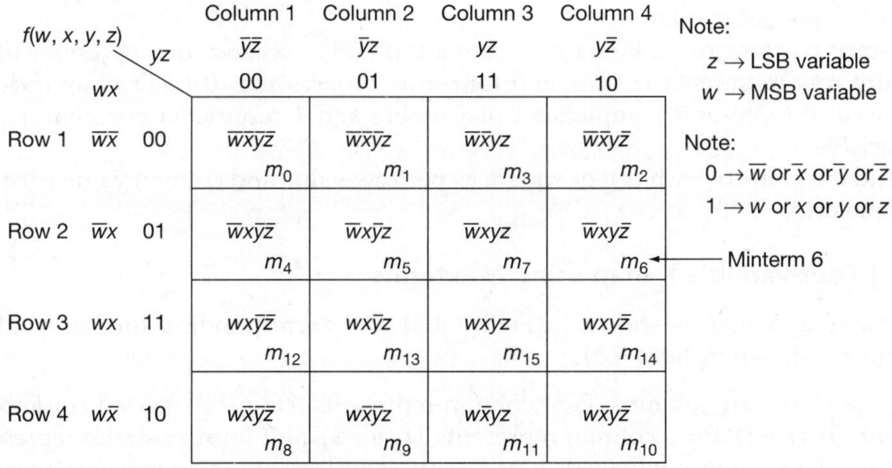

FIGURE 4.30 | Four-variable K-map set-up

TABLE 4.21 | Fundamental/Standard product for four inputs

Decimal Equivalent	Inputs w	x	y	z	Fundamental/ Standard Product	Minterms	Fundamental/ Standard Sum	Maxterms
0	0	0	0	0	$\bar{w}\bar{x}\bar{y}\bar{z}$	m_0	$w+x+y+z$	M_0
1	0	0	0	1	$\bar{w}\bar{x}\bar{y}z$	m_1	$w+x+y+\bar{z}$	M_1
2	0	0	1	0	$\bar{w}\bar{x}y\bar{z}$	m_2	$w+x+\bar{y}+z$	M_2
3	0	0	1	1	$\bar{w}\bar{x}yz$	m_3	$w+x+\bar{y}+\bar{z}$	M_3
4	0	1	0	0	$\bar{w}x\bar{y}\bar{z}$	m_4	$w+\bar{x}+y+z$	M_4
5	0	1	0	1	$\bar{w}x\bar{y}z$	m_5	$w+\bar{x}+y+\bar{z}$	M_5
6	0	1	1	0	$\bar{w}xy\bar{z}$	m_6	$w+\bar{x}+\bar{y}+z$	M_6
7	0	1	1	1	$\bar{w}xyz$	m_7	$w+\bar{x}+\bar{y}+\bar{z}$	M_7
8	1	0	0	0	$w\bar{x}\bar{y}\bar{z}$	m_8	$\bar{w}+x+y+z$	M_8
9	1	0	0	1	$w\bar{x}\bar{y}z$	m_9	$\bar{w}+x+y+\bar{z}$	M_9
10	1	0	1	0	$w\bar{x}y\bar{z}$	m_{10}	$\bar{w}+x+\bar{y}+z$	M_{10}
11	1	0	1	1	$w\bar{x}yz$	m_{11}	$\bar{w}+x+\bar{y}+\bar{z}$	M_{11}
12	1	1	0	0	$wx\bar{y}\bar{z}$	m_{12}	$\bar{w}+\bar{x}+y+z$	M_{12}
13	1	1	0	1	$wx\bar{y}z$	m_{13}	$\bar{w}+\bar{x}+y+\bar{z}$	M_{13}
14	1	1	1	0	$wxy\bar{z}$	m_{14}	$\bar{w}+\bar{x}+\bar{y}+z$	M_{14}
15	1	1	1	1	$wxyz$	m_{15}	$\bar{w}+\bar{x}+\bar{y}+\bar{z}$	M_{15}

- K-map has four columns. First column represents 00 ($\bar{y}\,\bar{z}$), second column represents 01 ($\bar{y}\,z$), third column represents 11 (yz), and fourth column represents 10 ($y\,\bar{z}$). Gray code is followed. 0 represents complement of variable and 1 represents non-complemented variable.
- K-map has four rows. First row represents 00 ($\bar{w}\,\bar{x}$), second row represents 01 ($\bar{w}\,x$), third row represents 11 (wx), and fourth row represents 10 ($w\,\bar{x}$). Gray code is followed. 0 represents complement of variable and 1 represents non-complemented variable.
- Find the value of each cell or square as per row value and column value (like matrix element).

4.4.1.6 | Four-variable K-map using Maxterms

A four-variable K-map is shown in Figure 4.31 and corresponding fundamental terms maxterms are shown in Table 4.21.

- K-map has four columns. First column represents 00 ($y + z$), second column represents 01 ($y+\bar{z}$), third column represents 11 ($\bar{y}+\bar{z}$), and fourth column represents 10 ($\bar{y}+z$). Gray code is followed. 1 represents complement of variable and 0 represents non-complemented variable.

Note: $z \rightarrow$ LSB variable, $w \rightarrow$ MSB variable

Note: $1 \rightarrow \overline{w}$ or $\overline{x}$ or $\overline{y}$ or $\overline{z}$; $0 \rightarrow w$ or x or y or z

$f(w, x, y, z)$ — wx \ yz	Column 1 $y+z$ 00	Column 2 $y+\overline{z}$ 01	Column 3 $\overline{y}+\overline{z}$ 11	Column 4 $\overline{y}+z$ 10
Row 1 $w+x$ 00	$w+x+y+z$ M_0	$w+x+y+\overline{z}$ M_1	$w+x+\overline{y}+\overline{z}$ M_3	$w+x+\overline{y}+z$ M_2
Row 2 $w+\overline{x}$ 01	$w+\overline{x}+y+z$ M_4	$w+\overline{x}+y+\overline{z}$ M_5	$w+\overline{x}+\overline{y}+\overline{z}$ M_7	$w+\overline{x}+\overline{y}+z$ M_6
Row 3 $\overline{w}+\overline{x}$ 11	$\overline{w}+\overline{x}+y+z$ M_{12}	$\overline{w}+\overline{x}+y+\overline{z}$ M_{13}	$\overline{w}+\overline{x}+\overline{y}+\overline{z}$ M_{15}	$\overline{w}+\overline{x}+\overline{y}+z$ M_{14}
Row 4 $\overline{w}+x$ 10	$\overline{w}+x+y+z$ M_8	$\overline{w}+x+y+\overline{z}$ M_9	$\overline{w}+x+\overline{y}+\overline{z}$ M_{11}	$\overline{w}+x+\overline{y}+z$ M_{10}

Maxterm 8 → M_8

FIGURE 4.31 | Four-variable K-map set-up

- K-map has four rows. First row represents 00 $(w+x)$, second row represents 01 $(w+\overline{x})$, third row represents 11 $(\overline{w}+\overline{x})$, and fourth row represents 10 $(\overline{w}+x)$. Gray code is followed. 0 represents complement of variable and 1 represents non-complemented variable.
- Find the value of each cell or square as per row value and column value (like matrix element).

4.4.2 | Mapping of 0's and 1's in the Karnaugh Map

Some rules are elaborated to map 0's and 1's in the K-map

- The 1's are placed in the squares of a K-map that represent minterms of a Boolean function in canonical SOP (Sum-of-Minterm) form. The number of l's in the K-map is equal to the number of product terms in the Boolean function, The cells that do not have a 1 are the cells for which the expression is 0. (see Figure 4.32)

$$f(x, y, z) = \overline{x}\,\overline{y}\,\overline{z} + \overline{x}\,\overline{y}\,z + x\,y\,\overline{z} + x\,\overline{y}\,\overline{z}$$

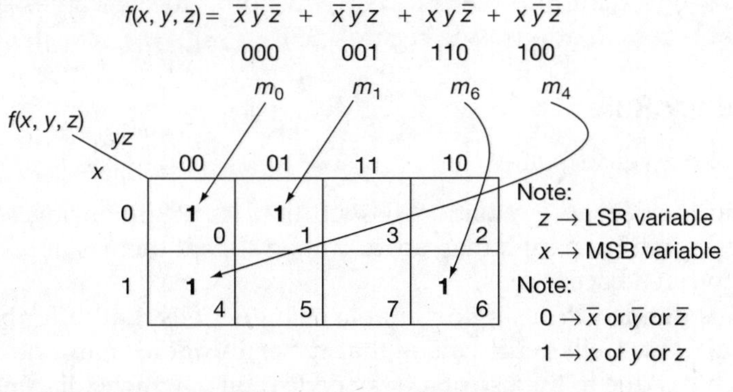

FIGURE 4.32 | Three-variable K-map set-up for SOP

- The 0's are placed in the squares of a K-map that represents the maxterms of a Boolean function in canonical POS (Product-of-Maxterm) form, The number of 0's in the K-map is equal to the number of sum terms in the Boolean function. The cells that do not have a 0 are the cells for which the expression is 1. (see Figure 4.33)
- A Boolean expression in standard (SOP or POS) form should be converted first into its canonical (Sum of Minterm or Product of Maxterm) expression before mapping the 1's (or 0's). (see Figure 4.34)

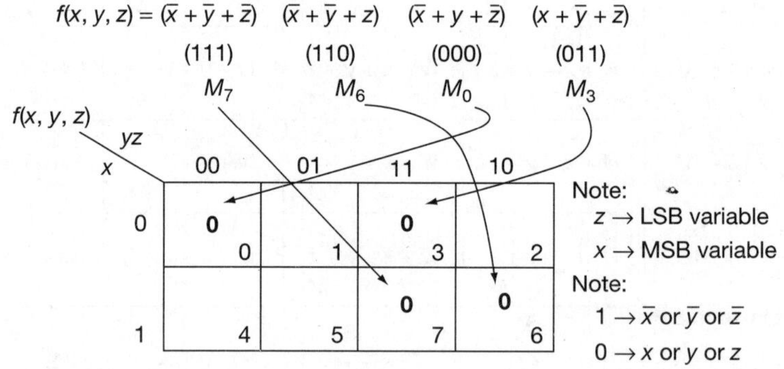

FIGURE 4.33 | Three-variable K-map set-up for POS

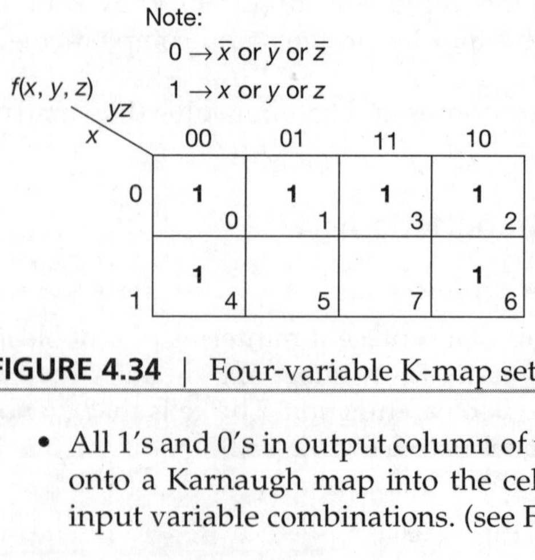

$$f(x,y,z) = \bar{x} + x\bar{y} + xy\bar{z}$$

$$\begin{array}{lll} 0XX & 10X & 110(m_6) \\ 000(m_0) & 100(m_4) & \\ 001(m_1) & 101(m_3) & \\ 010(m_2) & & \\ 011(m_3) & & \end{array}$$

FIGURE 4.34 | Four-variable K-map set-up for SOP

- All 1's and 0's in output column of the truth table (Table 4.22) can be mapped directly onto a Karnaugh map into the cells corresponding to the values of the associated input variable combinations. (see Figure 4.35)

4.4.3 | Adjacency Rule

Adjacency rules are given as below

- Each cell in the K-map is positioned such that its neighbouring cells are adjacent to it. The cells in a Karnaugh map are arranged so that there is only a single-variable change between adjacent cells.
- Adjacency is defined by a single-variable change. Cells that differ by only one variable are adjacent. Cells with values that differ by more than one variable are not adjacent. This is due to the use of a Gray code (one in which adjacent numbers differ in only one position) to label the edges of a map.

TABLE 4.22 | Truth table

Minterm	*x*	*y*	*z*	*f*(*x*, *y*, *z*)
m_0	0	0	0	1
m_1	0	0	1	0
m_2	0	1	0	0
m_3	0	1	1	0
m_4	1	0	0	1
m_5	1	0	1	0
m_6	1	1	0	1
m_7	1	1	1	1

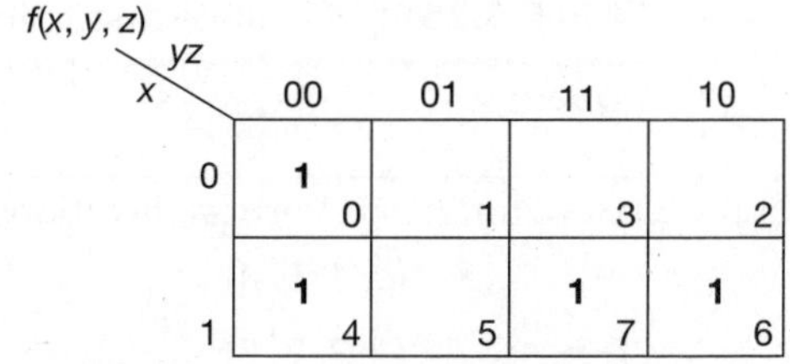

FIGURE 4.35 | Three-variable K-map set-up for SOP

- Cells those are side by side in the horizontal and vertical directions (but not diagonal) are adjacent cells.
- For a map row: the leftmost cell and the rightmost cell are adjacent cells.
- For a map column: the topmost cell and the bottom most cell are adjacent cells.
- For a four-variable map: cells occupying the four corners of the map are adjacent cells.

Table 4.23 gives possible pair(s) formed by grouping two adjacent cells/squares side by side in the horizontal and vertical directions for two-variable K-map. Decimal numbers 0, 1, 2 and 3 are representing minterms or maxterms.

TABLE 4.23 | Possible pairs for two-variable K-map and derived term

Pair (Horizontal Direction)	Derived Product Term	Derived Sum Term	Pair (Vertical Direction)	Derived Product Term	Derived Sum Term
(0, 1)	$\bar{x}$	x	(0, 2)	$\bar{y}$	y
(2, 3)	x	$\bar{x}$	(1, 3)	y	$\bar{y}$

Table 4.24 gives possible pair(s) formed by grouping two adjacent cells/squares side by side in the horizontal and vertical directions for three-variable K-map. Decimal numbers 0, 1, 2, 3, 4, 5, 6 and 7 are representing minterms or maxterms.

TABLE 4.24 | Possible pairs for three-variable K-map and their input literals

Pair (Horizontal Direction)	Derived Product Terms	Derived Sum Terms	Pair (Vertical Direction)	Derived Product Terms	Derived Sum Terms	Pair (On Rolling Over K-map)	Derived Product Terms	Derived Sum Terms
(0, 1)	$\bar{x}\bar{y}$	$x+y$	(0, 4)	$\bar{y}\bar{z}$	$y+z$	(0, 2)	$\bar{x}\bar{z}$	$x+z$
(1, 3)	$\bar{x}z$	$x+\bar{z}$	(1, 5)	$\bar{y}z$	$y+\bar{z}$	(4, 6)	$x\bar{z}$	$\bar{x}+z$
(3, 2)	$\bar{x}y$	$x+\bar{y}$	(3, 7)	yz	$\bar{y}+\bar{z}$			
(4, 5)	$x\bar{y}$	$\bar{x}+y$	(2, 6)	$y\bar{z}$	$\bar{y}+z$			
(5, 7)	xz	$\bar{x}+\bar{z}$						
(7, 6)	xy	$\bar{x}+\bar{y}$						

Table 4.25 gives possible quad(s) formed by grouping four adjacent cells/squares side by side in the horizontal and vertical directions for three-variable K-map.

TABLE 4.25 | Possible quads for three-variable K-map and derived terms

Quad	Derived Product Term	Derived Sum Term
Horizontal and Vertical directions		
(0, 1, 5, 4)	$\bar{y}$	y
(1, 3, 7, 5)	z	$\bar{z}$
(3, 2, 6, 7)	y	$\bar{y}$
On rolling over K-map		
(0, 2, 6, 4)	$\bar{z}$	z

Table 4.26 gives possible pairs formed by grouping four adjacent cells/squares side by side in the horizontal and vertical directions for four-variable K-map. Decimal numbers 0, 1, 2, 3, 4, 5, 6, 7, 8, 9, 10, 11, 12, 13, 14 and 15 are representing minterm or maxterm terms.

TABLE 4.26 | Possible pairs for four-variable K-map and their input literals

Pair (Horizontal Direction)	Derived Product Term	Derived Sum Term	Pair (Vertical Direction)	Derived Product Term	Derived Sum Term	Pair (On Rolling Over K-map)	Derived Product Term	Derived Sum Term
(0, 1)	$\bar{w}\bar{x}\bar{y}$	$w+x+y$	(0, 4)	$\bar{w}\bar{y}\bar{z}$	$w+y+z$	Horizontal direction		
(1, 3)	$\bar{w}\bar{x}z$	$w+x+\bar{z}$	(1, 5)	$\bar{w}\bar{y}z$	$w+y+\bar{z}$	(0, 2)	$\bar{w}\bar{x}\bar{z}$	$w+x+z$
(3, 2)	$\bar{w}\bar{x}y$	$w+x+\bar{y}$	(3, 7)	$\bar{w}yz$	$w+\bar{y}+\bar{z}$	(4, 6)	$\bar{w}x\bar{z}$	$w+\bar{x}+z$
(4, 5)	$\bar{w}x\bar{y}$	$w+\bar{x}+y$	(2, 6)	$\bar{w}y\bar{z}$	$w+\bar{y}+z$	(12, 14)	$wx\bar{z}$	$\bar{w}+\bar{x}+z$
(5, 7)	$\bar{w}xz$	$w+\bar{x}+\bar{z}$	(4, 12)	$x\bar{y}\bar{z}$	$\bar{x}+y+z$	(8, 10)	$w\bar{x}\bar{z}$	$\bar{w}+x+z$
(7, 6)	$\bar{w}xy$	$w+\bar{x}+\bar{y}$	(5, 13)	$x\bar{y}z$	$\bar{x}+y+\bar{z}$	Vertical direction		
(12, 13)	$wx\bar{y}$	$\bar{w}+\bar{x}+y$	(7, 15)	xyz	$\bar{x}+\bar{y}+\bar{z}$	(0, 8)	$\bar{x}\bar{y}\bar{z}$	$x+y+z$
(13, 15)	wxz	$\bar{w}+\bar{x}+\bar{z}$	(6, 14)	$xy\bar{z}$	$\bar{x}+\bar{y}+z$	(1, 9)	$\bar{x}\bar{y}z$	$x+y+\bar{z}$
(15, 14)	wxy	$\bar{w}+\bar{x}+\bar{y}$	(12, 8)	$w\bar{y}\bar{z}$	$\bar{w}+y+z$	(3, 11)	$\bar{x}yz$	$x+\bar{y}+\bar{z}$
(8, 9)	$w\bar{x}\bar{y}$	$\bar{w}+x+y$	(13, 9)	$w\bar{y}z$	$\bar{w}+y+\bar{z}$	(2, 10)	$\bar{x}y\bar{z}$	$x+\bar{y}+z$
(9, 11)	$w\bar{x}z$	$\bar{w}+x+\bar{z}$	(15, 11)	wyz	$\bar{w}+\bar{y}+\bar{z}$			
(11, 10)	$w\bar{x}y$	$\bar{w}+x+\bar{y}$	(14, 10)	$wy\bar{z}$	$\bar{w}+\bar{y}+z$			

Table 4.27 gives possible quads(s) formed by grouping four adjacent cells/squares side by side in the horizontal and vertical directions for four-variable K-map.

TABLE 4.27 | Possible quads for four-variable K-map and their input literals

Quad	Derived Product Term	Derived Sum Term	Quad	Derived Product Term	Derived Sum Term
Horizontal direction			Horizontal and vertical directions		
(0, 1, 3, 2)	$\bar{w}\bar{x}$	$w + x$	(0, 1, 5, 4)	$\bar{w}\bar{y}$	$w + y$
(4, 5, 7, 6)	$\bar{w}x$	$w + \bar{x}$	(1, 3, 7, 5)	$\bar{w}z$	$w + \bar{z}$
(12, 13, 15, 14)	wx	$\bar{w} + \bar{x}$	(3, 2, 6, 7)	$\bar{w}y$	$w + \bar{y}$
(8, 9, 11, 10)	$w\bar{x}$	$\bar{w} + x$	(4, 5, 13, 12)	$x\bar{y}$	$\bar{x} + y$
Vertical directions			(5, 7, 15, 13)	xz	$\bar{x} + \bar{z}$
(0, 4, 12, 8)	$\bar{y}\bar{z}$	$y + z$	(7, 6, 14, 15)	xy	$\bar{x} + \bar{y}$
(1, 5, 13, 9)	$\bar{y}z$	$y + \bar{z}$	(12, 13, 9, 8)	$w\bar{y}$	$\bar{w} + y$
(3, 7, 15, 11)	yz	$\bar{y} + \bar{z}$	(13, 15, 11, 9)	wz	$\bar{w} + \bar{z}$
(2, 6, 14, 10)	$y\bar{z}$	$\bar{y} + z$	(15, 14, 10, 11)	wy	$\bar{w} + \bar{y}$
On rolling over K-map					
(0, 4, 2, 6)	$\bar{w}\bar{z}$	$w + z$			
(4, 12, 6, 14)	$x\bar{z}$	$\bar{x} + z$			
(12, 8, 14, 10)	$w\bar{z}$	$\bar{w} + z$			
(0, 1, 8, 9)	$\bar{x}\bar{y}$	$x + y$			
(1, 3, 9, 11)	$\bar{x}z$	$x + \bar{z}$			
(3, 2, 11, 10)	$\bar{x}y$	$x + \bar{y}$			
(0, 2, 10, 8)	$\bar{x}\bar{z}$	$x + z$			

Table 4.28 gives possible octet(s) formed by grouping eight adjacent cells/squares side by side in the horizontal and vertical directions for four-variable K-map.

TABLE 4.28 | Possible octets for four-variable K-map and their input literals

Octet Grouping	Derived Product Term	Derived Sum Term
Horizontal direction		
(0, 1, 3, 2, 4, 5, 7, 6)	$\bar{w}$	w
(4, 5, 7, 6, 12, 13, 15, 14)	x	$\bar{x}$
(12, 13, 15, 14, 8, 9, 11, 10)	w	$\bar{w}$
Vertical direction		
(0, 4, 12, 8, 1, 5, 13, 9)	$\bar{y}$	y
(1, 5, 13, 9, 3, 7, 15, 11)	z	$\bar{z}$
(3, 7, 15, 11, 2, 6, 14, 10)	y	$\bar{y}$
On rolling over K-map		
(0, 1, 3, 2, 8, 9, 11, 10)	$\bar{x}$	x
(0, 4, 12, 8, 2, 6, 14, 10)	$\bar{z}$	z

4.4.4 | Grouping of 0's and 1's

Group the 1's (or 0's) on the Karnaugh map according to the rules given below. The objective is to maximize the number of elements in a group and to minimize the number of groups.

(a) Always group adjacent cell containing 1's (or 0's) in powers of 2 {1, 2, 4, 8, 16, 32 or 64 1s (or 0s) in a group}. Start by combining the maximum number of adjacent cell containing 1's (or 0's).

(b) Always include the largest possible number of 1's (or 0's) in a group in accordance with rule (a) to reduce the number of literals in a term.

(c) Ensure that each 1's (or 0's) are covered when combining the squares. The 1's (or 0's) already in a group can be included in another group as long as the overlapping groups include non-common 1's (or 0's).

(d) Minimize the number of groups to reduce the number of terms in the simplified function. Avoid redundant grouping.

(e) Cells may only be grouped/looped together in twos, fours, or eights. As few groups as possible must be formed. Groups may overlap one another and may contain only one cell.

(f) The larger the number of 1s grouped/looped together in a group the simpler is the product term that the group represents (Table 4.29).

TABLE 4.29 | Number of input literals to represent the group on K-map

No. of Adjacent Cells Group	Named	Number of Input Literals to Represent the Group			
		Two-variable, K-map	Three-variable, K-map	Four-variable, K-map	Five-variable K-map
1	Single	2	3	4	5
2	Pair	1	2	3	4
4	Quad	$f = 1$ or 0	1	2	3
8	Octet		$f = 1$ or 0	1	2
16	—			$f = 1$ or 0	1
32	—				$f = 1$ or 0

Prime-implicants A prime implicant is a group of minterms that cannot be combined with any other minterm or groups.

Essential prime-implicants An essential prime implicant is a prime-implicant that covers the output of the function but no other prime-implicant is able to cover. A prime implicant is essential prime implicant if it covers a minterm that cannot be covered by any other prime implicant. Essential prime implicant is one in which one or more minterms terms are unique.

Redundant group or non-essential prime implicant Group is known as redundant group if all the 1's (or 0's) forming group already have been participated in other groups. In other words all the participating minters are not unique.

4.4.5 | Determination of Simplified Boolean Function in SOP and POS Form

In determining the simplified Boolean function in SOP form, the following steps are followed:

(a) Obtain a simplified product term for each group of cell containing 1's by considering only variables that occur in common (either un-complemented or complemented) within the group. Different variables that occur both un-complemented and complemented within each group are eliminated.
(b) When the entire simplified product terms are derived from the Karnaugh map, they are summed to form the simplified Boolean function in SOP form.
(c) Table 4.23 gives simplified derived product terms obtained after forming groups, i.e. pair for two-variable K-map.
(d) Tables 4.24 and 4.25 give simplified derived product terms obtained after forming groups, i.e. pair, and quad for three-variable K-map.
(e) Tables 4.26, 4.27 and 4.28 give simplified derived product terms obtained after forming groups, i.e. pair, quad, octet, etc. for four-variable K-map.

In determining the simplified Boolean function in POS form, the following steps are done:

(a) Obtain a simplified sum term for each group of cell containing 0's by considering only variables that occur in common (either un-complemented or complemented) within the group. Different variables that occur both un-complemented and complemented within each group are eliminated.
(b) When all the simplified sum terms are derived from the Karnaugh map, AND all the simplified sum terms to form the simplified Boolean function in POS form.
(c) Table 4.23 gives simplified derived sum terms obtained after forming groups, i.e. pair for two-variable K-map.
(d) Tables 4.24 and 4.25 give simplified derived sum terms obtained after forming groups, i.e. pair, and quad for three-variable K-map.
(e) Tables 4.26, 4.27 and 4.28 give simplified derived sum terms obtained after forming groups, i.e. pair, quad, octet, etc. for four-variable K-map.

EXAMPLE 4.27 Simplify $f(x, y, z) = \sum m(0, 1, 2, 3, 4, 6)$ using Karnaugh Map.

SOLUTION Set up K-map as given in Figure 4.36. Fill 1 for each minterm. There are two quads. One quad is formed by grouping (0, 1, 3, 2) cells and on rolling over K-map another quad is formed by grouping (0, 4, 2, 6) cells/square. These quads are known as prime-implicants as well as essential prime-implicants.

Overlapping: There is overlapping of Cell 0 and 2 in forming quad (0, 4, 2, 6).

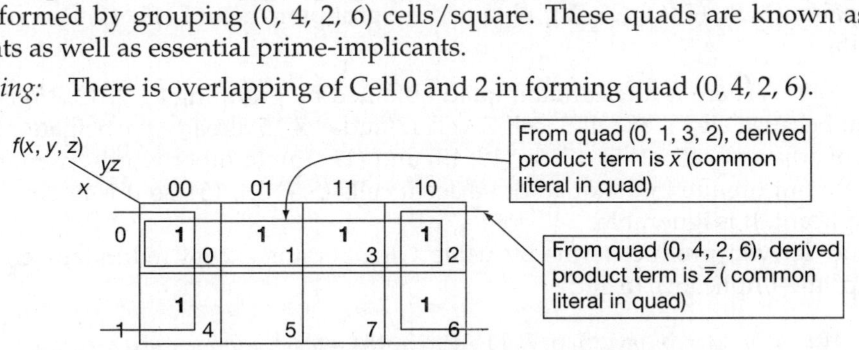

FIGURE 4.36 | Three-variable K-map set-up for SOP

Combine derived literal by using OR operator, to get:

$$f(x, y, z) = \sum m(0, 1, 2, 3, 4, 6) = \overline{x} + \overline{z} \quad \textbf{Ans.}$$

EXAMPLE 4.28 Simplify $f(x,y,z) = \sum m(2, 3, 5, 7)$ using Karnaugh Map.

SOLUTION Set up K-map as given in Figure 4.37. Fill 1 for each minterm. There are three possible pairs. One pair is formed by grouping (3, 2) cells. Second pair is formed by grouping/looping (5, 7) cells. Third pair is formed by grouping/looping (3, 7) cells/squares.

Redundant pair: There is a redundant pair obtained by grouping (3, 7) cells. It is redundant because cell 3 and cell 7 already participated in forming pairs (3, 2) and (5, 7) cells. In other words, minterms 3 and 7 are not unique in the pair (3, 7) so it is not an essential prime-implicant. It is known as redundant group.

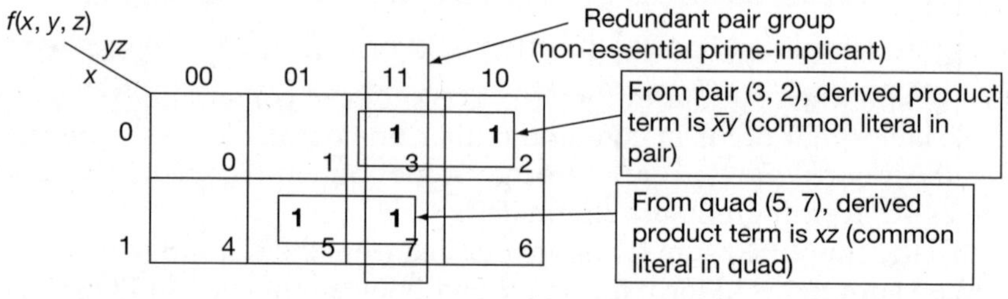

FIGURE 4.37 | Three-variable K-map set-up for SOP

Combine derived literal by using OR operator except redundant group (non-essential prime-implicant), to get:

$$f(x,y,z) = \sum m(2, 3, 5, 7) = \overline{x}y + xz$$

EXAMPLE 4.29 Simplify $f(w,x,y,z) = \sum m(1,5,6,7,11,12,13,15)$ using Karnaugh Map.

SOLUTION Set up K-map as given in Figure 4.38. Fill 1 for each minterm. There are four possible pairs; (1, 5), (7, 6), (12, 13) obtained by grouping 1s and (15, 11) and one quad; (5, 7, 13, 15).

Redundant pair: There is a redundant quad obtained by grouping (5, 7, 13, 15) cells. It is redundant because all 1's of cell 5, cell 7, cell 13 and cell 15 already participated in forming pairs of adjacent cell: (1, 5), (7, 6), (12, 13) and (11, 15). In other words, minterms 5, 7, 13 and 7 are not unique in the quad of adjacent cells (5, 7, 13, 15), so it is not an essential prime-implicant. It is ignorable.

Combine derived product terms by using OR operator except redundant group (not essential prime-implicant), to get:

$$f(w, x, y, z) = \sum m(1, 5, 6, 7, 11, 12, 13, 15) = \overline{w}\,\overline{y}\,z + \overline{w}\,x\,y + w\,y\,z + w\,x\,\overline{y}$$

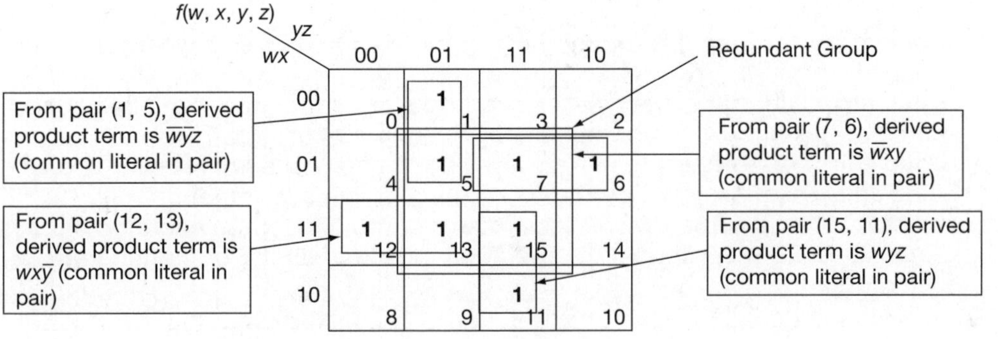

FIGURE 4.38 | Four-variable K-map for SOP

EXAMPLE 4.30 Simplify $f(w,x,y,z) = \prod M(1, 5, 6, 7, 11, 12, 13, 15)$ using Karnaugh Map.

SOLUTION Set up K-map as given in Figure 4.39. Fill 0 for each maxterm. There are four possible pair of adjacent cells; (1, 5), (7, 6), (12, 13) obtained by grouping 0s and (15, 11) and one quad of adjacent cells; (5, 7, 13, 15) which forms redundant group.

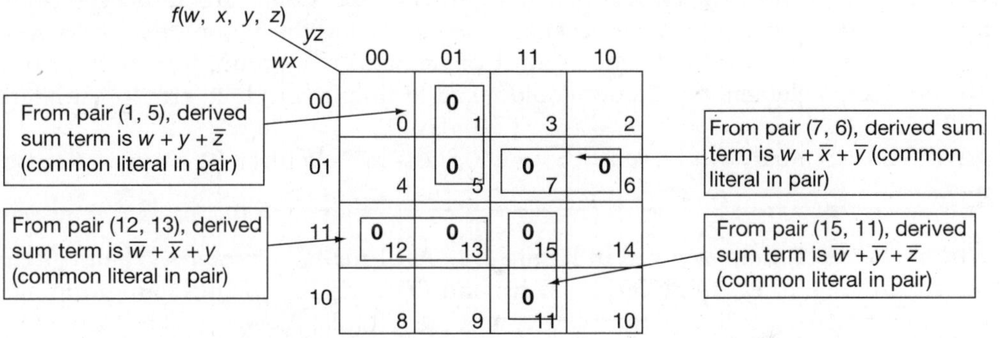

FIGURE 4.39 | Four-variable K-map for POS

Combine derived sum terms using AND operator except redundant group (not essential prime-implicant), to get:

$$f(w,x,y,z) = \prod m(1,5,6,7,11,12,13,15) = (w+y+\overline{z})(w+\overline{x}+\overline{y})(\overline{w}+\overline{y}+\overline{z})(\overline{w}+\overline{x}+y)$$

Complete simplification process

1. Construct the K-map and place 1s and 0s in the squares according to the truth table.
2. Group the isolated 1's or 0's which are not adjacent to any other 1's or 0's. (single loops)
3. Group any pair which contains a 1or 0 adjacent to only one other 1or 0. (double loops)

(Continued)

4. Group any octet even if it contains one or more 1's or 0's that have already been grouped.
5. Group any quad that contains one or more 1's or 0's that have not already been grouped, making sure to use the minimum number of groups.
6. Group any pairs necessary to include any 1's or 0's that have not yet been grouped being adjacent, making sure to use the minimum number of groups.
7. In case of SOP, form the OR operation of all the product terms generated by each group. In case of POS, form AND operation of sum of all the terms generated by each group.

4.5 | KARNAUGH MAP WITH 'DON'T CARE' CONDITIONS

Sometimes a situation arises in which some input variable combinations are not allowed. For example, in the BCD code, there are six invalid combinations these are: 1010, 1011, 1100, 1101, 1110 and 1111. Since these un-allowed states will never occur in an application involving the BCD code, they can be treated as 'don't care' terms with respect to their effect on the output. That is, for these 'does not care' terms either a 1 or a 0 may be assigned to the output; it really does not matter since they will never occur.

For each 'don't care' term, an X is placed in the cell. The 'don't care' terms can be used as an advantage when simplifying because when grouping the 1s (or 0's), the Xs can be treated as 1s (or 0's) to maximize the number of elements in a group. Note that maximizing the number of element per group would result to simpler product term (or sum term).

EXAMPLE 4.31 Simplify $f(w,x,y,z) = \sum m(2,3,4,5,14,15) + d(10,11,12,13)$ using Karnaugh Map.

SOLUTION Set up K-map as given in Figure 4.40. Fill 1 for each minterm. There are three possible quads; (4, 5, 12, 13), (15, 14, 11, 10) and (3, 2, 11, 10) obtained by grouping 1s. Groups are formed considering don't care conditions. At least one '1' must be there to form group. Don't care conditions can be overlapped to form another group.

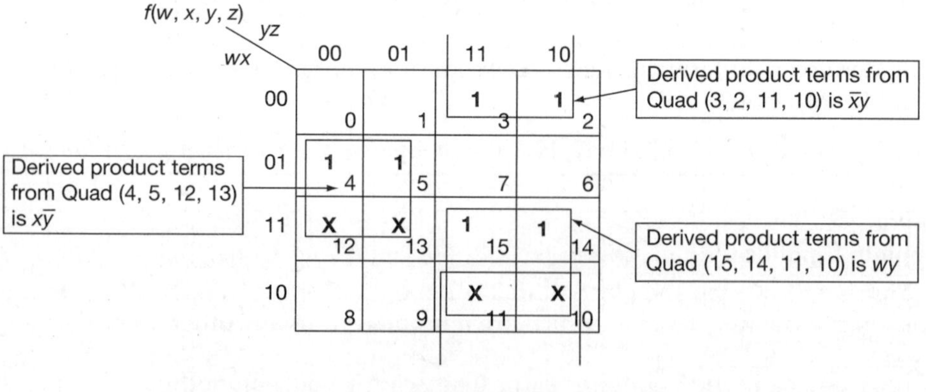

FIGURE 4.40 | Four-variable K-map set-up for SOP

Combine derived literal by using OR operator except redundant group, to get:

$$f(w,x,y,z)=\sum m(2,3,4,5,14,15)+d(10,11,12,13)=x\overline{y}+wy+\overline{x}y$$

EXAMPLE 4.32 Simplify $f(w,x,y,z)=\prod m(2,3,4,5,14,15)\cdot d(10,11,12,13)$ using Karnaugh Map.

SOLUTION Set up K-map as given in Figure 4.41. Fill 0 for each maxterm. There are three possible quads; (4, 5, 12, 13), (15, 14, 11, 10) and (3, 2, 11, 10) obtained by grouping 0s. Groups are formed considering don't care conditions. At least one '0' must be there to form group. Don't care conditions can be overlapped to form another group.

Combine derived literal by using AND operator except redundant group, to get:

$$f(w,x,y,z)=\prod M(2,3,4,5,14,15)+d(10,11,12,13)=(\overline{x}+y)(\overline{w}+\overline{y})(x+\overline{y})$$

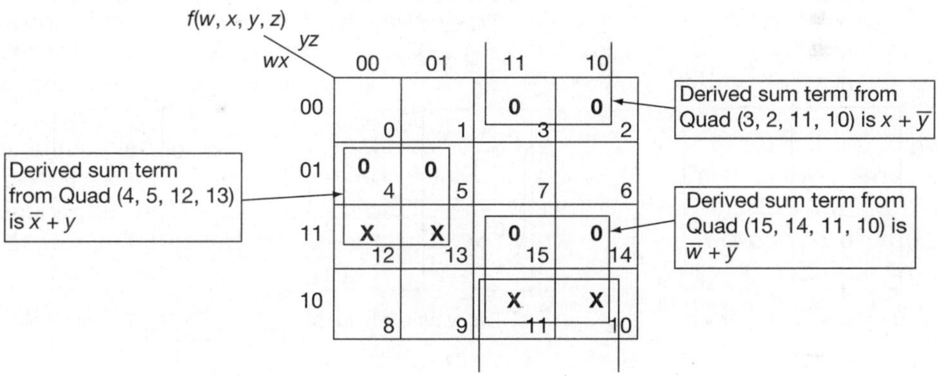

FIGURE 4.41 | Four-variable K-map set-up for POS

Some examples to simplify two-variable Boolean expressions in SOP form are given in Figure 4.42. Some examples to simplify two-variable Boolean expressions in POS form are given in Figure 4.43.

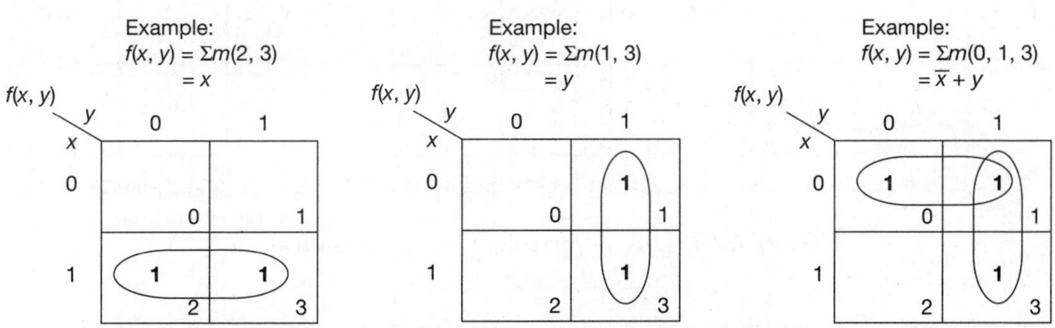

(a) Two-variable K-map (b) Two-variable K-map (c) Two-variable K-map

FIGURE 4.42 | *(Continued)*

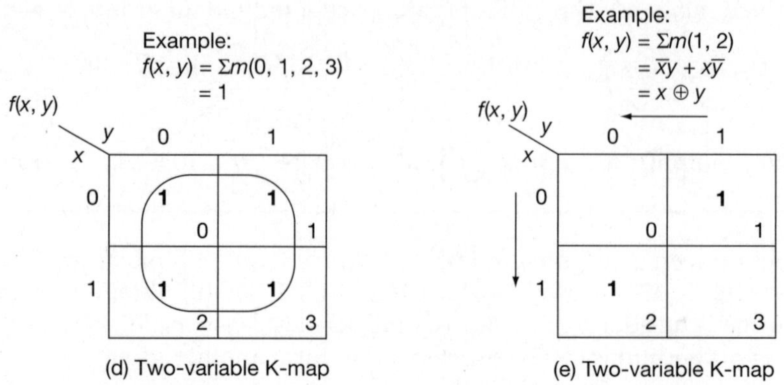

FIGURE 4.42 | Two-variable, K-Map simplification

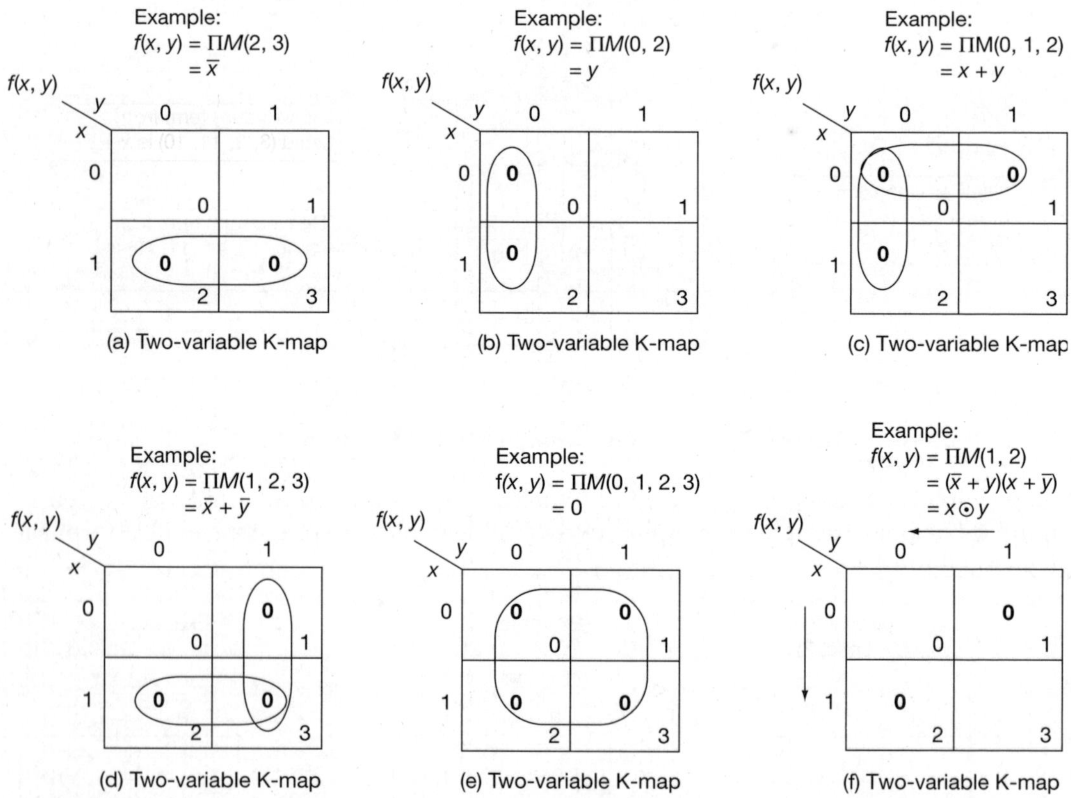

FIGURE 4.43 | Two-variable, K-map simplification

Some examples to simplify three-variable Boolean expressions in SOP form are given in Figure 4.44. Some examples to simplify three-variable Boolean expressions in POS form are given in Figure 4.45.

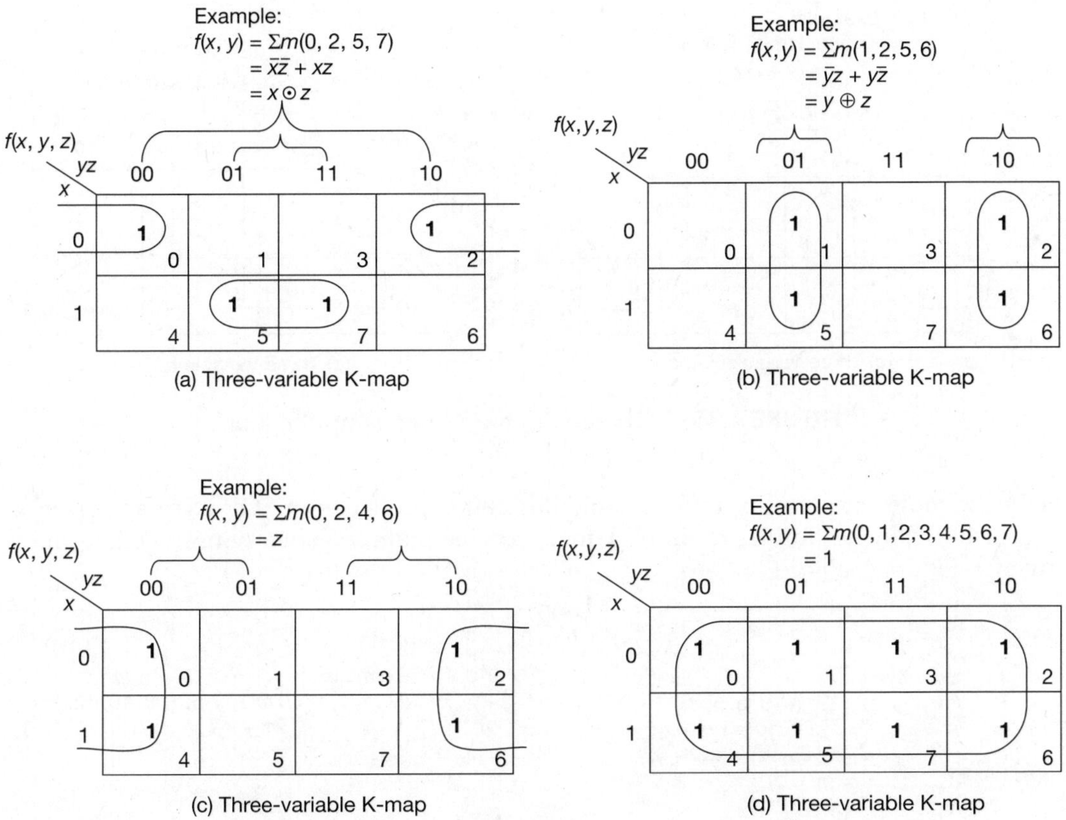

FIGURE 4.44 | Three-variable, K-map simplification

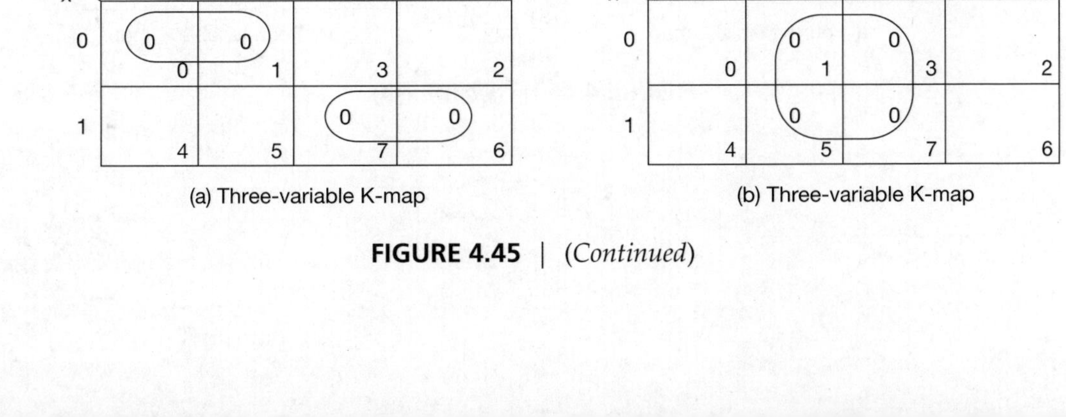

FIGURE 4.45 | *(Continued)*

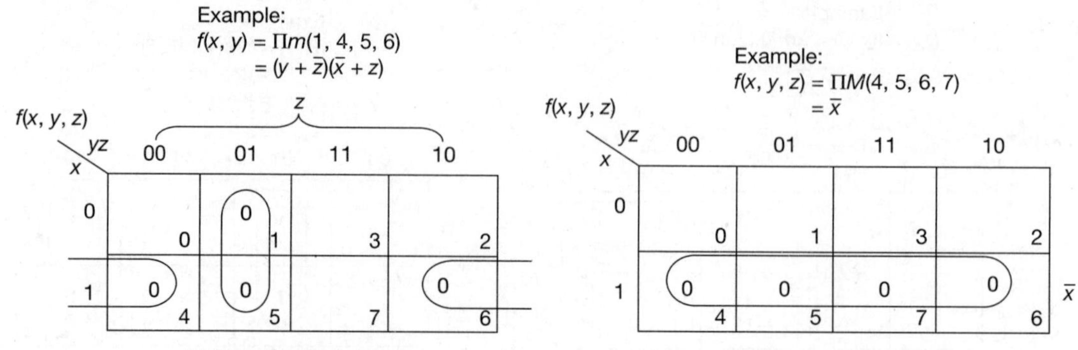

(c) Three-variable K-map

(d) Three-variable K-map

FIGURE 4.45 | Three-variable K-map simplification

Some examples to simplify four-variable Boolean expressions in SOP form are given in Figure 4.46. Some examples to simplify four-variable Boolean expressions in POS form are given in Figure 4.47.

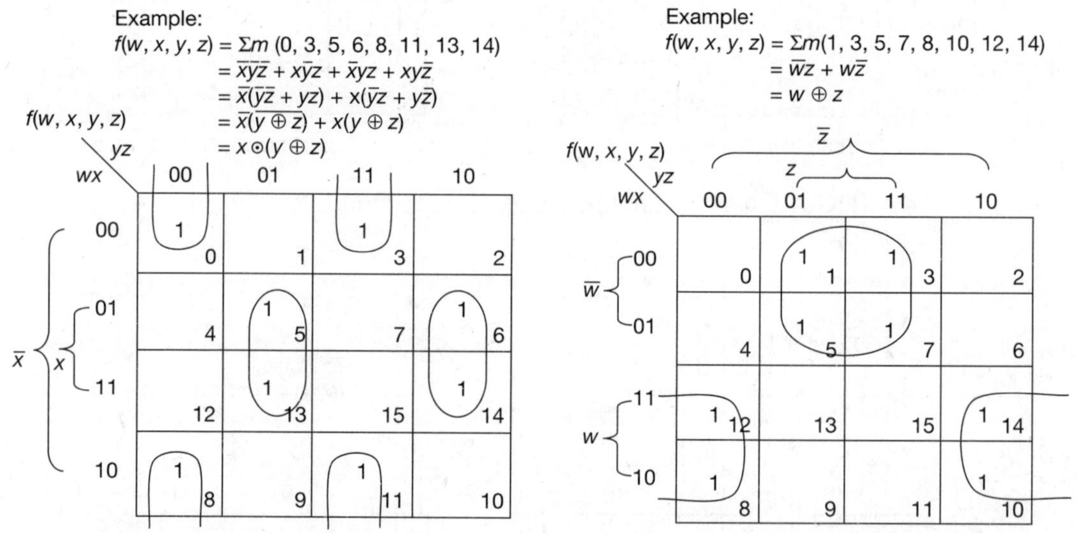

(a) Four-variable K-map

(b) Four-variable K-map

FIGURE 4.46 | *(Continued)*

Example:

$f(w, x, y, z) = \Sigma m(1, 3, 4, 6, 9, 11, 12, 14)$
$= \bar{x}z + x\bar{z}$
$= x \oplus z$

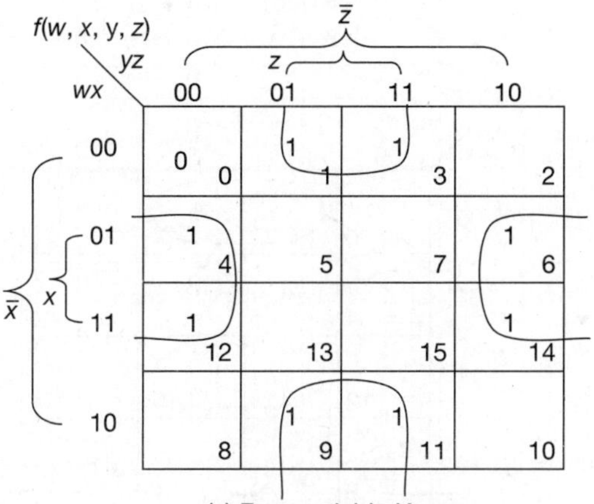

(c) Four-variable K-map

Example:

$f(w, x, y, z) = \Sigma m(0, 2, 5, 7, 8, 10, 13, 15)$
$= \bar{x}\bar{z} + xz$
$= \overline{x \oplus z} = x \odot z$

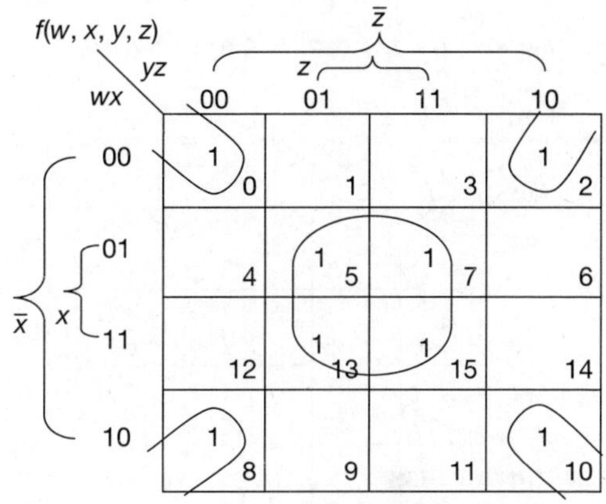

(d) Four-variable K-map

Example:

$f(w, x, y, z) = \Sigma m(0, 1, 2, 3, 8, 9, 10, 11)$
$= \bar{x}$

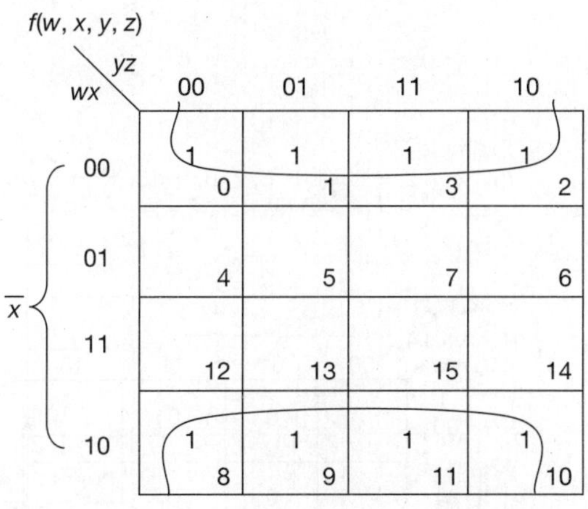

(e) Four-variable K-map

Example:

$f(w, x, y, z) = \Sigma m(4, 5, 6, 7, 12, 13, 14, 15)$
$= x$

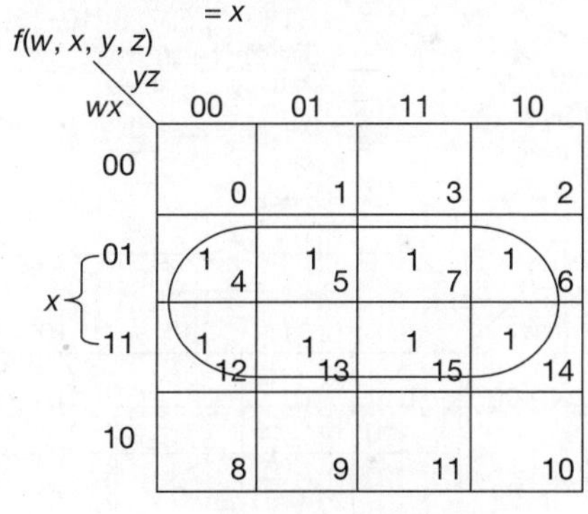

(f) Four-variable K-map

FIGURE 4.46 | *(Continued)*

Example:

$f(w, x, y, z) = \Sigma m\,(0, 2, 4, 6, 8, 10, 12, 14)$
$= \bar{z}$

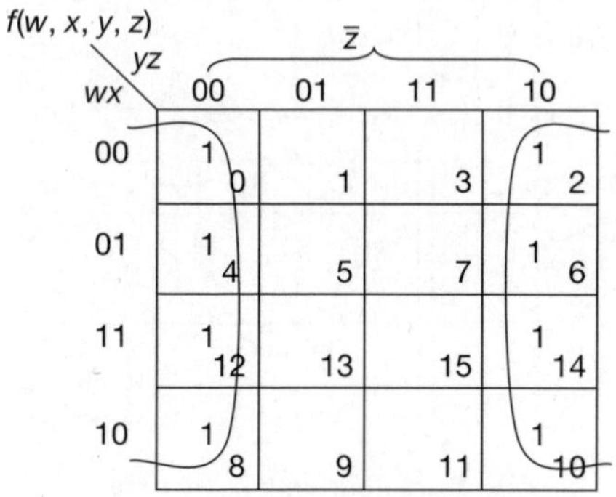

(g) Four-variable K-map

Example:

$f(w, x, y, z) = \Sigma m(1, 3, 4, 6, 8, 10, 13, 15)$
$= \overline{w}\overline{x}z + \overline{w}x\overline{z} + wxz + w\overline{x}\overline{z}$
$= \overline{w}(\overline{x}z + x\overline{z}) + w(xz + \overline{x}\overline{z})$
$= \overline{w}(x \oplus z) + w(\overline{x \oplus z})$
$= w \oplus (x \oplus z)$

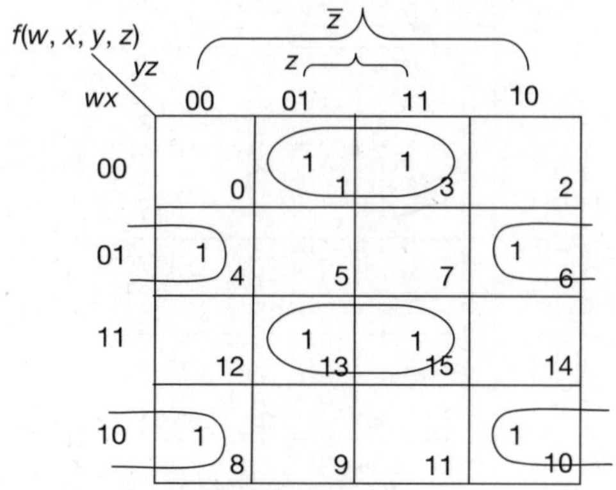

(h) Four-variable K-map

FIGURE 4.46 | Four-variable K-map simplification

Example:

$f(w, x, y, z) = \Pi M(0, 2, 5, 7, 8, 10, 13, 15)$
$= (\bar{x} + \bar{z})\,(x + z)$
$= x\bar{x} + x\bar{z} + z\bar{x} + z\bar{z}$
$= 0 + x\bar{z} + x\bar{z} + 0$
$= x\bar{z} + \bar{x}z$
$= x \oplus z$

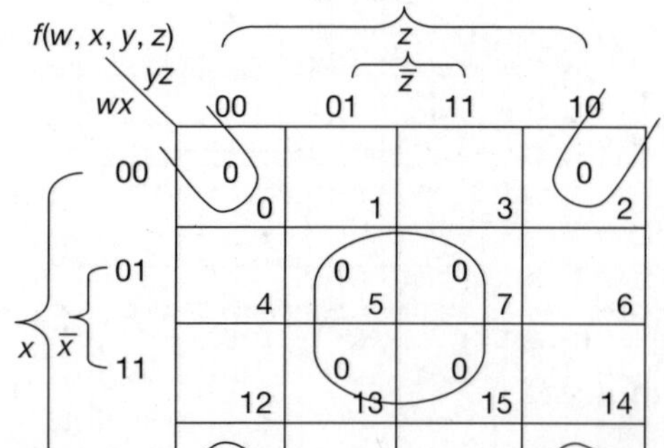

(a) Four-variable K-map

Example:

$f(w, x, y, z) = \Sigma M(1, 3, 5, 7, 8, 9, 10, 12, 13, 14)$
$= (w + \bar{z})\,(\overline{w} + z)\,(\overline{w} + y)$

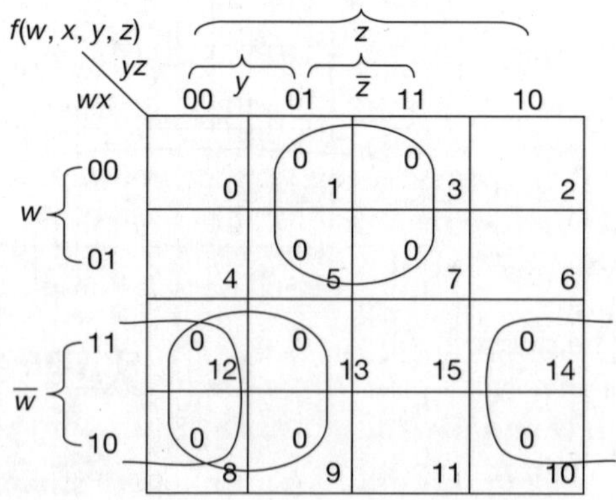

(b) Four-variable K-map

FIGURE 4.47 | *(Continued)*

Example:
$f(w, x, y, z) = \Pi M(0, 1, 4, 5, 7, 10, 11, 13, 14, 15)$
$= (\bar{x} + \bar{z})(w + y)(\bar{w} + \bar{y})$

Example:
$f(w, x, y, z) = \Pi M(0, 2, 4, 5, 6, 7, 8, 10, 12, 13, 14, 15)$
$= \bar{x}.z$

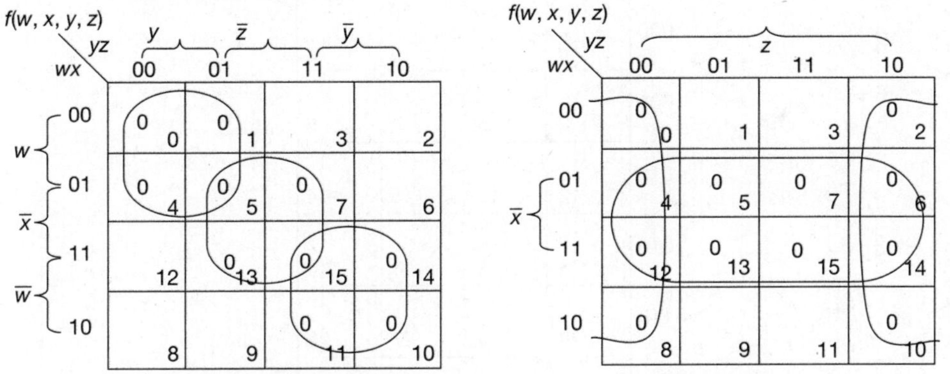

(c) Four-variable K-map

(d) Four-variable K-map

FIGURE 4.47 | Four-variable K-map simplification

4.6 | FIVE-VARIABLE KARNAUGH MAP (K-MAP)

Five-variable map is constructed by laying out two variables horizontally and three variables vertically.

Number of cells/squares in the map = $2^5 = 32$.

Cells are numbered ranging from 0 to 31 ($2^5 - 1$). The numbers are calculated as given below.

Rows of K-map			Columns of K-map	
MSB			LSB	
2^4	2^3	2^2	2^1	2^0
16	8	4	2	1
v	*w*	*x*	*y*	*z*
1	0	0	1	1 = 19

The decimal number gives the minterm, m_{19} represented by that cell (row 8 and column 3).

Five-variable map is arranged by two methods to follow adjacency; (i) mirror image view and (ii) stacked view.

Mirror image view: Gray code is followed to arrange the cells/squares horizontally and vertically. Figure 4.48 shows the set of five-variable K-map.

f(v, w, x, y, z)

vwx \ yz	00	01	11	10
000	0	1	3	2
001	4	5	7	6
011	12	13	15	14
010	8	9	11	10
110	24	25	27	26
111	28	29	31	30
101	20	21	23	22
100	16	17	19	18

FIGURE 4.48 | Four-variable K-map set-up

Stack view: Without losing the adjacency of cells/squares, two four-variable K-maps are stacked. The number of cells/squares remains the same.

Most significant variable v is fixed either to 0 or to 1 for a single four-variable K-map. Figure 4.49 shows the stack view of two four-variable K-map. This view can be shown by Figure 4.50. In figures quad formed by m_0, m_2, m_8, m_{10} of K-map-I is adjacent to quad formed by m_{16}, m_{18}, m_{24}, m_{26} of K-map-II. It is little confusing in Figure 4.49 and is very much visible in Figure 4.50. Lines drawn show this adjacency. Moreover, m_0 is adjacent to m_{16} and m_2 is adjacent to m_{18}. But m_2 is not adjacent to m_{16}.

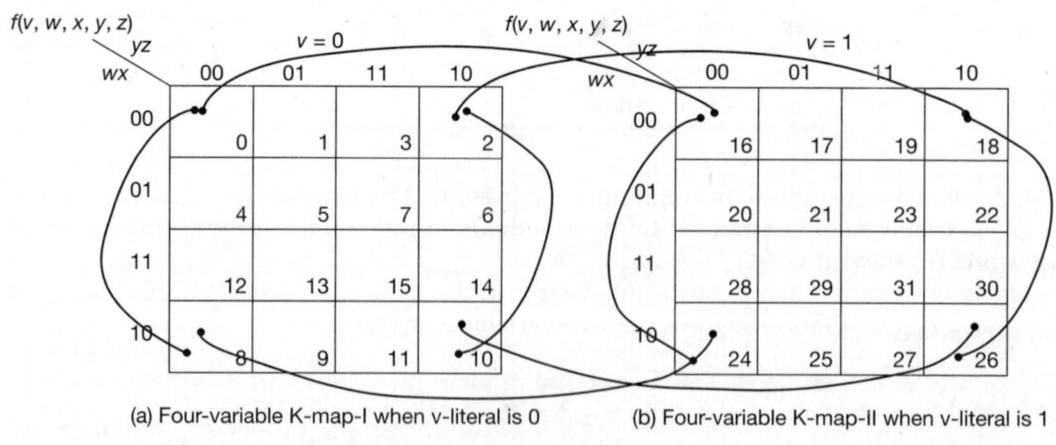

(a) Four-variable K-map-I when v-literal is 0 (b) Four-variable K-map-II when v-literal is 1

FIGURE 4.49 | Five-variable K-map set-up following stack

Grouping can be of 16 cells, 8 cells, 4 cells and 2 cells, which will reduce the number of literals 4, 3, 2 and 1, respectively. Common literals in group are retained whereas varying literals are eliminated.

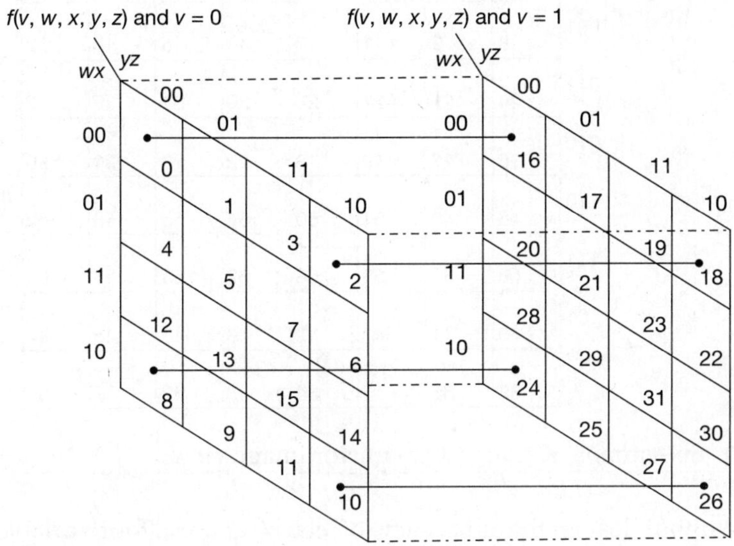

FIGURE 4.50 | Five-variable K-map set-up following stacked view

4.7 | SIX-VARIABLE KARNAUGH MAP (K-MAP)

Six-variable map is constructed by laying out three variables horizontally and three variables vertically.

Number of cells/squares in the map = 2^6 = 64.

Cells are numbered ranging from 0 to 63 (2^6 – 1). The numbers are calculated as given below.

Rows of K-map			Columns of K-map		
MSB			LSB		
2^5	2^4	2^3	2^2	2^1	2^0
32	16	8	4	2	1
u	*v*	*w*	*x*	*y*	*z*
1	1	0	0	0	0 = 48

The decimal number gives the minterm, m_{48} represented by that cell (row 5 and column 0).

Six-variable map is arranged by two methods to follow adjacency; (i) mirror image view and (ii) stacked view.

Mirror image view: Gray code is followed to arrange the cells/squares horizontally and vertically. Figure 4.51 shows the set of six-variable K-map.

f(u, v, w, x, y, z)

uvw \ xyz	000	001	011	010	110	111	101	100
000	0	1	3	2	6	7	5	4
001	8	9	11	10	14	15	13	12
011	24	25	27	26	30	31	29	28
010	16	17	19	18	22	23	21	20
110	48	49	51	50	54	55	53	52
111	56	57	59	58	62	63	61	60
101	40	41	43	42	46	47	45	44
100	32	33	35	34	38	39	37	36

FIGURE 4.51 | Six-variable K-map set-up mirror image view

Stack view: Without losing the adjacency of cells/squares, four-variable K-maps are stacked. The number of cells/squares remains same. Variables u and x are fixed either to 0 or to 1 for a single four-variable K-map giving rise to four possible combinations, i.e. 2^2. This view can be shown by Figure 4.52. In figures quad formed by m_0, m_2, m_{18}, m_{16} of K-map-I, quad formed by m_4, m_6, m_{22}, m_{20} of K-map-II, quad formed by $m_{32}, m_{34}, m_{50}, m_{48}$ of K-map-III and quad formed by $m_{36}, m_{38}, m_{54}, m_{52}$ of K-map-IV are adjacent to each other. Lines drawn show this adjacency of stacked four-variable K-maps. Moreover, m_0 is adjacent to m_4, m_4 is adjacent to m_{32}, m_{32} is adjacent to m_{36} and m_{36} is adjacent to m_0. In other words m_0, m_4, m_{32} and m_{36} are adjacent to each other. But m_2 is not adjacent to m_4 or m_{36} or m_{32}.

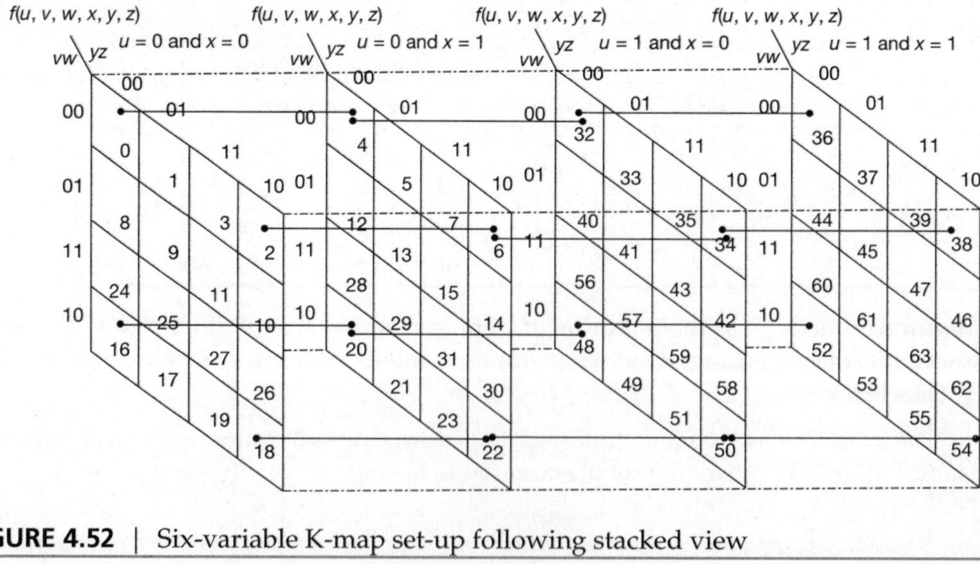

FIGURE 4.52 | Six-variable K-map set-up following stacked view

Grouping can be of 32 cells, 16 cells, 8 cells (Octet), 4 cells (quad), 2 cells (pair), which will reduce the number of literals 5, 4, 3, 2, and 1, respectively. Common literals in group are retained whereas varying literals are eliminated.

EXAMPLE 4.33 Simplify the given function:

$$f(v,w,x,y,z) = \sum m(0,2,5,7,8,10,13,15,16,18,21,23,24,26,29,31)$$
$$= \bar{x}\,\bar{z} + xz$$
$$= x \odot z$$

Explanation:

Follow Figure 4.53.

(i) Quad (m_0, m_2, m_{10}, m_8) is adjacent to Quad $(m_{16}, m_{18}, m_{26}, m_{24})$

(ii) Quad $(m_5, m_7, m_{15}, m_{13})$ is adjacent to Quad $(m_{21}, m_{23}, m_{31}, m_{29})$

Two adjacent quads form octet and reduce three variables.

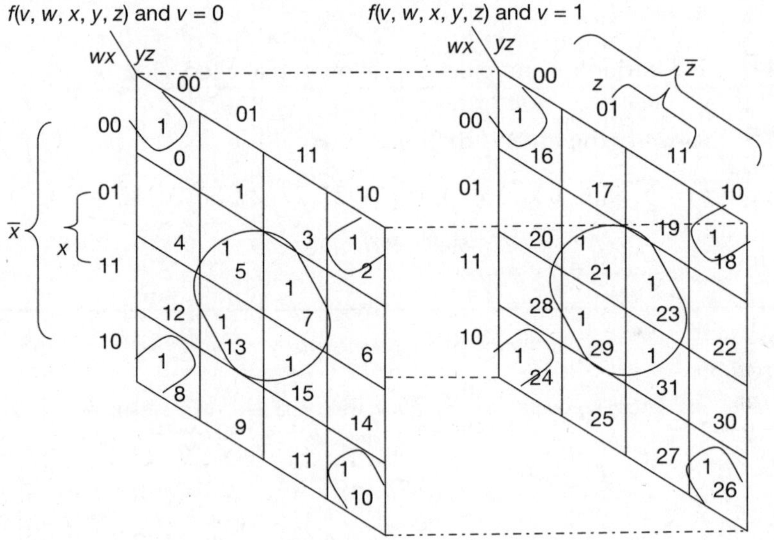

FIGURE 4.53 | Five-variable K-map set-up following stacked view

EXAMPLE 4.34 Simplify the given function:

$$f(v,w,x,y,z) = \prod M(0, 1, 2, 3, 5, 7, 8, 10, 13, 15, 16, 17, 18, 19, 21, 23, 24, 25, 26, 27, 29, 31)$$
$$= (\bar{x} + \bar{z})(x + z)(w + x)(\bar{v} + \bar{w} + x)$$

Explanation:

Follow Figure 4.54.

(i) Quad (m_0, m_2, m_{10}, m_8) is adjacent to Quad $(m_{16}, m_{18}, m_{26}, m_{24})$

(ii) Quad $(m_5, m_7, m_{15}, m_{13})$ is adjacent to Quad $(m_{21}, m_{23}, m_{31}, m_{29})$

(iii) Quad (m_0, m_1, m_3, m_2) is adjacent to Quad $(m_{16}, m_{17}, m_{19}, m_{18})$

Only Quad $(m_{24}, m_{25}, m_{27}, m_{26})$

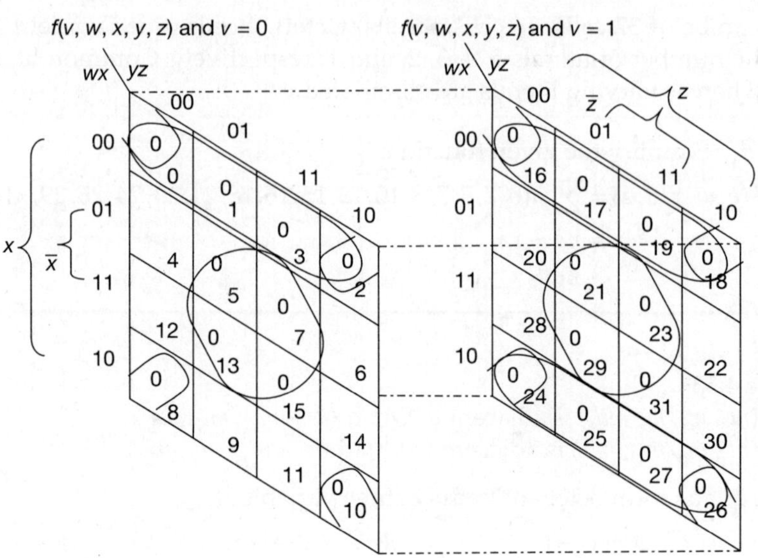

FIGURE 4.54 | Five-variable K-map set-up following stacked view

EXAMPLE 4.35 Simplify the given function:

$$f(u,v,w,x,y,z) = \sum m(8, 9, 11, 13, 14, 15, 24, 25, 27, 29, 30, 31, 40, 41, 43, 45, 46, 47, 56, 57, 59, 61, 62, 63)$$
$$= wz + w\overline{x}\,\overline{y} + wxy$$

Explanation:
Follow Figure 4.55.

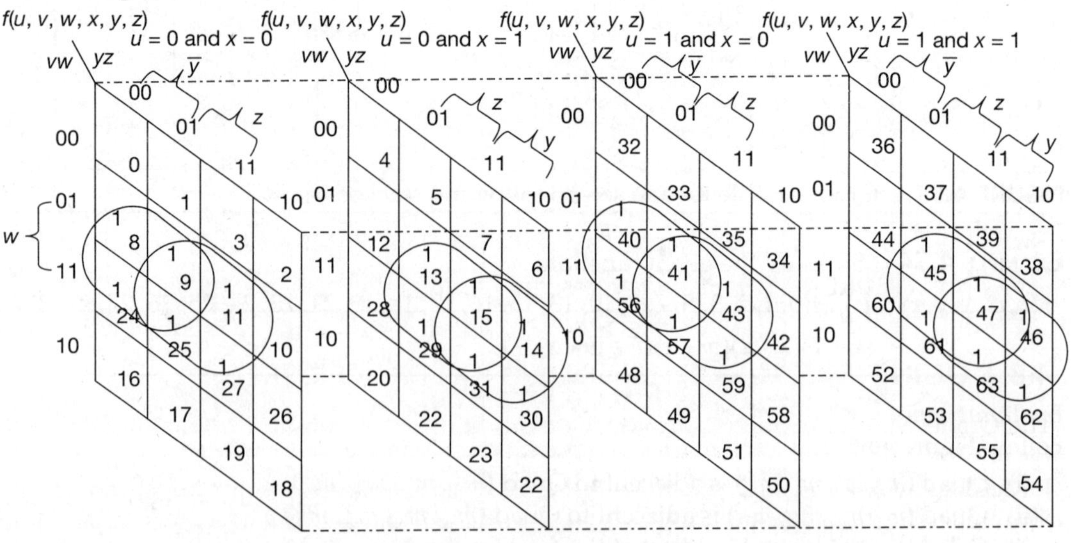

FIGURE 4.55 | Six-variable K-map set-up following stacked view

Quad (m_9, m_{11}, m_{25}, m_{27}), (m_{13}, m_{15}, m_{29}, m_{31}), (m_{41}, m_{43}, m_{57}, m_{59}) and (m_{45}, m_{47}, m_{61}, m_{63}) are adjacent to each other. Quad (m_8, m_9, m_{24}, m_{25}) is adjacent to Quad (m_{40}, m_{41}, m_{56}, m_{57}) Quad (m_{15}, m_{14}, m_{31}, m_{30}) is adjacent to Quad (m_{47}, m_{46}, m_{63}, m_{62})

EXAMPLE 4.36 Simplify the given function:

$$f(u,v,w,x,y,z) = \sum m(7,12,22,23,28,34,37,38,40,42,44,46,56,58,60,62)$$
$$= uw\bar{z} + u\bar{v}y\bar{z} + wx\bar{y}\,\bar{z} + \bar{u}\,v\,\bar{w}xy + \bar{u}\bar{w}xyz + u\bar{v}\bar{w}x\bar{y}z$$

Explanation:

Follow Figure 4.56.

(i) Quad (m_{40}, m_{56}, m_{42}, m_{58}) and Quad (m_{44}, m_{60}, m_{46}, m_{62}) are adjacent.
(ii) Pair (m_{12}, m_{28}) is adjacent to Pair (m_{44}, m_{60}).
(iii) Pair (m_{34}, m_{42}) is adjacent to Pair(m_{38}, m_{46}).
(iv) Pair (m_7, m_{23}).
(v) Pair (m_{22}, m_{23}) and single m_{37}.

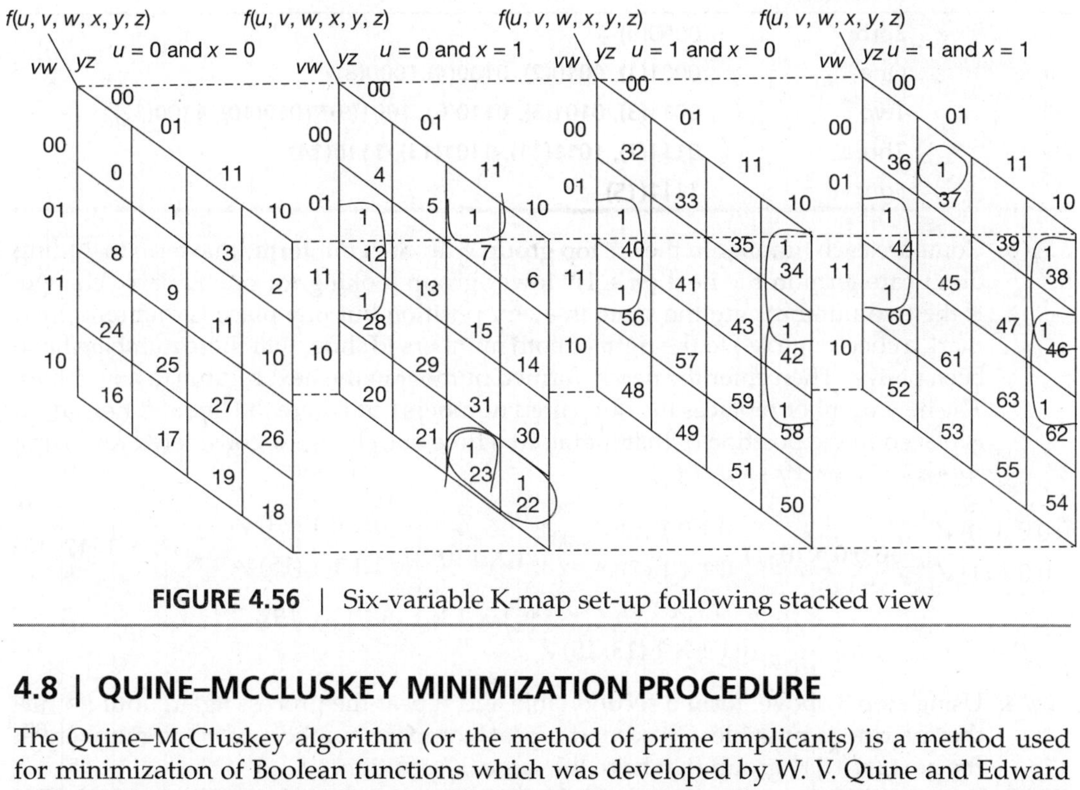

FIGURE 4.56 | Six-variable K-map set-up following stacked view

4.8 | QUINE–MCCLUSKEY MINIMIZATION PROCEDURE

The Quine–McCluskey algorithm (or the method of prime implicants) is a method used for minimization of Boolean functions which was developed by W. V. Quine and Edward J. McCluskey in 1956. It is functionally identical to Karnaugh mapping, but the tabular form makes it more efficient for use in computer algorithms, and it also gives a deterministic way to check that the minimal form of a Boolean function has been reached. It is also known as the tabulation method.

The method involves two steps:

- Finding all prime-implicants of the function.
- Finding the essential prime-implicants of the function, as well as other prime implicants that are necessary to cover the function using prime-implicant.

This is basically a tabular method of minimization and as much it is suitable for computer applications. The procedure for minimization as follows:

Step 1: Describe individual minterms/maxterms and don't care terms of the given Boolean expression by their equivalent binary numbers. Differentiate don't care terms using * symbol.

Step 2: Form a table by grouping numbers representing minterms/maxterms having equivalent number of 1's and arrange them in ascending order, i.e. first numbers having no 1's, then numbers having one 1, and then numbers having two 1's, etc. (see Table 4.30)

TABLE 4.30 | Grouping based on number of 1's present in minterm represented by binary number

Group	Number of 1's in Minterm	Four-bit Binary Numbers Representing Minterm
I	Zero	0000(0)
II	One	0001(1), 0010(2), 0100(4),1000(8)
III	Two	0011(3), 0101(5), 0110(6), 1001(9), 1010(10), 1100(12)
IV	Three	0111(7), 1011(11), 1101(13), 1110(14)
V	Four	1111(15)

Step 3: Compare each number in the mth top group with each minterm/maxterm including don't care term in the next $(m + 1)^{th}$ lower group looking for one variable change. If the two numbers are the same in every position but one place is different, then mark a check sign (✓) to the right of both numbers to show that these numbers have been paired. Then enter the newly formed number in the next column (a new table). The new number replaces the old paired numbers but where the literal differ, an 'X' is placed in the position of that literal. At a time one change is considered following $xy + x\bar{y} = x$, or $\bar{x}y + \bar{x}\,\bar{y} = \bar{x}$

0 0 0 0 (0) ✓
0 0 0 1 (1) ✓ ⇨ 0 0 0 X (0, 1)

0 1 0 1 (5) ✓
0 1 1 1 (7) ✓ ⇨ 0 1 X 1 (0, 3)

1 1 0 1 (13) ✓
1 1 1 1 (15) ✓ ⇨ 1 1 X 1 (13, 15)

0 1 X 1 (5, 7) ✓
1 1 X 1 (13, 15) ✓ ⇨ X 1 X 1 (5, 7, 13, 15)

Step 4: Using step 3 above, form a second table and repeat the process again until no further pairing is possible. On second repeat, compare numbers of mth to numbers in the next $(m + 1)^{th}$ group that have the same 'X' position.

Step 5: Terms which were not covered are the prime-implicants and are ORed and ANDed together to form final function. Name prime-implicants as P_j, where j is integer number.

Step 6: Construct the prime-implicant map and determine essential prime-implicants. Don't considers don't care term part of group.

Select the row corresponding to column that has one 'X' and mark the minterms included by that row and columns. Repeat the process of selection of rows and columns till all the minterms groped by all the prime-implicants are marked.

4.8.1 | Reduction Techniques

Another simplest procedure is given below

1. Check for essential columns and remove them;
2. Check for row dominance and remove all dominating rows;
3. Check for column dominance and remove all dominated columns;
4. Repeat 1, 2, 3 if there is any removal occurs.
   - *Essential Columns:* A column is essential if it covers one 1-entry that cannot be covered by any other columns.
   - Row r_i dominates row r_j if r_i has all the 1-entries in r_j. r_i is dominating and r_j is dominated.

Column p_i dominates p_j if p_i has all the 1-entries in p_j and p_i costs (e.g., number of literals) no more than p_j.

> ***Note:*** The procedure above gives the prime implicant but not **essential** prime-implicant.

EXAMPLE 4.37 Simplify the given Boolean expression using Quine–McCluskey procedure

$$f(w,x,y,z) = \bar{w}\bar{x}\bar{y}\bar{z} + \bar{w}x\bar{y}z + \bar{w}xy\bar{z} + \bar{w}xyz + w\bar{x}\bar{y}z + w\bar{x}y\bar{z} + wx\bar{y}z + wxy\bar{z} + wxyz$$

SOLUTION Table 4.31 assigns the group number.

TABLE 4.31 | Group number assignment

Minterm	Binary Representation	Minterm Number	Number of 1's in Minterm	Assign Group Number
$\bar{w}\bar{x}\bar{y}\bar{z}$	0000	m_0	0	1
$\bar{w}x\bar{y}z$	0101	m_5	2	2
$\bar{w}xy\bar{z}$	0110	m_6	2	2
$\bar{w}xyz$	0111	m_7	3	3
$w\bar{x}\bar{y}z$	1001	m_9	2	2
$w\bar{x}y\bar{z}$	1010	m_{10}	2	2
$wx\bar{y}z$	1101	m_{13}	3	3
$wxy\bar{z}$	1110	m_{14}	3	3
$wxyz$	1111	m_{15}	4	4

Step 1: Form a table of functions of minterms according to the number of 1's in each minterm as shown in Table 4.32.

TABLE 4.32 | Groups in ascending order

Group	Step II *wxyz*
1	0000(0)
2	0101(5)
2	0110(6)
2	1001(9)
2	1010(10)
3	0111(7)
3	1101(13)
3	1110(14)
4	1111(15)

Step 2:

- Start pairing off each element of first group with the next, however since m_0 has no 1's, it and the next group of numbers with one 1's are missing, therefore they cannot be paired off.
- Start by pairing elements of m_5 with m_7, m_{13}, m_{14}, and m_6 with m_7, m_{13}, m_{14}, and so on.
- If these pair off, write them in a separate table and ✓ the minterms of that pair.
- m_5 (0101) and m_7 (0111) pair off to produce 01X1, so in the next Table 4.33 under 'minterm paired' Enter (5, 7) and under '*wxyz*' enter '01X1' and place a ✓ sign in front of 5 and 7 in Table 4.33.

TABLE 4.33 | Finding the prime-implicants

Group	Step-1 Minterm *wxyz*	Step-2 Minterms Paired
1	0000(0) P_1	
2	0101(5) ✓	01X1(5,7)
2	0110(6) ✓	X101(5,13)
2	1001(9) ✓	011X(6,7)
2	1010(10) ✓	X110(6,14)
2		1X01(9,13)
2		1X10(10,14)
3	0111(7) ✓	X111(7,15)
3	1101(13) ✓	11X1(13,15)
3	1110(14) ✓	111X(14,15)
4	1111(15) ✓	

Step 3: Now repeat the same procedure by pairing off each element of a group with the elements of the next group for elements that have 'X' in the same position.

Let (5, 7) matches (13, 15) to produce X1X1. These elements are placed in Table 4.34 as shown, and the elements in stage-II Table 4.33 are ✓ checked. (The elements that produce the same *wxyz* pattern are eliminated.)

TABLE 4.34 | Prime-implicants

Group	Step-1	Step-2	Step-3
	wxyz	*wxyz*	*wxyz*
1	0000(0) P_1		
2	0101(5) ✓	01X1(5, 7) ✓	X1X1(5, 7, 13, 15) P_4
2	0110(6) ✓	X101(5, 13) ✓	X11X(6, 7, 14, 15) P_5
2	1001(9) ✓	011X(6, 7) ✓	
2	1010(10) ✓	X110(6, 14) ✓	
2		1X01(9, 13) P_2	
2		1X10(10, 14) P_3	
3	0111(7) ✓	11X1(13, 15) ✓	
3	1101(13) ✓	111X(14, 15) ✓	
3	1110(14) ✓		
4	1111(15) ✓		

Since (9, 13) and (10, 14) in Table 4.34 do not pair off, these are prime-implicants. m_0, from Table 4.33 is prime-implicant. (5, 7, 13, 15) and (6, 7, 14, 15) from Table 4.34 are unpaired individuals. P_1, P_2, P_3, P_4 and P_5 are marked as prime-implicates and are given Table 4.35.

TABLE 4.35 | Minimal prime-implicants

Prime-implicant		Decimal	0	5	6	7	9	10	13	14	15
P_1	$\bar{w}\bar{x}\bar{y}\bar{z}$	(0)	X								
P_2	$w\bar{y}z$	(9, 13)					X		X		
P_3	$wy\bar{z}$	(10, 14)						X		X	
P_4	xz	(5, 7, 13, 15)		X		X			X		X
P_5	xy	(6, 7, 14, 15)			X	X				X	X
			✓	✓	✓		✓	✓	✓	✓	✓

Select row-wise and column-wise so that all the minterm represented by prime-implicants are covered (Table 4.35). Select columns 0, 5, 6, 9 and 10 because there is only X in these columns. Select P_1, P_4, P_5, P_2 and P_3 rows corresponding to 0, 5, 6, 9 and 10 columns, respectively. Therefore, write the minimized SOP as given below.

$$F(w,x,y,z) = \bar{w}\bar{x}\bar{y}\bar{z} + w\bar{y}z + wy\bar{z} + xz + xy$$

EXAMPLE 4.38 Simplify the given Boolean expression using Quine–McCluskey procedure

$$F(A,B,C) = \sum m(0,1,2,5,6,7)$$

SOLUTION Minterms are grouped in various groups and are given in Table 4.36. Stages are given in Table 4.37 to find the prime-implicants.

TABLE 4.36 | Grouping of minterms

Group	Minterm	*A*	*B*	*C*	Marks
1	0	0	0	0	✓
2	1	0	0	1	✓
2	2	0	1	0	✓
3	5	1	0	1	✓
3	6	1	1	0	✓
4	7	1	1	1	✓

TABLE 4.37 | Prime-implicants

Group	Step-0	Step-2			
	Minterm	*A*	*B*	*C*	
1	(0, 1)	0	0	X	P_1 $\bar{A}\bar{B}$
1	(0, 2)	0	X	0	P_2 $\bar{A}\bar{C}$
2	(1, 5)	X	0	1	P_3 $\bar{B}C$
2	(2, 6)	X	1	0	P_4 $B\bar{C}$
3	(5, 7)	1	X	1	P_5 AC
4	(6, 7)	1	1	X	P_6 AB

P_1, P_2, P_3, P_4 and P_5 are marked as prime-implicates because these are not further pair off and are written in Table 4.38. 1 denotes the normal (un-complemented) variable and 0 denotes the complement of the variable.

TABLE 4.38 | Minimal prime-implicants map

Prime-implicant	Decimal	Minterms	0	1	2	5	6	7
P_1	(0, 1)	$\bar{A}\bar{B}$	X	X				
P_2	(0, 2)	$\bar{A}\bar{C}$	X		X			
P_3	(1, 5)	$\bar{B}C$		X		X		
P_4	(2, 6)	$B\bar{C}$			X		X	
P_5	(5, 7)	AC				X		X
P_6	(6, 7)	AB					X	X
			✓	✓	✓	✓	✓	✓

Select P_2 and P_3 rows in Table 4.38. Mark the columns having X. Complete the marking of all columns by selecting P_6 row. The Boolean expression is obtained by applying OR operator to prime-implicants.

$$F = \overline{A}\overline{C} + \overline{B}C + AB$$

Another possibility can also be explored as given in Table 4.39. 1 denotes the normal (un-complemented) variable and 0 denotes the complement of the variable.

TABLE 4.39 | Minimal prime-implicants map

Prime-implicant	Decimal	Minterms	0	1	2	5	6	7
P_1	(0, 1)	$\overline{A}\overline{B}$	X	X				
P_2	(0, 2)	$\overline{A}\overline{C}$	X		X			
P_3	(1, 5)	$\overline{B}C$		X		X		
P_4	(2, 6)	$B\overline{C}$			X		X	
P_5	(5, 7)	AC				X		X
P_6	(6, 7)	AB					X	X
			✓	✓	✓	✓	✓	✓

Select P_1 and P_4 rows in Table 4.39. Mark the columns having X. Complete the marking of all columns by selecting P_5 row. The Boolean expression is obtained by applying OR operator to prime-implicants. 1 denotes the normal (un-complemented) variable and 0 denotes the complement of the variable.

$$F = \overline{A}\overline{B} + B\overline{C} + AC$$

So, there is need to find the procedure that gives all minimal possibilities. Petrick's method is one of them.

4.8.2 | Petrick's Method

S. R. Petrick has submitted a technical report AFCRC-TR-56-110, Air Force Cambridge Research Center (AFCRC), Cambridge, MA, USA, April, 1956, to determine directly the irredundant forms of a Boolean function from the set of prime-implicants. *A* prime-implicant chart, determines all minimum SOP solutions. Consider the prime-implicants given in Table 4.40 that are obtained in Example 4.38.

TABLE 4.40 | Prime-implicant chart

			0	1	2	5	6	7
P_1	(0, 1)	$\overline{A}\overline{B}$	X	X				
P_2	(0, 2)	$\overline{A}\overline{C}$	X		X			
P_3	(1, 5)	$\overline{B}C$		X		X		
P_4	(2, 6)	$B\overline{C}$			X		X	
P_5	(5, 7)	AC				X		X
P_6	(6, 7)	AB					X	X

Step 1: Label all the rows in the chart.
Step 2: Form a logic function P with the logic variables P_1, P_2, P_3 that is true when all the minterms in the chart are covered.

The first column has an X in rows P_1 and P_2. Therefore one of these rows is included in order to cover minterm m_0. Thus the following term must be in P, i.e. $(P_1 + P_2)$. Similarly determine P as given below:

$$P = (P_1 + P_2)(P_1 + P_3)(P_2 + P_4)(P_3 + P_5)(P_4 + P_6)(P_5 + P_6)$$

Apply commutative law so that distributive law can be applied

$$P = (P_1 + P_2)(P_1 + P_3)(P_4 + P_2)(P_5 + P_3)(P_4 + P_6)(P_5 + P_6)$$

Rearrange them

$$P = (P_1 + P_2)(P_1 + P_3)(P_4 + P_2)(P_4 + P_6)(P_5 + P_3)(P_5 + P_6)$$

Apply distributive law

$$P = (P_1 + P_2P_3)(P_4 + P_2P_6)(P_5 + P_3P_6)$$

Rewriting above equation

$$P = (P_1P_4 + P_1P_2P_6 + P_2P_3P_4 + P_2P_3P_6)(P_5 + P_3P_6)$$

$$P = P_1P_4P_5 + P_1P_2P_5P_6 + P_2P_3P_4P_5 + P_2P_3P_5P_6 + P_1P_3P_4P_6 + P_1P_2P_3P_6 + P_2P_3P_4P_6 + P_2P_3P_6$$

$F = \bar{A}\bar{B} + B\bar{C} + AC$ or $F = \bar{A}\bar{C} + \bar{B}C + AB$. This expression explains that to cover all the minterms there are eight possible ways. $P_1P_4P_5$ means line P_1, line P_4 and line P_5, cover all columns. Similarly, $P_1P_2P_5P_6$ means line P_1, line P_2, line P_5, and line P_6 covers all columns. So, find minimal forms of the function considering that all the terms P_1, P_2, … have same cost. The two minimal forms are $P_1P_4P_5$ and $P_2P_3P_6$:

$$F = \bar{A}\bar{B} + B\bar{C} + AC \text{ or } F = \bar{A}\bar{C} + \bar{B}C + AB$$

EXAMPLE 4.39 Simplify the given Boolean expression using Quine–McCluskey procedure

$$F(A,B,C,D) = \sum m(4,5,6,8,9) + \sum d(0,7,15)$$

SOLUTION Don't care minterms are included in the groups in stage-I to find prime implicants. Prime-implicants are obtained in Step 1, Step 2 and Step 3 are given in Tables 4.41(a), 4.41(b) and 4.41(c), respectively.

TABLE 4.41(a) | Prime-implicants formation (Step-1)

Group	Minterm	Variable				Mark
		A	B	C	D	
1	(0)	0	0	0	0	✓
2	(4)	0	1	0	0	✓
2	(8)	1	0	0	0	✓

(*Continued*)

TABLE 4.41(a) | (Continued)

Group	Minterm	A	B	C	D	Mark
		Variable				
3	(5)	0	1	0	1	✓
3	(6)	0	1	1	0	✓
3	(9)	1	0	0	1	✓
3	(10)	1	0	1	0	✓
4	(7)	0	1	1	1	✓
4	(13)	1	1	0	1	✓
4	(15)	1	1	1	1	✓

TABLE 4.41(b) | Prime-implicants formation (Step-2)

Group	Minterm	A	B	C	D	Mark
		Variable				
1	(0, 4)	0	X	0	0	P_1, $\bar{A}\bar{C}\bar{D}$
1	(0, 8)	X	0	0	0	P_2 $\bar{B}\bar{C}\bar{D}$
2	(4, 5)✓	0	1	0	X	✓
2	(4, 6)✓	0	1	X	0	✓
2	(8, 9)	1	0	0	X	P_3, $A\bar{B}\bar{C}$
2	(8, 10)	1	0	X	0	P_4, $A\bar{B}\bar{D}$
3	(5, 7)✓	0	1	X	1	✓
3	(5, 13)✓	X	1	0	1	✓
3	(6, 7)✓	0	1	1	X	✓
3	(9, 13)	1	X	0	1	P_5, $A\bar{C}D$
4	(7, 15)✓	X	1	1	1	✓
4	(13, 15)✓	1	1	X	1	✓

TABLE 4.41(c) | Prime-implicants formation (Step-3)

Group	Minterm	A	B	C	D	Mark
		Variable				
2	(4, 5, 6, 7)	0	1	X	X	P_6, $\bar{A}B$
2	(5, 7, 13, 15)	X	1	X	1	P_7, BD

Don't care minterms 0, 7 and 15 are not considered as columns to find essential prime-implicants as given in Table 4.42. Columns 6 and 10 have only one cross (X) as shown in Table 4.42. So, select P_4 and P_6 rows. Select the columns corresponding to P_4 and P_6 rows. Only columns 9 and 13 are not selected. So, select row P_5 to complete the selection of columns.

TABLE 4.42 | Minimal prime-implicants map

Prime-implicant		Decimal	4	5	6	8	9	10	13
P_1	$\bar{A}\bar{C}\bar{D}$	(0, 4)	X						
P_2	$\bar{B}\bar{C}\bar{D}$	(0, 8)				X			
P_3	$A\bar{B}\bar{C}$	(8, 9)				X	X		
P_4	$A\bar{B}\bar{D}$	(8, 10)				X		X	
P_5	$A\bar{C}D$	(9, 13)					X		X
P_6	$\bar{A}B$	(4, 5, 6, 7)	X	X	X				
P_7	BD	(5, 7, 13, 15)		X					X
Selection of minterm			✓	✓	✓	✓	✓	✓	✓

The Boolean expression is obtained by applying OR operator to prime-implicants. 1 denotes the normal (un-complemented) variable and 0 denotes the complement of the variable.

$$F(A,B,C,D) = A\bar{B}\bar{D} + A\bar{C}D + \bar{A}B$$

EXAMPLE 4.40 Simplify the given Boolean expression using Quine–McCluskey procedure

$$F(A,B,C,D) = \prod m(4,5,6,8,9) \cdot \prod d(0,7,15)$$

SOLUTION POS prime-implicants are of sum form rather than product form. Don't care maxterms are included in the groups in stage-I to find prime implicants. Prime-implicants are obtained in Step 1, Step 2 and Step 3 are given Tables 4.43(a), 4.43(b) and 4.43(c), respectively.

TABLE 4.43(a) | Prime-implicants formation (Step-1)

Group	Minterm	Variable				Mark
		A	*B*	*C*	*D*	
1	(0)	0	0	0	0	✓
2	(4)	0	1	0	0	✓
2	(8)	1	0	0	0	✓
3	(5)	0	1	0	1	✓
3	(6)	0	1	1	0	✓
3	(9)	1	0	0	1	✓
3	(10)	1	0	1	0	✓
4	(7)	0	1	1	1	✓
4	(13)	1	1	0	1	✓
4	(15)	1	1	1	1	✓

TABLE 4.43(b) | Prime-implicants formation (Step-2)

Group	Minterm	Variable				Mark
		A	*B*	*C*	*D*	
1	(0, 4)	0	X	0	0	P_1, $\bar{A}\bar{C}\bar{D}$
1	(0, 8)	X	0	0	0	P_2, $\bar{B}\bar{C}\bar{D}$
2	(4, 5)✓	0	1	0	X	✓
2	(4, 6)✓	0	1	X	0	✓
2	(8, 9)	1	0	0	X	P_3, $A\bar{B}\bar{C}$
2	(8, 10)	1	0	X	0	P_4, $A\bar{B}\bar{D}$
3	(5, 7)✓	0	1	X	1	✓
3	(5, 13)✓	X	1	0	1	✓
3	(6, 7)✓	0	1	1	X	✓
3	(9, 13)	1	X	0	1	P_5, $A\bar{C}D$
4	(7, 15)✓	X	1	1	1	✓
4	(13, 15)✓	1	1	X	1	✓

TABLE 4.43(c) | Prime-implicants formation (Step-3)

Group	Minterm	Variable				Mark
		A	*B*	*C*	*D*	
2	(4, 5, 6, 7)	0	1	X	X	P_6, $\bar{A}B$
2	(5, 7, 13, 15)	X	1	X	1	P_7, BD

Don't care maxterms 0, 7 and 15 are not considered as columns to find essential prime-implicants as given in Table 4.44. Columns 6 and 10 have only one cross (X) as shown in Table 4.44. So, select P_4 and P_6 rows. Mark the columns covered by P_4 and P_6 rows. Only columns 9 and 13 are not selected. So, select row P_5 to complete the selection of columns.

TABLE 4.44 | Minimal prime-implicants map

Prime-implicant		Decimal	4	5	6	8	9	10	13
P_1	$A+C+D$	(0, 4)	X						
P_2	$B+C+D$	(0, 8)				X			
P_3	$\bar{A}+B+C$	(8, 9)				X	X		
P_4	$\bar{A}+B+D$	(8, 10)				X		X	
P_5	$\bar{A}+C+\bar{D}$	(9, 13)					X		X
P_6	$A+\bar{B}$	(4, 5, 6, 7)	X	X	X				
P_7	$\bar{B}+\bar{D}$	(5, 7, 13, 15)		X					X
Selection of minterm			✓	✓	✓	✓	✓	✓	✓

The Boolean expression is obtained by applying AND operator to prime-implicants. 0 is considered as normal variable and 1 denotes the complement of the variable in prime-implicants.

$$F(A,B,C,D) = (\bar{A}+B+D)(\bar{A}+C+\bar{D})(A+\bar{B})$$

4.9 | MAP-ENTERED VARIABLE METHOD

Karnaugh mapping is the best manual technique for Boolean equation simplification yet when the map size exceed five or six variables it becomes difficult. The technique called map-entered variable (MEVs) increases the effective size of K-map, allowing a smaller map to handle a greater number of variables. Normally K-map dimensions and the number of problem variables are related by $2^n = m$, where n is the number of variables and m gives number of cells/squares in K-map. Using MEV techniques number of variables is reduced by 1 or 2.

The procedure for getting the MEV for MEV K-map is given as below for given truth table (Table 4.45): Let XY literals form minterm for MEV, K-map. Z-literal is used as map-entered variable (MEP).

TABLE 4.45 | Truth table for Z as MEV

Minterms		Binary Variable			Output	MEV
Two-literal K-map	Three-literal K-map	*X*	*Y*	*Z*(MEV)	*F*	K-map
0	0	0	0	0	1	1
0	1	0	0	1	1	
1	2	0	1	0	0	0
1	3	0	1	1	0	
2	4	1	0	0	1	$\bar{Z}$
2	5	1	0	1	0	
3	6	1	1	0	0	Z
3	7	1	1	1	1	

- In case, the output variable, *F* is a 0 for both minterms including (Z and $\bar{Z}$) covered by a MEV K-map square, then a 0 is written in that MEV K-map square

MEV Minterm	Actual Minterm	*X*	*Y*	MEV (*Z*)	*F*	Value Written in MEV K-map
1	2	0	1	0	0	0
1	3	0	1	1	0	

- In case, the output variable is a 1 for both minterms including (Z and $\bar{Z}$) covered by a MEV K-map square, then a 1 is written in that MEV K-map square.

MEV Minterm	Actual Minterm	X	Y	MEV (Z)	F	Value Written in MEV K-map
0	0	0	0	0	1	1
0	1	0	0	1	1	

- In case, the output variable is the same value for minterms covered by a MEV K-map square, as the MEV (Z and $\bar{Z}$) then the MEV (Z) is written into MEV K-map square:

MEV Minterm	Actual Minterm	X	Y	MEV (Z)	F	Value Written in MEV K-map
3	6	1	1	0	0	Z
3	7	1	1	1	1	

- In case, the output variable is the complement value for minterms covered by a MEV K-map square, as the complement of MEV (Z and $\bar{Z}$) then the complement of MEV ($\bar{Z}$) is written into MEV K-map square:

MEV Minterm	Actual Minterm	X	Y	MEV (Z)	F	Value Written in MEV K-map
2	4	1	0	0	1	$\bar{Z}$
2	5	1	0	1	0	

- In case, the output variable is a don't care term, for minterms covered by a MEV K-map square, write a don't care symbol (X) into the MEV K-map.

MEV Minterm	Actual Minterm	X	Y	MEV (Z)	F	Value Written in MEV K-map
2	4	1	0	0	X	X
2	5	1	0	1	X	

- In case, the output variable is a don't care term in one case and a 0 in the other for minterms covered by a MEV K-map square, write 0 in the appropriate square:

MEV Minterm	Actual Minterm	X	Y	MEV (Z)	F	Value Written in MEV K-map
2	4	1	0	0	0	0
2	5	1	0	1	X	

- In case, the output variable is a don't care term in one case and a 1 in the other for minterms covered by a MEV K-map square, write 1 in the appropriate square:

MEV Minterm	Actual Minterm	X	Y	MEV (Z)	F	Value Written in MEV K-map
2	4	1	0	0	1	1
2	5	1	0	1	X	

For a Boolean expression given below, MEV Boolean expression for MEV K-map can be obtained as follows.

Let

$$F(X,Y,Z) = \sum m(0,1,4,7)$$

$$F(X,Y,Z) = \overline{X}\overline{Y}\overline{Z} + \overline{X}\overline{Y}Z + X\overline{Y}\overline{Z} + XYZ$$

$$F(X,Y,Z) = \overline{X}\overline{Y}\left(\overline{Z} + Z\right) + X\overline{Y}\overline{Z} + XYZ$$

$$F(X,Y,Z) = m_0\left(\overline{Z} + Z\right) + m_2\overline{Z} + m_3 Z$$

$$F(X,Y,Z) = m_0 + m_2\overline{Z} + m_3 Z \tag{4.33}$$

where m_0, m_1 and m_3 are the minterms of two variables X and Y.

Truth table for Z as MEV is given in Table 4.45.

To the simplify function from a MEV K-map, following given steps are to be followed:

- Determine the groups consisting of only 1s along with any don't care terms that may exist. In other words group the 1's in the K-map.
- Consider the 1's as don't care terms once step 1 is completed, because all of the 1s have been previously covered.
- Group all identical MEV terms with 1's or don't care terms to maximize the group size. Any minterms that are not contained in the MEV essential prime-implicants are considered, to be 0s. So, cover all the MEVs in the K-map.
- Determine the MEV essential prime-implicants by reading the K-map in the normal fashion. Then AND the MEV variable or expression with remaining map variables.

MEV K-map considering truth table given as Table 4.45 has been set up and is shown in Figure 4.57.

- 1 in m_0 has no grouping so is written in reduced form and gives $\overline{X}\overline{Y}$.
- m_0 and m_2, forms pair with and gives $\overline{Y}\overline{Z}$.
- m_3 has no grouping gives $XY(Z)$.

Use OR operator to combine.

$$F(X,Y,Z) = \overline{X}\overline{Y} + \overline{Y}\overline{Z} + XY(Z)$$

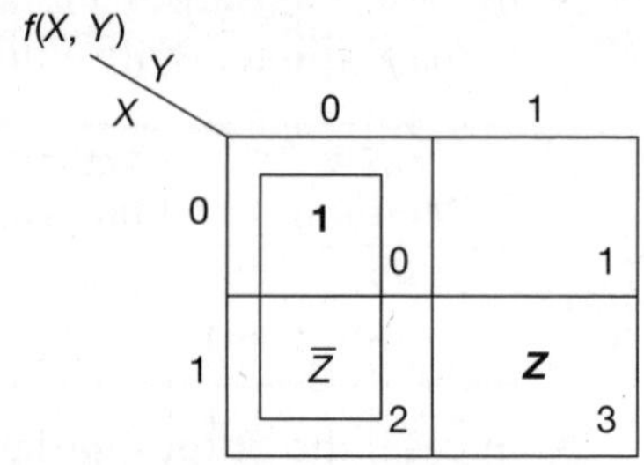

FIGURE 4.57 | MEV K-map

EXAMPLE 4.41 Simplify $F(W,X,Y,Z) = \sum m(2, 4, 5, 10, 11, 14) + \sum d(7, 8, 9, 12, 13, 15)$ using MEV K-map method.

SOLUTION Let Z be the variable to be entered.

$$F(W,X,Y,Z) = \left(\overline{W}\overline{X}Y\overline{Z} + \overline{W}X\overline{Y}\overline{Z} + \overline{W}X\overline{Y}Z + W\overline{X}Y\overline{Z} + W\overline{X}YZ + WXY\overline{Z}\right)$$
$$+ d(\overline{W}XYZ + W\overline{X}\overline{Y}\overline{Z} + W\overline{X}\overline{Y}Z + WX\overline{Y}\overline{Z} + WX\overline{Y}Z + WXYZ)$$

$$F(W,X,Y,Z) = \left(\overline{W}\overline{X}Y\overline{Z} + \overline{W}X\overline{Y}\left(\overline{Z} + Z\right) + W\overline{X}Y\left(\overline{Z} + Z\right) + WXY\overline{Z}\right)$$
$$+ d(\overline{W}XYZ + W\overline{X}\overline{Y}\left(\overline{Z} + Z\right) + WX\overline{Y}\left(\overline{Z} + Z\right) + WXYZ)$$

$$F(W,X,Y,Z) = \left(m_1\bar{Z} + m_2\left(\bar{Z}+Z\right) + m_5\left(\bar{Z}+Z\right) + m_7\bar{Z}\right)$$
$$+ d\left(m_3Z + m_4\left(\bar{Z}+Z\right) + m_6\left(\bar{Z}+Z\right) + m_7Z\right)$$

m_3Z is considered 0 being only don't care term.

m_7Z don't care term is combined with $m_7\bar{Z}$ and treated as 1

$$F(W,X,Y,Z) = \left(m_1\bar{Z} + m_2 + m_5 + m_7(\bar{Z}+Z)\right) + d\left(m_4 + m_6\right)$$

$$F(W,X,Y,Z) = \left(m_1\bar{Z} + m_2 + m_5 + m_7\right) + d\left(m_4 + m_6\right)$$

Same Boolean function can be considered using truth Table 4.46.

TABLE 4.46 | Truth table

Minterms		Binary Variables				Output	
Three-literal K-map	**Four-literal K-map**	***W***	***X***	***Y***	***Z* (MEV)**	***F***	**MEV K-map**
0	0	0	0	0	0	0	0
0	1	0	0	0	1	0	
1	2	0	0	1	0	1	$\bar{Z}$
1	3	0	0	1	1	0	
2	4	0	1	0	0	1	1
2	5	0	1	0	1	1	
3	6	0	1	1	0	0	0
3	7	0	1	1	1	X	
4	8	1	0	0	0	X	X
4	9	1	0	0	1	X	
5	10	1	0	1	0	1	1
5	11	1	0	1	1	1	
6	12	1	1	0	0	X	X
6	13	1	1	0	1	X	
7	14	1	1	1	0	1	1
7	15	1	1	1	1	X	

Set up the three-variable K-map in Figure 4.58.

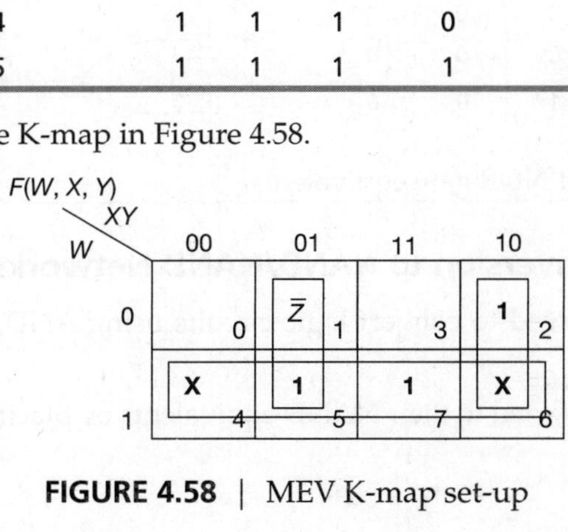

FIGURE 4.58 | MEV K-map set-up

- Quad of m_0, m_5, m_6 and m_7 is formed to get W.
- m_1 and m_5 forms pair with and gives $\bar{X}Y(\bar{Z})$.
- m_2 and m_6 forms pair and gives $X\bar{Y}$.

Use OR operator to combine.

$$F(X,Y,Z) = W + X\bar{Y} + \bar{X}Y(\bar{Z}) \text{ Ans.}$$

4.10 | REALIZATION OF CIRCUIT USING NAND/NOR GATES ONLY

NAND/NOR gates are the universal gates. Universal gates are easier to fabricate with electronic components. Basic gates AND, OR and NOT gates can be designed using universal gates. Only one type of gates is required to realize the logic circuit. SOP logic circuits can be designed using AND, OR and NOT gates known as AOI realization. Whereas POS (POS) logic circuits can be designed using OR, AND and NOT gates. SOP and POS logic circuits can easily be converted using NAND or NOR gates only. De Morgan's theorem can be applied for such conversion. OR gate equivalent to bubbled NAND gate. NAND gate is equivalent to bubbled OR gate (Figure 4.59). AND gate equivalent to bubbled NOR gate. NOR gate is equivalent to bubbled AND gate (Figure 4.60).

A	$\bar{A}$	B	$\bar{B}$	$A+B$	$\overline{\bar{A}\bullet\bar{B}}$	$\bar{A}+\bar{B}$	$\overline{A\bullet B}$
0	1	0	1	0	0	1	1
0	1	1	0	1	1	1	1
1	0	0	1	1	1	1	1
1	0	1	0	1	1	0	0

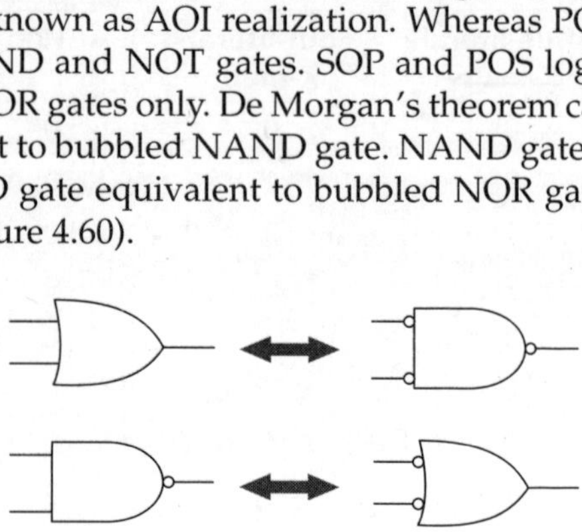

FIGURE 4.59 | OR and NAND gate equivalents

A	$\bar{A}$	B	$\bar{B}$	$A\bullet B$	$\overline{\bar{A}+\bar{B}}$	$\bar{A}\bullet\bar{B}$	$\overline{A+B}$
0	1	0	1	0	0	1	1
0	1	1	0	0	0	0	0
1	0	0	1	0	0	0	0
1	0	1	0	1	1	0	0

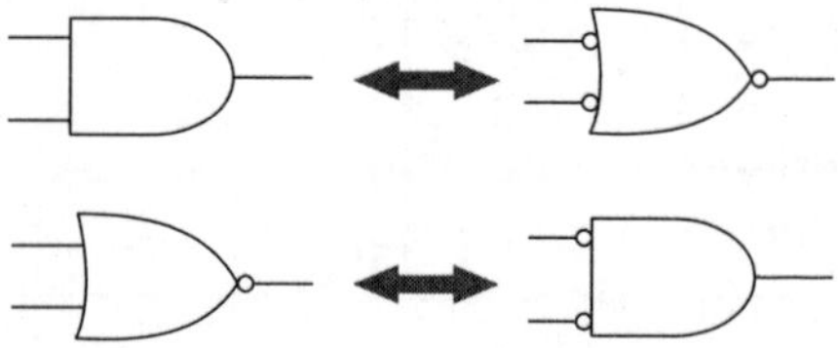

FIGURE 4.60 | AND and NOR gate equivalents

4.10.1 | AND/OR Conversion to NAND/NAND Networks

Following steps are followed to convert logic circuits using AND/OR into logic circuits using NAND gates

- AND gates are converted to their NAND equivalents by placing bubble at the output of gate.

- Complement the inputs to OR gates to conserve the circuit's logic function. In other words Convert OR gate to bubbled OR gate which is equivalent to NAND gate (De Morgan Theorem-I).
- Bubble or inversion is conserved as and when a new bubble or inversion is introduced as input or output of gate. So, it must be balanced by a complementary inversion.
- OR gate with complemented inputs is replaced by NAND gate.

The procedure to implement Logic Circuit using NAND gate is explained in Figures 4.61(a), 4.61(b) and 4.61(c).

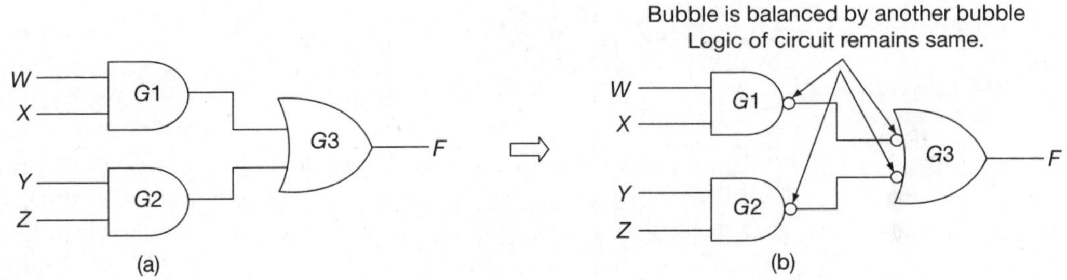

FIGURE 4.61 | (a) Logic circuit using AND-OR gates; (b) Logic circuit

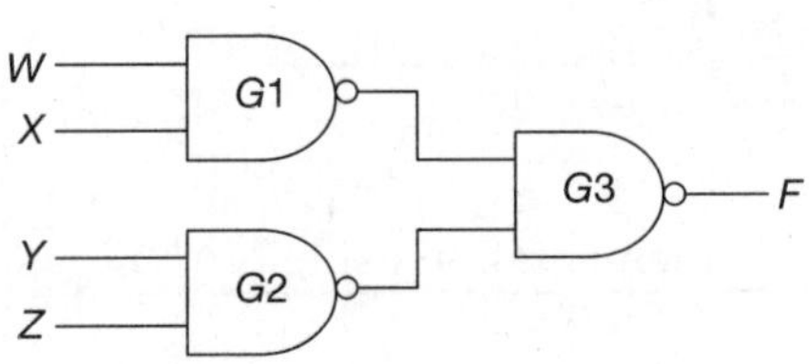

FIGURE 4.61(c) | Logic circuit using NAND gates

Multilevel conversion to NAND gates Multilevel conversion to NAND gates has been implemented for the following logic function $F = A(B + CD) + BC$.

The conversion procedure is explained in Figure 4.62.

4.10.2 | AND/OR Conversion to NOR/NOR Networks

Following steps are followed to convert logic circuits using AND/OR into logic circuits using NOR gates

- Complemented inputs are added at the two AND gates. Convert AND gate to bubbled-AND gate which is equivalent to NOR gate (De Morgan Theorem-II).
- AND gates with complemented inputs are replaced by NOR gates
- OR gate is converted to NOR gate after introducing a matching inverter

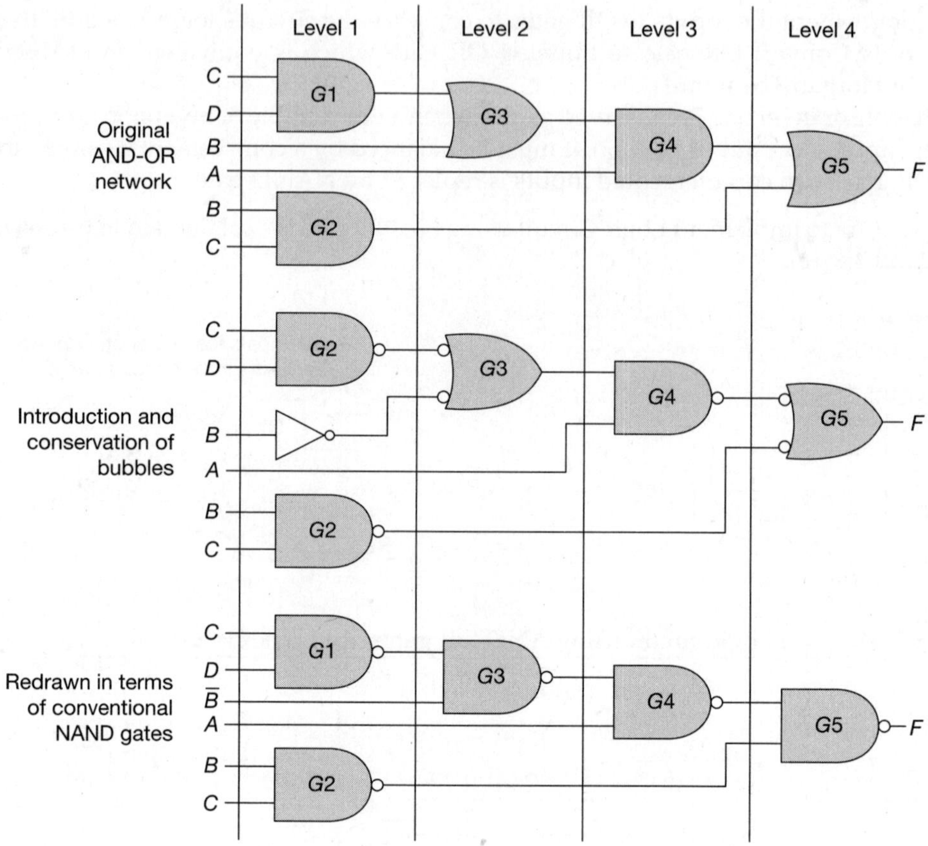

FIGURE 4.62 | Logic circuit implementation using NAND gate

- Bubble or inversion is conserved as and when a new bubble or inversion is introduced as input or output of gate. So, it must be balanced by a complementary inversion

The procedure to implement logic circuit using NOR gate is explained in Figures 4.63(a), 4.63(b) and 4.63(c).

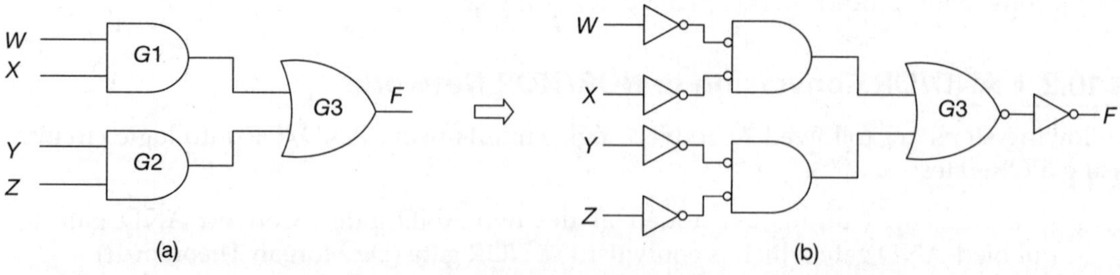

FIGURE 4.63 | (a) Logic circuit using AND-OR gates; (b) Logic circuit

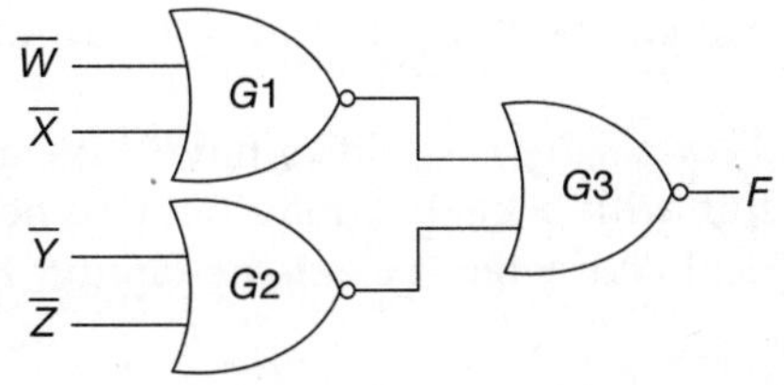

FIGURE 4.63(c) | Logic circuit using NOR gates

Multilevel conversion to NOR gates Multilevel conversion to NOR gate has been implemented for the following logic function $F = A(B+CD)+BC$.

The conversion procedure is explained in Figure 4.64.

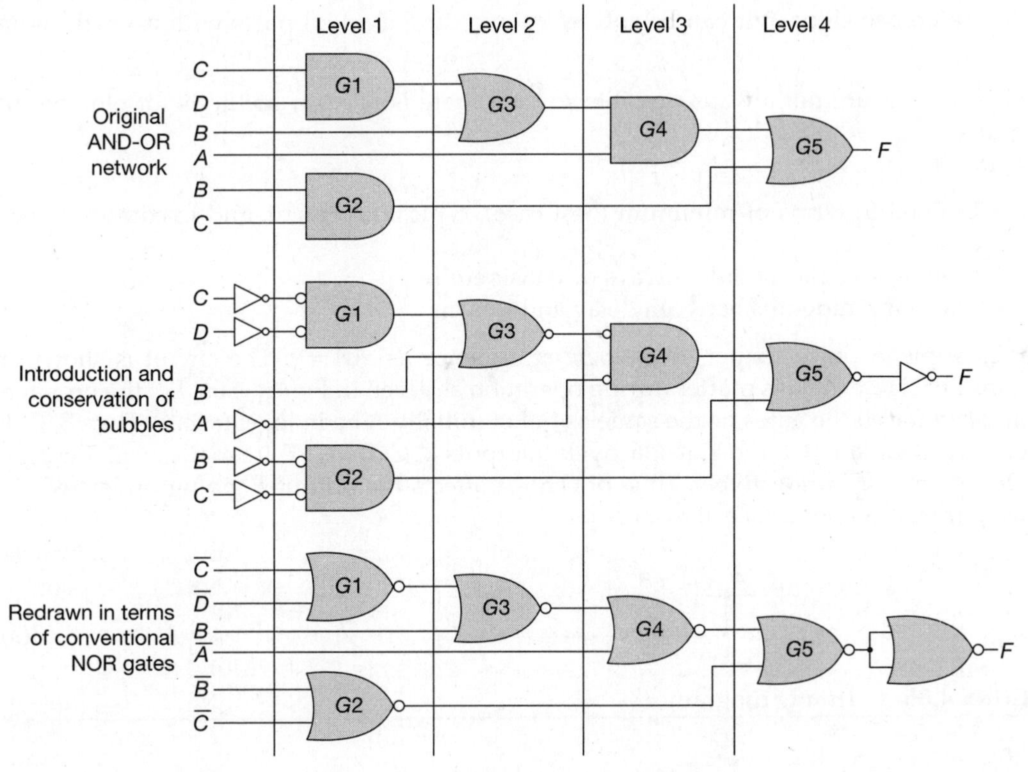

FIGURE 4.64 | Logic circuit implementation using NOR gate

4.11 | HAZARDS

Boolean functions used to provide description of the operation of circuits. Minimization of Boolean expressions is done to design simpler circuits to implement the task in question. Under certain circumstances, a minimal gate implementation of logic function may not be satisfactory solution of a design problem. Time response tells about a circuit's dynamic behaviour.

A glitch or logic-spike is an unwanted pulse at the output of a combinational logic network.

A circuit with the potential for a glitch is said to have a hazard. A hazard is something intrinsic about a circuit. Circuit with a hazard may or may not glitch depending on the input patterns and the electrical characteristics of the circuit. Hazards are a problem for digital systems in two cases

1. Time sensitive logic makes a decision based on the output of a function without allowing the output to settle to a final steady-state value. It can be resolved by increasing the interval between the time when inputs first begin to change and the time when outputs are examined by the decision making logic, i.e. increase the system clock period.
2. Hazardous circuits when connected to a component with asynchronous inputs (inputs take effect as soon as they change rather than when sampled with as reference signal). It can be solved by avoiding clocked parts with asynchronous inputs.

Both solutions are not always possible so then there is need of methods for eliminating hazards.

Gate Delays

- Defined in terms of minimum (best case), typical (average), and maximum (worst case) times.
- Worst-case delay should always be considered.
- There are trade-offs between delay and power.

For example, a pulse shaper gives logic at a glance as $A\bar{A} = 0$. The circuit is shown in Figure 4.65. Here, delays matter during operation as given in Figure 4.66. Let the propagation delay for all the gates be the same as t_p. Let initially A be in logic-0 state, Then B, C, D and F are at logic 1, 0, 1 and 1, respectively. Outputs B, C, D and F are made available time, t_p, $2t_p$, $3t_p$ and $4t_p$, respectively. Time of change affects the output F giving undesired out during transition period, i.e. 0 to 4 t_p.

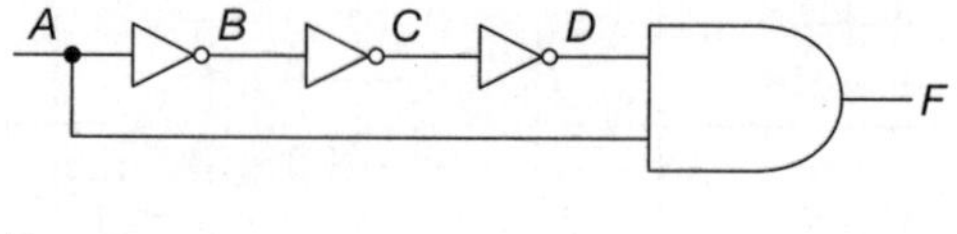

FIGURE 4.65 | Timing diagram

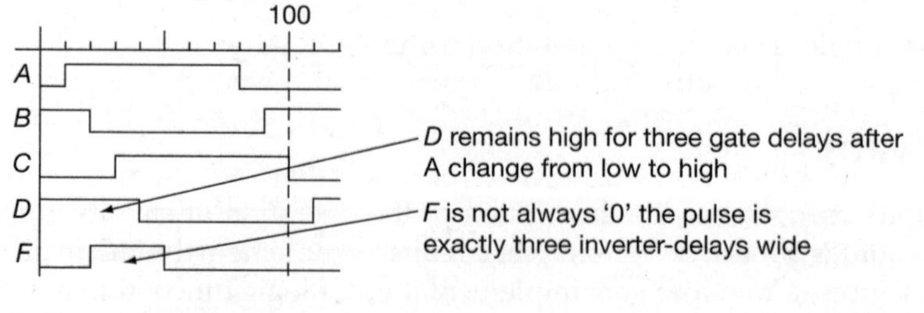

FIGURE 4.66 | Timing diagram

Analysis of a pulse-shaper circuit is given in Figure 4.67. Timing diagram is shown in Figure 4.68 without any hazards. Static input, A controls the change.

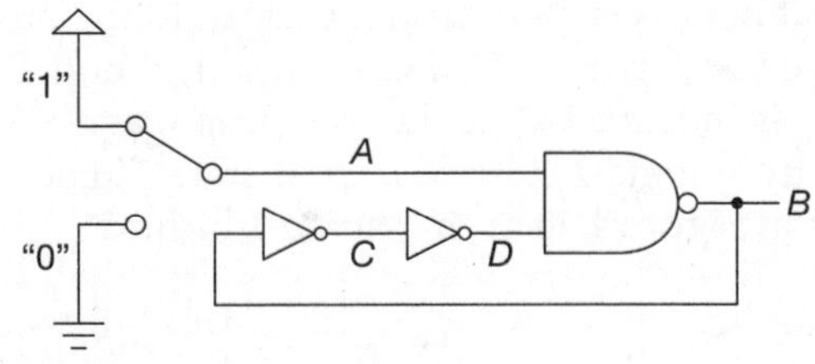

FIGURE 4.67 | Timing diagram

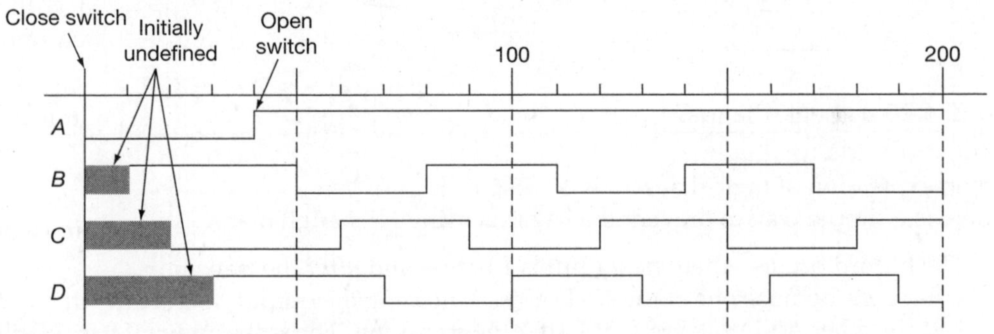

FIGURE 4.68 | Timing diagram

The design of hazard free circuitry is very important. Hazards are classified in two groups (1) Static and (2) dynamic.

4.11.1 | Static Hazards

Static hazards occur when it is possible for an output to undergo a momentary transition when it is expected to remain unchanged.

Static-1 hazard occurs when the output momentarily goes to 0 when it should remain a 1 (Figure 4.69a).

Static-0 hazard occurs when the output momentarily goes to 1 when it should remain a 0 (Figure 4.69b).

Dynamic hazard occurs when the output signal has the potential to change more than once when it is expected to make a single transition form $0 \rightarrow 1$ or $1 \rightarrow 0$. It occurs when there are multiple paths with different delays from the inputs to the outputs (Figure 4.69c).

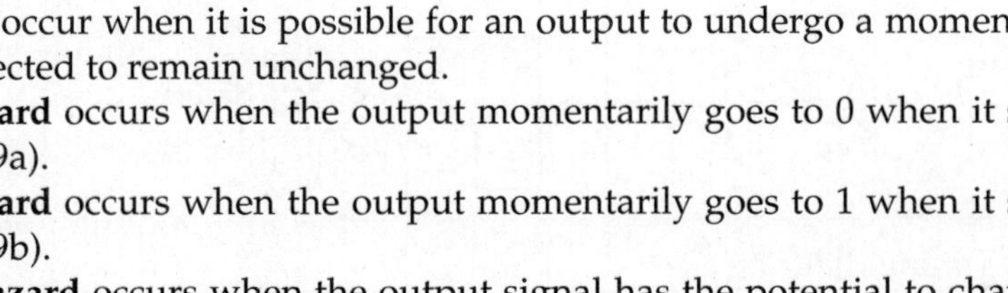

FIGURE 4.69 | Hazard types

4.11.2 | Static-1 Hazards

Without the delay in the gates, the output Z would be constant at 1. The spike is due to the gate delay and the fact that not every path from on input to the output is the same length. Consider logic circuit given in Figure 4.70. Let inputs A, S and B are initially in logic-1, respectively. The output Z is in logic-1 state. Let the propagation delay for all the gates be the same as t_p. S changes from logic-1 state to logic-0 state. Z goes to logic-0 due to delay. Timing diagram is shown in Figure 4.71 giving static-1 hazard.

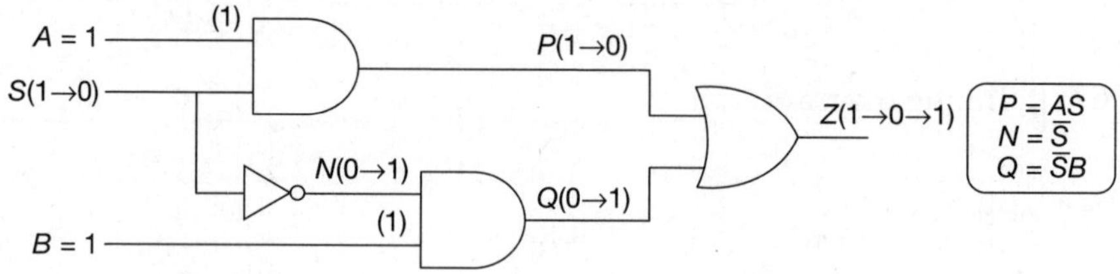

FIGURE 4.70 | Logic diagram

Boolean expression of logic diagram is $Z = SA + \bar{S}B$

There is a simple way to prevent the hazards using Karnaugh maps.

- The hazard occurs when the output is 1 before and after the transition.
- It happens because one of the AND gates is holding the output 1 before the transition and the other AND gate holds it 1 after the transition. The spike occurs if it is possible for the first to turn off before the second turns on.

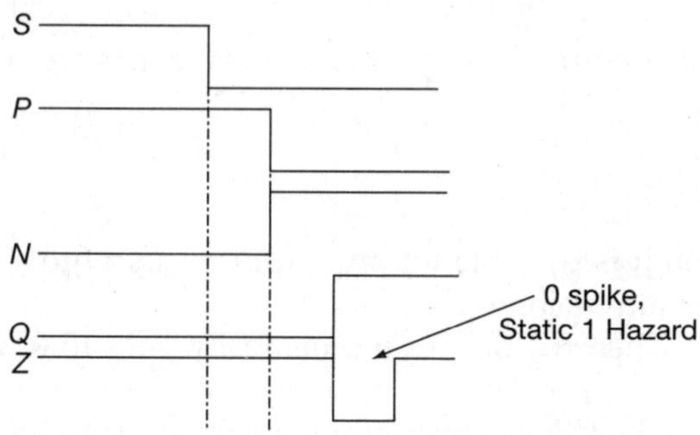

FIGURE 4.71 | Timing diagram

- This can happen anytime there are to adjacent squares on the Karnaugh map that are both 1 and not both covered by a common loop. The hazard occurs in making a transition from one square to the other.
- The way to prevent it is to put in an extra product term that is 1 on both sides of the transition. On the Karnaugh map, a loop covering the two adjacent one squares is formed.
- The loop added to prevent the hazard is redundant and not needed to realize the logical expression. Its only purpose is to prevent the hazard.

- This technique will avoid all static 1 hazards if only one variable is changed at a time.
- It is not possible to get a 1 spike from a 2-level AND–OR network by changing only one variable.

$$Z(S,A,B) = SA\overline{B} + SAB + \overline{S}\,\overline{A}B + \overline{S}AB$$

$$Z = m_6 + m_7 + m_1 + m_3$$

On simplification, Boolean expression is obtained using K-map (Figure 4.72) as $Z = SA + \overline{S}\,B + AB$ that eliminates the hazard. Logic diagram is shown in Figure 4.73.

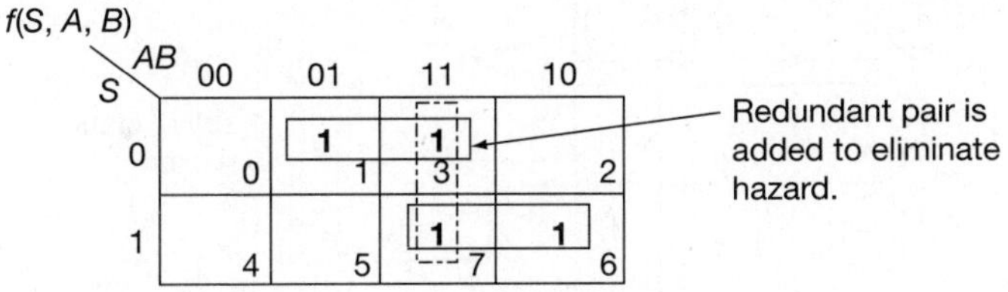

FIGURE 4.72 | 3-variable K-map

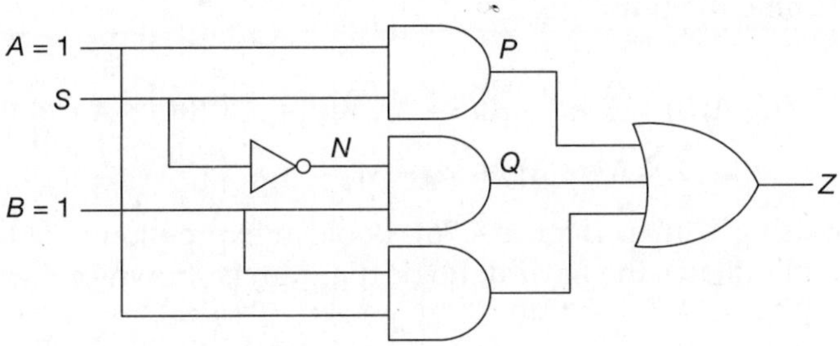

FIGURE 4.73 | Logic diagram

4.11.3 | Static-0 Hazard

Consider logic circuit given in Figure 4.74. Let inputs A, S and B are initially in logic-0, respectively. The output Z is in logic-0 state. Let the propagation delay for all the gates be same as t_p. S changes from logic-0 state to logic-1 state. Z goes to logic-1 due to delay. Timing diagram is shown in Figure 4.75 giving static 0 hazard.

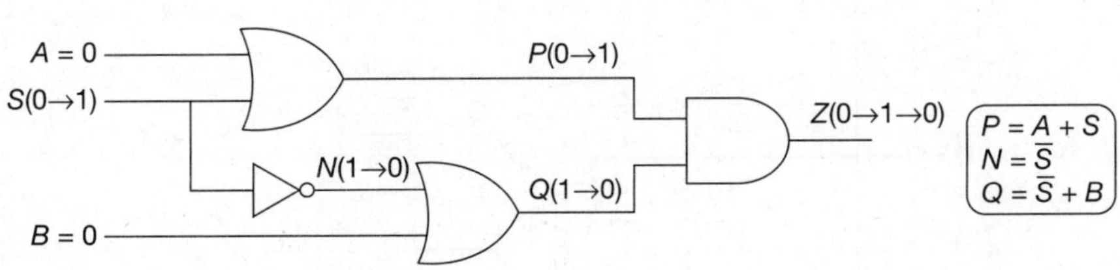

FIGURE 4.74 | Logic diagram

- The static 0 hazard is caused by two OR gates where one holds the output zero before the transition and the other after the transition, and it is possible for them to both be 1 during the transition.
- The hazard is eliminated by putting in an extra OR gate to hold the output zero during the transition.

Boolean expression of logic diagram is $Z = (S + A)(\overline{S} + B)$

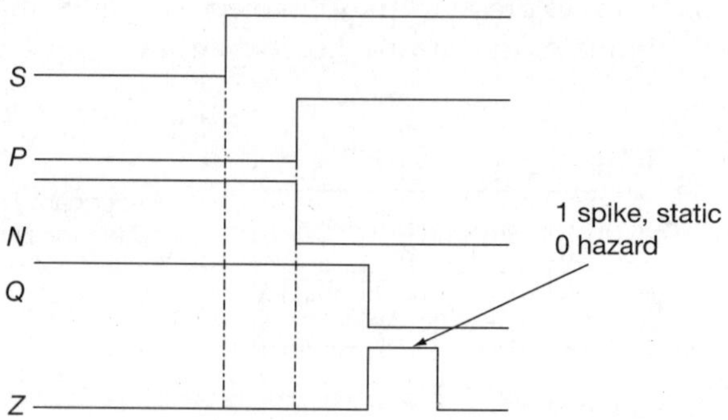

FIGURE 4.75 | Timing diagram

$$Z(S, A, B) = (S + A + \overline{B})(S + A + B)(\overline{S} + \overline{A} + B)(\overline{S} + A + B)$$

$$Z = M_1 + M_0 + M_6 + M_4$$

On simplification using K-map (Figure 4.76), Boolean expression is obtained $Z = (S + A)$ $(\overline{S} + B)(A + B)$ that eliminates the hazard, Logic diagram is shown in Figure 4.77.

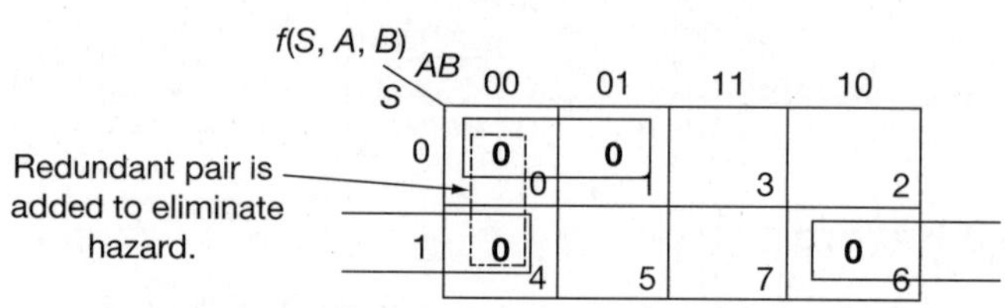

FIGURE 4.76 | Three-variable K-map

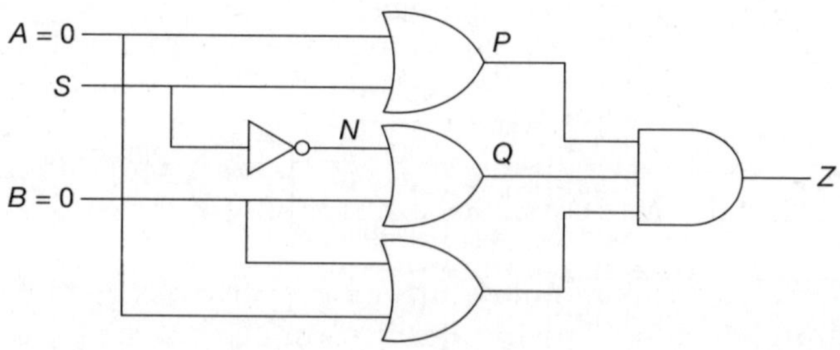

FIGURE 4.77 | Logic diagram

Hazards can frequently be ignored. If the signal Z above is not used during the transition, then a spike is considered don't care.

When a signal is used as the input to a flip-flop, it is only sampled with the clock makes a transition. Therefore, if it is ensured that any spikes occur at some other time, these will not create a problem.

Static hazards are eliminated as follows: It is assumed that only one input changes

1. The function expanded using canonical expansion.
2. Mapped onto a Karnaugh Map.
3. Minterms of formed, to specify the minimal solution.
4. Minterms that do not overlap any other minterm groups are identified as potential hazards.
5. Additional minterms are introduced crossing over the existing minterms so that no minterm group is not overlapped by another group.

4.11.4 | Dynamic Hazard

A dynamic hazard is the possibility to change the output more than once as the result of a single input transition. Multiple transitions occur if there are multiple paths with different delays from changing input to changing output.

Consider a circuit shown in Figure 4.78.

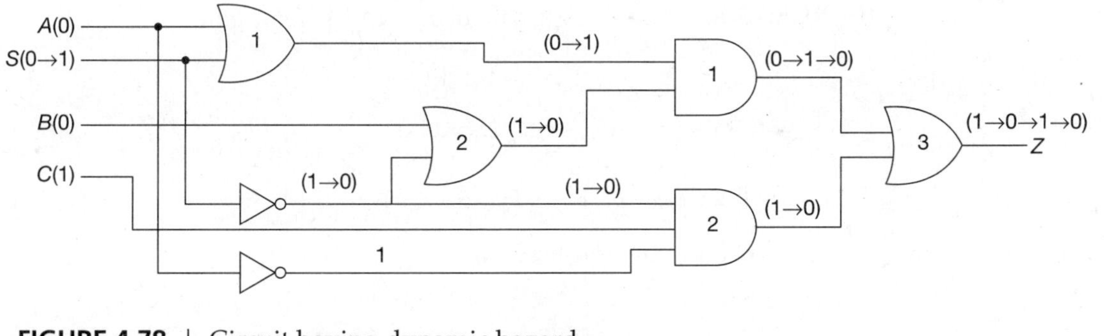

FIGURE 4.78 | Circuit having dynamic hazard

$$Z = (A+S)(\overline{S}+B)+\overline{A}\,\overline{S}C$$

Let inputs of the circuit A, S, B and C are in logic 0, 0, 0 and 1 states, respectively, and then output Z is in logic-1. Let propagation delay is same for types of gates. AND gate numbered 1 gives output after 3 times propagation delay, t_p whereas AND gate numbered 2 gives output after two times propagation delay, t_p. For gate 3, inputs are available at two different time interval causes dynamic hazard as shown the variation at the output of the gate.

Generally, dynamic hazard does not occur in a properly designed two level AND-OR or OR-AND circuit in which no variable and its complement are connected to the same first-level.

EXAMPLE 4.42 Implement basic logic functions using NAND gates.

SOLUTION

(i) *NOT gate:*

$$\overline{x.x} = \overline{x} \qquad (x.x = x) \tag{4.34a}$$

The circuit diagram is given in Figure 4.79(a)

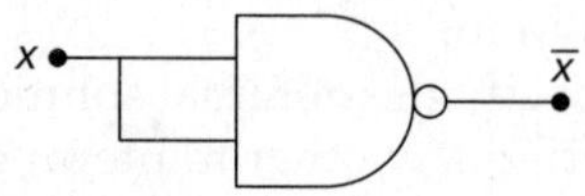

FIGURE 4.79(a) | NOT logic gate using NAND logic gate

(ii) *AND gate:*

$$xy = \overline{\overline{xy}} \qquad (\overline{\overline{x}} = x) \tag{4.34b}$$

The circuit diagram is given in Figure 4.79(b)

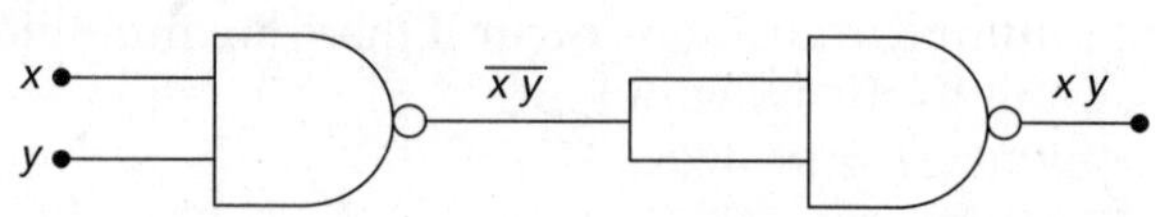

FIGURE 4.79(b) | AND logic gate using NAND logic gate

(iii) *OR gate:*

$$x + y = \overline{\overline{x + y}} \qquad (\overline{\overline{x}} = x)$$

$$x + y = \overline{\overline{x}\,\overline{y}} \qquad (\overline{x + y} = \overline{x}\,\overline{y}) \tag{4.34c}$$

The circuit diagram is given in Figure 4.79(c)

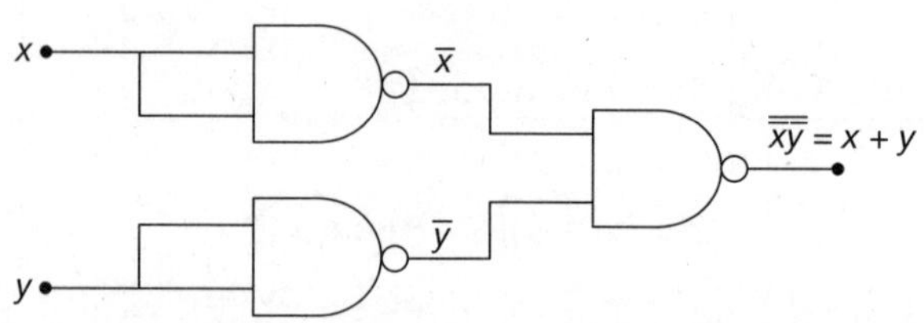

FIGURE 4.79(c) | OR logic gate using NAND logic gate

(iv) *NOR gate:*

$$x + y = \overline{\overline{x + y}} \qquad (\overline{\overline{x}} = x)$$

$$x + y = \overline{\overline{x}\,\overline{y}} \qquad (\overline{x + y} = \overline{x}\,\overline{y})$$

$$\overline{x + y} = \overline{\overline{x}\,\overline{y}} \tag{4.34d}$$

The circuit diagram is given in Figure 4.79(d)

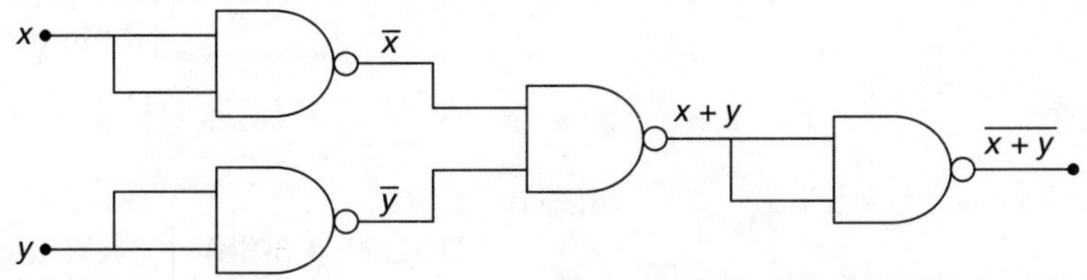

FIGURE 4.79(d) | NOR logic gate using NAND logic gate

(v) *XOR gate:*

$$x \oplus y = x\overline{y} + \overline{x}y \qquad (\because x + 0 = x \text{ and } x\overline{x} = 0)$$

$$x \oplus y = x\overline{x} + x\overline{y} + \overline{x}y + y\overline{y}$$

$$x \oplus y = x(\overline{x} + \overline{y}) + y(\overline{x} + \overline{y}) \qquad (\because \overline{x} + \overline{y} = \overline{xy}\text{, De Morgan's theorem})$$

$$x \oplus y = x(\overline{xy}) + y(\overline{xy})$$

$$x \oplus y = \overline{\overline{x(\overline{xy}) + y(\overline{xy})}} \qquad (\because \overline{\overline{x}} = x)$$

$$x \oplus y = \overline{\left(\overline{x(\overline{xy})}\right)\left(\overline{y(\overline{xy})}\right)} \qquad (\because \overline{x + y} = \overline{x}\,\overline{y}) \qquad (4.34e)$$

The circuit diagram is given in Figure 4.79(e)

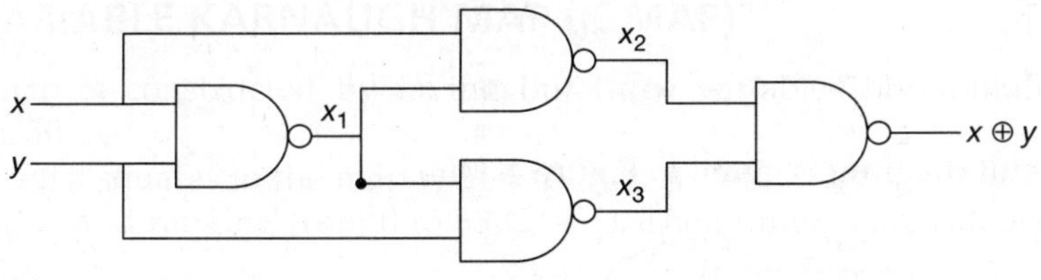

FIGURE 4.79(e) | XOR gate using NAND gates ($X_2 = \overline{xX_1}$, $X_3 = \overline{yX_1}$ and $X_1 = \overline{xy}$)

(vi) *XNOR gate:*

$$x \odot y = \overline{x \oplus y} \qquad (4.34f)$$

Complement of Eq. (4.33e), gives the XNOR gate
The circuit diagram is given in Figure 4.79(f)

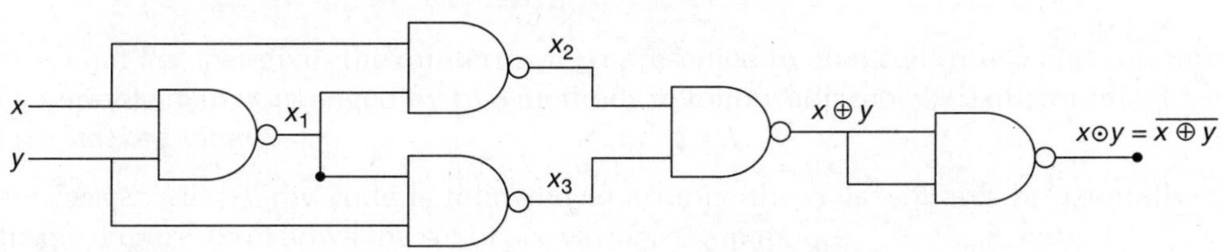

FIGURE 4.79(f) | X-NOR gate using NAND gates ($X_2 = \overline{xX_1}$, $X_3 = \overline{yX_1}$ and $X_1 = \overline{xy}$)

EXAMPLE 4.43 Implement basic logic functions using NOR gates.

SOLUTION

(i) *NOT gate:*

$$\overline{x+x}=\overline{x} \qquad (x+x=x) \qquad (4.35a)$$

The circuit diagram is given in Figure 4.80(a).

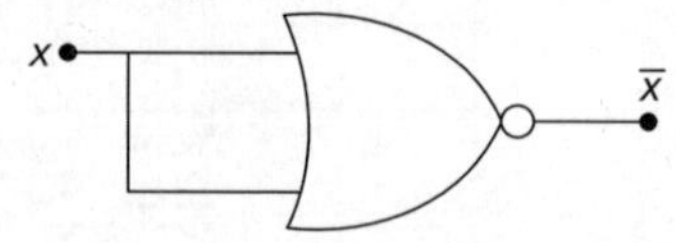

FIGURE 4.80(a) | NOT logic gate using NOR logic gate

(ii) *OR gate:*

$$x+y=\overline{\overline{x+y}} \qquad \left(\overline{\overline{x}}=x\right) \qquad (4.35b)$$

The circuit diagram is given in Figure 4.80(b).

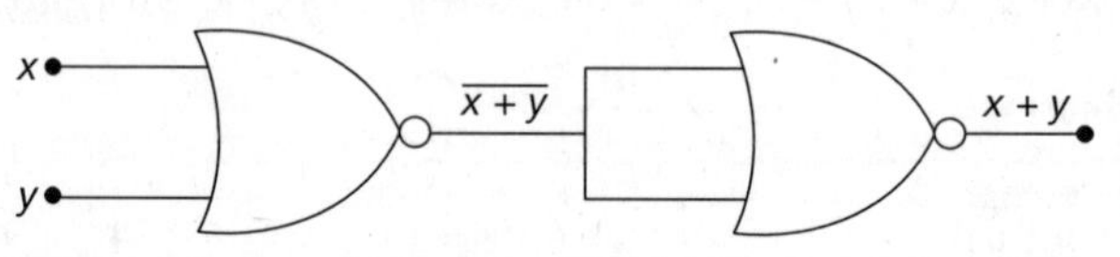

FIGURE 4.80(b) | OR logic gate using NOR logic gates

(iii) *AND gate:*

$$xy=\overline{\overline{xy}} \qquad \left(\overline{\overline{x}}=x\right)$$

$$xy=\overline{\overline{x}+\overline{y}} \qquad \left(\overline{xy}=\overline{x}+\overline{y}\right) \qquad (4.35c)$$

The circuit diagram is given in Figure 4.80(c).

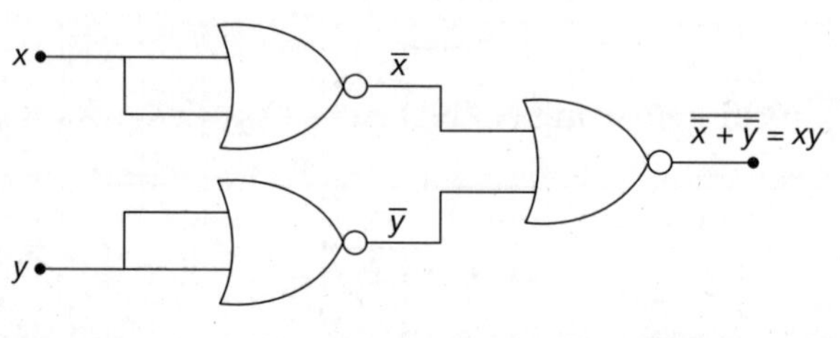

FIGURE 4.80(c) | AND logic gate from NOR logic gates

(iv) *NAND gate:*

$$xy=\overline{\overline{xy}} \qquad \left(\overline{\overline{x}}=x\right)$$

$$xy=\overline{\overline{x}+\overline{y}} \qquad \left(\overline{xy}=\overline{x}+\overline{y}\right)$$

$$\overline{x.y}=\overline{\overline{\overline{x}+\overline{y}}} \qquad (4.35d)$$

The circuit diagram is given in Figure 4.80(d).

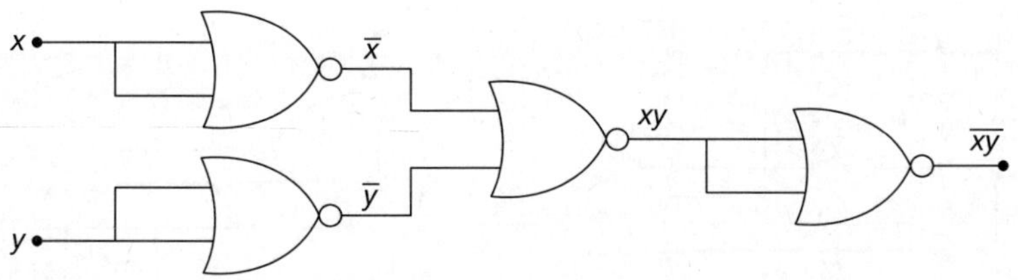

FIGURE 4.80(d) | NAND logic gate from NOR logic gates

(v) *XOR gate:*

$$x \oplus y = x\overline{y} + \overline{x}y$$

$$x \oplus y = x\overline{x} + x\overline{y} + \overline{x}y + y\overline{y} \qquad (\because x + 0 = x \text{ and } x\overline{x} = 0)$$

$$x \oplus y = \overline{x}(x+y) + \overline{y}(x+y)$$

$$x \oplus y = (x+y)(\overline{x}+\overline{y})$$

$$x \oplus y = \overline{\overline{(x+y)(\overline{x}+\overline{y})}}$$

$$x \oplus y = \overline{\overline{(x+y)} + \overline{(\overline{x}+\overline{y})}} \qquad (\because \overline{xy} = \overline{x}+\overline{y}) \qquad (4.35e)$$

The combinational circuit diagram is given in Figure 4.80(e).

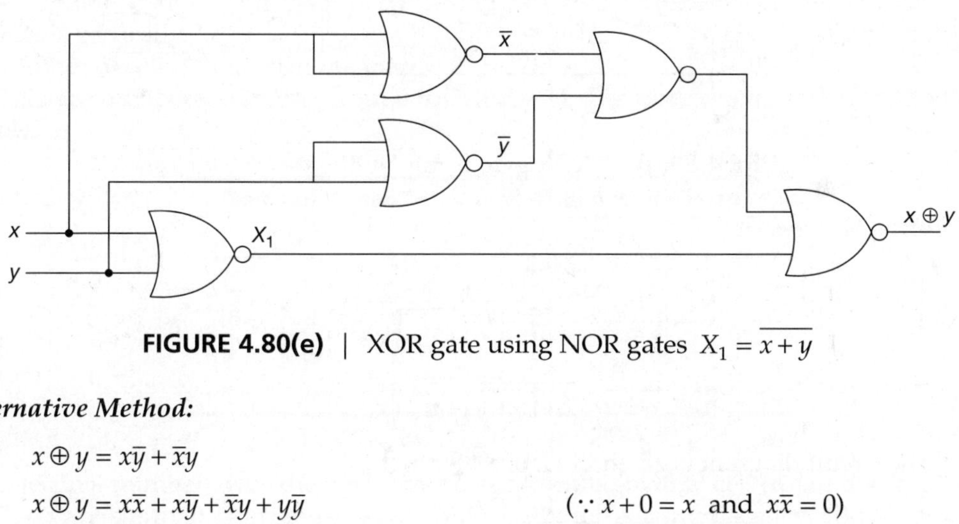

FIGURE 4.80(e) | XOR gate using NOR gates $X_1 = \overline{x+y}$

Alternative Method:

$$x \oplus y = x\overline{y} + \overline{x}y$$

$$x \oplus y = x\overline{x} + x\overline{y} + \overline{x}y + y\overline{y} \qquad (\because x + 0 = x \text{ and } x\overline{x} = 0)$$

$$x \oplus y = \overline{x}(x+y) + \overline{y}(x+y)$$

$$x \oplus y = \overline{x + \left(\overline{x+y}\right)} + \overline{y + \left(\overline{x+y}\right)}$$

$$x \oplus y = \overline{\overline{\overline{x + \left(\overline{x+y}\right)} + \overline{y + \left(\overline{x+y}\right)}}} \qquad (4.35f)$$

The combinational circuit diagram is given in Figure 4.80(f).

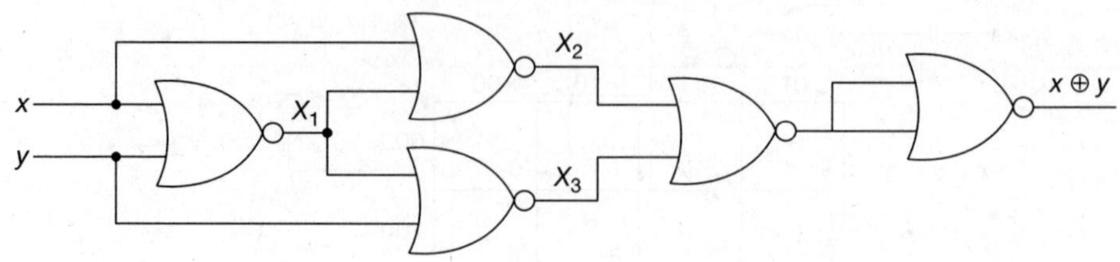

FIGURE 4.80(f) | XOR gate using NOR gates $X_1 = \overline{x+y}$, $X_2 = \overline{x+X_1}$ and $X_3 = \overline{y+X_1}$

(vi) *XNOR gate:*

$$x \odot y = \overline{x \oplus y}$$

$$\overline{x \oplus y} = \overline{\overline{\overline{(x+y)}+\overline{(\bar{x}+\bar{y})}}} \qquad (4.35g)$$

Complement of Eq. (4.35e), gives the X-NOR gate
The circuit diagram is given in Figure 4.80(g).

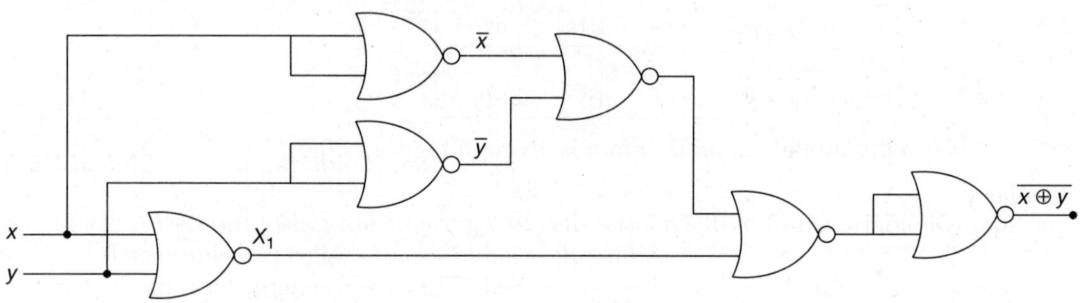

FIGURE 4.80(g) | XOR gate using NOR gates $X_1 = \overline{x+y}$

Alternative Method:

$$x \odot y = \overline{x \oplus y}$$

$$\overline{x \oplus y} = \overline{\overline{\overline{\overline{x+\overline{(x+y)}}+\overline{y+\overline{(x+y)}}}}}$$

$$\overline{x \oplus y} = \overline{\overline{x+\overline{(x+y)}}+\overline{y+\overline{(x+y)}}}$$

The circuit diagram is given in Figure 4.80(h).

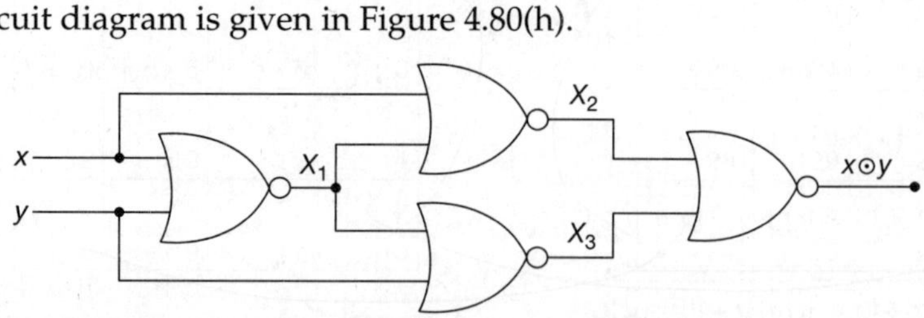

FIGURE 4.80(h) | XOR gate using NOR gates $X_1 = \overline{x+y}$, $X_2 = \overline{x+X_1}$ and $X_3 = \overline{y+X_1}$

EXAMPLE 4.44 Prove the following identities:

(a) $\overline{A \oplus B} = A \odot B$

(b) $A \odot B = \overline{A \oplus B} = \bar{A} \oplus B = A \oplus \bar{B}$

(c) $A(B \oplus C) = AB \oplus AC$

(d) $(A \oplus B \oplus AB) = A + B = A \oplus \bar{A}B$

SOLUTION

(a) $\overline{A \oplus B} = \overline{\bar{A}B + A\bar{B}}$

$= \left(\overline{\bar{A}B}\right)\left(\overline{A\bar{B}}\right)$

$= \left(A + \bar{B}\right)\left(\bar{A} + B\right)$

$= AB + \bar{A}\bar{B}$ (Consensus theorem)

$$\overline{A \oplus B} = A \odot B \tag{4.36a}$$

(b) $\bar{A} \oplus B = \bar{A}\bar{B} + \bar{\bar{A}}B$

$= \bar{A}\bar{B} + AB$

$$= A \odot B \tag{4.36b}$$

Similarly

$A \oplus \bar{B} = A\bar{\bar{B}} + \bar{A}\bar{B}$

$= AB + \bar{A}\bar{B}$

$$= A \odot B \tag{4.36c}$$

Compare Eqs. (4.36a), (4.36b) and (4.36c) to get

$\overline{A \oplus B} = A \odot B = \bar{A} \oplus B = A \oplus \bar{B}$ Proved

(c) LHS $= A(B \oplus C)$

$= A\left(B\bar{C} + \bar{B}C\right)$

$$= AB\bar{C} + A\bar{B}C \tag{4.36d}$$

RHS $= AB \oplus AC$

$= (AB)\left(\overline{AC}\right) + \left(\overline{AB}\right)(AC)$

$= (AB)\left(\bar{A} + \bar{C}\right) + \left(\bar{A} + \bar{B}\right)(AC)$

$= \left(AB\bar{A}\right) + \left(AB\bar{C}\right) + \left(\bar{A}AC + AC\bar{B}\right)$

$$= AB\bar{C} + AC\bar{B} \tag{4.36e}$$

LHS = RHS. Hence proved.

(d) $(A \oplus B \oplus AB) = A \oplus (B \oplus AB)$

$$= A \oplus \left(B\left(\overline{AB}\right) + \bar{B}(AB)\right)$$

$$= A \oplus \left(B\left(\bar{A} + \bar{B}\right) + A\bar{B}B\right)$$

$$= A \oplus \left(\bar{A}B + B\bar{B}\right)$$

$$= A \oplus \bar{A}B \tag{4.36f}$$

$$A \oplus \bar{A}B = A\left(\overline{\bar{A}B}\right) + \bar{A}\left(\bar{A}B\right)$$

$$= A\left(A + \bar{B}\right) + \bar{A}\left(\bar{A}B\right)$$

$$= A + A\bar{B} + \bar{A}B$$

$$= A\left(1 + \bar{B}\right) + \bar{A}B$$

$$= A + \bar{A}B$$

$$= \left(A + \bar{A}\right)(A + B)$$

$$= (A + B) \tag{4.36g}$$

EXAMPLE 4.45 Simplify applying De-Morgan's theorem, the following Boolean expressions:

(a) $\overline{\bar{A}B\bar{C}}$ (b) $\overline{\bar{A} + \bar{B}C}$ (c) $\overline{AB\overline{CD}}$

(d) $\overline{A\left(\overline{\bar{B} + C}\right)D}$ (e) $\overline{\left(M + \bar{N}\right)\left(\bar{M} + N\right)}$

SOLUTION

(a) $\overline{\bar{A}B\bar{C}} = \bar{\bar{A}} + \bar{B} + \bar{\bar{C}}$ $\left(\because \overline{AB} = \bar{A} + \bar{B}\right)$

$= A + \bar{B} + C$ **Ans.** $\left(\because \bar{\bar{A}} = A\right)$

(b) $\overline{\bar{A} + \bar{B}C} = \bar{\bar{A}} \cdot \overline{\bar{B}C}$ $(\because \overline{A + B} = \bar{A} \cdot \bar{B})$

$= A\left(\bar{\bar{B}} + \bar{C}\right)$ $(\because \bar{\bar{A}} = A)$

$= A\left(B + \bar{C}\right)$ $(\because \bar{\bar{B}} = B)$

$= AB + A\bar{C}$ **Ans.**

(c) $\overline{AB\overline{CD}} = \bar{A} + \bar{B} + \overline{\overline{CD}}$ $\quad\left(\because \overline{AB} = \bar{A} + \bar{B}\right)$

$= \bar{A} + \bar{B} + CD$ **Ans.** $\quad(\because \bar{\bar{A}} = A)$

(d) $\overline{A\left(\overline{B+C}\right)D} = \overline{A\left(\bar{B}\bar{C}\right)D}$ $\quad\left(\because \overline{A+B} = \bar{A}\cdot\bar{B}\right)$

$= \bar{A} + \bar{\bar{B}} + \bar{\bar{C}} + \bar{D}$ $\quad\left(\because \overline{AB} = \bar{A} + \bar{B}\right)$

$= \bar{A} + B + C + \bar{D}$ **Ans.**

(e) $\overline{\left(M+\bar{N}\right)\left(\bar{M}+N\right)} = (\overline{M+\bar{N}}) + (\overline{\bar{M}+N})$ $\quad\left(\because \overline{MN} = \bar{M} + \bar{N}\right)$

$= \left(\bar{M}\cdot\bar{\bar{N}}\right) + \left(\bar{\bar{M}}\cdot\bar{N}\right)$ $\quad\left(\because \overline{M+N} = \bar{M}\cdot\bar{N}\right)$

$= \bar{M}N + M\bar{N}$

$= M \oplus N$ **Ans.**

EXAMPLE 4.46 Simplify the following Boolean expressions:

(a) $\bar{A}\bar{B}C + \left(\overline{A+B+C}\right) + \bar{A}\bar{B}\bar{C}D$ (b) $(B+BC)\left(B+\bar{B}C\right)(B+D)$

(c) $ABCD + AB\left(\overline{CD}\right) + \overline{AB}CD$ (d) $ABC\left[AB+\bar{C}(BC+AC)\right]$

SOLUTION

(a) $\bar{A}\bar{B}C + \left(\overline{A+B+C}\right) + \bar{A}\bar{B}\bar{C}D$

$= \bar{A}\bar{B}C + \bar{A}\bar{B}C + \bar{A}\bar{B}C$ $\quad\left(\because \overline{A+B} = \bar{A}\cdot\bar{B}\right)$

$= \bar{A}\bar{B}\left(C + C + \bar{C}D\right)$

$= \bar{A}\bar{B}\left(C + \bar{C}D\right)$ $\quad(\because C + C = C)$

$= \bar{A}\bar{B}\left(C+\bar{C}\right)(C+D)$ $\quad\left(\because C + \bar{C}D = (C+\bar{C})(C+D)\right)$

$= \bar{A}\bar{B}(C+D)$ $\quad\left(\because C + \bar{C} = 1\right)$

$= \bar{A}\bar{B}C + \bar{A}\bar{B}D$ **Ans.**

(b) $(B+BC)\left(B+\bar{B}C\right)(B+D)$

$= B(1+C)\left[\left(B+\bar{B}\right)(B+C)\right](B+D)$ $\quad\left(\because B + \bar{B}C = \left(B+\bar{B}\right)(B+C)\right)$

$= B(B+C)(B+D)$ $\quad\left(\because 1+C = 1, B+\bar{B} = 1, 1\cdot B = B, 1\cdot(B+C) = B+C\right)$

$= B(B+CD)$ $\qquad (\because (B+C)(B+D) = B+CD)$

$= BB+BCD$ $\qquad (\because BB = B)$

$= B+BCD$

$= B(1+CD)$

$= B$ **Ans.**

(c) $ABCD+AB(\overline{CD})+\overline{AB}CD$

$= AB(CD+\overline{CD})+\overline{AB}CD$ $\qquad (\because CD+\overline{CD}=1)$

$= AB+\overline{AB}CD$ $\qquad (\because A+\bar{A}B=(A+\bar{A})(A+B))$

$= (AB+\overline{AB})(AB+CD)$ $\qquad (\because AB+\overline{AB}=1)$

$= AB+CD$ **Ans.**

(d) $ABC[AB+\bar{C}(BC+AC)]$

$= ABC[AB+\bar{C}C(B+A)]$

$= ABC[AB+0\cdot(B+A)]$ $\qquad (\because \bar{C}C=0)$

$= ABC[AB+0]$ $\qquad (\because 0\cdot A=0)$

$= ABC[AB]$ $\qquad (\because 0+A=A)$

$= ABCAB$

$= ABC$ **Ans.** $\qquad (\because AA=A, BB=B)$

EXAMPLE 4.47 Prove $\overline{\overline{A(\bar{B}+BC)}+A(B+A)}=0.$

SOLUTION LHS $= \overline{\overline{A(\bar{B}+BC)}+AB+AA}$ $\qquad (\because \bar{B}+BC=\bar{B}+C)$

$= \overline{\bar{A}+\overline{\bar{B}+C}+A(1+B)}$ $\qquad (\because AA=A, \overline{AB}=\bar{A}+\bar{B})$

$= \overline{\bar{A}+B\bar{C}+A}$ $\qquad (\because 1+B=1)$

$= \bar{\bar{A}}\cdot\overline{B\bar{C}}\cdot\bar{A}$ $\qquad (\because \overline{A+B}=\bar{A}+\bar{B})$

$= A\cdot\bar{A}\cdot\overline{B\bar{C}}$ $\qquad (\because A\cdot\bar{A}=0)$

$= 0\cdot\overline{B\bar{C}}$ $\qquad (\because 0\cdot A=0)$

$= 0 =$ RHS

EXAMPLE 4.48 Given $A\bar{B}+\bar{A}B=C$

Prove $A\bar{C}+\bar{A}C=B$.

SOLUTION Given $A\bar{B}+\bar{A}B=C$

Substitute in the expression $A\bar{C}+\bar{A}C$

$$A\bar{C}+\bar{A}C = A\overline{(A\bar{B}+\bar{A}B} + \bar{A}\left(A\bar{B}+\bar{A}B\right)$$

$$= A\left(AB+\bar{A}\bar{B}\right) + \bar{A}A\bar{B}+\bar{A}\bar{A}B) \qquad (\because \overline{A\oplus B}=A\odot B,\ \overline{A\bar{B}+\bar{A}B}=AB+\bar{A}\bar{B})$$

$$= AAB+A\bar{A}\bar{B} + 0 + \bar{A}B \qquad (\because A\bar{A}=0, \bar{A}\bar{A}=\bar{A})$$

$$= AB+0+\bar{A}B \qquad (\because 0+A=A)$$

$$= AB+\bar{A}B$$

$$= (A+\bar{A})B \qquad (\because A+\bar{A}=1)$$

$$= 1\cdot B$$

$= B$ Hence proved.

EXAMPLE 4.49 Convert the canonical form to standard SOP (Minterms).

(a) $f(A,B,C)=A+\bar{B}\bar{C}$

(b) $f(A,B,C,D)=ABC+AB+DC+D$

SOLUTION

(a) $f(A,B,C)=A+\bar{B}\bar{C}$

$$A = AXX \qquad \text{(let } A=1\text{)}$$

$$= 1XX$$

$$= 100+101+110+111$$

$$= m_4+m_5+m_6+m_7$$

$$\bar{B}\bar{C} = X\bar{B}\bar{C} \qquad \left(\text{let } \bar{B}=0, \bar{C}=0\right)$$

$$= X00$$

$$= 000+100$$

$$= m_0+m_4$$

$$f(A,B,C) = (m_4+m_5+m_6+m_7)+(m_0+m_4)$$

$$= m_0+m_4+m_5+m_6+m_7$$

$= \sum m(0,4,5,6,7)$ **Ans.**

$= \bar{A}\bar{B}\bar{C}+A\bar{B}\bar{C}+A\bar{B}C+AB\bar{C}+ABC$ **Ans.**

(b) $f(A,B,C,D) = ABC + AB + DC + D$

$$ABC = ABCX$$
$$= 111X \quad (\text{let } A = 1, B = 1, C = 1)$$
$$= 1110 + 1111$$
$$= m_{14} + m_{15}$$
$$AB = ABXX$$
$$= 11XX$$
$$= 1100 + 1101 + 1110 + 1111$$
$$= m_{12} + m_{13} + m_{14} + m_{15}$$
$$DC = XXCD \quad (\text{let } C = 1, D = 1)$$
$$= XX11$$
$$= 0011 + 0111 + 1011 + 1111$$
$$= m_3 + m_7 + m_{11} + m_{15}$$
$$D = XXXD \quad (\text{let } D = 1)$$
$$= 0001 + 0011 + 0101 + 0111 + 1001 + 1011 + 1101 + 1111$$
$$= m_1 + m_3 + m_5 + m_7 + m_9 + m_{11} + m_{13} + m_{15}$$
$$f(A,B,C,D) = (m_{14} + m_{15}) + (m_{12} + m_{13} + m_{14} + m_{15}) + (m_3 + m_7 + m_{11} + m_{15})$$
$$+ (m_1 + m_3 + m_5 + m_7 + m_9 + m_{11} + m_{13} + m_{15})$$

Remove duplicate terms.

$$f(A,B,C,D) = m_1 + m_3 + m_5 + m_7 + m_9 + m_{11} + m_{12} + m_{13} + m_{14} + m_{15}$$
$$= \sum m(1, 3, 5, 7, 9, 11, 12, 13, 14, 15) \quad \textbf{Ans.}$$
$$f = \bar{A}\bar{B}\bar{C}D + \bar{A}\bar{B}CD + \bar{A}B\bar{C}D + \bar{A}BCD + A\bar{B}\bar{C}D + A\bar{B}CD$$
$$+ AB\bar{C}\bar{D} + AB\bar{C}D + ABC\bar{D} + ABCD \quad \textbf{Ans.}$$

EXAMPLE 4.50 Convert the canonical form to standard POS form (maxterm) for the following:

(a) $f(A,B,C,D) = (A + \bar{B})(\bar{A} + D)$

(b) $f(A,B,C,D) = (A + B + \bar{D})(\bar{A} + C + D)$

SOLUTION

(a) $f(A,B,C,D) = (A + \bar{B})(\bar{A} + D)$

$$A + \bar{B} = A\bar{B}XX \quad (\text{let } A = 0, \bar{B} = 1)$$
$$= 0100$$

$$= (0100)(0101)(0110)(0111)$$
$$= M_4 \cdot M_5 \cdot M_6 \cdot M_7$$

$$\bar{A} + D = \bar{A}XXD$$
$$= 1XX0 \qquad \left(\text{let } \bar{A} = 1,\ D = 0\right)$$
$$= (1000)(1010)(1100)(1110)$$
$$= M_8 \cdot M_{10} \cdot M_{12} \cdot M_{14}$$

$$\left(A+\bar{B}\right)\left(\bar{A}+D\right) = \left(M_4 \cdot M_5 \cdot M_6 \cdot M_7\right)\left(M_8 \cdot M_{10} \cdot M_{12} \cdot M_{14}\right)$$
$$= \prod M(4,5,6,7,8,10,12,14) \quad \textbf{Ans.}$$

(b) $f(A,B,C,D) = \left(A+B+\bar{D}\right)\left(\bar{A}+C+D\right)$

$$A + B + \bar{D} = ABX\bar{D}$$
$$= 00X1 \qquad \left(\text{let } A = 0,\ B = 0,\ \bar{D} = 1\right)$$
$$= (0001)(0011)$$
$$= M_0 \cdot M_3$$

$$\bar{A} + C + D = \bar{A}XCD$$
$$= 1X00 \qquad \left(\text{let } \bar{A} = 1,\ C = 0,\ D = 0\right)$$
$$= (1000)(1100)$$
$$= M_8 \cdot M_{12}$$

$$\left(A+B+\bar{D}\right)\left(\bar{A}+C+D\right) = M_0 \cdot M_3 \cdot M_8 \cdot M_{12}$$
$$= \prod M(0,3,8,12) \quad \textbf{Ans.}$$

$$\left(A+B+\bar{D}\right)\left(\bar{A}+C+D\right) = (A+B+C+D)\left(A+B+\bar{C}+\bar{D}\right)\left(\bar{A}+B+C+D\right)\left(\bar{A}+\bar{B}+C+D\right) \quad \textbf{Ans.}$$

EXAMPLE 4.51 Reduce Boolean expression using K-map $AB + A\bar{B}C + \bar{A}B\bar{C} + B\bar{C}$.

SOLUTION There are three variables *A*, *B* and *C*. A is MSB.

$$AB = ABX$$
$$= 11X$$
$$= 110 + 111$$
$$= m_6 + m_7$$

$$B\bar{C} = XB\bar{C}$$
$$= X10$$
$$= 010 + 110$$
$$= m_2 + m_6$$

$$A\bar{B}C = 101 = m_5$$
$$\bar{A}B\bar{C} = 010 = m_2$$

$$f(A,B,C) = m_6 + m_7 + m_2 + m_6 + m_5 + m_2$$
$$= m_2 + m_5 + m_6 + m_7$$
$$= \sum m(2,5,6,7)$$

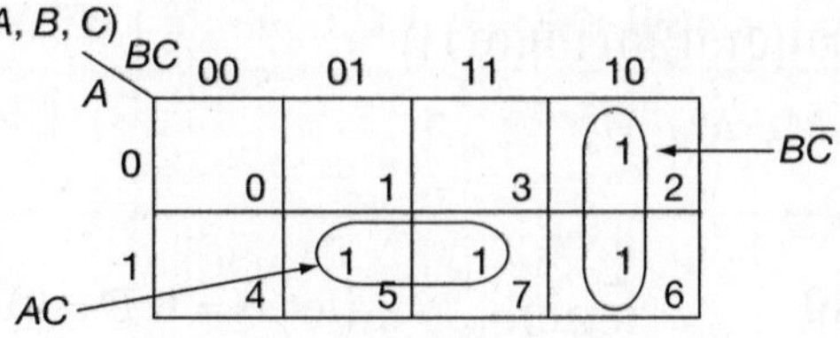

K-map for Example 4.51

$$f(A,B,C) = AC + B\overline{C} \quad \textbf{Ans.}$$

EXAMPLE 4.52 Reduce using K-map $(A+C)(A+\overline{B}+C)(A+C+\overline{D})(A+B+C+\overline{D})$.

SOLUTION There are four variables, A is MSB

$$f(A,B,C,D) = (A+C)(A+\overline{B}+C)(A+C+\overline{D})(A+B+C+\overline{D})$$

$$A+C = 0X0X = (0000)\cdot(0001)\cdot(0100)\cdot(0101) = M_0 \cdot M_1 \cdot M_4 \cdot M_5$$

$$A+\overline{B}+C = 010X = (0100)\cdot(0101) = M_4 \cdot M_5$$

$$A+C+\overline{D} = 0X01 = (0001)\cdot(0101) = M_1 \cdot M_5$$

$$(A+B+C+\overline{D}) = 0001 = M_1$$

$$f(A,B,C,D) = M_0 \cdot M_1 \cdot M_4 \cdot M_5 = \prod M(0,1,4,5)$$

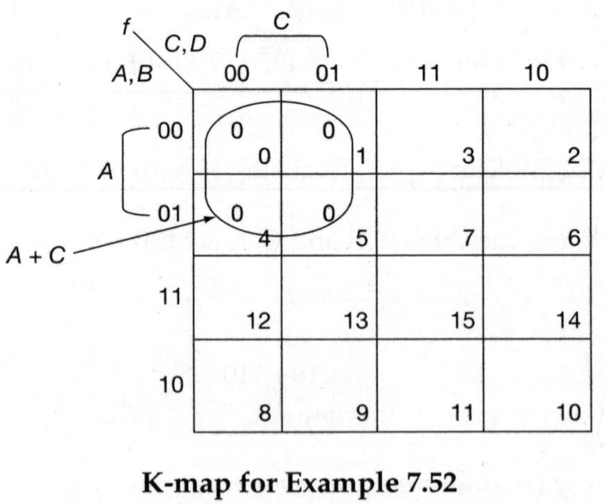

K-map for Example 7.52

$$f(A,B,C,D) = A + C \quad \textbf{Ans.}$$

EXAMPLE 4.53 Reduce using K-map $f(A,B,C,D) = \prod M(1,2,3,5,6,7,8,9,12,13)$.

SOLUTION

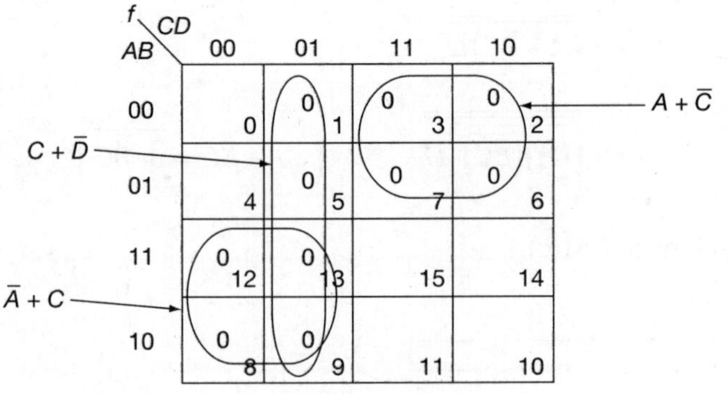

K-map for Example 4.53

$$f(A,B,C,D) = (\bar{A}+C)(C+\bar{D})(A+\bar{C}) \quad \textbf{Ans.}$$

EXAMPLE 4.54 Implement the Boolean logic using NAND gate $f = (A + \bar{B}C)D$.

SOLUTION Draw the logic diagram as shown in Figures 4.81(a). Figure 4.81(b) shows the conversion of AND gate(s) to NAND gate(s) and OR gate to bubbled OR gate.

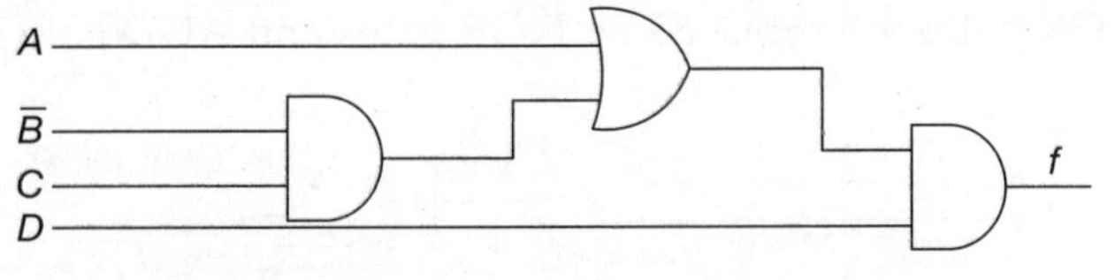

FIGURE 4.81(a) | Logic circuit diagram

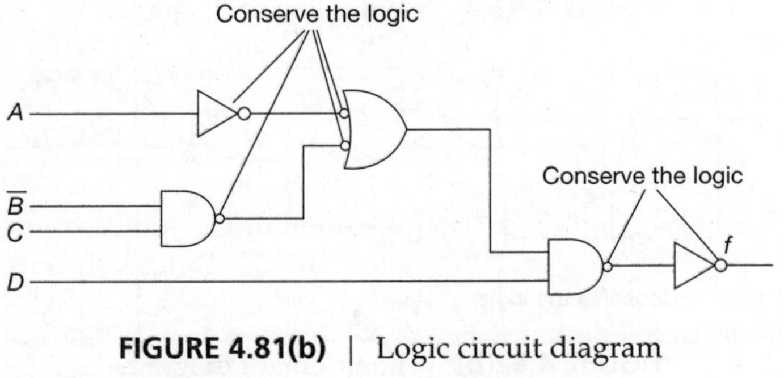

FIGURE 4.81(b) | Logic circuit diagram

Alternative Method:

$$f = \left(A + \bar{B}C\right)D$$

Taking double complement and De Morganize

$$f = \overline{\overline{\left(A + \bar{B}C\right)D}}$$

$$f = \overline{\left[\overline{\left(\bar{A}\right)\left(\overline{\bar{B}C}\right)}\right]D} \qquad \left(\because A + \bar{B}C = \overline{\bar{A} \cdot \overline{\bar{B}C}}\right)$$

Follow the drawn Figure 4.81(c).

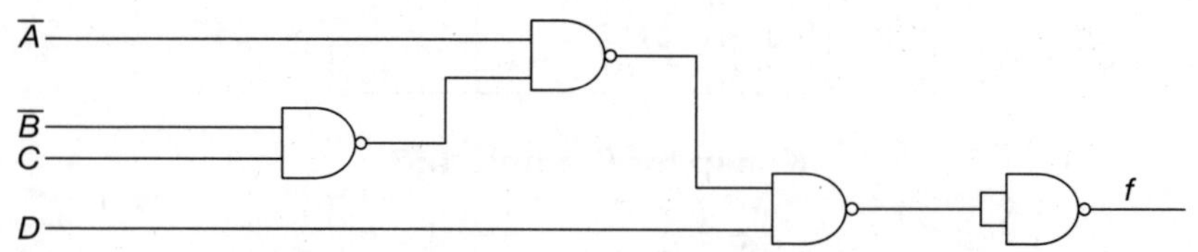

FIGURE 4.81(c) | Logic circuit diagram

EXAMPLE 4.55 Implement the logic expression using NOR gate only.

$$f = (X + Y) + XY$$

SOLUTION Draw the logic circuit diagram as shown in Figure 4.82(a), Figure 4.82(b) gives diagram in which OR gates are replaced by NOR gates and AND gate is replaced to bubbled AND gates.

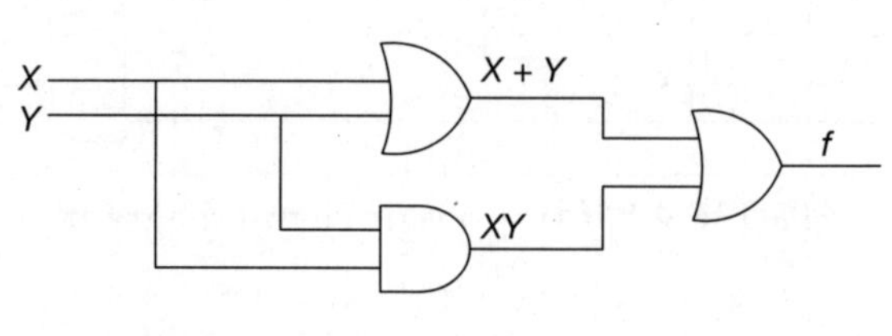

FIGURE 4.82(a) | Logic circuit diagram

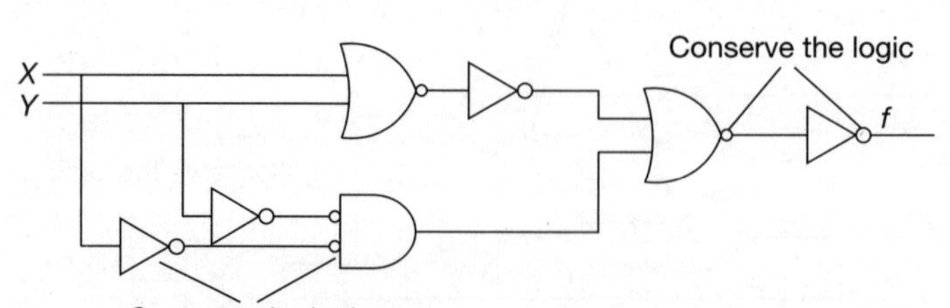

FIGURE 4.82(b) | Logic circuit diagram

Algebraic expression

$$f = (X+Y)+XY$$
$$= (X+Y)+\left(\overline{\overline{X}+\overline{Y}}\right) \qquad \left(\because XY = \overline{\overline{X}+\overline{Y}}\right)$$
$$= \left(\overline{\overline{X+Y}+\overline{\overline{X}+\overline{Y}}}\right)$$

Take double negation

$$f = \overline{\overline{\left(\overline{\overline{X+Y}}\right)+\left(\overline{\overline{X}+\overline{Y}}\right)}}$$

Diagram is given in Figure 4.82(c).

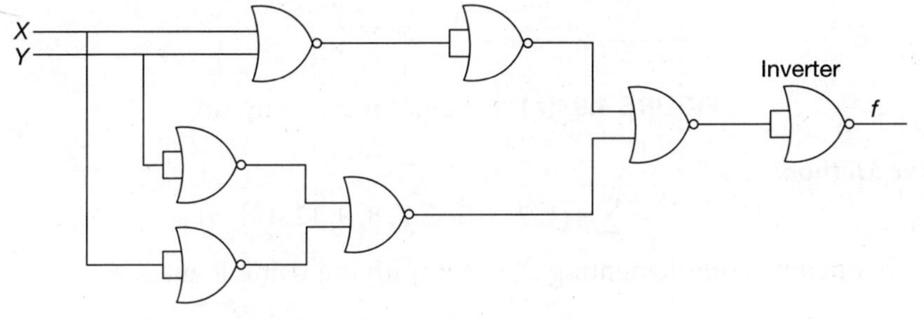

FIGURE 4.82(c) | Logic circuit diagram

Example 4.56 Implement the Boolean expression using (i) NOR gate and (ii) NAND gate

$$f(a,b,c,d) = \prod M(1,2,3,5,6,7,8,9,12,13)$$

Solution

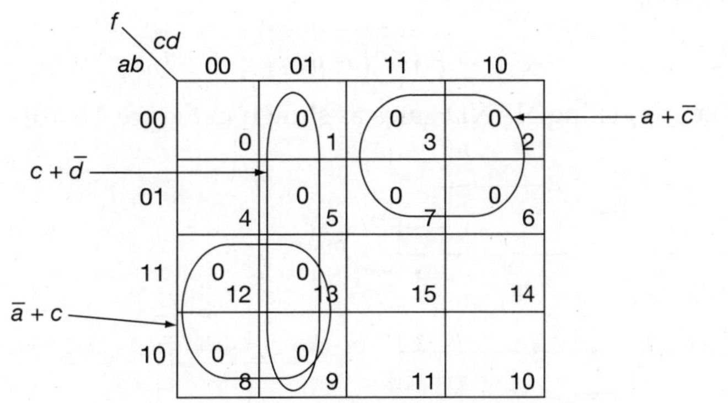

K-map for Example 4.56

$$f(a,b,c,d) = (a+\overline{c})(\overline{a}+c)(c+\overline{d})$$

Taking double negation

$$f(a,b,c,d) = \overline{\overline{(a+\bar{c})(\bar{a}+c)(c+\bar{d})}}$$

$$= \overline{\overline{(a+\bar{c})}+\overline{(\bar{a}+c)}+\overline{(c+\bar{d})}}$$

Draw the logic diagram as shown in Figure 4.83(a)

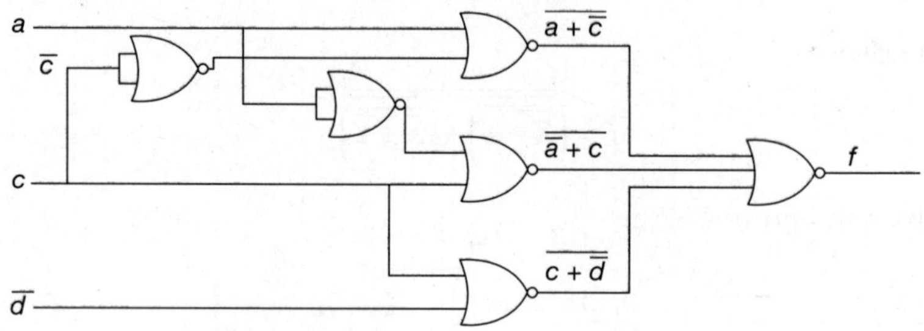

FIGURE 4.83(a) | Logic circuit diagram

Alternative Method:

$$\bar{f} = \sum m(1, 2, 3, 5, 6, 7, 8, 9, 12, 13)$$

∵ SOP is obtained by complementing POS. Simplifying using K-map.

$$\bar{f} = \bar{a}c + a\bar{c} + \bar{c}d$$

Complement both sides

$$f = \overline{\bar{a}c + a\bar{c} + \bar{c}d}$$

Apply De Morgan Theorem

$$f = (\overline{\bar{a}c})(\overline{a\bar{c}})(\overline{\bar{c}d})$$

Taking double negation

$$f = \overline{\overline{(\overline{\bar{a}c})(\overline{a\bar{c}})(\overline{\bar{c}d})}}$$

Draw the logic diagram using NAND gates as shown in Figure 4.83(b).

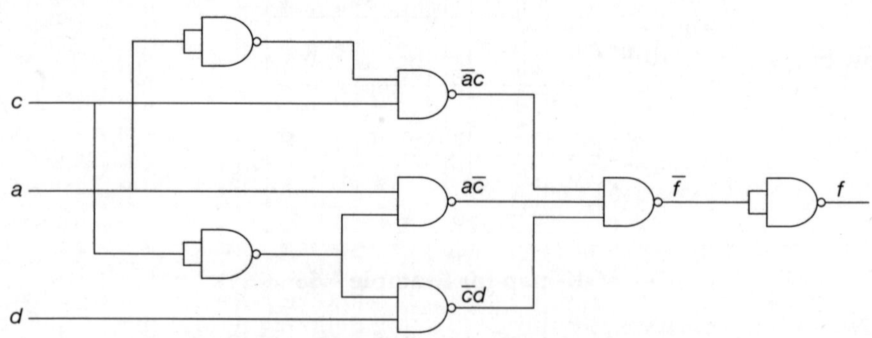

FIGURE 4.83(b) | Logic circuit diagram

EXAMPLE 4.57 Implement the Boolean expression free from hazards if any.

$$f(a,b,c,d)=\sum m(1,3,4,5,7,10,11,12,14,15)$$

SOLUTION

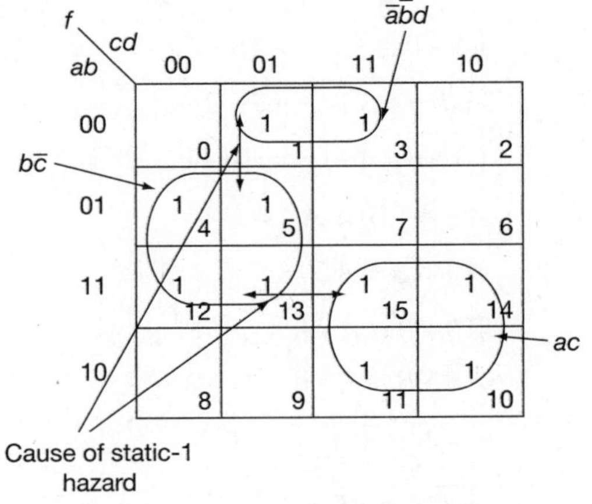

K-map for Example 4.57

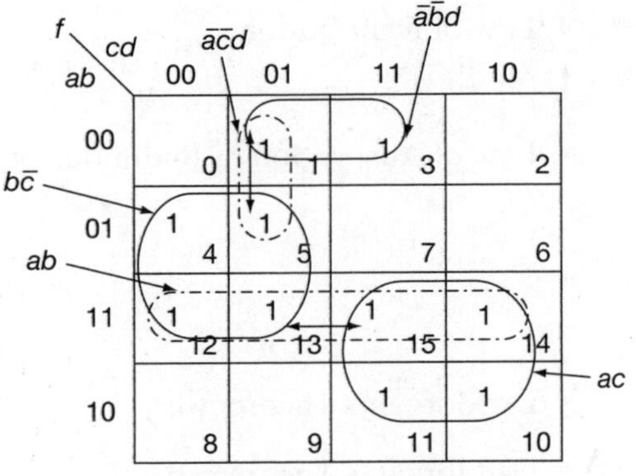

K-map for Example 4.57

The following logic gives static-1 hazard

$$f = ac + b\bar{c} + \bar{a}\bar{b}d$$

The following logic is free from any hazard

$$f = ac + b\bar{c} + \bar{a}\bar{b}d + \bar{a}\bar{c}d + ab$$

SUMMARY

- Law of Intersection: $A\cdot 1 = A$, $A\cdot 0 = 0$
- Law of Union: $A+1=1$, $A+0=A$
- Law of Identity: $A\cdot 1 = A$, $A+1=1$
- Law of Null: $A\cdot 0 = 0$, $A+0=A$
- Law of Tautology or Idempotence: $A\cdot A = A$, $A+A=A$
- Law of Complement or Negation: $A\cdot\bar{A}=0$, $A+\bar{A}=1$

- Law of Double Negation or Involution: $\bar{\bar{A}} = A$
- Law of Commutation: $A + B = B + A$
  $A \cdot B = B \cdot A$
- Law of Association: $(A \cdot B) \cdot C = A \cdot (B \cdot C)$
  $(A + B) + C = A + (B + C)$
- Law of Distribution: $A \cdot B + A \cdot C = A \cdot (B + C)$
  $(A + B)(A + C) = A + BC$
- Law of Absorption or Redundancy: $A \cdot (A + B) = A$
  $A + AB = A$
  $AB + \bar{B} = A + \bar{B}$
  $A\bar{B} + B = A + B$
- De Morgan's Theorem-I: $\overline{A + B} = \bar{A} \cdot \bar{B}$
- De Morgan's Theorem-II: $\overline{A \cdot B} = \bar{A} + \bar{B}$
- Consensus Theorem: $AB + \overline{A}C + BC = AB + \bar{A}C$
  $(A + B)(\overline{A} + C)(B + C) = (A + B)(\bar{A} + C)$
- Transposition Theorem: $AB + \overline{A}C = (A + C)(\bar{A} + B)$
- **Term:** A term is a collection of variables, e.g. ABCD.
- **Constant:** A constant is a value or quantity which has a fixed meaning. In Boolean algebra, there are only two possible constants, one and zero. These two constants are used to describe true and false, up and down, go and not go, etc.
- **Variable:** A variable is a quantity which changes by taking on the value of any constant in the algebraic system. At any one time the variable has a particular value of constant. There are only two values of constants in the system-therefore a variable can only be zero or one. Variables are denoted by letters.
- **Literal:** A literal is a variable or its complement $A, \bar{A}, B, \bar{B}, C, \bar{C}, D, \bar{D}$.
- **Minterm:** Minterm is known as the standard product or canonic product term. This is a term such as $A\bar{B}CD$ or $\bar{A}BC\bar{D}$ or $AB\bar{C}\bar{D}$, etc., where each variable is used once and once only.
- **Maxterm:** Also known as the standard sum or canonic sum term. This is a term such as $(A + \bar{B} + C + D)$ or $(\bar{A} + B + C + \bar{D})$ or $(A + B + \bar{C} + \bar{D})$, etc., where each variable is used once and once only.
- **Standard sum-of-products (SOP) form:** Standard SOP is known as the minterm canonic form or canonic sum function. A function in the form of the 'sum' (OR) of minterms, e.g.: $f(A, B, C, D) = A\bar{B}CD + \bar{A}BC\bar{D} + AB\bar{C}\bar{D}$.

- **Standard product-of-sums (POS) form:** Standard POS is known as the maxterm canonic form or canonic product function. A function in the form of the 'product' (AND) of maxterms, e.g.: $f(A,B,C,D)=(A+\bar{B}+C+D)(\bar{A}+B+C+\bar{D})(A+B+\bar{C}+\bar{D})$
- **Sum-of-products (SOP):** SOP is known as the normal sum function. A function in the form of the 'sum'(OR) of normal product (AND) terms, e.g.: $f(A,B,C,D)=\bar{A}\bar{D}+\bar{A}CD+BC+B\bar{C}\bar{D}$.
- **Product-of-sums (POS):** POS is known as the normal product function. A function in the form of the 'product' (AND) of normal sum (OR) terms, e.g.: $f(A,B,C,D)=(B+\bar{C})(\bar{A}+C+D)(B+D)(\bar{A}+B)$
- **Normal (general) sum term:** A term such as $(A+\bar{C})$ or $(B+D)$ or $(\bar{B}+C)$, etc.
- **Normal (general) product term:** A term such as AC or $\bar{C}D$ or $\bar{A}BD$, etc.
- **Truth table:** The name 'truth table' comes from a similar table used in symbolic logic, in which the truth or falsity of a statement is listed for all possible proposition conditions. The truth table consists of two parts; one part comprising all combinations of values of the input variables in a statement (or algebraic expression), the other part contains the output values of the statement for each combination. The truth table is useful in that it can be used to verify Boolean identities.
- **Prime-implicants:** It is an implicant of a function which does not imply any other implicant of the function.
- **Prime-implicant chart:** The chart is used to remove redundant prime implicants. A grid is prepared having all the prime-implicants listed down the left and all the minterms of the function along the top. Each minterm covered by a given prime-implicant is marked in the appropriate position.
- **A glitch or logic-spike** is an unwanted pulse at the output of a combinational logic network.
- A circuit with the potential for a glitch is said to have a hazard.
- **Static-1 hazard** occurs when the output momentarily goes to 0 when it should remain a 1.
- **Static-0 hazard** occurs when the output momentarily goes to 1 when it should remain a 0.
- **Dynamic hazard** occurs when the output signal has the potential to change more than once when it is expected to make a single transition form $0 \rightarrow 1$ or $1 \rightarrow 0$.
- The number of cell/square in a K-map is equal to 2^n, where n is the number of input variables.
- The map is drawn to show the relationship between squares and input variables, Variables are assigned to row and column. Binary marking are placed in each row and column using reflected (Gray) code sequence.
- A Karnaugh map comprises a square (or cell or box) for every line in the truth table; the binary value for each square is the binary value of the input terms in the corresponding table row.
- One's are placed in the squares of a K-map to represents minterms of a Boolean function in canonical SOP (Sum-of-Minterm) form. The number of l's in the K-map is equal to the number of product terms in the Boolean function, The cells that do not have a 1 are the cells for which the expression is 0.

- Zeros are placed in the squares of a K-map represents the maxterms of a Boolean function in canonical POS (Product-of-Maxterm) form, The number of 0's in the K-map is equal to the number of sum terms in the Boolean function. The cells that do not have a 0 are the cells for which the expression is 1.
- Grouping can be of $8(2^3)$-cells (Octet), $4(2^2)$-cells (quad), $2(2^1)$-cells (pair) which will reduce the number of literals by 3, 2, and 1 literal, respectively. Common literals in group are retained whereas varying literals are eliminated.

MULTIPLE CHOICE QUESTIONS

4.1 Number of squares in K-map of m-variables is

(a) $2m$ (b) 2

(c) 2^m (d) 2^{m-1}

4.2 The minterm designator for the term $WX\bar{Y}\bar{Z}$ is

(a) 12
(b) 15
(c) 03
(d) None of the above

4.3 In a K-map of four-variable if eight-minterms are combined they will reduce to a term which will have number of variables appearing

(a) 1 (b) 2

(c) 3 (d) 4

4.4 If function A, B, C and D are as follows:

$A = R + \bar{P}Q + \bar{R}S$, $B = PQ\bar{R}\bar{S} + \bar{P}\bar{Q}\bar{R}\bar{S} + P\bar{Q}\bar{R}\bar{S}$, $C = RS + \overline{PR + P\bar{Q} + \bar{P}\bar{Q}}$ and $D = R + S + \overline{PQ + \bar{P}\bar{Q}\bar{R} + P\bar{Q}\bar{S}}$

Choose the correct Boolean relation from the following:

(a) $A = D,\ B = \bar{D}$
(b) $A = D,\ B = C$
(c) $A = C$
(d) $A = \bar{D}$

4.5 Don't care condition is

(a) Logic 0
(b) Logic 1
(c) Logic 0 or Logic 1 where ever required
(d) None of above

4.6 The simplified expression in SOP form $f(A, B, C, D) = \sum m\,(1, 2, 3, 5, 7, 9, 10, 11, 13, 15)$

(a) $D(\bar{A} + B) + \bar{B}(C + AD)$
(b) $D + \bar{B}C$
(c) $D + \bar{B}C\bar{D}$
(d) None of above

4.7 Karnaugh map method is not convenient to use for more than

(a) three variables
(b) four variables
(c) five variables
(d) six variables

4.8 To construct a K-map, the order is crucial and is obtained by using

(a) Binary code progression
(b) BCD code
(c) Gray code progression
(d) Excess-3 code

4.9 For a function $f = \sum m\,(0, 1, 2, 3, 4, 6, 7)$, the minimal expression obtained will be

(a) $\bar{A}\bar{B} + \bar{A}B(\bar{C}\bar{D} + C)$
(b) $\bar{A}\bar{B} + \bar{A}C + \bar{A}\bar{D}$
(c) $\bar{A}C + \bar{A}\bar{C}\bar{D} + \bar{A}\bar{B}\bar{C}\bar{D}$
(d) None of above

4.10 The AND operation can be produced with

(a) two NAND gates
(b) three NAND gates
(c) one NOR gate
(d) two NOR gates

4.11 The OR operation can be produced with
(a) two NOR gates
(b) three NAND gates
(c) four NAND gates
(d) both answers (a) and (b)

4.12 When using dual symbols in a logic diagram,
(a) bubble outputs are connected to bubble inputs
(b) the NAND symbols produce the AND operations
(c) the negative-OR symbols produce the OR operations
(d) all of these answers are true

4.13 A logic circuit with an output $X = A\bar{B}C + A\bar{C}$ consists of
(a) two AND gates and one OR gate
(b) two AND gates, one OR gate, and two inverters
(c) two OR gates, one AND gate, and two inverters
(d) two AND gates, one OR gate, and one inverter

4.14 To implement the expression $\bar{A}BCD + A\bar{B}CD + AB\bar{C}\bar{D}$ it takes one OR gate and
(a) one AND gate
(b) three AND gates
(c) two AND gates and two inverters
(d) three AND gates and three inventers

4.15 An exclusive-OR function is expression as
(a) $\bar{A}\bar{B} + AB$
(b) $\bar{A}B + A\bar{B}$
(c) $(\bar{A}+B)(A+\bar{B})$
(d) $(\bar{A}+\bar{B})+(A+B)$

Answers

4.1 (a)	4.2 (a)	4.3 (a)	4.4 (a)	4.5 (c)	4.6 (b)	4.7 (d)
4.8 (c)	4.9 (b)	4.10 (a)	4.11 (d)	4.12 (d)	4.13 (b)	4.14 (d)
4.15 (b)						

QUESTIONS

4.1 What is K-map? Explain K-map method to simplify Boolean expression.

4.2 What are prime implicants and explain their role in Boolean function representation in its minimal form?

4.3 How don't care conditions are implemented in K-map? Explain with example.

4.4 What is Quine–McCluskey method and why it is used?

4.5 Why gray code is used to label the Karnaugh Map?

4.6 Explain minterm, maxterm and don't care conditions.

4.7 What do you mean by adjacent squares of a K-map?

4.8 Write a procedure to reduce Boolean expression using K-maps;

4.9 What do you mean by a real minimal expression?

4.10 What is the main criterion for the design of digital circuits?

4.11 How do you compare the cost of realizing a circuit from the Boolean expressions?

4.12 What do you mean by two-level logic? What is its main advantage?

4.13 What is hybrid logic? What are its main advantages and disadvantages? What is the criterion for the minimization of multiple output switching functions?

4.14 What is variable mapping? What are its advantages?

4.15 What is meant by index of a term in the tabular method?

4.16 When can two minterms or maxterms be combined?

4.17 In the tabular method, why two terms whose codes differ by a power of 2 but have the same index cannot be combined?

4.18 What are (a) a prime-implicant chart and (b) a reduced prime-implicant chart?

4.19 What are dominating rows and dominating columns? Does elimination of dominating columns and dominated rows end the search for minimal expression?

4.20 How many variables are eliminated by (i) a 2-square, (ii) a 4-square, (iii) an 8-square and (iv) a 16-square?

PROBLEMS

4.1 Obtain the simplified expressions in SOP for the following Boolean functions:

(a) $f(x,y,z)=\sum m(2,3,6,7)$

(b) $f(A,B,C,D)=\sum m(7,13,14,15)$

(c) $f(A,B,C,D)=\sum m(4,6,7,15)$

(d) $f(w,x,y,z)=\sum m(2,3,12,13,14,15)$

4.2 Obtain the simplified expressions in SOP for the following Boolean functions:

(a) $xy+\overline{x}y\overline{z}+\overline{x}\overline{y}\overline{z}$

(b) $\overline{A}B+B\overline{C}+\overline{B}\overline{C}$

(c) $\overline{a}\overline{b}+bc+\overline{a}b\overline{c}$

(d) $D(\overline{A}+B)+\overline{B}(C+AD)$

(e) $\overline{k}l\overline{m}+\overline{k}\overline{m}n+kl\overline{m}\overline{n}+lm\overline{n}$

4.3 Obtain the simplified expressions in POS:

(a) $f(x,y,z)=\prod M(0,1,4,5)$

(b) $f(a,b,c,d)=\prod M(0,1,2,3,4,10,11)$

(c) $f(w,x,y,z)=\prod M(1,3,5,7,13,15)$

4.4 Simplify the Boolean function in SOP using don't condition:

(a) $f = \overline{y} + \overline{xz}$

(b) $d = yz + xy$

(c) $f = \overline{b}\overline{c}\overline{d} + bc\overline{d} + abc\overline{d}$

(d) $d = \overline{b}c\overline{d} + \overline{a}b\overline{c}d$

(e) $f = \overline{w}\left(\overline{x}y + \overline{x}\overline{y} + xyz\right) + \overline{x}\overline{z}\left(y + w\right)$

(f) $d = \overline{w}x\left(\overline{y}z + y\overline{z}\right) + wyz$

4.5 Implement the following function with either NAND or NOR gates:

(a) $f = \overline{w}xz + \overline{w}yz + \overline{x}y\overline{z} + wx\overline{y}z$

(b) $d = wyz$

4.6 Simplify the following Boolean function by means of the tabulation method:

(a) $f = \sum m(0,1,4,5,6,7,8,12,13)$

(b) $f = \sum m(1,3,5,7,13,15)$

$d = \sum m(0,2,6)$

(c) $f = \prod M(0,1,2,3,5,6,10,11)$

$d = \prod M(13,14,15)$

where d is don't care condition.

5

Logic Circuit Design: Arithmetic Operation

CHAPTER OBJECTIVES

The main goal of this chapter is to impart knowledge about the digital logic circuit design of arithmetic operation. Readers will be able to discuss the following aspects in this chapter:

- Design of combinational circuits
- Design of single-bit binary adder and subtractor
- Design of four-bit binary parallel adder and look-ahead carry adder
- Design for combinational circuit for complement
- Design of binary parallel subtractor using parallel adder
- Design of binary-coded decimal (BCD) adder and subtractor
- Design of multiplier and divider
- Design of Excess-3 adder and subtractor
- Design of magnitude comparator
- Design of parity generator and checker
- Design of code conversion circuits
- Arithmetic logic unit design

5.1 | COMBINATIONAL CIRCUITS

A rule or logic is to denote a relationship of binary input and output. A tabular representation of the combinations that a group of binary variables can assume is called a truth table. An output variable X is said to be a function of a set of input variables, A, B and C. Mathematically it is represented as $X = f(A,B,C)$. X is binary output of logical function. A, B and C are the binary input variables. Function can be AND. OR, NOT, NAND, NOR, EX-OR, and EX-NOR logical functions or combination of these gates. Binary input and output variables can have either zero (0) or one (1) value. The circuit to represent the logic function is known as combinational circuit. Propagation delay in the gates exists. When input is fed at a time, gate(s) produce output after a delay. It is known as propagation delay.

Let x_1, x_2, x_3 and x_4 are input binary variables. y_1, y_2 and y_3 are output binary variables. The logic function, f gives output binary variables (y_1, y_2 and y_3) based on input binary variables (x_1, x_2, x_3 and x_4). Logic function is a combination of gates is known as combinational circuit that gives output (y_1, y_2 and y_3) when input (x_1, x_2, x_3 and x_4) is fed at a given time; t. Combinational circuit is a feed forward circuit and has no feedback (Figure 5.1).

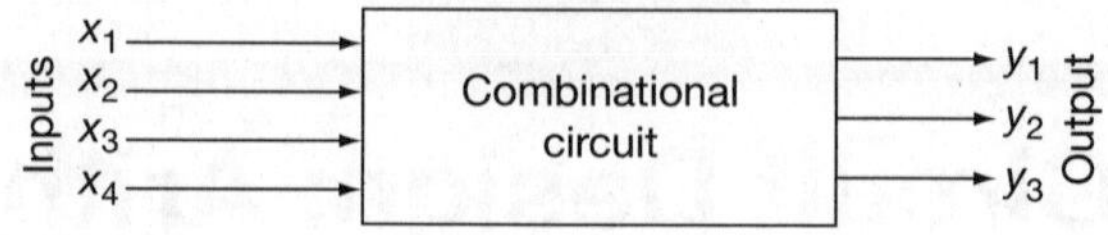

FIGURE 5.1 | Combinational circuit

Mathematical, logic functions of the combinational circuit are given below as:

$$y_1(t) = f(x_1(t), x_2(t), x_3(t), x_4(t)) \quad (5.1)$$

$$y_2(t) = f(x_1(t), x_2(t), x_3(t), x_4(t)) \quad (5.2)$$

$$y_3(t) = f(x_1(t), x_2(t), x_3(t), x_4(t)) \quad (5.3)$$

> ***Note:*** Combinational circuit is a feed forward circuit that consists of combination of gates.

To design a combinational circuit, the following steps are followed.

1. Problem statement is given.
2. Find number of input binary variables and output binary variables required.
3. Assign mnemonics or variable name (alphabet) to input binary variables and output binary variables.
4. Find the number of minterms and maxterms.
   For *n* input variables there are 2^n binary input possibilities giving minterms or maxterms.
5. Construct a truth table for all possible input combinations or minterms or maxterms.
6. Find simplified Boolean (logical) expression using K-map or Quine McCluskey method.
7. Draw the logic diagram known as combinational circuit.

EXAMPLE 5.1 Design a combinational circuit that has three binary inputs and generate output as 1 if and only if (i) most significant bit is a 0 or any one of other bits is 1, (ii) second bit is 0 or any one of other bits is 1 and (iii) all the bits are 1.

SOLUTION

Inputs: There are three-binary input variables (literals): x_2, x_1 and x_0. x_0 is considered least significant bit (LSB) and x_2 is considered as most significant bit (MSB).

Output(s): One binary output variable, *Y*.

Since there are three inputs so there are $2^3 = 8$ possible combinations that give eight-minterms. Truth table that gives relation between input variables and output is depicted in Table 5.1. K-map is setup as shown in Figure 5.2.

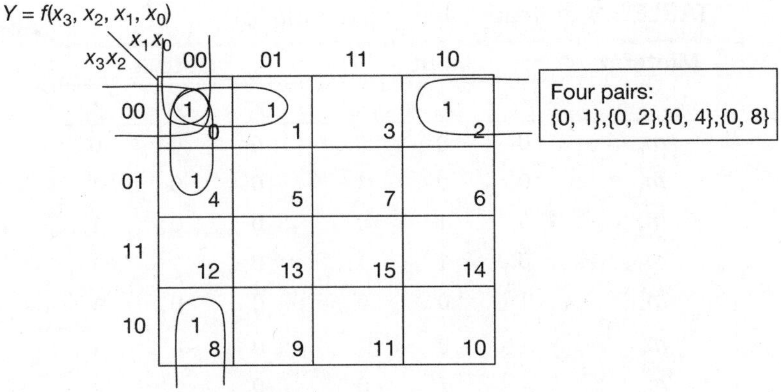

FIGURE 5.4 | K-map of Example 5.2

The combinational circuit is given in Figure 5.5.

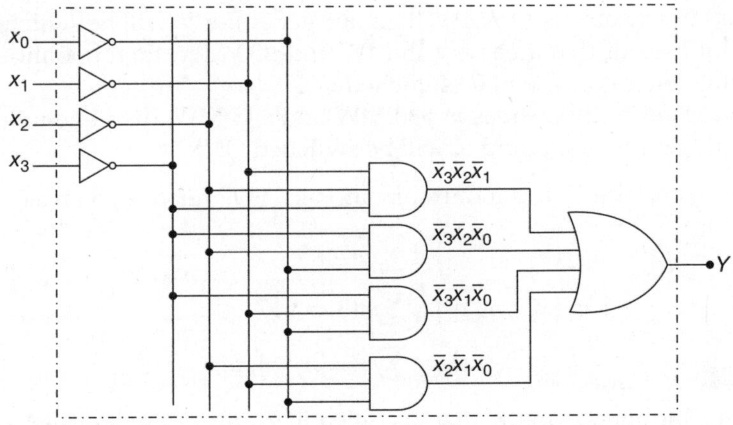

FIGURE 5.5 | Combinational circuit of Example 5.2

EXAMPLE 5.3 Three generators are to be switched on as per the load put on them. The generator can supply maximum of 15 MW at a time. The generators are supplying three areas. Each area has maximum of 15 MW load. Design a combinational circuit to provide the switching pattern.

SOLUTION

Inputs: There are three binary inputs variables: x_0, x_1 and x_2 representing load areas. x_0 is considered LSB and x_2 is considered as MSB.

Outputs: There are three binary output variables; y_0, y_1 and y_2. y_0 is considered LSB and y_2 is considered as MSB.

There are $2^3 = 8$ possible combinations representing eight minterms. Truth table is shown in Table 5.3.

TABLE 5.3 | Truth table of Example 5.3

Minterm	Input			Output		
	x_2	x_1	x_0	y_2	y_1	y_0
m_0	0	0	0	0	0	0
m_1	0	0	1	0	0	1
m_2	0	1	0	0	0	1
m_3	0	1	1	0	1	1
m_4	1	0	0	0	0	1
m_5	1	0	1	0	1	1
m_6	1	1	0	0	1	1
m_7	1	1	1	1	1	1

Logic followed is elaborated below:

- If the load of any area is ≤15 MW, then one generator y_0 will be switched ON.
- If the total load of two areas is >15 MW and ≤30 MW, then one more generator y_1 along with generator, y_0 will be switched ON.
- If the total load of three areas is >30 MW and ≤45 MW, then one more generator y_2 along with generators, y_0 and y_1 will be switched ON.

Considering truth Table 5.3, Standard or canonical SOP form is given as

$$y_2(x_2, x_1, x_0) = \sum m(7)$$

$$y_1(x_2, x_1, x_0) = \sum m(3, 5, 6, 7)$$

$$y_0(x_2, x_1, x_0) = \sum m(1, 2, 3, 4, 5, 6, 7)$$

Output, y_0 is simplified using K-map as given in Figure 5.6(a). Output, y_1 is simplified using K-map as given in Figure 5.6(b). One minterm (m_7) gives logic 1 for y_2, so no simplification is required. In Figure 5.6(a), three-quads are formed. Three quads give three product terms having one literal each. Two literals are reduced by forming pair. The simplified logic function, y_0 is given below:

$$y_0 = x_2 + x_1 + x_0 \tag{5.6}$$

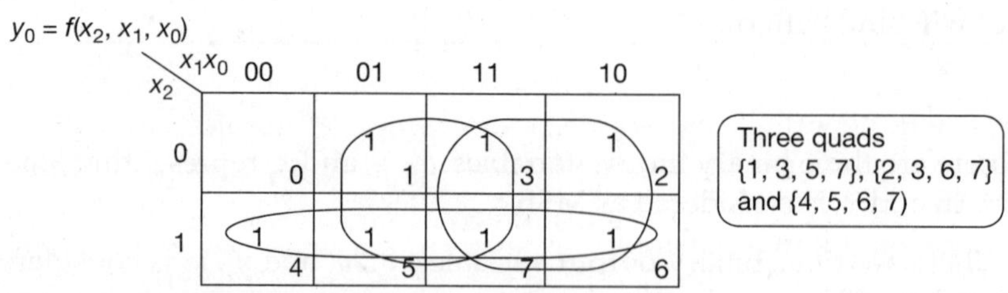

(a)

FIGURE 5.6(a) | K-map for Example 5.3

In Figure 5.6(b), three-pairs are formed. Three pairs give three product terms. Each term is of two literals. The simplified logic function y_1 is given below:

$$y_1 = x_1x_0 + x_2x_0 + x_2x_1 \tag{5.7}$$

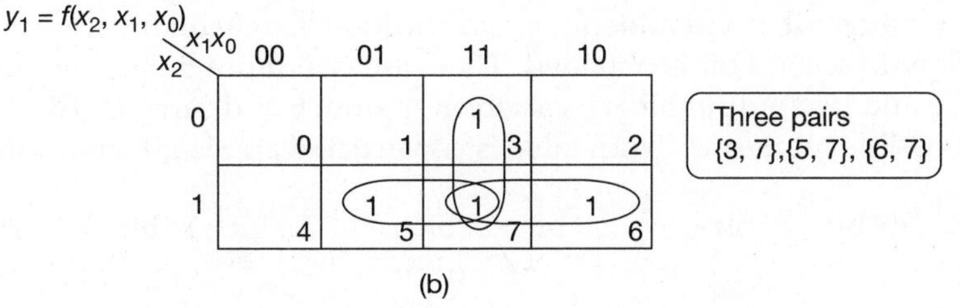

FIGURE 5.6(b) | K-map for y_1 of Example 5.3

The logic function y_2 is given below:

$$y_2 = x_2x_1x_0 \tag{5.8}$$

Logic diagram is shown in Figure 5.7.

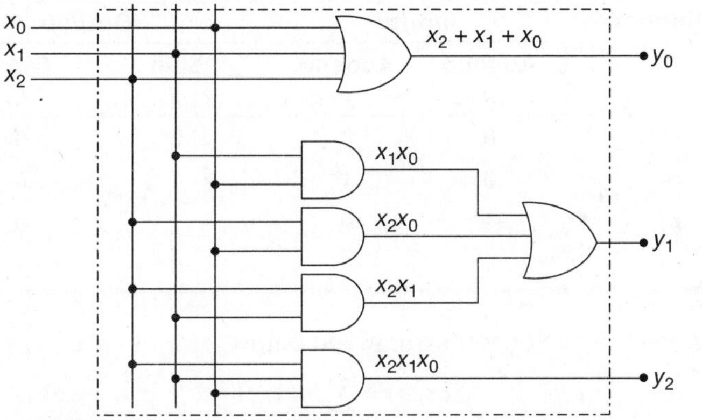

FIGURE 5.7 | Combinational circuit of Example 5.3

5.2 | BINARY ADDER

To process the numerical binary data, arithmetic circuits are necessary. Arithmetic circuits perform arithmetic operations like addition, subtraction, multiplication, division, comparison and parity calculations. Adders are the basic building blocks of all arithmetic circuits. Adder adds two binary numbers and gives out sum and carry-out as outputs. Adder circuits are divided into two categories

1. Half-adder
2. Full-adder

> ***Note:*** Single *bit* means binary digit.

5.2.1 | Half-Adder

Two single bits (binary digits) are added to produce sum and carry bits. Any overflow from first bit into second bit is known as carry. Half-adder circuit performs addition of two single bit values and produces single bit sum and a carry.

Let single-bit augend, x and addend, y. On addition, single bit binary sum, S and carry-out an overflow in second bit is obtained. There are two input binary variables; augend, x and addend, y and two output binary variables as sum, S and carry, C. Two input variables give 2^2 (=4) possible minterms. Truth table is constructed and is given in Table 5.4.

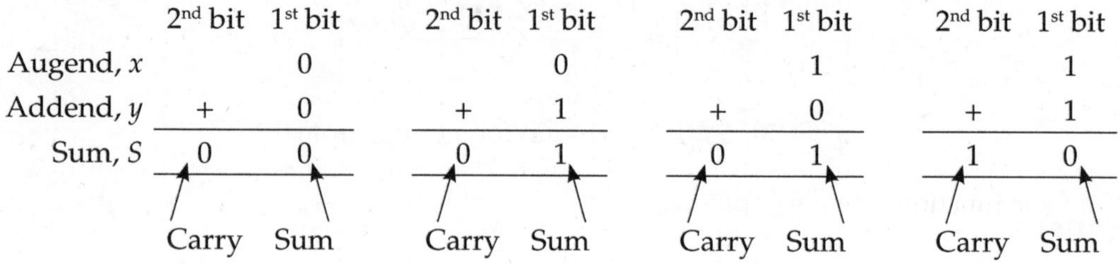

TABLE 5.4 | Truth table for half-adder

Minterm	Inputs		Outputs	
	Augend x	Addend y	Sum S	Carry C
m_0	0	0	0	0
m_1	0	1	1	0
m_2	1	0	1	0
m_3	1	1	0	1

Boolean equations for two outputs are given below:

$$S(x,y)=\sum m(1,2) \tag{5.9}$$

$$C(x,y)=\sum m(3) \tag{5.10}$$

Two variable K-maps are drawn to simply the output Boolean functions. K-maps are given in Figure 5.8(a) for sum, S and Figure 5.8(b) for carry-out, C. Function cannot be simplified as no pair is formed

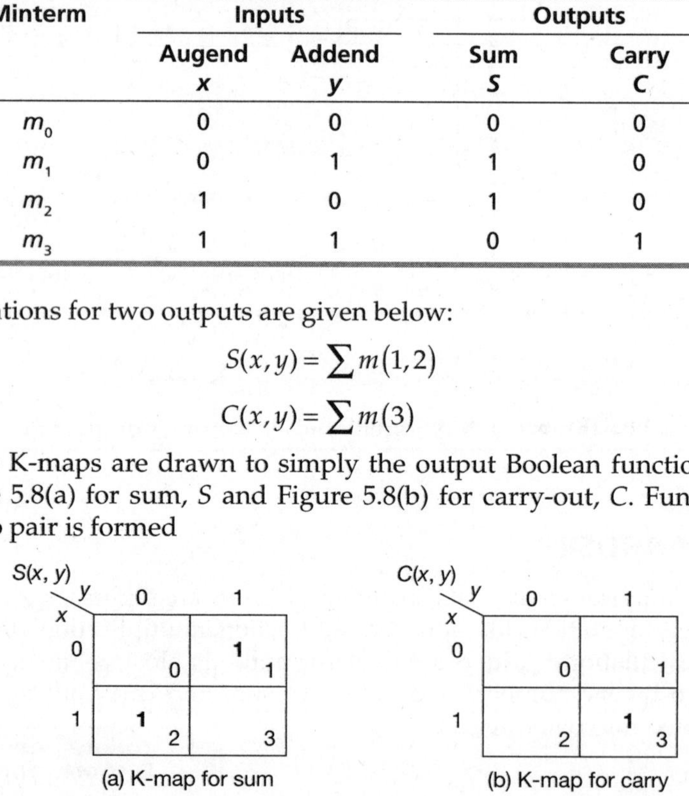

FIGURE 5.8 | K-map simplification of half adder

Logic expressions sum, S and carry, C for half-adder are given below.

$$S(x,y) = \bar{x}y + x\bar{y} \tag{5.11}$$

$$S(x,y) = x \oplus y \tag{5.12}$$

$$C(x,y) = xy \tag{5.13}$$

Logic diagram for half-adder is shown in Figure 5.9 and block diagram is given in Figure 5.10.

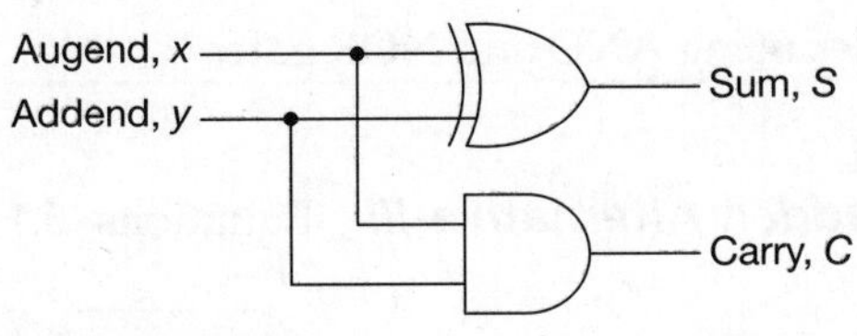

FIGURE 5.9 | Logic diagram of half-adder

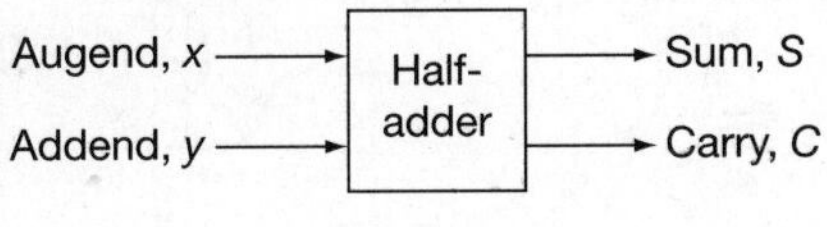

FIGURE 5.10 | Block diagram of half-adder

Logic diagram of half-adder: Alternative-I Logic diagram of half-adder can be obtained using AND-OR-INVERTER (AOI) fundamental gates. Figure 5.11 shows the half-adder using AOI gates to implement Boolean function given by Eq. (5.11) and Eq. (5.13).

Logic diagram of half-adder: Alternative-II Equation (5.11) is implemented by Exclusion-OR gate. Exclusive-OR logic function is equivalent to inverted exclusive-NOR function and is given below

$$S(x,y) = \overline{xy + \bar{x}\,\bar{y}} \quad \because (x \oplus y = x\bar{y} + \bar{x}y = \overline{xy + \bar{x}\,\bar{y}} = x \odot y) \tag{5.14}$$

$$C(x,y) = xy \tag{5.15}$$

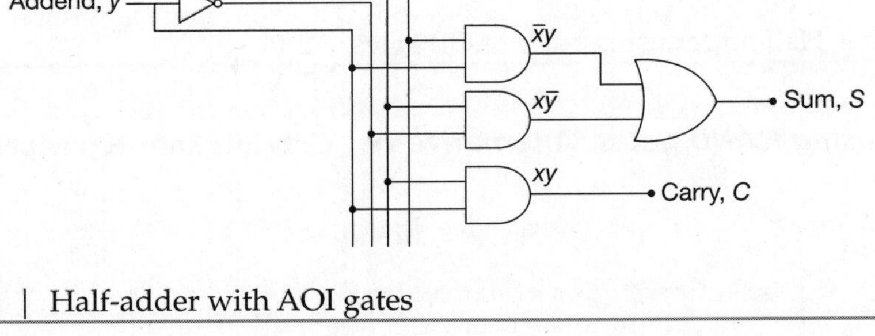

FIGURE 5.11 | Half-adder with AOI gates

One AND gate is saved. The logic diagram is shown in Figure 5.12.

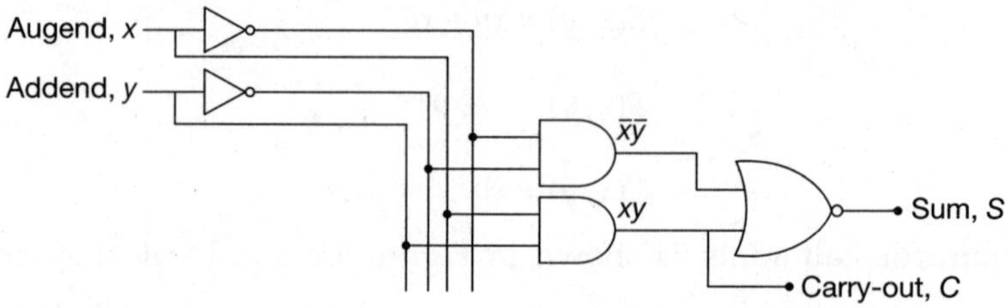

FIGURE 5.12 | Half-adder using AND and NOR gates

Logic diagram of half-adder: Alternative-III Equations (5.11) and (5.14) give the sum logic.

$$S(x,y) = \overline{xy + \bar{x}\,\bar{y}}$$

> **Half-adder:**
> Sum = Augend ⊕ Addend
> Carry Out = (Augend) · (Addend)

Applying De Morgan's theorem: $\overline{x + y} = \bar{x}\,\bar{y}$, to get

$$S(x,y) = \overline{(xy)}\left(\overline{\bar{x}\,\bar{y}}\right)$$

Again apply De Morgan's theorem: $\overline{x\,y} = \bar{x} + \bar{y}$, to get

$$S(x,y) = (\bar{x} + \bar{y})(x + y) \tag{5.16}$$

Double negation never changes the function value ($\bar{\bar{x}} = x$). So take double negation to get

$$C(x,y) = \overline{\overline{xy}}$$

Apply De Morgan's theorem: $\overline{x\,y} = \bar{x} + \bar{y}$ to get

$$C(x,y) = (\overline{\bar{x} + \bar{y}}) \tag{5.17}$$

The logic diagram is shown in Figure 5.13.

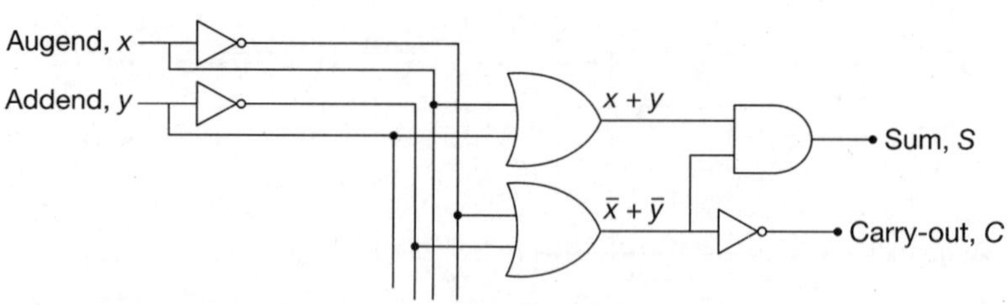

FIGURE 5.13 | Half-adder using basic (AOI) gates

Half-adder using NAND gates: Alternative-IV Consider sum, S function as given in Eq. (5.11)

$$S(x,y) = x\bar{y} + \bar{x}y$$

Performing OR operation of x binary variable with 0-logic does not change the variable value.

$$S(x,y) = 0 + x\overline{y} + \overline{x}y + 0 \qquad (\because x + 0 = x)$$

Zero-logic can be obtained from AND operation of binary variable with its complement.

$$S(x,y) = x\overline{x} + x\overline{y} + \overline{x}y + y\overline{y} \qquad (\because\ x\overline{x} = 0 \text{ and } y\overline{y} = 0)$$

On simplification

$$S(x,y) = x\left(\overline{x} + \overline{y}\right) + y\left(\overline{x} + \overline{y}\right) \quad (\because \overline{x} + \overline{y} = \overline{xy}, \text{ De Morgan's theorem})$$

Apply De Morgan's theorem: $\overline{x\,y} = \overline{x} + \overline{y}$ to get

$$S(x,y) = x\left(\overline{xy}\right) + y\left(\overline{xy}\right)$$

Double negation does not change the function value ($\overline{\overline{x}} = x$). So, take double negation to get

$$S(x,y) = \overline{\overline{x(\overline{xy}) + y(\overline{xy})}} \qquad (\because \overline{\overline{x}} = x)$$

Applying De Morgan's theorem: $\overline{x + y} = \overline{x}\,\overline{y}$, to get

$$S(x,y) = \overline{\left(\overline{x(\overline{xy})}\right)\left(\overline{y(\overline{xy})}\right)} \quad (\because \overline{x + y} = \overline{x}\,\overline{y}) \tag{5.18}$$

$$C(x,y) = xy$$

Take double negation to get

$$C(x,y) = \overline{\overline{xy}} \tag{5.19}$$

The logic diagram using only NAND gates is shown in Figure 5.14 to implement half-adder using Eq. (5.18) and Eq. (5.19).

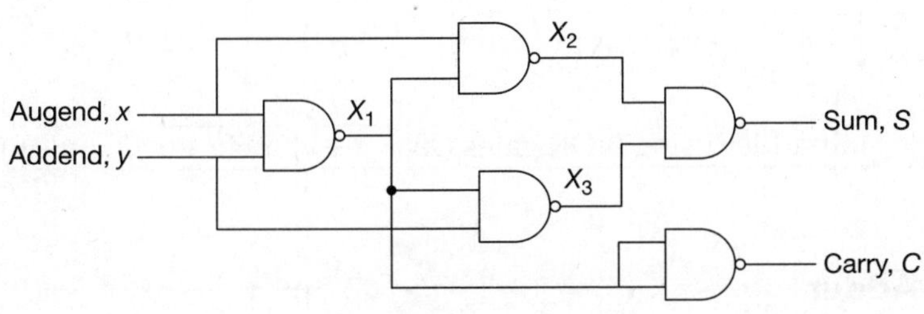

FIGURE 5.14 | Half-adder using NAND gates ($X_2 = \overline{xX_1}$, $X_3 = \overline{yX_1}$ and $X_1 = \overline{xy}$).

Half-adder using NOR gates: Alternative-V Consider sum, S function as given in Eq. (5.11)

$$S(x,y) = x\overline{y} + \overline{x}y$$

Rewrite above equation to get

$$S(x,y) = x\overline{x} + x\overline{y} + \overline{x}y + y\overline{y} \quad (\because x + 0 = x \text{ and } x\overline{x} = 0)$$

Rewrite above equation to get

$$S(x,y) = \bar{x}(x+y) + \bar{y}(x+y)$$

$$S(x,y) = (x+y)(\bar{x}+\bar{y})$$

Take double negation to get

$$S(x,y) = \overline{\overline{(x+y)(\bar{x}+\bar{y})}}$$

De Morganize the above function to get

$$S(x,y) = \overline{\overline{(x+y)} + \overline{(\bar{x}+\bar{y})}} \quad (\because \overline{xy} = \bar{x}+\bar{y}) \tag{5.20}$$

Take double negation of Eq. (5.13) to get

$$C(x,y) = \overline{\overline{xy}} \quad (\because \bar{\bar{x}} = x)$$

De Morganize the above function to get

$$C(x,y) = \overline{\bar{x}+\bar{y}} \quad (\because \overline{xy} = \bar{x}+\bar{y}) \tag{5.21}$$

The logic diagram using only NOR gates is shown in Figure 5.15 to implement half-adder using Eq. (5.20) and Eq. (5.21).

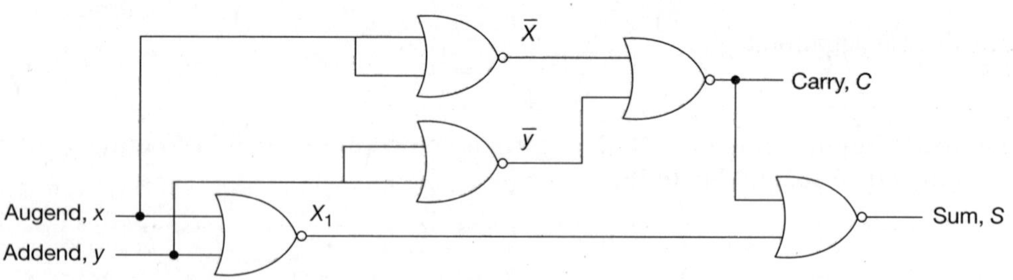

FIGURE 5.15 | Half-adder using NOR gates ($X_1 = \overline{x+y}$, $\bar{x} = \overline{x+x}$, $\bar{y} = \overline{y+y}$).

5.2.2 | Full-Adder

Two single binary digits (bits) along with carry bit are added to produce two outputs, a sum and a carry. Any overflow from first bit into second bit is known as carry out. Full-adder circuit performs addition of two single bit binary values along with carry to produce single bit binary sum and a carry-out.

Let single-bit binary augend, x, addend, y and previous carry, z. On addition single bit binary sum, S and carry, C an overflow in second bit is obtained. Full-adder can be designed following two ways

1. Cascading half-adders.
2. Full-adder design

5.2.2.1 | Full-Adder by Cascading Half-Adders

Full-adder is designed using two half-adders and one OR gate. Augend, x and addend, y are added using first half-adder to get partial sum S_1 and partial carry C_1. Second half-adder adds partial sum, S_1 and previous carry, z to find sum, S and partial carry C_2. Either carry C_1 or carry C_2 gives carry-out, C. Full-adder diagram is shown in Figure 5.16 using half-adders and OR gate.

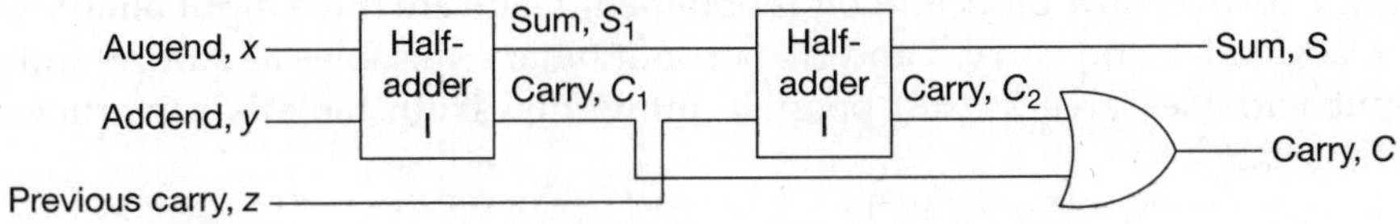

FIGURE 5.16 | Full-adder from half-adders and OR gate

The combinational circuit diagram of full-adder is given in Figure 5.17 using half-adder circuit diagram given in Figure 5.9.

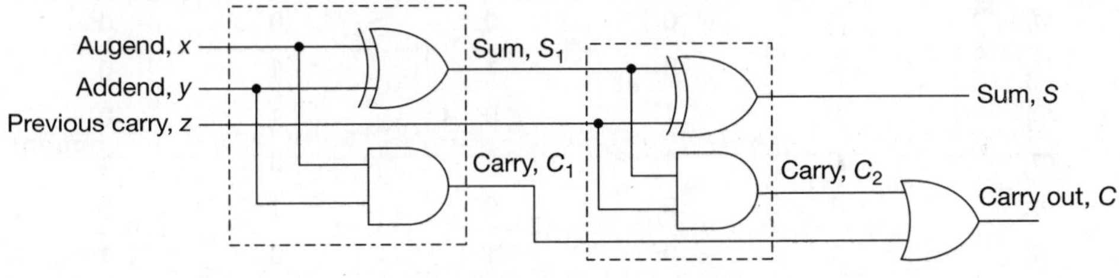

FIGURE 5.17 | Full-adder from half-adders and OR gate

Outputs from first half-adder are sum S_1, and carry-out, C_1 and are given below:

$$S_1 = x \oplus y$$

$$C_1 = xy$$

Outputs from second half-adder are obtained as sum, S_2 and carry-out, C_2 by adding S_1 and previous carry. Boolean equations are given below:

$$S = S_1 \oplus z$$

$$C_2 = S_1 z$$

Substitute S_1 in above equations, to get

$$S = (x \oplus y) \oplus z \tag{5.22}$$

$$C_2 = (x \oplus y) z$$

Either half-adder may produce carry-out to consider. So carry-out, C is obtained as

$$C = C_1 + C_2$$

Substitute the values of C_1 and C_2 in above equation to get

$$C = xy + (x \oplus y) z \tag{5.23}$$

5.2.2.2 | Full-Adder Design

Three single-binary digit (bit) values consisting augend, addend and previous carry are added to produce sum and carry-out. Any overflow from first bit into second bit is known as carry. Full-adder circuit performs addition of three single-bit values consisting augend, addend and previous carry and produces single-bit sum and a carry.

Let single-bit augend, x, addend, y and previous carry, z. On addition, single bit sum, S and carry, C an overflow in second bit is obtained. There are three input binary variables; augend, x, addend, y and carry, z and two output binary variables as sum, S and carry, C. Three input variables give 2^3 (=8) possible minterms. Truth table is constructed and is given in Table 5.5.

TABLE 5.5 | Truth table for full-adder

Minterm	Inputs			Outputs	
	Augend x	Addend y	Carry-in z	Sum S	Carry-out C
m_0	0	0	0	0	0
m_1	0	0	1	1	0
m_2	0	1	0	1	0
m_3	0	1	1	0	1
m_4	1	0	0	1	0
m_5	1	0	1	0	1
m_6	1	1	0	0	1
m_7	1	1	1	1	1

	2nd bit	Ist bit	2nd bit	Ist bit	2nd bit	Ist bit	2nd bit	Ist bit
Carry, z		0		1		1		1
Augend, x	+	1	+	0		1	+	1
Addend, y	+	1	+	1	+	0	+	1
Sum, S	1	0	1	0	1	0	1	1

Carry

Boolean equations for two outputs are given below:

$$S(x, y, z) = \sum m(1, 2, 4, 7) \tag{5.24}$$

$$C(x, y, z) = \sum m(3, 5, 6, 7) \tag{5.25}$$

Three variables K-maps are drawn to simplify the output Boolean function. K-maps are given in Figure 5.18(a) for sum, S and Figure 5.18(b) for carry-out, C. Sum, S function cannot be simplified as no pair is formed. Carry function forms three pairs {3, 7}, {5, 7} and {6, 7}.

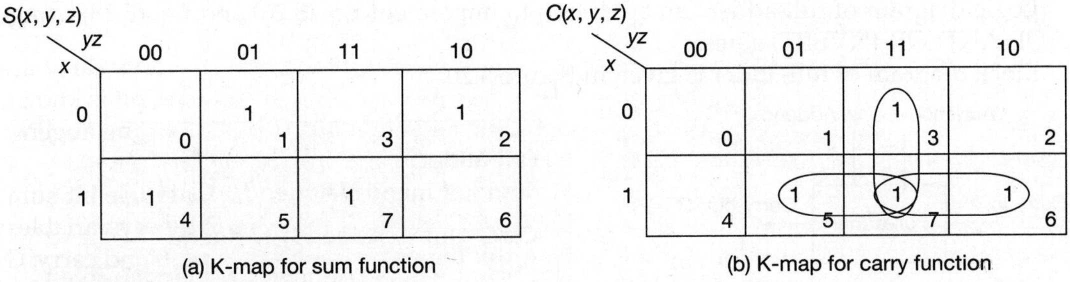

FIGURE 5.18 | K-map for full-adder

Logic expression of sum, S for full-adder is given below.

$$S(x,y,z)=\bar{x}\bar{y}z+\bar{x}y\bar{z}+x\bar{y}\bar{z}+xyz \qquad (5.26)$$

Rewriting Eq. (5.26),

$$S=\bar{x}\left(\bar{y}z+y\bar{z}\right)+x\left(\bar{y}\bar{z}+yz\right)$$

Rewriting above equation,

$$S=\bar{x}\left(y\oplus z\right)+x\left(y\odot z\right) \quad \because \left(y\oplus z\right)=\left(\bar{y}z+y\bar{z}\right)\text{and}\left(y\odot z\right)=\left(\bar{y}\bar{z}+yz\right)$$

Rewriting above equation,

$$S=\bar{x}\left(y\oplus z\right)+x\left(\overline{y\oplus z}\right) \quad \because \left(y\odot z\right)=\left(\overline{y\oplus z}\right)$$

Rewriting above equation as

$$S=x\oplus\left(y\oplus z\right)$$

Rewriting above equation as

$$S=x\oplus y\oplus z \qquad (5.27)$$

Using K-map (Figure 5.18(b)), simplified logic expression for carry-out, C of full-adder is given below.

$$C(x,y,z)=xz+yz+xy \qquad (5.28)$$

Logic diagram for full-adder is shown in Figure 5.19 to implement Eq. (5.27) and Eq. (5.28).

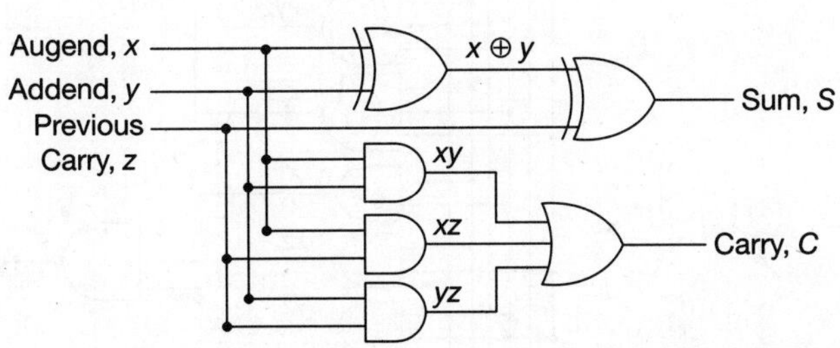

FIGURE 5.19 | Logic diagram of full-adder

Logic diagram of full-adder can be drawn to implement Eq. (5.26) and Eq. (5.28) using AOI (AND-OR-INVERT) gates.

Block diagram of full-adder is given in Figure 5.20.

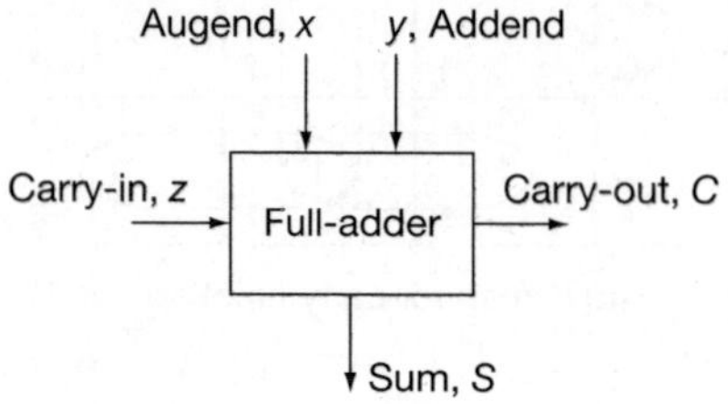

Full-adder:
Sum = Augend ⊕ Addend ⊕ Carry-in
Carry-Out = (Augend) · (Addend) + (Augend + Addend)Carry-in

FIGURE 5.20 | Block diagram of adder

Implementing full-adder using NAND gates Taking double negation, Eq. (5.26) remains the same

$$S(x,y,z)=\overline{\overline{\bar{x}\bar{y}z+\bar{x}y\bar{z}+x\bar{y}\bar{z}+xyz}}$$

Apply De Morgan's Theorem ($\because \overline{x+y}=\bar{x}\,\bar{y}$) to implement using NAND gates (5.29)

$$S(x,y,z)=\overline{\left(\overline{\bar{x}\bar{y}z}\right)\left(\overline{\bar{x}y\bar{z}}\right)\left(\overline{x\bar{y}\bar{z}}\right)\left(\overline{xyz}\right)}$$

Taking double negation, Eq. (5.28) remains the same

Apply De Morgan's Theorem ($\because \overline{x+y}=\bar{x}\,\bar{y}$) to implement using NAND gates

$$C(x,y,z)=\overline{\left(\overline{xz}\right)\left(\overline{yz}\right)\left(\overline{xy}\right)} \tag{5.30}$$

Logic diagram of full-adder is shown in Figure 5.21 to implement Eq. (5.29) and Eq. (5.30) using NAND gate.

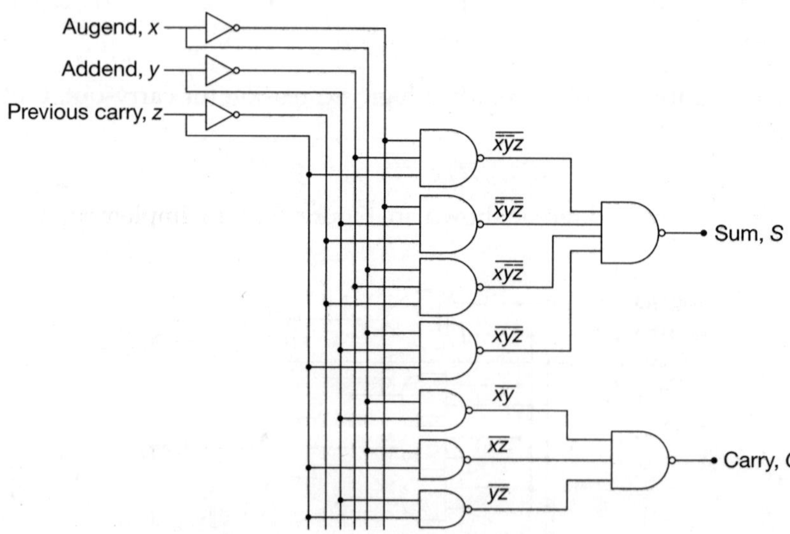

FIGURE 5.21 | Combination circuit of full-adder with fundamental (AOI) gates

5.3 | BINARY SUBTRACTOR

To process the numerical data, arithmetic operations are necessary. To perform arithmetic operations, arithmetic circuits are required. To process the binary data, subtract operation needs to study. In the proceeding section, half and full-subtractor circuits are discussed. Subtractor circuits are divided into two categories

1. Half-subtractor
2. Full-subtractor

> ***Note:*** Single *bit* means binary digit.

5.3.1 | Half-Subtractor

Single-bit value is subtracted from another single bit value to produce difference and borrow bits. Any borrow from second bit into first bit is known as borrow taken. Half-subtractor circuit performs subtraction of single bit value from another single bit value and produces single bit difference and borrows.

Let single-bit minuend, x and subtrahend, y. On performing subtraction single bit binary difference, D and borrow, B from second bit is obtained. There are two input binary variables; minuend, x and subtrahend, y and two output binary variables as difference, D and borrow, B. Two input variables give 2^2 (=4) possible minterms. Truth table is constructed and is given in Table 5.6.

TABLE 5.6 | Truth table half-subtractor

Minterm	**Inputs**		**Outputs**	
	Minuend x	**Subtrahend** y	**Difference** D	**Borrow** B
m_0	0	0	0	0
m_1	0	1	1	1
m_2	1	0	1	0
m_3	1	1	0	0

	2nd bit	1st bit	2nd bit	1st bit	2nd bit	1st bit	2nd bit	1st bit
Minuend		0		1		1		0
Subtrahend	–	0	–	1	–	0	–	1
Difference	0	0	0	0	0	1	1	1
	Borrow	Difference	Borrow	Difference	Borrow	Difference	Borrow	Difference

Boolean equations for two outputs are given below:

$$D(x,y) = \sum m(1,2) \tag{5.31}$$

$$B(x,y) = \sum m(1) \tag{5.32}$$

Two variable K-maps are drawn to simply the output Boolean function. K-maps are given in Figure 5.22(a) for difference, D and Figure 5.22(b) for borrow, B. Function cannot be simplified as no pair is formed

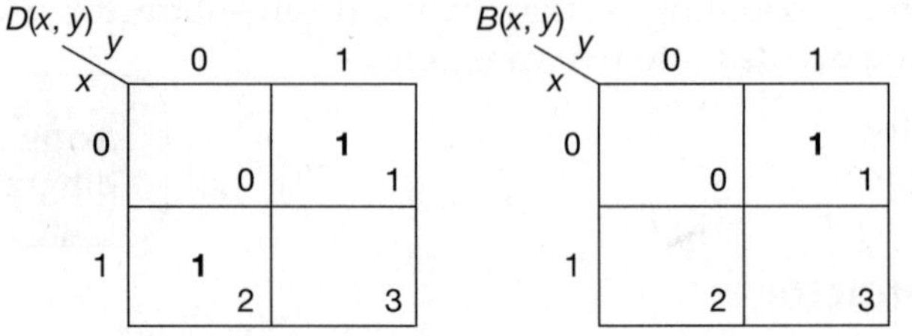

(a) K-map for difference (b) K-map for borrow

FIGURE 5.22 | K-map for half-subtractor

Logic expressions difference, D and borrow, B for half-subtractor are given below.

$$D(x,y) = \overline{x}y + x\overline{y} \tag{5.33}$$

$$D(x,y) = x \oplus y \tag{5.34}$$

$$B(x,y) = \overline{x}y \tag{5.35}$$

Logic diagram for half-subtractor is shown in Figure 5.23 and block diagram is given in Figure 5.24.

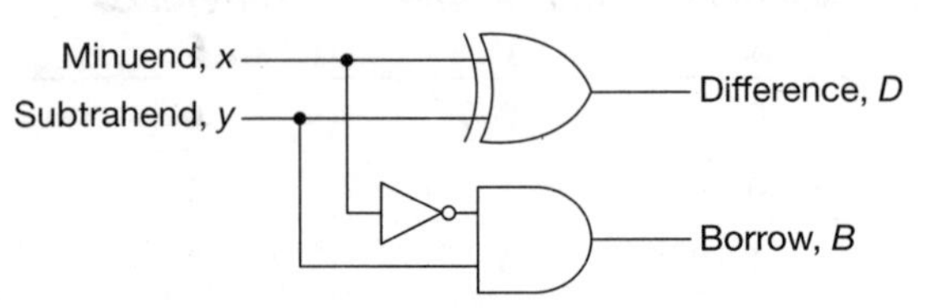

FIGURE 5.23 | Logic diagram of half-subtractor

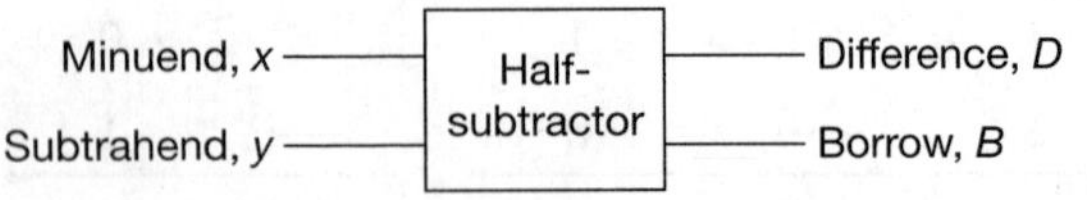

> **Half-subtractor:**
> Difference = Minuend ⊕ Subtrahend
> Borrow = $(\overline{\text{Minuend}}) \cdot (\text{Subtrahend})$

FIGURE 5.24 | Block diagram of half-subtractor

Logic diagram of half-subtractor using AOI gates Logic diagram of half-subtractor can be obtained using AOI fundamental gates. Figure 5.25 shows the half-subtractor using AOI gates to implement Boolean function given by Eq. (5.33) and Eq. (5.35).

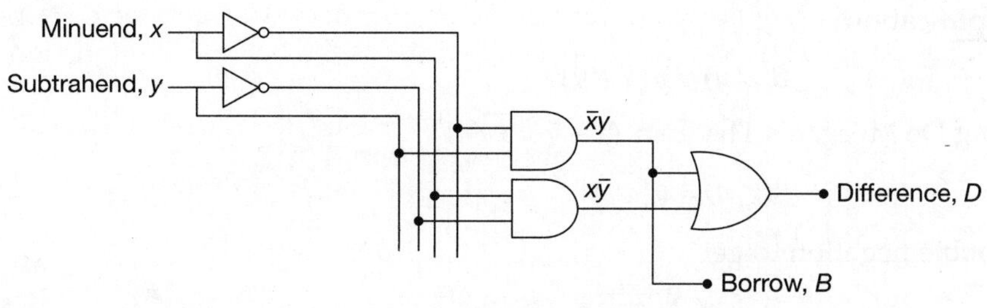

FIGURE 5.25 | Half-subtractor with basic AOI gates

Half-subtractor using NAND gates Consider difference, D function as given in Eq. (5.33)

$$D(x,y) = x\overline{y} + \overline{x}y$$

Performing OR operation of x binary variable with 0-logic does not change the variable value.

$$D(x,y) = 0 + x\overline{y} + \overline{x}y + 0 \qquad (\because x + 0 = x)$$

Zero-logic can be obtained from AND operation of binary variable with its complement.

$$D(x,y) = x\overline{x} + x\overline{y} + \overline{x}y + y\overline{y} \qquad (\because\ x\overline{x} = 0 \text{ and } y\overline{y} = 0)$$

On simplification

$$D(x,y) = x(\overline{x} + \overline{y}) + y(\overline{x} + \overline{y}) \qquad (\because \overline{x} + \overline{y} = \overline{xy}, \text{ De Morgan's theorem})$$

Apply De Morgan's theorem: $\overline{x\,y} = \overline{x} + \overline{y}$ to get

$$D(x,y) = x(\overline{xy}) + y(\overline{xy})$$

Double negation never changes the function value ($\overline{\overline{x}} = x$). So, take double negation to get

$$D(x,y) = \overline{\overline{x(\overline{xy}) + y(\overline{xy})}} \qquad (\because \overline{\overline{x}} = x)$$

Applying De Morgan's theorem: $\overline{x + y} = \overline{x}\,\overline{y}$, to get

$$D(x,y) = \overline{\left(\overline{x(\overline{xy})}\right)\left(\overline{y(\overline{xy})}\right)} \qquad (\because \overline{x + y} = \overline{x}\,\overline{y}) \tag{5.36}$$

Consider Eq. (5.35) of borrow, B

$$B(x,y) = \overline{x}y$$

Performing OR operation with 0-logic does not change the equation

$$B(x,y) = \overline{x}y + 0$$

Zero-logic can be obtained from AND operation of binary variable with its complement.

$$B(x,y) = \overline{x}y + \overline{y}y$$

On simplification

$$B(x,y) = y(\bar{x} + \bar{y})$$

Applying De Morgan's Theorem $(\bar{x} + \bar{y} = \overline{xy})$

$$B(x,y) = y(\overline{xy})$$

Take double negation to get

$$B(x,y) = \overline{\overline{y(\overline{xy})}} \tag{5.37}$$

The logic diagram using only NAND gates is shown in Figure 5.26 for Boolean Eq. (5.36) and Eq. (5.37) to implement half-subtractor.

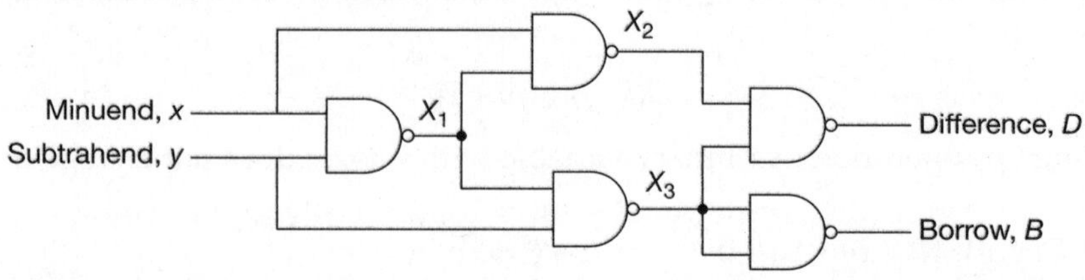

FIGURE 5.26 | Half-subtractor using NAND gates ($X_2 = \overline{xX_1}$, $X_3 = \overline{yX_1}$ and $X_1 = \overline{xy}$)

Half-subtractor using NOR gates Consider difference, D function as given in Eq. (5.33)

$$D(x,y) = x\bar{y} + \bar{x}y$$

Above equation can be rewritten as

$$D(x,y) = x\bar{x} + x\bar{y} + \bar{x}y + y\bar{y} \qquad (\because x + 0 = x \text{ and } x\bar{x} = 0)$$

Rearranging the above equation,

$$D(x,y) = x\bar{x} + \bar{x}y + x\bar{y} + y\bar{y}$$

Consider common terms

$$D(x,y) = \bar{x}(x+y) + \bar{y}(x+y)$$

Performing double negation,

$$D(x,y) = \overline{\overline{\bar{x}(x+y) + \bar{y}(x+y)}}$$

Applying De Morgan's Theorem $(\overline{xy} = \bar{x} + \bar{y})$

$$D(x,y) = \overline{\left(\overline{\bar{x}(x+y)}\right)\left(\overline{\bar{y}(x+y)}\right)}$$

Applying De Morgan's Theorem $(\overline{xy} = \bar{x} + \bar{y})$

$$D(x,y) = \overline{\left(x + (\overline{x+y})\right)\left(y + (\overline{x+y})\right)}$$

Applying De Morgan's Theorem $(\overline{xy} = \bar{x} + \bar{y})$

$$D(x,y) = \left(\overline{x + (\overline{x+y})}\right) + \left(\overline{y + (\overline{x+y})}\right)$$

Performing double negation

$$D(x,y)=\overline{\overline{\left(\overline{x+\overline{(x+y)}}\right)+\left(\overline{y+\overline{(x+y)}}\right)}} \tag{5.38}$$

Consider Eq. (5.35) of borrow, B

$$B(x,y)=\bar{x}y$$

Performing OR operation with 0-logic does not change the equation

$$B(x,y)=\bar{x}y+0$$

Zero logic can be obtained from AND operation of binary variable with its complement.

$$B(x,y)=\bar{x}y+\bar{x}x$$

Consider common term

$$B(x,y)=\bar{x}(y+x)$$

Take double negation to get

$$B(x,y)=\overline{\overline{\bar{x}(y+x)}}$$

Applying De Morgan Theorem $(\overline{xy}=\bar{x}+\bar{y})$

$$B(x,y)=\overline{x+\overline{(y+x)}} \tag{5.39}$$

The logic diagram using only NOR gates is shown in Figure 5.27 for Boolean Eq. (5.38) and Eq. (5.39) to implement half-subtractor.

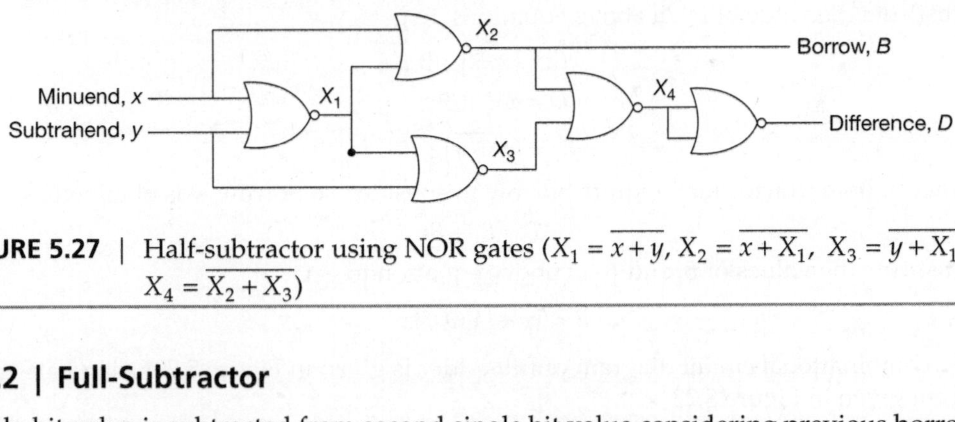

FIGURE 5.27 | Half-subtractor using NOR gates ($X_1=\overline{x+y}$, $X_2=\overline{x+X_1}$, $X_3=\overline{y+X_1}$ and $X_4=\overline{X_2+X_3}$)

5.3.2 | Full-Subtractor

Single-bit value is subtracted from second single bit value considering previous borrow to produce difference and borrow bits. Any borrow from second bit into first bit is known as borrow. Full-subtractor circuit performs subtraction of single bit from another single-bit value considering borrow and produces single-bit difference and borrow.

Let single-bit binary minuend, x, subtrahend, y and previous borrow, z. On performing subtraction single-bit binary difference, D and borrow, B is obtained. Full-subtractor can be designed following two ways

1. Cascading half-subtractor.
2. Full-subtractor design

5.3.2.1 | Full-subtractor by Cascading Half-subtractors

Full-subtractorcan be designed using two half-subtractors and one OR gate. Subtrahend, y is subtracted from Minuend, x using first half-subtractor to get partial difference D_1 and partial borrow B_1. Second half-subtractor subtracts previous borrow z from partial difference, D_1 to find difference, D and partial borrow B_2. Either borrow B_1 or borrow B_2 gives borrow, B. Full-subtractor diagram is shown in Figure 5.28 using half-adders and OR gate.

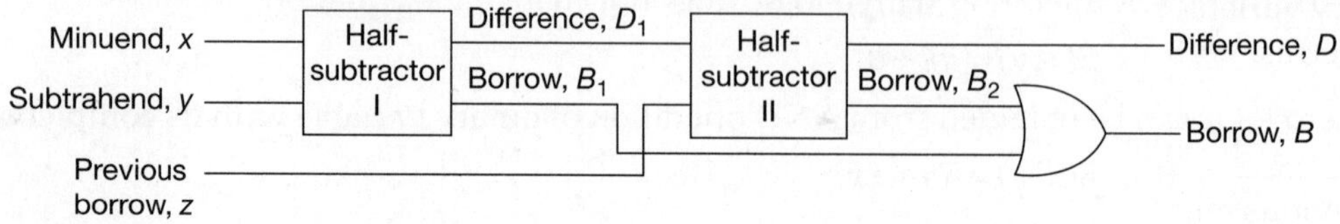

FIGURE 5.28 | Block diagram of full-subtractor using half-subtractors

Outputs from first half-subtractor are difference D_1, and borrow B_1 as given as follows:

$$D_1 = x \oplus y$$
$$B_1 = \bar{x}y$$

Outputs from second half-subtractor are obtained as difference, D_2 and borrow B_2 by subtracting previous borrow, z from D_1. Boolean equations are given as follows:

$$D = D_1 \oplus z$$
$$B_2 = \bar{D}_1 z$$

Substitute the value of D_1 in above equations, to get

$$D = (x \oplus y) \oplus z$$
$$D = x \oplus y \oplus z \tag{5.40}$$
$$B_2 = \left(\overline{x \oplus y}\right) z$$

Either half-subtractor may require borrow to consider, so borrow, B is obtained as

$$B = B_1 + B_2$$

Substitute the values of B_1 and B_2 in above equation to get

$$B = \bar{x}y + \left(\overline{x \oplus y}\right) z \tag{5.41}$$

The combinational circuit diagram of full-adder is given in Figure 5.29 using half-adder diagram given in Figure 5.23.

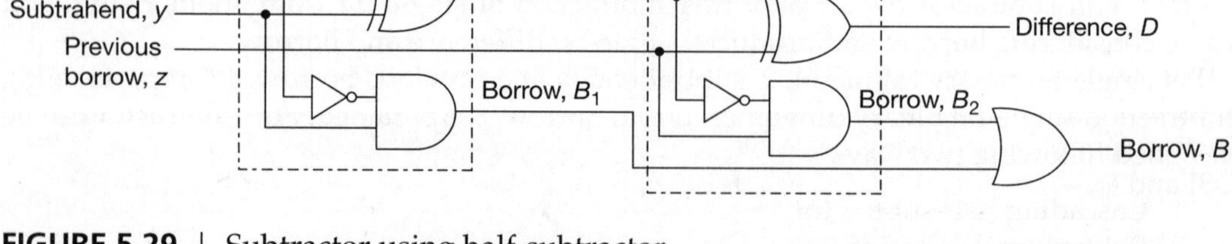

FIGURE 5.29 | Subtractor using half-subtractor

5.3.2.2 | Full-subtractor Design

Full-subtractor circuit performs subtraction of subtrahend single bit from minuend single bit value considering previous borrow, *z* and produces single bit difference, *D* and borrow, *B*.

Let single-bit minuend, *x*, subtrahend, *y* and previous borrow, *z*. On performing subtraction of single bit operands difference, *D* and borrow from second bit is obtained. There are three input binary variables; minuend, *x*, subtrahend, *y* and borrow, *z* and two output binary variables as difference, *D* and borrow, *B*. Three input variables give 2^3 (=8) possible minterms. Truth table is constructed and is given in Table 5.7.

Borrow, *B*	1	10	1	10	1	10	1	10
Previous borrow, *z*	−	1	−	1	−	1	−	1
Minuend, *x*	+	0	+	0	+	0	+	1
Subtrahend, *y*	−	0	−	1	−	0	−	1
Difference, *D*	0	1	1	0		1	1	1

TABLE 5.7 | Truth table of full-subtractor

Minterm	Inputs			Outputs	
	Minuend *x*	Subtrahend *y*	Previous Borrow *z*	Difference *D*	Borrow *B*
m_0	0	0	0	0	0
m_1	0	0	1	1	1
m_2	0	1	0	1	1
m_3	0	1	1	0	1
m_4	1	0	0	1	0
m_5	1	0	1	0	0
m_6	1	1	0	0	0
m_7	1	1	1	1	1

Boolean equations for two outputs are given below:

$$D(x, y, z) = \sum m(1, 2, 4, 7) \tag{5.42}$$

$$B(x, y, z) = \sum m(1, 2, 3, 7) \tag{5.43}$$

Two variable K-maps are drawn to simply the output Boolean function. K-maps are given in Figure 5.30(a) for difference, *D* and Figure 5.30(b) for borrow, *B*. Difference, *D* function cannot be simplified as no pair is formed. Borrow function forms three pairs {1, 3}, {2, 3} and {3, 7}.

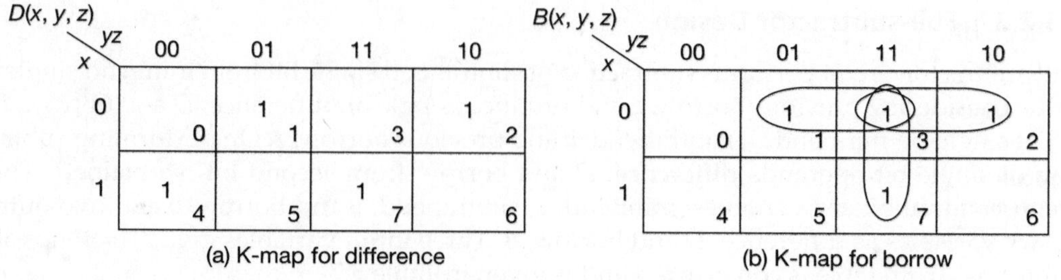

FIGURE 5.30 | K-map for full-subtractor

Logic expression difference, S for full-subtractor is given below.

$$D(x,y,z)=\bar{x}\bar{y}z+\bar{x}y\bar{z}+x\bar{y}\bar{z}+xyz \tag{5.44}$$

Rewriting above equation by considering common terms,

$$D=\bar{x}(\bar{y}z+y\bar{z})+x(\bar{y}\bar{z}+yz)$$

Rewriting above equation, we have

$$D=\bar{x}(y\oplus z)+x(y\odot z)\qquad \because (y\oplus z)=(\bar{y}z+y\bar{z})\text{and}(y\odot z)=(\bar{y}\bar{z}+yz)$$

Rewriting above equation as

$$D=\bar{x}(y\oplus z)+x(\overline{y\oplus z})\qquad \because (y\odot z)=(\overline{y\oplus z})$$

Rewriting above equation as

$$D=x\oplus(y\oplus z)$$

Rewriting above equation as

$$D=x\oplus y\oplus z \tag{5.45}$$

Using K-map given in Figure 5.30(b), simplified logic expression borrow, B for full-subtractor is given below.

$$B(x,y,z)=\bar{x}z+yz+\bar{x}y \tag{5.46}$$

Logic diagram for full-subtractor is shown in Figure 5.31 to implement Eq. (5.45) and Eq. (5.46)

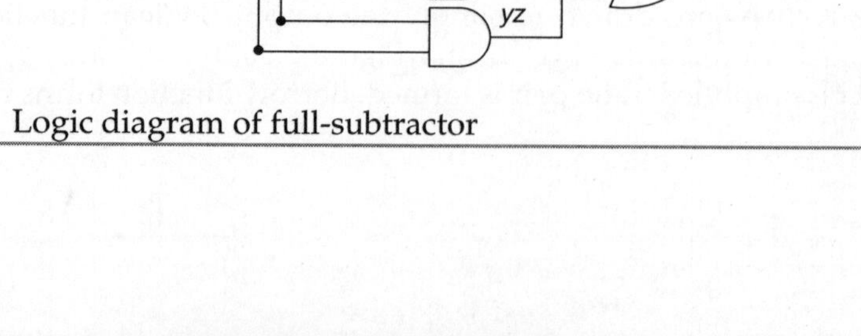

FIGURE 5.31 | Logic diagram of full-subtractor

Block diagram of full-subtractor is given in Figure 5.32.

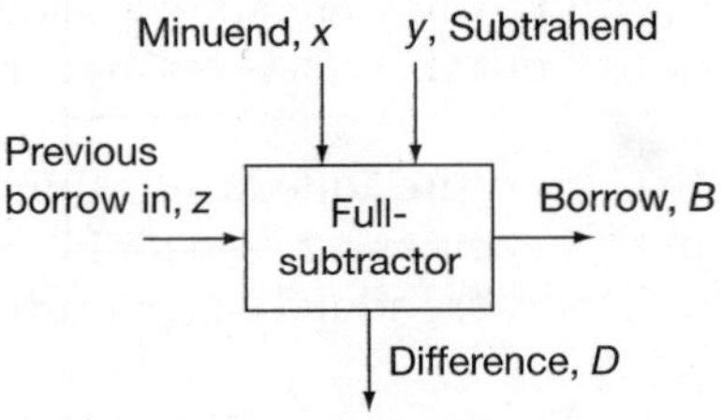

Full-subtractor:
Difference = Minuend ⊕ Subtrahend ⊕ Previous Borrow

Borrow = $(\overline{\text{Minuend}})\cdot(\text{Subtrahend}) + (\overline{\text{Minuend}} + \text{Subtrahend})\,\text{Prev. Borrow}$

FIGURE 5.32 | Block diagram of subtractor

Taking double negation of Eq. (5.44) remains the same.

$$S(x,y,z) = \overline{\overline{\bar{x}\bar{y}z + \bar{x}y\bar{z} + x\bar{y}\bar{z} + xyz}}$$

Applying De Morgan's Theorem to implement using NAND gates,

$$S(x,y,z) = \overline{\left(\overline{\bar{x}\bar{y}z}\right)\left(\overline{\bar{x}y\bar{z}}\right)\left(\overline{x\bar{y}\bar{z}}\right)\left(\overline{xyz}\right)} \tag{5.47}$$

Taking double negation, Eq. (5.46) remains the same.

$$B(x,y,z) = \overline{\overline{\bar{x}z + yz + \bar{x}y}}$$

Applying De Morgan's Theorem to implement using NAND gates,

$$B(x,y,z) = \overline{\left(\overline{\bar{x}z}\right)\left(\overline{yz}\right)\left(\overline{\bar{x}y}\right)} \tag{5.48}$$

Combinational circuit for full-subtractor using NAND gates to implement Eq. (5.47) and Eq. (5.48).

5.4 | BINARY PARALLEL ADDER

To add more bits, there is need of more than one full-adder. Let four-bit numbers, augend (X) and addend, (Y) with carry-in as zero are added to generate four-bit sum, S and carry out in fourth bit.

	Bit 4	Bit 3		Bit 2		Bit 1		Bit 0
Previous carry		C_3		C_2		C_1		C_0
Augend		x_3		x_2		x_1		x_0
Addend		y_3		y_2		y_1		y_0
Sum	C_4	S_3	C_3	S_2	C_2	S_1	C_1	S_0

- Firstly, bit 0 operands, augend, x_0 and addend, y_0 with c_0 carry are added using first full-adder and produces output, sum S_0 and carry-out C_1. C_1 acts as carry-in for bit 1.
- Secondly, bit 1 operands, augend, x_1 and addend y_1 with C_1 carry are added using second full-adder and produces output, sum S_1 and carry-out C_2. C_2 acts as carry-in for bit 2.
- Thirdly, bit 2 operands, augend, x_2 and addend y_2 with C_2 carry are added using third full-adder and produces output, sum S_2 and carry-out C_3. C_3 acts as carry-in for bit 3.
- Finally, bit 3 operands, augend, x_3 and addend y_3 with C_3 carry are added using fourth full-adder and produces output, sum S_3 and carry-out C_4.
- C_4 acts as carry-out.

Figure 5.33 gives the cascading of four full-adders to get four-bit operation. Carry-out of full-adder-I is connected to carry-in of full-adder-II and so on.

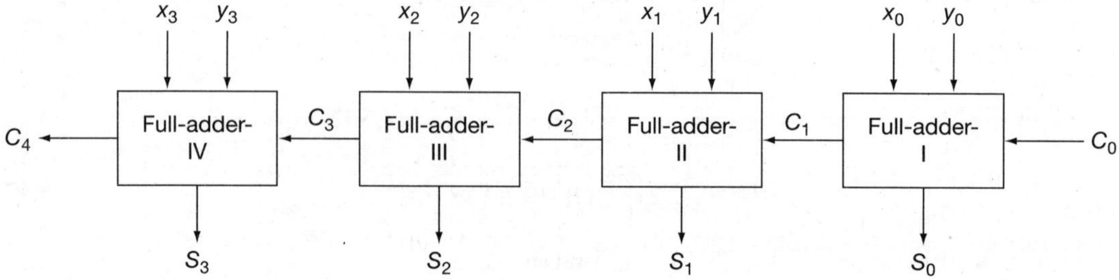

FIGURE 5.33 | Four-bit binary parallel adder

Consider Eq. (5.28) to generate carry for single bit.

$$C(x,y,z) = xy + (x+y)z$$

Represent carry in z as C_{in} and carry out C as C_{out}.

$$C_{out} = xy + (x+y)C_{in}$$

To generalize, full-adder equations can be represented for i^{th} bit where C_i is carry-in and will generate carry-out C_{i+1}.

$$S_i = x_i \oplus y_i \oplus C_i \quad (i = 0,1,2,\ldots,n) \tag{5.49}$$

$$C_{i+1} = x_i y_i + (x_i + y_i)C_i \quad (i = 0,1,\ldots,n) \tag{5.50}$$

Block diagram of four-bit parallel adder is shown in Figure 5.34. Parallel adder is also known as ripple carry adders. The main disadvantage is that it takes four times propagation delay of full-adder to complete four-bit addition because higher order bit operation can be performed only if carry-out from lower order bit occurs.

Inputs:

- C_0 acts as carry-in as input.
- Four-bit augend and four-bit addend operands are required as inputs.

Outputs:

- Four-bit sum will be output.
- C_4 as single bit carry-out will be output.

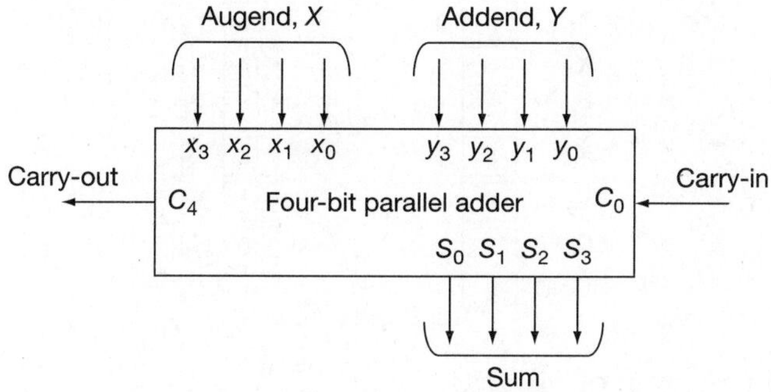

FIGURE 5.34 | Block diagram of four-bit parallel adder

Two four-bit parallel adders can be cascaded to get 1 byte addition. Carry-out of lower nibble adder is given to carry-in of higher nibble adder. Combination of four-bits is known as 1 nibble. Combination of eight-bits is known as 1 byte.

	bit 4	Higher nibble		Lower nibble
Carry		$C_7C_6C_5C_4$		$C_3C_2C_1C_0$
Augend		$X_7X_6X_5X_4$		$X_3X_2X_1X_0$
Addend		$Y_7Y_6Y_5Y_4$		$Y_3Y_2Y_1Y_0$
Sum	C_8	$S_7S_6S_5S_4$	C_4	$S_3S_2S_1S_0$

> ***Note:*** 1 nibble is a combination of four bits.

> ***Note:*** 1 byte is a combination of eight bits.

Block diagram of 1 byte adder is given in Figure 5.35.

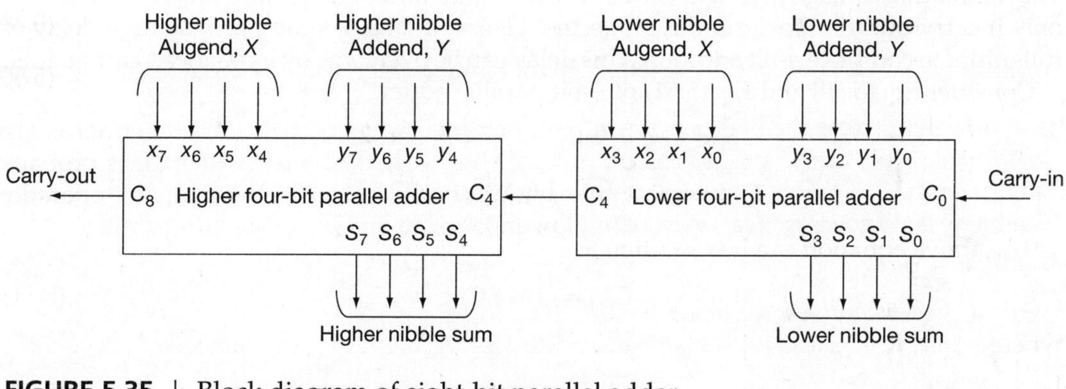

FIGURE 5.35 | Block diagram of eight-bit parallel adder

Pin diagram of 16-pin, 74LS283 IC is given in Figure 5.36 and logic symbol is given Figure 5.37.

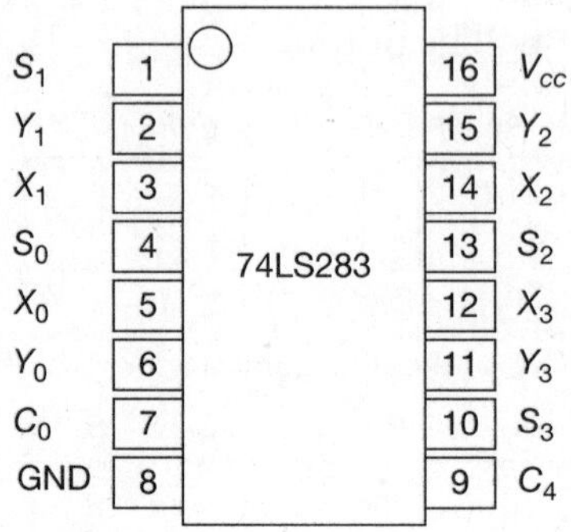

FIGURE 5.36 | Pin diagram of 74LS283

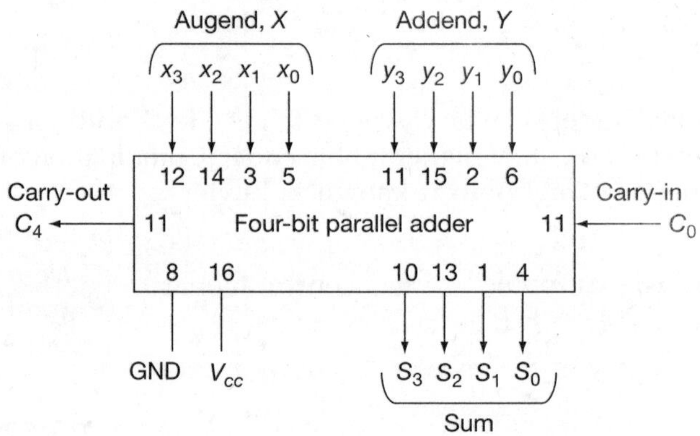

FIGURE 5.37 | 74LS283 logic symbol

5.5 | THE LOOK-AHEAD CARRY BINARY ADDERS

The main disadvantage of four-bit adder is that higher order bit operation can be performed only if carry-out from lower order bit occurs. Hence, it takes n-time's propagation delay of full-adder to complete n-bit addition. This delay can be overcome by look-ahead carry adder.

Consider Eq. (5.50) and Eq. (5.51) of n-bit parallel adder.

$$S_i = x_i \oplus y_i \oplus C_i \qquad (i = 0, 1, 2, \ldots, n)$$

$$C_{i+1} = x_i y_i + (x_i + y_i)C_i \quad (i = 0, 1, \ldots, n)$$

The above equation can be rewritten as

$$C_{i+1} = G_i + P_i C_i \tag{5.51}$$

where

$$G_i = x_i y_i \tag{5.52}$$

It is known as carry generator. When both input bits are 1, then carry is generated by full-adder.

$$P_i = x_i + y_i \tag{5.53}$$

It is known as carry propagation. An input carry is propagated by full-adder when either or both input bits are 1. Augend and addend bits are input bits.

Let $i = 0$. Equation (5.51) can be written as

$$C_1 = G_0 + P_0C_0 \tag{5.54}$$

Let $i = 1$. Equation (5.51) can be written as

$$C_2 = G_1 + P_1C_1$$

Substituting the value of carry generated, C_1 by Eq. (5.54),

$$C_2 = G_1 + P_1(G_0 + P_0C_0)$$

$$C_2 = G_1 + P_1G_0 + P_1P_0C_0 \tag{5.55}$$

Let $i = 2$. Equation (5.51) can be written as

$$C_3 = G_2 + P_2C_2$$

Substitute the value of carry generated, C_2 by Eq. (5.55).

$$C_3 = G_2 + P_2(G_1 + P_1G_0 + P_1P_0C_0)$$

$$C_3 = G_2 + P_2G_1 + P_2P_1G_0 + P_2P_1P_0C_0 \tag{5.56}$$

Let $i = 3$. Equation (5.51) can be written as

$$C_4 = G_3 + P_3C_3$$

Substitute the value of carry generated, C_3 by Eq. (5.56).

$$C_4 = G_3 + P_3(G_2 + P_2G_1 + P_2P_1G_0 + P_2P_1P_0C_0)$$

$$C_4 = G_3 + P_3G_2 + P_3P_2G_1 + P_3P_2P_1G_0 + P_3P_2P_1P_0C_0 \tag{5.57}$$

Carry generator is generalized as given below.

$$C_{i+1} = \sum_{k=0}^{i}\left\{\left(\prod_{j=k+1}^{i} P_j\right)G_k\right\} + \left(\prod_{j=0}^{i} P_j\right)C_0 \quad (i = 0, 1, \ldots, n) \tag{5.58}$$

Combinational circuit for four-bit look-ahead carry adder is shown in Figure 5.38.

5.6 | COMBINATIONAL CIRCUIT FOR COMPLEMENTS

Complements are used in digital computers for simplifying the subtraction operation and to represent signed numbers. There are two types of complements for each binary system.

1. One's complement or radix-minus-one complement
2. Two's complement or true complement.

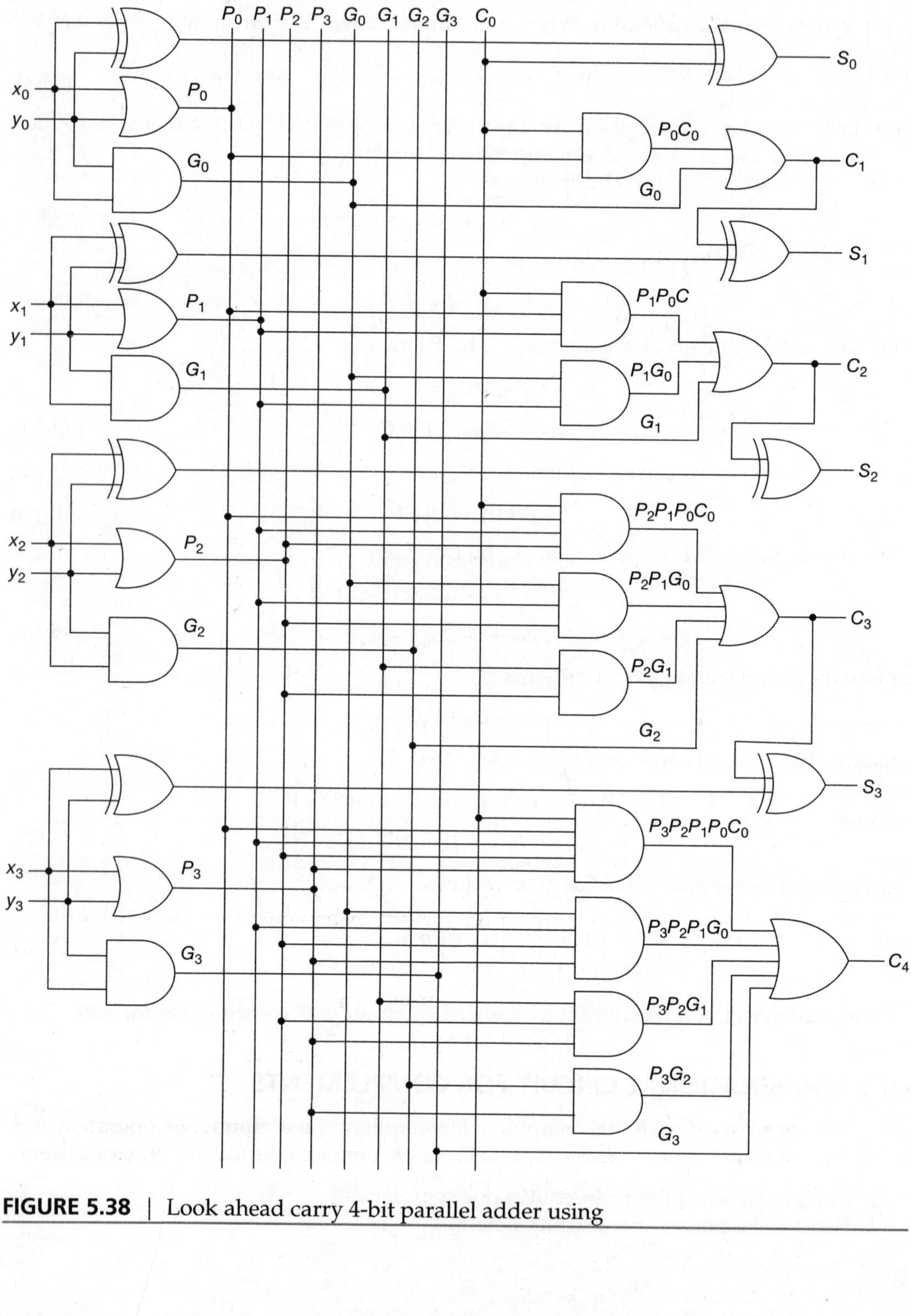

FIGURE 5.38 | Look ahead carry 4-bit parallel adder using

5.6.1 | One's Complement

Single bit can be inverted by using inverter, NAND gate, NOR gate, exclusive-OR (XOR) or exclusive-NOR (XNOR) gates as shown in Figure 5.39(a), Figure 5.39(b), Figure 5.39(c), Figure 5.39(d) and Figure 5.39(e), respectively.

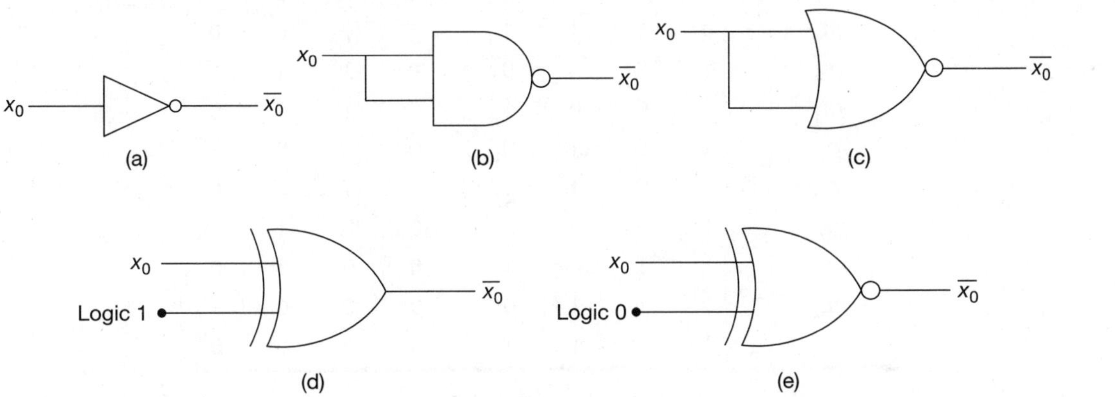

FIGURE 5.39 | (a) Complement using inverter gate; (b) Complement using NAND gate; (c) Complement using NOR gate; (d) Complement using XOR gates; (e) Complement using XNOR gates

5.6.1.1 | One's Complement of Four-bit Binary Number

Let four-bit binary number whose 1's complement can be obtained.

Number of inputs binary variables (or literals) = 4
Number of minterms = $2^4 (= 16)$
Number of output binary variables = 4

Let x_3, x_2, x_1 and x_0 are three single-bit variables. LSB is x_0 and MSB is x_3. Output is represented by three single-bit variables y_3, y_2, y_1 and y_0. For output LSB is y_0 and MSB is y_3. The truth table is shown in Table 5.8. Logic: Convert binary bit 0 to 1 and 1 to 0 to find 1's complement.

TABLE 5.8 | 1's complement of four-bit number

Minterm	Input				Output			
	x_3	x_2	x_1	x_0	y_3	y_2	y_1	y_0
m_0	0	0	0	0	1	1	1	1
m_1	0	0	0	1	1	1	1	0
m_2	0	0	1	0	1	1	0	1
m_3	0	0	1	1	1	1	0	0
m_4	0	1	0	0	1	0	1	1
m_5	0	1	0	1	1	0	1	0

(Continued)

TABLE 5.8 | (Continued)

Minterm	Input				Output			
	x_3	x_2	x_1	x_0	y_3	y_2	y_1	y_0
m_6	0	1	1	0	1	0	0	1
m_7	0	1	1	1	1	0	0	0
m_8	1	0	0	0	0	1	1	1
m_9	1	0	0	1	0	1	1	0
m_{10}	1	0	1	0	0	1	0	1
m_{11}	1	0	1	1	0	1	0	0
m_{12}	1	1	0	0	0	0	1	1
m_{13}	1	1	0	1	0	0	1	0
m_{14}	1	1	1	0	0	0	0	1
m_{15}	1	1	1	1	0	0	0	0

Boolean equations of outputs y_0, y_1, y_2 and y_3

$$y_0(x_3, x_2, x_1, x_0) = \sum m(0, 2, 4, 6, 8, 10, 12, 14) \tag{5.59}$$

$$y_1(x_3, x_2, x_1, x_0) = \sum m(0, 1, 4, 5, 8, 9, 12, 13) \tag{5.60}$$

$$y_2(x_3, x_2, x_1, x_0) = \sum m(0, 1, 2, 3, 8, 9, 10, 11) \tag{5.61}$$

$$y_3(x_3, x_2, x_1, x_0) = \sum m(0, 1, 2, 3, 4, 5, 6, 7) \tag{5.62}$$

K-map simplification method is adopted. Figure 5.40 shows K-map for output y_0, y_1, y_2 and y_3.

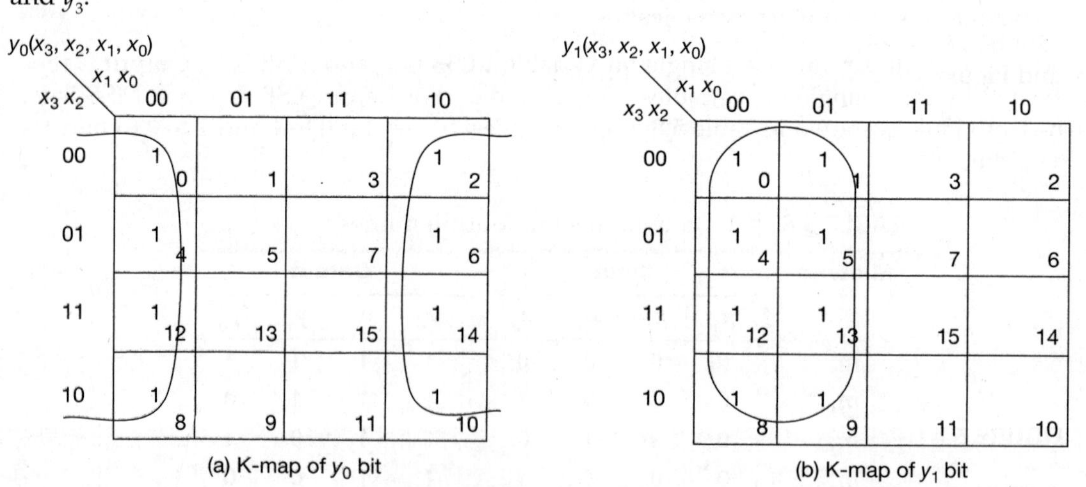

FIGURE 5.40 | *(Continued)*

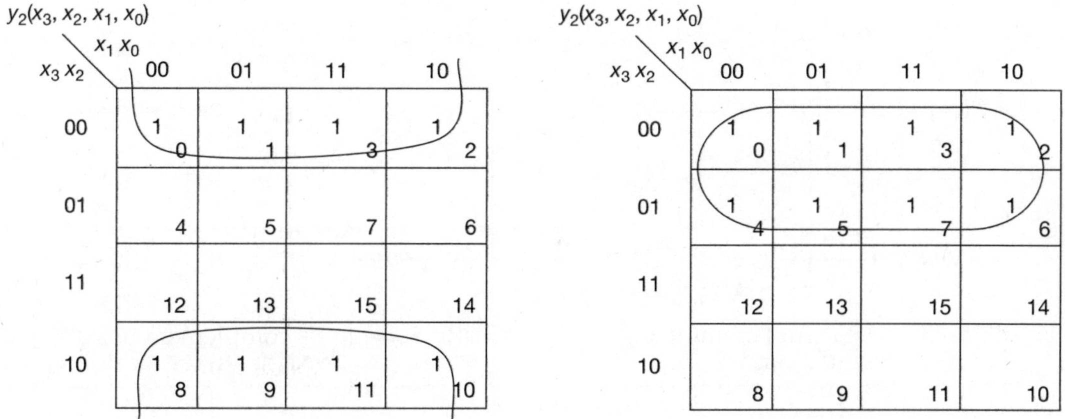

(c) K-map of y_2 bit

(d) K-map of y_3 bit

FIGURE 5.40 | K-map for 1's complement of 4-bit binary number

Figure 5.40(a): one octet {0, 2, 4, 6, 8, 10, 12, 14} gives simplified term $\overline{x}_0$. Figure 5.40(b): one octet {0, 1, 4, 5, 8, 9, 12, 13} gives simplified term $\overline{x}_1$. Figure 5.40(c): one octet {0, 1, 2, 3, 8, 9, 10, 11} gives simplified term $\overline{x}_2$. Figure 5.40(c): one octet {0, 1, 2, 3, 4, 5, 6, 7} gives simplified term $\overline{x}_3$. Simplified output expressions are given as follows:

$$y_0(x_3, x_2, x_1, x_0) = \overline{x}_0 \tag{5.63}$$

$$y_1(x_3, x_2, x_1, x_0) = \overline{x}_1 \tag{5.64}$$

$$y_2(x_3, x_2, x_1, x_0) = \overline{x}_2 \tag{5.65}$$

$$y_3(x_3, x_2, x_1, x_0) = \overline{x}_3 \tag{5.66}$$

Combinational circuits for 1's complement are shown in Figure 5.41, Figure 5.42, Figure 5.43 and Figure 5.44 using Inverter, NAND, NOR, XOR and XNOR gates, respectively.

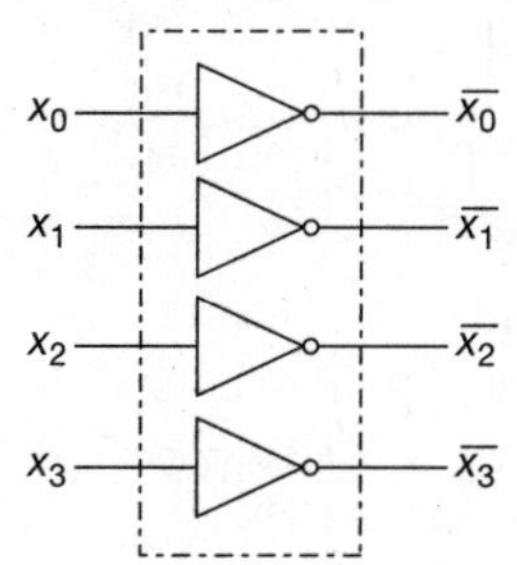

FIGURE 5.41 | 1's complement using inverters

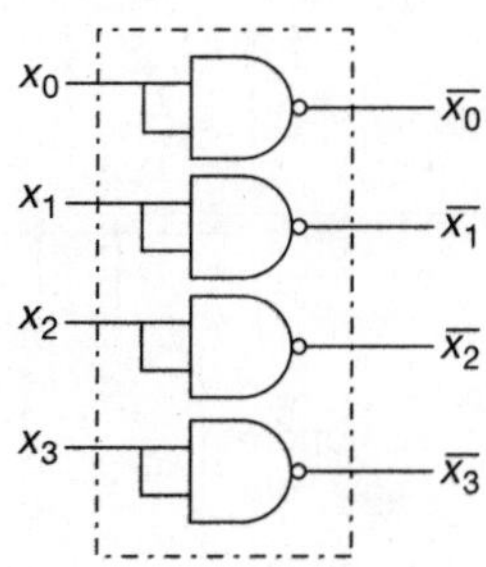

FIGURE 5.42 | 1's complement using NAND gates

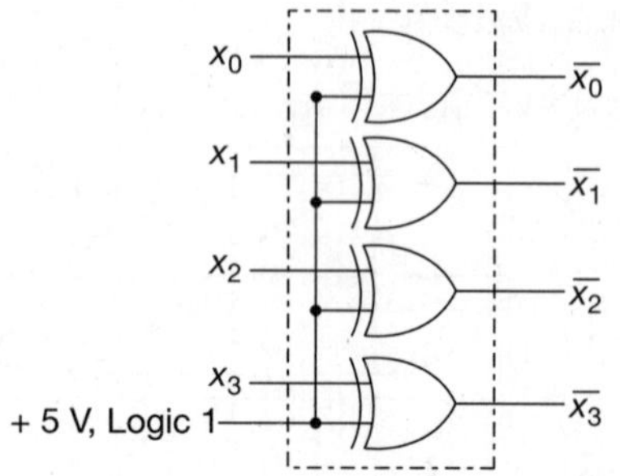

FIGURE 5.43 | 1′s complement using XOR gates

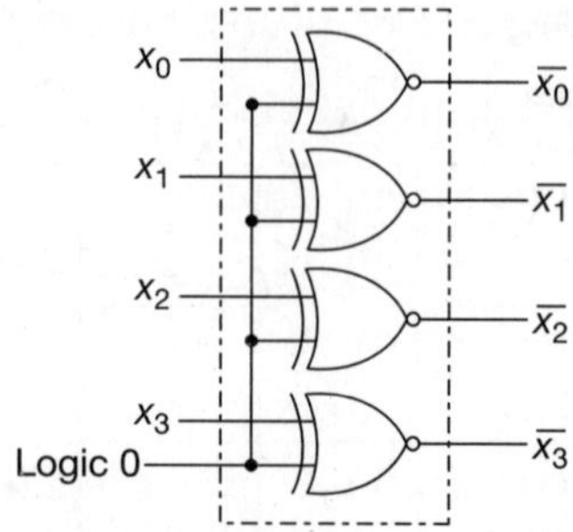

FIGURE 5.44 | 1′s complement using XNOR gates

5.6.1.2 | One's Complement using Binary Parallel Adder

One's complement can be obtained by performing following operations on four-bit parallel adder

- Take one's complement using any circuit given in Figure 5.43 or Figure 5.44.
- Perform addition with addend as (0000) without carry in will transfer the data without any change.

Circuit diagram shown in Figure 5.45 gives ones complement of four-bit number
Consider following example to get 1′s complement

Let $N =$		$(b_3b_2b_1b_0) =$		1001
1′s complement of $N =$		$(y_3y_2y_1y_0) =$		0110
Add zero	+	$(0\,0\,0\,0) =$		0000
Add 0 as carry (to LSB)	+	0	+	0
1′s complement of $N =$		$(y_3y_2y_1y_0)$		0110

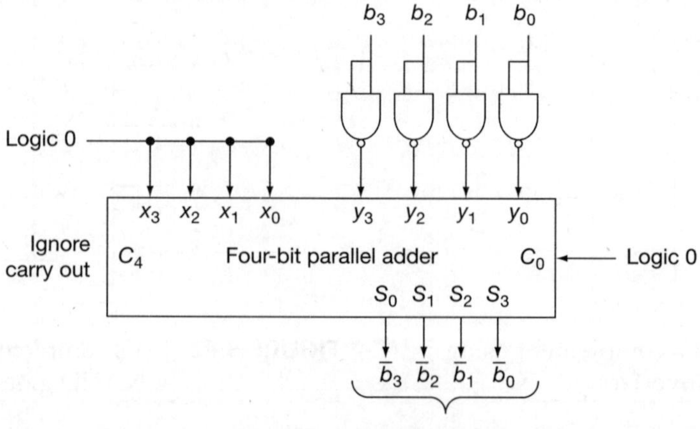

FIGURE 5.45 | 1′s complement from 4-bit binary adder

5.6.2 | Two's Complement using Binary Parallel Adder

Two's complement can be obtained by performing following operations

- Take one's complement
- Add one to LSB

Two's complement can be obtained by performing following operations on four-bit parallel adder

- Take one's complement using any circuit given in Figure 5.43 or Figure 5.44.
- Perform addition with addend as (0000) with carry as 1 to get 2's complement.

Circuit diagram shown in Figure 5.46 gives 2's complement of four-bit number. Example 5.4 explains operation of the circuit.

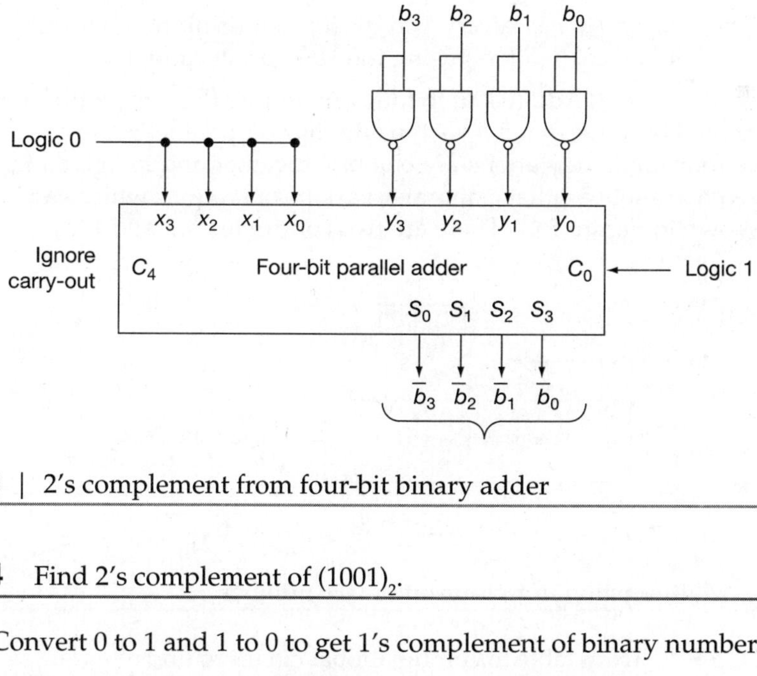

FIGURE 5.46 | 2's complement from four-bit binary adder

EXAMPLE 5.4 Find 2's complement of $(1001)_2$.

SOLUTION Convert 0 to 1 and 1 to 0 to get 1's complement of binary number.

Add 1 to the 1's complement of number to get 2's complement.

Let $N =$		$(b_3 b_2 b_1 b_0) =$	1001
1's complement of $N =$		$(y_3 y_2 y_1 y_0) =$	0110
Add Zero	+	$(0\,0\,0\,0) =$	0000
1's complement of $N =$		$(y_3 y_2 y_1 y_0) =$	0110
Add 1 as carry (to LSB)	+	$(0\,0\,0\,1) =$	0001
2's complement of $N =$		$(\bar{b}_3 \bar{b}_2 \bar{b}_1 \bar{b}_0) =$	0111

5.6.3 | Multifunction from Binary Parallel Adder

Parallel four-bit adder can be used to perform following function

(i) To transfer data (data buffer)
(ii) Increment data by 1
(iii) One's complement of data
(iv) Two's complement of data

Data transfer: XOR gate can be used to transfer the data because: $x \oplus 0 = x$. Figure 5.44 gives implementation of the data transfer using XOR gate, when logic 0 (0 V) is given instead of logic 1(+5 V).

Inverter: XOR gate can be used to find 1's complement because: $x \oplus 1 = \bar{x}$. Figure 5.44 gives implementation to find 1's complement of data using XOR gate.

Data Transfer using four-bit parallel adder: Add (0000) to four-bit number $(b_3b_2b_1b_0)$ without carry to get same number, i.e. $(b_3b_2b_1b_0)$ using four-bit parallel adder.

Increment four-bit datum by 1: Add (0000) to four-bit number $(b_3b_2b_1b_0)$ with carry as 1 to get number incremented by 1, i.e. $(b_3b_2b_1b_0) + 1$, using four-bit parallel adder.

All the above four implementations are clubbed together and are given in Figure 5.47. Truth table is given in Table 5.9 that elaborates various operations which can be performed on the circuit shown in Figure 5.47. There are two control lines, z and C_0.

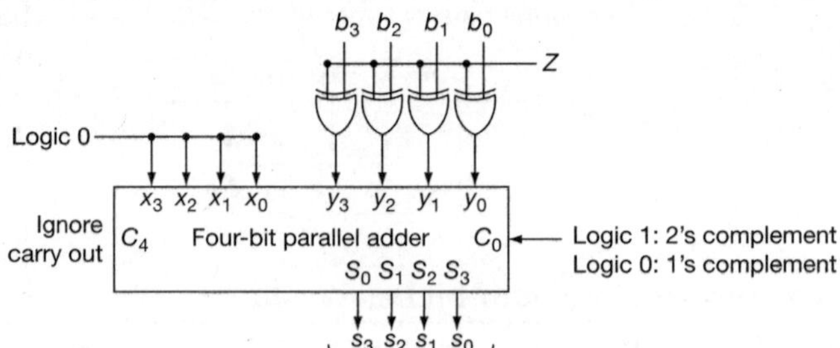

FIGURE 5.47 | Multifunction from four-bit binary number

TABLE 5.9 | Truth table to combinational circuit to find complement of four-bit binary number

Control Lines		Input	Output	Function
Z	C_0	$b_3b_2b_1b_0$	$y_3y_2y_1y_0$	
0	0	$b_3b_2b_1b_0$	$b_3b_2b_1b_0$	Data transfer
0	1	$b_3b_2b_1b_0$	$(b_3b_2b_1b_0) + 1$	Increment by 1
1	0	$b_3b_2b_1b_0$	$(\bar{b}_3\bar{b}_2\bar{b}_1\bar{b}_0)$	1's Complement
1	1	$b_3b_2b_1b_0$	$(\bar{b}_3\bar{b}_2\bar{b}_1\bar{b}_0)+1$	2's Complement

When Z = 1, XOR gate acts as inverter. So four-bit data is inverted or 1's complement is obtained as

$$y_3y_2y_1y_0 = \bar{b}_3\bar{b}_2\bar{b}_1\bar{b}_0 \tag{5.67}$$

When Z = 0, XOR gate does not act as inverter. So four-bit data is transferred without any change as

$$y_3y_2y_1y_0 = b_3b_2b_1b_0 \quad (5.68)$$

5.7 | BINARY SUBTRACTOR USING PARALLEL ADDER

Complements are used to perform subtraction and to represent signed numbers. Signed numbers can be positive or negative numbers. Subtraction is performed by doing addition using complement. There are two methods

1. Subtraction with 1's complement
2. Subtraction with 2's complement

5.7.1 | Subtraction with One's Complement

To perform subtraction of $(y_3y_2y_1y_0 - z_3z_2z_1z_0)$, the steps are given below. Let $(y_3y_2y_1y_0)_2$ and $(z_3z_2z_1z_0)_2$ both are positive binary numbers.

- Take 1's complement of four-bit subtrahend, $(z_3z_2z_1z_0)$ as $(x_3x_2x_1x_0)$ using combinational circuit given in Figure 5.47.
- Four-bit parallel adder-I adds minuend $(y_3y_2y_1y_0)_2$ and 1's complement of the subtrahend. $(x_3x_2x_1x_0)_2$ to get $(S_3S_2S_1S_0)_2$ and carry out either 0 or 1.
- Check carry out, C_4 from four-bit parallel binary adder (stage-I).
  - If carry occurs (C_4= 1), then add '1' to the four-bit sum, $(S_3S_2S_1S_0)_2$. Four-bit binary parallel adder (stage-II) adds (0000) to sum, $(S_3S_2S_1S_0)_2$ with carry in (C_0) as 1 to get $(d_3d_2d_1d_0)_2$. Result is positive.
  - If carry does not occur (C_4 = 0) then sum $(d_3d_2d_1d_0)_2$ gives negative value. Negative number is in 1's complement form.

Circuit diagram is given in Figure 5.48. The circuit is able to perform addition as well as subtraction. If data is not complemented and second stage addition is performed with zero without any carry. C_0 is fed after performing AND operation of Z and C_4. Z is 1's complement control line. C_4 is carry out from stage-I four-bit binary parallel adder.

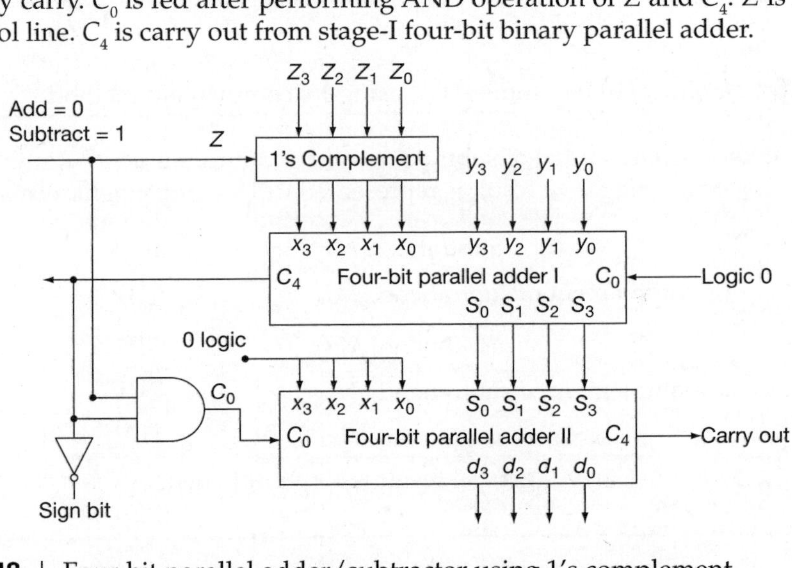

FIGURE 5.48 | Four-bit parallel adder/subtractor using 1's complement

Truth table is given is given in Table 5.10.

TABLE 5.10 | Truth table of Figure 5.64

Input	Output	Input	Operation	
Z	C_4	$C_0 = ZC_4$		
0	0	0	Add	$(y_3y_2y_1y_0)_2 + (z_3z_2z_1z_0)_2$
0	1	0		
1	0	0	Subtract	$(y_3y_2y_1y_0)_2 - (z_3z_2z_1z_0)_2$
1	1	1		

EXAMPLE 5.5 Subtract $(0111)_2$ from $(1001)_2$ using 1's complement method.

SOLUTION Take 1's complement of subtrahend and add it to minuend. Add 1 if carry occurs.

Subtrahend N =		0111
1's complement of subtrahend, $\bar{N}$ =		1000
Minuend, M =		1001
1's complement of subtrahend, $\bar{N}$ =	+	1000
$M + \bar{N}$ =	1	0001
$M + \bar{N}$ + Carry(1)	+	0001
$(M)_2 - (N)_2$ =	+	0010

EXAMPLE 5.6 Subtract $(1001)_2$ from $(0111)_2$ using 1's complement method.

SOLUTION Take 1's complement of subtrahend and add it to minuend. Result is negative if carry does not occur. Negative result is represented in 1's complement form.

Subtrahend N =		1001
1's complement of subtrahend, $\bar{N}$ =		0110
Minuend, M =		0111
1's complement of subtrahend, $\bar{N}$ =	+	0110
Carry is 0, so $M + \bar{N}$ =	0	1101

Carry is zero so result is 1's complement

5.7.2 | Subtraction with Two's Complement

To perform subtraction of $(Y_3Y_2Y_1Y_0{-}X_3X_2X_1X_0)$, the steps are given below. Let $(Y_3Y_2Y_1Y_0)_2$ and $(X_3X_2X_1X_0)_2$ both are positive binary numbers.

- Find 2's complement of four-bit subtrahend $(X_3X_2X_1X_0)_2$ using combinational circuit given in Figure 5.47.
- Add four-bit minuend $(Y_3Y_2Y_1Y_0)_2$ and 2's complement of the subtrahend. $(X_3X_2X_1X_0)_2$ using parallel adder and get $(S_3S_2S_1S_0)_2$ and carry out either 0 or 1.
- Check carry-out, C_4 from four-bit parallel binary adder.
  - If carry occurs ($C_4 = 1$), then ignore it. Result is $(S_3S_2S_1S_0)_2$ and is positive.
  - If carry does not occur ($C_4 = 0$), then sum $(S_3S_2S_1S_0)_2$ gives negative value. Negative number is in 2's complement form.

Circuit diagram is given in Figure 5.49.

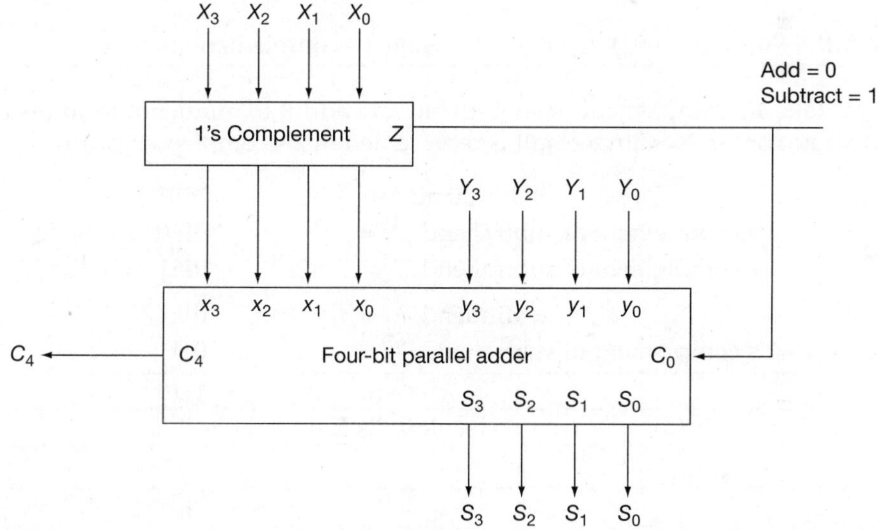

FIGURE 5.49 | Subtractor using 2's complement

Truth table is given is given in Table 5.11.

TABLE 5.11 | Truth table of Figure 5.65

Input		Operation	
Z	C_0		
0	0	Add	$(y_3y_2y_1y_0)_2 + (z_3z_2z_1z_0)_2$
1	1	Subtract	$(y_3y_2y_1y_0)_2 - (z_3z_2z_1z_0)_2$

EXAMPLE 5.7 Subtract $(0111)_2$ from $(1001)_2$ using 2's complement method.

SOLUTION Take 1's complement of subtrahend and add it to minuend. Ignore carry if it occurs.

Subtrahend $N =$		0111
1's complement of subtrahend, $\bar{N} =$		1000
2's complement of subtrahend, $\bar{N}+1 =$		1001
Minuend, $M =$		1001
2's complement of subtrahend, $\bar{N} =$	+	1001
$M+\bar{N} =$	1	0010
$(M)_2 - (N)_2 =$	+	0010

EXAMPLE 5.8 Subtract $(1001)_2$ from $(0111)_2$ using 2's complement method.

SOLUTION Take 1's complement of subtrahend and add it to minuend. Result is negative if carry does not occur. Negative result is represented in 2's complement form.

Subtrahend $N =$		1001
1's complement of subtrahend, $\bar{N} =$		0110
1's complement of subtrahend, $\bar{N} =$		0111
Minuend, $M =$		0111
1's complement of subtrahend, $\bar{N} =$		0111
$M+\bar{N} =$	0	1110

Carry is zero, so result is 2's complement.

5.8 | BINARY MULTIPLIER

Multiplication table of binary digits is given by:

$$0 \times 0 = 0$$
$$1 \times 1 = 1$$
$$0 \times 1 = 0$$
$$1 \times 0 = 0$$

Basically, multiplication is AND logic of two bits. So, the binary multiplication is performed by doing two-basic operations.

- AND operation of each bit of multiplicand with the multiplier bit.
- Multiplier bits are selected with reference to LSB.
- Addition of two operands is obtained after AND operation of multiplicand with two multiplier bits.

Consider two-bit multiplicand $(X_1X_0)_2$ and two-bit multiplier $(Y_1Y_0)_2$. Multiplication process is elaborated below.

Multiplicand (two-bit):			X_1	X_0
Multiplier (two-bit):			Y_1	Y_0
			X_1Y_0	X_0Y_0
		X_1Y_1	X_0Y_1	
Product (four-bit):	P_3	P_2	P_1	P_0

At a time two operands are added using one-bit half-adder. Four-bit product is obtained when two-bit multiplicand is multiplied with two-bit multiplier. The AND logic and ADD operation is performed at two-different stages. The multiplication operation can be elaborated as below.

Multiplicand (two-bit)			X_1	X_0
Multiplier (two-bit)			Y_1	Y_0
Stage-I:				X_0Y_0
AND logic of multiplicand X_1 and multiplier bit Y_0 (Augend)			X_1Y_0	↓
AND logic of multiplicand X_0 and multiplier bit Y_1 (Addend)			X_0Y_1	↓
Sum and Carry of single-bit half-adder I		C_{11}	S_{10}	↓
Stage-II:			↓	↓
Carry of stage-I half-adder (Augend)		C_{11}	↓	↓
AND logic of multiplicand X_1 and multiplier bit Y_1 (Addend)		X_1Y_1	↓	↓
Sum and Carry of one-bit half-adder II	C_{22}	S_{20}	↓	↓
Product (eight-bit)	P_3	P_2	P_1	P_0

Circuit diagram of two-bit multiplier is given in Figure 5.50 as per the aforementioned explained process.

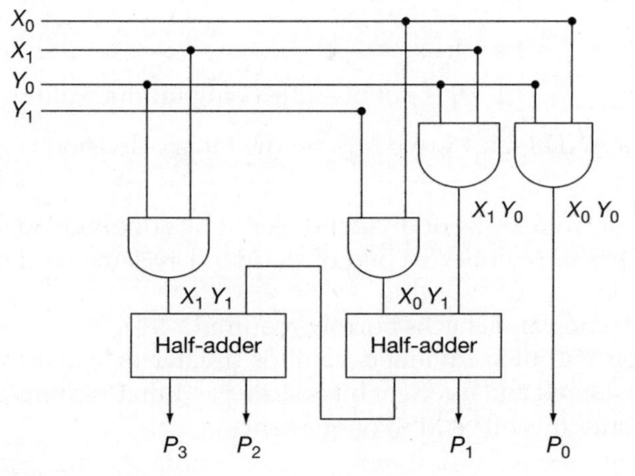

FIGURE 5.50 | 2-bit Multiplier

EXAMPLE 5.9 Multiply$(1101)_2$ by $(1110)_2$.

SOLUTION

Multiplicand					1	1	0	1	
Multiplier				×	1	1	1	0	
					0	0	0	0	
				0	0	0	0		
				1	1	0	1		
			0	1	1	0	1		
			0	1	1	0			
			1	1	0	1			
		1	0	0	1	1			
		1	0	0	1				
		1	1	0	1				
	1	0	0	1	0				
Product	1	0	0	1	0	1	1	0	**Ans.**

5.9 | BINARY DIVIDER

The procedure of binary division is sequenced for successive subtraction of divisor from dividend and giving the left shift to dividend. If subtraction is performed, then 1 is added to quotient otherwise zero is added.

Binary division is performed in the same manner as is done in decimal system. Division table of binary digits is given by:

$0 \div 0 = 0$

$1 \div 1 = 1$

$0 \div 1 = 0$

$1 \div 0$ is not possible being infinite value.

Basically, division is AND logic of two bits. So the binary division is performed by doing two basic operations

- Subtract divisor from MSBs of dividend. Select bits of dividend starting from higher place value (MSB). Number of bits of dividend is same as the number of bits of divisor.
- Perform subtraction and checks borrow required.
  - If borrow bit is 0 then dividend value is greater or equal to divisor. Difference value is considered and quotient bit is stored as 1 that is complement of the borrow value. Consider this bit as MSB of quotient.

- ○ If borrow bit is 1 then dividend value is lesser than divisor. Original dividend value is retained. Quotient bit is stored as 0 that is complement of the borrow value.
- Concatenate next LSB of dividend with the resulting value after subtraction at the LSB place. Repeat the subtraction till LSB of dividend is used.

Figure 5.51 gives the combinational circuit that implements the logic of division and is known as divider cell. The circuit constitutes full-subtractor circuit and a multiplexer. Dividend, *x* acts as minuend and divisor, *y* acts as subtrahend. Borrow, *z* is used for i^{th}-bit subtraction. So, full-subtractor subtracts divisor, *y* from dividend, *x*, considering previous borrow *z* and gives difference value *D* and borrow, *B*. Multiplexer selects either difference, *D* or dividend, *x* based on the borrow, *B* value.

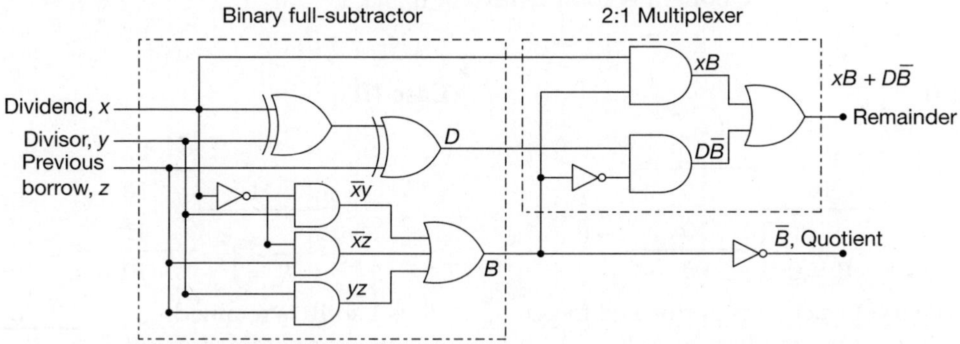

FIGURE 5.51 | Binary divider cell

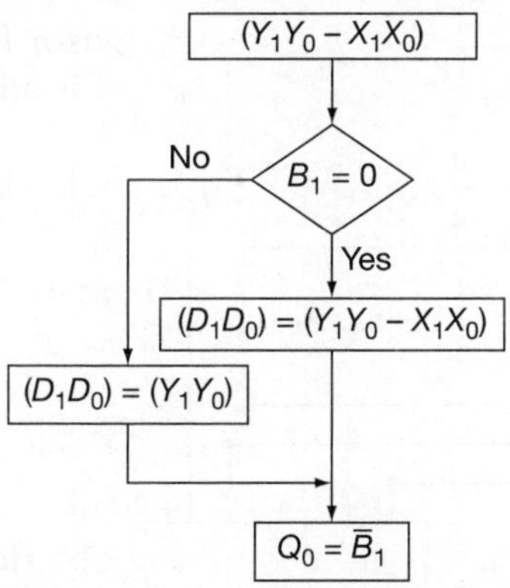

FIGURE 5.52 | Flow chart for division

If borrow is 0, then *D* is selected and acts as remainder.

If borrow is 1, then dividend value, *x* is selected.

When dividend is less than divisor, borrow occurs ($B = 1$). On the other hand if dividend is greater or equal than divisor than borrow does not occur ($B = 0$).

The circuit is extended to perform two-bit division. Consider two-bit divisor $(X_1X_0)_2$ and two-bit dividend $(Y_1Y_0)_2$. Division is elaborated below. Flow chart to perform the division is given in Figure 5.52.

Q_0 ⟵ Quotient

Divisor ⟶ X_1X_0) Y_1Y_0 ⟵ Dividend

X_1X_0

B_1 D_1D_0 ⟵ Remainder

Combinational circuit for two-bit division is given in Figure 5.53. End borrow B_1 is considered either to select the two-bit difference (D_1D_0) or two-bit dividend, (Y_1Y_0) is retained. Three possibilities of two-bit division are given below.

Case-I

```
         1
  01 ) 10
       01
      ----
B1 = 0   01
```

B_1 is 0 so subtraction is performed to get 00.
Quotient is complement of B_1, i.e. 1

Case-II

```
         1
  10 ) 10
       10
      ----
B1 = 0   00
```

B_1 is 0 so subtraction is performed to get 00. Quotient is complement of B_1

Case-III

```
         0
  11 ) 10
       11
      ----
B1 = 1   10
```

B_1 is 1 so 10 is retained.
Quotient is complement of B_1, i.e. 0

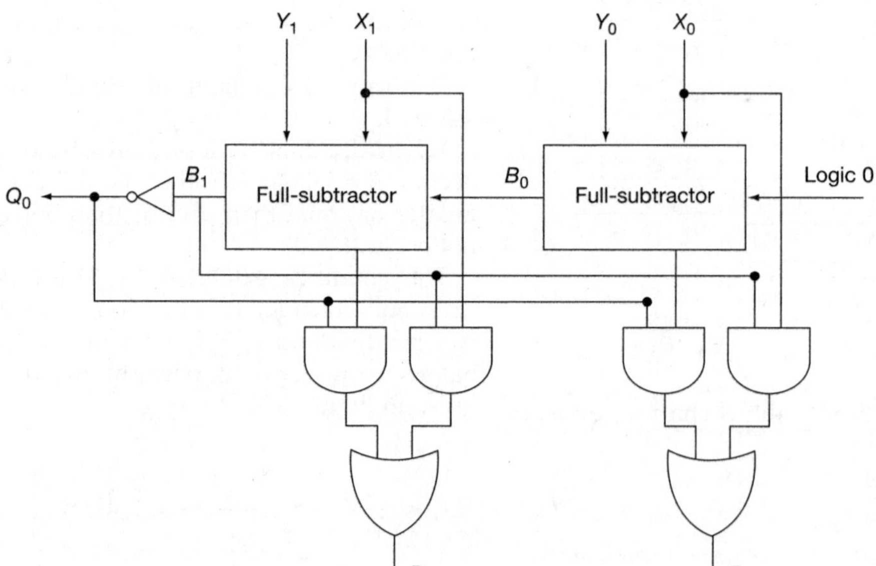

FIGURE 5.53 | Two-bit divider

EXAMPLE 5.10 Divide $(110)_2$ by $(10)_2$.

SOLUTION

			1	1	0		← Quotient
Divisor →	1	0	1	1	0	0	← Dividend
			1	0			
$B=0$			0	1	0	0	
				1	0	0	
$B=0$				0	0	0	
					1	0	
$B=1$					0	0	← Remainder

$$(1100)_2 \times (10)_2 = (110)_2 \times (10)_2 + (000)_2 \quad \textbf{Ans.}$$

5.10 | BCD ADDER

In calculators, games and digital instruments to feed inputs and get outputs in decimal, binary-coded decimal (BCD) code is used to represent decimal numbers. BCD code is also known as 8421 code and is a four-bit code. It is a weighted code. The decimal digit is translated directly to its four-bit BCD equivalent or vice-versa (Table 5.12).

TABLE 5.12 | Sum of BCD number resulting from four-bit binary numbers

Decimal Value	Sum in Binary					Sum in BCD					Z
	S_4	S_3	S_2	S_1	S_0	S_4	S_3	S_2	S_1	S_0	
0	0	0	0	0	0	0	0	0	0	0	0
1	0	0	0	0	1	0	0	0	0	1	0
2	0	0	0	1	0	0	0	0	1	0	0
3	0	0	0	1	1	0	0	0	1	1	0
4	0	0	1	0	0	0	0	1	0	0	0
5	0	0	1	0	1	0	0	1	0	1	0
6	0	0	1	1	0	0	0	1	1	0	0
7	0	0	1	1	1	0	0	1	1	1	0
8	0	1	0	0	0	0	1	0	0	0	0
9	0	1	0	0	1	0	1	0	0	1	0
10	0	**1**	0	**1**	0	1	0	0	0	0	1
11	0	**1**	0	**1**	1	1	0	0	0	1	1
12	0	**1**	**1**	0	0	1	0	0	1	0	1
13	0	**1**	**1**	0	1	1	0	0	1	1	1

(Continued)

TABLE 5.12 | (Continued)

Decimal Value	Sum in Binary					Sum in BCD					Z
	S_4	S_3	S_2	S_1	S_0	S_4	S_3	S_2	S_1	S_0	
14	0	1	1	1	0	1	0	1	0	0	1
15	0	1	1	1	1	1	0	1	0	1	1
16	1	0	0	0	0	1	0	1	1	0	1
17	1	0	0	0	1	1	0	1	1	1	1
18	1	0	0	1	0	1	1	0	0	0	1
19	1	0	0	1	1	1	1	0	0	1	1

BCD code represents only 10 digits varying from 0 to 9. On combining four-bit binary numbers, 2^4 (16) possible values are obtained. Range of data obtained by combining four-bits is from 0 to 15(2^4-1) as given in Table 5.12. Six values represented by four-bit are 1010, 1011, 1100, 1101, 1110 and 1111 and gives 2-digit decimal values 10, 11, 12, 13, 14 and 15, respectively. These six values are not used in BCD code to represent 1-digit decimal number. So BCD code is obtained by adding $(0110)_2$ to four-bit binary value if it is more than $(1001)_2$. Six is added to skip 6 four-bit binary numbers: 1010, 1011, 1100, 1101, 1110 and 1111.

BCD addition procedure is given below:

- Add augend and addend digit pair. BCD is a four-bit number
- Add 0110 $(6)_{10}$ to sum if
  - there is carry over after addition OR sum is more than 1001 $(9)_{10}$
- No need to add 0110 $(6)_{10}$ to sum if
  - there is no carry over after addition AND sum is not more than 1001 $(9)_{10}$
- Propagate carry to next digit if there occurs after adding $(0110)_2$.

	Decimal		BCD code		
Digits	2nd	1st	2nd	1st	**Explanation:**
Carry		0	0 [0]		• First BCD digit is not more than 1001 and carry is 0 so add (0000).
Augend		3		0011	
Addend		6		0110	
Sum	0	9	0	1001	
BCD				9	

	Decimal		BCD code		
Digits	2nd	1st	2nd	1st	**Explanation:**
Carry		0	0 [0]		• 1st BCD digit is more than 1001 but carry is 0 so add (0110)
Augend		5		0101	
Addend		6		0110	
Sum	1	1	0	1011	

Carry	1	
		1011
Add 6	+	0110
Sum		0001
BCD	1	1

Possible sum values after adding two digits following BCD code is give in Table 5.14. The possible binary caused after addition in BCD code is also given. Binary is divided into 3-categories as given in Table.

(i) $(0000)_2$ to $(1001)_2$: No need to add $(0110)_{BCD}$ as binary value is not more than $(1001)_{BCD}$.
(ii) $(1010)_2$ to $(1111)_2$: Need to add $(0110)_{BCD}$ as binary value is more than $(1001)_{BCD}$.
(iii) $(10000)_2$ to $(10011)_2$: Need to add $(0110)_{BCD}$ because S_4 is 1. S_4 is carry from four-bit addition.

Z variable is set to 1 when there is need to add $(0110)_{BCD}$.

$$Z(S_3, S_2, S_1, S_0) = \sum m(10, 11, 12, 13, 14, 15)$$

K-map is used to find the Boolean logic. K-map is given in Figure 5.54. Logic is given in Eq. (5.69).

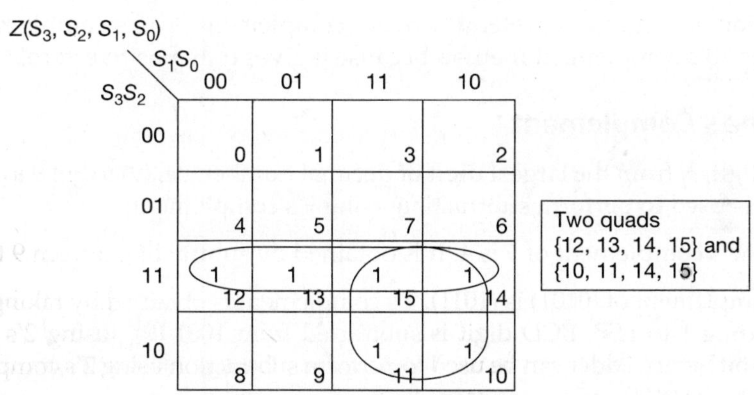

FIGURE 5.54 | K-map to convert binary to BCD

$$Z = S_3S_2 + S_3S_1 \tag{5.69}$$

There is need to add $(0110)_{BCD}$ because S_4 is 1. The above is rewritten as

$$Z = S_4 + S_3S_2 + S_3S_1$$

Combinational circuit is given in Figure 5.55 to add the 2-digits using BCD code. Two digits $(X_3X_2X_1X_0)$ and $(Y_3Y_2Y_1Y_0)$ in BCD code are added using 4-binary digit (bit) parallel adder in fist stage. Then six (0110) or (0000) is added based on the value of Z is 1 or 0, respectively, in second stage using four-bit parallel adder. Z sets the x_2 and x_1 bits to form one operand either 0110 or 0000.

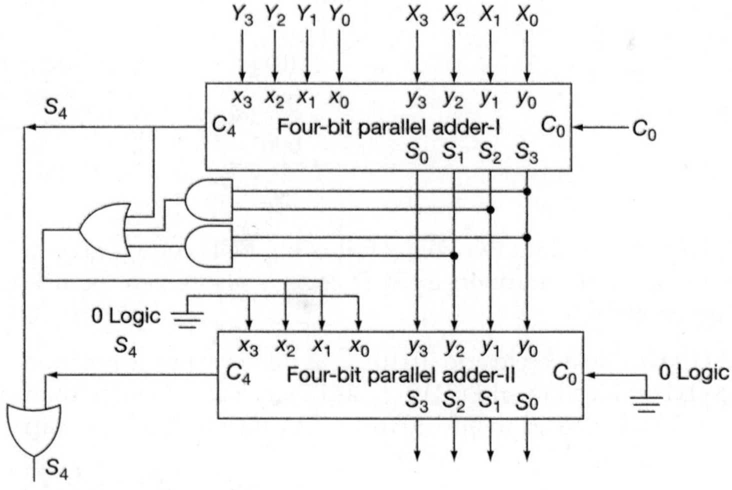

FIGURE 5.55 | Single digit BCD adder

5.11 | BCD SUBTRACTOR USING BCD ADDER

BCD subtraction is performed preferably by 9's complement or 10's complement method. Best method is 10's complement method because it gives only positive zero.

5.11.1 | Nine's Complement

Subtract the digit, N from the largest digit of decimal number, i.e. (9) to get 9's complement. Binary adder is used to perform subtraction using 9's complement.

In decimal: Nine's complement of 5 is 4. It is obtained by subtracting 5 from 9 to get 4.

In BCD: 2's complement of (0101) is (1011). 2's complement is obtained by taking 1's complement, and adding 1 to LSB. BCD digit is subtracted from 1001 $(9)_{10}$ using 2's complement method. Four-bit binary adder can be used to perform subtraction using 2's complement. Add 1001 and 1011 to get 0100 and ignore carry 1.

Operation	C_4	d_3	d_2	d_1	d_0	Remarks
N		0	1	0	1	; Given number N is 5
$\bar{N}$		1	0	1	0	; 1's complement of N. Change 0 to 1 and 1 to 0.
+1	+				1	: Add 1 to 1's complement
$\bar{N}+1$		1	0	1	1	; 2's complement of N
+9	+	1	0	0	1	; Add 9 to 2's complement of N
$(\bar{N}+10)$	1	0	1	0	0	; 9's complement of 5 is 4.

So, 9's complement of BCD digit is obtained by adding (1010) to 1's complement of BCD digit.

Figure 5.56 gives 9's complement of BCD digit. Exclusive OR gate gives complement of bit ($\bar{d}_i = d_i \oplus 1$) to get

$$(y_3y_2y_1y_0) = (\bar{d}_3\bar{d}_2\bar{d}_1\bar{d}_0)$$

2's complement is obtained by adding 1 to 1's complement. Four-bit adder gives sum which is equivalent to the 9's complement of BCD digit.

$$(S_3S_2S_1S_0) = (1001) + ((\bar{d}_3\bar{d}_2\bar{d}_1\bar{d}_0) + 1)$$

$$(S_3S_2S_1S_0) = (1001 + 1) + (\bar{d}_3\bar{d}_2\bar{d}_1\bar{d}_0)$$

$$(S_3S_2S_1S_0) = (1010) + (\bar{d}_3\bar{d}_2\bar{d}_1\bar{d}_0)$$

Carry is ignored. 9's complement of BCD digit gives a positive digit always.

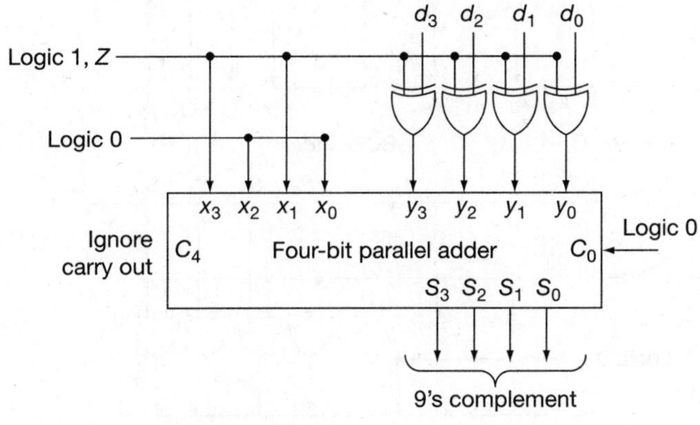

FIGURE 5.56 | 9's complement of BCD digit

5.11.2 | Subtractor using Nine's Complement

To perform subtraction of ($X_3X_2X_1X_0 - Y_3Y_2Y_1Y_0$) following BCD code, the steps are given below. Let $(X_3X_2X_1X_0)_{BCD}$ and $(Y_3Y_2Y_1Y_0)_{BCD}$ both are positive binary numbers.

- Take nine's complement of subtrahend, $(Y_3Y_2Y_1Y_0)_{BCD}$ as $(y_3y_2y_1y_0)$ using combinational circuit given in Figure 5.56.
- BCD adder-I adds minuend $(X_3X_2X_1X_0)_{BCD}$ and 9's complement of the subtrahend. $(Y_3Y_2Y_1Y_0)$ to get $(S_3S_2S_1S_0)_{BCD}$ and carry out either 0 or 1.
- Check carry out, C_4 from BCD adder.
  - If carry occurs ($C_4 = 1$), then add '1' to the BCD sum, $(S_3S_2S_1S_0)$. Four-bit binary parallel adder (stage-II) adds (0000) to sum, $(S_3S_2S_1S_0)_2$ with carry in (C_0) as 1 to get $(D_3D_2D_1D_0)_2$. Result is positive.
  - If carry does not occur ($C_4 = 0$) then sum $(d_3d_2d_1d_0)_2$ gives negative value. Negative number is in 9's complement form. 9's complement is obtained by adding 1010 to 1's complement of sum.

Circuit diagram is given in Figure 5.57. The circuit is able to perform addition as well as subtraction. If data is not complemented and second stage addition is performed with zero without any carry. C_0 is fed after performing AND operation of Z and C_4. Z is 1's complement control line. C_4 is carry out from stage-I four-bit binary parallel adder.

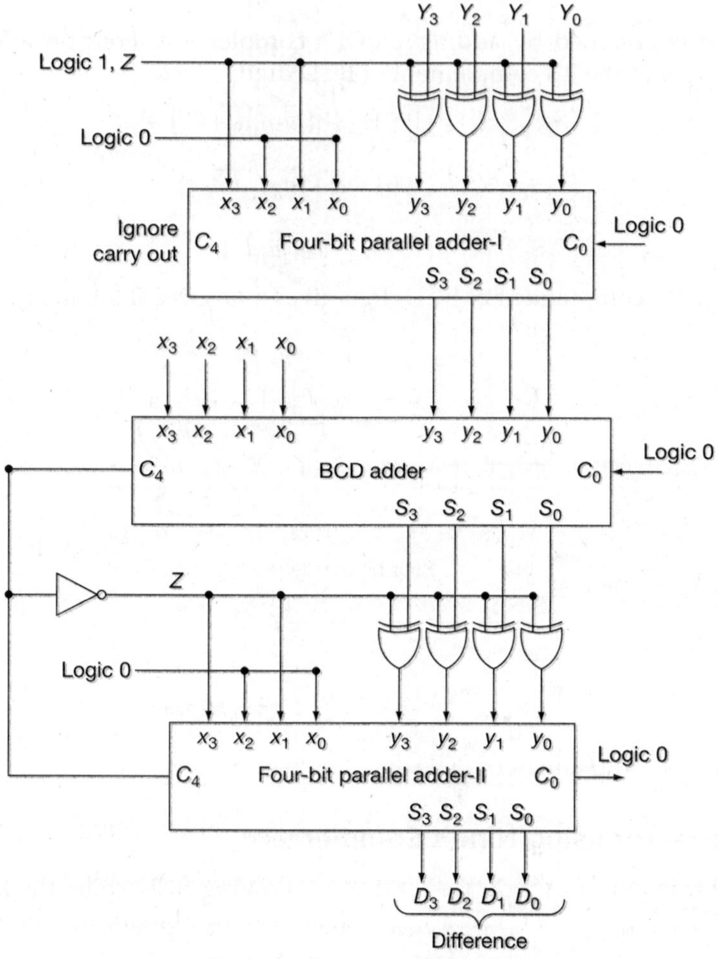

FIGURE 5.57 | BCD subtractor digit using 9's complement

EXAMPLE 5.11 Subtract 4 from 9 using BCD code and 9's complement.

SOLUTION Take 9's complement of $(N)_{BCD}$ = 0100 and add it to $(M)_{BCD}$ = 1001

1's complement of $\bar{N}$ = 1011
9's complement of 5 = 1010 + 1011
BCD code of 9's complement of 5 = 0101

		0		0	0
(*M*) =		9	⇔		1001
9′s complement of (*N*) =		5			0101
(*M*) + (9′s complement of (*N*)) =	1	4		0	1110
					1110
Convert binary into BCD					0110
				1	0100

End carry is one, so add 1 to the LSD and place positive sign

(*M*) + (9′s complement of (*N*)) =		4				0100	
Add 1 to LSD =	+	1			+	1	
(*M*) − (*N*) =	+	5	**Ans.**	⇔	+	0101	**Ans.**
						(+)5	

EXAMPLE 5.12 Subtract 9 from 4 using BCD code and 9′s complement.

SOLUTION Take 9′s complement of $(N)_{BCD} = 1001$ and add it to $(M)_{BCD} = 0100$

1′s complement of $\bar{N} = 0110$
9′s complement of 5 = 1010 + 0110
BCD code of 9′s complement of 9 = 0000

		0		0	0
(*M*) =		4	⇔		0100
9′s complement of (*N*) =		0			0000
(M) + (9′s complement of (*N*)) =	0	4		0	0100

End carry is zero, so find 9′s complement of result and place negative sign
1′s complement of 0100 = 1011
9′s complement of 0100 = 1010 + 1011
BCD code of 9′s complement of 0100 = (−)0101 **Ans.**

5.11.3 | Ten′s Complement

Subtract the digit, *N* from the largest digit of decimal number, i.e. (9) to get 9′s complement. Add 1 to 9′s complement to get 10′s complement. In other words subtract number from 10 to get 10′s complement. Binary adder is used to perform 2′s complement subtraction.

In decimal: 10′s complement of 6 is 4. Subtract 6 from 9 to get 3. 3 is 9′s complement of 6. Add 1 to 3 to get 4. 4 is 10′s complement of 6. So, 10′s complement can be obtained by subtracting 6 from 10.

In BCD: 2's complement of (0110) is (1010). Two's complement is obtained by taking 1's complement, and adding 1 to LSB. Four-bit binary adder can be used to perform subtraction using 2's complement. Add 1001 and 1010 to get 0011 and ignore carry 1. Add 0011 and 0001 to get 10's complement 0100.

Operation	C_4	d_3	d_2	d_1	d_0	Remarks
N		0	1	1	0	; Given number N is 5
$\bar{N}$		1	0	0	1	; 1's complement of N. Change 0 to 1 and 1 to 0.
+1	+				1	: Add 1 to 1's complement
$\bar{N}+1$		1	0	1	0	; 2's complement of N
+10	+	1	0	1	0	; Add (9+1) to 10's complement of N
$(\bar{N}+11)$	1	0	1	0	0	; 10's complement of 5 is 4.

So, 10's complement of BCD digit is obtained by adding (1011) to 1's complement of BCD digit.

Figure 5.58 gives 10's complement of BCD digit. Exclusive OR gate gives complement of bit ($\bar{d}_i = d_i \oplus 1$) to get

$$(y_3 y_2 y_1 y_0) = (\bar{d}_3 \bar{d}_2 \bar{d}_1 \bar{d}_0)$$

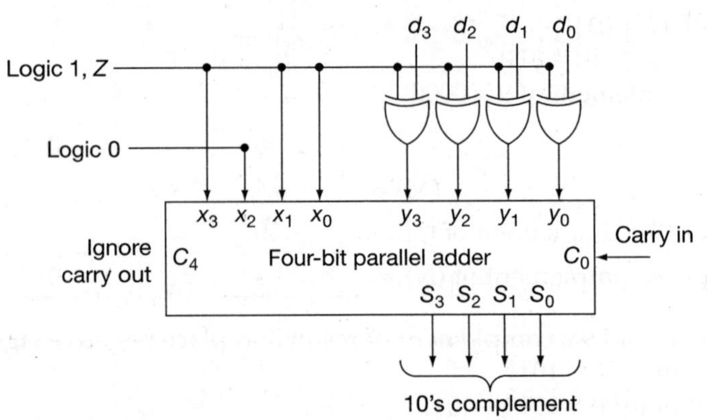

FIGURE 5.58 | Ten's complement of BCD digit

Two's complement is obtained by adding 1 to 1's complement. Four-bit adder gives sum which is equivalent to the 9's complement of BCD digit.

$$9\text{'s complement of digit} = (1001) + \left(\left(\bar{d}_3 \bar{d}_2 \bar{d}_1 \bar{d}_0\right) + 1\right)$$

$$9\text{'s complement of digit} = (1001 + 1) + \left(\bar{d}_3 \bar{d}_2 \bar{d}_1 \bar{d}_0\right)$$

$$9\text{'s complement of digit} = (1010) + \left(\bar{d}_3 \bar{d}_2 \bar{d}_1 \bar{d}_0\right)$$

Ten's complement of BCD digit is obtained by adding 1 to 9's complement

$$10\text{'s complement of digit} = (1010) + \left(\bar{d}_3\bar{d}_2\bar{d}_1\bar{d}_0\right) + 1$$

$$10\text{'s complement of digit} = (1011) + \left(\bar{d}_3\bar{d}_2\bar{d}_1\bar{d}_0\right)$$

Carry is ignored. Ten's complement of BCD digit gives a positive digit always.

5.11.4 | Subtractor using Ten's Complement

To perform subtraction of $(X_3X_2X_1X_0 - Y_3Y_2Y_1Y_0)$ in BCD code, the steps are given below. Let $(X_3X_2X_1X_0)_{BCD}$ and $(Y_3Y_2Y_1Y_0)_{BCD}$ both are positive binary numbers.

- Take 10's complement of subtrahend, $(Y_3Y_2Y_1Y_0)_{BCD}$ as $(y_3y_2y_1y_0)$ using combinational circuit given in Figure 5.59.

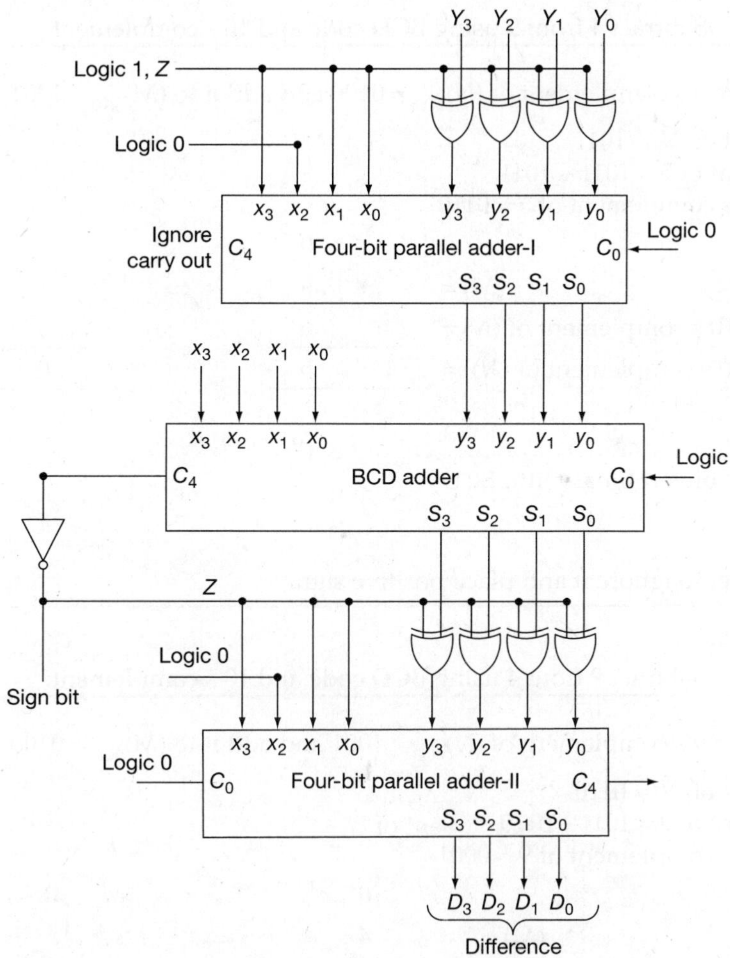

FIGURE 5.59 | BCD subtractor digit using 10's complement

- BCD adder-I adds minuend $(X_3X_2X_1X_0)_{BCD}$ and 10's complement of the subtrahend. $(Y_3Y_2Y_1Y_0)$ to get $(S_3S_2S_1S_0)_{BCD}$ and carry out either 0 or 1.
- Check carry out, C_4 from BCD adder.
  - If carry occurs ($C_4 = 1$), then ignore it. BCD sum, $(S_3S_2S_1S_0)$ is the result. Four-bit binary parallel adder (stage-II) adds (0000) to sum, $(S_3S_2S_1S_0)_2$ with carry in (C_0) as 0 to get $(D_3D_2D_1D_0)_2$. Result is positive.
  - If carry does not occur ($C_4 = 0$) then sum $(d_3d_2d_1d_0)_2$ gives negative value. Negative number is in 10's complement form. Ten's complement is obtained by adding 1011 to 1's complement of sum.

Circuit diagram is given in Figure 5.59. The circuit is able to perform addition as well as subtraction. If data is not complemented and second stage addition is performed with zero without any carry. C_0 is fed after performing AND operation of Z and C_4. Z is 1's complement control line. C_4 is carry out from stage-I four-bit binary parallel adder.

EXAMPLE 5.13 Subtract 4 from 9 using BCD code and 10's complement.

SOLUTION Take 9's complement of $(N)_{BCD} = 0100$ and add it to $(M)_{BCD} = 1001$

1's complement of $\bar{N} = 1011$
10's complement of 5 = 1011 + 1011
BCD code of 9's complement of 5 = 0110

		0		0	0
(M) =		9	⇔		1001
10's complement of (N) =		6			0110
(M) + (10's complement of (N)) =	1	5		0	1111
					1111
Convert binary into BCD					0110
				1	0101

End carry is one, so ignore it and place positive sign.

EXAMPLE 5.14 Subtract 9 from 4 using BCD code and 10's complement.

SOLUTION Take 9's complement of $(N)_{BCD} = 1001$ and add it to $(M)_{BCD} = 0100$

1's complement of $\bar{N} = 0110$
10's complement of 9 = 1011 + 0110
BCD code of 9's complement of 9 = 0001

		0		0	0
(M) =		4	⇔		0100
9's complement of (N) =		1			0001
(M) + (9's complement of (N)) =	0	5		0	0101

End carry is zero, so find 10's complement of result and place negative sign
1's complement of 0101 = 1010
10's complement of 0100 = 1011 + 1010
BCD code of 10's complement of 0101 = (–)0101 **Ans.**

5.12 | EXCESS-3 (XS-3) CODE ADDERS

Excess-3 (XS-3) code is a kind of BCD code. Each decimal digit is coded into four-bit binary code. The XS-3 code for each digit is obtained by adding decimal 3 to the natural BCD code of the digit.

For example,

Decimal number	=	4
BCD code	=	0100
BCD code + 3	=	(0100 + 0011)
Excess-3 (XS-3) code	=	0111

As XS-3 code can be obtained by adding 3 (0011) to BCD digit, so four-bit parallel binary adder is used. The combinational circuit diagram is given in Figure 5.60.

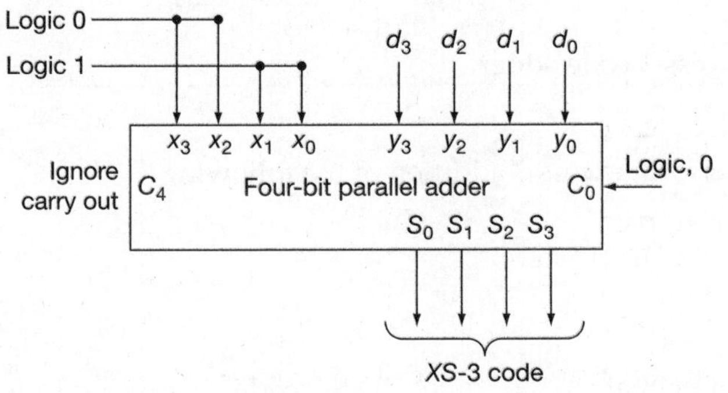

FIGURE 5.60 | Excess-3 code converter

Excess-3 code is not a weighted code. Excess-3 is self-complementary code. It means 1's complement of excess-3 code of a digit gives 9's complement of the digit itself.

- Addition is performed by adding augend and addend digits. Carry occurred after addition of every digit pair is observed. Binary addition is performed and binary result will occur. Binary number should be converted into XS-3 code.
- If there is carry over after the addition then add (0011) to the resulting sum.
- If there is no carry over after the addition then subtract (0011) from the resulting sum.
- Subtraction is performed by 2's complement method. One's complement of (0011) is 1100 and 2's complement is 1101.

$$(S_3S_2S_1S_0)-(0011)=(S_3S_2S_1S_0)+2\text{'s complement of }(0011)$$
$$(S_3S_2S_1S_0)-(0011)=(S_3S_2S_1S_0)+(1101)$$

The complement circuit is given in Figure 5.61 to implement the above logic to add two XS-3 digits.

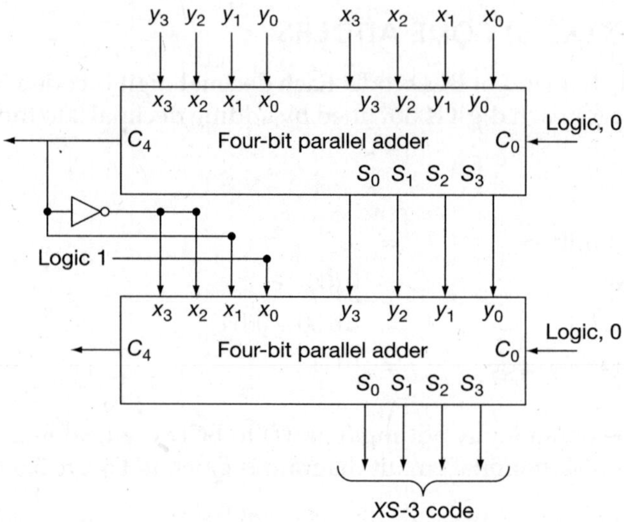

FIGURE 5.61 | Excess-3 code adder

EXAMPLE 5.15 Perform excess-3 addition of the following

(a) 5 + 6 (b) 2 + 7

SOLUTION

(a)

	Decimal		XS-3 code		
Digits	2nd	1st	2nd	1st	**Explanation:**
Carry	1	0	1		• 1 as carry after addition of 1st digit pair so add (0011) to sum (0001)
Augend		5		1000	
Addend		6	0	1001	
Sum	1	1	1	0001	
Carry					
			1	0001	
				+ 0011	
			1	0100	
Digit			1	1	

(b)

	Decimal		XS-3 code	
Digits	2nd	1st	2nd	1st
Carry		0	0	
Augend		2		0101
Addend		7		1010
Sum	0	9	0	1111
Carry				
			0	1111
				+1101
			1	1100
Digit			Ignore carry	9

Explanation:

- 0 as carry after addition subtract (0011) to sum (0001). Subtraction can be achieved by adding 1101 to sum.

5.13 | EXCESS-3 (XS-3) CODE SUBTRACTOR

Excess-3 (XS-3) code subtraction is performed preferably by 9's complement or 10's complement method. Best method is 10's complement method because it gives only zero. 10's complement for subtraction is discussed here.

Subtraction using 10's complement method in XS-3 code

- Compute 10's complement of the subtrahend, $(Y_3Y_2Y_1Y_0)$ by taking 2's complement of XS-3 code of the number.
- Add minuend $(X_3X_2X_1X_0)$ and 10's complement of the subtrahend (N). Addition is performed using four-bit parallel adder (stage-I).
- To convert binary number into XS-3 code, check carry over after addition and perform Stage-II addition using four-bit parallel adder.
  - o If there is carry over after the addition. Then add $(0011)_2$ to the resulting sum.
  - o If there is no carry over after the addition. Then subtract $(0011)_2$ from the resulting sum. Subtraction is performed using 2's complement method. So, add 2's complement of 0011, i.e. 1101 to the sum and ignore resulting carry.
  - o To complete 10's complement subtraction, check end carry obtained after stage I.
    - – If an end carry occurs, then discard it and place positive sign (+ve).
    - – If an end carry does not occur then take 2's complement of XS-3 code of the number and place negative sign (−ve).

The complement circuit is given in Figure 5.62.

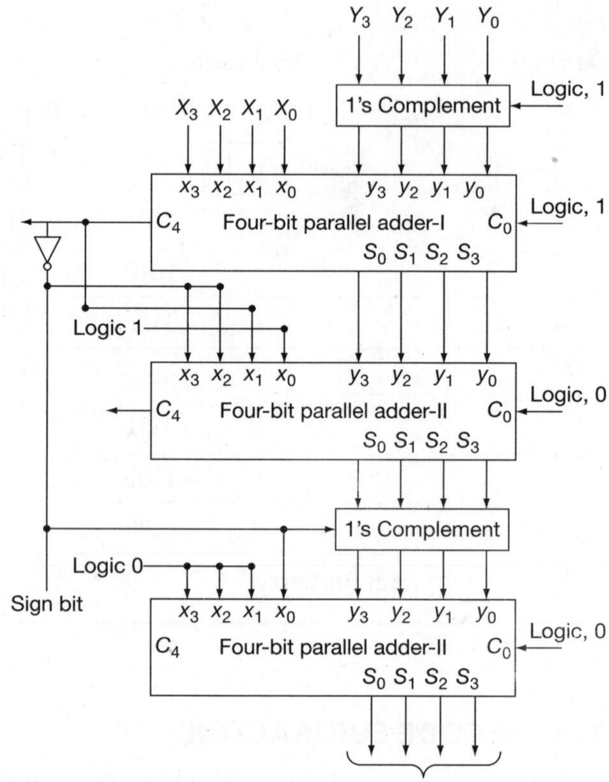

FIGURE 5.62 | Excess-3 code subtractor

Combinational circuit diagram to implement above logic is given in Figure 5.62.

EXAMPLE 5.16 Perform subtraction of XS-3 numbers using 10's complement.

(a) (6 – 2) (b) (4 – 8)

SOLUTION (a) Take 10's complement of (*N*) = 2 and add it to (*M*) = 6

XS-3 code of 10's complement of 2 = 1011 (2's complement of 0101)

		0	1	0
(*M*) =		6		1001
10's complement of (*N*) =		8		1011
(*M*) + (10's complement of (*N*)) =	1	4	1	0100
				0100
				+ 0011
			1	0111

End carry is one, so ignore it and consider positive sign

(b) Take 10′s complement of $(N) = 8$ and add it to $(M) = 4$
10′s complement of 8 in XS-3 code = 0101 (2′s complement of 1011)

		0	0	0
$(M) =$		4		0111
10′s complement of $(N) =$		2		0101
(M) + (10′s complement of (N)) =	0	6	**0**	1100
				1100
				+ 1101
Carry is 0 so compute 10′s complement	(–)	4	1	1001

End carry is zero, so compute 10′s complement and place negative sign
10′s complement of 1001 in XS-3 code = (–) 0111 (2′s complement of 1001) **Ans.**

5.14 | COMPARATOR

Comparator is a device that compares two binary bits. Output of device gives logic 1 if one bit is equal or less than or greater than other bit. Let x binary digit is compared with y binary digit. Truth table for single-bit comparator is given in Table 5.13.

TABLE 5.13 | Comparison of single bit binary numbers

Minterms	Input		Output		
	x	y	f_1 $(x > y)$	f_2 $(x = y)$	f_3 $(x < y)$
m_0	0	0	0	1	0
m_1	0	1	0	0	1
m_2	1	0	1	0	0
m_3	1	1	0	1	0

Boolean logic for $(x > y)$, $(x = y)$, and $(x < y)$ are given below

$$f_{x>y}(x, y) = \sum m(2) \tag{5.70}$$

$$f_{x=y}(x, y) = \sum m(0, 3) \tag{5.71}$$

$$f_{x<y}(x, y) = \sum m(1) \tag{5.72}$$

K-maps for above functions are given in Figures 5.63(a), 5.63(b) and 5.63(c). No pair is possible. The logic equations are given below.

$$f_{x>y}(x, y) = x\overline{y} \quad \text{or} \quad f_{x>y}(x, y) = 1 \text{ then } x > y$$

$$f_{x=y}(x, y) = xy + \overline{x}\,\overline{y} \Rightarrow f_{x=y}(x, y) = \overline{x \oplus y} \quad \text{or} \quad \text{if } f_{x=y}(x, y) = 1 \text{ then } x = y$$

$$f_{x<y}(x, y) = \overline{x}\,y \quad \text{or} \quad \text{if } f_{x<y}(x, y) = 1 \text{ then } x < y$$

Generalized logic is given below.

$$f(x,y)=\begin{cases} x\overline{y}=1 & ;x>y \\ \overline{\overline{x}y+x\overline{y}}=1 & ;x=y \\ \overline{x}y=1 & ;x<y \end{cases} \tag{5.73}$$

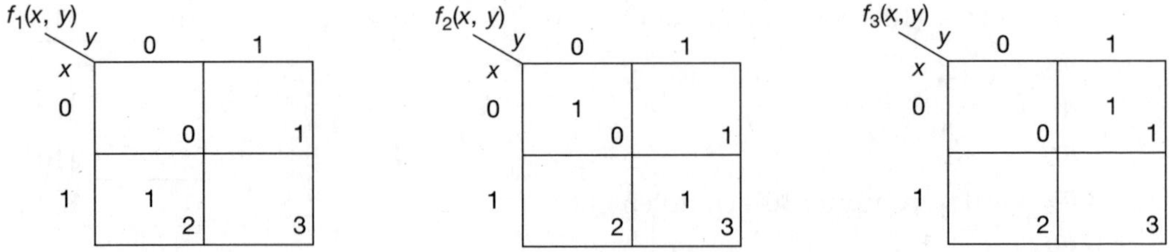

(a) K-map for x > y (b) K-map for x = y (c) K-map for x < y

FIGURE 5.63 | K-map for comparison of single-bit binary numbers

Combination circuit is given in Figure 5.64. It gives possible three outputs based on two inputs. If two inputs are equal then ($x = y$) becomes high and others become low. Block diagram is given in Figure 5.65.

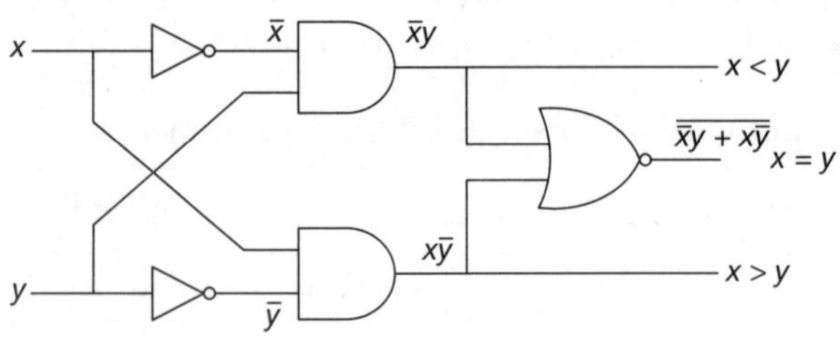

FIGURE 5.64 | Logic diagram of comparison of single-bit binary numbers

The logic can be extended for *i*-th bit as given below.

$$E_i(x_i,y_i)=\overline{\overline{x}_i y_i + x_i\overline{y}_i}=\overline{x}_i\overline{y}_i + x_i y_i \tag{5.74}$$

$$G_i(x_i,y_i)=\overline{x}_i y_i \tag{5.75}$$

$$L_i(x_i,y_i)=x_i\overline{y}_i \tag{5.76}$$

Comparison of two-bit numbers Comparison of two-bit numbers is discussed here. Truth table is given in Table 5.14. y_1y_0 and x_1x_0 are compared. y_1 is MSB and y_0 is LSB. x_1 is MSB and x_0 is LSB. There are four inputs and three outputs. Sixteen ($=2^4$) minterms are tabulated in the truth table.

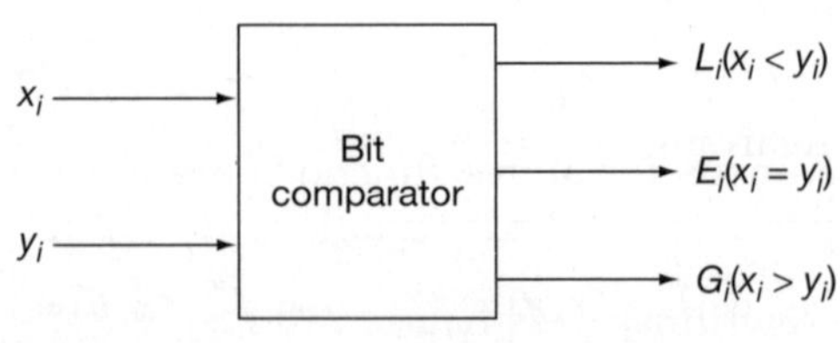

FIGURE 5.65 | Block diagram of comparison of *i*-th bit

TABLE 5.14 | Comparison of two-bit binary numbers

Minterm	Input				Output		
	Y		*X*		*Y* = *X*	*Y* > *X*	*Y* < *X*
	y_1	y_0	x_1	x_0			
m_0	0	0	0	0	1	0	0
m_1	0	0	0	1	0	0	1
m_2	0	0	1	0	0	0	1
m_3	0	0	1	1	0	0	1
m_4	0	1	0	0	0	1	0
m_5	0	1	0	1	1	0	0
m_6	0	1	1	0	0	0	1
m_7	0	1	1	1	0	0	1
m_8	1	0	0	0	0	1	0
m_9	1	0	0	1	0	1	0
m_{10}	1	0	1	0	1	0	0
m_{11}	1	0	1	1	0	0	1
m_{12}	1	1	0	0	0	1	0
m_{13}	1	1	0	1	0	1	0
m_{14}	1	1	1	0	0	1	0
m_{15}	1	1	1	1	1	0	0

Boolean equations are given below corresponding to the comparison of two-bit numbers.

$$f_{Y=X}(y_1,y_0,x_1,x_0)=\sum m(0,5,10,15) \tag{5.77}$$

$$f_{Y>X}(y_1,y_0,x_1,x_0)=\sum m(4,8,9,12,13,14) \tag{5.78}$$

$$f_{Y<X}(y_1,y_0,x_1,x_0)=\sum m(1,2,3,6,7,11) \tag{5.79}$$

Four-variable K-maps are given in Figure 5.66(a), Figure 5.66(b) and Figure 5.66(c) for $f_{Y=X}, f_{Y>X}$ and $f_{Y<X}$, respectively.

In K-map given in Figure 5.66(a), no pair, quad or octet may be formed. So logic function is written as

$$f_{X=Y}(y_1y_0x_1x_0)=\overline{y}_1\overline{y}_0\overline{x}_1\overline{x}_0+y_1\overline{y}_0x_1\overline{x}_0+\overline{y}_1y_0\overline{x}_1x_0+y_1y_0x_1x_0$$

Rearranging the logic equation

$$f_{X=Y}(y_1y_0x_1x_0)=\overline{y}_1\overline{y}_0\overline{x}_1\overline{x}_0+\overline{y}_1y_0\overline{x}_1x_0+y_1\overline{y}_0x_1\overline{x}_0+y_1y_0x_1x_0$$

Rewriting the equation

$$f_{X=Y}(y_1y_0x_1x_0)=\overline{y}_1\overline{x}_1(\overline{y}_0\overline{x}_0+y_0x_0)+y_1x_1(\overline{y}_0\overline{x}_0+y_0x_0)$$

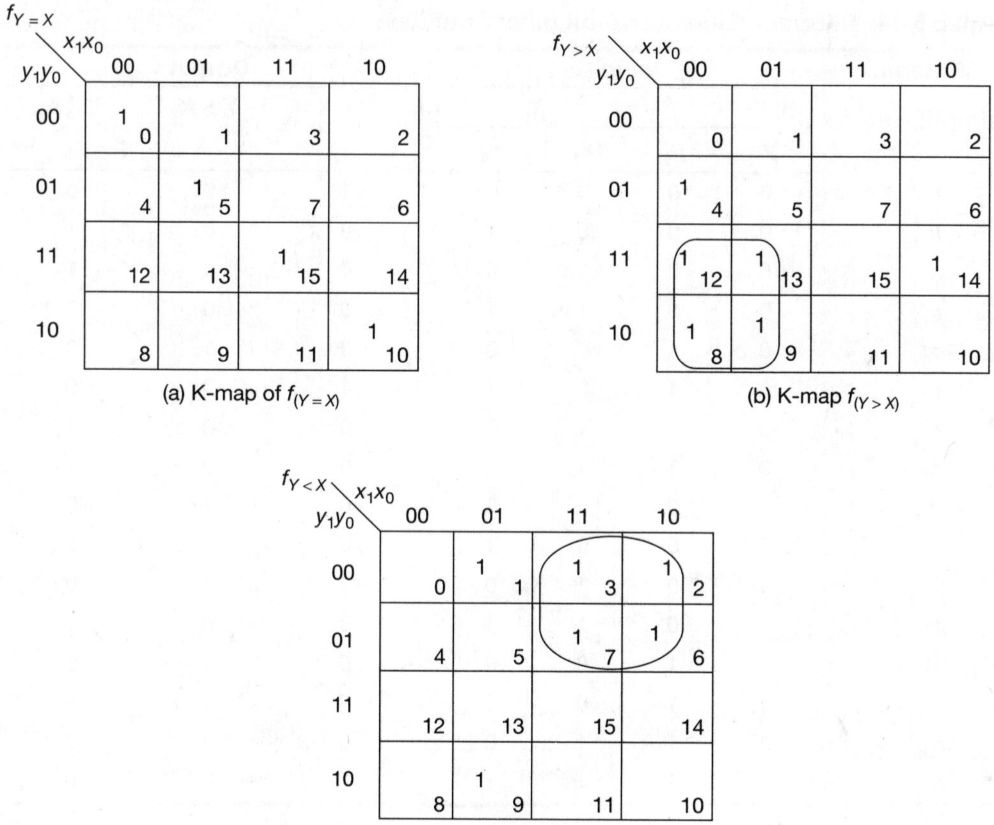

(a) K-map of $f_{(Y=X)}$ (b) K-map $f_{(Y>X)}$ (c) K-map of $f_{(Y<X)}$

FIGURE 5.66 | K-map to compare 2-bit binary numbers

Equation is revised as

$$f_{X=Y}(y_1y_0x_1x_0) = (\bar{y}_1\bar{x}_1 + y_1x_1)(\bar{y}_0\bar{x}_0 + y_0x_0) \quad (5.80)$$

Using Eq. (5.74), the above equation can be written as

$$f_{X=Y} = (E_1)(E_0) \quad (5.81)$$

In K-map given in Figure 5.66(b), one quad {8, 9, 12, 1 3} is formed. Logic function is written as

$$f_{Y>X}(y_1y_0x_1x_0) = y_1\bar{x}_1 + \bar{y}_1y_0\bar{x}_1\bar{x}_0 + y_1y_0x_1\bar{x}_0$$

On simplification

$$f_{Y>X}(y_1y_0x_1x_0) = y_1\bar{x}_1 + y_0\bar{x}_0(\bar{y}_1\bar{x}_1 + y_1x_1) \quad (5.82)$$

Using Eq. (5.75), the above equation can be written as

$$f_{Y>X} = G_1 + G_0(E_1) \quad (5.83)$$

In K-map given in Figure 5.66(b), one quad {2, 3, 6, 7} is formed. Logic function is written as

$$f_{Y<X}(y_1y_0x_1x_0) = \overline{y}_1x_1 + (\overline{y}_1\overline{x}_1\overline{y}_0x_0 + y_1x_1\overline{y}_0x_0)$$

On simplification

$$f_{Y<X}(y_1y_0x_1x_0) = \overline{y}_1x_1 + \overline{y}_0x_0(\overline{y}_1\overline{x}_1 + y_1x_1) \qquad (5.84)$$

Using Eq. (5.76), the above equation can be written as

$$f_{Y<X} = L_1 + L_0(E_1) \qquad (5.85)$$

Based on Eq. (5.81), Eq. (5.83) and Eq. (5.85), combination circuit diagram is given in Figure 5.67.

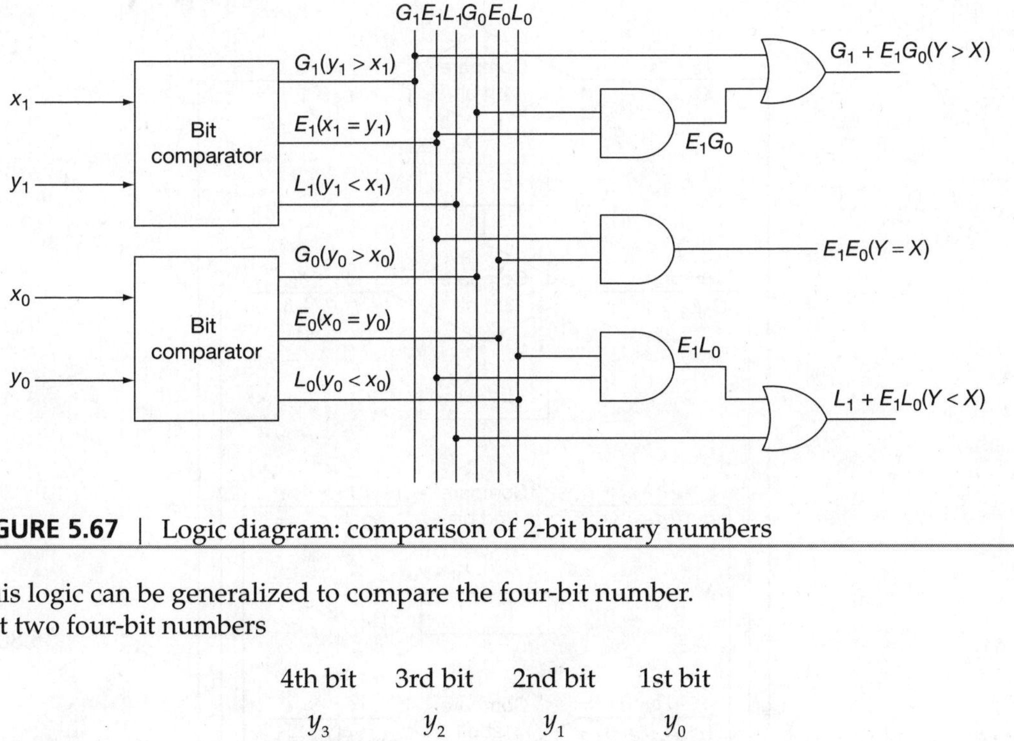

FIGURE 5.67 | Logic diagram: comparison of 2-bit binary numbers

This logic can be generalized to compare the four-bit number.
Let two four-bit numbers

4th bit	3rd bit	2nd bit	1st bit
y_3	y_2	y_1	y_0
x_3	x_2	x_1	x_0

y_3 is MSB and y_0 is LSB. x_3 is MSB and x_0 is LSB.

Logic given in Eq. (5.81), Eq. (5.83) and Eq. (5.85) can be extended to compare four-bit number. MSB y_3 is compared with MSB x_3.

- If $y_3 > x_3$, the $y_3y_2y_1y_0$ binary number is greater than $x_3x_2x_1x_0$ binary number. If $y_3 < x_3$, the $y_3y_2y_1y_0$ binary number is lesser than $x_3x_2x_1x_0$ binary number.
- In case $y_3 = x_3$, then compare y_2 and x_2 bits of the numbers. If $y_2 > x_2$, then $y_3y_2y_1y_0$ binary number is greater than $x_3x_2x_1x_0$ binary number. If $y_2 < x_2$ then $y_3y_2y_1y_0$ binary number is lesser than $x_3x_2x_1x_0$ binary number.

- In case $y_2 = x_2$, then compare y_1 and x_1 bits of the numbers. If $y_1 > x_1$, then the $y_3y_2y_1y_0$ binary number is greater than $x_3x_2x_1x_0$ binary number. If $y_1 < x_1$, the $y_3y_2y_1y_0$ binary number is lesser than $x_3x_2x_1x_0$ binary number.
- In case $y_1 = x_1$, then compare y_0 and x_0 bits of the numbers. If $y_0 > x_0$ the $y_3y_2y_1y_0$ binary number is greater than $x_3x_2x_1x_0$ binary number. If $y_0 < x_0$, the $y_3y_2y_1y_0$ binary number is lesser than $x_3x_2x_1x_0$ binary number.
- In case $y_0 = x_0$, then $y_3y_2y_1y_0$ binary number is equal to $x_3x_2x_1x_0$ binary number.

For equal all the bits are compared. If all bits are equal then the number is same. The complete operation is depicted in the flow chart given in Figure 8.68.

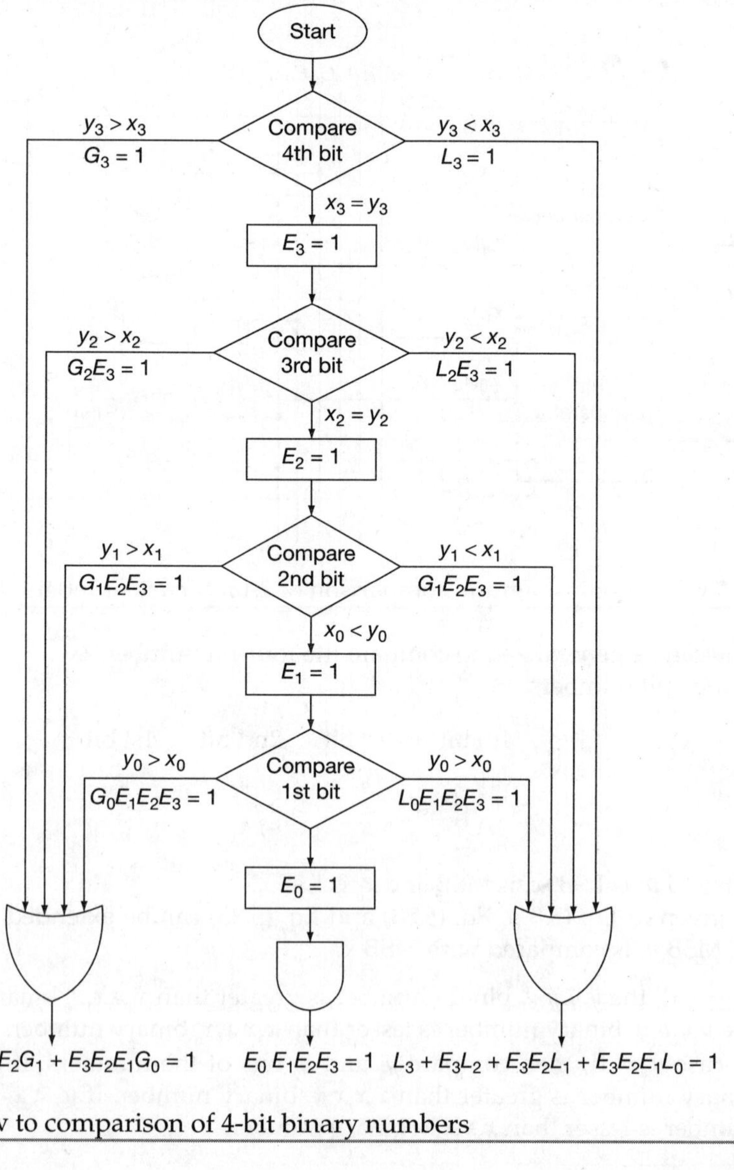

FIGURE 5.68 | Flow to comparison of 4-bit binary numbers

Equation (5.81) can be extended for four-bit comparison

$$f_{X=Y} = (E_3)(E_2)(E_1)(E_0) \tag{5.86}$$

Equation (5.83) can be extended for four-bit comparison

$$f_{Y>X} = G_3 + E_3G_2 + E_3E_2G_1 + E_3E_2E_1G_0 \tag{5.87}$$

Equation (5.83) can be extended for four-bit comparison

$$f_{Y>X} = L_3 + E_3L_2 + E_3E_2L_1 + E_3E_2E_1L_0 \tag{5.88}$$

Combination circuit is given in Figure 5.69 that compares four-bit numbers. Equations (5.86), (5.87) and (5.88) can be generalized for n-bits. Cascading is preferred.

$$\prod_{i=0}^{n-1} E_i = 1 \quad \text{for} \quad f_{Y=X} \tag{5.89}$$

$$\sum_{k=0}^{n-1}\left[\left(\prod_{i=k+1}^{n-1} E_i\right)G_k\right] = 1 \quad \text{for} \quad f_{Y>X} \tag{5.90}$$

$$\sum_{k=0}^{n-1}\left[\left(\prod_{i=k+1}^{n-1} E_i\right)L_k\right] = 1 \quad \text{for} \quad f_{Y<X} \tag{5.91}$$

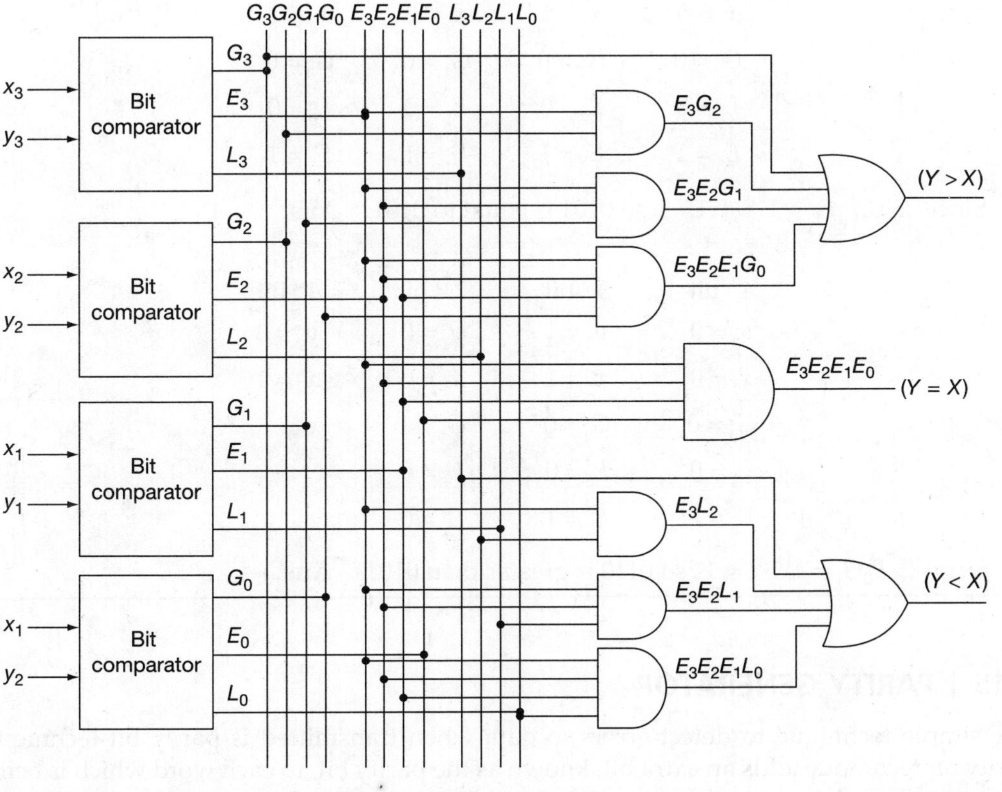

FIGURE 5.69 | Logic diagram to compare 4-bit binary numbers

EXAMPLE 5.17 Perform comparison of the following binary numbers:

(a) 0111 and 1000 (b) 0101 and 0101 (c) 0110 and 0101

SOLUTION

(a)

4th bit	3rd bit	2nd bit	1st bit
$y_3 = 0$	$y_2 = 1$	$y_1 = 1$	$y_0 = 1$
$x_3 = 1$	$x_2 = 0$	$x_1 = 0$	$x_0 = 0$
$G_3 = 0$			
$L_3 = 1$			
$E_3 = 0$			

Since $L_3 = 1$ so 0111 is lesser than 1000. **Ans.**

(b)

4th bit	3rd bit	2nd bit	1st bit
$y_3 = 0$	$y_2 = 1$	$y_1 = 0$	$y_0 = 1$
$x_3 = 0$	$x_2 = 1$	$x_1 = 0$	$x_0 = 1$
$G_3 = 0$	$G_2 = 0$	$G_1 = 0$	$G_0 = 0$
$L_3 = 0$	$L_2 = 0$	$L_1 = 0$	$L_0 = 0$
$E_3 = 1$	$E_2 = 1$	$E_1 = 1$	$E_0 = 1$

Since $E_3E_2E_1E_0 = 1 \cdot 1 \cdot 1 \cdot 1 = 1$ so 0101 is equal to 0101. **Ans.**

(c)

4th bit	3rd bit	2nd bit	1st bit
$y_3 = 0$	$y_2 = 1$	$y_1 = 0$	$y_0 = 0$
$x_3 = 0$	$x_2 = 1$	$x_1 = 0$	$x_0 = 1$
$G_3 = 0$	$G_2 = 0$	$G_1 = 1$	
$L_3 = 0$	$L_2 = 0$	$L_1 = 0$	
$E_3 = 1$	$E_2 = 1$	$E_1 = 0$	

Since $E_3E_2G_1 = 1 \cdot 1 \cdot 1 = 1$, so 0110 is greater than 0101. **Ans.**

5.15 | PARITY GENERATOR

The simple technique to detect errors in data when transmitted is parity bit technique. Parity bit technique adds an extra bit, known as the parity bit, to each word which is being transmitted. There are two types of parity—odd parity and even parity.

Odd parity The parity bit is set to a 0 or a 1 at the sender such that the total number of 1 bit in the word including the parity bit is an odd number.

Even parity The parity bit is set to a 0 or a 1 at the sender such that the total number of 1 bit in the word including the parity bit is an even number.

Odd parity is preferred than even parity because even parity does not detect the situation where all 0s are created by short-circuit or some other fault condition.

5.15.1 | Even-Parity Generator

Table 5.15 shows the even-parity bits to be added to transmit to four-bit binary data. x_3 is considered MSB and x_0 is considered as LSB

TABLE 5.15 | Even-parity generator

Minterm	Input				Output
	x_3	x_2	x_1	x_0	Even-parity Bit, P_{even}
m_0	0	0	0	0	0
m_1	0	0	0	1	1
m_2	0	0	1	0	1
m_3	0	0	1	1	0
m_4	0	1	0	0	1
m_5	0	1	0	1	0
m_6	0	1	1	0	0
m_7	0	1	1	1	1
m_8	1	0	0	0	1
m_9	1	0	0	1	0
m_{10}	1	0	1	0	0
m_{11}	1	0	1	1	1
m_{12}	1	1	0	0	0
m_{13}	1	1	0	1	1
m_{14}	1	1	1	0	1
m_{15}	1	1	1	1	0

Boolean equation is given below

$$P_{\text{even}}(x_3,x_2,x_1,x_0)=\sum m(1,2,4,7,8,11,13,14) \tag{5.92}$$

K-map is used to simplify the logic as given in Figure 5.70. No pair, quad or octet is formed.

Logic equations are given below.

$$P_{\text{even}}=\overline{x}_3\overline{x}_2(x_1\oplus x_0)+x_3x_2(x_1\oplus x_0)+\overline{x}_3x_2(\overline{x_1\oplus x_0})+x_3\overline{x}_2(\overline{x_1\oplus x_0})$$

Rewriting above equation,

$$P_{\text{even}}=(\overline{x}_3\overline{x}_2+x_3x_2)(x_1\oplus x_0)+(\overline{x}_3x_2+x_3\overline{x}_2)(\overline{x_1\oplus x_0})$$

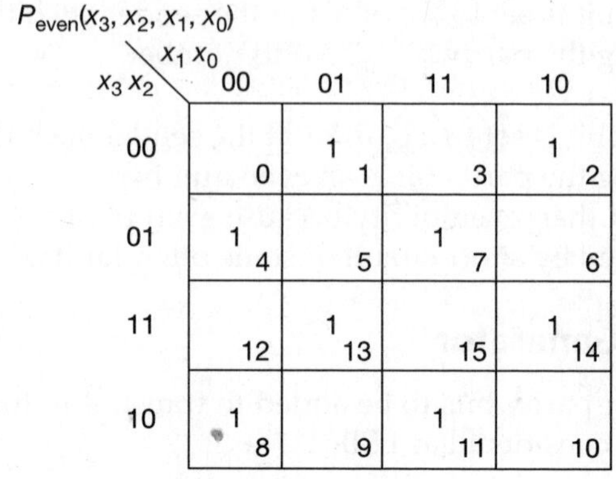

FIGURE 5.70 | K-map for even-parity generator

Rewriting above equation,

$$P_{\text{even}} = \overline{(x_3 \oplus x_2)}(x_1 \oplus x_0) + (x_3 \oplus x_2)\overline{(x_1 \oplus x_0)}$$

Rewriting above equation,

$$P_{\text{even}} = (x_3 \oplus x_2) \oplus (x_1 \oplus x_0). \tag{5.93}$$

Combinational circuit is given in Figure 5.71.

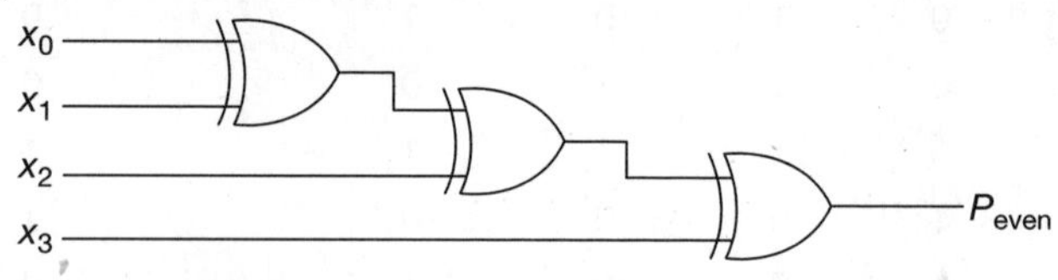

FIGURE 5.71 | Even-parity generator

5.15.2 | Odd-Parity Generator

Table 5.16 shows the odd-parity bits to be added to transmit to four-bit binary data. x_3 is considered MSB and x_0 is considered as LSB.

TABLE 5.16 | Odd-parity generator

Minterm	Input				Output
	x_3	x_2	x_1	x_0	Odd-parity Bit, P_{odd}
m_0	0	0	0	0	1
m_1	0	0	0	1	0
m_2	0	0	1	0	0
m_3	0	0	1	1	1
m_4	0	1	0	0	0

(Continued)

TABLE 5.16 | (Continued)

Minterm	Input				Output
	x_3	x_2	x_1	x_0	Odd-parity Bit, P_{odd}
m_5	0	1	0	1	1
m_6	0	1	1	0	1
m_7	0	1	1	1	0
m_8	1	0	0	0	0
m_9	1	0	0	1	1
m_{10}	1	0	1	0	1
m_{11}	1	0	1	1	0
m_{12}	1	1	0	0	1
m_{13}	1	1	0	1	0
m_{14}	1	1	1	0	0
m_{15}	1	1	1	1	1

Boolean equation is given below

$$P_{\text{odd}}(x_3, x_2, x_1, x_0) = \sum m(0, 3, 5, 6, 9, 10, 12, 15) \tag{5.94}$$

K-map is used to simplify the logic as given in Figure 5.72. No pair, quad or octet is formed.

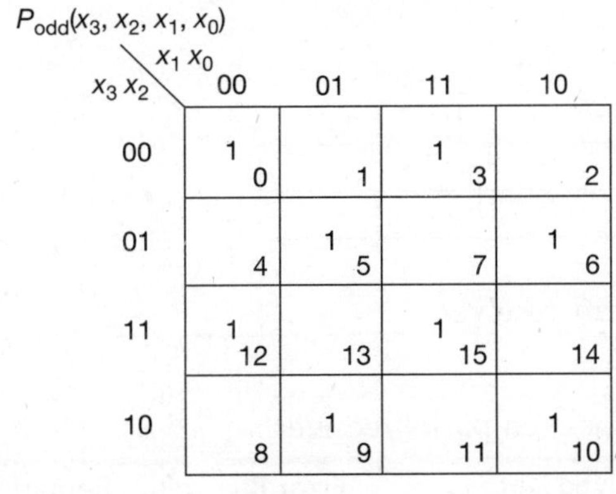

FIGURE 5.72 | K-map for odd-parity generator

Logic equations are given below.

$$P_{\text{odd}} = \bar{x}_3\bar{x}_2\overline{(x_1 \oplus x_0)} + x_3x_2\overline{(x_1 \oplus x_0)} + \bar{x}_3x_2(x_1 \oplus x_0) + x_3\bar{x}_2(x_1 \oplus x_0)$$

Rewriting above equation,

$$P_{\text{odd}} = (\bar{x}_3\bar{x}_2 + x_3x_2)\overline{(x_1 \oplus x_0)} + (\bar{x}_3x_2 + x_3\bar{x}_2)(x_1 \oplus x_0)$$

Rewriting above equation,

$$P_{\text{odd}} = (\overline{x_3 \oplus x_2})(\overline{x_1 \oplus x_0}) + (x_3 \oplus x_2)(x_1 \oplus x_0)$$

Rewriting above equation,

$$P_{\text{odd}} = \overline{(x_3 \oplus x_2) \oplus (x_1 \oplus x_0)} \qquad (5.95)$$

Combinational circuit is given in Figure 5.73.

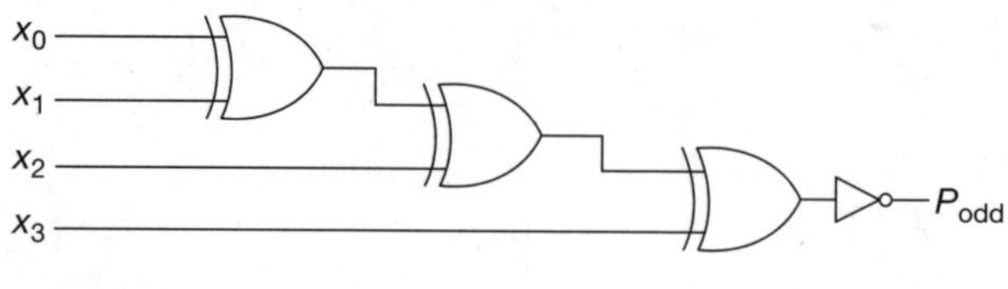

FIGURE 5.73 | Odd-parity generator

5.15.3 | Even-Parity Bit Receiver

On receiving the digital data, a parity checking circuit generates an error signal if the total number of 1's is even in an even-parity system. The even-parity checker detects a single-bit error.

The total number of 1 bit in the word including the even-parity bit should be even. Combinational circuit is given in Figure 5.74. Table 5.17 gives the error bit for even-parity checker.

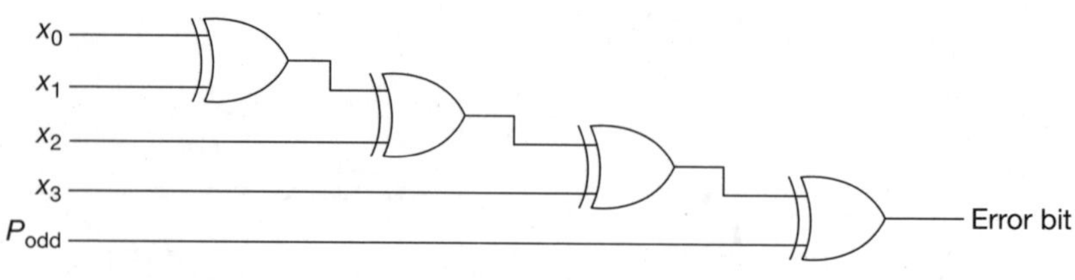

FIGURE 5.74 | Odd-parity receiver

TABLE 5.17 | Error for even-parity receiver

Number of 1's in Received Data	Error Bit	Remarks
Even	0	No error in received data
Odd	1	Error in received data

5.15.4 | Odd-Parity Bit Receiver

On receiving the digital data, a parity checking circuit generates an error signal if the total number of 1's is odd in an odd-parity system. The odd-parity checker detects a single-bit error. It cannot detect two or more errors within the data word.

The total number of 1 bit in the word including the odd-parity bit should be odd. Combinational circuit is given in Figure 5.75. Table 5.18 gives the error bit for odd-parity checker.

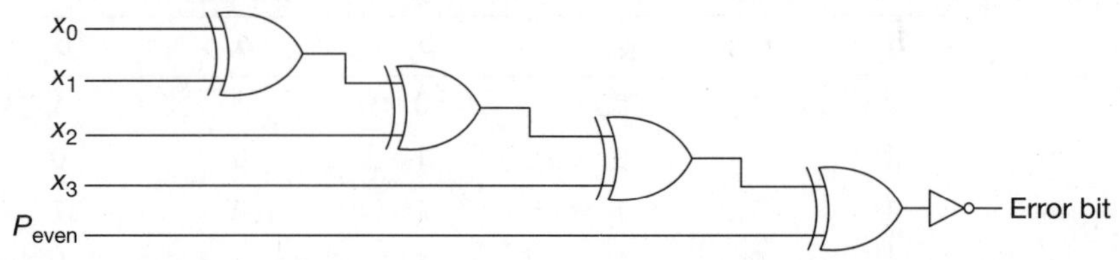

FIGURE 5.75 | Odd-parity receiver

TABLE 5.18 | Error for odd-parity receiver

Number of 1's in Received Data	Error Bit	Remarks
Even	1	Error in received data
Odd	0	No error in received data

5.16 | CODE CONVERTER

Digital circuits process data in the binary format. Different codes are used to represent the data. Data may be numeric, alphanumeric or special characters. Numeric codes used to represent decimal digits are called BCD codes. There are a large number of BCD codes. In order to represent decimal digits 0, 1, 2, …, 9, it is necessary to use a sequence of at least four binary digits. Combinational circuits are required those convert one form of code to another form. Few examples are discussed below to design code converters.

EXAMPLE 5.18 Design a combinational circuit that converts three-bit gray number into binary code.

SOLUTION

Inputs: Three-binary digits $g_2g_1g_0$ represent gray number. g_2 gives MSB and g_0 gives LSB.

Outputs: Three-binary digits $b_2b_1b_0$ represent binary number. b_2 is considered MSB and b_0 is considered as LSB.

Truth table: truth table is given in Table 5.19. Three-bit gray code gives minterms which are not in sequence.

TABLE 5.19 | Gray to Binary code conversion of three-bit number

Minterm	Input			Output		
	g_2	g_1	g_0	b_2	b_1	b_0
m_0	0	0	0	0	0	0
m_1	0	0	1	0	0	1
m_3	0	1	1	0	1	0

(Continued)

TABLE 5.19 | (Continued)

Minterm	Input			Output		
	g_2	g_1	g_0	b_2	b_1	b_0
m_2	0	1	0	0	1	1
m_6	1	1	0	1	0	0
m_7	1	1	1	1	0	1
m_5	1	0	1	1	1	0
m_4	1	0	0	1	1	1

Boolean equation is given below

$$b_0(g_2, g_1, g_0) = \sum m(1, 2, 4, 7) \tag{5.96}$$

$$b_1(g_2, g_1, g_0) = \sum m(2, 3, 4, 5) \tag{5.97}$$

$$b_2(g_2, g_1, g_0) = \sum m(4, 5, 6, 7) \tag{5.98}$$

K-maps are used to simplify the logic functions. Figure 5.76(a) gives three-variable K-map for function b_0. No pair is formed. The logic expression is given below.

$$b_0(g_2, g_1, g_0) = \bar{g}_2\bar{g}_1 g_0 + \bar{g}_2 g_1 \bar{g}_0 + g_2\bar{g}_1\bar{g}_0 + g_2 g_1 g_0$$

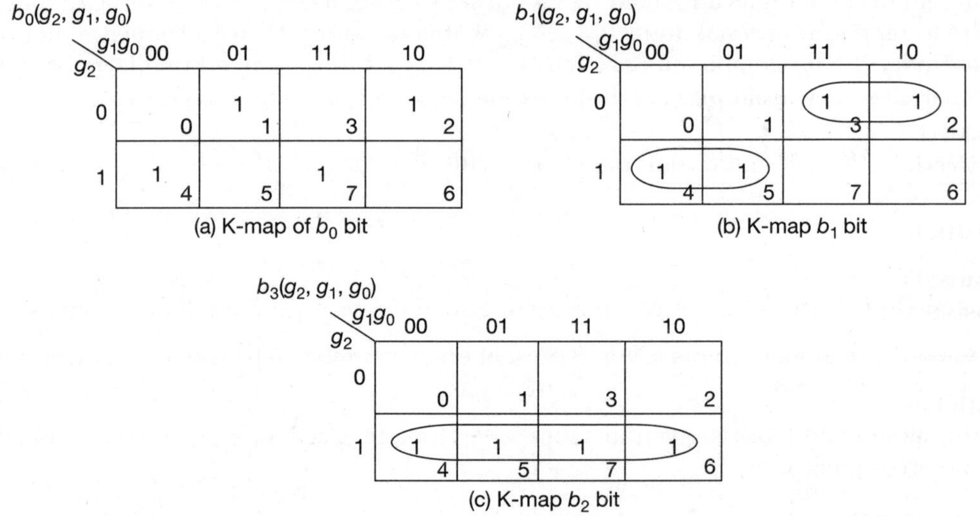

FIGURE 5.76 | K-map to convert 3-bit gray number into binary number

Rewriting above equation,

$$b_0(g_2, g_1, g_0) = \bar{g}_2(\bar{g}_1 g_0 + g_1\bar{g}_0) + g_2(\bar{g}_1\bar{g}_0 + g_1 g_0)$$

Rewriting above equation,

$$b_0(g_2, g_1, g_0) = \bar{g}_2(g_1 \oplus g_0) + g_2(\overline{g_1 \oplus g_0})$$

Rewriting above equation,

$$b_0(g_2, g_1, g_0) = g_2 \oplus (g_1 \oplus g_0) \tag{5.99}$$

Figure 5.76(b) gives three-variable K-map for function b_1. Two pairs are formed: {2, 3} and {4, 5}. The logic expression is given below.

$$b_1(g_2, g_1, g_0) = \bar{g}_2 g_1 + g_2 \bar{g}_1$$

Rewriting above equation,

$$b_1(g_2, g_1, g_0) = g_2 \oplus g_1 \tag{5.100}$$

Figure 5.76(c) gives three-variable K-map for function b_2. One quad is formed: {4, 5, 6, 7}. The logic expression is given below.

$$b_2(g_2, g_1, g_0) = g_2 \tag{5.101}$$

Combinational circuit is given in Figure 5.77.

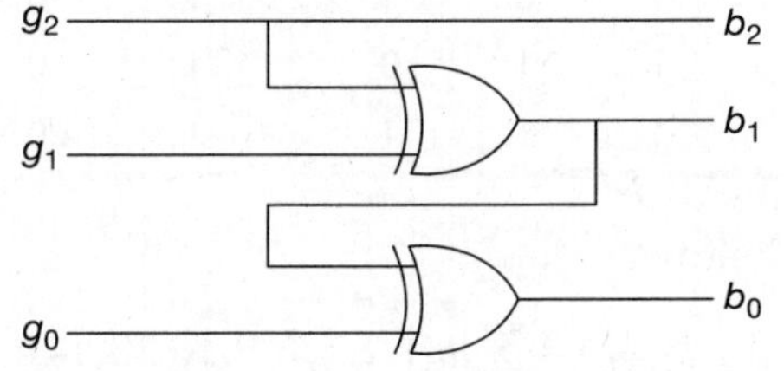

FIGURE 5.77 | Logic diagram for 3-bit gray to binary code converter

EXAMPLE 5.19 Design a combinational circuit that converts four-bit binary number into gray code.

SOLUTION

Inputs: Four-binary digits $b_3b_2b_1b_0$ represent binary number. b_3 is considered MSB and b_0 is considered as LSB.

Outputs: Four- binary digits $g_3g_2g_1g_0$ represent gray number. g_3 gives MSB and g_0 gives LSB.

Truth table: truth table is given in Table 5.20.

TABLE 5.20 | Binary to Gray code conversion of four-bit number

Minterm	Input				Output			
	b_3	b_2	b_1	b_0	g_3	g_2	g_1	g_0
m_0	0	0	0	0	0	0	0	0
m_1	0	0	0	1	0	0	0	1
m_2	0	0	1	0	0	0	1	1
m_3	0	0	1	1	0	0	1	0

(*Continued*)

TABLE 5.20 | (Continued)

Minterm	Input				Output			
	b_3	b_2	b_1	b_0	g_3	g_2	g_1	g_0
m_4	0	1	0	0	0	1	1	0
m_5	0	1	0	1	0	1	1	1
m_6	0	1	1	0	0	1	0	1
m_7	0	1	1	1	0	1	0	0
m_8	1	0	0	0	1	1	0	0
m_9	1	0	0	1	1	1	0	1
m_{10}	1	0	1	0	1	1	1	1
m_{11}	1	0	1	1	1	1	1	0
m_{12}	1	1	0	0	1	0	1	0
m_{13}	1	1	0	1	1	0	1	1
m_{14}	1	1	1	0	1	0	0	1
m_{15}	1	1	1	1	1	0	0	0

Boolean equation is given below

$$g_0(b_3,b_2,b_1,b_0)=\sum m(1,2,5,6,9,10,13,14) \tag{5.102}$$

$$g_1(b_3,b_2,b_1,b_0)=\sum m(2,3,4,5,10,11,12,13) \tag{5.103}$$

$$g_2(b_3,b_2,b_1,b_0)=\sum m(4,5,6,7,8,9,10,11) \tag{5.104}$$

$$g_3(b_3,b_2,b_1,b_0)=\sum m(8,9,10,11,12,13,14,15) \tag{5.105}$$

K-maps are used to simplify the logic functions. Figure 5.78(a) gives four-variable K-map for function g_0. Two quads are formed: {1, 5, 9, 13} and {2, 6, 10, 14}. The logic expression is given below.

$$g_0(b_3,b_2,b_1,b_0)=\overline{b}_1 b_0+b_1\overline{b}_0$$

Rewriting above equation

$$g_0(b_3,b_2,b_1,b_0)=b_1\oplus b_0 \tag{5.106}$$

Figure 5.78(b) gives four-variable K-map for function g_1. Two quads are formed: {2, 3, 10, 11} and {4, 5, 12, 13}. The logic expression is given below.

$$g_1(b_3,b_2,b_1,b_0)=\overline{b}_2 b_1+b_2\overline{b}_1$$

Rewriting above equation,

$$g_1(b_3,b_2,b_1,b_0)=b_2\oplus b_1 \tag{5.107}$$

Figure 5.78(c) gives four-variable K-map for function g_2. Two quads are formed: {4, 5, 6, 7} and {8, 9, 10, 11}. The logic expression is given below.

$$g_2(b_3,b_2,b_1,b_0)=\overline{b}_3 b_2+b_3\overline{b}_2$$

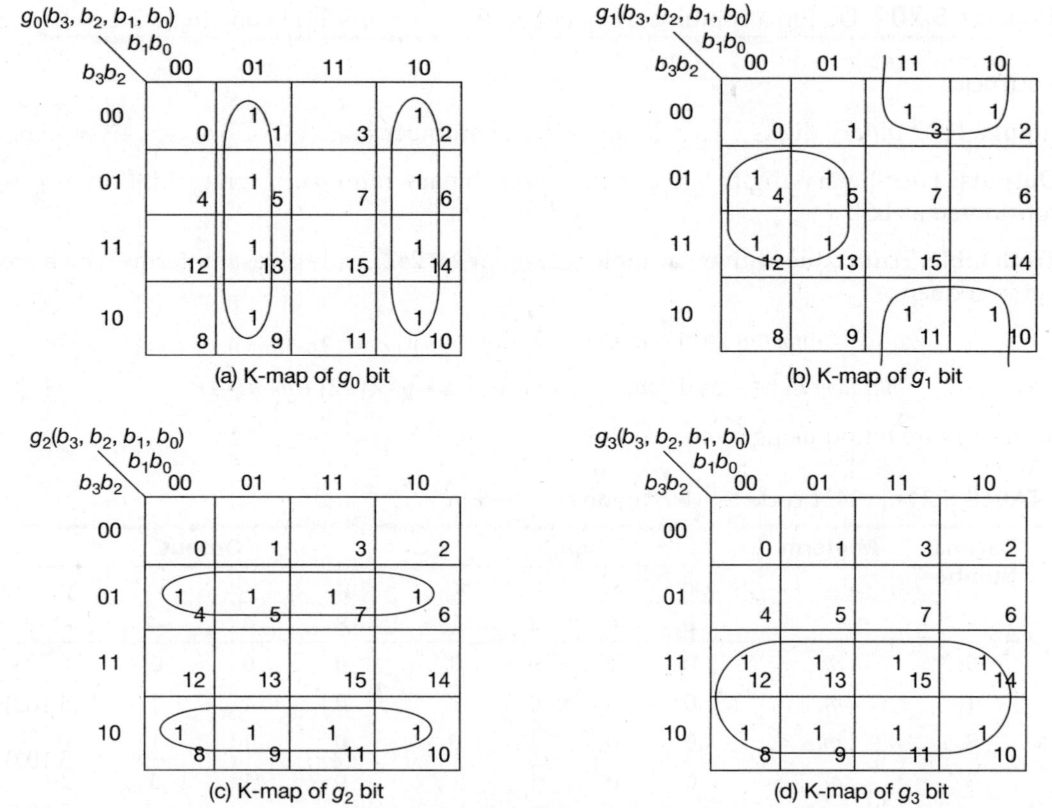

FIGURE 5.78 | K-map to convert 4-bit binary number into gray number

Rewriting above equation,

$$g_2(b_3, b_2, b_1, b_0) = b_3 \oplus b_2 \tag{5.108}$$

Figure 5.78(d) gives four-variable K-map for function g_3. One octet is formed: {8, 9, 10, 11, 12, 13, 14, 15}. The logic expression is given below.

$$g_3(b_3, b_2, b_1, b_0) = b_3 \tag{5.109}$$

Combinational circuit is given in Figure 5.79.

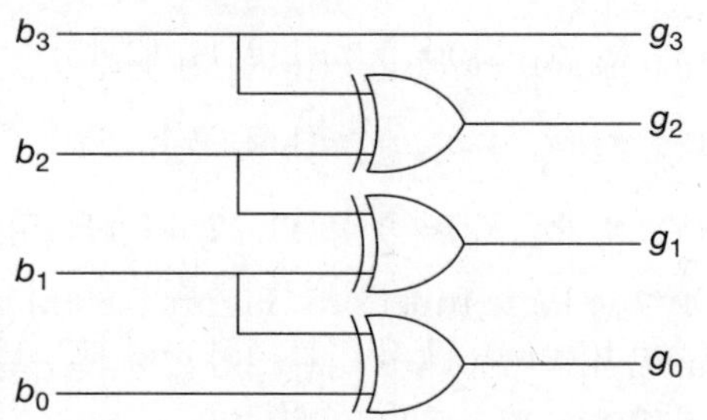

FIGURE 5.79 | Logic diagram of 4-bit binary to gray code converter

EXAMPLE 5.20 Design a combinational circuit that converts 2421 code to 84-2-1 code.

SOLUTION

Inputs: Four-binary digits $x_3x_2x_1x_0$ represent gray number. x_3 gives MSB and x_0 gives LSB.

Outputs: Four-binary digits $y_3y_2y_1y_0$ represent binary number. y_3 gives MSB and y_0 is considered as LSB.

Truth table: Truth table is given in Table 5.21. Four-bit 2421 code gives minterms which are not in sequence.

$$\text{Number in 2421 code} = x_3 \times 2 + x_2 \times 4 + x_1 \times 2 + x_0 \times 1$$

$$\text{Number in } 84-2-1 \text{ code} = y_3 \times 8 + y_2 \times 4 + y_1 \times (-2) + y_0 \times (-1)$$

Minterms are found using binary coding.

TABLE 5.21 | 2421 code to 84-2-1code conversion

Decimal Number	Minterm	Input				Output			
		x_3	x_2	x_1	x_0	y_3	y_2	y_1	y_0
		2	4	2	1	8	4	−2	−1
0	m_0	0	0	0	0	0	0	0	0
1	m_1	0	0	0	1	0	1	1	1
2	m_2	0	0	1	0	0	1	1	0
3	m_3	0	0	1	1	0	1	0	1
4	m_4	0	1	0	0	0	1	0	0
5	m_{11}	1	0	1	1	1	0	1	1
6	m_{12}	1	1	0	0	1	0	1	0
7	m_{13}	1	1	0	1	1	0	0	1
8	m_{14}	1	1	1	0	1	0	0	0
9	m_{15}	1	1	1	1	1	1	1	1

Boolean equations are given below for output functions.

$$y_0(x_3, x_2, x_1, x_0) = \sum m(1, 3, 11, 13, 15) \tag{5.110}$$

$$y_1(x_3, x_2, x_1, x_0) = \sum m(1, 2, 11, 12, 15) \tag{5.111}$$

$$y_2(x_3, x_2, x_1, x_0) = \sum m(1, 2, 3, 4, 15) \tag{5.112}$$

$$y_3(x_3, x_2, x_1, x_0) = \sum m(11, 12, 13, 14, 15) \tag{5.113}$$

K-maps are used to simplify the logic functions. Figure 5.80(a) gives four-variable K-map for function y_0. Three pair are formed; {1, 3}, {11, 15} and {13, 15}. The logic expression is given below.

$$y_0(x_3, x_2, x_1, x_0) = \overline{x}_3\overline{x}_2x_0 + x_3x_2x_0 + x_3x_1x_0 \tag{5.114}$$

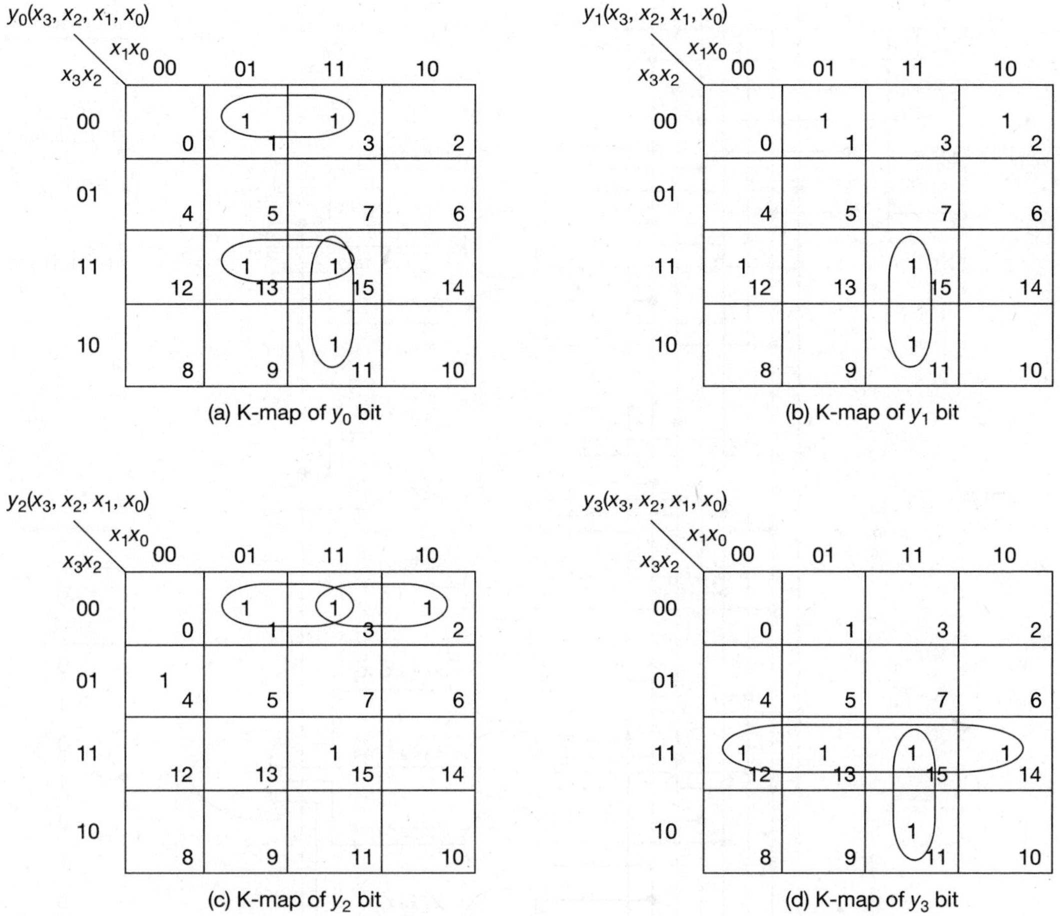

FIGURE 5.80 | K-map to convert 2421 code into 84-2-1 code

Figure 5.80(b) gives four-variable K-map for function y_1. One pair is formed: {11, 15}. The logic expression is given below.

$$y_1(x_3,x_2,x_1,x_0) = \overline{x}_3\overline{x}_2\overline{x}_1x_0 + \overline{x}_3\overline{x}_2x_1\overline{x}_0 + x_3x_2\overline{x}_1\overline{x}_0 + x_3x_1x_0 \tag{5.115}$$

Figure 5.80(c) gives four-variable K-map for function y_2. Two pairs are formed: {1, 3}, {2, 3}. The logic expression is given below.

$$y_2(x_3,x_2,x_1,x_0) = \overline{x}_3\overline{x}_2x_0 + \overline{x}_3\overline{x}_2x_1 + \overline{x}_3x_2\overline{x}_1\overline{x}_0 + x_3x_2x_1x_0 \tag{5.116}$$

Figure 5.80(d) gives four-variable K-map for function y_3. One pair and one quad are formed: {11, 15}, {12, 13, 14, 15}. The logic expression is given below.

$$y_3(x_3,x_2,x_1,x_0) = x_3x_2 + x_3x_1x_0 \tag{5.117}$$

Combinational circuit is given in Figure 5.81.

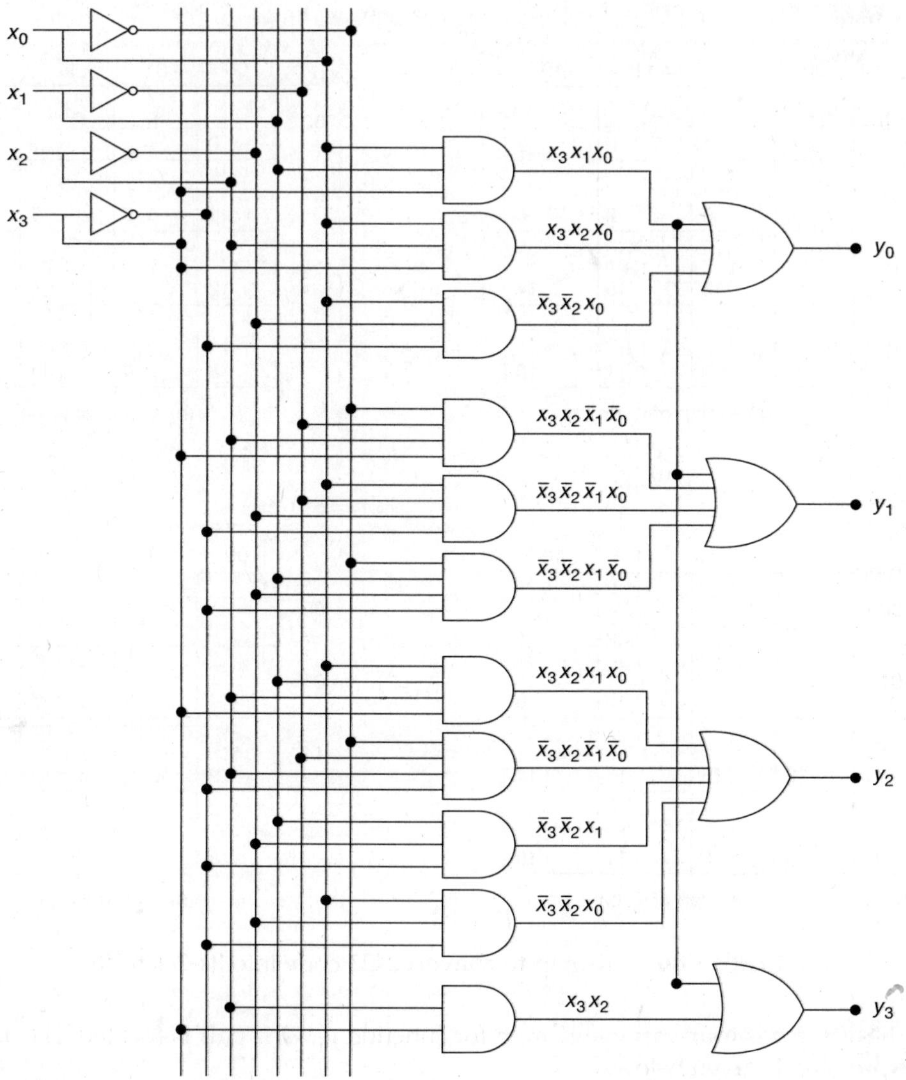

FIGURE 5.81 | Code converter: 2421 to 84-2-1

EXAMPLE 5.21 Design a combinational circuit that converts BCD code to Excess-3 code.

SOLUTION

Inputs: Four binary digits $d_3d_2d_1d_0$ represent gray number. d_3 gives MSB and d_0 gives LSB.

Outputs: Four-binary digits $y_3y_2y_1y_0$ represent binary number. y_3 gives MSB and y_0 is considered as LSB.

Truth table: Truth table is given in Table 5.22. BCD code is used for numeric digits. Four-bit binary values form more values can be used as don't care conditions.

TABLE 5.22 | BCD to Excess-3 code conversion

Minterm	Input				Output			
	d_3	d_2	d_1	d_0	y_3	y_2	y_1	y_0
m_0	0	0	0	0	0	0	1	1
m_1	0	0	0	1	0	1	0	0
m_2	0	0	1	0	0	1	0	1
m_3	0	0	1	1	0	1	1	0
m_4	0	1	0	0	0	1	1	1
m_5	0	1	0	1	1	0	0	0
m_6	0	1	1	0	1	0	0	1
m_7	0	1	1	1	1	0	1	0
m_8	1	0	0	0	1	0	1	1
m_9	1	0	0	1	1	1	0	0

Boolean equations are given below for output functions

$$y_0(d_3, d_2, d_1, d_0) = \sum m(0, 2, 4, 6, 8) + \sum d(10, 11, 12, 13, 14, 15) \tag{5.118}$$

$$y_1(d_3, d_2, d_1, d_0) = \sum m(0, 3, 4, 7, 8) + \sum d(10, 11, 12, 13, 14, 15) \tag{5.119}$$

$$y_2(d_3, d_2, d_1, d_0) = \sum m(1, 2, 3, 4, 9) + \sum d(10, 11, 12, 13, 14, 15) \tag{5.120}$$

$$y_3(d_3, d_2, d_1, d_0) = \sum m(5, 6, 7, 8, 9) + \sum d(10, 11, 12, 13, 14, 15) \tag{5.121}$$

K-maps are used to simplify the logic functions. Figure 5.82(a) gives four-variable K-map for function y_0. One octet is formed: {0, 4, 8, 12, 2, 6, 10, 14}. The logic expression is given below.

$$y_0(d_3, d_2, d_1, d_0) = d_0 \tag{5.122}$$

Figure 5.82(b) gives four-variable K-map for function y_1. Two quads are formed: {0, 4, 8, 12}, {3, 7, 11, 15}. The logic expression is given below.

$$y_1(d_3, d_2, d_1, d_0) = \bar{d}_1\bar{d}_0 + d_1 d_0 \tag{5.123}$$

Figure 5.82(c) gives four-variable K-map for function y_2. One pair and two quads are formed: {4, 12}, {1, 3, 9, 11}, {2, 3, 10, 11}. The logic expression is given below.

$$y_2(d_3, d_2, d_1, d_0) = \bar{d}_2 d_0 + \bar{d}_2 d_1 + d_2\bar{d}_1\bar{d}_0 \tag{5.124}$$

Figure 5.82(d) gives four-variable K-map for function y_3. Two quads and one octet are formed: {5, 7, 13, 15}, {6, 7, 14, 15} and {8, 9, 10, 11, 12, 13, 14, 15}. The logic expression is given below.

$$y_3(d_3, d_2, d_1, d_0) = d_3 + d_2 d_0 + d_2 d_1 \tag{5.125}$$

Combinational circuit is given in Figure 5.83.

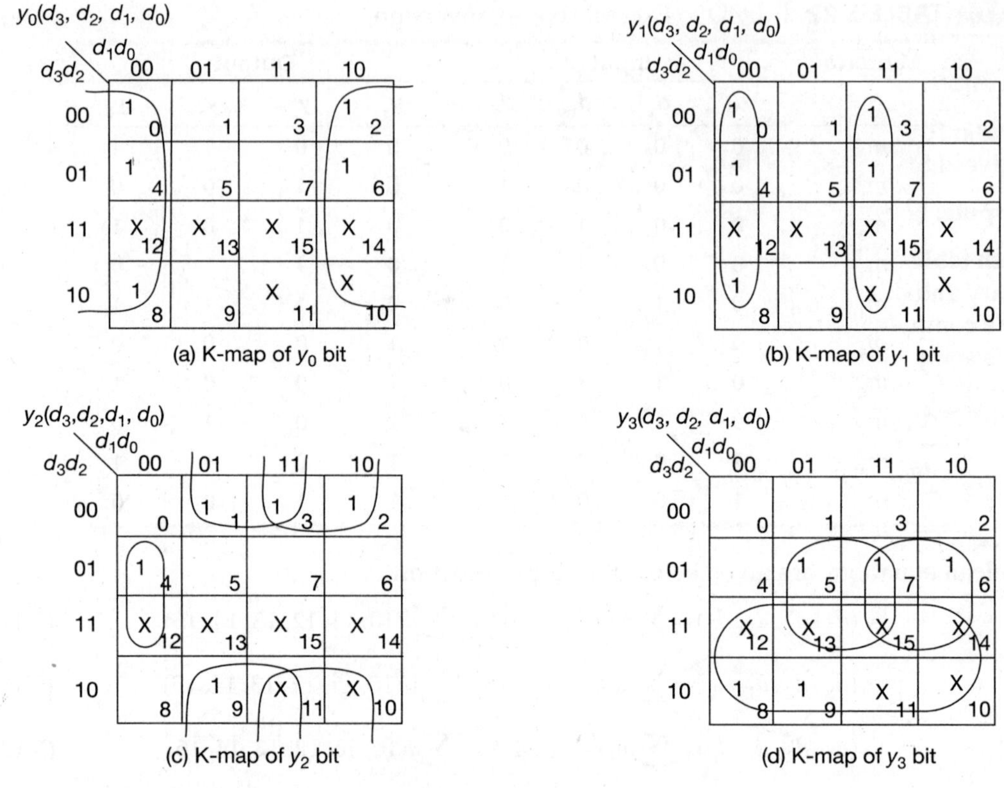

FIGURE 5.82 | K-map to convert four-bit BCD into excess-3 code

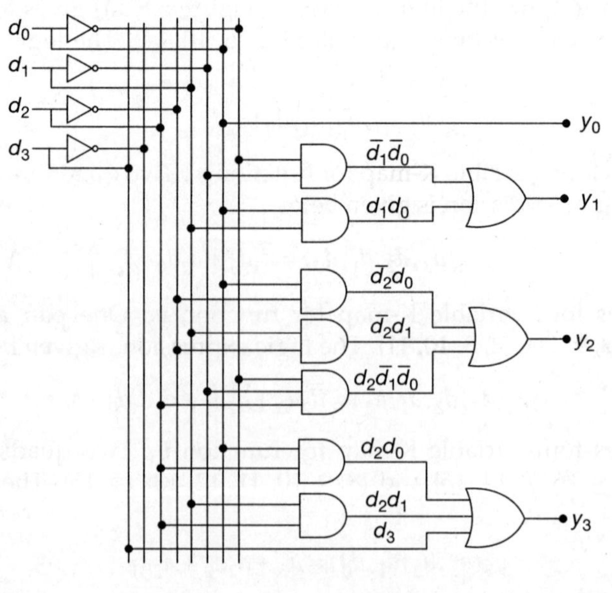

FIGURE 5.83 | Logic diagram to convert BCD code to excess-3 code

EXAMPLE 5.22 Design a combinational circuit that finds 9's complement of BCD digit

SOLUTION

Inputs: Four-binary digits $d_3d_2d_1d_0$ represent BCD code of numeric digit. d_3 gives MSB and d_0 gives LSB.

Outputs: Four-binary digits $y_3y_2y_1y_0$ represent gray number. y_3 gives MSB and y_0 gives LSB.

Truth table: Truth table is given in Table 5.23. BCD code is used for numeric digits. Four-bit binary values form more values can be used as don't care conditions.

9's complement of 0 $(0000)_{BCD}$ is 9 $(1001)_{BCD}$, i.e. 9-0
9's complement of 3 $(0000)_{BCD}$ is 6 $(0110)_{BCD}$, i.e. 9-3

TABLE 5.23 | 9's complement of BCD number

Minterm	Input				Output			
	d_3	d_2	d_1	d_0	y_3	y_2	y_1	y_0
m_0	0	0	0	0	1	0	0	1
m_1	0	0	0	1	1	0	0	0
m_2	0	0	1	0	0	1	1	1
m_3	0	0	1	1	0	1	1	0
m_4	0	1	0	0	0	1	0	1
m_5	0	1	0	1	0	1	0	0
m_6	0	1	1	0	0	0	1	1
m_7	0	1	1	1	0	0	1	0
m_8	1	0	0	0	0	0	0	1
m_9	1	0	0	1	0	0	0	0

Boolean equations are given below for output functions.

$$y_0(d_3, d_2, d_1, d_0) = \sum m(0, 2, 4, 6, 8) + \sum d(10, 11, 12, 13, 14, 15) \quad (5.126)$$

$$y_1(d_3, d_2, d_1, d_0) = \sum m(2, 3, 6, 7) + \sum d(10, 11, 12, 13, 14, 15) \quad (5.127)$$

$$y_2(d_3, d_2, d_1, d_0) = \sum m(2, 3, 4, 5) + \sum d(10, 11, 12, 13, 14, 15) \quad (5.128)$$

$$y_3(d_3, d_2, d_1, d_0) = \sum m(0, 1) + \sum d(10, 11, 12, 13, 14, 15) \quad (5.129)$$

K-maps are used to simplify the logic functions. Figure 5.84(a) gives four-variable K-map for function y_0. One octet is formed: {0, 4, 8, 12, 2, 6, 10, 14}. The logic expression is given below.

$$y_0(d_3, d_2, d_1, d_0) = \bar{d}_0 \quad (5.130)$$

Figure 5.84(b) gives four-variable K-map for function y_1. One octet is formed: {2, 3, 6, 7, 10, 11, 14, 15}. The logic expression is given below.

$$y_1(d_3, d_2, d_1, d_0) = d_1 \tag{5.131}$$

Figure 5.84(c) gives four-variable K-map for function y_2. Two quads are formed: {2, 3, 10, 11}, {4, 5, 12, 13}. The logic expression is given below.

$$y_2(d_3, d_2, d_1, d_0) = d_2\bar{d}_1 + \bar{d}_2 d_1 \tag{5.132}$$

Figure 5.84(d) gives four-variable K-map for function y_3. One pair is formed: {0, 1}. The logic expression is given below.

$$y_3(d_3, d_2, d_1, d_0) = \bar{d}_3\bar{d}_2\bar{d}_1 \tag{5.133}$$

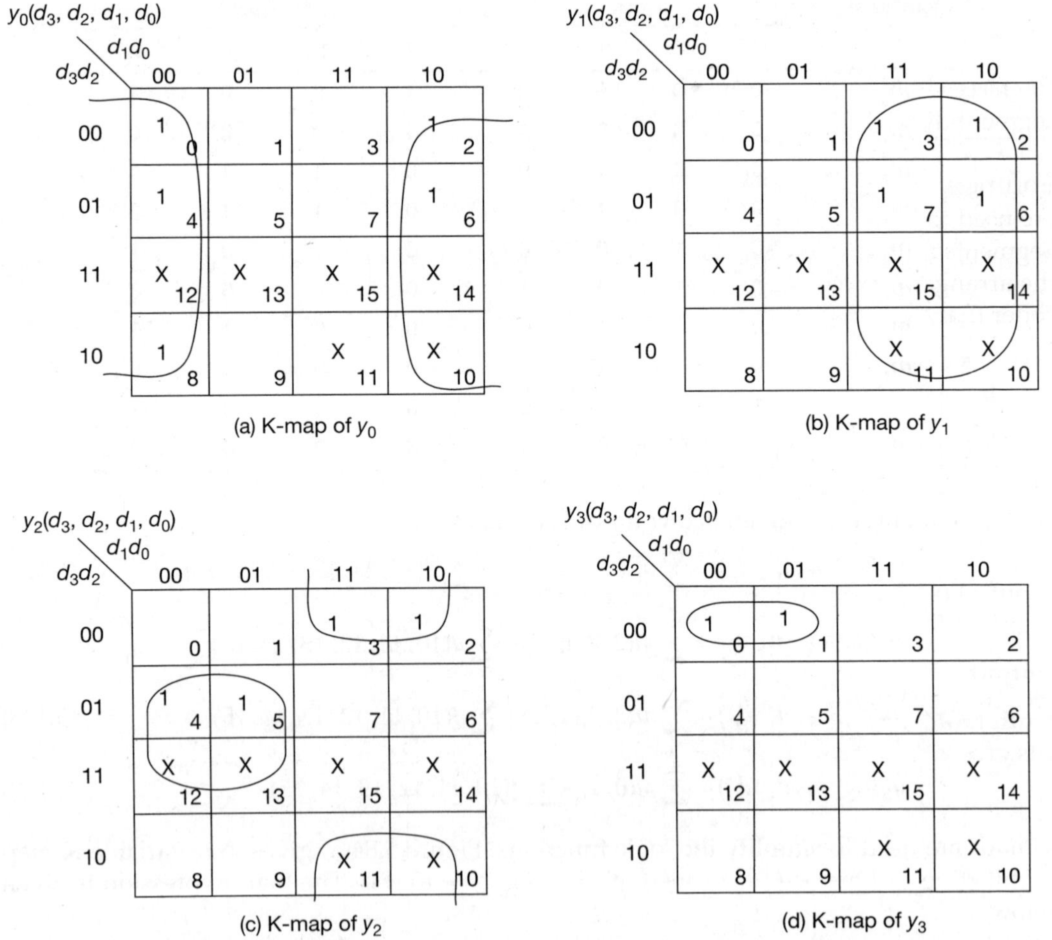

FIGURE 5.84 | K-map for 9's complement of BCD digit

Combinational circuit is given in Figure 5.85.

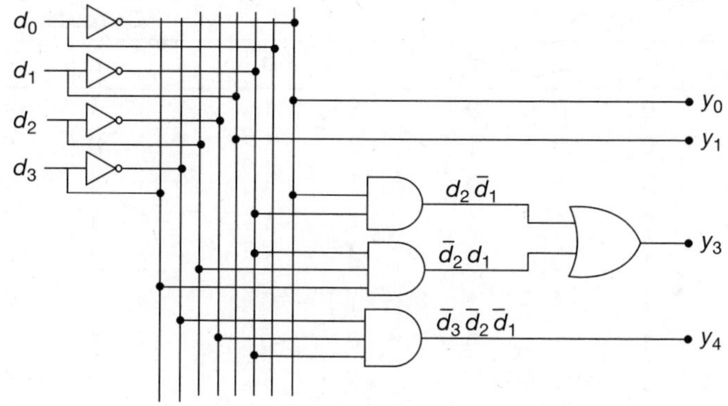

FIGURE 5.85 | Logic diagram of 9's complement of BCD digit

EXAMPLE 5.23 Design a combinational circuit that displays decimal digit to seven-segment display.

SOLUTION Seven-segment display consists of seven light emitting diodes (LEDs). These are used to display numeric digits. Segments are illuminated to display a particular digit. Segments are named a through g. Illuminated segmented is treated as 1. Figure 5.86 gives the arrangement depicting illuminated segments of 0 to 9 digits. The aim is to design decoder that converts BCD code to seven segment code as shown in Figure 5.87.

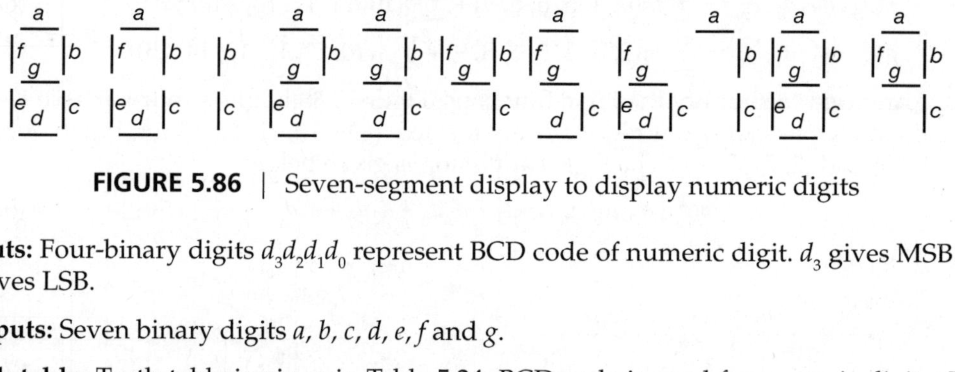

FIGURE 5.86 | Seven-segment display to display numeric digits

Inputs: Four-binary digits $d_3d_2d_1d_0$ represent BCD code of numeric digit. d_3 gives MSB and d_0 gives LSB.

Outputs: Seven binary digits *a*, *b*, *c*, *d*, *e*, *f* and *g*.

Truth table: Truth table is given in Table 5.24. BCD code is used for numeric digits. Four bits give values larger than 9 those can be used as don't care conditions.

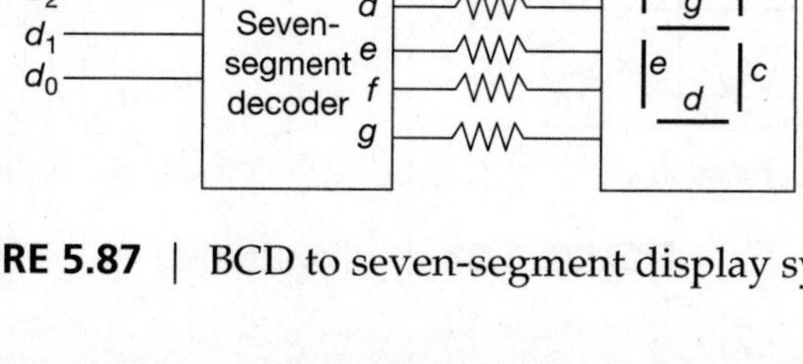

FIGURE 5.87 | BCD to seven-segment display system

TABLE 5.24 | Truth table BCD to seven-segment decoder

Decimal Digit	Inputs				Outputs						
	d_3	d_2	d_1	d_0	a	b	c	d	e	f	g
0	0	0	0	0	1	1	1	1	1	1	0
1	0	0	0	1	0	1	1	0	0	0	0
2	0	0	1	0	1	1	0	1	1	0	1
3	0	0	1	1	1	1	1	1	0	0	1
4	0	1	0	0	0	1	1	0	0	1	1
5	0	1	0	1	1	0	1	1	0	1	1
6	0	1	1	0	0	0	1	1	1	1	1
7	0	1	1	1	1	1	1	0	0	0	0
8	1	0	0	0	1	1	1	1	1	1	1
9	1	0	0	1	1	1	1	0	0	1	1

Boolean equations are given below for output functions

$$a(d_3, d_2, d_1, d_0) = \sum m(0, 2, 3, 5, 7, 8, 9) + \sum d(10, 11, 12, 13, 14, 15) \quad (5.134)$$

$$b(d_3, d_2, d_1, d_0) = \sum m(0, 1, 2, 3, 4, 7, 8, 9) + \sum d(10, 11, 12, 13, 14, 15) \quad (5.135)$$

$$c(d_3, d_2, d_1, d_0) = \sum m(0, 1, 3, 4, 5, 6, 7, 8, 9) + \sum d(10, 11, 12, 13, 14, 15) \quad (5.136)$$

$$d(d_3, d_2, d_1, d_0) = \sum m(0, 2, 3, 5, 6, 8) + \sum d(10, 11, 12, 13, 14, 15) \quad (5.137)$$

$$e(d_3, d_2, d_1, d_0) = \sum m(0, 2, 6, 8) + \sum d(10, 11, 12, 13, 14, 15) \quad (5.138)$$

$$f(d_3, d_2, d_1, d_0) = \sum m(0, 4, 5, 6, 8, 9) + \sum d(10, 11, 12, 13, 14, 15) \quad (5.139)$$

$$g(d_3, d_2, d_1, d_0) = \sum m(2, 3, 4, 5, 6, 8, 9) + \sum d(10, 11, 12, 13, 14, 15) \quad (5.140)$$

K-maps are used to simplify the logic functions. Figure 5.88(a) gives four-variable K-map for function a. One octet and three quads are formed: {8, 9, 10, 11, 12, 13, 14, 15}, {0, 2, 8, 10}, {3, 7, 11, 15} and {5, 7, 13, 15}. The logic expression is given below.

$$a(d_3, d_2, d_1, d_0) = d_3 + \bar{d}_2\bar{d}_0 + d_2 d_0 + d_1 d_0 \quad (5.141)$$

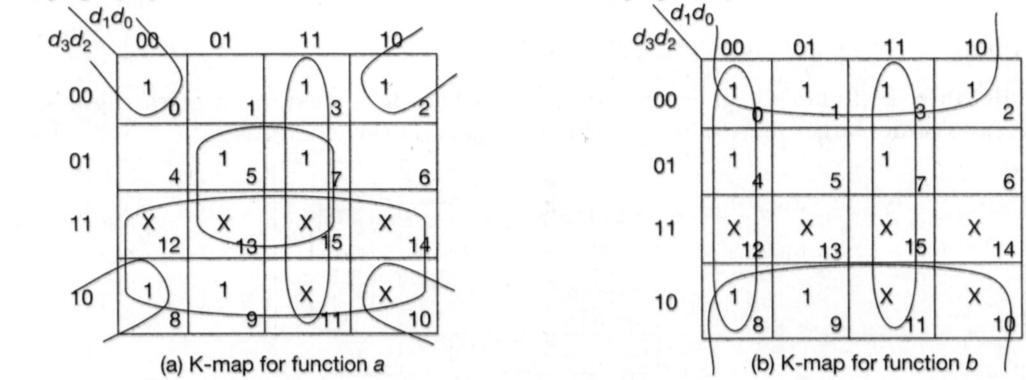

(a) K-map for function a

(b) K-map for function b

FIGURE 5.88 | (Continued)

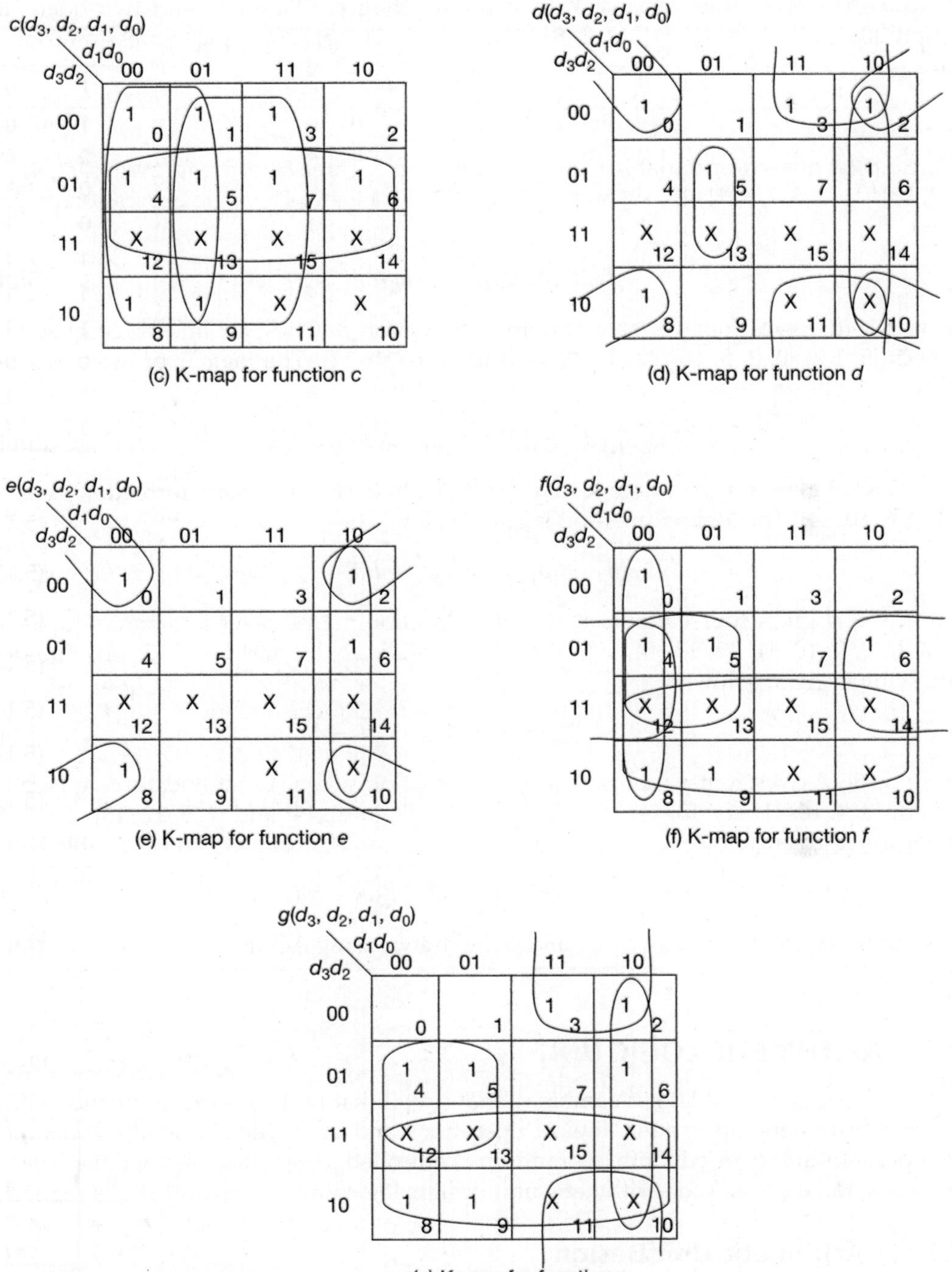

(c) K-map for function c

(d) K-map for function d

(e) K-map for function e

(f) K-map for function f

(g) K-map for function g

FIGURE 5.88 | K-map for BCD to seven-segment display decoder

Figure 5.88(b) gives four-variable K-map for function *b*. One octet and two quads are formed: {0, 1, 2, 3, 8, 9, 10, 11}, {0, 4, 8, 12} and {3, 7, 11, 15}. The logic expression is given below.

$$b(d_3, d_2, d_1, d_0) = \bar{d}_2 + \bar{d}_1\bar{d}_0 + d_1 d_0 \tag{5.142}$$

Figure 5.88(c) gives four-variable K-map for function *c*. Three octet are formed: {0, 1, 4, 5, 8, 9, 12, 13}, {1, 3, 5, 7, 9, 11, 13, 15} and {4, 5, 6, 7, 12, 13, 14, 15}. The logic expression is given below.

$$c(d_3, d_2, d_1, d_0) = \bar{d}_1 + d_0 + d_2 \tag{5.143}$$

Figure 5.88(d) gives four-variable K-map for function *d*. One pair and three quads are formed: {5, 13}, {0, 2, 8, 10}, {2, 3, 10, 11} and {2, 6, 10, 14}. The logic expression is given below.

$$d(d_3, d_2, d_1, d_0) = \bar{d}_2\bar{d}_0 + \bar{d}_2 d_1 + d_1\bar{d}_0 + d_2\bar{d}_1 d_0 \tag{5.144}$$

Figure 5.88(e) gives four-variable K-map for function *e*. Two quads are formed: {0, 2, 8, 10} and {2, 6, 10, 14}. The logic expression is given below.

$$e(d_3, d_2, d_1, d_0) = \bar{d}_2\bar{d}_0 + d_1\bar{d}_0 \tag{5.145}$$

Figure 5.88(f) gives four-variable K-map for function *f*. One octet and three quads are formed: {8, 9, 10, 11, 12, 13, 14, 15}, {0, 4, 8, 12}, {4, 5, 12, 13} and {4, 6, 12, 14}. The logic expression is given below.

$$f(d_3, d_2, d_1, d_0) = d_3 + d_2\bar{d}_0 + d_2\bar{d}_1 + \bar{d}_1\bar{d}_0 \tag{5.146}$$

Figure 5.88(g) gives four-variable K-map for function *g*. One octet and three quads are formed: {8, 9, 10, 11, 12, 13, 14, 15}, {2, 3, 10, 11}, {2, 6, 10, 14} and {4, 5, 12, 13}. The logic expression is given below.

$$g(d_3, d_2, d_1, d_0) = d_3 + \bar{d}_2 d_1 + d_2\bar{d}_1 + d_1\bar{d}_0 \tag{5.147}$$

Using above equations, circuit diagram can be drawn using AOI gates.

5.17 | ARITHMETIC LOGIC UNIT

Arithmetic logic unit (ALU) is a combinational circuit that performs multi-functions. ALU performs arithmetic operations, logical operations and gives the status after operation. Two operands are required. Both operands are controlled to get various functions. The following section discusses the arithmetic unit design, logic unit design and status register.

5.17.1 | Arithmetic Unit Design

Four-bit parallel adder is the main circuit of ALU. Four bit parallel adder is constructed by cascading four full-adder circuits. Figure 5.89 gives the representation of four-bit data as operands to four-bit parallel adder. Figure 5.90 gives simple circuit of four-bit parallel adder.

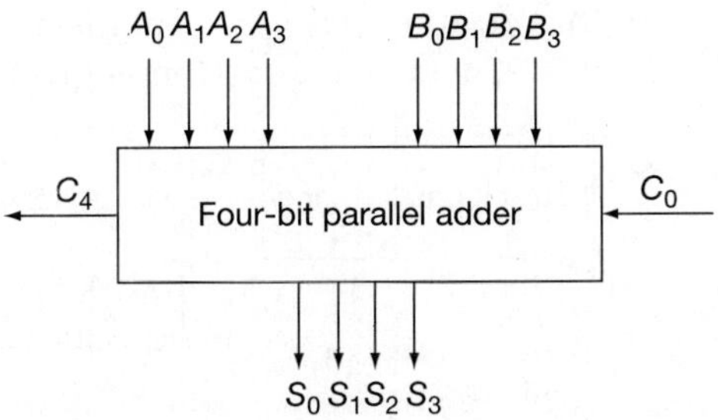

FIGURE 5.89 | Four-bit binary adder

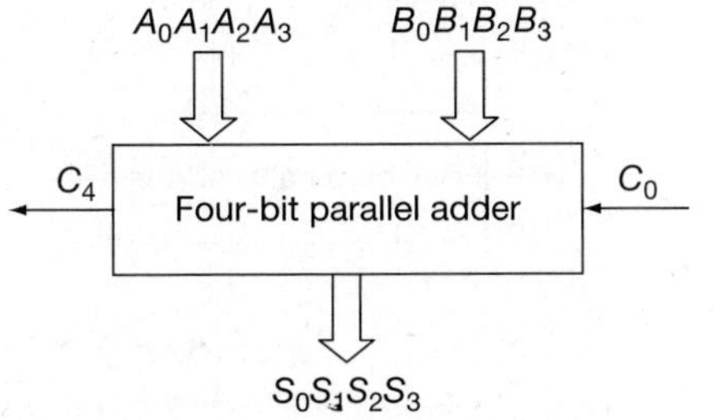

FIGURE 5.90 | Four-bit binary adder

Various types of arithmetic operations can be obtained by controlling one operand externally. Moreover carry-in plays a role in arithmetic operations. Figure 5.91 gives various arithmetic operations obtained by controlling one operand and carry-in.

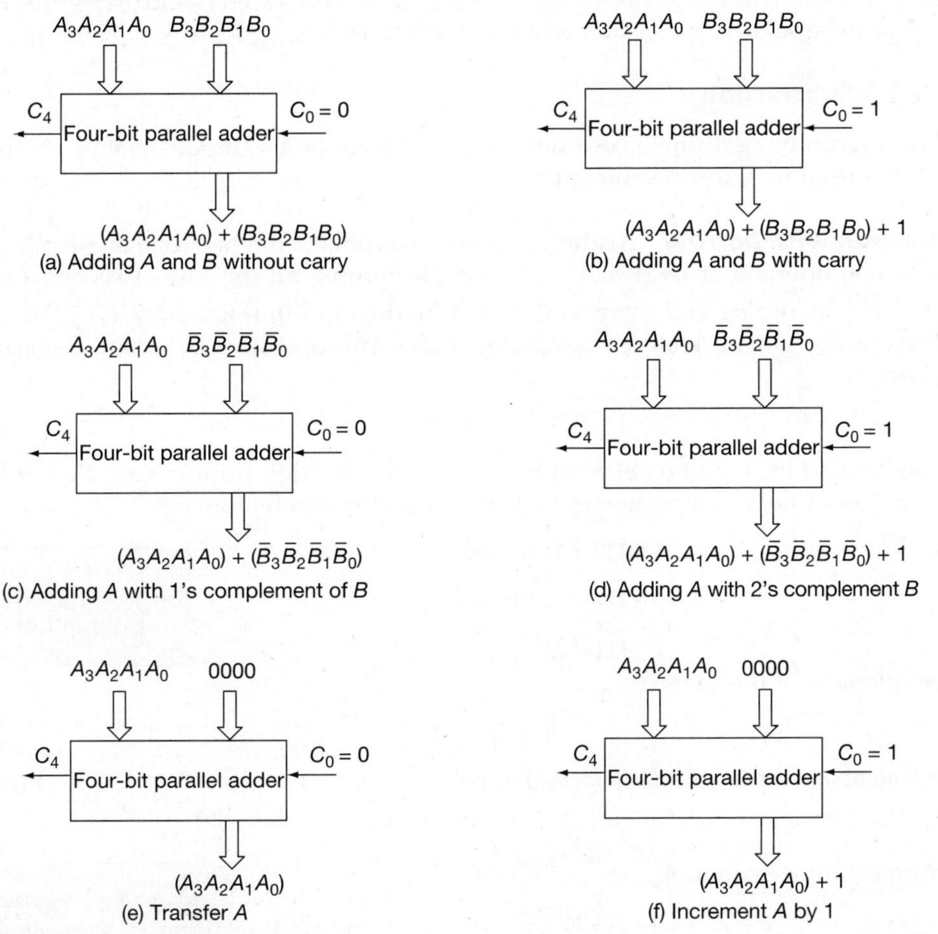

FIGURE 5.91 | (Continued)

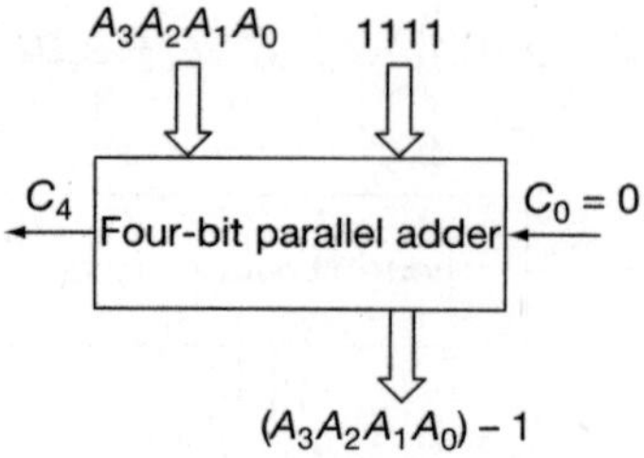

(g) Decrement A by 1

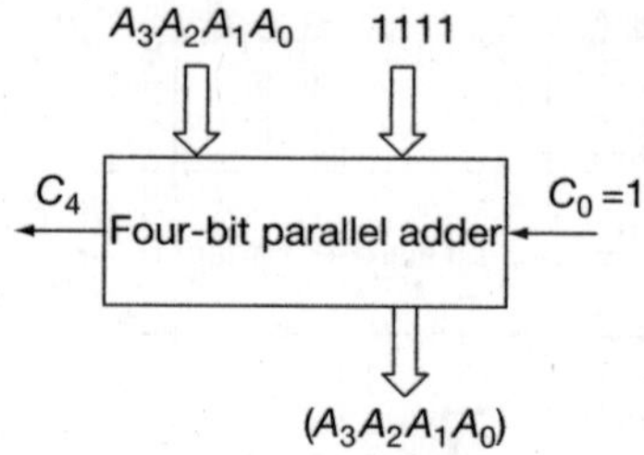

(h) Transfer A

FIGURE 5.91 | Addition by varying operand B

5.17.1.1 | Addition

Arithmetic addition is obtained when four-bit parallel adder receives two operands A ($A_3A_2A_1A_0$) and B ($B_3B_2B_1B_0$) as inputs and carry-in (C_0) is cleared (set to 0) (Figure 5.91(a)). Carry-in can be set to 1 to consider addition with carry (Figure 5.91(b)).

5.17.1.2 | Subtraction

Subtraction can be performed consider borrow. Moreover, 1′s complement or 2′s complement for is used to perform subtraction.

Subtraction with borrow Arithmetic addition is obtained when four-bit parallel adder receives one operand A ($A_3A_2A_1A_0$) and complementing all the bits of second operand $B\left(\bar{B}_3\bar{B}_2\bar{B}_1\bar{B}_0\right)$ as inputs and carry-in (C_0) is cleared (set to 0) (Figure 5.91(c)). The output produced is the sum of A and 1′s complement of B. This operation is similar to subtraction with borrow.

$$S = A + \bar{B} \tag{5.148}$$

1′s complement of B can be obtained by subtracting the digit from largest digit of binary number. The value 2^4 is represented by 1 0000. 1 is followed by four 0′s.

So $$1\,0000 = 1111 + 1$$

Or $$1111 = 1\,0000 - 1$$

$$1111 = 2^4 - 1$$

> 2^n is carryover of n-bit number

1′s complement of B is given by

$$\bar{B} = (2^4 - 1) - B \tag{5.149}$$

Substitute above equation in Eq. (5.148) to get

$$S = A + 2^4 - 1 - B$$

Rearranging above equation,

$$S = 2^4 + (A - B - 1) \tag{5.150}$$

> $(2^n - 1)_{10} - (B)_2$ is 1′s complement of B

The operation gives the subtraction of A and B with borrow.

EXAMPLE 5.24 Subtract 6 from 9

SOLUTION

$B =$		0110	$= (6)_{10}$	
$2^4 =$	1	0000	$= (16)_{10}$	
$2^4 - 1 =$		1111	$= (15)_{10}$	
$2^4 - 1 - B =$		1001	$= (15 - 6)_{10}$	1's complement of B
$A =$		1001	$= (9 + 15 - 6)_{10}$	
$2^4 + A - B - 1 =$	1	0010	$= (16 + 2)_{10}$	Ignore carry

Let the minuend, A is larger than subtrahend, B, so

$$A > B$$

Rewriting above equation,

$$A - B > 0$$

Adding $(2^4 - 1)$ on both sides,

$$(2^4 - 1) + (A - B) > (2^4 - 1)$$

$$2^4 + (A - B - 1) > (2^4 - 1) \tag{5.151}$$

Number is greater than $(2^4 - 1)$, so there is carry. Ignoring the carry, the result is subtraction with borrow and is stated below

$$S = (A - B - 1)$$

EXAMPLE 5.25 Subtract 9 from 6

SOLUTION

$B =$		1001	$= (9)_{10}$	
$2^4 =$	1	0000	$= (16)_{10}$	
$2^4 - 1 =$		1111	$= (15)_{10}$	
$2^4 - 1 - B =$		0110	$= (15 - 9)_{10}$	1's complement of B
$A =$		0110	$= (6 + 15 - 9)_{10}$	
$(2^4 - 1) - (B - A) =$	0	1100	$= (15 - 3)_{10}$	1's complement of $(B - A)$

Let the minuend, A is lesser than or equal to subtrahend, B, so

$$A \leq B$$

Rewriting above equation,

$$A - B \leq 0$$

Adding $(2^4 - 1)$ on both sides

$$(2^4 - 1) + (A - B) \le (2^4 - 1)$$

$$2^4 + (A - B - 1) \le (2^4 - 1) \tag{5.152}$$

Number is less than $(2^4 - 1)$, so there is no carry.

$$F = (2^4 - 1) - (B - A) \tag{5.153}$$

The result is equivalent to 1's complement of $(B - A)$.

$$F = (\overline{B - A}) \tag{5.154}$$

Subtraction using 2's complement Arithmetic addition is obtained when four-bit parallel adder receives one operand A $(A_3A_2A_1A_0)$ and complementing all the bits of second operand B $(\bar{B}_3\bar{B}_2\bar{B}_1\bar{B}_0)$ as inputs and carry-in (c_0) is set to 1 (Figure 5.91(d)). The output produced is the sum of A and 2's complement of B. This operation is similar to subtraction with borrow. Carry-in can be set to 1 to consider sum of A and 2's complement of B (Figure 91(d)). This operation is similar to subtraction and output carry is discarded if occurs.

1's complement of B can be obtained by subtracting the digit from largest digit of binary number. The value 2^4 is represented by 1 0000. 1 is followed by four 0's.

So, $$1\,0000 = 1111 + 1$$

Or $$1111 = 1\,0000 - 1$$

$$1111 = 2^4 - 1$$

2's complement of B can be obtained by taking 1's complement and adding 1 to it. Two's complement of B is given by

$$\bar{B} = (2^4 - 1) - B + 1$$

$$\bar{B} = (2^4 - B) \tag{5.155}$$

Substitute above equation in Eq. (5.148) to get

$$S = A + 2^4 - B$$

> $(2^n)_{10} - (B)_2$ is 2's complement of B

Rearranging above equation,

$$S = 2^4 + (A - B) \tag{5.156}$$

The operation gives the subtraction of A and B using 2's complement method.

Example 5.26 Subtract 6 from 9

Solution

B =	0110	$= (6)_{10}$	
2^4 =	1 0000	$= (16)_{10}$	
$2^4 - B$ =	1010	$= (16 - 6)_{10}$	2's complement of B
A =	1001	$= (9 + 16 - 6)_{10}$	
$2^4 + A - B - 1$ =	1 0011	$= (16 + 3)_{10}$	Ignore carry

Let the minuend, A is larger than subtrahend, B, so

$$A > B$$

Rewriting above equation,

$$A - B > 0$$

Adding (2^4) on both sides,

$$2^4 + (A - B) > 2^4 \tag{5.157}$$

Number is greater than (2^4), so there is carry. Ignoring the carry, the result is subtraction and is stated below

$$S = (A - B) \tag{5.158}$$

EXAMPLE 5.27 Subtract 9 from 6

SOLUTION

$B =$		1001	$= (9)_{10}$	
$2^4 =$	1	0000	$= (16)_{10}$	
$2^4 - B =$		0111	$= (16 - 9)_{10}$	2's complement of B
$A =$		0110	$= (6 + 16 - 9)_{10}$	
$2^4 + A - B - 1 =$	0	1101	$= (16 - 3)_{10}$	2's complement of $B - A$

Let the minuend, A is less than or equal to subtrahend, B, so

$$A \leq B$$

Rewriting above equation,

$$A - B \leq 0$$

Adding (2^4) on both sides, we get

$$2^4 + (A - B) \leq 2^4$$

$$2^4 + (A - B) \leq 2^4 \tag{5.159}$$

Number is less than (2^4), so there is no carry.

$$F = 2^4 - (B - A) \tag{5.160}$$

The result is equivalent to 2's complement of ($B - A$).

$$F = \left(\overline{B - A}\right) \tag{5.161}$$

5.17.1.3 | Increment by One

Arithmetic addition is obtained when four-bit parallel adder receives one operand A ($A_3A_2A_1A_0$) and all the bits of second operand B is cleared to 0 (0000) as inputs and carry-in (c_0) is cleared (set to 0) (Figure 5.91(e)). This operation transfers A operand. Carry-in can be set to 1 to consider increment A by 1 (Figure 5.91(f)).

5.17.1.4 | Decrement by One

Arithmetic addition is obtained when four-bit parallel adder receives one operand A $(A_3A_2A_1A_0)$ and all the bits of second operand B is set to 1(1111) as inputs and carry-in (c_0) is cleared (set to 0) (Figure 5.91(g)). This operation is similar to decrement the operand, A by 1. Carry-in can be set to 1 to consider the transfer operation of A (Figure 5.91(h)).

EXAMPLE 5.28 Subtract 9 from $(2^4 - 1)$

SOLUTION

$$\begin{array}{llll} A = & & 1001 & = (9)_{10} \\ 2^4 = & 1 & 0000 & = (16)_{10} \\ 2^4 - 1 = & & 1111 & = (15)_{10} \\ A + 2^4 - 1 & 1 & 1000 & = (16 + 8)_{10} \\ 2^4 + (A - 1) = & 1 & 1000 & \end{array}$$

It can be written that

$$1111 = 2^4 - 1$$

Adding $(2^4 - 1)$ to A to get

$$S = A + 2^4 - 1$$

$$S = 2^4 + (A - 1) \tag{5.162}$$

Number is greater than (2^4), so there is carry. Ignoring the carry, the result is equal to decrement A by 1.

All above operations are enlisted in truth Table 5.25.

TABLE 5.25 | Arithmetic operations from four-bit adder

Status Lines			Operand 1	Operand 2	Carry	Function	Action
S_2	S_1	S_0	X $(X_3X_2X_1X_0)$	Y $(Y_3Y_2Y_1Y_0)$	Z_0	$X + Y + Z_0$	
					$C_0 \leftarrow 0$	$X + Y$	
					$C_0 \leftarrow 1$	$X + Y + 1$	
0	0	0	$(A_3A_2A_1A_0)$	(0000)	0	A	Transfer data
					1	$A + 1$	Increment by 1
0	0	1	$A_3A_2A_1A_0$	$(B_3B_2B_1B_0)$	0	$A + B$	Add without carry
					1	$A + B + 1$	Add with carry
0	1	0	$A_3A_2A_1A_0$	$\bar{B}_3\bar{B}_2\bar{B}_1\bar{B}_0$	0	$A + \bar{B}$	Subtract without borrow
					1	$A + \bar{B} + 1$	Subtract with borrow
0	1	1	$A_3A_2A_1A_0$	(1111)	0	$A - 1$	Decrement by 1
					1	$A - 1 + 1$	Data transfer

$A \rightarrow (A_3A_2A_1A_0)$, $B \rightarrow (B_3B_2B_1B_0)$ and $\bar{B} \rightarrow (\bar{B}_3\bar{B}_2\bar{B}_1\bar{B}_0)$

Operand B can be controlled by the following logic obtained from variable K-map given in Figure 5.92.

$$Y(s_1, s_0) = s_0 B + s_1 \overline{B} \tag{5.163}$$

The combinational circuit diagram is given below in Figure 5.93 and block diagram is given in Figure 5.94. This single-bit diagram can be extended for four bits as shown in Figure 5.93. Four-bit adder circuit given in Figure 5.95 is modified to have controlled second operand. Circuit diagram is given in Figure 5.96.

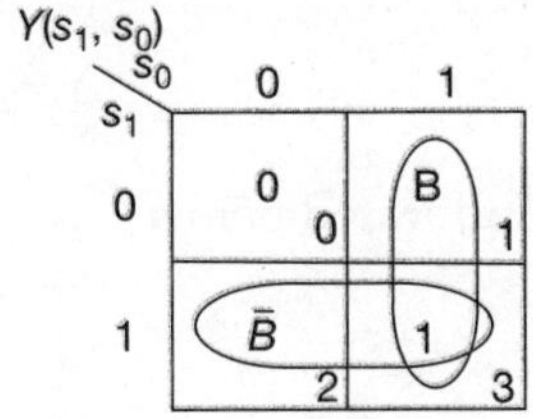

FIGURE 5.92 | Variable map

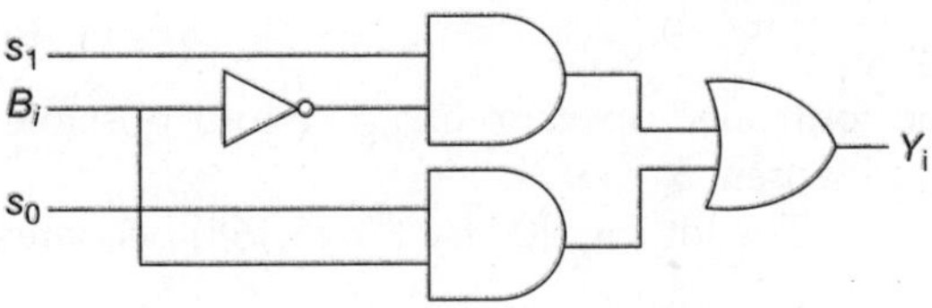

FIGURE 5.93 | Logic circuit

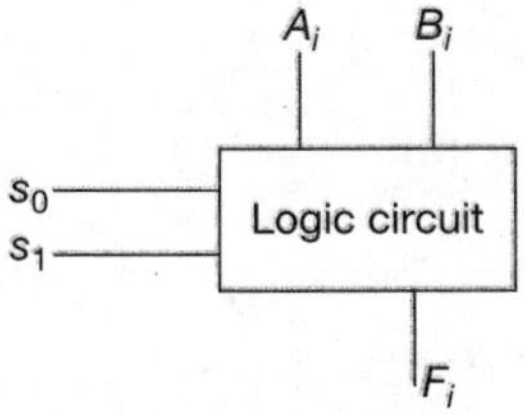

FIGURE 5.94 | Block diagram of 1-bit logic circuit

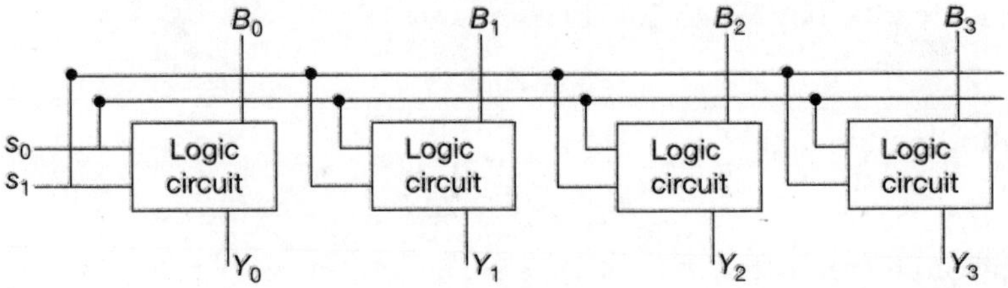

FIGURE 5.95 | Four-bit logic circuit

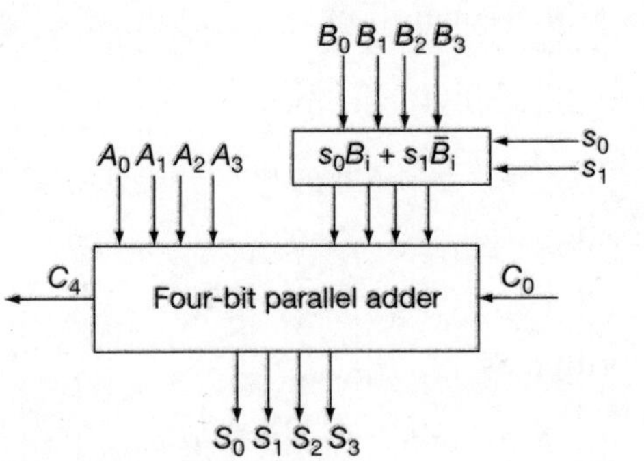

FIGURE 5.96 | Four-bit arithmetic unit

5.17.2 | Logic Unit Design

Basic logic operations are AND, OR and NOT. Adder circuit is designed using X-OR gate. In full-adder circuit if carry-in is always cleared to zero. Sum logic for full-adder circuit is given by

$$S_i = A_i \oplus (B_i \oplus C_i)$$

Clear carry-in bit, $C_i \to 0$. Full-adder logic becomes half-adder sum logic.

$$S_i = A_i \oplus (B_i \oplus 0)$$

$$S_i = A_i \oplus B_i \tag{5.164}$$

One controlled operand can give four possible operations as elaborated below.

Case-I: when $B_i \to 0$

The logic given by Eq. (5.164) becomes

$$S_i = A_i \oplus 0$$

Above logic can be rewritten as

$$S_i = A_i \tag{5.165}$$

This operation transfers the bit.

Case-II: when $B_i \to B_i$

The logic given by Eq. (5.164) becomes

$$S_i = A_i \oplus B_i \tag{5.166}$$

The operation is X-OR operation.

Case-III: when $B_i \to \overline{B}_i$

The logic given by Eq. (5.164) becomes

$$S_i = A_i \oplus \overline{B}_i$$

Expanding the logic

$$S_i = \overline{A}_i \overline{B}_i + A_i B_i$$

Logic can be rewritten as

$$S_i = \overline{A_i \oplus B_i} \tag{5.167}$$

The operation is X-NOR operation.

Case-III: when $B_i \to 1$

The logic given by Eq. (5.164) becomes

$$S_i = A_i \oplus 1$$

Expanding the logic

$$S_i = A_i \overline{1} + \overline{A}_i 1$$

Logic can be rewritten as

$$S_i = A_i \cdot 0 + \overline{A}_i 1$$

Logic can be rewritten as

$$S_i = \overline{A}_i \tag{5.168}$$

The logic gives the NOT operation.

The truth table is given Table 5.26.

TABLE 5.26 | Logical bit operations

Status Lines		Operand 1	Operand 2	Carry	Function	Action
s_1	s_0	X_i	Y_i	$C_0 \leftarrow 0$	$X_i + Y_i$	
0	0	A_i	0	0	A_i	Transfer data
0	1	A_i	B_i	0	$A_i \oplus B_i$	Exclusive OR operation
1	0	A_i	$\overline{B}_i$	0	$\overline{A_i \oplus B_i}$	Exclusive NOR operation
1	1	A_i	1	0	$\overline{A}_i$	Inverter

Transfer and Ex-NOR operations are not required. By applying some logic, transfer logic is modified to OR logic and Ex-NOR logic is modified to AND logic. It is achieved by controlling the other operand, i.e. A. For this operation it becomes necessary to select same operand for Y_i.

5.17.2.1 | OR Operation

For OR logic modify one operand as $A_i + B_i$ and second operand as 0 which is selected when s_0 and s_1 are 0 and 0, respectively. Equation (5.108) becomes

$$S_i = (A_i + B_i) \oplus 0$$

Above logic can be rewritten as

$$S_i = (A_i + B_i) \tag{5.169}$$

5.17.2.2 | AND Operation

For AND logic modify one operand as $A_i + \overline{B}_i$ and second operand as $\overline{B}_i$ which is selected when s_0 and s_1 are 0 and 1, respectively. Equation (5.108) becomes

$$S_i = (A_i + \overline{B}_i) \oplus \overline{B}_i$$

Expanding the Ex-OR logic in AND-OR-NOT logics

$$S_i = \overline{(A_i + \overline{B}_i)}\,\overline{B}_i + (A_i + \overline{B}_i)\,\overline{\overline{B}}_i$$

Simplifying and applying DeMorgan's theorem to get

$$S_i = (\overline{A}_i \overline{\overline{B}}_i)\,\overline{B}_i + (A_i + \overline{B}_i)\,B_i$$

On simplification,

$$S_i = (\overline{A}_i B_i)\,\overline{B}_i + A_i B_i + \overline{B}_i B_i$$

Further, simplification gives

$$S_i = \overline{A}_i B_i \overline{B}_i + A_i B_i + 0 \qquad (\because \overline{B}_i B_i = 0)$$

$$S_i = \bar{A}_i 0 + A_i B_i \qquad (\because A_i B_i + 0 = A_i B_i)$$
$$S_i = 0 + A_i B_i \qquad (\because \bar{A}_i 0 = 0)$$
$$S_i = A_i B_i \tag{5.170}$$

Truth table given by Table 5.26 is revised to get truth table, Table 5.27.

TABLE 5.27 | Logical operations

Status Lines		Operand 1	Operand 2	Carry	Function	Action
s_1	s_0	X_i	Y_i	Z_i	$X_i \oplus Y_i$	
0	0	$A_i + B_i$	0	0	$A_i + B_i$	OR
0	1	A_i	B_i	0	$A_i \oplus B_i$	XOR
1	0	$A_i + \bar{B}_i$	$\bar{B}_i$	0	$A_i B_i = (A_i + \bar{B}_i) \oplus \bar{B}_i$	AND
1	1	A_i	1	0	$\bar{A}_i = A_i \oplus 1$	Complement

$$X_i(s_1, s_0) = A_i + \bar{s}_1 \bar{s}_0 B_i + s_1 \bar{s}_0 \bar{B}_i \tag{5.171}$$
$$Y(s_1, s_0) = s_1 B + s_0 \bar{B} \tag{5.172}$$

Combinational circuit to implement the truth table is given in Figure 5.97. Block diagram is given in Figure 5.98.

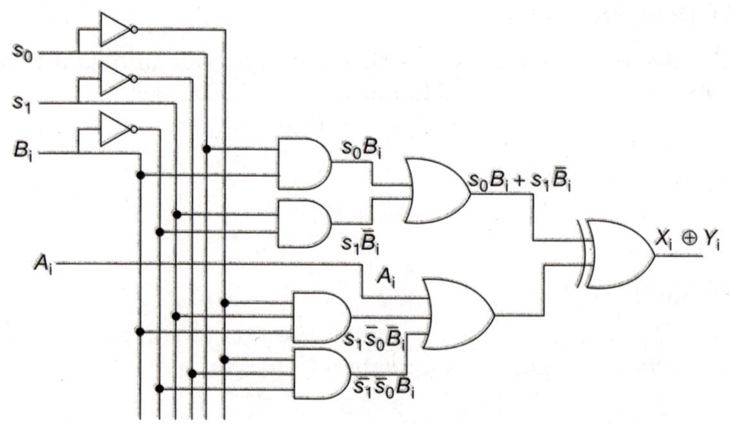

FIGURE 5.97 | Logic circuit for a bit

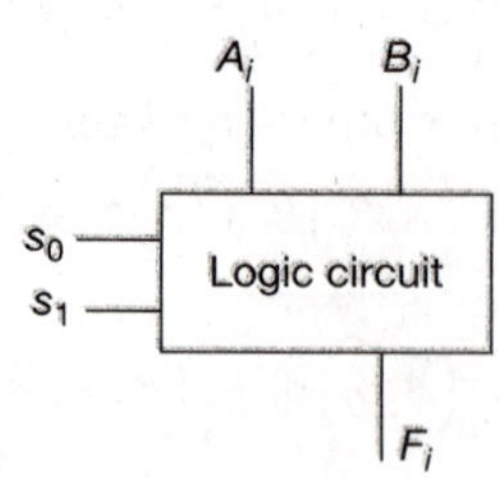

FIGURE 5.98 | Block diagram of 1-bit logic circuit

Single bit logic is extended for 1-nibble (four bits) operation and combinational circuit is given in Figure 5.99. Selection lines, s_1 and s_0 are common to select the bits. Block diagram of four-bit logic circuit is given in Figure 5.100.

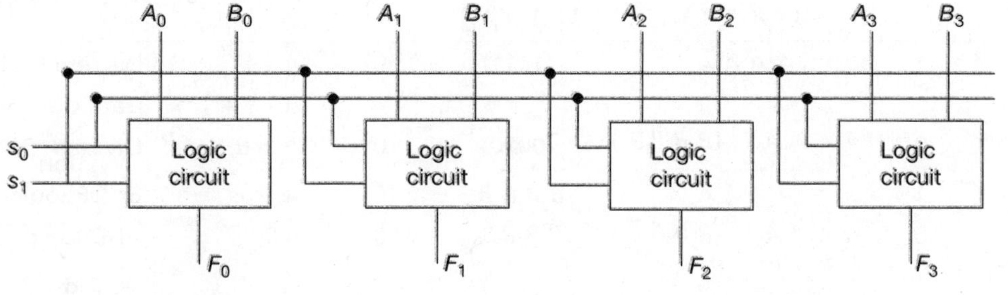

FIGURE 5.99 | Four-bit logic circuit

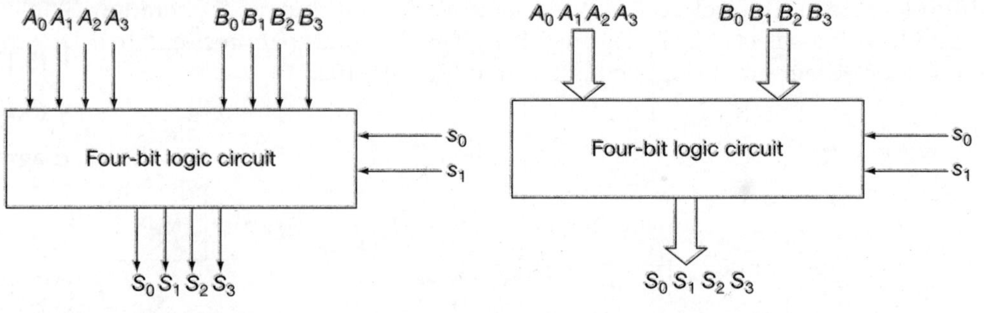

FIGURE 5.100 | Block diagram of four-bit logic circuit

Truth table for ALU is given in Table 5.28. To select arithmetic or logic operations, s_2 line is used. Arithmetic unit is selected, when s_2 is 0. Logic unit is selected, when s_2 is 1.

TABLE 5.28 | Arithmetic and logical operations from four-bit adder

Status Lines			Operand 1	Operand 2	Carry	Function	Action
s_2	s_1	s_0	X $(X_3X_2X_1X_0)$	Y $(Y_3Y_2Y_1Y_0)$	Z_0	$X + Y + Z_0$	
0	0	0	$A_3A_2A_1A_0$	(0000)	0	A	Transfer data
					1	$A + 1$	Increment by 1
0	0	1	$A_3A_2A_1A_0$	$B_3B_2B_1B_0$	0	$A + B$	Add without carry
					1	$A + B + 1$	Add with carry
0	1	0	$A_3A_2A_1A_0$	$\bar{B}_3\bar{B}_2\bar{B}_1\bar{B}_0$	0	$A+\bar{B}$	Subtract without borrow
					1	$A+\bar{B}+1$	Subtract with borrow

(*Continued*)

TABLE 5.28 | (Continued)

Status Lines			Operand 1	Operand 2	Carry	Function	Action
s_2	s_1	s_0	X $(X_3X_2X_1X_0)$	Y $(Y_3Y_2Y_1Y_0)$	Z_0	$X+Y+Z_0$	
0	1	1	$A_3A_2A_1A_0$	(1111)	0	$A-1$	Decrement by 1
					1	$A-1+1$	Data transfer
1	0	0	$(A_3A_2A_1A_0)+(B_3B_2B_1B_0)$	(0000)	0	A_i+B_i	OR Logic
1	0	1	$A_3A_2A_1A_0$	$B_3B_2B_1B_0$	0	$A_i \oplus B_i$	Ex-OR logic
1	1	0	$(A_3A_2A_1A_0)+(\bar{B}_3\bar{B}_2\bar{B}_1\bar{B}_0)$	$\bar{B}_3\bar{B}_2\bar{B}_1\bar{B}_0$	0	A_iB_i	AND logic
1	1	1	$A_3A_2A_1A_0$	(1111)	0	$\bar{A}_i$	NOT logic
			$A \to (A_3A_2A_1A_0),\ B \to (B_3B_2B_1B_0)\bar{B} \to (\bar{B}_3\bar{B}_2\bar{B}_1\bar{B}_0)$				

Multiplexer is used to select the logic operation, F_i or arithmetic S_i operation. Single bit, s_2 is used to either select S_i or F_i. Figure 5.101 gives the combination circuit to select *i*th bit of S or F. Block diagram of single bit is given in Figure 5.102.

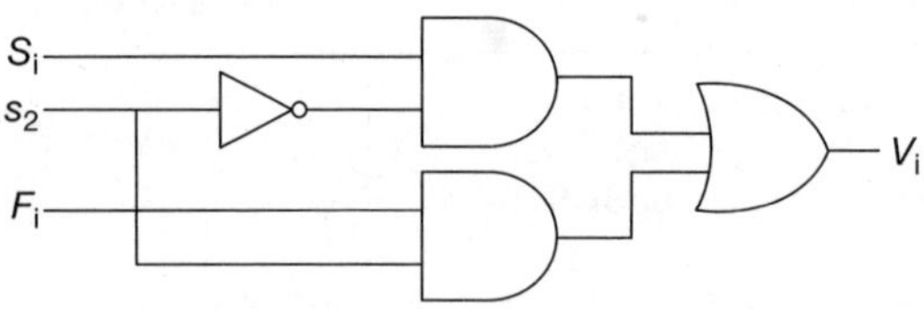

FIGURE 5.101 | Single-bit multiplexer

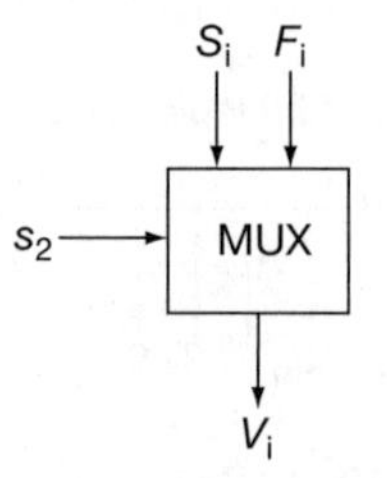

FIGURE 5.102 | Single-bit multiplexer

Single-bit multiplexer is used to select four bits simultaneously whose circuit diagram is given in Figure 5.103 and its block diagram are given in Figure 5.104. Combinational circuits of arithmetic unit given in Figure 5.96, logic circuit given in Figure 5.99 are combined and output is selected with four-bit multiplexer given in Figure 5.104. ALU is given in Figure 5.105.

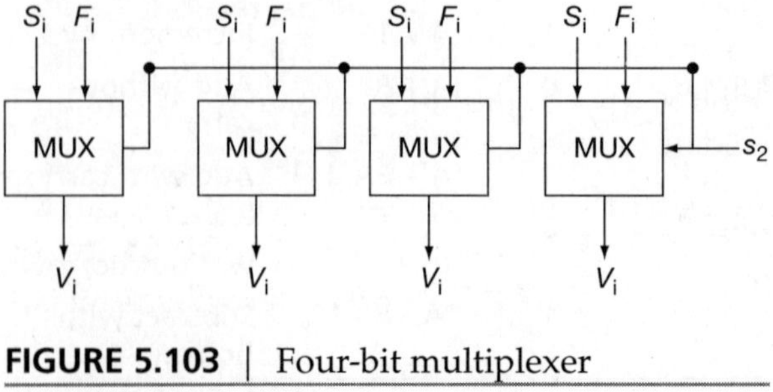

FIGURE 5.103 | Four-bit multiplexer

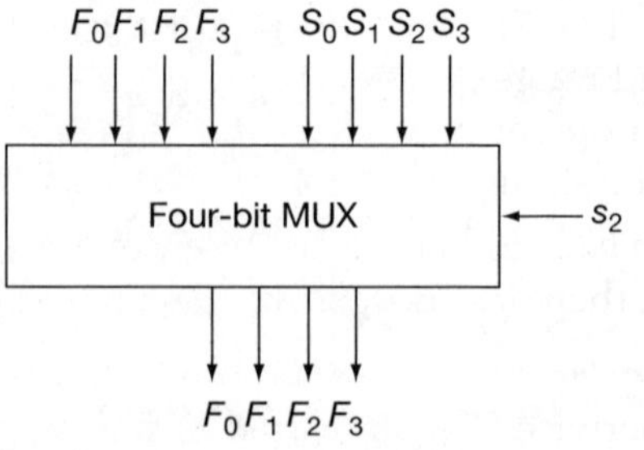

FIGURE 5.104 | Block diagram of four-bit multiplexer

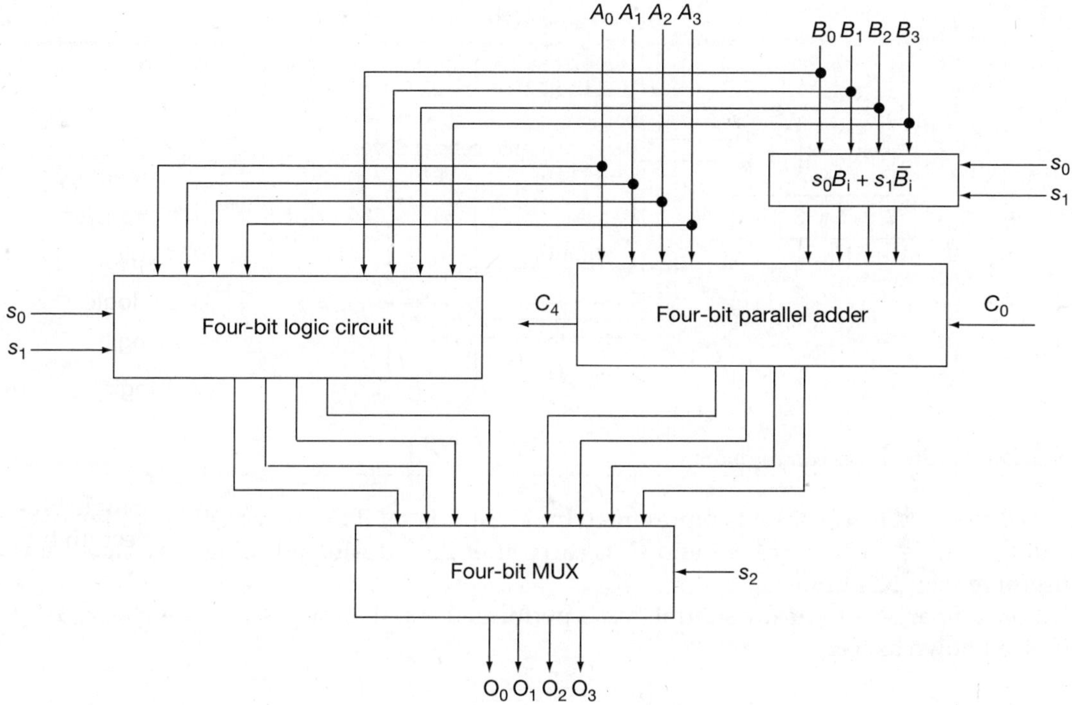

FIGURE 5.105 | Arithmetic and logic operations

5.17.3 | Status Register

Arithmetic operation includes addition and subtraction to perform. Unsigned or signed are considered as operands. If one operand *A* is greater than operand, *B* then result is positive. On the other hand, operand *A* is lesser than operand *B* then result is negative. Signed numbers are represented in complement form either 1's complement or 2's complement. The generally byte operations are performed. Valid range of unsigned eight-bit data in decimal is 0 to $2^8 - 1$ (255). Valid range of signed eight-bit data in decimal is -2^7 (−128) to $2^7 - 1$ (127). Similarly, valid range of unsigned four-bit data in decimal is 0 to $2^4 - 1$ (15). Valid range of signed four-bit data in decimal is -2^3 (−8) to $2^3 - 1$ (7). The resulting value after operation is valid or it is invalid. So, status of result is given by status register.
In the status register, there are four bits those give the status.

Sign bit: Sign bit is represented by *S*. The MSB of data is considered as sign bit. If sign bit is 0, then data is positive. If sign bit is 1, then data is negative.

Carry bit: Carry bit is represented by C. If the output is more than $2^4 - 1$ then there is carry. Otherwise *C* bit is cleared to zero.

Zero bit: Zero bit is represented by Z. In case the result after operation is zero then *Z* bit is set to 1 else it is cleared to 0.

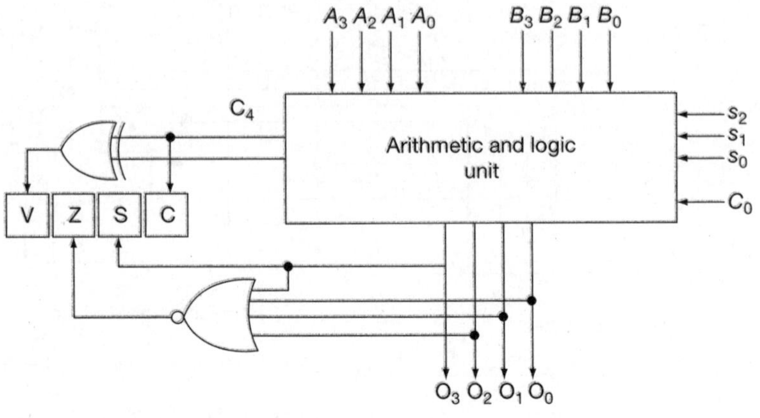

FIGURE 5.106 | Status register

Overflow bit: Overflow bit is represented by V. Bit V is set if exclusive OR operation of C_4 and C_3 is 1. C_4 carry after bit 4 and C_3 is carry after bit 3 during addition. Overflow bit is useful for signed numbers.

For comparison, basically subtraction is performed and then status of bits is checked. Bit is also known as flag.

$$A \lesseqgtr B = \begin{cases} C \text{ is 1 and } Z \text{ is 0} & ;A < B \\ Z \text{ is 1} & ;A = B \\ C \text{ is 0 and } Z \text{ is 0} & ;A > B \end{cases} \tag{5.173}$$

Commonly used ICs are enlisted in Table 5.29.

TABLE 5.29 | Commonly used IC type numbers used for arithmetic operations

IC Type Number	Function and Logic Family
7483	Four-bit full adder, TTL
7485	Four-bit magnitude comparator, TTL
74181	Four-bit ALU and function generator TTL
74182	Look-ahead carry generator TTL
74183	Dual carry save full-adder TTL
74283	Four-bit full binary adder TTL
74885	Eight-bit magnitude comparator TTL
4008	Four-bit binary full-adder CMOS
4527	BCD rate multiplier CMOS
4585	Four-bit magnitude comparator CMOS
40181	Four-bit arithmetic logic unit CMOS
40182	Look-ahead carry generator CMOS
10179	Look-ahead carry block ECL
10180	Dual high-speed two-bit-adder/subtractor ECL
10181	Four-bit arithmetic logic unit/function generator ECL
10182	Four-bit arithmetic logic unit/function generator ECL
10183	4 × 2 multiplier ECL

SUMMARY

- A half-adder is an arithmetic circuit that adds two single binary digits: augend and addend binary digits.
- A full-adder is an arithmetic circuit. It adds three single binary digits; augend, addend and carry binary digits.
- Carry is overflow in second binary digit resulting from addition of two single binary digits.
- 2^n is the carry of n-bit binary number.
- A parallel adder adds two numbers in parallel form and produces the sum bits in parallel form.
- A ripple carry adder is a parallel adder in which the carry-out of each full-adder is the carry-in to the next most significant adder.
- The look-ahead carry adder speeds up the process by eliminating the ripple carry.
- A half-subtractor is an arithmetic circuit that subtracts one binary digit from another.
- A half-subtractor is an arithmetic circuit that subtracts one binary digit from another including borrows.
- Unsigned numbers are positive numbers.
- Signed binary numbers are numbers having positive or negative value.
- Signed binary numbers are represented in one's or two's complement form.
- Two's complement form is better than one's complement form because two's complement of zero is zero or one's complement has one value.
- Sign bit is set to 1 if the binary number is negative.
- Sign bit is set to 0 if the binary number is negative.
- Most significant binary digit (MSB) is the signed bit.
- One's complement is obtained from exclusive OR gate by controlling one bit as 1.
- Two's complement is obtained from parallel adder when one operand is zero and other operand is controlled 1's complement and carry in is set to 1.
- Subtraction is performed by adding minuend and 1's complement or 2's complement of subtrahend.
- Signed BCD numbers are represented in nine's or ten's complement form.
- Ten's complement form is better than nine's complement form because ten's complement of zero is zero.
- Ten's complement is nine's complement plus one.
- Subtraction of BCD number is performed using nine's complement or Ten's complement.
- Minimum and maximum range of unsigned n-bit binary number is 0 and 2^{n-1}, respectively giving value in decimal.
- Minimum and maximum range of signed n-bit binary number is $-2^{(n-1)}$ to $2^{(n-1)} - 1$, respectively, giving value in decimal.
- Overflow arising from n-bit binary numbers is exclusive-OR of carry after $(n - 1)^{\text{th}}$ bit and carry after n^{th} bit.
- A comparator is a logic circuit that compares the magnitudes of two binary numbers. The output produced is logic to indicate that one binary number is equal to, greater than or less than other binary number.
- Code converters are logic circuits whose inputs are bit patterns representing numbers or characters in one code and whose outputs are the corresponding representation in a different code.

- Arithmetic logic unit is a combinational circuit that performs multi-functions like arithmetic and logical.
- A decoder is a logic circuit that converts an n-input binary code into a corresponding single numeric output code.
- Enable inputs are used to control the operation of the decoder.
- LED displays are of two types-common anode type and common cathode type.
- Incandescent displays are used in cash registers and other line-operated devices, where power consumption is not critical.
- An LCD does not emit light energy. So, it cannot be seen in the dark like an LED.
- LCDs consume much less power than LED displays do, and are, therefore, widely used in battery-powered devices such as calculators and watches.
- In LCDs, the frequency of the AC signal should not be less than 25 Hz, because it produces visible flicker.

MULTIPLE CHOICE QUESTIONS

5.1 A half-adder is characterized by
(a) two inputs and two outputs
(b) three inputs and two outputs
(c) two inputs and three outputs
(d) two inputs and one output

5.2 A full-adder is characterized by
(a) two inputs and two outputs
(b) three inputs and two outputs
(c) two inputs and three outputs
(d) two inputs and one output

5.3 The inputs to a full-adder are $A = 1$, $B = 1$. C_{in}: 0. The outputs are
(a) Sum is 1, Carry out is 1
(b) Sum is 1, Carry out is 0
(c) Sum is 0, Carry out is 1
(d) Sum is 0, Carry out is 0

5.4 A four-bit parallel adder can add
(a) two 4-bit binary numbers
(b) two 2-bit binary numbers
(c) four bits at a time
(d) four bits in sequence

5.5 To expand a four-bit parallel adder to an eight-bit parallel adder, you must
(a) use four 4-bit adders with no interconnections
(b) use two 4-bit adders and connect the sum outputs of one to the bit inputs of the other.
(c) use eight 4-bit adders with no interconnections
(d) use two 4-bit adders with the carry output of one connected to the carry inputs of the other

5.6 If a 74LS85 magnitude comparator has $A = 1011$ and $B = 1001$ on its inputs, the outputs are
(a) $A > B = 0, A < B = 0, A > B = 0$
(b) $A > B = 1, A < B = 0, A > B = 0$
(c) $A > B = 1, A < B = 1, A > B = 0$
(d) $A > B = 0, A < B = 0, A > B = 1$

5.7 On adding two BCD digits there is need to add 6 if
(a) Result is more than 9
(b) Only carry occurs
(c) Result is less than 9 and carry occurs
(d) (a) or (c) above

5.8 A BCD-to-7 segment decoder has 0100 on its inputs. The active outputs are
(a) a, c, f, g
(b) b, c, f, g
(c) b, c, e, f
(d) b, d, e, g

5.9 Which of the following codes exhibit even parity?
(a) 10011000
(b) 01111000
(c) 11111111
(d) both answers (b) and (c)

5.10 An n-bit parallel adder consists of
(a) $(n + 1)$ full-adders
(b) $(n - 1)$ full-adders
(c) n-full-adders
(d) $2n$ full-adders

5.11 In which of the following adder circuits, the carry look ripple delay is eliminated
(a) Full-adder
(b) Parallel adder
(c) Serial adder
(d) Carry look-ahead adder

5.12 A carry look-ahead adder is frequently used for addition because it
(a) is faster
(b) is more accurate
(c) uses fewer gates
(d) its cost is less

5.13 A full-adder can be implemented with half-adders and OR gates. A four-bit parallel full-adder without any initial carry requires
(a) 8 half-adder, 4 OR gates
(b) 8 half-adder, 3 OR gates
(c) 7 half-adder, 4 OR gates
(d) 7 half-adder, 3 OR gates

5.14 Binary subtraction using one's complements method
(a) Ignores carry if any
(b) Adds carry if any to sum
(c) Takes one's complement if carry occurs
(d) All of above

5.15 Binary subtraction using one's complements method considers result in one's complement form if
(a) carry is 1
(b) carry is 0
(c) carry 0 after adding it to result
(d) all of above

5.16 Binary subtraction using two's complements method
(a) Ignores carry if any
(b) Adds carry if any
(c) Takes one's complement if carry occurs
(d) All of above

5.17 Binary subtraction using two's complements method considers result positive if
(a) carry is 1
(b) carry is 0
(c) after taking two's complement
(d) All of above

5.18 BCD subtraction using nine's complements method
(a) Ignores carry if any
(b) Adds carry if any to result
(c) Takes one's complement if carry occurs
(d) All of above

5.19 BCD subtraction using nine's complements method considers result in nine's complement form if
(a) carry is 0
(b) carry is 1
(c) carry 0 after adding it to result
(d) All of above

5.20 BCD subtraction using ten's complements method
(a) Ignores carry if any
(b) Adds carry if any
(c) Takes nine's complement if carry occurs
(d) All of above

Answers

5.1 (a)	5.2 (b)	5.3 (c)	5.4 (a)	5.5 (d)	5.6 (d)	5.7 (d)
5.8 (b)	5.9 (d)	5.10 (c)	5.11 (d)	5.12 (a)	5.13 (b)	5.14 (b)
5.15 (b)	5.16 (a)	5.17 (a)	5.18 (a)	5.19 (a)	5.20 (a)	

QUESTIONS

5.1 How do you characterize or define a combinational circuit? How does it differ from a sequential circuit? Give two examples each of combinational and sequential logic devices.

5.2 Beginning with the statement of the problem, outline different steps involved in the design of a suitable combinational logic circuit to implement the hardware required to solve the given problem.

5.3 Write down Boolean expressions representing the SUM and CARRY outputs in terms of three input binary variables to be added. Design a suitable combinational circuit to hardware-implement the design using NAND gates only.

5.4 Draw the truth table of a full-subtractor circuit. Write a minterm Boolean expression for DIFFERENCE and BORROW outputs in terms of minuend variable, subtrahend variable and BORROW-IN. Minimize the expressions and implement them in hardware.

5.5 Draw the logic diagram of a three-digit BCD adder and briefly describe its functional principle.

5.6 Briefly describe the concept of look-ahead carry generation with respect to its use in adder circuits. What is its significance while implementing hardware for addition of binary numbers of longer lengths?

5.7 With the help of a block schematic of the logic circuit, briefly describe how individual four-bit magnitude comparators can be used in a cascade arrangement to perform magnitude comparison of binary numbers of longer lengths.

5.8 Describe the operations performed by half-adder arithmetic circuits.

5.9 Describe the operations performed by full-adder arithmetic circuits:

5.10 Describe the operations performed by half subtractor arithmetic circuits.

5.11 Describe the operations performed by full-subtractor arithmetic circuits:

5.12 Describe the (a) Parallel adder (b) Ripple carry adder and (c) Look-ahead carry adder.

5.13 How do the look-ahead carry adder speed up the addition process?

5.14 When is a carry generated and when is a carry propagated?

5.15 What is cascading of parallel adders? Why is it required?

5.16 What is a parity bit generator?

5.17 What is (a) an odd-parity generator and (b) an even-parity generator?

5.18 What is a code converter? List some of the code converters.

5.19 Describe the operations performed by the comparator.

5.20 Which LED segments will be ON for a decoder/driver input of 1001?

PROBLEMS

5.1 Design an 8421-to-2421 BCD code converter.

5.2 Find the simplest logic for a 2421-to-51111 BCD code converter.

5.3 Design a logic diagram for 2421 to decimal converter.

5.4 Design the following, code converters: (a) 5211 to 2421 (b) four-bit BCD to gray code

5.5 Draw a logic circuit that generates an even-parity bit for the 2421 BCD code.

5.6 Draw a logic circuit that generates an odd-parity bit for the 8421 BCD code.

5.7 A, B, B_{in}, D and B_{out} are, respectively, the minuend, the subtrahend, the BORROW-IN, the DIFFERENCE output and the BORROW-OUT in the case of a full-subtractor. Determine the bit status of D and B_{out} for the following values of A, B and B_{in}: (a) $A = 0, B = 1, B_{in} = 1$, (b) $A = 1, B = 1, B_{in} = 0$, (c) $A = 1, B = 1, B_{in} = 1$ and (d) $A = 0, B = 0, B_{in} = 1$

5.8 Determine the number of half and full-adder circuit blocks required to construct a 64-bit binary parallel adder. Also, determine the number and type of additional logic gates needed to transform this 64-bit adder into a 64-bit adder–subtractor.

5.9 If the minuend, subtrahend and BORROW-IN bits are, respectively, applied to the augend, addend and the CARRY-IN inputs of a full-adder, prove that the SUM output of the full-adder will produce the correct DIFFERENCE output.

5.10 The objective is to design a BCD adder circuit using four-bit binary adders and additional combinational logic. If the decimal numbers to be added can be anywhere in the range from 0 to 9999, determine the number of four-bit binary adder circuit blocks of type IC 7483 required to do the job.

6

Logic Circuit Design: Data Processing

CHAPTER OBJECTIVES

The main goal of this chapter is to impart knowledge about the digital logic circuits for data processing. Readers will be able to discuss the following aspects in this chapter:

- Decoder
- BCD to decimal decoder
- Implementing logic on decoders
- Cascading of decoder
- Seven-segment decoder
- Encoder
- Priority encoders
- Multiplexer (MUX)
- Implementing logic on multiplexer
- Cascading of multiplexers
- Demultiplexer (DeMUX)
- Cascading of demultiplexers

6.1 | INTRODUCTION

Day-to-day work follows decimal system. Digital system works on binary system. Binary codes are used to represent alphanumeric characters. A simple digital system of hand held calculator is shown in Figure 6.1. In a digital system, the decimal input pressed from the keyboard is converted into binary form. This process is accomplished by the digital device known as encoder. Translation from a signal representing decimal to its binary equivalent is known as encoding. Central processing unit (CPU) outputs binary data after processing it. Then the binary data is converted into decimal so that it can be displayed on seven-segment display unit. This process is accomplished by the digital device known as decoder. The translation from binary to its signal representing decimal equivalent is known as decoding. The device is called decoder. The encoder and decoder in this system are electronic code translators.

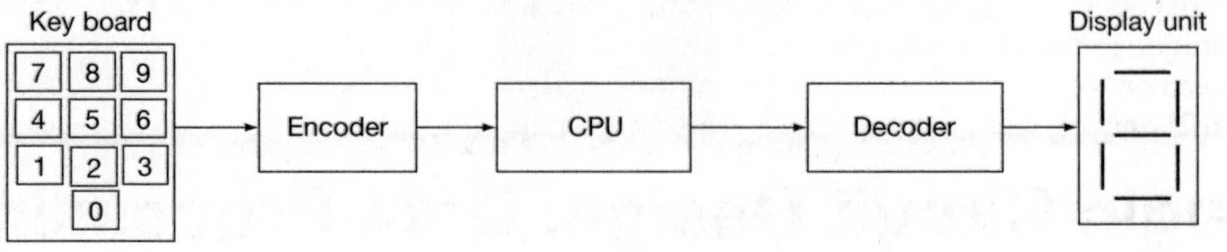

FIGURE 6.1 | Simplified blocked diagram of calculator

Generally parallel binary data is transmitted at short distances. At long distances, data is transmitted serially. The process of converting parallel data to serial data is accomplished by the device known as multiplexer (MUX) (Figure 6.2). Similarly, the process of converting serial data to parallel data is accomplished by the device known as demultiplexer (DeMUX). Moreover, multiplexer selects data from many lines and places on single line. Demultiplexer places datum on the selected line available from single line. The line, where datum is placed, is selected from many lines.

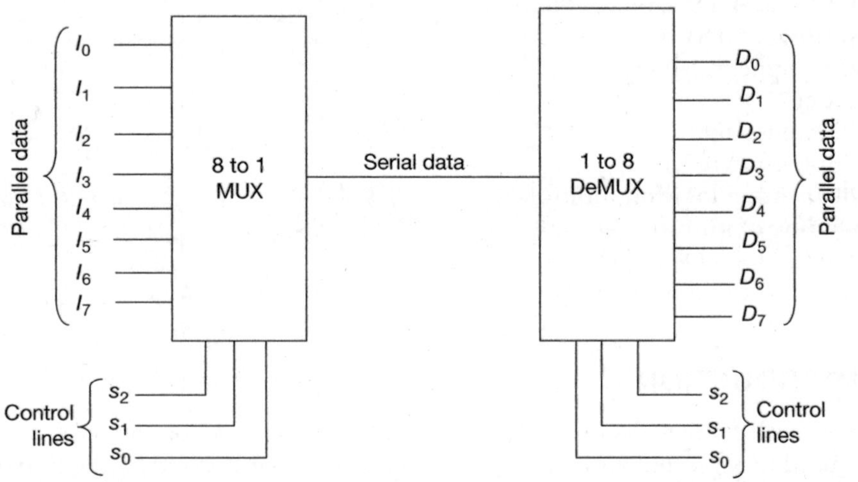

FIGURE 6.2 | Parallel-to-serial-to-parallel data transfer using MUX and DEMUX

This chapter takes a comprehensive look on another class of building blocks used to design more complex combinational circuits, and covers building blocks such as decoders, encoders, multiplexers and demultiplexers and other derived devices such as adder and code converters. Particular emphasis is given to the operational basics and use of these devices to design more complex combinational circuits.

6.2 | DECODERS

Digital system works on discrete quantities of information. Discrete information is represented by binary codes. A binary code of n-bits represents 2^n distinct possible decimal values of the coded information. For instance 3-bit binary code gives $8(2^3)$ distinct decimal values varying from 0 to 7.

A decoder is a combinational circuit that converts n-bit binary data to a maximum of 2^n unique output lines. The unique lines are numbered from 0 to $2^n - 1$. If there are some unused or 'don't care' combinations in the *n*-bit binary data (code), then there will be lesser than 2^n output lines. The decoder is called *n*-to-*m* line decoder, where $m \le 2^n$.

6.2.1 | One-to-Two Line Decoder

A 1-to-2 line decoder is a combinational circuit that converts 1-bit binary data into 2 (2^1) unique outputs. The truth table is given in Table 6.1. In case, b_0 input is 0 then D_0 line is set to 1 and D_1 line is cleared to 0. In case, b_0 input is 1 then D_0 line is cleared to 0 and D_1 line is set to 1. This logic can be obtained from minterms given below:

$$D_0 = \overline{b}_0 \tag{6.1}$$

$$D_1 = b_0 \tag{6.2}$$

TABLE 6.1 | One-to-two line decoder

Input	Output	
b_0	D_0	D_1
0	1	0
1	0	1

Combinational circuit of 1-to-2 line decoder is given in Figure 6.3 based on Boolean expression shown in Eqs. (6.1) and (6.2). The block diagram of 1-to-2 line decoder is given in Figure 6.4.

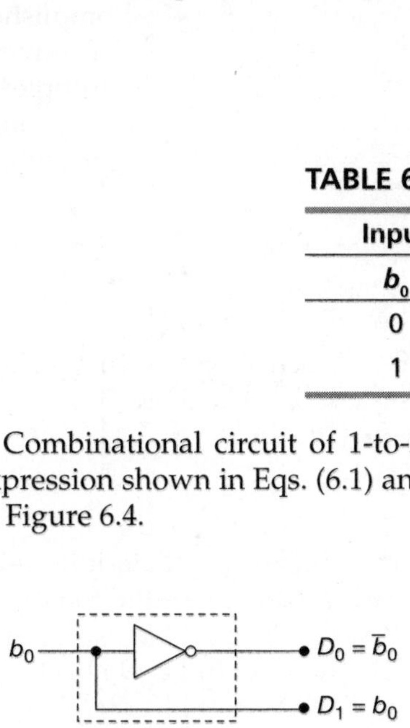

FIGURE 6.3 | 1-to-2 line decoder

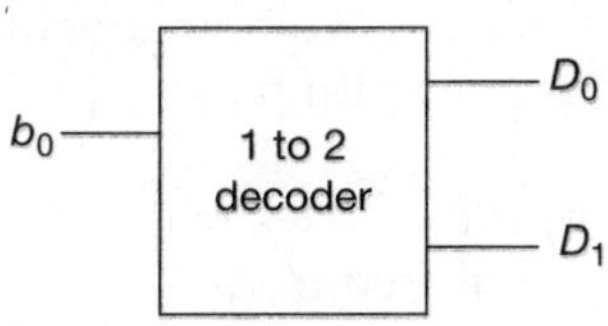

FIGURE 6.4 | Block diagram of 1-to-2 line decoder

Many devices can be cascaded. A device is selected so that output is obtained from device. Device is made to work. The signal required to select the device is known as enable signal. Table 6.2 shows the truth table of 1-to-2 line decoder with enable signal.

TABLE 6.2 | One-to-two line decoder with enable

Input		Output	
$\overline{E}$	b_0	D_0	D_1
0	0	1	0
0	1	0	1
1	X	0	0

0 means logic-0 or LOW state.

1 means logic-1 or HIGH state.

When enable signal, $\bar{E}$ is at logic 1 then output lines D_0 and D_1 are cleared to 0. In case, enable signal, $\bar{E}$ is at logic 0 and b_0 input is 0 then D_0 line is set to 1 and D_1 line is cleared to 0. When enable signal, $\bar{E}$ is at logic 0, b_0 input is 1 then D_0 line is cleared to 0 and D_1 line is set to 1. This logic can be obtained following minterms along with enable signal as given below:

$$D_0 = \bar{E}\bar{b}_0 \tag{6.3}$$

$$D_1 = \bar{E}b_0 \tag{6.4}$$

Considering above Boolean expressions, combinational circuit of 1-to-2 line decoder is given in Figure 6.5. The block diagram of 1-to-2 line decoder is shown in Figure 6.6.

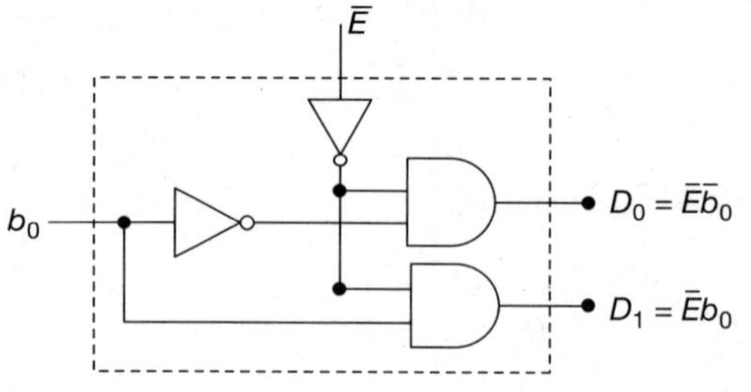

FIGURE 6.5 | A 1-to-2 line decoder with enable signal

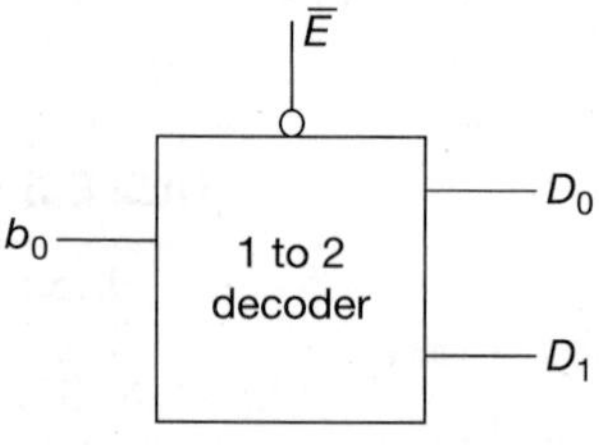

FIGURE 6.6 | Block diagram of 1-to-2 line decoder with enable

6.2.2 | Two-to-Four Line Decoder

A 2-to-4 line decoder is combinational circuit that converts 2-bits binary data into 4 (2^2) unique outputs. The truth table is given in Table 6.3. The truth table gives the following information

- When inputs b_1 and b_0 are 0 and 0, respectively, then D_0 line is set to 1 and D_1, D_2 and D_3 lines are cleared to 0.
- When inputs b_1 and b_0 are 0 and 1, respectively, then D_1 line is set at 1 and D_0, D_2 and D_3 lines are cleared to 0.
- When inputs b_1 and b_0 are 1 and 0, respectively, then D_2 line is set to 1 and D_0, D_1 and D_3 lines are cleared to 0.
- When inputs b_1 and b_0 are 1 and 1, respectively, then D_3 line is set to 1 and D_0, D_1 and D_2 lines are cleared to 0.

TABLE 6.3 | Two-to-four line decoder

Input		Output			
b_1	b_0	D_0	D_1	D_2	D_3
0	0	1	0	0	0
0	1	0	1	0	0
1	0	0	0	1	0
1	1	0	0	0	1

This logic can be obtained from minterms given below:

$$D_0 = \overline{b}_1\overline{b}_0 \quad (6.5)$$

$$D_1 = \overline{b}_1 b_0 \quad (6.6)$$

$$D_2 = b_1\overline{b}_0 \quad (6.7)$$

$$D_3 = b_1 b_0 \quad (6.8)$$

Combinational circuit of 2-to-4 line decoder is given in Figure 6.7 based on Boolean expression shown in Eqs. (6.5) to (6.8). The block diagram of 1-to-2 line decoder is given Figure 6.8.

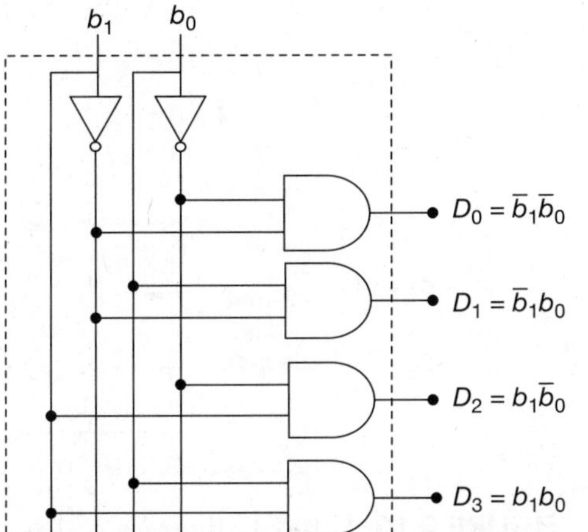

FIGURE 6.7 | A 2-to-4 line decoder

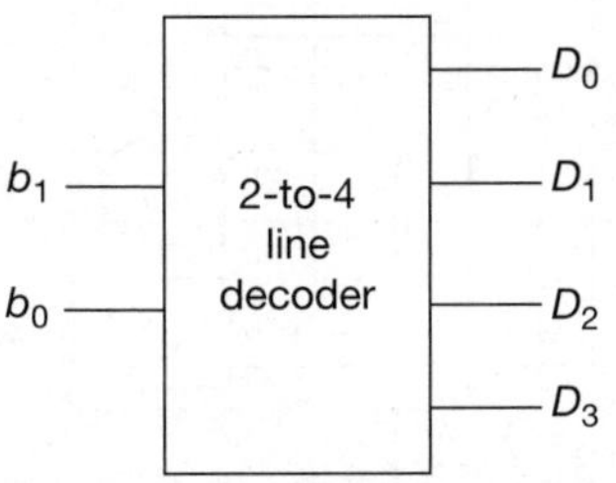

FIGURE 6.8 | Block diagram of 2-to-4 line decoder

The signal required to select the device is known as enable signal. The device works when enable signal is active otherwise device does not work. Table 6.4 shows the truth table of 2-to-4 line decoder with enable signal.

TABLE 6.4 | Two-to-four line decoder with enable

Input			Output			
$\overline{E}$	b_1	b_0	D_0	D_1	D_2	D_3
0	0	0	1	0	0	0
0	0	1	0	1	0	0
0	1	0	0	0	1	0
0	1	1	0	0	0	1
1	X	X	0	0	0	0

When enable signal, $\overline{E}$ is at logic 1 then output lines D_0, D_1, D_2 and D_3 are cleared to logic 0. In case, enable signal, $\overline{E}$ is at logic 0 then decoder gives the output as explained in Table 6.4. This logic is the minterms ANDed along with enable signal and is given below:

$$D_0 = \bar{E}\bar{b}_1\bar{b}_0 \tag{6.9}$$

$$D_1 = \bar{E}\bar{b}_1 b_0 \tag{6.10}$$

$$D_2 = \bar{E}b_1\bar{b}_0 \tag{6.11}$$

$$D_3 = \bar{E}b_1 b_0 \tag{6.12}$$

Combinational circuit of 2-to-4 line decoder with enable signal is given in Figure 6.9. The block diagram of 2-to-4 line decoder with enable signal is given in Figure 6.10.

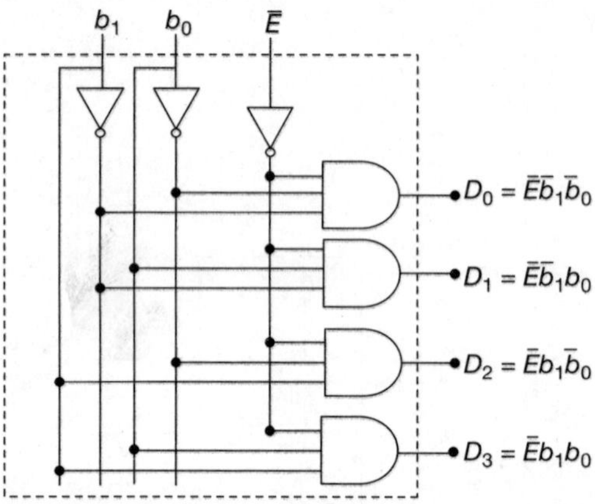

FIGURE 6.9 | A 2-to-4 line decoder with enable

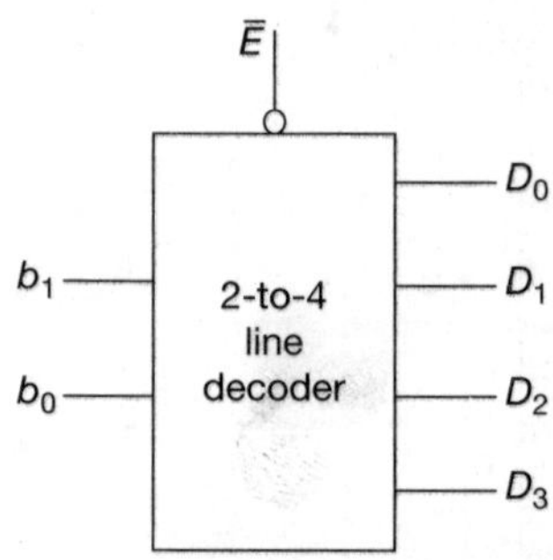

FIGURE 6.10 | Block diagram of 2-to-4 line decoder with enable

6.2.3 | Three-to-Eight Line Decoder

A 3-to-8 line decoder is a combinational circuit that converts 3-bits binary data into 8 (2^3) unique outputs. The truth table is given in Table 6.5 considering enable signal. The truth table gives the following information

TABLE 6.5 | Three-to-eight line decoder with enable

Input				Output							
$\bar{E}$	b_2	b_1	b_0	D_0	D_1	D_2	D_3	D_4	D_5	D_6	D_7
0	0	0	0	1	0	0	0	0	0	0	0
0	0	0	1	0	1	0	0	0	0	0	0
0	0	1	0	0	0	1	0	0	0	0	0
0	0	1	1	0	0	0	1	0	0	0	0
0	1	0	0	0	0	0	0	1	0	0	0

(Continued)

TABLE 6.5 | (Continued)

Input				Output							
$\bar{E}$	b_2	b_1	b_0	D_0	D_1	D_2	D_3	D_4	D_5	D_6	D_7
0	1	0	1	0	0	0	0	0	1	0	0
0	1	1	0	0	0	0	0	0	0	1	0
0	1	1	1	0	0	0	0	0	0	0	1
1	X	X	X	0	0	0	0	0	0	0	0

- When $\bar{E}$ is at logic 0 and inputs b_2, b_1 and b_0 are 0, 0 and 0, respectively, then D_0 line is set to 1 and D_1, D_2, D_3, D_4, D_5, D_6 and D_7 lines are cleared to 0.
- When $\bar{E}$ is at logic 0 inputs b_2, b_1 and b_0 are 0, 0 and 1, respectively, then D_1 line is set to 1 and D_0, D_2, D_3, D_4, D_5, D_6 and D_7 lines are cleared to 0.
- When $\bar{E}$ is at logic 0 inputs b_2, b_1 and b_0 are 0, 1 and 0, respectively, then D_2 line is set to 1 and D_0, D_1, D_3, D_4, D_5, D_6 and D_7 lines are cleared to 0.
- When $\bar{E}$ is at logic 0 inputs b_2, b_1 and b_0 are 0, 1 and 1, respectively, then D_3 line is set to 1 and D_0, D_1, D_2, D_4, D_5, D_6 and D_7 lines are cleared to 0.
- When $\bar{E}$ is at logic 0 inputs b_2, b_1 and b_0 are 1, 0 and 0, respectively, then D_4 line is set to 1 and D_0, D_1, D_2, D_3, D_5, D_6 and D_7 lines are cleared to 0.
- When $\bar{E}$ is at logic 0 inputs b_2, b_1 and b_0 are 1, 0 and 1, respectively, then D_5 line is set to 1 and D_0, D_1, D_2, D_3, D_4, D_6 and D_7 lines are cleared to 0.
- When $\bar{E}$ is at logic 0 inputs b_2, b_1 and b_0 are 1, 1 and 0, respectively, then D_6 line is set to 1 and D_0, D_1, D_2, D_3, D_4, D_5 and D_7 lines are cleared to 0.
- When $\bar{E}$ is at logic 0 inputs b_2, b_1 and b_0 are 1, 1 and 1, respectively, then D_7 line is set to 1 and D_0, D_1, D_2, D_3, D_4, D_5 and D_6 lines are cleared to 0.
- When enable signal, $\bar{E}$ is at logic 1 then output lines D_0, D_1, D_2, D_3, D_4, D_5, D_6 and D_7 are cleared to 0.

This logic is the minterms ANDed along with enable signal and is given below:

$$D_0 = \bar{E}\bar{b}_2\bar{b}_1\bar{b}_0 \quad (6.13)$$

$$D_1 = \bar{E}\bar{b}_2\bar{b}_1 b_0 \quad (6.14)$$

$$D_2 = \bar{E}\bar{b}_2 b_1\bar{b}_0 \quad (6.15)$$

$$D_3 = \bar{E}\bar{b}_2 b_1 b_0 \quad (6.16)$$

$$D_4 = \bar{E}b_2\bar{b}_1\bar{b}_0 \quad (6.17)$$

$$D_5 = \bar{E}b_2\bar{b}_1 b_0 \quad (6.18)$$

$$D_6 = \bar{E}b_2 b_1\bar{b}_0 \quad (6.19)$$

$$D_7 = \bar{E}b_2 b_1 b_0 \quad (6.20)$$

Combinational circuit of 3-to-8 line decoder with enable signal is given in Figure 6.11. The block diagram of 3-to-8 line decoder with enable signal is given in Figure 6.12.

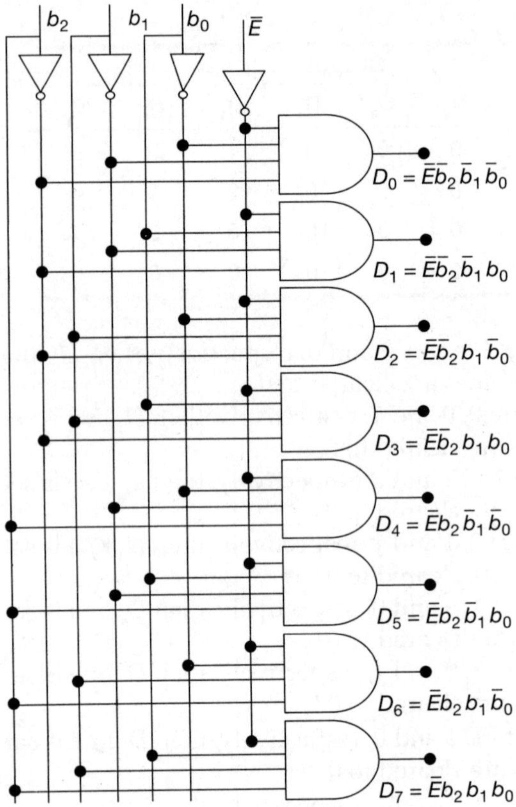

FIGURE 6.11 | A 3-to-8 line decoder with enable signal

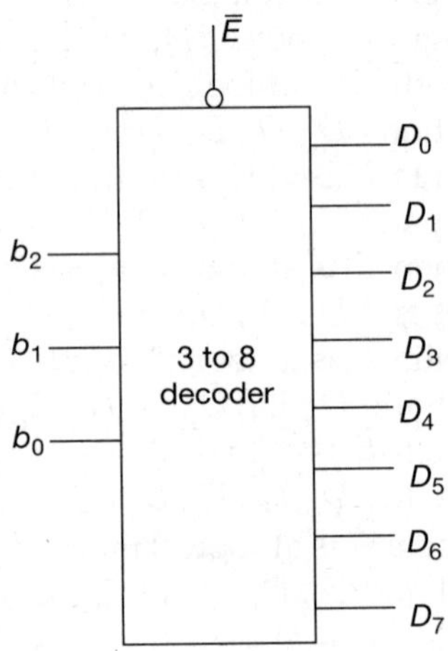

FIGURE 6.12 | Block diagram of a 3-to-8 line decoder with enable

6.2.4 | BCD-to-Decimal Decoder

A BCD decoder is a combinational circuit that converts 4-bits binary data into 10 ($\leq 2^4$) unique outputs. Since, there are some unused or 'don't care' combinations giving 10 to 15 decimal numbers. The truth table is given in Table 6.6 considering enable signal.

TABLE 6.6 | BCD-to-decimal decoder with enable

Input					Output									
$\bar{E}$	b_3	b_2	b_1	b_0	D_0	D_1	D_2	D_3	D_4	D_5	D_6	D_7	D_8	D_9
0	0	0	0	0	1	0	0	0	0	0	0	0	0	0
0	0	0	0	1	0	1	0	0	0	0	0	0	0	0
0	0	0	1	0	0	0	1	0	0	0	0	0	0	0
0	0	0	1	1	0	0	0	1	0	0	0	0	0	0
0	0	1	0	0	0	0	0	0	1	0	0	0	0	0
0	0	1	0	1	0	0	0	0	0	1	0	0	0	0

(Continued)

TABLE 6.6 | (Continued)

Input					Output									
$\bar{E}$	b_3	b_2	b_1	b_0	D_0	D_1	D_2	D_3	D_4	D_5	D_6	D_7	D_8	D_9
0	0	1	1	0	0	0	0	0	0	0	1	0	0	0
0	0	1	1	1	0	0	0	0	0	0	0	1	0	0
0	1	0	0	0	0	0	0	0	0	0	0	0	1	0
0	1	0	0	1	0	0	0	0	0	0	0	0	0	1
1	X	X	X	X	0	0	0	0	0	0	0	0	0	0

When enable signal, $\bar{E}$ is at logic 1 then all the output lines are cleared to logic 0. When $\bar{E}$ is at logic 0, then a line, corresponding the binary inputs b_3, b_2, b_1 and b_0, is set to logic-1 and rest other lines are cleared to logic-0. As don't care terms are involved and each binary combination represents minterm, so the K-map is used to simplify the expression. In K-map shown in Figure 6.13, form possible quad or pair of minterm giving decimal number with don't care minterms only. Minterm representing D_2 forms pair with don't care minterm 10. Similarly, minterms representing D_3, D_4, D_5, D_6 and D_7, forms pair with don't care minterms 11, 12, 13, 14 and 15, respectively. Minterms D_8 forms quad with don't care minterms 12, 10 and 14. Minterms D_9 forms quad with don't care minterms 11, 13 and 15.

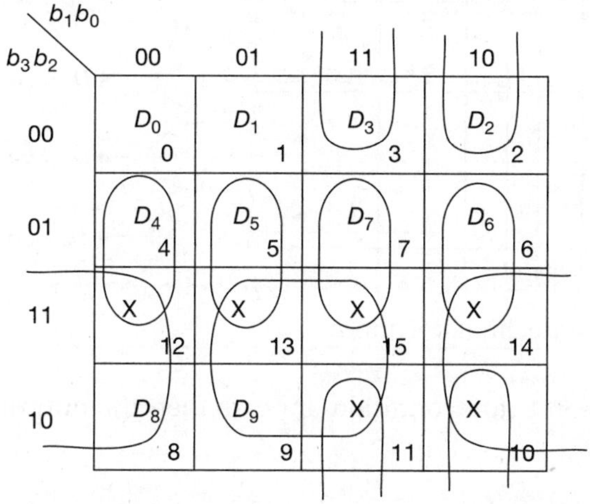

FIGURE 6.13 | K-map simplification

The logic expressions are represented below along with enable signal.

$$D_0 = \bar{E}\bar{b}_3\bar{b}_2\bar{b}_1\bar{b}_0 \quad (6.21)$$

$$D_1 = \bar{E}\bar{b}_3\bar{b}_2\,\bar{b}_1 b_0 \quad (6.22)$$

$$D_2 = \bar{E}\bar{b}_2 b_1\bar{b}_0 \quad (6.23)$$

$$D_3 = \bar{E}\bar{b}_2 b_1 b_0 \quad (6.24)$$

$$D_4 = \bar{E}b_2\bar{b}_1\bar{b}_0 \quad (6.25)$$

$$D_5 = \bar{E}b_2\bar{b}_1 b_0 \quad (6.26)$$

$$D_6 = \bar{E}b_2 b_1\bar{b}_0 \quad (6.27)$$

$$D_7 = \bar{E}b_2 b_1 b_0 \quad (6.28)$$

$$D_8 = \bar{E}b_3\bar{b}_0 \quad (6.29)$$

$$D_9 = \bar{E}b_3 b_0 \quad (6.30)$$

Combination circuit is given below in Figure 6.14.

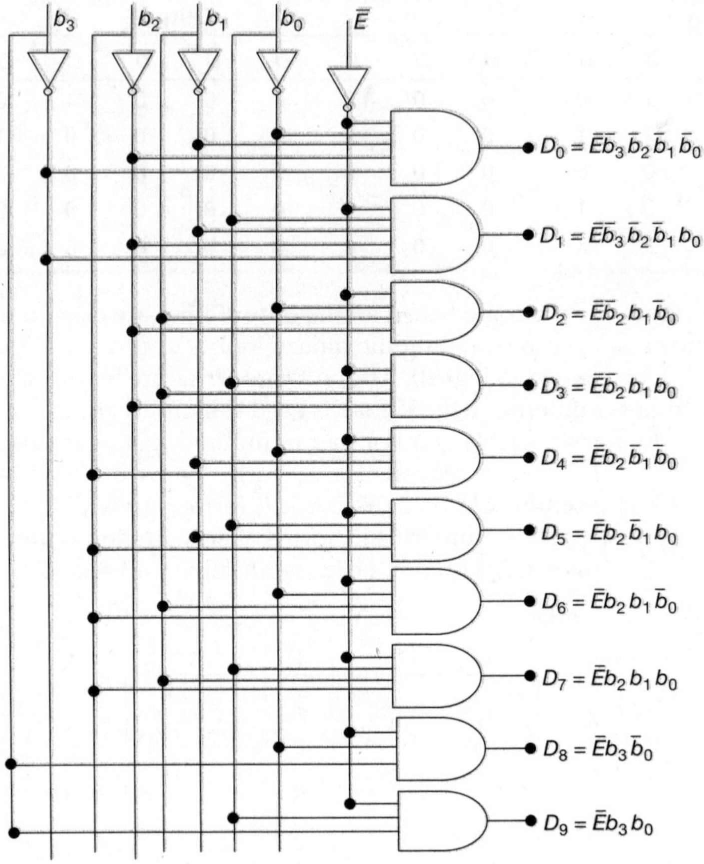

FIGURE 6.14 | BCD-to-decimal decoder

Limitation: If any don't care condition appears then the output produced is given in Table 6.7.

TABLE 6.7 | BCD-to-decimal decoder with enable

Input					Output									
$\bar{E}$	b_3	b_2	b_1	b_0	D_0	D_1	D_2	D_3	D_4	D_5	D_6	D_7	D_8	D_9
0	1	0	1	0	0	0	1	0	0	0	0	0	1	0
0	1	0	1	1	0	0	0	1	0	0	0	0	0	1
0	1	1	0	0	0	0	0	0	1	0	0	0	1	0
0	1	1	0	1	0	0	0	0	0	1	0	0	0	1
0	1	1	1	0	0	0	0	0	0	0	1	0	1	0
0	1	1	1	1	0	0	0	0	0	0	0	1	0	1

6.2.5 | Combinational Circuit using Decoder

A combinational circuit can be implemented using decoder and OR gate. The procedure is outlined below for SOP function:

Sum-of-product (SOP) function is given:

1. Express the Boolean function in standard or canonical sum-of-product (SOP) form.
2. Choose the decoder depending on the number of input variables.
3. Select OR-gates depending on output functions.
4. Input the OR gates with the minterms giving Logic 1 as output.

EXAMPLE 6.1 Implement full-adder circuit using decoder and OR gates.

SOLUTION Consider the truth table of full-adder given in Table 5.6 (Chapter 5). Augend, addend and carry in are represented by x, y and z binary variables, respectively. Sum-of-product (SOP) in canonical form is given by

$$S(x,y,z)=\sum m(1,2,4,7) \tag{6.31}$$

$$C(x,y,z)=\sum m(3,5,6,7) \tag{6.32}$$

A 3-to-8 line decoder is required to implement the full-adder. Two OR gates are required. The circuit is given in Figure 6.15. Output of minterms 1, 2, 4 and 7 are given to 4-inputs OR-gate to get sum, S. Output of minterms 3, 5, 6 and 7 are given to 4-inputs OR-gate to get carry out, C.

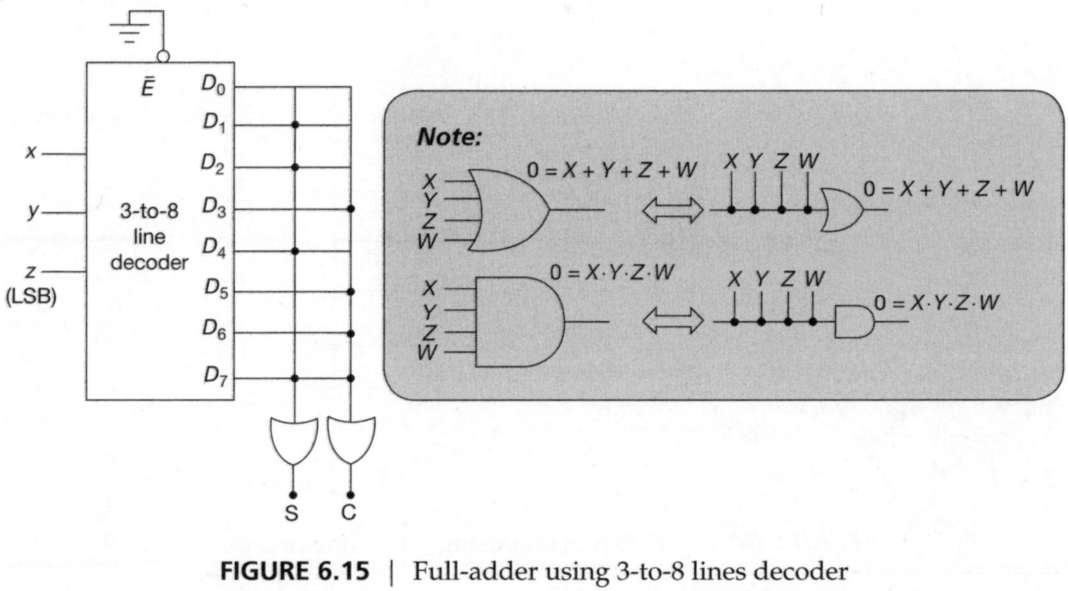

FIGURE 6.15 | Full-adder using 3-to-8 lines decoder

Product-of-sum (POS) function is given:

1. Express the Boolean function in standard POS form.
2. Convert standard POS form into its equivalent standard SOP form by taking complement of the function.
3. Choose the decoder depending on the number of input variables.
4. Select NOR-gates depending on output functions.
5. Input the NOR gates with the minterms giving Logic 1 as output.

EXAMPLE 6.2 Implement $f(x,y,z)=\prod M(0,1,3,5,7)$ using decoder and NOR gates.

SOLUTION Product-of-sum (POS) is given by

$$f(x,y,z)=\prod M(0,1,3,5,7)$$

Complementing the function, sum-of-product (SOP) in canonical form is obtained.

$$\bar{f}(x,y,z)=\sum m(0,1,3,5,7)$$

A 3-to-8 line decoder is required to implement the above function. One NOR gate is required. The circuit is given in Figure 6.16. Output of minterms 0, 1, 3, 5 and 7 are given to NOR-gate to get complemented function.

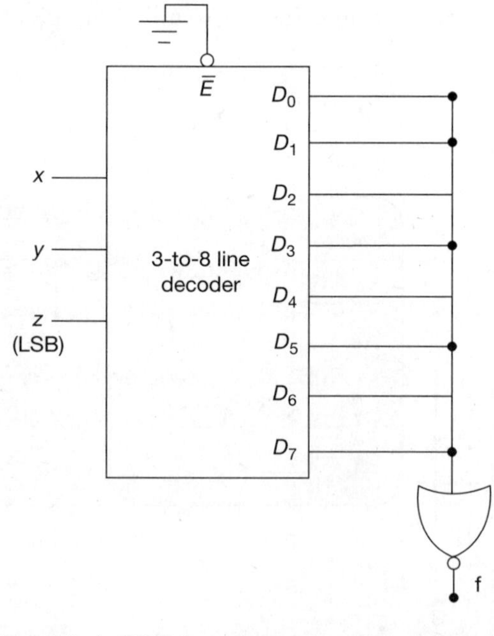

FIGURE 6.16 | POS function using 3-to-8 line decoder

EXAMPLE 6.3 Implement full-subtractor circuit using decoder and OR gates.

SOLUTION Consider the truth table of full-subtractor given in Table 5.8 (Chapter 5). Minuend, subtrahend and previous borrow are represented by x, y and z binary variables, respectively. Sum-of-product (SOP) in canonical form is given by

$$D(x,y,z)=\sum m(1,2,4,7) \tag{6.33}$$

$$B(x,y,z)=\sum m(1,2,3,7) \tag{6.34}$$

A 3-to-8 line decoder is required to implement the full-subtractor. Two OR gates are required. The circuit is given in Figure 6.17. Output of minterms 1, 2, 4 and 7 are given to 4-imputs OR-gate to get difference, D. Output of minterms 1, 2, 3 and 7 is given to 4-inputs OR-gate to get borrow, B.

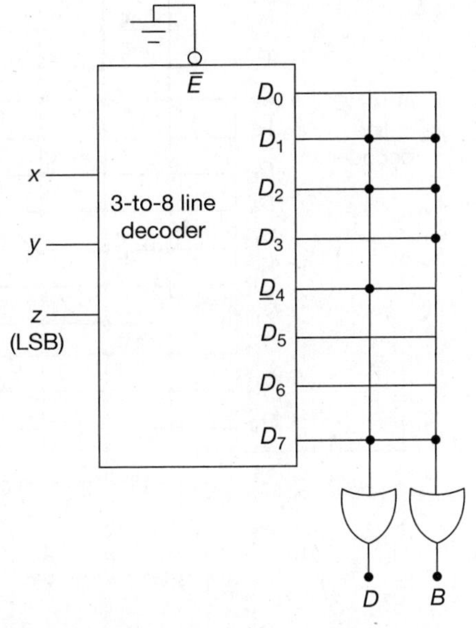

FIGURE 6.17 | Full-subtractor using 3-to-8 line decoder

EXAMPLE 6.4 Implement 4-bit binary to gray code converter circuit using decoder and OR gates.

SOLUTION Consider the truth table of 4-bit binary to gray code converter given in Table 5.23 in Chapter 5. Binary variable b_0 represents LSB and b_3 represents MSB.

Sum-of-product (SOP) in canonical form to convert 4-bit binary code to 4-bit gray code is given by Eqs. (6.35) to (6.38).

$$g_0(b_3,b_2,b_1,b_0)=\sum m(1,2,5,6,9,10,13,14) \tag{6.35}$$

$$g_1(b_3,b_2,b_1,b_0)=\sum m(2,3,4,5,10,11,12,13) \quad (6.36)$$

$$g_2(b_3,b_2,b_1,b_0)=\sum m(4,5,6,7,8,9,10,11) \quad (6.37)$$

$$g_3(b_3,b_2,b_1,b_0)=\sum m(8,9,10,11,12,13,14,15) \quad (6.38)$$

A 4-to-16 line decoder is required to implement 4-bit binary to gray code converter circuit. Four 8-inputs OR gates are required. The circuit is given in Figure 6.18.

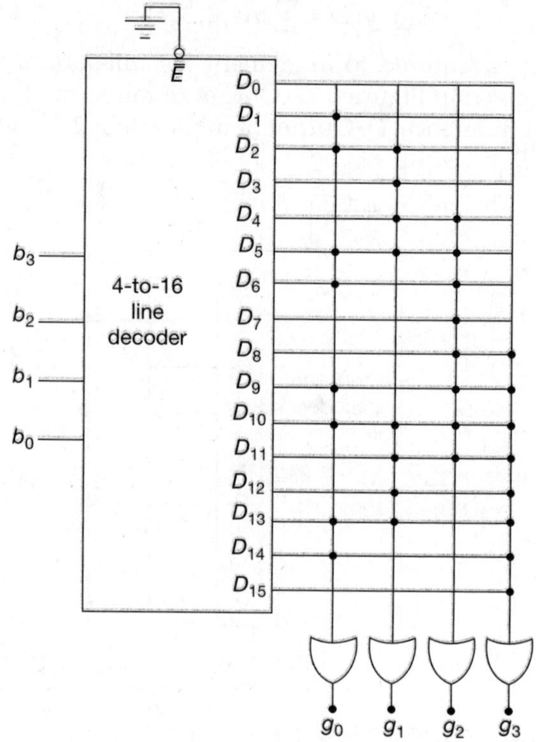

FIGURE 6.18 | Binary to gray code converter using 4-to-16 line decoder

EXAMPLE 6.5 Implement BCD to seven-segment decoder circuit using decoder and OR gates.

SOLUTION Consider the truth table of BCD to seven-segment display code converter given in Table 5.30 (see Example 5.28 in Chapter 5). Binary variable d_0 represents LSB and d_3 represents as MSB. Boolean equations to get seven-segment displays are given by Equations (6.39) to (6.45).

$$a(d_3,d_2,d_1,d_0)=\sum m(0,2,3,5,7,8,9) \quad (6.39)$$

$$b(d_3,d_2,d_1,d_0)=\sum m(0,1,2,3,4,7,8,9) \quad (6.40)$$

$$c(d_3,d_2,d_1,d_0)=\sum m(0,1,3,4,5,6,7,8,9) \tag{6.41}$$

$$d(d_3,d_2,d_1,d_0)=\sum m(0,2,3,5,6,8) \tag{6.42}$$

$$e(d_3,d_2,d_1,d_0)=\sum m(0,2,6,8) \tag{6.43}$$

$$f(d_3,d_2,d_1,d_0)=\sum m(0,4,5,6,8,9) \tag{6.44}$$

$$g(d_3,d_2,d_1,d_0)=\sum m(2,3,4,5,6,8,9) \tag{6.45}$$

A BCD-to-decimal line decoder is required to implement 4-bit binary to seven-segment display circuit. Seven OR gates are required. Input to gates varies from 4 to 9. The circuit is given in Figure 6.19.

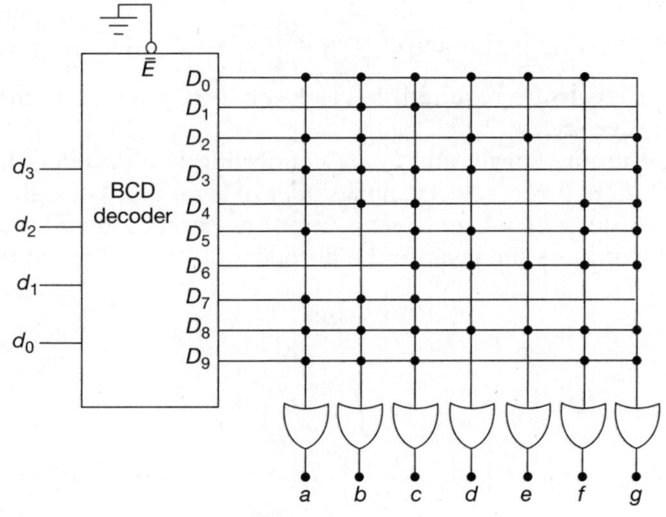

FIGURE 6.19 | Seven-segment decoder using BCD-to-decimal decoder

6.2.6 | Cascading of Decoders

Number of similar decoders are cascaded (combined together) to provide the required higher range decoder by selecting the decoder through its enable signal. The procedure is outlined below.

- Compute the number of decoders required. A number of decoders are arranged in the different levels.
  - o Let n is output lines of decoder to be designed and m is output lines of the available decoder.
  - o Number of decoders required in Level 1:

$$n_1 = \frac{n(\text{Output lines of decoder to be designed})}{m(\text{Output lines of available decoder})}$$

- If $n_i \geq m$ then number of decoders required in Level $l + 1$:
- $n_{1+1} = \dfrac{n_1(\text{Number of decoders in Level 1})}{m(\text{Output lines of available decoder})} \quad (l = 1, 2, 3, \ldots.)$

- Inputs lines are connected commonly. Level 1 has lowest significant input lines. Level 2 will have next higher significant input lines and so on.
- Output lines of decoder are in parallel.
- Output lines of (l + 1) level are connected to enable signal of lth level decoders to select them.

EXAMPLE 6.6 Construct 3-to-8 line decoder with 2-to-4 line decoders.

SOLUTION Let $n = 8$ (output lines of decoder to be constructed) and $m = 4$ (output lines of available decoder)

Number of decoders required in Level 1: $n_1 = \dfrac{8}{4} = 2$

Since 2 < 4 so 2 levels are required, Level 1 connects 2 decoders in parallel. Level 2 requires 1-to-2 line decoder.

Least significant inputs lines b_1 and b_0 are connecting two decoders I and II of level I. Higher significant input line b_2 selects one decoder of level II. A 1-to-2 line decoder is used to select the level II decoder where b_2 act as input to decoder. Inverter acts as 1-to-2 line decoder. In other words, b_2 line enables the decoders of level I. Circuit diagram is shown in Figure 6.20.

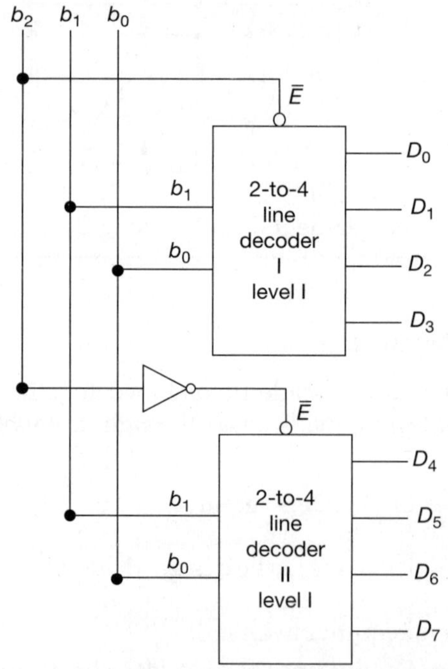

FIGURE 6.20 | A 3-to-8 line decoder using 2-to-4 line decoders

Table 6.8 explains the working of 3-to-8 line decoder after cascading. When Input line, b_2 is cleared to 0, decoder I of level I is selected to get out D_0 to D_3 by feeding input to b_1 and b_0 lines. When input line, b_2 is set to 1, it selects decoder II of level I to get output D_4 to D_7 by feeding input to b_1 and b_0 lines.

TABLE 6.8 | A 3-to-8 line decoder with 2-to-4 line decoders

Level II Input Line	Enables	Level I Input Lines		Output	
b_2		b_1	b_0		
0	Decoder-I	0	0	D_0	Decoder-I
		0	1	D_1	
		1	0	D_2	
		1	1	D_3	
1	Decoder-II	0	0	D_4	Decoder-II
		0	1	D_5	
		1	0	D_6	
		1	1	D_7	

EXAMPLE 6.7 Construct 4-to-16 line decoder when 2-to-4 line decoders are available in abundance.

SOLUTION Let n = 16 (output lines of decoder to be constructed) and m = 4 (output lines of available decoder)

Number of decoders required in Level I: $n_1 = \dfrac{16}{4} = 4$

Number of decoders required in Level II: $n_2 = \dfrac{4}{4} = 1$

Five (4 + 1) 2-to-4 line decoders are required and arranged in 2 levels. Least significant input lines b_1 and b_0 are connected to four decoders I, II, III and IV of level I. Higher significant input lines b_3 and b_2 are given to decoder I of level II. Output of decoder I of level II selects the decoders in Level I through enable line of that decoder. Circuit diagram is shown in Figure 6.21.

When input lines, b_3 and b_2 are set to 0 and 0, respectively, of decoder I of level II, decoder I of level I is selected to get output D_0 to D_3 by feeding proper inputs to b_1 and b_0 lines.

When input lines, b_3 and b_2 are set to 0 and 1, respectively, of decoder I of level II, decoder II of level I is selected to get output D_4 to D_7 by feeding proper inputs to b_1 and b_0 lines.

When input lines, b_3 and b_2 are fed 1 and 0, respectively, to decoder I of level II, decoder III of level I is selected to get outputs D_8 to D_{11} by feeding proper inputs to b_1 and b_0 lines.

When input lines, b_3 and b_2 are fed 1 and 1, respectively, to decoder I of level II, decoder IV of level I is selected to get outputs D_{12} to D_{15} by feeding proper inputs to b_1 and b_0 lines.

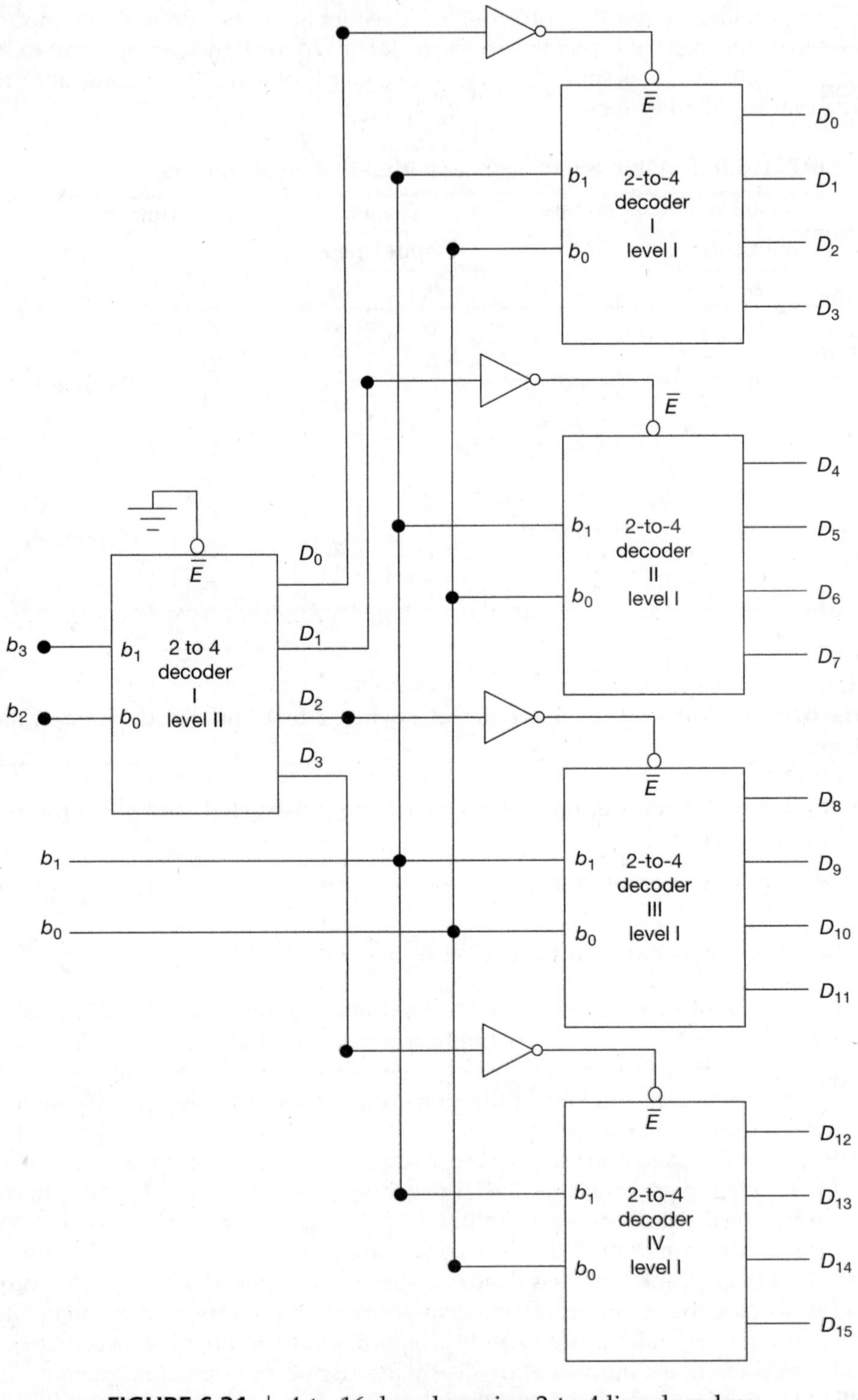

FIGURE 6.21 | 4-to-16 decoder using 2-to-4 line decoders

EXAMPLE 6.8 How many 2-to-4 line decoders are required to construct 6-to-64 line decoder?

SOLUTION Let n = 64 (output lines of decoder to be constructed) and m = 4 (output lines of available decoder)

Number of decoders required in Level I: $n_1 = \frac{64}{4} = 16$

Number of decoders required in Level II: $n_2 = \frac{16}{4} = 4$

Number of decoders required in Level III: $n_3 = \frac{4}{4} = 1$

Total number of 2-to-4 line decoders required = 16 + 4 + 1 = 21 **Ans.**

Level III Inputs		Level II Inputs		Level I Inputs	
b_5	b_4	b_3	b_2	b_1	b_0
To 1 decoder		To 4 decoders		To 16 Decoders	

6.3 | ENCODERS

An encoder is a combinational logic function that has 2^n (or fewer) input lines and n output lines. The n output lines generate the binary code for the possible 2^n input lines. Consider the case of a 2-to-1 line binary encoder. Such an encoder has two (2^1) input lines and one output lines representing the single-bit binary equivalent. The truth table of such an encoder is given in Table 6.9. In the truth table, I_0 and I_1 represent two digits 0 and 1. b_0 represents the output binary digit.

TABLE 6.9 | Two-to-one line encoder

Decimal Value	Input		Output
	I_1	I_0	b_0
0	0	1	0
1	1	0	1

The output is the same as given by input I_1. The expression is given by

$$b_0 = I_1 \tag{6.46}$$

The logic is depicted in Figure 6.22.

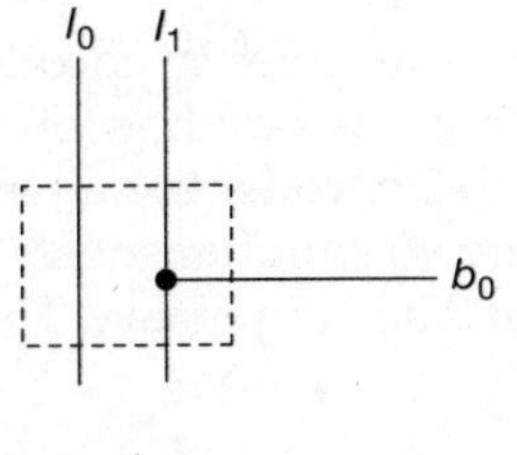

FIGURE 6.22 | 2-to-1 line binary encoder

6.3.1 | Four-to-Two Line Binary Encoder

An encoder is a combinational logic function that has 2^2 input lines and 2 output lines. The 2 output lines generate the binary code for the possible 2^2 input lines. A 4-to-2 line encoder has four input lines, and two output lines representing the two-bit binary equivalent. The truth table of such an encoder is given in Table 6.10. In the truth table, I_0 to I_3 represent four digits 0 to 3. b_1 and b_0 represent the binary digits.

TABLE 6.10 | Four-to-two line encoder

Decimal Value	Input				Output	
	I_3	I_2	I_1	I_0	b_1	b_0
0	0	0	0	1	0	0
1	0	0	1	0	0	1
2	0	1	0	0	1	0
3	1	0	0	0	1	1

The output, b_0 is 1 when I_3 or I_1 input is 1. The output, b_1 is 1 when I_3 or I_2 input is 1. The logical expressions are given below:

$$b_0 = I_3 + I_1 \tag{6.47}$$

$$b_1 = I_3 + I_2 \tag{6.48}$$

Combinational circuit is shown in Figure 6.23. Block diagram of 4-to-2 line encoder is given by Figure 6.24.

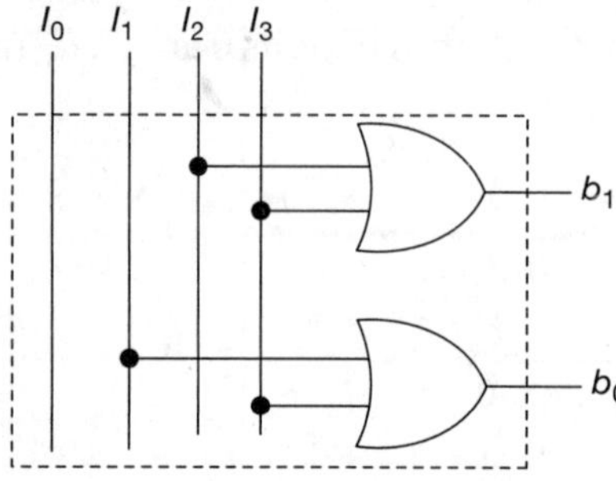

FIGURE 6.23 | 4-to-2 line encoder

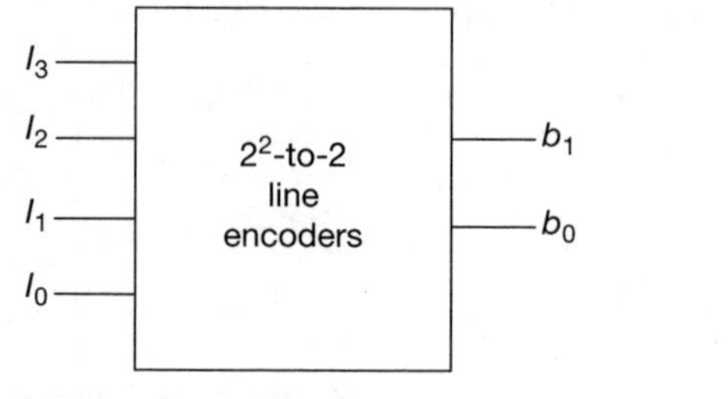

FIGURE 6.24 | 4-to-2 line encoder

6.3.2 | Four-to-Two Line Priority Encoder

The encoders available in IC form are all priority encoders. A priority is assigned to each input so that, when more than one input is simultaneously active, the input with the highest priority is encoded. Consider 4-to-2 encoder has an input priority for higher-order digits. In case input lines I_0, I_2 and I_3 are all simultaneously in logic-1 state. In that case, only I_3 will be encoded and the output will be 11 in binary. The truth table of 4-to-2 line priority encoder is shown in Table 6.11.

TABLE 6.11 | Four-to-Two priority encoder

Decimal Value	Input				Output	
	I_3	I_2	I_1	I_0	b_1	b_0
0	0	0	0	1	0	0
1	0	0	1	X	0	1
2	0	1	X	X	1	0
3	1	X	X	X	1	1

Looking at the last row of the table, it implies that, if I_3 is 1, then, irrespective of the logic status of other inputs, the output is 11 in binary as only I_3 is encoded. I_0, I_1 and I_2 act as don't care conditions. Output b_0 is in logic-1 state when minterms formed from 001X and 1XXX give the output b_0 as 1. In SOP form, the logic function is given by

$$b_0(I_3, I_2, I_1, I_0) = \sum m(2,3,8,9,10,11,12,13,14,15) \tag{6.49}$$

Output b_1 is in logic-1 state when minterms formed from 01XX and 1XXX give the output b_1 as 1. In standard SOP form, the logic function is given by

$$b_1(I_3, I_2, I_1, I_0) = \sum m(4,5,6,7,8,9,10,11,12,13,14,15) \tag{6.50}$$

To simplify the logic function, K-map is used as shown in Figure 6.25 and Figure 6.26.

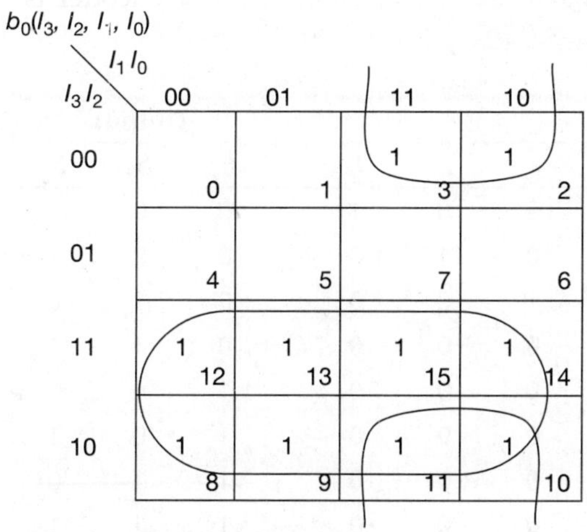

FIGURE 6.25 | K-map for output b_0

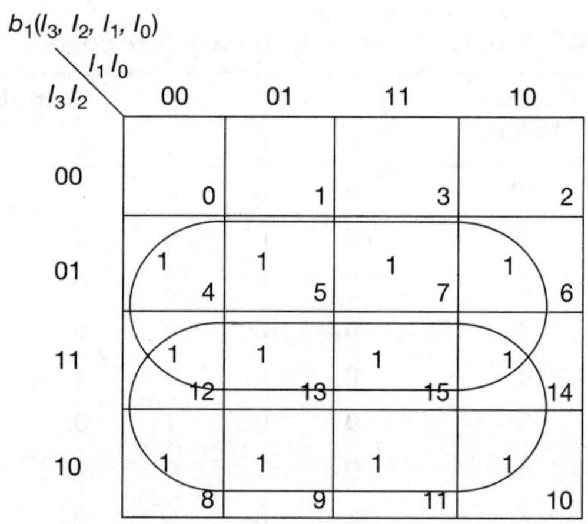

FIGURE 6.26 | K-map for output b_1

Simplified logic expressions are given below:

$$b_0 = I_3 + \bar{I}_2 I_1 \tag{6.51}$$

$$b_1 = I_3 + I_2 \tag{6.52}$$

The combinational circuit is shown in Figure 6.27.

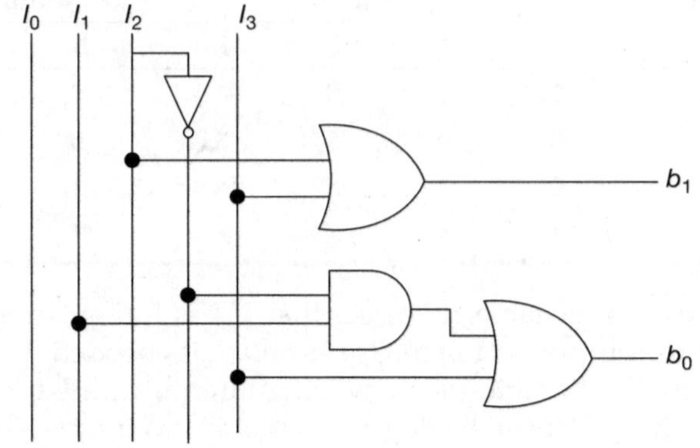

FIGURE 6.27 | 4-to-2 priority encoder

6.3.3 | Octal-to-Binary Encoder

Octal-to-binary encoder has eight (2^3) input lines, each representing an octal digit, and three output lines representing the three-bit binary equivalent. The truth table of such an encoder is given in Table 6.12. In the truth table, I_0 to I_7 represent octal digits 0 to 7. The output lines b_2, b_1 and b_0 represent the binary digits.

TABLE 6.12 | Octal-to-binary line encoder

Decimal Value	Input								Output		
	I_7	I_6	I_5	I_4	I_3	I_2	I_1	I_0	b_2	b_1	b_0
0	0	0	0	0	0	0	0	1	0	0	0
1	0	0	0	0	0	0	1	0	0	0	1
2	0	0	0	0	0	1	0	0	0	1	0
3	0	0	0	0	1	0	0	0	0	1	1
4	0	0	0	1	0	0	0	0	1	0	0
5	0	0	1	0	0	0	0	0	1	0	1
6	0	1	0	0	0	0	0	0	1	1	0
7	1	0	0	0	0	0	0	0	1	1	1

The output, b_0 is 1 when I_7, I_5, I_3 or I_1 input is 1. The output, b_1 is 1 state when I_7, I_6, I_3 or I_2 input is 1. The output, b_2 is 1 when I_7, I_6, I_5 or I_4 input is 1. The logic expressions are given below:

$$b_0 = I_7 + I_5 + I_3 + I_1 \tag{6.53}$$

$$b_1 = I_7 + I_6 + I_3 + I_2 \tag{6.54}$$

$$b_2 = I_7 + I_6 + I_5 + I_4 \tag{6.55}$$

Combinational circuit is shown in Figure 6.28. Block diagram of 4-to-2 line encoder is given by Figure 6.29.

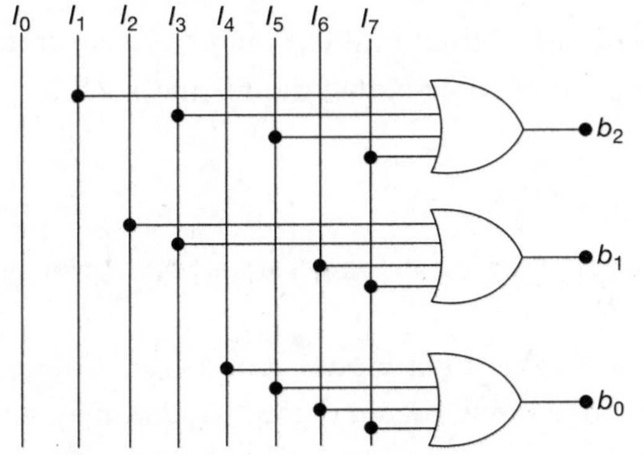

FIGURE 6.28 | Octal-to-binary encoder

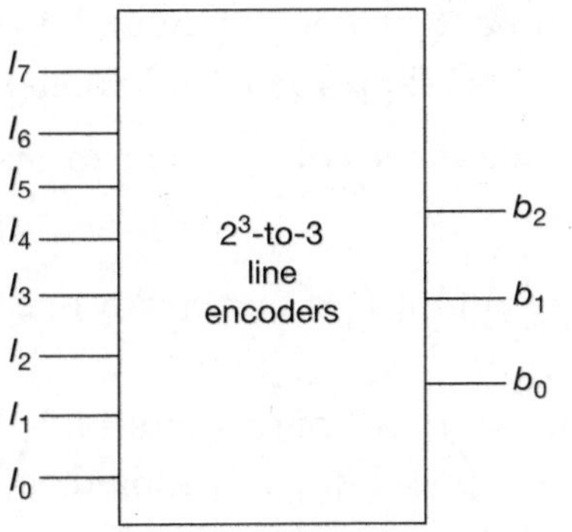

FIGURE 6.29 | 8-to-3 line encoder

6.3.4 | Octal-to-Binary Priority Encoder

Octal-to-binary priority has an input priority for higher-order digits. In case input lines I_0, I_5 and I_6 are all simultaneously in logic-1 state, then only I_6 will be encoded and the output will be 110 in binary. A priority is assigned to each input so that, when more than one input is simultaneously active, the input with the highest priority is encoded. The truth table of 8-to-3 line priority encoder is shown in Table 6.13.

TABLE 6.13 | Octal-to-binary priority encoder

Decimal Value	Input								Output		
	I_7	I_6	I_5	I_4	I_3	I_2	I_1	I_0	b_2	b_1	b_0
0	0	0	0	0	0	0	0	1	0	0	0
1	0	0	0	0	0	0	1	X	0	0	1
2	0	0	0	0	0	1	X	X	0	1	0
3	0	0	0	0	1	X	X	X	0	1	1
4	0	0	0	1	X	X	X	X	1	0	0
5	0	0	1	X	X	X	X	X	1	0	1
6	0	1	X	X	X	X	X	X	1	1	0
7	1	X	X	X	X	X	X	X	1	1	1

Binary output, b_0 is in logic 1 state when any one of inputs I_1, I_3, I_5 or I_7 is set to logic 1. In priority encoder, binary output b_0 will be 1 only if no higher order digit input is high except the inputs of digits activate the b_0 to 1 like I_3, I_5 and I_7. So, in priority encoder,

- output b_0 is 1 only if I_1 is 1 and other higher digit inputs, I_2, I_4 and I_6 are ensured to 0. So, expression $\bar{I}_6\bar{I}_4\bar{I}_2I_1$ should be 1 whereas remaining higher digits I_3, I_5 or I_7 may set b_0 to 1.

- output b_0 is 1 only if input, I_3 is 1 and other higher digit inputs, I_4 and I_6 are ensured to 0. So, expression $\bar{I}_6\bar{I}_4I_3$ should be 1 whereas remaining higher digits I_5 or I_7 may set b_0 to 1.
- output b_0 is 1 only if input, I_5 is 1 and other higher digit inputs, I_4 and I_6 are ensured to 0. So, expression $\bar{I}_6I_5$ should be 1 whereas remaining higher digit I_7 may set b_0 to 1.

Combine above with OR gate to get

$$b_0 = I_7 + \bar{I}_6I_5 + \bar{I}_6\bar{I}_4I_3 + \bar{I}_6\bar{I}_4\bar{I}_2I_1 \tag{6.56}$$

Binary output, b_1 is 1 when any one of inputs I_2, I_3, I_6 or I_7 is in logic-1 state. So, in priority encoder,

- output b_1 is 1 only if input I_2 is 1 and other higher digit inputs, I_4 and I_5 are 0 state. So, expression $\bar{I}_5\bar{I}_4I_2$ should be 1 whereas remaining higher digits I_3, I_6 or I_7 may set b_1 to 1.
- output b_1 is 1only if input, I_3 is 1 and other higher digit inputs, I_4 and I_5 are 0. So, expression $\bar{I}_5\bar{I}_4I_3$ should be 1 whereas remaining higher digits I_6 or I_7 may set b_1 to 1.

Combine above with OR gate to get

$$b_1 = I_7 + I_6 + \bar{I}_5\bar{I}_4I_3 + \bar{I}_5\bar{I}_4I_2 \tag{6.57}$$

Binary output, b_2 is in logic-1 state when any one of inputs I_4, I_5, I_6 or I_7 is in logic-1 state. In priority encoder, no other higher digit other than I_4 may be high.

$$b_2 = I_7 + I_6 + I_5 + I_4 \tag{6.58}$$

The combinational circuit of octal-to-binary encoder is shown in Figure 6.30.

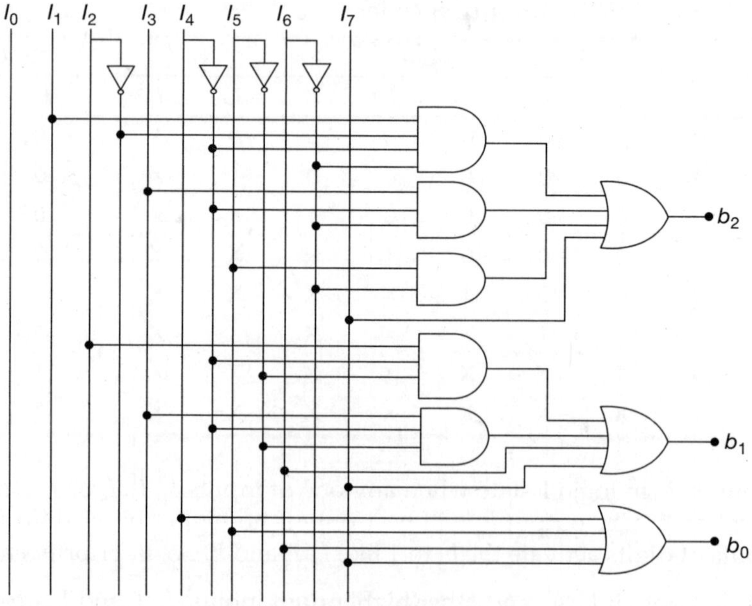

FIGURE 6.30 | Octal-to-binary priority encoder

6.3.5 | Decimal-to-BCD Encoder

An encoder is a combinational logic function that has 10 (< 2^n) input lines and 4 output lines. The 4 output lines generate the BCD code for the possible 10 decimal input lines. The truth table of decimal-to-BCD encoder is given in Table 6.14.

TABLE 6.14 | Decimal-to-BCD encoder

Decimal Value	Input										Output			
	I_9	I_8	I_7	I_6	I_5	I_4	I_3	I_2	I_1	I_0	d_3	d_2	d_1	d_0
0	0	0	0	0	0	0	0	0	0	1	0	0	0	0
1	0	0	0	0	0	0	0	0	1	0	0	0	0	1
2	0	0	0	0	0	0	0	1	0	0	0	0	1	0
3	0	0	0	0	0	0	1	0	0	0	0	0	1	1
4	0	0	0	0	0	1	0	0	0	0	0	1	0	0
5	0	0	0	0	1	0	0	0	0	0	0	1	0	1
6	0	0	0	1	0	0	0	0	0	0	0	1	1	0
7	0	0	1	0	0	0	0	0	0	0	0	1	1	1
8	0	1	0	0	0	0	0	0	0	0	1	0	0	0
9	1	0	0	0	0	0	0	0	0	0	1	0	0	1

Binary output, d_0 is in logic 1 state when any one of inputs I_1, I_3, I_5, I_7 or I_9 is 1.

$$d_0 = I_9 + I_7 + I_5 + I_3 + I_1 \tag{6.59}$$

Binary output, d_1 is in logic 1 state when any one of inputs I_2, I_3, I_6 or I_7 is 1.

$$d_1 = I_7 + I_6 + I_3 + I_2 \tag{6.60}$$

Binary output, d_2 is in logic 1 state when any one of inputs I_4, I_5, I_6 or I_7 is 1.

$$d_2 = I_7 + I_6 + I_5 + I_4 \tag{6.61}$$

Binary output, d_3 is in logic 1 state when any one of inputs I_8 or I_9 is

$$d_3 = I_9 + I_8 \tag{6.62}$$

The combinational circuit of decimal-to-BCD encoder is shown in Figure 6.31.

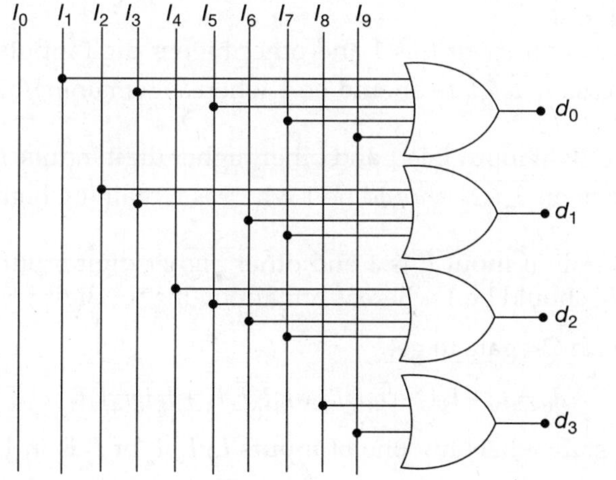

FIGURE 6.31 | Decimal-to-BCD encoder

6.3.6 | Decimal-to-BCD Priority Encoder

An encoder is a combinational logic function that has 10 ($< 2^n$) input lines and 4 output lines. The 4 output lines generate the BCD code for the possible 10 decimal input lines. A priority is assigned to each input so that, when more than one input is simultaneously active, the input with the highest priority is encoded. The truth table of decimal-to-BCD priority encoder is given in Table 6.15.

TABLE 6.15 | Decimal-to-BCD Priority encoder

Decimal Value	Input										Output			
	I_9	I_8	I_7	I_6	I_5	I_4	I_3	I_2	I_1	I_0	b_3	b_2	b_1	b_0
0	0	0	0	0	0	0	0	0	0	1	0	0	0	0
1	0	0	0	0	0	0	0	0	1	X	0	0	0	1
2	0	0	0	0	0	0	0	1	X	X	0	0	1	0
3	0	0	0	0	0	0	1	X	X	X	0	0	1	1
4	0	0	0	0	0	1	X	X	X	X	0	1	0	0
5	0	0	0	0	1	X	X	X	X	X	0	1	0	1
6	0	0	0	1	X	X	X	X	X	X	0	1	1	0
7	0	0	1	X	X	X	X	X	X	X	0	1	1	1
8	0	1	X	X	X	X	X	X	X	X	1	0	0	0
9	1	X	X	X	X	X	X	X	X	X	1	0	0	1

Binary output, d_0 is in logic 1 state when any one of inputs I_1, I_3, I_5, I_7 or I_9 is 1. In priority encoder, binary output b_0 will be 1 only if no higher order digit input is high except the inputs of digits activate the b_0 to 1 like I_3, I_5 and I_7. So, in priority encoder,

- output, d_0 is 1 only if input I_1 is 1 and other higher digit inputs, I_2, I_4, I_6 and I_8 are in 0 state. So, expression $\overline{I}_8\overline{I}_6\overline{I}_4\overline{I}_2I_1$ should be 1 whereas remaining higher digits I_3, I_5, I_7 or I_9 may set d_0 to 1.
- output, d_0 is 1 only if input I_3 is 1 and other higher digit inputs, I_4, I_6 and I_8 are in 0 state. So, expression $\overline{I}_8\overline{I}_6\overline{I}_4I_3$ should be 1 whereas remaining higher digits I_5, I_7 or I_9 may set d_0 to 1.
- output, d_0 is 1 only if input I_5 is 1 and other higher digit inputs, I_6 and I_8 are in logic-0 state. So, expression $\overline{I}_8\overline{I}_6I_5$ should be 1 whereas remaining higher digits I_7 or I_9 may set d_0 to 1.
- output, d_0 is 1 only if input I_7 is 1 and other higher digit input, I_8 are in 0 state. So, expression $\overline{I}_8I_7$ should be 1 whereas remaining higher digit I_9 may set d_0 to 1.

Combine above with OR gate to get

$$d_0 = I_9 + \overline{I}_8I_7 + \overline{I}_8\overline{I}_6I_5 + \overline{I}_8\overline{I}_6\overline{I}_4I_3 + \overline{I}_8\overline{I}_6\overline{I}_4\overline{I}_2I_1 \tag{6.63}$$

Output, d_1 is in 1 state when any one of inputs I_2, I_3, I_6 or I_7 is in 1 state. So, in priority encoder,

- output, d_1 is 1 only if input I_2 is 1 and other higher digit inputs, I_4, I_5, I_8 and I_9 are in 0 state. So, expression $\bar{I}_9\bar{I}_8\bar{I}_5\bar{I}_4I_2$ should be 1 whereas remaining higher digits I_3, I_6 or I_7 may set d_1 to 1.
- output, d_1 is 1 only if input I_3 is 1 and other higher digit inputs, I_4, I_5, I_8 and I_9 are in 0 state. So, expression $\bar{I}_9\bar{I}_8\bar{I}_5\bar{I}_4I_3$ should be 1 whereas remaining higher digits I_6 or I_7 may set d_1 to 1.
- output, d_1 is 1 only if input I_6 is 1 and other higher digit inputs, I_8 and I_9 are in 0 state. So, expression $\bar{I}_9\bar{I}_8I_6$ should be 1 whereas remaining higher digit I_7 may set d_1 to 1.
- output, d_1 is 1 only if input I_7 is 1 and other higher digit inputs, I_8 and I_9 are in 0 state. So, expression $\bar{I}_9\bar{I}_8I_7$ should be 1.

Combine above with OR gate to get

$$d_1 = \bar{I}_9\bar{I}_8I_7 + \bar{I}_9\bar{I}_8I_6 + \bar{I}_9\bar{I}_8\bar{I}_5\bar{I}_4I_3 + \bar{I}_9\bar{I}_8\bar{I}_5\bar{I}_4I_2 \tag{6.64}$$

d_2 is in logic 1 state when any one of I_4, I_5, I_6 or I_7 is in logic-1 state.

- output, d_2 is 1 only if input I_4 is 1 and other higher digit inputs, I_8 and I_9 are in logic-0 state. So, expression $\bar{I}_9\bar{I}_8I_4$ should be 1 whereas remaining higher digits I_5, I_6 or I_7 may set d_2 to 1.
- output, d_2 is 1 only if input I_5 is 1 and other higher digit inputs, I_8 and I_9 are in logic-0 state. So, expression $\bar{I}_9\bar{I}_8I_5$ should be 1 whereas remaining higher digits I_6 or I_7 may set d_2 to 1.
- output, d_2 is 1 only if input I_6 is 1 and other higher digit inputs, I_8 and I_9 are in logic-0 state. So, expression $\bar{I}_9\bar{I}_8I_6$ should be 1 whereas remaining higher digit I_7 may set d_2 to 1.
- output, d_2 is 1 only if input I_7 is 1 and other higher digit inputs, I_8 and I_9 are in logic-0 state. So, expression $\bar{I}_9\bar{I}_8I_7$ should be 1.

Combine above with OR gate to get

$$d_2 = \bar{I}_9\bar{I}_8I_7 + \bar{I}_9\bar{I}_8I_6 + \bar{I}_9\bar{I}_8I_5 + \bar{I}_9\bar{I}_8I_4 \tag{6.65}$$

d_3 is in logic 1 state when any one of I_8 or I_9 is in logic-1 state.

$$d_3 = I_9 + I_8 \tag{6.66}$$

The combinational circuit of decimal-to-BCD encoder is shown in Figure 6.32.

6.4 | MULTIPLEXERS

A multiplexer is a combinational circuit with more than one input lines, one output line and one or more selection lines. A multiplexer selects binary information present on any one of the input lines, depending upon the logic status of the selection inputs, and sends it to the output line. If there are n selection lines, then the number of maximum possible input lines is 2^n and such multiplexer is referred to as a 2^n-to-1 multiplexer or $2^n \times 1$ multiplexer. A multiplexer or MUX, also called a data selector. There are some multiplexer ICs those give complementary outputs. Multiplexers have an enable or strobe input, which requires being active for the multiplexer to perform its intended function.

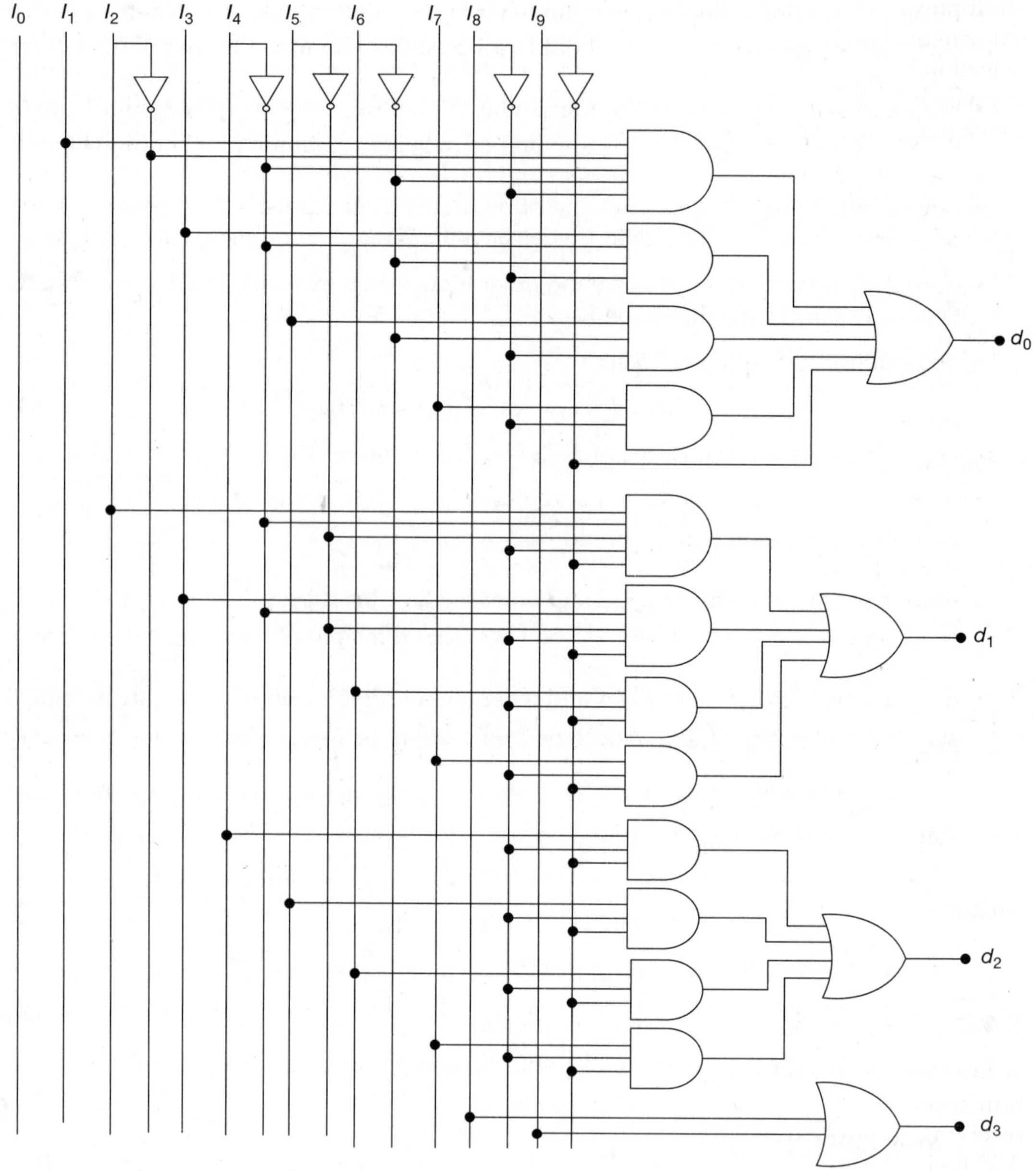

FIGURE 6.32 | Decimal to BCD priority encoder

6.4.1 | Two-to-One Multiplexer

A two-to-one (2^1-to-1) multiplexer (MUX) is a combinational circuit with two input lines, I_0 and I_1 one output line, O and one selection line, S_0. On receiving active low enable input, multiplexer is able to perform its intended function. A 2^1-to-1 multiplexer selects binary information present on either I_0 or I_1 input line, depending upon the logic status of the selection input, S_0, and sends it to the output line, O. Truth table of the multiplexer action is given in Table 6.16. Combinational circuit of 2^1-to-1 MUX is shown in Figure 6.33 and its block diagram is given in Figure 6.34.

TABLE 6.16 | Two-to-one MUX

Input		Output
$\bar{E}$	S_0	O
1	X	0
0	0	I_0
0	1	I_1

The logic expression for the 2-to-1 MUX is given in Eq. (6.67).

$$O = \bar{E}\bar{S}_0 I_0 + \bar{E} S_0 I_1 \quad (6.67)$$

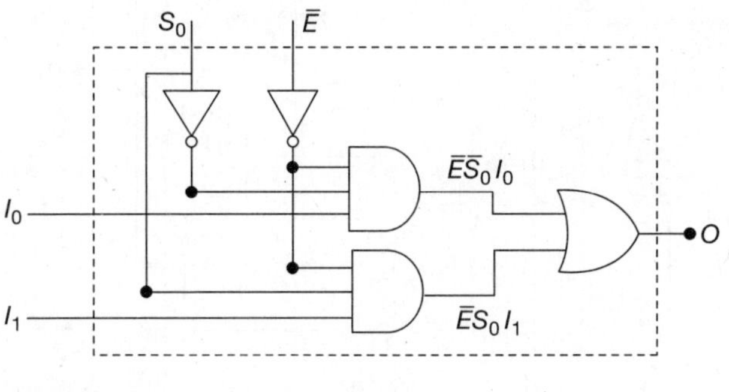

FIGURE 6.33 | 2-to-1 MUX

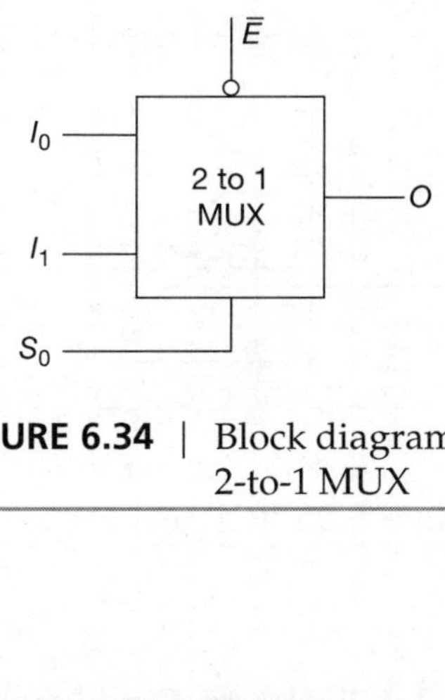

FIGURE 6.34 | Block diagram of 2-to-1 MUX

6.4.2 | Four-to-one Multiplexer

A four-to-one (or 2^2-to-1or 4-to-1) multiplexer (MUX) is a combinational circuit with four input lines, I_0, I_1, I_2 and I_3 one output line, O and two selection lines, S_0 and S_1. On receiving active low enable input, multiplexer is able to perform its intended function. A 2^2-to-1 (or 4-to-1) multiplexer selects binary information present on any one of I_0, I_1, I_2 or I_3 input line, depending upon the logic status of two selection inputs, S_0 and S_1, and sends it to the output line, O. Truth table of the multiplexer action is given in Table 6.17. Combinational circuit of 2^2-to-1 MUX is shown in Figure 6.35 and its block diagram is given in Figure 6.36.

The logic expression for the 2^2-to-1 MUX is given in Eq. (6.68).

$$O = \bar{E}\bar{S}_1\bar{S}_0 I_0 + \bar{E}\bar{S}_1 S_0 I_1 + \bar{E} S_1 \bar{S}_0 I_2 + \bar{E} S_1 S_0 I_3 \quad (6.68)$$

TABLE 6.17 | 2^2- to- 1 (4-to-1) MUX

Input			Output
$\overline{E}$	S_1	S_0	O
1	X	X	0
0	0	0	I_0
0	0	1	I_1
0	1	0	I_2
0	1	1	I_3

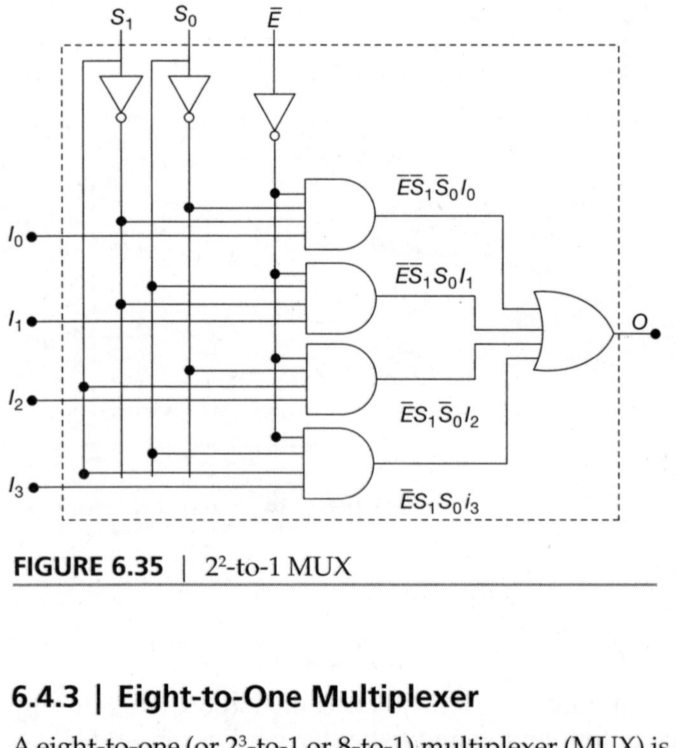

FIGURE 6.35 | 2^2-to-1 MUX

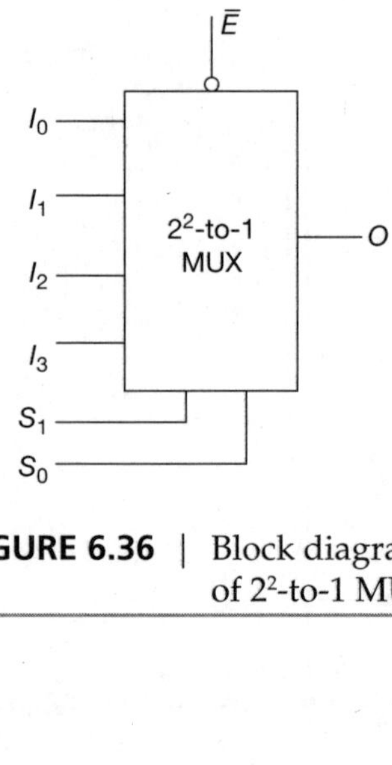

FIGURE 6.36 | Block diagram of 2^2-to-1 MUX

6.4.3 | Eight-to-One Multiplexer

A eight-to-one (or 2^3-to-1 or 8-to-1) multiplexer (MUX) is a combinational circuit with eight input lines, I_0, I_1, I_2, I_3, I_4, I_5, I_6 and I_7, one output line, O and three selection lines, S_2, S_1 and S_0. On receiving active low enable input, multiplexer is able to perform its intended function. A 2^3-to-1 (or 8-to-1 line) multiplexer selects binary information present on any one of I_0, I_1, I_2, I_3, I_4, I_5, I_6 or I_7 input line, depending upon the logic status of three selection inputs, S_2, S_1 and S_0, and sends it to the output line, O. Truth table of the multiplexer action is given in Table 6.18. Combinational circuit of 2^3-to-1 MUX is shown in Figure 6.37 and its block diagram is given in Figure 6.38.

$$O=\overline{E}\overline{S}_2\overline{S}_1\overline{S}_0I_0+\overline{E}\overline{S}_2\overline{S}_1S_0I_1+\overline{E}\overline{S}_2S_1\overline{S}_0I_2+\overline{E}\overline{S}_2S_1S_0I_3+\overline{E}S_2\overline{S}_1\overline{S}_0I_4+\overline{E}S_2\overline{S}_1S_0I_5+\overline{E}S_2S_1\overline{S}_0I_6+\overline{E}S_2S_1S_0I_7 \tag{6.69}$$

TABLE 6.18 | 2^3-to-1 MUX

Input				Output
$\bar{E}$	S_2	S_1	S_0	O
1	X	X	X	0
0	0	0	0	I_0
0	0	0	1	I_1
0	0	1	0	I_2
0	0	1	1	I_3
0	1	0	0	I_4
0	1	0	1	I_5
0	1	1	0	I_6
0	1	1	1	I_7

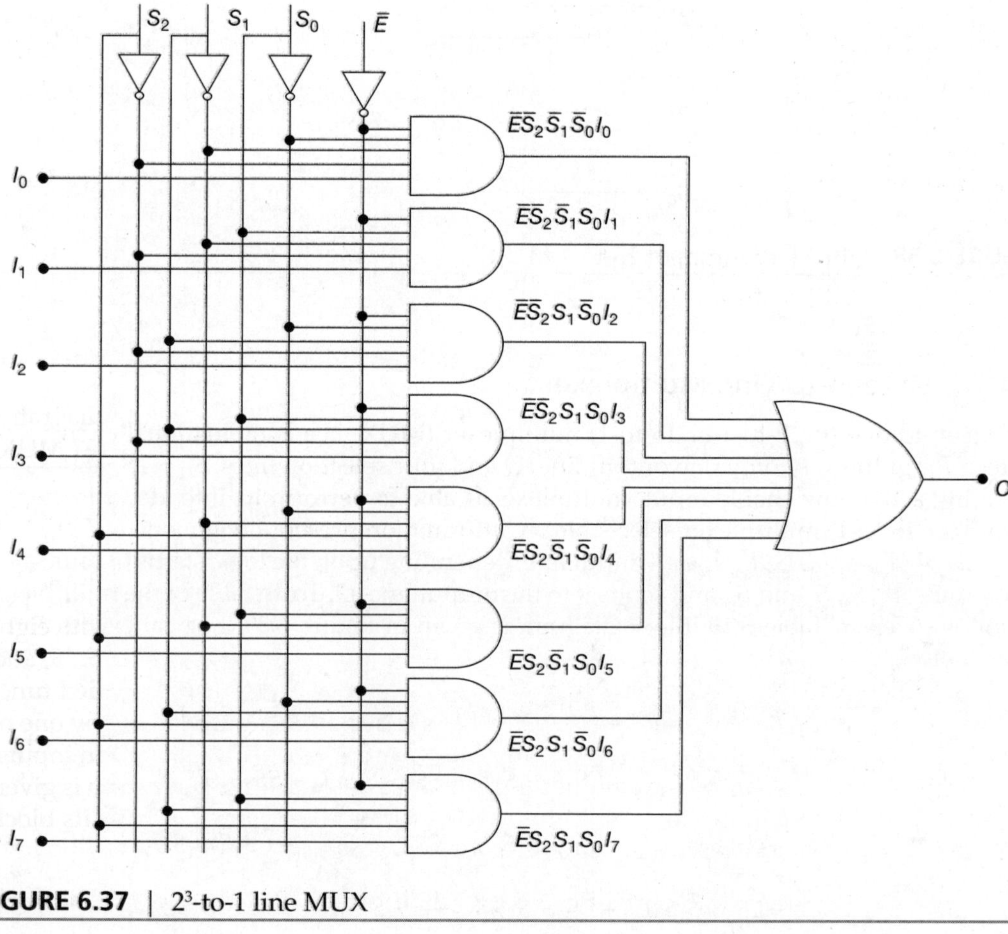

FIGURE 6.37 | 2^3-to-1 line MUX

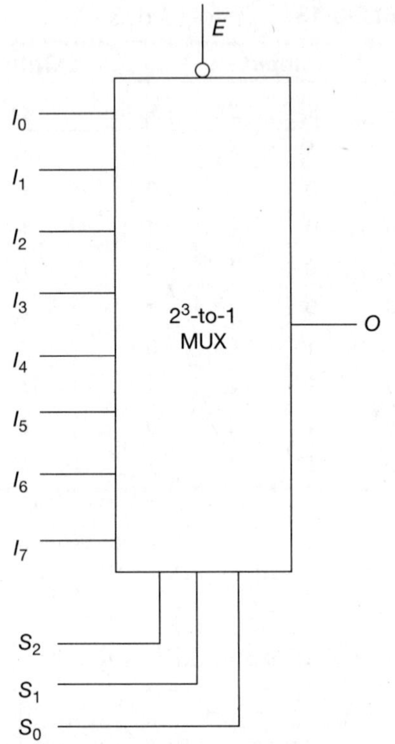

FIGURE 6.38 | Block diagram of 23-to-1 MUX

6.4.4 | Sixteen-to-One Multiplexer

A sixteen-to-one (or 2^4-to-1 or 16-to-1) multiplexer (MUX) is a combinational circuit with sixteen input lines, I_0 to I_{15}, one output line, O and four selection lines, S_3, S_2, S_1 and S_0. On receiving active low enable input, multiplexer is able to perform its intended function. A 2^4-to-1 (or 16-to-1) multiplexer selects binary information present on any one of I_0, I_1, I_2, I_3, I_4, I_5, I_6, I_7, I_8, I_9, I_{10}, I_{11}, I_{12}, I_{13}, I_{14} or I_{15} input line, depending upon the logic status of four selection inputs, S_3, S_2, S_1 and S_0, and sends it to the output line, O. Truth table of the multiplexer action is given in Table 6.19. Block diagram is given in Figure 6.39. Logical expression is given below:

$$
\begin{aligned}
O = {} & \bar{E}\bar{S}_3\bar{S}_2\bar{S}_1\bar{S}_0 I_0 + \bar{E}\bar{S}_3\bar{S}_2\bar{S}_1 S_0 I_1 + \bar{E}\bar{S}_3\bar{S}_2 S_1\bar{S}_0 I_2 + \bar{E}\bar{S}_3\bar{S}_2 S_1 S_0 I_3 \\
& + \bar{E}\bar{S}_3 S_2\bar{S}_1\bar{S}_0 I_4 + \bar{E}\bar{S}_3 S_2\bar{S}_1 S_0 I_5 + \bar{E}\bar{S}_3 S_2 S_1\bar{S}_0 I_6 + \bar{E}\bar{S}_3 S_2 S_1 S_0 I_7 \\
& + \bar{E} S_3\bar{S}_2\bar{S}_1\bar{S}_0 I_8 + \bar{E} S_3\bar{S}_2\bar{S}_1 S_0 I_9 + \bar{E} S_3\bar{S}_2 S_1\bar{S}_0 I_{10} + \bar{E} S_3\bar{S}_2 S_1 S_0 I_{11} \\
& + \bar{E} S_3 S_2\bar{S}_1\bar{S}_0 I_{12} + \bar{E} S_3 S_2\bar{S}_1 S_0 I_{13} + \bar{E} S_3 S_2 S_1\bar{S}_0 I_{14} + \bar{E} S_3 S_2 S_1 S_0 I_{15}
\end{aligned}
\tag{6.70}
$$

TABLE 6.19 | 2^4-to-1 MUX

Input					Output
$\bar{E}$	S_3	S_2	S_1	S_0	O
1	X	X	X	X	0
0	0	0	0	0	I_0
0	0	0	0	1	I_1
0	0	0	1	0	I_2
0	0	0	1	1	I_3
0	0	1	0	0	I_4
0	0	1	0	1	I_5
0	0	1	1	0	I_6
0	0	1	1	1	I_7
0	1	0	0	0	I_8
0	1	0	0	1	I_9
0	1	0	1	0	I_{10}
0	1	0	1	1	I_{11}
0	1	1	0	0	I_{12}
0	1	1	0	1	I_{12}
0	1	1	1	0	I_{14}
0	1	1	1	1	I_{15}

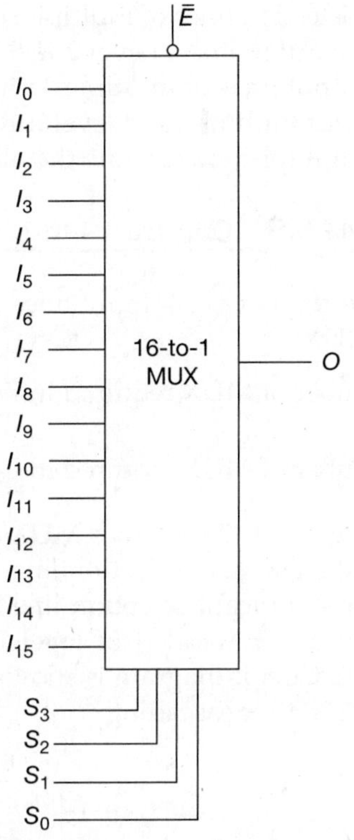

FIGURE 6.39 | Block diagram of 16-to-1 MUX

6.4.5 | Cascading of Multiplexers

Number of similar multiplexers are cascaded (combined together) to construct the required multiplexer of higher rating by selecting the multiplexer through its enable signal. The procedure is outlined below.

- Compute the number of multiplexers required. Required number is arranged in the different levels.
  - Let n is input lines of multiplexer to be designed and m is input lines of the available multiplexer.
  - Number of multiplexers required in Level 1:

$$n_1 = \frac{n(\text{Input lines of multiplexer to be designed})}{m(\text{Input lines of available multiplexer})}$$

  - If $n_i \geq m$ then number of multiplexer required in Level $l + 1$:
  - $n_{l+1} = \dfrac{n_l(\text{Number of multiplexer in Level l})}{m(\text{Input lines of available multiplexer})} \quad (l = 1, 2, 3, \ldots.)$

- Selection lines of multiplexer are connected commonly. Level 1 has lowest significant selection lines. Level 2 will have next higher significant selection lines and so on.
- Input lines of multiplexer are in parallel.
- Output lines of lth level multiplexers are connected to input signal of $(l+1)^{th}$ level of multiplexers to route the output.

EXAMPLE 6.9 Construct 4-to-1 lines multiplexer using 2-to-1 line multiplexer.

SOLUTION Let $n = 4$ (Input lines of MUX to be constructed) and $m = 2$ (Input lines of available MUX)

Number of MUX required in Level I: $n_1 = \frac{4}{2} = 2$

Number of MUX required in Level II: $n_2 = \frac{2}{2} = 1$

Three (2 + 1) 2-to-1 lines MUX are required and arranged in two levels.

Least significant selection line S_0 is connected to two MUX numbered I and II of level I. Higher significant selection line S_1 is given to MUX numbered III of level II. Outputs of MUX number I and II of level I are connected to input lines of MUX numbered III of Level II. Circuit diagram is shown in Figure 6.40. Table 6.20 explains the working of 4-to-1 line MUX after cascading.

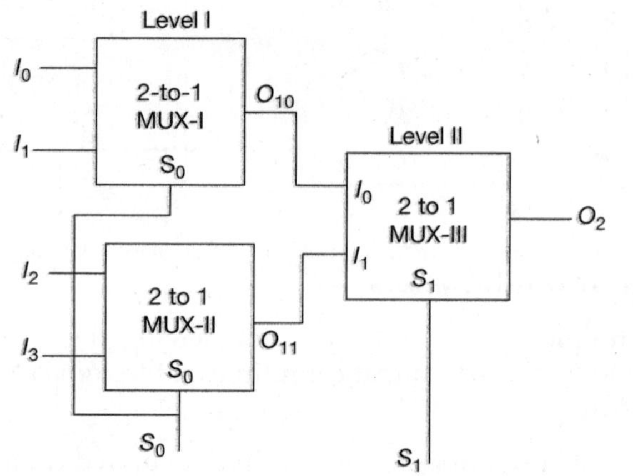

FIGURE 6.40 | Block diagram of $4(2^2)$-to-1 MUX

TABLE 6.20 | Operation of expanded MUX

Level I				Level II			
MUX	**Input**	**Selection**	**Output**	**MUX**	**Input**	**Selection**	**Output**
I	I_0	0	O_{10}	III	O_{10}	0	O_2
	I_1	1					
II	I_2	0	O_{11}		O_{11}	1	
	I_3	1					

Let selection line, S_0 is set to 1.
Input, I_1 of MUX-I is sent to output O_{10}, i.e. I_1.
Input, I_3 of MUX-II is sent to output O_{11}, i.e. I_3.
Let selection line S_1 is set to 1.
Input, O_{11}, i.e. I_3 of MUX-III is sent to output O_2, i.e. I_3.
When selection lines, S_1 and S_0 are 1 and 1, respectively, I_3 is routed to O_2.

EXAMPLE 6.10 Construct 8-to-1 line multiplexer using 2-to-1 line multiplexer.

SOLUTION Let $n = 8$ (Input lines of MUX to be constructed) and $m = 2$ (Input lines of available MUX)

Number of MUX required in Level I: $n_1 = \frac{8}{2} = 4$

Number of MUX required in Level II: $n_2 = \frac{4}{2} = 2$

Number of MUX required in Level III: $n_3 = \frac{2}{2} = 1$

7(4 + 2 + 1) 2-to-1 line MUX are required and arranged in three levels.

Least significant selection line S_0 is connected to four MUX numbered I, II, III and IV of level I. Higher significant selection line S_1 is given to MUX numbered V and VI of level II. Most significant selection line S_2 is given to MUX numbered VII of level III.

Outputs of MUX number I and II of level I are connected to input lines of MUX numbered V of Level II. Outputs of MUX number III and IV of level I are connected to input lines of MUX numbered VI of Level II. Outputs of MUX number V and VI of level II are connected to input lines of MUX numbered VII of Level III. Circuit diagram is shown in Figure 6.41.

Let selection line, S_0 is 0.
MUX-I sends input I_0 to output line O_{10}.
MUX-II sends input I_2 to output line O_{11}.
MUX-III sends input I_4 to output line O_{12}.
MUX-IV sends input I_6 to output line O_{13}.
Let selection line, S_1 is set to 1.
MUX-V sends input O_{11} to output line O_{20}, i.e. I_2.
MUX-VI sends input O_{13} to output line O_{21}, i.e. I_6.
Let selection line, S_2 is cleared to 0.
MUX-VII sends input O_{20} to output line O_3, i.e. I_2.
Output of MUX-VII is O_3 is O_{20} or O_{11} or I_2.
When selection lines, S_2, S_1 and S_0 are 0, 1 and 0, respectively, input I_2 is routed to O_3.

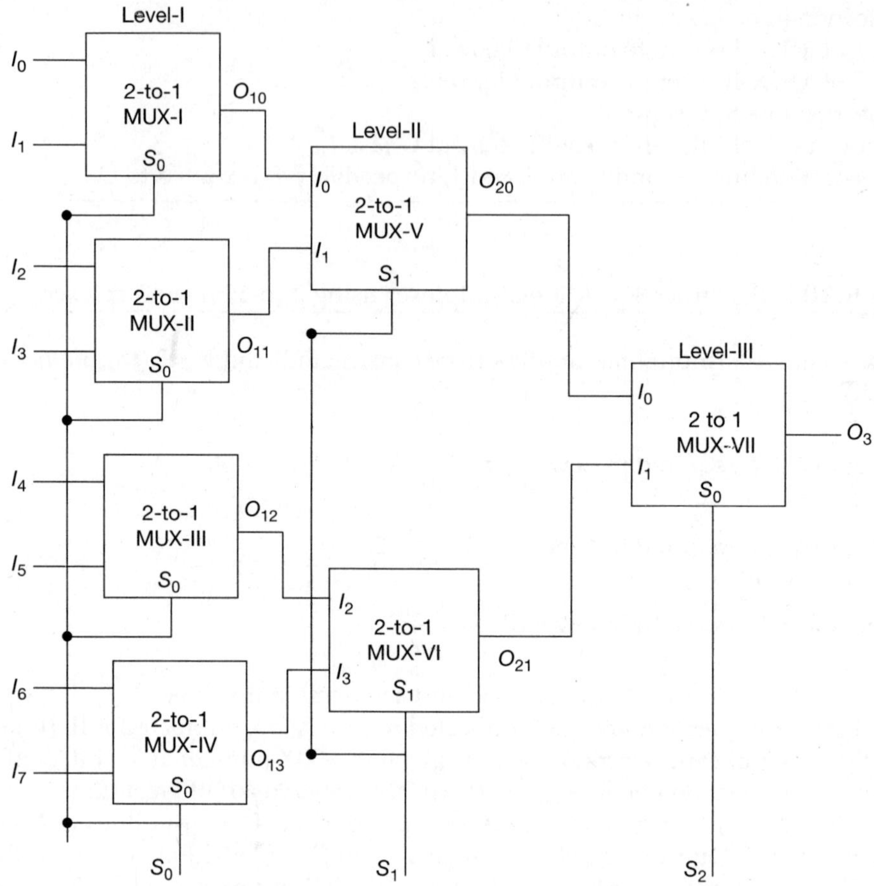

FIGURE 6.41 | Block diagram of 8(2^3) to 1 MUX

EXAMPLE 6.11 Construct 16-to-1 line multiplexer using 4-to-1 line multiplexer.

SOLUTION Let $n = 16$ (Input lines of MUX to be constructed) and $m = 4$ (Input lines of available MUX)

Number of MUX required in Level I: $n_1 = \frac{16}{4} = 4$

Number of MUX required in Level II: $n_2 = \frac{4}{4} = 1$

5(4 + 1) 4-to-1 line MUX are required and arranged in 2 levels.

Least significant selection line S_0 and S_1 is connected to four MUX numbered I, II, III and IV of level I. Higher significant selection line S_2 and S_3 are given to MUX numbered V of level II. Outputs of MUX number I, II, III and IV of level I are connected to input lines of MUX numbered V of Level II. Circuit diagram is shown in Figure 6.42.

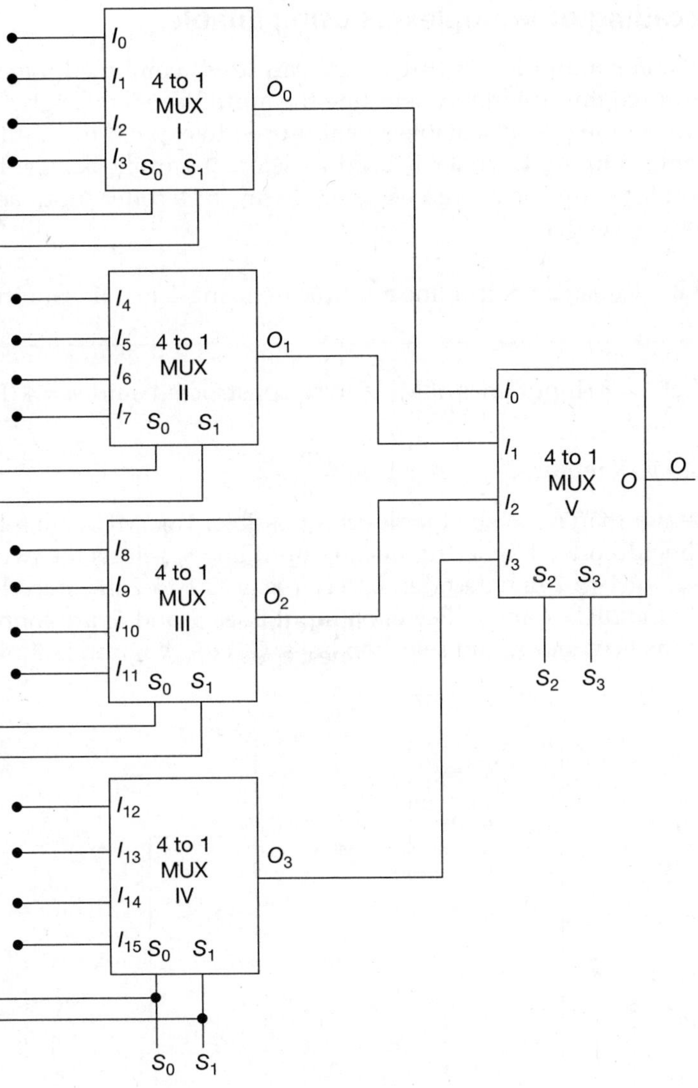

FIGURE 6.42 | 16-to-1 MUX

Let selection lines, S_1 and S_0 are 0 and 1, respectively.
Input, I_1 of MUX-I is sent to output line, O_0, i.e. I_1
Input, I_5 of MUX-II is sent to output line, O_1, i.e. I_5
Input, I_9 of MUX-III is sent to output line, O_2, i.e. I_9
Input, I_{13} of MUX-IV is sent to output line, O_3, i.e. I_{13}
Let selection lines, S_3 and S_2 are 1 and 0, respectively.
Input, O_2, i.e. I_9 of MUX-V is sent to output line, O, i.e. I_9
When S_3, S_2, S_1 and S_0 are 1, 0, 0 and 1, I_9 is routed to O.

6.4.6 | Cascading of Multiplexers using Enable

Number of similar multiplexers (MUX) are cascaded (combined together) to construct/ design the required multiplexer by selecting the multiplexers through its enable signal. OR gate is used to get the output. Enable signal is used to expand the multiplexers in various rows and single column. Decoder is used to select the multiplexers. Decoder input and MUX selection lines are considered as selection lines. Decoder input acts most significant selection lines of decoder.

EXAMPLE 6.12 Construct 8-to-1 line multiplexer using 4-to-1 line multiplexer with enable lines.

SOLUTION Let $n = 8$ (Input lines of MUX to be constructed) and $m = 4$ (Input lines of available MUX)

Number of MUX required in Level I: $n_1 = \frac{8}{4} = 2$

n_1 is not greater than m, so no other level is possible. Two MUX are selected using enable using 1-to-2 line decoder. Higher significant input line S_2 selects the two MUX arranged in one column using 1-to-2 line decoder. Inverter acts as 1-to-2 line decoder. In other words, S_2 line enables the MUX I and II. Selection input lines S_1 and S_0 are connected to two MUX which will act as least significant selection lines. Circuit diagram is shown in Figure 6.43.

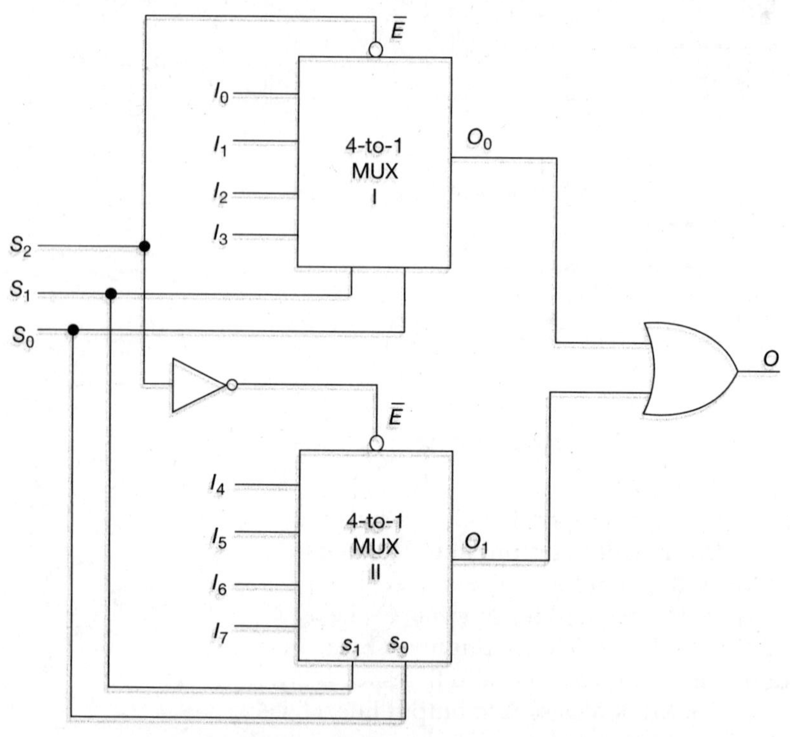

FIGURE 6.43 | Block diagram of 8-to-1 MUX

Operation of the expanded MUX is explained below

When S_2 is 0, MUX I is selected and

S_1 and S_0 selection lines are 0 and 0, respectively, then input, I_0 is sent to output line, O_0.
S_1 and S_0 selection lines are 0 and 1, respectively, then input, I_1 is sent to output line, O_0.
S_1 and S_0 selection lines are 1 and 0, respectively, then input, I_2 is sent to output line, O_0.
S_1 and S_0 selection lines are 1 and 1, respectively, then input, I_3 is sent to output line, O_0.

When S_2 is 1, MUX II is selected, and

S_1 and S_0 selection lines are 0 and 0, respectively, then input, I_4 is sent to output line, O_1.
S_1 and S_0 selection lines are 0 and 1, respectively, then input, I_5 is sent to output line, O_1.
S_1 and S_0 selection lines are 1 and 0, respectively, then input, I_6 is sent to output line, O_1.
S_1 and S_0 selection lines are 1 and 1, respectively, then input, I_7 is sent to output line, O_1.

EXAMPLE 6.13 Construct 16-to-1 line multiplexer using 4-to-1 line multiplexer and 2-to-4 lines decoder.

SOLUTION Let $n = 16$ (Input lines of MUX to be constructed) and $m = 4$ (Input lines of available MUX)

Number of MUX required in Level I: $n_1 = \dfrac{16}{4} = 4$

n_1 is not greater than m, so no other level is possible. Four MUX are selected using enable lines provided by 2-to-4 line decoder. Higher significant input line S_3 and S_2 selects four MUX arranged in one column using 2-to-4 line decoder. Selection input lines S_1 and S_0 are connected to four MUX those act as least significant selection lines. Circuit diagram is shown in Figure 6.44.

Operation of the expanded MUX is explained below:

When S_3 and S_2 are 0 and 0, respectively, decoder sends high signal on D_0 line of decoder. MUX-I is selected.

S_1 and S_0 selection lines are 0 and 0, respectively, then input, I_0 is sent to output line, O_0.
S_1 and S_0 selection lines are 0 and 1, respectively, then input, I_1 is sent to output line, O_0.
S_1 and S_0 selection lines are 1 and 0, respectively, then input, I_2 is sent to output line, O_0.
S_1 and S_0 selection lines are 1 and 1, respectively, then input, I_3 is sent to output line, O_0.

When S_3 and S_2 are 0 and 1, respectively, decoder sends high signal on D_1 line of decoder. MUX-II is selected.

S_1 and S_0 selection lines are 0 and 0, respectively, then input, I_4 is sent to output line, O_1.
S_1 and S_0 selection lines are 0 and 1, respectively, then input, I_5 is sent to output line, O_1.
S_1 and S_0 selection lines are 1 and 0, respectively, then input, I_6 is sent to output line, O_1.
S_1 and S_0 selection lines are 1 and 1, respectively, then input, I_7 is sent to output line, O_1.

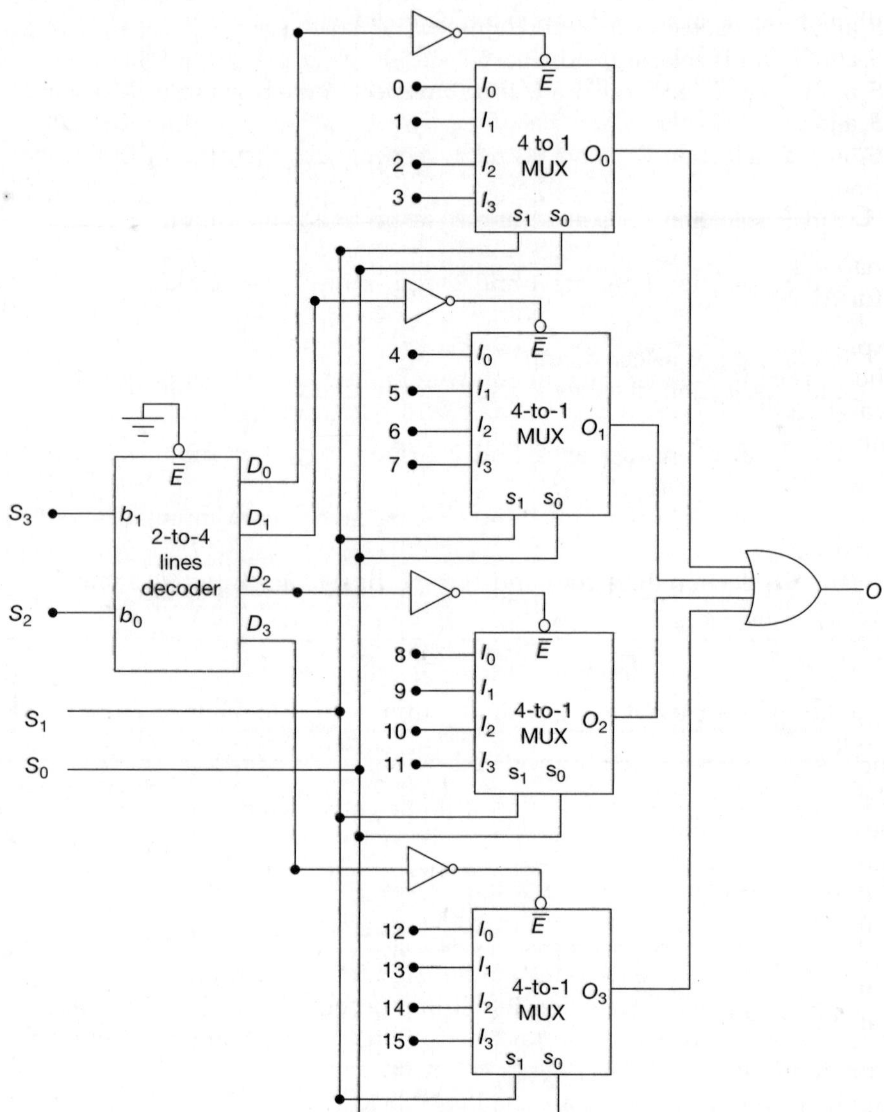

FIGURE 6.44 | 16-to-1 line MUX using 2-to-4 line decoders and 4-to-1 line MUX

When S_3 and S_2 are 1 and 0, respectively, decoder sends high signal on D_2 line of decoder. MUX-III is selected.

S_1 and S_0 selection lines are 0 and 0, respectively, then input, I_8 is sent to output line, O_2.
S_1 and S_0 selection lines are 0 and 1, respectively, then input, I_9 is sent to output line, O_2.
S_1 and S_0 selection lines are 1 and 0, respectively, then input, I_{10} is sent to output line, O_2.
S_1 and S_0 selection lines are 1 and 1, respectively, then input, I_{11} is sent to output line, O_2.

When S_3 and S_2 are 1 and 1, respectively, decoder sends high signal on D_3 line of decoder. MUX-IV is selected.

S_1 and S_0 selection lines are 0 and 0, respectively, then input, I_{12} is sent to output line, O_3.
S_1 and S_0 selection lines are 0 and 1, respectively, then input, I_{13} is sent to output line, O_3.
S_1 and S_0 selection lines are 1 and 0, respectively, then input, I_{14} is sent to output line, O_3.
S_1 and S_0 selection lines are 1 and 1, respectively, then input, I_{15} is sent to output line, O_3.

6.4.7 | Combinational Circuit using Multiplexer

A combinational circuit can be implemented using MUX. The procedure is outlined below for SOP functions:

1. Express the Boolean function in standard SOP.
2. Choose the MUX depending on number of input variables minus one.
3. Draw a table by separating variable which represents most significant literal of minterm. Binary code is followed.

yz / x		00	01	10	11
		$\bar{y}\bar{z}$	$\bar{y}z$	$y\bar{z}$	yz
		I_0	I_1	I_2	I_3
0	$\bar{x}$	0	1	2	3
1	x	4	5	6	7
Input to MUX					

4. Encircle the minterms having output 1.
5. Check each column having encircled minterms to decide the input to MUX. Follow the given procedure
   - If both the minterms in a column are not encircled then consider 0.
   - If both the minterms in a column are encircled then consider 1.
   - If the minterm of first row in a column is encircled then consider complement variable.
   - If the minterm of second row in a column is encircled then consider uncomplement variable.

The procedure is outlined below for POS functions:

1. Express the Boolean function in standard POS.
2. Choose the MUX depending on number of input variables minus one.
3. Draw a table by separating variable which represents most significant literal of maxterm. Binary code is followed.

yz / x		00	01	10	11
		$y+z$	$y+\bar{z}$	$\bar{y}+z$	$\overline{y+z}$
		I_0	I_1	I_2	I_3
0	x	0	1	2	3
1	$\bar{x}$	4	5	6	7
Input to MUX					

4. Encircle the maxterms having 0 output.
5. Check each column having encircled maxterms to decide the input to MUX. Follow the given procedure
   - If both the maxterm in a column are not encircled then consider 1.
   - If both the maxterm in a column are encircled then consider 0
   - If the maxterm of first row in a column is encircled then consider un-complemented variable.
   - If the maxterm of second row in a column is encircled then consider complement.

EXAMPLE 6.14 Implement $f(b_2, b_1, b_0) = \sum m(0,2,4,6)$ using MUX.

SOLUTION There are three variables. One $2^{(3-1)}$-to-1 line MUX is required to implement given Boolean expression. Table 6.21 gives the input to MUX using minterms. Variable representing most signification digit gives two rows, one with 0 or complement and other with 1 or uncomplemented to decide the input to MUX. Encircle minterms 0, 2, 4 and 6.

The circuit diagram is given in Figure 6.45.

- Both minterms, m_0 and m_4 in first column are encircled, so input to MUX, I_0 is 1.
- Both minterms, m_1 and m_5 in second column are not encircled, so input to MUX, I_1 is 0.
- Both minterms, m_2 and m_6 in third column are encircled, so input to MUX, I_2 is 1.
- Both minterms, m_3 and m_7 in fourth column are not encircled, so input to MUX, I_3 is 0.

TABLE 6.21 | MUX input table of minterms for Example 6.14

b_2 \ b_1b_0		00	01	10	11
		$\bar{b}_1\bar{b}_0$	$\bar{b}_1 b_0$	$b_1\bar{b}_0$	$b_1 b_0$
		I_0	I_1	I_2	I_3
0	$\bar{b}_2$	(0)	1	(2)	3
1	b_2	(4)	5	(6)	7
Input to MUX		1	0	1	0

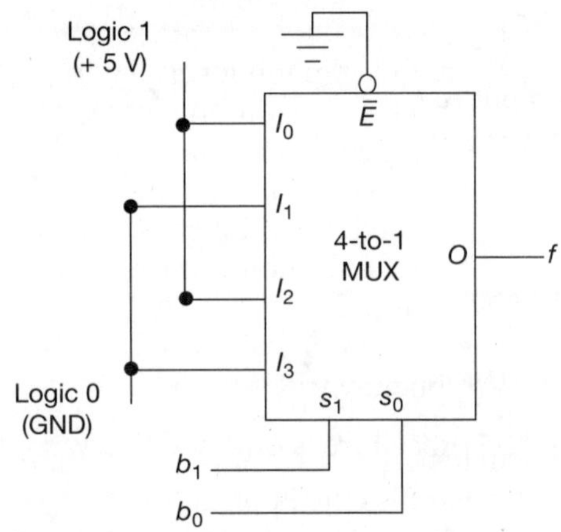

FIGURE 6.45 | 4-to-1 line MUX

EXAMPLE 6.15 Implement $f(b_2, b_1, b_0) = \sum m(1,3,4,6)$ using MUX.

SOLUTION There are three variables. One $2^{(3-1)}$-to-1 line MUX (i.e. 4-to-1 line MUX) is required to implement the given Boolean logic. Table 6.22 gives the input to MUX using minterms. Variable representing most signification digit gives two rows, one with 0 or complement and other with 1 or uncomplemented to decide the input to MUX. Encircle minterms 1, 3, 4 and 6.

- One minterm, m_4 in first column is encircled, so input to MUX, I_0 is b_2.
- One minterm, m_1 in second column is encircled, so input to MUX, I_1 is $\bar{b}_2$.
- One minterm, m_6 in third column is encircled, so input to MUX, I_2 is b_2.
- One minterm, m_3 in fourth column is encircled, so input to MUX, I_3 is $\bar{b}_2$.

The circuit diagram is given in Figure 6.46.

TABLE 6.22 | MUX input table of minterms for Example 6.15

b_2 \ b_1b_0		00	01	10	11
		$\bar{b}_1\bar{b}_0$	$\bar{b}_1b_0$	$b_1\bar{b}_0$	b_1b_0
		I_0	I_1	I_2	I_3
0	$\bar{b}_2$	0	(1)	2	(3)
1	b_2	(4)	5	(6)	7
Input to MUX		b_2	$\bar{b}_2$	b_2	$\bar{b}_2$

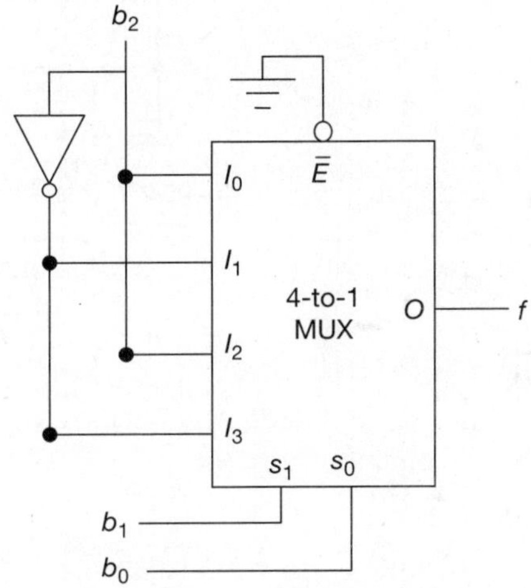

FIGURE 6.46 | 4-to-1 line MUX

EXAMPLE 6.16 Implement $f(b_3, b_2, b_1, b_0) = \sum m(2,4,5,7,9,14)$ using MUX.

SOLUTION There are four variables. $2^{(4-1)}$-to-1 line MUX (i.e. 8-to-1 line MUX) is required to implement the given Boolean logic. Table 6.23 gives the input to MUX using minterms. Encircle minterms 2, 4, 5, 7, 9 and 14. The circuit diagram is given in Figure 6.47.

TABLE 6.23 | MUX input table of minterms for Example 6.16

b_3 \ $b_2b_1b_0$		000	001	010	011	100	101	110	111
		$\bar{b}_2\bar{b}_1\bar{b}_0$	$\bar{b}_2\bar{b}_1b_0$	$\bar{b}_2b_1\bar{b}_0$	$\bar{b}_2b_1b_0$	$b_2\bar{b}_1\bar{b}_0$	$b_2\bar{b}_1b_0$	$b_2b_1\bar{b}_0$	$b_2b_1b_0$
		I_0	I_1	I_2	I_3	I_4	I_5	I_6	I_7
0	$\bar{b}_3$	0	1	(2)	3	(4)	(5)	6	(7)
1	b_3	8	(9)	10	11	12	13	(14)	15
Input to MUX		0	b_3	$\bar{b}_3$	0	$\bar{b}_3$	$\bar{b}_3$	b_3	$\bar{b}_3$

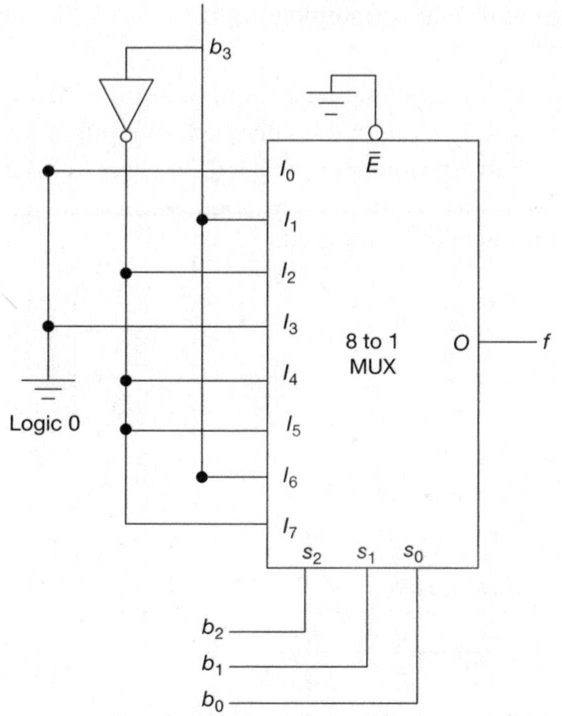

FIGURE 6.47 | 8-to-1 line MUX

EXAMPLE 6.17 Implement $f(x,y,z,w)=\sum m(1,3,5,10,11,13,14)+d(0,6)$ using MUX.

SOLUTION There are four variables. $2^{(4-1)}$-to-1 line MUX (i.e. 8-to-1 line MUX) is required to implement the given Boolean logic. Table 6.24 gives the input to MUX using minterms. Encircle minterms 2, 4, 5, 7, 9 and 14. Also encircle don't care minterms 0 and 6.

TABLE 6.24 | MUX input table of minterms for Example 6.17

yzw \ x		000	001	010	011	100	101	110	111
		$\bar{y}\bar{z}\bar{w}$	$\bar{y}\bar{z}w$	$\bar{y}z\bar{w}$	$\bar{y}zw$	$y\bar{z}\bar{w}$	$y\bar{z}w$	$yz\bar{w}$	yzw
x		I_0	I_1	I_2	I_3	I_4	I_5	I_6	I_7
0	$\bar{x}$	0	1	2	3	4	5	6	7
1	x	8	9	10	11	12	13	14	15
Input to MUX		$\bar{x}$	$\bar{x}$	x	1	0	1	1	0

The circuit diagram is given in Figure 6.48.

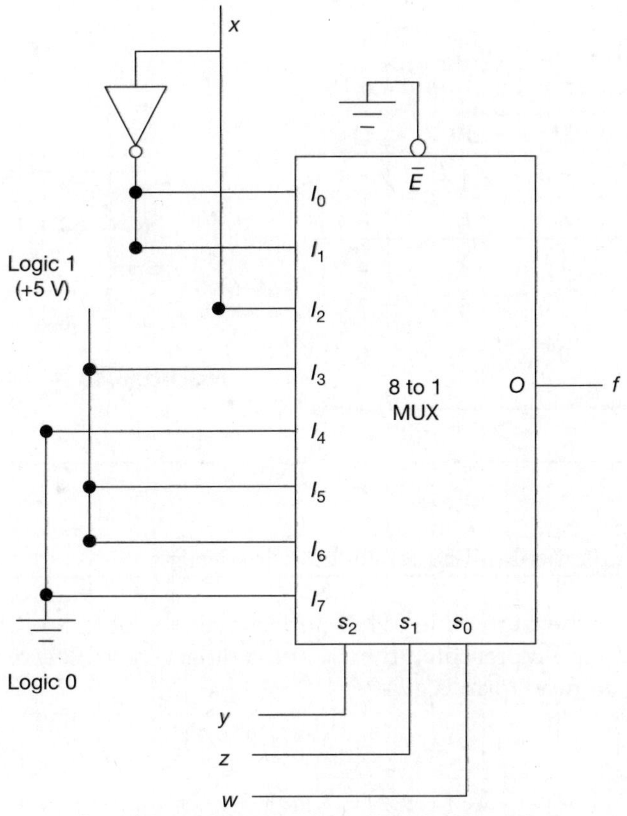

FIGURE 6.48 | 8-to-1 line MUX

EXAMPLE 6.18 Implement $f(x,y,z) = \prod M(0,1,3,5,7)$ using MUX.

SOLUTION There are four variables. $2^{(3-1)}$-to-1 line MUX (i.e. 4-to-1 line MUX) is required to implement the given Boolean logic. Table 6.25 gives the input to MUX using maxterms.

Encircle maxterms 0, 1, 3, 5, and 7.

- One maxterms, M_0 in first column is encircled, so input to MUX, I_0 is x.
- Both maxterms M_1 and M_5 in second column are encircled, so input to MUX, I_1 is 0.
- Both maxterms m_2 and m_6 in third column are not encircled, so input to MUX, I_2 is 1.
- Both maxterms m_3 and m_7 in fourth column are not encircled, so input to MUX, I_3 is 0.

The circuit diagram is given in Figure 6.49.

TABLE 6.25 | MUX input table of maxterms for Example 6.18

x \ yz		00	01	10	11
		$y+z$	$y+\bar{z}$	$\bar{y}+z$	$\bar{y}+\bar{z}$
		I_0	I_1	I_2	I_3
0	x	(0)	(1)	2	(3)
1	$\bar{x}$	4	(5)	6	(7)
Input to Mux		x	0	1	0

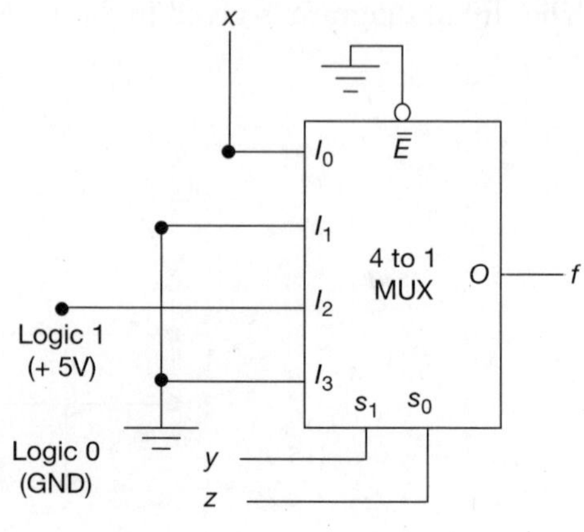

FIGURE 6.49 | 4-to-1 line MUX

EXAMPLE 6.19 Implement full-adder circuit using MUX.

SOLUTION Consider the truth table of full-adder given in Table 5.6 (Chapter 5). Augend, addend and carry in are represented by x, y and z binary variables, respectively. Sum-of-product (SOP) in canonical form is given by

$$S(x,y,z)=\sum m(1,2,4,7) \tag{6.71}$$

$$C(x,y,z)=\sum m(3,5,6,7) \tag{6.72}$$

Two $2^{(3-1)}$-to-1 line MUX are required to implement the full-adder.

Table 6.26 gives the input to MUX using minterms for sum of full-adder. Encircle maxterms 1, 2, 4 and 7. Table 6.27 gives the input to MUX using minterms for carry out of full-adder. Encircle maxterms 3, 5, 6 and 7.

TABLE 6.26 | MUX input table of minterms for sum

x \ yz		00	01	10	11
		$\bar{y}\bar{z}$	$\bar{y}z$	$y\bar{z}$	yz
		I_0	I_1	I_2	I_3
0	$\bar{x}$	0	(1)	(2)	3
1	x	(4)	5	6	(7)
Input to MUX		x	$\bar{x}$	$\bar{x}$	x

TABLE 6.27 | MUX input table of minterms for carry in

x \ yz		00	01	10	11
		$\bar{y}\bar{z}$	$\bar{y}z$	$y\bar{z}$	yz
		I_0	I_1	I_2	I_3
0	$\bar{x}$	0	1	2	(3)
1	x	4	(5)	(6)	(7)
Input to MUX		0	x	x	1

The circuit diagram is given in Figure 6.50.

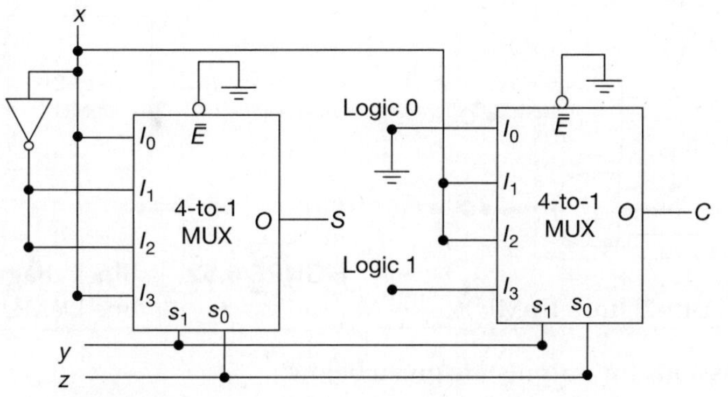

FIGURE 6.50 | Full-adder using 4-to-1 line MUX

6.5 | DEMULTIPLEXERS

A demultiplexer is a combinational circuit with one input line, two or more output lines and one or more selection lines. A demultiplexer, routes binary information present on the input line, to any one of the output lines. The output line that receives the information present on the input line is selected by the bit status of the selection lines. A demultiplexer is a combinational logic circuit with an input line, 2^n output lines and n select lines. A decoder is a kind of a demultiplexer without the input line. If there are n selection lines, then the number of maximum possible output lines is 2^n and such demultiplexer is referred to as a 1-to-2^n demultiplexer or 1×2^n demultiplexer. A demultiplexer or DeMUX is also called a data distributor. Demultiplexers (DeMUX) have an enable or strobe input, which must be active for the demultiplexer to perform its intended function.

6.5.1 | One-to-Two Line Demultiplexer

A one-to-two (1-to-2^1) line demultiplexer (DeMUX) is a combinational circuit with one input line, I two output lines, O_0 and O_1 and one selection line, S_0. On receiving active low enable input, demultiplexer is able to perform its intended function. A 1-to-2^1 demultiplexer places binary information available on input line, *I* on either O_0 or O_1 output line. The output line is selected based on the logic status of the selection input, S_0. Truth table of the demultiplexer (DeMUX) action is given in Table 6.28. Combinational circuit of 1-to-2^1 DeMUX is shown in Figure 6.51 and its block diagram is given in Figure 6.52.

TABLE 6.28 | 1-to-2^1 line DeMUX

Input		Output	
$\bar{E}$	S_0	O_0	O_1
1	X	0	0
0	0	I	0
0	1	0	I

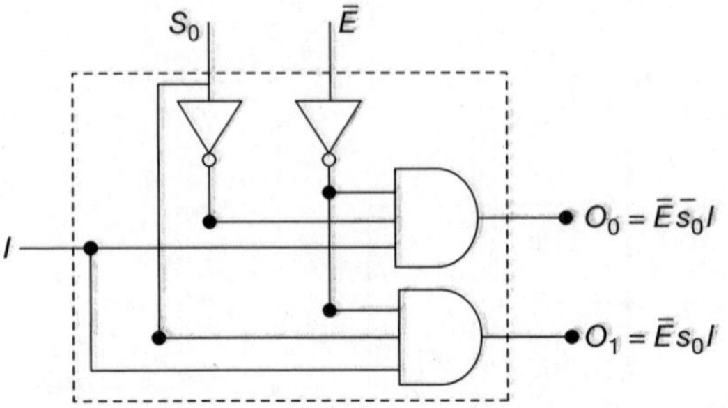

FIGURE 6.51 | 1-to-2^1 lines DeMUX

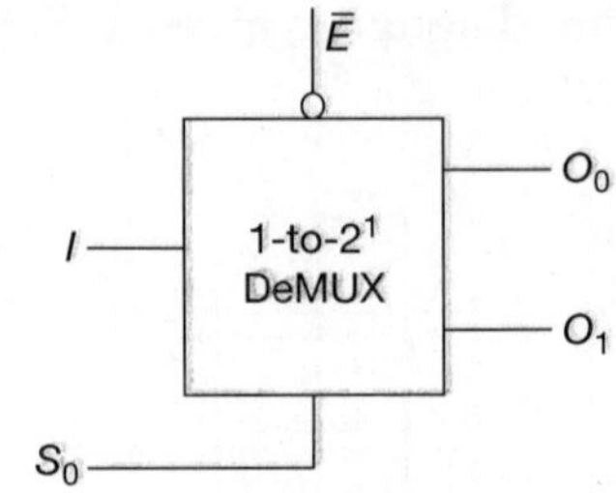

FIGURE 6.52 | Block diagram of 1-to-2^1 line DeMUX

Boolean expressions for outputs are given below.

$$O_0 = \overline{E}\overline{S}_0 I \tag{6.73}$$

$$O_1 = \overline{E}S_0 I \tag{6.74}$$

6.5.2 | One-to-Four Line Demultiplexer

A one-to-four (or 1-to-2^2) line demultiplexer (DeMUX) is a combinational circuit with one input line, I four output lines, O_0, O_1, O_2, and O_3 and two selection lines, S_1 and S_0.

On receiving active low enable input, demultiplexer is able to perform its intended function. A 1-to-2^2 line demultiplexer places binary information available on input line, I, on any one O_0, O_1, O_2 or O_3 output line. The output line is selected based on the logic status of the selection inputs, S_1 and S_0. Truth table of the demultiplexer (DeMUX) action is given in Table 6.29. Combinational circuit of 1-to-2^2 line (1-to-4 lines) DeMUX is shown in Figure 6.53 and its block diagram is given in Figure 6.54. This logic is the minterms ANDed along with enable signal and is given below:

$$O_0 = \overline{E}\overline{S}_1\overline{S}_0 I \tag{6.75}$$

$$O_1 = \overline{E}\overline{S}_1 S_0 I \tag{6.76}$$

$$O_2 = \overline{E}S_1\overline{S}_0 I \tag{6.77}$$

$$O_3 = \overline{E}S_1 S_0 I \tag{6.78}$$

TABLE 6.29 | 1-to-4 lines DeMUX

Input			Output			
$\bar{E}$	S_1	S_0	D_0	D_1	D_2	D_3
1	X	X	0	0	0	0
0	0	0	I	0	0	0
0	0	1	0	I	0	0
0	1	0	0	0	I	0
0	1	1	0	0	0	I

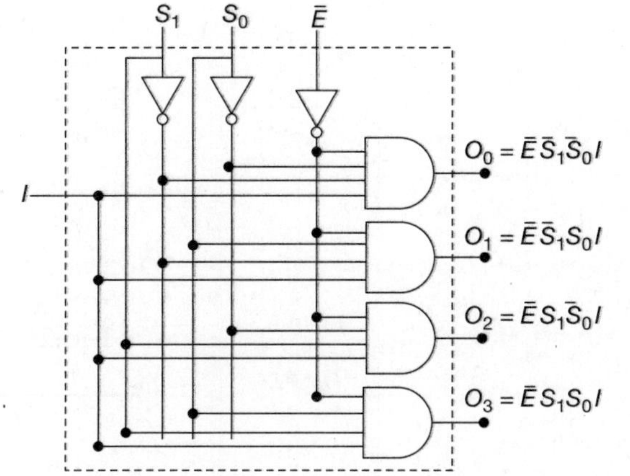

FIGURE 6.53 | One-to-four lines DeMux

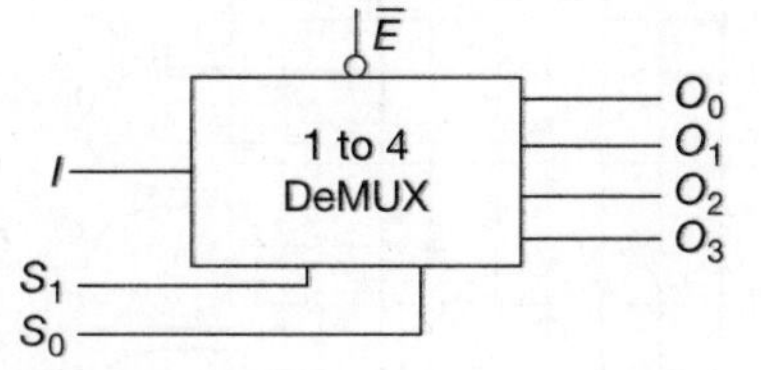

FIGURE 6.54 | Block diagram of 1-to-4 lines DeMux

6.5.3 | One-to-Eight Line Demultiplexer

A one-to-eight (or 1-to-2^3 i.e. 1-to-8) line demultiplexer (DeMUX) is a combinational circuit with one input line, I eight output lines, O_0, O_1, O_2, O_3, O_4, O_5, O_6 and O_7 and three selection lines, S_2, S_1 and S_0. On receiving active low enable input, demultiplexer is able to perform its intended function. A 1-to-2^3 line demultiplexer places binary information available on input line, *I*, on any one of O_0 to O_7 output lines. The output line is selected based on the logic status of the selection inputs, S_2, S_1 and S_0. Truth table of the demultiplexer (DeMUX) action is given in Table 6.30. Combinational circuit of 1-to-2^3 line (1-to-8 lines) DeMUX is shown in Figure 6.55 and its block diagram is given in Figure 6.56.

TABLE 6.30 | 1-to-8 line DeMUX

Input				Output							
$\bar{E}$	S_2	S_1	S_0	D_0	D_1	D_2	D_3	D_4	D_5	D_6	D_7
1	X	X	X	0	0	0	0	0	0	0	0
0	0	0	0	I	0	0	0	0	0	0	0
0	0	0	1	0	I	0	0	0	0	0	0
0	0	1	0	0	0	I	0	0	0	0	0
0	0	1	1	0	0	0	I	0	0	0	0
0	1	0	0	0	0	0	0	I	0	0	0
0	1	0	1	0	0	0	0	0	I	0	0
0	1	1	0	0	0	0	0	0	0	I	0
0	1	1	1	0	0	0	0	0	0	0	I

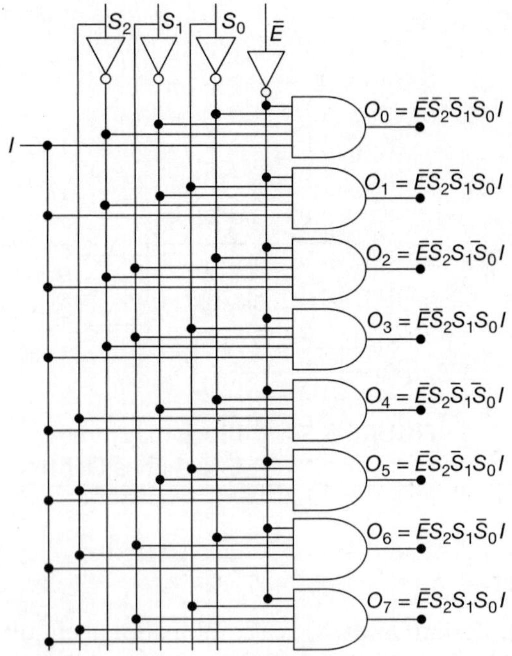

FIGURE 6.55 | 1- to- 8 line DeMUX

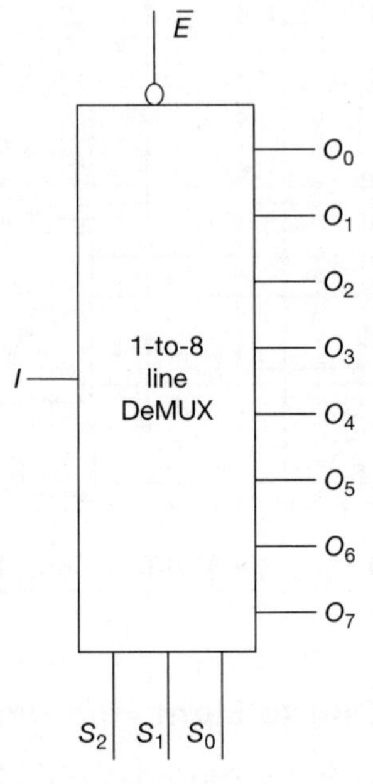

FIGURE 6.56 | Block diagram of 1-to-8 line DeMUX

This logic is the minterms ANDed along with enable and input signal and is given below:

$$O_0 = \overline{E}\overline{S}_2\overline{S}_1\overline{S}_0I \quad (6.79)$$

$$O_1 = \overline{E}\overline{S}_2\overline{S}_1S_0I \quad (6.80)$$

$$O_2 = \overline{E}\overline{S}_2S_1\overline{S}_0I \quad (6.81)$$

$$O_3 = \overline{E}\overline{S}_2S_1S_0I \quad (6.82)$$

$$O_4 = \overline{E}S_2\overline{S}_1\overline{S}_0I \quad (6.83)$$

$$O_5 = \overline{E}S_2\overline{S}_1S_0I \quad (6.84)$$

$$O_6 = \overline{E}S_2S_1\overline{S}_0I \quad (6.85)$$

$$O_7 = \overline{E}S_2S_1S_0I \quad (6.86)$$

6.5.4 | Cascading of Demultiplexers

Number of similar demultiplexers are cascaded (combined together) to construct/design the required higher rating demultiplexer by selecting the demultiplexer through its enable signal. The procedure is outlined below.

- Compute the number of demultiplexers required. Required number is arranged in the different levels.
  - Let n is output lines of demultiplexer to be designed and m is output lines of the available demultiplexer.
  - Number of demultiplexers required in Level 1:

$$n_1 = \frac{n(\text{Output lines of demultiplexer to be designed})}{m(\text{Output lines of available demultiplexer})}$$

- If $n_i \geq m$ then number of demultiplexer required in Level $l + 1$:
- $n_{l+1} = \dfrac{n_1(\text{Number of demultiplexer in Level l})}{m(\text{output lines of available demultiplexer})} \quad (l = 1,2,3,\ldots.)$

- Selection lines of demultiplexer are connected commonly. Level 1 has most significant selection lines. Level 2 will have next lower significant selection lines and so on.
- Input lines of demultiplexer are connected commonly.
- Output lines of lth level demultiplexers are connected to input signal of $(l - 1)^{th}$ level of demultiplexers to be routed to output.

EXAMPLE 6.20 Construct 1-to-4 line demultiplexer using 1-to-2 line demultiplexer.

SOLUTION Let $n = 4$ (Output lines of DeMUX to be constructed) and $m = 2$ (Output lines of available DeMUX)

Number of DeMUX required in Level I: $n_1 = \dfrac{4}{2} = 2$

Number of DeMUX required in Level II: $n_2 = \dfrac{2}{2} = 1$

3 (2 + 1) 1-to-2 line DeMUX are required and arranged in two levels.

Most significant selection line S_1 is connected to DeMUX-I of level II. Least significant selection line S_0 is given to DeMUX numbered II and III of level I. Outputs of DeMUX-I of level II are connected to input lines of DeMUX numbered II and III of Level I. Circuit diagram is shown in Figure 6.57. Table 6.31 explains the working of 1-to-4 line DeMUX after cascading.

Let selection line, S_1 is set to 1 for DeMUX-I.

Input, I of DeMUX-I is sent to output line, O_{11} that is input I_3 for DeMUX-III

Let selection line S_0 is set to 1 for DeMUX numbered II and III.

Input, I_3 of DeMUX-III is sent to output line, O_3.

When selection lines, S_1 and S_0 are 1 and 1, respectively, I is sent to O_3

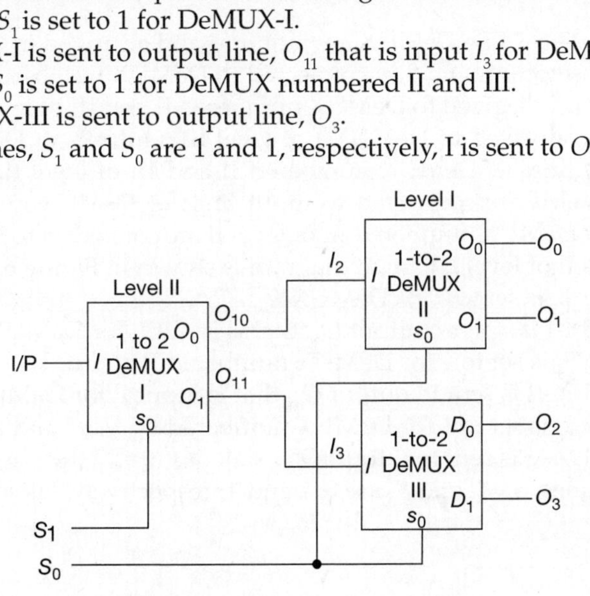

FIGURE 6.57 | 1-to-4 line DeMUX

TABLE 6.31 | Operation of expanded DeMUX

Level II			Level I			
DeMUX	Selection Line, S_1	Output O	DeMUX	Input	Selection Line, S_0	Output
I	0	O_{10}	II	I_2	0	O_0
					1	O_1
	1	O_{11}	III	I_3	0	O_2
					1	O_3

EXAMPLE 6.21 Construct 1-to-8 line demultiplexer using 1-to-2 line demultiplexer.

SOLUTION Let $n = 8$ (Output lines of DeMUX to be constructed) and $m = 2$ (Output lines of available DeMUX)

Number of DeMUX required in Level I: $n_1 = \frac{8}{2} = 4$

Number of DeMUX required in Level II: $n_2 = \frac{4}{2} = 2$

Number of DeMUX required in Level III: $n_3 = \frac{2}{2} = 1$

Seven (4 + 2 + 1) 1-to-2 line DeMUX are required and arranged in three levels.

Least significant selection line S_0 is connected to DeMUX numbered IV, V, VI and VII of level I. Selection line S_1 is given to DeMUX numbered II and III of level II. Most significant selection line S_2 is connected to DeMUX-I of level III. Outputs of DeMUX-I of level I are connected to input lines of DeMUX numbered II and III of level II. Outputs of DeMUX numbered II of level II are connected to input lines of DeMUX numbered IV and V of level III. Outputs of DeMUX numbered III of level II are connected to input lines of DeMUX numbered VI and VII of level I. Circuit diagram is shown in Figure 6.58.

Let selection line, S_2 is set to 0 for DeMUX-I.
Input, I of DeMUX-I is sent to output O_{20} that is input I_2 for DeMUX-II
Let selection line S_1 is set to 1 for DeMUX numbered II and III.
Input, I_2 of DeMUX-II is sent to output O_{11} that is input I_5 for DeMUX-V
Let selection line S_0 is set to 1 for DeMUX numbered IV, V, VI and III.
Input, I_5 of DeMUX-V is sent to output O_3.
When selection lines, S_2, S_1 and S_0 are 0, 1 and 1, respectively, I is sent to O_3.

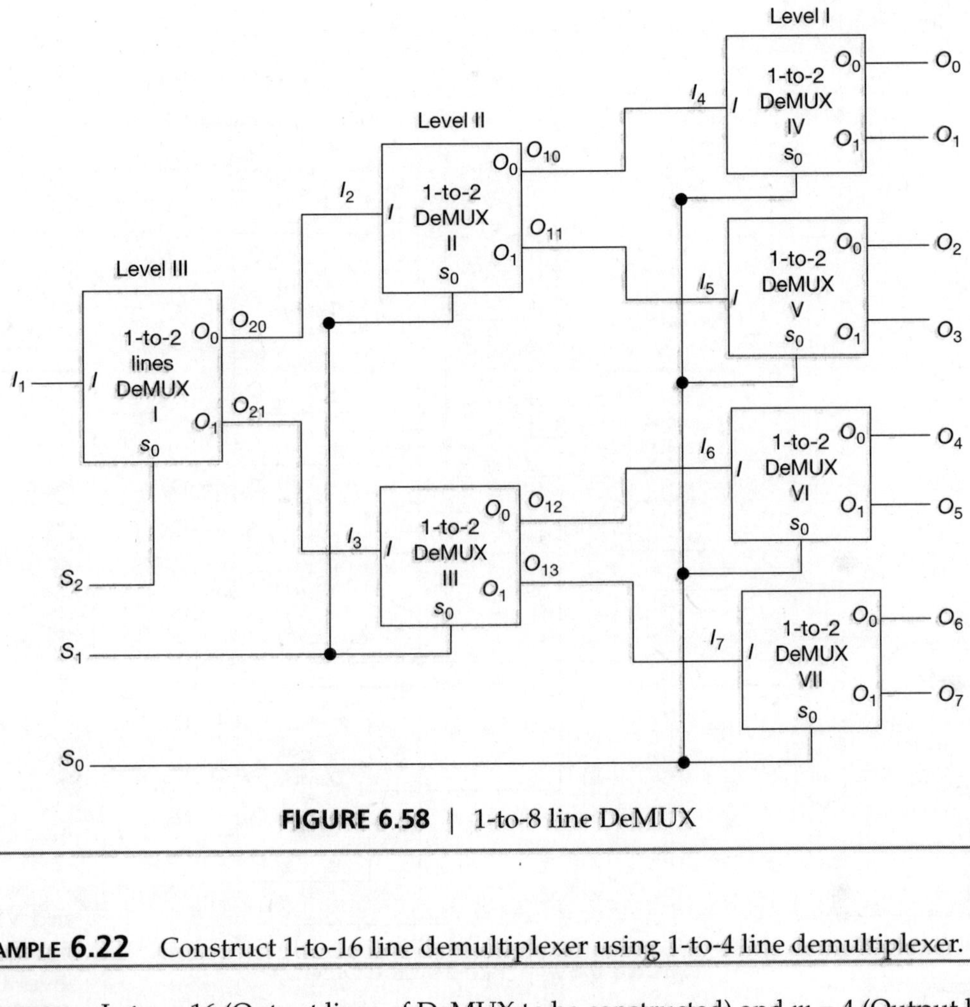

FIGURE 6.58 | 1-to-8 line DeMUX

EXAMPLE 6.22 Construct 1-to-16 line demultiplexer using 1-to-4 line demultiplexer.

SOLUTION Let $n = 16$ (Output lines of DeMUX to be constructed) and $m = 4$ (Output lines of available DeMUX)

Number of DeMUX required in Level I: $n_1 = \dfrac{16}{4} = 4$

Number of DeMUX required in Level II: $n_2 = \dfrac{4}{4} = 1$

Five (4 + 1) 1-to-4 line DeMUX are required and arranged in two levels.

Most significant selection lines S_3 and S_2 are connected to DeMUX-I of level II. Least significant selection lines S_1 and S_0 are given to DeMUX numbered II, III, IV and V of level I. Outputs of DeMUX-I of level II are connected to input lines of DeMUX numbered II, III, IV and V of Level I. Circuit diagram is shown in Figure 6.59.

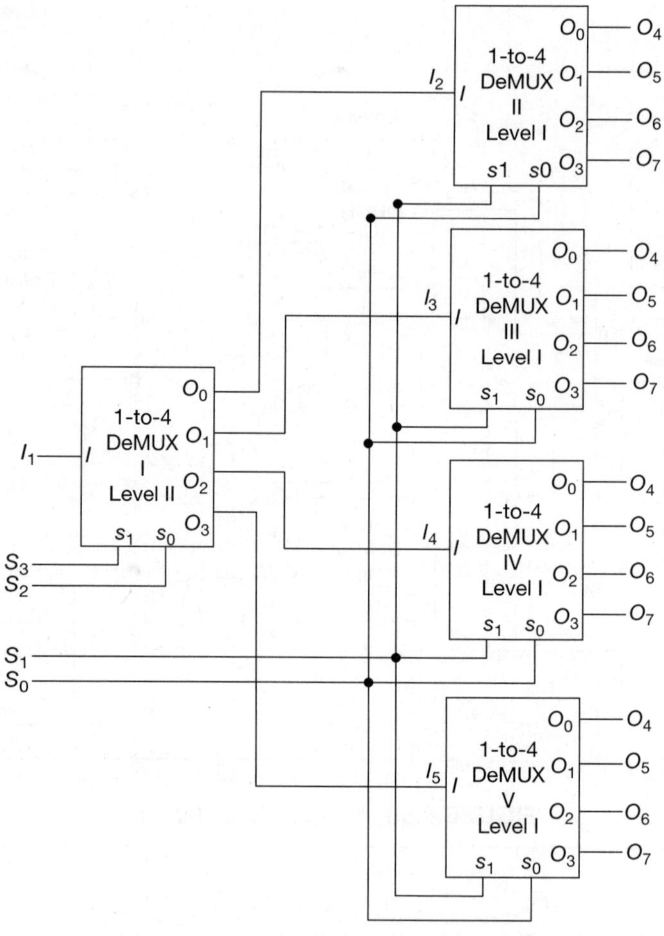

FIGURE 6.59 | 1-to-16 Line DeMUX

Let selection line, S_3 and S_2 is set to 0 and 1, respectively, to DeMUX-I.
Input, I_1 of DeMUX-I is sent to output O_{11} that becomes input I_3 for DeMUX-III
Let selection line $S_1$and S_0 is set to 1 and 0 for DeMUX numbered II, III, IV and V.
Input, I_3 of DeMUX-II is sent to third line of output O_6.
When selection lines, S_3, S_2, S_1 and S_0 are 0, 1, 1 and 0, respectively, I is sent to O_6.

6.5.5 | Cascading of Demultiplexers using Enable

Number of similar demultiplexers (DeMUX) are cascaded (combined together) to construct the required demultiplexer by selecting the available demultiplexers through their enable signals. Enable signal is used to expand the demultiplexers in various rows and single column. Decoder is used to select the demultiplexers. Decoder input and DeMUX selection lines are considered as selection lines. Decoder input acts most significant selection lines. Decoder output is used to activate the enable signal of deMUX.

EXAMPLE 6.23 Construct 1-to-8 line multiplexer using 1-to-2 line multiplexer with enable lines.

SOLUTION Let n = 8 (Output lines of DeMUX to be constructed) and m = 4 (Output lines of available DeMUX)

Number of DeMUX required in Level I: $n_1 = \frac{8}{4} = 2$

n_1 is not greater than m, so one level is possible. Two MUX are selected using enable using 1-to-2 line decoder. Higher significant input line S_2 selects the two DeMUX arranged in single column using 1-to-2 line decoder. Inverter acts as 1-to-2 line decoder. In other words S_2 line enables the MUX-I and-II. Selection input lines S_1 and S_0 are connected to two DeMUX and act as least significant selection lines. Circuit diagram is shown in Figure 6.60.

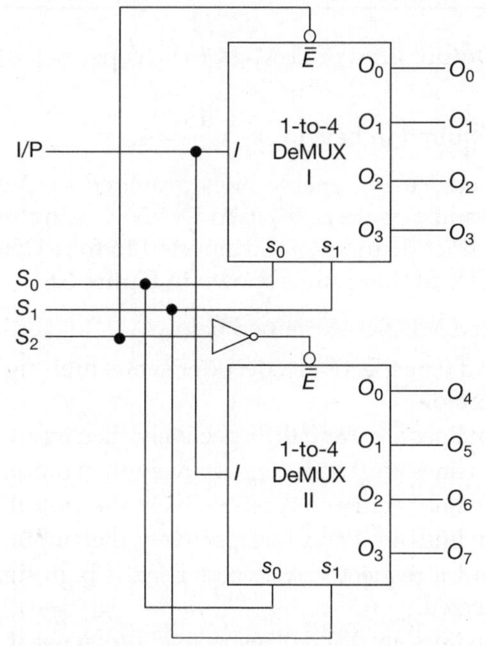

FIGURE 6.60 | 1-to-8 line DeMux

Operation of the expanded DeMUX is explained below.

When S_2 is 0, DeMUX-I is selected and

S_1 and S_0 selection lines are 0 and 0, respectively, then input, I of DeMUX-I is sent to output line, O_0.

S_1 and S_0 selection lines are 0 and 1, respectively, then input, I of DeMUX-I is sent to output line, O_1.

S_1 and S_0 selection lines are 1 and 0, respectively, then input, I of DeMUX-I is sent to output line, O_2.

S_1 and S_0 selection lines are 1 and 1, respectively, then input, I of DeMUX-I is sent to output line, O_3.

When S_2 is 1, DeMUX-II is selected, and
S_1 and S_0 selection lines are 0 and 0, respectively, then input, I of DeMUX-II is sent to output line, O_4.
S_1 and S_0 selection lines are 0 and 1, respectively, then input, I of DeMUX-II is sent to output line, O_5.
S_1 and S_0 selection lines are 1 and 0, respectively, then input, I of DeMUX-II is sent to output line, O_6.
S_1 and S_0 selection lines are 1 and 1, respectively then input, I of DeMUX-II is sent to output line, O_7.

EXAMPLE 6.24 Construct 1-to-16 line multiplexer using 1-to-4 line multiplexer and 2-to-4 line decoder.

SOLUTION Let $n = 16$ (Output lines of DeMUX to be constructed) and $m = 4$ (Output lines of available DeMUX)

Number of DeMUX required in Level I: $n_1 = \frac{16}{4} = 4$

Four DeMUX are selected using enable lines provided by 2-to-4 line decoder. Higher significant input line S_3 and S_2 selects the two DeMUX using output lines of 2-to-4 line decoder. Selection input lines S_1 and S_0 are connected to four DeMUX and act as least significant selection lines. Circuit diagram is shown in Figure 6.61.

Operation of the expanded MUX is explained below:

When S_3 and S_2 are 0 and 0, respectively, decoder sends high signal on D_0 line of decoder. DeMUX-I is selected.
S_1 and S_0 selection lines are 0 and 0, respectively, then input, I is sent to output line, O_0.
S_1 and S_0 selection lines are 0 and 1, respectively, then input, I is sent to output line, O_1.
S_1 and S_0 selection lines are 1 and 0, respectively, then input, I is sent to output line, O_2.
S_1 and S_0 selection lines are 1 and 1, respectively, then input, I is sent to output line, O_3.

When S_3 and S_2 are 0 and 1, respectively, decoder sends high signal on D_1 line of decoder. DeMUX-II is selected.
S_1 and S_0 selection lines are 0 and 0, respectively, then input, I is sent to output line, O_4.
S_1 and S_0 selection lines are 0 and 1, respectively, then input, I is sent to output line, O_5.
S_1 and S_0 selection lines are 1 and 0, respectively, then input, I is sent to output line, O_6.
S_1 and S_0 selection lines are 1 and 1, respectively, then input, I is sent to output line, O_7.

When S_3 and S_2 are 1 and 0, respectively, decoder sends high signal on D_2 line of decoder. DeMUX-III is selected.
S_1 and S_0 selection lines are 0 and 0, respectively, then input, I is sent to output line, O_8.
S_1 and S_0 selection lines are 0 and 1, respectively, then input, I is sent to output line, O_9.
S_1 and S_0 selection lines are 1 and 0, respectively, then input, I is sent to output line, O_{10}.
S_1 and S_0 selection lines are 1 and 1, respectively, then input, I is sent to output line, O_{11}.

When S_3 and S_2 are 1 and 1, respectively, decoder sends high signal on D_3 line of decoder. DeMUX-IV is selected.
S_1 and S_0 selection lines are 0 and 0, respectively, then input, I is sent to output line, O_{12}.
S_1 and S_0 selection lines are 0 and 1, respectively, then input, I is sent to output line, O_{13}.

S_1 and S_0 selection lines are 1 and 0, respectively, then input, I is sent to output line, O_{14}.
S_1 and S_0 selection lines are 1 and 1, respectively, then input, I is sent to output line, O_{15}.

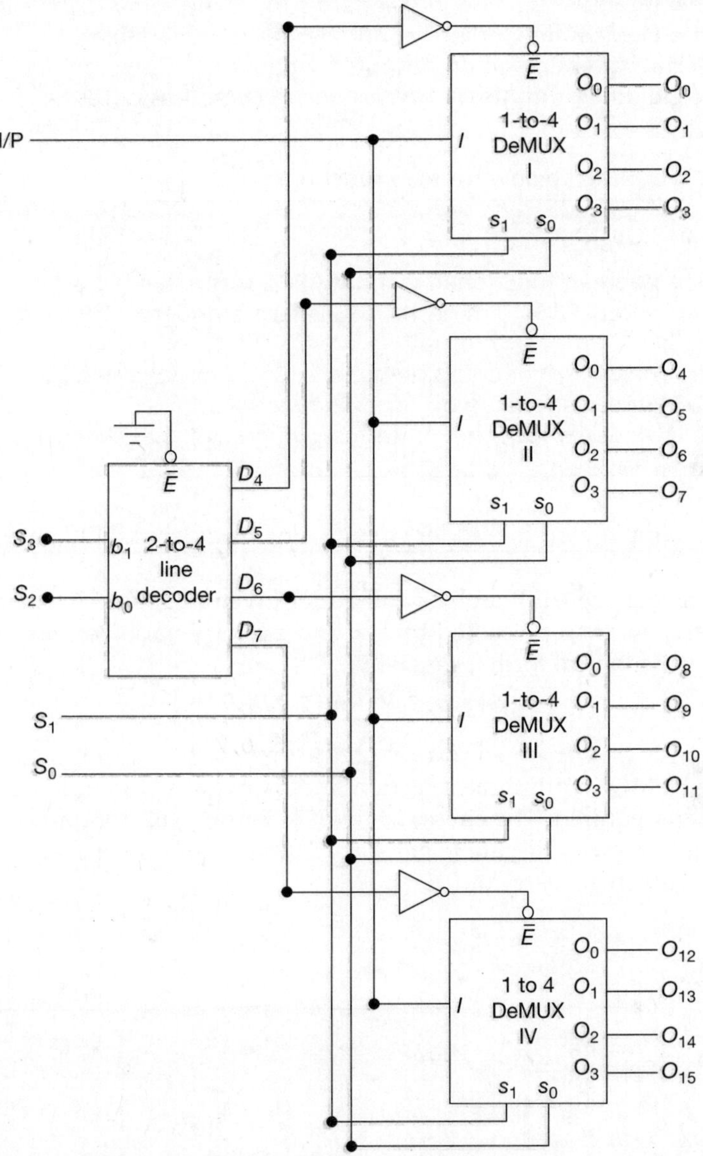

FIGURE 6.61 | 1-to-16 DeMUX using 2-to-4 line decoders and 1-to-4 DeMUX

6.5.6 | Combinational Circuit using Demultiplexer

A combinational circuit can be implemented using DeMUX. The procedure is outlined below for SOP functions:

SOP function is given:

1. Express the Boolean function in standard sum of minterms (SOP) form.
2. Choose the DeMUX depending on number of input variables.
3. Select OR-gates depending on outputs functions.
4. Input the OR gates with the minterms giving Logic 1 as output.

The procedure is outlined below for POS functions:

POS function is given:

1. Express the Boolean function in standard POS form.
2. Convert standard POS form to its equivalent standard SOP form by taking its complement.
3. Choose the DeMUX depending on number of input variables.
4. Select NOR-gates depending on outputs functions.
5. Input the NOR gates with the minterms giving Logic 1 as output.

EXAMPLE 6.25 Implement full-adder circuit using decoder and OR gates.

SOLUTION Consider the truth table of full-adder given in Table 5.6 (Chapter 5). Augend, addend and carry in are represented by x, y and z binary variables, respectively. Sum-of-product (SOP) in canonical form is given by

$$S(x,y,z)=\sum m(1,2,4,7) \tag{6.87}$$

$$C(x,y,z)=\sum m(3,5,6,7) \tag{6.88}$$

A 1-to-8 line DeMUX with three selection lines is required to implement the full-adder. Two OR gates are required. The circuit is given in Figure 6.62. Output of minterms 1, 2, 4 and 7 is given to 4-inputs OR-gate to get sum, S. Output of minterms 3, 5, 6 and 7 is given to 4-inputs OR-gate to get carry out, C.

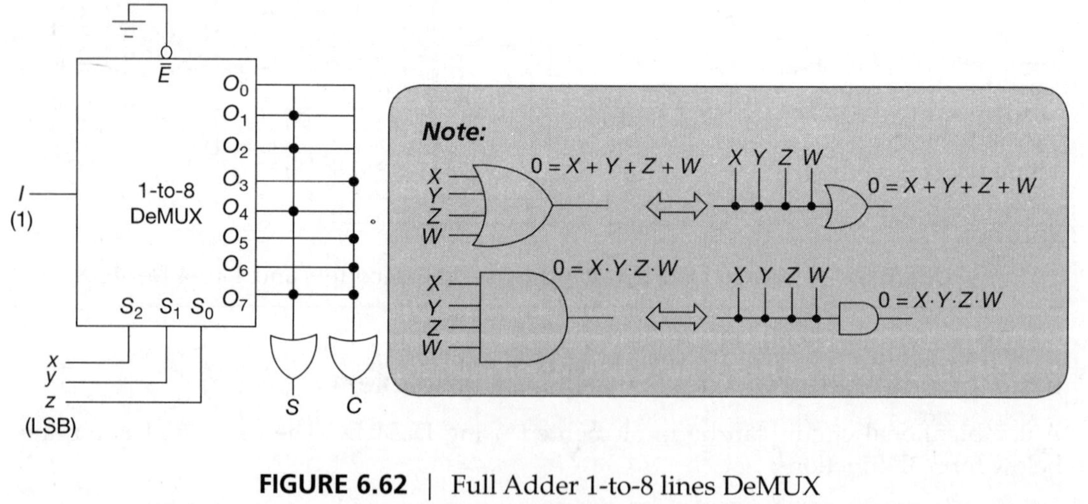

FIGURE 6.62 | Full Adder 1-to-8 lines DeMUX

EXAMPLE 6.26 Implement full-subtractor circuit using decoder and OR gates.

SOLUTION Consider the truth table of full-adder given in Table 5.8 (Chapter 5). Minuend, subtrahend and previous borrow are represented by x, y and z binary variables, respectively. Sum-of-product (SOP) in canonical form is given by

$$D(x,y,z) = \sum m(1,2,4,7) \quad (6.89)$$

$$B(x,y,z) = \sum m(1,2,3,7) \quad (6.90)$$

A 3-to-8 line decoder is required to implement the full-subtractor. Two OR gates are required. The circuit is given in Figure 6.63. Output of minterms 1, 2, 4 and 7 is given to 4-imputs OR-gate to get difference, D. Output of minterms 1, 2, 3 and 7 is given to 4-inputs OR-gate to get borrow, B.

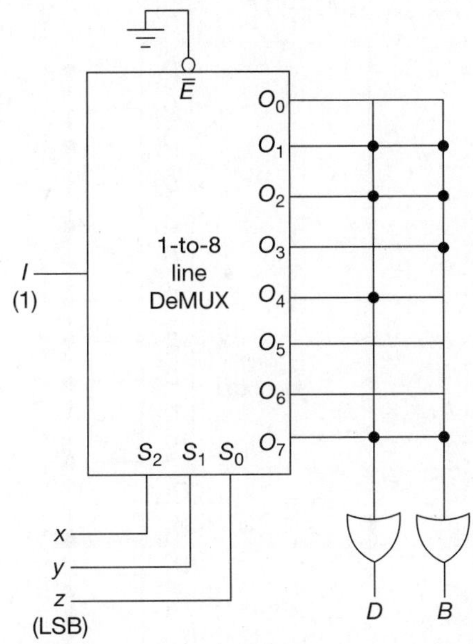

FIGURE 6.63 | Full-subtractor using 1-to-8 line DeMUX

EXAMPLE 6.27 A combinational circuit is given by $f(w,x,y,z) = xy + wxy + \overline{z}$. Design a circuit using decoder and OR gates.

SOLUTION The function is having four w, x, y and z binary variables, respectively. Sum-of-product (SOP) in canonical form is given by

$$f(w,x,y,z) = xy + wxy + \overline{z}$$

$$xy = \text{X11X} \qquad \text{(Let } x = 1 \text{ and } y = 1\text{)}$$

$$= 0110 + 0111 + 1110 + 1111$$

$$= m_6 + m_7 + m_{14} + m_{15}$$

$$wxy = 111X \qquad \text{(Let } w = 1, x = 1 \text{ and } y = 1\text{)}$$

$$= 1110 + 1111$$

$$= m_{14} + m_{15}$$

$$\overline{z} = XXX0 \qquad \text{(Let } \overline{z} = 0\text{)}$$

$$= 0000 + 0010 + 0100 + 0110 + 1000 + 1010 + 1100 + 1110$$

$$= m_0 + m_2 + m_4 + m_6 + m_8 + m_{10} + m_{12} + m_{14}$$

$$f(w,x,y,z) = \sum m(0,2,4,6,7,8,10,12,14,15)$$

The circuit is given in Figure 6.64.

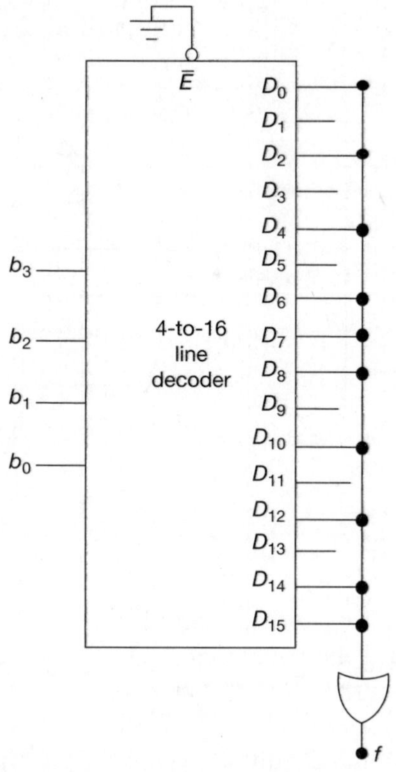

FIGURE 6.64 | Logic implementation using 4-to-16 lines decoder

EXAMPLE 6.28 A combinational circuit is given by $f(w,x,y,z) = xy + \overline{z}$. Design a circuit using MUX.

SOLUTION The function is having four w, x, y and z binary variables, respectively. $2^{(4-1)}$-to-1 line MUX (i.e. 8-to-1 line MUX) is required to implement the given Boolean logic. Sum-of-product (SOP) in canonical form is given by

$$f(w,x,y,z) = xy + wxy + \overline{z}$$

$$xy = \text{X11X} \qquad \text{(Let } x = 1 \text{ and } y = 1)$$

$$= 0110 + 0111 + 1110 + 1111$$

$$= m_6 + m_7 + m_{14} + m_{15}$$

$$\overline{z} = \text{XXX0} \qquad (\text{Let } \overline{z} = 0)$$

$$= 0000 + 0010 + 0100 + 0110 + 1000 + 1010 + 1100 + 1110$$

$$= m_0 + m_2 + m_4 + m_6 + m_8 + m_{10} + m_{12} + m_{14}$$

$$f(w,x,y,z) = \sum m(0,2,4,6,7,8,10,12,14,15)$$

Table 6.32 gives the input to MUX using minterms. Encircle minterms 0, 2, 4, 6, 7, 8, 10, 12, 14 and 15.

TABLE 6.32 | MUX input table of minterms for Example 6.28

w \ xyz		000	001	010	011	100	101	110	111
		$\overline{y}\overline{z}\overline{w}$	$\overline{y}zw$	$\overline{y}z\overline{w}$	$\overline{y}zw$	$y\overline{z}\overline{w}$	$y\overline{z}w$	$yz\overline{w}$	yzw
		I_0	I_1	I_2	I_3	I_4	I_5	I_6	I_7
0	$\overline{x}$	(0)	1	(2)	3	(4)	5	(6)	(7)
1	x	(8)	9	(10)	11	(12)	13	(14)	(15)
Input to MUX		1	0	1	0	1	0	1	0

Circuit diagram is given in Figure 6.65.

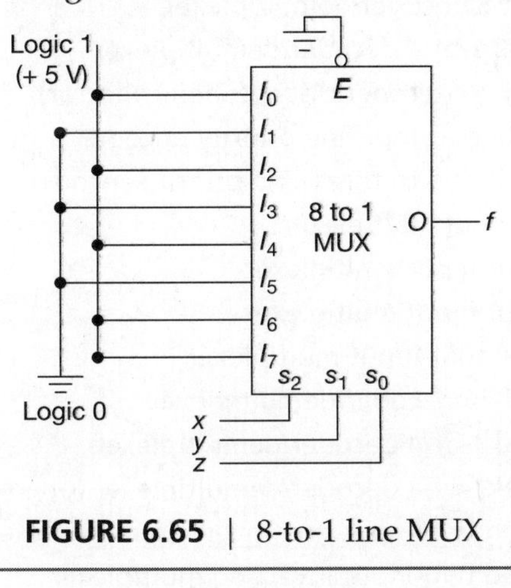

FIGURE 6.65 | 8-to-1 line MUX

6.6 | LIST OF IC's

A list of commonly used IC type numbers used as multiplexers, encoders, demultiplexers and decoders belonging to CMOS, TTL and ECL logic families is given in Tables 6.33–6.35. Information regarding the pin connection diagram, truth table, etc., in respect of these IC numbers is given in the handbook or companion website.

TABLE 6.33 | IC of CMOS logic family

IC Number	Function of IC
4019	Quad two-input multiplexer
4028	1-of-10 decoder
40147	10-line to four-line BCD priority encoder
4511	BCD to seven-segment latch/decoder/driver
4512	Eight-input three-state multiplexer
4514	1-of-16 decoder/demultiplexer with input latch
4515	1-of-16 decoder/demultiplexer with input latch
4532	Eight-line to three-line priority encoder
4539	Dual four-input multiplexer
4543	BCD to seven-segment latch/decoder/driver for LCD displays
4555	Dual 1-of-4 decoder/demultiplexers
4556	Dual 1-of-4 decoder/demultiplexers
4723	Dual four-bit addressable latch
4724	Eight-bit addressable latch

TABLE 6.34 | IC of TTL logic family

IC Number	Function of IC
7442	1-of-10 decoder
74138	1-of-8 decoder/demultiplexer
74139	Dual 1-of-4 decoder/demultiplexer
74145	1-of-10 decoder/driver (open collector)
74147	10-line to four-line priority encoder
74148	Eight-line to three-line priority encoder
74150	16-input multiplexer
74151	Eight-input multiplexer
74152	Eight-input multiplexer
74153	Dual four-input multiplexer
74154	4-of-16 decoder/demultiplexer
74155	Dual 1-of-4 decoder/demultiplexer
74156	Dual 1-of-4 decoder/demultiplexer (open collector)
74157	Quad two-input non-inverting multiplexer
74158	Quad two-input inverting multiplexer
74247	BCD to seven-segment decoder/driver (open collector)
74248	BCD to seven-segment decoder/driver with Pull-ups
74251	Eight-input three-state multiplexer
74253	Dual four-input three-state multiplexer
74256	Dual four-bit addressable latch

(Continued)

TABLE 6.34 | (Continued)

IC Number	Function of IC
74257	Quad two-input non-inverting three-state multiplexer
74258	Quad two-input inverting three-state multiplexer
74259	Eight-bit addressable latch
74298	Dual two-input multiplexer with output latches
74348	Eight-line to three-line priority encoder (three-state)
74353	Dual four-input multiplexer
74398	Quad two-input multiplexer with output register
74399	Quad two-input multiplexer with output register

TABLE 6.35 | IC of ECL logic family

IC Number	Function of IC
10132	Dual two-input multiplexer with latch and common reset
10134	Dual multiplexer with latch
10158	Quad two-input multiplexer (non-inverting)
10159	Quad two-input multiplexer (inverting)
10161	3-to-8 line decoder (LOW)
10162	3-to-8 line decoder (HIGH)
10164	Eight-line multiplexer
10165	Eight-input priority encoder
10171	Dual 2-to-4 line decoder (LOW)
10172	Dual 2-to-4 line decoder (HIGH)
10173	Quad two-input multiplexer/latch
10174	Dual 4-to-1 multiplexer

SUMMARY

- **Decoder:** A decoder is a combinational circuit that converts n-bit binary data to a maximum of 2^n unique output lines. The unique lines are numbered from 0 to $2^n - 1$. If there are some unused or 'don't care' combinations in the n-bit binary data (code), then there will be lesser than 2^n output lines. The decoder is called n-to-m line decoder, where $m \leq 2^n$.
- **Encoder:** An encoder is a combinational logic function that has 2^n (or fewer) input lines and n output lines. The n output lines generate the binary code for the possible 2^n input lines.
- **Multiplexer:** A multiplexer is a combinational circuit with more than one input lines, one output line and one or more selection lines.
- A multiplexer selects binary information present on any one of the input lines, depending upon the logic status of the selection inputs, and sends it to the output line.
- A multiplexer has n selection lines, then the number of maximum possible input lines is 2^n and such multiplexer is specified as a 2^n-to-1 multiplexer (or $2^n \times 1$ multiplexer).

- A multiplexer or MUX is also called a data selector.
- **Demultiplexer:** A demultiplexer is a combinational circuit with one input line, two or more output lines and one or more selection lines.
- A demultiplexer, routes binary information present on the input line, to any one of the output lines. The output line that receives the information present on the input line is selected by the bit status of the selection lines.
- A demultiplexer is a combinational logic circuit with an input line, 2^n output lines and n selection lines.
- If there are n selection lines, then the number of maximum possible output lines is 2^n and such demultiplexer is specified as a 1-to-2^n demultiplexer or 1×2^n multiplexer.
- A demultiplexer or DeMUX is also called a data distributor.
- A decoder is a kind of a demultiplexer without the input line.

MULTIPLE CHOICE QUESTIONS

6.1 A multiplexer has
   (a) One datum input line, several data output lines and selection inputs.
   (b) One datum input line, one datum output line and selection inputs.
   (c) Several data input lines, several data output lines and selection inputs.
   (d) Several data input line, one datum output line and selection inputs.

6.2 A demultiplexer has
   (a) One datum input line, several data output lines and selection inputs.
   (b) One datum input line, one datum output line and selection inputs.
   (c) Several data input lines, several data output lines and selection inputs.
   (d) Several data input line, one datum output line and selection inputs.

6.3 Data selectors are basically the same as
   (a) Decoder
   (b) Demultiplexer
   (c) Multiplexer
   (d) Priority encoder

6.4 Data distributers are basically the same as
   (a) Decoder
   (b) Demultiplexer
   (c) Multiplexer
   (d) Priority encoder

6.5 The operation of multiplexer is same as
   (a) OR-AND operation
   (b) AND-OR operation
   (c) OR-NAND operation
   (d) NOR-AND operation

6.6 What is the number of selection lines required in a single input, n-output demultiplexer?
   (a) 2 (b) n
   (c) 2^n (d) $\log_2 n$

6.7 The logic circuit used to change a BCD number into an equivalent decimal number
   (a) Decoder (b) Encoder
   (c) MUX (d) Code

6.8 What is the number of selection lines required for a 64-to-1 line output multiplexer?
   (a) 2 (b) 4
   (c) 5 (d) 6

6.9 ABCD decoder has
(a) Four input lines and 16 output lines
(b) Four selection lines, one input line and 16 output lines
(c) Sixteen input lines and four output lines
(d) Four input lines and ten output lines

6.10 Number of 2-to-4 line decoders needed to design 6-to-64 line decoder.
(a) 3
(b) 16
(c) 21
(d) none of the above

6.11 Size of decoder needed to design 16-to-1 line multiplexer.
(a) 2-to-4 line decoder
(b) 3-to-8 line decoder
(c) 4-to-16 line decoder
(d) 5-to-32 line decoder

6.12 Number of 1-to-4 line demultiplexer needed to design 1-to-256 line demultiplexer.
(a) 64 (b) 85
(c) 128 (d) 256

6.13 Number of 4-to-1 line multiplexer needed to design 128-to-1 line multiplexer.
(a) 32 (b) 42
(c) 43 (d) 64

6.14 The following switching function is to be implemented using decoder

$$f = \sum m(1,2,4,8,14)$$

What is the size of decoder?
(a) 2-to-4 line
(b) 3-to-8 line
(c) 4-to-16 line
(d) 5-to-32 line

6.15 The following switching function is to be implemented using multiplexer

$$f = \sum m(1,2,4,8,14,45)$$

What is the size of multiplexer?
(a) 8-to-1 line
(b) 16-to-1 line
(c) 32-to-1 line
(d) 64-to-1 line

Answers

6.1 (d) 6.2 (a) 6.3 (c) 6.4 (b) 6.5 (b) 6.6 (d) 6.7 (b)
6.8 (d) 6.9 (d) 6.10 (c) 6.11 (c) 6.12 (b) 6.13 (b) 6.14 (c)
6.15 (d)

QUESTIONS

6.1 What is a multiplexer circuit? Briefly describe one or two applications of a multiplexer?
6.2 Is it possible to enhance the capability of an available multiplexer in terms of the number of input lines it can handle by using more than one device? If yes, briefly describe the procedure to do so, with the help of an example.
6.3 What is an encoder? How does a priority encoder differ from a conventional encoder? With the help of a truth table, briefly describe the functioning of a 10-to-4 line priority encoder with active LOW inputs and outputs and priority assigned to the higher-order inputs.
6.4 What is a demultiplexer and how does it differ from a decoder? Can a decoder be used as a demultiplexer? If yes, from where do we get the required input line?

6.5 Briefly describe how decoder is used to implement a given Boolean function? Illustrate the answer with the help of an example.
6.6 Design a BCD decoder.
6.7 Design a hexadecimal to binary encoder.
6.8 Design a binary to octal decoder.
6.9 Design octal to one multiplexer.
6.10 Draw truth tables for the following:
(a) an 8-to-1 multiplexer with active LOW inputs and an active LOW ENABLE input;
(b) a four-line to 16-line decoder with active HIGH inputs and active LOW outputs and an active LOW ENABLE input;
(c) an eight-line to three-line priority encoder with active LOW inputs and outputs and an active LOW ENABLE input.

PROBLEMS

6.1 Implement the following functions by means of the decoder and fundamental gates
(a) $F = \overline{y} + \overline{x}\,\overline{z}$
(b) $F = bc + \overline{a}b\overline{c}$
(c) $F = \prod M(0,1,4,5)$
(d) $F = (\overline{A}+\overline{B}+D)(A+\overline{B}+\overline{C})(\overline{A}+B+\overline{D})(B+\overline{C}+\overline{D})$

6.2 Implement the following functions by means of the demultiplexer and fundamental gates
(a) $F = \overline{y} + \overline{x}\,\overline{z}$
(b) $F = bd + \overline{a}b\overline{c}$
(c) $F = \prod M(0,1,3,5)$
(d) $F = (\overline{A}+\overline{B}+D)(A+\overline{B}+\overline{C})(\overline{A}+B+\overline{D})(B+\overline{C}+\overline{D})$

6.3 Implement the following functions by means of the multiplexer.
(a) $F = \overline{y} + \overline{x}\,\overline{z}$
(b) $F = bd + \overline{a}b\overline{c}$
(c) $F = \prod M(0,1,3,5)$
(d) $F = (\overline{A}+\overline{B}+D)(A+\overline{B}+\overline{C})(\overline{A}+B+\overline{D})(B+\overline{C}+\overline{D})$

6.4 Design a 5-to-32 line decoder with 3-to-8 line decoder and one 2-to-4 line decoder.
6.5 Design a 64-to-1 line multiplexer using 4-to-1 line multiplexer.
6.6 Design a 32-to-1 MUX using 16-to-1 MUX and 2-to-1 MUX.
6.7 Design a 1-to-64 line demultiplexer using 1-to-4 lines demultiplexer.
6.8 Design a 1-to-32 DeMUX using 1-to-16 DeMUX and 1-to-2 DeMUX.
6.9 Design a circuit to find 9's complement using MUX.
6.10 Design a circuit to convert 4-bit gray code to 4-bit binary code using demultiplexer and OR gates.

7

Flip-Flops

CHAPTER OBJECTIVES

The main goal of this chapter is to impart knowledge about the flip-flops being memory element of sequential circuits. Readers will be able to discuss the following aspects in this chapter:

- Latches and flip-flops
- SR flip-flop
- Gated flip-flop
- Edge-triggered D flip-flop
- Edge-triggered JK flip-flop
- Edge-triggered T flip-flop
- Operating characteristics of flip-flop
- Master-slave flip-flop
- Conversion of flip-flop from SR to JK, SR to D, JK to T, D to JK and T to D-flip-flop

7.1 | INTRODUCTION

Logic circuits are classified into combinational logic circuits and sequential logic circuits. A combinational circuit consists of logic gates whose outputs at any time are determined directly from present inputs without considering the previous inputs. The combinational logic circuits consist of input variables, logic gates and output variables. The combinational logic circuit (Figure 7.1) has no memory to save the previous states. In some circuits, some of outputs are explicitly or implicitly delayed and is connected as inputs. Such circuits have a feedback from the output to the inputs. The output depends not only on the present state but also on the past history of the inputs. So it has memory.

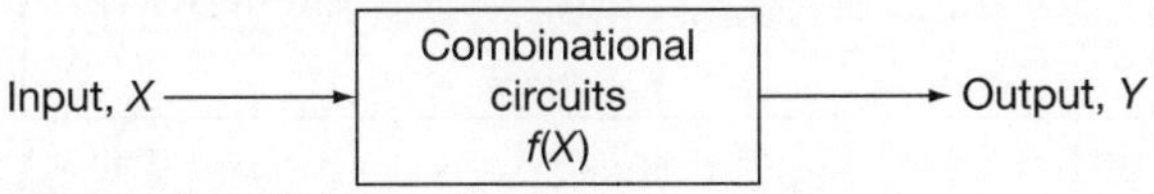

FIGURE 7.1 | Combinational circuit

$$Y = f(X) \tag{7.1}$$

where X is the input variable
Y is the output variable

Excitation

$$E = f(S(t), X) \tag{7.2a}$$

Next state

$$S(t+1) = f(E, S(t)) \tag{7.2b}$$

Output

$$Y = f(X, S(t)) \tag{7.2c}$$

where X is the input variable
E is excitation for the memory device
$S(t)$ is the previous output state (at time t) of memory device
Y is the output variable

Input variables: Variables fed to sequential circuit are known as input variables.

Output variables: Variables given out by the sequential circuit are called as output variables.

State variables: The output of memory (flip-flop) defines the state of the sequential circuit. State variables are outputs of flip-flop.

Excitation variables: Excitation excites the memory to change its state. So, excitation variables are input to memory device (flip-flop). State variables are functions of excitation variables.

State: The state of sequential circuit is defined by the contents of memory. $S(t)$ is the state before the arrival of clock. $S(t + 1)$ is state after arrival of clock.

Previous state: The status of all state variables at some time, t before the next clock triggering represents condition called present state.

Next state: The status of all state variables at some time, $t + 1$ after the clock triggering represents condition called next state.

State diagram: State diagram is a graphical representation of sequential circuit (Figure 7.2) where individual states are represented by a circle with an identifying symbol located inside the circle.

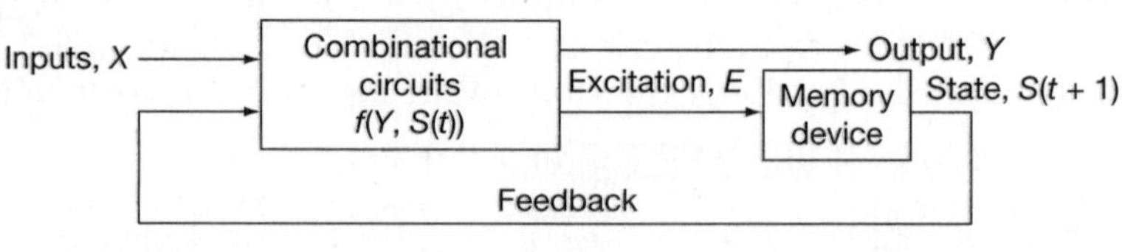

FIGURE 7.2 | Sequential circuit

Sequential circuits are classified into two categories, known as asynchronous and synchronous sequential circuits. A sequential circuit whose behaviour depends on the change of input signal at any time is referred to as asynchronous sequential circuits. A combinational circuit with feedback is termed as as asynchronous sequential circuit. A sequential circuit, whose behaviour is defined from the knowledge of its signal at discrete instants of time, is

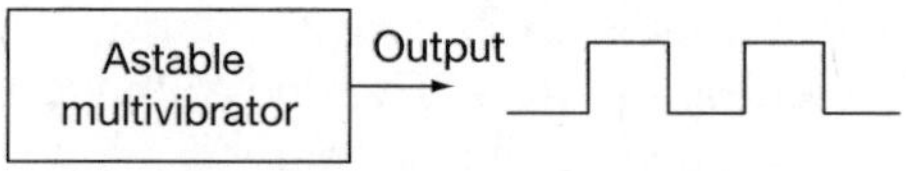

FIGURE 7.3 | Bistable multivibrators

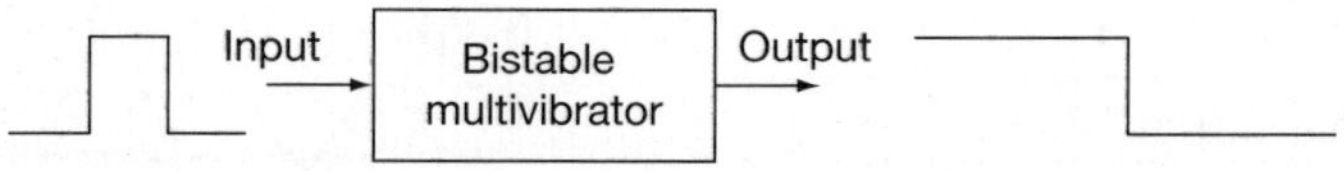

FIGURE 7.4 | Bistable multivibrators

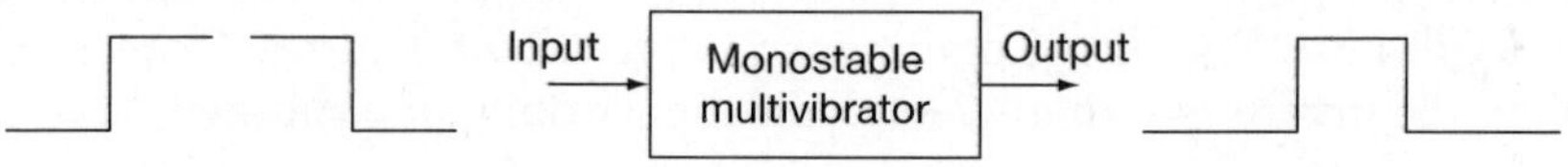

FIGURE 7.5 | Monostable multivibrators

referred as synchronous sequential circuits. The synchronization is achieved by a timing device known as a system clock which generates a train of clock pulses. Synchronous circuits are also known as clocked-sequential circuit.

The basic building block of the sequential logic circuit is the flip-flop circuit. Sequential logic circuits are valuable because of their memory characteristic. Flip-flops are also called latches or bistable multi-vibrators, or binaries. Flip-flops are interconnected to form sequential logic circuits for data storage, timing, counting, and sequencing. A multivibrator is a pulse generator circuit which produces a rectangular-wave output. Multivibrators are classified as astable, bistable, or monostable. An astable multivibrator is also called a free-running multivibrator. The astable multivibrator generates a continuous flow of pulses as depicted in Figure 7.3. A bistable multivibrator is also called a flip-pop. The bistable multivibrator gives one out of two stable states (set or reset). The basic of a bistable multivibrator is depicted in Figure 7.4, where the input pulse triggers a change in output from LOW to HIGH. A monostable multivibrator is also called a one-shot multivibrator. When the one-shot is triggered, as shown in Figure 7.5, the multivibrator generates a single short pulse.

7.2 | BASIC BISTABLE ELEMENT

The circuit that has two stable states is known as bistable circuit. A switch has two stable states that are logic-0 and logic-1 states. The switch is said to have memory since it remains as ON (set) or OFF (clear) until someone changes its position. A Flip-flop is a bistable electronic circuit. So, flip-flop has two stable states. Flip-flop is regarded as memory because its output remains in the same state until someone changes its input. When, the output of flip-flop is 1, it is considered as SET state. When the output of flip-flop is 0, it is considered as RESET state. The flip-flop is sometimes known as latch because it holds or locks in either state.

Two inverters, connected in series with feedback (Figure 7.6), are used to produce two stable states i.e. either logic-0 or logic-1 state. The line connecting the output of inverter-II back to the input of inverter-I is called feedback line.

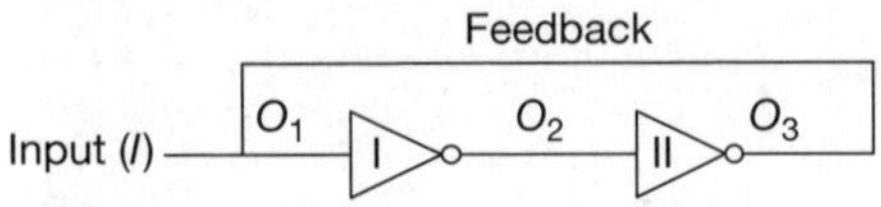

FIGURE 7.6 | Basic latch

TABLE 7.1 | Truth table

I	O_1	O_2	O_3
0	0	1	0
1	1	0	1

Let Input, *I* is 0 then output of inverter-I, O_2 becomes logic-0 (Table 7.1). Output, O_2 acts as input for inverter-II and produces logic-1 state. Since, output, O_3 of inverter-II acts input, O_1, of inverter-I. So, output, O_2 of inverter-I and output, O_3 of inverter-II remains same. Two stable states logic-1 and logic-0 are maintained in the circuit. Figure 7.7 can be redrawn by cross-coupling the inverters. Table 7.2 explains the working of basic latch.

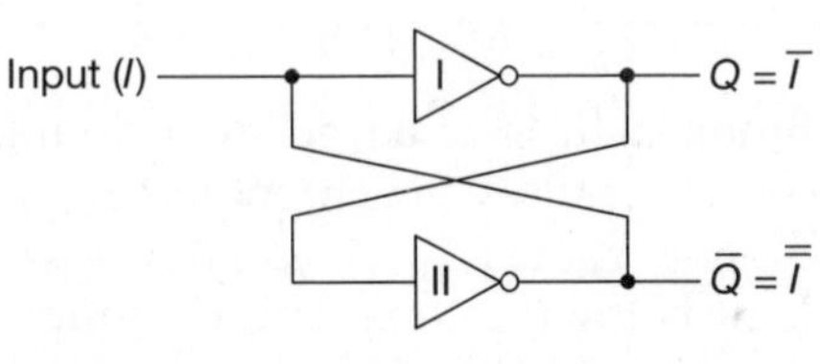

FIGURE 7.7 | Basic latch

TABLE 7.2 | Truth table of latch

Input	Output		Remarks
I	*Q*	$\bar{Q}$	
0	1	0	Set state as Q is in Logic 1 state. It stores binary 1 value.
1	0	1	Reset state as Q is in logic 0 state. It stores binary 0 value.

The main drawback of the latch circuit is that it is difficult to feed the desired binary digit (datum or information). The drawback can be overcome by cross coupling NOR or NAND gates.

7.3 | SR LATCH

SR latch can be designed using universal gates; NOR gates or NAND gates. The proceeding sections explain the design SR latch using universal gates. Latch is also known as transparent because is any change in input is transmitted immediately to the output lines.

7.3.1 | SR Latch using NOR Gates

Two NOR gates are connected in series with feedback (Figure 7.8), are used to construct SR latch. Latch is bi-stable circuit with two stable states (0 and 1). Figure 7.9 is drawn to achieve cross coupling to construct SR latch using NOR gates. The output of SR latch is Q that is output of NOR gate whose one of the inputs is R.

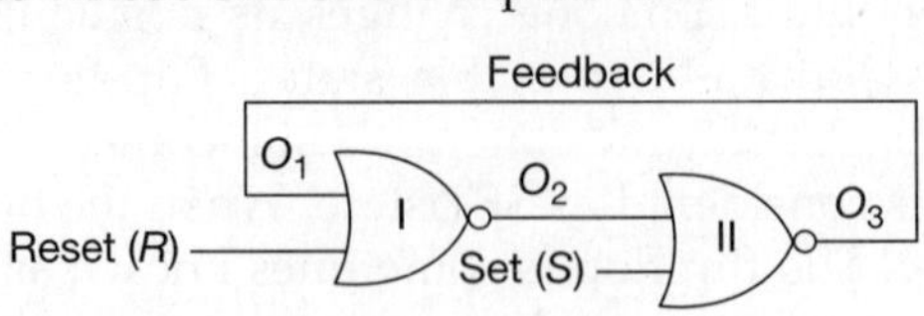

FIGURE 7.8 | SR latch

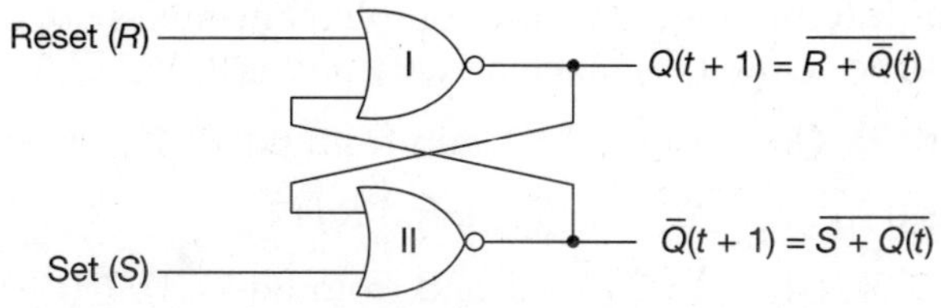

FIGURE 7.9 | SR latch

The output, $Q(t+1)$ of NOR-I gate depends on previous output, $\bar{Q}(t)$ of NOR-II gate and input, R. The Boolean expression for NOR-I gate is given below:

$$Q(t+1) = \overline{R+\bar{Q}(t)} \tag{7.3}$$

The output, $\bar{Q}(t+1)$ of NOR-II gate depends on previous output, $Q(t)$ of NOR-I gate and input, S. The Boolean expression for NOR-II gate is given below:

$$\bar{Q}(t+1) = \overline{S+Q(t)} \tag{7.4}$$

Following Boolean expressions given in Eq. (7.3) and Eq. (7.4), the characteristic truth table is obtained and is given in Table 7.3. Characteristic table gives relation between input and output. Outputs of latch give two stable states following the complement rule of each other.

TABLE 7.3 | Characteristics table of SR latch

Inputs			Outputs		Remarks
$Q(t)$	S	R	$Q(t+1)$	$\bar{Q}(t+1)$	
0	0	0	0	1	Previous state 0
0	0	1	0	1	Reset
0	1	0	1	0	Set
0	1	1	0	0	(Indeterminate)*
1	0	0	1	0	Previous state 1
1	0	1	0	1	Reset
1	1	0	1	0	Set
1	1	1	0	0	(Indeterminate)*

* Complement rule is not followed.

Explanation:

Case I: Let $Q(t) = 0$ so that $\bar{Q}(t) = 1$. S is 0 and R is 0.

From Eq. (7.3), $Q(t+1) = \overline{0+1} = 0$ and from Eq. (7.4), $\bar{Q}(t+1) = \overline{0+0} = 1$

Case II: Let $Q(t)$ is 0 so that $\bar{Q}(t) = 1$. S is 0 and R is 1.

From Eq. (7.3), $Q(t+1) = \overline{1+1} = 0$ and from Eq. (7.4), $\bar{Q}(t+1) = \overline{0+0} = 1$

Case III: Let $Q(t)$ is 0 so that $\bar{Q}(t) = 1$. S is 1 and R is 0.

From Eq. (7.3), $Q(t+1) = \overline{0+1} = 0$ and from Eq. (7.4), $\bar{Q}(t+1) = \overline{1+0} = 0$

During this intermediate state the complement rule is not followed. So, wait for stable state. So, $\bar{Q}(t+1)$ is taken as $\bar{Q}(t)$ and is 0.

From Eq. (7.3), $Q(t+1)=\overline{0+0}=1$ and from Eq. (7.4), $\bar{Q}(t+1)=\overline{1+1}=0$

Case IV: Let $Q(t)$ is 0 so that $\bar{Q}(t)=1$. S is 1 and R is 1.

From Eq. (7.3), $Q(t+1)=\overline{1+1}=0$ and from Eq. (7.4), $\bar{Q}(t+1)=\overline{1+0}=0$

It is considered as indeterminate state because output $Q(t+1)$ and $\bar{Q}(t+1)$ are same.

Case V: Let $Q(t)$ is 1 so that $\bar{Q}(t)=0$. S is 0 and R is 0.

From Eq. (7.3), $Q(t+1)=\overline{0+0}=1$ and from Eq. (7.4), $\bar{Q}(t+1)=\overline{0+1}=0$

Case VI: Let $Q(t)$ is 1 so that $\bar{Q}(t)=0$. S is 0 and R is 1.

From Eq. (7.3), $Q(t+1)=\overline{1+0}=0$ and from Eq. (7.4), $\bar{Q}(t+1)=\overline{0+1}=0$

During this intermediate state the complement rule is not followed. So, wait for state to stabilize. Hence, $Q(t+1)$ is taken as $Q(t)$ and is 0.

From Eq. (7.3), $Q(t+1)=\overline{1+0}=0$ and from Eq. (7.4), $\bar{Q}(t+1)=\overline{0+0}=1$

Case VII: Let $Q(t)$ is 1 so that $\bar{Q}(t)=0$. S is 1 and R is 0.

From Eq. (7.3), $Q(t+1)=\overline{0+0}=1$

From Eq. (7.4), $\bar{Q}(t+1)=\overline{1+1}=0$

Case VIII: Let $Q(t)$ is 1 so that $\bar{Q}(t)=0$. S is 1 and R is 1.

From Eq. (7.3), $Q(t+1)=\overline{1+1}=0$ and from Eq. (7.4), $\bar{Q}(t+1)=\overline{1+0}=0$

It is considered as indeterminate state because outputs $Q(t+1)$ and $\bar{Q}(t+1)$ are same.

7.3.2 | Gated SR Latch using NOR Gates

The output of latch changes its state as and when the inputs of latch are changed. So, the latch is called asynchronous latch. The change of output state of latch is controlled with an additional signal known as clock along with the change of inputs of latch. So, the change of output state of latch is synchronized with the clock. Such, latch is known as gated or clocked latch. High level clock is considered as active clock for the change. Figure 7.10 is drawn to construct SR latch using NOR gates when clock pulse (C_p) controls the inputs with AND logic. The output of NOR is Q that has input, R having AND operation with clock pulse.

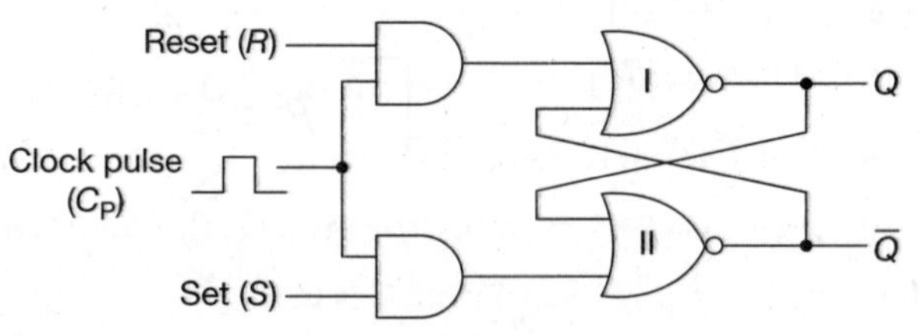

FIGURE 7.10 | SR flip-flop with clock

The output $Q(t+1)$ of NOR-I gate depends on previous output, $\bar{Q}(t)$ of NOR-II gate, input, R and active part of clock pulse, (C_p). The Boolean expression of NOR-I gate is given below.

$$Q(t+1) = \overline{RC_P + \bar{Q}(t)} \tag{7.5}$$

The output $\bar{Q}(t+1)$ of NOR-II gate depends on previous output, $Q(t)$ of NOR-I gate input, R and active part of clock pulse, (C_p). The Boolean expression of NOR-II gate is given below.

$$\bar{Q}(t+1) = \overline{SC_P + Q(t)} \tag{7.6}$$

The truth table of gated SR latch is given in Table 7.4.

TABLE 7.4 | Characteristic table of gated SR latch

Inputs			Outputs	Remarks
S **(Set)**	***R*** **(Reset)**	C_P **(Clock Pulse)**	$Q(t+1)$	
0	0	1	$Q(t)$	Previous state or hold
0	1	1	0	Set
1	0	1	1	Clear (Reset)
1	1	1	?	Indeterminate (It does not follow Complement rule)
X	X	0	$Q(t)$	Hold state

7.3.3 | SR Latch using NAND Gates

In Figure 7.8, NOR gates are replaced with bubbled AND gate following De Morgan theorem $\left(\overline{A+B} = \bar{A}\,\bar{B}\right)$ to get Figure 7.11. Inverters are conserved and SR latch is shown in Figure 7.12.

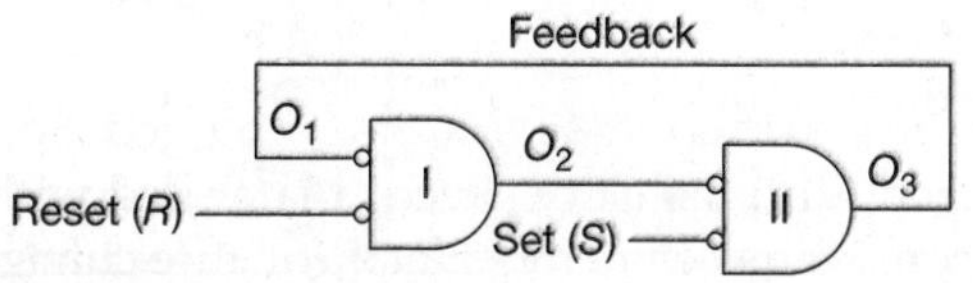

FIGURE 7.11 | SR latch

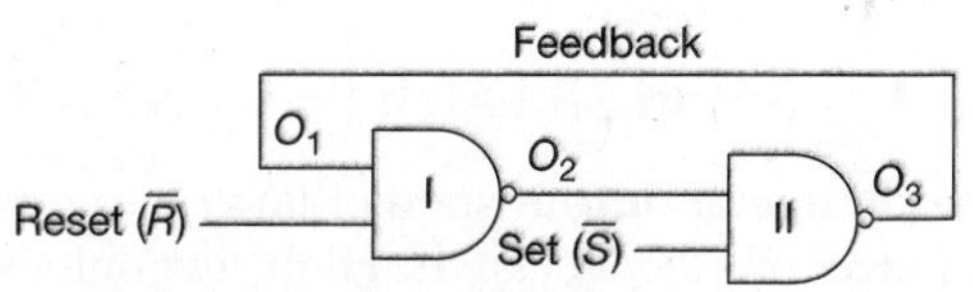

FIGURE 7.12 | SR latch

Figure 7.13 is redrawn to achieve cross coupling to construct SR latch using NAND gates. Inputs S and R are inverted. The output of NAND-II gate is Q that has one inverted input, S. The truth table of latch is given Table 7.5.

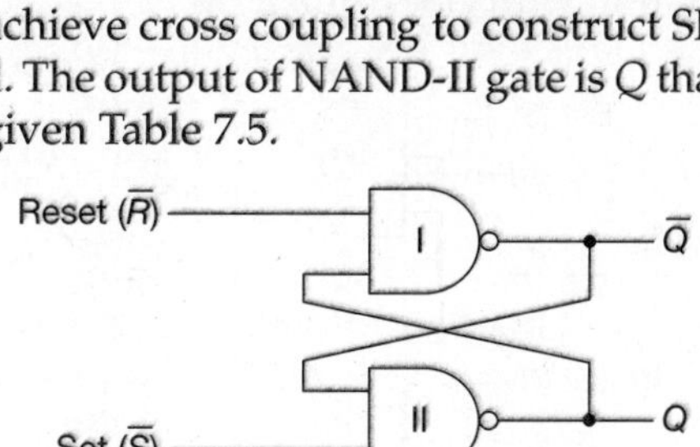

FIGURE 7.13 | SR latch

The output, $\bar{Q}(t+1)$ of NAND-I gate depends on previous output, $\bar{Q}(t)$ of NAND-II gate and inverted input, *R*. The Boolean expression for NAND-I gate is given below:

$$\bar{Q}(t+1) = \overline{\bar{R}Q(t)} \tag{7.7}$$

The output, $Q(t+1)$ of NAND-II gate depends on previous output, $Q(t)$ of NAND-I gate and inverted input, *S*. The Boolean expression for NAND-II gate is given below:

$$Q(t+1) = \overline{\bar{S}\bar{Q}(t)} \tag{7.8}$$

Characteristic truth table is given in Table 7.5. Characteristic table gives relation between inputs and outputs. Outputs of latch give two stable states following the complement rule of each other.

TABLE 7.5 | Characteristics table of SR-flip-flop with NAND gates

Inputs			Outputs		Remarks
$Q(t)$	$\bar{S}$	$\bar{R}$	$Q(t+1)$	$\bar{Q}(t+1)$	
0	0	0	0	1	Previous state 0
0	0	1	0	1	Reset
0	1	0	1	0	Set
0	1	1	0	0	(Indeterminate)*
1	0	0	1	0	Previous state 1
1	0	1	0	1	Reset
1	1	0	1	0	Set
1	1	1	0	0	(Indeterminate)*

* Complement rule is not followed.

7.3.4 | Gated SR Latch using NAND Gates

The change of output state of latch is synchronized with the clock. Such, a latch is known as gated or clocked latch. High level clock is considered as active clock for the change. Figure 7.14 is drawn to construct SR latch using NAND gates when clock pulse (C_P) controls the inputs with NAND logic using non-inverted inputs. The output of NAND is *Q* that has input, *S* having NAND operation with clock pulse.

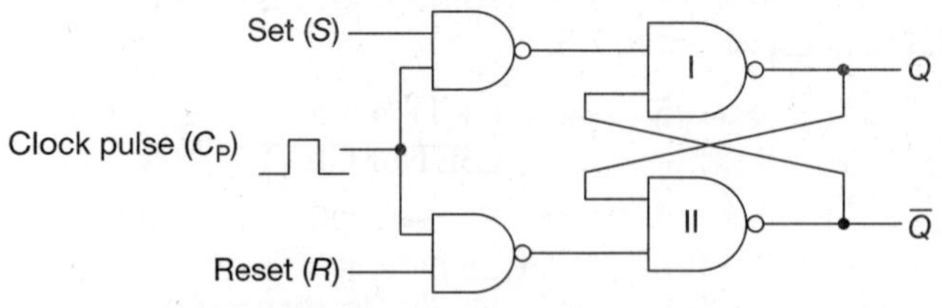

FIGURE 7.14 | SR flip-flop with edge triggered

The output, $\bar{Q}(t+1)$ of NAND-I gate depends on previous output, $\bar{Q}(t)$ of NAND-II gate, inverted input, R and input clock pulse, C_P. The Boolean expression for NAND-I gate is given below:

$$\bar{Q}(t+1)=\overline{\left(\overline{RC_P}\right)Q(t)} \tag{7.9}$$

The output, $Q(t+1)$ of NAND-II gate depends on previous output, Q_t of NAND-I gate, inverted input, S and input clock pulse, C_P. The Boolean expression for NAND-II gate is given below:

$$Q(t+1)=\overline{\left(\overline{SC_P}\right)\left(\overline{Q(t)}\right)} \tag{7.10}$$

Characteristic truth table is given in Table 7.4.

7.3.5 | Characteristic Equation of SR-Latch

Characteristic Table 7.3 of SR latch is considered where $Q(t)$, S and R are considered as inputs and $Q(t+1)$ is taken as output. Three-variable K-map is set up to find the simplified Boolean expressions giving characteristic equation. Figure 7.15 shows K-map, which is set up using truth table given in Table 7.3. One pair {4, 6} and one quad {2, 3, 6, 7} are formed. Indeterminate state give don't care minterms.

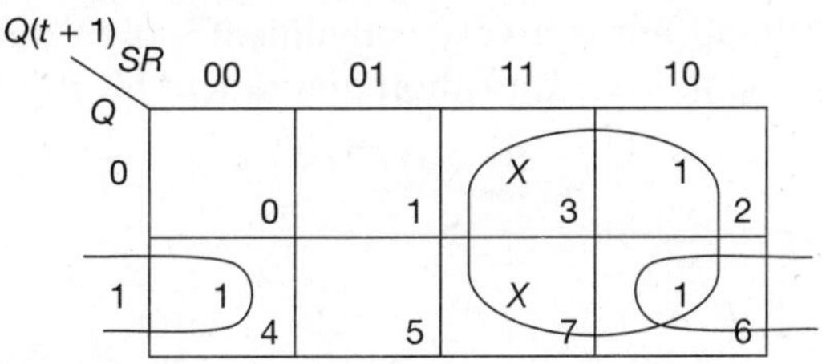

FIGURE 7.15 | K-map for SR latch

The characteristic equations are given below for SR latch.

$$Q(t+1)=S+\bar{R}Q(t) \tag{7.11}$$

$$SR=0 \tag{7.12}$$

Equation (7.12) avoids the indeterminate states when S and R are 1.

Observations from characteristic Table 7.3

- When both S and R are 0, the output, $Q(t+1)$ is same as that of previous output, $Q(t)$.
- When S is 0 and R is 1, the output $Q(t+1)$ is 0 irrespective to previous output, $Q(t)$. As output is 0 so it is considered as RESET or CLEAR state.
- When S is 1 and R is 0, the output $Q(t+1)$ is 1 irrespective to previous output, $Q(t)$. As output is 1 so it is considered as SET state.
- When both S and R are 1, the output, $Q(t+1)$ and $\bar{Q}(t+1)$ are same irrespective to previous output, $Q(t)$. This state is known as indeterminate state. Output does not complement of the other.

The characteristic table in simplified form is given in Table 7.6.

TABLE 7.6 | Characteristics table of SR-flip-flop

Inputs		Outputs	Remarks
S	*R*	*Q*(*t* + 1)	
0	0	*Q*(*t*)	Previous state
0	1	0	Reset
1	0	1	Set
1	1	?	(Indeterminate)*

* Complement rule is not followed.

7.3.6 | State Transition Diagram of SR Latch

In a sequential logic circuit, the memory elements at a given time define the state of that circuit at that time. Finite state offers an alternative to truth table in understanding progress of sequential logic with time. For a complex circuit, a truth table is difficult to adopt due to its large size. The functional behaviour of the circuit is explained using finite number of states. State transition diagram is a very convenient tool to describe finite state of the flip-flop. Figure 7.16 represents state transition diagram of SR-flip-flops.

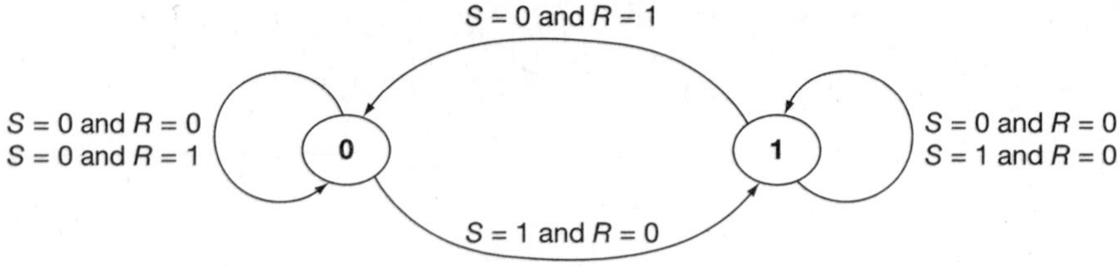

FIGURE 7.16 | State transition diagram of SR flip-flop

Output of the SR-flip-flop goes from previous state of logic 0 to next state of logic-1 (Set state) when *S* is set to 1 and *R* is cleared to 0.

Output of the SR-flip-flop goes from previous state of logic 1 to next state of logic-0 (Reset state) when *S* is cleared to 0 and *R* is set to 1.

Output of the SR-flip-flop remains in state of logic 0 (Reset state) when *S* is cleared to logic-0 and *R* is either clear to 0 or set to 1.

Output of the SR-flip-flop remains in state of logic 1 (Set state) when *S* is either cleared to logic-0 or set to 1 and *R* is cleared to 0.

7.3.7 | Excitation Table of SR-Latch

To analyse the sequential circuit, there is need of truth table (or Characteristic Table). To synthesis or design sequential circuits excitation of flip-flop or latch is required. The importance of excitation tables of flip-flop is analogous to that of truth table useful while analysing problem. Characteristic table of flip-flop/latch gives the relation between inputs and output where inputs *S* and *R* of flip-flop are given and output *Q*(*t* + 1) can be obtained.

Output of flip-flop depends on inputs of flip-flop. Excitation table of a flip-flop gives the relation between output and inputs where output transition $Q(t) \rightarrow Q(t + 1)$ of flip-flop is given and input of flip-flop can be found. Flip-flop input is presented as a dependent function of transition $Q(t) \rightarrow Q(t + 1)$ and comes later in the table. This is derived from flip-flop truth table or characteristic equation but more directly from its state transition diagram. Table 7.7 presents the excitation table of SR the flip-flop.

TABLE 7.7 | Excitation table SR-flip-flop

Outputs		Inputs	
$Q(t)$	$Q(t + 1)$	S (Set)	R (Reset)
0	0	0	X
0	1	1	0
1	0	0	1
1	1	X	0

7.3.8 | SR-Flip-Flop with Asynchronous Inputs

On switching ON the flip-flop, the output of flip-flop is uncertain. So, two more inputs are added to flip-flops, called preset (Pr) and clear (Cr). Such inputs are known as asynchronous inputs or direct inputs. Clocked SR-flip-flop with preset and clear inputs are given in Figure 7.17 using NOR gates. Characteristic table with preset and clear inputs is given as Table 7.8.

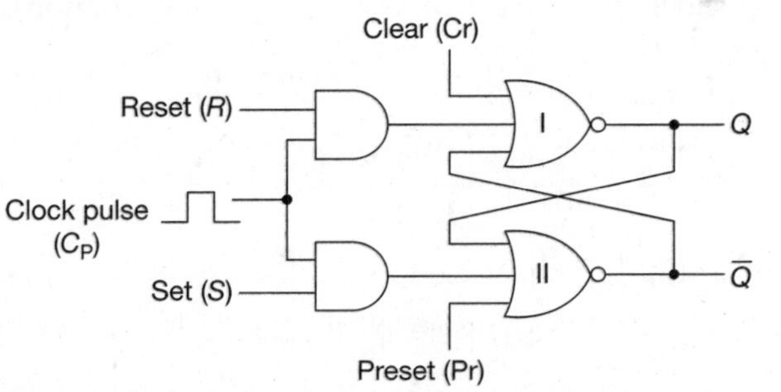

FIGURE 7.17 | Clocked SR flip-flop with preset and clear

TABLE 7.8 | Characteristics of SR flip-flop with preset and clear inputs

Inputs					Outputs	Remarks
S (Set)	R (Reset)	C_P (Clock Pulse)	Pr (Preset)	Cr (Clear)	$Q(t + 1)$	
0	0	1	0	0	$Q(t)$	Previous state
0	1	1	0	0	0	Set
1	0	1	0	0	1	Clear (Reset)

(Continued)

TABLE 7.8 | (Continued)

Inputs					Outputs	Remarks
S (Set)	*R* (Reset)	C_P (Clock Pulse)	Pr (Preset)	Cr (Clear)	*Q*(*t* + 1)	
1	1	1	0	0	?	Indeterminate*
X	X	0	1	0	1	Preset
X	X	0	0	1	0	Clear

* Complement rule is not followed.

The output, $Q(t+1)$ of NOR-I gate depends on previous output, $\bar{Q}(t)$ of NOR-II gate, input, R, non-active part of clock pulse, ($C_P = 0$) and Clear, Cr input. The Boolean expression of NOR-I gate is given below

$$Q(t+1) = \overline{RC_P + Cr + \bar{Q}(t)}$$

For non-active part of clock pulse, ($C_P = 0$), $RC_P = 0$. Above equation is rewritten as

$$Q(t+1) = \overline{Cr + \bar{Q}(t)} \tag{7.13}$$

The output, $\bar{Q}(t+1)$ of NOR-II gate depends on previous output, $Q(t)$ of NOR-I gate input, S, non-active part of clock pulse, ($C_P = 0$) and preset input, Pr. The Boolean expression of NOR-II gate is given below

$$\bar{Q}(t+1) = \overline{SC_P + Pr + Q(t)}$$

For non-active part of clock pulse, ($C_P = 0$), $SC_P = 0$. Above equation is rewritten as

$$\bar{Q}(t+1) = \overline{Pr + Q(t)} \tag{7.14}$$

Let Pr is 1 and Cr is 0. From Eq. (7.14), Output, $\bar{Q}(t+1)$ is cleared to 0.

$$\bar{Q}(t+1) = \overline{1 + Q(t)} \Rightarrow 0 \qquad (\because 1 + x = 1)$$

Output, $\overline{Q(t+1)}$ acts as input in Eq. (7.13) to set the output $Q(t+1)$ as 1.

$$Q(t+1) = \overline{0+0} \Rightarrow 1$$

Let Pr is 0 and Cr is 1. From Eq. (7.13), output $Q(t+1)$ is cleared to 0 (Reset)

$$Q(t+1) = \overline{1 + \bar{Q}(t)} \Rightarrow 0 \quad (\because 1 + x = 1)$$

Output, $\bar{Q}(t+1)$ acts as input in Eq. (7.14) to set the output $\bar{Q}(t+1)$ as 1.

$$\bar{Q}(t+1) = \overline{0+0} \Rightarrow 1$$

When C_P is active ($C_P = 1$), Cr is 0 and Pr is 0 then

$$Q(t+1) = \overline{RC_P + \bar{Q}(t)} \text{ and } \bar{Q}(t+1) = \overline{SC_P + Q(t)}$$

These equations are similar to gated SR latch Eq. (7.5) and Eq. (7.6). SR latch function normally as given in Table 7.8.

Clocked SR-flip-flop with preset and clear inputs are given in Figure 7.18 using NAND gates. Characteristic table with preset and clear inputs is given as Table 7.9.

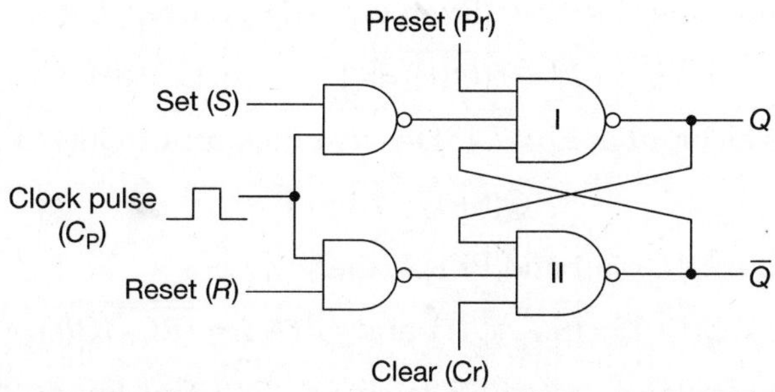

FIGURE 7.18 | Clocked SR flip-flop with preset and clear

TABLE 7.9 | Characteristics of SR-flip-flop with preset and clear inputs

Inputs					Outputs	Remarks
R **(Set)**	***S*** **(Reset)**	**CP (Clock Pulse)**	**Pr (Preset)**	**Cr (Clear)**	***Q*(*t* + 1)**	
0	0	1	1	1	*Q*(*t*)	Previous state
0	1	1	1	1	0	Set
1	0	1	1	1	1	Clear (Reset)
1	1	1	1	1	?	Indeterminate state
X	X	0	0	1	1	Output is set to 1
X	X	0	1	0	0	Output is clear to 0

The output, $Q(t+1)$ of NAND-I gate depends on previous output, $\bar{Q}(t)$ of NAND-II gate, input, S, non-active input clock pulse, $(C_P = 0)$ and preset input, Pr. Boolean expression for NAND-I gate is given below:

$$Q(t+1) = \overline{(\overline{SC_P})(Pr)\bar{Q}(t)}$$

For non-active part of clock pulse, $(C_P = 0)$, $\overline{SC_P} = 1$. Above equation is rewritten as

$$Q(t+1) = \overline{(Pr)\bar{Q}(t)} \tag{7.15}$$

The output, $\bar{Q}(t+1)$ of NAND-II gate depends on previous output, $Q(t)$ of NAND-I gate, inverted input, R, non-active input clock pulse, $(C_P = 0)$. Boolean expression for NAND-II gate is given below:

$$\bar{Q}(t+1) = \overline{(\overline{RC_P})Cr.(Q(t))}$$

For non-active part of clock pulse, $(C_P = 0)$, $\overline{RC_P} = 1$. Above equation is rewritten as

$$\bar{Q}(t+1) = \overline{Cr.(Q(t))} \tag{7.16}$$

Let Pr is 0 and Cr is 1. From Eq. (7.15), Output, $Q(t+1)$ is set to 1.

$$Q(t+1) = \overline{(0)(\overline{Q(t)})} \Rightarrow 1 \qquad (\because 0.x = 0)$$

Output, $Q(t+1)$ acts as input in Eq. (7.16) to clear the output $\bar{Q}(t+1)$ at logic-0.

$$\bar{Q}(t+1) = \overline{1.1} \Rightarrow 0$$

Let Pr is at logic-1 and Cr is at logic-0. From Eq. (7.16), output $\bar{Q}(t+1)$ is set to 1

$$\bar{Q}(t+1) = \overline{0.(Q(t))} \Rightarrow 1 \qquad (\because 0.x = 0)$$

Output, $\overline{Q(t+1)}$ acts as input in Eq. (7.15) to clear the output $Q(t+1)$ at logic-0.

$$\bar{Q}(t+1) = \overline{1.1} \Rightarrow 0$$

When C_P is active ($C_P = 1$), Cr is 1 and Pr is 1, then

$$Q(t+1) = \overline{(\overline{SC_P})\bar{Q}(t)} \text{ and } \bar{Q}(t+1) = \overline{(\overline{RC_P})Q(t)}$$

These equations are similar to gated SR latch Eq. (7.9) and Eq. (7.10). SR latch function normally as given in Table 7.9.

7.4 | TRIGGERING OF LATCHES

The output of latch changes its state as and when the inputs of latch are changed. So, the latches are called asynchronous latches. The change of output state of latch is controlled with an additional signal known as clock along with the change of inputs of latch. So, the change of output state of latch is synchronized with the clock. Such, latch is known as gated or clocked latch.

A pulse may be a positive pulse (Figure 7.19) or a negative pulse (Figure 7.20). A single positive pulse is generated when a normally LOW voltage goes to its HIGH level and then returns to its normal LOW level. A single negative pulse is generated when a normally HIGH voltage goes to its LOW level and then returns to its normal HIGH level.

A pulse has LOW and HIGH level states. High level of clock is known as positive level. Low level of clock is known as negative level. A pulse also has two edges: a rising (or leading) edge and a falling (or trailing) edge. For a positive pulse, the rising (or leading) edge is a positive going transition that changes from low voltage level to high voltage level and the falling (or trailing) edge is a negative going transition that changes from HIGH voltage level to LOW voltage level. For a negative pulse, the rising (or leading) edge is a negative going transition and the trailing edge is a positive going transition. The pulses are ideal, because the rising and falling edges change instantaneously, that is in zero time (Figure 7.21). Practical pulses do not change instantaneously from LOW to HIGH voltage or from HIGH to LOW voltage.

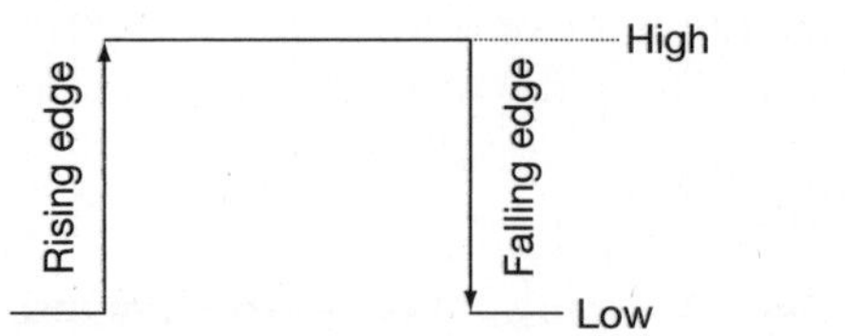

FIGURE 7.19 | Positive pulse

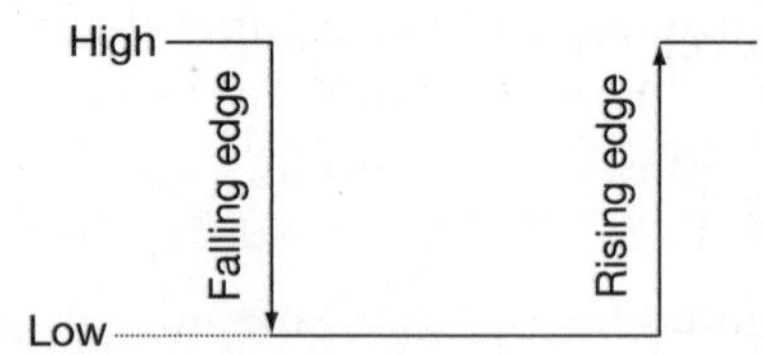

FIGURE 7.20 | Negative pulse

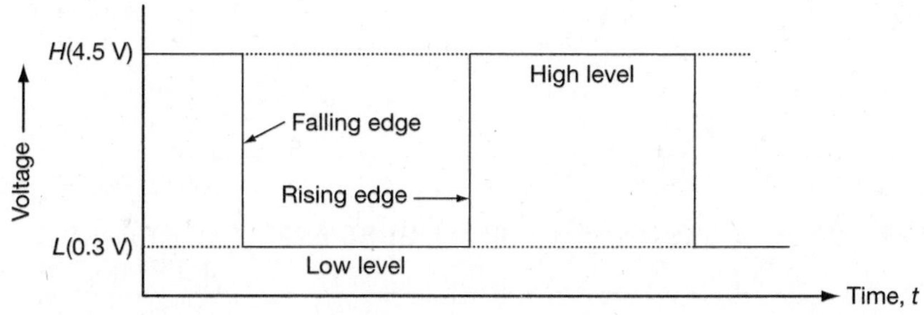

FIGURE 7.21 | Ideal pulse and its transition

The state of a flip-flop is changed by a change in the input signal. This change caused is said to trigger the flip-flop. A clocked flip-flop is triggered by level or edge triggering. The gated latch is said to be level-triggered circuit, if the circuit responds to the changes in input only when the clock pulse is in HIGH level or LOW level state. The output will change as per the changes in enable input as long as the active level is maintained at the enable input. The gated latch is said to be edge-triggered circuit, if the circuit responds to the changes in input only when the clock pulse is changes either from high to low (or rising edge) or low to high (or falling edge) transition. Edge triggering is also known as dynamic triggering. Triggering of latches is classified as follows.

- Level triggering
  - Positive (or high) level Triggering
  - Negative (or low) level Triggering
- Edge triggering
  - Positive (or rising) edge triggering
  - Negative (or falling) edge triggering

7.4.1 | Positive (or high) Level Triggering

The gated latch is said to be positive level-triggered circuit, if the circuit responds to the changes in input only when the clock input (C_P) is in high state. The output will change as long as the active high level is maintained at the clock pulse input. Block diagram of positive level-triggered flip-flop is shown in Figure 7.22.

7.4.2 | Negative (or low) Level Triggering

The gated latch is said to be negative level-triggered circuit, if the circuit responds to the changes in input only when the clock pulse (C_P) is in low state. The output will change as long as the active low level is maintained at the clock input. Block diagram of positive level-triggered flip-flop is shown in Figure 7.23.

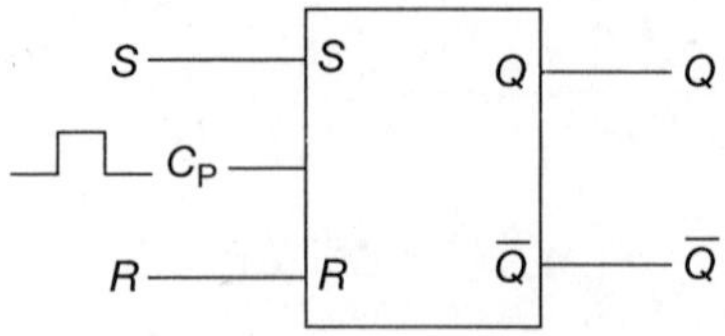

FIGURE 7.22 | Positive level triggered SR flip-flop

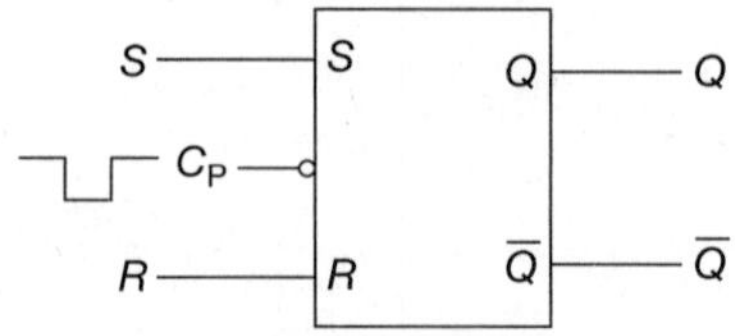

FIGURE 7.23 | Negative level triggered SR flip-flop

7.4.3 | Positive (or leading or rising) Edge Triggering

The gated latch is said to be positive edge-triggered circuit, if the circuit responds to the changes in input only when the clock input (C_P) goes from high to low level state. The rising (or leading) edge is a positive going transition that changes from low voltage level to high voltage level. The output will change as and when the transition from high level to low level of the clock pulse takes place. Block diagram of positive edge-triggered flip-flop is shown in Figure 7.24.

7.4.4 | Negative (or low) Level Triggering

The gated latch is said to be negative edge-triggered circuit, if the circuit responds to the changes in input only when the clock input (C_P) goes from low to high level state. The falling (or trailing) edge is a negative going transition that changes from high voltage level to low voltage level. The output will change as and when the transition from low level to high level of the clock pulse takes place. Block diagram of positive edge-triggered flip-flop is shown in Figure 7.25.

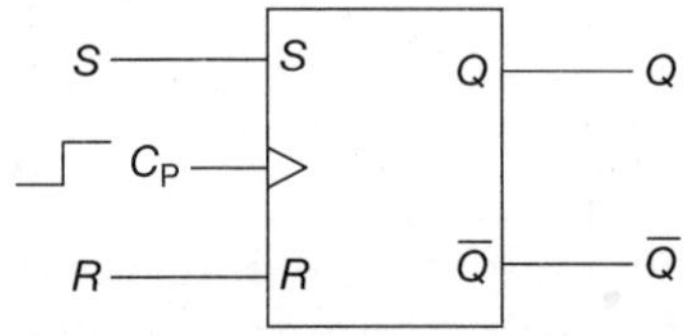

FIGURE 7.24 | Positive edge triggered SR flip-flop

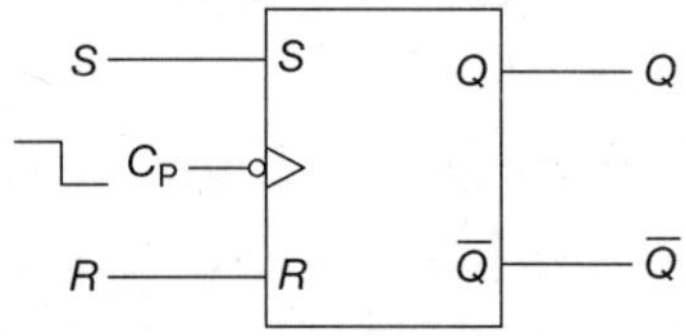

FIGURE 7.25 | Negative edge triggered SR flip-flop

7.4.5 | Generation of Spikes

It is essential to apply the clock signal (timing signal) in the form of sharp positive and negative spikes instead of in the form of a pulse train, for the edge-triggered flip-flops. These spikes are generated from the rectangular clock pulses with the help of a passive differentiator circuit shown in Figure 7.26. Placing diode in series with the differentiator circuit will act as half rectifier. The diode blocks negative edge spike and allows the positive edge spike as shown in Figure 7.27.

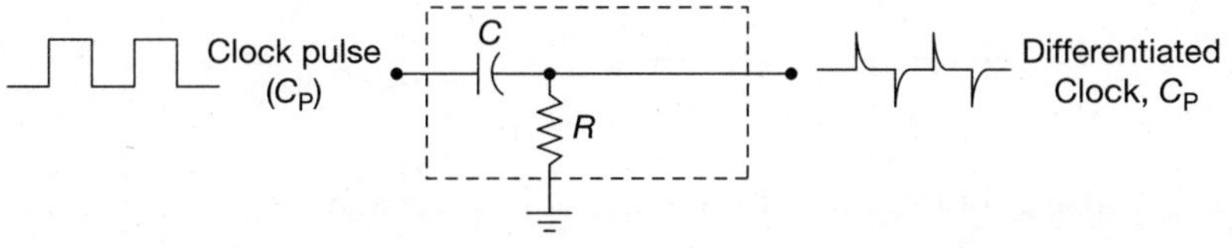

FIGURE 7.26 | Generation of positive and negative spikes of clock pulse

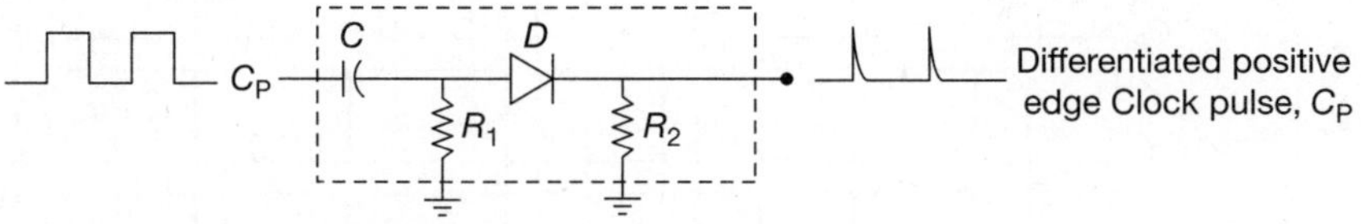

FIGURE 7.27 | Generation of positive edge clock pulse

7.4.6 | Generation of Pulse at Rising Edge of Clock Pulse

A positive pulse is generated at rising edge of clock pulse using an inverter and AND gate as shown in Figure 7.28. The timing diagram is shown in Figure 7.29. Inverter produces delay of nanoseconds. AND gate produces output as high for the delayed duration by inverter after rising edge of clock pulse, C_P.

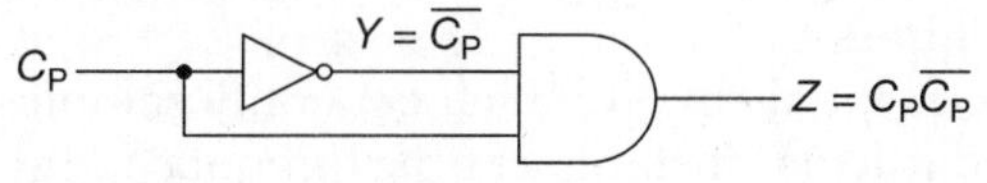

FIGURE 7.28 | Pulse generator at rising edge of clock pulse

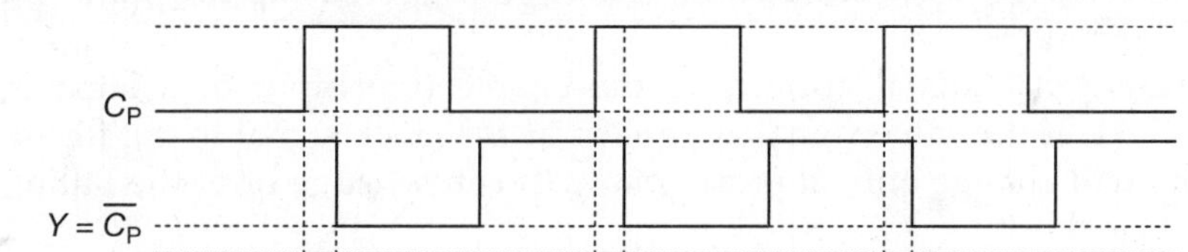

FIGURE 7.29 | Timing diagram of positive pulse generator

7.4.7 | Generation of Pulse at Falling Edge of Clock Pulse

A positive pulse is generated at falling edge of clock pulse using an inverter and NOR gate as shown in Figure 7.30. The timing diagram is shown in Figure 7.31. Inverter produces delay of nanoseconds. NOR gate produces output as high for the delayed duration by inverter after falling edge of clock pulse, C_P.

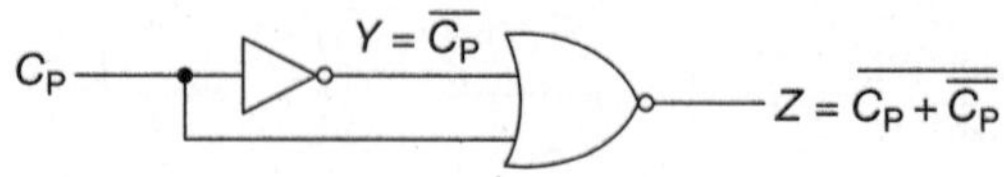

FIGURE 7.30 | Pulse generator at falling edge of clock pulse

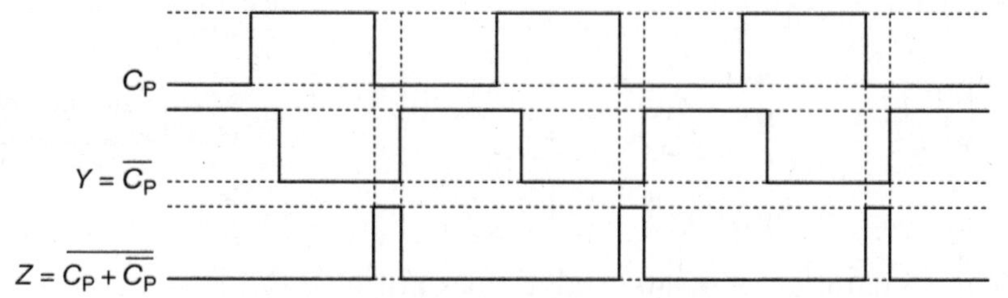

FIGURE 7.31 | Timing diagram of negative pulse generator

7.4.8 | Latch Versus Flip-Flop

- Latches and flip-flops both are basically the bistable multivibrators.
- A latch has an enable input signal known as clock. As long as, it is active, the latch output will keep changing according to the changes in its input. In other words latch is a level-triggered flip-flop.
- But, flip-flop is a sequential circuit which generally samples its inputs and changes its outputs only at particular instants of time but not continuously.
- The flip-flops are therefore said to be edge sensitive or edge triggered rather than being level triggered like latches. The edge-triggered flip-flops are also called as clocked flip-flops.
- The rectangular signal applied to the enable (C_p) input of a latch is shown in Figure 7.31. If the same signal is applied at the clock signal to the flip-flop, then its outputs will change only at either rising (positive) edge or at the falling (negative) edge of clock.

7.5 | D-FLIP-FLOP

The main disadvantage of SR-flip-flop is its indeterminate (forbidden) state. D-flip-flop is designed to overcome the disadvantage of SR-flip-flop and avoids the indeterminate state. Inverted set (*S*) input is given to reset (*R*) input of the flip-flop so that flip-flop has only one input and two outputs which are compliment to each other. D-flip-flop has the ability to store the data into the flip-flop. D-flip-flop is also called as data flip-flop. D-flip-flop is used to design registers. Figure 7.32 shows diagram of D-flip-flop using SR-flip-flop. Figure 7.33 shown the diagram of D-flip-flop using NOR gate. Flip-flop is triggered using positive level clock. Figure 7.34 shows diagram of D-flip-flop using NAND gate. D-flip-flop is triggered using positive edge transition clock.

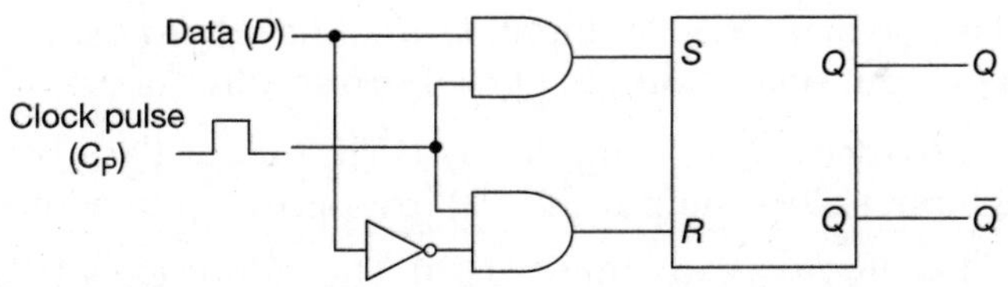

FIGURE 7.32 | Clocked D flip-flop

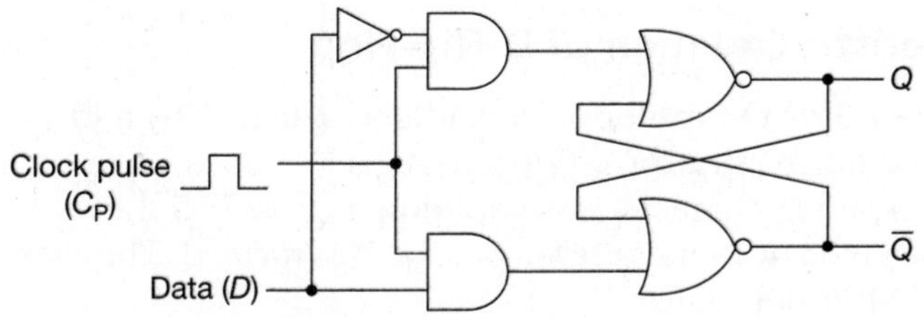

FIGURE 7.33 | D-flip-flop with clock using NOR gate

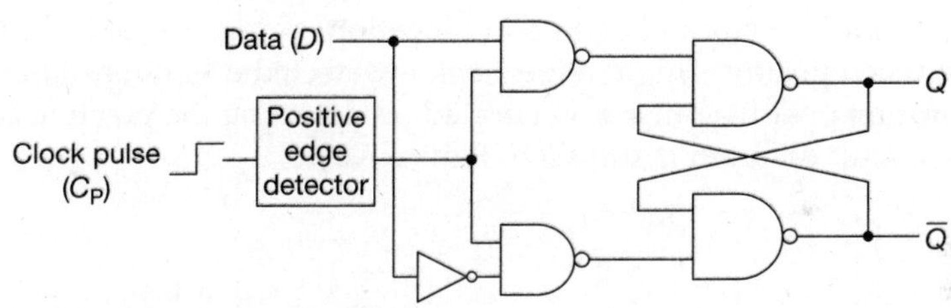

FIGURE 7.34 | D-flip-flop with clock using NAND gate

Characteristic table of D-flip-flop is given in Table 7.10. D and previous output of flip-flop $Q(t)$ is acting as input. $Q(t + 1)$ and its inverted signal acts as outputs.

TABLE 7.10 | Characteristics of D-flip-flop

Inputs		Outputs	
$Q(t)$	D	$Q(t+1)$	$\bar{Q}(t+1)$
0	0	0	1
0	1	1	0
1	0	0	1
1	1	1	0

Case I: Let $Q_t = 0$ is 0 so that $\bar{Q}(t) = 1$. Input, D is 0. The output $Q(t+1)$ is 0 and $\bar{Q}(t+1)$ is 1. It saves input state as output $Q(t + 1)$ irrespective of previous state Q(t).

Case II: Let $Q(t)$ is 0 so that $\bar{Q}(t) = 1$. Input, D is 1. The output $Q(t+1)$ is 1 and $\bar{Q}(t+1)$ is 0. It saves input state as output $Q(t + 1)$ irrespective of previous state $Q(t)$.

Case III: Let $Q(t)$ is 1 so that $\bar{Q}(t) = 0$. Input D is 0. The output $Q(t+1)$ is 0 and $\bar{Q}(t+1)$ is 1. It saves input state as output $Q(t + 1)$ irrespective of previous state $Q(t)$.

Case IV: Let $Q(t)$ is 1 so that $\bar{Q}(t) = 0$. Input, D is 1. The output $Q(t+1)$ is 1 and $\bar{Q}(t+1)$ is 0. It saves input state as output $Q(t + 1)$ irrespective of previous state $Q(t)$.

7.5.1 | Characteristic Equation of D-Flip-Flop

Characteristic Table 7.8 of D-flip-flop is considered where $Q(t)$ and D are considered as inputs and $Q(t + 1)$ is taken as output. Two variables K-map is set up to find the simplified Boolean expressions giving characteristic equation. Figure 7.35 shows K-map, which is set up using truth table given in Table 7.8. One pair {1, 3} is formed. The characteristic equation is given below for D-flip-flop/latch.

$$Q(t+1) = D \tag{7.17}$$

Characteristic table of clocked or gated D-flip-flop is given in Table 7.11. The flip-flop changes its state when positive (high) level clock is given, otherwise it holds (saves) the data. Characteristic table of clocked or gated D-flip-flop is given in Table 7.12. The flip-flop changes its state when positive (high) edge clock is detected otherwise it holds (saves) the data. A short duration positive pulse is generated on detecting the positive edge transition of the clock. The circuit diagram is shown in Figure 7.28.

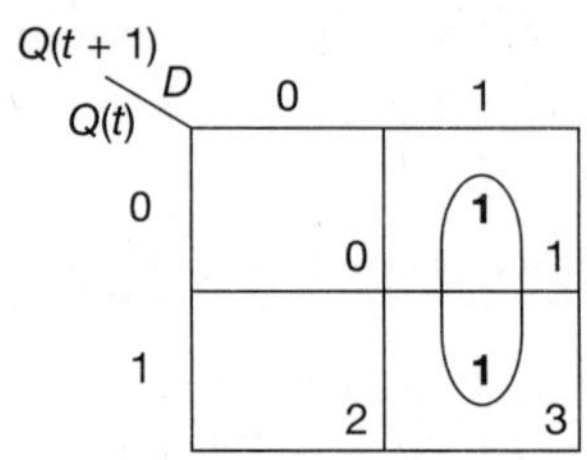

FIGURE 7.35 | K-map D latch

TABLE 7.11 | Characteristics of gated or clocked D-flip-flop/latch

Inputs		Outputs	Remarks
***D* (Data)**	**CP (Positive Level)**	***Q*(*t* + 1)**	
0	1	0	Save 0-state
1	1	1	Save 1-state
X	0	*Q*(*t*)	No change or holds data

TABLE 7.12 | Characteristics of gated or clocked D-flip-flop/latch

Inputs		Outputs	Remarks
D (Data)	CP (Positive Edge)	*Q*(*t* + 1)	
0	↑	0	Save 0-state
1	↑	1	Save 1-state
X	otherwise	*Q*(*t*)	No change or holds data

7.5.2 | State Transition Diagram of D-Flip-Flop

The functional behaviour of the circuit is explained using finite number of states. State transition diagram is a convenient tool to describe finite state of the flip-flop. Figure 7.36 represents state transition diagram of D-flip-flops.

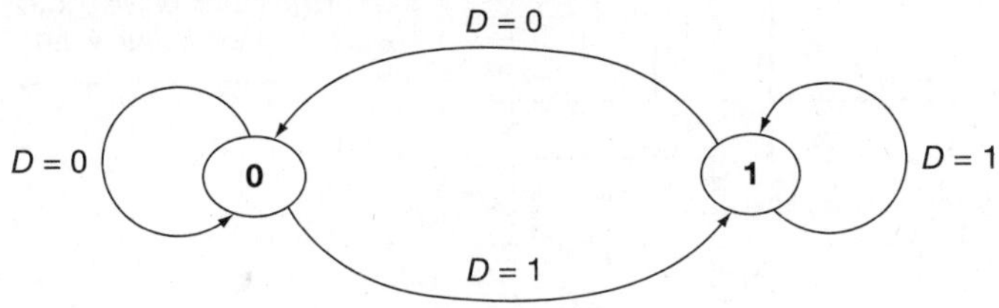

FIGURE 7.36 | State transition diagram of D flip-flop

Output of the D-flip-flop goes from previous state of 0 to next state 1 when input *D* is set to 1. Output of the D-flip-flop goes from previous state 1 to next state 0 when input *D* is cleared to 0. Output of the D-flip-flop remains in state 0, when input *D* is having 0 as input. Output of the D-flip-flop remains in state 1, when *D* is having 1 as input.

7.5.3 | Excitation Table of D-Flip-Flop

Excitation table of a flip-flop gives the relation between output and inputs where output transition $Q(t) \rightarrow Q(t+1)$ of flip-flop is given and input, *D* of flip-flop can be found. Flip-flop input is presented as a dependent function of transition $Q(t) \rightarrow Q(t+1)$ and comes later in the table. This is derived from flip-flop truth table or characteristic equation but more directly from its state transition diagram. Table 7.13 presents the excitation table of D-flip-flop.

TABLE 7.13 | Excitation table D-flip-flop

Outputs		Inputs
Q(*t*)	*Q*(*t* + 1)	*D* (Data)
0	0	0
0	1	1
1	0	0
1	1	1

7.6 | FLIP-FLOP TIMING

Diode and transistor takes time to switch states. To turn on or off, Diode takes small amount of time in nanoseconds. Similarly, transistor takes time to switch from saturation to cut-off. So, propagation signal from input to output lines is due to switching time. Data must be present before the clock edge transition is detected. It is required to stray capacitance at input lines. In case of edge-triggered flip-flop, the transfer of information from flip-flop inputs to the output of the flip-flop only during the brief interval of time; when the clock signal is making a transition from the 1 to 0 (↓) state or from 0 to 1 (↑) state and in this the data lock out occurs at end of the edge. The timing specifications of an edge-triggered flip-flop are illustrated in Figure 7.37.

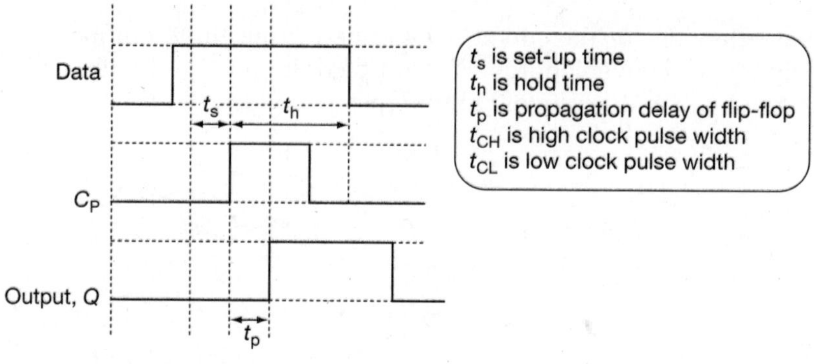

FIGURE 7.37 | Timing diagram

Set-up Time: Set-up time (t_s) is the minimum time interval that the data present immediately before the active transition of the clock signal during which the input (data input) must be maintained at the proper level. Undesired output is obtained, if time interval t is less than setup time, t_s.

Hold Time: Hold time (t_h) is the minimum time interval immediately after the active transition of the clock signal during which the control input (data input) must be maintained at the level. Undesired output is obtained, if time interval t is less than hold time, t_h.

Propagation Delay: The flip-flop output changes exactly on the active transition of the clock signal, which in fact does not happen practically. Propagation delay (t_p) is amount of time taken for the output of flip-flop to change states after an input trigger is received.

High Clock Pulse Width (t_{CH}): It is the minimum time for which the clock remains high.

Low Clock Pulse Width (t_{CL}): It is the minimum time for which the clock remains low.

Maximum Frequency: Maximum frequency of clock signal is the highest frequency at which flip-flop performs proper and stable operation.

$$f_{max} = \frac{1}{T} \tag{7.18}$$

$$T \geq t_s + t_p + t_{ns}$$

where

t_s = Set-up time

t_p = Propagation delay of flip-flop

t_{ns} = Propagation delay of next state decoder

Power Dissipation: The power dissipation of a flip-flop is the total power consumption of the device. Power dissipation is product of supply voltage (V_{CC}) and current drawn from supply (I_{CC}) by the device.

$$P = V_{CC}I_{CC} \tag{7.19}$$

where

V_{CC} is the supply voltage.

I_{CC} is the current drawn from supply.

The power dissipation is usually in milliwatts (mW).

7.7 | JK-FLIP-FLOP

The main disadvantage of SR-flip-flop is its indeterminate (forbidden) state when both S and R are 1. JK-flip-flop behaves like SR-flip-flop and permits to have valid operation when both J and K are 1.

The output, $Q(t)$ is fed back as input to AND gate along with inputs K and clock. Complement of $Q(t)$ is fed back as input to AND gate along with inputs J and clock. Figure 7.38 shows diagram of JK-flip-flop designed from SR-flip-flop. Flip-flop is triggered using positive level clock. Figure 7.39 shows diagram of JK-flip-flop using NAND gate and feedback of outputs as NAND inputs. JK-flip-flop is triggered using positive edge transition clock.

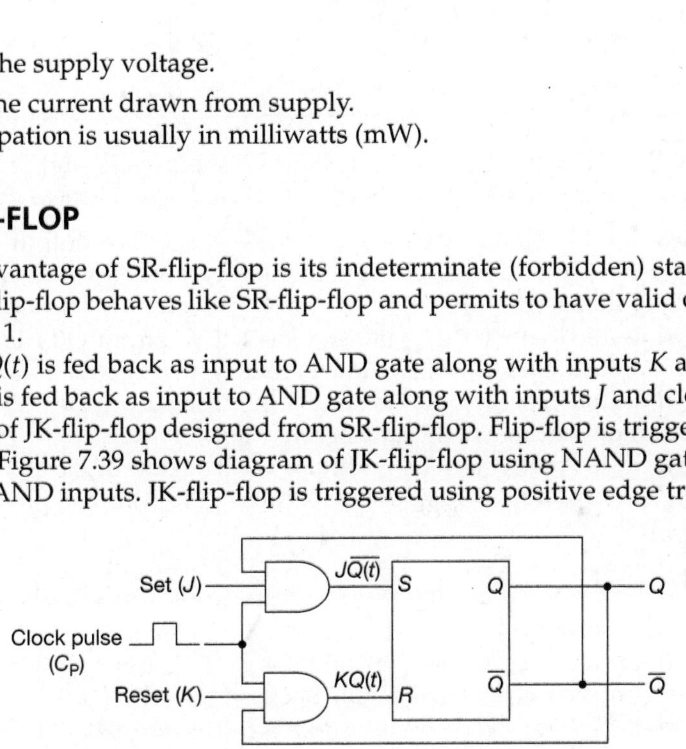

FIGURE 7.38 | Clocked JK latch

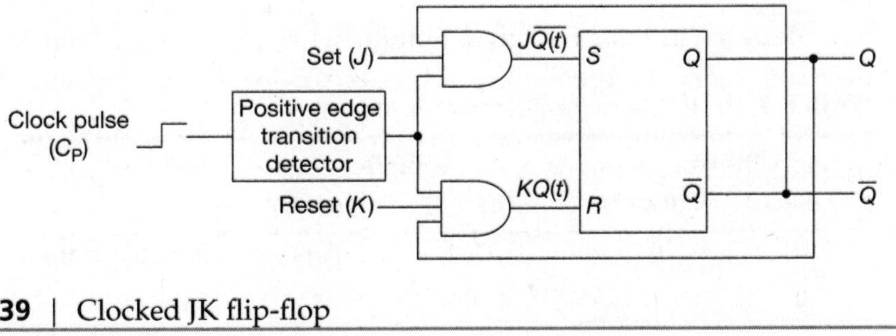

FIGURE 7.39 | Clocked JK flip-flop

TABLE 7.14 | Characteristics of JK-flip-flop

Inputs			Outputs	
$Q(t)$	J	K	$Q(t+1)$	$\overline{Q(t+1)}$
0	0	0	0	1
0	0	1	0	1
0	1	0	1	0
0	1	1	1	0
1	0	0	1	0
1	0	1	0	1
1	1	0	1	0
1	1	1	0	1

Characteristic table of JK-flip-flop is given in Table 7.14. Inputs JK and previous output of flip-flop $Q(t)$ are acting as inputs. $Q(t + 1)$ and its inverted signal acts as outputs. When

Case I: Let $Q(t) = 0$ is 0 so that $\bar{Q}(t) = 1$. J is 0 and K is 0. The output $Q(t+1)$ is 0 and $\bar{Q}(t+1)$ is 1. It is considered as hold state because $Q(t + 1)$ is same as that of previous state $Q(t)$.

Case II: Let $Q(t)$ is 0 so that $\bar{Q}(t) = 1$. J is 0 and K is 1. The output $Q(t+1)$ is 0 and $\bar{Q}(t+1)$ is 1. It is considered as RESET state because $Q(t + 1)$ is 0.

Case III: Let $Q(t)$ is 0 so that $\bar{Q}(t) = 1$. J is 1 and K is 0. The output $Q(t+1)$ is 1 and $\bar{Q}(t+1)$ is 0. It is considered as SET state because $Q(t + 1)$ is 1.

Case IV: Let $Q(t)$ is 0 so that $\bar{Q}(t) = 1$. J is 1 and K is 1. The output $Q(t+1)$ is 1 and $\bar{Q}(t+1)$ is 0. It is considered as Toggle state because $Q(t + 1)$ is complement of previous state $Q(t)$.

Case V: Let $Q(t)$ is 1 so that $\bar{Q}(t) = 0$. J is 0 and K is 0. The output $Q(t+1)$ is 1 and $\bar{Q}(t+1)$ is 0. It is considered as hold state because $Q(t + 1)$ is same as that of previous state $Q(t)$, i.e. logic-1.

Case VI: Let $Q(t)$ is 1 so that $\overline{Q(t)} = 0$. J is 0 and K is 1. The output $Q(t+1)$ is 0 and $\overline{Q(t+1)}$ is 1. It is considered as RESET state because $Q(t + 1)$ is 0.

Case VII: Let $Q(t)$ is 1 so that $\overline{Q(t)} = 0$. J is 1 and K is 0. The output $Q(t+1)$ is 1 and $\overline{Q(t+1)}$ is 0. It is considered as SET state because $Q(t + 1)$ is 1.

Case VIII: Let $Q(t)$ is 1 so that $\overline{Q(t)} = 0$. J is 1 and K is 1. The output $Q(t+1)$ is 1 and $\overline{Q(t+1)}$ is 0. It is considered as Toggle state because $Q(t + 1)$ is complement of previous state $Q(t)$.

Characteristic table given in Table 7.15 gives outputs when inputs J and K are known.

TABLE 7.15 | Characteristics of J-K-flip-flop

Inputs		Outputs		Remarks
J (Set)	K (Reset)	$Q(t+1)$	$\bar{Q}(t+1)$	
0	0	$Q(t)$	$\bar{Q}(t)$	Previous state
0	1	0	1	Set

(Continued)

TABLE 7.15 | (Continued)

Inputs		Outputs		Remarks
J (Set)	*K* (Reset)	$Q(t+1)$	$\bar{Q}(t+1)$	
1	0	1	0	Clear (Reset)
1	1	$\bar{Q}(t)$	$Q(t)$	Toggle state

7.7.1 | Characteristic Equation of JK-Flip-Flop

Characteristic Table 7.12 of JK-flip-flop is considered where $Q(t)$, J and K are considered as inputs and $Q(t + 1)$ is taken as output. Three variables K-map is set up to find the simplified Boolean expressions giving characteristic equation. Figure 7.40 shows K-map, which is set up using truth table given in Table 7.12. Two pairs {2, 3} and {1, 6} are formed.

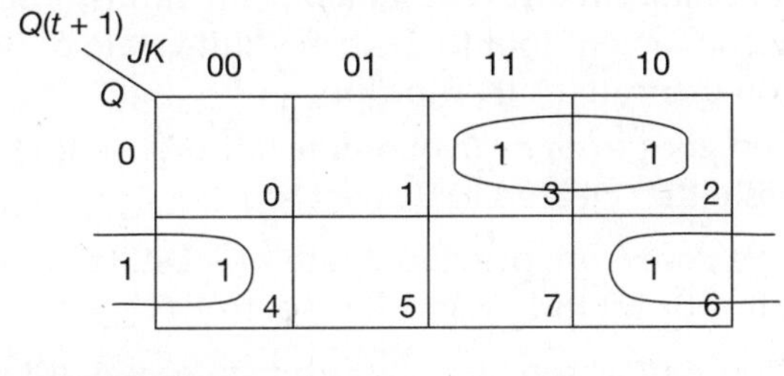

FIGURE 7.40 | K-map JK Flip-flop

The characteristic equations are given below for D-flip-flop.

$$Q(t+1) = J\bar{Q}(t) + \bar{K}Q(t) \tag{7.20}$$

Characteristic table of clocked or gated JK-flip-flop is given in Table 7.16. The flip-flop changes its state when positive (high) level clock is given otherwise it holds (saves) the data. Characteristic table of clocked or gated JK-flip-flop is given in Table 7.17. The flip-flop changes its state when positive (high) edge clock is detected otherwise it holds (saves) the data. A short duration positive pulse is generated on detecting the positive edge transition of the clock. The circuit diagram is shown in Figure 7.28.

TABLE 7.16 | Characteristics of J-K-flip-flop

Inputs			Outputs	Remarks
J (Set)	*K* (Reset)	CP (Clock Pulse)	$Q(t+1)$	
0	0	1	$Q(t)$	Previous state
0	1	1	0	Set
1	0	1	1	Clear (Reset)
1	1	1	$\overline{Q(t)}$	Toggle state
X	X	0	$Q(t)$	No change

TABLE 7.17 | Characteristics of J-K-flip-flop

Inputs			Outputs	Remarks
J (Set)	*K* (Reset)	CP (Positive Edge)	*Q*(*t* + 1)	
0	0	↑	*Q*(*t*)	Previous state
0	1	↑	0	Set
1	0	↑	1	Clear (Reset)
1	1	↑	$\bar{Q}(t)$	Toggle state
X	X	Otherwise	*Q*(*t*)	No change

7.7.2 | State Transition Diagram of JK-Flip-Flop

The functional behaviour of the circuit is explained using finite number of states. State transition diagram is a very convenient tool to describe finite state of the flip-flop. Figure 7.41 represents state transition diagram of JK-flip-flop.

Output of the J-K-flip-flop goes from previous state 0 (Reset state) to next state 1 (Set state) when *J* is set to 1 and *K* is either cleared to 0 or set to 1.

Output of the J-K-flip-flop goes from previous state 1 (Set state) to next state 0 (Reset state) when *J* is either cleared to 0 or set to 1 to and *K* is set to 1.

Output of the J-K-flip-flop remains in state 0 (Reset state) when *J* is cleared to 0 and *K* is either clear to 0 or set to 1.

Output of the J-K-flip-flop remains in state 1 (Set state) when *J* is either cleared to 0 or set to 1 and *K* is cleared to 0.

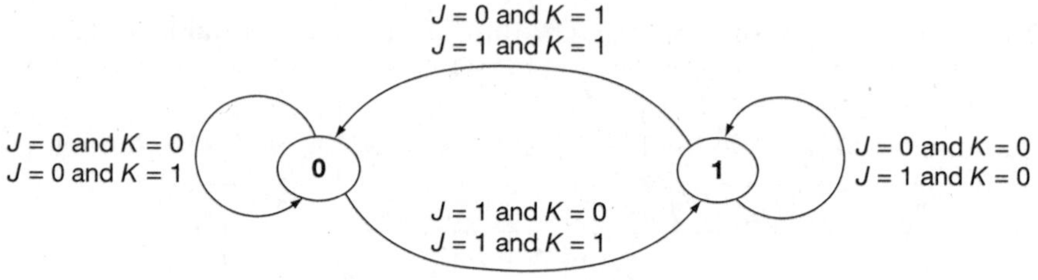

FIGURE 7.41 | State transition diagram of J-K flip-flop

7.7.3 | Excitation Table of JK-Flip-Flop

Excitation table of a flip-flop gives the inputs where output transition $Q(t) \rightarrow Q(t+1)$ of flip-flop is given. Inputs, *J* and *K* of flip-flop can be found when output and previous outputs are known. Flip-flop input is presented as a dependent function of transition $Q(t) \rightarrow Q(t+1)$ and comes later in the table. This is derived from flip-flop truth table or characteristic equation but more directly from its state transition diagram. Table 7.18 presents the excitation table of D-flip-flop.

TABLE 7.18 | Excitation table J-K-flip-flop

Outputs		Inputs	
$Q(t)$	$Q(t+1)$	J (Set)	K (Reset)
0	0	0	X
0	1	1	X
1	0	X	1
1	1	X	0

7.8 | T-FLIP-FLOP

Toggle (*T*) flip-flop is constructed from JK flip when set (*J*) input is given to reset (*K*) input. T-flip-flop has only one input and two outputs which are compliment to each other. T-flip-flop has the ability to store the data and toggle (change) the data. T-flip is used to design counters. T-flip-flop is also called as toggle flip-flop. Figure 7.42 shows diagram of T-flip-flop using JK-flip-flop. Flip-flop is triggered using positive level clock.

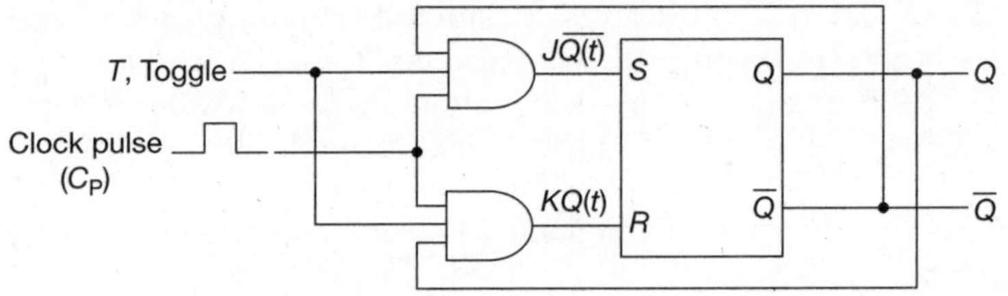

FIGURE 7.42 | Clocked T-flip-flop

Characteristic table of *T*-flip-flop is given in Table 7.19. Input *T* and previous output of flip-flop $Q(t)$ are acting as inputs. $Q(t+1)$ and its inverted signal acts as outputs. When

TABLE 7.19 | Characteristics of T-flip-flop

Inputs		Outputs	
$Q(t)$	T	$Q(t+1)$	$\overline{Q(t+1)}$
0	0	0	1
0	1	1	0
1	0	1	0
1	1	0	1

Case I: Let $Q_t = 0$ is 0 so that $\bar{Q}(t) = 1$. Input, *T* is 0. The output $Q(t+1)$ is 0 and $\bar{Q}(t+1)$ is 1. It is considered as hold state because $Q(t+1)$ is same as that of previous state $Q(t)$.

Case II: Let $Q(t)$ is 0 so that $\bar{Q}(t) = 1$. Input, *T* is 1. The output $Q(t+1)$ is 1 and $\bar{Q}(t+1)$ is 0. It is considered as Toggle state because $Q(t+1)$ is complement of previous state $Q(t)$.

Case III: Let $Q(t)$ is 1 so that $\bar{Q}(t) = 0$. Input T is 0. The output $Q(t+1)$ is 1 and $\bar{Q}(t+1)$ is 0. It is considered as hold state because $Q(t + 1)$ is same as that of previous state $Q(t)$ i.e. logic-1.

Case IV: Let $Q(t)$ is 1 so that $\bar{Q}(t) = 0$. Input, T is 1. The output $Q(t+1)$ is 1 and $\bar{Q}(t+1)$ is 0. It is considered as Toggle state because $Q(t + 1)$ is complement of previous state $Q(t)$.

Characteristic table given in Table 7.20 gives outputs when inputs J and K are known.

TABLE 7.20 | Characteristics of T-flip-flop

Inputs	Outputs		Remarks
T (Toggle)	**$Q(t + 1)$**	**$\bar{Q}(t + 1)$**	
0	$Q(t)$	$\bar{Q}(t)$	Save state
1	$\bar{Q}(t)$	$Q(t)$	Inverted state

7.8.1 | Characteristic Equation of T-Flip-Flop

Characteristic Table 7.21 of T-flip-flop is considered where $Q(t)$ and T are considered as inputs and $Q(t + 1)$ is taken as output. Two variables K-map is set up to find the simplified Boolean expressions giving characteristic equation. Figure 7.43 shows K-map, which is set up using truth table given in Table 7.21. No pair is formed.

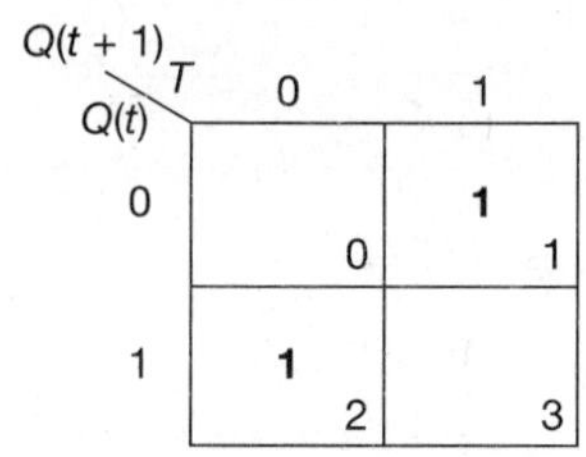

FIGURE 7.43 | K-map T-flip-flop

The characteristic equations are given below for T-flip-flop.

$$Q(t+1) = T\bar{Q}(t) + \bar{T}Q(t) \tag{7.21}$$

Characteristic table of clocked or gated T-flip-flop is given in Table 7.18. The flip-flop changes its state when positive (high) level clock is given otherwise it holds (saves) the data.

TABLE 7.21 | Characteristics of T-flip-flop

Inputs		Outputs	Remarks
T (Toggle)	**C_P (Clock Pulse)**	**$Q(t + 1)$**	
0	1	$Q(t)$	Save or hold state
1	1	$\bar{Q}(t)$	Toggle state
X	0	$Q(t)$	No change

7.8.2 | State Transition Diagram of T-Flip-Flop

The functional behaviour of the circuit is explained using finite number of states. State transition diagram is a very convenient tool to describe finite state of the flip-flop. Figure 7.44 represents state transition diagram of T-flip-flops.

Output of the T-flip-flop goes from previous state 0 to next state 1 when input T is set to 1. Output of the T-flip-flop goes from previous state 1 to next state 0 when input T is set to 1. Output of the T-flip-flop remains in state 0, when input T is cleared to 0. Output of the T-flip-flop remains in state 1, when T is cleared to 0.

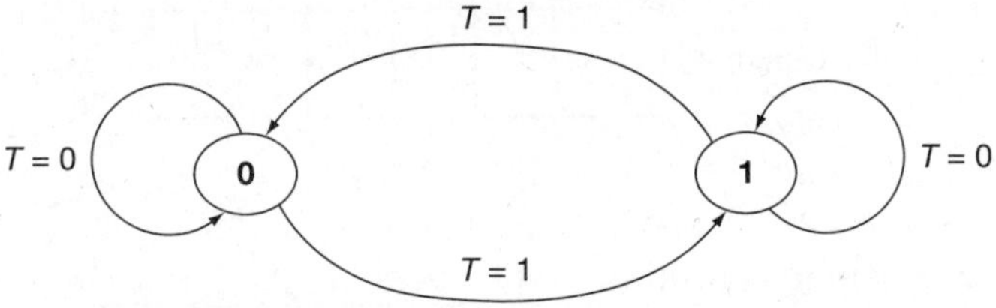

FIGURE 7.44 | State transition diagram of T-flip-flop

7.8.3 | Excitation Table of T-Flip-Flop

Excitation table of a flip-flop gives the inputs where output transition $Q(t) \rightarrow Q(t + 1)$ of flip-flop is given. Input, T of flip-flop can be found when output and previous outputs are known. Flip-flop input is presented as a dependent function of transition $Q(t) \rightarrow Q(t + 1)$ and comes later in the table. This is derived from T-flip-flop truth table or characteristic equation but more directly from its state transition diagram. Table 7.22 presents the excitation table of T-flip-flop.

TABLE 7.22 | Excitation table T-flip-flop

Outputs		Inputs
$Q(t)$	$Q(t+1)$	T (Toggle)
0	0	0
0	1	1
1	0	1
1	1	0

7.9 | RACE AROUND CONDITION

Let both the inputs of JK-flip-flop are 1 and the previous output, $Q(t)$ is logic-0 then the output of JK-flip-flop will toggle (change) the previous output after a propagation time, t_P giving output as logic-1.

Positive level clock is fed. Let the time duration of clock pulse is large than the propagation delay, t_P. After propagation time delay, the output will change from logic-1 to logic-0 (Figure 7.45). Output will oscillate from 0 to 1 state and 1 to 0 state. The output of flip-flop becomes uncertain. This situation is known as race around condition. Race around condition can be avoided if the time duration of active positive level clock pulse, t is reduced than the propagation delay, t_P.

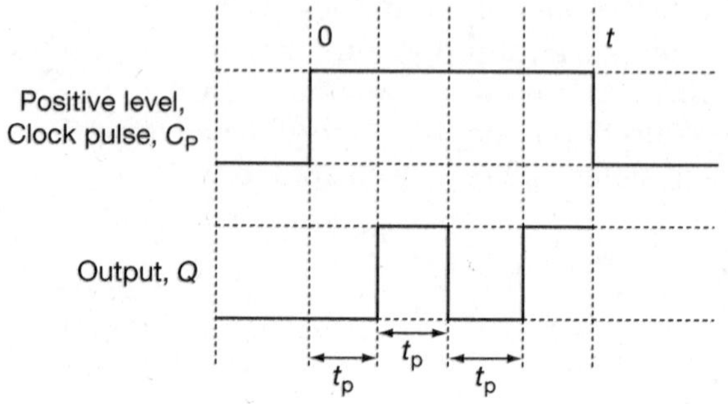

FIGURE 7.45 | Race around condition

7.10 | MASTER-SLAVE FLIP-FLOP

A better way to solve feedback timing problem is to make the flip-flop sensitive to the pulse transition rather than the pulse duration. In case, a flip-flop respond to the positive or negative edge transition only, instead of the entire pulse duration then the multiple transition problem can be eliminated. Edge triggering is achieved by designing the flip-flop to neglect one spike and trigger on the occurrence of the other spike. Two flip-flops are cascaded to form master-slave flip-flops. One flip-flop performs operation during positive edge triggering and SR-flip-flop copy the operation during negative edge triggering.

A master-slave flip-flop is constructed from same type of flip-flops. One flip-flop serves as master flip-flop and other flip-flop acts as slave flip-flop. Master flip-flop performs operation during positive level or edge triggering and slave flip-flop performs the operation during negative level or edge triggering while sharing the same clock pulse.

Figure 7.46 shows master-slave SR-flip-flop. To understand the operation let S is 1 and R is 0 with any output of the master and slave flip-flops.

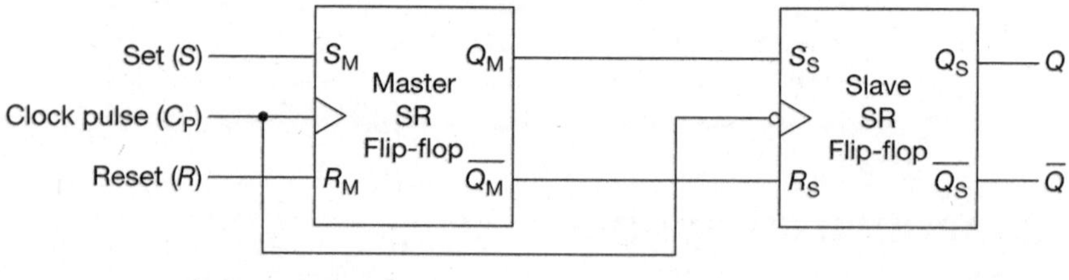

FIGURE 7.46 | Master-slave SR flip-flop

During positive edge triggering when clock goes from low to high level, master SR-flip-flop changes its state so Q_M is set to 1 state because S_M is 1 and R_M is 0. Slave SR-flip-flop remains active. Same state of Q_S is held.

During negative edge triggering when clock goes from high to low level, slave SR-flip-flop changes its state so Q_S is set to 1 state because S_S is 1 and R_S is 0. Master SR-flip-flop remains non-active. Same state of Q_M is held. Table 7.23 explains the truth table of the master-slave SR-flip-flop.

TABLE 7.23 | Characteristics of master and slave SR-flip-flop

Inputs			Outputs	Remarks
S (Set)	*R* (Reset)	CP	*Q*(*t* + 1)	
0	0	⎍	*Q*(*t*)	Previous state
0	1	⎍	0	Set
1	0	⎍	1	Clear (Reset)
1	1	⎍	?	Indeterminate state

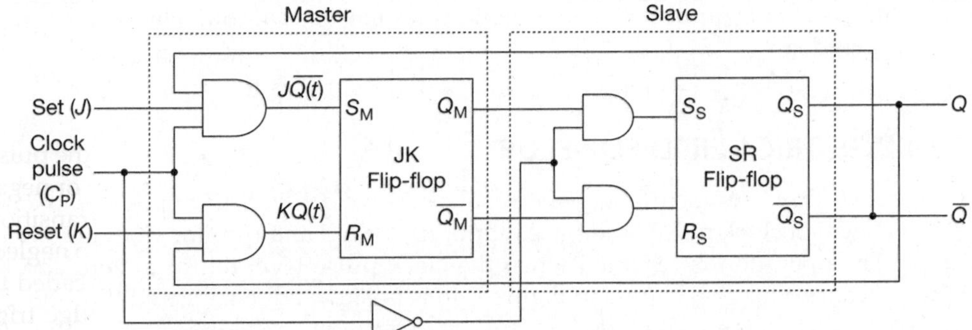

FIGURE 7.47 | Master-slave J-K flip-flop

Master-slave combination can be constructed for any type of flip-flop by adding a gated SR-flip-flop. Figure 7.47 explains the master-slave operation of JK-flip-flop. It requires two flip-flops where one is gated JK acts as master and other SR-flip-flop acts as slave. Output of slave flip-flop is fed as input to AND gate of master flip-flop which acts as JK-flip-flop. Information present at JK-flip-flop is transmitted to output of master flip-flop on positive level clock and is held there until negative level clock pulse appears. Table 7.24 explains the truth table of the master-slave JK-flip-flop. The truth table follows negative level triggering. This cascading avoids race around condition when J and K inputs are having logic-1 information.

TABLE 7.24 | Characteristics of master and slave J-K-flip-flop

Inputs			Outputs	Remarks
J (Set)	*K* (Reset)	CP	*Q*(*t* + 1)	
0	0	⎍⎍	*Q*(*t*)	Previous state
0	1	⎍⎍	0	Set
1	0	⎍⎍	1	Clear (Reset)
1	1	⎍⎍	$\bar{Q}(t)$	Toggle state

Let both inputs J and K are 1. Previous outputs of master and slave flip-flops are $Q_M = 0$ and $Q_S = 0$, respectively.

During high level clock pulse, information is transmitted to master flip-flop because $C_P \rightarrow 1$ for mater and slave flip-flop holds previous output because $C_P \rightarrow 0$. So outputs of master flip-flop are $Q_M \rightarrow 1$ and $\bar{Q}_M \rightarrow 0$ and outputs of slave flip-flop are $Q_S \rightarrow 0$ and $\bar{Q}_S \rightarrow 1$.

During low level clock pulse, information is transmitted to slave flip-flop because $C_P \rightarrow 1$ for slave and master flip-flop holds previous output because $C_P \rightarrow 0$. So outputs of master flip-flop are $Q_M \rightarrow 1$ and $\bar{Q}_M \rightarrow 0$ and outputs of slave flip-flop are $Q_S \rightarrow 1$ and $\bar{Q}_S \rightarrow 0$.

During next high level clock pulse, information is transmitted to master flip-flop because $C_P \rightarrow 1$ for master and slave flip-flop holds previous output because $C_P \rightarrow 0$ for slave. Input of master flip-flop is $S_M = JC_P\bar{Q}_S \Rightarrow 1.1.0 \Rightarrow 0$ and $R_M = KC_PQ_S \Rightarrow 1.1.1 \Rightarrow 1$ will reset the master flip-flop.

During low level clock pulse, information is transmitted to slave flip-flop and master flip-flop holds previous output. Slave flip-flop copies master's output. Hence, race around situation is avoided.

7.11 | EDGE-TRIGGERED FLIP-FLOP

Edge-triggered flip-flop synchronizes state change with clock pulse. Output is transmitted at a specific level of clock pulse otherwise inputs are locked and flip-flop does not perform operation. The operation is resumed when the clock pulse level return to zero and high level pulse occurs. Flip-flop shown in Figure 7.48 is positive edge-triggered flip-flop. Flip-flop can be negative edge-triggered also. Truth table is given in Table 7.25.

Let D be 0 and clock pulse, CP $\rightarrow$ 0. NAND-II and NAND-III gates give outputs $R \rightarrow 1$ and $S \rightarrow 1$ as one input; C_P is 0. Outputs of NAND-I and NAND-IV gates are 0 and 1, respectively. The output will not change any state. So, the inputs are locked.

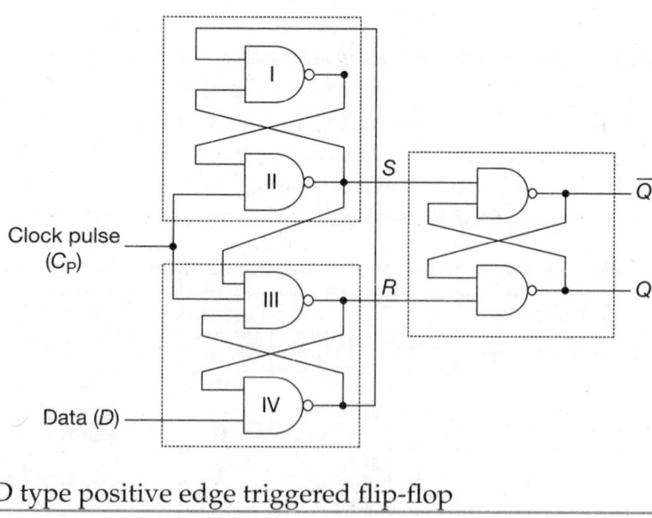

FIGURE 7.48 | D type positive edge triggered flip-flop

TABLE 7.25 | Characteristic table of gated SR-flip-flop

Inputs		Intermediate Output		Outputs	Remarks
D (Data)	C_P (Clock Pulse)	*S* (Set)	*R* (Reset)	*Q*(*t* + 1)	
0	1	1	0	0	Clear (Reset)
1	1	0	1	1	Set
0	0	1	1	*Q*(*t*)	Previous state or locked
1	0	1	1	*Q*(*t*)	Previous state or locked

Let D be 1 and clock pulse, CP $\rightarrow$ 0. NAND-II and NAND-III gates give outputs $R \rightarrow 1$ and $S \rightarrow 1$ as one input; C_P is 0. Outputs of NAND-I and NAND-IV gates are be 1 and 0, respectively. The output will not change any state. So, the inputs are locked.

Let D is 0 and clock pulse, $C_P \rightarrow 1$. Previous intermediate outputs are $R \rightarrow 1$ and $S \rightarrow 1$. As D is at logic-0 so output of NAND-IV gate is 1. Output of NAND-II gives output $S \rightarrow 0$. As all inputs of NAND-II gate are at logic-1. Both the inputs to NAND-I gate are logic-1 and gives output as logic-0. Output of NAND-II gate remains at logic-1. Finally, intermediate outputs are $R \rightarrow 0$ and $S \rightarrow 1$. Flip-flop is reset.

Let D is 1 and clock pulse, $C_P \rightarrow 1$. Previous intermediate outputs are $R \rightarrow 1$ and $S \rightarrow 1$. As D is 1 and $S \rightarrow 1$, so output of NAND-IV gate becomes 0. Outputs of NAND-I gate changes to 0 as all inputs are 1. Output of NAND-II gives output $S \rightarrow 0$. Inputs to NAND-III gate are 0 and 1 so outputs become 1. Finally, intermediate outputs are $R \rightarrow 1$ and $S \rightarrow 0$. Flip-flop is set as NAND gates are used for latch.

7.12 | CONVERSION OF FLIP-FLOPS

In the conversion from one type of flip-flop to another type of flip-flop, a combinational logic is designed based on the inputs of the flip-flop to be constructed (designed). The output of the designed combination logic is fed to the given flip-flop. So that output of constructed (designed) flip-flop can be achieved from the given flip-flop. Combinational logic is known as flip-flop conversion logic. Conversion of flip-flop process is shown in Figure 7.49. Steps involved are outlined below.

- Obtain the characteristic table of the flip-flop to be obtained (constructed or designed).
- Obtain the excitation table of the given flip-flop to find the input values.
- Find the relation between input of the given flip-flop based on the input and output of the obtained flip-flop.
- Any logic simplification technique can be applied to find the conversion logic
- Draw the diagram.
- Verify the characteristic table of obtained flip-flop.

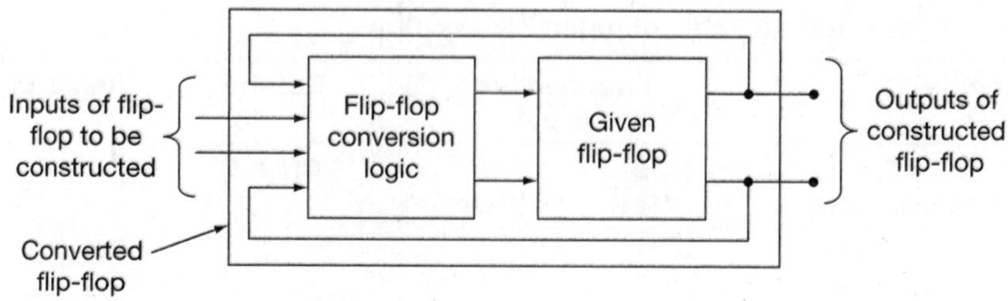

FIGURE 7.49 | Conversion of one flip-flop to another flip-flop

EXAMPLE 7.1 Convert SR-flip-flop to J-K-flip-flop.

SOLUTION J-K-flip-flop is to be constructed. SR-flip-flop is the given flip-flop. Figure 7.50 gives the conversion process. Characteristic table of J-K-flip-flop is given in Table 7.26. Based on $Q(t)$ and $Q(t + 1)$ outputs of J-K-flip-flop the excitation required for SR-flip-flop is obtained following excitation table of SR-flip-flop. SR excitation is given in Table 7.26.

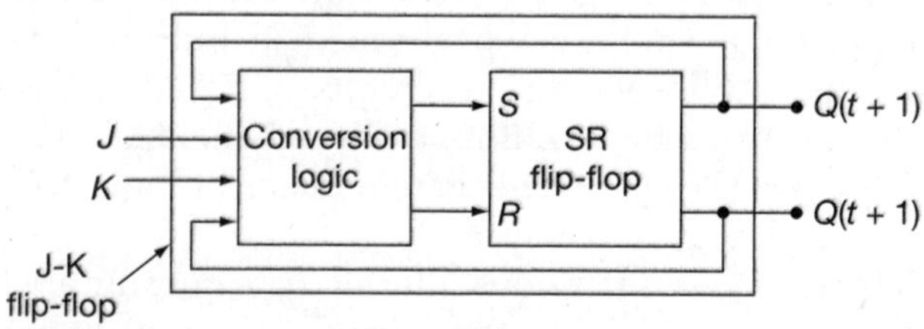

FIGURE 7.50 | Conversion of SR-flip-flop to JK-flip-flop

TABLE 7.26 | Conversion of SR-flip-flop to J-K-flip-flop

Minterm	Input of J-K-flip-flop		Output of J-K-flip-flop		Input of SR-flip-flop	
	J	K	$Q(t)$	$Q(t+1)$	S	R
m_0	0	0	0	0	0	X
m_1	0	0	1	1	X	0
m_2	0	1	0	0	0	X
m_3	0	1	1	0	0	1
m_4	1	0	0	1	1	0
m_5	1	0	1	1	X	0
m_6	1	1	0	1	1	0
m_7	1	1	1	0	0	1

Find the combination logic of $S(J, K, Q)$ and $R(J, K, Q)$ based on inputs J, K and $Q(t)$ using K-map simplification technique. Three-variable K-map is set up for $S(J, K, Q(t))$ and $R(J, K, Q(t))$ functions and is shown in Figures 7.51 and 7.52, respectively.

$$S(J,K,Q(t)) = \sum m(4,6) + \sum d(1,5)$$

$$R\left(J,K,Q(t)\right)=\sum m(3,7)+\sum d(0,2)$$

One pair {4, 6} is formed in Figure 7.51 and one pair {3, 7} is formed in Figure 7.52.

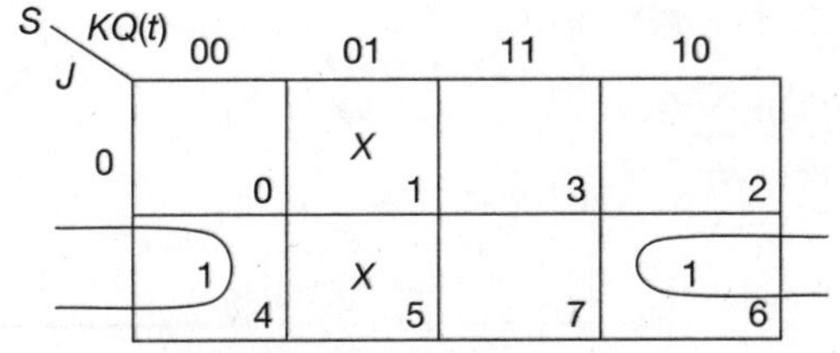

FIGURE 7.51 | K-map for S

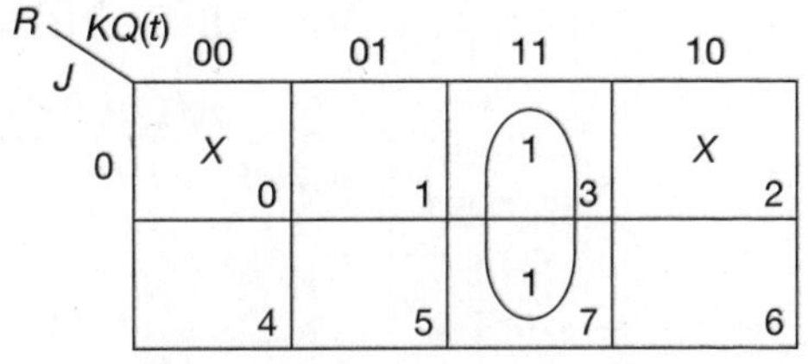

FIGURE 7.52 | K-map for R

The conversion logic is given below:

$$S=J\bar{Q}(t) \tag{7.22}$$

$$R=KQ(t) \tag{7.23}$$

The conversion diagram of JK-flip-flop from SR-flip-flop is given in Figure 7.53.

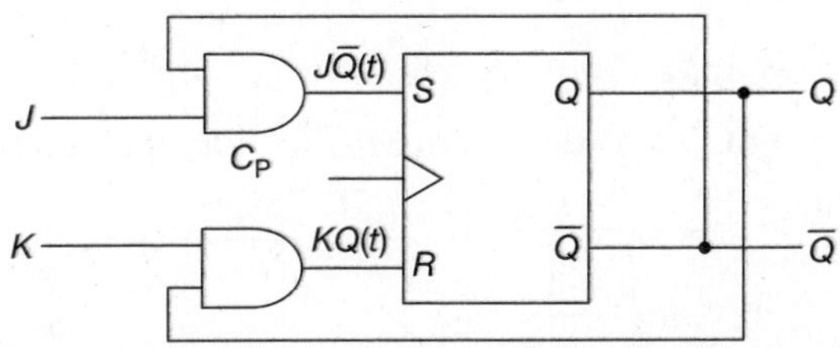

FIGURE 7.53 | J-K flip-flop from SR flip-flop

EXAMPLE 7.2 Convert SR-flip-flop to D-flip-flop.

SOLUTION D-flip-flop is to be constructed. SR-flip-flop is the given flip-flop. Characteristic table of D-flip-flop is given in Table 7.27. Based on $Q(t)$ and $Q(t+1)$ outputs of D-flip-flop the excitation required for SR-flip-flop is obtained following excitation table of SR-flip-flop. SR excitation is given in Table 7.27.

TABLE 7.27 | Conversion of SR-flip-flop to D-flip-flop

Minterm	Input of D-flip-flop	Output of SR-flip-flop		Input of SR-flip-flop	
	D	**Q(t)**	**Q(t + 1)**	**S**	**R**
m_0	0	0	0	0	X
m_1	0	1	0	0	1
m_2	1	0	1	1	0
m_3	1	1	1	X	0

Find the combination logic of $S(D, Q)$ and $R(D, Q)$ based on inputs D and $Q(t)$ using K-map simplification technique. Two-variable K-map is set up for $S(D, Q)$ and $R(D, Q)$ functions and is shown in Figures 7.54 and 7.55, respectively.

$$S(D,Q(t)) = \sum m(2) + \sum d(3)$$
$$R(D,Q(t)) = \sum m(1) + \sum d(0)$$

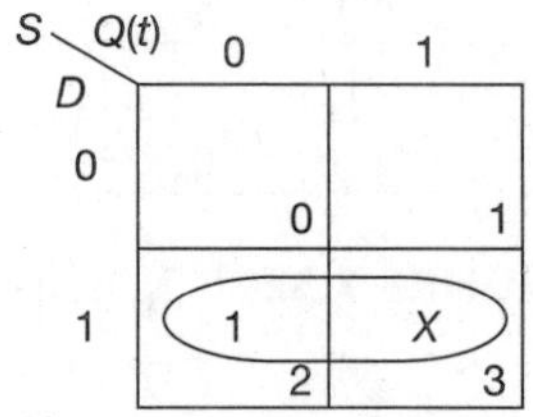

FIGURE 7.54 | K-map for S

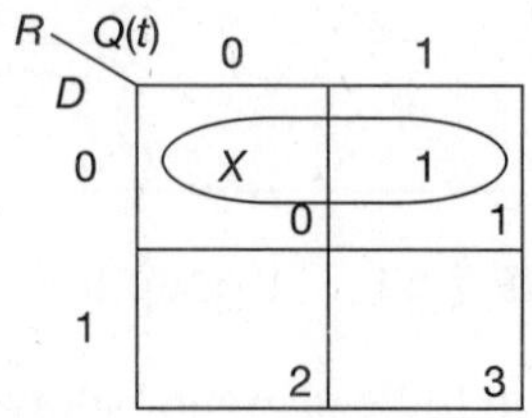

FIGURE 7.55 | K-map for R

One pair {2, 3} is formed in Figure 7.54 and one pair {0, 1} is formed in Figure 7.55. The conversion logic is given below.

$$S = D \tag{7.24}$$
$$R = \bar{D} \tag{7.25}$$

The conversion diagram of D-flip-flop from SR-flip-flop is given in Figure 7.56.

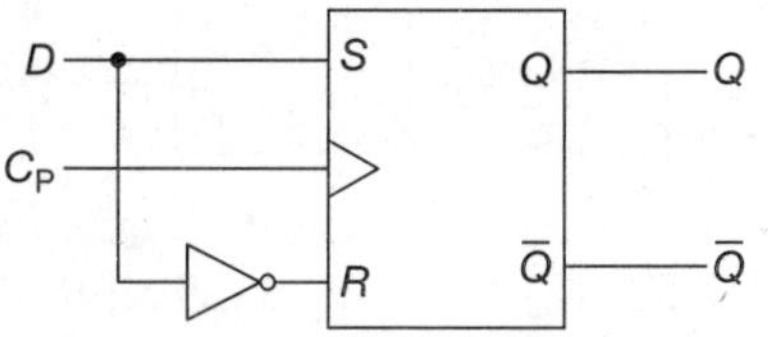

FIGURE 7.56 | D flip-flop from SR flip-flop

EXAMPLE 7.3 Convert J-K-flip-flop to T-flip-flop.

SOLUTION T-flip-flop is to be constructed. J-K-flip-flop is the given flip-flop. Characteristic table of T-flip-flop is given in Table 7.28. Based on $Q(t)$ and $Q(t+1)$ outputs of T-flip-flop the excitation required for J-K-flip-flop is obtained following excitation table of J-K-flip-flop. J-K excitation is given in Table 7.28.

TABLE 7.28 | Conversion of J-K-flip-flop to T-flip-flop

Minterm	Input of T-flip-flop	Output of T-flip-flop		Input of J-K-flip-flop	
	T	$Q(t)$	$Q(t+1)$	J	K
m_0	0	0	0	0	X
m_1	0	1	1	X	0
m_2	1	0	1	1	X
m_3	1	1	0	X	1

$$J(T,Q) = \sum m(2) + \sum d(1,3)$$

$$K(T,Q) = \sum m(3) + \sum d(0,2)$$

Find the combination logic of $J(T, Q)$ and $K(T, Q)$ based on inputs T and $Q(t)$ using K-map simplification technique. Two-variable K-maps are set up for $J(T, Q)$ and $K(T, Q)$ functions and are shown in Figures 7.57 and 7.58, respectively.

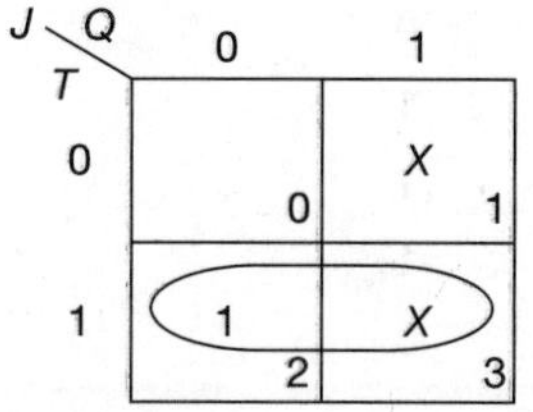

FIGURE 7.57 | K-map for J

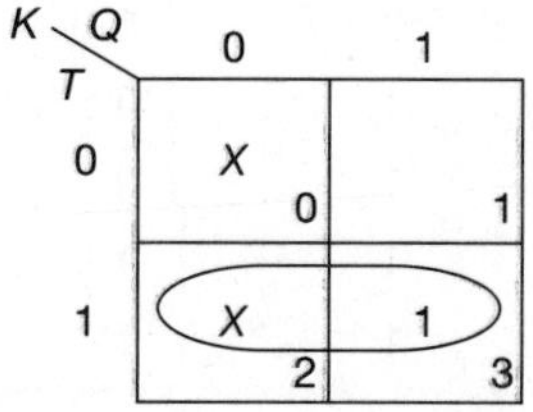

FIGURE 7.58 | K-map for K

One pair {2, 3} is formed in Figure 7.57 and {2, 3} are formed in Figure 7.58. The conversion logic is given below

$$J = T \tag{7.26}$$

$$K = T \tag{7.27}$$

The conversion diagram of J-K-flip-flop from T-flip-flop is given in Figure 7.59.

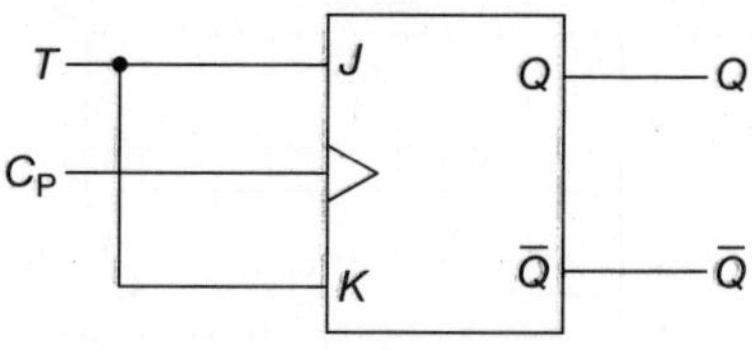

FIGURE 7.59 | T flip-flop from J-K flip-flop

EXAMPLE 7.4 Convert D-flip-flop to J-K-flip-flop.

SOLUTION SR-flip-flop is to be constructed. D-flip-flop is the given flip-flop. Characteristic table of J-K-flip-flop is given in Table 7.29. Based on $Q(t)$ and $Q(t + 1)$ outputs of J-K-flip-flop the excitation required for D flip-flop is obtained following excitation table of D-flip-flop. D excitation is given in Table 7.29. Find the combination logic of $D(J, K, Q)$ based on inputs J, K and $Q(t)$ using K-map simplification technique.

$$D(J,K,Q(t)) = \sum m(1,4,5,6)$$

Three-variable K-maps are set up for $D(J, K, Q)$ function and are shown in Figure 7.60. Two pairs {1, 5} and {4, 6} are formed in Figure 7.60. The conversion logic is given below:

$$D = J\overline{Q(t)} + \bar{K}Q(t) \tag{7.28}$$

The conversion diagram of J-K-flip-flop from D-flip-flop is given in Figure 7.61.

TABLE 7.29 | Conversion of D-flip-flop to J-K-flip-flop

Minterm	Input of J-K-flip-flop		Output of J-K-flip-flop		Input of D-flip-flop
	J	K	$Q(t)$	$Q(t+1)$	D
m_0	0	0	0	0	0
m_1	0	0	1	1	1
m_2	0	1	0	0	0
m_3	0	1	1	0	0
m_4	1	0	0	1	1
m_5	1	0	1	1	1
m_6	1	1	0	1	1
m_7	1	1	1	0	0

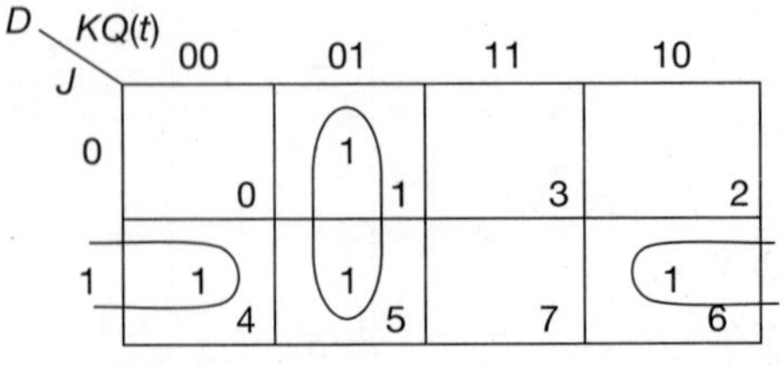

FIGURE 7.60 | K-map for D

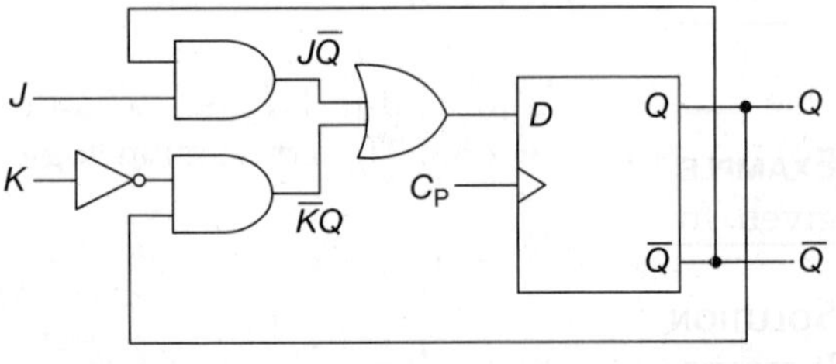

FIGURE 7.61 | J-K-flip-flop from D-flip-flop

EXAMPLE 7.5 Convert T-flip-flop to D-flip-flop.

SOLUTION D-flip-flop is to be constructed. T-flip-flop is the given flip-flop. Characteristic table of D-flip-flop is given in Table 7.30. Based on $Q(t)$ and $Q(t + 1)$ outputs of D-flip-flop the excitation required for T-flip-flop is obtained following excitation table of T-flip-flop. T-flip-flop excitation is given in Table 7.30. Find the combination logic of $D(T, Q)$ based on inputs T and $Q(t)$ using K-map simplification technique.

TABLE 7.30 | Conversion of D-flip-flop to T-flip-flop

Minterm	Input of D-flip-flop	Output of D-flip-flop		Input of T-flip-flop
	D	$Q(t)$	$Q(t+1)$	T
m_0	0	0	0	0
m_1	0	1	0	1
m_2	1	0	1	1
m_3	1	1	1	0

$$T(D,Q(t)) = \sum m(1,2)$$

Two-variable K-maps are set up for $T(D, Q)$ function and are shown in Figure 7.62. No pair is formed in Figure 7.62. The conversion logic is given below.

$$T = D\overline{Q(t)} + \bar{D}Q(t) \tag{7.29}$$

The conversion diagram of D-flip-flop from T-flip-flop is given in Figure 7.63.

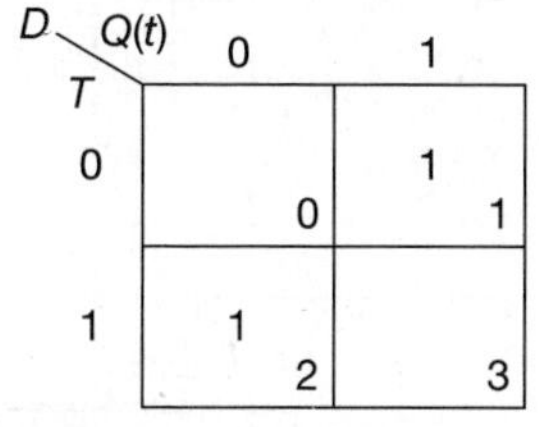

FIGURE 7.62 | K-map

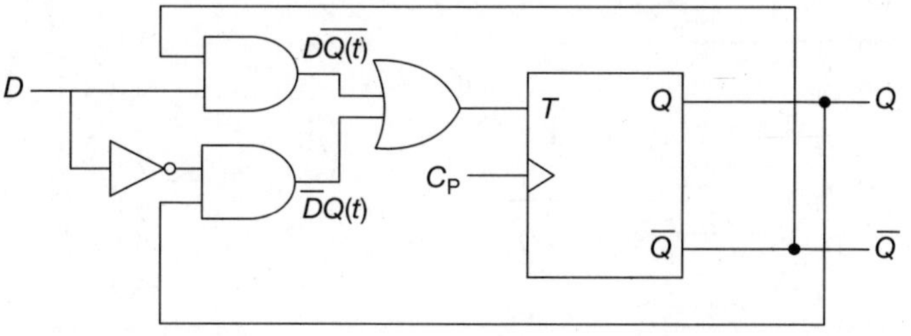

FIGURE 7.63 | D-flip-flop from T-flip-flop

EXAMPLE 7.6 Determine the output waveform of SR flip-flop when input waveform is given. Initial output is 0.

SOLUTION SR-flip-flop is shown in Figure 7.64. Flip-flop is positive edge triggered. It means the change in output takes place when pulse goes from 0 to 1 only. Output wave form is shown in Figure 7.65.

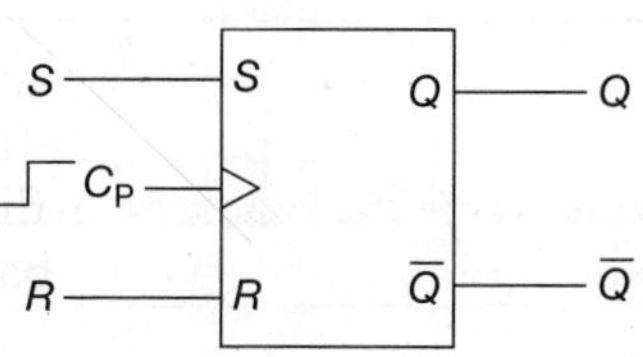

FIGURE 7.64 | Positive edge triggered SR flip-flop

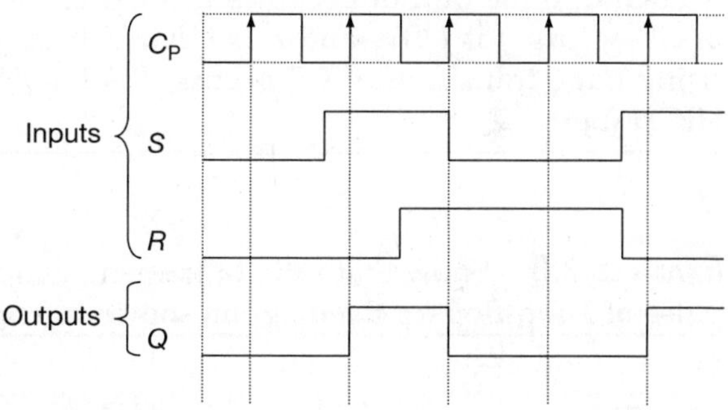

FIGURE 7.65 | Timing diagram of SR flip-flop

At first time when rising edge transition of CP occurs then S is LOW and R is low so Q will remain in LOW (unchanged). At second time, when rising edge transition of CP occurs, S is HIGH and R is LOW, so the output becomes HIGH (Set). At third time, when rising edge transition of CP occurs, S is LOW and R is HIGH, so the output becomes LOW (Reset). Fourth time when rising edge transition of CP occurs, S is LOW and R is HIGH so the output remains LOW (Reset). Fifth time rising edge transition of CP occurs, S is HIGH and R is LOW, so the output becomes HIGH (Set).

EXAMPLE 7.7 Determine the output waveform of J-K-flip-flop when input waveform is given. Initial output is 0.

SOLUTION J-K-flip-flop is shown in Figure 7.66. Flip-flop is positive edge triggered. It means the change in output will take place when pulse goes from state 0 to state 1 only. Output wave form is shown in Figure 7.67.

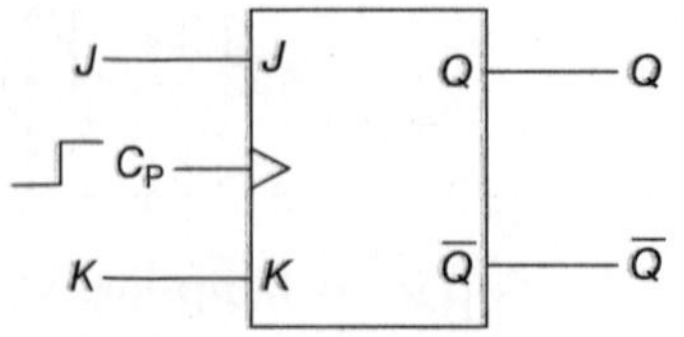

FIGURE 7.66 | Positive edge triggered J-K-flip-flop

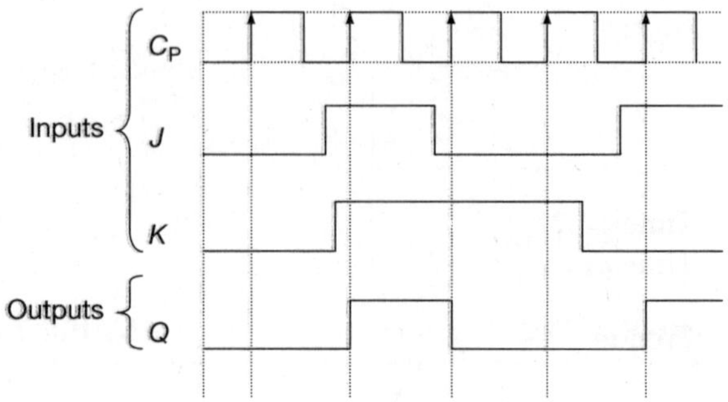

FIGURE 7.67 | Timing diagram of J-K flip-flop

At first time when rising edge transition of CP occurs then *J* is LOW and *K* is LOW, so *Q* will remain in LOW (unchanged). At second time, when rising edge transition of CP occurs, *J* is HIGH and *K* is HIGH, so the output becomes HIGH (Toggle) as previous output is LOW. At third time, when rising edge transition of CP occurs, *J* is LOW and *K* is HIGH, so the output becomes LOW (Reset). Fourth time when rising edge transition of CP occurs, *J* is LOW and *K* is HIGH, so the output remains LOW (Reset). Fifth time rising edge transition of CP occurs, *J* is HIGH and K is LOW, so the output becomes HIGH (Set).

EXAMPLE 7.8 Show that ratio of frequency input pulse is twice the frequency of output pulse of T-flip-flop for the diagram shown in Figure 7.68.

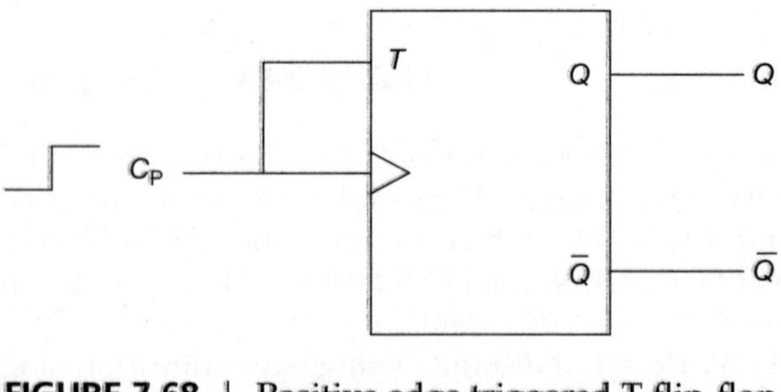

FIGURE 7.68 | Positive edge triggered T-flip-flop

SOLUTION Input of T-Flip-flop is same as CP. Timing diagram is given in Figure 7.69.

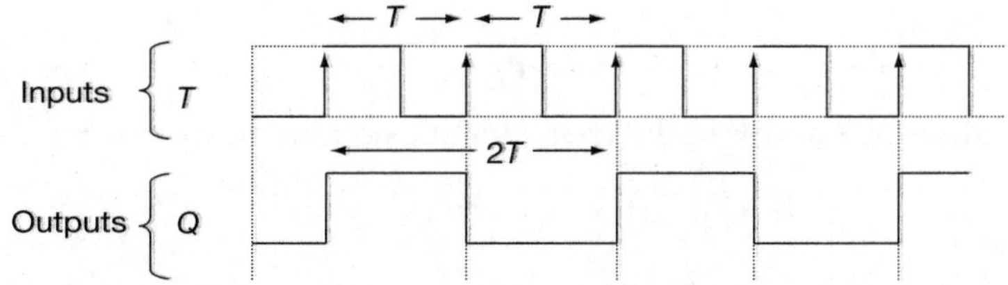

FIGURE 7.69 | Timing diagram of T-flip-flop

Time period of input clock of T-flip-flop = T
Time period of output clock of T-flip-flop = 2T

Frequency of input clock of T-flip-flop $f_{\text{input}} = \frac{1}{T}$

Frequency of output clock of T-flip-flop

$$f_{\text{output}} = \frac{1}{2T}$$

$$\frac{1}{T} = 2f_{\text{output}} \Rightarrow f_{\text{input}} = 2f_{\text{output}} = \frac{1}{T}$$

EXAMPLE 7.9 Write the characteristic equation of the flip-flop. The circuit diagram is shown in Figure 7.70.

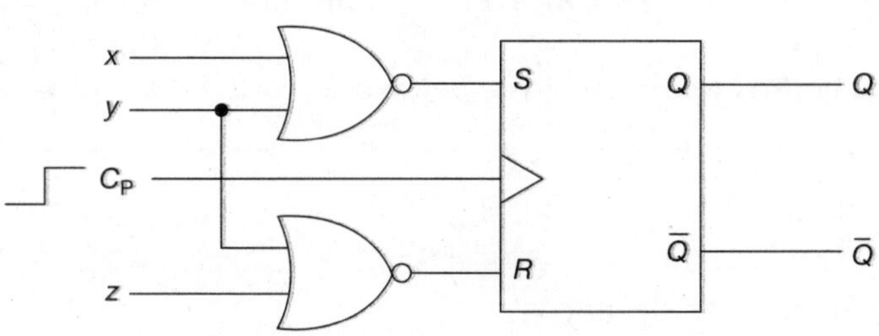

FIGURE 7.70 | SR flip-flop

SOLUTION Inputs of SR-flip-flop are given below:

$$S = \overline{x + y} \quad \text{(i)}$$

$$R = \overline{y + z} \quad \text{(ii)}$$

Characteristic equation of SR-flip-flop is

$$Q(t+1) = S + \bar{R}Q(t) \tag{iii}$$

and

$$SR = 0 \tag{iv}$$

Substitute the values of S and R in the characteristic equation given as Eq. (iii) and Eq. (iv)

$$Q(t+1) = \overline{x+y} + (\overline{\overline{y+z}})Q(t)$$

$$(\overline{x+y})(\overline{y+z}) = 0$$

On simplification,

$$Q(t+1) = \overline{x+y} + (y+z)Q(t) \quad \textbf{Ans.}$$

$$(\bar{x}\,\bar{y})(\bar{y}\,\bar{z}) = 0 \Rightarrow (\bar{x}\,\bar{y}\,\bar{z}) = 0 \quad \textbf{Ans.}$$

EXAMPLE 7.10 Obtain the characteristic table and equation of the flip-flop. The circuit diagram is shown in Figure 7.71.

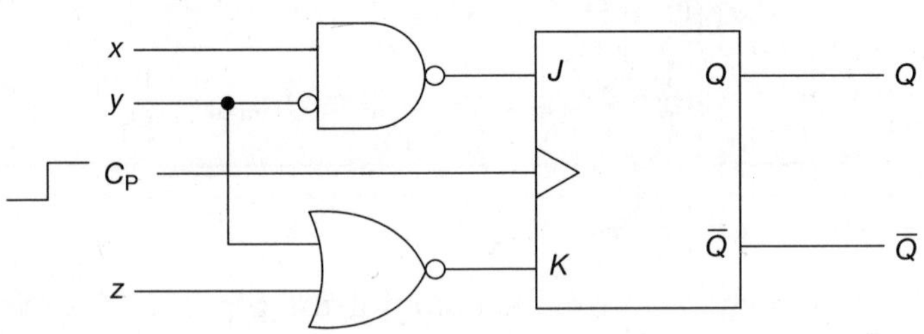

FIGURE 7.71 | JK flip-flop

SOLUTION Inputs of JK-flip-flop are given below:

$$J = \overline{x\bar{y}} \tag{i}$$

$$K = \overline{y+z} \tag{ii}$$

Characteristic equation of JK-flip-flop is

$$Q(t+1) = J\overline{Q(t)} + \bar{K}Q(t) \tag{iii}$$

Substituting the values of J and K in the characteristic equation given as Eq. (iii),

$$Q(t+1) = (\overline{x\bar{y}})\overline{Q(t)} + \left(\overline{y+z}\right)Q(t) \tag{iv}$$

Characteristic table for circuit shown in Figure 7.71 is given below in Table 7.31. Inputs for the truth table are x, y, z and $Q(t)$. Inputs for JK-flip-flop are J, K and $Q(t)$.

TABLE 7.31 | Characteristic table

Inputs			Inputs to JK-flip-flop			Output
x	y	z	$Q(t)$	J $(x\bar{y})$	K $\overline{(y+z)}$	$Q(t)$
0	0	0	0	1	1	1
0	0	0	1	1	1	0
0	0	1	0	1	0	1
0	0	1	1	1	0	1
0	1	0	0	1	0	1
0	1	0	0	1	0	1
0	1	1	1	1	0	1
0	1	1	0	1	0	1
1	0	0	0	0	1	0
1	0	0	1	0	1	0
1	0	1	0	0	0	0
1	0	1	1	0	0	1
1	1	0	0	1	0	1
1	1	0	0	1	0	1
1	1	1	1	1	0	1
1	1	1	0	1	0	1

EXAMPLE 7.11 Determine the output sequence for the given serial data fed to the flip-flop shown in Figure 7.72. Assume Q is at 0 logic initially. There is one clock pulse for each bit.

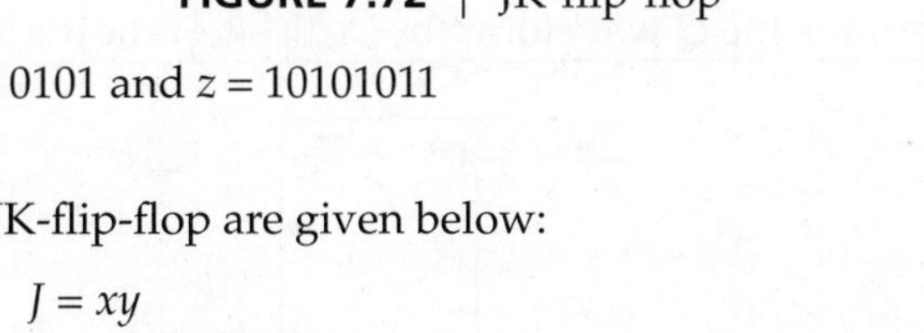

FIGURE 7.72 | JK-flip-flop

$x = 1010\ 0101$, $y = 1110\ 0101$ and $z = 10101011$

SOLUTION Inputs of JK-flip-flop are given below:

$$J = xy \quad \text{(i)}$$

$$J = (x_7x_6x_5x_4x_3x_2x_1x_0) \,.\, (y_7y_6y_5y_4y_3y_2y_1y_0)$$

Perform bitwise AND logical operation, e.g. x_0 *AND* y_0, x_1 *AND* y_1, and x_7 *AND* y_7

$$x = (1010\ 0101)$$

$$y = (1110\ 0101)$$

$$xy = (1010\ 0101)$$

$$K = y \oplus z$$

$$K = (x_7x_6x_5x_4x_3x_2x_1x_0) \oplus (y_7y_6y_5y_4y_3y_2y_1y_0) \quad \text{(ii)}$$

TABLE 7.32 | Truth table

Q(*t*)	*J*	*K*	*Q*(*t* + 1)
0 (Given)	1 (LSB)	0 (LSB)	1
1	0	1	0
0	1	1	1
1	0	1	0
0	0	0	0
0	1	0	1
1	0	1	0
0	1 (MSB)	0 (MSB)	1

Perform bitwise Exclusive OR logical operation. e.g. $x_0 \oplus y_0$, $x_1 \oplus y_1$ and $x_7 \oplus y_7$

$$y = (1110\ 0101)$$

$$z = (1010\ 1011)$$

$$y \oplus z = (0100\ 1110)$$

$$J = (1010\ 0101)$$

$$K = (0100\ 1110)$$

$$Q = (1010\ 0101) \quad \textbf{Ans.}$$

It is explained in truth Table 7.32. Output $Q(t + 1)$ becomes $Q(t)$ for next bit.

EXAMPLE 7.12 Determine the Q waveform for D-flip-flop shown in Figure 7.73.

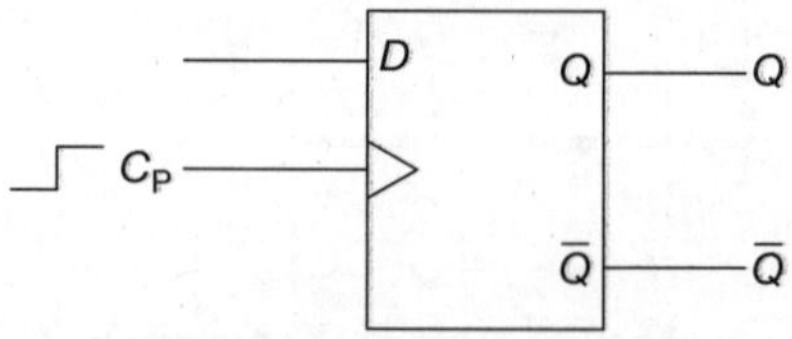

FIGURE 7.73 | D-flip-flop

SOLUTION Input of D-flip-flop is given in Figure 7.74. Output remains same till the positive edge is detected.

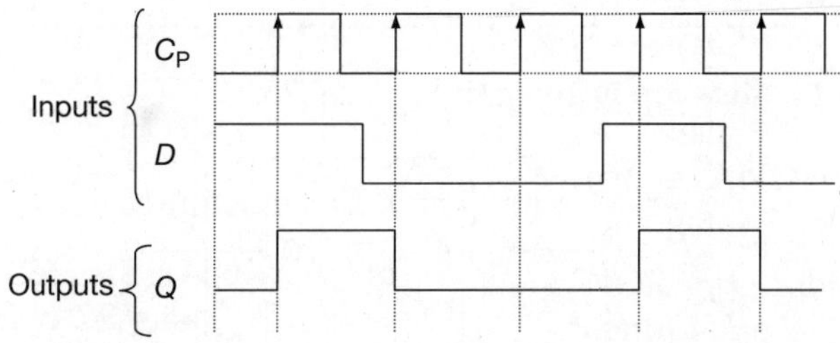

FIGURE 7.74 | Timing diagram of D-flip-flop

EXAMPLE 7.13 A sequential circuit using D-flip-flop is shown in Figure 7.75. Determine the circuit output.

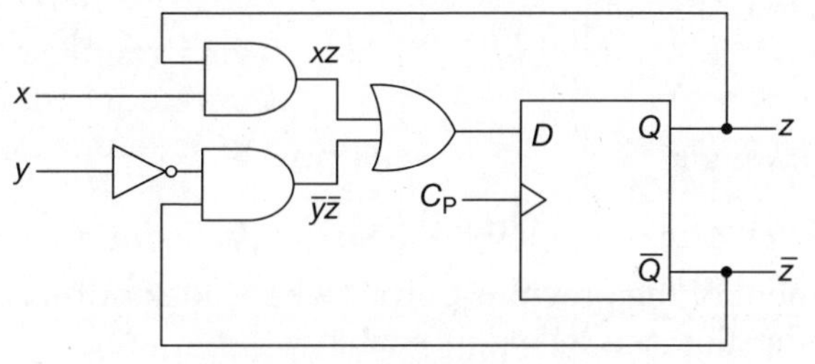

FIGURE 7.75 | D flip-flop

SOLUTION Input of D-flip-flop is given below which is based on x and y inputs of sequential circuit. Output of circuit is z.

$$D = xz + \bar{y}\,\bar{z}$$

The output of D-flip-flop is

$$Q(t+1) = D$$

$$Q(t+1) = xz + \bar{y}\,\bar{z} \tag{i}$$

When output is fed back to inputs, the characteristic equation can be equated to J-K-flip-flop. Characteristic equation of JK-flip-flop is

$$Q(t+1) = J\overline{Q(t)} + \bar{K}Q(t) \tag{ii}$$

Equation (i) can be rewritten as given below:

$$Q(t+1) = \bar{y}\,\bar{z} + xz \tag{iii}$$

Comparing Eq. (ii) with Eq. (iii), we get

$$J = \bar{y},\ K = x \text{ and } Q(t) = z.$$

So, above circuit acts as JK-flip-flop when excitation is given by $J = \bar{y},\ K = x$ to get output $Q(t) = z$.

EXAMPLE 7.14 Verify D-flip-flop acts as toggle switch when $\bar{Q}$ output is connected to input of D-flip-flop.

SOLUTION Circuit of D-flip-flop is given in Figure 7.76.

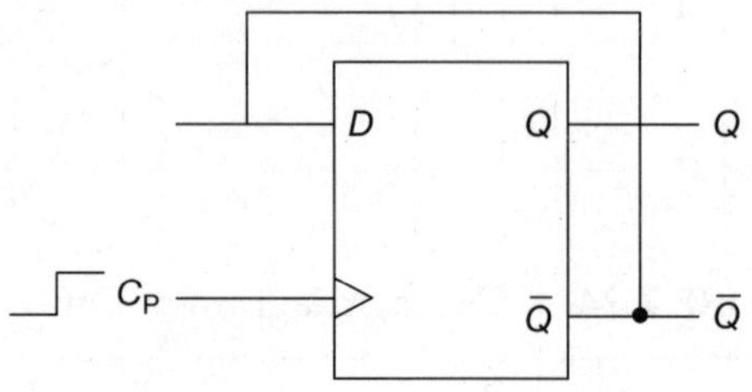

FIGURE 7.76 | D-flip-flop

The output of D-flip-flop is

$$Q(t+1) = D \tag{i}$$

But $D = \overline{Q(t)}$
Substituting in Eq. (i), we get

$$Q(t+1) = \overline{Q(t)} \tag{ii}$$

Output is complement of the previous output when clock pulse is positive edge transition. Hence, D-flip-flop acts as toggle switch.

EXAMPLE 7.15 Construct truth table for the sequential circuit shown in Figure 7.77. Show that circuit acts as T-flip-flop.

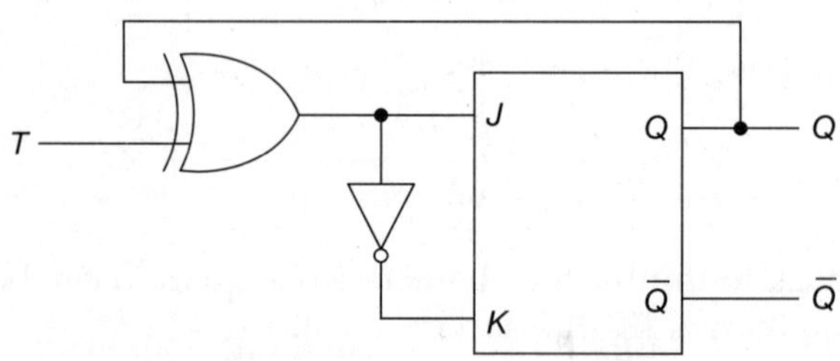

FIGURE 7.77 | JK flip-flop

SOLUTION Input to JK-flip-flop is given below.

$$J = T \oplus Q(t) \tag{i}$$

$$K = \overline{T \oplus Q(t)} \tag{ii}$$

Truth table for the circuit is given in Table 7.33.

TABLE 7.33 | Truth table

T	*Q(t)*	*J*	*K*	*Q(t + 1)*
0	0	0	1	0
0	1	1	0	1
1	0	1	0	1
1	1	0	1	0

Truth table gives the output equivalent to T-flip-flop.
Characteristic equation of JK-flip-flop is

$$Q(t+1) = J\overline{Q(t)} + \bar{K}Q(t) \qquad \text{(iii)}$$

Substituting the value of *J* and *K* in the above equation, we get

$$Q(t+1) = (T \oplus Q(t))\overline{Q(t)} + (\overline{\overline{T \oplus Q(t)}})Q(t) \qquad \text{(ii)}$$

On simplification,

$$Q(t+1) = (T \oplus Q(t))\overline{Q(t)} + (T \oplus Q(t))Q(t)$$
$$Q(t+1) = (T \oplus Q(t))\left[\overline{Q(t)} + Q(t)\right]$$
$$Q(t+1) = (T \oplus Q(t))$$

It is characteristic equation of T-flip-flop.

EXAMPLE 7.16 Obtain truth table and characteristic equation of the JK-flip-flop with an inverter between external input *X* and input of flip-flop *J*.

SOLUTION Input to JK-flip-flop is shown in Figure 7.78. Truth table is given in Table 7.34.

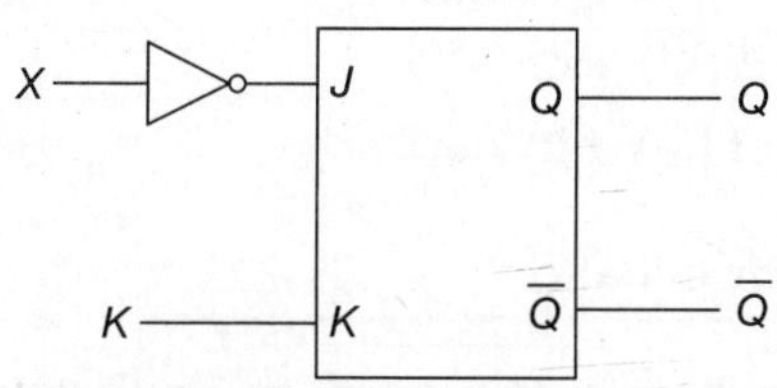

FIGURE 7.78 | JK flip-flop

TABLE 7.34 | Truth table

Q(t)	*X*	*K*	*J*	*Q(t + 1)*
0	0	0	1	1
0	0	1	1	1
0	1	0	0	0
0	1	1	0	0
1	0	0	1	1

(Continued)

TABLE 7.34 | (Continued)

Q(t)	X	K	J	Q(t + 1)
1	0	1	1	0
1	1	0	0	1
1	1	1	0	0

Characteristic equation is obtained using K-map given in Figure 7.79. Two pairs are formed {0, 1} and {4, 6}. The expression is given below.

$$Q(t+1) = \bar{X}\overline{Q(t)} + \bar{K}Q(t)$$

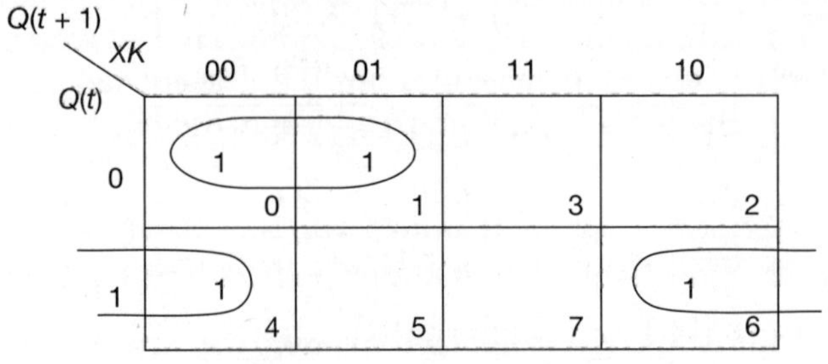

FIGURE 7.79 | K-map

EXAMPLE 7.17 A new clocked U-V-flip-flop is designed with two inputs U and V. The flip-flop functions are stated below:

- If U is 0 and V is 0, the flip-flop is set.
- If U is 0 and V is 1, the flip-flop toggles.
- If U is 1 and V is 0, the flip-flop holds.
- If U is 1 and V is 1, the flip-flop is reset.

(a) Obtain the truth table of U-V-flip-flop.
(b) Write excitation table,
(c) Obtain the characteristic equation.

SOLUTION Definition of flip-flop is tabulated as truth table in Table 7.35. Following Table 7.35, characteristic table is given in Table 7.36. Excitation is obtained from characteristic table and is given in Table 7.37.

TABLE 7.35 | Truth table

U	V	Q(t + 1)	Remarks
0	0	1	Set
0	1	$\bar{Q}(t)$	Toggle
1	0	$Q(t)$	Hold
1	1	0	Reset

TABLE 7.36 | Characteristic table

Minterm	Q(t)	U	V	Q(t + 1)
m_0	0	0	0	1
m_1	0	0	1	1
m_2	0	1	0	0
m_3	0	1	1	0
m_4	1	0	0	1
m_5	1	0	1	0
m_6	1	1	0	1
m_7	1	1	1	0

TABLE 7.37 | Excitation table

Q(t)	Q(t + 1)	U	V	Remarks
0	0	1	X	Consider minterms 2 and 3. *U* is 1 and *V* may be either 0 or 1
0	1	0	X	Consider minterms 0 and 1. *U* is 0 and *V* may be either 0 or 1
1	0	X	1	Consider minterms 5 and 7. *V* is 1 and *U* may be either 0 or 1
1	1	X	0	Consider minterms 4 and 6. *V* is 0 and *U* may be either 0 or 1

Characteristic equation is obtained from truth table giving output as 1. K-map is given in Figure 7.80.

$$Q(Q(t),U,V) = \sum m(0,1,4,6)$$

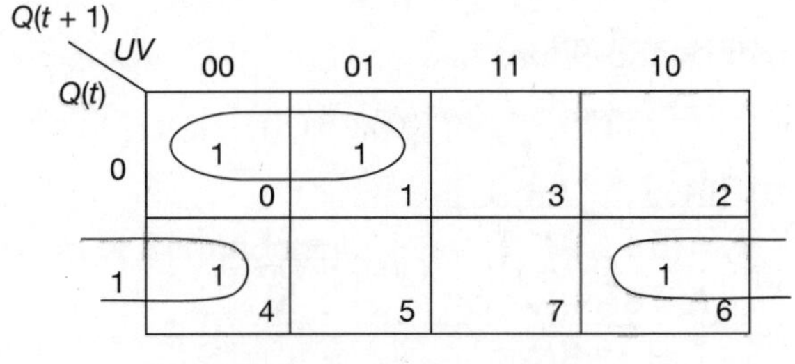

FIGURE 7.80 | K-map

Two pairs are formed {0, 1} and {4, 6}. The expression is given below.

$$Q(t+1) = \bar{U}\,\overline{Q(t)} + \bar{V}Q(t) \quad \text{(i)}$$

When output depends on previous output, flip-flop has J-K-flip-flop like characteristics. Characteristic equation of JK-flip-flop is

$$Q(t+1) = J\,\overline{Q(t)} + \bar{K}Q(t) \quad \text{(ii)}$$

Comparing Eq. (i) with Eq. (ii), we get

$$J = \bar{U} \text{ and } K = V$$

Circuit diagram of the UV-flip-flop is shown in Figure 7.81.

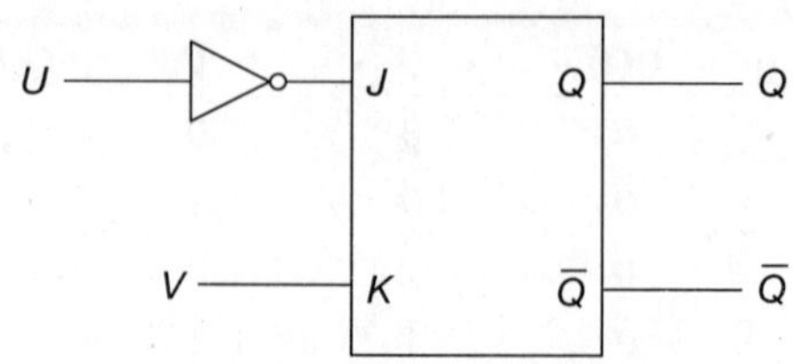

FIGURE 7.81 | UV-flip-flop

EXAMPLE 7.18 Write the characteristic equation of the flip-flop. The circuit diagram is shown in Figure 7.82.

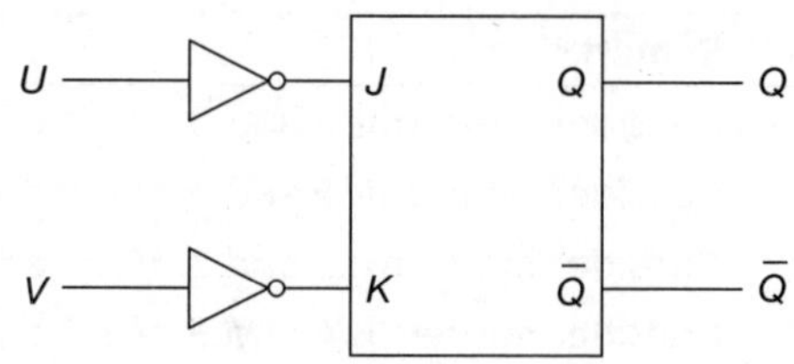

FIGURE 7.82 | UV-flip-flop

SOLUTION Inputs of SR-flip-flop are given below:

$$J = \bar{U} \quad \text{(i)}$$

$$K = \bar{V} \quad \text{(ii)}$$

Characteristic equation of JK-flip-flop is

$$Q(t+1) = J\overline{Q(t)} + \bar{K}Q(t) \quad \text{(iii)}$$

Substitute Eq. (i) and Eq. (ii) in Eq. (iii) to get

$$Q(t+1) = \bar{U}\overline{Q(t)} + \bar{\bar{V}}Q(t)$$

On simplification,

$$Q(t+1) = \bar{U}\overline{Q(t)} + VQ(t) \quad \textbf{Ans.}$$

7.13 | LIST OF FLIP-FLOP ICs

A list of commonly used IC type numbers used as flip-flop flops belonging to CMOS, TTL and ECL logic families is given in Tables 7.38–7.40. Information regarding the pin connection diagram, truth table, etc., in respect of these IC numbers is given in the handbook or companion website.

TABLE 7.38 | Flip-flop of CMOS logic family

IC Number	Function of IC
4013	Dual D-type flip-flop
4027	Dual J-K-flip-flop
4042	Quad D-type latch
4044	Quad SR latch with three-state output
4076	Quad D-type flip-flop with three-state output
40174	Hex D-type flip-flop
40175	Quad D-type flip-flop
4511	BCD to seven-segment latch/decoder/driver
4543	BCD to seven-segment latch/decoder/driver for LCD
4723	Dual four-bit addressable latch
4724	Eight-bit addressable latch

TABLE 7.39 | IC of TTL logic family

IC Number	Function of IC
54/7473	Dual J-K negative edge-triggered flip-flop with CLEAR
54/7474	Dual D-type positive edge-triggered flip-flop with PRESET and CLEAR
54/7475	Four-bit D-type latch
54/7476	Dual J-K-flip-flop with PRESET and CLEAR
54/7478	Dual J-K-flip-flop with PRESET and CLEAR
54/74107	Dual J-K-flip-flop with CLEAR
54/74109	Dual J-K-positive edge-triggered flip-flop with PRESET and CLEAR
54/74112	Dual J-K-negative edge-triggered flip-flop with PRESET and CLEAR
54/74113	Dual J-K-negative edge-triggered flip-flop with PRESET TTL
54/74114	Dual J-K-negative edge-triggered flip-flop with PRESET and CLEAR
54/74174	Hex D-type flip-flop with CLEAR
54/74175	Quad edge-triggered D-type flip-flop with CLEAR
54/74256	Dual four-bit addressable latch
54/74259	Eight-bit addressable latch
54/74273	Octal D-type flip-flop with MASTER RESET
54/74279	Quad SET/RESET latch
54/74373	Octal transparent latch (three-state)
54/74374	Octal D-type flip-flop (three-state)
54/74377	Octal D-type flip-flop with common ENABLE
54/74378	Hex D-type flip-flop with ENABLE

(*Continued*)

TABLE 7.39 | (Continued)

IC Number	Function of IC
54/74379	Four-bit D-type flip-flop with ENABLE
54/74533	Octal transparent latch (three-state)
54/74534	Octal D-type flip-flop (three-state)
54/74573	Octal D-type latch (three-state)
54/74574	Octal D-type flip-flop (three-state)

TABLE 7.40 | IC of ECL Logic Family

IC Number	Function of IC
MC10130	Quad D-type latch
MC10131	Dual D-type master/slave flip-flop
MC10133	Quad D-type latch (negative transition)
MC10135	Dual J-K master/slave flip-flop
MC10153	Quad latch (positive transition)
MC10168	Quad D-type latch
MC10176	Hex D-type master/slave flip-flop
MC10231	High-Speed dual D-type M/S flip-flop
MC1666	Dual clocked SR flip-flop
MC1668	Dual clocked latch
MC1670	D-type master/slave flip-flop

SUMMARY

- Asynchronous sequential circuit is a sequential circuit whose behaviour depends on the change of input signal at any time.
- A combinational circuit with feedback is termed as asynchronous sequential circuit.
- Synchronous sequential circuit is a sequential circuit, whose behaviour is defined from the knowledge of its signal at discrete instants of time.
- The synchronization is achieved by a timing device known as a system clock which generates a train of clock pulses.
- Synchronous circuits are also known as clocked-sequential circuit.
- A multivibrator is a pulse generator circuit which produces a rectangular-wave output.
- An astable multivibrator is also called a free-running multivibrator. The astable multivibrator generates a continuous flow of pulses.
- A bistable multivibrator is also called a flip-pop. The bistable multivibrator is gives one out of two stable states (set or reset).

- A monostable multivibrator is called a one-shot multivibrator. When the one-shot is triggered, the multivibrator generates a single short pulse.
- Clock: Signal generated by an oscillator that receives the flow of current carriers.
- Flip-flops are used for data storage, counting, frequency division, parallel-to-serial and serial-to-parallel data conversion, etc.
- An unclocked flip-flop is called a latch because the output of the flip-flop latches on to a I or a 0 immediately after the input is applied.
- Non-gated latches are called asynchronous latches and clocked (but not edge-triggered) latches are called synchronous latches.
- A latch is constructed using two cross-coupled N AND gates or NOR gates.
- **Latch:** Fundamental binary storage device and is called a flip-flop.
- **Flip-flop:** Basic sequential logic device having two stable states. It can serve as a memory device. Flip-flop is called a bistable multivibrator.
- The NOR gate SR latch is an active High SR latch.
- The NAND gate SR latch is an active-LOW SR latch.
- The clocked D latch is called a transparent D latch, because its output follows the input when the clock is HIGH.
- Flip-flops may be level-triggered or edge-triggered.
- The level-triggered flip-flops are those which respond to changes in inputs when the clock is HIGH.
- The edge-triggered flip-flops are those which respond only to inputs present at the transition of the clock pulse.
- Edge triggering is also called dynamic triggering.
- **SR-flip-flop:** Flip-flop with at least set, reset, and hold modes of operation. Fundamental latching (memory) circuit.
- **D-flip-flop:** Flip-flop with at least set and reset modes of operation. D-flip-flop is called a data or delay flip-flop.
- **J-K-flip-flop:** Flip-flop with at least set, reset, toggle and hold modes of operation. It is very adaptable.
- **T-flip-flop:** It is toggle-flip-flop. The output toggles to the opposite logic state on repeated clock pulse.
- Due to wiring between components, a clock pulse may undergo varying degrees of delay before it arrives at the flip-flops. It is called clock skew.
- **Master-slave flip-flop is made up of two flip-flops:** a master and a slave. It performs synchronous operation of flip-flop by overcoming the problem of race in clocked flip-flop.
- **Set-up Time:** Setup time (t_s) is the minimum time interval that the data present immediately before the active transition of the clock signal during which the input (data input) must be maintained at the proper level.
- Undesired output is obtained, if time interval t is less than setup time, t_s.
- **Hold Time:** Hold rime (t_h) is the minimum time interval immediately after the active transition of the clock signal during which the control input (data input) must be maintained at the level. Undesired output is obtained, if time interval t is less than hold time, t_h.

- **Propagation Delay:** The flip-flop output changes exactly on the active transition of the clock signal, which in fact does not happen practically. Propagation delay (t_p) is amount of time taken for the output of flip-flop to change states after an input trigger is received.
- **High Clock Pulse Width (t_{CH}):** It is the minimum time for which the clock remains high.
- **Low Clock Pulse Width (t_{CL}):** It is the minimum time for which the clock remains low.
- **Maximum Frequency:** Maximum frequency of clock signal is the highest frequency at which flip-flop performs proper and stable operation.
- **Power Dissipation:** The power dissipation of a flip-flop is the total power consumption of the device. Power dissipation is product of supply voltage and current drawn from supply by the device. The power dissipation is usually in mill watts (mW).

Characteristic equations of flip-flops:

SR-Flip-Flop: $Q(t+1) = S + \bar{R}Q(t)$ provided $SR = 0$

D-Flip-Flop: $Q(t+1) = D$

J-K-Flip-Flop: $Q(t+1) = J\bar{Q}(t) + \bar{K}Q(t)$

T-Flip-Flop: $Q(t+1) = T \oplus Q(t)$

Characteristic tables of flip-flops

SR-flip-flop			J-K-flip-flop			D-flip-flop		T-flip-flop	
Inputs		Output	Inputs		Output	Input	Output	Input	Output
S (Set)	*R* (Reset)	*Q*(*t* + 1)	*J* (Set)	*K* (Reset)	*Q*(*t* + 1)	*D* (Data)	*Q*(*t* + 1)	*T* (Toggle)	*Q*(*t* + 1)
0	0	$Q(t)$	0	0	$Q(t)$	0	0	0	$Q(t)$
0	1	0	0	1	0	1	1	1	$\bar{Q}(t)$
1	0	1	1	0	1				
1	1	?	1	1	$\bar{Q}(t)$				

Excitation of flip-flops

Outputs		SR Excitation		JK Excitation		D Excitation	T Excitation
Q(*t*)	*Q*(*t* + 1)	*S* (Set)	*R* (Reset)	*J* (Set)	*K* (Reset)	*D* (data)	*T* (Toggle)
0	0	0	X	0	X	0	0
0	1	1	0	1	X	1	1
1	0	0	1	X	1	0	1
1	1	X	0	X	0	1	0

MULTIPLE CHOICE QUESTIONS

7.1 A flip-flop has
(a) one input line and one output line.
(b) one input line and two output lines.
(c) two input lines and two output lines.
(d) two input lines and one output line.

7.2 A flip-flop has
(a) monostable multivibrator.
(b) astable multivibrator.
(c) bistable multivibrator.
(d) None of the above

7.3 When S and R are set to 1-logic, SR-flip-flop gives
(a) Hold state
(b) Set state
(c) Reset state
(d) Indeterminate state

7.4 When J and K are set to 1-logic and 0-logic, respectively, JK-flip-flop gives
(a) Hold state
(b) Set state
(c) Reset state
(d) Toggle state

7.5 When S and R are set to 0-logic and 1-logic, respectively, SR-flip-flop gives
(a) Hold state
(b) Set state
(c) Reset state
(d) Indeterminate state

7.6 When J and K are set to 1-logic, JK-flip-flop gives
(a) Hold state
(b) Set state
(c) Reset state
(d) Toggle state

7.7 Characteristic equation of T-flip-flop is
(a) $Q(t+1)=(T\oplus Q(t))$
(b) $Q(t+1)=Q(t)$
(c) $Q(t+1)=T$
(d) $Q(t+1)=TQ(t)+\bar{T}\bar{Q}(t)$

7.8 Characteristic equation of D-flip-flop is
(a) $Q(t+1)=(D\oplus Q(t))$
(b) $Q(t+1)=Q(t)$
(c) $Q(t+1)=D$
(d) $Q(t+1)=DQ(t)+\bar{D}\bar{Q}(t)$

7.9 Characteristic equation of J-K-flip-flop is
(a) $Q(t+1)=J\bar{Q}(t)+KQ(t)$
(b) $Q(t+1)=J\bar{Q}(t)+\bar{K}Q(t)$
(c) $Q(t+1)=\bar{J}\bar{Q}(t)+\bar{K}Q(t)$
(d) $Q(t+1)=JQ(t)+\bar{K}\bar{Q}(t)$

7.10 Indeterminate state in SR-flip-flop is avoided if
(a) $SR=0$
(b) $S\bar{R}=0$
(c) $SR=1$
(d) $\bar{S}R=1$

7.11 The race around condition can be avoided if
(a) $t>t_p<T$
(b) $t<t_p<T$
(c) $t>t_p>T$
(d) $t<t_p>T$

7.12 In master-slave flip-flop, when clock pulse is zero then
(a) Master flip-flop is enabled
(b) Slave flip-flop is enabled
(c) Both master and slave flip-flops are enabled
(d) None of the above

7.13 SR-flip-flop is converted to D-flip-flop when
(a) $S = D$ and $R = D$
(b) $S = D$ and $R = \bar{D}$
(c) $S = \bar{D}$ and $R = D$
(d) $S = \bar{D}$ and $R = \bar{D}$

7.14 JK-flip-flop is converted to T-flip-flop when
(a) $J = T$ and $K = T$
(b) $J = T$ and $K = \bar{T}$
(c) $J = \bar{T}$ and $K = T$
(d) $J = \bar{T}$ and $K = \bar{T}$

7.15 The race around condition occurs in J-K-flip-flop if
(a) $J = 0$ and $K = 0$
(b) $J = 0$ and $K = 1$
(c) $J = 1$ and $K = 0$
(d) $J = 1$ and $K = 1$

Answers

7.1 (c)	7.2 (c)	7.3 (d)	7.4 (b)	7.5 (c)	7.6 (d)	7.7 (a)
7.8 (c)	7.9 (b)	7.10 (a)	7.11 (b)	7.12 (b)	7.13 (b)	7.14 (a)
7.15 (d)						

QUESTIONS

7.1 Briefly describe the operational aspects of bistable, monostable and astable multivibrators. Which multivibrator closely resembles a flip-flop?

7.2 What is a flip-flop? Show the logic implementation of an SR flip-flop having active HIGH R and S inputs. Draw its truth table and mark the invalid entry.

7.3 With the help of the logic diagram, describe the operation of a clocked SR-flip-flop with active LOW R and S inputs. Draw the truth table of this flip-flop if it were negatively edge-triggered.

7.4 What is a clocked J-K-flip-flop? What improvement does it have over a clocked SR-flip-flop?

7.5 Differentiate between:
(a) synchronous and asynchronous inputs;
(b) level-triggered and edge-triggered flip-flops;
(c) active LOW and active HIGH inputs.

7.6 Briefly describe the following flip-flop timing parameters: (a) set-up time and hold time; (b) propagation delay; (c) maximum clock frequency.

7.7 Draw the truth table for the following types of flip-flop:
(a) a positive edge-triggered J-K-flip-flop with active HIGH J and K inputs and active LOW PRESET and CLEAR inputs;
(b) a negative edge-triggered J-K-flip-flop with active LOW J and K inputs and active LOW PRESET and CLEAR inputs.

7.8 What is meant by the race problem in flip-flops? How does a master-slave configuration help in solving this problem?

7.9 Differentiate between a D-flip-flop and a D latch.

7.10 Draw the function table for (a) a negative edge-triggered D-flip-flop and (b) a D latch with an active LOW ENABLE input.

7.11 With the help of a schematic arrangement, explain how a J-K-flip-flop can be used as a (a) a D-flip-flop and (b) a T-flip-flop.

7.12 With the help of a suitable circuit, briefly explain how a D-flip-flop can be used to detect the sequence of occurrence of edges of synchronous inputs.
7.13 What is a master-slave flip-flop? Discuss its working.
7.14 What is the major restriction when operating a pulse-triggered flip-flop?
7.15 What do you mean by: (a) clock skew and (b) time race?

PROBLEMS

7.1 A 100 kHz clock signal is applied to a J-K-flip-flop with $J = K = 1$.

(a) If the flip-flop has active HIGH J and K inputs and is negative edge-triggered, determine the frequency of the Q output.

(b) If the flip-flop has active LOW J and K inputs and is positive edge-triggered, what should the frequency of the Q output be? Assume that Q is initially '0'.

7.2 In a Schmitt trigger inverter circuit, the two trip points are observed to occur at 1.8 and 2.8 V. At what input voltage levels will this device make (a) HIGH-to-LOW transition and (b) LOW-to-HIGH transition?

7.3 In the case of a presettable, clearable J-K-flip-flop with active HIGH J and K inputs and active LOW PRESET and CLEAR inputs, what would the Q output logic status be for the following input conditions, assuming that Q is initially '0', immediately after it is clocked?

(a) $J = 1, K = 0$, PRESET = 1, CLEAR = 1;
(b) $J = 1, K = 1$, PRESET = 0, CLEAR = 1;
(c) $J = 0, K = 1$, PRESET = 1, CLEAR = 0;
(d) $J = K = 0$, PRESET = 0, CLEAR = 1.

7.4 Derive the expression for $Q(t + 1)$ in terms of $Q(t)$ and J and K inputs for a clocked J-K-flip-flop with active LOW J and K inputs. $Q(t)$ and $Q(t + 1)$ have the usual meaning.

7.5 A new clocked U-V flip-flop is designed with two inputs U and V. The flip-flop functions are stated below:

- If U is 0 and V is 0, the flip-flop is toggle.
- If U is 0 and V is 1, the flip-flop is set
- If U is 1 and V is 0, the flip-flop is reset.
- If U is 1 and V is 1, the flip-flop is hold.

(a) Obtain the truth table of U-V-flip-flop.
(b) Write excitation table.
(c) Obtain the characteristic equation.

7.6 Determine the output sequence for the given serial data fed to the flip-flop shown. Assume Q is at 0 logic initially. There is one clock pulse for each bit.

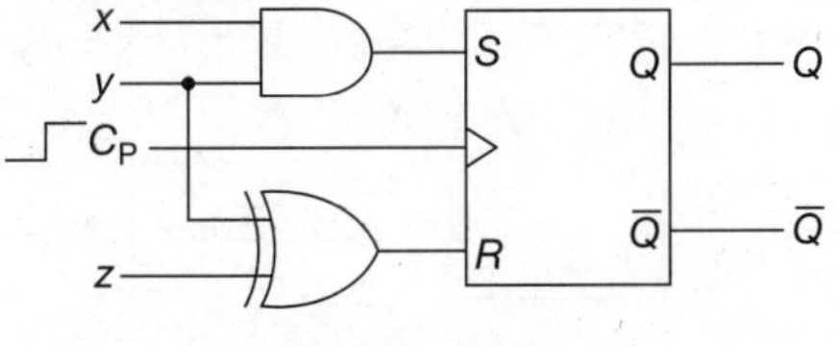

PROBLEM 7.6

$x = 1010\ 0101$, $y = 1110\ 0101$ and $z = 10101111$

7.7 Obtain truth table and characteristic equation of the J-K-flip-flop with an inverter between external input X and input of flip-flop K.

7.8 Construct truth table for the sequential circuit shown in the following figure.

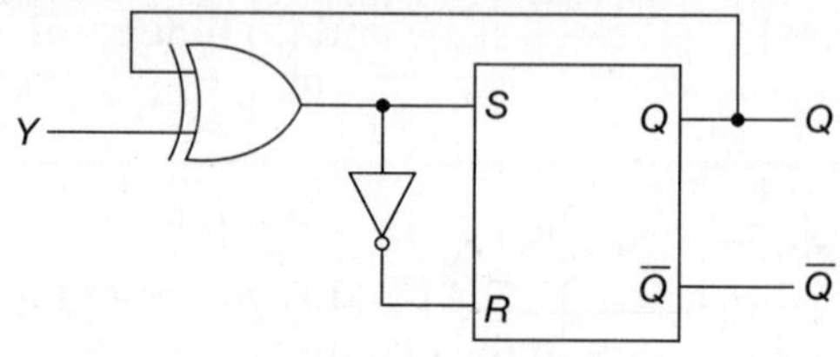

PROBLEM 7.8

7.9 A D-flip-flop has the following characteristic: set-up time is 10 ns, hold time is 15 ns and propagation time is 25 ns. (a) How far ahead of the triggering clock edge must the data be applied? (b) How long after the clock edge must the data be present to ensure correct storage? (c) What is the highest frequency of clock pulses for proper and stable operation if propagation of next state decoder t_{ns} is 20 ns.

7.10 Obtain the characteristic table and equation of the flip-flop. The circuit diagram is shown in the following figure.

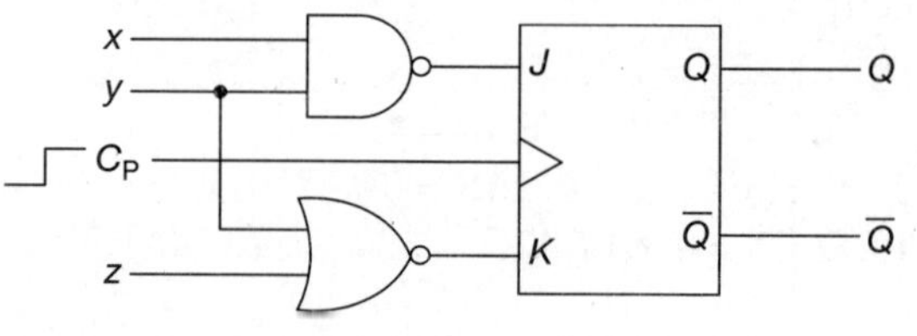

PROBLEM 7.10

8

Design of Sequential Circuits

CHAPTER OBJECTIVES

The main goal of this chapter is to impart knowledge about the design of sequential circuits. Readers will be able to discuss the following aspects in this chapter:

- Analysis of sequential circuits: Moore and Mealy sequential circuits
- Reduction of states and assignments
- Design of sequential circuits
- Analysis of asynchronous circuits
- Problems in asynchronous sequential circuits
- Design of asynchronous circuits
- Algorithmic state machines (ASMs)

8.1 | INTRODUCTION

Logic circuits are classified into combinational and sequential logic circuits. A combinational logic circuit consists of logic gates whose outputs at any time are determined directly from present inputs without considering the previous inputs. In some circuits, some of outputs are explicitly or implicitly delayed and are connected as inputs. Such circuits are considered to have a feedback from the output to the inputs. The output depends not only on the present state but also on the past history of the inputs. In other words, circuit has memory. Sequential circuits are classified into two categories, known as asynchronous and synchronous sequential circuits. A sequential circuit whose behaviour depends on the change of input signal at any time is referred to as asynchronous sequential circuits. A combinational circuit with feedback is termed as an asynchronous sequential circuit. A sequential circuit, whose behaviour is defined from the knowledge of its signal at discrete instants of time, is referred as synchronous sequential circuits. The synchronization is achieved by a timing device known as a system clock which generates a train of clock pulses. Synchronous circuits are also known as clocked-sequential circuit.

Sequential circuit design problem normally starts with a given word description of input output relationship and ends with a circuit diagram having sequential and combinatorial logic elements. The word description is first converted to a state transition diagram or algorithmic state machine (ASM) chart followed by preparation of state synthesis table. For flip-flop based implementation, excitation tables are used to generate design equations through Karnaugh Map the final circuit diagram is developed from these design equations. In read-only memory (ROM)-based implementation, excitation tables are not required however;

flip-flops are used as delay elements. On the availability of the sequential circuit, analysis of the circuit is performed by feeding input and to obtain the state table of state diagram.

There are two different approaches of state machine design called Moore model and Mealy model. In Moore model circuit outputs, also called primary outputs are generated solely from secondary outputs or memory values. In Mealy model circuit inputs, also known as primary inputs combine with memory elements to generate circuit output.

8.2 | NOTATIONS

Sequential circuits use different names for Boolean variables. The variables are named as these are generated in the sequential circuit.

Input variable(s): All variables that are required for the sequential circuit to work when input is fed, are known as input variables.

Output variables: All variables that are produced by the sequential circuit are said to be output variables.

State variable: The output of memory (flip-flops) defines the state of a sequential circuit (machine). Decoded state variables (along with- the input variables for a Mealy machine) produce the output variables. In other words, the state variables are the flip-flop outputs.

Excitation variable: Excitation variables are the inputs to memory (flip-flops). The name excitation is used because the variable excites the memory to change. When flip-flops are used for the system memory, the excitation variables are the inputs of SR-, JK-, D- and T-flip-flops, respectively. State variables are a function of excitation variables. Excitation variables are generated by the input combinational logic operating on the state variables and input variables.

State: The state of a sequential machine is defined by the content of memory. When memory is realized with-flip-flops, the machine state is defined by the outputs. Each state of a sequential machine must be unique and unambiguous.

State variables and states are related by the expression

$$2^x = y$$

where x = number of state variables like flip-flops
y = maximum number of states possible

For 4-state variables, maximum of 16 (2^4) states are possible.

Present state: The status of all state variables, $Q(t)$, at time, t, before the next clock edge, represents a condition called present state (Figure 8.1). The present state or status of sequential circuit is a reference point with respect to time.

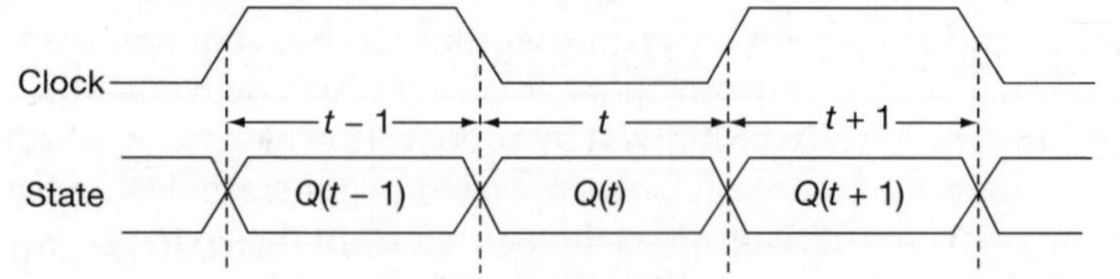

FIGURE 8.1 | Present and next state illustration

Next state: The status of all state variables, $Q(t + 1)$ at time, $t + 1$ represents a condition called next state (Figure 8.1). The next state of a sequential circuit is represented by the memory status a particular clock, t.

State table: The behaviour of a sequential circuit is determined from the inputs, outputs, and states of its flip-flops. Both the outputs and next state are functions of the inputs and the present states. A sequential circuit is given by a state table which relates outputs and next states as functions of the inputs and the present states. The present state column lists all the possible states in the sequential circuit. A series of next state columns exist, one for each input combination. The state table indicates the state transitions. In clocked sequential circuits, the transition from present to next state is activated by a clock pulse.

State diagram: Generally states are represented by their binary equivalent values. The information available in a state table is represented graphically in a state diagram. So, a state diagram is a graphical representation of states of a sequential circuit those it may undergo. A single state is represented by a circle with an identifying symbol located inside. Changes from present state to next state are indicated by directed lines or arcs. Input conditions that cause state changes to occur and the resulting output signals are written adjacent to the directed lines or arcs.

Let x is input variable. y is output variable. A and B are two different states. Figure 8.2 shows state diagram used for Mealy model of sequential circuit. The state A changes to state B when input, x is given and produces output, y. When input, $\bar{x}$ is given to the sequential circuit goes from state, A to the same state, A and produces output, y. Figure 8.3 also represents state diagram used for Moore model of sequential circuit. The state, A having output, y changes to state B having output, y when input, x is given.

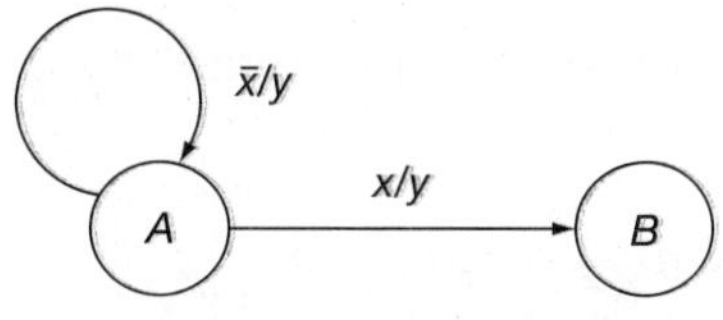

FIGURE 8.2 | State diagram

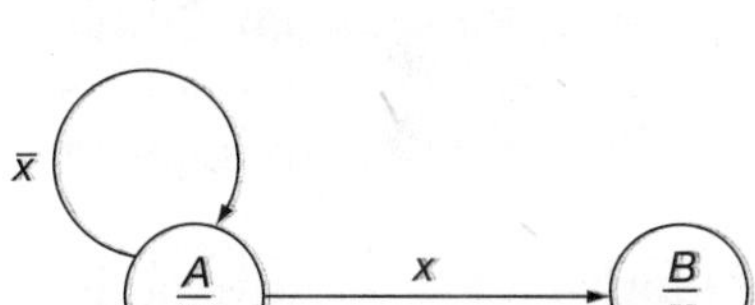

FIGURE 8.3 | State diagram

Transition table: The state diagram and state table represent states using symbols or names. Transition table is a kind of state table where states are represented by actual binary values. Assignment of binary values to states symbols is called state assignments.

8.3 | MOORE AND MEALY SEQUENTIAL CIRCUIT

Synchronous sequential circuit memory, usually consisting of flip-flops, update circuit states information. The logic functions are shown in the model diagrams translate input and flip-flop output information. The input logic produces the excitation inputs to flip-flops. The output logic converts input and flip-flop data represents states to satisfy the output variable requirements. The value of the next-state variables, $S(t + 1)$, is determined by the present value of the state variables, $Q(t + 1)$, and the input, X. $Q(t)$, and $Q(t + 1)$, represent the binary value of state variables, $Q(t)$, at different time, t. Memory data can be changed with time. The memory device has a synchronizing clock input, C_P. The clock input is a synchronizing train of pulses. Synchronous memory changes its data only at certain time intervals. The present state, $Q(t)$ is considered to be the flip-flop's value prior to the arrival of a clock pulse of interest. Based on the known present state, $Q(t)$ of memory and input, X conditions, next sate, $Q(t + 1)$ of memory is obtained.

Moore sequential circuit: The clocked synchronous circuit in which the output, Y depends only on the present state, $Q(t)$ of the flip-flops is known as Moore sequential circuit. It does not depend upon present inputs. The block diagram of Moore sequential circuit is shown in Figure 8.4.

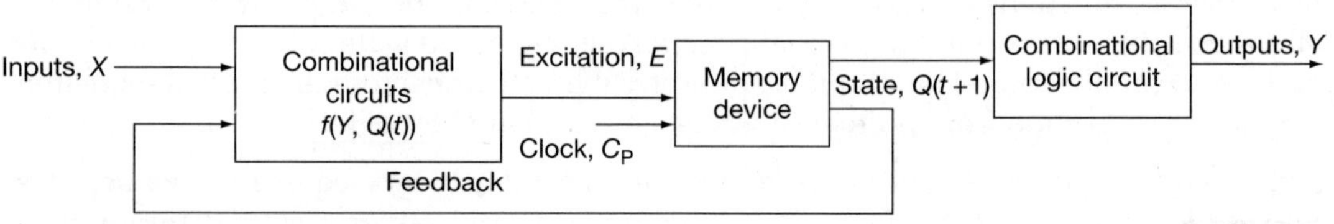

FIGURE 8.4 | Moore sequential circuit

Excitation is given by

$$E = f(Q(t), X) \tag{8.1}$$

Next state is determined by

$$Q(t+1) = f(E, Q(t)) \tag{8.2}$$

Output is given by

$$Y = f(Q(t)) \tag{8.3}$$

where X is the input variables
E is excitation for the memory device
$Q(t)$ is the present output state (at time t) of memory device.
Y is the output variables

Algorithm 8.1 | Analysis of clocked sequential circuits

1. Find input variables, output variables and state variables of the given sequential circuit.
2. Number of flip-flops in the circuit gives the number of state variables.
3. Find the number of possible states i.e. 2^n, where n is number of state variables or flip-flops.
4. Find excitation of each flip-flop in the given sequential circuit.
5. Use characteristic table and excitations of the flip-flop(s) to find next state(s) when present states are known.
6. Prepare characteristic table from characteristic equations of flip-flop and output logic. Prepare transition table from characteristic table. Draw the table which shows the present state, then next state when input is 0, the next state when input is 1, the output when input is 0, and the output when input is 1.
7. Assign symbols to state binary value like a, A, etc. Input value is represented by its variable. Assign $X \leftarrow 1$ and $\bar{X} \leftarrow 0$. Represent output value to its variable.
8. Prepare state stable and draw state diagram.

An example of Moore sequential circuit is shown in Figure 8.5. The output depends upon the output states of the JK-flip-flops.

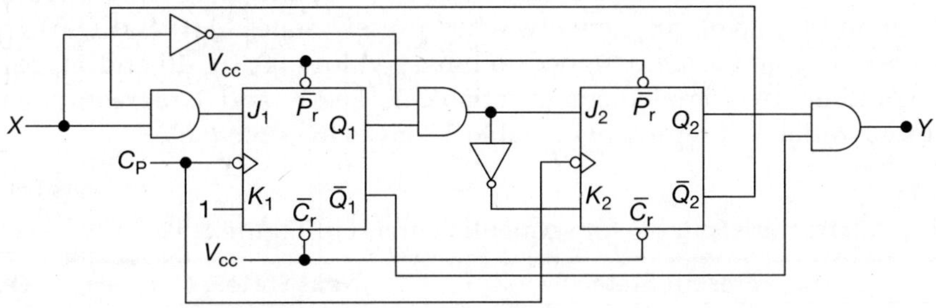

FIGURE 8.5 | Moore sequential circuit

Excitations of JK-flip-flops are given below:

$Q_1(t)$ is present state
$Q_1(t+1)$ is next state

$$J_1 = X\bar{Q}_2(t) \quad \text{and} \quad K_1 = 1$$

$$J_2 = \bar{X}Q_1(t) \quad \text{and} \quad K_2 = \bar{J}_2$$

Output of the sequential circuit is given below:

$$Y = \bar{Q}_1(t)Q_2(t) \tag{8.4}$$

Characteristic equation of JK-flip-flop-I is given below:

$$Q_1(t+1) = J_1\bar{Q}_1(t) + \bar{K}_1Q_1(t)$$

Substitute the excitations of JK-flip-flop in the above characteristic equation of JK-flip-flop to get

$$Q_1(t+1) = (X\bar{Q}_2(t))\,\bar{Q}_1(t) + \bar{1}Q_1(t)$$

Simplify the above characteristic equation

$$Q_1(t+1) = X\bar{Q}_1(t)\,\bar{Q}_2(t) \tag{8.5}$$

Characteristic equation of JK-flip-flop-II is given below:

$$Q_2(t+1) = J_2\bar{Q}_2(t) + \bar{K}_2Q_2(t)$$

Substitute the excitations of JK-flip-flop in the above characteristic equation of JK-flip-flop to get.

$$Q_2(t+1) = J_2\bar{Q}_2(t) + \bar{\bar{J}}_2Q_2(t)$$

$$Q_2(t+1) = J_2(\bar{Q}_2(t) + Q_2(t))$$

$$Q_2(t+1) = J_2$$

$$Q_2(t+1) = \bar{X}Q_1(t) \tag{8.6}$$

Characteristic table is prepared using Eq. (8.4), Eq. (8.5) and Eq. (8.6) and is given in Table 8.1. Transition table is given in Table 8.2. It is developed from the truth table given in Table 8.1. Transition table gives the next states $Q_1(t+1)$ and $Q_2(t+1)$ which are obtained from Eq. (8.5) and Eq. (8.6), respectively, when present states $Q_1(t)$ and $Q_2(t)$ are known. State symbols a, b, c and d are assigned to binary values 00, 01, 10 and 11, respectively. Output 0 represents by complemented variable Y, i.e. $\bar{Y}$ and 1 represents variable, Y. Similarly, input variable X represents 1. Table 8.3 gives the state table.

TABLE 8.1 | Characteristic table for sequential circuit of Figure 8.5

Input	Present State		Next State		Output
X	$Q_2(t)$	$Q_1(t)$	$Q_2(t+1)$	$Q_1(t+1)$	Y
0	0	0	0	0	0
0	0	1	1	0	0
0	1	0	0	0	1
0	1	1	1	0	0
1	0	0	0	1	0
1	0	1	0	0	0
1	1	0	0	0	1
1	1	1	0	0	0

TABLE 8.2 | Transition table for circuit of Figure 8.5

Present State		Next State				Output
		$X=0$		$X=1$		
$Q_2(t)$	$Q_1(t)$	$Q_2(t+1)$	$Q_1(t+1)$	$Q_2(t+1)$	$Q_1(t+1)$	Y
0	0	0	0	0	1	0
0	1	1	0	0	0	0
1	0	0	0	0	0	1
1	1	1	0	0	0	0

TABLE 8.3 | State table

Present State	Next State		Output
	$\bar{X}$	X	
a	a	b	$\bar{Y}$
b	c	a	$\bar{Y}$
c	a	a	Y
d	c	a	$\bar{Y}$

State diagram related to Table 8.3 is shown in Figure 8.6 and is explained as given below:

- When input X is given to sequential circuit, it changes its state 'a' with output $\overline{Y}$ to state 'b' with output $\overline{Y}$. State 'a' with output $\overline{Y}$ remains same when $\overline{X}$ input is given.
- When input $\overline{X}$ is given to sequential circuit, it changes its state 'b' with output $\overline{Y}$ to state 'c' with output Y. When input X is given to sequential circuit, it changes its state 'b' with output $\overline{Y}$ to state 'a' with output $\overline{Y}$.
- Sequential circuit undergoes change of its state 'c' with Y output to state 'a' with output $\overline{Y}$ when input $\overline{X}$ or X is given to the circuit.
- When input $\overline{X}$ is given to sequential circuit, it changes its state 'd' with output $\overline{Y}$ to state 'c' with output Y. When input X is given to sequential circuit, it changes its state 'd' with output $\overline{Y}$ to state 'a' with output $\overline{Y}$.

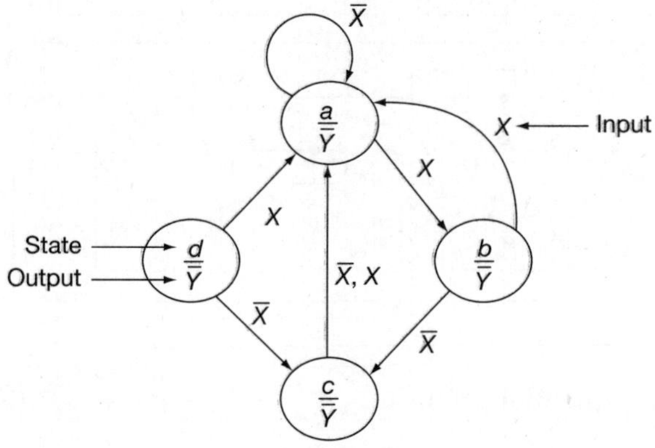

FIGURE 8.6 | State diagram

Mealy sequential circuit: The clocked synchronous circuit in which the output, Y depends only on the present state, $Q(t)$ of the flip-flops and inputs, X is known as Mealy sequential circuit. It depends upon present inputs. The block diagram of Mealy sequential circuit is shown in Figure 8.7.

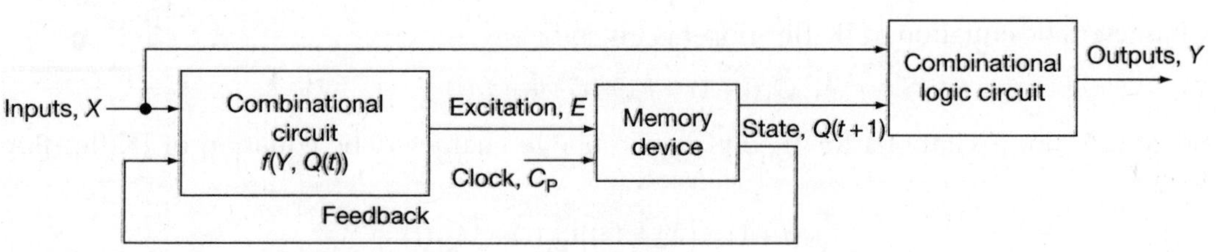

FIGURE 8.7 | Mealy circuit-based sequential circuit

Excitation is given by

$$E = f(Q(t), X) \tag{8.7}$$

Next state is determined as

$$Q(t+1) = f(E, Q(t)) \tag{8.8}$$

Output is given by

$$Y = f(X, Q(t)) \tag{8.9}$$

where X is the input variables
E is excitation for the memory device
$Q(t)$ is the present output state (at time t) of memory device.
Y is the output variables

An example of Mealy sequential circuit is shown in Figure 8.8. The output depends upon the output states of the JK-flip-flops and input X.

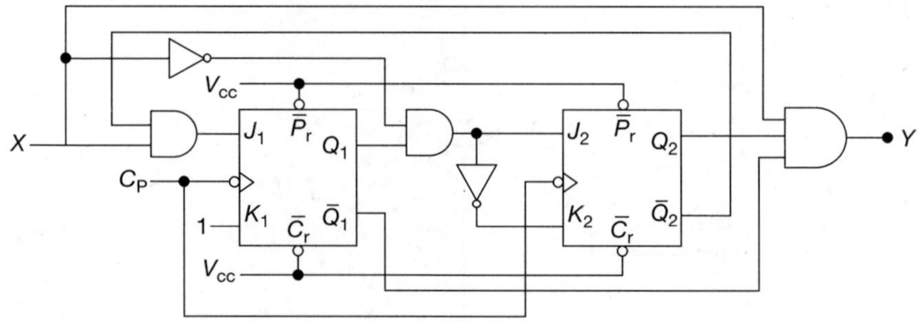

FIGURE 8.8 | Mealy sequential circuit

Excitations of JK-flip-flops are given below:

$$J_1 = X\bar{Q}_2(t) \quad \text{and} \quad K_1 = 1$$

$$J_2 = \bar{X}Q_1(t) \quad \text{and} \quad K_2 = \bar{J}_2$$

Output of the sequential circuit is given below:

$$Y = x\bar{Q}_1(t)Q_2(t) \tag{8.10}$$

Characteristic equation of JK-flip-flop-I is given below:

$$Q_1(t+1) = J_1\bar{Q}_1(t) + \bar{K}_1Q_1(t)$$

Substitute the excitations for JK-flip-flop in above characteristic equation of JK-flip-flop to get

$$Q_1(t+1) = \left(X\bar{Q}_2(t)\right)\bar{Q}_1(t) + \bar{1}Q_1(t)$$

Simplify the above characteristic equation

$$Q_1(t+1) = X\bar{Q}_1(t)\bar{Q}_2(t) \tag{8.11}$$

Characteristic equation of JK-flip-flop-II is given below:

$$Q_2(t+1) = J_2\bar{Q}_2(t) + \bar{K}_2Q_2(t)$$

Substitute the excitation for JK-flip-flop in above characteristic equation of JK-flip-flop to get

$$Q_2(t+1) = \bar{X}Q_1(t) \tag{8.12}$$

Characteristic table is prepared using Eq. (8.10), Eq. (8.11) and Eq. (8.12) and is given in Table 8.4. Transition table is given in Table 8.5. It is developed from the truth table given in Table 8.4. Transition table for sequential circuit gives the next states $Q_1(t+1)$ and $Q_2(t+1)$ which are obtained from Eq. (8.11) and Eq. (8.12), respectively, when present states $Q_1(t)$ and $Q_2(t)$ are known. Similarly, output is obtained from Eq. (8.10) for the known present states.

TABLE 8.4 | Characteristic table for sequential circuit of Figure 8.8

Input	Present State		Next State		Output
X	$Q_2(t)$	$Q_1(t)$	$Q_2(t+1)$	$Q_1(t+1)$	Y
0	0	0	0	0	0
0	0	1	1	0	0
0	1	0	0	0	0
0	1	1	1	0	0
1	0	0	0	1	0
1	0	1	0	0	0
1	1	0	0	0	1
1	1	1	0	0	0

TABLE 8.5 | Transition table for circuit of Figure 8.8

Present State		Next State				Output, Y	
		$X = 0$		$X = 1$			
$Q_2(t)$	$Q_1(t)$	$Q_2(t+1)$	$Q_1(t+1)$	$Q_2(t+1)$	$Q_1(t+1)$	$X = 0$	$X = 1$
0	0	0	0	0	1	0	0
0	1	1	0	0	0	0	0
1	0	0	0	0	0	0	1
1	1	1	0	0	0	0	0

State symbols A, B, C and D are assigned to binary values 00, 01, 10 and 11, respectively. Output 0 represents complemented variable Y, i.e. $\bar{Y}$ and 1 represents variable, Y. Similarly, input variable X represents 1. Table 8.6 gives the state table.

State diagram related to Table 8.4 is shown in Figure 8.9 and is explained as given below:

- When input X is given to sequential circuit, it changes its state 'A' to state 'B' and produces output $\bar{Y}$. State 'A' remains same when input, $\bar{X}$ is given and produces output $\bar{Y}$.

TABLE 8.6 | State table

Present State	Next State		Output, Y	
	$\bar{X}$	X	$\bar{X}$	X
A	A	B	$\bar{Y}$	$\bar{Y}$
B	C	A	$\bar{Y}$	$\bar{Y}$
C	A	A	$\bar{Y}$	Y
D	C	A	$\bar{Y}$	$\bar{Y}$

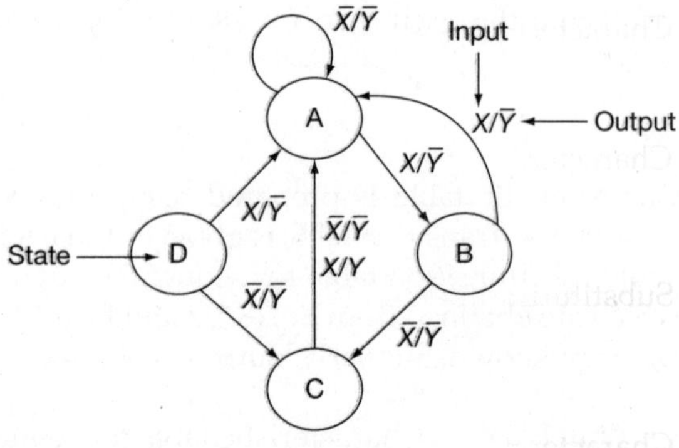

FIGURE 8.9 | State diagram

- When input $\bar{X}$ is given to sequential circuit, it changes its state 'B' to state 'C' and output $\bar{Y}$ is produced. When input X is given to sequential circuit, it changes its state 'B' to state 'A' and produces output $\bar{Y}$.
- Sequential circuit undergoes change of its state 'C' to state 'A' but produces output $\bar{Y}$ or Y when input $\bar{X}$ or X is given to the circuit, respectively.
- When input $\bar{X}$ is given to sequential circuit, it changes its state 'D' to state 'C' and produces output $\bar{Y}$. When input X is given to sequential circuit, it changes its state 'D' to state 'A' and produces output $\bar{Y}$.

EXAMPLE 8.1 Analyse the sequential circuit given in Figure 8.10 and draw the state diagram.

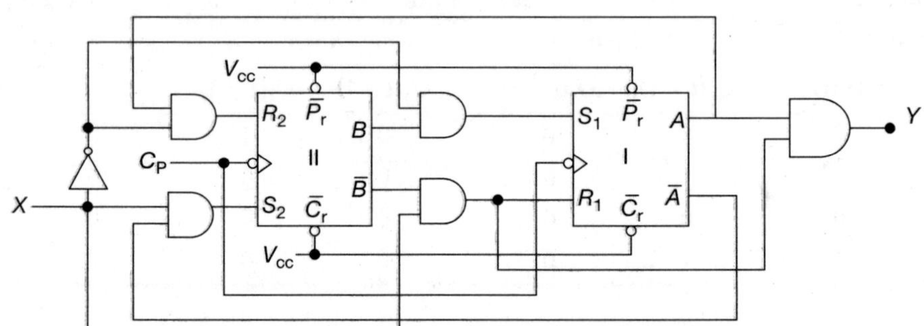

FIGURE 8.10 | Sequential circuit

SOLUTION Excitations of SR-flip-flops are given below:

$$S_1 = \bar{X}B(t) \quad \text{and} \quad R_1 = X\bar{B}(t)$$

$$R_2 = \bar{X}A(t) \quad \text{and} \quad S_2 = X\bar{A}(t)$$

Characteristic equation of SR-flip-flop is:

$$Q(t+1) = S + \bar{R}Q(t) \qquad \text{Provided } SR = 0$$

Characteristic equation for SR-flip-flop I is:

$$A(t+1) = S_1 + \bar{R}_1 A(t) \quad \text{and} \quad S_1 R_1 = 0$$

Substituting S_1 and R_1 excitations in above characteristic equation of SR-flip-flop to get

$$A(t+1) = \bar{X}B(t) + \left(\overline{X\overline{B(t)}}\right)A(t) \tag{8.13}$$

Characteristic equation for SR-flip-flop II is:

$$B(t+1) = S_2 + \bar{R}_2 B(t) \quad \text{and} \quad S_2 R_2 = 0$$

Substituting S_1 and R_1 excitations in above characteristic equation of SR-flip-flop to get

$$B(t+1) = X\bar{A}(t) + \left(\overline{\bar{X}A(t)}\right)B(t) \tag{8.14}$$

Output of the sequential circuit is

$$Y = XA(t)\bar{B}(t) \tag{8.15}$$

Figure 8.10 shows Mealy model of sequential circuit because output depends on input. Characteristic table is prepared using Eq. (8.13), Eq. (8.14) and Eq. (8.15) and is given in Table 8.7. Transition table is given in Table 8.8. It is developed from the truth table given in Table 8.7. Transition table for sequential circuit gives the next states $A(t + 1)$ and $B(t + 1)$ which are obtained from Eq. (8.13) and Eq. (8.14), respectively, when present states $A(t)$ and $B(t)$ are known. Similarly, output is obtained from Eq. (8.15) for the known present states.

TABLE 8.7 | Characteristic table for sequential circuit of Figure 8.10

Input	Present State		Next State		Output
X	$A(t)$	$B(t)$	$A(t+1)$	$B(t+1)$	Y
0	0	0	0	0	0
0	0	1	1	0	0
0	1	1	0	0	0
0	1	1	0	0	0
1	0	0	1	1	0
1	0	1	1	0	0
1	1	0	0	0	1
1	1	1	1	0	0

State symbols a, b, c and d are assigned to binary values 00, 01, 10 and 11, respectively. Output 0 represents complemented variable Y, i.e. $\bar{Y}$ and 1 represents variable, Y. Similarly, input variable X represents 1. Table 8.9 gives the state table. State diagram is shown in Figure 8.11.

TABLE 8.8 | Transition table for circuit of Figure 8.7

Present State		Next State				Output, Y	
		X = 0		X = 1			
A(t)	B(t)	A(t + 1)	B(t + 1)	A(t + 1)	B(t + 1)	X = 0	X = 1
0	0	0	0	0	1	0	0
0	1	1	1	0	1	0	0
1	0	1	0	0	0	0	1
1	1	1	0	1	1	0	0

TABLE 8.9 | State table

Present State	Next State		Output, Y	
	$\bar{X}$	X	$\bar{X}$	X
a	a	b	$\bar{Y}$	$\bar{Y}$
b	d	b	$\bar{Y}$	$\bar{Y}$
c	c	a	$\bar{Y}$	Y
d	c	d	$\bar{Y}$	$\bar{Y}$

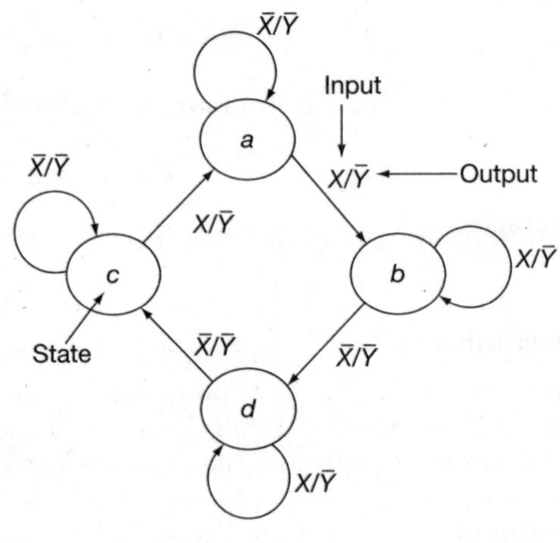

FIGURE 8.11 | State diagram

EXAMPLE 8.2 Analyse the sequential circuit given in Figure 8.12 and draw the state diagram.

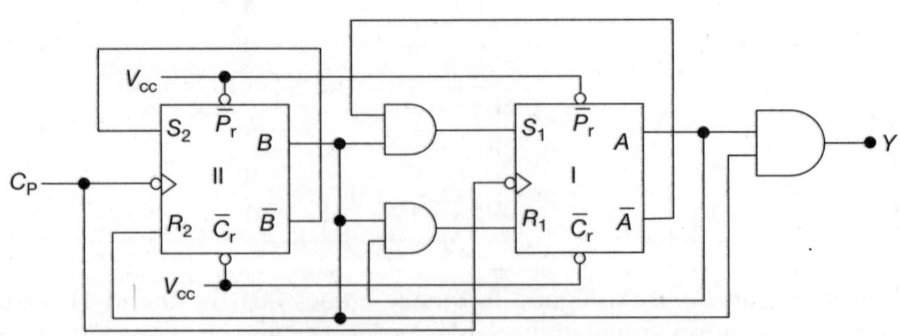

FIGURE 8.12 | Sequential circuit

SOLUTION Excitations of SR-flip-flops are given below:

$$S_1 = \bar{A}(t)B(t) \quad \text{and} \quad R_1 = A(t)B(t)$$

$$S_2 = \bar{B}(t) \quad \text{and} \quad R_2 = B(t)$$

Characteristic equation for SR-flip-flop I is

$$A(t+1) = S_1 + \bar{R}_1 A(t) \quad \text{and} \quad S_1 R_1 = 0$$

Substituting S_1 and R_1 excitations in above characteristic equation of SR-flip-flop to get

$$A(t+1) = \bar{A}(t)B(t) + \left(\overline{A(t)B(t)}\right)A(t)$$

$$A(t+1) = \bar{A}(t)B(t) + \left(\bar{A}(t) + \bar{B}(t)\right)A(t)$$

$$A(t+1) = \bar{A}(t)B(t) + A(t)\bar{B}(t) \qquad (\because A(t)\bar{A}(t) = 0)$$

$$A(t+1) = A(t) \oplus B(t) \tag{8.16}$$

Characteristic equation for SR-flip-flop II is:

$$B(t+1) = S_2 + \bar{R}_2 B(t) \quad \text{and} \quad S_2 R_2 = 0$$

Substituting S_2 and R_2 excitations in above characteristic equation of SR-flip-flop to get

$$B(t+1) = \bar{B}(t) + (\bar{B}(t))B(t)$$

$$B(t+1) = \bar{B}(t) \qquad (\because B(t)\bar{B}(t) = 0) \tag{8.17}$$

Output of the sequential circuit is

$$Y = A(t)B(t) \tag{8.18}$$

Figure 8.12 shows Moore model of sequential circuit because output does not depend on input. Transition table for sequential circuit shown in Figure 8.10, is given in Table 8.10. Next states $A(t + 1)$ and $B(t + 1)$ are obtained from Eq. (8.16) and Eq. (8.17), respectively, when present states $A(t)$ and $B(t)$ are known. Similarly, output is obtained from Eq. (8.18) for the known present states.

TABLE 8.10 | Transition table for circuit of Figure 8.12

Present State		Next State		Output
B(t)	**A(t)**	**B(t + 1)**	**A(t + 1)**	**Y**
0	0	0	1	0
0	1	1	0	0
1	0	1	1	0
1	1	0	0	1

Assign the states symbols *a*, *b*, *c* and *d* to binary values 00, 01, 10 and 11, respectively. Output 0 represents complemented variable *Y*, i.e. $\bar{Y}$ and 1 represents variable, *Y*.

Similarly, input variable X represents 1. Table 8.11 gives the state table. State diagram is drawn in Figure 8.13.

TABLE 8.11 | State table

Present State	Next State	Output Y
a	b	$\bar{Y}$
b	c	$\bar{Y}$
c	d	$\bar{Y}$
d	a	Y

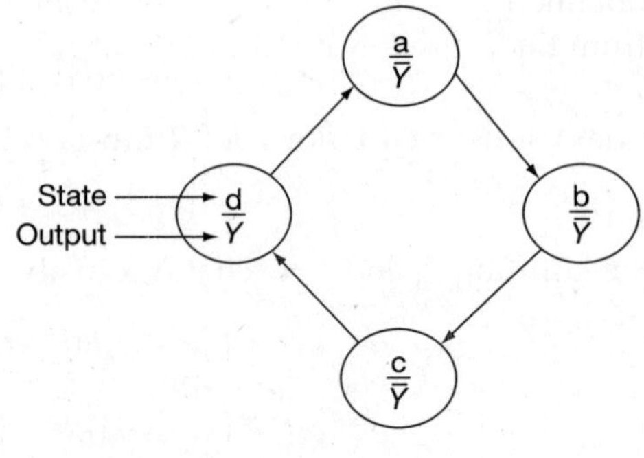

FIGURE 8.13 | State diagram

EXAMPLE 8.3 Analyse the sequential circuit given in Figure 8.14 and draw the state diagram.

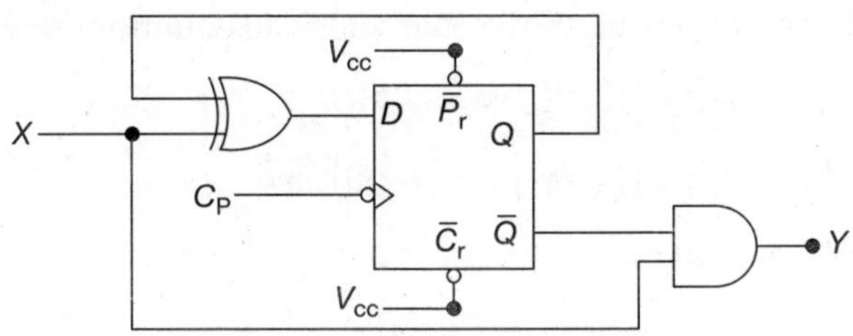

FIGURE 8.14 | Sequential circuit

SOLUTION Excitation of D-flip-flop is given below:

$$D = X \oplus Q(t)$$

Characteristic equation for D-flip-flop is:

$$Q(t+1) = D$$

Substitute D excitation in above characteristic equation of D-flip-flop to get

$$Q(t+1) = X \oplus Q(t) \tag{8.19}$$

Output of the sequential circuit is

$$Y = X\bar{Q}(t) \tag{8.20}$$

Figure 8.14 shows Mealy model of sequential circuit because output depends on input. Characteristic table is prepared using Eq. (8.19) and Eq. (8.20) and is given in

Table 8.12. Transition table is given in Table 8.13. It is developed from the truth table given in Table 8.12. Transition table for sequential circuit gives the next state $Q(t + 1)$ which is obtained from Eq. (8.19) when present state $Q(t)$ is known. Similarly, output is obtained from Eq. (8.20) for the known present states.

TABLE 8.12 | Characteristic table for sequential circuit of Figure 8.14

Input	Present State	Next State	Output
X	$Q(t)$	$Q(t+1)$	Y
0	0	0	0
0	1	1	0
1	0	1	1
1	1	0	0

TABLE 8.13 | Transition table for circuit of Figure 8.14

Present State	Next State, $Q(t+1)$		Output, Y	
$Q(t)$	$X = 0$	$X = 1$	$X = 0$	$X = 1$
0	0	1	0	1
1	1	0	0	0

State symbols a and b are assigned to binary values 0 and 1, respectively. Output 0 represents complemented variable Y and 1 represents variable, Y. Similarly, input variable X represents 1. Table 8.14 gives the state table. State diagram is drawn in Figure 8.15.

TABLE 8.14 | State table for circuit of Figure 8.12

Present State	Next State, $Q(t+1)$		Output, Y	
$Q(t)$	$\bar{X}$	X	$\bar{X}$	X
a	a	b	$\bar{Y}$	Y
b	b	a	$\bar{Y}$	$\bar{Y}$

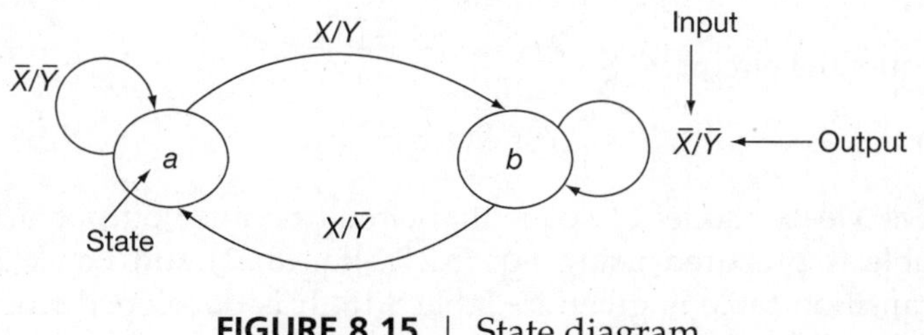

FIGURE 8.15 | State diagram

EXAMPLE 8.4 Analyse the sequential circuit given in Figure 8.16 and draw the state diagram.

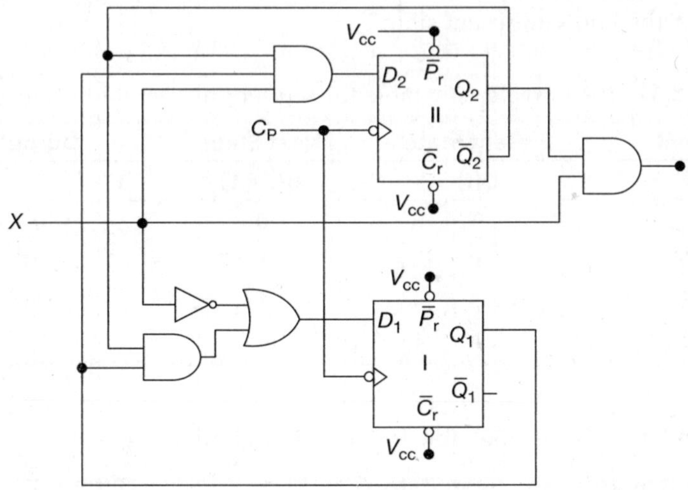

FIGURE 8.16 | Sequential circuit

SOLUTION Excitations of D-flip-flops are given below:

$$D_1 = \overline{X} + Q_1(t)\overline{Q}_2(t)$$

$$D_2 = XQ_1(t)\overline{Q}_2(t)$$

Characteristic equations of D-flip-flops are:

$$Q_1(t+1) = D_1$$

$$Q_2(t+1) = D_2$$

Substituting D_1 and D_2 excitations in above characteristic equations of D-flip-flops to get

$$Q_1(t+1) = \overline{X} + Q_1(t)\overline{Q}_2(t) \tag{8.21}$$

$$Q_2(t+1) = XQ_1(t)\overline{Q}_2(t) \tag{8.22}$$

Output of the sequential circuit is

$$Y = XQ_2(t) \tag{8.23}$$

Figure 8.16 shows Mealy model of sequential circuit because output depends on input. Characteristic table is prepared using Eq. (8.21), Eq. (8.22) and Eq. (8.23) and is given in Table 8.15. Transition table is given in Table 8.16. It is developed from the truth table given in Table 8.15. Transition table for sequential circuit gives the next states $Q_1(t + 1)$

and $Q_2(t + 1)$ which are obtained from Eq. (8.21) and Eq. (8.22), respectively, when present states $Q_1(t)$ and $Q_2(t)$ are known. Similarly, output is obtained from Eq. (8.23) for the known present states.

TABLE 8.15 | Characteristic table for sequential circuit of Figure 8.16

Input	Present State		Next State		Output
X	$Q_2(t)$	$Q_1(t)$	$Q_2(t+1)$	$Q_1(t+1)$	Y
0	0	0	0	1	0
0	0	1	0	1	0
0	1	0	0	1	0
0	1	1	0	1	0
1	0	0	0	0	0
1	0	1	1	1	1
1	1	0	0	0	1
1	1	1	0	0	0

TABLE 8.16 | Transition table for sequential circuit of Figure 8.16

Present State		Next State				Output, Y	
		$X = 0$		$X = 1$			
$Q_2(t)$	$Q_1(t)$	$Q_2(t+1)$	$Q_1(t+1)$	$Q_2(t+1)$	$Q_1(t+1)$	$X = 0$	$X = 1$
0	0	0	1	0	0	0	0
0	1	0	1	1	1	0	0
1	0	0	1	0	0	0	1
1	1	0	1	0	0	0	1

State symbols *A*, *B*, *C* and *D* are assigned to binary values 00, 01, 10 and 11, respectively. Output 0 represents complemented variable *Y* and 1 represents variable, *Y*. Similarly, input variable *X* represents 1. Table 8.17 gives the state table. State diagram is drawn in Figure 8.17.

TABLE 8.17 | State table for sequential circuit of Figure 8.16

Present State	Next State		Output	
	$\bar{X}$	X	$\bar{X}$	X
A	*B*	*A*	$\bar{Y}$	$\bar{Y}$
B	*B*	*D*	$\bar{Y}$	$\bar{Y}$
C	*B*	*A*	$\bar{Y}$	*Y*
D	*B*	*A*	$\bar{Y}$	*Y*

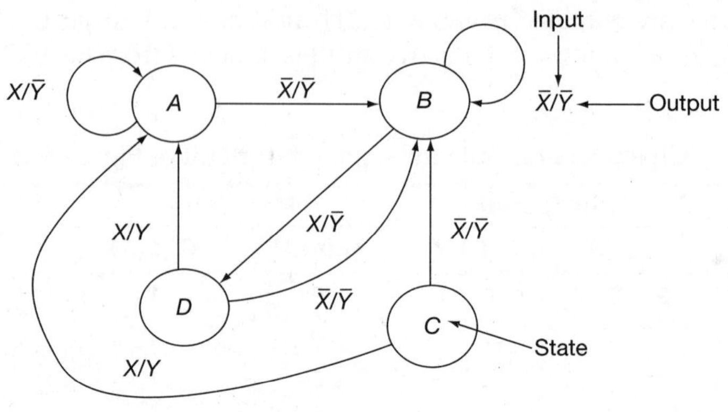

FIGURE 8.17 | State diagram

8.4 | STATE REDUCTION

Two approaches to state equivalency identification and reduction are presented that are equivalence group and the implication chart. The equivalence group algorithm partitions a set of states into groups based on the next state and output results for a given input. If group contains a single state, it means no equivalency or redundancy exists. The equivalent or redundant states are members of partitioned group. Implication chart is a graphical technique to find equivalent or redundant states. Each group is identified by a box in the implication chart.

8.4.1 | Equivalence Groups

A procedure for state classification in order to find *n*-equivalence is given below.

Algorithm 8.2 | Constructing the reduced state table using state equivalency

1. Partition the present states in a group in such a manner that input and output values of the next states are same.
2. Construct a table in which each next state is assigned the group to which the next state belongs.
3. Partition the present states in the subgroups in such a manner that each next state belong to same group for the given input.
4. Construct a table in which each next state is assigned the subgroup in which the next state belongs.
5. Step-3 and 4 are repeated till no further partitioning of subgroups is possible.
6. The present states within the partitioned subgroup/ group are equivalent or redundant.
7. Remove the equivalent present states to get reduced state table.

EXAMPLE 8.5 Reduce the state table given in Table 8.18.

TABLE 8.18 | State table

Present State	Next State		Output, Z	
	X = 0	X = 1	X = 0	X = 1
A	A	B	0	0
B	C	D	0	0
C	E	F	0	0
D	B	A	0	1
E	C	D	0	0
F	B	A	0	1

SOLUTION

Step 1: The present states are partitioned into two groups G_1 and G_2. Present states *A*, *B*, *C* and *E* have same outputs 0 and 0 when inputs are 0 and 1, respectively. Present states *D* and *F* have same outputs 0 and 1 when inputs are 0 and 1, respectively.

$$G_1 = (A,B,C,E)$$

$$G_2 = (D,F)$$

Step 2: Table 8.18(a) is constructed where next states are replaced by the corresponding group G_1 or G_2.

TABLE 8.18(a) | Partitioned groups table

Present State	Next State	
	X = 0	X = 1
A	G_1	G_1
B	G_1	G_2
C	G_1	G_2
E	G_1	G_2
D	G_1	G_1
F	G_1	G_1

Step 3: The present states of group G_1 are partitioned into two groups G_{11} and G_{12}. Present state *A* has next states which belong to group G_1 for input $X = 0$ and to group G_1 for input $X = 1$. Present states *B*, *C* and *E* have next states which belong to group G_1 for input $X = 0$ and to group G_2 for input $X = 1$. Group G_2 is not further partitioned but renamed as G_{21}.

$$G_{11} = (A)$$

$$G_{12} = (B, C, E)$$

$$G_{21} = (D, F)$$

Step 4: Table 8.18(b) is constructed where next states are replaced by the corresponding subgroups, G_{11}, G_{12} and G_{21}.

Step 5: No further partitioning of subgroups is possible. States of G_{12} subgroups are equivalent, so $B = C = E$. Similarly, States of G_{12} subgroups are equivalent, so $D = F$.

Step 6: Replace states C and E with state B and state F with state D in the given state Table 8.18 to get state Table 8.18(c).

TABLE 8.18(b) | Partitioned subgroups table

Present State	Next State	
	$X=0$	$X=1$
A	G_{11}	G_{12}
B	G_{12}	G_{21}
C	G_{12}	G_{21}
E	G_{12}	G_{21}
D	G_{12}	G_{11}
F	G_{12}	G_{11}

TABLE 8.18(c) | State table with equivalent states

Present State	Next State		Output, Z	
	$X=0$	$X=1$	$X=0$	$X=1$
A	A	B	0	0
B	~~C~~ B	D	0	0
~~C~~ B	~~E~~ B	~~F~~ D	0	0
D	B	A	0	1
~~E~~ B	~~C~~ B	D	0	0
~~F~~ D	B	A	0	1

Remove the duplicating states C, E and F to get the reduced state Table 8.18(d).

TABLE 8.18(d) | Reduced state table

Present State	Next State		Output, Z	
	$X=0$	$X=1$	$X=0$	$X=1$
A	A	B	0	0
B	B	D	0	0
D	B	A	0	1

EXAMPLE 8.6 Reduce the states given in the state Table 8.19 using equivalence groups.

TABLE 8.19 | State table

Present State	Next State		Output, Z	
	X = 0	X = 1	X = 0	X = 1
S_0	S_4	S_2	0	0
S_1	S_2	S_0	0	0
S_2	S_1	S_6	0	0
S_3	S_6	S_0	0	0
S_4	S_5	S_1	1	0
S_5	S_4	S_3	0	0
S_6	S_3	S_6	0	0

SOLUTION

Step 1: The present states are partitioned into two groups, G_1 and G_2 those have same outputs for both inputs and are given below:

$$G_1 = (S_0, S_1, S_2, S_3, S_5, S_6)$$

$$G_2 = (S_4)$$

Step 2: Table 8.19(a) is constructed where next states are replaced by the corresponding group G_1 or G_2.

Step 3: The present states of group G_1 are partitioned into two subgroups G_{11} and G_{12}. Group G_2 is not further partitioned but renamed as G_{21}. Table 8.19(b) is constructed where next states are replaced by the corresponding subgroups, G_{11}, G_{12} and G_{21}.

$$G_{11} = (S_0, S_5)$$

$$G_{12} = (S_1, S_2, S_3, S_6)$$

$$G_{21} = (S_4)$$

TABLE 8.19(a) | Partitioned state table into groups

Present State	Next State	
	X = 0	X = 1
S_0	G_2	G_1
S_1	G_1	G_1
S_2	G_1	G_1
S_3	G_1	G_1
S_5	G_2	G_1
S_6	G_1	G_1
S_4	G_1	G_1

TABLE 8.19(b) | Partitioned state table into subgroups

Present State	Next State	
	$X = 0$	$X = 1$
S_0	G_{21}	G_{12}
S_5	G_{21}	G_{12}
S_1	G_{12}	G_{11}
S_2	G_{12}	G_{12}
S_3	G_{12}	G_{11}
S_6	G_{12}	G_{12}
S_4	G_{11}	G_{12}

Step 4: The present states of subgroup G_{12} are partitioned into two subgroups G_{121} and G_{122}. Groups G_{11} and G_{21} are not further partitioned but renamed as G_{111} and G_{211}, respectively. Table 8.19(c) is constructed where next states are replaced by the corresponding subgroups, G_{111}, G_{121}, G_{122} and G_{211}.

$$G_{111} = (S_0, S_5)$$
$$G_{121} = (S_1, S_3)$$
$$G_{122} = (S_2, S_6)$$
$$G_{211} = (S_4)$$

TABLE 8.19(c) | Partitioned state table into sub-subgroups

Present State	Next State	
	$X = 0$	$X = 1$
S_0	G_{211}	G_{122}
S_5	G_{211}	G_{121}
S_1	G_{122}	G_{111}
S_3	G_{122}	G_{111}
S_2	G_{121}	G_{122}
S_6	G_{121}	G_{122}
S_4	G_{111}	G_{121}

TABLE 8.19(d) | Partitioned state table into sub-subgroups

Present State	Next State	
	$X = 0$	$X = 1$
S_0	G_{2111}	G_{1221}
S_5	G_{2111}	G_{1211}

(*Continued*)

TABLE 8.19(d) | (Continued)

Present State	Next State	
	$X = 0$	$X = 1$
S_1	G_{1221}	G_{1111}
S_3	G_{1221}	G_{1111}
S_2	G_{1211}	G_{1221}
S_6	G_{1211}	G_{1221}
S_4	G_{1112}	G_{1211}

Step 5: The present states of subgroup G_{111} are partitioned into two subgroups G_{1111} and G_{1112}. Groups G_{121}, G_{122} and G_{211} are not further partitioned but renamed as G_{1211}, G_{1221} and G_{2111}, respectively. Table 8.19(d) is constructed where next states are replaced by the corresponding subgroups, G_{1111}, G_{1112}, G_{1211}, G_{1221} and G_{2111}.

$$G_{1111} = (S_0)$$

$$G_{1112} = (S_5)$$

$$G_{1211} = (S_1, S_3)$$

$$G_{1221} = (S_2, S_6)$$

$$G_{2111} = (S_4)$$

Step 6: No further partitioning of sub subgroups is possible. States of G_{1211} and G_{1221} subgroups are equivalent that are $S_1 = S_3$ and $S_2 = S_6$.

Step 7: Replace states S_3 with state S_1 and state S_6 with S_2 in the given state Table 8.19 to get state Table 8.19(e).

TABLE 8.19(e) | State table with equivalent states

Present State	Next State		Output, Z	
	$X = 0$	$X = 1$	$X = 0$	$X = 1$
S_0	S_4	S_2	0	0
S_1	S_2	S_0	0	0
S_2	S_1	~~S_6~~ S_2	0	0
~~S_3~~ S_1	~~S_6~~ S_2	S_0	0	0
S_4	S_5	S_1	1	0
S_5	S_4	~~S_3~~ S_1	0	0
~~S_6~~ S_2	~~S_3~~ S_1	~~S_6~~ S_2	0	0

Remove the duplicating states S_3 and S_6 to get the reduced state Table 8.19(f).

TABLE 8.19(f) | Stable table with equivalent states

Present State	Next State		Output, Z	
	$X = 0$	$X = 1$	$X = 0$	$X = 1$
S_0	S_4	S_2	0	0
S_1	S_2	S_0	0	0
S_2	S_1	S_2	0	0
S_4	S_5	S_1	1	0
S_5	S_4	S_1	0	0

EXAMPLE 8.7 Reduce the states given in the state Table 8.20.

TABLE 8.20 | State table

Present State	Next State		Output, Z	
	$X = 0$	$X = 1$	$X = 0$	$X = 1$
A	B	D	0	1
B	E	F	0	1
C	B	D	0	1
D	E	B	0	1
E	E	F	0	1
F	F	C	0	0

SOLUTION

Step 1: The present states are partitioned into two groups, G_1 and G_2 those have same outputs for both inputs and are given below:

$$G_1 = (A, B, C, D, E)$$

$$G_2 = (F)$$

Step 2: Table 8.20(a) is constructed where next states are replaced by the corresponding group G_1 or G_2.

TABLE 8.20(a) | Partitioned state table into groups

Present State	Next State	
	$X = 0$	$X = 1$
A	G_1	G_1
B	G_1	G_2
C	G_1	G_1
D	G_1	G_1
E	G_1	G_2
F	G_2	G_1

Step 3: The present states of group G_1 are partitioned into two subgroups G_{11} and G_{12}. Group G_2 is not further partitioned but renamed as G_{21}. Table 8.20(b) is constructed where next states are replaced by the corresponding subgroups, G_{11}, G_{12} and G_{21}.

$$G_{11} = (A, C, D)$$

$$G_{12} = (B, E)$$

$$G_{21} = (F)$$

TABLE 8.20(b) | Partitioned state table into subgroups

Present State	Next State	
	$X = 0$	$X = 1$
A	G_{12}	G_{11}
C	G_{12}	G_{11}
D	G_{12}	G_{12}
B	G_{12}	G_{21}
E	G_{12}	G_{21}
F	G_{21}	G_{11}

Step 4: The present states of subgroup G_{11} are partitioned into two subgroups G_{111} and G_{112}. Groups G_{12} and G_{21} are not further partitioned but renamed as G_{121} and G_{311}, respectively. Table 8.20(c) is constructed where next states are replaced by the corresponding subgroups, G_{111}, G_{112}, G_{121} and G_{211}, respectively.

$$G_{111} = (A, C)$$

$$G_{112} = (D)$$

$$G_{121} = (B, E)$$

$$G_{211} = (F)$$

TABLE 8.20(c) | Partitioned state table into sub-subgroups

Present State	Next State	
	$X = 0$	$X = 1$
A	G_{121}	G_{112}
C	G_{121}	G_{112}
D	G_{121}	G_{121}
B	G_{121}	G_{211}
E	G_{121}	G_{211}
F	G_{211}	G_{111}

TABLE 8.20(d) | State table with equivalent states

Present State	Next State		Output, Z	
	$X = 0$	$X = 1$	$X = 0$	$X = 1$
A	B	D	0	1
B	~~E~~ B	F	0	1
~~C~~ A	B	D	0	1
D	~~E~~ B	B	0	1
~~E~~ B	~~E~~ B	F	0	1
F	F	~~C~~ A	0	0

Step 5: No further partitioning of sub-subgroups is possible. States of G_{111} and G_{121} subgroups are equivalent that are $A = C$ and $B = E$.

Step 6: Replace states C with state A and state B with state D in the given state Table 8.20 to get state Table 8.20(d).

Remove the duplicating states to get the reduced state Table 8.20(e).

TABLE 8.32(e) | Reduced state table

Present State	Next State		Output, Z	
	$X = 0$	$X = 1$	$X = 0$	$X = 1$
A	B	D	0	1
B	B	F	0	1
D	B	B	0	1
F	F	A	0	0

8.4.2 | Implication Chart

The implication chart uses a grid array to list the possible state equivalencies for different inputs. The chart is constructed by listing all of the states except the last along the horizontal axis and all the states except the first along the vertical axis. The intersection of the horizontal and vertical state spaces indicates a possible state equivalency. A procedure for state classification in order to find n-equivalence using implication method is given below.

Algorithm 8.3 | Constructing the reduced states using implication chart method

1. Construct Implication chart: Implication chart has $(n - 1)$ horizontal rows and $(n - 1)$ vertical columns for n number of states forming $\sum_{i=1}^{n-1}(n-i)$ number of non-repetitive grids. Counting is started from top. First horizontal row corresponds to second state and first vertical column correspond first state. Intersection of these forms a square and represents the possibility of second and first state pair equivalency. First horizontal row has one square. Second horizontal row contains two squares and intersects the first and second columns.
2. Fill the squares for state pairs which have different outputs: Place cross (X) in the squares when the state pairs have different outputs.
3. Fill the squares for state pairs which may have same outputs: Compare present states, S_i and S_j having next states S_l and S_m, respectively, for a given input. If next states S_l and S_m for the given input become equal then the present states may be equal that is $S_i = S_j$ so fill the pair of next states (S_l, S_m). Such pairs depend on number of inputs.
4. Apply equivalency test of each square by checking implied equivalent state pairs written in square. Put cross (X) in the square if implied equivalent state pairs are not equal.

EXAMPLE 8.8 Reduce the states given in the state Table 8.21 using implication chart.

TABLE 8.21 | State table

Present State	Next State		Output, Z	
	$X = 0$	$X = 1$	$X = 0$	$X = 1$
A	A	B	0	0
B	C	D	0	0
C	E	F	0	0
D	B	A	0	1
E	C	D	0	0
F	B	A	0	1

SOLUTION

Step 1: Set up implication chart. Implication chart is set up as shown in Figure 8.18(a). Total number of states is 6 that are A, B, C, D, E and F. Number of horizontal rows is 5. Number of vertical columns is 5. Total numbers of intersecting grids is $\sum_{i=1}^{6-1}(6-i) = 5+4+3+2+1 = 15$.

Top row count start from second state, i.e. *B*. Rest rows count to down as third, *C*, fourth *D*, fifth *E* and last sixth state *F*. Left most column count starts from first state, i.e. *A*. Rest columns count towards right as second, *B*, third, *C*, fourth *D* and right most fifth state *E*.

Step 2: Fill the squares for state pairs which have different outputs: Figure 8.18(b) has the crossed squares those cannot become equal.

Output of present *A* state is not equal to output of present state *D* when input $X = 1$. So, cross the square intersecting states *A* and *D*.

Output of present *A* state is not equal to output of present state *F* when input $X = 1$. So, cross the square intersecting states *A* and fifth (bottom) row from top marked as state *F*.

Output of present *B* state is not equal to output of present state *D* when input $X = 1$. So, cross the square intersecting the second column, marked for state *B* and third row from top marked as state *D*.

Output of present *B* state is not equal to output of present state *F* when input $X = 1$. So, cross the square intersecting the second column, marked for state *B* and fifth (bottom) row from top marked as state *F*.

Output of present *C* state is not equal to output of present state *D* when input $X = 1$. So, cross the square intersecting the third column, marked for state *C* and third row from top marked as state *D*.

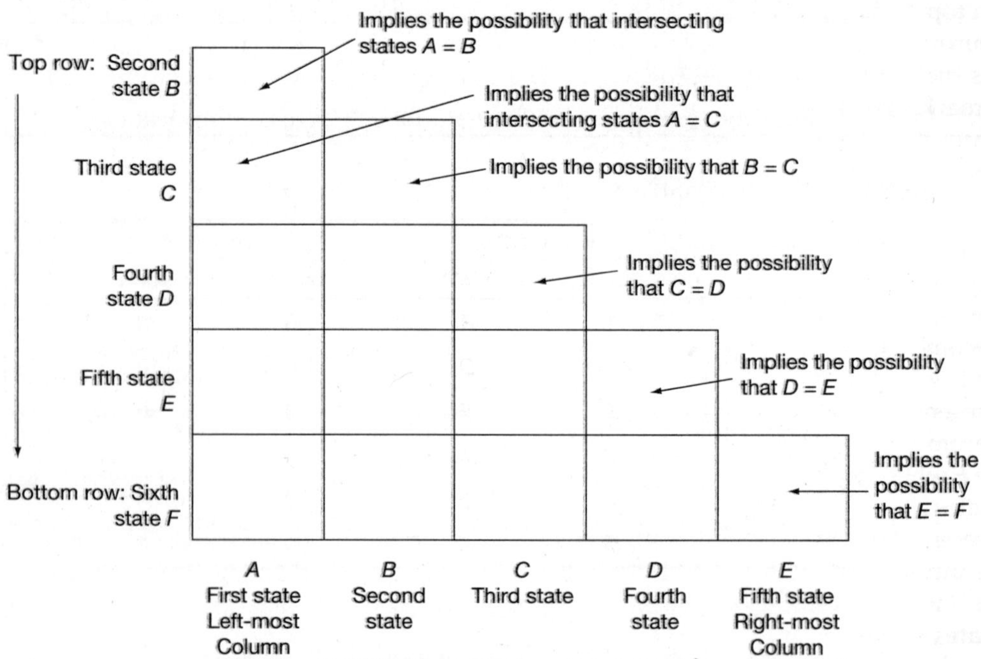

FIGURE 8.18(a) | Implication chart

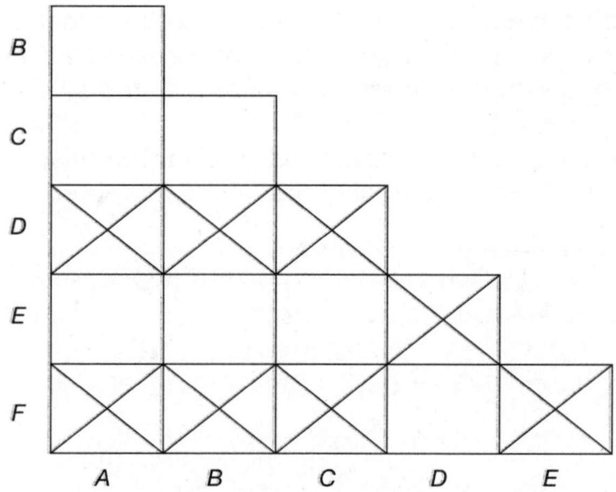

FIGURE 8.18(b) | Implication chart

Output of present C state is not equal to output of present state F when input $X = 1$. So, cross the square intersecting the third column, marked for state C and fifth (bottom) row from top marked as state F.

Output of present D state is not equal to output of present state E when input $X = 1$. So, cross the square intersecting the fourth column, marked for state D and fourth row from top marked as state E.

Output of present E state is not equal to output of present state F when input $X = 1$. So, cross the square intersecting the fifth column, marked for state E and fifth (bottom) row from top marked as state F.

Step 3: Fill the squares for state pairs which may have same outputs

Figure 8.18(c) shows the filled squares for state pair that have same outputs.

Present states A and B having equal outputs, so can become equal if and only if next state A becomes equal to next state C when input $X = 0$ and next state B becomes equal to next state D when input $X = 1$. Write (A, C and B, D) in intersecting square of states A and B.

Present states A and C having equal outputs, so can become equal if and only if next state A becomes equal to next state E when input $X = 0$ and next state B becomes equal to next state F when input $X = 1$. So, write equivalency pair (A, E) and (B, F) in intersecting square of states A and C.

Present states A and E having equal outputs, so can become equal if and only if next state A becomes equal to next state C when input $X = 0$ and next state B becomes equal to next state D when input $X = 0$. So, write equivalency pair (A, C) and (B, D) in intersecting square of states A and E.

Present states B and C have equal outputs and can become equal if and only if next state C becomes equal to next state E when input $X = 0$ and next state D becomes equal to next state F when input $X = 1$. So, write equivalency pair (C, E) and (D, F) in intersecting square of states B and C.

Present states B and E have equal outputs and are equal because next states are C and D when input $X = 0$ and $X = 1$. So mark double tick (✓) because both next states are same for both inputs.

Present states C and *E* have equal outputs and can become equal if and only if next state *E* becomes equal to next state *C* when input $X = 0$ and next state *F* becomes equal to next state *D* when input $X = 1$. So, write equivalency pair (*C*, *E*) and (*D*, *F*) in intersecting square of states *C* and *E*.

Present states *D* and *F* have equal outputs and are equal because next states are *B* and *A* when input $X = 0$ and $X = 1$. So mark double tick (✓) because both next states are same for both inputs.

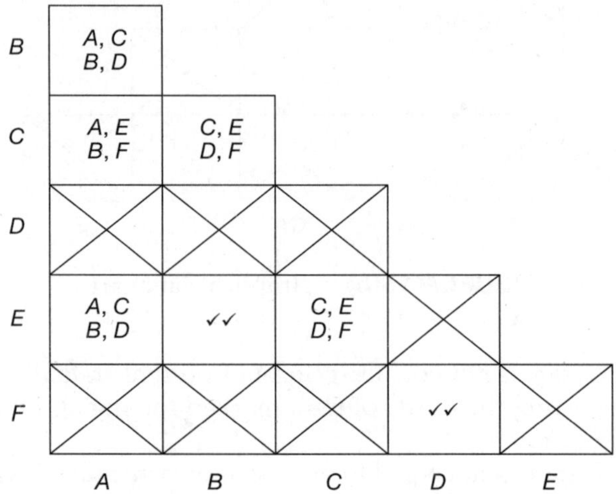

FIGURE 8.18(c) | Implication chart

Step 4: Apply equivalency test of each square

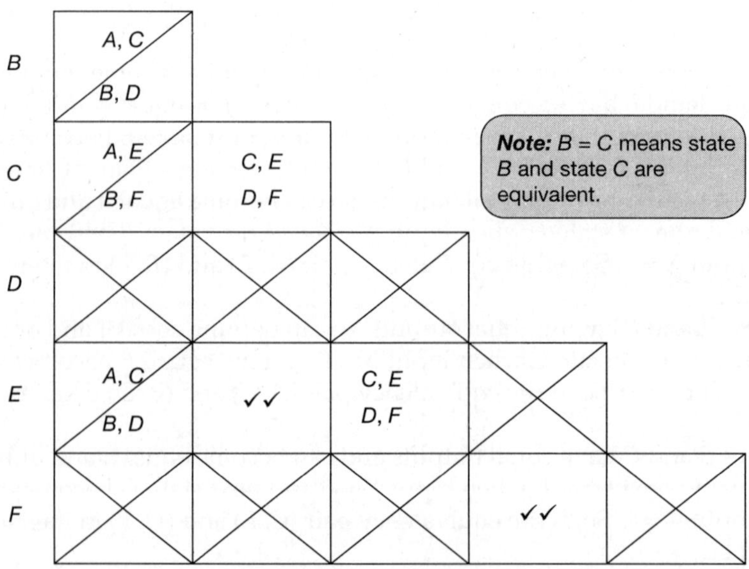

FIGURE 8.18(d) | Implication chart

Start inspecting from column marked as state *E*.

E, *F* are not equivalent ($E \neq F$), because there is a cross.

E: *E*

D, *F* are equivalent so $D = F$ (Check ✓✓ the square).

D: *E*(DF)

C, *E* are equivalent if and only if *D*, *F* are equivalent but *D*, *F* are equivalent so *C*, *E* are also equivalent. So, do not cross the intersecting square.

C: *E*(DF)(CE) or (CE)(DF)

B, *E* are equivalent so $B = E$ (Check ✓✓ the square).

B, *C* are equivalent if and only if *C*, *E* and *D*, *F* are equivalent but *D*, *F* and *C*, *E* are equivalent so *B*, *C* are also equivalent. So do not cross the intersecting square.

B: (CE)(DF)(BE)(BC) or (BCE)(DF) ($B = C = E$ because $C = E$, $B = E$ and $B = C$)

A, *B* are equivalent if and only if *A*, *C* and *B*, *D* are equivalent but *B*, *D* are not equivalent so *A*, *B* cannot be equivalent. So cross the intersecting square.

A, *C* are equivalent if and only if *A*, *E* and *B*, *F* are equivalent but *B*, *F* are not equivalent so *A*, *C* cannot be equivalent. So cross the intersecting square.

A, *E* are equivalent if and only if *B*, *D* are equivalent but *B*, *D* are not equivalent so *A*, *E* cannot be equivalent. So cross the intersecting square. Figure 8.18(d) shows the revised implication chart.

So $B = C = E$ and $D = F$. Delete *C*, *E* and *F* rows in Table 8.33 and replace states *C* and *E* with state *B* and state *F* with state *D* to get state Table 8.22.

TABLE 8.22 | Reduced state table

Present State	Next State		Output, *Z*	
	$X = 0$	$X = 1$	$X = 0$	$X = 1$
A	*A*	*B*	0	0
B	*B*	*D*	0	0
D	*B*	*A*	0	1

EXAMPLE 8.9 Reduce the states given in the state Table 8.23 using implication chart.

TABLE 8.23 | State table

Present State	Next State		Output, *Z*	
	$X = 0$	$X = 1$	$X = 0$	$X = 1$
A	*D*	*B*	0	0
B	*E*	*A*	0	1
C	*G*	*F*	0	1
D	*A*	*D*	1	0
E	*A*	*D*	1	0
F	*C*	*B*	0	0
G	*A*	*E*	1	0

Solution

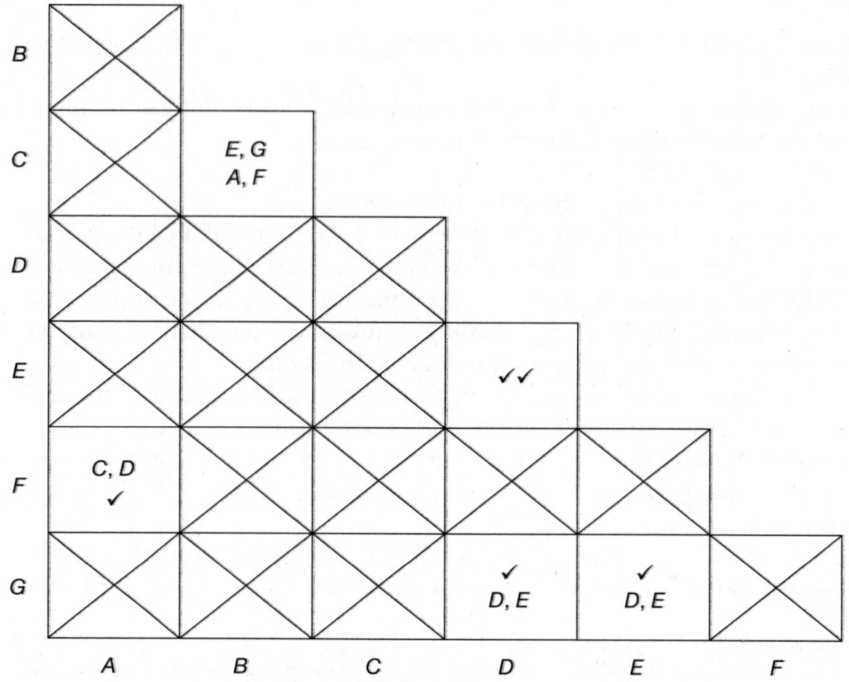

FIGURE 8.19(a) | Implication chart

Step 1: Implication chart is set up as shown in Figure 8.19(a). It has seven rows and seven columns. Total number of intersecting grids is $\sum_{i=1}^{7-1}(7-i) = 6+5+4+3+2+1 = 21.$

Step 2: Fill the intersecting squares for present state pairs with cross (X) which have different outputs. Implication chart is revised as shown in Figure 8.19(a).

Step 3: Fill the intersecting squares of present state pairs which have same outputs and may become equal if and only if next states are equated. Fill such intersecting squares with the condition of equality. Implication chart is revised as shown in Figure 8.19(b).

Step 4: Apply equivalency test for each square. Start inspecting from column marked as state *F*.

$G \neq F$ because there is cross.

F: *F*

$E = G$ iff $D = E$. $D = E$ because intersecting square has double tick. So, $E = G$.

E: (*EG*)*F*

$D = E$ because intersecting square has double tick. $D = E = G$ because $D = E$ and $E = G$.

D: (*DE*)(*EG*)*F* = (*DEG*)*F*

C: *C*(*DEG*)*F*

$B = C$ iff $A = F$ and $E = G$. A can be equal to F iff $C = D$ but $C \neq D$ (X against intersecting square). So $B \neq C$. Put cross

B: $BC(DEG)F$

$A = F$ iff $C = D$ but $C \neq D$ (X against intersecting square). So $A \neq F$. Put cross

A: ABC $(DEG)F$

So, $D = E = G$. Delete E and G rows and replace states G and E with state D in the given state Table 8.24 to get state Table 8.38.

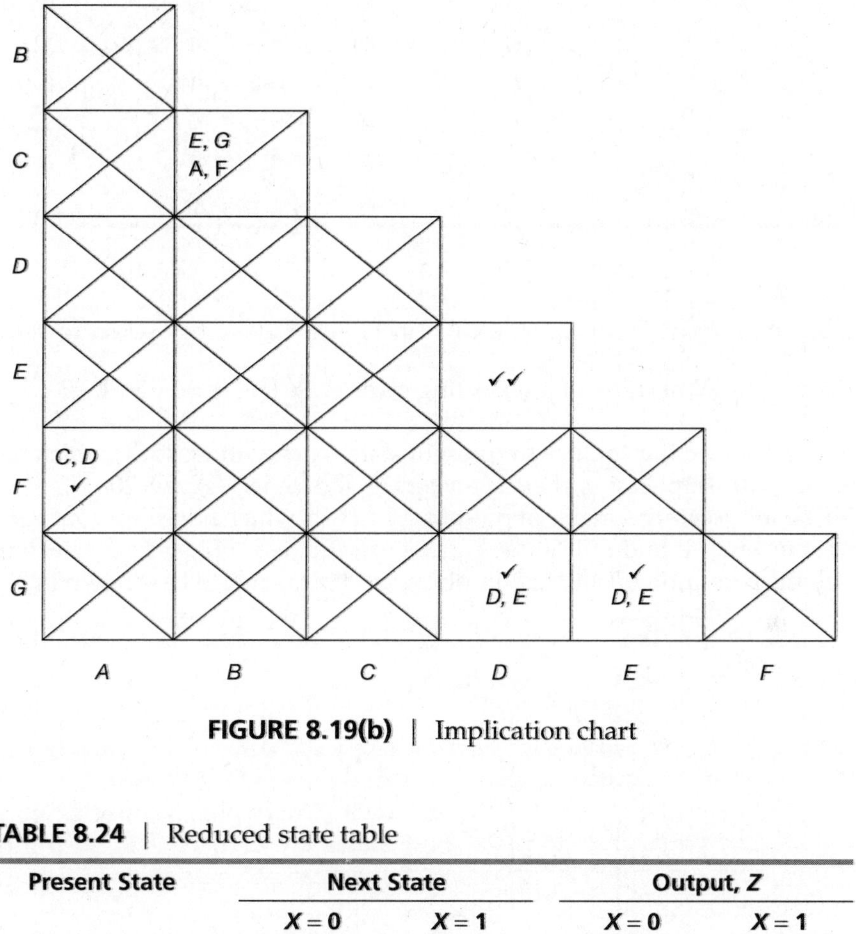

FIGURE 8.19(b) | Implication chart

TABLE 8.24 | Reduced state table

Present State	Next State		Output, Z	
	$X = 0$	$X = 1$	$X = 0$	$X = 1$
A	D	B	0	0
B	D	A	0	1
C	D	F	0	1
D	A	D	1	0
F	C	B	0	0

EXAMPLE 8.10 Reduce the states given in the state Table 8.25 using implication chart.

TABLE 8.25 | State table

Present State	Next State		Output, Z	
	$X=0$	$X=1$	$X=0$	$X=1$
A	*C*	*B*	0	0
B	*D*	*C*	0	0
C	*G*	*D*	1	1
D	*E*	*F*	1	0
E	*F*	*A*	0	1
F	*G*	*F*	1	0
G	*F*	*A*	0	1

SOLUTION

Step 1: Implication chart is set up as shown in Figure 8.20. It has seven rows and seven columns. Total number of intersecting grids is $\sum_{i=1}^{7-1}(7-i)=6+5+4+3+2+1=21.$

Step 2: Fill the intersecting squares for present state pairs with cross (X) which have different outputs. Implication chart is revised as shown in Figure 8.20.

Step 3: Fill the intersecting squares of present state pairs which have same outputs and may become equal if and only if next states are equated. Fill such intersecting squares with the condition of equality. Implication chart is revised as shown in Figure 8.20.

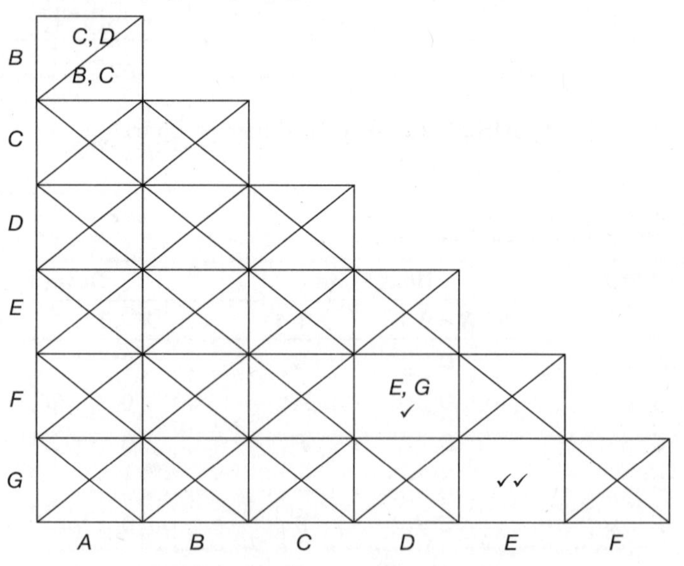

FIGURE 8.20 | Implication chart

Step 4: Apply equivalency test for each square. Start inspecting from column marked as state *F*.

$F \neq G$ (*X* is found)

F: *F*

$E = G$ because double tick is found

$E \neq F$ (*X* is present)

E: *F* (*EG*)

$D = F$ iff $E = G$. But $E = G$ so $D = F$

D: *F* (*EG*)(*DF*)

C: *F* (*EG*)(*DF*) *C*

B: *F* (*EG*)(*DF*) *C B*

$A = B$ iff $C = D$ and $B = C$. But $C \neq D$ and $B \neq C$ so $A \neq B$

A: *F* (*EG*)(*DF*) *C B A*

So, $E = G$ and $D = F$. Delete *G* and *F* rows and replace states *G* and *F* with state *E* and *D*, respectively, in the given state Table 8.25 to get reduced state Table 8.26.

TABLE 8.26 | State table with equivalent states

Present State	Next State		Output, Z	
	X = 0	X = 1	X = 0	X = 1
A	C	B	0	0
B	D	C	0	0
C	E	D	1	1
D	E	D	1	0
E	D	A	0	1

8.5 | STATE ASSIGNMENT

The clocked sequential circuit is defined in terms of present state, next state and the output for different input conditions. The input for flip-flops are calculated to get the desired next state for the present state and input(s), hence there is a need to specify the present state and next state in binary format. In case the present state and next state are specified by alphabets, then binary values are assigned for the alphabets. The number of bits used to assign binary values to alphabets depends on the number of alphabets used. So for M number of alphabets, n-bits are required to represent the alphabets in binary, such that $2^n > M$. The selection of value of n should be minimum possible because n numbers of flip-flops are used to design such circuit. So, minimum numbers of flip-flops are used to design a system. Since one flip-flop has two states, therefore, the number of flip-flops required can be determined from the number of states in the circuit. The redundant states are eliminated by state reduction method.

The following equation gives the number of possible state assignments, *S*.

$$S = \frac{2^n\,!}{(2^n - M)!}$$

where n is number state variables (flip-flops).

M is number of states.

E.g. $n = 2$ and $M = 4$. The number of possible state assignments is given by

$$S = \frac{2^2!}{(2^2 - 4)!} = \frac{4!}{0!} = \frac{4\times3\times2\times1}{1} = 24$$

All the possible assignments are not worth considering. Some assignments produce the same results. Such assignments are said to be equivalent. It happens when one state assignment is compliment of other assignment. Two state assignments are equivalent if the place values of binary value assignments are interchanged. So, non-equivalent assignments are required to be found. The following equation gives the number of possible non-equivalent state assignments, S.

$$S = \frac{(2^n - 1)!}{(2^n - M)!n!}$$

where n is number state variables (flip-flops). M is number of states.

Let $n = 2$ and $M = 4$. The number of possible non-equivalent state assignments, S is

$$S = \frac{(2^2 - 1)!}{(2^2 - 4)!2!} = \frac{3!}{0!2!} = 3$$

Let $n = 3$ and $M = 5$. The number of possible non-equivalent state assignments, S is

$$S = \frac{(2^3 - 1)!}{(2^3 - 5)!3!} = \frac{7!}{3!3!} = \frac{7\times6\times5\times4}{3\times2\times1} = 140$$

The number of possible non-equivalent state assignments increases drastically if number of state variables, n and number of states, M increase. So, hit-and-trial attempt for state assignments is not practical. A systematic state assignment approach is to be followed. No general solution to state assignment problem is possible; however, a reasonable solution can be obtained using following rules.

Rules:

1. Adjacent binary codes should be assigned to present states having the same next state, for a given input value. Adjacent codes can be grouped to form into a prime implicant.
2. Adjacent binary codes should be assigned to next state(s) of single present state.
3. Adjacent binary codes should be assigned to states those have the same outputs for given inputs.

EXAMPLE 8.11 Partial state diagram of a sequential machine is shown in Figure 8.21. Make suitable state assignment for obtaining minimal logic expression. Assume the machine with a single input and the complete state diagram contains seven states.

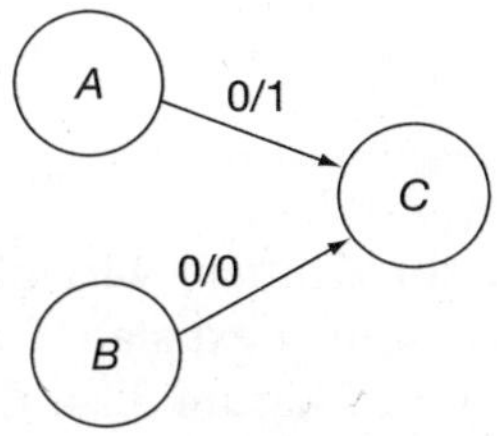

FIGURE 8.21 | Partial state diagram

SOLUTION

Number of states given is 7.
Number of flip-flops required = $\log_2 7 = 3$.
So, three binary bits are assigned to three binary variables.
A and *B* present states have the same next state *C* for input 0.
Set a three-variable K-map as shown in Figure 8.22. Consider two adjacent cells to fill the square with present states.

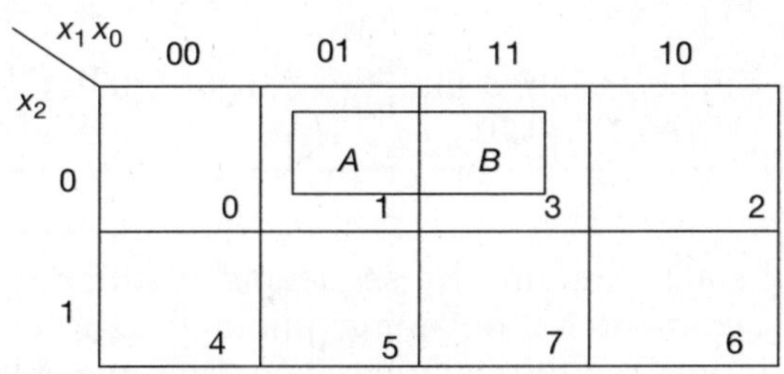

FIGURE 8.22 | K-map

Two adjacent values can be assigned like $A \rightarrow 001$ and $B \rightarrow 011$ or $A \rightarrow 000$ and $B \rightarrow 001$ or $A \rightarrow 011$ and $B \rightarrow 010$ or $A \rightarrow 100$ and $B \rightarrow 101$ or $A \rightarrow 101$ and $B \rightarrow 111$ or $A \rightarrow 111$ and $B \rightarrow 110$.

EXAMPLE 8.12 Partial state diagram of a sequential machine is shown in Figure 8.23. Make suitable state assignment for obtaining minimal logic expression. Assume the machine with a double inputs and the complete state diagram contains seven states.

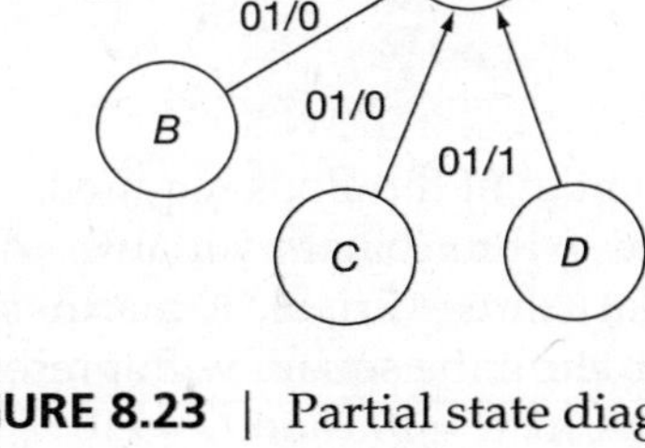

FIGURE 8.23 | Partial state diagram

Solution

Number of states given is 7.
Number of flip-flops required = $\log_2 7 = 3$.
So, three binary bits are assigned to three binary variables.
A, B, C and D present states have the same next state E for an input 01.
Set a three-variable K-map as shown in Figure 8.24. Consider four cells adjacent to one another to fill the square with present states.

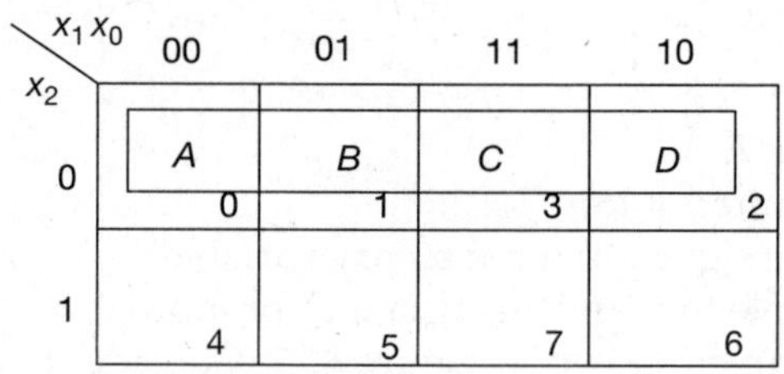

FIGURE 8.24 | K-map

Two adjacent values can be assigned like $A \rightarrow 001$, $B \rightarrow 001$, $C \rightarrow 011$ and $D \rightarrow 010$ or $A \rightarrow 100$, $B \rightarrow 101$, $C \rightarrow 111$ and $D \rightarrow 110$.

Example 8.13 Partial state diagram of a sequential machine is shown in Figure 8.25. Make suitable state assignment for obtaining minimal logic expression. Assume the machine with a double inputs and the complete state diagram contains seven states.

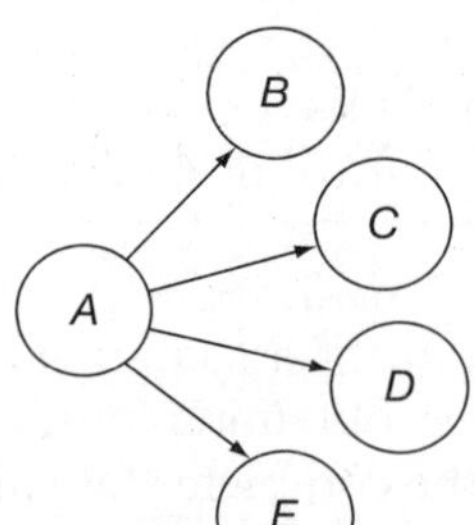

FIGURE 8.25 | Partial state diagram

Solution

Number of states given is 7. Number of flip-flops required = $\log_2 7 = 3$.
So, three binary bits are assigned to three binary variables. A and B present states have the same next state C for a input 0. Set a three-variable K-map as shown in Figure 8.26. Consider four cells adjacent to each other to fill the square with present states. Two adjacent values can be assigned like $A = 000$, $B = 010$, $C = 100$ and $D = 110$.

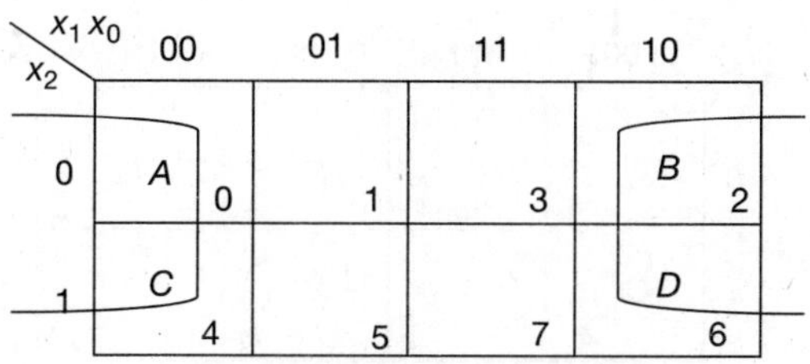

FIGURE 8.26 | K-map

EXAMPLE 8.14 State table of a sequential machine is shown in Table 8.27. Make suitable state assignment for obtaining minimal logic expression.

TABLE 8.27 | State table

Present State	Next State		Output, Z	
	X = 0	X = 1	X = 0	X = 1
A	B	D	0	1
B	C	D	1	0
C	E	F	0	1
D	F	E	1	0
E	G	A	0	0
F	A	G	0	0
G	B	D	0	0

SOLUTION

Number of states given is 7.

Number of flip-flops required = $\log_2 7 = 3$.

So, three binary bits are assigned to three binary variables.

Rule 1: Find present states having the same next state, for a given input value.

Present states *A* and *G* move to next state *B* when input *X* = 0. Present states *A* and *G* move to next state *D* when input *X* = 1.

Targeted adjacent states are [*A*, *G*]

Rule 2: Find next states arrived from single present state.

Present state *A* moves to *B* and *D* next states. Present state *B* moves to *C* and *D* next states. Present state *C* and *D* moves to next states *E* and *F*. Present states *E* and *F* move to next states *A* and *G*. Present state *F* moves to *B* and *D* next states.

Targeted adjacent states are [*B*, *D*], [*C*, *D*], [*E*, *F*] and [*A*, *G*].

Rule 3: Find states those have the same outputs.

States *A* and *C* have same outputs for given input. States *B* and *D* have same outputs. States *E*, *F* and *G* have same outputs for given input.

Targeted adjacent states are [*A*, *C*], [*B*, *D*] and [*E*, *F*, *G*]

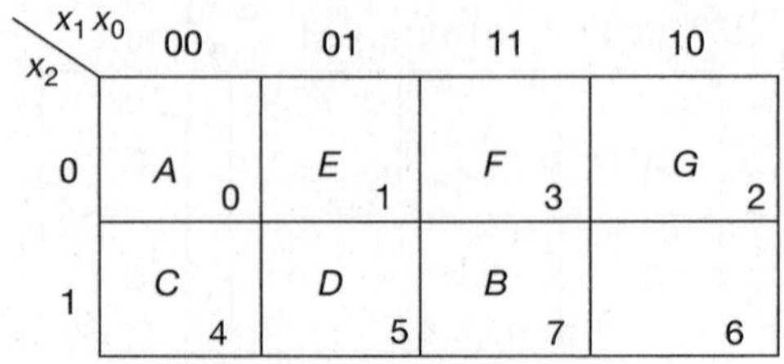

FIGURE 8.27 | K-map

Set a three-variable K-map as shown in Figure 8.27. Try to adjust as many as possible targeted adjacent states. Give top priority to Rule 1 targeted adjacent state pairs then to Rule 2 and Rule 3. [*A*, *G*] states are targeted two times, so have top priority. One of the possible state assignments is given as below.

[*A*, *G*], [*C*, *D*], [*E*, *F*], [*A*, *C*], [*B*, *D*] and [*E*, *F*, *G*] are adjusted in K-map to follow adjacent assignments.

So, $A \rightarrow 000$, $B \rightarrow 111$, $C \rightarrow 100$, $D \rightarrow 101$, $E \rightarrow 001$, $F \rightarrow 011$ and $G \rightarrow 010$.

Example 8.15 State table of a sequential machine is shown in Table 8.28. Make suitable state assignment for obtaining minimal logic expression.

TABLE 8.28 | State table

Present State	Next State		Output, *Z*	
	$X=0$	$X=1$	$X=0$	$X=1$
A	*A*	*C*	0	0
B	*C*	*D*	0	1
C	*A*	*F*	1	1
D	*B*	*E*	0	0
E	*C*	*C*	1	0
F	*D*	*E*	1	0

Solution

Number of states given is 6.

Number of flip-flops required $= \log_2 6 = 3$.

So, three binary bits are assigned to three binary variables.

Rule 1: Find present states having the same next state, for a given input value.

Present states *A* and *C* move to next state *A* when input $X = 0$. Present states *B* and *E* move to next state *C* when input $X = 0$. Present states *A* and *E* move to next state *C* when input $X = 1$. Present states *D* and *F* move to next state *E* when input $X = 1$. Targeted adjacent states are [*A*, *C*], [*B*, *E*], [*A*, *E*], [*D*, *F*]

Rule 2: Find next states arrived from single present state.
Present state *A* moves to *A* and *C* next states. Present state *B* moves to next states *C* and *D*. Present state *C* moves to next states *A* and *F*. Present state *D* moves to *B* and *E* next states. Present state *F* moves to *D* and *E* next states.
Targeted adjacent states are [*A*, *C*], [*C*, *D*], [*A*, *F*], [*B*, *E*] and [*D*, *E*].

Rule 3: Find states those have the same outputs.
States *A* and *D* have same outputs for given input. States *E* and *F* have same outputs.
Targeted adjacent states are [*A*, *D*] and [*E*, *F*]

Set a three-variable K-map as shown in Figure 8.28. Try to adjust as many as possible targeted adjacent states. Give top priority to Rule 1 targeted adjacent state pairs then to Rule 2 and Rule 3. [*A*, *C*] and [*B*, *E*] states are targeted two times so have top priority. One of the possible state assignments is given below.

x_2 \ x_1x_0	00	01	11	10
0	*A* 0	1	*B* 3	*E* 2
1	*C* 4	5	*F* 7	*D* 6

FIGURE 8.28 | K-map

[*A*, *C*], [*B*, *E*], [*A*, *E*], [*D*, *F*], [*C*, *D*] and [*D*, *E*] are adjusted in K-map to follow adjacent assignments where states [*A*, *F*], [*A*, *D*] and [*E*, *F*] are ignored.

So, $A \rightarrow 000$, $B \rightarrow 011$, $C \rightarrow 100$, $D \rightarrow 110$, $E \rightarrow 010$ and $F \rightarrow 111$.

8.6 | DESIGN OF CLOCK SEQUENTIAL CIRCUIT

The clocked sequential circuit is given or defined by a set of statements or state table or state diagram. The stepwise procedure is given in Algorithm 8.4.

Algorithm 8.4 | Design of clocked sequential circuits

1. Prepare a state table from the given set of statements or from state diagram.
2. Using the state reduction technique, reduce the number of states.
3. If the states are not given in binary, assign binary values to each state in the state table.
4. Determine the number of flip-flops required based on number of state variables to implement the given state table.
5. Select the flip-flop(s).
6. Draw the excitation table for the circuit as per the state table and.
7. Using K-map, simplify and write the Boolean equations for the input and output of flip-flops.
8. Draw the logic diagram.

Example 8.16 Design a sequential circuit whose state diagram is shown Figure 8.29.

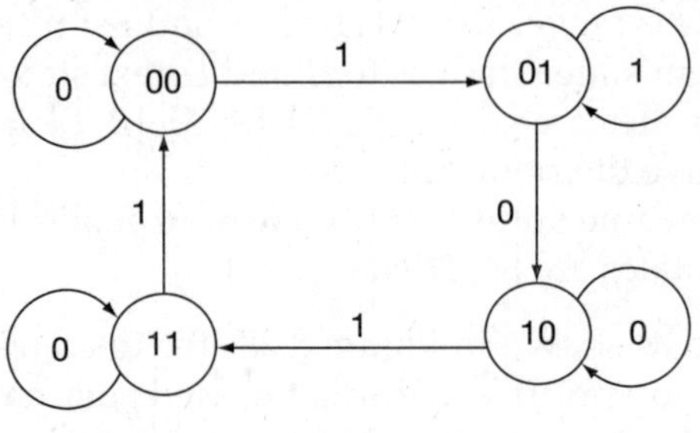

FIGURE 8.29 | State diagram

Solution

Step 1: Prepare a state table (Table 8.29) from state diagram shown in Figure 8.29.

TABLE 8.29 | State table

Present State		Next State			
		Input $X = 0$		Input $X = 1$	
A_1	A_0	A_1	A_0	A_1	A_0
0	0	0	0	0	1
0	1	1	0	0	1
1	0	1	0	1	1
1	1	1	1	0	0

Step 2: No further state reduction is possible.
Step 3: Binary values are already assigned to states in the diagram.
Step 4: Find number of flip-flops
Number of states = 4.
State variables = 2.
Flip-flops required = 2 (2^2 or no. of variables)
Step 5: JK-flip flop is selected.
Step 6: Excitation table is prepared and is given as Table 8.30.

Excitation of JK-flip-flop

$Q(t+1)$	$Q(t)$	J	K
0	0	0	X
0	1	1	X
1	0	X	1
1	1	X	0

TABLE 8.30 | Excitation table

Input	Present State		Next State		Excitation of Flip-flop			
X	A_1	A_0	A_1	A_0	JA_1	KA_1	JA_0	KA_0
0	0	0	0	0	0	X	0	X
0	0	1	1	0	1	X	X	1

(Continued)

TABLE 8.30 | (Continued)

Input	Present State		Next State		Excitation of Flip-flop			
X	A_1	A_0	A_1	A_0	JA_1	KA_1	JA_0	KA_0
0	1	0	1	0	X	0	0	X
0	1	1	1	1	X	0	X	0
1	0	0	0	1	0	X	1	X
1	0	1	0	1	0	X	X	0
1	1	0	1	1	X	0	1	X
1	1	1	0	0	X	1	X	1

Excitations of JK-flip-flops are given below in SOP form

$$JA_1(X, A_1, A_0) = \sum m(1) + \sum d(2,3,6,7) \quad (8.24(a))$$

$$KA_1(X, A_1, A_0) = \sum m(7) + \sum d(0,1,4,5) \quad (8.24(b))$$

$$JA_0(X, A_1, A_0) = \sum m(4,6) + \sum d(1,3,5,7) \quad (8.24(c))$$

$$KA_0(X, A_1, A_0) = \sum m(1,7) + \sum d(0,2,4,6) \quad (8.24(d))$$

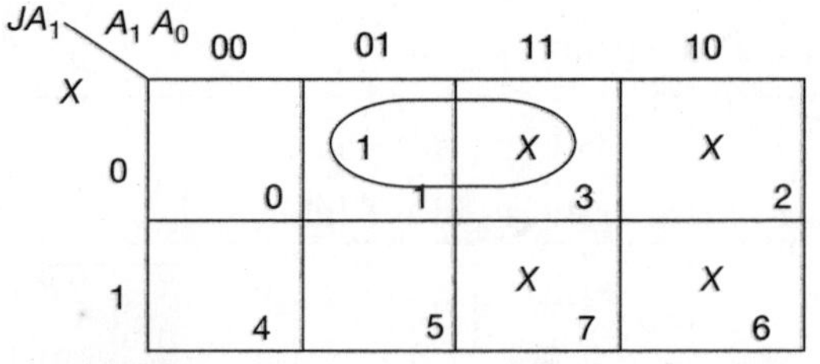

FIGURE 8.30(a) | K-map for JA_1

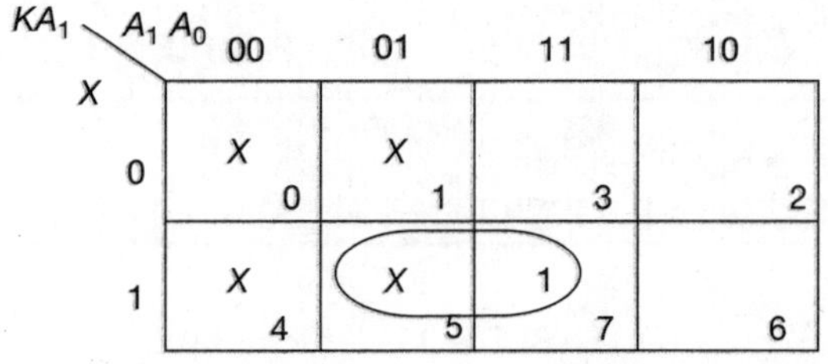

FIGURE 8.30(b) | K-map for KA_1

Step 7: Simplified Boolean expression for the excitations of JK-flip-flops using K-maps shown in Figures 8.30(a)–8.30(d), are given below:

$$JA_1(X, A_1, A_0) = \bar{X}A_0 \quad (8.25(a))$$

$$KA_1(X, A_1, A_0) = XA_0 \quad (8.25(b))$$

$$JA_0(X, A_1, A_0) = X \quad (8.25(c))$$

$$KA_0(X, A_1, A_0) = \bar{X}\bar{A}_1 + XA_1 \quad (8.25(d))$$

$$KA_0(X, A_1, A_0) = X \odot A_1 \quad (8.25(e))$$

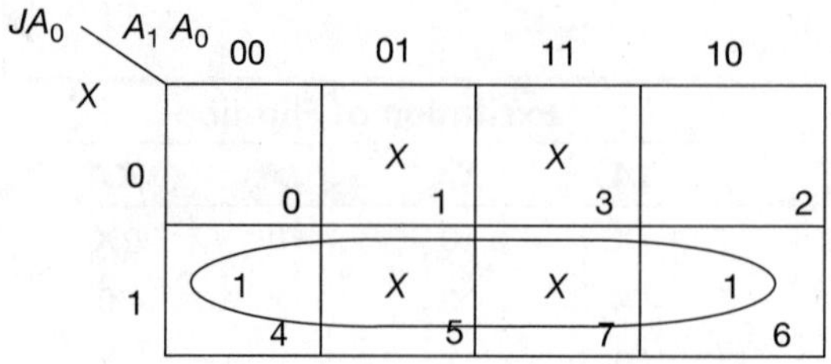

FIGURE 8.30(c) | K-map for JA_0

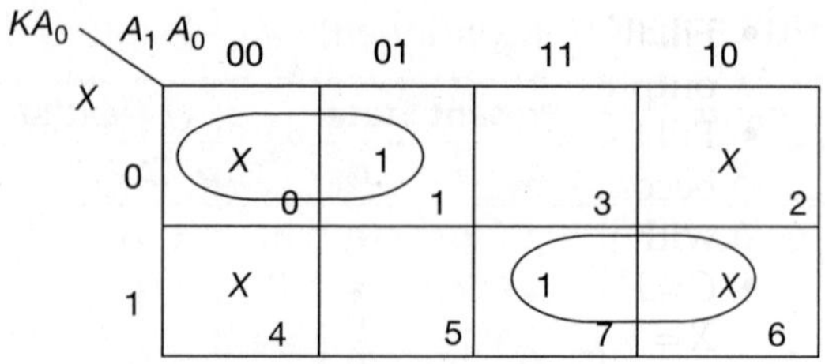

FIGURE 8.30(d) | K-map for KA_0

Step 8: Circuit diagram is drawn and is shown in Figure 8.31 considering excitation expression.

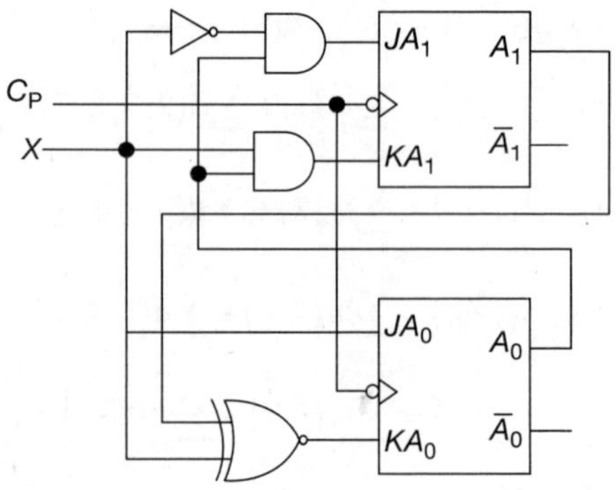

FIGURE 8.31 | Circuit diagram

EXAMPLE 8.17 Design a sequential circuit whose state table is shown Table 8.31.

TABLE 8.31 | State table

Present State	Next State		Output, Z	
	X = 0	X = 1	X = 0	X = 1
A	A	B	0	0
B	C	B	0	1
C	D	A	1	1
D	D	A	1	1

SOLUTION

Step 1: Perform reduction of states.

- Implication chart is set up as shown in Figure 8.32. It has three rows and three columns. Total numbers of intersecting grids are $\sum_{i=1}^{4-1}(4-i) = 3+2+1 = 6$.

- Fill the intersecting squares for present state pairs with cross (X) which have different outputs. Implication chart is shown in Figure 8.32.
- Fill the intersecting squares of present state pairs which have same outputs and may become equal if and only if next states are equated. Fill such intersecting squares with the condition of equality.
- $C = D$ if and only if $D = D$ and $A = A$ for $X = 0$ and $X = 1$, respectively. It is represented by double ✓ because next state is same. It is shown in Implication chart depicted in Figure 8.32.
- Apply equivalency test for each square. Start inspecting from column marked as state C.
  $C = D$ (double ✓ is found)
  C: (CD)
  $B \neq D$ and $B \neq C$ (Square is crossed)
  B: $B(CD)$
  $A \neq D$, $A \neq C$ and $A \neq B$ (Square is crossed)
  A: $AB(CD)$
  So, $C = D$.
  Reduced state table is given in Table 8.32.

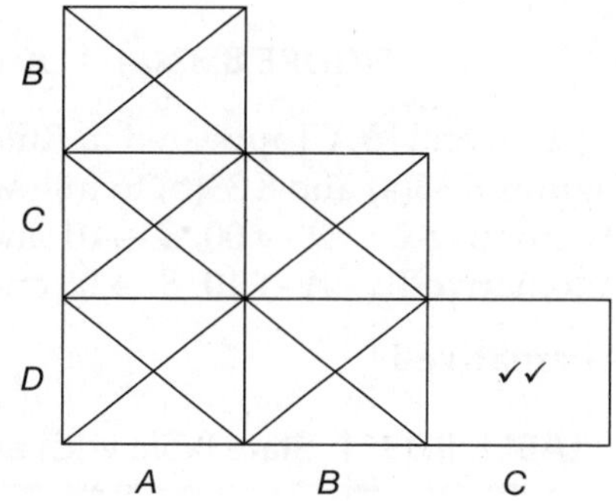

FIGURE 8.32 | Implication chart

TABLE 8.32 | Reduced state table

Present State	Next State		Output, Z	
	$X = 0$	$X = 1$	$X = 0$	$X = 1$
A	A	B	0	0
B	C	B	0	1
C	C	A	1	1

Step 2: Assign binary values to states.
Number of states given are 3.
Number of flip-flops required = $\log_2 3 = 2$.
So, two binary bits are assigned to two binary variables.

Rule 1: Find present states having the same next state, for a given input value.
Present states B and C move to next state B when input $X = 0$. Present states A and B move to next state B when input $X = 1$.
Targeted adjacent states are $[A, B]$ and $[B, C]$

Rule 2: Find next states arrived from single present state.
Present state A moves to A and B next states. Present state B moves to C and B next states. Present state C moves to next states C and A.
Targeted adjacent states are $[A, B]$, $[B, C]$ and $[C, A]$.

Rule 3: Find states those have the same outputs.
No two states have same outputs.

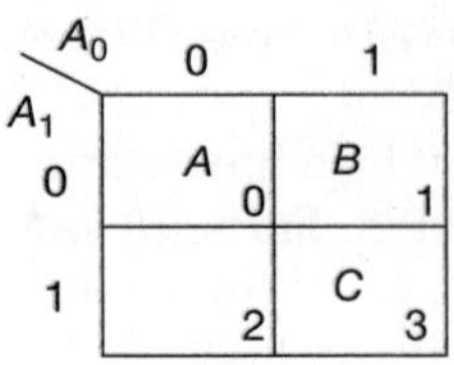

FIGURE 8.33(a) | K-map

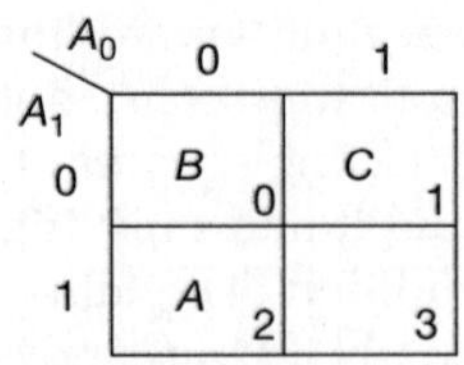

FIGURE 8.33(b) | K-map

[A, B] and [B, C] appeared in Rule-I and Rule-II so these are adjusted in K-map shown in Figures 8.33(a) and 8.33(b) to follow adjacent assignments

Alternative-I: $A \rightarrow 00$, $B \rightarrow 01$ and $C \rightarrow 11$

Alternative-II: $A \rightarrow 10$, $B \rightarrow 00$ and $C \rightarrow 01$.

Alternative-I:

TABLE 8.33 | State table with assigned binary values

Present State		Next State				Output	
		Input $X = 0$		Input $X = 1$		Input $X = 0$	Input $X = 1$
A_1	A_0	A_1	A_0	A_1	A_0		
0	0	0	0	0	1	0	0
0	1	1	1	0	1	0	1
1	1	1	1	0	0	1	1

T-flip-flops are selected. Excitation table is prepared and is given as Table 8.34.

TABLE 8.34 | Excitation table

Input	Present State		Next State		Excitation of Flip-flop		Output
X	A_1	A_0	A_1	A_0	TA_1	TA_0	Z
0	0	0	0	0	0	0	0
0	0	1	1	1	1	0	0
0	1	1	1	1	0	0	1
1	0	0	0	1	0	1	0
1	0	1	0	1	0	0	1
1	1	1	0	0	1	1	1

Excitations of T-flip-flops and output Z are given in standard sum-of-product form as given below.

$$TA_1(X, A_1, A_0) = \sum m(4,7) + \sum d(2,6) \qquad (8.26(a))$$

$$TA_0(X, A_1, A_0) = \sum m(1,7) + \sum d(2,6) \qquad (8.26(b))$$

$$Z(X, A_1, A_0) = \sum m(3,5,7) + \sum d(2,6) \qquad (8.26(c))$$

Simplified Boolean expressions for excitations of T-flip-flops are obtained using K-maps shown in Figures 8.34(a), 8.34(b) and 8.34(c).

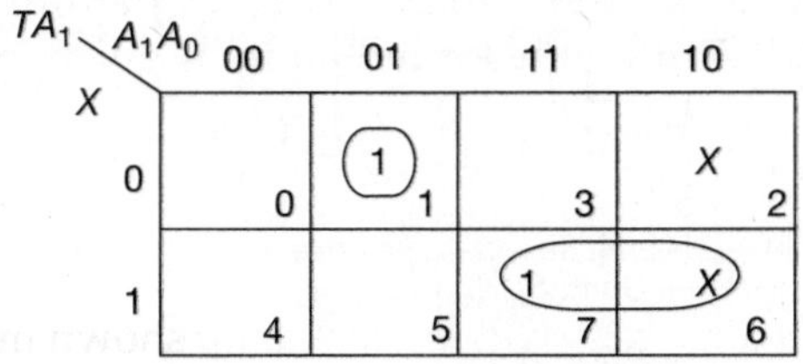

FIGURE 8.34(a) | K-map for TA_1

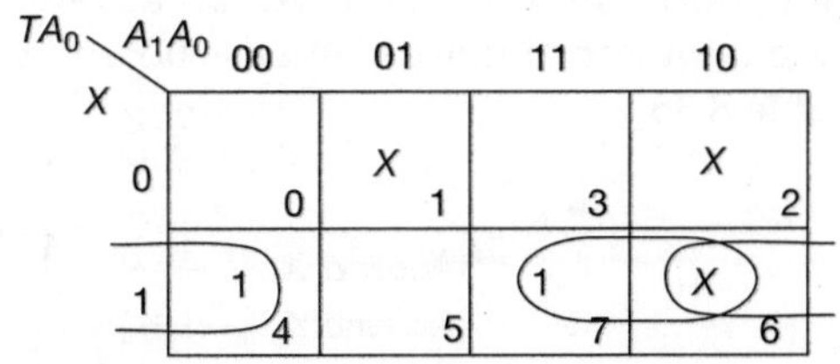

FIGURE 8.34(b) | K-map for TA_0

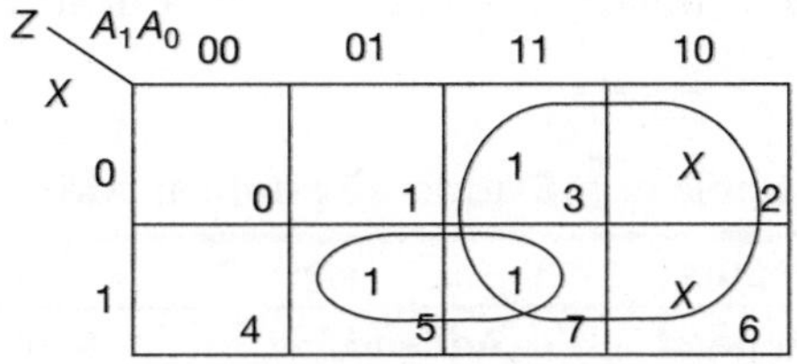

FIGURE 8.34(c) | K-map for Z

$$TA_1(X, A_1, A_0) = XA_1 + \bar{X}\bar{A}_1A_0 \tag{8.27(a)}$$

$$TA_0(X, A_1, A_0) = X\bar{A}_0 + XA_1 \tag{8.27(b)}$$

$$TA_0(X, A_1, A_0) = X(\bar{A}_0 + A_1) \tag{8.27(c)}$$

$$Z(X, A_1, A_0) = XA_0 + A_1 \tag{8.27(d)}$$

Circuit diagram is drawn and is shown in Figure 8.35 using excitation expressions of T-flip-flop and output is given in Eqs. (8.30)–(8.33).

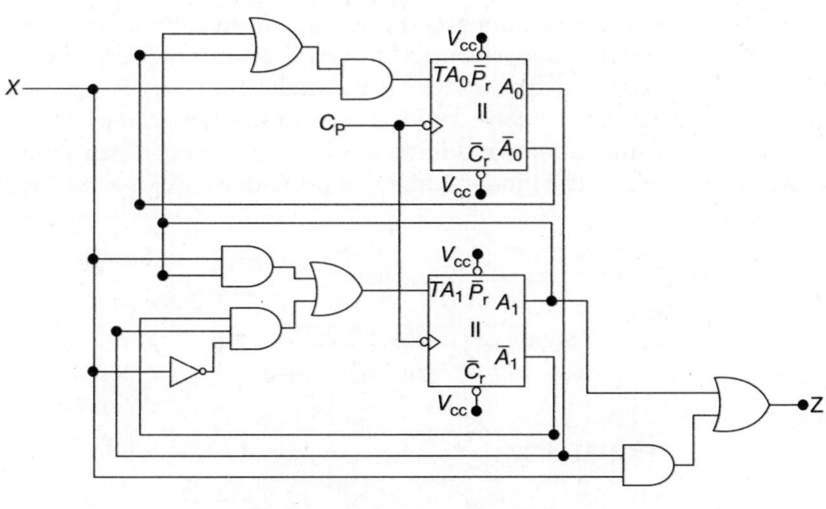

FIGURE 8.35 | Sequential circuit

EXAMPLE 8.18 Design a serial full adder using Mealy model and clocked D-flip-flops.

SOLUTION Full adder requires augend, X_i, addend, Y_i and carries in, C_i to perform single-bit addition for i^{th} bit as shown in Figure 8.36. Truth table for full-adder is given in Table 8.35.

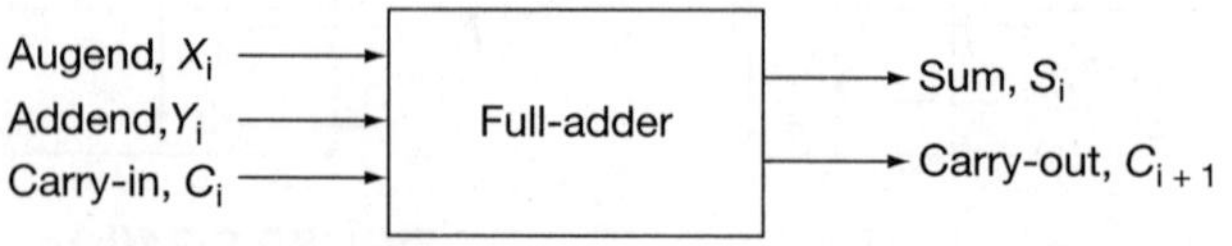

FIGURE 8.36 | Binary full-adder to perform addition of ith bit

TABLE 8.35 | Truth table of full-adder to perform addition of ith bit

Inputs			Outputs	
Carry-in C_i	Augend X_i	Addend Y_i	Sum S_i	Carry-out C_{i+1}
0	0	0	0	0
0	0	1	1	0
0	1	0	1	0
0	1	1	0	1
1	0	0	1	0
1	0	1	0	1
1	1	0	0	1
1	1	1	1	1

To perform 4-bit or multi-bit additions, carry out, C_{i+1} from i^{th} bit is propagated for next $(i+1)^{th}$ bit addition. Sum of bits is represented by sum, S_i as output. Carry-in and carry-out can be 0 or 1, so there is need of two states for Mealy model. So, Carry is represented by two states $A(0)$ and $B(1)$. Carry in (C_i) is considered as present state and carry out C_{i+1} is taken as next state. Figure 8.36 of full adder is modified as Figure 8.37 to perform serial full binary adder for i^{th} bit. Truth table of full binary adder is modified and is given in Table 8.36.

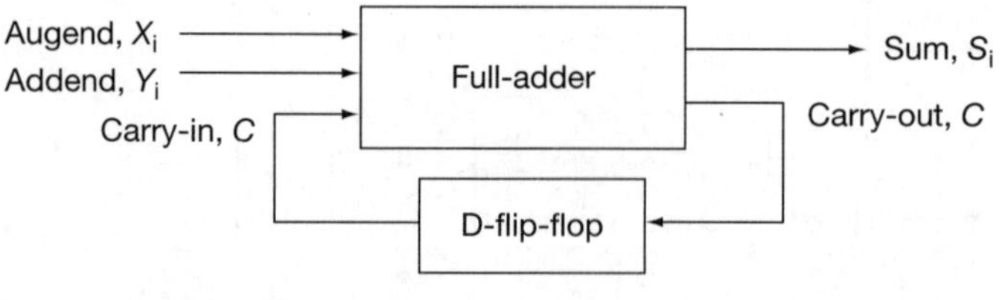

FIGURE 8.37 | Synchronous serial adder

TABLE 8.36 | State truth table of serial adder to perform addition of ith bit

Inputs			Outputs	
Present State (Carry-in) *A*	**Augend X_i**	**Addend Y_i**	**Sum S_i**	**Next State (Carry-out) *B***
0	0	0	0	0
0	0	1	1	0
0	1	0	1	0
0	1	1	0	1
1	0	0	1	0
1	0	1	0	1
1	1	0	0	1
1	1	1	1	1

Present state A *(Figure 8.38(a))*

When present state (carry-in) is *A*(0) and inputs, X_iY_i is 00, the next state (carry-out) is *A*(0) and output, S_i is 0. For inputs, X_iY_i is 01, the next state (carry-out) is *A*(0) and output, S_i is 1. For inputs, X_iY_i is 10, the next state (carry-out) is *A*(0) and output, S_i is 1. For inputs, X_iY_i is 11, the next state (carry-out) is *B* (1) and output, S_i is 0.

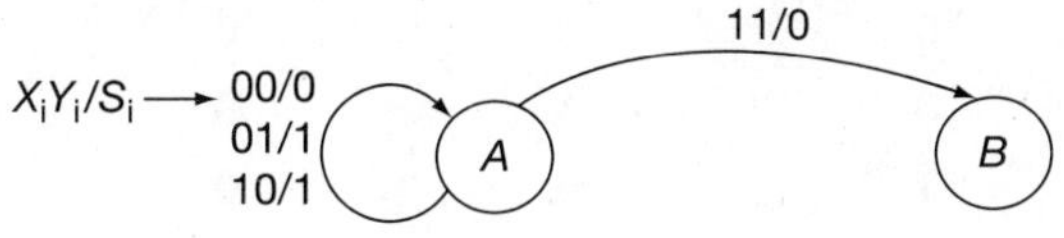

FIGURE 8.38(a) | Partial state diagram of serial adder as Mealy model

Present state B (Figure 8.38(b))

When present state (carry-in) is *B* (1) and inputs, X_iY_i is 00, the next state (carry-out) is *A*(0) and output, S_i is 1. For inputs, X_iY_i is 01, the next state (carry-out) is *B* (1) and output, S_i is 0. For inputs, X_iY_i is 10, the next state (carry-out) is *B* (1) and output, S_i is 0. For inputs, X_iY_i is 11, the next state (carry-out) is *B* (1) and output, S_i is 1.

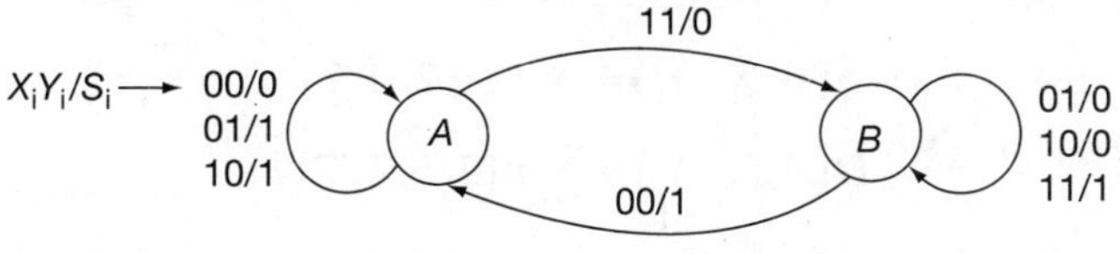

FIGURE 8.38(b) | State diagram of serial adder as Mealy model

State table is obtained from state diagram shown in Figure 8.38(b) is given in Table 8.37. Table 8.38 is shown when '0' is assigned to *A* variable and '1' is assigned to *B* variable.

TABLE 8.37 | State table of serial binary adder

Carry as Present State	Carry as Next State				Sum as Output, S_i			
	$X_iY_i = 00$	$X_iY_i = 01$	$X_iY_i = 10$	$X_iY_i = 11$	$X_iY_i = 00$	$X_iY_i = 01$	$X_iY_i = 10$	$X_iY_i = 11$
A	*A*	*A*	*A*	*B*	0	1	1	0
B	*A*	*B*	*B*	*B*	1	0	0	1

TABLE 8.38 | State assignment table of serial binary adder

Carry as Present State	Carry as Next State				Sum as Output, S_i			
	$X_iY_i = 00$	$X_iY_i = 01$	$X_iY_i = 10$	$X_iY_i = 11$	$X_iY_i = 00$	$X_iY_i = 01$	$X_iY_i = 10$	$X_iY_i = 11$
0	0	0	0	1	0	1	1	0
1	0	1	1	1	1	0	0	1

One D-flip-flop is selected to design the synchronous circuit. Excitation table is given in Table 8.39.

TABLE 8.39 | Excitation table for full binary adder

Present State *A*	Inputs X_i	Y_i	Next State *B*	Excitation of D-flip-flop	Output S_i
0	0	0	0	0	0
0	0	1	0	0	1
0	1	0	0	0	1
0	1	1	1	1	0
1	0	0	0	0	1
1	0	1	1	1	0
1	1	0	1	1	0
1	1	1	1	1	1

Excitations of D-flip-flop and output, S_i are given in standard sum-of-product form as given below.

$$S(A, X_i, Y_i) = \sum m(1, 2, 4, 7)$$

$$D(A, X_i, Y_i) = \sum m(3, 5, 6, 7)$$

Using K-map (Figures 8.39(a) and 8.39(b)), simplified logic expression for sum, S_i and excitation of *D* is given below.

$$S = X_i \oplus Y_i \oplus A \tag{8.28(a)}$$

$$D = X_iA + Y_iA + X_iY_i \tag{8.28(b)}$$

Circuit diagram of serial adder is given in Figure 8.40.

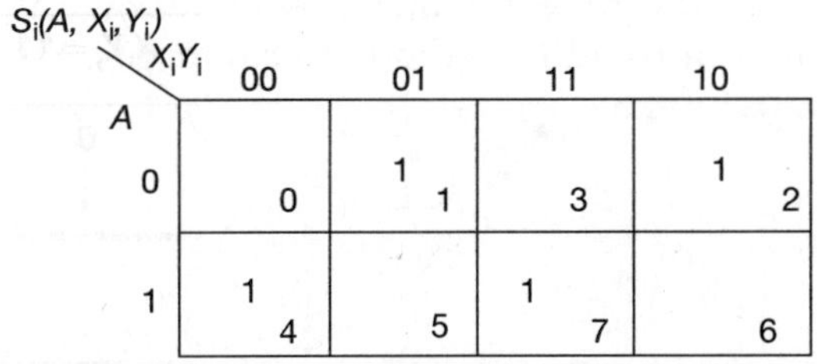

FIGURE 8.39(a) | K-map for sum of serial adder

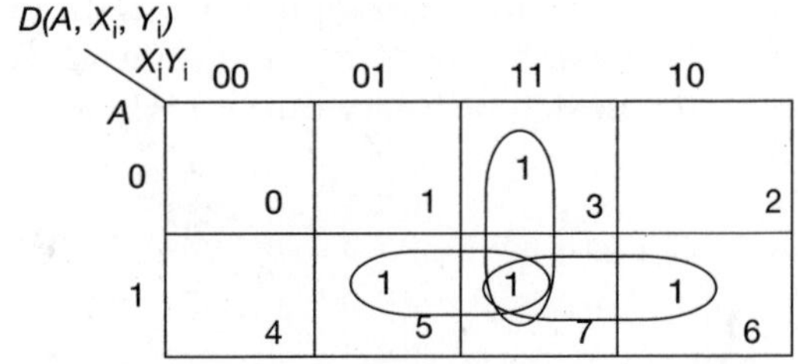

FIGURE 8.39(b) | K-map for D-flip-flop excitation

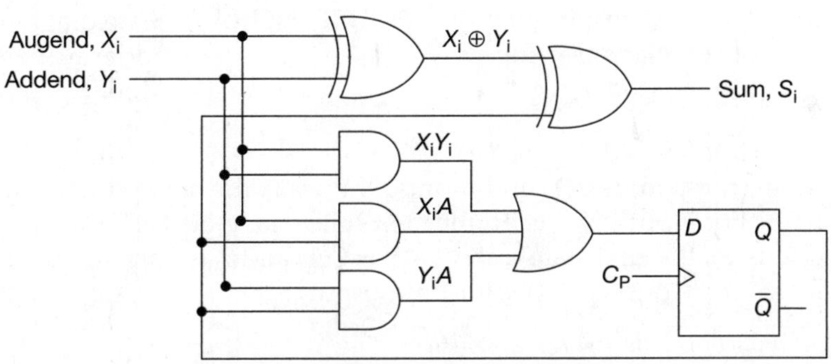

FIGURE 8.40 | Circuit diagram of serial adder using Mealy model

EXAMPLE 8.19 Design a serial full-adder using Moore model and clocked D-flip-flops.

SOLUTION Full-adder requires augend, X_i, addend, Y_i and carries in, C_i to perform single bit addition for i^{th} bit as shown in Figure 8.36. Truth table for full-adder is given in Table 8.59. To perform 4-bit or multi-bit additions, carry out, C_{i+1} from i^{th} bit is propagated for next $(i + 1)^{th}$ bit addition. Sum of bits is represented by sum, S_i as output. Carry-in and carry-out can be 0 or 1. There is need of four states for Moore model because there are two states and two outputs. So, carry is represented by two states *A*(0) and *B* (1) and outputs are also 0 or 1. There are *A*/0, *B*/1, *C*/0 and *D*/1 are possible states. When carry is 0 and output is 0 the state is *A*/0. When carry is 0 and output is 1 the state is *B*/1. When carry is 1 and output is 0, the state is *C*/0. When carry is 1 and output is 1, the state is *D*/1.

Present state A and output, S_i is 0 (Figure 8.41(a))

When present state (carry-in) is *A*(0) and inputs, X_iY_i is 00, the next state (carry-out) is *A*(0) with 0 as output, S_i. For inputs, X_iY_i is 01, the next state (carry-out) is *B* (0) with 1 as output, S_i. For inputs, X_iY_i is 10, the next state (carry-out) is *B*(0) and 1 as output, S_i. For inputs, X_iY_i is 11, the next state (carry-out) is C (1) with 0 as output, S_i.

Present state B and output, S_i is 1 (Figure 8.41(b))

When present state (carry-in) is *B* (0) and inputs, X_iY_i is 00, the next state (carry-out) is *A*(0) with 0 as output, S_i. For inputs, X_iY_i is 01, the next state (carry-out) is *B* (0) with 1 as output, S_i. For inputs, X_iY_i is 10, the next state (carry-out) is *B*(0) and 1 as output, S_i. For inputs, X_iY_i is 11, the next state (carry-out) is *C* (1) with 0 as output, S_i.

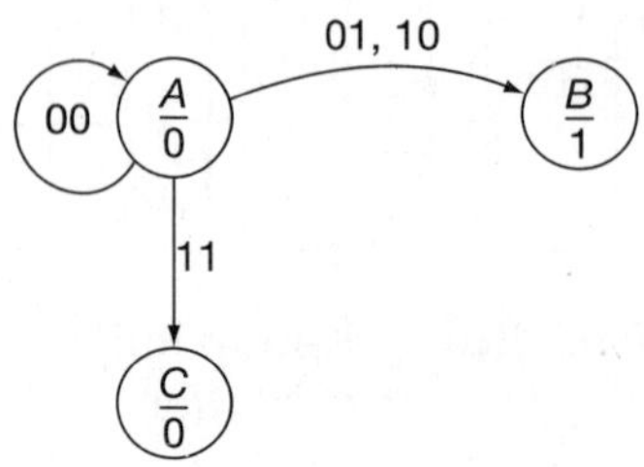

FIGURE 8.41(a) | Partial state diagram of serial adder using Moore Model

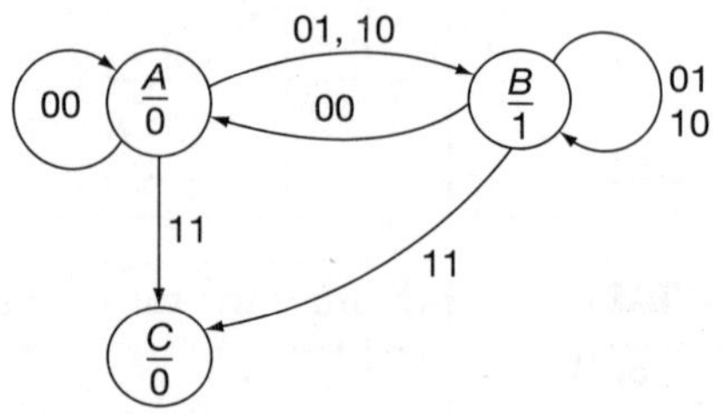

FIGURE 8.41(b) | State diagram of serial adder using Moore Model

Present state C and output, S_i is 0 (Figure 8.41(c))

When present state (carry-in) is *C*(1) and inputs, X_iY_i is 00, the next state (carry-out) is *B*(0) with 1 as output, S_i. For inputs, X_iY_i is 01, the next state (carry-out) is *C* (1) with 0 as output, S_i. For inputs, X_iY_i is 10, the next state (carry-out) is *C*(1) and 0 as output, S_i. For inputs, X_iY_i is 11, the next state (carry-out) is *D*(1) with 1 as output, S_i.

Present state D and output, S_i is 1 (Figure 8.41(d))

When present state (carry-in) is *D*(1) and inputs, X_iY_i is 00, the next state (carry-out) is *B*(0) with 1 as output, S_i. For inputs, X_iY_i is 01, the next state (carry-out) is *C*(1) with 0 as output, S_i. For inputs, X_iY_i is 10, the next state (carry-out) is *C*(1) and 0 as output, S_i. For inputs, X_iY_i is 11, the next state (carry-out) is *D*(1) with 1 as output, S_i.

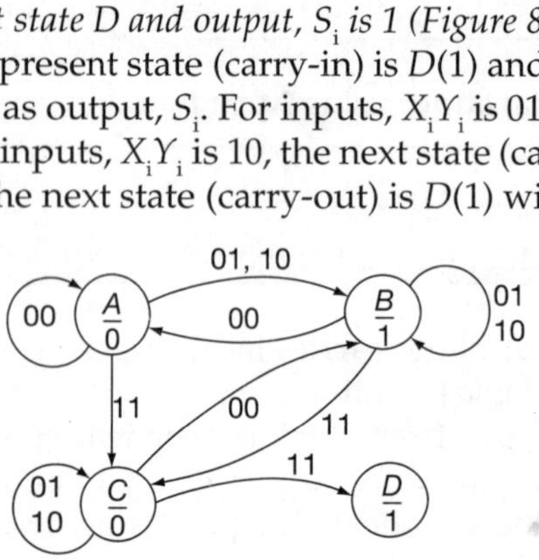

FIGURE 8.41(c) | Partial state diagram of serial adder using Moore model

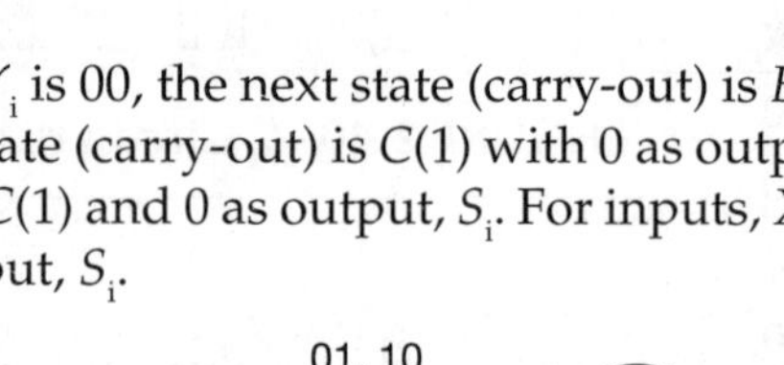

FIGURE 8.41(d) | State diagram of serial adder using Moore model

State table is obtained from sate diagram shown in Figure 8.41(d) is given in Table 8.40. Table 8.41 is shown when '00' is assigned to *A* state variable, '01' is assigned to b state variable '10' is assigned to C state variable and '11' is assigned to state *D*-state variable. Two D-flip-flops are selected. Excitations D_A and D_B of flip-flop is given in excitation Table 8.42 is derived from state assignment Table 8.41.

TABLE 8.40 | State table of serial binary adder

Present State (Carry-in)	Next State (Carry-out)				Sum as Output, S_i
	$X_iY_i = 00$	$X_iY_i = 01$	$X_iY_i = 10$	$X_iY_i = 11$	
A	A	B	B	C	0
B	A	B	B	C	1
C	B	C	C	D	0
D	B	C	C	D	1

TABLE 8.41 | State assignment table of serial binary adder

Present State (Carry-in)	Next State (Carry-out)				Sum as Output, S_i
	$X_iY_i = 00$	$X_iY_i = 01$	$X_iY_i = 10$	$X_iY_i = 11$	
00	00	01	01	10	0
01	00	01	01	10	1
10	01	10	10	11	0
11	01	10	10	11	1

TABLE 8.42 | Excitation table for serial adder

Inputs		Present State		Next State		Excitation		Output
X_i	Y_i	A	B	A	B	D_A	D_B	S_i
0	0	0	0	0	0	0	0	0
0	0	0	1	0	0	0	0	1
0	0	1	0	0	1	0	1	0
0	0	1	1	0	1	0	1	1
0	1	0	0	0	1	0	1	0
0	1	0	1	0	1	0	1	1
0	1	1	0	1	0	1	0	0
0	1	1	1	1	0	1	0	1
1	0	0	0	0	1	0	1	0
1	0	0	1	0	1	0	1	1
1	0	1	0	1	0	1	0	0
1	0	1	1	1	0	1	0	1
1	1	0	0	1	0	1	0	0
1	1	0	1	1	0	1	0	1
1	1	1	0	1	1	1	1	0
1	1	1	1	1	1	1	1	1

Excitations of D-flip-flop and output, S_i are given in standard sum-of-product form as given below.

$$D_A(X_i, Y_i, A, B) = \sum m(6, 7, 10, 11, 12, 13, 14, 15)$$
$$D_B(X_i, Y_i, A, B) = \sum m(2, 3, 4, 5, 8, 9, 14, 15)$$
$$X_i(A, B) = \sum m(1, 3)$$

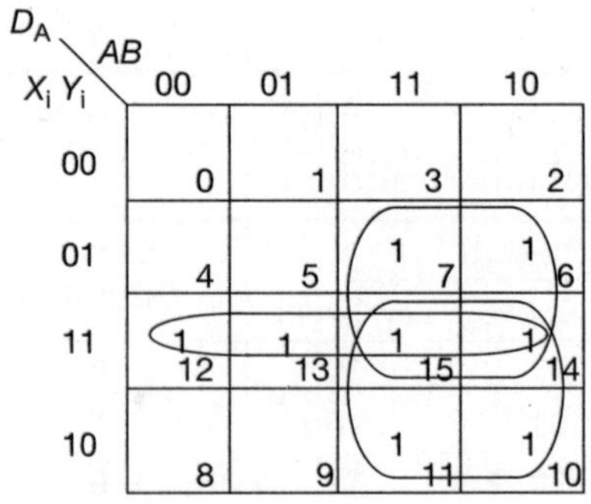

FIGURE 8.42(a) | K-map

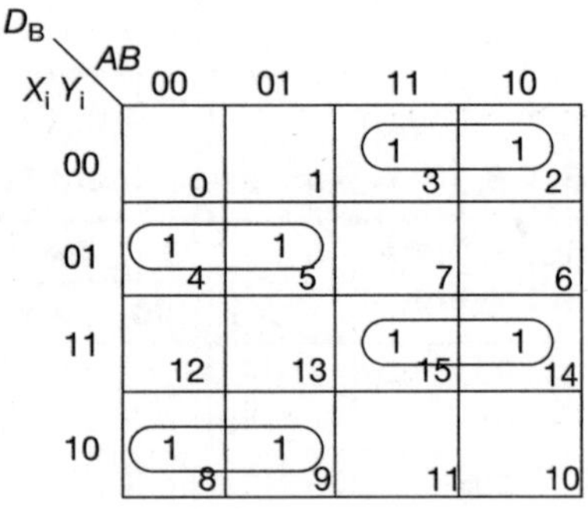

FIGURE 8.42(b) | K-map

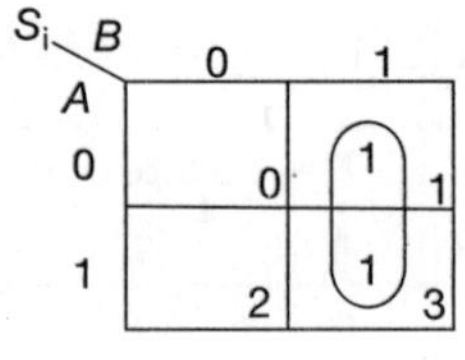

FIGURE 8.42(c) | K-map

Using K-map (Figures 8.42(a) and 8.42(b)), simplified logic expression for sum, S_i and excitation of D is given below.

$$D_B = X_i \oplus Y_i \oplus A \quad (8.29(a))$$
$$D_A = (X_iA + Y_iA + X_iY_i) \quad (8.29(b))$$
$$S_i = AB \quad (8.29(c))$$

Circuit diagram of serial adder is given in Figure 8.43.

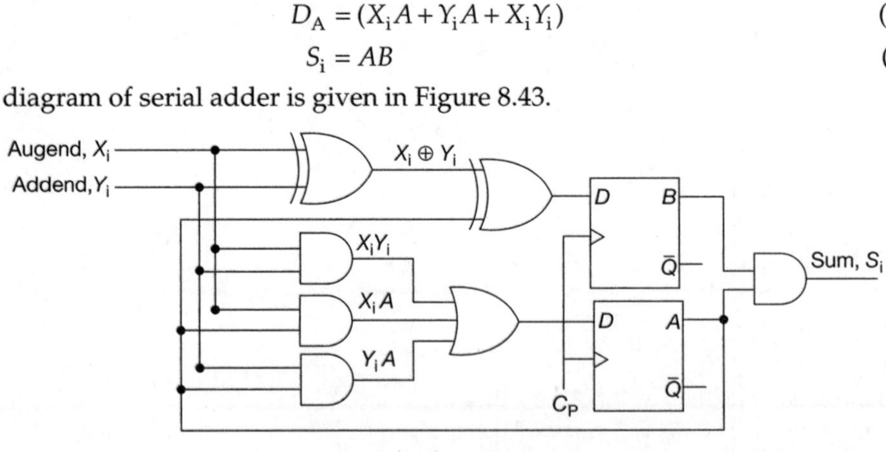

FIGURE 8.43 | Circuit diagram of serial adder using Moore model

EXAMPLE 8.20 Design even-parity generator using clocked JK-flip-flops.

SOLUTION A serial-parity generator a circuit that receives 3-bit message and adds a parity bit to every 3-bit message. Fourth bit is set as blank. For even parity, a parity bit is inserted as 1 when the proceeding number of 1's in a string is odd.

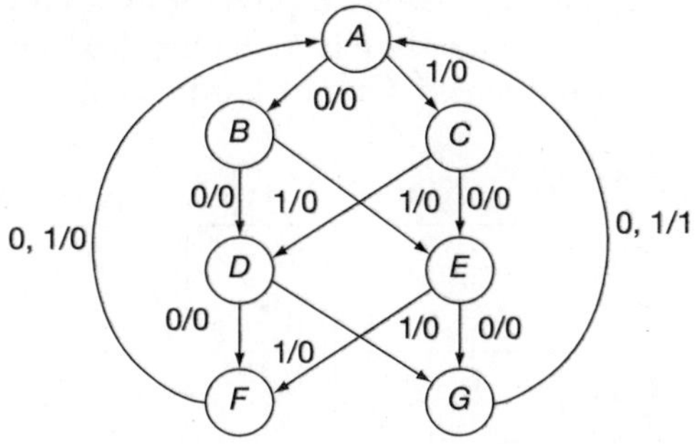

FIGURE 8.44 | State diagram

The state diagram of even-parity generator is shown in Figure 8.44. As and when any bit enters as 0 or 1 then the state generates output as 1 for even parity and 0 for odd parity. States lead from $A \to B$, $B \to D$ and $D \to F$ when inputs are entered as 0, 0 and 0, respectively. Next input may any 0 or 1 being a blank bit and gives output as 0 as even parity being the received string is 000b. States lead from $A \to C$, $C \to E$ and $E \to G$ when inputs are entered as 1, 0 and 0, respectively. Next input may be 0 or 1 and gives output as 1 as odd parity as the received string is 100b. All possible combinations are taken care in the state diagram. State table of even-parity generator is derived from Figure 8.44 is given in Table 8.43.

TABLE 8.43 | State table of even-parity generator

Present State	Next State		Output, Z	
	X = 0	X = 1	X = 0	X = 1
A	B	C	0	0
B	D	E	0	0
C	E	D	0	0
D	F	G	0	0
E	G	F	0	0
F	A	A	0	0
G	A	A	1	1

No further reduction of states is possible.

State Assignment:

Rule 1: Find present states having the same next state, for a given input value.

Present states *F* and *G* moves to next state *A* when input X = 0 and X = 1.

Targeted adjacent states are [*F*, *G*].

Rule 2: Find next states arrived from single present state.

Present state *A* moves to next states *B* and *C*. Present states *B* and *C* move to *D* and *E* next states. Present states *D* and *E* move to next states *F* and *G*.

Targeted adjacent states are [*B*, *C*], [*D*, *E*] and [*F*, *G*].

Rule 3: Find states those have the same outputs.

States *A*, *B*, *C*, *D*, *E* and *F* have same outputs for given input.

Set a three-variable K-map as shown in Figure 8.45. Try to adjust as many as possible targeted adjacent states. One of the possible state assignments is given below

x_2 \ x_1x_0	00	01	11	10
0	A 0	1	C 3	B 2
1	F 4	G 5	E 7	D 6

FIGURE 8.45 | K-map

So, state assignments are $A \rightarrow 000$, $C \rightarrow 011$, $B \rightarrow 010$, $D \rightarrow 110$, $E \rightarrow 111$, $F \rightarrow 100$ and $G \rightarrow 101$.

State assignment table of even-parity generator is given in Table 8.44.

TABLE 8.44 | State assignment table of even-parity generator

Present State			Next State						Output, *Z*	
A_3	A_2	A_1	*X* = 0			*X* = 1			*X* = 0	*X* = 1
			A_3	A_2	A_1	A_3	A_2	A_1		
0	0	0	0	1	0	0	1	1	0	0
0	1	0	1	1	0	1	1	1	0	0
0	1	1	1	1	1	1	1	0	0	0
1	1	0	1	0	0	1	0	1	0	0
1	1	1	1	0	1	1	0	0	0	0
1	0	0	0	0	0	0	0	0	0	0
1	0	1	0	0	0	0	0	0	1	1

Three JK-flip-flops are selected and excitation table for flip-flops is given in Table 8.45.

TABLE 8.45 | Excitation table for JK-flip-flop of even-parity generator

Input	Present State			Next State			Excitations of Flip-flops						Output
X	A_2	A_1	A_0	A_2	A_1	A_0	JA_2	KA_2	JA_1	KA_1	JA_0	KA_0	*Z*
0	0	0	0	0	1	0	0	X	1	X	0	X	0
0	0	1	0	1	1	0	1	X	X	0	0	X	0
0	0	1	1	1	1	1	1	X	X	0	X	0	0
0	1	1	0	1	0	0	X	0	X	1	0	X	0

(Continued)

TABLE 8.45 | (Continued)

Input	Present State			Next State			Excitations of Flip-flops						Output
X	A_2	A_1	A_0	A_2	A_1	A_0	JA_2	KA_2	JA_1	KA_1	JA_0	KA_0	Z
0	1	1	1	1	0	1	X	0	X	1	X	0	0
0	1	0	0	0	0	0	X	1	0	X	0	X	0
0	1	0	1	0	0	0	X	1	0	X	X	1	1
1	0	0	0	0	1	1	0	X	1	X	1	X	0
1	0	1	0	1	1	1	1	X	X	0	1	X	0
1	0	1	1	1	1	0	1	X	X	0	X	1	0
1	1	1	0	1	0	1	X	0	X	1	1	X	0
1	1	1	1	1	0	0	X	0	X	1	X	1	0
1	1	0	0	0	0	0	X	1	0	X	0	X	0
1	1	0	1	0	0	0	X	1	0	X	X	1	1

Excitations of JK-flip-flop and output, Z are given in standard sum-of-product form as given below.

$$JA_2(X,A_2,A_1,A_0)=\sum m(2,3,10,11)+\sum d(1,4,5,6,7,9,12,13,14,15)$$
$$KA_2(X,A_2,A_1,A_0)=\sum m(4,5,12,13)+\sum d(0,1,2,3,8,9,10,11)$$
$$JA_1(X,A_2,A_1,A_0)=\sum m(0,8)+\sum d(1,2,3,6,7,9,10,11,14,15)$$
$$KA_1(X,A_2,A_1,A_0)=\sum m(6,7,14,15)+\sum d(0,1,4,5,8,9,12,13)$$
$$JA_0(X,A_2,A_1,A_0)=\sum m(8,10,14)+\sum d(1,3,5,7,9,11,13,15)$$
$$KA_0(X,A_2,A_1,A_0)=\sum m(5,11,13,15)+\sum d(0,1,2,4,6,8,9,10,12,14)$$
$$Z(X,A_2,A_1,A_0)=\sum m(5,13)+\sum d(1,9)$$

Using K-map (Figures 8.46(a) to 8.46(g)), simplified logic expressions for excitations of JK-flip-flop and output, Z are given below.

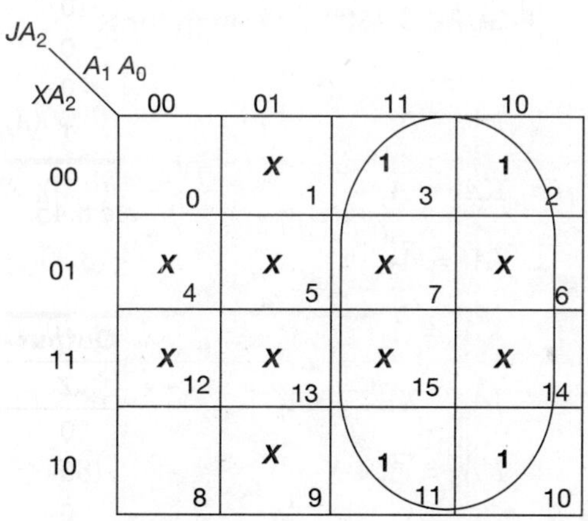

FIGURE 8.46(a) | K-map for JA_2

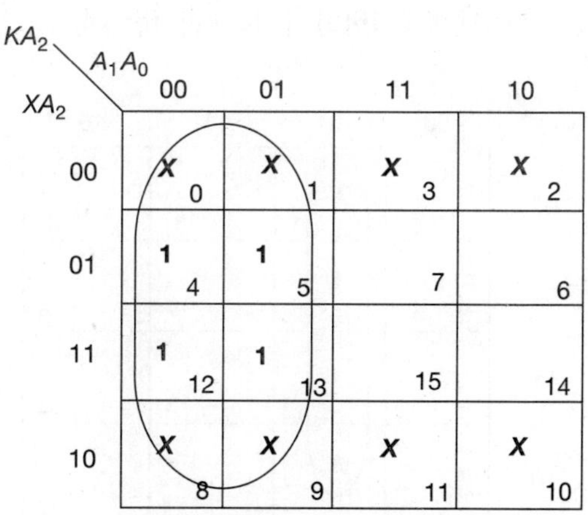

FIGURE 8.46(b) | K-map for KA_2

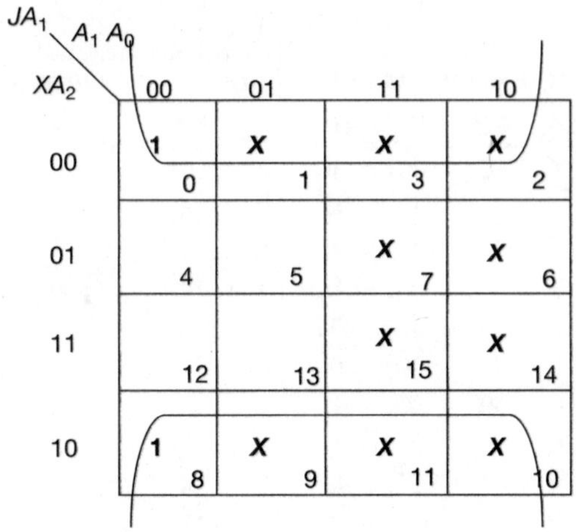

FIGURE 8.46(c) | K-map for JA_1

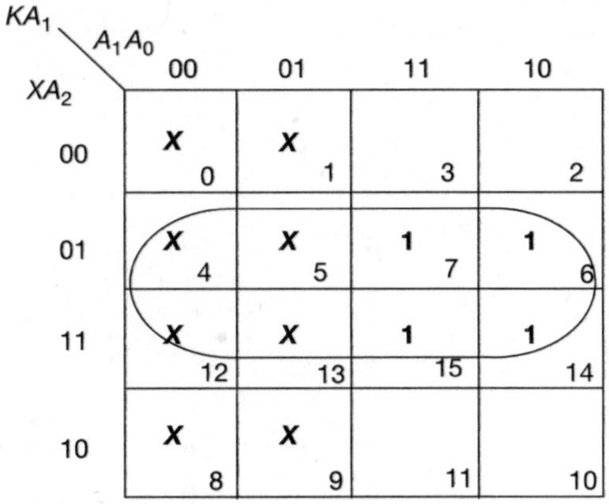

FIGURE 8.46(d) | K-map for KA_1

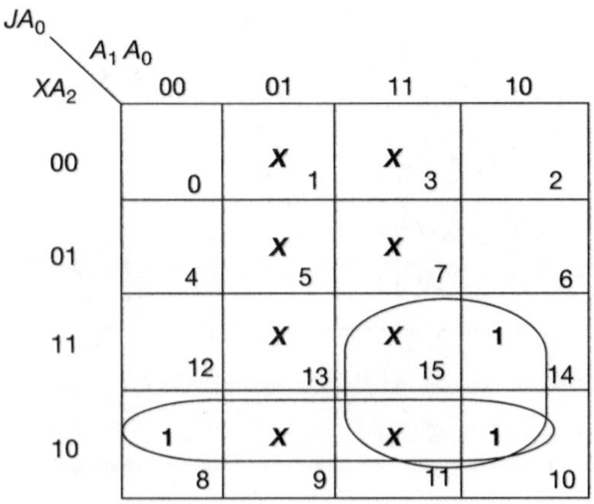

FIGURE 8.46(e) | K-map for JA_0

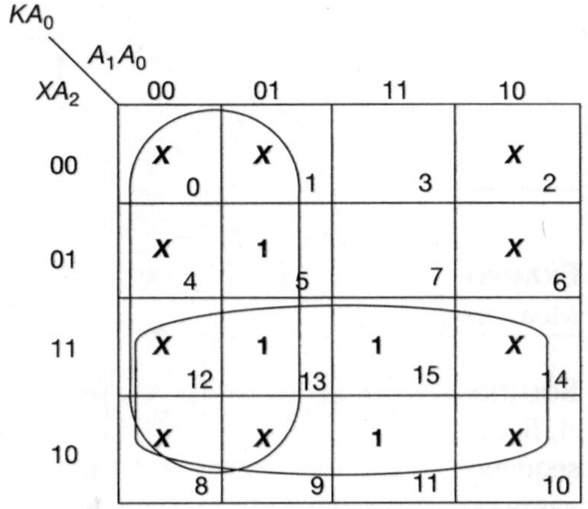

FIGURE 8.46(f) | K-map for KA_0

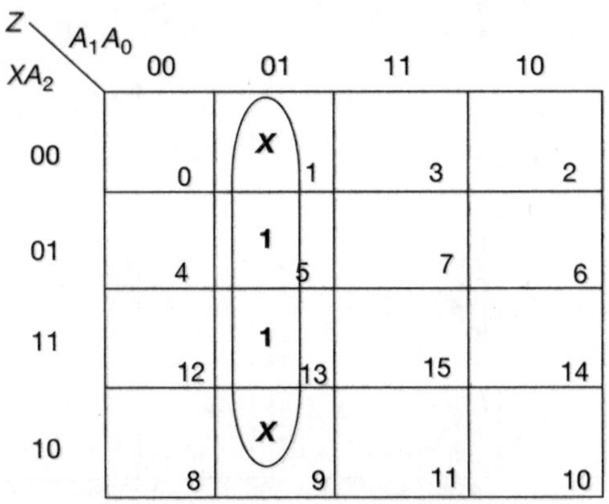

FIGURE 8.46(g) | K-map for Z

$$JA_2 = A_1 \quad (8.30(a))$$

$$KA_2 = \bar{A}_1 \quad (8.30(b))$$

$$JA_1 = \bar{A}_2 \quad (8.30(c))$$

$$KA_1 = A_2 \quad (8.30(d))$$

$$JA_0 = X\bar{A}_2 + XA_1 \quad (8.30(e))$$

$$KA_0 = \bar{A}_1 + X \quad (8.30(f))$$

$$Z = \bar{A}_1 A_0 \quad (8.30(g))$$

Circuit diagram is given in Figure 8.47.

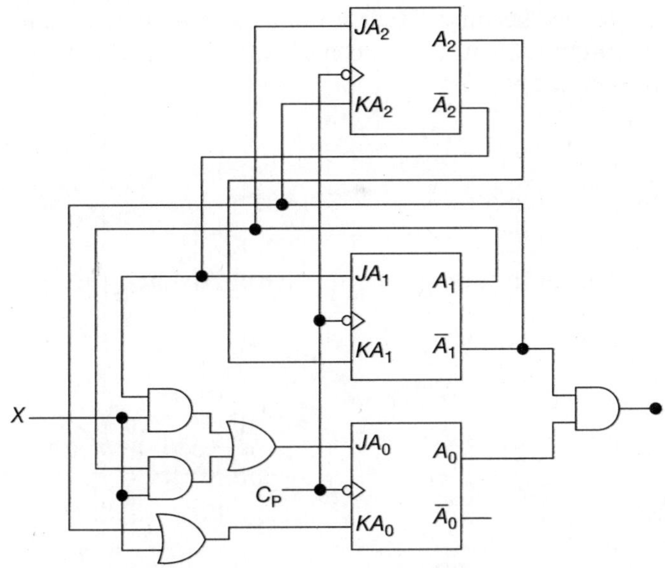

FIGURE 8.47 | Circuit diagram

EXAMPLE 8.21 Design and implement sequence detector 1101001 using D-flip-flops and Mealy model.

SOLUTION The number of bits is 7 in the given sequence. So, seven states are required. Let *A*, *B*, *C*, *D*, *E*, *F* and *G* are states. Figure 8.48(a) shows the required sequence. When complete sequence is detected, the output is 1 otherwise 0. However, unexpected bit can be available. So, every possibility while receiving the bit is explored and is shown in Figures 8.48(b) to 8.48(h).

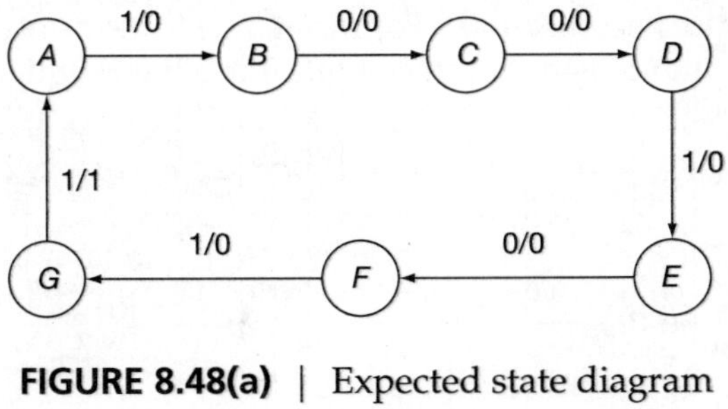

FIGURE 8.48(a) | Expected state diagram

Present state A (Figure 8.48(b))

When input, *X* is 1, the next state is *B* and output, *Z* is 0 because the first bit of sequence is 1 and complete sequence is not detected. If input, *X* is 0 then next state is *A* and output, *Z* is 0 because first bit of sequence is not correctly detected.

Present state B (Figure 8.48(c))
When input, *X* is 0, the next state is *C* and output, *Z* is 0 because the second bit of sequence is 0. The received sequence becomes 10. If input, *X* is 1 then next state is *B* and output, *Z* is 0 because second bit of sequence is not correctly detected. Out of received sequence 11, 1 is accepted and next state becomes *B*.

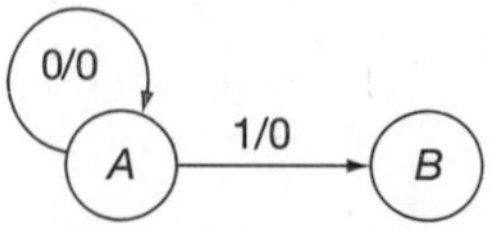

FIGURE 8.48(b) | Partial state diagram

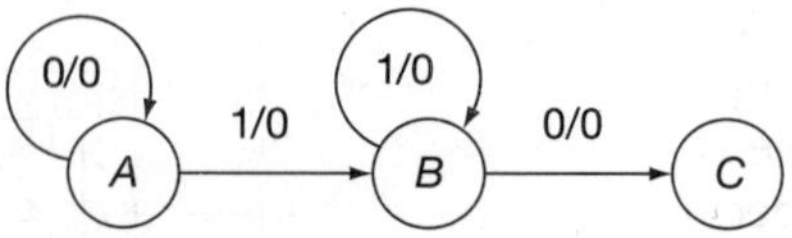

FIGURE 8.48(c) | Partial state diagram

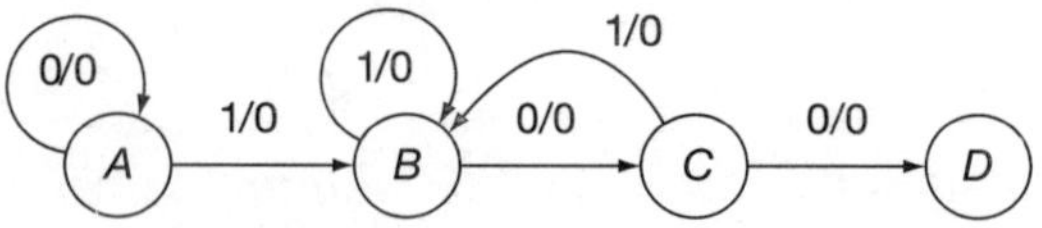

FIGURE 8.48(d) | Partial state diagram

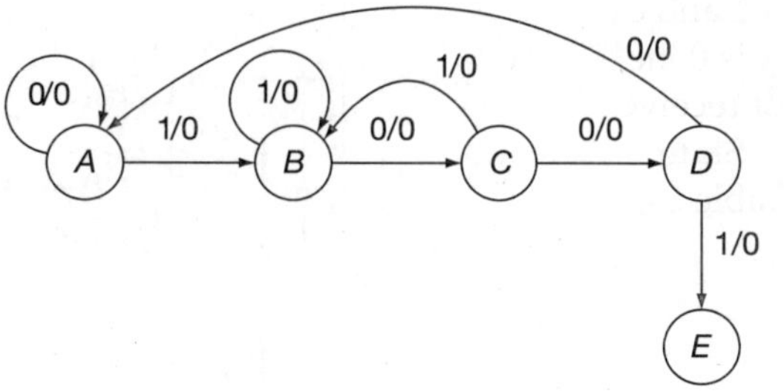

FIGURE 8.48(e) | Partial state diagram

Present state C (Figure 8.48(d))
When input, *X* is 0, the next state is *D* and output, *Z* is 0 because the third bit of sequence is 0. The received sequence becomes 100. For input, *X* is 1, next state is *B* and output, *Z* is 0 because third bit of sequence is not correct. Out of received sequence 101, 1 is accepted and *B* state is considered.

Present state D (Figure 8.48(e))
When input, *X* is 1, the next state is *E* and output, *Z* is 0 because the fourth bit of sequence is 1. The received sequence becomes 1001. For input, *X* is 0, next state is *A* and output, *Z* is 0 because fourth bit of sequence is not correct. Out of received sequence 1000, 0 is accepted and *A* state is considered.

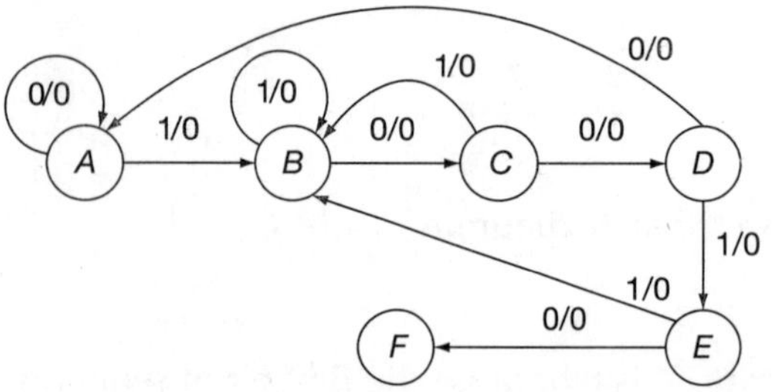

FIGURE 8.48(f) | Partial state diagram

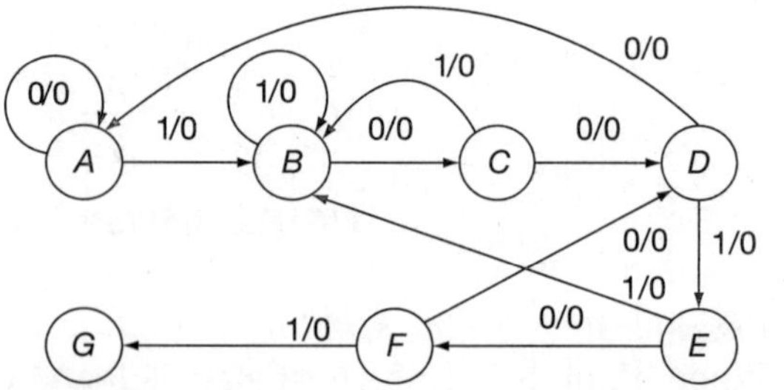

FIGURE 8.48(g) | Partial state diagram

Present state E (Figure 8.48(f))
When input, X is 0, the next state is F and output, Z is 0 because the fifth bit of sequence is 0. The received sequence becomes 10010. For input, X is 1, next state is C and output, Z is 0 because fifth bit of sequence is not correct. Out of received sequence 10011, 1 is accepted and B state is considered.

Present state F (Figure 8.48(g))
When input, X is 1, the next state is G and output, Z is 0 because the sixth bit of sequence is 1. The received sequence becomes 100101. For input, X is 0, next state is D and output, Z is 0 because sixth bit of sequence is not correct. Out of received sequence 100100, 100 is accepted and D state is considered.

Present state G (Figure 8.48(h))
When input, X is 1, the next state is A and output, Z is 1 because the seventh bit of sequence is 1 and complete sequence is accepted. The received sequence becomes 1001011. For input, X is 0, next state is C and output, Z is 0 because seventh bit of sequence is not correct. Out of received sequence 1001010, 10 is accepted and C state is considered.

State table is constructed from state diagram shown in Figure 8.48(h) and is given in Table 8.46.

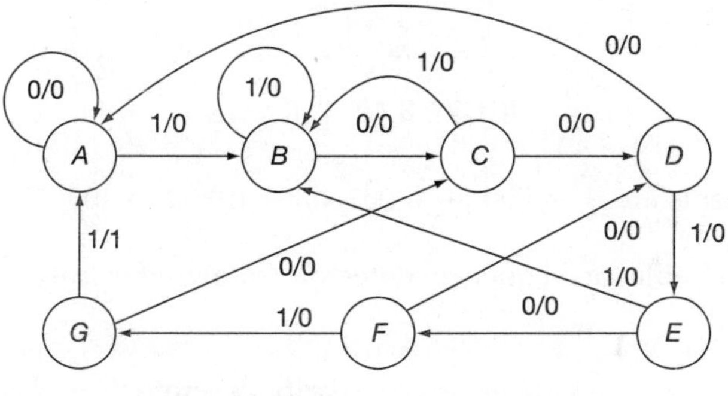

FIGURE 8.48(h) | State diagram

TABLE 8.46 | State table of sequence detector

Present State	Next State		Output, Z	
	$X = 0$	$X = 1$	$X = 0$	$X = 1$
A	A	B	0	0
B	C	B	0	0
C	D	B	0	0
D	A	E	0	0
E	F	B	0	0
F	D	G	0	0
G	C	A	0	1

No further reduction of states is possible.
State assignment:

Rule 1: Find present states having the same next state, for a given input value.
Targeted adjacent states are [*A*, *B*, *C*, *E*], [*A*, *D*] [*B*, *G*] and [*C*, *F*]

Rule 2: Find next states arrived from single present state.
Targeted adjacent states are [*B*, *D*], [*B*, *F*], [*A*, *E*], [*A*, *C*] and [*D*, *G*].

Rule 3: Find states those have the same outputs.
States *A*, *B*, *C*, *D*, *E* and *F* have same outputs for given input.
Set a three variable K-map as shown in Figure 8.49. Try to adjust as many as possible targeted adjacent states. Priority is given to Rule-I. One of the possible state assignments is given below.

x_2 \ x_1x_0	00	01	11	10
0	*A* (0)	*B* (1)	*E* (3)	*C* (2)
1	*D* (4)	*G* (5)	(7)	*F* (6)

FIGURE 8.49 | K-map

So, state assignments are $A \to 000$, $B \to 001$, $C \to 010$, $D \to 100$, $E \to 011$, $F \to 110$ and $G \to 101$.

State assignment table of sequence detector following Mealy model is given in Table 8.47.

TABLE 8.47 | State assignment table of even-parity generator

	Present State			Next State						Output, *Z*	
				X = 0			*X* = 1				
	A_3	A_2	A_1	A_3	A_2	A_1	A_3	A_2	A_1	*X* = 0	*X* = 1
A	0	0	0	0	0	0	0	0	1	0	0
B	0	0	1	0	1	0	0	0	1	0	0
C	0	1	0	1	0	0	0	0	1	0	0
E	0	1	1	1	1	0	0	0	1	0	0
D	1	0	0	0	0	0	0	1	1	0	0
G	1	0	1	0	1	0	0	0	0	0	1
F	1	1	0	1	0	0	1	0	1	0	0

Three D-flip-flops are selected and excitation table for flip-flops is given in Table 8.48.

TABLE 8.48 | Excitation table for D-flip-flop of even-parity generator

Input	Present State			Next State			Excitations of Flip-flops			Output
X	A_2	A_1	A_0	A_2	A_1	A_0	DA_2	DA_1	DA_0	Z
0	0	0	0	0	0	0	0	0	0	0
0	0	0	1	0	1	0	0	1	0	0
0	0	1	0	1	0	0	1	0	0	0
0	0	1	1	1	1	0	1	1	0	0
0	1	0	0	0	0	0	0	0	0	0
0	1	0	1	0	1	0	0	1	0	0
0	1	1	0	1	0	0	1	0	0	0
1	0	0	0	0	0	1	0	0	1	0
1	0	0	1	0	0	1	0	0	1	0
1	0	1	0	0	0	1	0	0	1	0
1	0	1	1	0	0	1	0	0	1	0
1	1	0	0	0	1	1	0	1	1	0
1	1	0	1	0	0	0	0	0	0	1
1	1	1	0	1	0	1	1	0	1	0

Excitations of JK-flip-flop and output, Z are given in standard sum-of-product form as given below.

$$DA_0(X, A_2, A_1, A_0) = \sum m(8,9,10,11,12,14) + \sum d(7,15)$$

$$DA_1(X, A_2, A_1, A_0) = \sum m(1,3,5,12) + \sum d(7,15)$$

$$DA_2(X, A_2, A_1, A_0) = \sum m(2,3,6,14) + \sum d(7,15)$$

$$Z(X, A_2, A_1, A_0) = \sum m(13) + \sum d(7,15)$$

Using K-map (Figures 8.50(a) to 8.50(d)), simplified logic expressions for excitations of D-flip-flop and output, Z are given below.

$$DA_0 = X\bar{A}_2 + X\bar{A}_0 \tag{8.31(a)}$$

$$DA_1 = XA_2\bar{A}_1\bar{A}_0 + \bar{X}A_0 \tag{8.31(b)}$$

$$DA_2 = A_2A_1\bar{A}_0 + \bar{X}A_1 \tag{8.31(c)}$$

$$Z = XA_2A_0 \tag{8.31(d)}$$

Circuit diagram for the sequence detector using Mealy model is given in Figure 8.51.

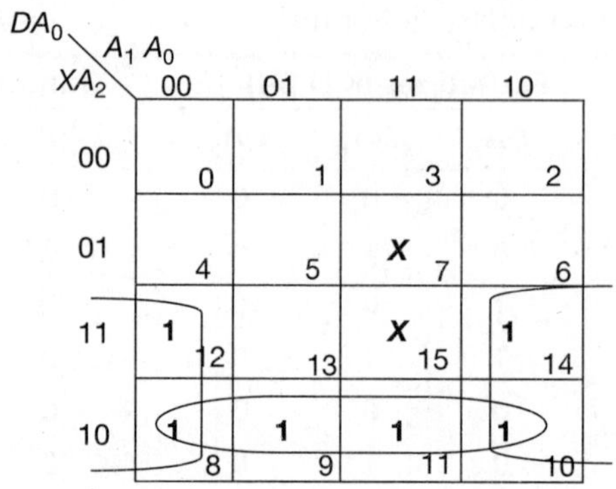

FIGURE 8.50(a) | K-map for DA_0

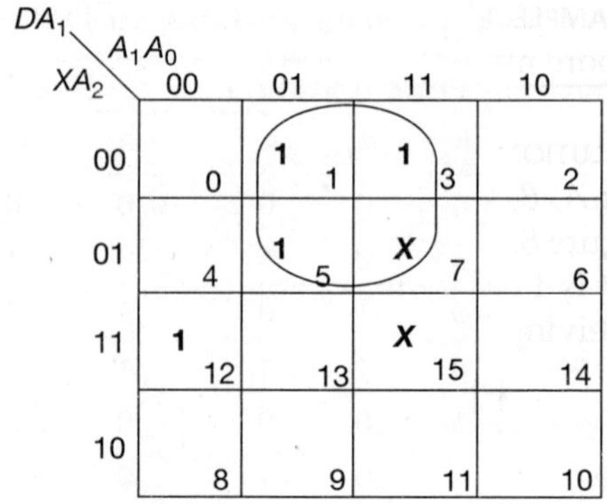

FIGURE 8.50(b) | K-map for DA_1

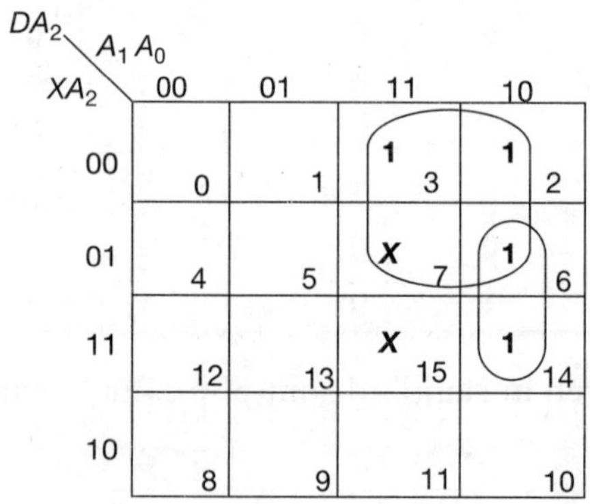

FIGURE 8.50(c) | K-map for DA_2

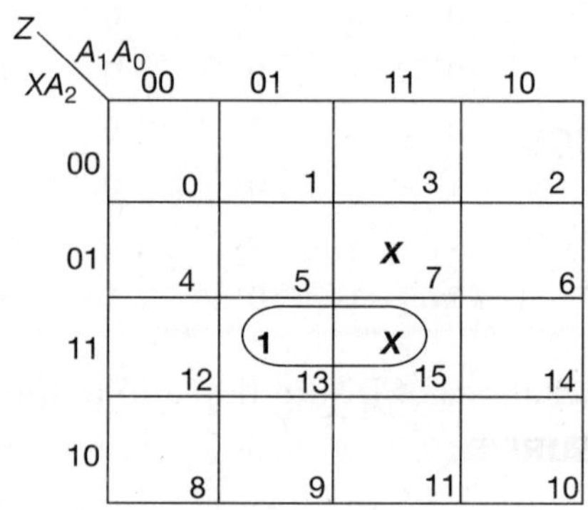

FIGURE 8.50(d) | K-map for Z

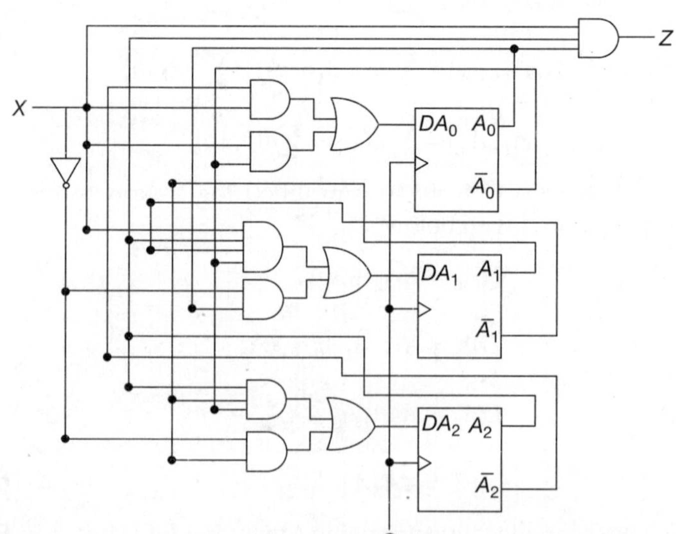

FIGURE 8.51 | Circuit diagram of sequence detector using Mealy model

EXAMPLE 8.22 Design and implement sequence detector 1101001 using D-flip-flops and Moore model.

SOLUTION The number of bits is 7 in the given sequence. So, eight states are required. Let *A*, *B*, *C*, *D*, *E*, *F*, *G* and *H* are states with outputs 0, 0, 0, 0, 0, 0, 0 and 1, respectively. Figure 8.52(a) shows the required sequence. When complete sequence is detected, the output is 1 otherwise 0. However unexpected bit can be received. So, every possibility while receiving the bit is explored and is shown in Figures 8.52(b) to 8.52(i).

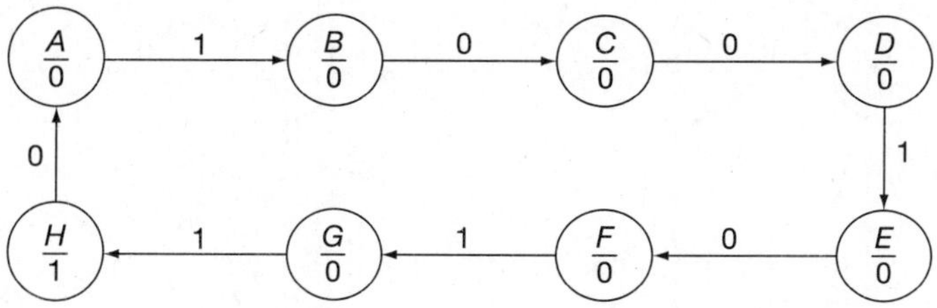

FIGURE 8.52(a) | Required state diagram of sequence detector using Moore model

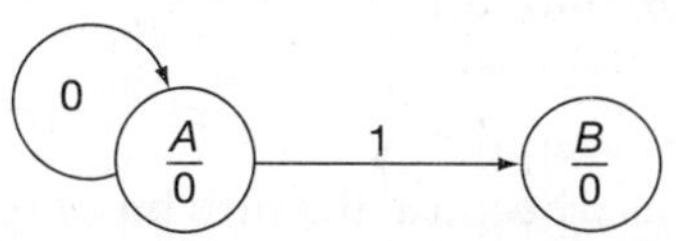

FIGURE 8.52(b) | Partial state diagram

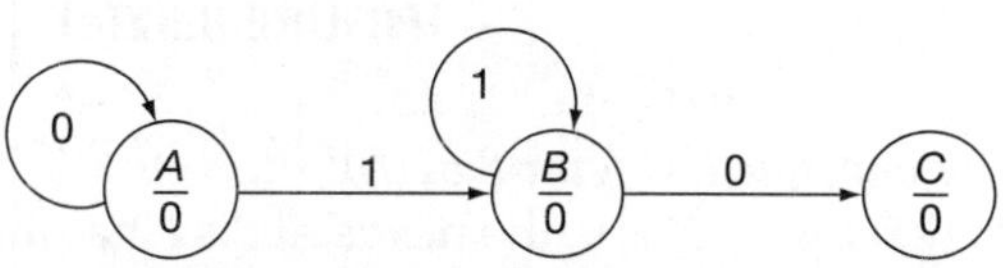

FIGURE 8.52(c) | Partial state diagram

Present state A (Figure 8.52(b))
When input, *X* is 1, the next state is *B* having output, *Z* as 0 because the first bit of sequence is 1. For input, *X* is 0, next state is *A*, because first bit of sequence is not correctly detected.

Present state B (Figure 8.52(c))
When input, *X* is 0, the next state is *C* having output, *Z* as 0 because the second bit of sequence is 0. The received sequence becomes 10. If input, *X* is 1 then next state is *B* having output, *Z* as 0 because second bit of sequence is not correctly detected. Out of received sequence 11, 1 is accepted and next state becomes *B*.

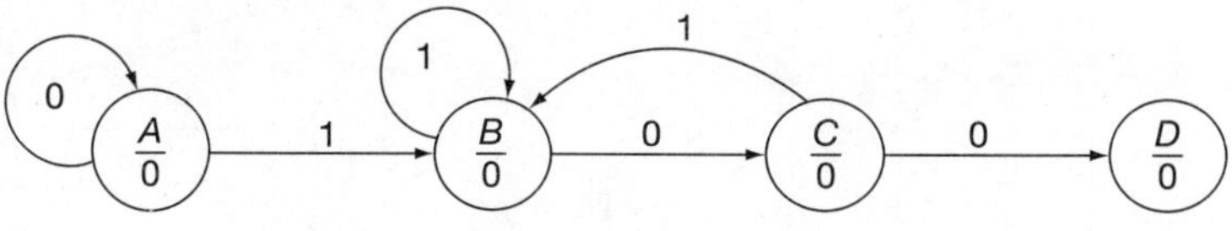

FIGURE 8.52(d) | Partial state diagram

Present state C (Figure 8.52(d))
When input, *X* is 0, the next state is *D* having output, *Z* as 0 because the third bit of sequence is 0 and complete sequence 100 is detected and next sequence is expected. If input, *X* is 1 then next state is *B* having output, *Z* as 0 because third bit of sequence is not correctly detected. Out of received sequence 101, 1 is accepted hence next state becomes *B*.

Present state D (Figure 8.52(e))

When input, X is 1, the next state is E having output, Z as 0 because the fourth bit of sequence is 1 and complete sequence 1001 is detected and next sequence is expected. If input, X is 0 then next state is A having output, Z as 0 because fourth bit of sequence is not correctly detected. Out of received sequence 1000, 0 is accepted hence next state becomes A.

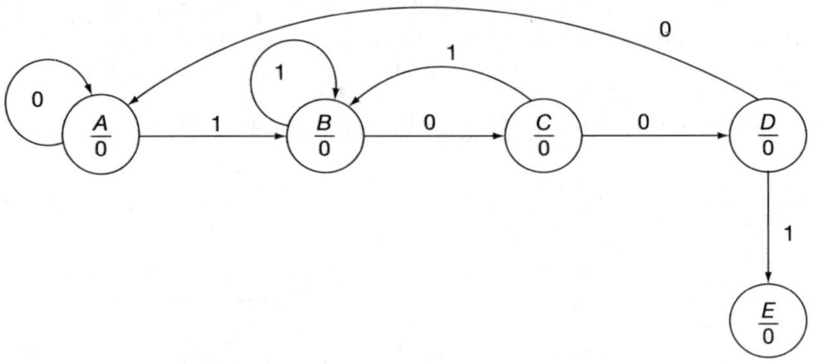

FIGURE 8.52(e) | Partial state diagram

Present state E (Figure 8.52(f))

When input, X is 0, the next state is F having output, Z as 0 because the fifth bit of sequence is 0 and complete sequence 10010 is detected and next sequence is expected. If input, X is 1 then next state is B having output, Z as 0 because fifth bit of sequence is not correctly detected. Out of received sequence 10011, 1 is accepted hence next state becomes B.

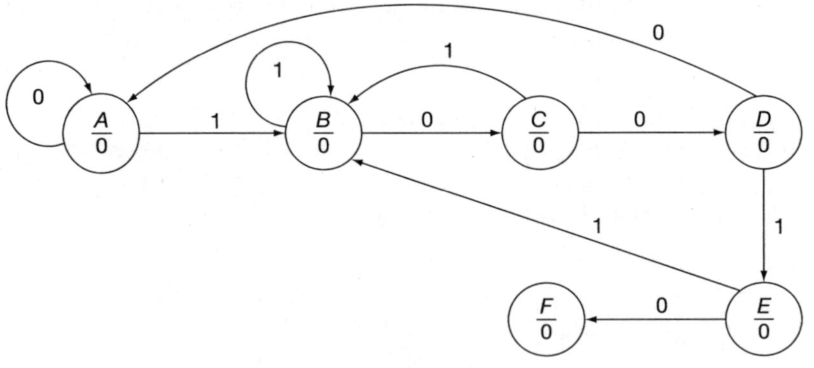

FIGURE 8.52(f) | Partial state diagram

Present state F (Figure 8.52(g))

When input, X is 1, the next state is G having output, Z as 0 because the sixth bit of sequence is 1 and complete sequence 100101 is detected and next sequence is expected. If input, X is 0 then next state is D having output, Z as 0 because sixth bit of sequence is not correctly detected. Out of received sequence 100100, 100 is accepted hence next state becomes D.

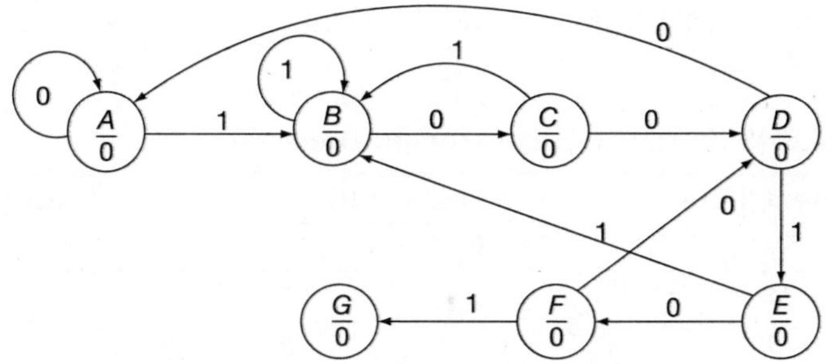

FIGURE 8.52(g) | Partial state diagram

Present state G (Figure 8.52(h))

When input, X is 1, the next state is H having output, Z as 1 because the seventh bit of sequence is 1 and complete sequence 1001011 is detected and correct sequence is detected. If input, X is 0 then next state is C having output, Z as 0 because seventh bit of sequence is not correctly detected. Out of received sequence 1001010, 10 is accepted hence next state becomes C.

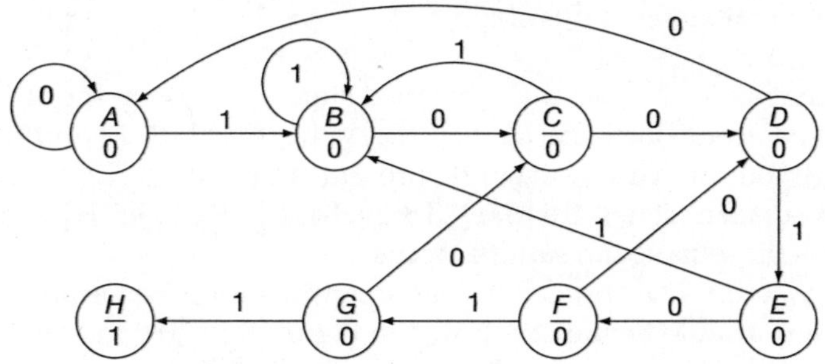

FIGURE 8.52(h) | Partial state diagram

Present state H (Figure 8.52(i))

When input, X is 0, the next state is A having output, Z as 01 because the first bit of next sequence is 0 and complete sequence 0 is detected and next bit of sequence is expected. If input, X is 1 then next state is B having output, Z as 0 because first bit of sequence is 1. Out of received sequence 1 leads to next state B.

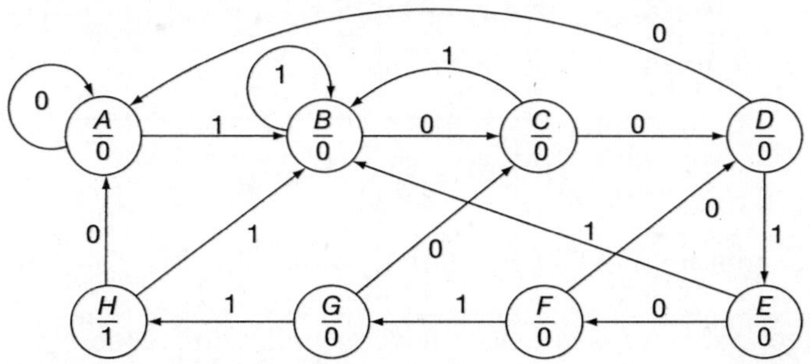

FIGURE 8.52(i) | State diagram of sequence detector using Moore model

State table is constructed from state diagram shown in Figure 8.43(i) and is given in Table 8.49.

TABLE 8.49 | State table of sequence detector

Present State	Next State		Output, Z
	X = 0	X = 1	
A	A	B	0
B	C	B	0
C	D	B	0
D	A	E	0
E	F	B	0
F	D	G	0
G	C	H	0
H	A	B	1

No further reduction of states is possible.

State Assignment:

Rule 1: Find present states having the same next state, for a given input value.
Targeted adjacent states are [*A*, *B*, *C*, *E*, *H*], [*A*, *D*, *H*] [*B*, *G*] and [*C*, *F*]

Rule 2: Find next states arrived from single present state.
Targeted adjacent states are [*B*, *D*], [*B*, *F*], [*A*, *E*], [*B*, *F*], [*C*, *H*] and [*D*, *G*].

Rule 3: Find states those have the same outputs.
States *A*, *B*, *C*, *D*, *E*, *F* and *G* have same outputs for given input.
Set a three variable K-map as shown in Figure 8.53. Try to adjust as many as possible targeted adjacent states. Priority is given to Rule-I. One of the possible state assignments is given below

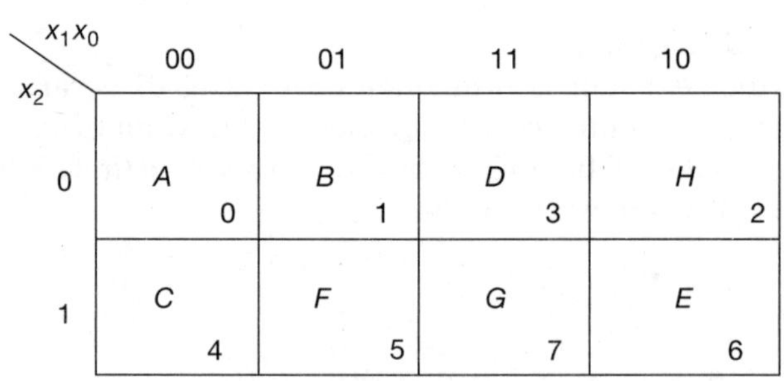

FIGURE 8.53 | K-map

So, state assignments are $A \rightarrow 000, B \rightarrow 001, C \rightarrow 100, D \rightarrow 011, E \rightarrow 110, F \rightarrow 101, G \rightarrow 111$ and $H \rightarrow 010$.

State assignment table of sequence detector following Moore model is given in Table 8.50.

TABLE 8.50 | State assignment table of even-parity generator

	Present State			Next State						Output, Z
	A_3	A_2	A_1	X = 0			X = 1			X = 0
				A_3	A_2	A_1	A_3	A_2	A_1	
A	0	0	0	0	0	0	0	0	1	0
B	0	0	1	1	0	0	0	0	1	0
H	0	1	0	0	0	0	0	0	1	1
D	0	1	1	0	0	0	1	1	0	0
C	1	0	0	0	1	1	0	0	1	0
F	1	0	1	0	1	1	1	1	1	0
E	1	1	0	1	0	1	0	0	1	0
G	1	1	1	1	0	0	0	1	0	0

Three *D*-flip-flops are selected and excitation table for flip-flops is given in Table 8.51.

Excitations of JK-flip-flop and output, *Z* are given in standard sum-of-product form as given below.

$$DA_0(X, A_2, A_1, A_0) = \sum m(4,5,6,8,9,10,12,13,14)$$

$$DA_1(X, A_2, A_1, A_0) = \sum m(4,5,11,13,15)$$

$$DA_2(X, A_2, A_1, A_0) = \sum m(1,6,7,11,13)$$

$$Z(A_2, A_1, A_0) = \sum m(2)$$

TABLE 8.51 | Excitation table for JK-flip-flop of even-parity generator

Input	Present State			Next State			Excitations of Flip-flops			Output
X	A_2	A_1	A_0	A_2	A_1	A_0	DA_2	DA_1	DA_0	Z
0	0	0	0	0	0	0	0	0	0	0
0	0	0	1	1	0	0	1	0	0	0
0	0	1	0	0	0	0	0	0	0	1
0	0	1	1	0	0	0	0	0	0	0
0	1	0	0	0	1	1	0	1	1	0
0	1	0	1	0	1	1	0	1	1	0
0	1	1	0	1	0	1	1	0	1	0
0	1	1	1	1	0	0	1	0	0	0
1	0	0	0	0	0	1	0	0	1	0
1	0	0	1	0	0	1	0	0	1	0
1	0	1	0	0	0	1	0	0	1	1
1	0	1	1	1	1	0	1	1	0	0

(*Continued*)

TABLE 8.51 | (Continued)

Input	Present State			Next State			Excitations of Flip-flops			Output
X	A_2	A_1	A_0	A_2	A_1	A_0	DA_2	DA_1	DA_0	Z
1	1	0	0	0	0	1	0	0	1	0
1	1	0	1	1	1	1	1	1	1	0
1	1	1	0	0	0	1	0	0	1	0
1	1	1	1	0	1	0	0	1	0	0

Using K-map (Figures 8.54(a) to 8.54(c)), simplified logic expressions for excitations of D-flip-flop and output, Z are given below.

$$DA_0 = A_2\bar{A}_0 + A_2\bar{A}_1 + X\bar{A}_0 + X\bar{A}_1 \quad (8.32(a))$$

$$DA_1 = \bar{X}A_2\bar{A}_1 + XA_2A_0 + XA_1A_0 \quad (8.32(b))$$

$$DA_2 = \bar{X}A_2A_1 + \bar{X}\bar{A}_2\bar{A}_1A_0 + XA_2\bar{A}_1A_0 + X\bar{A}_2A_1A_0 \quad (8.32(c))$$

$$Z = \bar{A}_2A_1\bar{A}_0 \quad (8.32(d))$$

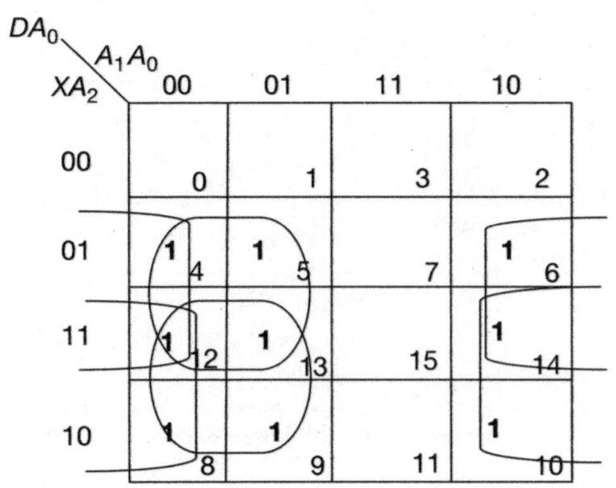

FIGURE 8.54(a) | K-map for DA_0

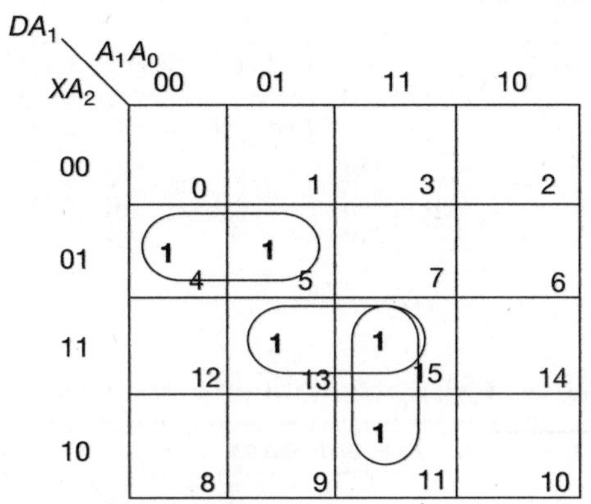

FIGURE 8.54(b) | K-map for DA_1

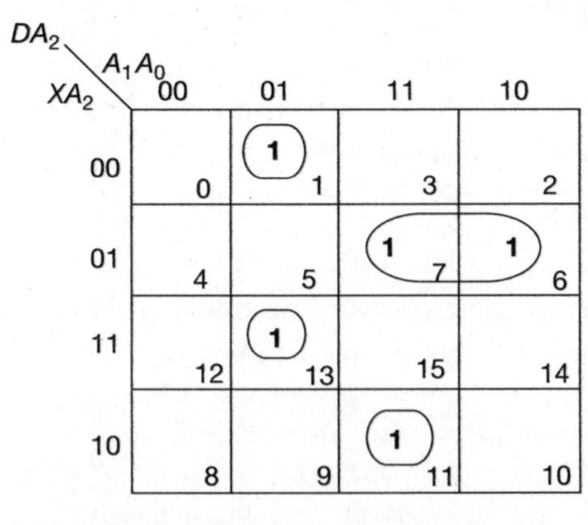

FIGURE 8.54(c) | K-map for DA_2

Circuit diagram for the sequence detector using Moore model is given in Figure 8.55.

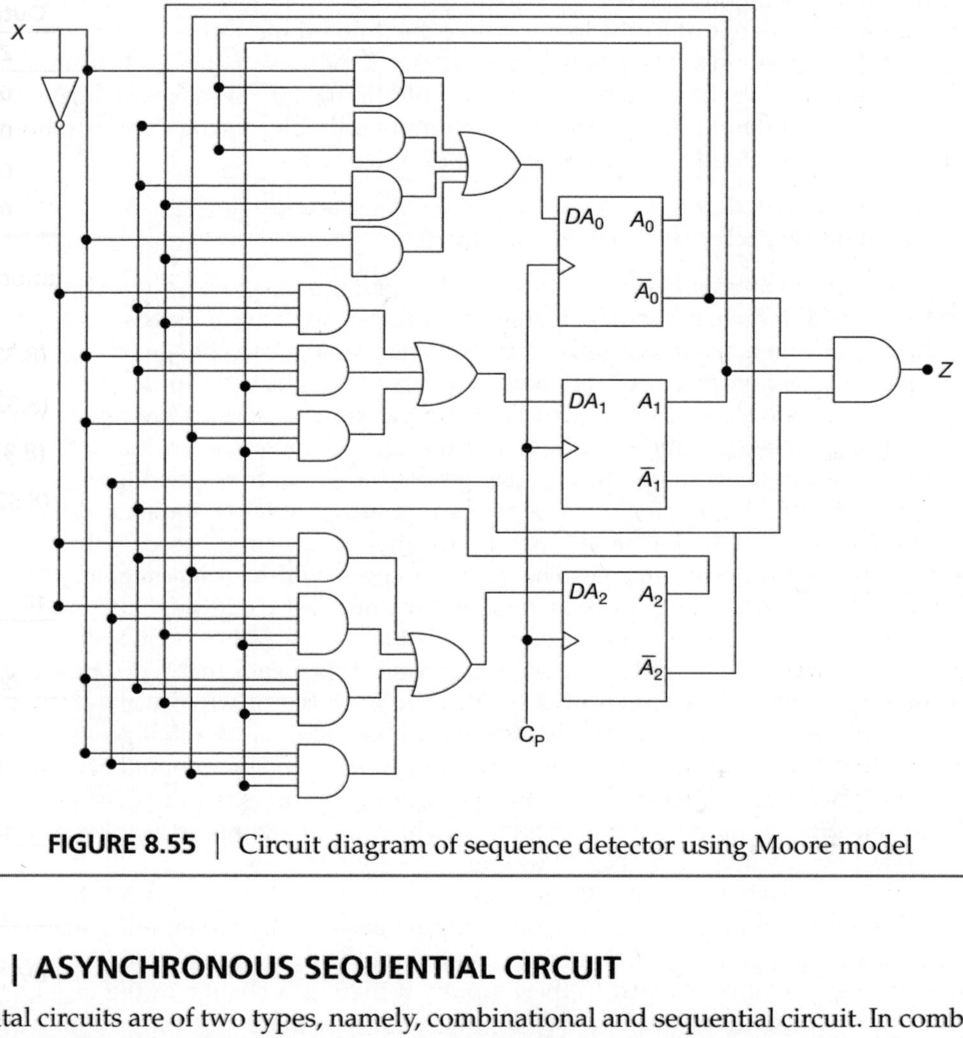

FIGURE 8.55 | Circuit diagram of sequence detector using Moore model

8.7 | ASYNCHRONOUS SEQUENTIAL CIRCUIT

Digital circuits are of two types, namely, combinational and sequential circuit. In combinational circuit, output depends on input. In sequential circuits, the outputs depend upon the present inputs as well as the past inputs and outputs. A sequential circuit can be classified in to synchronous and asynchronous circuit.

In, a sequential circuit, a change of state occurs only in response to a synchronized clock pulse. All the flip-flops are clocked simultaneously by a common clock pulse. Input changes are assumed to occur between clock pulses. The circuit must be in the stable state before next clock pulse arrives. The speed of operation depends on the maximum allowed clock frequency. The memory elements are clocked flip-flops. Any number of inputs can change simultaneously, during the absence of the clock.

In an asynchronous sequential circuit the state of the circuit can change immediately when an input change occurs. Input changes occur only when the circuit is in a stable state. Asynchronous sequential circuits do not require clock pulses and these can change state with the change of input. The asynchronous sequential circuits are faster than the synchronous sequential circuits. The memory elements are either unclocked flip-flops (latches) or gate circuits with feedback producing the effect of latch operation. Only one input is allowed to change at a time in the case of the level inputs and only one pulse input is allowed to be present in the case of the pulse inputs. If more than one level inputs change simultaneously or more than one pulse input is present, the circuit makes erroneous state transitions due to different delay paths for each input variable. Asynchronous sequential circuits are classified based on the type of input into two types:

- Fundamental mode asynchronous sequential circuits and
- Pulse mode asynchronous sequential circuits.

In ***fundamental mode asynchronous sequential circuits***, inputs and outputs are represented by levels. Fundamental mode operation assumes that the input signals change only when the circuit is in a stable state and that only one variable can change at a given time. Fundamental mode asynchronous sequential circuits use unclocked flip-flop or latches.

In ***pulse mode asynchronous sequential circuits***, inputs and outputs are represented by pulses. In pulse mode operation, the width of the of clock pulses is critical to the circuit operation. The input pulse should be long enough for the circuit to respond to the input but it should not be longer so that new state is reached. In such a situation the state of the circuit goes for another transition. The minimum pulse width requirement is based on the propagation delay through the next-state logic. The maximum pulse width is determined by the total propagation delay through the next state logic and the memory elements. Pulse mode asynchronous sequential circuits use unclocked flip-flop or latches. In pulse-mode operation, only one input is allowed to have pulse present at any time. This means that when pulse occurs on any one input, while the circuit is in stable state, pulse must not arrive at any other input.

Since, synchronous sequential circuits respond immediately to the change in any input variable without waiting for a clock pulse. So, an asynchronous sequential circuit is preferred over synchronous circuit when high speed of operation is required Asynchronous sequential circuits are useful in applications in which input signals changes at any time. Asynchronous sequential circuits cost less than the sequential circuits.

In the design of asynchronous circuits, it is assumed that a change occurs in only one of the input and no change occurs in any other inputs until the circuit enters a stable state. In an asynchronous sequential circuit, the circuit is said to be in stable state if the outputs of memory devices become equal to their inputs. If there is a change in the inputs, then the combinational circuit produces a new set of excitation variables, the inputs of memory devices and the circuit enters into an unstable state. When the outputs of memory devices become equal to their inputs, the circuit is said to enter the next stable state.

8.8 | ANALYSIS OF ASYNCHRONOUS SEQUENTIAL CIRCUIT

When input is known, the output can be obtained. Fundamental mode can be constructed without using latches as well as using latches. The circuit may have explicit memory or not.

8.8.1 | Fundamental Mode Asynchronous Sequential Circuit without Latches

Ensuing examples elaborate the analysis of asynchronous sequential circuit without latches. All logical gates are involved in the circuit.

EXAMPLE 8.23 Analyse the two input AND gate with output feedback being asynchronous sequential circuit.

SOLUTION The circuit diagram is shown in Figure 8.56. X is input. Output A^+ is fed back to AND gate, A.

Logic expression is stated as below:

$$A^+ = AX \tag{8.33}$$

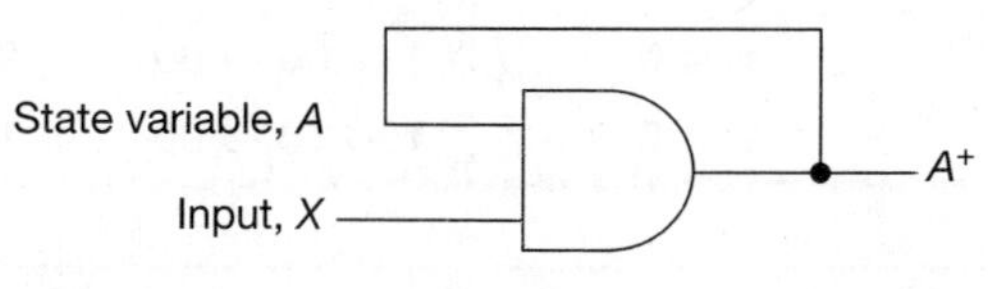

FIGURE 8.56 | AND Gate with feedback

State table is given in Table 8.52. The state is said to be stable if and only if A and A^+ are same after propagation of signal. Otherwise state is unstable. Transition takes place to attain stable state.

TABLE 8.52 | State table of AND gate

State Variable A	Input X	State Variable A^+	Remarks
0	0	0	Stable
0	1	0	Stable
1	0	0	Unstable
1	1	1	Stable

The transition table from state table of Table 8.76 is shown in Figure 8.57.

Input remains same: Transition takes place to attain stable state in case state is unstable if input remains same. $AX = 10$ attains a stable state 00. Let $AX = 10$, the state A^+ changes to 0 after finite time t_d, which is unstable state. After propagation delay time t_d, AX becomes 00 and leads to state A^+ to 0 which is a stable state.

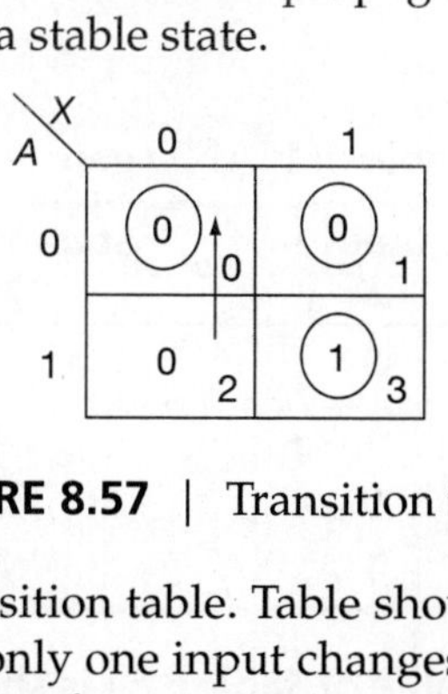

FIGURE 8.57 | Transition table

Input changes: Table 8.53 is a transition table. Table shows the changes, the circuit undergoes as input changes. At a time, only one input changes. Let, the circuit is at $A = 1$, $X = 1$, a stable state. Now, input, X changes from 1 to 0 ($1 \rightarrow 0$). So, $A = 1$, $X = 0$. After finite time

t_d, $A+ = 0$. After finite time t_d, $A = 0$, $X = 0$ and $A+ = 0$, a stable state. In transition table, AX changes from $11 \rightarrow 10 \rightarrow 00$.

TABLE 8.53 | Transition table of AND gate

State Variable A	Input X	State Transition AX	Output A^+
0	$0 \rightarrow 1$	$00 \rightarrow 01 \rightarrow 00$	$0 \rightarrow 0 \rightarrow 0$
1	$0 \rightarrow 1$	$10 \rightarrow 11 \rightarrow 11$	$1 \rightarrow 1 \rightarrow 1$
0	$1 \rightarrow 0$	$01 \rightarrow 00 \rightarrow 00$	$0 \rightarrow 0 \rightarrow 0$
1	$1 \rightarrow 0$	$11 \rightarrow 10 \rightarrow 00$	$1 \rightarrow 0 \rightarrow 0$

EXAMPLE 8.24 Analyse the two input NAND gate with output feedback being asynchronous sequential circuit.

SOLUTION The circuit diagram is shown in Figure 8.58. X is input. Output A^+ is fed back to NAND gate, A.

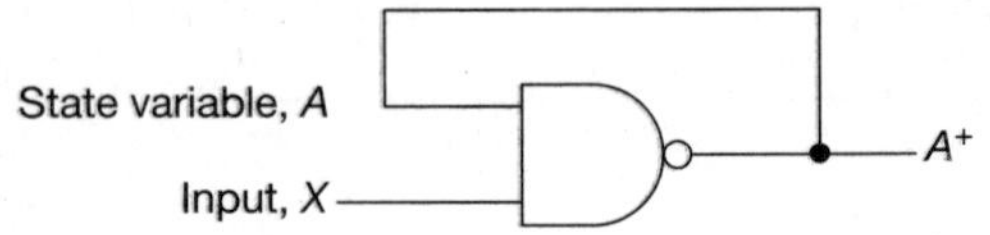

FIGURE 8.58 | NAND gate with feedback

Logic expression is stated as below:

$$A^+ = \overline{AX} \tag{8.34}$$

State table is given in Table 8.54. The state is said to be stable if and only if A and A^+ are same after propagation of signal. Otherwise state is unstable. Transition takes place to attain stable state

TABLE 8.54 | State table of NAND gate

State Variable A	Input X	State Variable A^+	Remarks
0	0	1	Unstable
0	1	1	Unstable
1	0	1	Stable
1	1	0	Unstable

The transition table from state table of Table 8.54 is shown in Figure 8.59.

Input remains same: Transition takes place to attain stable state in case state is unstable if input remains same. $AX = 00$ attains a stable state 10. Let $AX = 00$, the state A^+ changes to 1 after finite time t_d, which is unstable state. After propagation delay time t_d, AX becomes 10 and leads to state A^+ to 1, which is a stable state. Further, let $AX = 01$, the state A^+ changes to 1 after finite time t_d, which is unstable state. After propagation delay time t_d, AX becomes 11 and leads to state A^+ to 0, an unstable state. AX toggles $01 \rightarrow 11 \rightarrow 01 \rightarrow 11$ and so on.

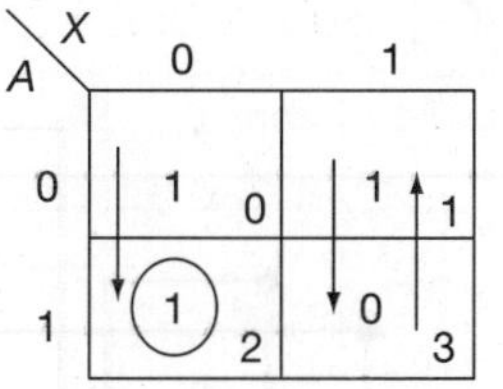

FIGURE 8.59 | Transition table

TABLE 8.55 | Transition table of NAND gate

State Variable A	Input X	State Transition AX	Output A^+	Remarks
0	$0 \rightarrow 1$	$00 \rightarrow 01 \rightarrow 11 \rightarrow 01$	$0 \rightarrow 0 \rightarrow 1$	Unstable
1	$0 \rightarrow 1$	$10 \rightarrow 11 \rightarrow 01 \rightarrow 11$	$1 \rightarrow 1 \rightarrow 0$	Unstable
0	$1 \rightarrow 0$	$01 \rightarrow 00 \rightarrow 10$	$0 \rightarrow 0 \rightarrow 1$	Unstable
1	$1 \rightarrow 0$	$11 \rightarrow 10 \rightarrow 10$	$1 \rightarrow 1 \rightarrow 1$	Stable

Input changes: Table 8.55 is a transition table. Transition table shows the changes, the circuit undergoes as input changes. At a time, only one input changes. Let, the circuit is at $A = 1, X = 1$, a stable state. Now, input, X changes from 1 to 0 ($1 \rightarrow 0$). So, $A = 1$ and $X = 0$. After finite time t_d, $A = 1$, $X = 0$ and $A^+ = 1$, a stable state. In transition table, AX changes from $11 \rightarrow 10 \rightarrow 10$. Other possibilities are also given in the transition table.

EXAMPLE 8.25 Analyse the two input OR gate with output feedback being asynchronous sequential circuit.

SOLUTION The circuit diagram is shown in Figure 8.60. X is input. Output A^+ is fed back to OR gate, A.

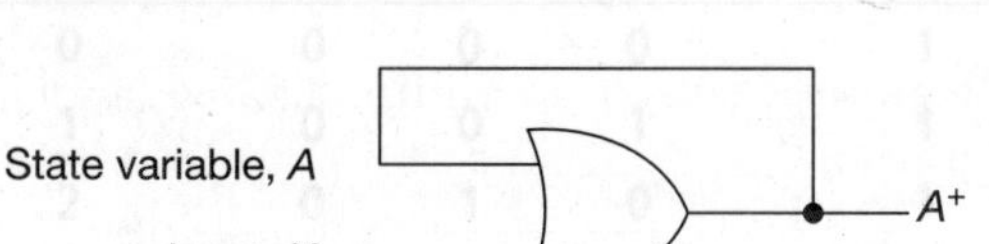

FIGURE 8.60 | OR gate with feedback

Logic expression is stated as below:

$$A^+ = A + X \tag{8.35}$$

State table is given in Table 8.56. The state is said to be stable if and only if A and A^+ are same after propagation of signal. Otherwise state is unstable. Transition takes place to attain stable state.

TABLE 8.56 | State table of OR gate

State Variable A	Input X	State Variable A^+	Remarks
0	0	0	Stable
0	1	1	Unstable
1	0	1	Stable
1	1	1	Stable

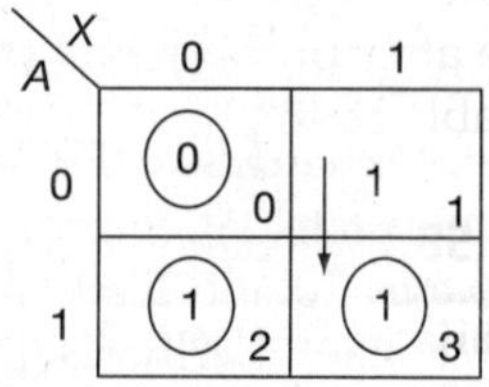

FIGURE 8.61 | Transition table

The transition table from state table of Table 8.56 is shown in Figure 8.61.

Input remains same: Transition takes place to attain stable state in case state is unstable if input remains same. $AX = 00$ attains a stable state 10. Let $AX = 01$, the state A^+ changes to 1 after finite time t_d, which is unstable state. After propagation delay time t_d, AX becomes 11 and leads to state A^+ to 1 which is a stable state. AX undergoes $01 \rightarrow 11 \rightarrow 11$.

Input changes: Table 8.57 is a transition table. Transition table shows the changes, the circuit undergoes as input changes. At a time, only one input changes. Let, the circuit is at $A = 0$, $X = 0$, a stable state. Now, input X changes from 0 to 1 $(0 \rightarrow 1)$. So, $A = 0$, $X = 1$. After finite time t_d, $A^+ = 1$. After finite time t_d, $A = 1$, $X = 1$ and $A^+ = 1$, a stable state. In transition table, AX changes from $00 \rightarrow 01 \rightarrow 11$. Other possibilities are also given in the transition table.

TABLE 8.57 | Transition table of OR gate

State Variable A	Input X	State Transition AX	Output A^+	Remarks
0	$0 \rightarrow 1$	$00 \rightarrow 01 \rightarrow 11$	$0 \rightarrow 0 \rightarrow 1$	Unstable
1	$0 \rightarrow 1$	$10 \rightarrow 11 \rightarrow 11$	$1 \rightarrow 1 \rightarrow 1$	Stable
0	$1 \rightarrow 0$	$01 \rightarrow 00 \rightarrow 00$	$0 \rightarrow 0 \rightarrow 0$	Stable
1	$1 \rightarrow 0$	$11 \rightarrow 10 \rightarrow 10$	$1 \rightarrow 1 \rightarrow 1$	Stable

EXAMPLE 8.26 Analyse the NAND latch as asynchronous sequential circuit.

SOLUTION The circuit diagram is shown in Figure 8.62. X_1 and X_0 is input. Output A^+ is fed back to NAND gate, A.

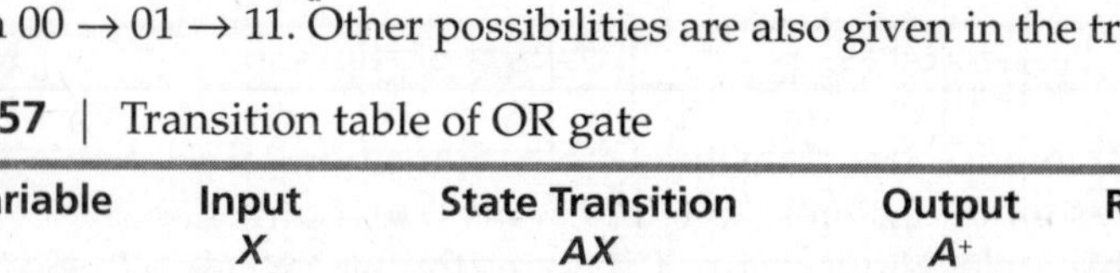

FIGURE 8.62 | NAND latch

Logic expressions are stated as below:

$$Z = \overline{AX_0} \quad (8.36)$$

$$A^+ = \overline{X_1(\overline{AX_0})} \quad (8.37)$$

State table is given in Table 8.58. The state is said to be stable if and only if A and A^+ are same after propagation of signal. Otherwise state is unstable. Transition takes place to attain stable state

TABLE 8.58 | State table of NAND latch

State Variable A	Input X_1	X_0	Intermediate Output, Z	State Variable A^+	Remarks
0	0	0	1	1	Unstable
0	0	1	1	1	Unstable
0	1	0	1	0	Stable
0	1	1	1	0	Stable
1	0	0	1	1	Stable
1	0	1	0	1	Stable
1	1	0	1	0	Unstable
1	1	1	0	1	Stable

The transition table from state table of Table 8.84 is shown in Figure 8.63.

Input remains same: Transition takes place to attain stable state in case state is unstable if input remains same. $AX_1X_0 = 000$ attains a stable state 100. $AX_1X_0 = 001$ attain stable state as 101 and $AX_1X_0 = 110$ attains stable state as 010. Explanation is given below:

When AX_1X_0 is 000, the state A^+ changes to 1 after finite time t_d, which is unstable state. After finite time t_d, AX_1X_0 becomes 100 and leads to state A^+ to 1 which is a stable state. AX_1X_0 transitions 000 → 100.

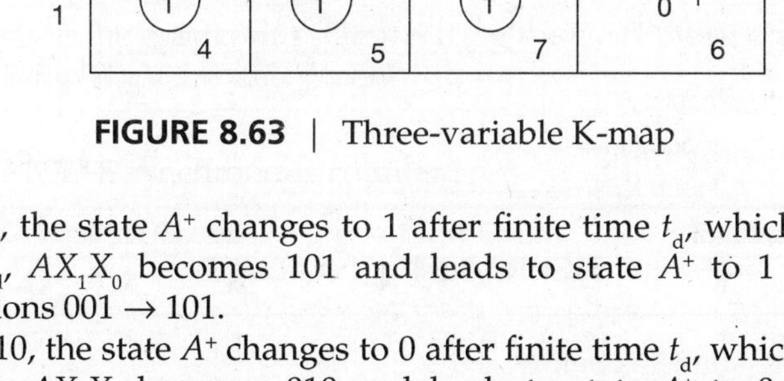

FIGURE 8.63 | Three-variable K-map

When AX_1X_0 is 001, the state A^+ changes to 1 after finite time t_d, which is unstable state. After finite time t_d, AX_1X_0 becomes 101 and leads to state A^+ to 1 which is a stable state. AX_1X_0 transitions 001 → 101.

When AX_1X_0 is 110, the state A^+ changes to 0 after finite time t_d, which is unstable state. After finite time t_d, AX_1X_0 becomes 010 and leads to state A^+ to 0 which is a stable state. AX_1X_0 transitions 110 → 010.

When AX_1X_0 is 011, the state A^+ changes to 0 after finite time t_d, which is stable state. No transition takes place.

Input changes: Table 8.59 is a transition table. Transition table shows the changes, the circuit undergoes as input changes. At a time, only one input changes. Let, the circuit is at $AX_1X_0 = 011$ and $AX_1X_0 = 111$. These are the stable state (Figure 8.63). Transition of X_1X_0, $00 \rightarrow 01$, reaches to $AX_1X_0 = 101$ state. So, Transition of X_1X_0, $00 \rightarrow 10$, reaches to $AX_1X_0 = 010$ state. Other possibilities are explained in Table 8.59.

TABLE 8.59 | Transition table of NAND latch

State Variable A	Input X_1X_0	State Transition AX_1X_0	Output A^+	Remarks
0	00 → 01	000 → 001 → 101	0 → 1 → 1	
1	00 → 01	100 → 101	1 → 1	Stable
0	00 → 10	000 → 010	0 → 0	Stable
1	00 → 10	100 → 110 → 010	1 → 1 → 0	
0	01 → 00	001 → 000 → 100	0 → 0 → 1	
1	01 → 00	101 → 100	1 → 1	Stable
0	01 → 11	001 → 011	0 → 0	Stable
1	01 → 11	101 → 111	1 → 1	Stable
0	10 → 00	010 → 000 → 100	0 → 0 → 1	
1	10 → 00	110 → 100	1 → 1	Stable
0	10 → 11	010 → 011	0 → 0	Stable
1	10 → 11	110 → 111	1 → 1	Stable
0	11 → 01	011 → 001 → 101	0 → 0 → 1	
1	11 → 01	111 → 101	0 → 0 → 1	
0	11 → 10	011 → 010	0 → 0	Stable
1	11 → 10	111 → 110 → 010	1 → 0 → 0	

EXAMPLE 8.27 Analyse the NOR latch as asynchronous sequential circuit.

SOLUTION The circuit diagram is shown in Figure 8.64. X_1 and X_0 is input. Output A^+ is fed back to NOR gate, A.

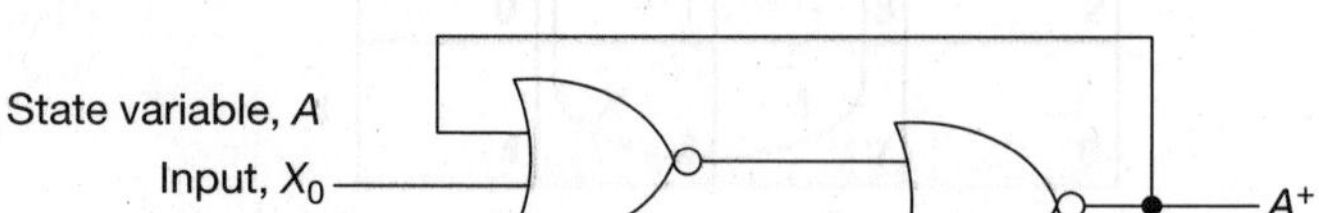

FIGURE 8.64 | NOR latch

Logic expressions are stated as below:

$$Z = \overline{A + X_0} \tag{8.38}$$

$$A^+ = \overline{X_1 + (\overline{A + X_0})} \tag{8.39}$$

State table is given in Table 8.60. The state is said to be stable if and only if A and A^+ are same after propagation of signal. Otherwise, state is unstable. Transition takes place to attain stable state.

TABLE 8.60 | State table of NOR latch

State Variable A	Input X_1	Input X_0	Intermediate Output	State Variable A^+	Remarks
0	0	0	1	0	Stable
0	0	1	0	1	Unstable
0	1	0	1	0	Stable
0	1	1	0	0	Stable
1	0	0	0	1	Stable
1	0	1	0	1	Stable
1	1	0	0	0	Unstable
1	1	1	0	0	Unstable

The transition table from state table of Table 8.86 is shown in Figure 8.65.

Input remains same: Transition takes place to attain stable state in case state is unstable if input remains same. $AX_1X_0 = 001$ attains a stable state 101. $AX_1X_0 = 110$ attain stable state as 010 and $AX_1X_0 = 111$ attains stable state as 011. Explanation is given below:

When AX_1X_0 is 001, the state A^+ changes to 1 after finite time t_d, which is unstable state. After finite time t_d, AX_1X_0 becomes 101 and leads to state A^+ as 1, which is a stable state. AX_1X_0 transitions 001 → 101.

When AX_1X_0 is 110, the state A^+ changes to 0 after finite time t_d, which is unstable state. After finite time t_d, AX_1X_0 becomes 010 and leads to state A^+ as 0, which is a stable state. AX_1X_0 transitions 110 → 010.

When AX_1X_0 is 111, the state A^+ changes to 0 after finite time t_d, which is unstable state. After finite time t_d, AX_1X_0 becomes 011 and leads to state A^+ as 1, which is a stable state. AX_1X_0 transitions 111 → 011.

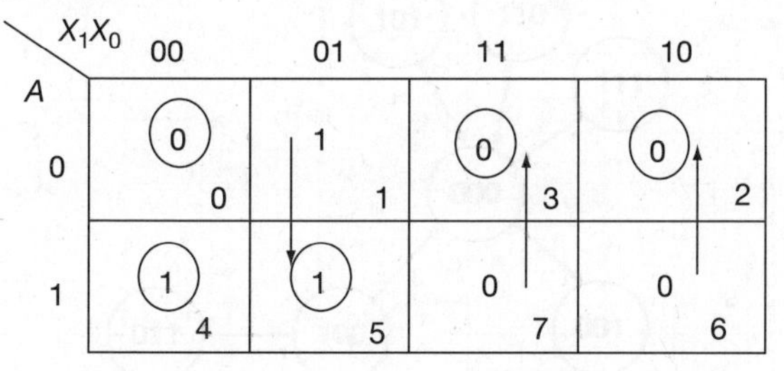

FIGURE 8.65 | Three-variable K-map

Input changes: Table 8.61 is a transition table. Transition table shows the changes, the circuit undergoes as input changes. At a time, only one input changes. Let the circuit be at $AX_1X_0 = 000$ and $AX_1X_0 = 100$. These are the stable state (Figure 8.65). Transition of X_1X_0, 00 → 01, reaches to $AX_1X_0 = 101$ state. So, Transition of X_1X_0, 00 → 10, reaches to $AX_1X_0 = 010$ state. Other possibilities are explained in Table 8.61.

TABLE 8.61 | Transition table of NOR latch

State Variable A	Input X_1X_0	State Transition AX_1X_0	Output A^+	Remarks
0	00 → 01	000 → 001 → 101	0 → 1 → 1	
1	00 → 01	100 → 101	1 → 1	Stable
0	00 → 10	000 → 010	0 → 0	Stable
1	00 → 10	100 → 110 → 010	1 → 1 → 0	
0	01 → 00	001 → 000	0 → 0	Stable
1	01 → 00	101 → 100	1 → 1	Stable
0	01 → 11	001 → 011	0 → 0	Stable
1	01 → 11	101 → 111 → 011	1 → 1 → 0	
0	10 → 00	010 → 000 → 100	0 → 0 → 1	
1	10 → 00	110 → 100	1 → 1	Stable
0	10 → 11	010 → 011	0 → 0	Stable
1	10 → 11	110 → 111 → 011	1 → 1 → 0	
0	11 → 01	011 → 001 → 101	0 → 0 → 1	
1	11 → 01	111 → 101	0 → 0 → 1	
0	11 → 10	011 → 010	0 → 0	Stable
1	11 → 10	111 → 110 → 010	1 → 0 → 0	

EXAMPLE 8.28 Analyse the fundamental mode of asynchronous sequential circuit given in Figure 8.66.

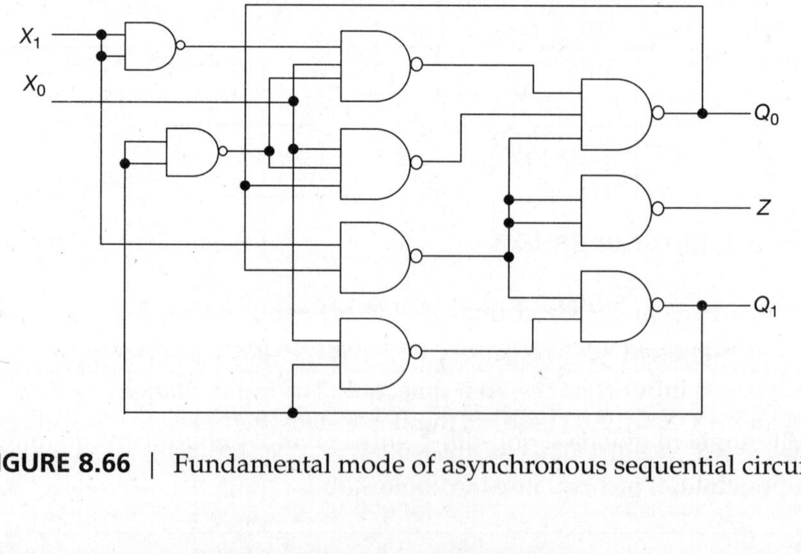

FIGURE 8.66 | Fundamental mode of asynchronous sequential circuit

SOLUTION The combination of the outputs of memory elements are known as secondary or internal states and these variables are known as internal or secondary state variables. Here, Q_1 and Q_0 are the internal state variables since no explicit elements are present. The combination of both, input state and the secondary state (Q_1, Q_0, X_1 and X_0) is known as the total state. Z is the output variable.

X_1 and X_0 are two inputs. Q_1 and Q_0 are two state variables. Logic expression to evaluate next state variable are given below:

$$Q_0^+ = \overline{(\overline{Q_0 X_1})(\overline{\bar{Q}_1 Q_0 X_0})(\overline{\bar{Q}_1 \bar{X}_1 X_0})} \tag{8.40}$$

$$Q_1^+ = \overline{(\overline{Q_0 X_1})(\overline{Q_1 X_0})} \tag{8.41}$$

Output of the circuit is given below.

$$Z = \overline{(\overline{Q_0 X_1})} \tag{8.42}$$

State variables Q_1, Q_0 and inputs X_1 and X_0 are the inputs required to find the next state variables. Table 8.62 is the state table expresses next states for 2^4 possible combinations. Next states are stable if these are same the state variable.

TABLE 8.62 | State table

State Variable		Input		Next State Variable		State	Output
Q_1	Q_0	X_1	X_0	Q_1	Q_0		Z
0	0	0	0	0	0	Stable	0
0	0	0	1	0	1	Unstable	0
0	0	1	0	0	0	Stable	0
0	0	1	1	0	0	Stable	0
0	1	0	0	0	0	Unstable	0
0	1	0	1	0	1	Stable	0
0	1	1	0	1	1	Unstable	1
0	1	1	1	1	1	Unstable	1
1	0	0	0	0	0	Unstable	0
1	0	0	1	1	0	Stable	0
1	0	1	0	0	0	Unstable	0
1	0	1	1	1	0	Stable	0
1	1	0	0	0	0	Unstable	0
1	1	0	1	1	0	Unstable	0
1	1	1	0	1	1	Stable	1
1	1	1	1	1	1	Stable	1

Transition table is drawn in Figure 8.67 from state Table 8.62. The unstable states attain the stable states after transitions, show by arrows.

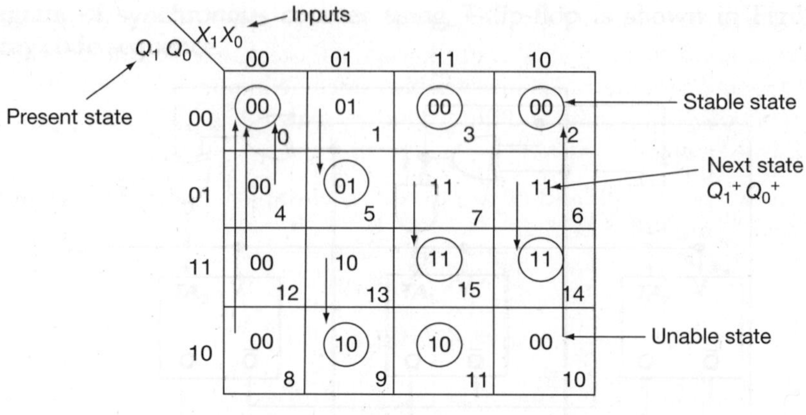

FIGURE 8.67 | Transition table

Let secondary state is referred as *A* for state variables $Q_1Q_0 = 00$. Secondary state is referred as *B* for state variables $Q_1Q_0 = 01$. Secondary state is referred as C for state variables $Q_1Q_0 = 11$ and secondary state is referred as *D* for state variables $Q_1Q_0 = 10$.

A flow table is basically similar to a transition table except that the internal states are represented symbolically rather than by binary states. The column headings are the input combinations and the entries are the next states, and outputs. The state changes occur with change of inputs (one input change at a time) and next state logic propagation delay. The flow table is obtained from transition table is given in Figure 8.68.

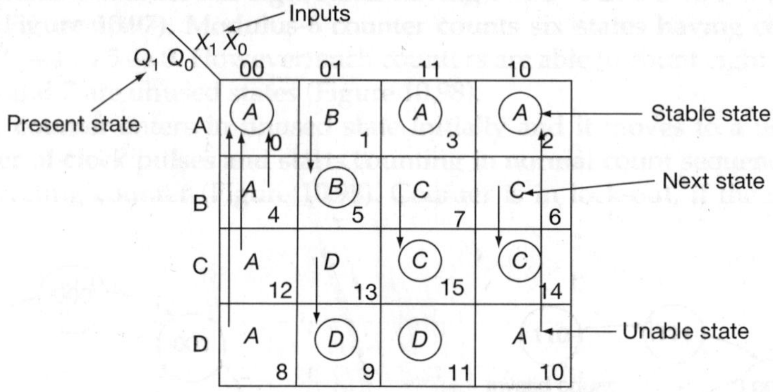

FIGURE 8.68 | Transition table

8.8.2 | Pulse Mode Asynchronous Sequential Circuit with Latches

In a pulse mode asynchronous sequential circuit, an Input pulse is permitted to occur only when the circuit is in stable state and there is no pulse present on any other input. When an input pulse arrives, it triggers the circuit and causes a transition from one stable state to another stable state so as to enable the circuit to receive another input pulse. In this mode of operation critical race is avoided. To keep the circuit stable between two pulses, flip-flops, whose outputs are levels, are used as memory elements.

For the analysis of pulse-mode circuits, the model used for the fundamental-mode circuits is not followed since the circuit is stable when there are no inputs and the absence of a pulse conveys no information.

In pulse mode asynchronous sequential circuits, the number of columns in the next state table is equal to the number of inputs.

EXAMPLE 8.29 Analyse the pulse mode of asynchronous sequential circuit given in Figure 8.69.

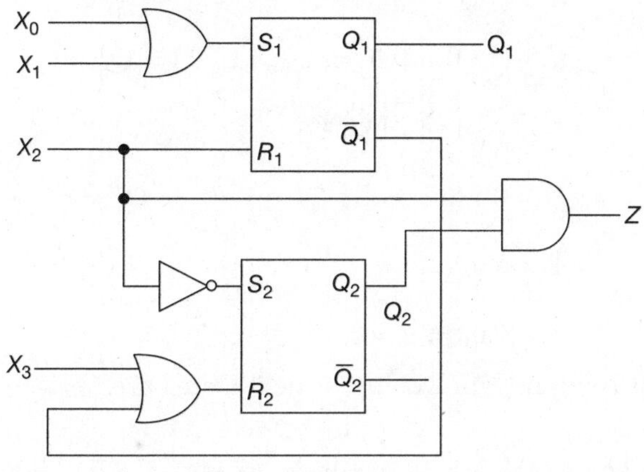

FIGURE 8.69 | Pulsed mode asynchronous sequential circuit

SOLUTION The combination of the outputs of memory elements are known as secondary or internal states and these variables are known as internal or secondary state variables. Here, Q_1 and Q_0 are the internal state variables since no explicit elements are present. The combination of both, input state and the secondary state (Q_1, Q_0, X_1 and X_0) is known as the total state. Z is the output variable.

The excitations and output expressions are:

$$S_1 = X_0 + X_1$$

$$R_1 = X_2$$

$$S_2 = \bar{X}_2$$

$$R_2 = X_3 + \bar{Q}_1$$

$$Z = X_2 Q_2 \tag{8.43}$$

The next state equations are obtained by using the excitations and characteristic equations of latch. The next state equations are:

$$Q_1^+ = S_1 + \bar{R}_1 Q_1$$

$$Q_1^+ = (X_0 + X_1) + \bar{X}_2 Q_1 \tag{8.44}$$

$$Q_2^+ = S_2 + \bar{R}_2 Q_2$$

$$Q_2^+ = \bar{X}_2 + (\overline{X_3 + \bar{Q}_1})Q_2 \quad (8.45)$$

To fill the cell of first row and first column, next states are obtained by considering the values as:

$$Q_2Q_1 = 00, X_0 = 1, X_1 = 0, X_2 = 0 \text{ and } X_3 = 0$$

$$Q_1^+ = (X_0 + X_1) + \bar{X}_2 Q_1$$

$$Q_1^+ = (1+0) + \bar{0}.0 \Rightarrow Q_1^+ = 1+0 \Rightarrow Q_1^+ = 1$$

$$Q_2^+ = \bar{X}_2 + (\overline{X_3 + \bar{Q}_1})Q_2$$

$$Q_2^+ = \bar{0} + (\overline{0+\bar{0}})0 \Rightarrow Q_2^+ = 1+0 \Rightarrow Q_2^+ = 1$$

$$Z = X_2Q_2$$

$$Z = 0.0 \Rightarrow Z = 0$$

To fill the cell of first row and third column, next states are obtained by considering the values as:

$$Q_2Q_1 = 00, X_0 = 0, X_1 = 0, X_2 = 1 \text{ and } X_3 = 0$$

$$Q_1^+ = (0+0) + \bar{1}.0 \Rightarrow Q_1^+ = 0+0 \Rightarrow Q_1^+ = 0$$

$$Q_2^+ = \bar{1} + (\overline{0+\bar{0}})0 \Rightarrow Q_2^+ = 0+0 \Rightarrow Q_2^+ = 0$$

Similarly, other cells are filled and transition table is given in Figure 8.70.

Flow table is shown in Figure 8.71 obtained from transition table where binary values are assigned with variables.

Q_2Q_1	X_0	X_1	X_2	X_3
00	11/0	11/0	00/0	01/0
01	11/0	11/0	00/1	01/0
11	11/0	11/0	01/1	11/0
10	11/0	11/0	00/0	11/0

FIGURE 8.70 | Transition table

Q_2Q_1	X_0	X_1	X_2	X_3
S_0	S_2/0	S_2/0	S_0/0	S_1/0
S_1	S_2/0	S_2/0	S_0/1	S_1/0
S_2	S_2/0	S_2/0	S_1/1	S_2/0
S_3	S_2/0	S_2/0	S_0/0	S_2/0

FIGURE 8.71 | Flow table

8.9 | PROBLEMS IN ASYNCHRONOUS SEQUENTIAL CIRCUIT

The asynchronous sequential circuits have three major problems: cycles, races and hazards.

8.9.1 | Cycles

Cycles are created due to the feedback in asynchronous sequential circuits. When the input changes, it induces a feedback transition through more than one unstable state, and such a situation is called a cycle. The circuit goes through a unique sequence of unstable states because of an input change. When a cycle exists in a state table of an asynchronous circuit, care must be taken to ensure that the circuit terminates in a stable state. Otherwise, the circuit goes form one unstable state to another until a new change occurs in the input.

8.9.2 | Races

In case of cycles, only one feedback variable is unstable at any time due to the change in an input variable. When two or more feedback variables change value in response to a change in an input variable with unequal propagation delay, then a RACE condition exists in an asynchronous sequential circuit. Races are of two types:

- Critical races and
- Non-critical races

Critical races should be eliminated in a circuit, whereas non-critical races may be tolerated.

8.9.3 | Critical Races

Critical races may always go to some wrong state. It is because of different propagation delays. Identical circuits with different propagation delays may go to different wrong states. Consider the present state/next state table shown in Table 8.63.

TABLE 8.63 | State table

Present State	Next State Q_1Q_0			
Q_1Q_0	$X_1X_0 = 00$	$X_1X_0 = 01$	$X_1X_0 = 11$	$X_1X_0 = 10$
00	11 (1)	00 (3)	11 (9)	11 (13)
01	01 (4)	11 (6)	11 (10)	00 (15)
10	10 (2)	00 (7)	11 (11)	10 (16)
11	10 (5)	11 (8)	11 (12)	10 (14)

Assume that the circuit is in stable state 3 (i.e. $Q_1Q_0 = 00$ corresponding to $X_1X_0 = 01$) and input X_0 changes from 1 to 0. Now the next state variable Q_1Q_0 begins to switch from 00 to 11. But due to an unequal propagation delay, one of the next state variables becomes 1 and other variables do not change from the value 0, and hence, $Q_1Q_0 = 01$. Then the circuit goes to stable state 4. If $Q_1Q_0 = 10$, then the circuit goes to stable state 2. Therefore, for an input change i.e. from $X_1X_0 = 01$ to 00, depending on the relative switching speed of the next state variables Q_1Q_0, the final state will be either stable state 2 or 4. But both final states 2, and 4, are wrong for the given input change. This situation is called a critical race, which must be avoided in an asynchronous circuit.

8.9.4 | Non-critical Races

A non-critical race problem may always go to a correct final stable state, after its transition through unstable states. It is also because of different propagation delay, Consider Table 8.63 and assume that the circuit is in stable state 3 (i.e. $Q_1Q_0 = 00$). Now, if the input changes from $X_1X_0 = 01$ to 11, then the next state variables (Q_1Q_0) of the circuit are supposed to switch from 00 to 11. Again, due to propagation delay Q_1Q_0 becomes either 01 or 10. Therefore, the circuit will either go to state 10 or 11 or finally reach the stable state 12. Such a situation where the correct final stable state is reached after a transition through unstable states is called a non-critical race. Non-critical races can be tolerated in an asynchronous circuit.

8.9.5 | Hazards

Combinational circuits used in asynchronous sequential circuits may have unequal propagation delays. Hazard is an unwanted transient that is spike or glitch that occurs due to unequal propagation delays through a combinational circuit. There are two types of hazards: static hazard and dynamic hazard.

Static hazard is a condition which results in a single momentary incorrect output due to the change, in an input variable when the output is expected to remain in the same state. Static hazard is of two types: Static-0 hazard and Static-1 hazard.

In case the output momentarily goes to state 1 due to change in input, when the output is expected to remain in state 0 as per the steady state analysis, such hazard is known as static-0 hazard. It can be removed by forming a group of adjacent cells.

Similarly, if the output momentarily goes to state 0 due to a change in input, when the output is expected to remain in state 1 as per the steady state analysis, such hazard is known as static-1 hazard. It can be removed by forming a group of adjacent cells.

Multiple glitch situations at the output due a change from 0 to 1 (or form 1 to 0) is known as dynamic hazard. It can be removed by forming a group of adjacent cells as in static hazard.

8.9.6 | Essential Hazards

It is a type of hazard that exists only in asynchronous sequential circuits with two or more feedbacks. Essential hazard occurs normally in toggling type circuits. It is an operational error generally caused by an excessive delay to a feedback variable in response to an input change leading to a transition to an improper state. Essential hazards cannot be eliminated by adding redundant gates, same as static hazards.

8.10 | ASYNCHRONOUS SEQUENTIAL CIRCUIT DESIGN

Design of asynchronous sequential circuits is more difficult than that of synchronous sequential circuits because of the timing problems involved in these circuits. Designing an asynchronous sequential circuit requires to obtain logic diagram for the given design specifications. Usually the design problem is specified in the form of statements of the desired circuit performance precisely specifying the circuit operation for every applicable input sequence.

- Formulate the circuit conditions have to follow, from the problem definition.
- Develop a primitive state table and specify the outputs that are associated with the stable state.
- Minimize the primitive state table.
- Assign state variables (i.e. secondary variables) to the rows of reduced primitive state table and obtain the present state/next state and output table. The outputs assigned to the unstable states vary according to the design requirements.
- Decide the memory element to be used and obtain the excitation and output table.
- Obtain the simplified expression of the excitation and output function.
- Draw the schematic circuit diagram.

EXAMPLE 8.30 Design a logic circuit using SR latch for the state diagram is given in Figure 8.72.

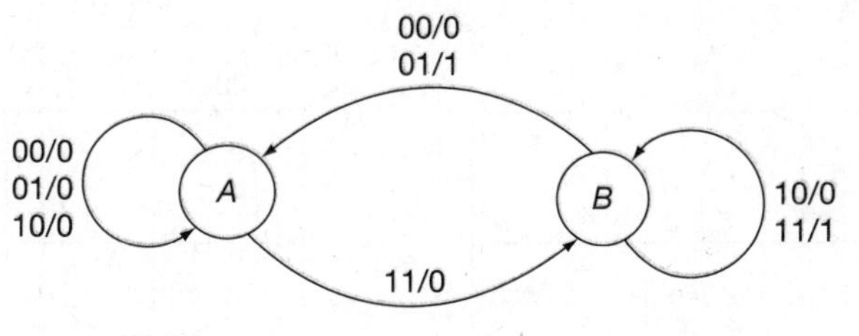

FIGURE 8.72 | State transition diagram

SOLUTION Flow table is obtained from state transition diagram as Figure 8.73. Present and next states are same considered as stable state. Let us assume there are two inputs X_1X_0, one output, Z and one state variable, Q. For inputs 00, 01 and 10, the machine achieves stable state A. When state moves from A to B or B to A, the state is said to be lead to unstable state.

Q \ X_1X_0	00	01	11	10
A	(A)/0	(A)/0 1	B/0 3	(A)/0 2
B	A/0 4	A/1 5	(B)/1 7	(B)/0 6

FIGURE 8.73 | Flow table

Q \ X_1X_0	00	01	11	10
0	(0)/0	(0)/0 1	1/0 3	(0)/0 2
1	0/0 4	0/1 5	(1)/1 7	(1)/0 6

FIGURE 8.74 | Transition table

There are two states. States are assigned binary value as 0 and 1. The transition table is given in Figure 8.74. SR flip is used. To determine excitations of SR, excitation table is given in Table 8.64. For $X_1X_0 = 00$, state Q moves from 0 to 0 then excitation $S = 0$ and $R = X$ (0 or 1). Similarly, for $X_1X_0 = 00$, state Q moves from 1 to 0 then excitation $S = 0$ and $R = 1$. Excitation table to excite S of SR latch is given Figure 8.75. Excitation table to excite R of SR latch is given Figure 8.76.

Logical expressions to excite *S* and *R* are given below by performing simplification.

$$S = X_1 X_0 \tag{8.46}$$

$$R = \bar{X}_1 \tag{8.47}$$

TABLE 8.64 | Excitation table SR-flip-flop

Outputs		Inputs	
***Q*(*t*)**	***Q*(*t* + 1)**	***S* (Set)**	***R* (Reset)**
0	0	0	X
0	1	1	0
1	0	0	1
1	1	X	0

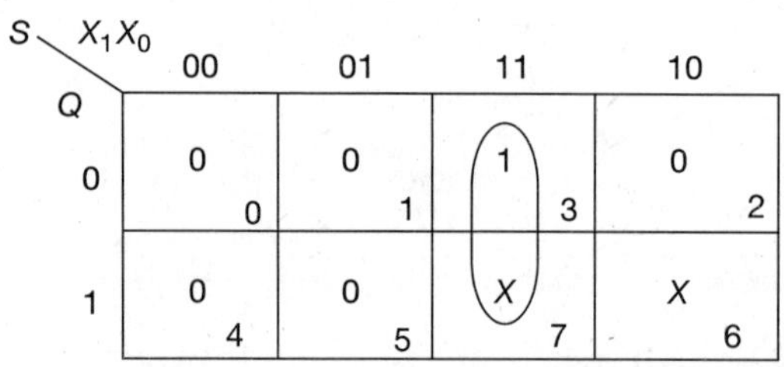

FIGURE 8.75 | Excitation table for *S*

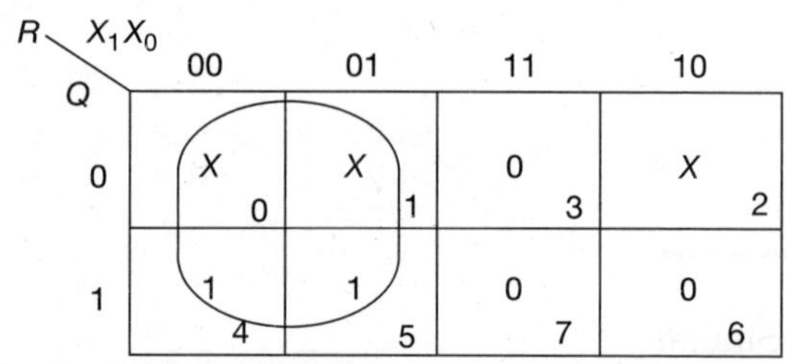

FIGURE 8.76 | Excitation table for *R*

Simplified logical expression of output, *Z* is obtained using K-map shown in Figure 8.77.

$$Z = QX_0 \tag{8.48}$$

The logic diagram is drawn in Figure 8.78 following excitation expressions of SR-flip-flop and output.

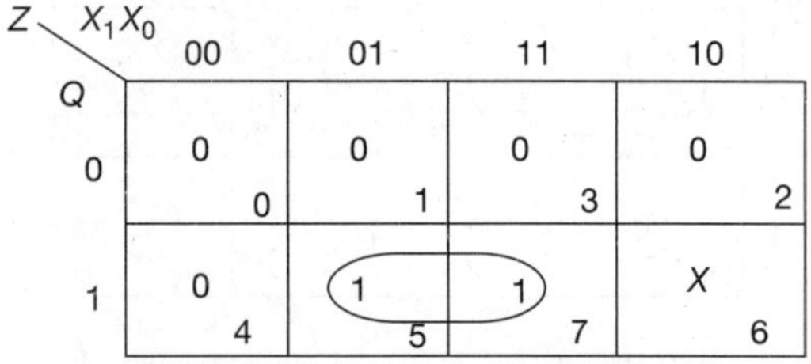

FIGURE 8.77 | Excitation table for *Z*

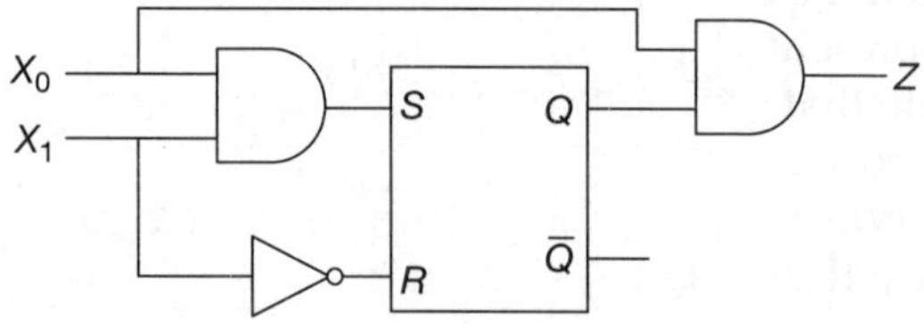

FIGURE 8.78 | Asynchronous sequential circuit

EXAMPLE 8.31 Design an asynchronous sequential circuit with two inputs, X_1 and X_0 and one output, *Z*. Initially, both inputs are equal to 0. When input X_1 changes from 0 to 1, output *Z* becomes 1. When X_2 changes from 0 to 1, *Z* becomes 0 otherwise *Z* is 0. Realize the circuit using S-R-flip flop.

SOLUTION The primitive table developed as per statement is given in Table 8.65.

Initially, both inputs are equal to 0 (i.e. $X_1 = X_2 = 0$), the circuit is said to be in a stable state1 and the output is 0, represented as stable *A*, 0.

When X_0 remains same at 0 and X_1 becomes 1, the circuit enters a stable state *B* and its output is 1. This is represented as encircled *B*, 1 in the second row. In the first row, *B* is entered to indicate that a transition to stable *B* state will occur as a result of the input change from 00 to 10. *B* represents an unstable state.

TABLE 8.65 | Primitive state table

Present State	Next State			
	$X_1X_0 = 00$	$X_1X_0 = 01$	$X_1X_0 = 11$	$X_1X_0 = 10$
A	(A),0	C, –	–	B, –
B	A, –	–	D, –	(B), 1
C	A, –	(C), 0	E, –	–
D	–	C, –	(D), 0	F, –
E	–	C, –	(E), 1	F, –
F	A, –	–	D, –	(F), 0

Similarly, when X_1 remains same at 0 and X_0 becomes 1, the circuit enters a stable state *C* and its output is 0. This is represented as encircled *C* with 0 outputs in the third row. In the first row, *C* is entered to indicate that a transition to stable state, *C* will occur as a result of the input change from 00 to 01. *C* represent as an unstable state.

When the circuit is in stable *B* state and the other input also becomes 1($X_1X_0 = 11$), then the circuit enters stable *D* with output 0. This is represented as encircled *D*, 0 in the fourth row. In the second row, an unstable *D* state is entered to indicate that a transition to stable *D* will occur as a result of the input change from 10 to 11.

When the circuit is in stable *C* state and the other input also becomes 1($X_1X_0 = 11$), then the circuit enters stable *E* state with output 1 because X_1 changes from 0 to 1. This is represented as stable *E*, 1 in the fifth row. In the third row, an unstable *E* is entered to indicate that a transition to stable state *E* state will occur as a result of the input change from 01 to 11.

When the circuit is in stable *D* or stable *E* and if X_1 remains same at 1 and X_0 becomes 0, then the circuit enters a stable state *F* and its output is 0. This is represented as encircled *F*, 0 in the sixth row. In the fourth and fifth rows, an unstable *F* state is entered to indicate that a transition to stable *F* state will occur as a result of the input change from 11 to 10.

Reduction of States

A primitive table is never completely specified. Some states and outputs are not specified in it as shown in Table 8.65 by dashes. Therefore, the concept of equivalent states cannot be used for reduction of states. However, incompletely specified states can be combined to reduce the number of states in the flow table or primitive table. Two incompletely specified states can be combined if these are compatible.

Two states are compatible if and only if, for even possible input sequence both produce the same output sequence whenever both outputs are specified and their next states are compatible. The unspecified outputs and states shown as dashes in the flow table or primitive table have no effect for compatible states. Draw the implication chart as shown in Figure 8.79.

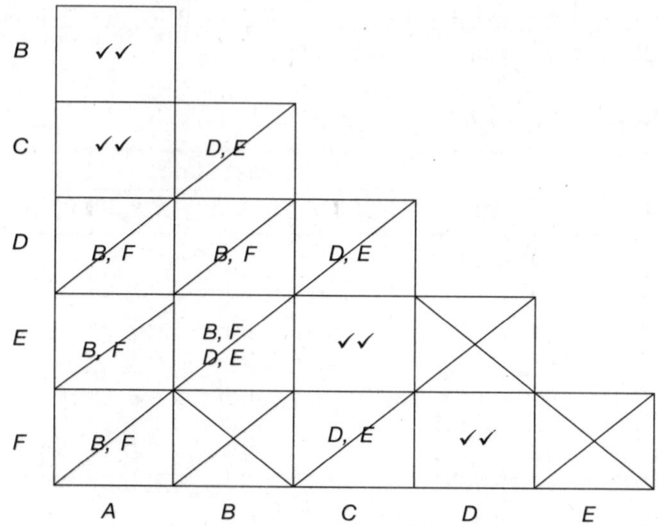

FIGURE 8.79 | Implication chart

Consider State *A* (row-1)

A and *B* are compatible.

(Row 1 and Row 2 compared. Column-1 stable *A* with 1 output is compatible with unstable *A* with unspecified output. Column 2, *C* state is compatible with unspecified state having unspecified outputs. Similarly other columns are compatible)

A and *C* are compatible

A and *D* are compatible iff *B*, *F* are compatible.

A and *E* are compatible iff *B*, *F* are compatible.

A and *F* are compatible iff *B*, *F* are compatible.

Consider State *B* (row-2)

B and *C* are compatible iff *D*, *E* are compatible.

B and *D* are compatible iff *B*, *F* are compatible.

B and *E* are compatible iff *B*, *F* and *D*, *E* are compatible.

B and *F* are not compatible. Having different output for $X_1X_0 = 10$ (Cross in square)

Consider State *C* (Row-3)

C and *D* are compatible iff *D*, *E* are compatible.

C and *E* are compatible.

C and *F* are compatible iff *D*, *E* are compatible.

Consider State *D* (Row-4)

D and *E* are not compatible (Cross in square).

D, *F* are compatible.

Consider State D (Row-5)

E and F are not compatible (Cross in square).

D and E and B and F are not compatible so cross the conditional squares like (A, D), (A, E) and (A, F) are also not compatible.

The following are the compatible states those can be merged. So, (A, B) (A, C)(C, E) (D, F) Since (A, C) is covered by (A, B) and (C, E) can be ignored. So, compatible states are (A, B)(C, E) (D, F) which are to be merged. The merged primitive table is given in Table 8.66.

TABLE 8.66 | Primitive state table

Present State	Next State			
	$X_1X_0 = 00$	$X_1X_0 = 01$	$X_1X_0 = 11$	$X_1X_0 = 10$
A	(A), 0	C, 0	D, 0	A, 1
C	A, 0	(C), 0	(C), 1	D, 0
D	A, 0	C, 0	(D), 0	(D), 0

Since there are three rows in the merged state table, 2-bit variables can be used, 00 assigned to row 1, 01 assigned to row 2, and 10 assigned to Row 3. The present state/next state and output table are given in Table 8.67. The unstable states are assigned 00/01/10 according to their stable state.

TABLE 8.67 | State table

Present State	Next State			
	$X_1X_0 = 00$	$X_1X_0 = 01$	$X_1X_0 = 11$	$X_1X_0 = 10$
00	00, 0	01, 0	10, 0	00, 1
01	00, 0	01, 0	01, 1	10, 0
10	00, 0	01, 0	10, 0	10, 0

TABLE 8.68 | State table

Present State	Next State			
	$X_1X_0 = 00$	$X_1X_0 = 01$	$X_1X_0 = 11$	$X_1X_0 = 10$
Q_1Q_0	$(S_1, R_1)(S_0, R_0)$, Z	$(S_1, R_1)(S_0, R_0)$, Z	$(S_1, R_1)(S_0, R_0)$, Z	$(S_1, R_1)(S_0, R_0)$, Z
00	(0, X)(0, X), 0	(0, X)(1, 0), 0	(1, 0)(0, X), 0	(0, X)(0, X), 1
01	(0, X)(0, 1), 0	(0, X)(X, 0), 0	(0, X)(X, 0),1	(1, 0)(0, 1), 0
10	(0, 1)(0, X), 0	(0, 1)(1, 0), 0	(X, 0)(0, X), 0	(X, 0)(0, X), 0

SR-flip-flops are used, and then an excitation and output table is required as given in Table 8.68. The excitation function (S_1, R_1), (S_0, R_0), and Z can be simplified using K-maps as shown in Figures 8.80(a) to 8.80(e).

$$S_1 = \bar{Q}_0 X_1 X_0 + Q_0 X_1 \bar{X}_0 \qquad (8.49(a))$$

$$R_1 = \bar{X}_1 \tag{8.49(b)}$$

$$S_0 = \bar{X}_1 X_0 \tag{8.49(c)}$$

$$R_1 = \bar{X}_0 \tag{8.49(d)}$$

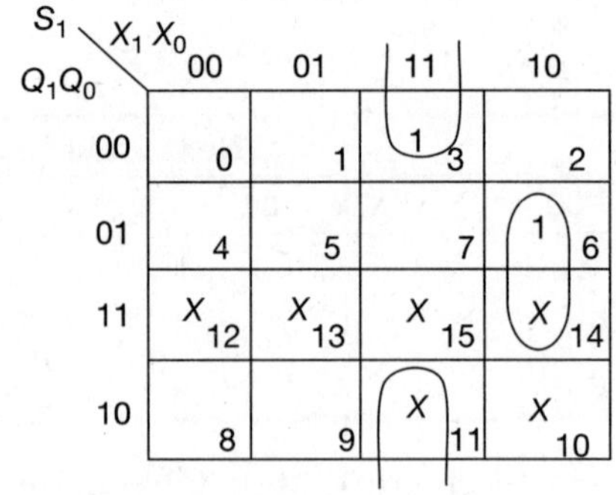

FIGURE 8.80(a) | K-map for S_1

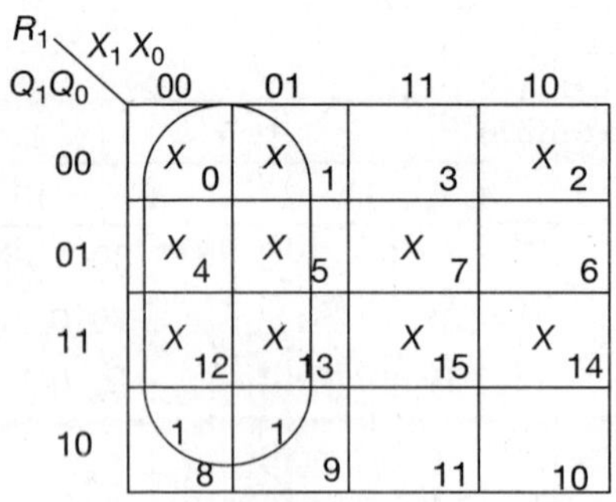

FIGURE 8.80(b) | K-map for R_1

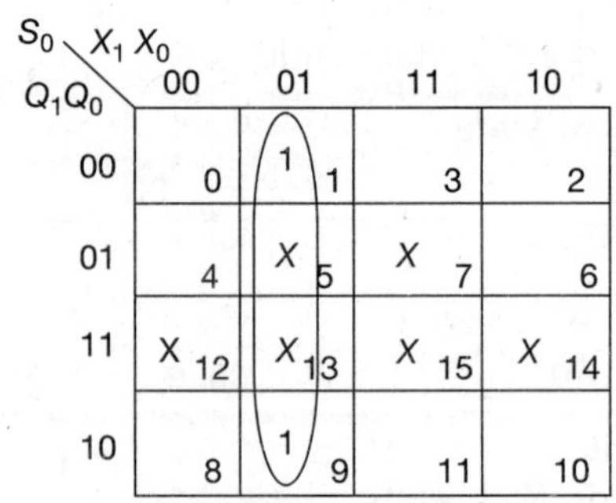

FIGURE 8.80(c) | K-map for S_0

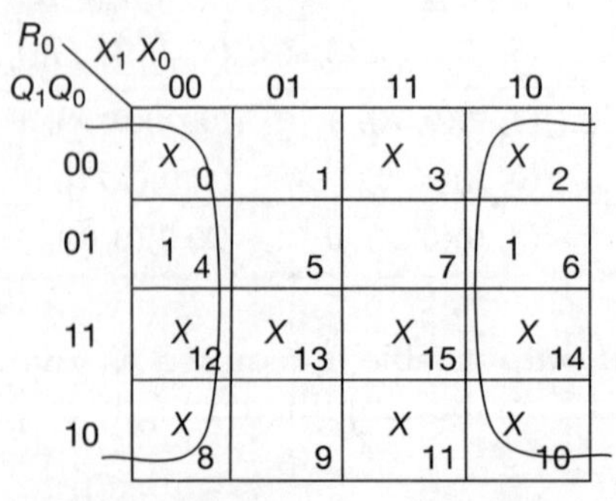

FIGURE 8.80(d) | K-map for R_0

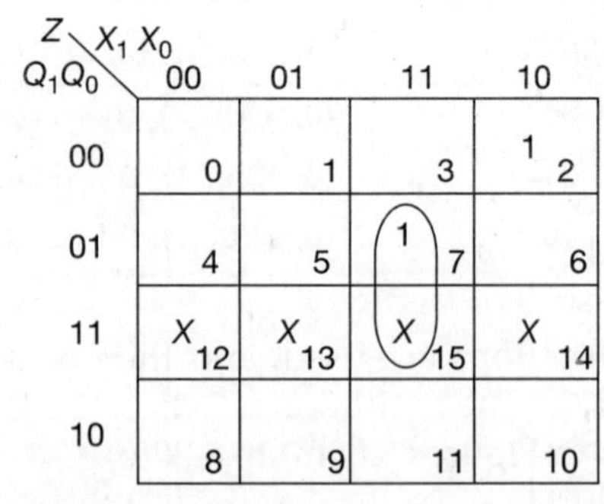

FIGURE 8.80(e) | K-map for Z

$$Z = \bar{Q}_1\bar{Q}_0X_1\bar{X}_0 + Q_0X_1X_0 \tag{8.50}$$

Using excitation and output logical expressions given in Eqs. (8.49(a))–(8.49(d)) and Eq. (8.50), the circuit diagram for the given asynchronous circuit is drawn, as shown in Figure 8.81.

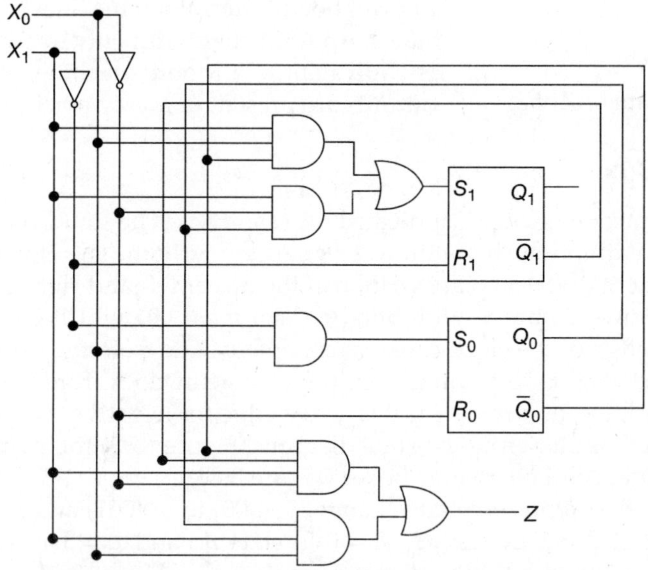

FIGURE 8.81 | Asynchronous sequential circuit

8.11 | ALGORITHMIC STATE MACHINES

The digital system design generally has two distinct parts. One part deals with data processing operations and is represented as data section. The second part deals with control signals generation and supervision and is represented as the control section. The control section is basically a sequential circuit which generates control signals to control data processing operations by considering the status signals and external signals at any given time. Thus, depending upon the status and external input signals, the sequential circuit goes to the next state to initiate further operations. ASM chart is one of the methods to represent hardware algorithm of control section and data sections.

ASM or simply state machine is another name used for a sequential circuit. These names are used when the sequential circuit is used to control a digital system that carries out a step-by-step procedure or algorithm. A special type of flowchart is called a state machine flowchart or SM chart or ASM chart. ASM chart is basically composed of three basic elements: (i) state box, (ii) decision box and (iii) conditional box.

An ASM chart is constructed from an ASM block. It is a structure which represents one state and contains diamond-shaped decision boxes and oval-shaped conditional boxes associated with it. An ASM block has one entrance path and one or more exit paths. Unconditional outputs are the functions of parent states only, and written inside the state box;

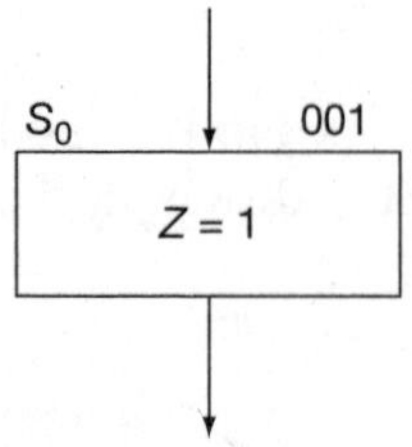

FIGURE 8.82 | State box of S_0

whereas conditional outputs depend upon the present state and the decision box path associated with it. The ASM chart notation is similar to a state diagram, where one state block is equivalent to a state in the sequential circuit. Each decision box is equivalent to the binary information written along the lines connecting the two states. An ASM chart with only unconditional outputs is equivalent to a Moore machine, where conditional outputs are present in Mealy machines.

8.11.1 | State Box

Any state in the control sequence is indicated by a state box. The state box is represented by rectangular shape within which the output signals are written. The name of the state and the binary assignment of the step are written at the upper left and right corner of the state box. Figure 8.82 shows a state S_0 with binary assignment 001 which generates an output $Z = 1$. Each active edge of the clock causes a change of state from the previous state to the next state in a synchronous sequential circuit. A change in clock from 0 to 1 is active edge for positive trigger logic, and from 1 to 0 is active edge for negative trigger logic.

The state changes, as shown in the state diagram (Figure 8.83) following Moore model, are represented using ASM chart state boxes (Figure 8.84).

Figures 8.83 and 8.84 represent state change, S_0 (000) to S_1 (001) and S_1 (001) to S_2 (010), producing an output $Z = 1$ in state S_0, $Z = 0$ in state S_1, and $Z = 1$ in state S_2. The state changes are represented with timing diagram, as shown in Figure 8.85.

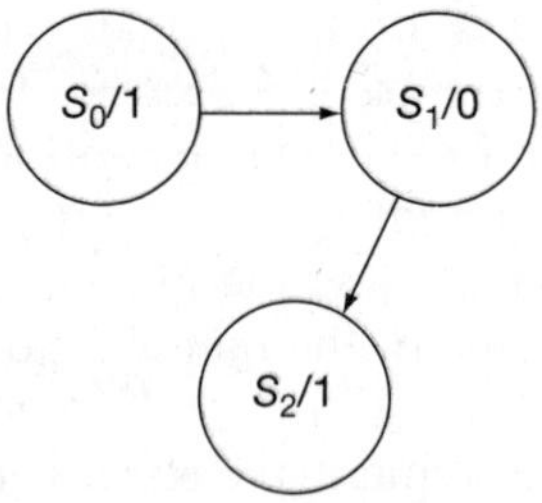

FIGURE 8.83 | State diagram

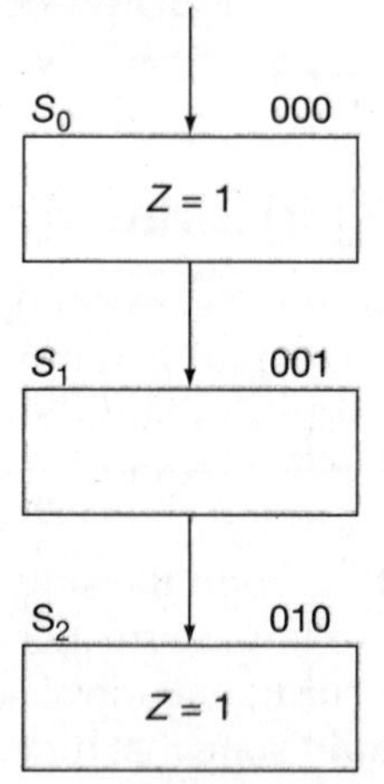

FIGURE 8.84 | State box using ASM chart

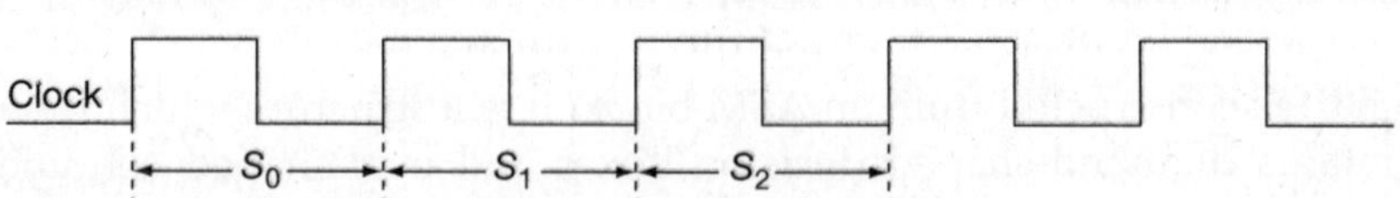

FIGURE 8.85 | Timing diagram representing state changes

8.11.2 | Decision Box

A decision box is a diamond-shaped box with two or more exit paths, as Figure 8.86. It expresses the conditions on the inputs which decide the state. The input condition, which is to be tested, is written inside the box. Generally condition written inside the box is a Boolean expression that is evaluated to find which decision to take for action. The symbol shown in Figure 8.86 is incorporated in the ASM chart by adding to a state box. Then the present state and the condition of the input select the corresponding exit path and decide the next state. The state diagram with a condition is shown in Figure 8.87 and its ASM representation with the use of decision box is shown in Figure 8.88. The state changes represented with timing diagram for input $X = 0$ or $X = 1$ is shown in Figure 8.89.

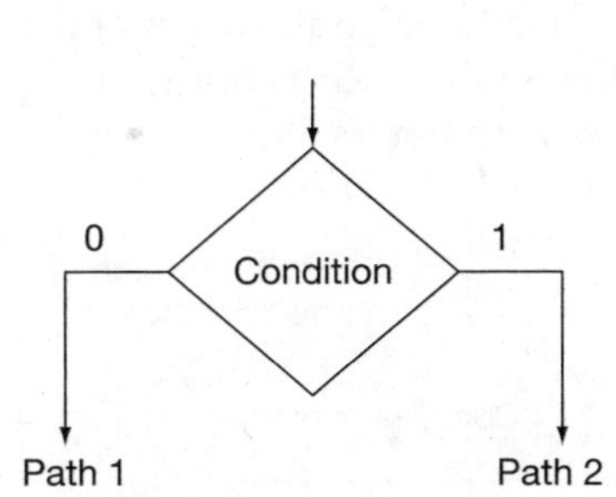

FIGURE 8.86 | Decision box

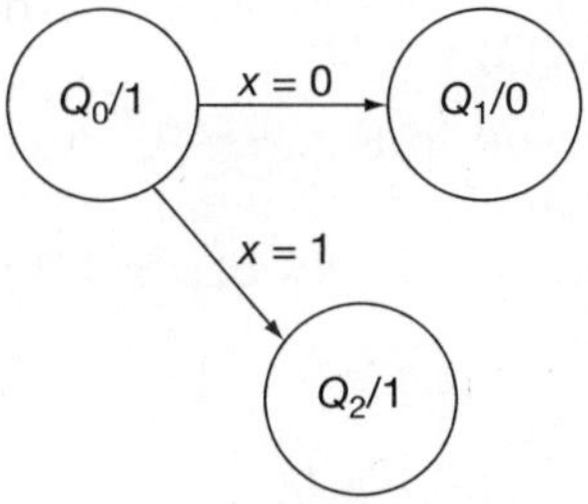

FIGURE 8.87 | State diagram

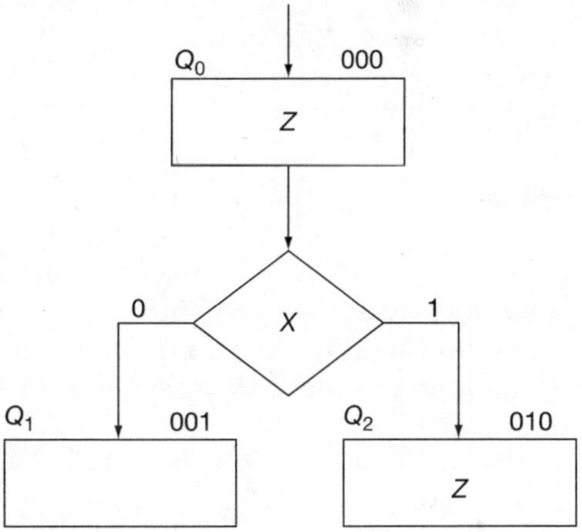

FIGURE 8.88 | ASM chart for Figure 8.87

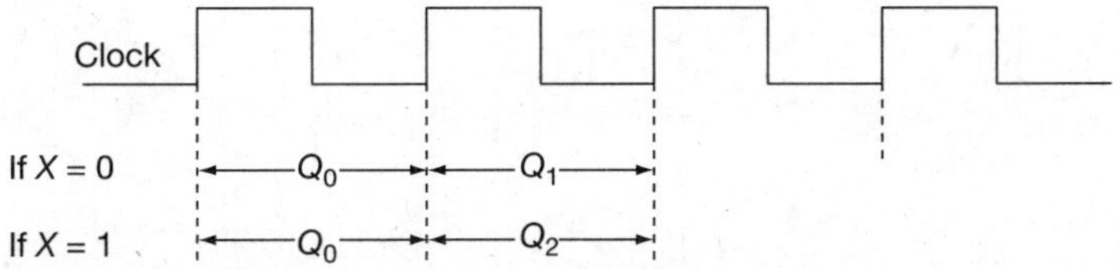

FIGURE 8.89 | Timing diagram representing state changes

8.11.3 | Conditional Box

A conditional box is an oval-shaped box which is used to represent the output produced during a state only when a certain input condition is satisfied. So, the input path to the conditional box should come from one of the exit paths of a decision box. Figure 8.90 represents the conditional output generation using a conditional box. The state diagram following Mealy model with a conditional output is shown in Figure 8.91 and its ASM chart representation with the use of decision box and conditional box is shown in Figure 8.92.

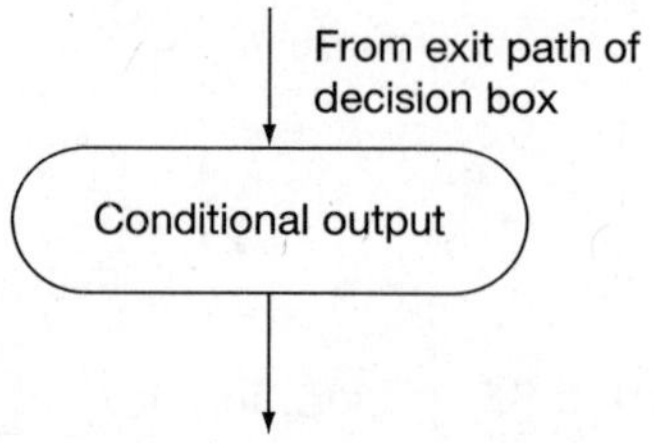

FIGURE 8.90 | Conditional box

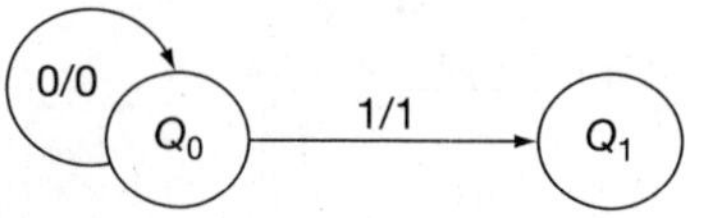

FIGURE 8.91 | State diagram

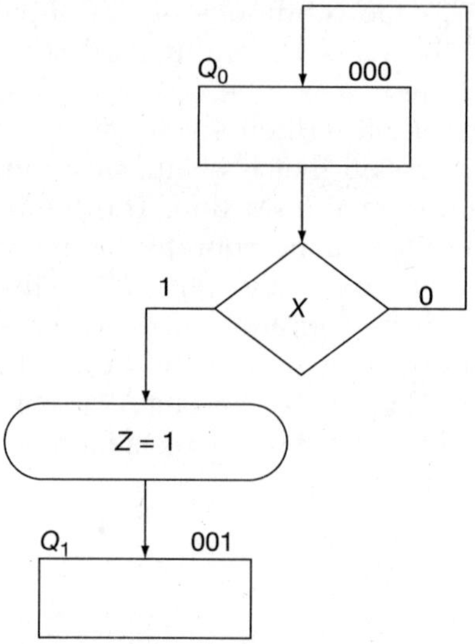

FIGURE 8.92 | ASM chart for Figure 8.91

Let another state diagram with a conditional output shown in Figure 8.93. ASM chart representation with the use of decision box and conditional box is shown in Figure 8.94 corresponding to state diagram shown in Figure 8.93. Decision boxes depend upon the inputs X and Y.

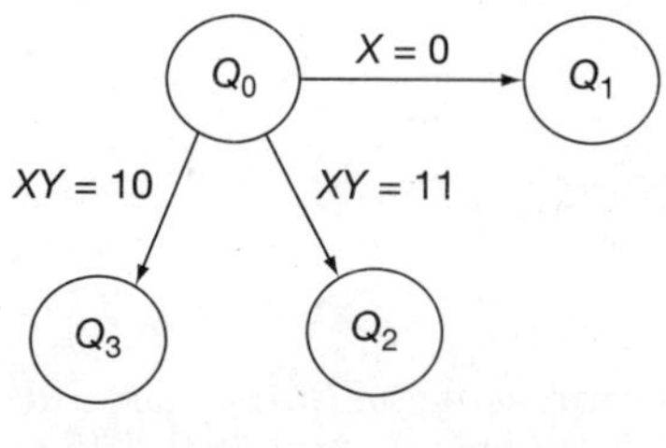

FIGURE 8.93 | State diagram

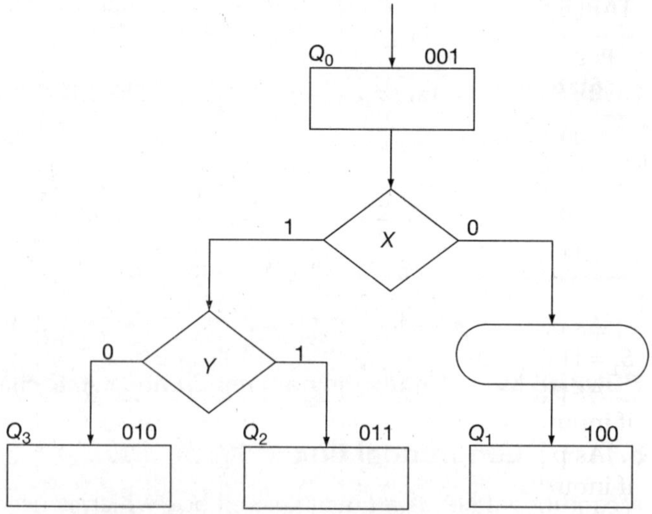

FIGURE 8.94 | ASM chart for Figure 8.93

EXAMPLE 8.32 Draw an ASM chart for a modulo-4 UP/DOWN counter.

SOLUTION The state table developed for MOD-4 binary Up/Down counter is given in Table 8.69. Counter moves to UP count when input $X = 0$ and moves to DOWN count when input $X = 1$. Counter has four (2^2) states $S_0 = 00$, $S_1 = 01$, $S_2 = 10$ and $S_3 = 11$.

TABLE 8.69 | State table

Present State	Next State	
	$X = 0$	$X = 1$
00	01	11
01	10	00
10	11	01
11	00	10

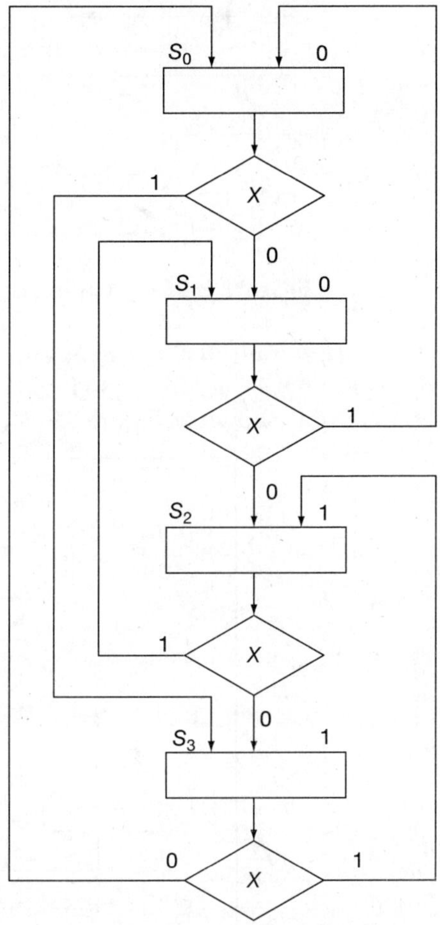

FIGURE 8.95 | ASM chart for UP/DOWN MOD-4 binary counter

As per input, initial state goes from $S_0 = 00$ to $S_1 = 01$ if input $X = 0$ and state $S_0 = 00$ to $S_1 = 11$ if input $X = 1$.

As per input, state goes from $S_1 = 01$ to $S_2 = 10$ if input $X = 0$ and state $S_1 = 01$ to $S_0 = 00$ if input $X = 1$.

As per input, state goes from $S_2 = 10$ to $S_3 = 11$ if input $X = 0$ and state $S_2 = 10$ to $S_1 = 01$ if input $X = 1$.

As per input, state goes from $S_3 = 11$ to $S_0 = 00$ if input $X = 0$ and state $S_3 = 11$ to $S_2 = 10$ if input $X = 1$.

ASM chart is shown in Figure 8.95.

EXAMPLE 8.33 Develop an ASM chart for a controllable wave generator that outputs any one of the four waveforms, as shown in Figure 8.96.

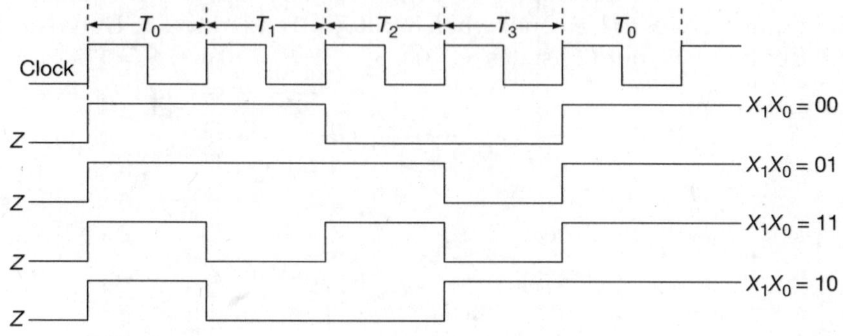

FIGURE 8.96 | Timing diagram representing state changes

SOLUTION The controllable waveform generator generates one of the waveforms, as shown in Figure 8.96, according to inputs X_1, X_0. The block diagram of a controllable wave form generator is shown in Figure 8.97. It has two inputs X_1, X_0 and one output, Z.

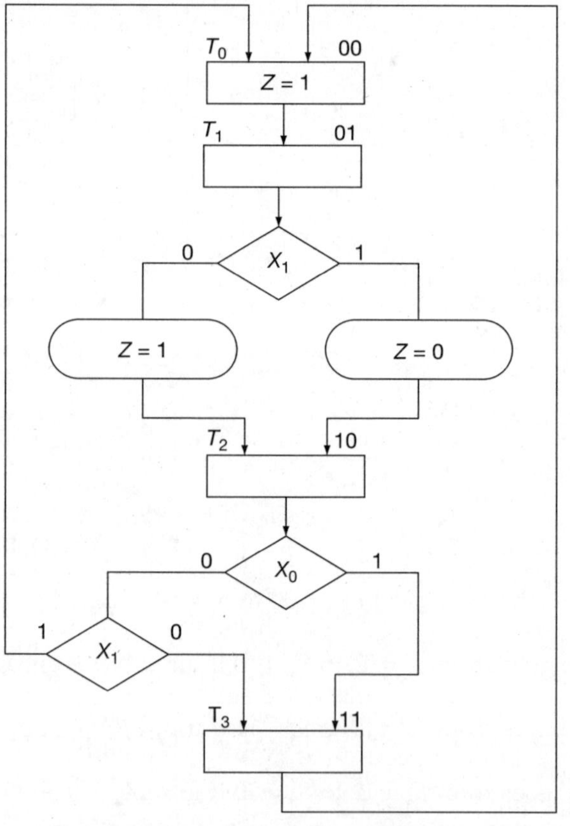

FIGURE 8.97 | ASM chart for Figure 8.96

(i) *State T_0*: In state S_0 corresponding to the first clock pulse, all the waveforms are high. Hence, the output of the waveform generator (Z) should be 1 irrespective of the values of X_1 and X_0.
(ii) *State T_1*: This state corresponds to the second clock pulse. Figure 8.96 shows that the output $Z = 1$ for the input when X_1 is 0 and the output $Z = 0$ for the input when X_1 is 1.
(iii) *State T_2*: This state corresponds to the third clock pulse. Figure 8.96 shows that the output $Z = 1$ for the input when X_0 is 1 and the output $Z = 0$ for the input when X_0 is 0.
(iv) *State T_3*: This state corresponds to the fourth clock pulse. Figure 8.96 shows that the output $Z = 1$ for the input when X_1X_0 is 10 and the output $Z = 0$ for other inputs.

ASM chart is shown in Figure 8.97.

8.12 | REALIZATION OF ASM CHARTS

An ASM chart provides all the information necessary to design a digital system. The requirements of the design of the data section are given inside the state and conditional boxes (outputs). The control logic is determined from the decision boxes and the state transitions. The different techniques used for synthesizing the circuits are described using ASM charts. The set techniques are listed as follows:

- Traditional synthesis using flip-flops and decoders
- Multiplexer controller method for synthesis
- One shot method of synthesis
- ASM synthesis using ROM/PLA devices

For simple controllers the traditional method is suitable, whereas larger designs are synthesized using one of the other three methods. In the following section, the first two methods are discussed while the other two are beyond the scope of this book.

8.12.1 | Traditional Synthesis from an ASM Chart

The ASM charts resemble a state diagram where each state box represents a state. The state diagram can be converted in to a state table from the sequential circuit of the controller.

The steps followed in the traditional synthesis are same as that of a sequential circuit. These steps are:

1. Assign binary values to each state in the ASM chart.
2. Obtain the state table from the ASM chart in a suitable form (considering don't-care input conditions).
3. Determine the number of flip-flops required and assign a letter symbol to each.
4. Choose the type of flip-flop. D-flip-flop is convenient to use.
5. Using map or other simplification method, derive the flip-flop excitation functions and circuit output functions.
6. Draw the logic diagram.

EXAMPLE 8.34 Realize the ASM chart shown in Figure 8.98 using D-flip-flops and combinational logic.

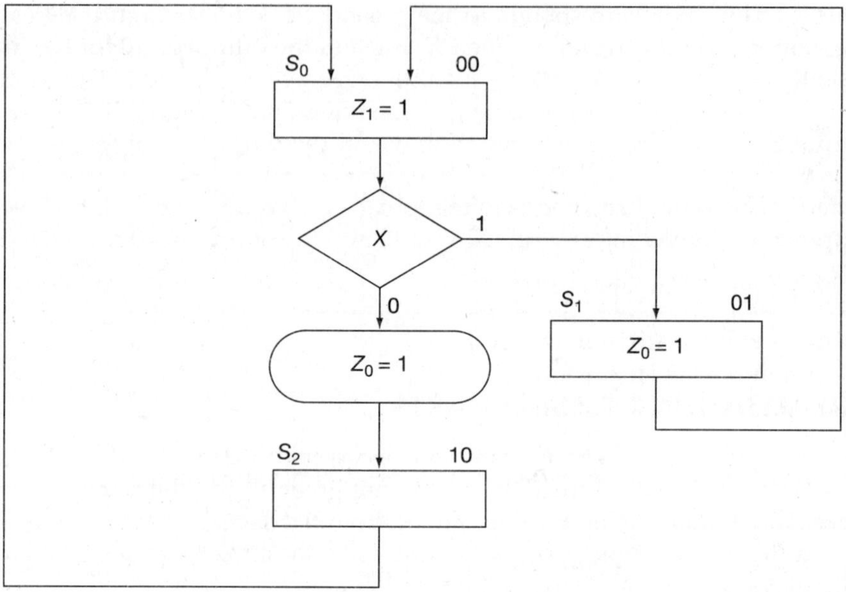

FIGURE 8.98 | ASM chart

SOLUTION The state table is constructed as Table 8.70 from the ASM chart given in Figure 8.98. There are three states so two D-flip-flops are selected. Two outputs Z_1 and Z_0 are obtained. One input, X is given.

TABLE 8.70 | State table

Input	Present State		Next State		Output		Excitations	
X	Q_1	Q_0	Q_1	Q_0	Z_1	Z_0	D_1	D_0
0	0	0	1	0	1	1	1	0
1	0	0	0	1	1	0	0	1
X	0	1	0	0	0	1	0	0
X	1	0	0	0	0	0	0	0

Excitations of D-flip-flops are given below:

$$D_1 = \bar{X}\bar{Q}_1\bar{Q}_0 \tag{8.51(a)}$$

$$D_0 = X\bar{Q}_1\bar{Q}_0 \tag{8.51(b)}$$

Outputs are given below in the following standard SOP form:

$$Z_1(X,Q_1,Q_0) = \sum m(0,4)$$

$$Z_0(X,Q_1,Q_0) = \sum m(0,1,5)$$

Outputs are simplified using K-maps given in Figures 8.99(a) and 8.99(b) and are given below

$$Z_1 = \bar{Q}_1\bar{Q}_0 \quad (8.52(a))$$

$$Z_0 = \bar{X}\bar{Q}_1 + \bar{Q}_1 Q_0 \quad (8.52(b))$$

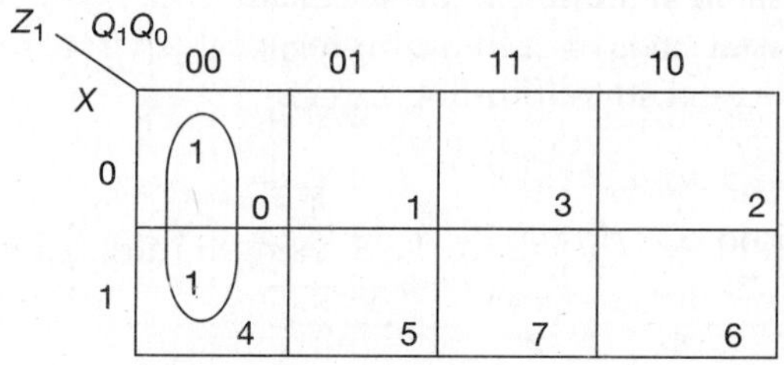

FIGURE 8.99(a) | K-map

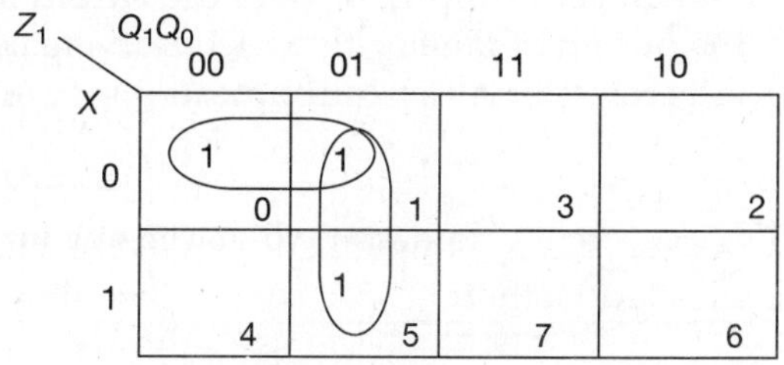

FIGURE 8.99(b) | K-map

The logic circuit diagram is given in Figure 8.100.

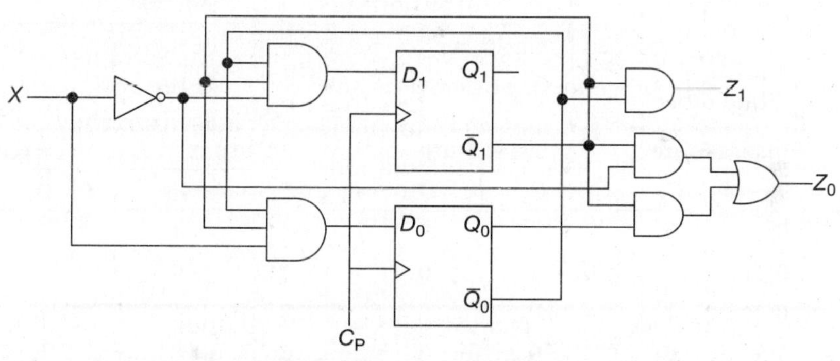

FIGURE 8.100 | Logic diagram

8.12.2 | Multiplexer Controller Method

This is a simple method of synthesizing the combinational logic for any controller having the following desirable features:

1. It produces a design which has a direct correspondence with the algorithm.
2. It is a standard method which can be applied to any ASM chart.
3. It needs the complete ASM chart for its realization.
4. It is very easy to learn and implement.

In this method, the design requires a regular pattern of three levels of components.

1. The first level contains multiplexers which determine the next state by replacing the discrete gates used in the traditional method.
2. The second level consists of registers to hold the present state. It is similar to flip-flops used in the traditional method.
3. The third level has decoders to provide outputs for each control state by replacing the combinational circuit using gates in the traditional method.

This method uses a multiplexer for the input to each state flip-flop. Each multiplexer produces a new input to its state flip-flop. The set of multiplexers, each producing an input to its respective flip-flop, gives the output for the next state. The present state code, which is the output of the flip-flops, feeds the select input lines of each multiplexer which in turn selects the appropriate multiplexer input for the next state (flip-flop inputs).

EXAMPLE 8.35 Implement the ASM chart shown in Figure 8.98 using multiplexer controlled method.

SOLUTION The state table is constructed as Table 8.71 from the ASM chart given in Figure 8.98. There are three states so two D-flip-flops and two multiplexers are required. As there are three states, so four-line to single-line multiplexers are required. Two outputs Z_1 and Z_0 are obtained. One input, X is given.

TABLE 8.71 | State table

Input	Present State		Next State		Output		Excitations	
X	Q_1	Q_0	Q_1	Q_0	Z_1	Z_0	D_1	D_0
0	0	0	0	1	1	1	0	1
1	0	0	1	0	1	0	1	0
X	0	1	0	0	0	1	0	0
X	1	0	0	0	0	0	0	0

TABLE 8.72 | MUX input table of next states Q_1

X \ Q_1Q_0		00	01	10	11
		$\bar{Q}_1\bar{Q}_0$	$\bar{Q}_1Q_0$	$Q_1\bar{Q}_0$	Q_1Q_0
		I_0	I_1	I_2	I_3
0	$\bar{X}$	(0)	1	2	3
1	X	4	5	6	7
Input to MUX		$\bar{X}$	0	0	0

TABLE 8.73 | MUX input table of next states Q_0

X \ Q_1Q_0		00	01	10	11
		$\bar{Q}_1\bar{Q}_0$	$\bar{Q}_1Q_0$	$Q_1\bar{Q}_0$	Q_1Q_0
		I_0	I_1	I_2	I_3
0	$\bar{X}$	0	1	2	3
1	X	(4)	5	6	7
Input to MUX		X	0	0	0

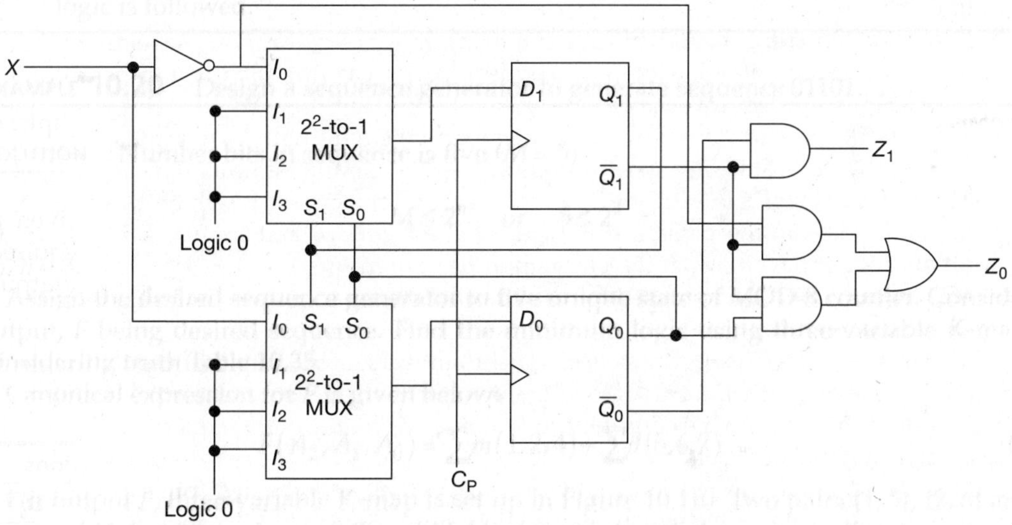

FIGURE 8.101 | Logic diagram

Inputs for multiplexer for giving next state, Q_1 is given in Table 8.72 for input, X and selection lines are represented by present states, Q_1 and Q_0. Similarly, Inputs for Multiplexer for giving next state, Q_0 is given in Table 8.73 for input, X and selection lines are represented by present states, Q_1 and Q_0. Circuit diagram is given in Figure 8.101.

SUMMARY

- An un-clocked flip-flop is called a latch because the output of the flip-flop latches on to a 1 or a 0 immediately after the input is applied.
- *Asynchronous sequential circuit:* It is a sequential circuit whose behaviour depends on the change of input signal at any time. A combinational circuit with feedback is termed as asynchronous sequential circuit.
- *Synchronous sequential circuit:* It is a sequential circuit, whose behaviour is defined from the knowledge of its signal at discrete instants of time. The synchronization is achieved by a timing device known as a system clock which generates a train of clock pulses.

- ***Critical race:*** In an asynchronous sequential circuit when two or more internal state variables have to change simultaneously due to change in input, then due to different delay paths, it is very difficult to predict the state variable which will change first. Due to this, race condition is said to exist and the circuit may go to different next-states depending upon the delays involved in different paths.
- ***Cycle:*** A cycle occurs is an asynchronous sequential circuit when multiple state transitions occur without changing the inputs, taking the circuit from one unstable state to another unstable state and finally reaching a stable state.
- Flip-flops are used for data storage, counting, frequency division, parallel-to-serial and serial-to-parallel data conversion, etc.
- ***Flow table:*** A tabular method for showing state transitions in asynchronous sequential circuits.
- ***Essential hazard:*** Inherent hazard occurring in an asynchronous sequential circuit due to different delay in different feedback paths, causing incorrect transition for a change in single input variable.
- ***Fundamental mode:*** An asynchronous sequential circuit with level inputs.
- ***Internal state:*** A state in asynchronous sequential circuit. Same as secondary state.
- ***Mealy Sequential Circuit:*** The clocked synchronous circuit in which the output, Y depends only on the present state, $Q(t)$ of the flip-flops and inputs, X is known as Mealy sequential circuit. It depends upon present inputs.
- ***Merger diagram:*** A graph used for the possible merging of states in an asynchronous sequential circuit.
- ***Moore Sequential Circuit:*** The clocked synchronous circuit in which the output, Y depends only on the present state, $Q(t)$ of the flip-flops is known as Moore sequential circuit.
- ***Non-critical race:*** A race condition in which an asynchronous sequential circuit will reach the same stable state.
- Non-gated latches are called asynchronous latches and clocked (but not edge triggered) latches are called synchronous latches.
- ***Primitive flow table:*** A flow table in which every row has only one stable state.
- ***Pulse mode:*** An asynchronous sequential circuit with pulse inputs.
- ***Race:*** A condition occurring in an asynchronous sequential circuit due to different path delays when transition takes place requiring more than one state to change simultaneously.
- ***State assignment:*** Assigning binary values to states in a sequential circuit.
- ***State diagram:*** It is a graphical representation of a sequential circuit indicating the states, input values, output values, and directed arcs for state transitions.
- ***State reduction:*** Process of reducing the states by eliminating the redundant states in sequential circuits.
- ***State table (or State transition table):*** A tabular form of representing the behaviour of a sequential circuit.
- Synchronous circuits are also known as clocked-sequential circuit.
- ***Transition table:*** A tabular form of representing the behaviour of a sequential circuit.

Characteristic equations of flip-flops:

SR-Flip-Flop: $Q(t+1) = S + \bar{R}Q(t)$ provided $SR = 0$

D-Flip-Flop: $Q(t+1) = D$

J-K-Flip-Flop: $Q(t+1) = J\bar{Q}(t) + \bar{K}Q(t)$

T-Flip-Flop: $Q(t+1) = T \oplus Q(t)$

Excitation of flip-flops

Outputs		SR-excitation		JK-excitation		D-excitation	T-excitation
Q(*t*)	*Q*(*t* + 1)	*S* (Set)	*R* (Reset)	*J* (Set)	*K* (Reset)	*D* (Data)	*T* (Toggle)
0	0	0	X	0	X	0	0
0	1	1	0	1	X	1	1
1	0	0	1	X	1	0	1
1	1	X	0	X	0	1	0

MULTIPLE CHOICE QUESTIONS

8.1 A flip-flop is a
(a) combinational circuit
(b) memory element
(c) arithmetic element
(d) memory and arithmetic element

8.2 In a sequential circuit the output at any instant depends on
(a) present state
(b) past inputs only
(c) past outputs only
(d) past output and present inputs

8.3 A combinational circuit with feedback is termed as
(a) asynchronous sequential circuit
(b) combinational circuit
(c) synchronous sequential circuit
(d) none of the above

8.4 A sequential circuit whose behaviour is defined from the knowledge of its signal at discrete instants of time is known as
(a) asynchronous
(b) synchronous
(c) fundamental mode of asynchronous
(d) none of the above

8.5 An asynchronous sequential circuit with level inputs is known as
(a) pulse mode
(b) fundamental mode
(c) synchronous mode
(d) none of the above

8.6 An asynchronous sequential circuit with pulse inputs is known as
(a) pulse mode
(b) fundamental mode
(c) synchronous mode
(d) none of the above

8.7 The clocked synchronous circuit in which the output depends only on the present state of the flip-flops and inputs is known as
(a) Mealy model
(b) Moore model
(c) Pulse mode asynchronous
(d) None of the above

8.8 A race condition in which an asynchronous sequential circuit reaches the same stable state.
(a) Critical race
(b) Non-critical race
(c) Cycle
(d) Static hazard

8.9 A flow table in which every row has only one stable state is known as
(a) primitive flow table
(b) transition table
(c) state table
(d) none of the above

8.10 The clocked synchronous circuit in which the output depends only on the present state of the flip-flops is known as
(a) Mealy model
(b) Moore model
(c) Pulse mode asynchronous
(d) None of the above

Answers

8.1 (b)	8.2 (d)	8.3 (a)	8.4 (b)	8.5 (b)	8.6 (a)	8.7 (a)
8.8 (b)	8.9 (a)	8.10 (b)				

QUESTIONS

8.1 Differentiate synchronous and asynchronous sequential circuits.
8.2 Explain the fundamental and pulse mode asynchronous sequential circuits.
8.3 Describe the design procedure for asynchronous sequential circuits.
8.4 Explain the problems in asynchronous circuits.
8.5 Describe cycles in asynchronous sequential circuits.
8.6 What do you mean by critical and non-critical races?
8.7 What are the different types of hazards in asynchronous circuits?
8.8 Differentiate static-0, static-1 and dynamic hazards with waveform.
8.9 Define races in asynchronous sequential circuits.

PROBLEMS

8.1 Design a clocked synchronous sequential circuit that checks a serial data line for odd parity. The circuit has two inputs, SYNC and DATA output ERROR using D-flip-flops.

8.2 Design a sequential circuit using D-flip-flops to find each of the sequence in a serial input signal (I). Whenever the sequence is found the output (Z) should be 1. (a) 11 0 (b) 0011 (c) 1010

8.3 Design an asynchronous sequential circuit with two inputs, I_1 and I_2, and two outputs, Z_1 and Z_2, for the following specifications:
When $I_1I_2 = 00$, the output $Z_1Z_2 = 00$. When $I_1 = 0$ and I_2, changes from 0 to 1 the output $Z_1Z_2 = 01$. When $I_2 = 1$ and I_1 changes from 0 to 1, the output $Z_1Z_2 = 10$. Otherwise, the output is constant.

8.4 Obtain a static hazard-free asynchronous circuit for the followings switching functions

$f = E(0, 2, 4, 5, 8, 10, 14)$

8.5 For the state diagram given in following, design the circuit using
   (a) T-flip-flops,
   (b) SR-flip-flops and
   (c) JK-flip-flops

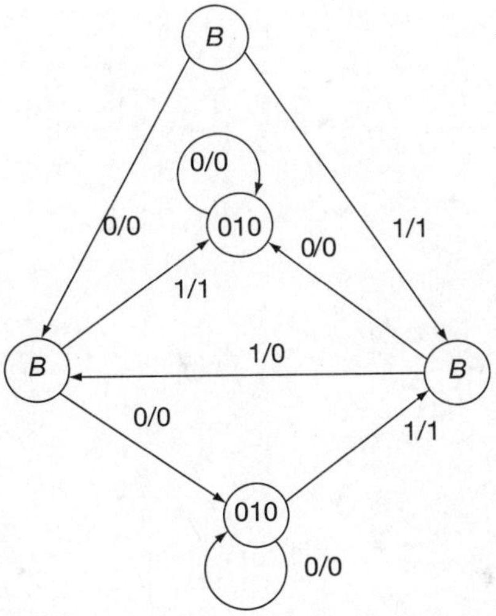

PROBLEM 8.5 | State diagram

9

Registers

CHAPTER OBJECTIVES

The main goal of this chapter is to impart knowledge about the registers. Readers will be able to discuss the following aspects in this chapter:

- Registers, registers with parallel load
- Serial-in, serial-out shift register
- Serial-in, parallel-out shift register
- Parallel-in, parallel-out shift register
- Parallel-in, serial-out shift register
- Bidirectional serial shift register
- Universal registers
- Ring counter, Johnson counter
- Serial adder
- Sequence generator and detector

9.1 | INTRODUCTION

In a computer, a register is a smallest data holding component that is a part of a computer processor. A register may hold a computer instruction, a storage address, or any kind of data (such as a bit sequence or individual characters). A, high-speed storage area within the central processing unit (CPU) is called register. Datum is stored in a register before it can be processed. For instance, for the addition of two numbers, both numbers are stored in registers, and the result is also placed in a register. The number of registers that a digital processing unit has and the size of each register that is number of bits determine the power and speed of a digital processor. For example, a 32-bit central processing unit (CPU) is one in which each register is of 32-bits size. Therefore, each CPU instruction is able to process 32-bits of data. Usually, the movement of data in and out of registers is completely transparent to users, and even to programmers. However, only assembly language programs can manipulate registers. In high-level languages, the compiler translates high-level operations into low-level operations that access registers.

Registers are data storage devices. A register is a group of binary cells (D-flip-flop) suitable for holding binary information. A group of cascaded flip-flops used to store related bits of information is known as a register. Registers can process parallel and serial types of digital information. Parallel registers take in multiple bits at a time. A register is constructed from multiple flip-flops connected to each other used to store multiple bits of data. Having '*n*' flip-flops, '*n*' bits of data can be stored. So, the size of register is defined by the number of bits can be stored in it. Shift Registers are the simplest serial interfaces. A serial shift register is able to shift the data it contains by one bit either to the left or to the right. Shift registers are used to transform serial data to parallel, and parallel to serial data. Shift Registers are used to make larger state machines like counters and universal asynchronous receiver/transmitter (UART). Registers process information in first-in-first-out (FIFO) and last-in-first-out (LIFO) ways. The simplest register is a FIFO. It is only one level deep and one bit wide. It is basically a single D-type flip-flop. A FIFO (first-in-first-out) is digital device that moves data in the same way as is followed in a queue. A LIFO (last-in-first-out) is like a FIFO only the data comes out in the reverse order it came in. Data is processed like a stack data structure. Imagine putting items onto a stack, the first item taken off would be at the top of the stack. Thus, the last item in (item at the top of the stack), would be the first item out.

9.2 | REGISTERS

Register is a group of binary cells suitable to hold binary information. Register is a group of flip-flops. Each flip-flop is capable of storing one bit of information. An n-bit register consists of a group of n-flip-flops. Figure 9.1 shows a four-bit register using D-flip-flop. This type of register is also known as buffer register. The common clock pulse, C_P enables all the four flip-flops to transfer input to flip-flops. The output lines are used to obtain output of register. The transfer of new binary information into a register is termed as loading the register. If all the bits of the register are loaded simultaneously with a single clock pulse, then the loading is done in parallel.

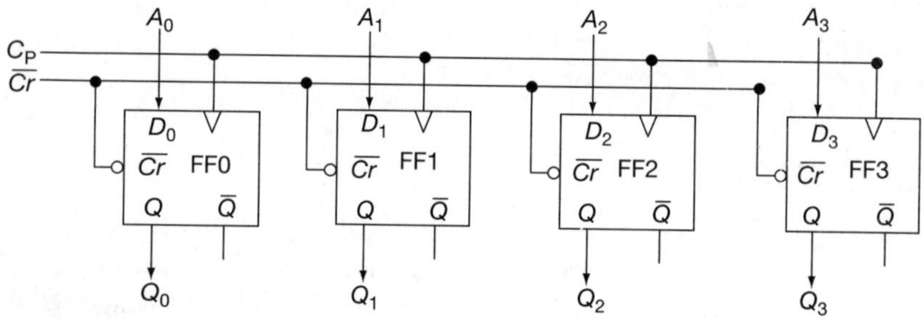

FIGURE 9.1 | Four-bit register

9.2.1 | Four-bit Latch

Triggering of latch in the register plays an important role. A group of latches is called four-bit latch if it is sensitive to pulse duration. The characteristic table of gated D-latch is given in Table 9.1.

Working of D-flip-flops is explained below:

- For a 1 (High) clock pulse, input information is stored and is available to output lines.
- For a 0 (Low) clock pulse, output does not change. No change in output happens if clock pulse changes from 0 to 1 or 1 to 0.
- When clear signal is active 0, the register is cleared to all 0's prior to its gated operation. Clear is asynchronous input.

The flip-flops are sensitive to the pulse duration, and the-register is enabled as for as long as the clock, C_P is 1 (High). A register that responds to the pulse duration is commonly

TABLE 9.1 | Characteristics of gated D-latch

Inputs			Outputs	Remarks
C_r Clear	D (Data)	C_P (Gated)	$Q(t+1)$	
1	0	High (1)	0	Input, *D* is stored and available as output, *Q*
1	1	High (1)	1	Input, *D* is stored and available as output, *Q*
1	X	Low (0)	$Q(t)$	No change in output, *Q* or holds data
0	X	Low (0) or High (1)	0	Cleared to zero (or reset to zero)

called a gated latch. Latches are used as temporary storage of binary information that is to be transferred to an external destination.

Figure 9.2 shows a four-bit buffer register using D-latches with load control. This type of register is also known as latch. The common clock pulse, C_P enables all the four latches to save inputs to latches. The output lines are used to obtain output of latch.

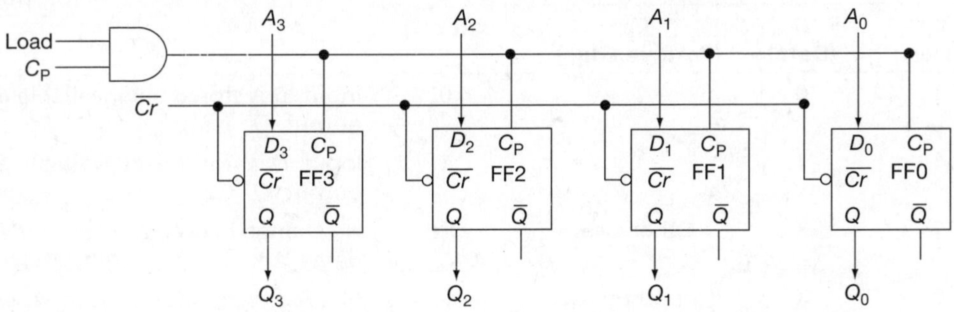

FIGURE 9.2 | Four-bit parallel-in-parallel-out buffer with load control

Table 9.2 gives the output obtained from four-bit data buffer. When clock is gated high then input of each latch is stored simultaneously and is available as output of buffer. The action is known as loading the buffer. The load operation is control by inhibiting the clock pulse using AND gate. The register is cleared by applying asynchronous clear $\overline{Cr}$ signal as active low. The change in buffer depends on clock pulse duration.

Limitation: Such latches are not used in the design of sequential circuits that have feedback connections.

9.2.2 | Register

A group of flip-flops is called register if it is sensitive to pulse transition. A flip-flop can be used in the design of clocked sequential circuits provided it is sensitive to the pulse transition rather than the pulse duration. So, the flip-flops in the register should be of the edge-triggered or master-slave type. A group of flip-flops sensitive to pulse transition is

TABLE 9.2 | Four-bit latch (buffer)

Control			Input of Latches				Output of Latches				Action
			FF0	FF1	FF2	FF3	FF0	FF1	FF2	FF3	
C_p, Clock	Load	$\overline{Cr}$	D_0	D_1	D_2	D_3	Q_0	Q_1	Q_2	Q_3	
High	1	1	A_0	A_1	A_2	A_3	A_0	A_1	A_2	A_3	Load
Low	1	1	X	X	X	X	A_0	A_1	A_2	A_3	No change in previous output
High	0	1	X	X	X	X	A_0	A_1	A_2	A_3	No change in output
X	X	0	X	X	X	X	0	0	0	0	Clear

called a register. A register can replace a latch. But latch can replace register by ensuring that outputs from a latch never go to other flip-flop inputs that are triggered with the same common clock pulse. The characteristic table of clocked D-flip-flop is given in Table 9.3.

TABLE 9.3 | Characteristics of clocked D-flip-flop

Inputs			Outputs	Remarks
C_r Clear	D (Data)	C_P (Positive Edge)	$Q(t+1)$	
1	0	↑	0	Input, *D* is stored and available as output, *Q*
1	1	↑	1	Input, *D* is stored and available as output, *Q*
1	X	Otherwise	$Q(t)$	No change in output, *Q* or holds data
0	X	↑ or otherwise	0	Cleared to zero (or reset to zero)

Working of D-flip-flop is explained below:

- As clock changes from 0 to 1 (rising or positive edge transition), input information is stored and is available to output lines.
- As clock changes from 1 to 0 (falling or negative edge transition), output does not change. Output does not change also if clock remains either at 0 (low level) or 1 (high level).
- When clear signal is active 0, the register is cleared to all 0's prior to its clocked operation. Clear is asynchronous input.

The output obtained from four-bit register is given in Table 9.4. When clock pulse is positive edge going then input of each flip-flop is stored and is available as output of register. The action is known as loading the register. The register is cleared by applying asynchronous clear $\overline{Cr}$ signal as active low. The change in register depends on clock pulse.

Limitation: The input is stored at every rising edge transition of pulse train. Then there is need to control the load operation of registers. The clock pulse is inhibited from the clock pulse, C_P terminal if the content of the register are not allowed to change (Figure 9.2 by

TABLE 9.4 | Four-bit register

Control		Input of Flip-flop				Output of Flip-flops				Action
		I	II	III	IV	I	II	III	IV	
Clock	$\overline{Cr}$	D_0	D_1	D_2	D_3	Q_0	Q_1	Q_2	Q_3	
↑	1	A_0	A_1	A_2	A_3	A_0	A_1	A_2	A_3	Load
		X	X	X	X	A_0	A_1	A_2	A_3	No change in previous output
X	0	X	X	X	X	0	0	0	0	Clear

changing gated pulse to clocked pulse). By adding the gates between clock generator and clock inputs to flip-flop produces propagation delay.

9.3 | REGISTER WITH PARALLEL LOAD

On applying input clock pulse, C_P, the register (Figure 9.2) will load all four inputs in parallel. One method to load the register is to control the clock pulse which causes propagation delay. Another method to load the register is to control the load and to hold (retain) the previous output of D-flip-flop. There is need to select the one out of two inputs, i.e. input to register or previous stored value. A multiplexing circuit is added to input of D-flip-flop. A multiplexer (2-to-1 line) selects either the input to flip-flop to load or the previous output of flip-flop to retain (save). Figure 9.3 shows diagram of 2-to-1 line multiplexer using logic gates and Figure 9.4 shows the block diagram of 2-to-1 line multiplexer. Figure 9.5 shows the four-bit parallel register. Load line is acting as control line of multiplexer. Table 9.5 explains the operation of four-bit parallel register. When load signal, L is 0 (Low) then multiplexer selects the input of each flip-flop and is saved in them. For a 1 (High) load signal, L, multiplexer selects the output of each flip-flop to save.

Circuit diagram of four-bit register is redrawn in Figure 9.6 using AND-OR-NOT gates where two NOT gates are saved. Input to D-flip-flop-I is given as below:

$$D_i = A_i\overline{L} + Q_iL \quad (I = 0, 1, 2, 3) \tag{9.1}$$

When input load, L is 0 then $D_i = A_i$ and performs load operation. When input load, L is 1 then, $D_i = Q_i$, output of each flip-flop is hold.

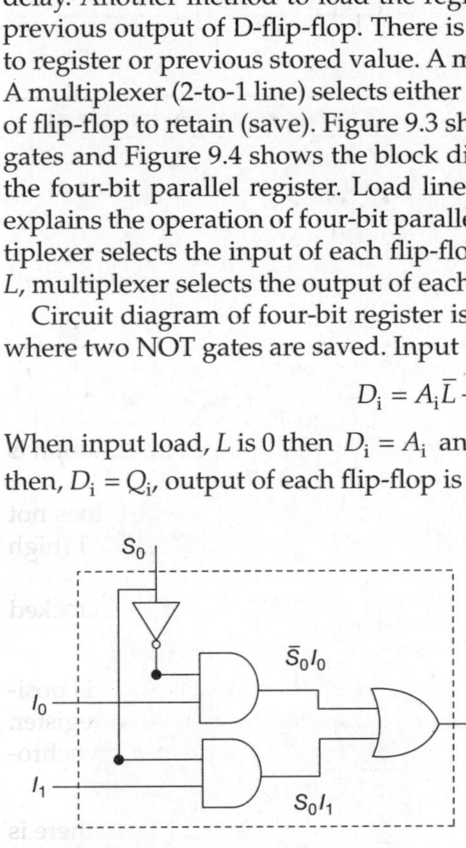

FIGURE 9.3 | Two-to-one MUX

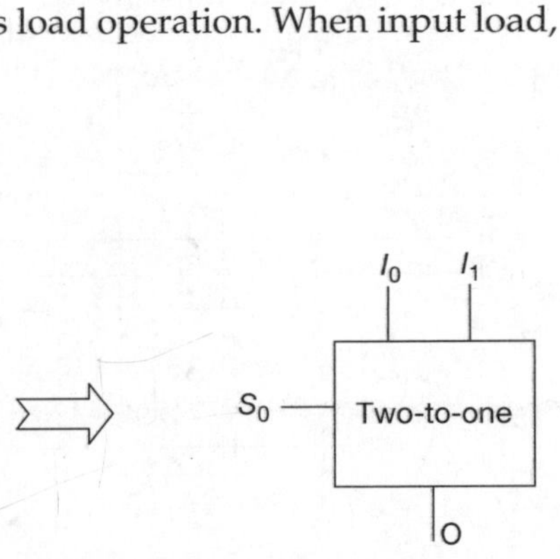

FIGURE 9.4 | Block diagram of 2-to-1 MUX

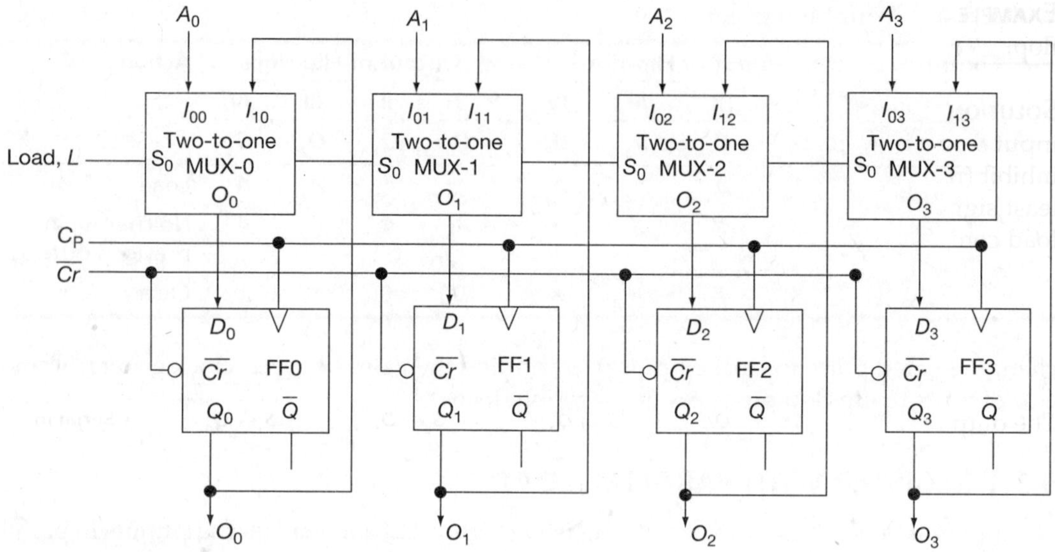

FIGURE 9.5 | Four-bit parallel buffer register

TABLE 9.5 | Parallel load register or parallel buffer register

Clock	Load	$\overline{Cr}$	Input	Output	Action
↑	0	1	$A_3A_2A_1A_0$	$A_3A_2A_1A_0$	Load
↑	1	1	$Q_3Q_2Q_1Q_0$	$Q_3Q_2Q_1Q_0$	Hold
X	X	0	XXXX	0000	Clear

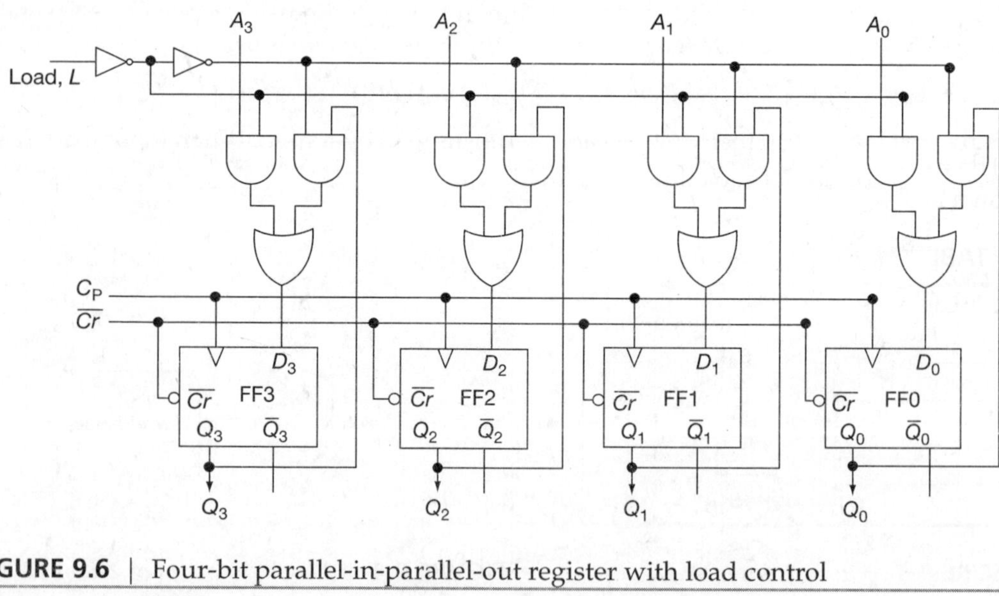

FIGURE 9.6 | Four-bit parallel-in-parallel-out register with load control

EXAMPLE 9.1 Draw a circuit diagram of four-bit register with load control using S-R-flip flop.

SOLUTION S-R-flip-flop is used as D-flip flop if input, *R* of SR-flip-flop is complement of input *S* of SR-flip-flop. Load control is having AND operation with input lines to select or inhibit (restrict). Diagram of four-bit register is shown in Figure 9.7. Flip-flop I is used for least significant bit (LSB) of register. Inputs of SR-flip-flops are given below having input load control, *L*.

$$R_i = L\bar{A}_i \ (i = 0, 1, 2, 3) \tag{9.2}$$

$$S_i = LA_i \ (i = 0, 1, 2, 3) \tag{9.3}$$

The output of SR-flip-flop is given in Table 9.6 in various conditions.

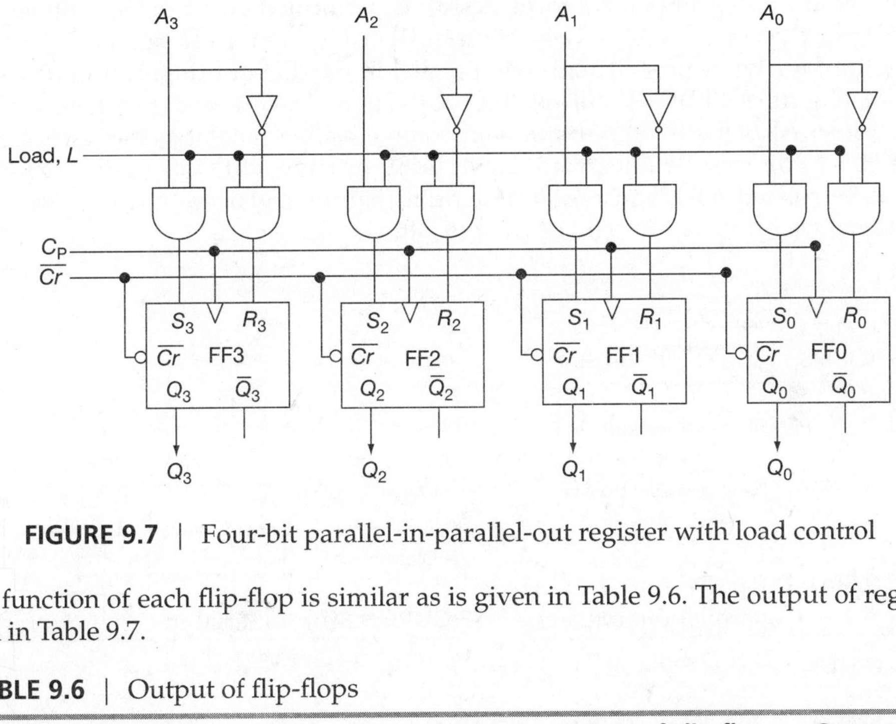

FIGURE 9.7 | Four-bit parallel-in-parallel-out register with load control

The function of each flip-flop is similar as is given in Table 9.6. The output of register is shown in Table 9.7.

TABLE 9.6 | Output of flip-flops

Load L	Clock Pulse C_P	Clear C_r	Input of Register A_i	Inputs of Flip-flop S_i	R_i	Output, Q_i
0	↑	1	X	0	0	Hold output
1	↑	1	0	0	1	Reset, 0
1	↑	1	1	1	0	Set, 1
X	otherwise	0	X	X	X	Clear to 0

TABLE 9.7 | Parallel load register or parallel buffer register

Clock	Load	$\overline{Cr}$	Input	Output	Action
↑	1	1	$A_3A_2A_1A_0$	$A_3A_2A_1A_0$	Load
↑	0	1	$Q_3Q_2Q_1Q_0$	$Q_3Q_2Q_1Q_0$	Hold
X	X	0	XXXX	0000	Clear

9.4 | SHIFT REGISTER

A number of flip-flops connected (cascaded) together in such a way that data can be shifted into and shifted out of them is called a shift register. It can be achieved if output of one flip-flop is connected to input of second flip-flop. Data can be shifted into or out of the register either in serial form or in parallel form. A shift is performed on occurrence of valid clock pulse. Register can be cleared to 0 asynchronously using input clear signal.

There are four types of shift registers: parallel-in-parallel-out (Figure 9.8(a)); serial-in-serial-out (Figure 9.8(b)); serial-in-parallel-out (Figure 9.8(c)); and parallel-in-serial-out (Figure 9.8(d)). All of these configurations are commercially available as transistor-transistor logic (TTL); medium-scale integrated circuit (MSI) or large-scale integrated circuits (LSI). Data can be rotated left (Figure 9.8(e)) or right (Figure 9.8(f)). Data can be shifted in a bidirectional way from left to right or right to left.

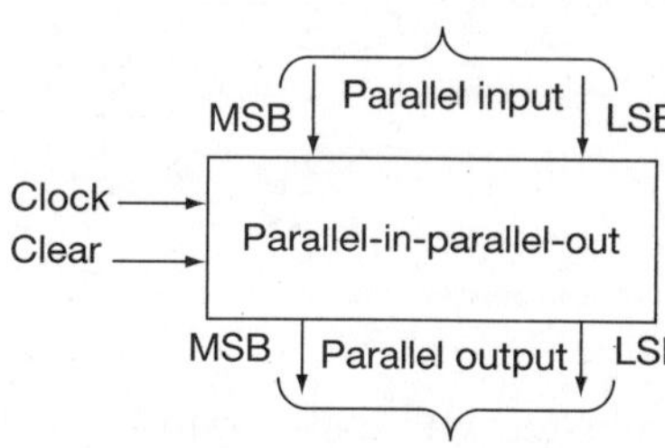

FIGURE 9.8(a) | Parallel-in-parallel-out register

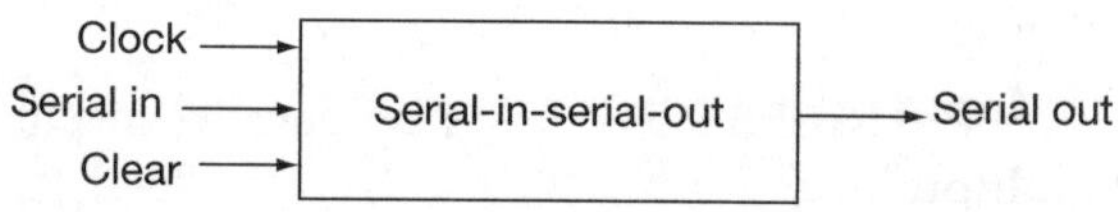

FIGURE 9.8(b) | Serial-in-serial-out register

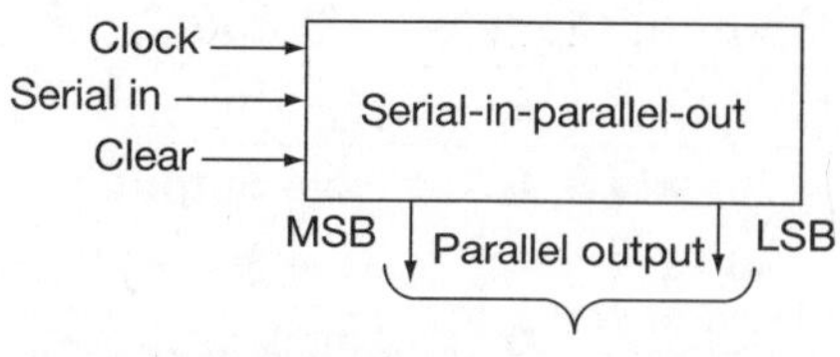

FIGURE 9.8(c) | Serial-in-parallel-out shift register

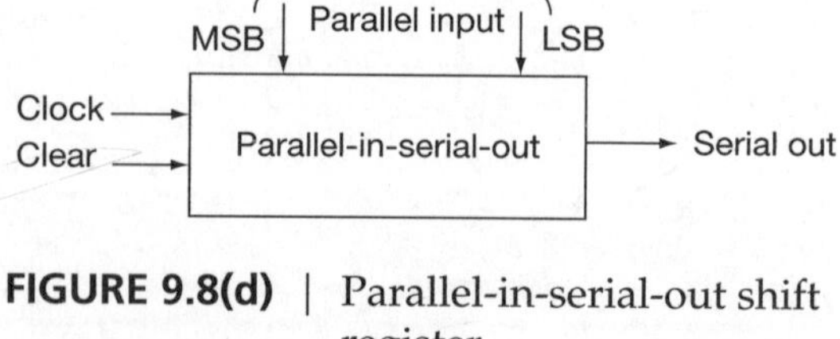

FIGURE 9.8(d) | Parallel-in-serial-out shift register

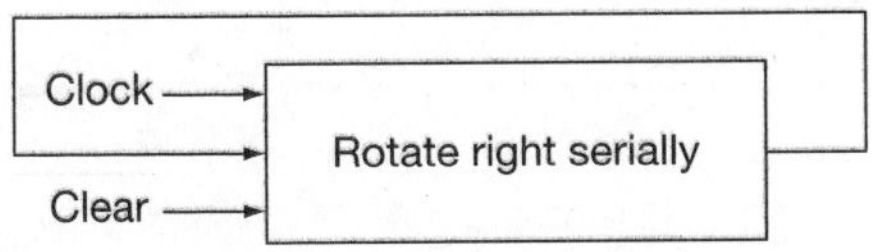

FIGURE 9.8(e) | Rotate right shift register

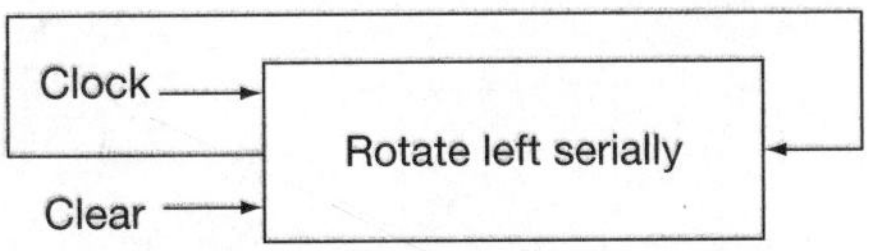

FIGURE 9.8(f) | Rotate left shift register

9.5 | SERIAL-IN, SERIAL-OUT SHIFT REGISTER

Binary data can be shifted serially towards left or right direction. Left shift multiplies the number by 2. Left shift performs shift from least significant bit (LSB) towards most significant bit (MSB) position. Bit having smallest, 2^0 weightage is called LSB. Bit having largest 2^n weightage is called MSB, where n is the number of bits in the binary number. Overflow in register occurs as per the size of register. Right shift divides the number by 2. Right-shift moves the bit from MSB to LSB position. Zero is fed as serial input.

9.5.1 | Left-shift Serial-in, Serial-out Register with D-flip-flop

Four D-flip-flops are cascaded serially by input. Output of one flip-flop is fed as input to another D-flip-flop. Figure 9.9 shows serial-in-serial-out (SISO) four-bit register that performs the shift towards left direction. Serial input, D_{in} is fed as input, D_0 of first flip-flop, FF0. Output Q_0 of first flip-flop, FF0 is fed as input D_1, to second flip-flop, FF1. Similarly, output Q_1 of second flip-flop, FF1 is fed as input D_2, to third flip-flop, FF2 and output Q_2 of third flip-flop, FF2 is fed as input D_3, to fourth flip-flop, FF3. Output, Q_3 of fourth flip-flop gives serial out bit, D_{out}. Q_0 is the LSB and Q_3 is the MSB. Lower weighted bit, Q_0 is shifted to the position of next higher weighted bit, Q_1.

Input of FF0 flip-flop is $D_0 = D_{in}$,

Input of FF1 flip-flop is $D_1 = Q_0$,

Input of FF2 flip-flop is $D_2 = Q_1$,

Input of FF3 flip-flop is $D_3 = Q_2$ and

Serial output is $D_{out} = Q_3$

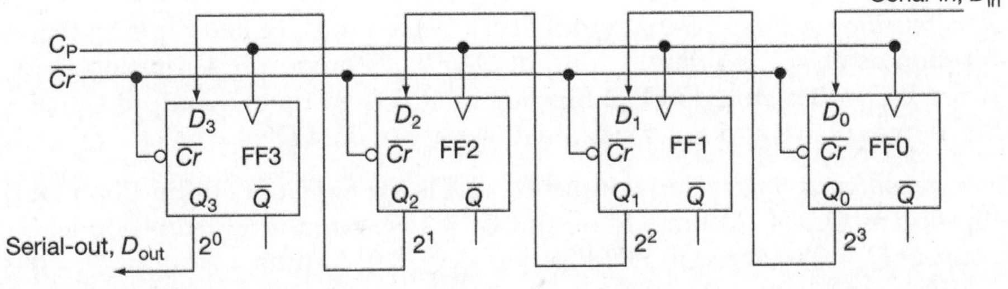

FIGURE 9.9 | Four-bit left serial-in-serial-out register

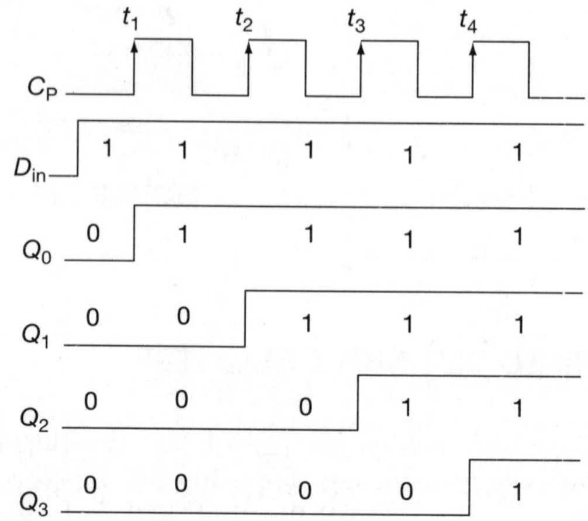

FIGURE 9.10 | Serial-in-serial-out left shift

All the flip-flops are initially cleared to 0 by giving active low logic to clear, $\overline{Cr}$ asynchronously. Timing diagram of shift-left operation is shown in Figure 9.10. Let the serial input, D_{in} fed to FF0-flip-flop is shown in timing diagram shown in Figure 9.10. Change in state of flip-flop takes place during rising edge transition of clock at time t_1, t_2, t_3 and t_4. Initially stored data in the register is $Q_3Q_2Q_1Q_0 = 0000$. The shifting operation serial in serial out to left is given in Table 9.8.

TABLE 9.8 | Shift-left operation

Time of Clock, C_P	$D_{out} \leftarrow Q_3$ (2^3)	$D_3 \leftarrow Q_2$ (2^2)	$D_2 \leftarrow Q_1$ (2^1)	$D_1 \leftarrow Q_0$ (2^0)	Serial Input $D_0 \leftarrow D_{in}$
t_0	0	0	0	0	–
t_1	0	0	0	1	1
t_2	0	0	1	1	1
t_3	0	1	1	1	1
t_4	1	1	1	1	1

Before rising edge transition at time t_1: The register stores $Q_3Q_2Q_1Q_0$ as 0000. LSB appears at Q_0.

Rising edge transition at time t_1: Serial input D_{in} as 1 is fed to D_0 of FF0-flip-flop that is stored in FF0-flip-flop as $Q_0 = 1$. At time, t_1, output $Q_0 = 0$ is stored in FF1-flip-flop as $Q_1 = 0$. At time, t_1, output $Q_1 = 0$ is stored in FF2-flip-flop as $Q_2 = 0$. At time, t_1, output $Q_2 = 0$ is stored in FF3-flip-flop as $Q_3 = 0$ and is considered as serial out data D_{out}.

Rising edge transition at time t_2: Serial input D_{in} as 1 is fed to D_0 of FF0-flip-flop that is stored in FF0-flip-flop as $Q_0 = 1$. At time, t_2, output $Q_0 = 1$ is stored in FF1-flip-flop as $Q_1 = 1$. At time, t_2, output $Q_1 = 0$ is stored in FF2-flip-flop as $Q_2 = 0$. At time, t_2, output $Q_2 = 0$ is stored in FF3-flip-flop as $Q_3 = 0$ and is considered as serial out data D_{out}.

Rising edge transition at time t_3: Serial input D_{in} as 1 is fed to D_0 of FF0-flip-flop that is stored in FF0-flip-flop as $Q_0 = 1$. At time, t_3, output $Q_0 = 1$ is stored in FF1-flip-flop as $Q_1 = 1$. At time, t_3, output $Q_1 = 1$ is stored in FF2-flip-flop as $Q_2 = 1$. At time, t_3, output $Q_2 = 0$ is stored in FF3-flip-flop as $Q_3 = 0$ and is considered as serial out data D_{out}.

Rising edge transition at time t_4: Serial input D_{in} as 1 is fed to D_0 of FF0-flip-flop that is stored in FFF0-flip-flop as $Q_0 = 1$. At time, t_3, output $Q_0 = 1$ is stored in FF1-flip-flop as $Q_1 = 1$. At time, t_3, output $Q_1 = 1$ is stored in FF2-flip-flop as $Q_2 = 1$. At time, t_3, output $Q_2 = 1$ is stored in FF3-flip-flop as $Q_3 = 1$ and is considered as serial out data D_{out}.

9.5.2 | Left-shift SISO Register with SR-flip-flop

Four SR-flip-flops are cascaded serially. Output of one flip-flop is fed as input to next flip-flop. Figure 9.11 shows the four-bit register using SR-flip-flop that performs the shift towards left direction. Input is given to SR-flip-flop, FF0 and output of SR-flip-flop, FF0 is given to SR-flip-flop, FF1 and so on. Active high logic is given to clear, $\overline{Cr}$ and preset, $\overline{Pr}$ asynchronously. Active low logic to clear, $\overline{Cr}$ clears the register and active low logic to preset, $\overline{Pr}$ sets the register to (1111).

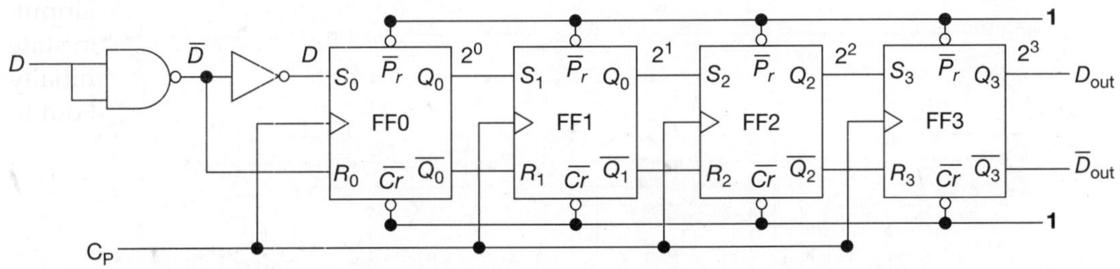

FIGURE 9.11 | SISO 4-bit register performs left shift

Input of SR-flip-flop, FF0 is $S_0 = \overline{\overline{\overline{D.D}}} \Rightarrow S_0 = \overline{\overline{D}} \Rightarrow S_0 = D$ and $R_0 = \overline{D.D} \Rightarrow R_0 = \overline{D}$

Input of SR-flip-flop, FF1 is $S_1 = Q_0$ and $R_1 = \overline{Q_0}$

Input of SR-flip-flop, FF2 is $S_2 = Q_1$ and $R_2 = \overline{Q_1}$

Input of SR-flip-flop, FF3 is $S_3 = Q_2$ and $R_3 = \overline{Q_2}$

Serial output is $D_{out} = Q_3$ and $\overline{D}_{out} = \overline{Q}_3$

Excitation Table 9.9 gives the inputs and outputs of SR-flip-flop, FF0. Excitation Table 9.10 gives the inputs and outputs of SR-flip-flop, FF1. Excitations of SR-flip-flops, FF2 and FF3 are similar to excitation of SR-II-flip-flop.

TABLE 9.9 | Truth table of FF0, flip-flop

Clock C_P	Input D	SR-flip-flop Inputs		Output of SR-flip-flop	
		S_0	R_0	Q_0	$\overline{Q_0}$
↑	1	1	0	1	0
↑	0	0	1	0	1

TABLE 9.10 | Truth table FF1-flip-flop

Clock C_P	SR-flip-flop, Inputs		Output of SR-flip-flop	
	$S_1 = Q_0$	$R_1 = \overline{Q_0}$	Q_1	$\overline{Q_1}$
↑	1	0	1	0
↑	0	1	0	1

9.5.3 | Left-shift SISO Register with Asynchronous Loading

Four D-flip-flops are cascaded serially. Output of one flip-flop is fed as input to next flip-flop. All the flip-flops are initially cleared to 0 by giving active low logic to clear, $\overline{Cr}$ asynchronously. Flip-flop is set to logic-1 through active low preset, $\overline{Pr}$ to flip-flop. The presetting of flip-flop is controlled with additional load, L input as high logic. Figure 9.12 gives the diagram of four-bit shift register with asynchronous loading. NAND gate is used to control preset, $\overline{Pr}$ input of flip-flop with load, L control line.

$$\overline{Pr} = \overline{LA_i} \quad (i = 0, 1, 2, 3) \tag{9.4}$$

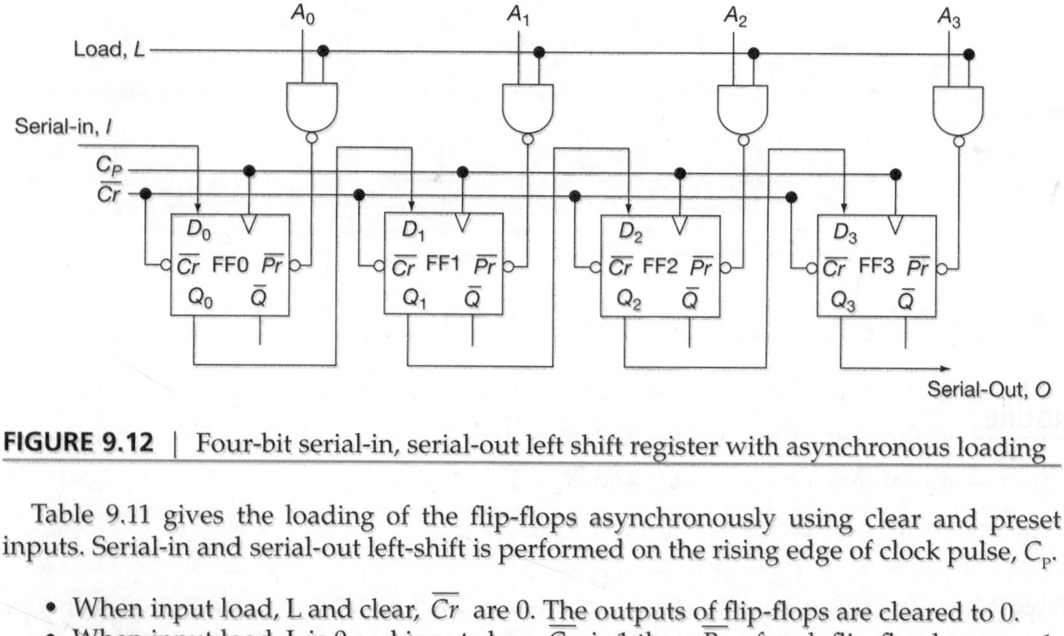

FIGURE 9.12 | Four-bit serial-in, serial-out left shift register with asynchronous loading

Table 9.11 gives the loading of the flip-flops asynchronously using clear and preset inputs. Serial-in and serial-out left-shift is performed on the rising edge of clock pulse, C_P.

- When input load, L and clear, $\overline{Cr}$ are 0. The outputs of flip-flops are cleared to 0.
- When input load, L is 0 and input clear, $\overline{Cr}$ is 1 then $\overline{Pr}$ of each flip-flop becomes 1. Input of each flip-flop is stored.
- When input load, L is 1 and input clear, $\overline{Cr}$ is 1 then preset, $\overline{Pr}$ of each flip-flop becomes 0. The outputs of flip-flops are preset to 1.

Alternatively, the flip-flops are loaded to 0 or 1 by controlling clear, $\overline{Cr}$ and preset, $\overline{Pr}$ inputs of flip-flops with load, *L*. Figure 9.13 gives the diagram of four-bit shift register with asynchronous loading. NAND gates are used to control the clear, $\overline{Cr}$ and preset, $\overline{Pr}$ inputs of flip-flops with load, *L* control line.

TABLE 9.11 | Truth table to load the flip flop

Load, L	Clear $\overline{Cr}$	D-flip-flop-FF0 A_0	$\overline{Pr}$	Q_0	D-flip-flop-FF1 A_1	$\overline{Pr}$	Q_1	D-flip-flop-FF2 A_2	$\overline{Pr}$	Q_2	D-flip-flop-FF3 A_3	$\overline{Pr}$	Q_3
0	0	0	1	0	0	1	0	0	1	0	0	1	0
0	0	1	1	0	1	1	0	1	1	0	1	1	0
0	1	0	1	D_0	0	1	D_1	0	1	D_2	0	1	D_3
0	1	1	1	D_0	1	1	D_1	1	1	D_2	1	1	D_3
1	1	0	1	D_0	0	1	D_1	0	1	D_2	0	1	D_3
1	1	1	0	1	1	0	1	1	0	1	1	0	1

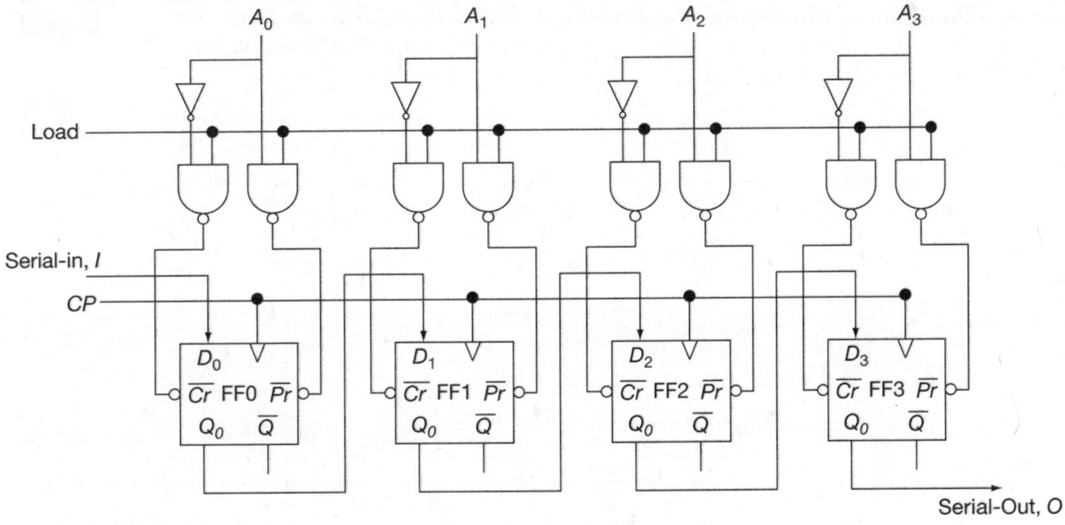

FIGURE 9.13 | Four-bit parallel-in-serial-out register giving left shift

$$\overline{Pr} = \overline{LA_i} \ (i = 0, 1, 2, 3)$$

$$\overline{Cr} = \overline{L\overline{A}_i} \ (i = 0, 1, 2, 3)$$

Table 9.12 gives the loading of the flip-flops asynchronously using clear and preset inputs.

- When input load, L is 0, clear, $\overline{Cr}$ and preset, $\overline{Pr}$ inputs become 1. Input of each flip-flop is stored for any input, A_i to NAND gates. The outputs of flip-flops are cleared to 0.
- When input load, L is 1 and inputs, A_i to NAND gates are 0 then clear, $\overline{Cr}$ input becomes 0 and preset, $\overline{Pr}$ input becomes 1 of each flip-flop. The outputs of flip-flops are cleared to 0.
- When load, L is 1 and inputs, A_i to NAND gates are 1 then clear, $\overline{Cr}$ inputs become 1 and preset, $\overline{Pr}$ inputs become 0 of each flip-flop. The outputs of flip-flops are set to 1.

Serial-in and serial-out left-shift is performed on the rising edge of clock pulse, C_P.

TABLE 9.12 | Truth table to load the flip flop

Load	D-flip-flop-FF0				D-flip-flop-FF1				D-flip-flop-FF2				D-flip-flop-FF3			
L	A_0	$\overline{Cr}$	$\overline{Pr}$	Q_0	A_1	$\overline{Cr}$	$\overline{Pr}$	Q_1	A_2	$\overline{Cr}$	$\overline{Pr}$	Q_2	A_3	$\overline{Cr}$	$\overline{Pr}$	Q_3
0	0	1	1	D_0	0	1	1	D_1	0	1	1	D_2	0	1	1	D_3
0	1	1	1	D_0	1	1	1	D_1	1	1	1	D_2	1	1	1	D_3
1	0	0	1	0	0	0	1	0	0	0	1	0	0	0	1	0
1	1	1	0	1	1	1	0	1	1	1	0	1	1	1	0	1

EXAMPLE 9.2 Draw a circuit diagram of four-bit SISO register to perform the left-shift using S-R-flip flop.

SOLUTION Figure 9.14 shows the four-bit register using SR-flip-flop. Input is given to SR-flip-flop and output of SR-I-flip-flop is given to SR-flip-flop-II and so on. 1010 is loaded to SR-flip-flops when load, L is 1.

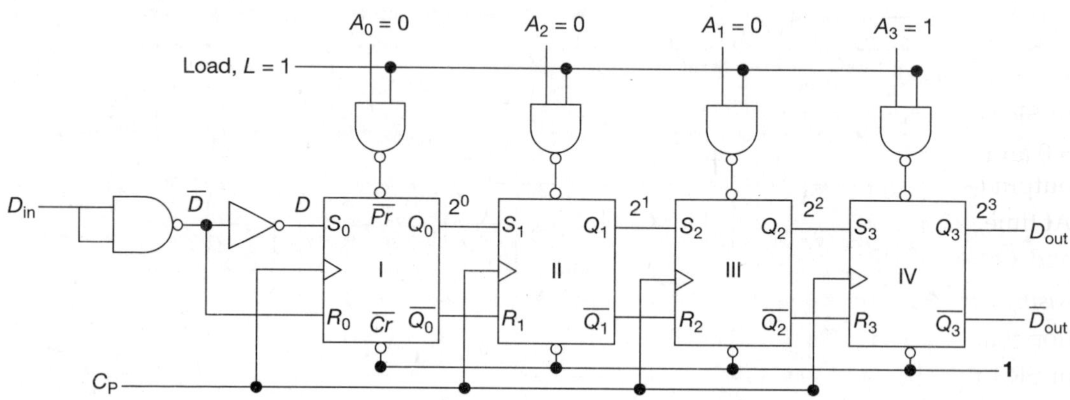

FIGURE 9.14 | SISO 4-bit register performs left shift

Let the serial input, D_{in} is fed to SR-I-flip-flop and is shown in timing diagram (Figure 9.15). Change in state of flip-flop takes place during rising edge transition of clock at time t_1, t_2, t_3 and t_4. Initially stored data in the register is $Q_3Q_2Q_1Q_0 = 1010$. The shifting is given in Table 9.13.

TABLE 9.13 | Shift-left operation

Time of Clock, C_P	$D_{out} \leftarrow Q_3$ (2^3)	$D_3 \leftarrow Q_2$ (2^2)	$D_2 \leftarrow Q_1$ (2^1)	$D_1 \leftarrow Q_0$ (2^0)	Serial in $D_0 \leftarrow D_{in}$
t_0	1	0	1	0	–
t_1	0 ←	1 ←	0 ←	1 ←	1
t_2	1 ←	0 ←	1 ←	0 ←	0
t_3	0 ←	1 ←	0 ←	0 ←	0
t_4	1 ←	0 ←	0 ←	1 ←	1

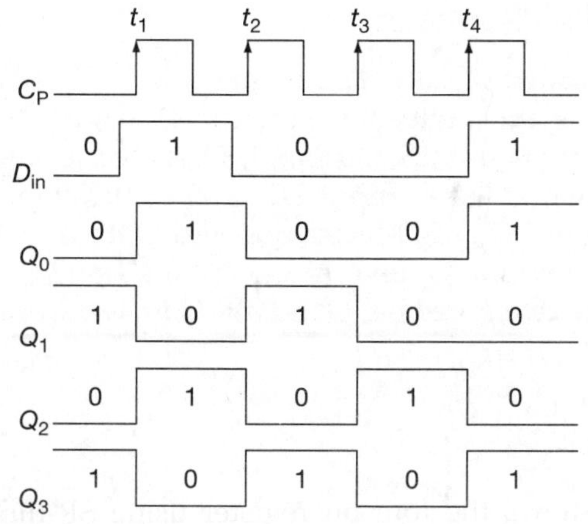

FIGURE 9.15 | Serial-in-serial-out left

Before rising edge transition at time t_1: The register stores $Q_3Q_2Q_1Q_0$ as 1010. LSB appears at Q_0.

Rising edge transition at time t_1: Serial input $D_{in} = 1$ and $\bar{D}_{in} = 0$ are fed to S_0 and R_0 inputs of SR-I-flip-flop and $Q_0 = 1$ and $\bar{Q}_0 = 0$ are stored in the flip-flop. At time, t_1, output $Q_0 = 0$ and $\bar{Q}_0 = 1$ of SR-I-flip-flop is stored in SR-II-flip-flop as $Q_1 = 0$ and $\bar{Q}_1 = 1$. At time, t_1, output $Q_1 = 1$ and $\bar{Q}_1 = 0$ of SR-II-flip-flop is stored in SR-III-flip-flop as $Q_2 = 1$ and $\bar{Q}_2 = 0$. At time, t_1, output $Q_2 = 0$ and $\bar{Q}_2 = 1$ of SR-III-flip-flop is stored in SR-IV-flip-flop as $Q_3 = 0$ and $\bar{Q}_3 = 1$ and is considered as serial out data D_{out}.

Rising edge transition at time t_2: Serial input D_{in} as 0 is fed to S_0 and R_0 inputs of SR-I-flip-flop that is stored in the flip-flop as $Q_0 = 0$ and $\bar{Q}_0 = 1$. At time, t_1, output $Q_0 = 1$ and $\bar{Q}_0 = 0$ of SR-I-flip-flop is stored in SR-II-flip-flop as $Q_1 = 1$ and $\bar{Q}_1 = 0$. At time, t_1, output $Q_1 = 0$ and $\bar{Q}_1 = 1$ of SR-II-flip-flop is stored in SR-III-flip-flop as $Q_2 = 0$ and $\bar{Q}_2 = 1$. At time, t_1, output $Q_2 = 1$ and $\bar{Q}_2 = 0$ of SR-III-flip-flop is stored in SR-IVI-flip-flop as $Q_3 = 1$ and $\bar{Q}_3 = 0$ and is considered as serial out data D_{out}.

Rising edge transition at time t_3: Serial input D_{in} as 0 is fed to S_0 and R_0 inputs of SR-I-flip-flop that is stored in the flip-flop as $Q_0 = 0$ and $\bar{Q}_0 = 1$. At time, t_1, output $Q_0 = 0$ and $\bar{Q}_0 = 1$ of SR-I-flip-flop is stored in SR-II-flip-flop as $Q_1 = 0$ and $\bar{Q}_1 = 1$. At time, t_1, output $Q_1 = 1$ and $\bar{Q}_1 = 0$ of SR-II-flip-flop is stored in SR-III-flip-flop as $Q_2 = 1$ and $\bar{Q}_2 = 0$. At time, t_1, output $Q_2 = 0$ and $\bar{Q}_2 = 1$ of SR-III-flip-flop is stored in SR-IV-flip-flop as $Q_3 = 0$ and $\bar{Q}_3 = 1$ and is considered as serial out data D_{out}.

Rising edge transition at time t_4: Serial input D_{in} as 1 is fed to S_0 and R_0 inputs of SR-I-flip-flop that is stored in the flip-flop as $Q_0 = 1$ and $\bar{Q}_0 = 0$. At time, t_1, output $Q_0 = 0$ and $\bar{Q}_0 = 1$ of SR-I-flip-flop is stored in SR-II-flip-flop as $Q_1 = 0$ and $\bar{Q}_1 = 1$. At time, t_1, output $Q_1 = 0$ and $\bar{Q}_1 = 1$ of SR-II-flip-flop is stored in SR-III-flip-flop as $Q_2 = 0$ and $\bar{Q}_2 = 1$. At time, t_1, output $Q_2 = 1$ and $\bar{Q}_2 = 0$ of SR-III-flip-flop is stored in SR-IV-flip-flop as $Q_3 = 1$ and $\bar{Q}_3 = 0$ and is considered as serial out data D_{out}.

9.5.4 | Right-Shift SISO Register

Four D-flip-flops are cascaded serially. The output of one flip-flop is fed as input to next flip-flop. Four D-flip-flops are numbered from I to IV. Figure 9.16 shows SISO four-bit register able to perform right-shift. Serial input, D_{in} is fed to input, D_3 of D-IV-flip-flop. Output Q_3 of D-IV-flip-flop is fed to input D_2, of D-III-flip-flop. Similarly, Output Q_2 of D-III-flip-flop is fed to input D_1, of D-II-flip-flop and Output Q_1 of D-I-flip-flop is fed to input D_0, of D-I-flip-flop. Output, Q_0 gives serial out bit, D_{out}. Q_0 is considered as LSB and Q_3 is taken as MSB. Higher weighted bit, Q_3 is shifted to next lower weighted bit, Q_2.

Input of D-I-flip-flop is $D_0 = Q_1$,
Input of D-II-flip-flop is $D_1 = Q_2$,
Input of D-III-flip-flop is $D_2 = Q_3$,
Input of D-IV-flip-flop is $D_3 = D_{in}$ and
Serial output is $D_{out} = Q_0$

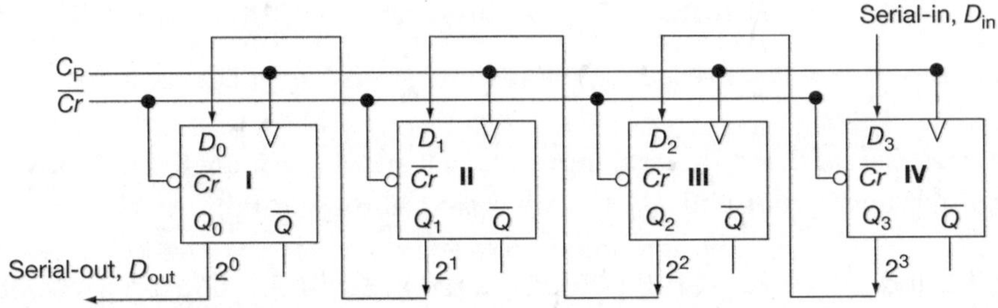

FIGURE 9.16 | Four-bit serial-in, serial-out right shift register

Timing diagram of shift-right operation is shown in Figure 9.17. Let the serial input, D_{in} be fed to D-IV flip is shown in timing diagram shown in Figure 9.17. Change in state of flip-flop takes place during rising edge transition of clock at time t_1, t_2, t_3 and t_4. Initially stored data in the register is $Q_3Q_2Q_1Q_0 = 0000$. The serial in serial out to right-shift operation is given in Table 9.14.

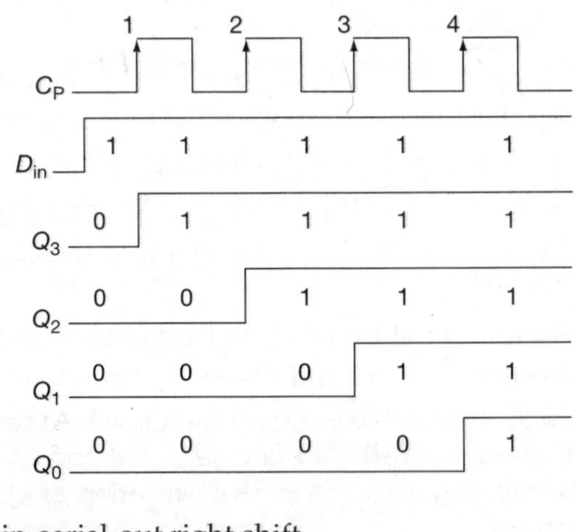

FIGURE 9.17 | Serial-in-serial-out right shift

TABLE 9.14 | Shift-right operation

Time of Clock, C_P	Serial Input $D_3 \leftarrow D_{in}$	$D_2 \leftarrow Q_3$ (2^3)	$D_1 \leftarrow Q_2$ (2^2)	$D_0 \leftarrow Q_1$ (2^1)	$D_{out} \leftarrow Q_0$ (2^0)
t_0		0	0	0	–
t_1	1	1	0	0	0
t_2	1	1	1	0	0
t_3	1	1	1	1	0
t_4	1	1	1	1	1

Before rising edge transition at time t_1: The register stores $Q_3Q_2Q_1Q_0$ as 0000. LSB appears at Q_0.

Rising edge transition at time t_1: Serial input D_{in} as 1 is fed to D_3 of D-IV-flip-flop that is stored in D-IV-flip-flop as $Q_3 = 1$. At time, t_1, output $Q_3 = 0$ of D-IV-flip-flop is stored in D-III-flip-flop as $Q_2 = 0$. At time, t_1, output $Q_2 = 0$ of D-III-flip-flop is stored in D-II-flip-flop as $Q_1 = 0$. At time, t_1, output $Q_1 = 0$ of D-II-flip-flop is stored in D-I-flip-flop as $Q_0 = 0$ and is considered as serial out data D_{out}.

Rising edge transition at time t_2: Serial input D_{in} as 1 is fed to D_3 of D-IV-flip-flop that is stored in D-IV-flip-flop as $Q_3 = 1$. At time, t_2, output $Q_3 = 1$ of D-IV-flip-flop is stored in D-III-flip-flop as $Q_2 = 1$. At time, t_2, output $Q_2 = 0$ of D-III-flip-flop is stored in D-II-flip-flop as $Q_1 = 0$. At time, t_2, output $Q_1 = 0$ of D-II-flip-flop is stored in D-I-flip-flop as $Q_0 = 0$ and is considered as serial out data D_{out}.

Rising edge transition at time t_3: Serial input D_{in} as 1 is fed to D_3 of D-IV-flip-flop that is stored in D-IV-flip-flop as $Q_3 = 1$. At time, t_3, output $Q_3 = 1$ of D-IV-flip-flop is stored in D-III-flip-flop as $Q_2 = 1$. At time, t_3, output $Q_2 = 1$ of D-III-flip-flop is stored in D-II-flip-flop as $Q_1 = 1$. At time, t_3, output $Q_1 = 0$ of D-II-flip-flop is stored in D-I-flip-flop as $Q_0 = 0$ and is considered as serial out data D_{out}.

Rising edge transition at time t_4: Serial input D_{in} as 1 is fed to D_3 of D-IV-flip-flop that is stored in D-IV-flip-flop as $Q_3 = 1$. At time, t_3, output $Q_3 = 1$ of D-IV-flip-flop is stored in D-III-flip-flop as $Q_2 = 1$. At time, t_3, output $Q_2 = 1$ of D-III-flip-flop is stored in D-II-flip-flop as $Q_1 = 1$. At time, t_3, output $Q_1 = 1$ of D-II-flip-flop is stored in D-I-flip-flop as $Q_0 = 1$ and is considered as serial out data D_{out}.

EXAMPLE 9.3 Draw a circuit diagram of four-bit SISO register to perform the left-shift using JK-flip-flop.

SOLUTION Figure 9.18 shows the four-bit register using JK-flip-flop. Input is given to JK-flip-flop and output of JK-IV-flip-flop is given to JK-III-flip-flop and so on.

Input of JK-IV-flip-flop is $J_3 = \overline{\overline{D}} \Rightarrow J_3 = D$ and $K_3 = \overline{D}$

Input of JK-III-flip-flop is $J_2 = Q_3$ and $K_2 = \overline{Q}_3$

Input of JK-II-flip-flop is $J_1 = Q_2$ and $K_1 = \overline{Q}_2$

Input of JK-I-flip-flop is $J_0 = Q_1$ and $K_0 = \overline{Q}_1$

Serial output from JK-I-flip-flop is $D_{out} = Q_0$ and $\overline{D}_{out} = \overline{Q}_0$

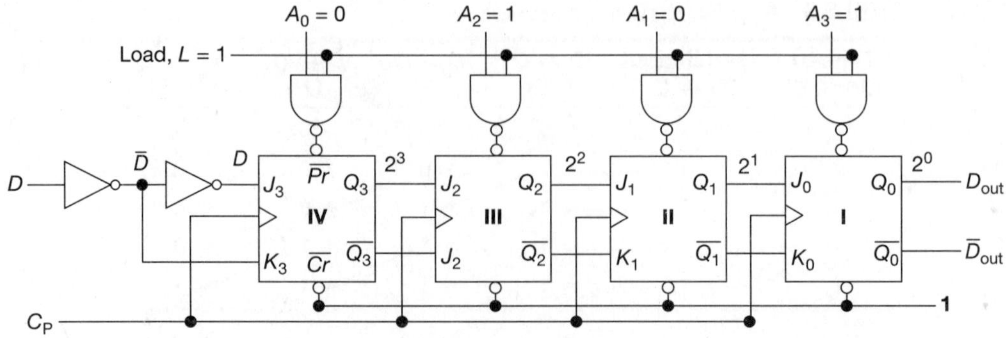

FIGURE 9.18 | Four-bit serial-in, serial-out right shift register

Truth Table 9.15 gives the excitation table of JK-IV-flip-flop. Excitation Table 9.16 gives the inputs and outputs of JK-III-flip-flop. Let the serial input, D_{in} fed to JK-IV flip is shown in timing diagram shown in Figure 9.19. Change in state of flip-flop takes place during rising edge transition of clock at time t_1, t_2, t_3 and t_4. Initially stored data in the register is $Q_3Q_2Q_1Q_0 = 1001$. The serial-in-serial-out right-shift operation is given in Table 9.17.

TABLE 9.15 | Truth table

Clock C_P	Input D	JK-IV-flip-flop Inputs J_3	K_3	Output of JK-IV-flip-flop Q_3	$\overline{Q_3}$
↑	1	1	0	1	0
↑	0	0	1	0	1

TABLE 9.16 | Truth table

Clock C_P	JK-III-flip-flop Inputs $J_2 = Q_3$	$K_2 = \overline{Q_3}$	Output of JK-III-flip-flop Q_2	$\overline{Q_2}$
↑	1	0	1	0
↑	0	1	0	1

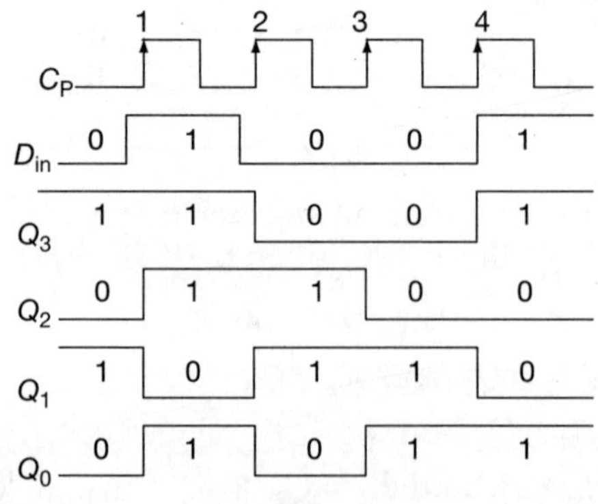

FIGURE 9.19 | Serial-in-serial-out right shift

TABLE 9.17 | Shift-right operation

Time of Clock, C_P	Serial Input $J_3 \leftarrow D$ $K_3 = \bar{D}$	$J_2 \leftarrow Q_3$ $K_2 = \bar{Q}_3$ (2^3)	$J_1 \leftarrow Q_2$ $K_1 = \bar{Q}_2$ (2^2)	$J_0 \leftarrow Q_1$ $K_0 = \bar{Q}_1$ (2^1)	$D_{out} \leftarrow Q_0$ (2^0)
t_0		1	0	1	0
t_1	1 →	1	1	0	1
t_2	0 →	0	1	1	0
t_3	0 →	0	0	1	1
t_4	1 →	1	0	0	1

Before rising edge transition at time t_1: The register stores $Q_3Q_2Q_1Q_0$ as 1010. LSB appears at Q_0.

Rising edge transition at time t_1: Serial input D as 1 is fed to J_3 and complement of D, 0 is fed to K_3 inputs of JK-IV-flip-flop that is stored in JK-IV-flip-flop as $Q_3 = 1$ and $\bar{Q}_3 = 0$. At time, t_1, output $Q_3 = 1$ and $\bar{Q}_3 = 0$ of JK-IV-flip-flop is stored in JK-III-flip-flop as $Q_2 = 1$ and $\bar{Q}_2 = 0$. At time, t_1, output $Q_2 = 0$ and $\bar{Q}_2 = 1$ of JK-III-flip-flop is stored in JK-II-flip-flop as $Q_1 = 1$ and $\bar{Q}_1 = 0$. At time, t_1, output $Q_1 = 1$ and $\bar{Q}_1 = 0$ of JK-II-flip-flop is stored in JK-I-flip-flop as $Q_0 = 1$ and $\bar{Q}_0 = 0$ and is considered as serial out data D_{out}.

Rising edge transition at time t_2: Serial input D as 0 is fed to J_3 and complement of D, 1 is fed to K_3 inputs of JK-IV-flip-flop that is stored in the JK-IV-flip-flop as $Q_3 = 0$ and $\bar{Q}_3 = 1$. At time, t_2, output $Q_3 = 1$ and $\bar{Q}_3 = 0$ of JK-IV-flip-flop is stored in JK-III-flip-flop as $Q_2 = 1$ and $\bar{Q}_2 = 0$. At time, t_2, output $Q_2 = 1$ and $\bar{Q}_2 = 0$ of JK-III-flip-flop is stored in JK-II-flip-flop as $Q_1 = 1$ and $\bar{Q}_1 = 0$. At time, t_2, output $Q_1 = 0$ and $\bar{Q}_1 = 1$ of JK-II-flip-flop is stored in JK-I-flip-flop as $Q_0 = 0$ and $\bar{Q}_0 = 1$ and is considered as serial out data D_{out}.

Rising edge transition at time t_3: Serial input D as 0 is fed to J_3 and complement of D, 1 is fed to K_3 inputs of JK-IV-flip-flop that is stored in JK-IV-flip-flop as $Q_3 = 0$ and $\bar{Q}_3 = 1$. At time, t_3, output $Q_3 = 0$ and $\bar{Q}_3 = 1$ of JK-IV-flip-flop is stored in JK-III-flip-flop as $Q_2 = 0$ and $\bar{Q}_2 = 1$. At time, t_3, output $Q_2 = 1$ and $\bar{Q}_1 = 0$ of JK-III-flip-flop is stored in JK-II-flip-flop as $Q_1 = 1$ and $\bar{Q}_1 = 0$. At time, t_3, output $Q_1 = 1$ and $\bar{Q}_1 = 0$ of JK-II-flip-flop is stored in JK-I-flip-flop as $Q_0 = 1$ and $\bar{Q}_0 = 0$ and is considered as serial out data D_{out}.

Rising edge transition at time t_4: Serial input D as 1 is fed to J_3 and complement of D, 0 is fed to K_3 inputs of JK-IV-flip-flop that is stored in the flip-flop as $Q_3 = 0$ and $\bar{Q}_3 = 1$. At time, t_4, output $Q_3 = 0$ and $\bar{Q}_3 = 1$ of JK-IV-flip-flop is stored in JK-III-flip-flop as $Q_2 = 0$ and $\bar{Q}_2 = 1$. At time, t_4, output $Q_2 = 0$ and $\bar{Q}_2 = 1$ of JK-III-flip-flop is stored in JK-II-flip-flop as $Q_1 = 0$ and $\bar{Q}_1 = 1$. At time, t_4, output $Q_1 = 1$ and $\bar{Q}_1 = 0$ of JK-II-flip-flop is stored in JK-I-flip-flop as $Q_0 = 1$ and $\bar{Q}_0 = 0$ and is considered as serial out data D_{out}.

9.5.5 | Bidirectional SISO Register

A unidirectional register performs shift operation on binary information only in one direction. So, it shifts binary data one bit at a time either to left or to right direction. A bidirectional register (Figure 9.20) performs controlled shift operation on binary data one bit at a time in both left and right directions. However, at a time it shifts binary data one bit either to left direction or to right direction.

To extend bidirectional shift operation to binary data, a control signal, C is added. Input to D-flip-flop is fed through 2-to-1 line multiplexer (MUX) to select the direction of shift operation. Right-shift operation is performed if control signal, C is 1 (Table 9.18) and left-shift operation is performed if control signal, C is 0.

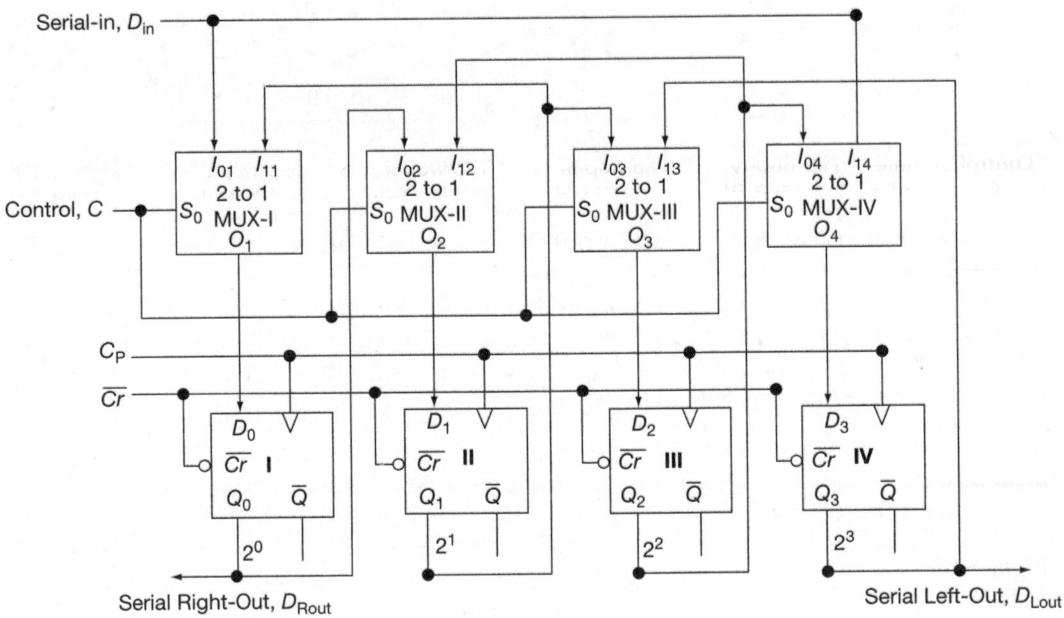

FIGURE 9.20 | Four-bit bidirectional serial register

TABLE 9.18 | Bidirectional serial register

Control, C	Action	Input of Flip Flop, D_i
0	Left-shift	I_{0i} of MUX
1	Right-shift	I_{1i} of MUX

Left-Shift: For left-shift operation, serial input is given to input, I_{01} of MUX-I. Output Q_0 of D-I-flip-flop is given to input, I_{02} of MUX-II. Output Q_1 of D-II-flip-flop is given to input, I_{03} of MUX-III. Output Q_2 of D-III-flip-flop is given to input, I_{04} of MUX-IV. Output Q_3 of D-IV-flip-flop is the serial out data. I_{0i} is selected by respective multiplexers (MUX) when control signal, C is 0. MUX output acts as input for the corresponding D-flip-flop. Table 9.19 explains the left- and right-shift operation when all the flip-flops are cleared initially by issuing asynchronous, $\overline{Cr}$ as low signal. Serial input 1 is given at t_1, t_2, t_3 and t_4.

TABLE 9.19 | Shift operation

Control, C	Time of Clock, C_P	Left-shift Out $D_{Lout} \leftarrow Q_3$ (2^3)	Flip-flop-IV and MUX-IV $D_3 \leftarrow I_{04} \leftarrow Q_2$ (2^2)	Flip-flop-III and MUX-III $D_2 \leftarrow I_{03} \leftarrow Q_1$ (2^1)	Flip-flop-II and MUX-II $D_1 \leftarrow I_{02} \leftarrow Q_0$ (2^0)	Flip-flop-I and MUX-I Serial in $D_0 \leftarrow I_{01} \leftarrow D_{in}$
0	t_0	0	0	0	0	-
	t_1	0	1	0	1	1
	t_2	1	0	1	0	1
	t_3	0	1	0	0	1
	t_4	1	0	0	1	1
Control, C	**Time of Clock, C_P**	**Flip-flop-IV and MUX-IV** Serial Input $D_3 \leftarrow I_{14} \leftarrow D_{in}$	**Flip-flop-III and MUX-III** $D_2 \leftarrow I_{13} \leftarrow Q_3$ (2^3)	**Flip-flop-II and MUX-II** $D_1 \leftarrow I_{12} \leftarrow Q_2$ (2^2)	**Flip-flop-I and MUX-I** $D_0 \leftarrow I_{11} \leftarrow Q_1$ (2^1)	**Right-shift Out** $D_{out} \leftarrow Q_0$ (2^0)
1	t_0	0	0	0	0	0
	t_1	1	1	1	0	1
	t_2	0	0	1	1	0
	t_3	0	0	0	1	1
	t_4	1	1	0	0	1

Right-Shift: For right operation, serial input is given to input, I_{14} of MUX-IV. Output Q_3 of D-IV-flip-flop is given to input, I_{13} of MUX-III. Output Q_2 of D-III-flip-flop is given to input, I_{12} of MUX-II. Output Q_1 of D-II-flip-flop is given to input, I_{11} of MUX-I. Output Q_0 of D-I-flip-flop is the serial out data. I_{1i} is selected by respective multiplexers (MUX) when control signal, C is 1. MUX output acts as input for the corresponding D-flip-flop. Table 9.19 explains the shift-right operation when all the flips are cleared initially by issuing asynchronous, $\overline{Cr}$ as active low. Serial input 1, 0, 1 and 1 is given at t_1, t_2, t_3 and t_4, respectively.

EXAMPLE 9.4 Draw a circuit diagram of four-bit bidirectional SISO register using SR-flip-flop and AND-OR-INVERTER (AOI) gates.

SOLUTION Figure 9.21 shows the four-bit register using SR-flip-flop. To perform right-shift, input is given to SR-flip-flop and output of SR-IV-flip-flop is given to SR-flip-flop-III and so on. To perform left shift, input is given to SR-flip-flop and output of SR-I-flip-flop is given to SR-flip-flop-II and so on. The left and right-shift operation is controlled with a Load, L signal given to inhibit or enable AND gates and any one output of AND gates is selected with OR gate. To construct D-flip-flop, complement of S is given to R.

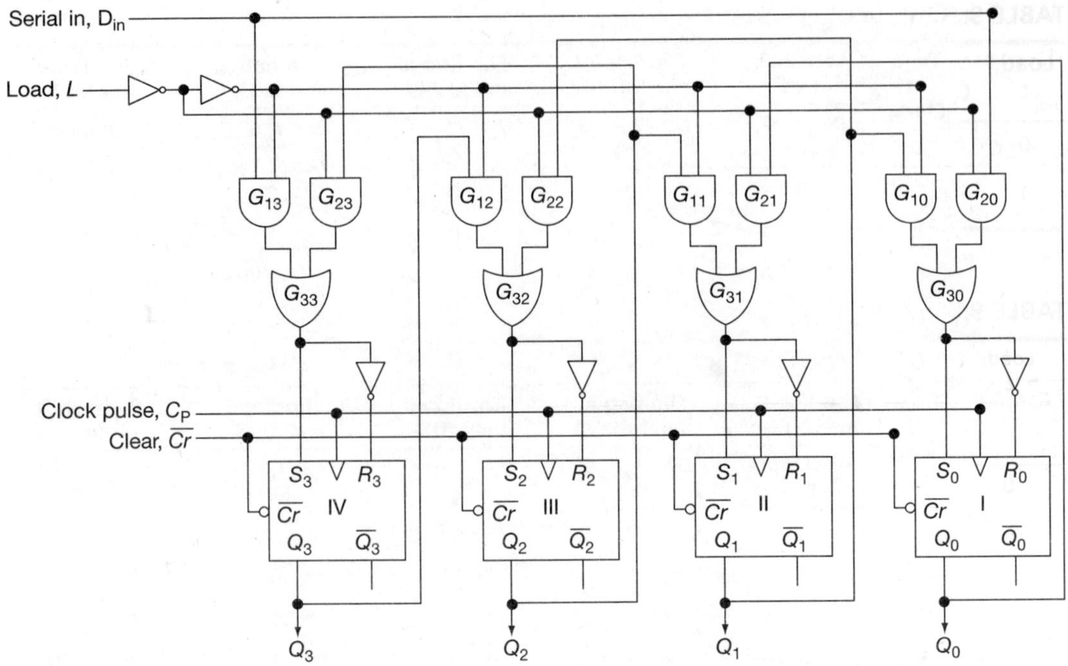

FIGURE 9.21 | Bidirectional serial shift register

Using two AND gates G_{10} and G_{20} and one OR gate G_{30}, inputs for SR-flip-flop, S_0 and R_0 are given below:

$$S_0 = \bar{L}D_{in} + LQ_1 \text{ and } R_0 = \overline{\bar{L}D_{in} + LQ_1}$$

Using two AND gates G_{11} and G_{21} and one OR gate G_{31}, inputs for SR-flip-flop, S_1 and R_1 are given below:

$$S_1 = \bar{L}Q_0 + LQ_2 \text{ and } R_1 = \overline{\bar{L}Q_0 + LQ_2}$$

Using two AND gates G_{12} and G_{22} and one OR gate G_{32}, inputs for SR-flip-flop, S_2 and R_2 are given below:

$$S_2 = \bar{L}Q_1 + LQ_3 \text{ and } R_2 = \overline{\bar{L}Q_1 + LQ_3}$$

Using two AND gates G_{13} and G_{23} and one OR gate G_{33}, inputs for SR-flip-flop, S_3 and R_3 are given below:

$$S_3 = \bar{L}Q_2 + LD_{in} \text{ and } R_3 = \overline{\bar{L}Q_2 + LD_{in}}$$

Table 9.20 gives the left and right shift operation when load L, is cleared (0) or set (1), respectively.

Table 9.21 explains the left right operation when all the flip-flops are cleared initially by issuing asynchronous, $\overline{Cr}$. Serial input 1 is given at t_1, t_2, t_3 and t_4. Table 9.21 explains the shift-right operation when all the flip-flops are cleared initially by issuing asynchronous, $\overline{Cr}$. Serial input 1, 0, 1 and 1 is given at t_1, t_2, t_3 and t_4, respectively.

TABLE 9.20 | Truth table for logic expressions

Load, L	Shift Direction	Flip-flop-I S_0	Flip-flop-I R_0	Flip-flop-II S_1	Flip-flop-II R_1	Flip-flop-III S_2	Flip-flop-III R_2	Flip-flop-IV S_3	Flip-flop-IV R_3	Output D_{out}
0	Left	D_{in}	$\bar{D}_{in}$	Q_0	$\bar{Q}_0$	Q_1	$\bar{Q}_1$	Q_2	$\bar{Q}_2$	Q_3
1	Right	Q_1	$\bar{Q}_1$	Q_2	$\bar{Q}_2$	Q_3	$\bar{Q}_3$	D_{in}	$\bar{D}_{in}$	Q_0

TABLE 9.21 | Shift operation

Load, L	Time of Clock, C_P	Left-shift Out $D_{Lout} \leftarrow Q_3$ (2^3)	Flip-flop-IV and MUX-IV $S_3 \leftarrow Q_2$ (2^2)	Flip-flop-III and MUX-III $S_2 \leftarrow Q_1$ (2^1)	Flip-flop-II and MUX-II $S_1 \leftarrow Q_0$ (2^0)	Flip-flop-I and MUX-I Serial in $S_0 \leftarrow D_{in}$
0	t_0	0	0	0	0	–
	t_1	0	1	0	1	1
	t_2	1	0	1	0	1
	t_3	0	1	0	0	1
	t_4	1	0	0	1	1
Load, L	**Time of Clock, C_P**	**Flip-flop-IV and MUX-IV** Serial Input $S_3 \leftarrow D_{in}$	**Flip-flop-III and MUX-III** $S_2 \leftarrow Q_3$ (2^3)	**Flip-flop-II and MUX-II** $S_1 \leftarrow Q_2$ (2^2)	**Flip-flop-I and MUX-I** $S_0 \leftarrow Q_1$ (2^1)	**Right-shift Out** $D_{out} \leftarrow Q_0$ (2^0)
1	t_0	0	0	0	0	0
	t_1	1	1	1	0	1
	t_2	0	0	1	1	0
	t_3	0	0	0	1	1
	t_4	1	1	0	0	1

9.6 | SERIAL-IN, PARALLEL-OUT SHIFT REGISTER

The serial-in, parallel-out (SIPO) shift register is similar to SISO shift register except that all the flip-flop outputs are externally accessible. Figure 9.22 shows the circuit of SIPO left-shift register. Basic circuit of SISO left-shift register is extended by four AND gates that allow the stored data in flip-flops to be read out. One clock pulse is required for each stage to load input data. So, four clock pulses are required for a four-bit register and only one pulse is required to output the parallel data. When a four-bit word is stored in the register, all the bits can be read out simultaneously by applying logic-1 to the read line, R.

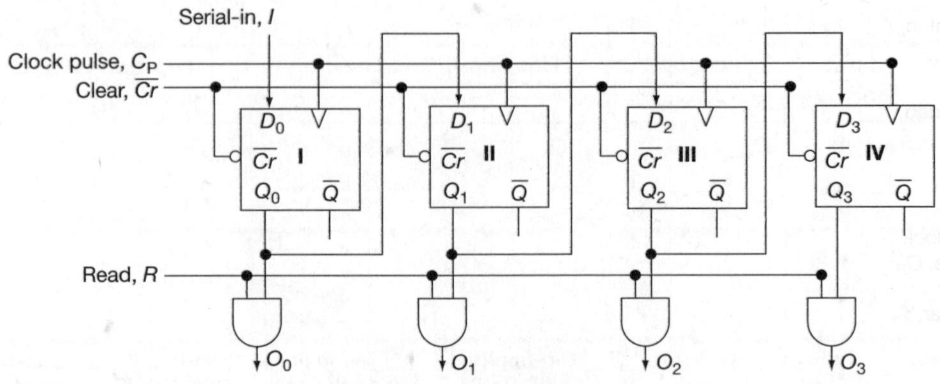

FIGURE 9.22 | Four-bit serial-in-parallel-out left shift register

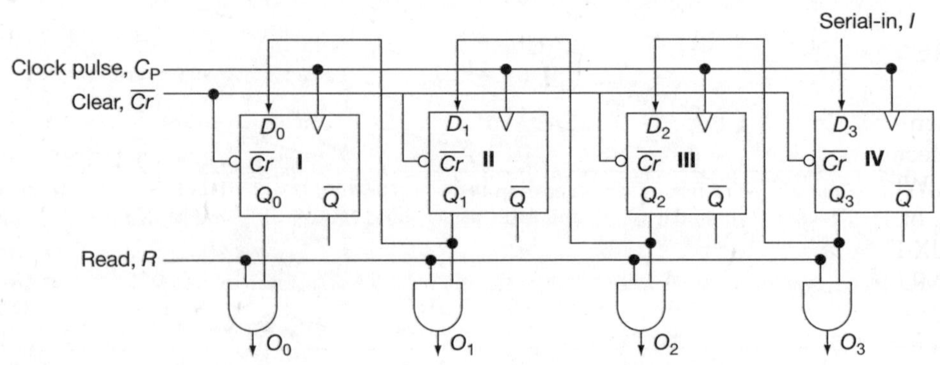

FIGURE 9.23 | Four-bit serial-in-parallel-out right shift register

Figure 9.23 shows the circuit of SIPO right-shift register. Basic circuit of SISO right-shift register is extended by four AND gates that allow the stored data in flip-flops to be read out. When a four-bit word is stored in the register, all the bits can be read out simultaneously by applying logic-1 to the read line, *R*.

9.7 | PARALLEL-IN, SERIAL-OUT, SHIFT REGISTER

The data bits are entered simultaneously into their respective flip-flops on lines simultaneously in a parallel-in, serial-out (PISO), shift register. Whereas, the data bits are transferred out of the register serially, that is on a bit-by-bit basis over a single line.

9.7.1 | PISO Left-Shift Register

Figure 9.24 illustrates a four-bit PISO, left-shift register using D-flip-flops. There are four data lines x_0, x_1, x_2, and x_3 through which the data is entered into the register simultaneously in parallel form. The input signal shift/load, *S* allows (i) the data to be entered in parallel form into the register and (ii) the data to be shifted out serially from terminal Q_3. These two operations are controlled by multiplexers.

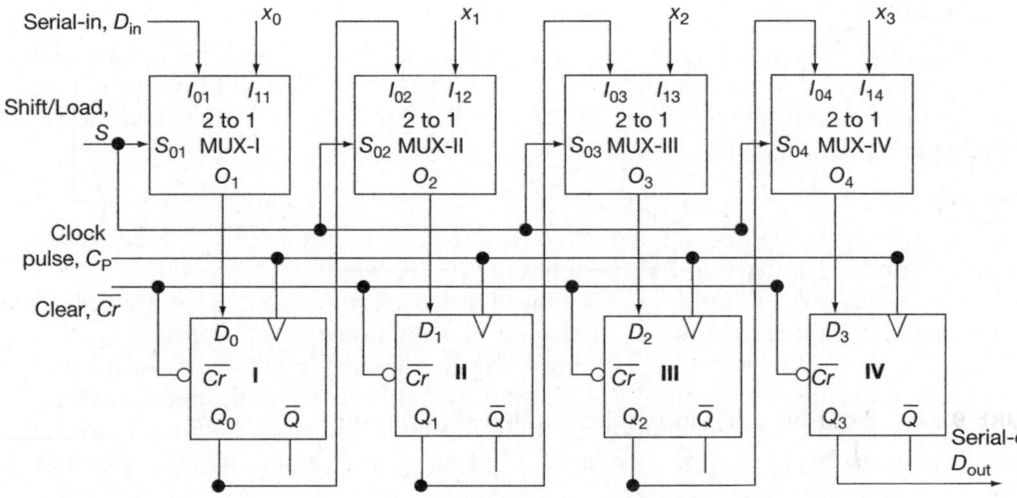

FIGURE 9.24 | Four-bit parallel-in, serial-out (left) shift register

When shift/load, S line is low (logic-0), inputs I_{01}, I_{02}, I_{03} and I_{04} of respective multiplexers are selected allowing the data bits to shift-left from flip-flop-I to the next flip-flop-II and so on. When a clock pulse is applied, D_{in} is fed to input, D_0 of D-I-flip-flop through input line I_{01} of MUX-I. The data bit Q_0 is shifted to input D_1 of flip-flop-II through input line I_{02} of MUX-II. Similarly, the data bit Q_1 is shifted to input D_2 of flip-flop-II through input line I_{03} of MUX-III. The data bit Q_2 is shifted to input, D_3 of flip-flop-IV through input line, I_{04} of MUX-IV.

When shift/load, S-line is high (logic-1), multiplexers inputs I_{10}, I_{11}, I_{12} and I_{13} are allowed as data inputs x_0, x_1, x_2, and x_3 to appear at inputs D_0, D_1, D_2, and D_3 of flip-flops-I, II, -III and -IV, respectively, those are subsequently stored in their respective flip-flops as Q_0, Q_1, Q_2, and Q_3. Table 9.22 explains the parallel loading of 1010 and performs four shifts when input 1 is fed at every clock pulse.

TABLE 9.22 | PISO shift register with parallel load

Shift/ Load, S	Action	Time of Clock, C_P	Flip-flop-IV and MUX-IV (2^3)	Flip-flop-III and MUX-III (2^2)	Flip-flop-II and MUX-II (2^1)	Flip-flop-I and MUX-I (2^0)	Serial Input
1	Load		$Q_3 \leftarrow D_3 \leftarrow I_{14} \leftarrow x_3$	$Q_2 \leftarrow D_2 \leftarrow I_{13} \leftarrow x_2$	$Q_1 \leftarrow D_1 \leftarrow I_{12} \leftarrow x_1$	$Q_0 \leftarrow D_0 \leftarrow I_{11} \leftarrow x_0$	–
1	Load	t_0	1	0	1	0	–
0	Shift-left		$Q_3 \leftarrow D_3 \leftarrow I_{04} \leftarrow Q_2$	$Q_2 \leftarrow D_2 \leftarrow I_{03} \leftarrow Q_1$	$Q_1 \leftarrow D_1 \leftarrow I_{02} \leftarrow Q_0$	$Q_0 \leftarrow D_0 \leftarrow I_{01} \leftarrow D_{in}$	D_{in}
0	Shift-left	t_1	0	1	0	1	1
0	Shift-left	t_2	1	0	1	1	1
0	Shift-left	t_3	0	1	1	1	1
0	Shift-left	t_4	1	1	1	1	1

Parallel load:

t_1; $(Q_3Q_2Q_1Q_0) \leftarrow (D_3D_2D_1D_0) \leftarrow (I_{14}I_{13}I_{12}I_{11}) \leftarrow (x_3x_2x_1x_0)$: (1010) as $(x_3x_2x_1x_0)$ appear simultaneously on $(I_{14}I_{13}I_{12}I_{11})$ of multiplexers I, II, III and IV, respectively. Multiplexers select $(I_{14}I_{13}I_{12}I_{11})$ and feed input to flip-flops $(D_3D_2D_1D_0)$. Finally $(D_3D_2D_1D_0)$ are stored in respective flip-flops as $(Q_3Q_2Q_1Q_0)$.

Serial left:

t_1; $(Q_0 \leftarrow D_0 \leftarrow I_{01} \leftarrow D_{in})$: D_{in} (1) is given to input, I_{01} of multiplexer-I. I_{01} is selected and appears as input, D_0 of flip-flop-I. D_0 is stored in flip-flop-I as Q_0.
t_2; $(Q_1 \leftarrow D_1 \leftarrow I_{02} \leftarrow Q_0)$: Output, Q_0 (0) of flip-flop-I is given to input, I_{02} of multiplexer-II and is selected to appear as input, D_1 of flip-flop-II. D_1 is stored in flip-flop-II as Q_1.
t_3; $(Q_2 \leftarrow D_2 \leftarrow I_{03} \leftarrow Q_1)$: Output, Q_1 (1) of flip-flop-II is given to input, I_{03} of multiplexer-III and is selected to appear on input, D_2 of flip-flop-III and is stored in flip-flop-III as Q_2.
t_4; $(Q_3 \leftarrow D_3 \leftarrow I_{04} \leftarrow Q_2)$ means output, Q_2 (0) of flip-flop-III is given to input, I_{04} of multiplexer-IV and is selected to appear on input, D_3 of flip-flop-IV and is stored in flip-flop-IV as Q_3.

9.7.2 | PISO, Right-Shift Register

Figure 9.25 shows a four-bit parallel-in, serial-out right-shift register using four D-flip-flops. There are four data lines x_0, x_1, x_2, and x_3 through which the data is entered into the register simultaneously in parallel form. The signal shift/load, S allows (i) the data to be entered in parallel form into the register and (ii) the data to be shifted out serially from terminal Q_3. The two operations are selected by multiplexers.

When shift/load, S line is high (logic-1) input, I_{0i} of its multiplexer is selected allowing the data bits to shift-right from one flip-flop-III to the next flip-flop-II and so on. When a clock pulse is applied, the data bits Q_3 is shifted to flip-flop-III input D_2 through MUX-III input line I_0. Similarly, the data bits Q_2 is shifted to flip-flop-II input D_1 through MUX-II input line I_0 and the data bits Q_1 is shifted to flip-flop-I input, D_1 through MUX-I input line, I_0.

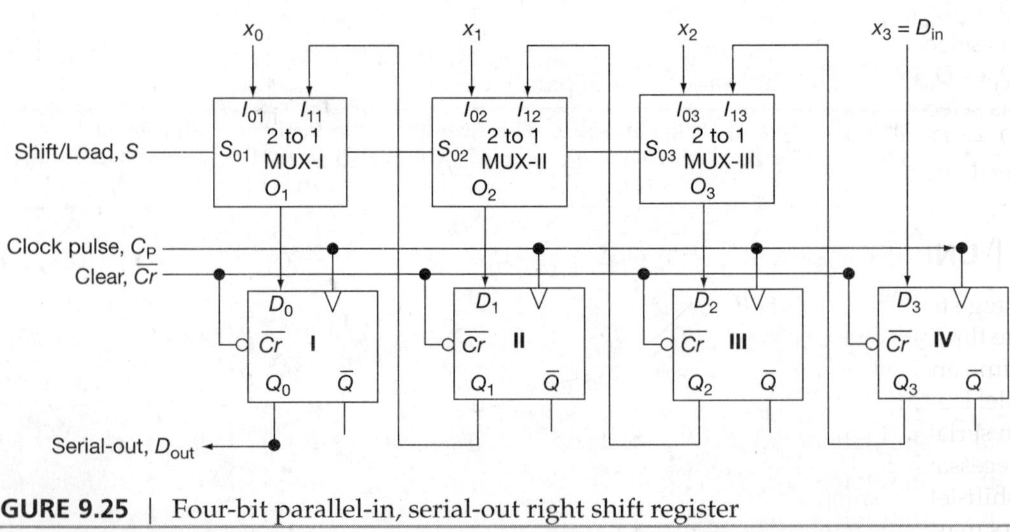

FIGURE 9.25 | Four-bit parallel-in, serial-out right shift register

When shift/load, S- line is low (logic-0), I_1 allowing the data inputs x_0, x_1, x_2, and x_3 to appear at inputs D_0, D_1, D_2, and D_3 of flip-flops-I, -II, -III and -IV, respectively. Flip-flop inputs are stored in the respective flip-flops as Q_0, Q_1, Q_2, and Q_3. x_0, x_1, and x_2 are selected through multiplexers as these are fed when I_{01}, I_{02} and I_{03} inputs of multiplexers I, II and II, respectively are selected. Table 9.23 explains the parallel loading of 1010 and performs four shifts when input 1 is fed at every clock pulse.

TABLE 9.23 | PISO shift register with parallel load

Shift/ Load, S	Action	Time of Clock, C_P	Serial Input x_3	Flip-flop-IV (2^3)	Flip-flop-III and MUX-III (2^2)	Flip-flop-II and MUX-II (2^1)	Flip-flop-I and MUX-I (2^0)
0	Load		–	$D_3 \leftarrow x_3$	$Q_2 \leftarrow D_2 \leftarrow I_{13} \leftarrow x_2$	$Q_1 \leftarrow D_1 \leftarrow I_{12} \leftarrow x_1$	$Q_0 \leftarrow D_0 \leftarrow I_{11} \leftarrow x_0$
0	Load	t_0	–	1	0	1	0
1	Shift-right		x_3	$Q_3 \leftarrow D_3 \leftarrow x_3$	$Q_2 \leftarrow D_2 \leftarrow I_{13} \leftarrow Q_3$	$Q_1 \leftarrow D_1 \leftarrow I_{12} \leftarrow Q_2$	$Q_0 \leftarrow D_0 \leftarrow I_{11} \leftarrow Q_1$
1	Shift-right	t_1	1	1	1	0	1
1	Shift-right	t_2	1	1	1	1	0
1	Shift-right	t_3	1	1	1	1	1
1	Shift-right	t_4	1	1	1	1	1

Parallel load:

t_1; $(Q_3Q_2Q_1Q_0) \leftarrow (D_3D_2D_1D_0) \leftarrow (x_3I_{03}I_{02}I_{01}) \leftarrow (x_3x_2x_1x_0)$: (1010) as $(x_3x_2x_1x_0)$ appear simultaneously on flip-flop-IV input D_3 and multiplexers inputs $(I_{03}I_{02}I_{01})$. Multiplexers select $(I_{03}I_{02}I_{01})$ and feed input to flip-flops $(D_2D_1D_0)$, respectively. Finally $(D_3D_2D_1D_0)$ are stored in respective flip-flops as $(Q_3Q_2Q_1Q_0)$.

Serial right:

t_1; $(Q_3 \leftarrow D_3 \leftarrow x_3)$: x_3 (1) is given to D_3 input of flip-flop-IV. D_3 is stored in flip-flop-IV as Q_3
t_2; $(Q_2 \leftarrow D_2 \leftarrow I_{13} \leftarrow Q_3)$: Output, Q_3 (1) of flip-flop-IV is given to input, I_{13} of multiplexer-III and is selected to appear as input, D_2 of flip-flop-III. D_2 is stored in flip-flop-III as Q_2.
t_3; $(Q_1 \leftarrow D_1 \leftarrow I_{12} \leftarrow Q_2)$: Output, Q_2 (0) of flip-flop-III is given to input, I_{12} of multiplexer-II and is selected to appear on input, D_1 of flip-flop-II and is stored in flip-flop-II as Q_1.
t_4; $(Q_0 \leftarrow D_0 \leftarrow I_{11} \leftarrow Q_1)$ means output, Q_1 (1) of flip-flop-II is given to input, I_{11} of multiplexer-I and is selected to appear on input, D_0 of flip-flop-I and is stored in flip-flop-I as Q_0.

9.8 | UNIVERSAL SHIFT REGISTER

Shift registers can be used for converting serial data to parallel data, and vice versa. To access all the flip-flop outputs of a shift register, information is entered serially by shifting one bit at a time and information can be taken out in parallel from the outputs of the flip-flops. If a parallel load capability is added to a shift register, then data entered in parallel can be taken out in serial fashion by shifting the data stored in the register. Some shift registers provide the necessary input and output terminals for transfer. They may also have both shift-right and shift-left capabilities. The most general shift register has all the capabilities listed below. Others may have only some of these functions, with at least one shift operation.

1. A clear control to clear the register to 0.
2. Input clock pulses, C_P to synchronize all operations.
3. A shift-right control to enable the shift-right operation and the serial input and output lines associated with the shift-right.
4. A shift-left control to enable the shift-left operation and the serial input and output lines associated with the shift-left.
5. A parallel-load control to enable a parallel transfer and the 'n', input lines associated with the parallel transfer.
6. 'n' parallel output lines.
7. A control state that leaves the information in the register unchanged even though clock pulses are continuously applied.

A register capable of shifting both right and left is called a bidirectional shift register. One that can shift in only one direction is called a unidirectional shift register. If the register has both shift and parallel-load capabilities, it is called a shift register with parallel load.

The diagram of a shift register that has all the capabilities listed above is shown in Figure 9.26. It consists of four D-flip-flops, although RS-flip-flops could be used provided an inverter is inserted between the S and R terminals. The four 4-to-1 line multiplexers (MUX) are part of the register and are shown here in block diagram form. The logic diagram of 4-to-1 multiplexer is shown in Figure 6.37 (Refer Chapter 6). The four multiplexers have two common selection lines, S_1, and S_0. Input I_{0i} in each MUX is selected when $S_1S_0 = 00$, input, I_{1i} is selected when $S_1S_0 = 01$, input, I_{2i} is selected when $S_1S_0 = 10$, and input, I_{3i} is selected when $S_1S_0 = 11$. Multiplexer logic expressions are given below

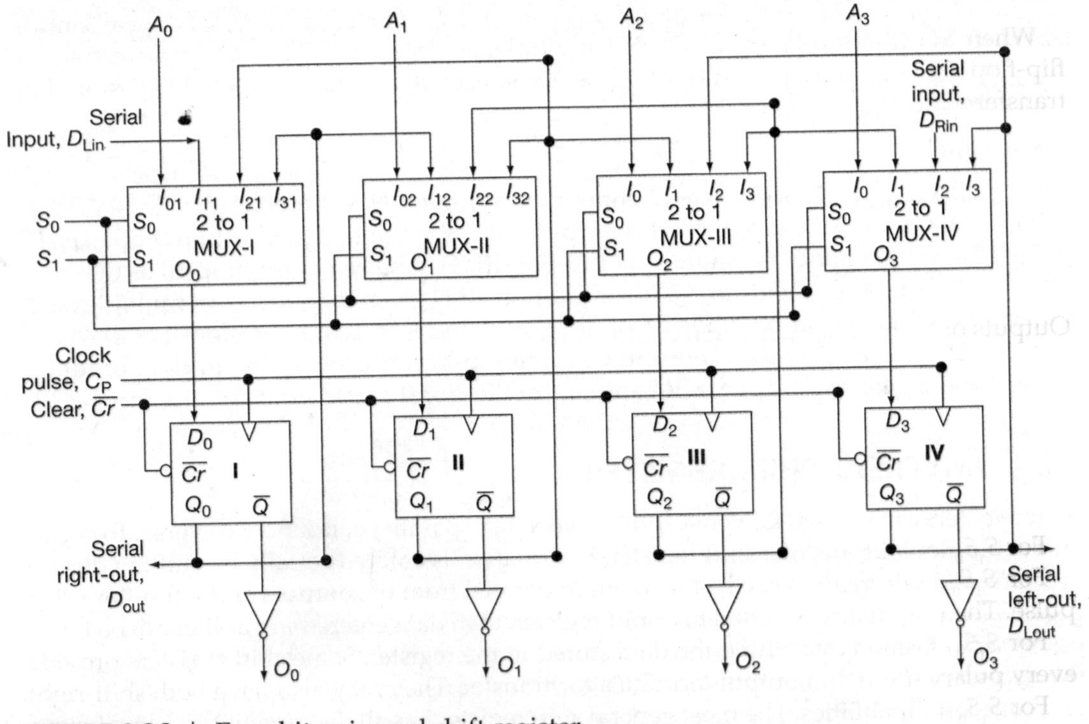

FIGURE 9.26 | Four-bit universal shift register

$$O_0 = S_1S_0Q_0 + S_1\bar{S}_0Q_1 + \bar{S}_1S_0D_{\text{Lin}} + \bar{S}_1\bar{S}_0A_0 \quad (9.5)$$

$$O_1 = S_1S_0Q_1 + S_1\bar{S}_0Q_2 + \bar{S}_1S_0Q_0 + \bar{S}_1\bar{S}_0A_1 \quad (9.6)$$

$$O_2 = S_1S_0Q_2 + S_1\bar{S}_0Q_3 + \bar{S}_1S_0Q_1 + \bar{S}_1\bar{S}_0A_2 \quad (9.7)$$

$$O_3 = S_1S_0Q_3 + S_1\bar{S}_0Q_{\text{Rin}} + \bar{S}_1S_0Q_2 + \bar{S}_1\bar{S}_0A_3 \quad (9.8)$$

The S_1, and S_0 inputs control the mode of operation of the register as specified in the function entries of Table 9.24. When $S_1S_0 = 11$, the present value of the register is applied to the D inputs of the flip-flops. This condition forms a path from the output of each flip-flop into the input of the same flip-flop. The next clock pulse transfers into each flip-flop the binary value it held previously, and no change of state occurs. When $S_1S_0 = 01$, terminals I_{1i} of the multiplexer inputs have a path to the D_i inputs of the flip-flops. This causes a shift-right operation, with the serial input transferred into flip-flop A_0.

TABLE 9.24 | Universal shift register

Selection Lines		Remarks	Flip-flop-IV	Flip-flop-III	Flip-flop-II	Flip-flop-I
S_1	S_0		D_3	D_2	D_1	D_0
0	0	Parallel load	A_3	A_2	A_1	A_0
0	1	Left-shift	Q_2	Q_1	Q_0	D_{Lin}
1	0	Right-shift	D_{Rin}	Q_3	Q_2	Q_1
1	1	Hold (No change)	Q_3	Q_2	Q_1	Q_0

When $S_1S_0 = 10$, a shift-left operation results, with the other serial input going into flip-flop A_3. Finally, when $S_1S_0 = 00$, the binary information on the parallel input lines is transferred into the register simultaneously during the next clock pulse.

$$D_0 = S_1S_0Q_0 + S_1\bar{S}_0Q_1 + \bar{S}_1S_0D_{\text{Lin}} + \bar{S}_1\bar{S}_0A_0 \quad (9.9)$$

$$D_1 = S_1S_0Q_1 + S_1\bar{S}_0Q_2 + \bar{S}_1S_0Q_0 + \bar{S}_1\bar{S}_0A_1 \quad (9.10)$$

$$D_2 = S_1S_0Q_2 + S_1\bar{S}_0Q_3 + \bar{S}_1S_0Q_1 + \bar{S}_1\bar{S}_0A_2 \quad (9.11)$$

$$D_3 = S_1S_0Q_3 + S_1\bar{S}_0Q_{\text{Rin}} + \bar{S}_1S_0Q_2 + \bar{S}_1\bar{S}_0A_3 \quad (9.12)$$

Outputs of D-flip-flops are given below:

$$Q_0(t+1) = D_0 \quad (9.13)$$

$$Q_1(t+1) = D_1 \quad (9.14)$$

$$Q_2(t+1) = D_2 \quad (9.15)$$

$$Q_3(t+1) = D_3 \quad (9.16)$$

For $S_1S_0 = 00$, 1111 is loaded into register.

For $S_1S_0 = 01$, Shift-right operation is performed when 0 is fed serially in after every pulse. The output becomes 0000 after four shifts.

For $S_1S_0 = 10$, shift-left operation is performed when 1, 0, 1 and 0 is fed serially in after every pulse. The output becomes 1010 after four shifts.

For $S_1S_0 = 11$, 1010 is retained in register.

9.9 | RING COUNTER

Ring counter is a kind of shift register in which output of last flip-flop is fed back to input of first flip-flop. Ring counter is also known as simple ring counter or basic ring counter. Figure 9.27 gives circuit diagram of ring counter using four D-flip-flops. Output, Q_3 of flip-flop-IV is given to flip-flop-III as input, D_2. Output, Q_2 of flip-flop-III is given to flip-flop-II as input, D_1. Output, Q_1 of flip-flop-II is given to flip-flop-I as input, D_0. Output, Q_0 of flip-flop-I is fed back to flip-flop-IV as input, D_3.

Let initial sequence $D_3D_2D_1D_0$ is 1000. Input-output relation of flip-flops is given below:

$$D_3 \rightarrow Q_3 \rightarrow D_2 \rightarrow Q_2 \rightarrow D_1 \rightarrow Q_1 \rightarrow D_0 \rightarrow Q_0 \rightarrow D_3 \rightarrow Q_3$$

Input D_3 of flip-flop-IV is stored as Q_3. Output, Q_3 of flip-flop-IV is feed forward to D_2 input of flip-flop-III and so on. Table 9.25 shows the output of flip-flops after each pulse. It is also represented by state diagram shown in Figure 9.28.

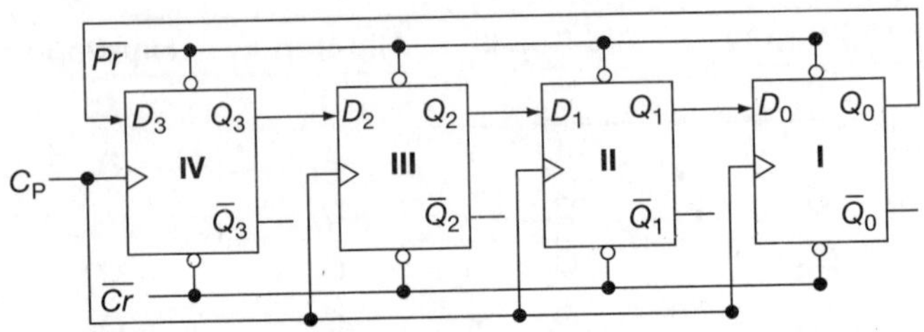

FIGURE 9.27 | Ring counter

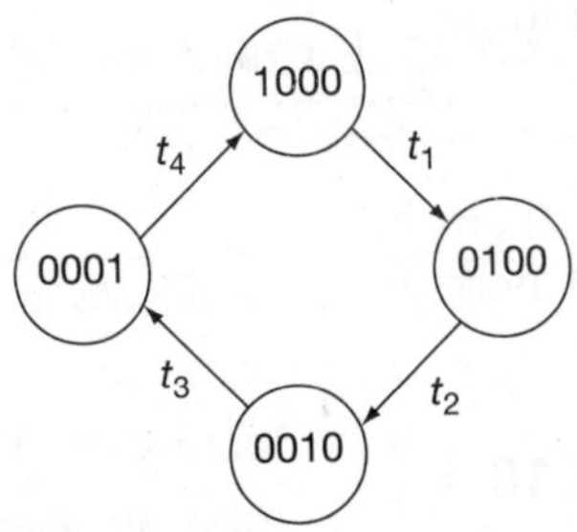

FIGURE 9.28 | State diagram

TABLE 9.25 | Sequence by ring counter

Clock Pulse, C_P	Flip-flop-IV D_3	Flip-flop-III D_2	Flip-flop-II D_1	Flip-flop-I D_0
t_0	1	0	0	0
t_1	0	1	0	0
t_2	0	0	1	0
t_3	0	0	0	1
t_4	1	0	0	0
t_5	0	1	0	0

The circuit is used to count the clock pulses. N-bit ring counter counts N clock pulses. The frequency of the pulses at LSB, Q_0 is the frequency of clock pulse, C_P divided by N. So, ring counter is also known as divided-by-N counter.

Self-Starting Ring Counter Ring counter sometimes suffers from lock-out problem. If a counter enters in unused state, it remains in moving from one unused state to another unused state and never finds its way to enter in used state. It can be overcome by

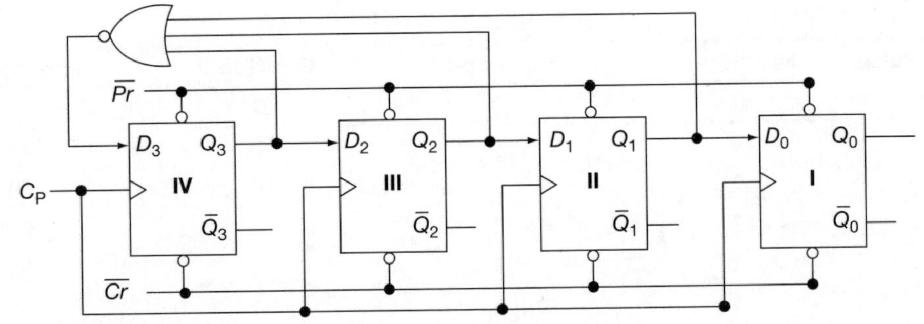

FIGURE 9.29 | Ring counter

self-starting sequence. Figure 9.29 shows a diagram with self-starting sequence. Irrespective of initial state, a single 1 will circulate ultimately. The logic applied is given below:

$$D_3 = \overline{Q_3 Q_2 Q_1} \tag{9.17}$$

If any of Q_3, Q_2 and Q_1 is 1, then D_3 becomes 0. If all the outputs Q_3, Q_2 and Q_1 are 0, then D_3 becomes 1 that is circulated.

9.10 | JOHNSON COUNTER

Johnson counter is a modified version of ring counter. Johnson counter is a kind of shift register in which inverted output of last flip-flop is fed back to input of first flip-flop. Johnson counter is also known as switch-tail ring counter or twisted-ring counter or Moebius counter.

Figure 9.30 gives circuit diagram of Johnson counter using four D-flip-flops. Output, Q_3 of flip-flop-IV is given to flip-flop-III as input, D_2. Output, Q_2 of flip-flop-III is given to flip-flop-II as input, D_1. Output, Q_1 of flip-flop-II is given to flip-flop-I as input, D_0. Inverted output, $\overline{Q}_0$ of flip-flop-I is fed back to flip-flop-IV as input, D_3. Table 9.26 shows the output of flip-flops after every pulse. Figure 9.31 shows the state diagram of the output.

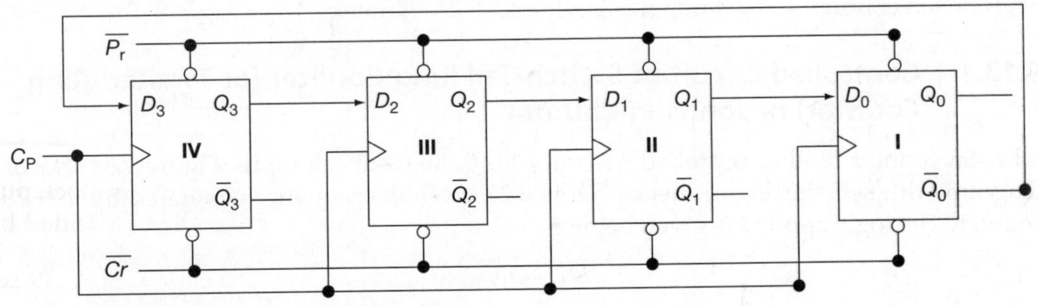

FIGURE 9.30 | Johnson counter

TABLE 9.26 | Sequence by twisted ring counter or Johnson counter

Clock Pulse, C_P	Flip-flop-IV D_3	Flip-flop-III D_2	Flip-flop-II D_1	Flip-flop-I D_0
t_0	0	0	0	0
t_1	1	0	0	0
t_2	1	1	0	0
t_3	1	1	1	0
t_4	1	1	1	1
t_5	0	1	1	1
t_6	0	0	1	1
t_7	0	0	0	1
t_8	0	0	0	0

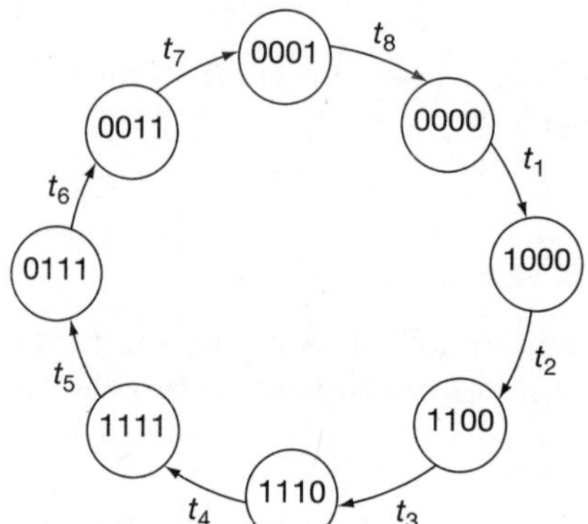

FIGURE 9.31 | State diagram

The circuit is used to count the clock pulses. *N*-bit ring counter counts 2*N* clock pulses. The frequency of the pulses at LSB, Q_0 is the frequency of clock pulse, C_P divided by 2*N*. So, Johnson counter is also known as divided-by-2*N* counter.

9.10.1 | Controlled Circuit of Switch-Tail Ring Counter (or Twisted-Ring Counter) or Johnson Counter

Johnson counter can be controlled by controlling the feedback input. Figure 9.32 shows a diagram with self-starting sequence. Table 9.27 gives the sequence generated by Johnson counter. The logic applied is given below:

$$D_3 = \bar{Q}_1\bar{Q}_0 \tag{9.18}$$

Both Q_1 and Q_0 are zero; then D_3 becomes 1, otherwise D_3 becomes 0. Subsequently, 1 is circulated.

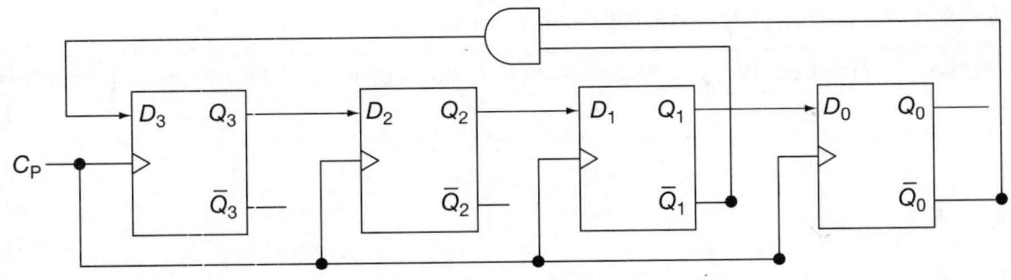

FIGURE 9.32 | Johnson counter to prevent lock out

TABLE 9.27 | Sequence by twisted ring counter or Johnson counter

Clock Pulse, C_P	Flip-flop-IV D_3	Flip-flop-III D_2	Flip-flop-II D_1	Flip-flop-I D_0
t_0	0	0	0	0
t_1	1	0	0	0
t_2	1	1	0	0
t_3	1	1	1	0
t_4	0	1	1	1
t_5	0	0	1	1
t_6	0	0	0	1
t_7	0	0	0	0

9.10.2 | Decoding Count of Johnson Counter

The count sequence of Johnson counter can be decoded using AND gate. Decoding logic is helpful in decoding the repeating logic. The logic diagram of a four-bit Johnson counter with decoding logic is shown in Figure 9.33 that decodes (0000) output of Johnson counter then Y becomes 1. Decoding logic generates

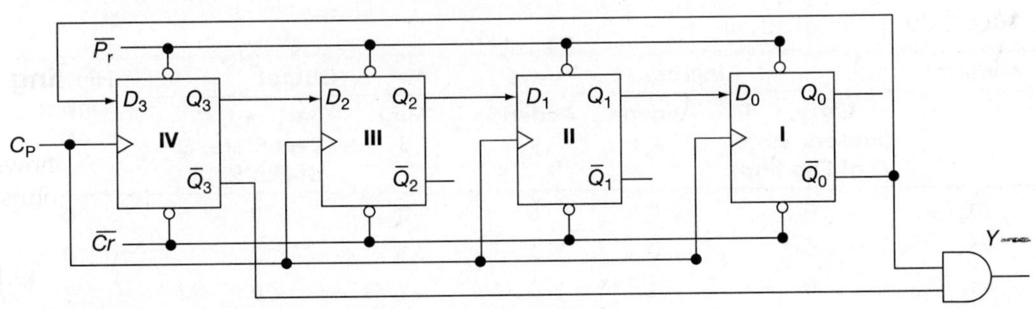

FIGURE 9.33 | Johnson counter with decoding logic

TABLE 9.28 | Decoding logic for Johnson counter

Clock Pulse, C_P	Flip-flop-IV D_3	Flip-flop-III D_2	Flip-flop-II D_1	Flip-flop-I D_0	Decoding Logic (Y)
t_0	0	0	0	0	$\bar{Q}_3\bar{Q}_0$
t_1	1	0	0	0	$Q_3\bar{Q}_2$
t_2	1	1	0	0	$Q_2\bar{Q}_1$
t_3	1	1	1	0	$Q_1\bar{Q}_0$
t_4	1	1	1	1	Q_3Q_0
t_5	0	1	1	1	$\bar{Q}_3Q_2$
t_6	0	0	1	1	$\bar{Q}_2Q_1$
t_7	0	0	0	1	$\bar{Q}_1Q_0$

$Y = 1$ only once during $2N$ clock cycle. Table 9.28 gives the Johnson counter sequence and corresponding decoding logic.

9.11 | SERIAL ADDER

Mostly, digital computers perform operation in parallel because it is a faster mode of operation. Serial operations are slower however requires less equipment. Design of a serial adder is presented here to demonstrate the serial mode of operation.

To add two binary numbers serially, two shift registers store these binary numbers. The serial outputs from the registers are designated by variables, x_i and y_i. The sequential circuit has two inputs, x_i and y_i, that provide a pair of significant bits, an output S_i that generates the sum bit, and flip-flop for storing the carry, C_{i+1}. The present state of flip-flop provides the present value of the carry, C_i. The clock pulse shifts the registers and enables flip-flop to load the next carry, C_{i+1}. This carry is used with the next pair of bits in x_{i+1} and y_{i+1}. The state table that specifies the sequential circuit is given in Table 9.29.

TABLE 9.29 | Serial adder

Minterm	Inputs			Output		Flip-flop Input
	Carry, C_i (Present State, Q of Flip-flop)	Augend x_i	Addend y_i	Sum S_i	Carry-out, C_{i+1} (Next State, Q of Flip-flop)	D
m_0	0	0	0	0	0	0
m_1	0	0	1	1	0	0
m_2	0	1	0	1	0	0
m_3	0	1	1	0	1	1

(Continued)

TABLE 9.29 | (Continued)

Minterm	Inputs			Output		Flip-flop Input D
	Carry, C_i (Present State, Q of Flip-flop)	Augend x_i	Addend y_i	Sum S_i	Carry-out, C_{i+1} (Next State, Q of Flip-flop)	
m_4	1	0	0	1	0	0
m_5	1	0	1	0	1	1
m_6	1	1	0	0	1	1
m_7	1	1	1	1	1	1

$$S_i = x_i \oplus y_i \oplus C \quad (i = 0,1,\ldots,7)$$

Next state of D-flip-flop

$$C_{i+1} = D \tag{9.19}$$

The two binary numbers to be added serially are stored in two shift registers. Bits are added one pair at a time, sequentially, through a single full-adder circuit, as shown in Figure 9.34.

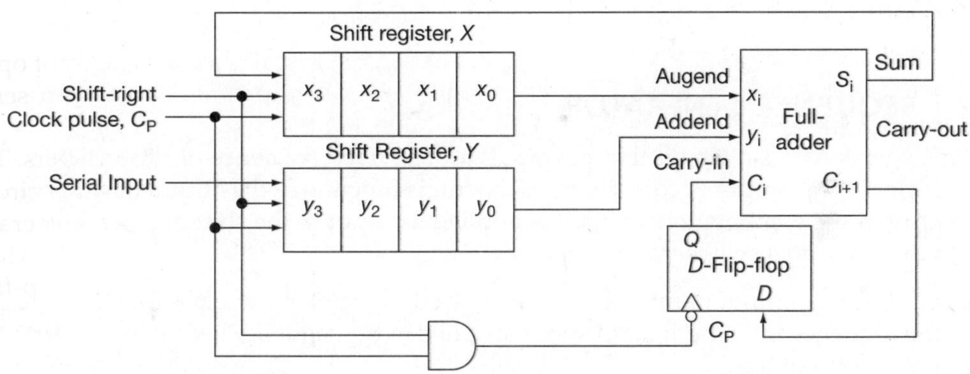

FIGURE 9.34 | Serial adder

The carry out, C_{i+1} of the full-adder is transferred to a D-flip-flop. The output of this flip-flop is used as an input carry, C_i for the next pair of significant bits. The two shift registers are shifted to the right for word-time period. The sum bits from the S_i output of the full-adder can be transferred into a third shift register. By shifting the sum into X register, it is used for storing both the augend and the sum bits. The serial input (SI) of register B is able to receive a new binary number while the addend bits are shifted out during the addition. The shift-right control enables the registers for a number of clock pulses equal to the number of bits in the registers. For each succeeding clock pulse, a new sum bit is transferred to X, a new carry is stored to Q, and both registers are shifted one time to the right. This process continues until the shift-right control is disabled. Thus, the addition is accomplished by passing each pair of bits together with the previous carry through a full-adder circuit and shifting the sum, one bit at a time, into register, X.

The operation of the serial adder is explained as follows. Initially, the X register holds the four-bit augend, the Y register holds four-bit addend, and the carry flip-flop is cleared to 0.

At time t_0, output, Q of the flip-flop gives the input carry at C_0. The serial outputs of shift registers x_0 and y_0 are added in full-adder to give S_0 and C_1.

At time t_1, as first shift pulse appears, both registers are shifted one time to the right and the sum bit, S_0 enters the left-most flip-flop of X as serial input. The output carry, C_1 is stored into D-flip-flop. Now, output Q of the D-flip-flop gives the input carry, C_1. The serial outputs of shift registers x_1 and y_1 are added in full-adder to give S_1 and C_2.

At time t_2, as first shift pulse appears, both registers are shifted one time to the right and the sum bit, S_0 enters the left-most flip-flop of X as serial input. The output carry, C_2 is stored into D-flip-flop. Now, output Q of the D-flip-flop gives the input carry, C_2. The serial outputs of shift registers x_2 and y_2 are added in full-adder to give S_2 and C_3.

At time t_3, as first shift pulse appears, both registers are shifted one time to the right and the sum bit, S_0 enters the left-most flip-flop of X as serial input. The output carry, C_3 is stored into D-flip-flop. Now, output Q of the D-flip-flop gives the input carry, C_3. The serial outputs of shift registers x_3 and y_3 are added in full-adder to give S_3 and C_4.

At time t_4, as first shift pulse appears both registers are shifted one time to the right and the sum bit, S_0 enters the left-most flip-flop of X as serial input. The output carry, C_4 is stored into D-flip-flop.

9.12 | SEQUENCE GENERATOR

Sequence generator is a circuit that generates a prescribed sequence of bits synchronized with the clock. The output of combinational circuit is function of the output of shift register. The output of the combinational circuit is applied as input to the shift register. The procedure to design sequence generator is given below:

1. Find the minimum number of flip-flops needed to generate sequence. The length of the sequence is the number of successive bits in the sequence before the repetition of the pattern of bits.
2. Minimum number of flip-flops, N can be determined to satisfy the following relation

$$L \le 2^N - 1 \text{ or } N \ge \log_{10}(L+1)/\log_{10} 2$$

3. Construct the truth table of N-bit shift register. The defined sequence is enlisted under Q_{N-1}.
   Check the repetition of states in the table. If repetition of states exists, it means N flip-flops are insufficient to generate sequence. Increase number of flip-flops as $N+1$ and repeat Step 2 till no repetition appears.
4. Set up K-map or tabular method to simplify the logic.
5. Draw the combination circuit.

The design of sequence generator is illustrated with example.

EXAMPLE 9.7 Design a sequence generator to generate the sequence 1011110...

SOLUTION Given sequence $S = 1011110$

Length of sequence, $L = 7$

$$7 \le 2^N - 1 \text{ or } 8 \le 2^N \text{ or } 8 \le 2^3 - 1$$

So, minimum number of flip-flop required = 3.

Table 9.30 gives truth table of pulse generator using 3-flip-flops. Assign sequence (1011110) to variable Q_3 (row-wise). Rotate the sequence by 1-bit to get (0111101) and assign it to variable Q_2. Similarly, variable Q_1 and output f are taken by rotating the sequence of Q_2 and Q_1, respectively by one bit. $Q_3 Q_2 Q_1$ represents number.

TABLE 9.30 | Truth table pulse generator

Minterm	Input			Output
	Q_3	Q_2	Q_1	Function, f
m_5	1	0	1	1
m_2	0	1	0	1
m_5	1	0	1	0
m_6	1	1	0	1
m_7	1	1	1	0
m_7	1	1	1	1
m_3	0	1	1	1

There is repetition of m_5 and m_7. So, number of flip-flops needed is 4 (3 + 1). Table 9.31 gives the truth table of pulse generator using 4-flip-flops.

TABLE 9.31 | Truth table pulse generator

Minterm	Input				Output
	Q_3	Q_2	Q_1	Q_0	Function, f
M_{11}	1	0	1	1	1
M_5	0	1	0	1	1
M_{10}	1	0	1	0	1
M_{13}	1	1	0	1	0
M_{14}	1	1	1	0	1
M_{15}	1	1	1	1	0
M_7	0	1	1	1	1

Canonical form of decoding logic is given below. Other minterms can be treated as don't care states.

$$f(Q_3, Q_2, Q_1, Q_0) = \sum m(5,7,10,11,14) + \sum d(0,1,2,3,4,6,8,9,12) \tag{9.20}$$

The logical function is simplified using four-variable K-map set-up and K-map set-up is shown in Figure 9.35. Three octets are formed. The simplified logical expression is given below.

$$F(Q_3,Q_2,Q_1,Q_0)=\bar{Q}_3+\bar{Q}_2+\bar{Q}_0 \tag{9.21}$$

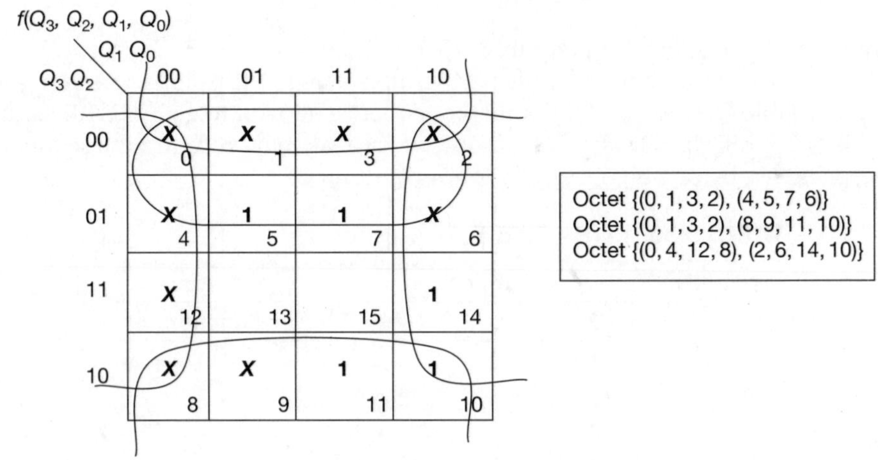

FIGURE 9.35 | K-map

The circuit diagram is shown in Figure 9.36.

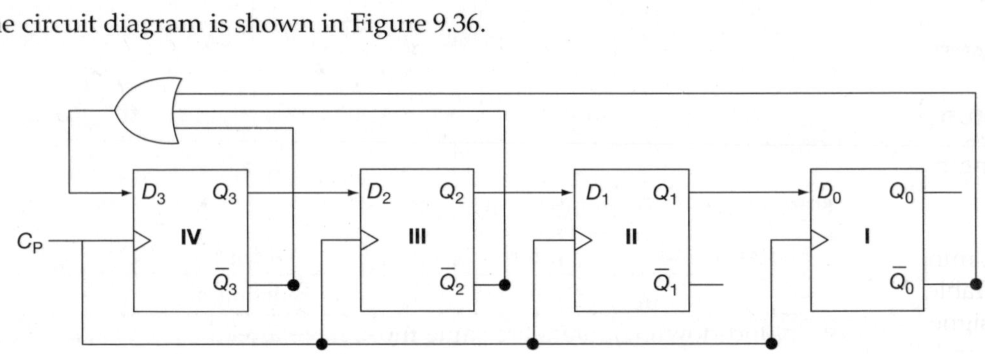

FIGURE 9.36 | Ring counter for sequence pulse generator

9.13 | SEQUENCE DETECTOR

A sequence detector circuit detects the correct sequence received by receiving which is transmitted by the transmitter. Shift register is used to detect the sequence. Shift register requires number of flip-flops that is equal to the length of the sequence. Figure 9.37 shows the sequence detector circuit. Sequence to be detected is stored in a register. The sequence can be modified also. Sequence is received by the shift register serially. The received bits are compared with the sequence bits to be detected using Exclusive-NOR gates. Output of flip-flops is compared with the corresponding desired bit. If received bit is equal to the desired sequence bit, then EX-NOR gates produce logic-0. OR gate produces logic-0, if received sequence is equal to the desired sequence then shift register will not receive bit because clock is inhibited to 0. Otherwise shift register receives another bit in sequence as MSB, Q_3 of sequence and LSB, Q_0 of sequence is lost is lost.

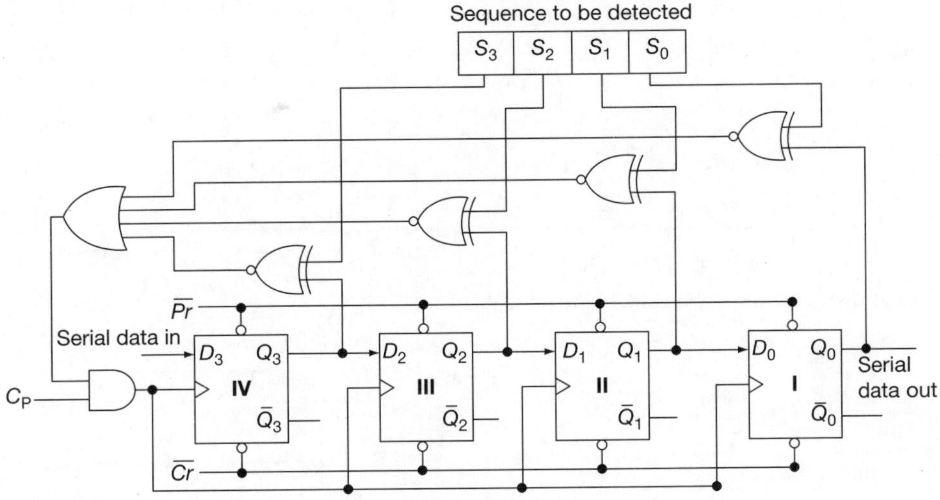

FIGURE 9.37 | Sequence detector

EXAMPLE 9.8 Design a sequence generator to generate the sequence 1001110...

SOLUTION Given sequence $S = 1001110$

Length of sequence, $L = 7$

$$7 \le 2^N - 1 \text{ or } 8 \le 2^N \text{ or } 8 \le 2^3 - 1$$

So, minimum number of flip-flop required = 3.

Table 9.32 gives the truth table of pulse generator. The sequence to be generated is assigned with variable, Q_3. Rotate the sequence by one bit and assign variable, Q_2. Similarly, Q_1 and output, f are taken by rotating the sequence by one bit. It ensures that no number is repeated when $Q_3Q_2Q_1$ represents number. Canonical form of decoding logic is given below. Other minterms can be treated as don't care states.

$$f(Q_3, Q_2, Q_1) = \sum m(2,3,4,5) + \sum d(0) \tag{9.22}$$

The logical function is simplified using three-variable K-map set-up and K-map set-up is shown in Figure 9.38. Two pairs are formed. The simplified logical expression is given below.

$$F(Q_3, Q_2, Q_1) = Q_3\bar{Q}_2 + \bar{Q}_3Q_2 \tag{9.23}$$

$$F(Q_3, Q_2, Q_1) = Q_3 \oplus Q_2 \tag{9.24}$$

TABLE 9.32 | Truth table pulse generator

Minterm	Input			Output
	Q_3	Q_2	Q_1	Function, f
m_5	1	0	1	1
m_2	0	1	0	1
m_1	0	0	1	0
m_4	1	0	0	1
m_6	1	1	0	0
m_7	1	1	1	0
m_3	0	1	1	1

The circuit diagram is shown in Figure 9.39.

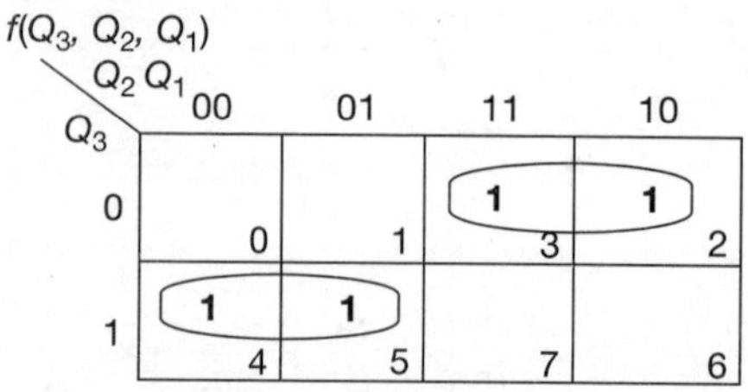

FIGURE 9.38 | Three-variable K-map

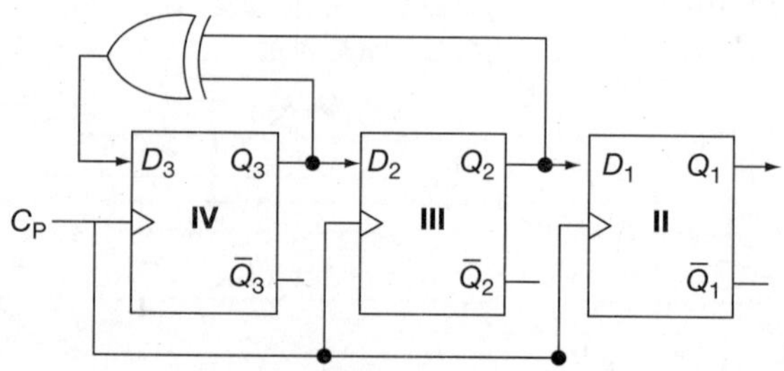

FIGURE 9.39 | Ring counter for sequence pulse generator

EXAMPLE 9.9 How much time delay is produced by eight-bit shift register operating on 2 MHz?

SOLUTION Time delay given by N-bit shift register $T_D = N \times \dfrac{1}{f}$

where N is the size of shift register and f is the clock frequency.
Here, $N = 8$ and $f = 10\,\text{MHz}$

$$T_D = 8 \times \frac{1}{10 \times 10^6} = 0.8\ \mu\text{s} \quad \textbf{Ans.}$$

EXAMPLE 9.10 How much time is required to shift hexadecimal number into four-bit serial shift register if operating clock frequency is 5 MHz?

SOLUTION Time required by *N*-bit shift register $T = N \times \frac{1}{f}$

where *N* is the size of shift register. *f* is the clock frequency.
Here, $N = 4$ and $f = 5\text{MHz}$

$$T = 4 \times \frac{1}{5 \times 10^6} = 0.8\ \mu\text{s} \quad \textbf{Ans.}$$

EXAMPLE 9.11 Eight-bit shift register and D-flip-flop is synchronized with same clock as shown in Figure 9.40. Initially D-flip-flop is in clear state and register has A9H. What will be the contents of register after four clock pulses?

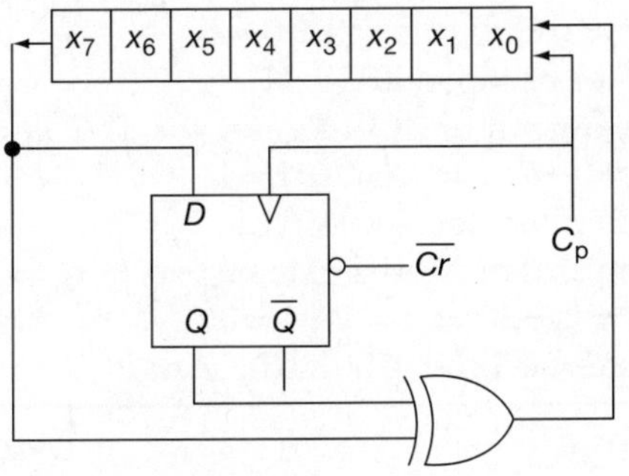

FIGURE 9.40 | Circuit diagram

SOLUTION Initial contents of register are A9H.

Q is 0 of D-flip-flop.

1	0	1	0	1	0	0	1

After first clock pulse: output of XOR gate, $y = x_7 \oplus Q \Rightarrow y = 1 \oplus 0 = 1$
So, contents of shift register become 01010011 and x_7 is stored in D-flip-flop, so $Q = 1$.
After second clock pulse: output of XOR gate, $y = x_6 \oplus x_7 \Rightarrow y = 0 \oplus 1 = 1$
So, contents of shift register become 10100111 and x_6 is stored in D-flip-flop, so $Q = 0$.
After third clock pulse: output of XOR gate, $y = x_5 \oplus x_6 \Rightarrow y = 1 \oplus 0 = 1$
So, contents of shift register become 01001111 and x_5 is stored in D-flip-flop, so $Q = 1$.
After fourth clock pulse: output of XOR gate, $y = x_4 \oplus x_5 \Rightarrow y = 0 \oplus 1 = 1$
So, contents of shift register become 10011111 and x_4 is stored in D-flip-flop, so $Q = 0$.

EXAMPLE 9.12 The initial contents of SIPO right-shift register shown in Figure 9.41 are 01101110. What will be the contents of shift register after three clock pulses?

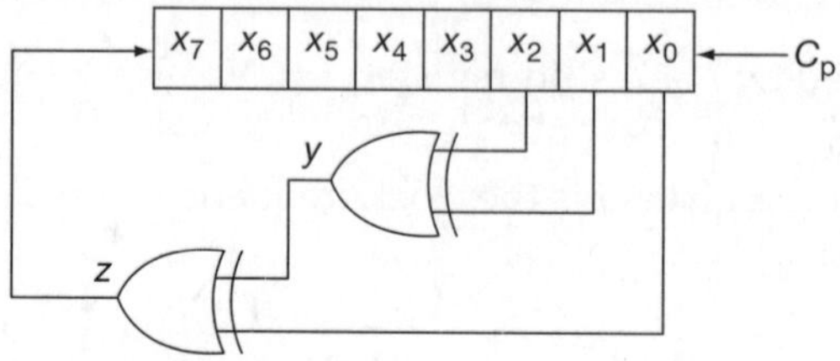

FIGURE 9.41 | Circuit diagram

SOLUTION Initial contents of register are

0	1	1	0	1	1	1	0

After first clock pulse: output of XOR-I gate, $y = x_2 \oplus x_1 \Rightarrow y = 1 \oplus 1 = 0$
Output of XOR-II gate, $z = y \oplus x_0 \Rightarrow z = 0 \oplus 0 = 0$
So, contents of shift register become 00110111.
After second clock pulse: output of XOR-I gate, $y = x_2 \oplus x_1 \Rightarrow y = 1 \oplus 1 = 0$
Output of XOR-II gate $z = y \oplus x_0 \Rightarrow z = 0 \oplus 1 = 1$
So, contents of shift register become 10011011.
After third clock pulse: output of XOR-I gate, $y = x_2 \oplus x_1 \Rightarrow y = 0 \oplus 1 = 1$
Output of XOR-II gate $z = y \oplus x_0 \Rightarrow z = 1 \oplus 1 = 0$
So, contents of shift register become 01001101 **Ans.**

9.14 | LIST OF SHIFT REGISTER ICs

A list of commonly used IC type numbers used as registers is provided in Table 9.33. Information regarding the pin connection diagram, truth table, etc., in respect of these IC numbers is given in the handbook or companion website.

TABLE 9.33 | Flip-flop of CMOS logic family

IC Number	Function of IC
74L91	Eight-bit SISO shift register
74164	Eight-bit SIPO shift register
74165	Eight-bit PISO shift register
74194	Eight-bit PIPO shift register
74198	Eight-bit PISO shift register
54L91	Eight-bit SIPO shift register
54164	Eight-bit SIPO shift register
54165	Eight-bit PISO shift register
54194	Four-bit PIPO shift register
54198	Eight-bit PIPO shift register

SUMMARY

- Bidirectional register is a shift register in which data can be shifted in both directions that is from left to right and right to left.
- Left-shift register is a shift register in which data gets shifted in the left direction in response to clock pulse.
- Lock out is a condition which may exist in a counter taking the counter from one unused state to another unused state and is not allowing it to come to any of the used states.
- Parallel data: All the data bits are available simultaneously on different data lines.
- Parallel-loading simultaneous loading of all the bits in a register or counter.
- Parallel-to-serial converter is a logic circuit that converts data from parallel to serial form.
- Right-shift register is a shift register in which the data gets shifted in the right direction (higher significant bit to lower significant bit) in response to clock pulses.
- Ring counter is a shift register whose output is fed back to the input that is the data circulates around the flip-flop in response to clock pulses.
- Sequence detector is a synchronous sequential circuit which can detect a sequence of binary input.
- Sequential circuit is a logic circuit whose outputs are determined by the sequence in which input signals are applied.
- Serial adder is an adder circuit in which addition is performed serially, that is bit by bit.
- Serial data is data arranged as one bit at a time at a regular interval starting from the LSB.
- Serial-to-parallel converter is a logic circuit that converts data from serial to parallel form.
- Shift register counters are shift registers with feedback that exhibit special sequences. Examples are Johnson counter and ring counter.
- Timing diagram is a graphical representation of various signals with reference to time.
- The Johnson counter has $2n$ states in its sequence, where n is number of stages (flip-flops).
- The ring counter has n-states in its sequence.
- Transition table is a tabular form of representing the behaviour of a sequential circuit.
- Twisted-ring counter or Johnson counter is a shift register with its complement output ($\bar{Q}$) of the last stage connected to the D-input of the first stage.

MULTIPLE CHOICE QUESTIONS

9.1 A stage in a shift register consists of
   (a) a latch
   (b) a flip-flop
   (c) a byte of storage
   (d) a nibble of storager

9.2 Dynamic shift registers are made up of
   (a) dynamic flip-flops
   (b) MOS inverters
   (c) MOS NAND gates
   (d) CMOS inverters

9.3 A PIPO register
(a) accepts serial input and transfers parallel data
(b) accepts parallel input and transfers parallel data
(c) accepts serial input and gives out serial data
(d) accepts parallel data and gives serial data

9.4 A universal register
(a) accepts serial input
(b) accepts parallel input
(c) gives serial and parallel outputs
(d) is capable of all of the above

9.5 A shift register accepts serial input and gives out serial data is known as
(a) SISO
(b) SIPO
(c) PIPO
(d) PISO

9.6 A pulse train can be delayed by a finite number of clock periods timing
(a) SISO
(b) SIPO
(c) PISO
(d) PIPO

9.7 Data can be changed from spatial code to temporal code and vice-versa using
(a) ADC and DAC
(b) Shift registers
(c) Timers
(d) Synchronous counters

9.8 A ring counter is same as
(a) up–down counter
(b) parallel counter
(c) shift register
(d) none of the above

9.9 How many flip flops are required for dividing the frequency by 64?
(a) 4 (b) 5
(c) 6 (d) 8

9.10 How much time is taken to shift a byte in eight-bit PIPO shift register with 1 MHz clock frequency?
(a) 1 μs
(b) 4 μs
(c) 8 μs
(d) none of the above

9.11 How much time is taken to shift a byte in eight-bit SISO shift register with 1 MHz clock frequency?
(a) 1 μs
(b) 4 μs
(c) 8 μs
(d) none of the above

9.12 How much time is taken to shift a nibble in four-bit SIPO shift register working on 1 MHz clock frequency?
(a) 1 μs
(b) 4 μs
(c) 8 μs
(d) none of the above

9.13 What is output of a 10-bit SISO right-shift register after three clock pulses if alternative 1 and 0 are fed in at MHz clock frequency? Initially register is cleared to zero.
(a) 1010000000
(b) 0000000101
(c) 1010101010
(d) none of the above

9.14 What is output of a 10-bit SISO left-shift register after three clock pulses if 1 is fed in at MHz clock frequency every time? Initially register is cleared to zero.
(a) 1110000000
(b) 0000000111
(c) 1111111111
(d) none of the above

9.15 How many flip flops are required to generate the sequence of binary bits as …1011110…?
(a) 3 (b) 4
(c) 5 (d) 6

9.16 To serially shift a byte of data into a shift register, there must be
 (a) one clock pulse
 (b) one load pulse
 (c) eight clock pulses
 (d) one clock pulse for each 1 in the data

9.17 To parallel load a byte of data into a shift register with a synchronous load, there must be
 (a) one clock pulse
 (b) one clock pulse for each 1 in the data
 (c) eight clock pulses
 (d) one clock pulse for each 0 in the data

9.18 The group of bits 10110101 is serially shifted (right-most bit first) into an eight-bit parallel output shift register with an initial state of 11100100. After two clock pulses, the register contains
 (a) 01011110
 (b) 10110101
 (c) 01111001
 (d) 00101101

9.19 With a 100 kHz clock frequency, eight bits can be serially entered into a shift register in
 (a) 80 μs (b) 8 μs
 (c) 80 ms (d) 10 μs

9.20 With a I MHz clock frequency, eight bits can be parallel entered into a shift register
 (a) in 8 μs
 (b) in the propagation delay time of eight flip-flops
 (c) in 1 μs
 (d) in the propagation delay time of one flip-flop

9.21 A modulus-10 Johnson counter requires
 (a) 10 flip-flops
 (b) 4 flip-flops
 (c) 5 flip-flops
 (d) 12 flip-flops

9.22 A modulus-10 ring counter requires a minimum of
 (a) 10 flip-flops
 (b) 4 flip-flops
 (c) 5 flip-flops
 (d) 12 flip-flops

9.23 When an eight-bit serial in/serial out shift register is used for a 24 μs time delay, the clock frequency must be
 (a) 41.67 kHz
 (b) 333 kHz
 (c) 125 kHz
 (d) 8 MHz

Answers

9.1 (b)	9.2 (b)	9.3 (b)	9.4 (d)	9.5 (a)	9.6 (a)	9.7 (b)
9.8 (c)	9.9 (c)	9.10 (a)	9.11 (c)	9.12 (b)	9.13 (a)	9.14 (b)
9.15 (b)	9.16 (c)	9.17 (a)	9.18 (c)	9.19 (a)	9.20 (d)	9.21 (c)
9.22 (a)	9.23 (b)					

QUESTIONS

9.1 What do you mean by serial data and parallel data?

9.2 What are the types of loading the registers? What is serial loading and parallel loading?

9.3 Discuss the applications of shift registers.

9.4 With neat diagrams explain the working of the following types of shift registers.
   (a) serial-in, serial-out
   (b) serial-in, parallel-out
   (c) parallel-in, serial-out
   (d) parallel-in, parallel-out
   (e) bidirectional

9.5 Design a four-bit universal shift register using JK-flip-flops and draw the circuit.

9.6 Design a sequence detector circuit.

9.7 Design a serial adder using JK-flip-flop.

9.8 Design a parallel-in parallel-out shift register using JK-flip-flops.

9.9 Design a serial-in parallel-out shift register using JK-flip-flops.

9.10 Design a parallel-in serial-out shift register using JK-flip-flops.

9.11 Design a bidirectional shift register using JK-flip-flop.

PROBLEMS

9.1 Design a sequence generator circuit to generate a sequence …101101100… using D-flip-flop

9.2 Design a sequence generator circuit to generate a sequence …1101000… using SR-flip-flop

9.3 Assume that 1011 input data pattern is loaded into a four-bit ring counter. Sketch the resulting flip-flop output waveform. Assume positive edge triggering

9.4 Assume that 0000 input data pattern is loaded into a four-bit Johnson counter. Sketch the resulting flip-flop output waveform. Assume positive edge triggering

9.5 Add 10110 and 01101 using serial adder. Explain each step.

10

Counters

CHAPTER OBJECTIVES

The main goal of this chapter is to impart knowledge about the counters. Readers will be able to discuss the following aspects in this chapter:

- Asynchronous counters: Modulus-4, -8, -10, -16 up-and-down counters
- Asynchronous counters with parallel load, decoding logic, modulus-M counters
- Synchronous counters: Modulus-4, -8, -10, -16 up-and-down counters
- Synchronous modulus counters with parallel load
- Self-correcting counters
- Cascading of counters: Modulus-6, -10
- Pulse Train/Sequence generator and detector

10.1 | INTRODUCTION

Logic circuits are classified into combinational logic circuits and sequential logic circuits. A logic where the output depends not only on the latest inputs, but also on the condition of earlier inputs, such circuits are known as sequential, and implicitly these contain memory elements. A memory stores data, usually one bit per element. A one bit memory is often called a bi-stable, i.e., it has 2 stable internal states. Flip-flops and latches are particular implementations of bi-stables. The sequential circuit is known as asynchronous sequential circuit, when the output state changes occur directly in response to changes in the inputs in a sequential circuit. So, asynchronous term refers to events that occur at different time or events do not have same time relationship with each other. The sequential circuit is known as synchronous sequential circuit, when the output of a sequential circuit is constrained to change only at a time specified by a global enabling signal. So, synchronous term refers to events that occur at same time or events have fixed time relationship with each other. This signal is generally known as the system clock. The clock is a square wave signal at a particular frequency.

- Counter is a clocked sequential circuit that goes through a predetermined sequence of states upon the application of input pulses.
- Counters are designed using T- or JK-flip flops because these flip-flops have toggle states.

- A counter follows binary sequence is called an n-bit binary counter.
- Binary counter has n-flip-flops and possible maximum number of states are 2^n, which are passed through in the order 0, 1, 2, …, $2^n - 1$ and repeats.
- Up binary counter increases the output count by one with the clock pulse. Up binary counter advances count upward through its sequence 0, 1, 2, …, $2^n - 1$ and repeats from 0.
- Down binary counter decreases the output count by one with the clock pulse. Down binary counter advances count downward through its sequence $2^n - 1$, $2^n - 2$, $2^n - 3$, …, 2, 1, 0 and repeats from $2^n - 1$.
- The modulus (MOD) of a counter is the number of unique states through which the counter sequences.
- For example MOD-128 progress 128 numbers of states. MOD-128 counter uses 7 ($2^7 = 128$) T- or JK-flip-flops.
- Binary counters can have truncated sequences. So, counters can have less than maximum of 2^n states.
- A n-bit counter has n-flip-flops and 2^n states and divides input frequency by 2^n or MOD-N. ($N \le 2^n$).
- Counters can be classified into three broad categories according to the way counters are clocked: (i) asynchronous, (ii) synchronous and (iii) hybrid.
- Asynchronous counter is one in which the flip-flops within the counter do not share the same clock pulse so the flip-flops do not change their states at the same time. Asynchronous counter is also known as ripple counter, serial or series counter.
- Synchronous counter is one in which all the flip-flops in the counter share same clock pulse, so the flip-flops change their states at the same time. Synchronous counters are also known as parallel counters.
- Hybrid counter combines asynchronous and synchronous counters.

Counter applications include counting, producing delays of a particular duration, sequencers for control logic in a processor and divide by m counter (a divider), as used in a digital watch.

10.2 | ASYNCRONOUS OR RIPPLE COUNTER

Asynchronous counter uses n number of T- or JK-flip-flops. First flip-flop is clocked by external clock pulse. After that output of each flip-flop is given to clock pulse input of each successive flip-flop. n-bit asynchronous counter counts 2^n clock pulses.

10.2.1 | Modulus-4 Asynchronous (Ripple) Up Counter

Modulus-4 (MOD-4) asynchronous or ripples up counter uses two number of T-flip-flops. First flip-flop, FF0 is clocked by external clock pulse, C_{P_0}. Output, Q_0 of flip-flop, FF0 is given to clock pulse of successive flip-flop, FF1. Inputs of T-flip-flops, FF0 and FF1 are:

$$T_0 \leftarrow 1 \text{ and } T_1 \leftarrow 1$$

Clock pulse of T-flip-flops, FF0 and FF1 are:

$$C_{P_0} \leftarrow C_P \quad \text{and} \quad C_{P_1} \leftarrow Q_0$$

Figure 10.1 shows the logic diagram of MOD-4 asynchronous or ripples up counter. MOD-4 asynchronous up counter counts 4 (2^2) clock pulses and counts sequence from 0, 1, 2, 3 and 0 as shown in Figure 10.2. The inputs of T-flip-flops are connected to 1-logic. The flip-flop changes its state when it follows negative clock pulse or falling edge clock pulse as shown in Table 10.1. Further, flip-flop changes its state when the output of flip-flop changes from 1 to 0-logic, because it is connected to input of next input clock of T-flip-flop. The counting sequence of MOD-4 ripples counter is given in Table 10.2 and the waveforms are shown in Figure 10.3.

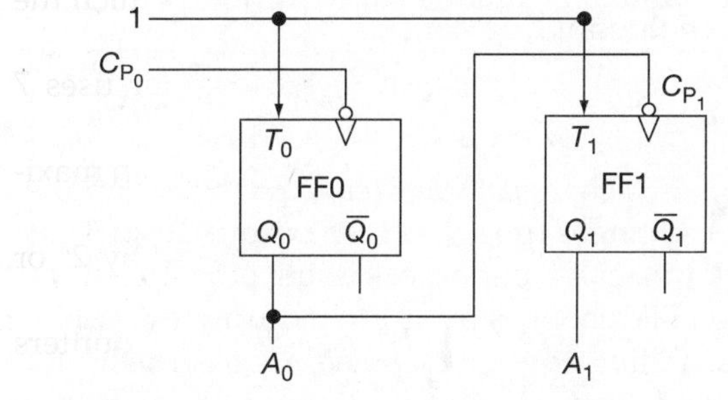

Output, Q_{n-1} of $(n-1)$th flip-flop is given to clock pulse, C_{P_n} of nth flip-flop.

FIGURE 10.1 | MOD-4 asynchronous up counter

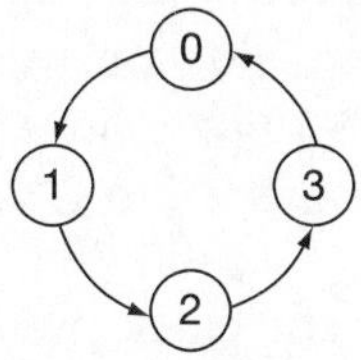

FIGURE 10.2 | Up count sequence

TABLE 10.1 | Characteristics of T-flip-flop

Input T (Toggle)	Output $Q(t+1)$	Remarks
0	$Q(t)$	Save state
1	$\overline{Q}(t)$	Inverted state

TABLE 10.2 | MOD-4 up asynchronous counter

Clock Count	Count Sequence A_1	A_0	Remarks
0	0	0	A_0 changes from 0 to 1. Since, A_0 changes from 0 to 1, so A_1 does not change.
1	0	1	A_0 changes from 1 to 0. Since, A_0 changes from 1 to 0, so A_1 changes from 0 to 1.
2	1	0	A_0 changes from 0 to 1. Since, A_0 changes from 0 to 1, so A_1 does not change.
3	1	1	A_0 changes from 1 to 0. Since, A_0 changes from 1 to 0, A_1 changes from 1 to 0.
0	0	0	

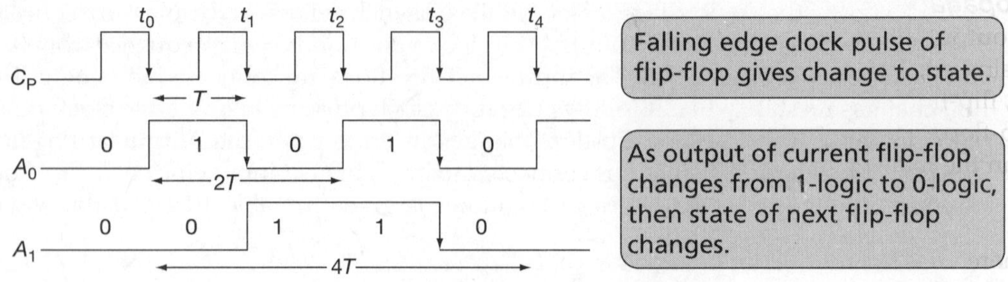

FIGURE 10.3 | Timing diagram of MOD-4 synchronous up counter

The working of counter is explained below.

- Input, T_0 is 1, so output, A_0 of T-flip-flop, FF0 changes from 0 to 1 during the applied falling edge of clock pulse, C_{P_0}. Since, A_0 changes from 0 to 1, so output, A_1 of T-flip-flop, FF1 does not change because clock pulse, C_{P_1} undergoes rising edge transition.
- Input, T_0 is 1, so output, A_0 of T-flip-flop, FF0 changes from 1 to 0 during the applied falling edge of clock pulse, C_{P_0}. It acts as falling edge clock pulse, C_{P_1} to FF1-flip-flop and input, T_1 is 1, so output, A_1 of T-flip-flop, FF1 changes from 0 to 1.
- Input, T_0 is 1, so output, A_0 of T-flip-flop, FF0 changes from 0 to 1 during the applied falling edge of clock pulse, C_{P_0}. Since, A_0 changes from 0 to 1, so output, A_1 of T-flip-flop, FF1 does not change because clock pulse, C_{P_1} undergoes rising edge transition.
- Input, T_0 is 1, so output, A_0 of T-flip-flop, FF0 changes from 1 to 0 during the applied falling edge of clock pulse, C_{P_0}. It acts as falling edge clock pulse, C_{P_1} to FF1-flip-flop and input, T_1 is 1, so output, A_1 of T-flip-flop, FF1 changes from 0 to 1.

Frequency Division Each flip-flop changes state whenever the clock pulse goes from 1 to 0 (falling edge transition).

A_0 changes state at the falling edge of each pulse. Time period of output, A_0 waveform is $2T$. So the output, A_0 waveform frequency is equal to one half of the clock frequency.

$$f_{A_0} = f/2$$

A_1 changes state each time when Q_0 changes from 1 to 0. Time period of output, A_1 waveform is $4T$. So the frequency of waveform of output, A_1 is equal to one half of the clock frequency, f_{A_0}.

$$f_{A_1} = \frac{f_{A_0}}{2} = \frac{f}{4}$$

It can be generalized as

$$f_{A_n} = \frac{f}{2^n} \Rightarrow$$

$$\text{Output frequency, } f_{\text{out}} = \frac{\text{Input frequency, } f}{\text{Modulus of counter, } 2^n} \tag{10.1}$$

where n is the number of flip-flops in MOD-2^n counter.

f_{A_n} is the frequency of waveform of output of nth flip-flop.

Propagation Delay In asynchronous counter, each flip-flop is triggered by the change in output of the proceeding flip-flop from 1 to 0 (falling edge clock pulse) or from 0 to 1 (rising edge clock pulse). Flip-flop has inherent propagation delay, t_{pd} to produce output. So, flip-flop, FF1 will respond after t_{pd} time on receiving the falling edge clock pulse by the flip-flop, FF0 (Figure 10.4). To avoid this problem, the period of input pulses made longer than the total propagation.

$$T_{\text{clock}} \geq nt_{\text{pd}} \tag{10.2}$$

where n is the number of flip flops in the ripples counter
t_{pd} is the inherent propagation delay of flip-flop.

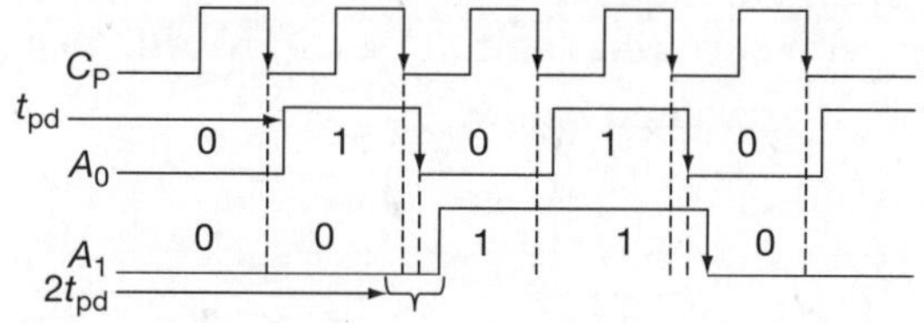

FIGURE 10.4 | Timing diagram of MOD-4 synchronous up counter with propagation delay

10.2.2 | Modulus-3 Asynchronous (Ripples) Up Counter with Decoded Output

A decoded logic is one which is connected to the outputs of a counter. Its output will be high only when counter state is equal to the given state. Decoding is used to truncate the sequence before reaching to maximum count, i.e. 3 ($2^2 - 1$). MOD-3 asynchronous counter counts three states that are 0, 1, 2. So $Y = \overline{A_1 \bar{A}_0}$ is used to clear the flip-flop outputs to 0. Table 10.3 shows the counting sequence. Figure 10.5 shows the asynchronous MOD-4 up counter with decoding logic using NAND gate to feed active low signal to $\overline{Cr}$ of flip-flops. Such count will count from 0 to 2 only as shown in Figure 10.6. The waveform counting sequence of MOD-3 ripples counter is shown in Figure 10.7.

TABLE 10.3 | MOD-4 up asynchronous counter

Clock Count	Count Sequence		Boolean Logic
	A_1	A_0	
0	0	0	$\bar{A}_1\bar{A}_0$
1	0	1	$\bar{A}_1 A_0$
2	1	0	$A_1\bar{A}_0$
3	1	1	$A_1 A_0$

At the decoder outputs false pulse of short duration are possible in any counter unless all flip-flops change state simultaneously. The false pulse is known as spike or glitch. This problem is overcome by using strobe or enables pulse input, so that counter is read only after spike is decayed and steady state is reached.

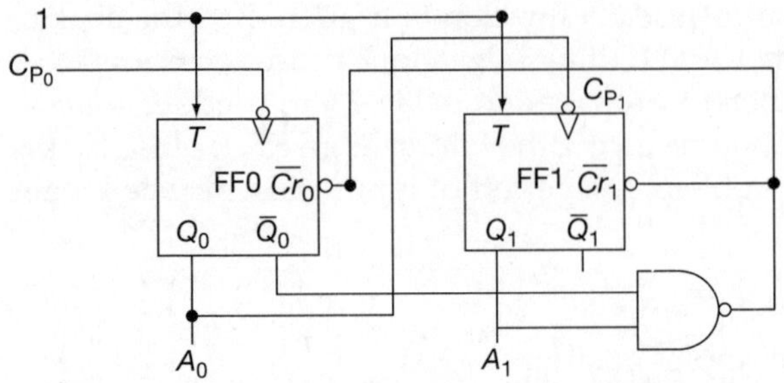

FIGURE 10.5 | MOD-4 asynchronous up counter with decoding logic

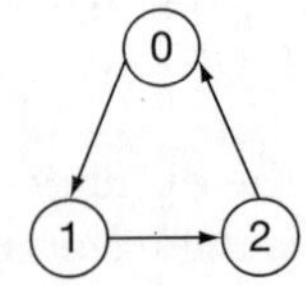

FIGURE 10.6 | Up count sequence

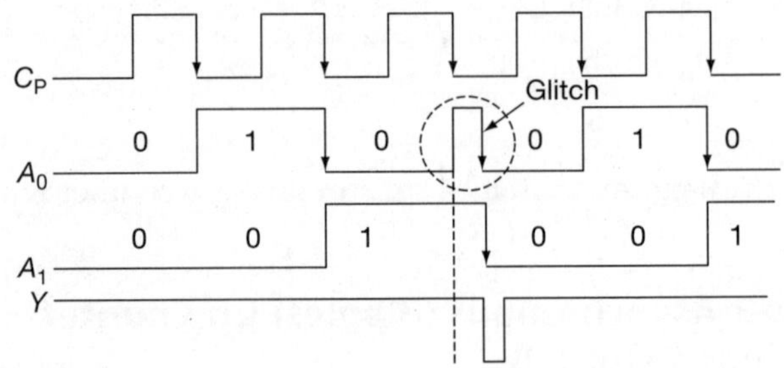

FIGURE 10.7 | Timing diagram of MOD-4 synchronous up counter

Then the frequency, f of clock pulses for reliable operation of counter is defined by

$$\frac{1}{f} \geq nt_{pd} + t_s$$

$$f \leq \frac{1}{nt_{pd} + t_s} \tag{10.3}$$

where n is the number of flip-flops in the asynchronous (ripples) counter
t_{pd} is inherent propagation delay of flip-flop
t_s is strobe pulse width

Total propagation delay in asynchronous counters become

$$T_{clock} \geq nt_{pd} + t_s \tag{10.4}$$

where n is number of flip flops in the ripples counter
t_{pd} is inherent propagation delay of flip-flop
t_s is strobe pulse width

10.2.3 | Modulus-4 Asynchronous (Ripples) Down Counter

Modulus-4 (MOD-4) asynchronous or ripples down counter uses two number of T-flip-flops. First T-flip-flop, FF0 is clocked by external clock pulse. Inverted output, $\bar{Q}_0$ of T-flip-flop, FF0 is given to clock pulse of successive T-flip-flop, FF1. Figure 10.8 shows the logic diagram of MOD-4 asynchronous or ripples down counter. MOD-4 asynchronous down counter counts 4 (2^2) clock pulses and counts sequence from 3, 2, 1, 0 and 3 as shown in Figure 10.9.

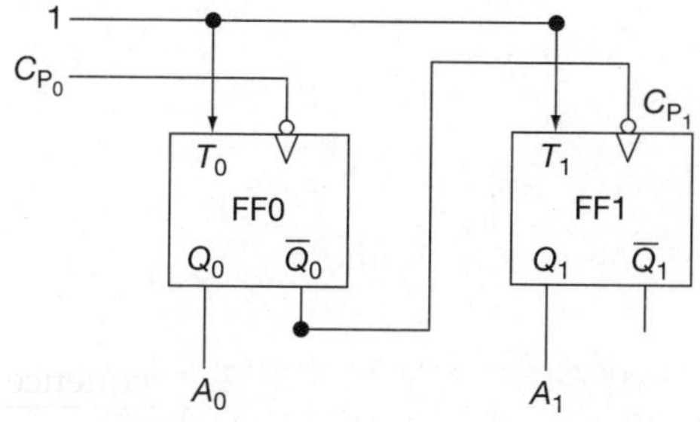

FIGURE 10.8 | MOD-4 asynchronous down counter

External clock pulse is given to first flip-flop.

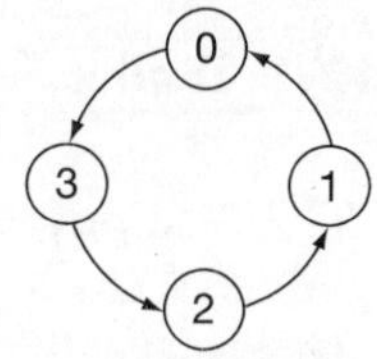

FIGURE 10.9 | Down count sequence

Inputs of T-flip-flops are:

$$T_0 = 1 \text{ and } T_1 = 1$$

Clock pulse of T-flip-flops, FF0 and FF1 are:

$$C_{P_0} \leftarrow C_P \text{ and } C_{P_1} = \bar{Q}_0$$

Complemented output, $\bar{Q}_{n-1}$ of $(n-1)$th flip-flop is given to clock pulse, C_{P_n} of nth flip-flop.

The inputs of T-flip-flops are connected to 1. The flip-flop changes its state when it follows negative clock pulse or falling edge clock pulse. Further, flip-flop changes its state when the output of flip-flop goes from 0 to 1 because inverted output goes from 1 to 0 that is connected to input of next input clock of T-flip-flop. The counting sequence of MOD-4 ripples down counter is given in Table 10.4 and the waveforms are shown in Figure 10.10.

TABLE 10.4 | MOD-8 asynchronous down counter

Clock Count	Countdown Sequence	
	A_1	A_0
3	1	1
2	1	0
1	0	1
0	0	0
3	1	1

Falling edge clock pulse of flip-flop gives change to state.

As output, Q of current flip-flop changes from 0-logic to 1-logic, then state of next flip-flop changes.

The working of counter is explained below.

- Input, T_0 is 1, so output, A_0 of T-flip-flop, FF0 changes from 1 to 0 during the applied falling edge of clock pulse, C_{P_0}. Since, A_0 changes from 1 to 0, inverted output, $\bar{Q}_0$ of T-flip-flop, FF0 changes from 0 to 1 which acts as clock pulse, C_{P_1} for FF1-flip-flop. Hence, A_1 of T-flip-flop, FF1 does not change.

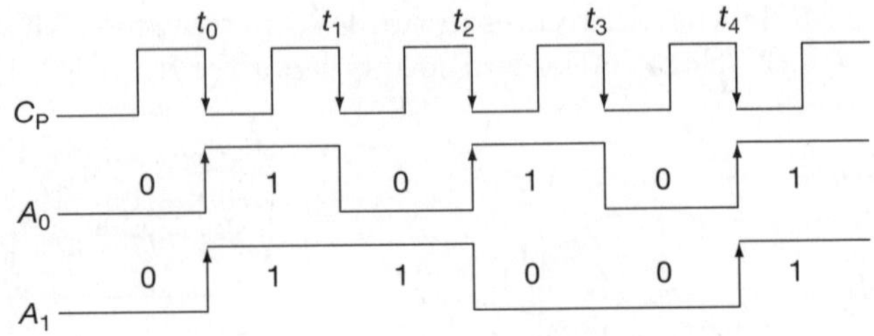

FIGURE 10.10 | Timing diagram of MOD-4 asynchronous down counter

- Input, T_0 is 1 so output A_0 of T-flip-flop, FF0 changes from 0 to 1 during the applied falling edge of clock pulse, C_{P_0}. Inverted output, $\bar{Q}_0$ of flip-flop FF0 changes from 1 to 0. It acts as falling edge clock pulse, C_{P_1} to T-flip-flop, FF1 and input T_1 is 1 so output, A_1 of T-flip-flop, FF1 changes from 1 to 0.
- Input, T_0 is 1, so output, A_0 of T-flip-flop, FF0 changes from 1 to 0 during the applied falling edge of clock pulse, C_{P_0}. Since, A_0 changes from 1 to 0. Inverted output, $\bar{Q}_0$ of T-flip-flop, FF0 changes from 0 to 1 which acts as clock pulse, C_{P_1} for FF1-flip-flop. Hence, A_1 of T-flip-flop, FF1 does not change.
- Input, T_0 is 1 so output A_0 of T-flip-flop, FF0 changes from 0 to 1 during the applied falling edge of clock pulse, C_{P_0}. Inverted output, $\bar{Q}_0$ of flip-flop FF0 changes from 1 to 0. It acts as falling edge clock pulse, C_{P_1} to T-flip-flop, FF1 and input T_1 is 1 so output, A_1 of T-flip-flop, FF1 changes from 1 to 0.

10.2.4 | Modulus-4 Asynchronous (Ripples) Up/Down Counter

The up-and-down counter counts the output increase and decrease, respectively with the clock pulse (Figure 10.11). In an up counter, the output count is increased with the clock pulse, as explained in ripple counter. In a down counter, the output count is decreased with the clock pulse. The up asynchronous counter is obtained by connecting the output, Q of previous flip-flop to the clock input of the next flip-flop. The down asynchronous counter is obtained by connecting the output, $\bar{Q}$ of a previous flip-flop to the clock input for the next flip-flop. The selection of output, Q or inverted output, $\bar{Q}$ can be selected using one control line. Multiplexer 2 to 1 line is used. The logic diagram of a MOD-4 up/down asynchronous counter is given in Figure 10.12 using multiplexer.

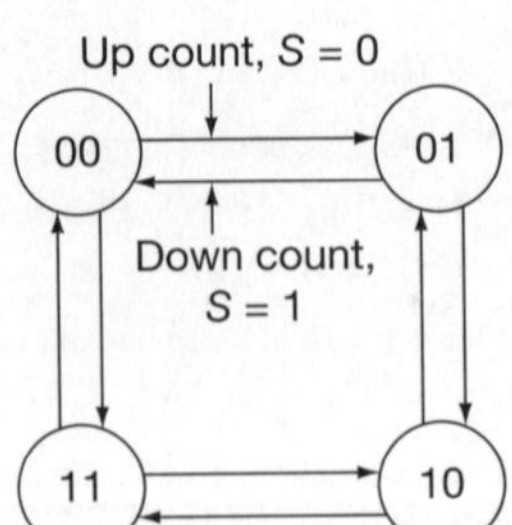

FIGURE 10.11 | Up/Down count sequence of MOD-4 counter

Input to both the T-flip-flop is set to 1.

$$T_0 = 1 \text{ and } T_1 = 1$$

Clock pulse of T-flip-flop, FF1 is given through the controlled logic given below. The logic can be implemented using 2-to-1 line multiplexer or AOI gates.

$$C_{P_1} = \bar{S}I_0 + SI_1 \tag{10.5}$$

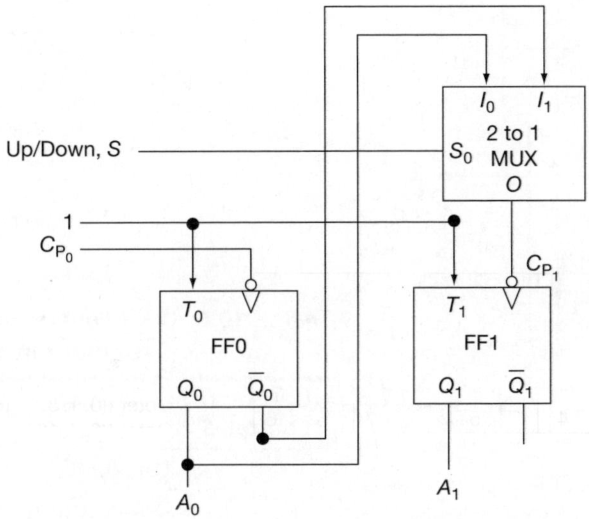

FIGURE 10.12 | MOD-4 asynchronous up/down counter

where output, Q_0 of FF0 is fed to input, I_0 of MUX an inverted output, $\bar{Q}_0$ of FF0 is fed to input, I_1 of MUX.

Table 10.5 gives the up-and-down function of asynchronous counter when up/down control line, S is 0 or 1, respectively.

TABLE 10.5 | MOD-4 up/down asynchronous counter function

Up/Down Control, S	Function	Clock Pulse of FF1 C_{P_1}
0	Up count	$I_0 \leftarrow Q_0$
1	Down count	$I_1 \leftarrow \bar{Q}_0$

The operations of up-and-down counters are same those have already been explained in Sections 10.2.1 and 10.2.3, respectively. The logic diagram of a MOD-4 up/down asynchronous counter is given in Figure 10.13 using AOI (AND, OR and INVERT) gates.

Input to both the T-flip-flop is set to 1.

$$T_0 = 1 \text{ and } T_1 = 1$$

Clock of T-flip-flop, FF1 is given through the controlled logic given below. The logic can be implemented using AOI gates.

$$C_{P_1} = \bar{S}Q_0 + S\bar{Q}_0 \tag{10.6}$$

Table 10.6 gives the up-and-down function of asynchronous counter when up/down control line, S is 0 or 1, respectively.

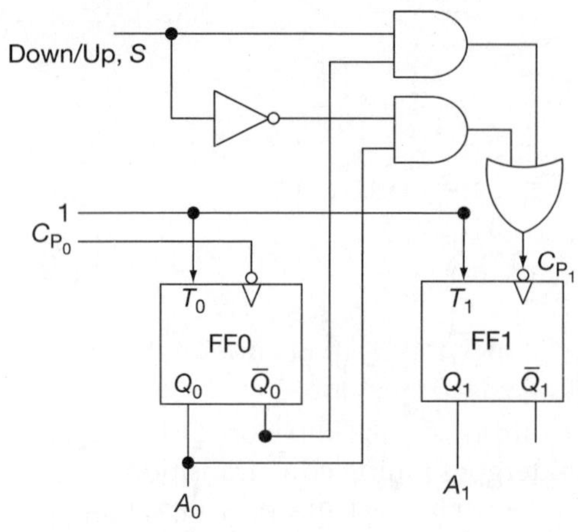

FIGURE 10.13 | MOD-4 asynchronous up/down counter

TABLE 10.6 | MOD-4 up/down asynchronous counter function

Up/Down Control, *S*	Function	Clock Pulse of FF1 C_{P_1}
0	Up count	Q_0
1	Down count	$\bar{Q}_0$

10.2.5 | Modulus-8 Asynchronous (Ripples) Up Counter

Modulus-8 (MOD-8) asynchronous or ripples up counter uses 3 $(\because 2^3 = 8)$ number of T-flip-flops. First T-flip-flop, FF0 is clocked by external clock pulse, C_{P_0}. Output, Q_0 of T-flip-flop, FF0 is given to clock pulse, C_{P_1} of successive T-flip-flop, FF1. Further, output, Q_1 of T-flip-flop, FF1 is given to clock pulse, C_{P_2} of successive flip-flop, FF2.

$$T_0 = 1,\ T_1 = 1 \text{ and } T_2 = 1$$

$$C_{P_1} = Q_0 \text{ and } C_{P_2} = Q_1$$

Figure 10.14 shows the logic diagram of MOD-8 asynchronous (or ripples) up counter. MOD-8 asynchronous up counter counts 8 (2^3) clock pulses and counts sequence from 0, 1, 2, 3, 4, 5, 6, 7 and 0 as shown in Figure 10.15. The inputs of T-flip-flops are connected to 1. The T-flip-flop changes its state when it follows negative clock pulse or falling edge clock pulse.

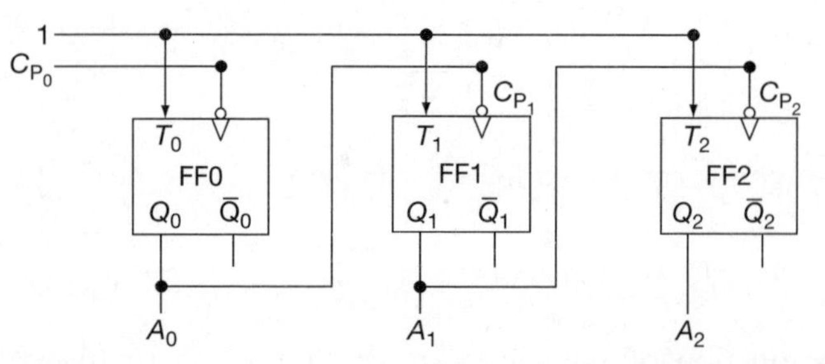

FIGURE 10.14 | MOD-8 synchronous up counter

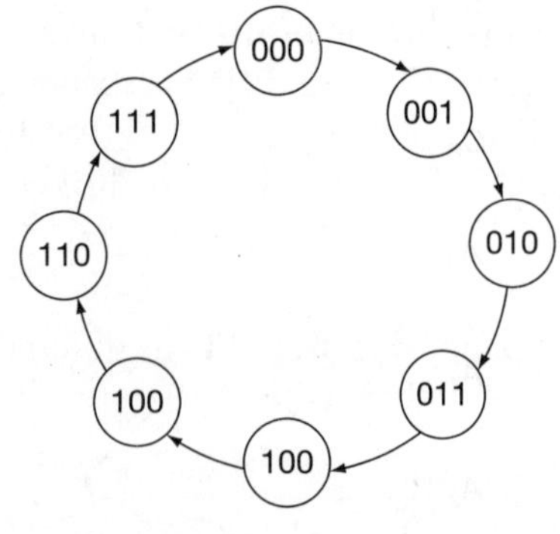

FIGURE 10.15 | Up count sequence

Further, flip-flop changes its state when the output of flip-flop changes from 1 to 0 logic because it is connected to next input clock of T-flip-flop. The counting sequence of MOD-4 ripples counter is given in Table 10.7 and the waveforms are shown in Figure 10.16. The working of counter is explained below.

- Input, T_0 is 1, so output, Q_0 of T-flip-flop, FF0 changes from 0 to 1 or 1 to 0 during the every applied falling edge of clock pulse, C_{P_0}.
- When output, Q_0 remains same either 0 or 1, the output, Q_1 of T-flip-flop, FF1does not change because clock pulse, C_{P_1} does not undergo falling edge transition.
- As output, Q_0 changes from 0 to 1, so output, Q_1 of T-flip-flop, FF1does not change because clock pulse, C_{P_1} does not undergo falling edge transition.
- As and when output, Q_0 changes from 1 to 0, so output, Q_1 of T-flip-flop, FF1 changes from 0 to 1 or 1 to 0 because clock pulse, C_{P_1} undergoes falling edge transition.
- As and when output, Q_1 changes from 1 to 0, so output, Q_2 of T-flip-flop, FF2 changes from 0 to 1 or 1 to 0 because clock pulse, C_{P_2} undergoes falling edge transition.

TABLE 10.7 | MOD-8 Asynchronous up counter

Clock Pulse Count	Count Sequence		
	A_2	A_1	A_0
0	0	0	0
1	0	0	1
2	0	1	0
3	0	1	1
4	1	0	0
5	1	0	1
6	1	1	0
7	1	1	1
0	0	0	0

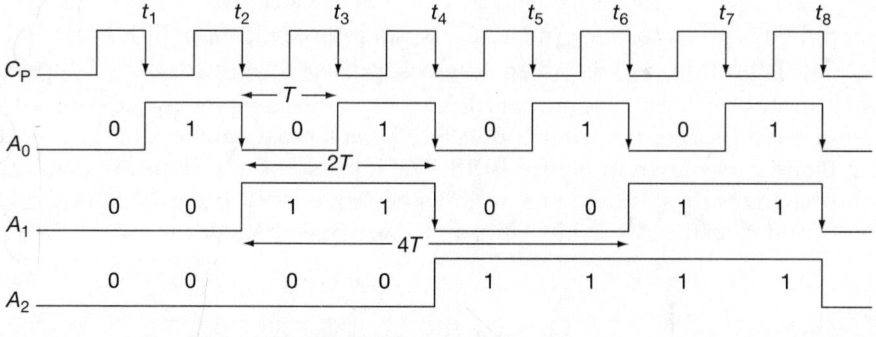

FIGURE 10.16 | Timing diagram of MOD-8 up synchronous counter

Frequency Division Each flip-flop changes state whenever the clock pulse goes from 1 to 0 (falling edge transition).

A_0 changes state at the falling edge of each pulse. Time period of output, A_0 waveform is $2T$. So the output, A_0 waveform frequency is equal to one half of the clock frequency.

$$f_{A_0} = f/2$$

A_1 changes state each time when Q_0 changes from 1 to 0. Time period of output, A_1 waveform is $4T$. So the frequency of waveform of output, A_1 is equal to one half of the clock frequency, f_{A_0}.

$$f_{A_1} = \frac{f_{A_0}}{2} = \frac{f}{4}$$

A_2 changes state each time when Q_1 changes from 1 to 0. Time period of output, A_2 waveform is $8T$. So, the frequency of waveform of output, A_2 is equal to one half of the clock frequency, f_{A_1}.

$$f_{A_2} = \frac{f_{A_1}}{2} = \frac{f_{A_0}}{4} = \frac{f}{8}$$

Propagation Delay In asynchronous counter, each flip-flop is triggered by the change in output of the proceeding flip-flop from 1 to 0 (falling edge clock pulse) or from 0 to 1 (rising edge clock pulse). Flip-flop has inherent propagation delay, t_{pd} to produce output. So, flip-flop, FF1 will respond after t_{pd} time on receiving the falling edge clock pulse by the flip-flop, FF0. So, flip-flop, FF2 will respond after t_{pd} time on receiving the falling edge clock pulse by the flip-flop, FF1. To avoid this problem, the period of input pulses made longer than the total propagation.

$$T_{\text{clock}} \geq 3t_{\text{pd}}$$

where n is number of flip flops in the ripples counter
t_{pd} is inherent propagation delay of flip-flop.

10.2.6 | Modulus-8 Asynchronous (Ripples) Down Counter

Modulus-8 (MOD-8) asynchronous or ripples down counter uses 3 $(\because 2^3 = 8)$ number of T-flip-flops. First flip-flop, FF0 is clocked by external clock pulse, C_{P_0}. Inverted output, $\bar{Q}_0$ of T-flip-flop, FF0 is given to clock pulse, C_{P_1} of successive flip-flop, FF1. Further, inverted output, $\bar{Q}_1$ of T-flip-flop, FF1 is given to clock pulse, C_{P_2} of successive T-flip-flop, FF2. Figure 10.17 shows the logic diagram of MOD-8 asynchronous (or ripples) down counter. MOD-8 asynchronous down counter counts 8 (2^3) clock pulses and counts sequence is 7, 6, 5, 4, 3, 2, 1, 0 and 7 as shown in Figure 10.18. The inputs of T-flip-flops are connected to 1. The flip-flop changes its state when it follows negative clock pulse or falling edge clock pulse. Inputs and clock pulses fed to T-flip-flops are given below:

$$T_0 = 1,\ T_1 = 1 \text{ and } T_2 = 1$$

$$C_{P_1} = \bar{Q}_0 \text{ and } C_{P_2} = \bar{Q}_1$$

Further, flip-flop changes its state when the inverted output of flip-flop changes from 0 to 1 logic because it is connected to next input clock of T-flip-flop. The counting sequence

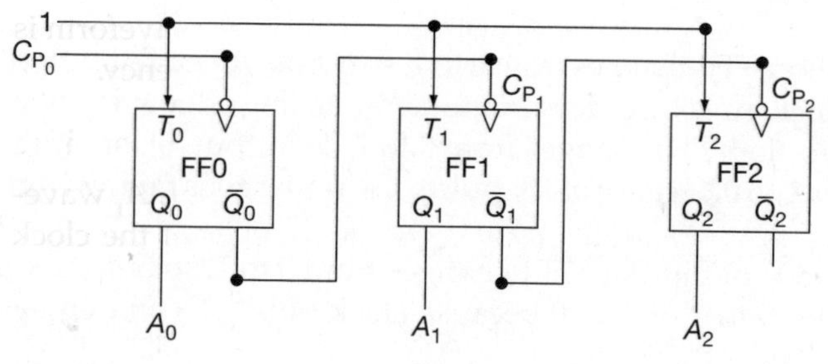

FIGURE 10.17 | MOD-8 asynchronous down counter

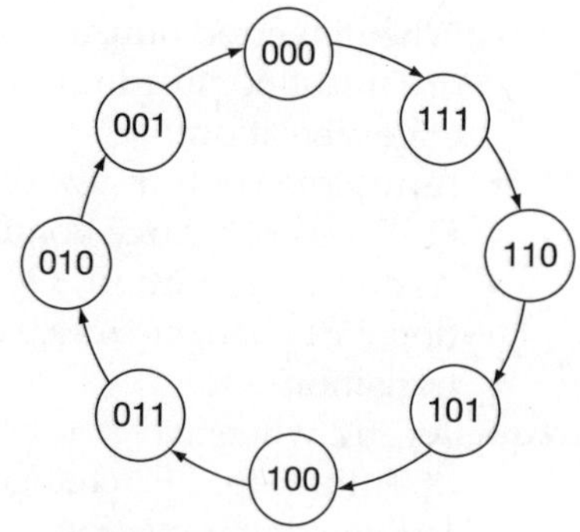

FIGURE 10.18 | Down count sequence

of MOD-8 ripples down counter is given in Table 10.8 and the waveforms are shown in Figure 10.19. The working of counter is explained below.

TABLE 10.8 | MOD-8 asynchronous down counter

Clock Pulse Count	Count Sequence A_2	A_1	A_0
0	1	1	1
1	1	1	0
2	1	0	1
3	1	0	0
4	0	1	1
5	0	1	0
6	0	0	1
7	0	0	0
0	1	1	1

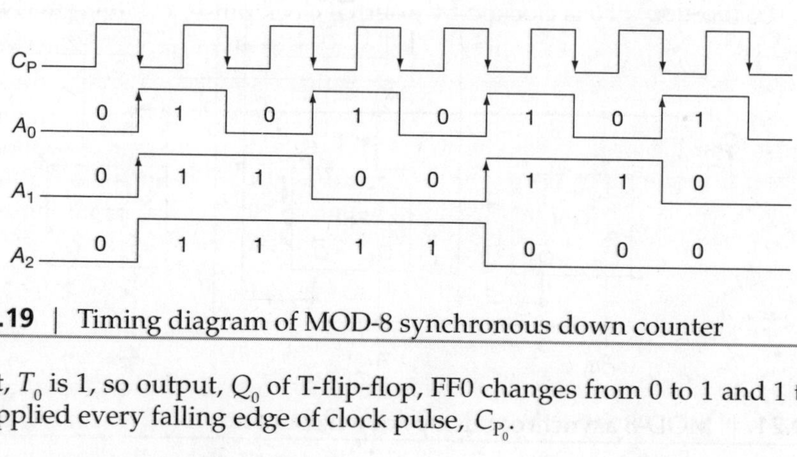

FIGURE 10.19 | Timing diagram of MOD-8 synchronous down counter

- Input, T_0 is 1, so output, Q_0 of T-flip-flop, FF0 changes from 0 to 1 and 1 to 0 during the applied every falling edge of clock pulse, C_{P_0}.

- When inverted output, $\bar{Q}_0$ of flip-flop, FF0 remains same either 0 or 1, the output, Q_1 of T-flip-flop, FF1does not change because clock pulse, C_{P_1} does not undergo falling edge transition.
- As inverted output, $\bar{Q}_0$ of flip-flop, FF0 changes from 0 to 1, so output, Q_1 of T-flip-flop, FF1does not change because clock pulse, C_{P_1} does not undergo falling edge transition.
- As and when output, $\bar{Q}_0$ of flip-flop, FF0 changes from 0 to 1, so output, Q_1 of T-flip-flop, FF1 changes from 0 to 1 or 1 to 0 because clock pulse, C_{P_1} undergoes falling edge transition.
- As and when inverted output, $\bar{Q}_1$ of flip-flop, FF1 changes from 0 to 1, so output, Q_2 of T-flip-flop, FF2 changes from 0 to 1 or 1 to 0 because clock pulse, C_{P_2} undergoes falling edge transition.

10.2.7 | Modulus-8 Asynchronous (Ripples) Up/Down Counter

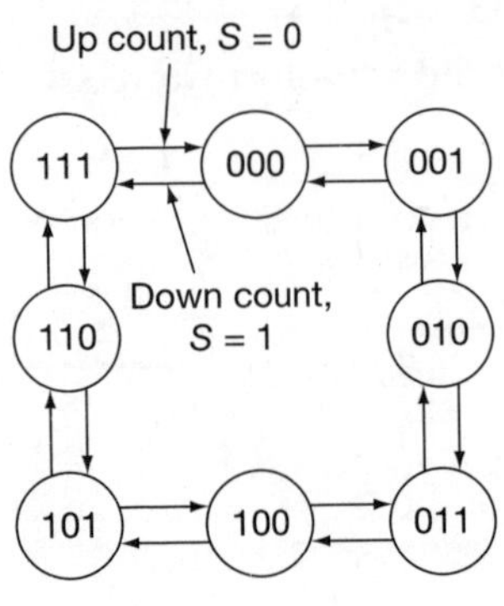

FIGURE 10.20 | Up/down count sequence

The up-and-down counter counts the output increase and decrease, respectively with the clock pulse (Figure 10.20). In an up counter, the output count is increased with the clock pulse, as explained in ripple counter. In a down counter, the output count is decreased with the clock pulse. The up asynchronous counter is obtained by connecting the output, Q of previous flip-flop to the clock input of the next flip-flop. The down asynchronous counter is obtained by connecting the output, $\bar{Q}$ of a previous flip-flop to the clock input for the next flip-flop. The selection of output, Q or inverted output $\bar{Q}$ can be selected using one control line. Multiplexer 2-to-1 line is used. The logic diagram of a 2-bit up/down asynchronous counter is given in Figure 10.21 using multiplexer.

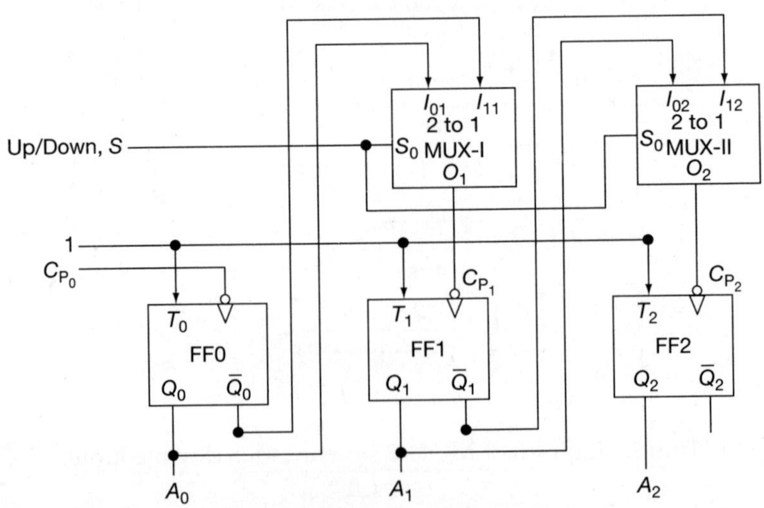

FIGURE 10.21 | MOD-8 asynchronous up/down counter

Input to both the T-flip-flop is set to 1.

$$T_0 = 1,\ T_1 = 1 \text{ and } T_2 = 1$$

Clock of T-flip-flop, FF1 is given through the controlled logic given below. The logic can be implemented using 2-to-1 line multiplexer

$$C_{P_1} = \bar{S}I_{01} + SI_{11} \tag{10.7}$$

where output, Q_0 of FF0 is fed to I_{01} of MUX-I an inverted output, $\bar{Q}_0$ of FF0 is fed to I_{11} of MUX-I.

Clock of T-flip-flop, FF2 is given through the controlled logic given below. The logic can be implemented using 2-to-1 line multiplexer.

$$C_{P_2} = \bar{S}I_{02} + SI_{12} \tag{10.8}$$

where output, Q_1 of FF1 is fed to I_{02} of MUX-II an inverted output, $\bar{Q}_1$ of FF1 is fed to I_{12} of MUX-II.

Table 10.9 gives the up-and-down function of asynchronous counter when up/down control line, S is 0 or 1, respectively.

TABLE 10.9 | Up/down asynchronous counter function

Up/Down Control, S	Function	Clock Pulse of FF1 C_{P_1}	Clock Pulse of FF2 C_{P_2}
0	Up count	$I_{01} \leftarrow Q_0$	$I_{02} \leftarrow Q_1$
1	Down count	$I_{11} \leftarrow \bar{Q}_0$	$I_{12} \leftarrow \bar{Q}_1$

The operations of up-and-down counters are same those have already been explained in section 10.2.5 and 10.2.6, respectively.

Circuit using AOI Gates

The logic diagram of a MOD-8 up/down asynchronous counter is given in Figure 10.22 using AOI (AND, OR and INVERT) gates. Input to both the T-flip-flop is set to 1.

$$T_0 = 1,\ T_1 = 1 \text{ and } T_2 = 1$$

Clock of T-flip-flop, FF1 is given through the controlled logic given below. The logic can be implemented using AOI gates.

$$C_{P_1} = \bar{S}Q_0 + S\bar{Q}_0 \tag{10.9}$$

Clock of T-flip-flop, FF2 is given through the controlled logic given below. The logic can be implemented using AOI gates.

$$C_{P_2} = \bar{S}Q_1 + S\bar{Q}_1 \tag{10.10}$$

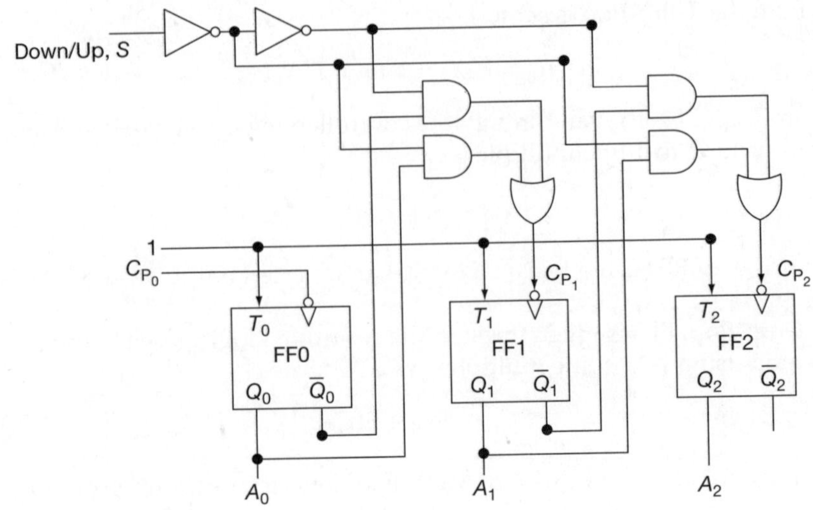

FIGURE 10.22 | MOD-8 asynchronous up/down counter

TABLE 10.10 | Up/down asynchronous counter function

Up/Down Control, S	Function	Clock Pulse of FF1 C_{P_1}	Clock Pulse of FF2 C_{P_2}
0	Up count	Q_0	Q_1
1	Down count	$\bar{Q}_0$	$\bar{Q}_1$

Table 10.10 gives the up-and-down function of asynchronous counter when up/down control line, S is 0 or 1, respectively.

10.2.8 | Modulus-16 Asynchronous (Ripples) Up/Down Counter

The up-and-down counter counts the output increase and decrease, respectively, with the clock pulse (Figure 10.23).

In an up counter, the output count is increased with the clock pulse, as explained in ripple counter. In a down counter, the output count is decreased with the clock pulse. The up asynchronous counter is obtained by connecting the output, Q of previous flip-flop to the clock input of the next flip-flop. The down asynchronous counter is obtained by connecting the output, $\bar{Q}$ of a previous flip-flop to the clock input for the next flip-flop. The selection of output, Q or inverted output, $\bar{Q}$ can be selected using one control line.
The logic diagram of a MOD-16 up/down asynchronous counter is given in Figure 10.23 using AOI (AND, OR and INVERT) gates. Input to both the T-flip-flop is set to 1.

$$T_0 = 1,\ T_1 = 1,\ T_2 = 1 \text{ and } T_3 = 1$$

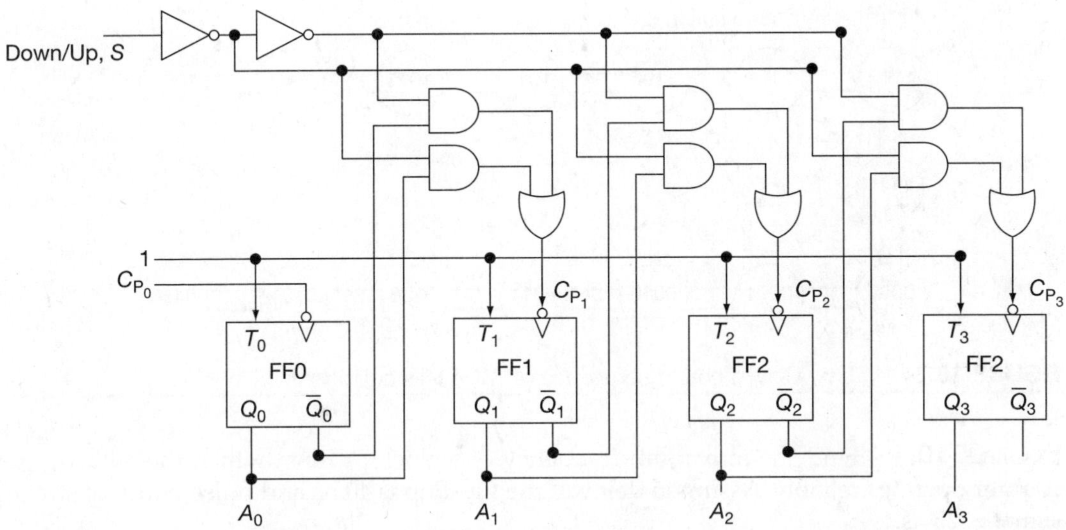

FIGURE 10.23 | MOD-8 asynchronous up/down counter

Clock of T-flip-flop, FF1 is given through the controlled logic given below. The logic can be implemented using AOI gates.

$$C_{P_1} = \bar{S}Q_0 + S\bar{Q}_0 \tag{10.11}$$

Clock of T-flip-flop, FF2 is given through the controlled logic given below. The logic is implemented using AOI gates.

$$C_{P_2} = \bar{S}Q_1 + S\bar{Q}_1 \tag{10.12}$$

Clock of T-flip-flop, FF3 is given through the controlled logic given below. The logic is implemented using AOI gates.

$$C_{P_3} = \bar{S}Q_2 + S\bar{Q}_2 \tag{10.13}$$

Table 10.11 gives the up-and-down function of asynchronous counter when up/down control line, S is 0 or 1, respectively, as shown in Figure 10.24.

TABLE 10.11 | Up/down asynchronous counter function

Up/Down Control, S	Function	Clock Pulse of FF1 C_{P_1}	Clock Pulse of FF2 C_{P_2}	Clock Pulse of FF2 C_{P_3}
0	Up count	Q_0	Q_1	Q_2
1	Down count	$\bar{Q}_0$	$\bar{Q}_1$	$\bar{Q}_2$

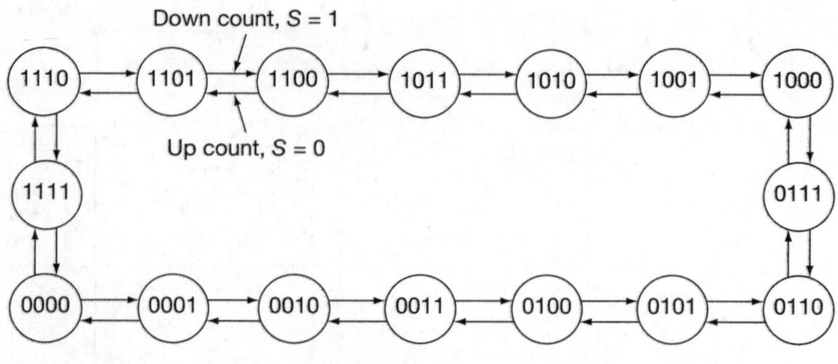

FIGURE 10.24 | Up/Down count sequence of MOD-16 counter

EXAMPLE 10.1 Find the maximum frequency of a clock pulse at which the 5-bit ripple counter operates reliably. Assumed delay of the flip-flop is 40 ns and pulse width of strobe signal is 25 ns.

SOLUTION Given: $n = 5$, $t_{pd} = 40$ ns, $t_s = 25$ ns.

$$f \le \frac{1}{nt_{pd} + t_s}$$

$$f \le \frac{1}{5\times(40\times10^{-9})+(25\times10^{-9})} \Rightarrow f \le \frac{1}{225\times10^{-9}} = 0.0044\times10^{9}\text{ Hz}$$

$$f \le 4.4\times10^{6}\text{ Hz} \le 4.4\text{ MHz}$$

Maximum frequency, $f = 4.4$ MHz **Ans.**

EXAMPLE 10.2 A binary ripple counter is required to count up to 8191_{10}. How many flip-flops are required? What is the frequency at the output of most significant bit (MSB) when clock frequency is 8.192 MHz?

SOLUTION Largest count is 8191.

or
$$2^n - 1 = 8191 \Rightarrow 2^n = 8191 + 1$$

Taking log on both sides

$$\log_{10} 2^n = \log_{10} 8192 \Rightarrow n\times\log_{10} 2 = \log_{10} 8192$$

$$n = \frac{\log_{10} 8192}{\log_{10} 2} = 13$$

Number of flip-flops required = 13 **Ans.**

Frequency at the output of MSB

$$f_{Q13} = \frac{f}{2^{13}} = \frac{8.192\text{ Mhz}}{8192} = \frac{8.192\times10^{6}}{8192}\text{Hz} = 1000\text{ Hz} = 1\text{ kHz}$$ **Ans.**

EXAMPLE 10.3 At what minimum propagation delay of each flip-flop, 5-bit asynchronous counter skips a count? The counter is clocked with 20 MHz.

SOLUTION Propagation delay of n flip-flop in asynchronous counter

$$T_{\text{clock}} = nt_{\text{pd}}$$

Frequency
$$f = \frac{1}{T_{\text{clock}}} = \frac{1}{nt_{\text{pd}}}$$

or
$$t_{\text{pd}} = \frac{1}{nf} = \frac{1}{5 \times 20\,\text{MHz}} = 0.01 \times 10^{-6}\,\text{s} = 10 \times 10^{-9}\,\text{s} = 10\,\text{ns} \quad \textbf{Ans.}$$

EXAMPLE 10.4 Design a MOD-5 up asynchronous counter from MOD-8 up asynchronous counter using decoding logic. Draw the timing diagram of the sequence.

SOLUTION MOD-8 up asynchronous counter is given in Figure 10.25. MOD-5 counts 0, 1, 2, 3 and 4 as shown in Figure 10.26. Decoding logic reset the flip-flops to 0. Inputs and clock pulses given to flip-flops are mentioned below:

$$T_0 = 1,\ T_1 = 1 \text{ and } T_2 = 1$$

$$C_{P_1} \leftarrow Q_0 \text{ and } C_{P_2} \leftarrow Q_1$$

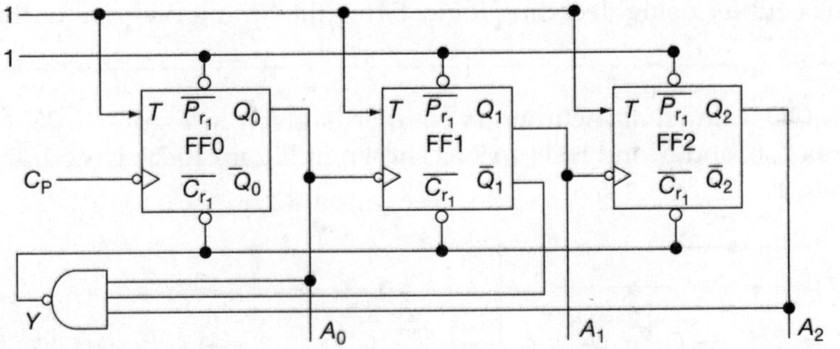

FIGURE 10.25 | MOD-5 asynchronous up counter with decoding logic

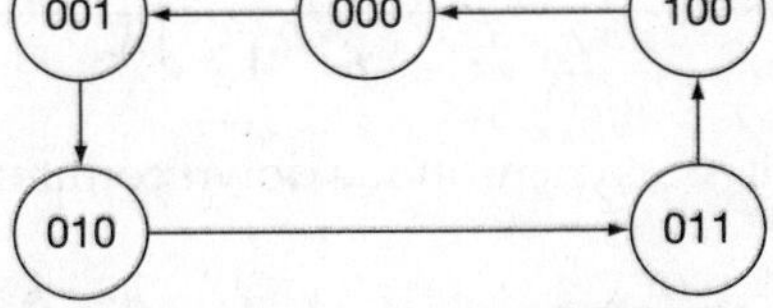

FIGURE 10.26 | Up/Down count sequence

The timing diagram of modulus-5 up counter is given in Figure 10.27. Input, T is 1, so output, Q_0 of T-flip-flop, FF0 changes from 0 to 1 or 1 to 0 during the every applied falling edge of clock pulse, C_{P_0}.

As and when output, Q_0 changes from 1 to 0, so output, Q_1 of T-flip-flop, FF1 changes from 0 to 1 or 1 to 0 because clock pulse, C_{P_1} undergoes falling edge transition. Similar operation is followed for next stages. When count reaches to 5, Y is cleared to 0-logic and is fed to clear, $\overline{Cr}$ the flip-flop outputs. The followed logic is given below:

$$Y = \overline{A_2 \bar{A}_1 A_0} \tag{i}$$

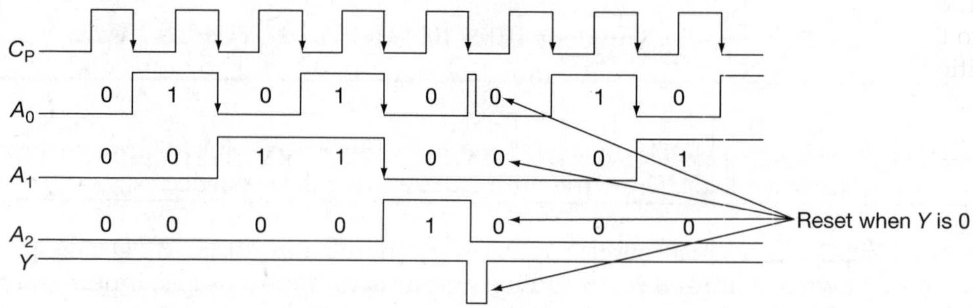

FIGURE 10.27 | Timing diagram of MOD-5 up synchronous counter

EXAMPLE 10.5 Design a MOD-3 down asynchronous counter from MOD-8 down asynchronous counter using decoding logic. Draw the timing diagram of the sequence of count.

SOLUTION MOD-8 down asynchronous counter is given in Figure 10.28. MOD-3 down counter counts 7, 6, and 5, and reset to 7 as shown in Figure 10.29. Decoding logic presets the flip-flops to 1.

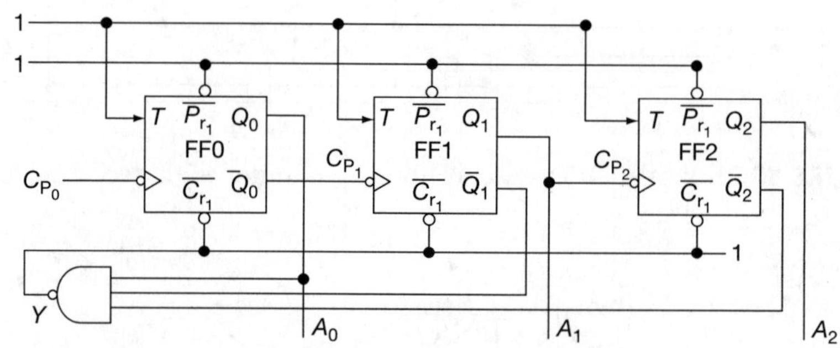

FIGURE 10.28 | MOD-8 asynchronous down counter with decoding logic

The inputs of T-flip-flops are connected to 1. The flip-flop changes its state when it follows negative clock pulse or falling edge clock pulse. Inputs and clock pulse for each flip-flop is given below:

$$T_0 = 1,\ T_1 = 1 \text{ and } T_2 = 1$$

$$C_{P_1} = \bar{Q}_0 \text{ and } C_{P_2} = \bar{Q}_1$$

The timing diagram is given in Figure 10.30. Input, T_0 is 1, so output, Q_0 of T-flip-flop, FF0 changes from 0 to 1 and 1 to 0 during the applied every falling edge of clock pulse, C_{P_0}.

As and when output, $\bar{Q}_0$ of flip-flop, FF0 changes from 1 to 0 or output, Q_0 of flip-flop, FF0 changes from 0 to 1, output, Q_1 of T-flip-flop, FF1 changes from 0 to 1 or 1 to 0 because clock pulse, C_{P_1} undergoes falling edge transition.

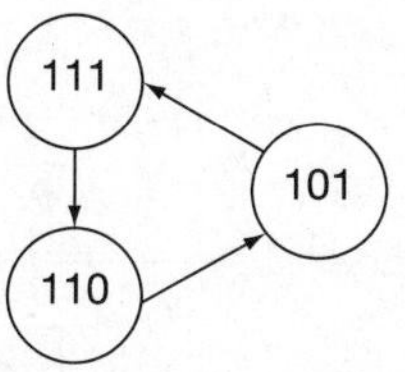

FIGURE 10.29 | Down count sequence

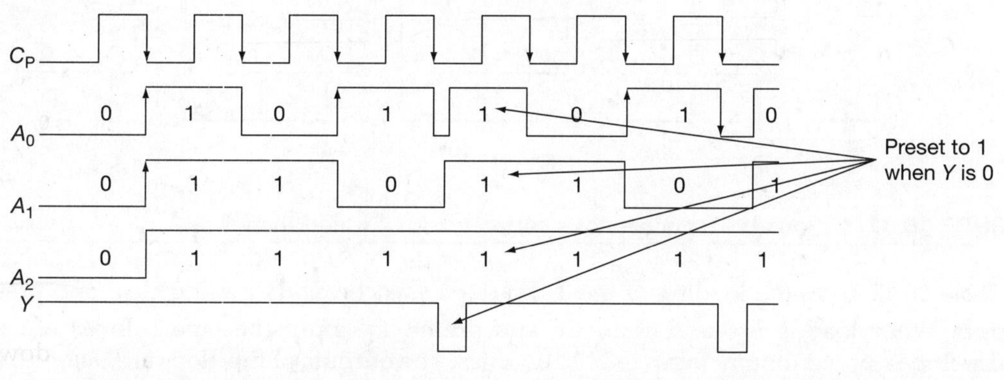

FIGURE 10.30 | Timing diagram of MOD-8 asynchronous down counter

Similar operation is followed for next stages. When count reaches to 4, Y is cleared to 0-logic and is fed to preset, $\overline{Pr}$ the flip-flop outputs to 1. The logic is given below:

$$Y = \overline{A_2 \bar{A}_1 \bar{A}_0} \qquad \text{(i)}$$

10.3 | ASYNCHRONOUS COUNTER WITH PARALLEL LOAD

Four T-flip-flops are cascaded serially. Output of one flip-flop is fed as clock pulse to another flip-flop. The T-flip-flops are loaded to 0 or 1 by controlling clear, $\overline{Cr}$ and preset, $\overline{Pr}$ inputs of flip-flops with load, L control line. Figure 10.31 gives the diagram of 4-bit asynchronous up counter having asynchronous loading. NAND gates are used to control the clear, $\overline{Cr}$ and preset, $\overline{Pr}$ inputs of flip-flops with load, L control line.

$$Cr_0 = \overline{L\bar{A}_0} \text{ and } Pr_0 = \overline{LA_0}$$

$$Cr_1 = \overline{L\bar{A}_1} \text{ and } Pr_1 = \overline{LA_1}$$

$$Cr_2 = \overline{L\bar{A}_2} \text{ and } Pr_2 = \overline{LA_2}$$

$$Cr_3 = \overline{L\bar{A}_3} \text{ and } Pr_3 = \overline{LA_3}$$

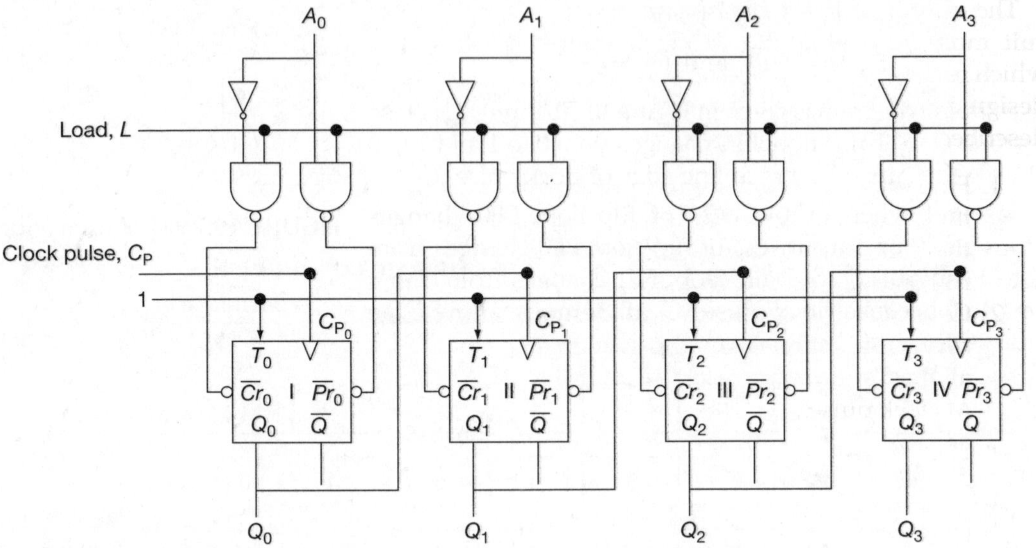

FIGURE 10.31 | Four-bit asynchronous counter with parallel load

Table 10.12 gives the loading of the T-flip-flops asynchronously using clear and preset inputs. When load, L is 0 and clear, $\overline{Cr}$ and preset, $\overline{Pr}$ inputs become 1. Input of each T-flip-flop is stored for any input to NAND gates. The outputs of flip-flops are cleared to 0. When load, L is 1 and inputs to NAND gates are 0 then clear, $\overline{Cr}$ input becomes 0 and preset, $\overline{Pr}$ input becomes 1 of each T-flip-flop. The outputs of flip-flops are set to 1. When load, L is 1 and inputs to NAND gates are 1 then clear, $\overline{Cr}$ inputs become 1 and preset, $\overline{Pr}$ inputs become 0 of each flip-flop. Asynchronous counting operation of counter is as usual is performed on the rising edge of clock pulse, Cp.

TABLE 10.12 | Truth table to load the T-flip-flop

Load	T-Flip-flop-I				T-Flip-flop-II				T-Flip-flop-III				T-Flip-flop-IV			
L	A_0	$\overline{Cr_0}$	$\overline{Pr_0}$	C_{P_0}	A_1	$\overline{Cr_1}$	$\overline{Pr_1}$	C_{P_1}	A_2	$\overline{Cr_2}$	$\overline{Pr_2}$	C_{P_2}	A_3	$\overline{Cr_3}$	$\overline{Pr_3}$	C_{P_3}
0	0	1	1	C_P	0	1	1	Q_1	0	1	1	Q_2	0	1	1	Q_3
0	1	1	1	C_P	1	1	1	Q_1	1	1	1	Q_2	1	1	1	Q_3
1	0	0	1	C_P	0	0	1	0	0	0	1	0	0	0	1	0
1	1	1	0	C_P	1	1	0	1	1	1	0	1	1	1	0	1

10.4 | MODULUS-M ASYNCHRONOUS COUNTER

It is observed that, n-bit asynchronous counters count $N = 2^n$ clock pulses. A 3-bit asynchronous (ripples) counter counts $2^3 = 8$ clock pulses and the count is 0 to 7. To count M clock pulses which is less than $N = 2^n$, reset terminal is used. The combinational circuit is designed such that all the flip-flops are reset after the count M. The block diagram of modulus-M counter is shown in Figure 10.32.

The output of a combinational logic circuit must be 0 after the M clock pulses, which resets the flip-flops. The procedure to design a MOD-M asynchronous counter is described below:

1. Find the minimum number of flip-flops required. For a MOD-M counter, the minimum number of flip-flops required is n so that $M \le 2^n$.
2. Construct the sequence and design the combinational circuit such that all the flip-flops are reset/set after the M clock pulse.
   a. Prepare the truth table of an asynchronous (ripples) counter with the output of a combinational logic circuit, such that $Y = 1$ for the used or valid state and $Y = 0$ for the unused or invalid state.
   b. Set up the K-map for output Y and simplify.

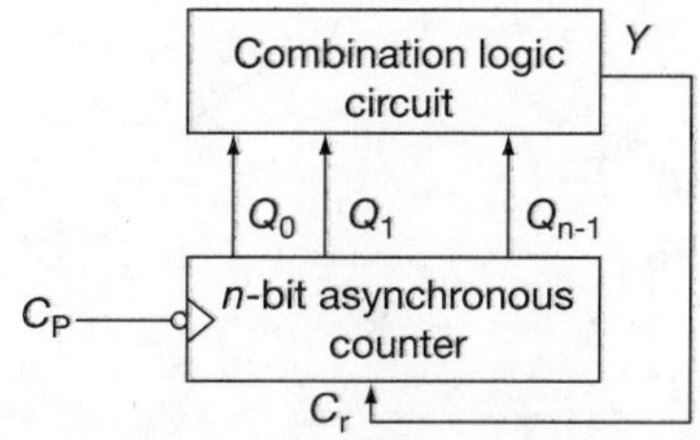

FIGURE 10.32 | Block diagram of MOD-M counter design

EXAMPLE 10.6 Design a MOD-5 up asynchronous counter. Draw the timing diagram of the sequence.

SOLUTION MOD-5 up counter counts 0, 1, 2, 3, 4 and reset to 0

Minimum number of flip-flops required is

$5 > 2^2$ $n = 2$ is not sufficient.

$5 \le 2^3$, so $n = 3$

Design of combinational circuit should reset the flip-flops after 5 clock pulses. Truth table of 3-bit asynchronous counter is given Table 10.13. Three-variable K-map is set up in Figure 10.33 for truth Table 10.13. Two quads are formed {0, 1, 3, 2} and {0, 2, 4, 6}.

TABLE 10.13 | Truth table of MOD-5 asynchronous counter

Clock Pulses	Input for Combinational Circuit			Output
	Q_2	Q_1	Q_0	Y
0	0	0	0	1
1	0	0	1	1
2	0	1	0	1
3	0	1	1	1
4	1	0	0	1
5	1	0	1	0
6	1	1	0	0
7	1	1	1	0

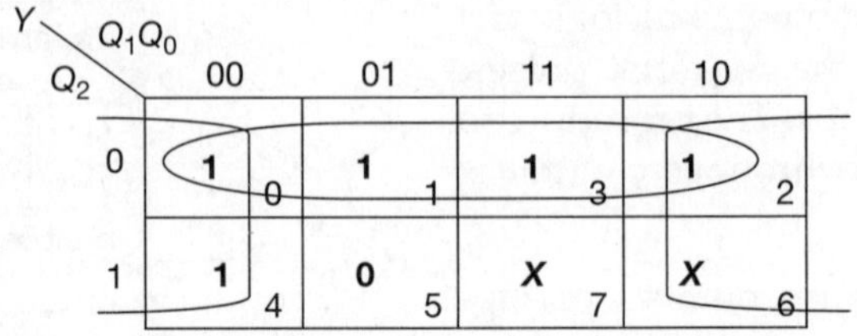

FIGURE 10.33 | K-map

Simplified logic expression is given below:

$$Y = \bar{Q}_2 + \bar{Q}_0 \quad \text{(i)}$$

The circuit diagram of MOD-5 up asynchronous counter is given in Figure 10.34. The timing diagram is same as given in Figure 10.27.

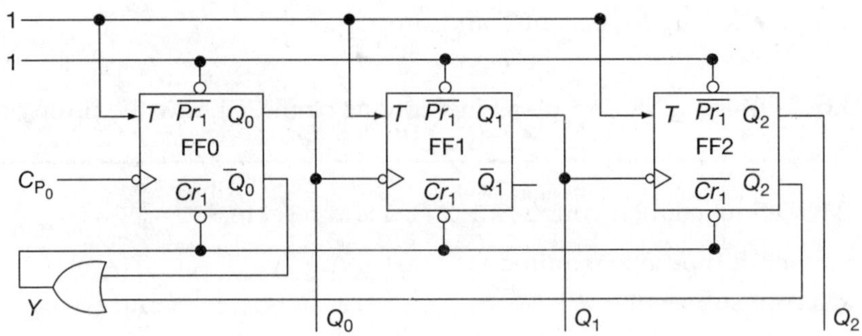

FIGURE 10.34 | MOD-5 asynchronous up counter with decoding logic

EXAMPLE 10.7 Design a BCD up asynchronous counter.

SOLUTION BCD up asynchronous counter is a MOD-10 up counter that counts 0, 1, 2, 3, 4, 5, 6, 7, 8, and 9, and reset to 0.

Minimum number of flip-flops required is $10 \le 2^4$, so $n = 4$.

Design of combinational circuit should reset the flip-flops after 10 clock pulses. Truth table of 4-bit asynchronous counter is given Table 10.14. Three-variable K-map is set up in Figure 10.35 for truth Table 10.14. Two octets are formed {0, 1, 3, 2, 4, 5, 7, 6} and {0, 1, 4, 5, 12, 13, 8, 9}

Simplified logic expression is given below:

$$Y = \bar{Q}_3 + \bar{Q}_1 \quad \text{(i)}$$

The circuit diagram of MOD-10 up asynchronous counter is given in Figure 10.36 and the timing diagram is given in Figure 10.37.

TABLE 10.14 | Truth table of MOD-10 (BCD) asynchronous counter

Clock Pulses	Input for Combinational Circuit				Output
	Q_3	Q_2	Q_1	Q_0	Y
0	0	0	0	0	1
1	0	0	0	1	1
2	0	0	1	0	1
3	0	0	1	1	1
4	0	1	0	0	1
5	0	1	0	1	1
6	0	1	1	0	1
7	0	1	1	1	1
8	1	0	0	0	1
9	1	0	0	1	1
10	1	0	1	0	0
11	1	0	1	1	X
12	1	1	0	0	X
13	1	1	0	1	X
14	1	1	1	0	X
15	1	1	1	1	X

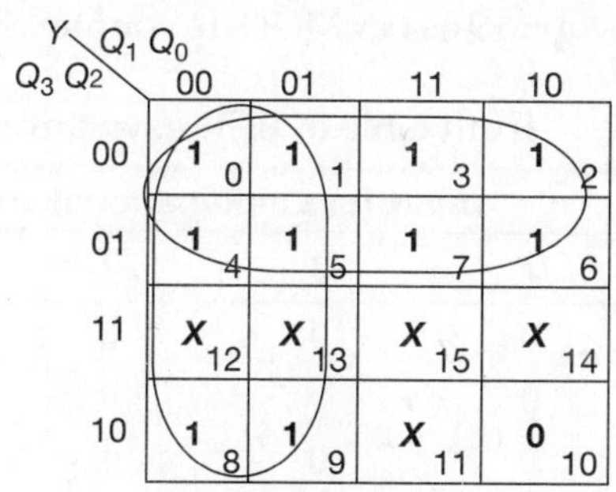

FIGURE 10.35 | K-map

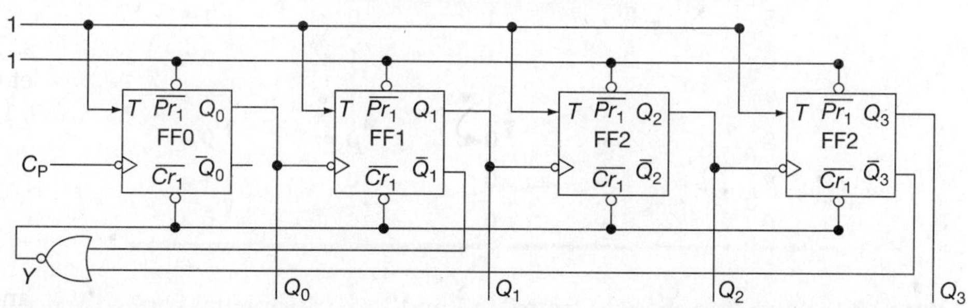

FIGURE 10.36 | MOD-10 asynchronous up counter with decoding logic

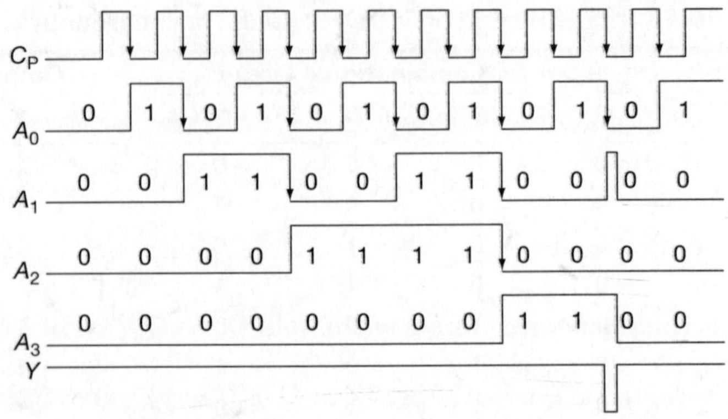

FIGURE 10.37 | Timing diagram of MOD-10 up asynchronous counter

EXAMPLE 10.8 Design a BCD up asynchronous counter.

SOLUTION BCD up asynchronous counter is a MOD-10 up counter that counts 0, 1, 2, 3, 4, 5, 6, 7, 8, and 9. Minimum number of flip-flops required is

$$10 \le 2^4, \text{ so } n = 4$$

Truth table of 4-bit asynchronous counter is given Table 10.15. It can be observed from the truth table: The counter works up to 9 as per MOD-16 counter. The count is restricted up to 9.

TABLE 10.15 | Truth table of BCD asynchronous counter

Clock Pulses	Input for Combinational Circuit			
	Q_3	Q_2	Q_1	Q_0
0	0	0	0	0
1	0	0	0	1
2	0	0	1	0
3	0	0	1	1
4	0	1	0	0
5	0	1	0	1
6	0	1	1	0
7	0	1	1	1
8	1	0	0	0
9	1	0	0	1
0	0	0	0	0

- Output, Q_0 of flip-flop changes from 0 to 1 and 1 to 0 during the applied every falling edge of clock pulse. So, $T_0 \leftarrow 1$.

- Output, Q_1 of flip-flop is clubbed with Q_3.
    - o IF Q_3 is 0 and Q_0 goes from 1 to 0, Q_1 is complemented. But when T is 1 then flip-flop will toggle state.
    - o IF Q_3 is 1 and Q_0 goes from 1 to 0, Q_1 is cleared or remains same. When T is 0 then flip-flop will hold state.
    - o So, $T_1 \leftarrow \bar{Q}_3$ and $C_{P_1} \leftarrow Q_0$.
- Output, Q_2 of flip-flop is complemented when Q_1 goes from 1 to 0. So, $T_2 \leftarrow 1$ and $C_{P_2} \leftarrow Q_1$.
- Output, Q_3 is complemented from 0 when both Q_2 and Q_1 are at 1 and Q_1 goes from 1 to 0.
- Output, Q_3 is cleared to 0 when either Q_2 or Q_1 is 0 and Q_1 goes from 1 to 0.

  $T_3 \leftarrow Q_3 + Q_2Q_1$ and $C_{P_3} \leftarrow Q_0$.

The circuit diagram of MOD-10 up asynchronous counter is given in Figure 10.38. The timing diagram is given in Figure 10.39.

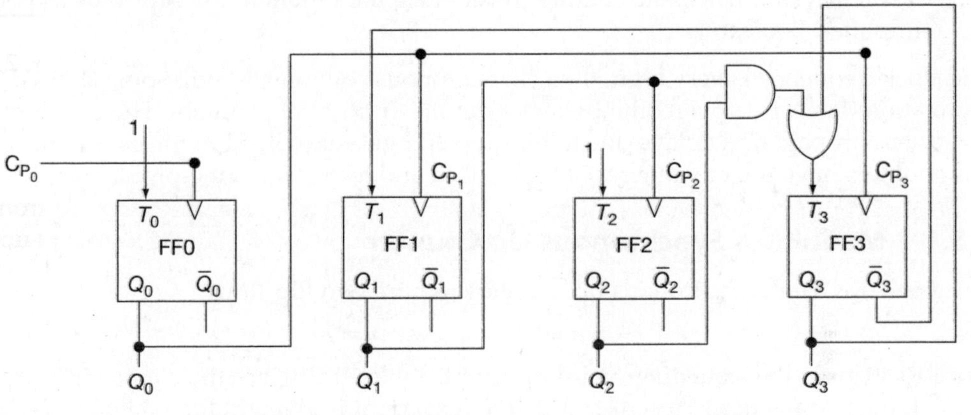

FIGURE 10.38 | BCD asynchronous up counter

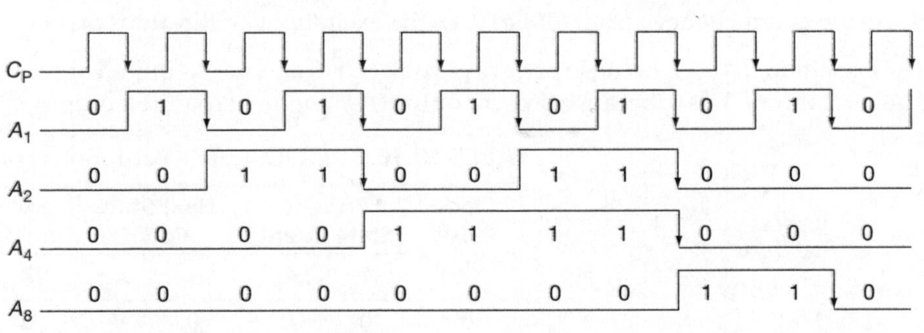

FIGURE 10.39 | Timing diagram of BCD asynchronous counter

Advantage of the design is that there is no glitch due to decoding or Boolean logic.

10.5 | SYNCHRONOUS COUNTER

In a synchronous counter, the clock pulse is given common to all the flip-flops and the transition at the output is in synchronism with the clock inputs. The procedure of synchronous counter is as follows:

1. *Number of flip-flops:* Find the minimum number of flip-flops required. For a MOD-M counter, the minimum number of flip-flops required is n so that $M \le 2^n$. It is noted that an n-bit counter is also known as a modulo 2^n counter.
2. *State diagram:* Draw the state diagram of counter that shows all states of counter progresses.
3. *Excitation table:* Write the given sequence in the form of presents state and next state. If 1 is present state for up counter is next state and for down counter 0 is next state. Determine excitations (inputs) of flip-flop for the given state transition following excitation table.
4. *Minimal expressions for excitations:* Set up K-map for each input of flip-flops in terms of output of the flip-flops and simplify.
5. *Circuit diagram:* Draw the circuits considering the inputs to the flip-flops as per the simplified Boolean.

If the clock frequency is very high, then asynchronous counter can skip some states due to propagation delay. It is overcome by synchronous or parallel counters. The synchronous counter has propagation delays due to flip-flop and gates involved in the circuit. Synchronous counters have the advantages of high speed and lesser decoding problems.

10.5.1 | Modulus-4 Synchronous Up Counter

Modulus 4 (MOD-4) synchronous up counters require two flip-flops.

$$2^n = 4 \Rightarrow 2^n = 2^2, \quad \text{so } n = 2$$

Figure 10.40 gives the state diagram of up count. Table 10.16 gives the count status during the clock pulse transition. Present count and next count is given in the table. Excitation of T-flip-flop is 0 for same present and next output states of flip-flop. Excitation of T-flip-flop is 1 for different present and next output states of flip-flop. When counter progresses from 00 to 01. As the count changes from (00) to (01), the excitations of flip-flops are:

- The excitation, TA_0 is 1, because present output, Q_0 is 0 and next required output, Q_0 is 1.
- The excitation TA_1 is 0, because present output, Q_1 and next required output, Q_1 are 0.

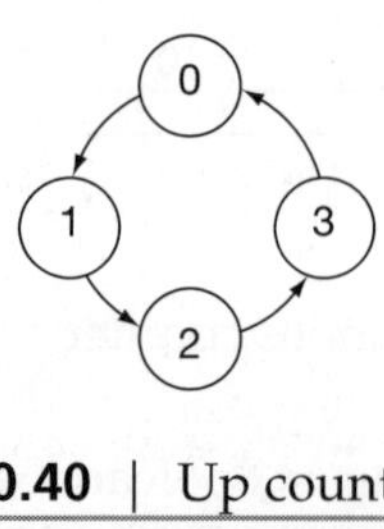

FIGURE 10.40 | Up count sequence

TABLE 10.16 | MOD-4 up synchronous counter

Clock Pulse Count	Present State Count		Next State Count		Excitation	
	A_1	A_0	A_1	A_0	TA_1	TA_0
0	0	0	0	1	0	1
1	0	1	1	0	1	1
2	1	0	1	1	0	1
3	1	1	0	0	1	1

Canonical expressions for excitations of T-flip-flops TA_0 and TA_1 are given below:

$$TA_0(A_1A_0) = \sum m(0,1,2,3) \tag{10.14}$$

$$TA_1(A_1A_0) = \sum m(1,3) \tag{10.15}$$

Two-variable K-maps are set up in Figures 10.41(a) and 10.41(b). In Figure 10.41(a), one quad is formed and all minterms are 1. In Figure 10.41(b), one pair {1, 3} is formed.

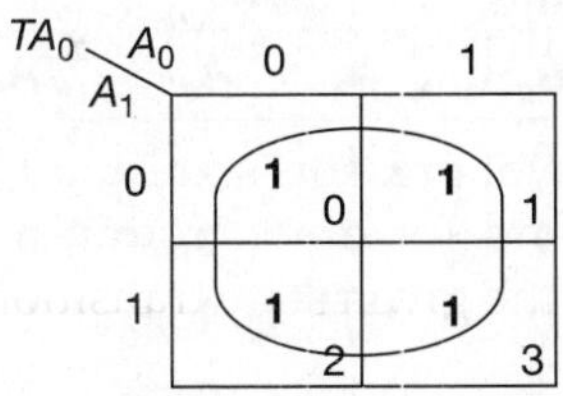

FIGURE 10.41(a) | K-map for TA_0

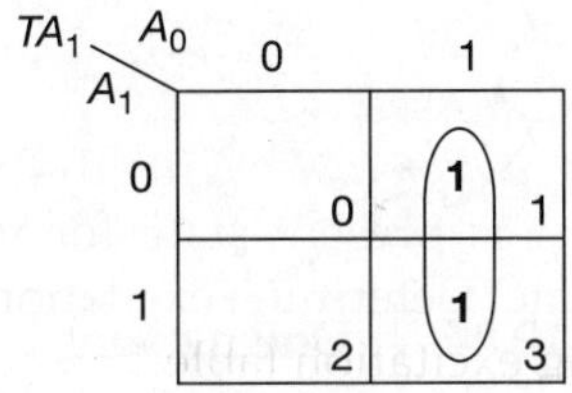

FIGURE 10.41(b) | K-map for TA_1

Simplified expressions for excitation of T-flip-flops are given below:

$$TA_0(A_1A_0) = 1 \tag{10.16}$$

$$TA_1(A_1A_0) = A_0 \tag{10.17}$$

Circuit diagram of MOD-4 up synchronous counter is shown in Figure 10.42 and timing diagram is shown in Figure 10.43.

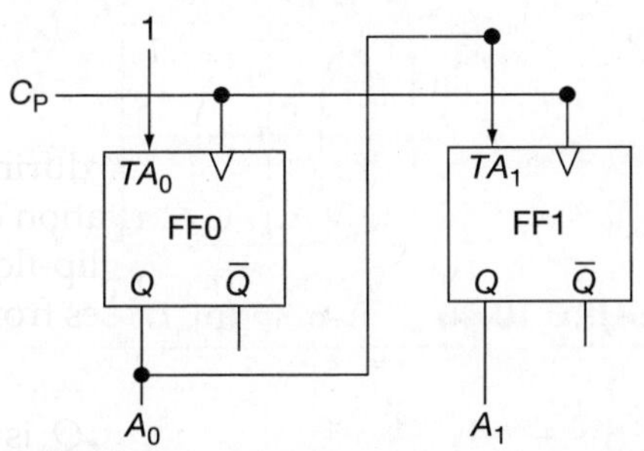

FIGURE 10.42 | MOD-4 up synchronous counter

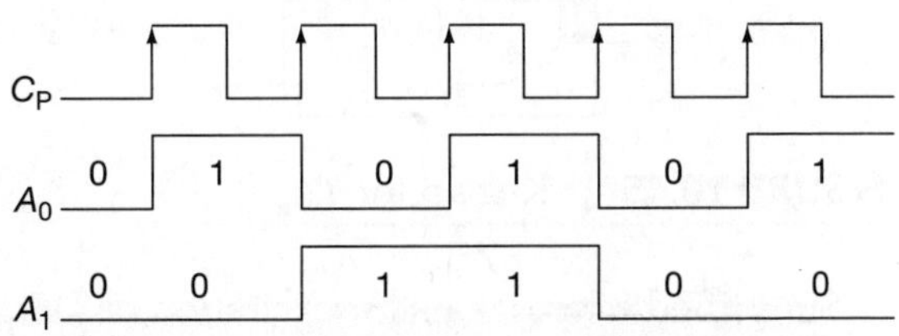

FIGURE 10.43 | Timing diagram of MOD-4 up synchronous counter

10.5.2 | Modulus-4 Synchronous Down Counter

Modulus 4 (MOD-4) synchronous counters require two flip-flops.

$$2^n = 4 \Rightarrow 2^n = 2^2, \quad \text{so } n = 2$$

Figure 10.44 gives the state diagram of up count. Table 10.17 gives the count status during the clock pulse transition. Present count and next count is given in the table. As the counts down, the count changes from (00) to (11), the excitations of flip-flops are:

- The excitation, TA_0 is 1, because present output, Q_0 is 0 and required next output, Q_0 is 1.
- The excitation, TA_1 is 1, because present output, Q_1 is 0 and next required output, Q_1 is 1.

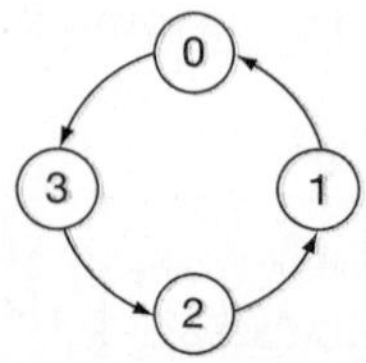

FIGURE 10.44 | Down count sequence

TABLE 10.17 | MOD-4 up synchronous counter

Clock Pulse Count	Present State Count		Next State Count		Excitation	
	A_1	A_0	A_1	A_0	TA_1	TA_0
0	0	0	1	1	1	1
1	0	1	0	0	0	1
2	1	0	0	1	1	1
3	1	1	1	0	0	1

Canonical expressions for excitations of T-flip-flops TA_0 and TA_1 are given below:

$$TA_0(A_1A_0) = \sum m(0,1,2,3) \tag{10.18}$$

$$TA_1(A_2A_1A_0) = \sum m(0,2) \tag{10.19}$$

Two-variable K-maps are set up in Figure 10.45 and Figure 10.46. In Figure 10.45, one quad {0, 1, 2, 3} is formed. All minterms are 1. In Figure 10.46, one pair {0, 2} is formed.

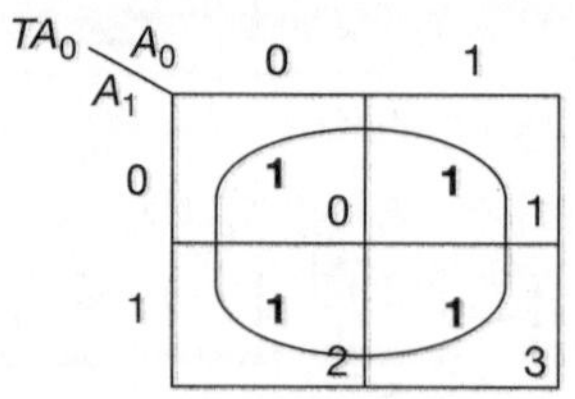

FIGURE 10.45 | K-map for TA_0

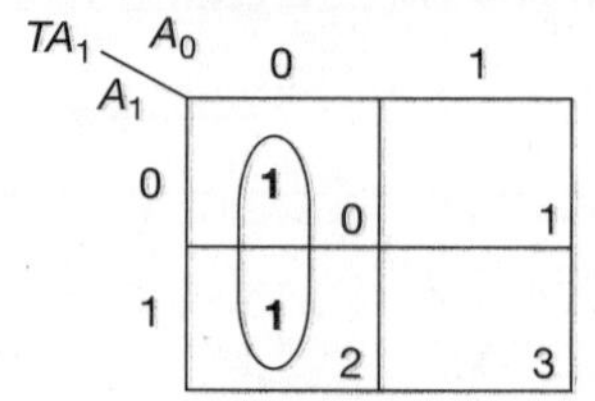

FIGURE 10.46 | K-map for TA_1

Simplified expressions for excitation of T-flip-flops are given below:

$$TA_0 = 1 \tag{10.20}$$

$$TA_1 = \bar{A}_0 \tag{10.21}$$

Circuit diagram of MOD-4 down synchronous counter is shown in Figure 10.47 and timing diagram is shown in Figure 10.48.

10.5.3 | Modulus-4 Synchronous UP/Down Counter

Modulus 4 (MOD-4) synchronous counters require two flip-flops. $\because 2^n = 4 \Rightarrow 2^n = 2^2$, so $n = 2$.

Figure 10.49 gives the state diagram of up count. Table 10.18 gives the count status during the clock pulse transition. One control variable is added to control the up-and-down

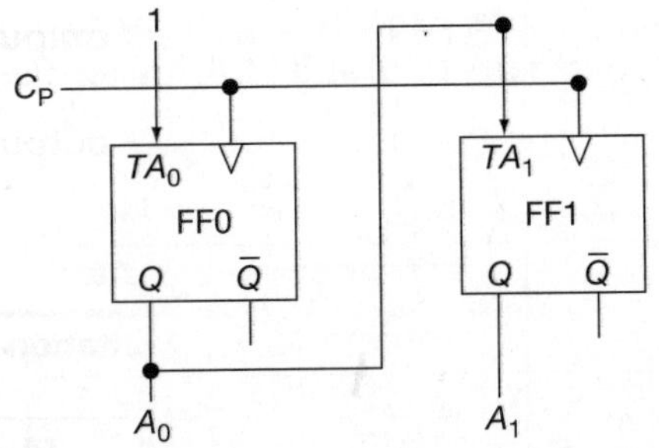

FIGURE 10.47 | MOD-4 down synchronous counter

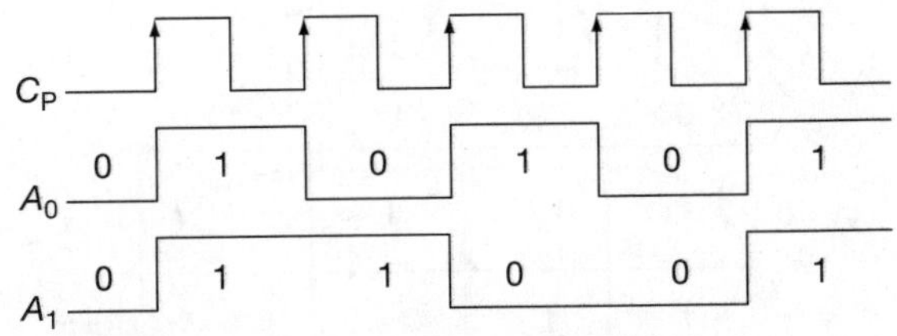

FIGURE 10.48 | Timing diagram of MOD-4 down synchronous counter

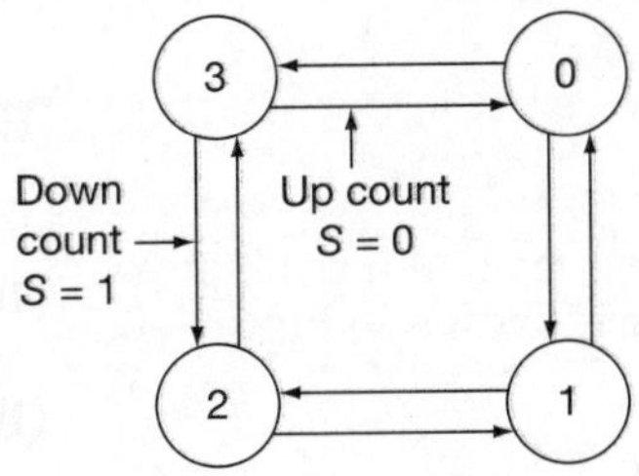

FIGURE 10.49 | UP/down count sequence

TABLE 10.18 | Up/down counter function

Control, S	Function
0	Up count
1	Down count

count. S is 0 for up count and S is 1 for down count (Table 10.18). Present count and next count is given in the table. T-flip-flop excitation of flip-flop is in given the Table 10.19.

Canonical expressions for TA_0 and TA_1 are given below:

$$TA_0(S, A_1, A_0) = \sum m(0,1,2,3,4,5,6,7) \tag{10.22}$$

$$TA_1(S, A_1, A_0) = \sum m(1,3,4,6) \tag{10.23}$$

TABLE 10.19 | MOD-4 up/down synchronous counter

Clock Pulse Count	Up/Down Control, S	Present Count		Next Count		Excitation of Flip-flop	
		A_1	A_0	A_1	A_0	TA_1	TA_0
0	0	0	0	0	1	0	1
1	0	0	1	1	0	1	1
2	0	1	0	1	1	0	1
3	0	1	1	0	0	1	1
0	1	0	0	1	1	1	1
1	1	0	1	0	0	0	1
2	1	1	0	0	1	1	1
3	1	1	1	1	0	0	1

Two-variable K-maps are set up in Figure 10.50 and Figure 10.51. In Figure 10.50, one Octet is formed covering all minterms. In Figure 10.51, two pairs {1, 5} and {2, 6} are formed.

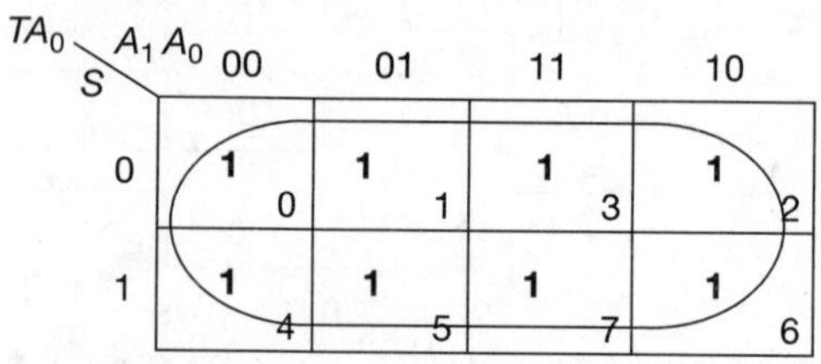

FIGURE 10.50 | K-map for TA_0

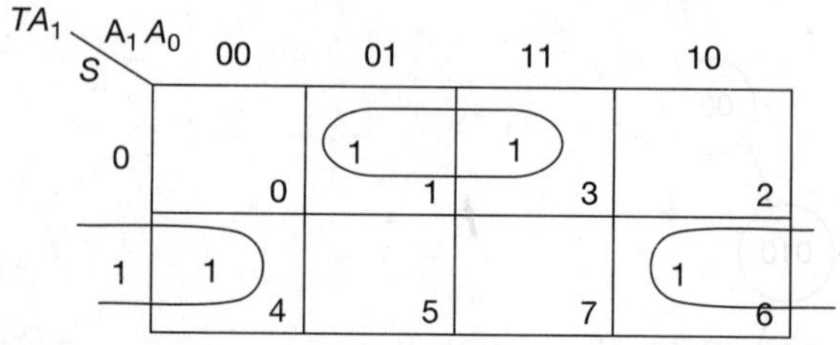

FIGURE 10.51 | K-map for TA_1

Simplified expressions are given below:

$$TA_0 = 1 \quad (10.24)$$

$$TA_1 = \bar{S}A_0 + S\bar{A}_0 \quad (10.25)$$

Circuit diagram of MOD-4 up/down synchronous counter is shown in Figure 10.52.

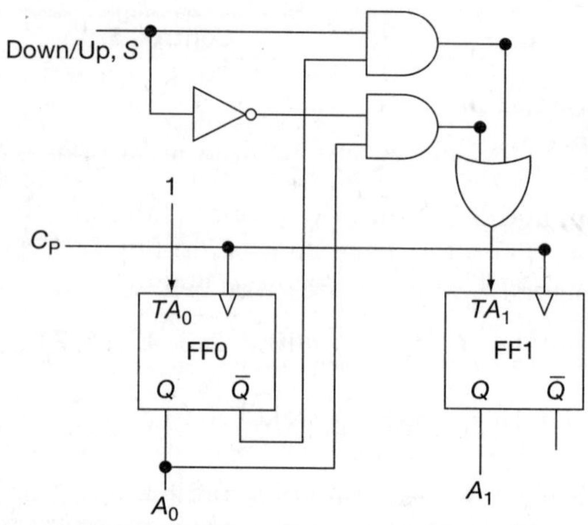

FIGURE 10.52 | MOD-9 up/down synchronous counter

10.5.4 | Modulus-8 Synchronous Up Counter

Modulus 8 (MOD-8) synchronous up counters require three flip-flops.

$$2^3 = 8 \Rightarrow 2^n = 2^3, \quad \text{so } n = 3$$

Figure 10.53 gives the state diagram of up count. Table 10.20 gives the count status during the clock pulse transition. Present count is (000) and next count is (001) and so on. When present count is (111), next count is (000). Excitation of T-flip-flop is 0 for same present and next output states of flip-flop. Excitation of T-flip-flop is 1 for different present and

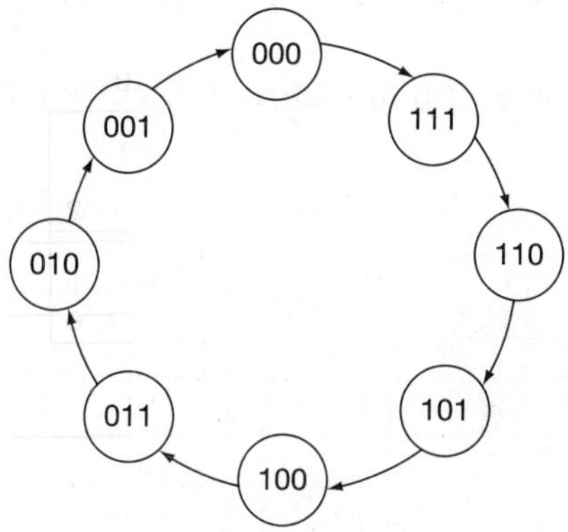

FIGURE 10.53 | Up count sequence of MOD-8 counter

TABLE 10.20 | MOD-8 synchronous up counter

Clock Pulse Count	Count Sequence			Excitation of Flip-flop		
	A_2	A_1	A_0	TA_2	TA_1	TA_0
0	0	0	0	0	0	1
1	0	0	1	0	1	1
2	0	1	0	0	0	1
3	0	1	1	1	1	1
4	1	0	0	0	0	1
5	1	0	1	0	1	1
6	1	1	0	0	0	1
7	1	1	1	1	1	1
	0	0	0			

next output states of flip-flop. As the count changes from (001) to (010) the excitations of flip-flops are:

- The excitation, TA_0 is 1, because present output, Q_0 is 1 and next required output, Q_0 is 0.
- The excitation TA_1 is 1 because present output, Q_1 is 0 and next required output, Q_1 is 1.
- The excitation TA_2 is 0 because present output, Q_2 and next required output, Q_2 are 0.

Canonical expression for excitations of T-flip-flops TA_0, TA_1 and TA_2 are given below:

$$TA_0(A_2,A_1,A_0)=\sum m(0,1,2,3,4,5,6,7) \quad (10.26)$$

$$TA_1(A_2,A_1,A_0)=\sum m(1,3,5,7) \quad (10.27)$$

$$TA_2(A_2,A_1,A_0)=\sum m(3,7) \quad (10.28)$$

For excitation TA_0 of T-flip-flop, three-variable K-map is set up in Figure 10.54. One octet is formed and all minterms give 1-logic.

$$TA_0 = 1 \quad (10.29)$$

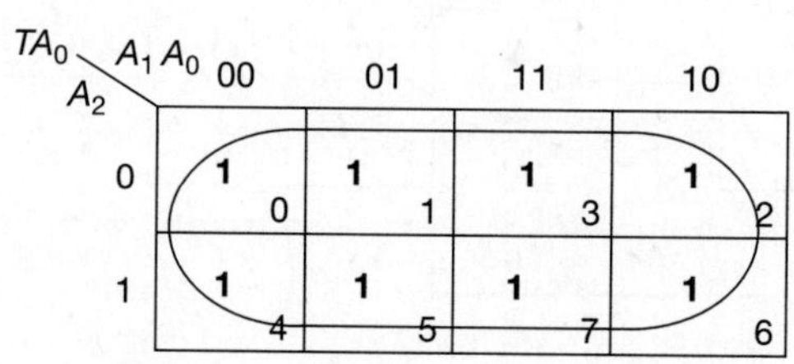

FIGURE 10.54 | K-map for TA_0

For excitation TA_1 of T-flip-flop, three-variable K-map is set up in Figure 10.55. One quad {1, 3, 5, 7} is formed that gives simplified logic as

$$TA_1 = A_0 \tag{10.30}$$

For excitation TA_2 of T-flip-flop, three-variable K-map is set up in Figure 10.56. One pair {3, 7} is formed that gives simplified logic as

$$TA_2 = A_1A_0 \tag{10.31}$$

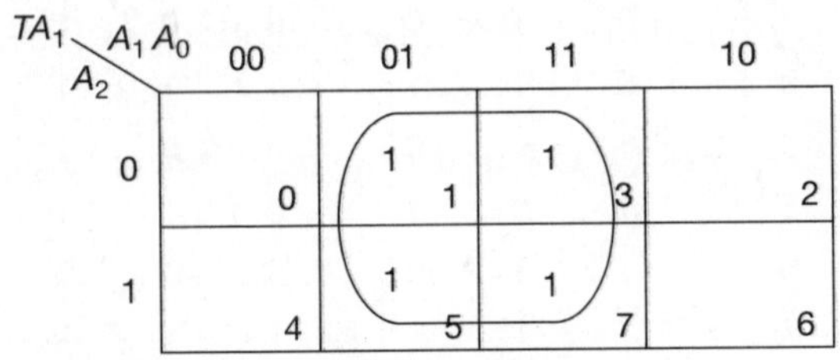

FIGURE 10.55 | K-map for TA_1

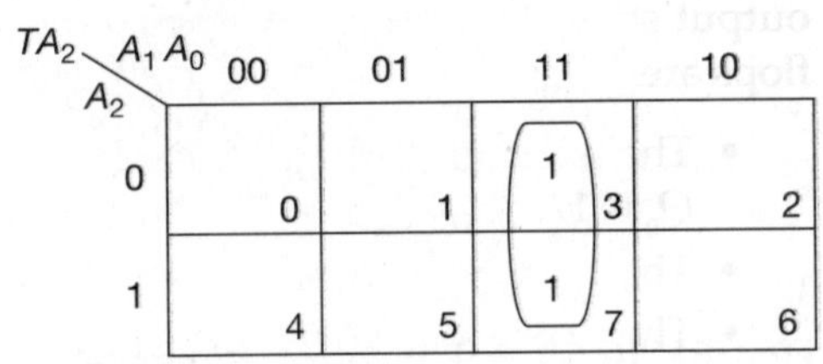

FIGURE 10.56 | K-map for TA_2

Circuit diagram of MOD-8 up synchronous counter is shown in Figure 10.57 and timing diagram is shown in Figure 10.58.

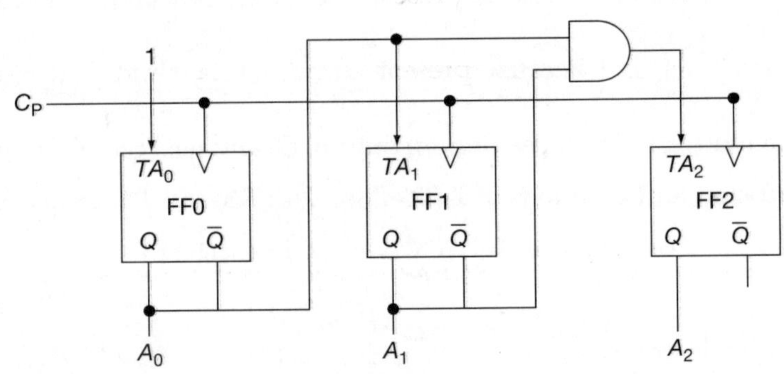

FIGURE 10.57 | MOD-8 up synchronous counter

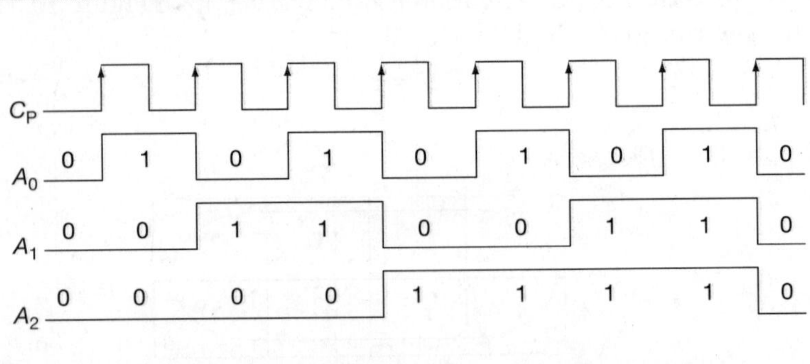

FIGURE 10.58 | Timing diagram of MOD-8 up synchronous counter

10.5.5 | Modulus-8 Synchronous Down Counter

Modulus 8 (MOD-8) synchronous up counters require three flip-flops.

$$2^3 = 8 \Rightarrow 2^n = 2^3, \text{ so } n = 3.$$

Figure 10.59 gives the state diagram of up count. Table 10.21 gives the count status during the clock pulse transition. Present count is (000) and next count is (111) and so on. When present count is (111), next count is (110). Excitation of T-flip-flop is 0 for same present and next output states of flip-flop. Excitation of T-flip-flop is 1 for different present and next output states of flip-flop. As the count changes from (110) to (101) the excitations of flip-flops are:

- The excitation, TA_0 is 1, because present output, Q_0 is 0 and next required output, Q_0 is 1.
- The excitation, TA_1 is 1, because present output, Q_1 is 1 and next required output, Q_1 is 0.
- The excitation, TA_2 is 0, because present output, Q_2 and next required output, Q_2 are 1.

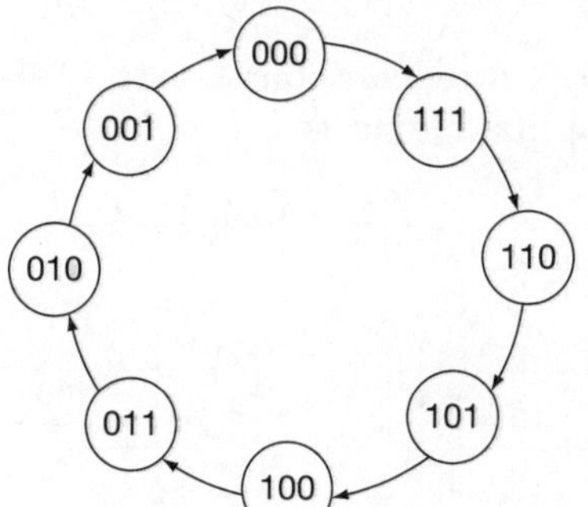

FIGURE 10.59 | Down count sequence of MOD-8 down counter

TABLE 10.21 | Down synchronous MOD-8 counter

Clock Pulse Count	Countdown Sequence			Excitations of Flip-flop		
	A_2	A_1	A_0	TA_2	TA_1	TA_0
7	1	1	1	0	0	1
6	1	1	0	0	1	1
5	1	0	1	0	0	1
4	1	0	0	1	1	1
3	0	1	1	0	0	1
2	0	1	0	0	1	1
1	0	0	1	0	0	1
0	0	0	0	1	1	1
7	1	1	1			

Canonical expressions for excitations of T-flip-flops TA_0, TA_1 and TA_2 are given below:

$$TA_0(A_2, A_1, A_0) = \sum m(0,1,2,3,4,5,6,7) \tag{10.32}$$

$$TA_1(A_2, A_1, A_0) = \sum m(0,2,4,6) \tag{10.33}$$

$$TA_2(A_2, A_1, A_0) = \sum m(0,2) \tag{10.34}$$

For excitation TA_0 of T-flip-flop, three-variable K-map is set up in Figure 10.60. One octet is formed and all minterms give 1-logic.

$$TA_0 = 1 \tag{10.35}$$

For excitation TA_1 of T-flip-flop, three-variable K-map is set up in Figure 10.61. One quad {0, 2, 4, 6} is formed that gives simplified logic as

$$TA_1 = \bar{A}_0 \tag{10.36}$$

For excitation TA_2 of T-flip-flop, three-variable K-map is set up in Figure 10.62. One pair {0, 4} is formed that gives simplified logic as

$$TA_2 = \bar{A}_1\bar{A}_0 \tag{10.37}$$

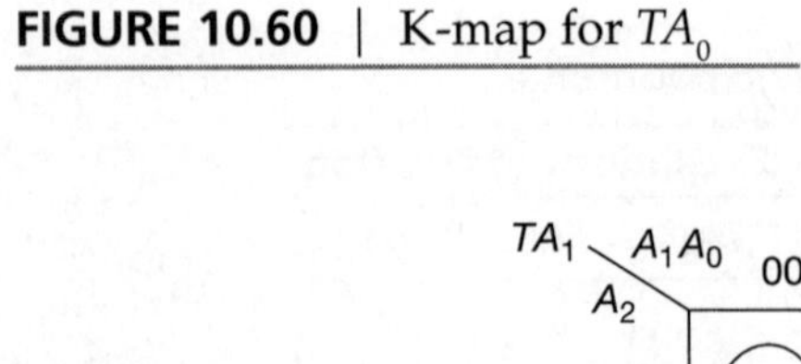

FIGURE 10.60 | K-map for TA_0

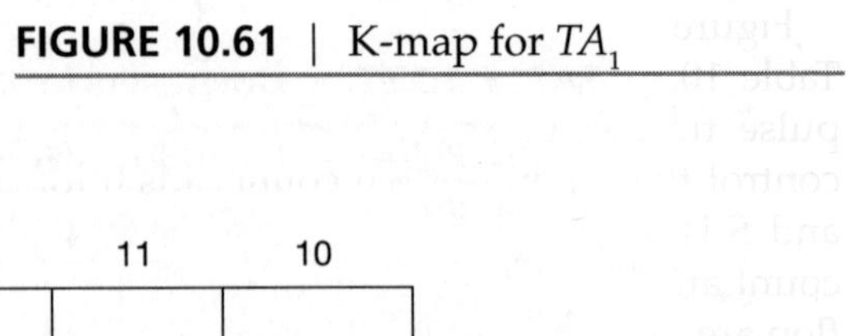

FIGURE 10.61 | K-map for TA_1

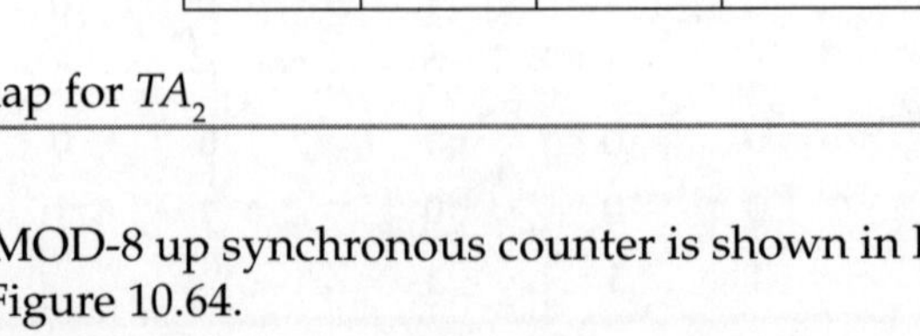

FIGURE 10.62 | K-map for TA_2

Circuit diagram of MOD-8 up synchronous counter is shown in Figure 10.63 and timing diagram is shown in Figure 10.64.

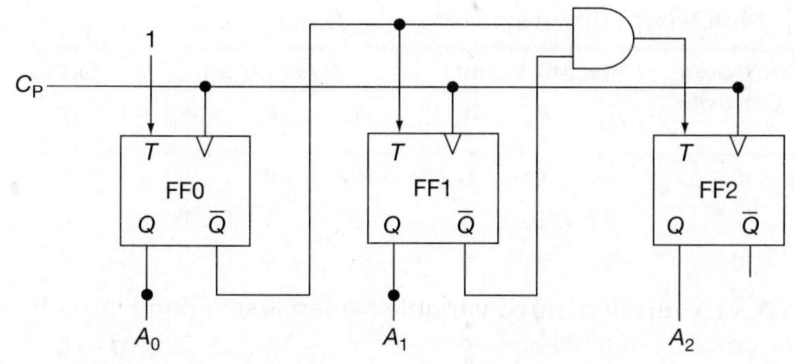

FIGURE 10.63 | MOD-8 down synchronous counter

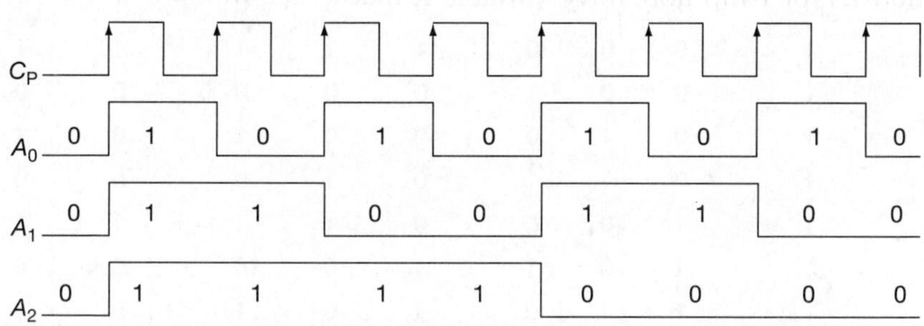

FIGURE 10.64 | Timing diagram of MOD-8 down synchronous counter

10.5.6 | Modulus-8 Synchronous UP/Down Counter

Modulus 8 (MOD-8) synchronous counters require three flip-flops. $\because 2^n = 8 \Rightarrow 2^n = 2^3$, so $n = 3$.

Figure 10.65 gives the state diagram of up count. Table 10.24 gives the count status during the clock pulse transition. One control variable is added to control the up-and-down count. S is 0 for up count and S is 1 for down count. Present count and next count are given in the Table 10.22. Excitations T-flip-flop are given in the Table 10.22.

As the count changes from (010) to (011) with S is 0 (up count), the excitations of flip-flops are:

- The excitation, TA_0 is 1, because present output, Q_0 is 0 and next output, Q_0 is 1.
- The excitation, TA_1 is 0, because present output, Q_1 is 0 and next required output, Q_1 is also 0.
- The excitation, TA_2 is 0, because present output, Q_2 and next required output, Q_2 are 0.

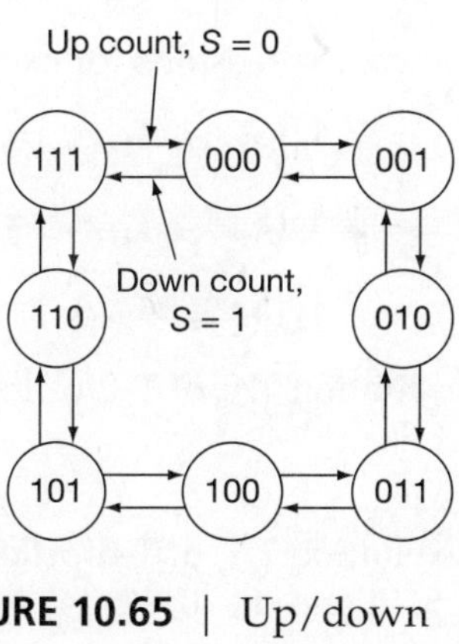

FIGURE 10.65 | Up/down count sequence

TABLE 10.22 | MOD-8 up/down synchronous counter

Clock Pulse Count	Up/Down Control S	Present Count			Next Count			Excitation of Flip-flop		
		A_2	A_1	A_0	A_2	A_1	A_0	TA_1	TA_1	TA_0
0	0	0	0	0	0	0	1	0	0	1
1	0	0	0	1	0	1	0	0	1	1
2	0	0	1	0	0	1	1	0	0	1
3	0	0	1	1	1	0	0	1	1	1
4	0	1	0	0	1	0	1	0	0	1
5	0	1	0	1	1	1	0	0	1	1
6	0	1	1	0	1	1	1	0	0	1
7	0	1	1	1	0	0	0	1	1	1
0	1	0	0	0	1	1	1	1	1	1
1	1	0	0	1	0	0	0	0	0	1
2	1	0	1	0	0	0	1	0	1	1
3	1	0	1	1	0	1	0	0	0	1
4	1	1	0	0	0	1	1	1	1	1
5	1	1	0	1	1	0	0	0	0	1
6	1	1	1	0	1	0	1	0	1	1
7	1	1	1	1	1	1	0	0	0	1

As the count changes from (101) to (100) with S is 1 (down count), the excitations of flip-flops are:

- The excitation, TA_0 is 1, because present output, Q_0 is 1 and next output, Q_0 is 0.
- The excitation TA_1 is 0 because present output, Q_1 is 0 and next required output, Q_1 is also 0.
- The excitation TA_2 is 0 because present output, Q_2 and next required output, Q_2 are 1.

Canonical expressions for excitations of T-flip-flops TA_0, TA_1 and TA_2 are given below:

$$TA_0(S, A_2, A_1, A_0) = \sum m(0, 1, 2, 3, 4, 5, 6, 7, 8, 9, 10, 11, 12, 13, 14, 15) \quad (10.38)$$

$$TA_1(S, A_2, A_1, A_0) = \sum m(1, 3, 5, 7, 8, 10, 12, 14) \quad (10.39)$$

$$TA_2(S, A_2, A_1, A_0) = \sum m(3, 7, 8, 12) \quad (10.40)$$

For excitation TA_0 of T-flip-flop, all the outputs are 1, so the excitation becomes:

$$TA_0 = 1 \quad (10.41)$$

For excitation TA_1 of T-flip-flop, three-variable K-map is set up in Figure 10.66. Two quads {1, 3, 5, 7} and {8, 10, 12, 14} are formed and give simplified logic as

$$TA_1 = \bar{S}A_0 + S\bar{A}_0 \quad (10.42)$$

For excitation TA_2 of T-flip-flop, three-variable K-map is set up in Figure 10.67. Two pairs {3, 7} and {8, 12} are formed those give simplified logic as

$$TA_2 = (\bar{S}A_0)A_1 + (S\bar{A}_0)\bar{A}_1 \tag{10.43}$$

Circuit diagram of MOD-8 up/down synchronous counter is shown in Figure 10.68.

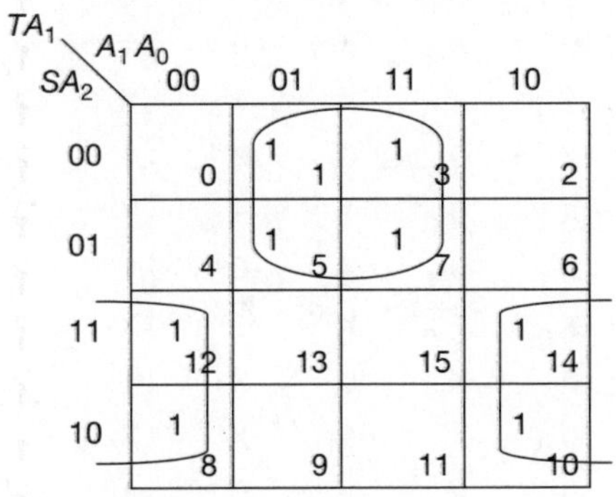

FIGURE 10.66 | K-map for TA_1

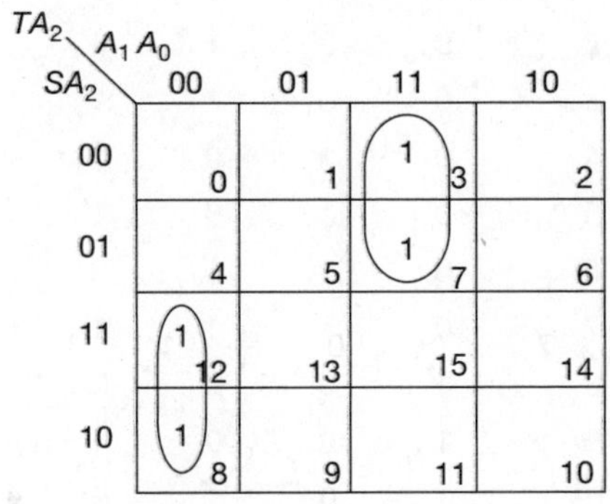

FIGURE 10.67 | K-map for TA_2

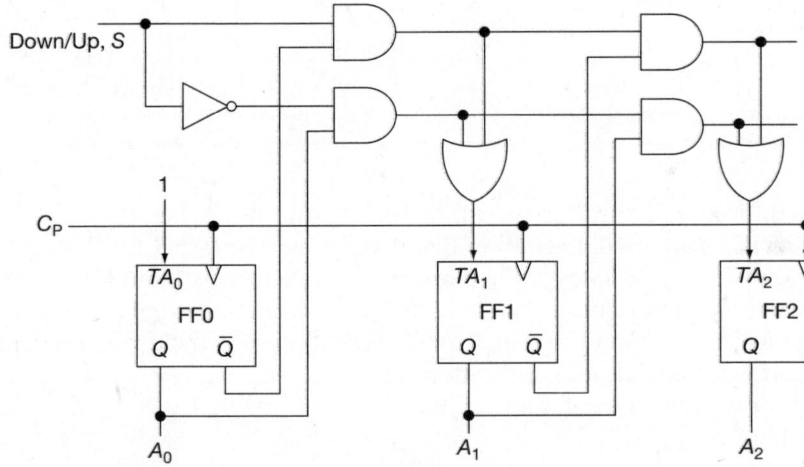

FIGURE 10.68 | MOD-8 up/down synchronous counter

10.6 | SYNCHRONOUS COUNTER WITH PARALLEL LOAD

JK-flip-flop can hold the output, store the input data and toggles the output. This property of JK-flip-flop is used to design a counter that can start counting with the given input. Let modulus-16 counter is considered for the design using JK-flip-flop. MOD-16 up synchronous counter counts 0, 1, 2, …, 14 and 15. Modulus 16 (MOD-16) synchronous up counter requires four flip-flops ($16 \le 2^4$). Table 10.23 gives the count status during the clock pulse

TABLE 10.23 | Up synchronous MOD-16 counter

Clock Pulse	Sequence of Count				Excitation							
	A_3	A_2	A_1	A_0	JA_3	KA_3	JA_2	KA_2	JA_1	KA_1	JA_0	KA_0
0	0	0	0	0	0	X	0	X	0	X	1	X
1	0	0	0	1	0	X	0	X	1	X	X	1
2	0	0	1	0	0	X	0	X	X	0	1	X
3	0	0	1	1	0	X	1	X	X	1	X	1
4	0	1	0	0	0	X	X	0	0	X	1	X
5	0	1	0	1	0	X	X	0	1	X	X	1
6	0	1	1	0	0	X	X	0	X	0	1	X
7	0	1	1	1	1	X	X	1	X	1	X	1
8	1	0	0	0	X	0	0	X	0	X	1	X
9	1	0	0	1	X	0	0	X	1	X	X	1
10	1	0	1	0	X	0	0	X	X	0	1	X
11	1	0	1	1	X	0	1	X	X	1	X	1
12	1	1	0	0	X	0	X	0	0	X	1	X
13	1	1	0	1	X	0	X	0	1	X	X	1
14	1	1	1	0	X	0	X	0	X	0	1	X
15	1	1	1	1	X	1	X	1	X	1	X	1
	0	0	0	0								

transition. Excitation J is 0 and K may be 0 or 1 for JK-flip-flop when the present and next output states of flip-flop are 0. Excitation K is 0 and J may be 0 or 1 for JK-flip-flop when the present and next output states of flip-flop are 1. Excitation J is 1 and K may be 0 or 1 for JK-flip-flop when the present output state of flip-flop is 0 and next output state of flip-flop is 1. Excitation K is 1 and J may be 0 or 1 for JK-flip-flop when the present output state of flip-flop is 1 and next output state of flip-flop is 0.

Canonical expressions for excitations of JK-flip-flops JA_0, KA_0, JA_1, KA_1, JA_2, KA_2, JA_3, and KA_3 are given below:

$$JA_0(A_3,A_2,A_1,A_0)=\sum m(0,2,4,6,8,10,12,14)+\sum d(1,3,5,7,9,11,13,15) \tag{10.44}$$

$$KA_0(A_3,A_2,A_1,A_0)=\sum m(1,3,5,7,9,11,13,15)+\sum d(0,2,4,6,8,10,12,14) \tag{10.45}$$

$$JA_1(A_3,A_2,A_1,A_0)=\sum m(1,5,9,13)+\sum d(2,3,6,7,10,11,14,15) \tag{10.46}$$

$$KA_1(A_3,A_2,A_1,A_0)=\sum m(3,7,11,15)+\sum d(0,1,4,5,8,9,12,13) \tag{10.47}$$

$$JA_2(A_3,A_2,A_1,A_0)=\sum m(3,11)+\sum d(4,5,6,7,12,13,14,15) \tag{10.48}$$

$$KA_2(A_3,A_2,A_1,A_0)=\sum m(7,15)+\sum d(0,1,2,3,8,9,10,11) \tag{10.49}$$

$$JA_3(A_3, A_2, A_1, A_0) = \sum m(7) + \sum d(8,9,10,11,12,13,14,15) \tag{10.50}$$

$$KA_3(A_3, A_2, A_1, A_0) = \sum m(15) + \sum d(0,1,2,3,4,5,6,7) \tag{10.51}$$

For excitation JA_0 and KA_0 of JK-flip-flop, all the outputs are 1 including don't care conditions and the excitation becomes:

$$JA_0 = 1 \text{ and } KA_0 = 1$$

For excitation JA_1 and KA_1 of JK-flip-flop, four-variable K-map is set up in Figures 10.69(a) and 10.94(b). One octet {1, 3, 5, 7, 9, 11, 13, 15} is formed in Figure 10.69(a) and Figure 10.69(a). Simplified logic for JK-flip-flop inputs are given as

$$JA_1 = A_0 \text{ and } KA_1 = A_0$$

For excitation JA_2 and KA_2 of JK-flip-flop, four-variable K-map is set up in Figures 10.69(c) and 10.94(d). One quad {3, 7, 11, 15} is formed in Figure 10.69(c) and Figure 10.69(d). Simplified logic for JK-flip-flop inputs are given as

$$JA_2 = A_1 A_0 \text{ and } KA_2 = A_1 A_0$$

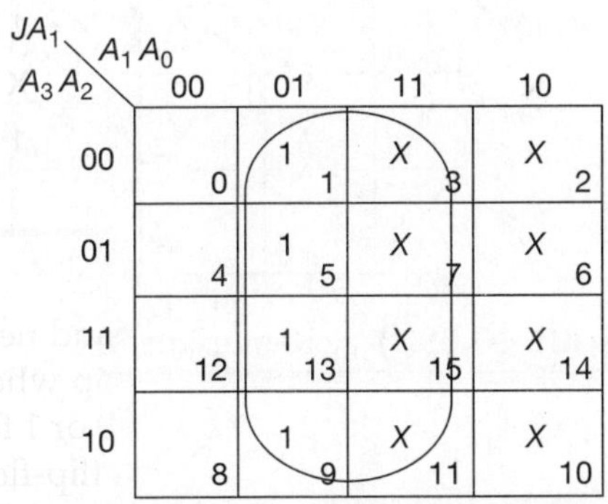

FIGURE 10.69(a) | K-map for JA_1

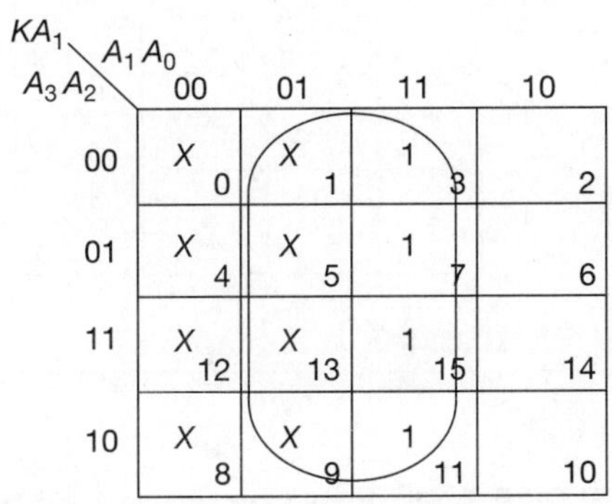

FIGURE 10.69(b) | K-map for KA_1

For excitation JA_3 and KA_3 of JK-flip-flop, four-variable K-map is set up in Figures 10.69(e) and 10.94(f). One quad {7, 15} is formed in Figure 10.69(e) and Figure 10.69(f). Simplified logic for JK-flip-flop inputs are given below as

$$JA_3 = A_2 A_1 A_0 \text{ and } KA_3 = A_2 A_1 A_0$$

Counters are loaded with an initial binary number prior the counting. The input load control, L is 1 then counter operation is disabled and causes a load operation. When the input load control, L is 0 and count input control, C is 1 then counter performs count operation. When the input load control, L and count input control, C are 0, then counter acts as register. Table 10.24 gives various operations to be performed. To store the value, the JK flip flop acts as D-flip-flop when K is complement of J.

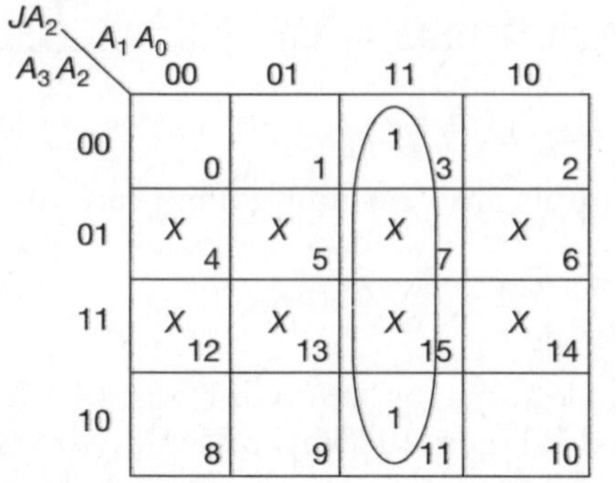

FIGURE 10.69(c) | K-map for JA_2

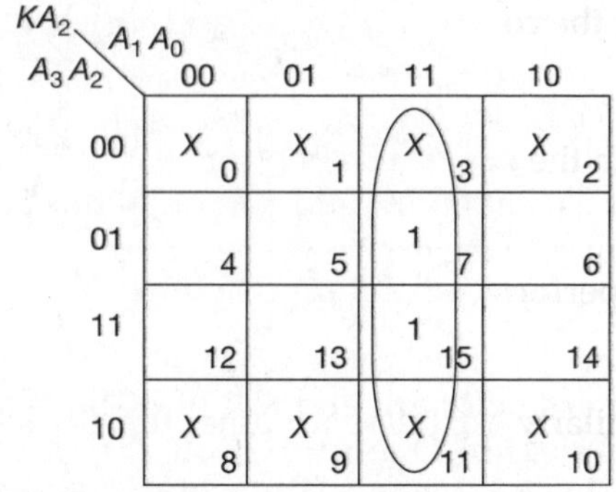

FIGURE 10.69(d) | K-map for KA_2

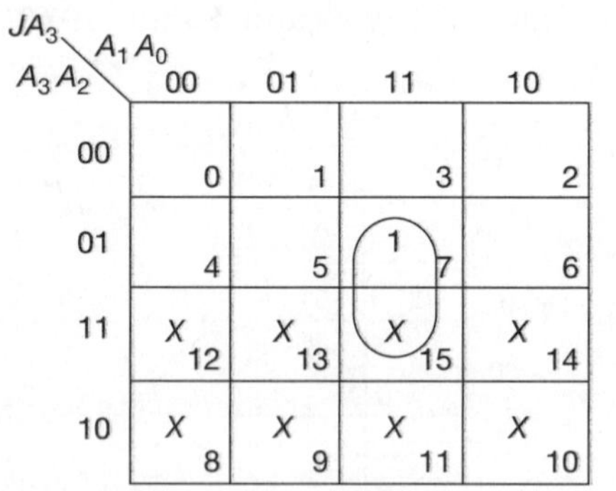

FIGURE 10.69(e) | K-map for JA_3

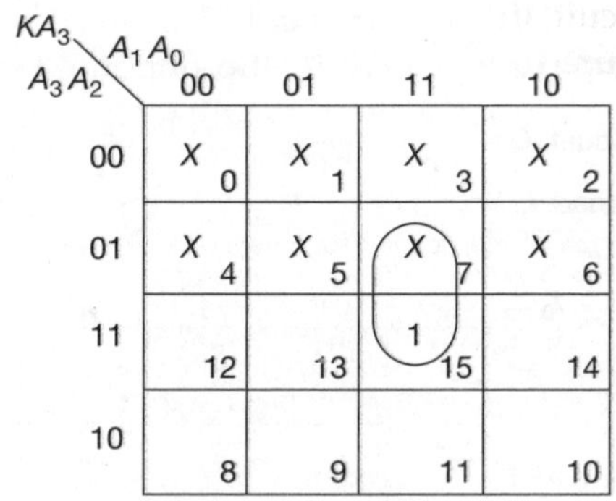

FIGURE 10.69(f) | K-map for KA_3

TABLE 10.24 | Function of the 4-bit binary count with parallel load

Clear (Cr)	Clock (C_p)	Load (L)	Count (C)	Function	J	K	J	K
0	X	X	X	Clear to 0	X	X	X	X
1	X	0	0	No Change (hold)	0	0	LI	$L\bar{I}$
1	↓	1	0	Load Inputs	I	$\bar{I}$	LI	$L\bar{I}$
1	↓	0	1	Count to next binary value	1	1	$\bar{L}C$	$\bar{L}C$

So $$J = I \text{ and } K = \bar{I}$$

To hold the output, J and K should be 0. So, the above expression is modified as

$$J = LI \text{ and } K = L\bar{I}$$

When L is 0 and I may be 0 or 1, then $J = 0$ and $K = 0$.

For the count operation, J and K should be 1. The flip-flop toggles when J and K are 1.

$$J = \bar{L}C \text{ and } K = \bar{L}C$$

Both the operations become:

$$J = LI + \bar{L}C \text{ and } K = L\bar{I} + \bar{L}C$$

To perform synchronous counter operation, the excitations become

$$JA_0 = LI + \bar{L}C \text{ and } KA_0 = L\bar{I} + \bar{L}C$$

Similarly excitation for other flip-flops become

$$JA_1 = LI + \bar{L}CA_0 \text{ and } KA_1 = LI + \bar{L}CA_0$$

$$JA_2 = LI + \bar{L}C\,A_1A_0 \text{ and } KA_2 = LI + \bar{L}C\,A_1A_0$$

$$JA_3 = LI + \bar{L}CA_2A_1A_0 \text{ and } KA_3 = LI + \bar{L}CA_2A_1A_0$$

Circuit diagram of MOD-16 up synchronous counter using JK-flip-flop is shown in Figure 10.70 and block diagram is given in Figure 10.71.

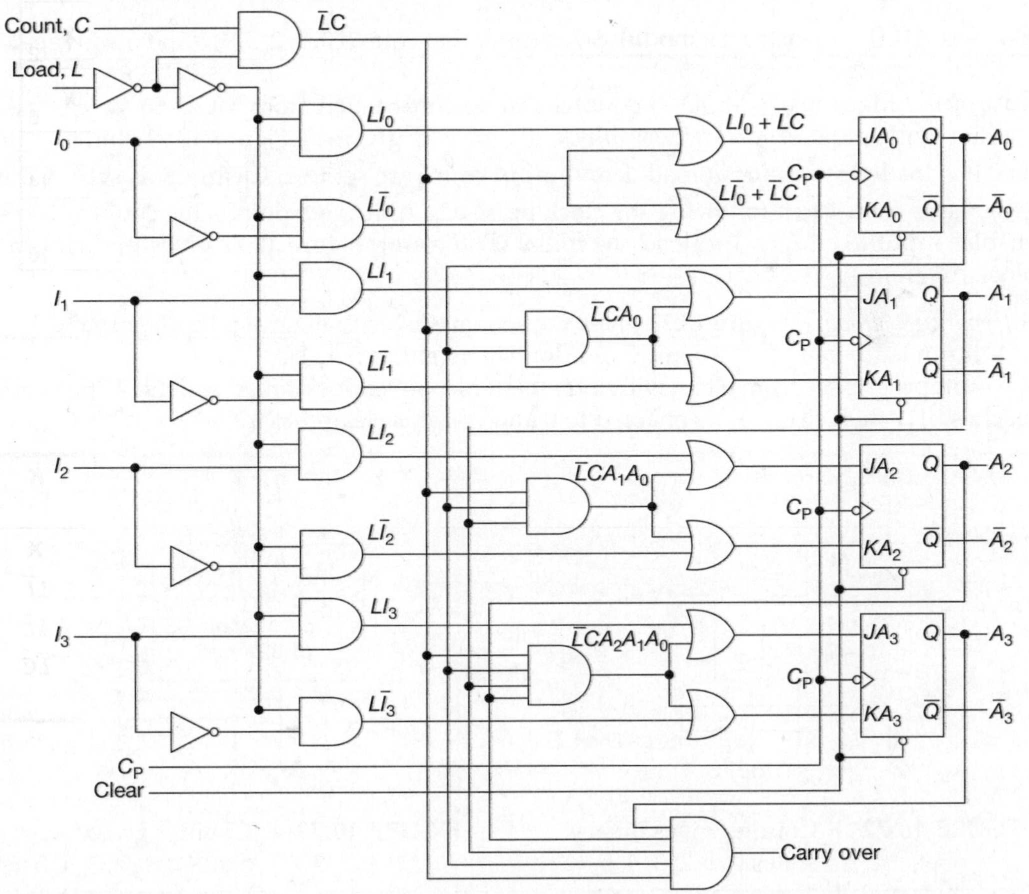

FIGURE 10.70 | Four-bit Binary counter with parallel load

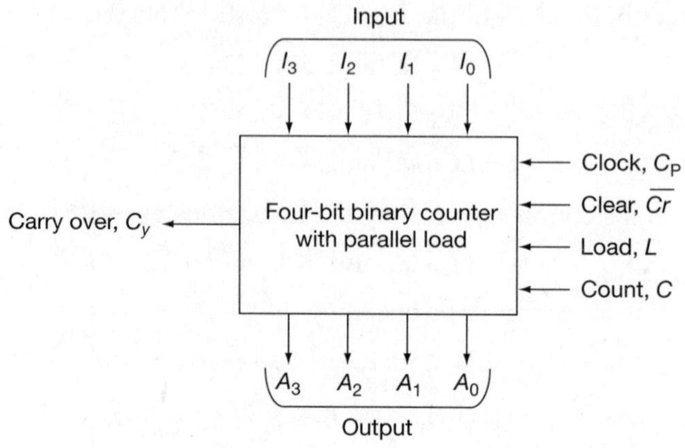

FIGURE 10.71 | Counter with parallel load

EXAMPLE 10.9 Construct a modulus-7 counter that counts 0, 1, 2, 3, 4, 5 and 6 and repeats.

SOLUTION Modulus-7 (MOD-7) counter can be constructed from MOD-16 synchronous counter with parallel load whose block diagram is given in Figure 10.71. Initial value (0000) is loaded when input load, L and input count are set to 1. Clear, $\overline{Cr}$ is set 1. Count progresses from 0 to 1 following the clock pulse, C_P. AND gate detects the count 0110 and enables input load, L as 1 to load the initial 0000 count. Figure 10.72 gives the schematic block diagram.

Alternative Method: Figure 10.73 gives the schematic block diagram. Input count, C is set to 1. Input load control, L is considered don't care (either 0 or 1).

Count progresses from 0 to 6 with increment of 1 on as clock pulse is given. When count reaches 0111 then clear, $\overline{Cr}$ is enabled to 0 and count is cleared to 0000.

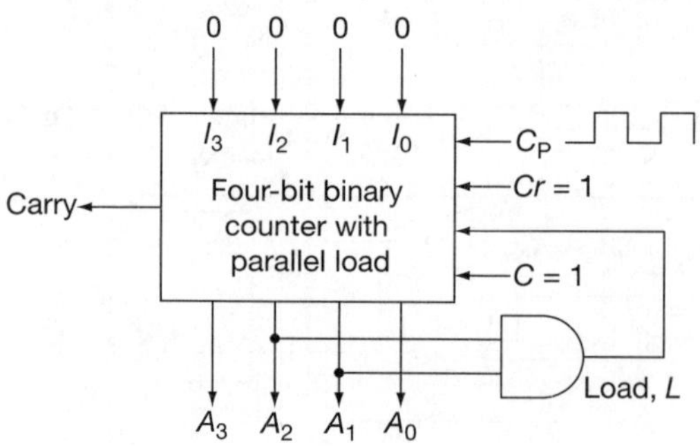

FIGURE 10.72 | Counter gives binary states 0, 1, 2, 3, 4, 5, 6

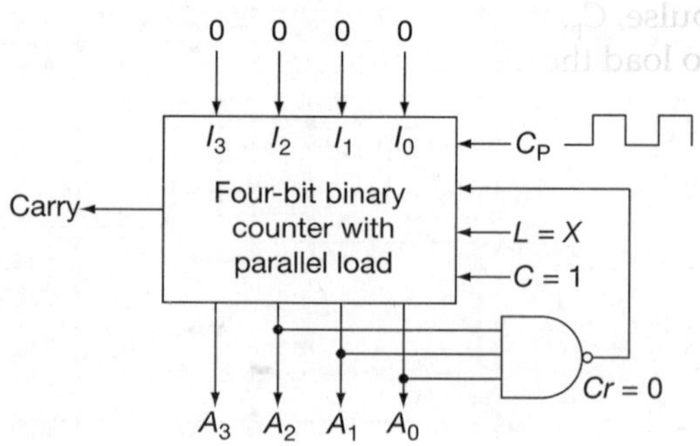

FIGURE 10.73 | Counter gives binary states 0, 1, 2, 3, 4, 5, 6

EXAMPLE 10.10 Construct a Modulus-7 counter that counts 9, 10, 11, 12, 13, 14 and 15 and repeats from 9.

SOLUTION Modulus-7 (MOD-7) counter can be constructed from MOD-16 synchronous counter with parallel load whose block diagram is given in Figure 10.71. Figure 10.74 gives the schematic block diagram. Initial value (1001) is loaded when input load, L is set to 1. Set the count input to 1. Clear, $\overline{Cr}$ is set 1. Count progresses from 9 to 15 following the clock pulse, C_P. Carry is generated when count reaches to 1111 that enables input load, L as 1 to load the initial 1001 count.

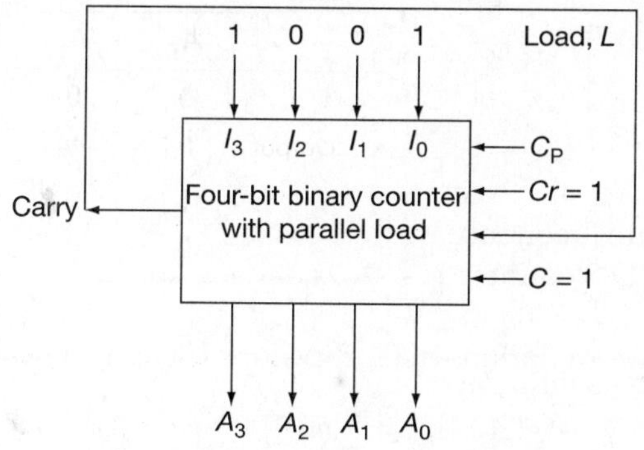

FIGURE 10.74 | Counter gives binary states 9, 10, 11, 12, 13, 14, 15

EXAMPLE 10.11 Construct a modulus-7 counter that counts 2, 3, 4, 5, 6, 7 and 8 and repeats from 2.

SOLUTION Modulus-7 (MOD-7) counter can be constructed from MOD-16 synchronous counter with parallel load whose block diagram is given in Figure 10.71. Figure 10.75 gives the schematic block diagram. Initial value (0010) is loaded when input load, L is set to 1. Set the count input to 1. Clear, $\overline{Cr}$ is set 1. Count progresses from 2 to 8 following the clock pulse, C_P. when reaches to 1000. As output at A_3 becomes 1 that enables input load, L to 1 to load the initial 0010 count.

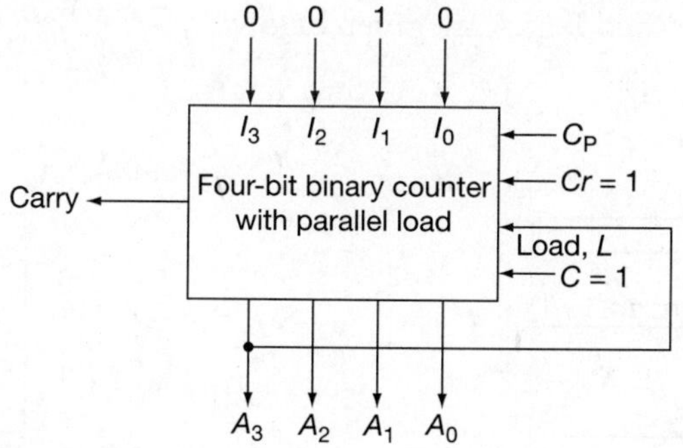

FIGURE 10.75 | Counter gives binary states 2, 3, 4, 5, 6, 7, 8

EXAMPLE 10.12 Design a modulus-3 (MOD-3) up synchronous counter using JK-flip-flop.

SOLUTION MOD-3 up synchronous counter counts 0, 1 and 2 (Figure 10.76). Modulus-3 (MOD-3) synchronous up counter requires two JK-flip-flops ($3 \le 2^2$). Table 10.25 gives the count status during the clock pulse transition. Excitation

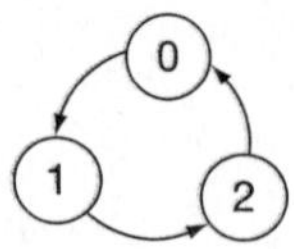

FIGURE 10.76 | Up count sequence of MOD-3 counter

TABLE 10.25 | MOD-3 synchronous counter

Clock Pulse	Sequence up Count		Excitation			
	A_1	A_0	JA_1	KA_1	JA_0	KA_0
0	0	0	0	X	1	X
1	0	1	1	X	X	1
2	1	0	X	1	0	X
	0	0	–		–	

Canonical expressions for excitations of JK-flip-flops JA_0, KA_0, JA_1, and KA_1 are given below:

$$JA_0(A_1, A_0) = \sum m(0) + \sum m(1,3) \tag{i}$$

$$KA_0(A_1, A_0) = \sum m(1) + \sum m(0,2,3) \tag{ii}$$

$$JA_1(A_1, A_0) = \sum m(1) + \sum d(2,3) \tag{iii}$$

$$KA_1(A_1, A_0) = \sum m(2) + \sum d(0,1,3) \tag{iv}$$

For excitation JA_0 and KA_0 of JK-flip-flop, two-variable K-map is set up in Figures 10.77(a) and 10.102(b). One pair {0, 1} is formed in Figure 10.77(a) and one quad {0, 1, 2, 3} is formed in Figure 10.77(b). Simplified logics are given below:

$$JA_0 = \bar{A}_1 \text{ and } KA_0 = 1$$

For excitation JA_1 and KA_1 of JK-flip-flop, two-variable K-map is set up in Figures 10.77(c) and 10.77(d). One pair {1, 3} is formed in Figure 10.77(c) and one quad {0, 1, 2, 3} is formed in Figure 10.77(d). Simplified logics are given below:

$$JA_1 = A_0 \text{ and } KA_1 = 1$$

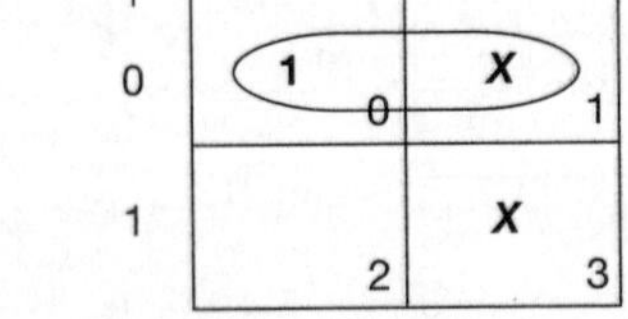

FIGURE 10.77(a) | K-map for JA_0

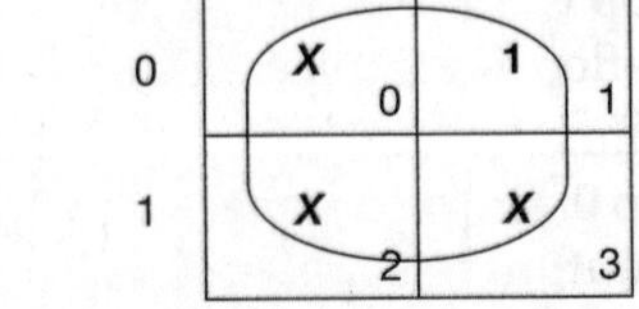

FIGURE 10.77(b) | K-map for KA_0

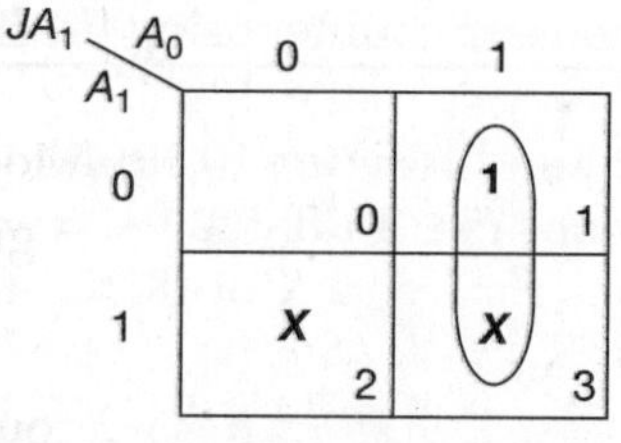

FIGURE 10.77(c) | K-map for JA_1

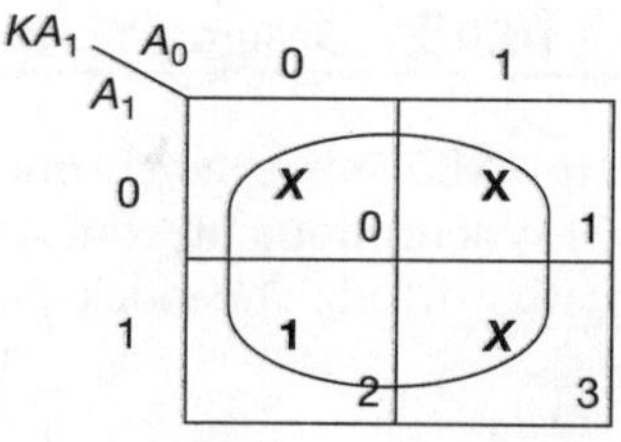

FIGURE 10.77(d) | K-map for KA_1

Circuit diagram of MOD-3 up synchronous counter using JK-flip-flop is shown in Figure 10.78 and timing diagram is given in Figure 10.79.

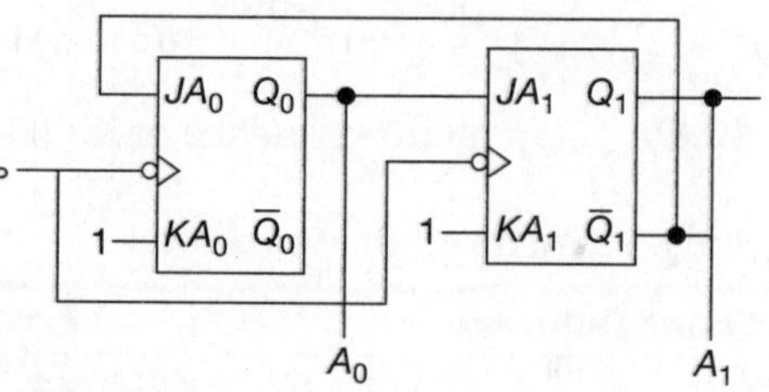

FIGURE 10.78 | MOD-3 synchronous up counter

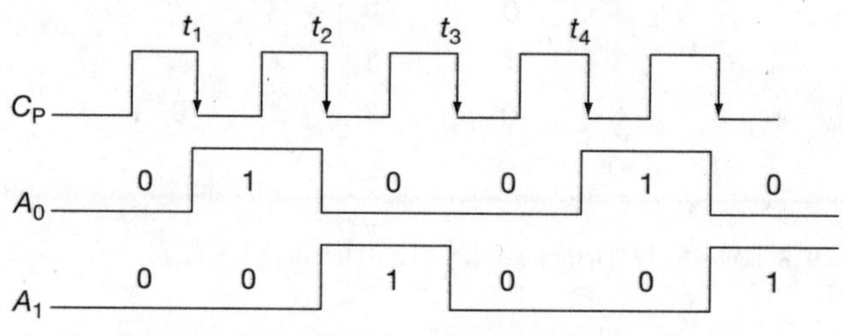

FIGURE 10.79 | Timing diagram of MOD-3 up synchronous counter

Let initial value of count is 00. Hence, A_0 is 0 and A_1 is 0.

At time t_1: As Q_1 is 0 and $\bar{Q}_1$ is 1, so JA_0 becomes 1 and KA_0 is 1 and the output Q_0 of flip flop is complemented to 1. For next stage, JA_1 is 0 and KA_1 is 1, so output, Q_1 of flip-flop is reset to 0. So, output of counter, A_0 is 1 and A_1 is 0.

At time t_2: As Q_1 is 0 and $\bar{Q}_1$ is 1, JA_0 becomes 1 and KA_0 are at 1 then the output, Q_0 of flip-flop is complemented and becomes 0. For next stage, JA_1 is 1 and KA_1 is 1, so output, Q_1 of flip-flop is complemented to 1. So, output of counter, A_0 is 0, and A_1 is 1.

At time t_3: As Q_1 is 1 and $\bar{Q}_1$ is 0, JA_0 is 0 and KA_0 is 1 then the output, Q_0 of flip-flop is reset to 0. For next stage, JA_1 is 1 and KA_1 is 1, so output, Q_1 of flip-flop is complemented to 0. So, output of counter, A_0 is 0 and A_1 is 0.

EXAMPLE 10.13 Design a modulus-5 (MOD-5) synchronous up counter using JK-flip-flop.

SOLUTION MOD-5 up synchronous counter counts 0, 1, 2, 3 and 4 (Figure 10.80). Modulus-3 (MOD-3) synchronous up counter requires three JK-flip-flops ($5 \le 2^3$). Table 10.26 gives the count status during the clock pulse transition. Excitations for J and K of JK-flip-flop are also given.

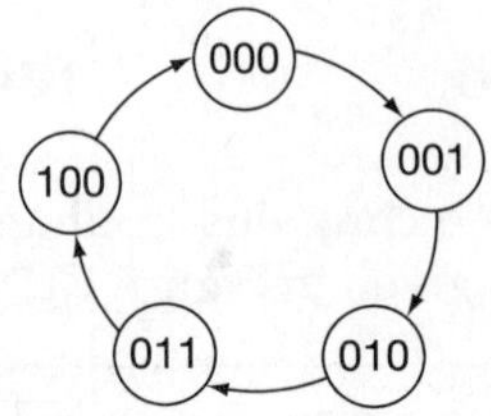

FIGURE 10.80 | Up count sequence of MOD-5 counter

TABLE 10.26 | MOD-5 synchronous up counter

Clock Pulse Count	Count Sequence			Excitation					
	A_2	A_1	A_0	JA_2	KA_2	JA_1	KA_1	JA_0	KA_0
0	0	0	0	0	X	0	X	1	X
1	0	0	1	0	X	1	X	X	1
2	0	1	0	0	X	X	0	1	X
3	0	1	1	1	X	X	1	X	1
4	1	0	0	X	1	0	X	0	X
0	0	0	0						

Canonical expressions for excitations of JK-flip-flops JA_0, KA_0, JA_1, KA_1, JA_2, and KA_2 are given below:

$$JA_0(A_2, A_1, A_0) = \sum m(0,2) + \sum d(1,3,5,6,7) \tag{i}$$

$$KA_0(A_2, A_1, A_0) = \sum m(1,3) + \sum d(0,2,4,5,6,7) \tag{ii}$$

$$JA_1(A_2, A_1, A_0) = \sum m(1) + \sum d(2,3,5,6,7) \tag{iii}$$

$$KA_1(A_2, A_1, A_0) = \sum m(3) + \sum d(0,1,4,5,6,7) \tag{iv}$$

$$JA_2(A_2, A_1, A_0) = \sum m(3) + \sum d(4,5,6,7) \tag{v}$$

$$KA_2(A_2, A_1, A_0) = \sum m(4) + \sum d(0,1,2,3,5,6,7) \tag{vi}$$

For excitation JA_0 and KA_0 of JK-flip-flop, two-variable K-map is set up in Figures 10.81(a) and 10.81(b). One quad {0, 1, 2, 3} is formed in Figure 10.81(a) and one octet {0, 1, 2, 3, 4, 5, 6, 7} is formed in Figure 10.81(b). Simplified logics are given below:

$$JA_0 = \bar{A}_2 \text{ and } KA_0 = 1$$

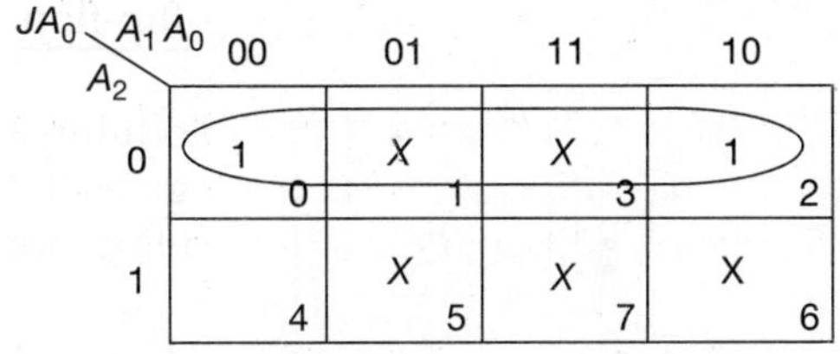

FIGURE 10.81(a) | K-map for JA_0

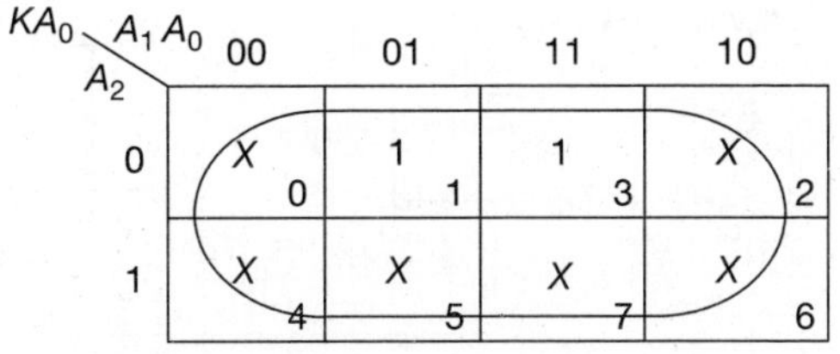

FIGURE 10.81(b) | K-map for KA_0

For excitation JA_1 and KA_1 of JK-flip-flop, two-variable K-map is set up in Figures 10.81(c) and 10.106(d). One quad {1, 3, 5, 7} is formed in Figure 10.81(c) and Figure 10.81(d). Simplified logics are given below:

$$JA_1 = A_0 \text{ and } KA_1 = A_0$$

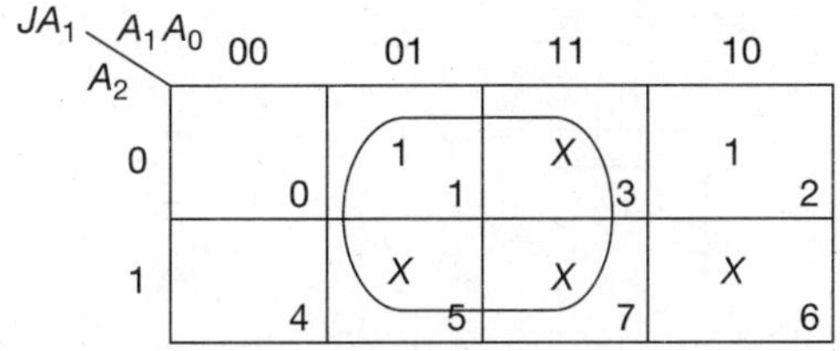

FIGURE 10.81(c) | K-map for JA_1

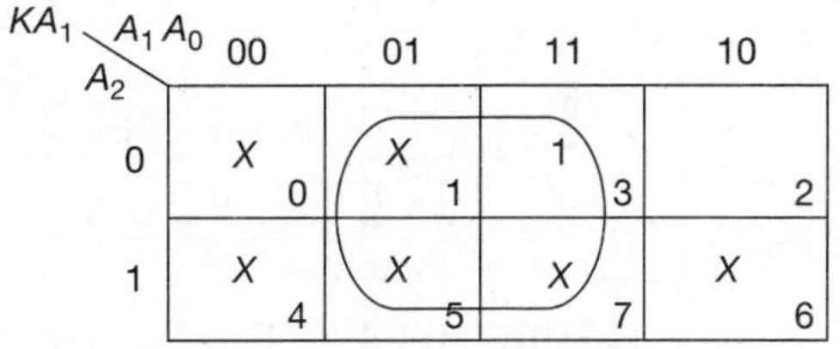

FIGURE 10.81(d) | K-map for KA_1

For excitation JA_1 and KA_1 of JK-flip-flop, two-variable K-map is set up in Figures 10.81(e) and 10.106(f). One pair {3, 7} is formed in Figure 10.81(e) and one octet {0, 1, 2, 3, 4, 5, 6, 7} is formed in Figure 10.81(f). Simplified logics are given below:

$$JA_2 = A_1A_0 \text{ and } KA_0 = 1$$

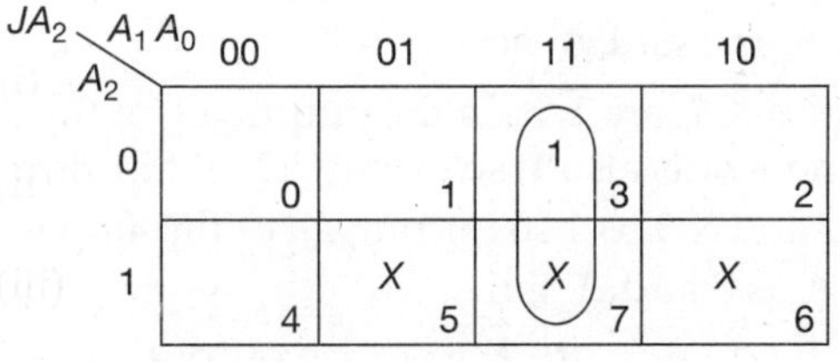

FIGURE 10.81(e) | K-map for JA_2

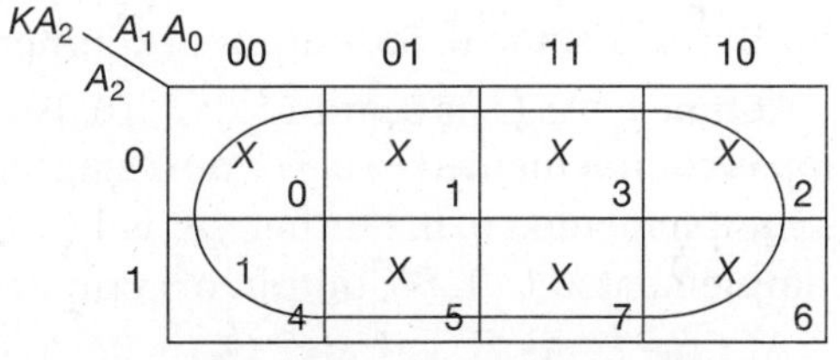

FIGURE 10.81(f) | K-map for KA_2

Circuit diagram of MOD-5 up synchronous counter using JK-flip-flop is shown in Figure 10.82 and timing diagram is given in Figure 10.83.

Let the initial value of count be 000. Hence, A_0 is 0, A_1 is 0 and A_2 is 0.

At time t_1: As Q_2 is 0 and $\bar{Q}_2$ is 1, so JA_0 becomes 1 and KA_0 is 1, and the output Q_0 of flip-flop is complemented to 1. For next stage, JA_1 is 0 and KA_1 is 1, so output, Q_1 of flip-flop is reset to 0. Further, JA_2 is 0 ($\because Q_0Q_1 = 0$) and KA_2 is 1 so output, Q_2 of flip-flop is reset to 0. So, output of counter, A_0 is 1, A_1 is 0 and A_2 is 0.

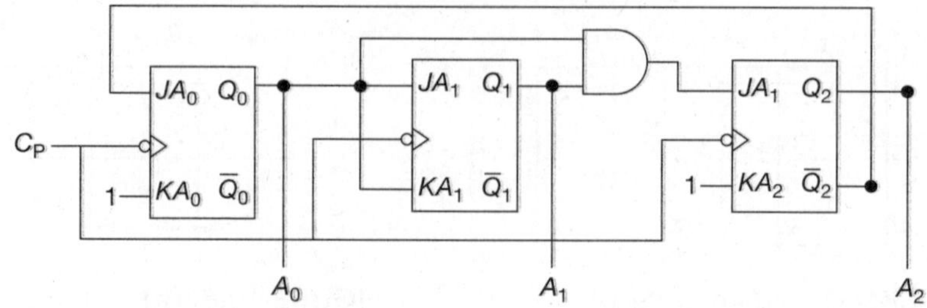

FIGURE 10.82 | MOD-5 synchronous up counter

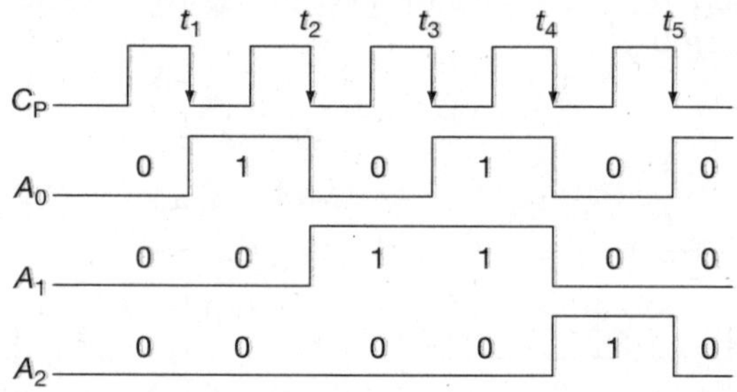

FIGURE 10.83 | Timing diagram of MOD-5 up synchronous counter

At time t_2: As Q_2 is 0 and $\bar{Q}_2$ is 1, JA_0 becomes 1 and KA_0 are at 1, then the output, Q_0 of flip-flop is complemented and becomes 0. For next stage, JA_1 is 1 and KA_1 is 1, so output, Q_1 of flip-flop is complemented to 1. Further, JA_2 is 0 ($\because Q_0Q_1 = 0$) and KA_2 is 1, so output, Q_2 of flip-flop is reset to 0. So, output of counter, A_0 is 0, A_1 is 1 and A_2 is 0.

At time t_3: As Q_2 is 0 and $\bar{Q}_2$ is 1, JA_0 becomes 1 and KA_0 are 1, then the output, Q_0 of flip-flop is complemented and becomes 1. For next stage, JA_1 is 0 and KA_1 is also 0, so output, Q_1 of flip-flop is holding as 1. Further, JA_2 is 0 ($\because Q_0Q_1 = 0$) and KA_2 is 1, so output, Q_2 of flip-flop is reset to 0. So, output of counter, A_0 is 1, A_1 is 1 and A_2 is 0.

At time t_4: As Q_2 is 0 and $\bar{Q}_2$ is 1, JA_0 becomes 1 and KA_0 are 1, then the output, Q_0 of flip-flop is complemented to 0. For next stage, JA_1 is 1 and KA_1 is also 1, so output, Q_1 of flip-flop is complemented to 0. Further, JA_2 is 1 ($\because Q_0Q_1 = 1$) and KA_2 is 1 so output, Q_2 of flip-flop is complemented to 1. So, output of counter, A_0 is 0, A_1 is 0 and A_2 is 1.

At time t_5: As Q_2 is 1 and $\bar{Q}_2$ is 0, JA_0 becomes 0 and KA_0 are 1, then the output, Q_0 of flip-flop is reset to 0. For next stage, JA_1 is 0 and KA_1 is also 0, so output, Q_1 of flip-flop is holding 0. Further, JA_2 is 0 ($\because Q_0Q_1 = 0$) and KA_2 is 1, so output, Q_2 of flip-flop is reset to 0. So, output of counter, A_0 is 0, A_1 is 0 and A_2 is 0.

10.7 | CASCADING OF COUNTERS

A counter can be synchronous, asynchronous or a combination of these types. An asynchronous or synchronous counters are cascaded to design hybrid counter. Asynchronous counter and synchronous counters can be cascaded. MOD-N_1 counter and MOD-N_2 counter are cascaded to form MOD-N counter where $N = N_1 \times N_2$. During cascading of counters, MOD-N_1 counter followed by MOD-N_2 counter is similar to MOD-N_2 counter followed by MOD-N_1 counter. The count sequence may differ. MOD-6 counter can be obtained by cascading MOD-3 synchronous counter and MOD-2 counter. Similarly, MOD-10 counter can be obtained by cascading MOD-5 synchronous counter and MOD-2 counter. Cascaded of counter saves the used counts by counter MOD-2^n to form MOD-M. To construct MOD-6, counter from MOD-8 counter wastes 2 counts. Whereas MOD-6 counter obtained by cascading MOD-3 synchronous counter and MOD-2 counter wastes only 1 count.

10.7.1 | Modulus-6 Counter

Modulus-6 (MOD-6) counter can be obtained by cascading MOD-3 synchronous counter and MOD-2 counter asynchronously as shown in Figure 10.84. Output of MOD-3 counter is used to trigger MOD-2 counter. In other words, flip-flop-III is triggered by the output of flip-flop-II. It does not follow binary sequence. The output frequency is 1/6 of input frequency. Figure 10.85 gives the waveform. Output A_2, of flip-flop-III changes when output A_1 of flip-flop-II changes from 1 to 0. It counts 0, 1, 2, 4, 5 and 6.

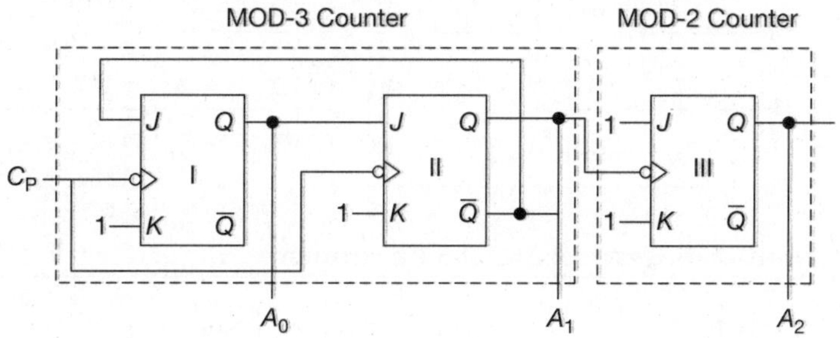

FIGURE 10.84 | MOD-6 counter

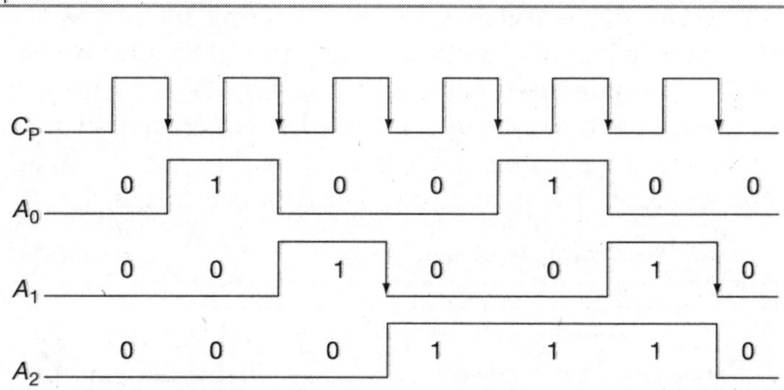

FIGURE 10.85 | Timing diagram of MOD-6 up counter

MOD-6 counter can be obtained by cascading MOD-2 counter and MOD-3 synchronous counter asynchronously as shown in Figure 10.86. Output of MOD-2 counter is used to trigger MOD-3 counter. In other words, flip-flop-II is triggered by the output of flip-flop-I. It follows binary sequence. The output frequency is 1/6 of input frequency. Figure 10.87

gives the waveform. Output A_1, of flip-flop-II changes when output A_0 of flip-flop-I changes from 1 to 0. Flip-flop II and III counts 0, 1, 2 and 0. The counter counts 0, 1, 2, 3, 4 and 5.

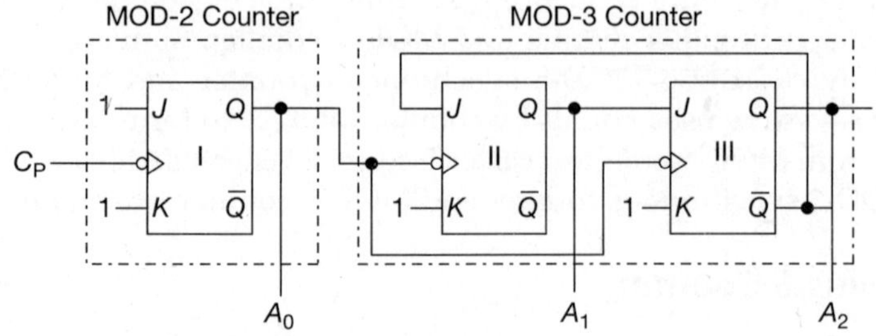

FIGURE 10.86 | MOD-6 counter

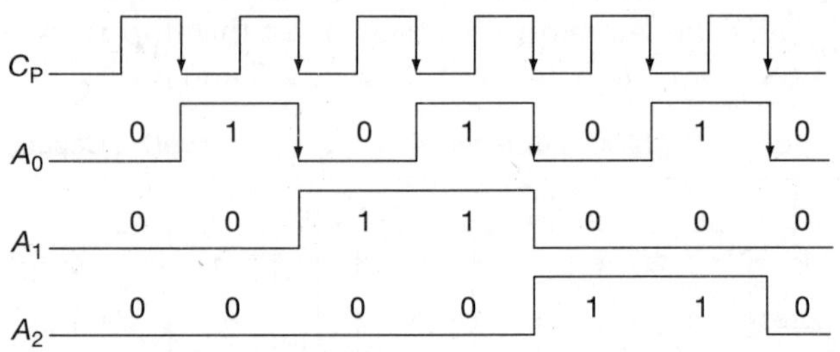

FIGURE 10.87 | Timing diagram of MOD-6 up counter

10.7.2 | Modulus-10 Counter

Modulus-10 (MOD-10) counter is constructed by cascading MOD-5 synchronous counter and MOD-2 counter asynchronously as shown in Figure 10.88. Output of MOD-5 counter is used to trigger MOD-2 counter. In other words, flip-flop-IV is triggered by the output of flip-flop-III. It does not follow binary sequence. The output frequency is 1/10 of input frequency. Figure 10.89 gives the waveform. Output A_3, of flip-flop-IV changes when output A_2 of flip-flop-III changes from 1 to 0. Such counter counts 0, 1, 2, 3, 4, 8, 9, 10, 11, 12.

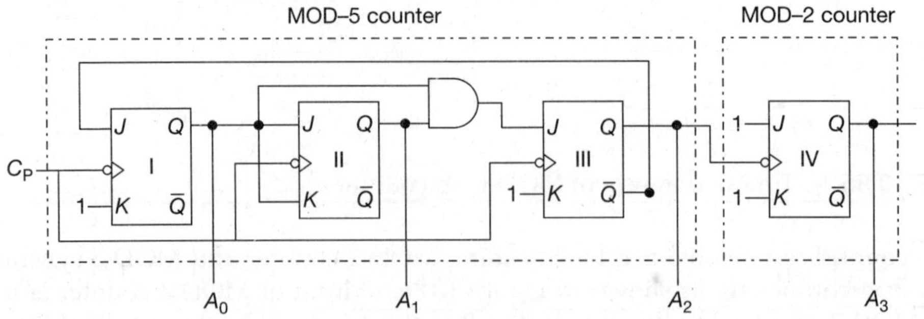

FIGURE 10.88 | MOD-10 counter

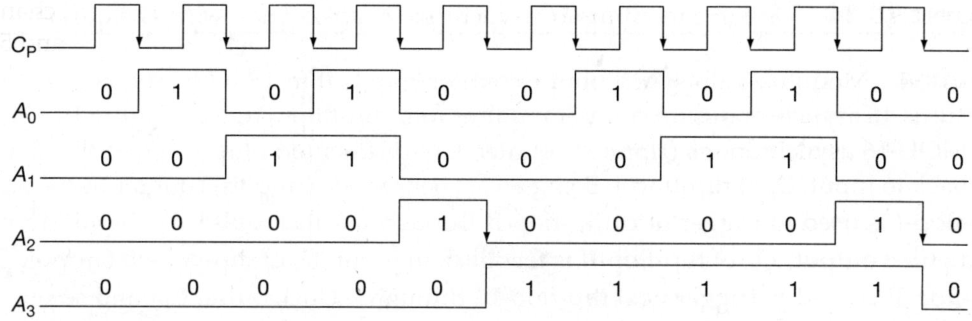

FIGURE 10.89 | Timing diagram of MOD-10 up counter

MOD-10 counter can be obtained by cascading MOD-2 counter and MOD-5 synchronous counter asynchronously as shown in Figure 10.90. Output of MOD-2 counter is used to trigger MOD-5 counter. In other words, flip-flop-II is triggered by the output of flip-flop-I. It follows binary sequence. The output frequency is 1/10 of input frequency. Figure 10.91 gives the waveform. Output A_1, of flip-II changes when output A_0 of flip-flop-I changes from 1 to 0. Flip-flop II and III counts 0, 1, 2, 3, 4 and 0. The counter counts 0, 1, 2, 3, 4, 5, 6, 7, 8, 9 and 0.

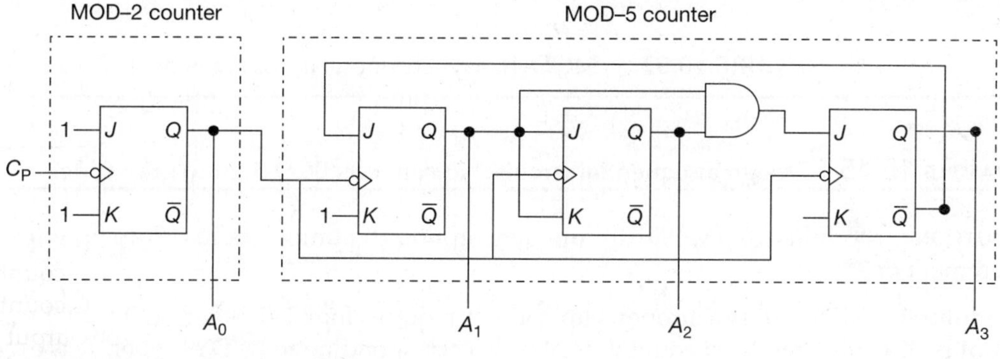

FIGURE 10.90 | MOD-10 counter

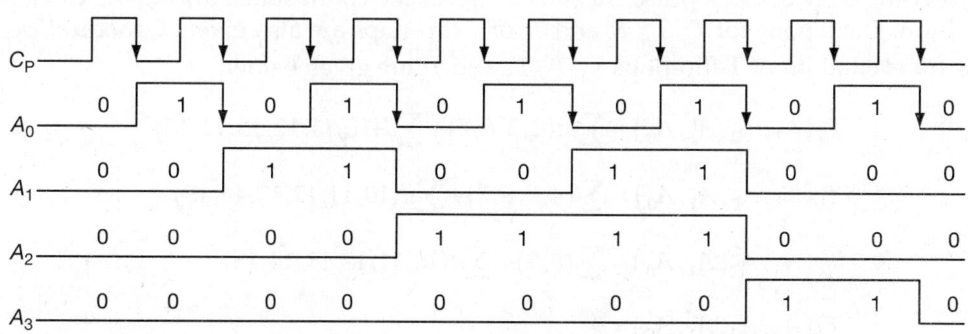

FIGURE 10.91 | Timing diagram of MOD-10 up counter

EXAMPLE 10.14 Design a modulus-16 asynchronous up counter using D-flip-flop.

SOLUTION Modulus-8 up synchronous counter counts 0 to 15 with an increment of 1. Modulus-16 asynchronous up counter requires four flip-flops ($16 \le 2^4$). Figure 10.92 gives the MOD-16 asynchronous (ripples) counter. Complemented output, $\bar{Q}_0$ of flip-flop-I is fed back to input, D_0 of flip-flop-I. Being an asynchronous (ripples) counter output, Q_0 of flip-flop-I is used to trigger next flip-flop-II through its clock pulse, C_{P_1}. Similarly, complemented output, $\bar{Q}_1$ of flip-flop-II is feedback to input, D_1 of flip-flop-II. Output, Q_1 of flip-flop-II is used to trigger next flip-flop-III through it clock pulse, C_{P_2} and so on.

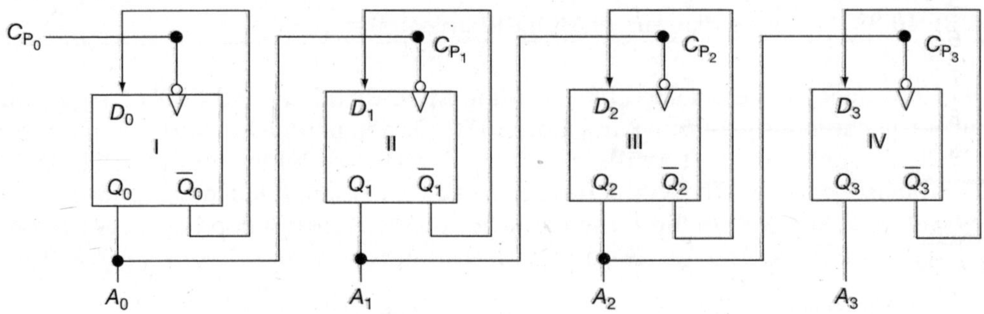

FIGURE 10.92 | MOD-16 asynchronous up counter

EXAMPLE 10.15 Design a sequential circuit that converts BCD digit to gray code.

SOLUTION Modulus-10 (MOD-10) up synchronous counter counts 0 to 9 with an increment of 1.

Modulus-10 (MOD-10) synchronous up counter requires four T-flip-flops ($10 \le 2^4$). Present count is BCD number. Next count is gray code corresponding to BCD number. T_0 is excitation of T-flip-flop giving output, g_0 when previous output is A_0. Previous output can be loaded by feeding loading input, L as 1. When loading input, L is 0 the count progresses on the rising edge of clock pulse. Table 10.27 gives the count status during the clock pulse transition. Excitations for T_0, T_1, T_2 and T_3 of T-flip-flops are also given. Canonical expressions for excitations of T-flip-flops T_0, T_1, T_2, and T_3 are given below:

$$T_0(A_3,A_2,A_1,A_0)=\sum m(2,3,6,7)+\sum d(10,11,12,13,14,15) \quad \text{(i)}$$

$$T_1(A_3,A_2,A_1,A_0)=\sum (4,5,6,7)+\sum d(10,11,12,13,14,15) \quad \text{(ii)}$$

$$T_2(A_3,A_2,A_1,A_0)=\sum (8,9)+\sum d(10,11,12,13,14,15) \quad \text{(iii)}$$

$$T_3(A_3,A_2,A_1,A_0)=0 \quad \text{(iv)}$$

TABLE 10.27 | Excitation table

Clock Pulse Count	Present Count				Next Count				Excitation of Flip-flop			
	A_2	A_2	A_1	A_0	g_3	g_2	g_1	g_0	T_3	T_2	T_1	T_0
0	0	0	0	0	0	0	0	0	0	0	0	0
1	0	0	0	1	0	0	0	1	0	0	0	0
2	0	0	1	0	0	0	1	1	0	0	0	1
3	0	0	1	1	0	0	1	0	0	0	0	1
4	0	1	0	0	0	1	1	0	0	0	1	0
5	0	1	0	1	0	1	1	1	0	0	1	0
6	0	1	1	0	0	1	0	1	0	0	1	1
7	0	1	1	1	0	1	0	0	0	0	1	1
8	1	0	0	0	1	1	0	0	0	1	0	0
9	1	0	0	1	1	1	0	1	0	1	0	0

For excitation T_0 of T-flip-flop, four-variable K-map is set up in Figure 10.93(a). One octet {2, 3, 6, 7, 10, 11, 14, 15} is formed. Simplified logics are given below:

$$T_0 = A_1 \quad \text{(v)}$$

For excitation T_1 of T-flip-flop, four-variable K-map is set up in Figure 10.93(b). One octet {4, 5, 6, 7, 12, 13, 14, 15} is formed. Simplified logics are given below:

$$T_1 = A_2 \quad \text{(vi)}$$

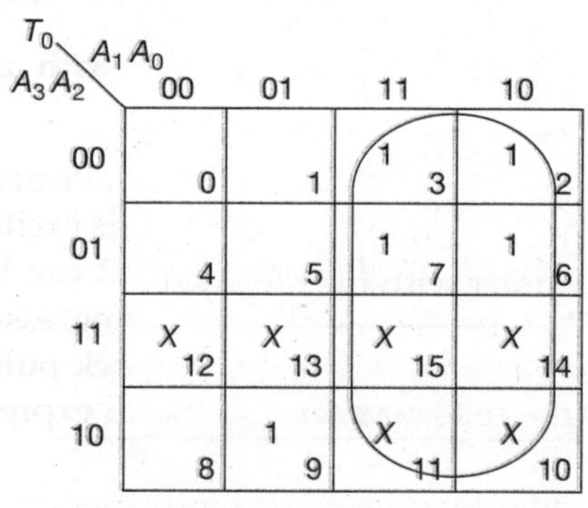

FIGURE 10.93(a) | K-map for T_0

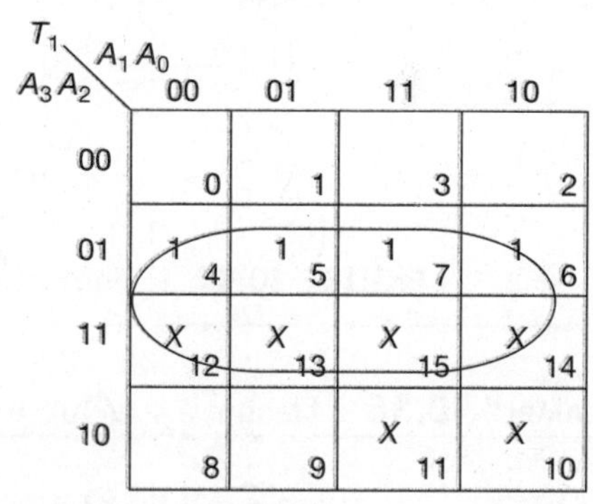

FIGURE 10.93(b) | K-map for T_1

For excitation T_2 of T-flip-flop, four-variable K-map is set up in Figure 10.93(c). One octet {8, 9, 10, 11, 12, 13, 14, 15} is formed. Simplified logics are given below:

$$T_2 = A_3 \quad \text{(vii)}$$

Circuit diagram of synchronous counter using T-flip-flop is shown in Figure 10.94 circuit that converts BCD digit to gray code. A_0, A_1, A_2 and A_3 are stored in T-flip-flop as Q_0, Q_1, Q_2 and Q_3, respectively.

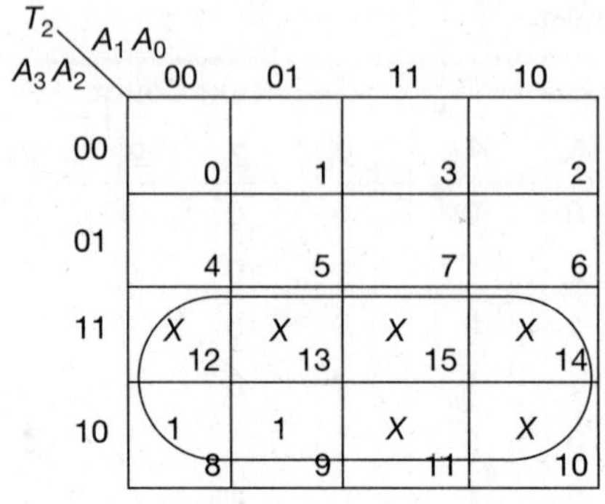

FIGURE 10.93(c) | K-map for T_2

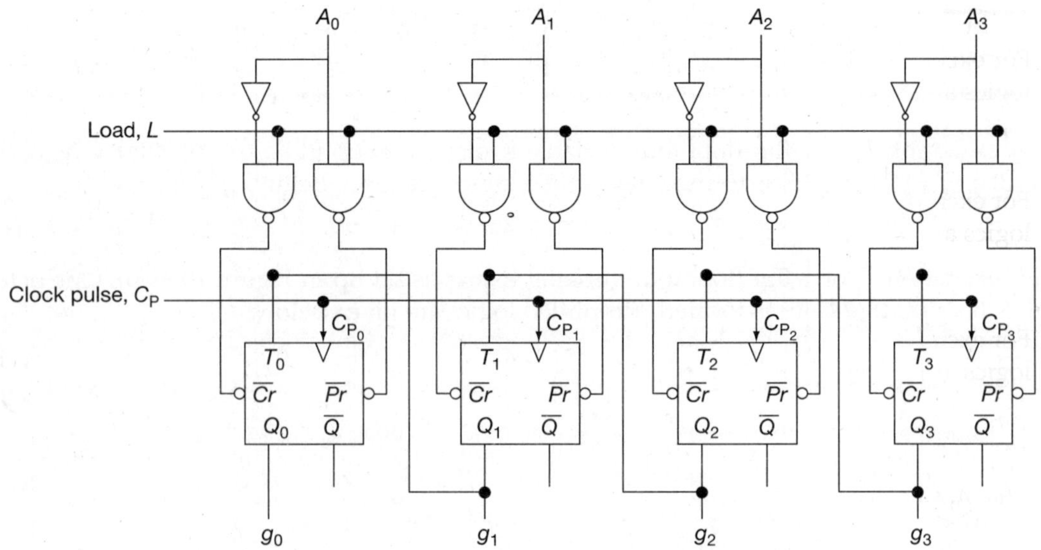

FIGURE 10.94 | Four-bit synchronous counter with parallel load

EXAMPLE 10.16 Design a synchronous Modulo-8 gray code counter.

SOLUTION Modulus-8 (MOD-8) synchronous up counter requires three flip-flops ($8 \le 2^3$). Table 10.28 gives the gray count status during the clock pulse transition. Excitation for T_0, T_1 and T_2 of T-flip-flops are also given.

Canonical expressions for excitations of T-flip-flops TA_0, TA_1 and TA_2 are given below:

$$T_0(A_2, A_1, A_0) = \sum m(0,3,5,6) \quad \text{(i)}$$

$$T_1(A_2, A_1, A_0) = \sum m(1,7) \quad \text{(ii)}$$

$$T_2(A_2, A_1, A_0) = \sum m(2,4) \quad \text{(iii)}$$

TABLE 10.28 | MOD-8 synchronous gray code counter

Clock Pulse Count	Minterm Number	Present Count			Next Count			Excitation of Flip-flop		
		A_2	A_1	A_0	A_2	A_1	A_0	T_2	T_1	T_0
0	0	0	0	0	0	0	1	0	0	1
1	1	0	0	1	0	1	1	0	1	0
2	3	0	1	1	0	1	0	0	0	1
3	2	0	1	0	1	1	0	1	0	0
4	6	1	1	0	1	1	1	0	0	1
5	7	1	1	1	1	0	1	0	1	0
6	5	1	0	1	1	0	0	0	0	1
7	4	1	0	0	0	0	0	1	0	0

For excitation T_0 of T-flip-flop, three-variable K-map is set up in Figure 10.95(a). Simplified logics are given below:

$$T_0 = \bar{A}_1\left(\overline{A_2 \oplus A_0}\right) + A_1\left(A_2 \oplus A_0\right) \text{ or } T_0 = \overline{A_1 \oplus \left(A_2 \oplus A_0\right)} \qquad \text{(iv)}$$

For excitation T_1 of T-flip-flop, three-variable K-map is set up in Figure 10.95(b). Simplified logics are given below:

$$T_1 = \left(\overline{A_2 \oplus A_1}\right)A_0 \qquad \text{(v)}$$

For excitation T_2 of T-flip-flop, three-variable K-map is set up in Figure 10.95(c). Simplified logics are given below:

$$T_2 = \left(A_2 \oplus A_1\right)\bar{A}_0 \qquad \text{(vi)}$$

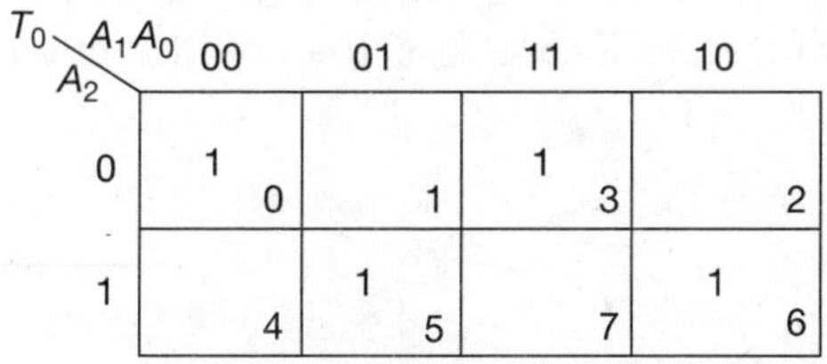

FIGURE 10.95(a) | K-map for T_0

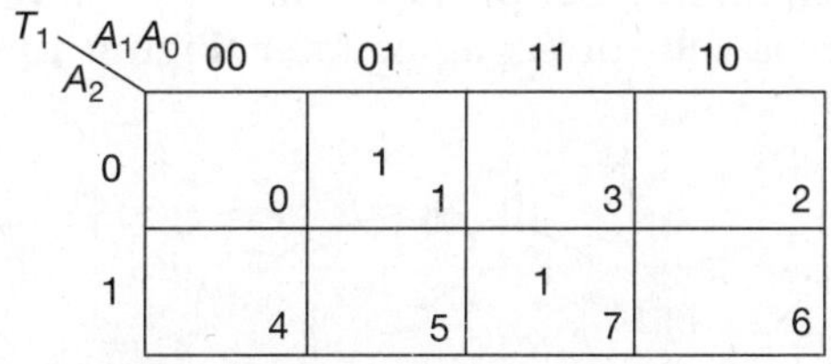

FIGURE 10.95(b) | K-map for T_1

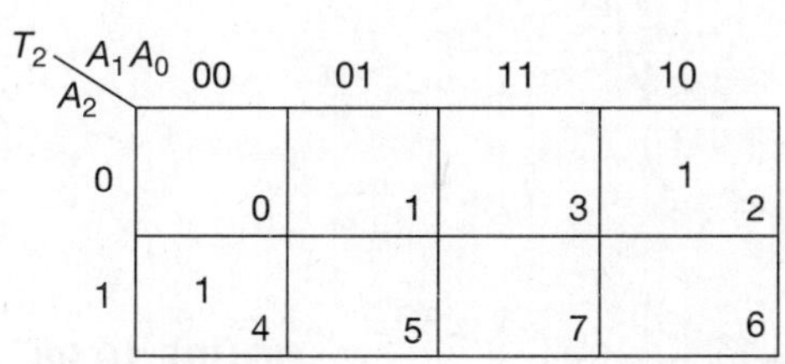

FIGURE 10.95(c) | K-map for T_2

Circuit diagram of synchronous counter using T-flip-flop is shown in Figure 10.96 to generate gray code sequence.

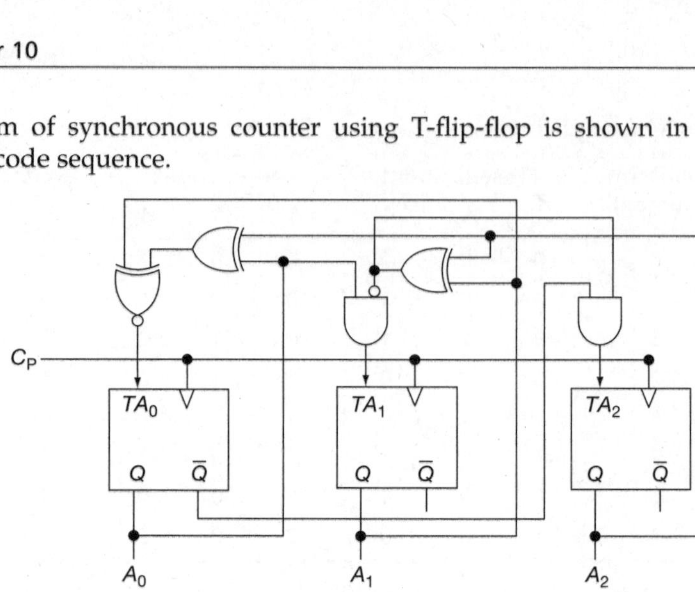

FIGURE 10.96 | MOD-8 synchronous gray code counter

10.8 | SELF-CORRECTING COUNTERS

Modulus-2^n counter has 0 to $2^n - 1$ count states. Modulus-M counter design has some unused states. Because $M \leq 2^n$, where n is number of flip-flop used to design counter. For example, Modulus-8 counter has eight states having $0 \rightarrow 1 \rightarrow 2 \rightarrow 3 \rightarrow 4 \rightarrow 5 \rightarrow 6 \rightarrow 7 \rightarrow 0$ count states (Figure 10.97). Modulus-6 counter counts six states having count sequence $0 \rightarrow 1 \rightarrow 2 \rightarrow 3 \rightarrow 4 \rightarrow 5 \rightarrow 0$. However, such counters are able to count eight states. So, two count states 6 and 7 are unused states (Figure 10.98).

Sometimes, counter enters in unused state initially and it moves to a used state after certain number of clock pulses and starts counting in normal count sequence. Such counter is self-correcting counter (Figure 10.98). Counter is in lock-out, if the counter enters

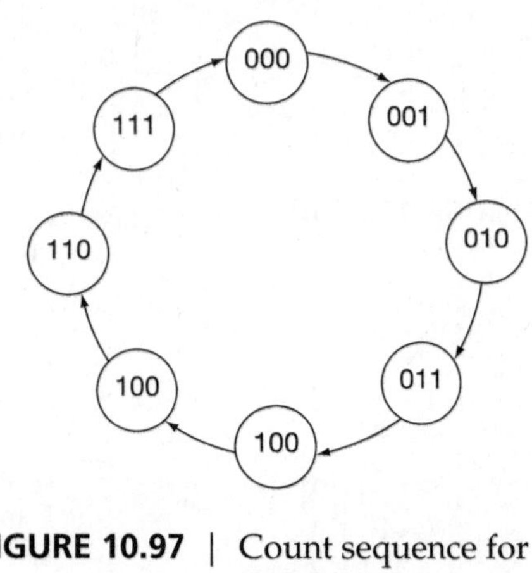

FIGURE 10.97 | Count sequence for MOD-8 counter

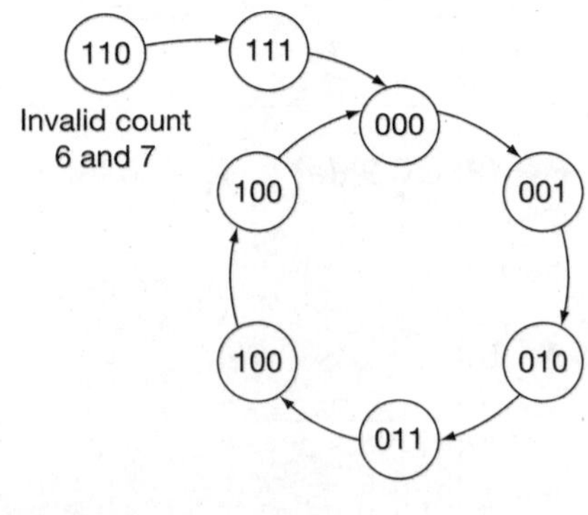

FIGURE 10.98 | Self-correcting MOD-6 counter

in unused state and unable to arrive at used state to follow the normal sequence. Let a modulus-4 (MOD-4) counter counts sequence $0 \rightarrow 3 \rightarrow 5 \rightarrow 6 \rightarrow 0$ as a normal sequence (Figure 10.99(a)). Count states 1, 2, 4 and 7 are invalid states (Figure 10.99(b)). Intially if counter enters in any of unused state and continues to remain in unused states like $1 \rightarrow 2 \rightarrow 4 \rightarrow 7 \rightarrow 1$ (Figure 10.99(b)). This is lock-out situation.

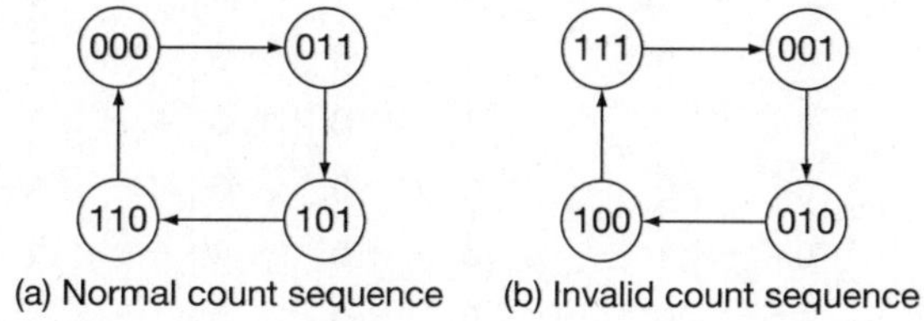

FIGURE 10.99 | Count sequence for MOD-4 counter

Self-correcting counters can be designed by the following two procedures:

(a) Redesign the counter presuming that counter enters to starting state whenever the counter goes to unused or invalid states. No don't care conditions should exist.
(b) Whenever counter enters in unused or invalid state, counter should reset to zero or preset to starting state.

EXAMPLE 10.17 Design a synchronous counter that goes through 0, 1, 2, 4, 0. Unused states must go to zero or to next state on next clock pulse.

SOLUTION State diagram is given in Figure 10.100. Such counter never enters in lock out state. It is also known as self-starting of counter. Modulus-8 (MOD-8) synchronous counter is required to count the sequence that goes up to 4 (100). Modulus-8 (MOD-8) synchronous counter requires three flip-flops ($8 \leq 2^3$). Table 10.29 gives the gray count status during the clock pulse transition. Excitations for T_0, T_1 and T_2 of T-flip-flops are also given.

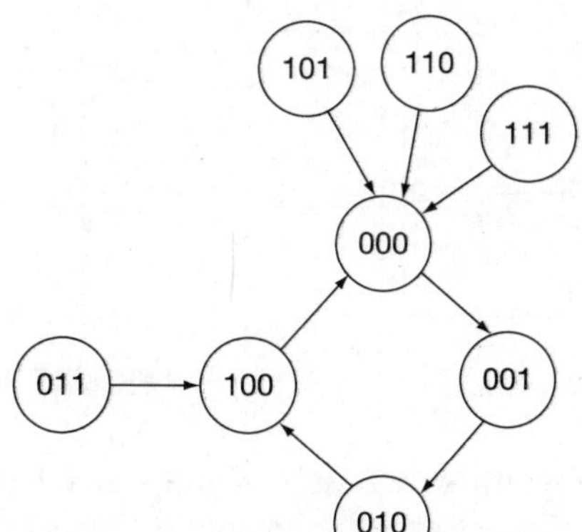

FIGURE 10.100 | State diagram

TABLE 10.29 | MOD-8 synchronous counter

Clock Pulse Count	Present Count			Next Count			Excitation of Flip-flop		
	A_2	A_1	A_0	A_2	A_1	A_0	T_2	T_1	T_0
0	0	0	0	0	0	1	0	0	1
1	0	0	1	0	1	0	0	1	1
2	0	1	0	1	0	0	1	1	0
3	0	1	1	1	0	0	1	1	1
4	1	0	0	0	0	0	1	0	0
5	1	0	1	0	0	0	1	0	1
6	1	1	0	0	0	0	1	1	0
7	1	1	1	0	0	0	1	1	1

Canonical expressions for excitations of T-flip-flops TA_0, TA_1 and TA_2 are given below:

$$T_0(A_2, A_1, A_0) = \sum m(0,1,3,5,7) \quad \text{(i)}$$

$$T_1(A_2, A_1, A_0) = \sum m(1,2,3,6,7) \quad \text{(ii)}$$

$$T_2(A_2, A_1, A_0) = \sum m(2,3,4,5,6,7) \quad \text{(iii)}$$

For excitation T_0 of T-flip-flop, three-variable K-map is set up in Figure 10.101(a). One pair {0, 1} and one quad {1, 3, 5, 7} are formed. Simplified logics are given below:

$$T_0 = \bar{A}_2\bar{A}_1 + A_0 \quad \text{(iv)}$$

For excitation T_1 of T-flip-flop, three-variable K-map is set up in Figure 10.101(b). One pair {1, 3} and one quad {2, 3, 6, 7} are formed. Simplified logics are given below:

$$T_1 = \bar{A}_2 A_0 + A_1 \quad \text{(v)}$$

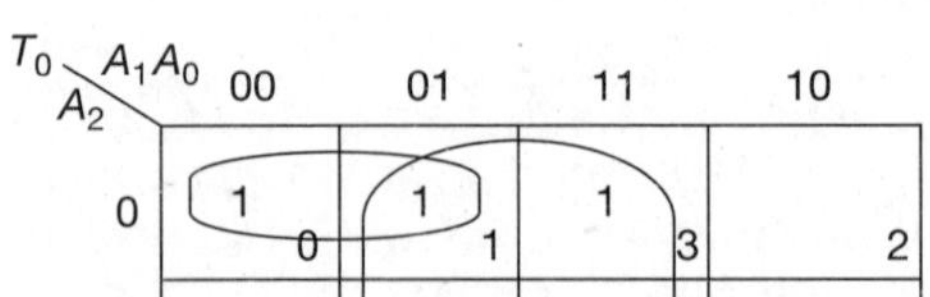

FIGURE 10.101(a) | K-map for T_0

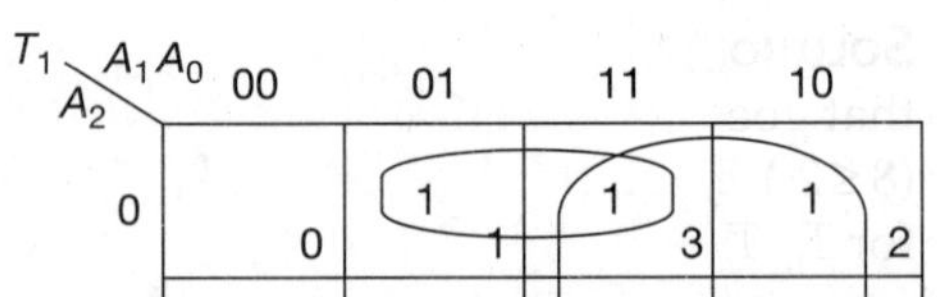

FIGURE 10.101(b) | K-map for T_1

For excitation T_2 of T-flip-flop, three-variable K-map is set up in Figure 10.101(c). Two quads {1, 3, 5, 7} and {2, 3, 6, 7} are formed. Simplified logics are given below:

$$T_2 = A_1 + A_2 \quad \text{(vi)}$$

Circuit diagram of synchronous counter using T-flip-flop is shown in Figure 10.102.

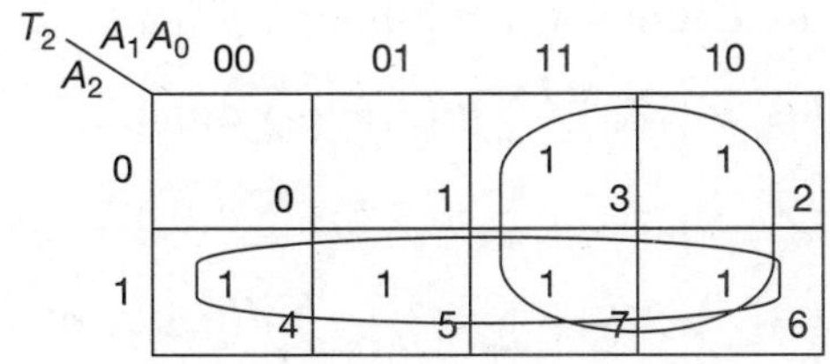

FIGURE 10.101(c) | K-map for T_2

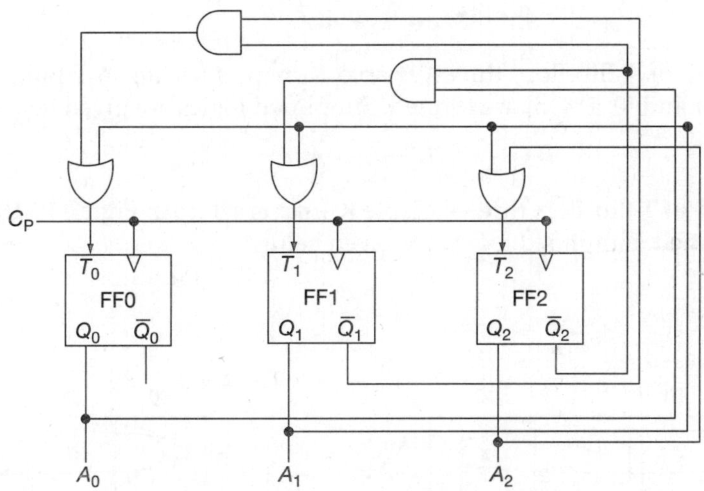

FIGURE 10.102 | Synchronous counter

EXAMPLE 10.18 Design a synchronous counter that goes through 3, 4, 6, 7, 3. Unused states must go to 3 on next clock pulse.

SOLUTION Modulus-8 (MOD-8) synchronous counter is required to count the sequence that goes up to 4 (100). Modulus-8 (MOD-8) synchronous counter requires three flip-flops ($8 \leq 2^3$). Table 10.30 gives the gray count status during the clock pulse transition. Excitations for T_0, T_1 and T_2 of T-flip-flops are also given.

TABLE 10.30 | MOD-8 synchronous counter

Clock Pulse Count	Present Count			Next Count			Excitation of Flip-flop		
	A_2	A_1	A_0	A_2	A_1	A_0	T_2	T_1	T_0
3	0	1	1	1	0	0	1	1	1
4	1	0	0	1	1	0	0	1	0
6	1	1	0	1	1	1	0	0	1
7	1	1	1	0	1	1	1	0	0

Canonical expressions for excitations of T-flip-flops T_0, T_1 and T_2 are given below:

$$T_0(A_2, A_1, A_0) = \sum m(3,6) + \sum d(0,1,2,5) \quad \text{(i)}$$

$$T_1(A_2, A_1, A_0) = \sum m(3,4) + \sum d(0,1,2,5) \quad \text{(ii)}$$

$$T_2(A_2, A_1, A_0) = \sum m(3,4 + \sum d(0,1,2,5) \quad \text{(iii)}$$

For excitation T_0 of T-flip-flop, three-variable K-map is set up in Figure 10.103(a). One pair {2, 6} and one quad {0, 1, 2, 3} are formed. Simplified logics are given below:

$$T_0 = \bar{A}_2 + A_1\bar{A}_0 \quad \text{(iv)}$$

For excitation T_1 of T-flip-flop, three-variable K-map is set up in Figure 10.103(b). Two quads {0, 1, 2, 3} and {0, 1, 4, 5} are formed. Simplified logics are given below:

$$T_1 = \bar{A}_2 + \bar{A}_1 \quad \text{(v)}$$

For excitation T_2 of T-flip-flop, three-variable K-map is set up in Figure 10.103(c). One quad {1, 3, 5, 7} is formed. Simplified logics are given below:

$$T_2 = A_0 \quad \text{(vi)}$$

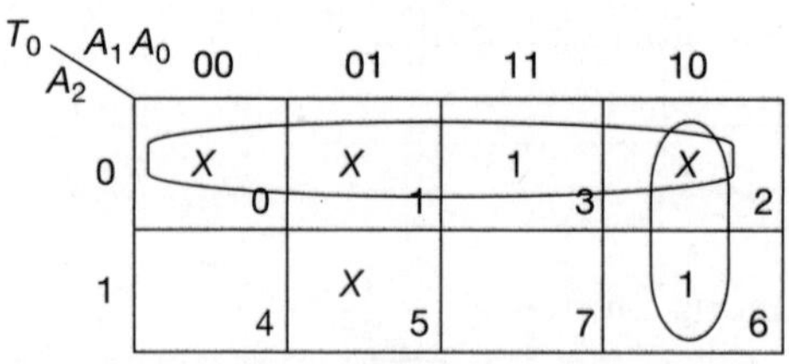

FIGURE 10.103(a) | K-map for T_0

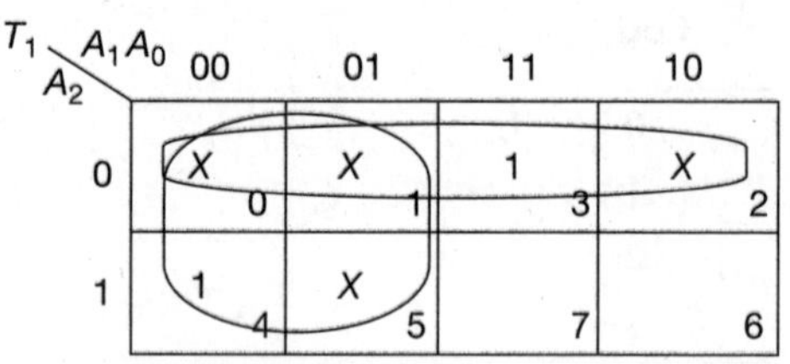

FIGURE 10.103(b) | K-map for T_1

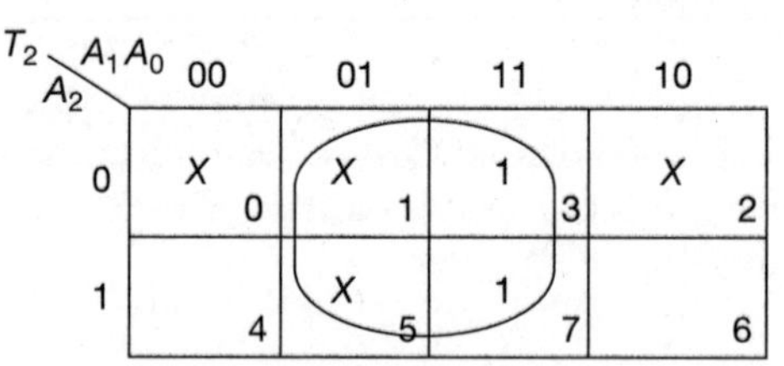

FIGURE 10.103(c) | K-map for T_2

Circuit diagram of synchronous counter using T-flip-flop is shown in Figure 10.104.

Table 10.31 gives the next state when counter goes to unused or invalid states 0, 1, 2 or 5. Counter moves from 0 to 3, 1 to 6, 2 to 1 and 5 to 3 state.

To obtain reset expression when counter goes to unused state. Table 10.32 is the truth table to find the logic.

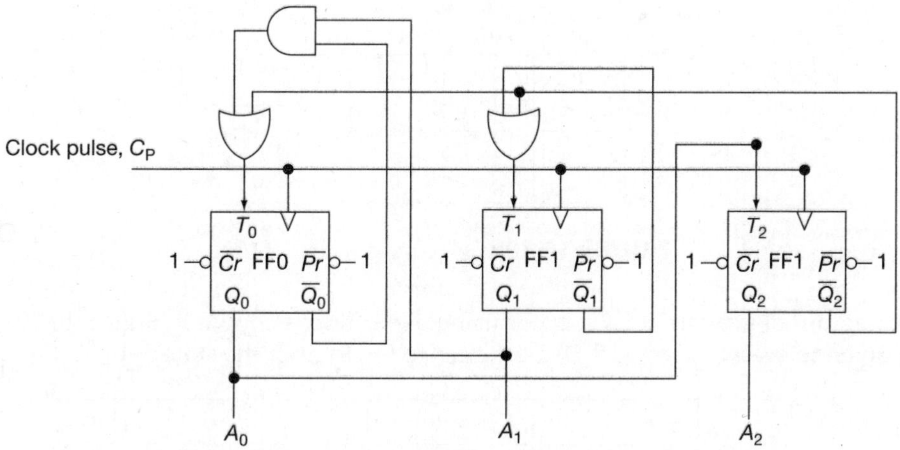

FIGURE 10.104 | Three-bit synchronous counter

TABLE 10.31 | Lock-out check

Clock Pulse Count	Present Count			Excitation of Flip-flop			Next Count		
	A_2	A_1	A_0	T_2	T_1	T_0	A_2	A_1	A_0
0	0	0	0	0	1	1	0	1	1
1	0	0	1	1	1	0	1	1	0
2	0	1	0	0	0	1	0	0	1
5	1	0	1	0	1	1	0	1	1

TABLE 10.32 | Truth table

Clock Pulse Count	Unused Count			Reset
	A_2	A_1	A_0	R
0	0	0	0	1
1	0	0	1	1
2	0	1	0	1
5	1	0	1	1

Canonical expression for reset expression is given below.

$$R(A_2, A_1, A_0) = \sum m(0,1,2,5) \quad \text{(vii)}$$

By setting up three-variable K-map (Figure 10.105) for reset expression, the following expression is obtained

$$R = \overline{A}_2\overline{A}_0 + \overline{A}_1 A_0 \quad \text{(viii)}$$

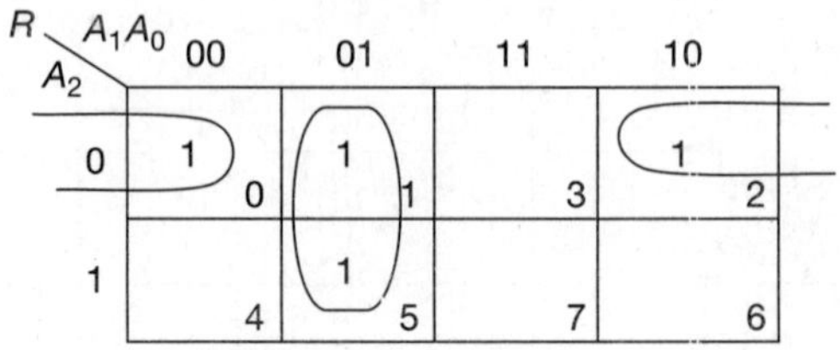

FIGURE 10.105 | K-map for reset, R

Circuit diagram of synchronous counter using T-flip-flop is shown in Figure 10.106 where reset is given to preset of FF_0 and FF_1 and clear to FF_2 to reset the state 011.

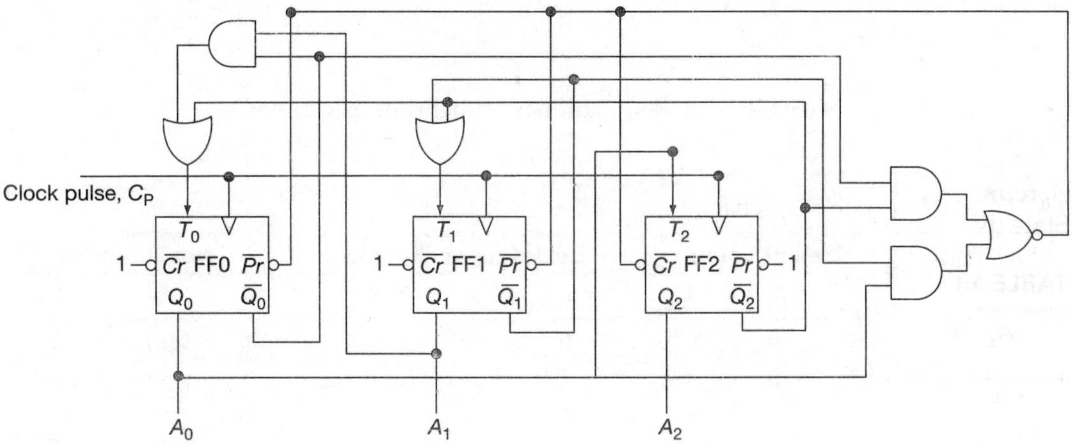

FIGURE 10.106 | Three-bit synchronous counter to go to state 3

10.9 | SEQUENCE GENERATOR

A sequence generator or pulse generator is a circuit which generates a desired sequence of binary digits (bits) in synchronism with *a*. These sequence generators are required to switch on the lights, switch off the lights, switch on the machine, open the valves etc. The procedure to design a sequence generator is as follows:

1. Find the minimum number of flip-flops required. The minimum number of flip-flops required depends on the nature of sequence. The required number of flip-flops is calculated from the equation

$$\text{Max}(N_0, N_1) < 2^{n-1}$$

where n is the number of flip-flops required, N_0 is the number of 0's in the given sequence and N_1 is the number of 1's in the given sequence.

2. Draw the state table as per the following procedure.
   (a) Assign the desired sequence (start from least significant bit (LSB) to the output of the flip flops, which produces the sequence as output.

(b) Assign the output to the other flip-flops 0 or 1, such that all the states are different from each other.

3. Draw the state diagram from the state table.
4. Prepare the excitation table on the basis of presents state and next state.
5. Draw the K-map and simplify the same.
6. Draw the logic diagram.

EXAMPLE 10.19 Design a sequence generator to generate sequence 0111.

SOLUTION Number of l's in the sequence is three ($N_1 = 3$)

Number of 0s in the sequence is one ($N_0 = 1$)

$$\text{Max}(1, 3) < 2^{n-1}$$

$$3 < 2^{3-1}$$

So, $n = 3$.

To generater the desired sequence, state table is given in Table 10.33. Least significant bit, A_0 represents the desired sequence. 00, 01, 10 or 11 are added to A_0 to get different states. State diagram is given in Figure 10.107.

TABLE 10.33 | State assignment

A_2	A_1	A_0 (LSB)	Decimal Equivalent
0	0	0	0
0	0	1	1
0	1	1	3
1	0	1	5

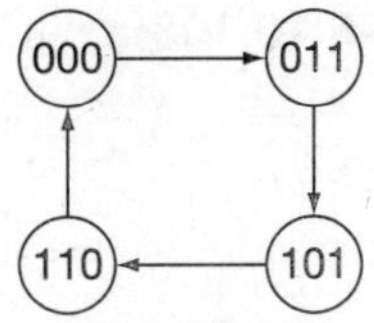

FIGURE 10.107 | State diagram

Excitation table for the state diagram is given in Table 10.34.

TABLE 10.34 | Excitation table for the state diagram (Figure 10.107)

Counter State	Present Count			Next Count			Excitation of Flip-flop		
	A_2	A_1	A_0	A_2	A_1	A_0	T_2	T_1	T_0
0	0	0	0	0	0	1	0	0	1
1	0	0	1	0	1	1	0	1	0
3	0	1	1	1	0	1	1	1	0
5	1	0	1	0	0	0	1	0	1

Canonical expressions for excitations of T-flip-flops T_0, T_1 and T_2 are given below:

$$T_0(A_2, A_1, A_0) = \sum m(0,5) + \sum d(2,4,6,7) \quad \text{(i)}$$

$$T_1(A_2, A_1, A_0) = \sum m(1,3) + \sum d(2,4,6,7) \quad \text{(ii)}$$

$$T_2(A_2,A_1,A_0)=\sum m(3,5)+\sum d(2,4,6,7) \quad \text{(iii)}$$

For excitation T_0 of T-flip-flop, three-variable K-map is set up in Figure 10.108(a). Two quads {0, 2, 4, 6} and {4, 5, 6, 7} are formed. Simplified logics are given below:

$$T_0 = A_2 + \bar{A}_0 \quad \text{(iv)}$$

For excitation T_1 of T-flip-flop, three-variable K-map is set up in Figure 10.108(b). Two quads {0, 2, 4, 6} and {2, 3, 6, 7} are formed. Simplified logics are given below:

$$T_1 = A_1 + \bar{A}_0 \quad \text{(v)}$$

For excitation T_2 of T-flip-flop, three-variable K-map is set up in Figure 10.108(c). Two quads {2, 3, 6, 7} and {4, 5, 6, 7} are formed. Simplified logics are given below:

$$T_1 = A_2 + A_1 \quad \text{(vi)}$$

Circuit diagram of synchronous counter using T-flip-flop is shown in Figure 10.109.

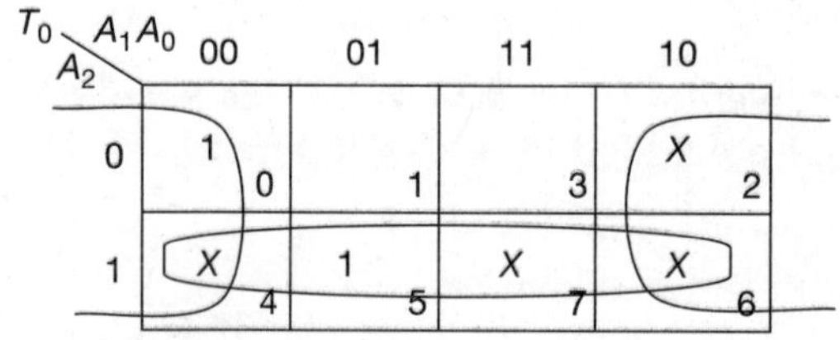

FIGURE 10.108(a) | K-map for T_0

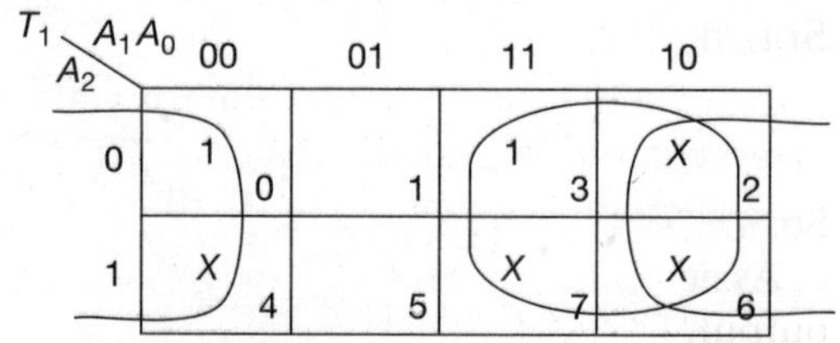

FIGURE 10.108(b) | K-map for T_1

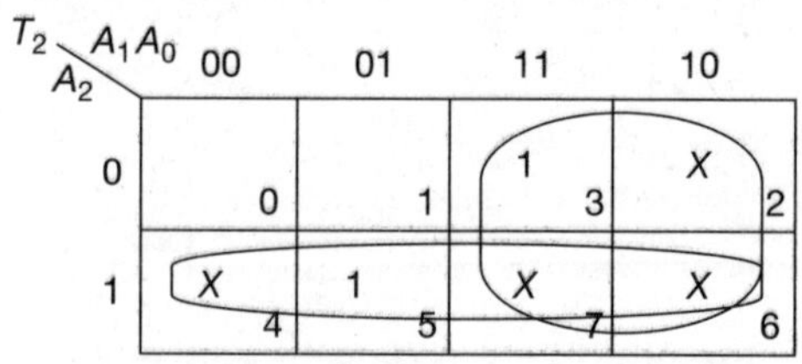

FIGURE 10.108(c) | K-map for T_2

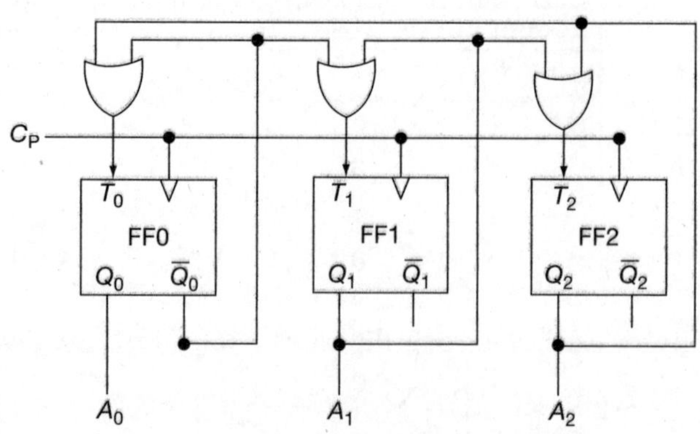

FIGURE 10.109 | Sequence generator

Alternative Method:

The procedure to design a sequence generator is as follows:

1. Find the minimum number of flip-flops required. The minimum number of flip-flops required depends on the nature of sequence. The required number of flip-flops is calculated from the equation

$$M < 2^n$$

where n is the number of flip-flops required, and M is the number of bits in sequence.
2. Design a MOD-M counter. Counter may be synchronous or asynchronous.
3. Find the decoding logic to output 1 from the states of the flip-flops. Minimization of logic is followed.

EXAMPLE 10.20 Design a sequence generator to generate sequence 01101.

SOLUTION Number bits in sequence is five ($M = 5$)

$$M \le 2^n \quad \text{or} \quad 5 \le 2^3$$

So, $n = 3$.

Assign the desired sequence generator to five unique state of MOD-8 counter. Consider output, F being desired sequence. Find the minimum logic using three-variable K-map, considering truth Table 10.35.

Canonical expression for F is given below:

$$F(A_2, A_1, A_0) = \sum m(1,2,4) + \sum d(5,6,7) \quad \text{(i)}$$

For output F, three-variable K-map is set up in Figure 10.110. Two pairs {1, 5}, {2, 6} and one quad {4, 5, 6, 7} are formed. Simplified logics are given below:

$$F = A_2 + A_1\bar{A}_0 + \bar{A}_1 A_0$$

$$F = A_2 + (A_1 \oplus A_0) \quad \text{(ii)}$$

TABLE 10.35 | MOD-8 counter states

A_2	A_1	A_0 (LSB)	F
0	0	0	0
0	0	1	1
0	1	0	1
0	1	1	0
1	0	0	1
1	0	1	X
1	1	0	X
1	1	1	X

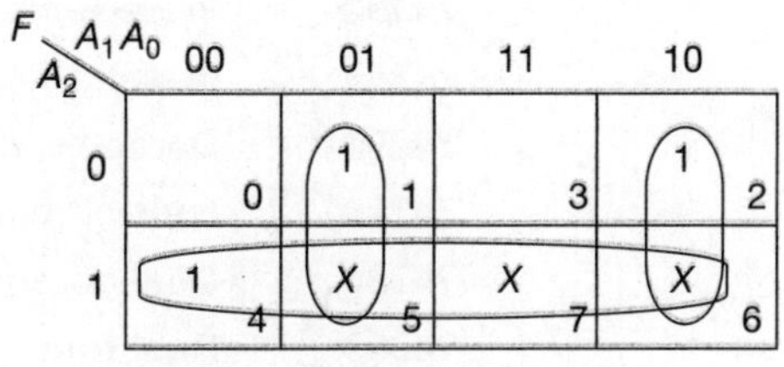

FIGURE 10.110 | K-map for F

Design a MOD-5 asynchronous up-counter and implement the logic to generate the required sequence. Circuit diagram is shown in Figure 10.111.

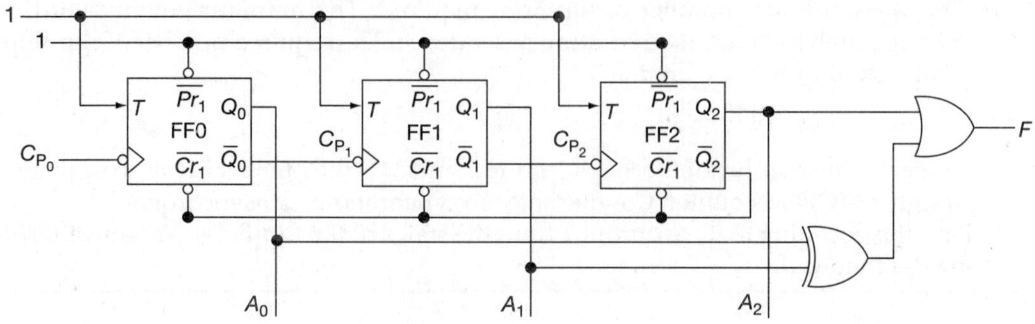

FIGURE 10.111 | Sequence generator using asynchronous MOD-5 up counter

10.10 | LIST OF COUNTER ICs

A list of commonly used IC type numbers used as counters belonging to CMOS, TTL and ECL logic families is given in Tables 10.36 to 10.38. Information regarding the pin connection diagram, truth table, etc., in respect of these IC numbers is given in the handbook or companion website.

TABLE 10.36 | Commonly used IC counters of TTL logic families

IC Number	Function of IC
7490	Decade counter
7493	Four-bit binary counter
74160	BCD decade counter with asynchronous CLEAR
74161	Four-bit binary counter with asynchronous CLEAR
74162	BCD decade counter with synchronous CLEAR
74163	Four-bit binary counter with synchronous CLEAR
74190	Presettable BCD decade UP/DOWN counter
74191	Presettable four-bit binary UP/DOWN counter
74192	Presettable BCD decade UP/DOWN counter
74193	Presettable four-bit binary UP/DOWN counter
74290	Decade counter
74293	Four-bit binary counter
74390	Dual decade counter
74393	Dual four-bit binary counter

TABLE 10.37 | Commonly used IC counters of CMOS logic families

IC Number	Function of IC
4029 B	Synchronous presettable four-bit UP/DOWN counter
40160 B	Decade counter with asynchronous CLEAR
40161 B	Binary counter with asynchronous CLEAR
40161 B	Decade counter
40163 B	Binary Counter
40192 B	Presettable BCD UP/DOWN counter
40193 B	Presettable Binary UP/DOWN counter
4510 B	Presettable UP/DOWN BCD counter
4518 B	Dual four-bit decade counter
4520 B	Dual four-bit binary counter
4522 B	Four-bit BCD programmable divide-by-N counter
4722 B	Programmable counter/timer

TABLE 10.38 | Commonly used IC counters of ECL logic families

IC Number	Function of IC
MC 10136	Universal hexadecimal counter
MC 10137	Universal decade counter
MC 10154	Binary counter (four-bit)
MC 10178	Four-bit binary counter

SUMMARY

- The different types of flip-flops are: (i) S-R-flip-flop, (ii) J-K flip-flop, (iii) Master-Slave J-K flip-flop, (iv) T-flip-flop, and (v) D-flip-flop.
- Preset and clear are the asynchronous inputs used to set the initial condition of the flip-flop.
- In a D-flip-flop, the output is equal to its input, but the output is delayed by the amount of clock.
- The maximum clock frequency that can be applied as a CLK input of flip-flop for proper and stable operation is given as;

$$f \leq \frac{1}{nt_{\text{pd}} + t_{\text{s}}}$$

where n is number of flip-flops in the asynchronous (ripples) counter. t_{pd} is inherent propagation delay of flip-flop. t_{s} is strobe pulse width

- Total propagation delay in asynchronous counters become

$$T_{\text{clock}} \geq nt_{\text{pd}} + t_{\text{s}}$$

where n is number of flip flops in the ripples counter. t_{pd} is inherent propagation delay of flip-flop. t_s is strobe pulse width.
- The sequential iogic circuit is a digital circuit, in which the output is a function of present inputs as well as past inputs and outputs.
- The N-bit ring counter counts N clock pulses. The frequency of the pulses obtained at least significant output, Q_0 is the frequency of the clock pulse divided by N. Hence, it is also known as divided-by-N counter.
- The n-bit ripples counter uses n-flip-flops, which counts 2^n clock pulses.
- In asynchronous counters, delay is more and speed is less. Applying a clock pulse to all the flip-flops of the counter increases the speed of operation.
- Counter is a clocked sequential circuit that goes through a predetermined sequence of states upon the application of input pulses.
- Counters are designed using T- or JK-flip flops.
- A counter follows binary sequence is called an n-bit binary counter.
- Binary counter has n-flip-flops (FFs) and possible maximum number of states are 2^n, which are passed through in the order 0, 1, 2, ..., $2^n - 1$ and repeats.
- Up binary counter increases the output count by one with the clock pulse. Up binary counter advances count upward through its sequence 0, 1, 2, ..., $2^n - 1$ and repeats from 0.
- Down binary counter decreases the output count by one with the clock pulse. Down binary counter advances count downward through its sequence $2^n - 1$, $2^n - 2$, $2^n - 3$, ..., 2, 1, 0 and repeats from $2^n - 1$.
- Binary counters can have truncated sequences. So, counters can have less than maximum of 2^n states.
- An n-bit counter has n flip-flops and 2^n states and divides input frequency by 2^N or MOD-N ($N \leq 2^n$).
- Counters can be classified into two broad categories according to the way counters are clocked: (i) asynchronous, (ii) synchronous and (iii) hybrid.
- Asynchronous counter is one in which the flip-flops within the counter do not share the same clock pulse so the flip-flops do not change their states at the same time. Asynchronous counter is also known as ripple counter, serial or series counter.
- Synchronous counter is one in which all the flip-flops in the counter share same clock pulse so the flip-flops change their states at the same time.
- Hybrid counter combines asynchronous and synchronous counters.
- BCD counter is a mod-10 counter, and it is a decade counter.
- The propagation delays owing to the ripple effect in asynchronous counters create transitional states in which the counter outputs change at slightly different times. These transitional states produce undesired voltage spikes of short duration called glitches.
- Elimination of decoding glitches by enabling the decoded outputs at a time after the glitches have had time to disappear is called strobing.

- Resetting the counter means bringing the counter to 0 state.
- Presetting the counter means bringing the counter to a specific state at the beginning.
- The circuit which generates the desired sequence of bits in synchronous mode with a clock is known as a sequence generator.
- The sequence detector is a logic circuit which detects the desired sequence.

MULTIPLE CHOICE QUESTIONS

10.1 How many states a 6-bit ripple counter can have?
(a) 6 (b) 12
(c) 32 (d) 64

10.2 How many flip-flops are needed to divide the input frequency by 40?
(a) 4 (b) 5
(c) 6 (d) 40

10.3 The number of flip-flops required for decade counter is
(a) 3 (b) 4
(c) 5 (d) 10

10.4 The minimum number of flip-flops required for a mod-12 ripple counter is
(a) 3 (b) 4
(c) 6 (d) 12

10.5 The maximum number that can be obtained by a ripple counter using five flip-flops is
(a) 5 (b) 16
(c) 31 (d) 32

10.6 In a 4-bit binary ripple counter, for every input clock pulse
(a) all the flip-flops get clocked simultaneously
(b) only one flip-flop gets clocked at a time
(c) two of the flip-flops get clocked at a time
(d) all the above statements are false

10.7 A 4-bit binary ripple counter uses flip-flops with propagation delay time of 25 ns each. The maximum possible time required for change of state will be
(a) 25 ns
(b) 50 ns
(c) 75 ns
(d) 100 ns

10.8 A 4-bit synchronous counter uses flip-flops with propagation delay time of 25 ns each. The maximum possible time required for change of state will be
(a) 25 ns
(b) 50 ns
(c) 75 ns
(d) 100 ns

10.9 A mod-2 counter followed by a mod-5 counter is
(a) the same a 5 a mod-5 counter followed by a mod-2 counter
(b) a decade counter
(c) a mod-7 counter
(d) none of the above

10.10 A 4-bit presettable up-counter has present input 0101. The presetting operation takes place as soon as the counter becomes maximum, i.e. 1111. The modulus of this counter is
(a) 5
(b) 10
(c) 11
(d) 15

10.11 The output frequency of a mod-16 counter, clocked from a 20-kHz clock input signal is
(a) 20 kHz
(b) 52 kHz
(c) 0.625 kHz
(d) 1.250 kHz

10.12 The output frequency of a mod-12 counter is 6 kHz. Its input frequency is
(a) 5000 Hz
(b) 6000 Hz
(c) 24,000 Hz
(d) 72,000 Hz

10.13 A 4-bit ripple counter uses a JK-flip-flop. If the propagation delay of each flip-flop is 50 ns, then the maximum clock frequency is equal to
(a) 20 MHz
(b) 10 MHz
(c) 5 MHz
(d) 4 MHz

10.14 A pulse train with a frequency of 1 MHz is counted using MOD-1024 ripples counter constructed with JK-flip-flop. For proper operation of the counter, the maximum permissible propagation delay per flip-flop stage is
(a) 100 ns
(b) 50 ns
(c) 20 ns
(d) 10 ns

10.15 A divide-by-78 counter can be realized by using
(a) 6 MOD-13 counters
(b) 16 MOD-6 counters
(c) 13 MOD-13 counters
(d) 1 MOD-13 and 1 MOD-6 counters

10.16 How many flip-flops are required for dividing the frequency by 64?
(a) 4
(b) 5
(c) 6
(d) 8

10.17 An n-stage ripple counter can count up to
(a) 2^n
(b) $2^n - 1$
(c) n
(d) 2^{n-1}

10.18 A count that counts 0 to 5 is called
(a) MOD-5 up counter
(b) MOD-5 down counter
(c) MOD-6 up counter
(d) MOD-6 down counter

10.19 Which of following has the highest speed?
(a) Asynchronous counter
(b) Synchronous counter
(c) Ring counter
(d) Ripples counter

10.20 Which of the following statements is correct? Asynchronous counter
(a) has flip-flops that change states simultaneously
(b) is capable of operating at high frequencies
(c) requires more logic elements
(d) has large carry propagation time

Answers

10.1 (d)	10.2 (c)	10.3 (b)	10.4 (d)	10.5 (b)	10.6 (d)	10.7 (d)
10.8 (a)	10.9 (b)	10.10 (b)	10.11 (d)	10.12 (d)	10.13 (c)	10.14 (a)
10.15 (d)	10.16 (c)	10.17 (b)	10.18 (c)	10.19 (b)	10.20 (d)	

QUESTIONS

10.1 Design a 4-bit binary up/down ripple counter.

10.2 Design a 3-bit binary up/down counter. Draw its timing diagram.

10.3 Design and implement a MOD-6 synchronous counter using J-K-flip-flop.

10.4 Design a pulse train generator using shift register for the pulse train 111010.

10.5 Design and implement a MOD-5 synchronous counter using J-K-flip-flop.

10.6 With general block diagrams, show how each of the following counters can be obtained usins flip-flop, a decade counter and a 4-bit binary counter or any combination of these.

(a) Divide-by-40 counter, (b) Divide-by-32 counter, (c) Divide-by-160 counter AND (d) Divide-by-320 counter

10.7 With general block diagrams, show how to obtain the following frequencies from a 1 MHz clock using single flip-flops, mod-5 counters and decade counters.

(a) 500 kHz, (b) 250 kHz, (c) 125 kHz, (d) 40 kHz, (e) 10 kHz

10.8 How counter is tested for lock-out of a counter? How lock-out problem is eliminated?

10.9 Write the design steps of synchronous counters.

10.10 What is a counter? Classify the counters. What is the difference between synchronous and asynchronous counters?

PROBLEMS

10.1 Design a MOD-7 asynchronous counter using J-K-flip-flop.

10.2 Design a MOD-12 asynchronous counter using J-K-flip-flop.

10.3 Design a MOD-9 synchronous counter using J-K-flip-flop.

10.4 Design a MOD-7 synchronous counter using S-R-flip-flop.

10.5 Design a MOD-12 synchronous counter using T-flip-flop.

10.6 Design a MOD-14 synchronous counter using D-flip-flop.

10.7 Design a J-K counter that goes through states 2, 4, 5, 1, 2, 4, … Is the counter self-startins? Modify the circuit such that whenever it goes to any invalid state it comes back to state 2.

10.8 Design a type-D counter that goes through states 0, 2, 4, 6, 0, …. The undesired states must alwavs go to a 0 on the next clock pulse.

10.9 Design a type-T counter that goes through states 6, 3, 7, 8, 2, 9, 1, 12, 14, 0, 6, 3, … Is the counter self-starting?

10.10 Design a type-T counter that goes through states 0, 5, 4, 2, 0, ... Is the counter self-starting?

10.11 Design a type-T counter that must go though states 0, 2, 4, 6, 0, ... if the control line is HIGH, and through states 0, 4, 2, 7,0, ... if the control line is LOW.

10.12 Design sequence generator to generate the sequence 1011010.

10.13 Design sequence generator to produce the pulse trains 10000111 and 10011010.

10.14 Generate the pulse train 100110 using indirect logic.

10.15 Design a BCD up/down counter using S-R-flip-flops.

10.16 Design an up/down counter using D-flip-flop to count 0, 3, 2, 6, 4, 0.

11

Memory

CHAPTER OBJECTIVES

The main goal of this chapter is to impart knowledge about the memory devices. Readers will be able to discuss the following aspects in this chapter:

- Read-only memory (ROM)
- Programmable logic device (PLD): programmable read-only memory (PROM), programmable array logic (PAL), programmable logic array (PLA)
- Complex-programmable logic devices (CPLDs)
- Field-programmable gate array (FPGA)
- Random-access memory (RAM): static random-access memory (SRAM), dynamic random-access memory (DRAM)
- Expanding word size and capacity
- Non-volatile read/write memory
- Flash memory
- First-in-first-out (FIFO) and last-in-first-out (LIFO) memories
- Content-addressable memory (CAM)
- Charge-coupled device (CCD) memory

11.1 | INTRODUCTION

As compared to analog systems, digital systems have the ability to store data in binary for short or long period of time. Information revolution is because of the availability of and use of memory or binary data storage devices. Internet is the media to transfer data from one storage device to another. Computers and telecommunication systems use large amount of binary data storage. Moreover, a digital data processing system stores unprocessed, partially processed and processed data. A system that stores unprocessed, partially processed and processed data is known as memory. So, memory is the physical device(s) used to store data and program temporarily or permanently for the use in a computing systems (Figure 11.1) or other digital electronic device.

A memory unit is a collection of storage cells like latches or flip-flops together with circuit needed to transfer information in and out of storage unit. The memory stores binary information in groups of binary digits (bits) called words. A memory word is a group of

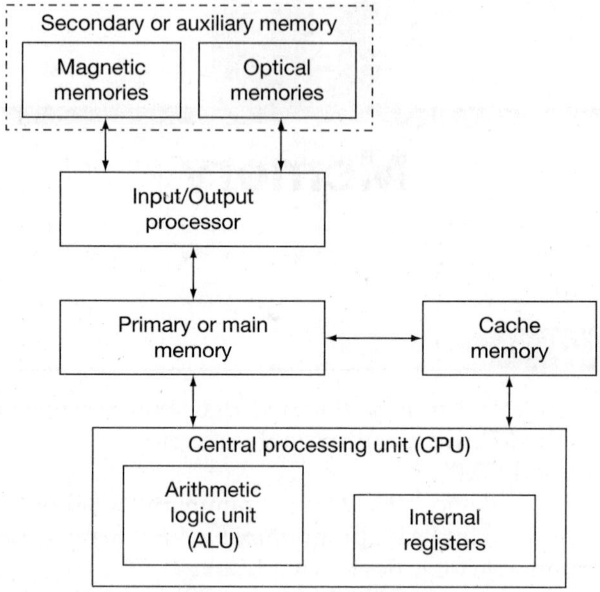

FIGURE 11.1 | Memory hierarchy in a computing system

1′s and 0′s. A number, an instruction code or alphanumeric characters or binary coded information is formed by combining binary bits. A group of eight bits is called byte. The 16-bit word contains 2 bytes. The 32-bit word contains 4 bytes. Total number of bytes can be stored give capacity of memory.

A memory unit is an essential part of digital processing unit and its primary purpose is to hold the program and data. The main goal of memory is speed to transfer data close to digital processing unit. The memory of a computing system is logically divided into four groups (Figure 11.2).

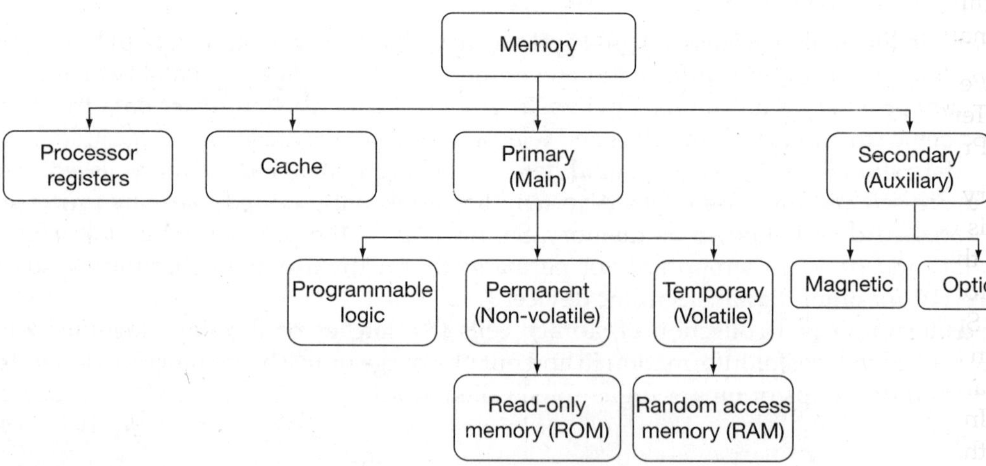

FIGURE 11.2 | Memory classification

- Processor memory or processor registers
- Cache memory
- Primary or main memory
- Secondary or auxiliary memory

Processor Memory: Processing unit internally has faster memory known as internal registers which use flip-flops (Figure 11.1). These registers are used to hold temporary results when a data processing is in progress. As cost involved is high, so small number of processor registers are used. The access time of processor memory is same as that of processor.

Cache: Cache memory is very small in size, expensive and less access time. Cache memory is placed between main memory and processor to improve the performance of processor (Figure 11.1). The active portions of program and data are placed in a fast and small memory, the average access time of memory is reduced. The cache is examined when *central processing unit* (CPU) needs to access memory. Availability of word in cache is called as cache hit. Non-availability of word in cache is considered as miss. During cache hit, processor reads the word from the cache. During the miss, the main memory is accessed by processor as word is not found in cache. The performance of cache is measured by hit ratio. Hit ratio is the ratio of the number of hits divided by total number of hits plus misses.

Cache memories are written using write-back and write-through methods. In the write-back method, a dirty bit is assigned to the cache word. Whenever the processor writes something into a cache word, the dirty word is first copied into the main memory before dirty word is replaced by the incoming new word. In the write-through method, the processor alters a cache address. The same alteration is made in the main memory by copying. So, altered cache address is copied to main memory address.

Primary Memory: Primary memory is the storage area in which all programs are executed. The processor directly accesses the program and data that are stored in primary memory (Figure 11.1). Hence, all programs and data must be available in the primary memory prior to execution. *Metal oxide semiconductor* (MOS) technology is used in primary memory design. Generally, the size of the primary memory is much larger than processor memory and its operating speed is slower than the processor registers.

Primary memory is further divided into three groups (Figure 11.2):

- Permanent or non-volatile memory
- Temporary or volatile memory
- Programmable memory

Primary memory normally includes *read-only memory* (ROM) or permanent memory. The ROM is non-volatile memory because on switching off the power, information is not lost. When the power is switch-on, the stored information is recalled. A ROM permits to read the information stored only. The contents are programmed by the manufacturer in the ROM. Such ROM is called mask-programmable ROM. ROM is inexpensive. ROM is built with an array of AND gates (AND-array) and an array of OR gates (OR-array). The links of AND-array and OR-array are fixed in ROM. ROM can be made programmable if fuses are used. Initially all the fuses are intact. Programming the ROM involves blowing the fuses along the paths as per requirement of logic function. The main disadvantage of a programmable read-only memory (PROM) is that it cannot be reprogrammed. PROM that allows reprogramming is called *erasable programmable read-only memory* (EPROM). Data are entered using electrical impulses and stored information is erased using ultraviolet (UV) ray. The process is done on separate socket of PROM. Stored information can be erased using electrical signal. Such ROM is known as *electrically erasable PROM* (EEPROM) or *electrically alterable PROM* (EAPROM). The ROM can be programmed while it is in the circuit board.

An *integrated circuit* (IC) that contains large numbers of gates and flip-flops, those can be configured by the user to perform different functions is called a *programmable logic device* (PLD). Either internal logic gates or connections or both of PLD can be changed (configured) by a programming process. One of the simplest programming methods is to use fuses. In the original state of the device, all the fuses are intact. Programming the device involves blowing those fuses along the paths that are required to obtain the particular configuration of the desired logic function. PLD is typically built with an array of AND gates (AND-array) and an array of OR gates (OR-array). There are three fundamental types of standard PLDs: PROM, *programmable array logic* (PAL), and *programmable logic array* (PLA). The main advantages of using PLDs are less board space, less access time (faster), lower power requirements, less costly assembly processes, higher reliability, easier troubleshooting and availability of design software.

Primary memory normally includes *random-access memory* (RAM) or temporary memory. RAM is volatile memory because information stored is lost when power is turned off. RAM can be backed up by batteries for a certain period of time and are known as *non-volatile RAM* (NVRAM). The semiconductor memory is designed using bipolar transistors and the information is stored in the form of voltage levels in flip-flops. The voltage levels do not change with time. Such semiconductor memories are known as static RAM because the information remains constant for some period of time. The semiconductor memory is also designed using MOS transistors and the information is stored in the form of electric charge in capacitors. The electric charge has tendency to leak out. So information changes from 1 to 0 with time. Such semiconductor memories are known as dynamic RAM because the information does not remain constant for some period of time. To prevent the loss, the dynamic RAM refreshes at regular intervals. Another read/write memories is the ferrite core memory where the reading process destroys the stored information. This property is called *destructive readout* (DRO) and it is unique with respect to ferrite cores. A semiconductor memory does not exhibit this property. So, solid-state memories are known as *non-destructive readout* (NDRO) memories. The information read from the ferrite core memory is first transferred to a buffer, and then it is rewritten into the location from where it was originally read. *Memory buffer register* (MBR) becomes a part of a RAM unit due to DRO phenomenon.

Secondary Memory: Secondary memory refers to the bulk storage medium comprising slow devices such as magnetic and optical. These devices are used to hold large information that is not needed by the processor frequently (Figure 11.1). Secondary memories are also referred to as auxiliary or backup storage. Floppy disk, hard disk (HD) and magnetic tapes come under magnetic storage devices. These devices store data in a thin coating of metal oxide on a flexible plastic disk. Alignment of domains in one direction or the other in metal oxide surface represent digital 0 or 1. A write head stores 0 and 1 on magnetic material while read head reads a 0 or 1 as magnetic material moves forward.

Common optical storage media includes *compact disk* (CD) and *digital versatile disk* (DVD). These are produced by industrial plastic injection equipment. Tiny pits and lands (no-pits) are moulded into shiny side of CD during manufacturing. The CD reader drive puts a laser beam at a track on the spinning CD. The reflected light bounced off the pits or lands are interpreted as 1 or 0. CD-writer drive is required to burn the CD-R (*recordable*) or CD-RW (*rewritable*).

Flash memory is very high density, low power, and non-volatile memory. It is rewriteable bit-by-bit memory in form of memory cards. It is very useful for portable systems those require low power consumption. Flash memory stores information in an array of memory cells made from floating-gate transistors. *Single-level cell* (SLC) device stores only one bit of information in each cell. *Multi-level cell* (MLC) devices, including *triple-level cell* (TLC) devices, can store more than one bit per cell by choosing between multiple levels of electrical charge to apply to the floating gates of its cells. The floating gate can be conductive typically polysilicon in most kinds of flash memory or non-conductive as in SONOS (silicon-oxide–nitride-oxide–silicon) flash memory.

11.2 | MEMORY BASICS

Basically, memories store data as 1-bit, 4-bit or 8-bit (Figure 11.3). The smallest unit of binary data is the bit (binary digit). Nibble is binary data obtained on combining 4 bits. Byte is binary data obtained on combining 8 bits. Generally, data are handled in a byte or in multiples of bytes. The byte can be split, into two nibbles. Bytes are grouped into words, i.e. 16-bit word, 32-bit word, 64-bit word, etc. In memories, single entity is stored as one memory location. Single entity is defined as a group of bits or bytes.

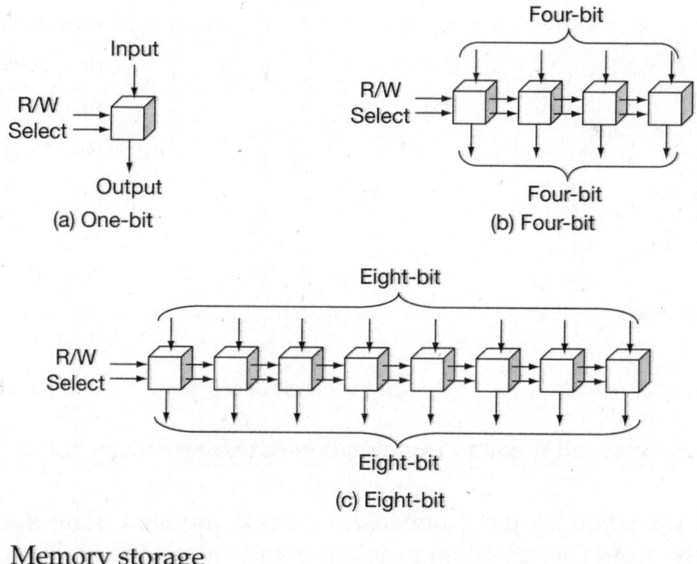

FIGURE 11.3 | Memory storage

Flip-flop is basic memory cell in some semiconductor memories. Registers and counters use flip-flop as memory unit. Register is small-scale memory. Flip-flop stores one binary bit (0 or 1). To store a nibble (4-bit), four flip-flops are required. To store a byte (8-bit), eight flip-flops are required. To store number of nibbles or bytes numbers of registers are required. Single register is termed as memory location. Number of memory locations where data is stored, gives the capacity of memory. Semiconductor memories are used because of their low cost, small size, better reliability and high speed. Moreover, available memory can be expanded to increase word length and word capacity.

11.2.1 | Memory Address

Each storage unit in a memory can retain either a 1 or a 0 and is called a cell. Memories are made up of arrays of cells, as illustrated in Figure 11.4 using 16-cells. Each block in the memory array represents one storage cell, and its location is identified by specifying a row and a column. The location of unit of data in a memory array called its address.

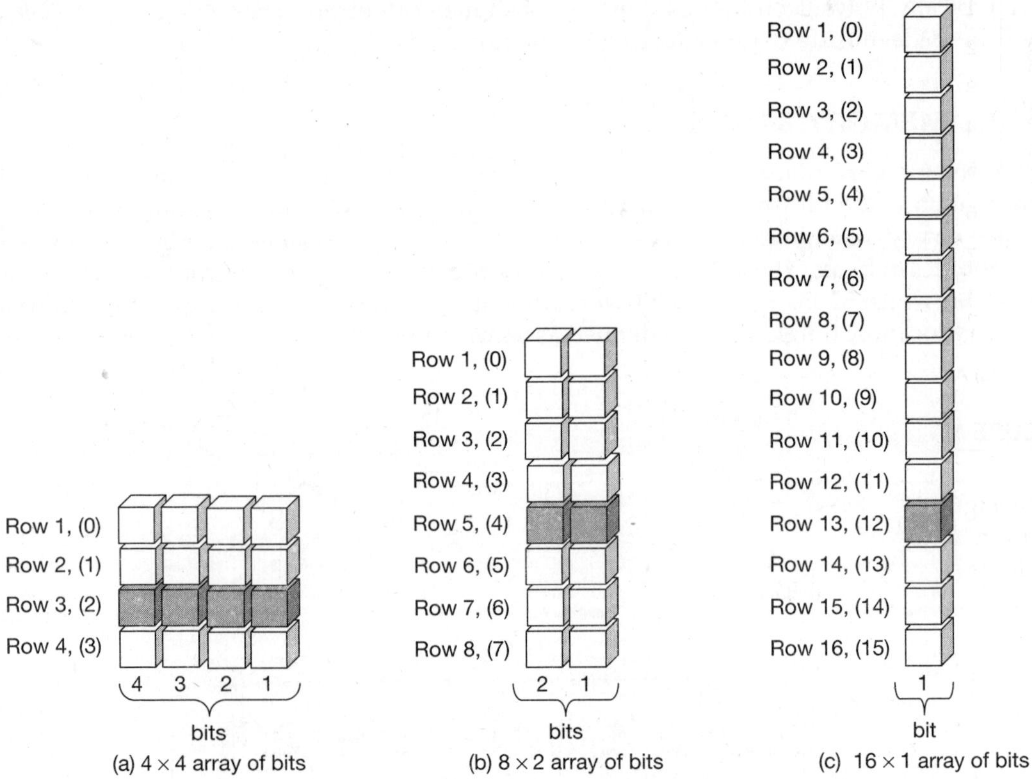

FIGURE 11.4 | Sixteen-cell memory organization in three different ways

A memory is identified by the number of words it can store. The size of each word remains same. The words are arranged in an order known as address. Figure 14.4(a) shows 4 × 4 array. This memory has 4-words and size of each word is 1-nibble. The number of addresses is 4. Address 0 is given to row 1, Address 1 is given to row 2 and so on. Figure 14.4(b) shows 8 × 2 array. This memory has 8-words and size of each word is 2 bits. The number of addresses is 8. Row 1 has address 0 and Row 8 has address 7. Figure 14.4(c) shows 16 × 1 array.

This memory has 16-words and size of each word is 1-bit. The number of addresses is 16. Row 1 has address 0 and Row 16 has address 15.

In Figure 11.5(a), the address of bit in two-dimensional array is specified by row and column as shown in 37th memory location (address). The size or unit of data is 1-bit. Sixty-four words of 1-bit size can be stored.

In Figure 11.5(b) the address is given by row. The size or unit of data is 8 bits. Eight number of words of 1-byte size can be stored. As shown in Figure 11.5(b), 1-byte word can be read from memory address 4. The smallest unit (size of word) of memory that can be addressed is byte.

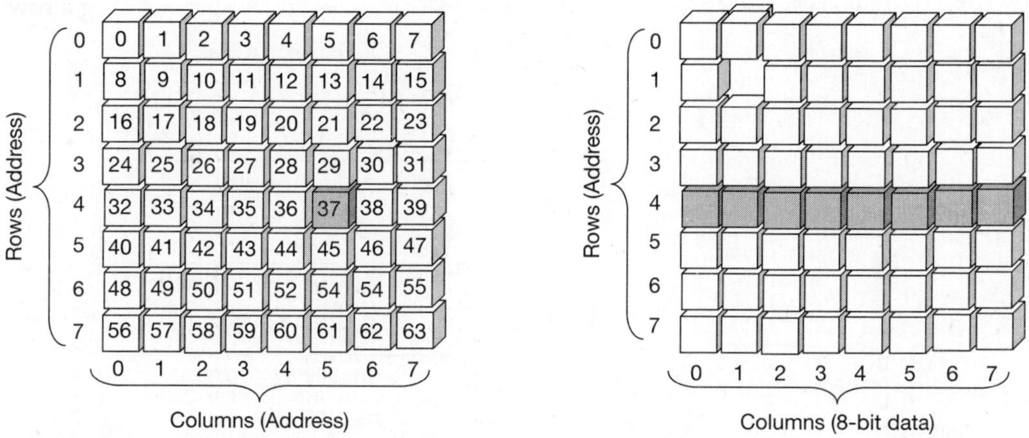

(a) Address of memory having 1-bit word

(b) Address of memory having 8-bit word

FIGURE 11.5 | Address of memory in two-dimensional array

In Figure 11.6, the address of nibble in three-dimensional array is given by row and column as shown in 39th memory location (address). Sixty-four numbers of words of 1-nibble size can be stored.

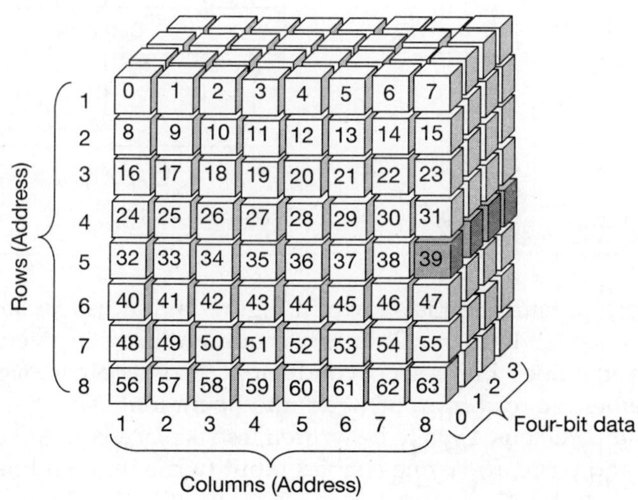

FIGURE 11.6 | Address of memory in three-dimensional array

The organization of 8 × 8 bits memory is shown in Figure 11.7 (Figure 11.5(b)). It has address 3 × 8 lines decoder that gives the 8 number of addresses from 3 address lines. The chip select ($\overline{CS}$) signal used as enable signal for decoder. The range of address is 0 to ($2^3 - 1$). Data can be stored at a particular memory location through input lines and enabling write signal ($\overline{WR}$). Data can be read from a particular memory location through output lines and enabling read signal ($\overline{RD}$). Address of particular memory location is accessed by feeding value on address lines. Input and output buffers hold data using tristate buffers controlled by write ($\overline{WR}$) or read ($\overline{RD}$) signal lines, respectively.

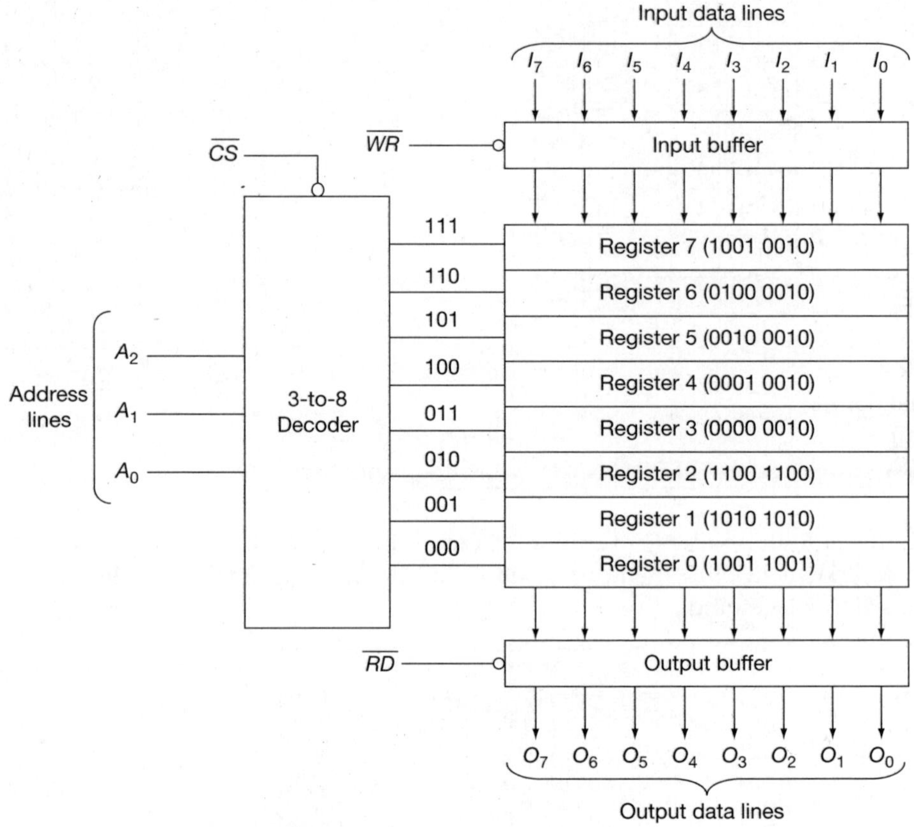

FIGURE 11.7 | 8 × 8 bits memory

Simplified memory organization is shown in Figure 11.8. It has *memory address register* (MAR) which acts as n-to-2^n lines ($n \times 2^n$ lines) decoder. Address is given through address bus. Address bus is on n-lines. Bus is set of conductors carrying binary signal. MBR is used to hold the data before the read from or write into operation. Data is read from is transferred to *input/output* (I/O) bus. Data to be written into is placed on I/O data bus. I/O bus is of m-lines. Read and write, $R/\overline{W}$ line enables input (write into) and output (read from) operations.

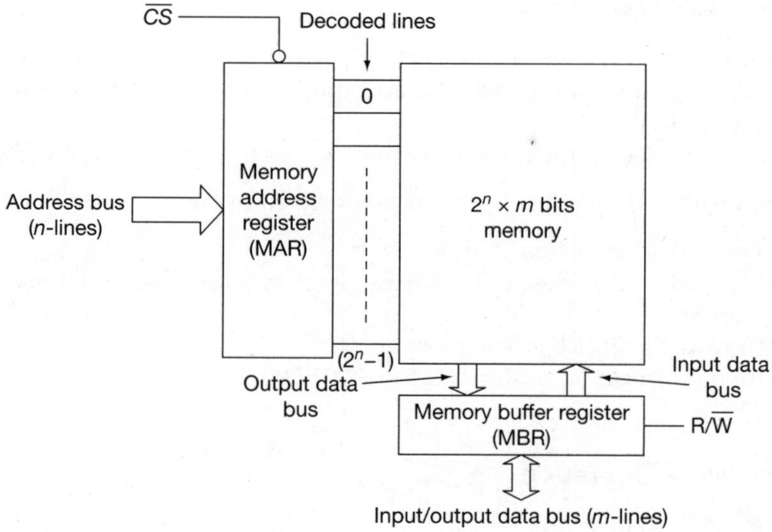

FIGURE 11.8 | Memory organization

Block diagram of memory is given in Figure 11.9. It has n-lines address bus. It is unidirectional. It has m-lines data bus. It is unidirectional for ROM and is bidirectional for RAM. Low chip select, $\overline{CS}$ signal is used to select the chip for operations. Active high read or low write $(R/\overline{W})$ signal is used to perform read from or write into data from memory.

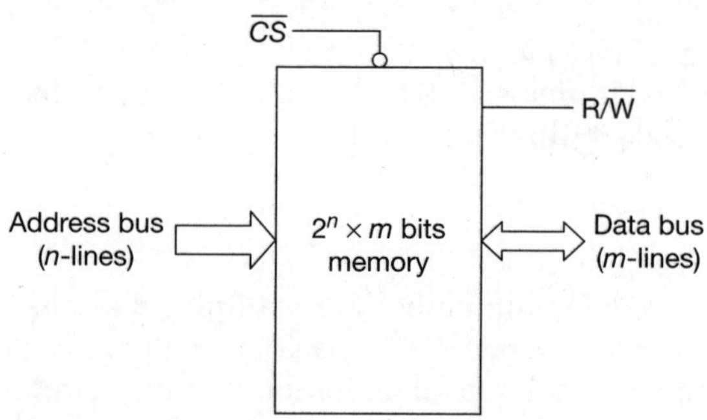

FIGURE 11.9 | Block diagram memory

11.2.2 | Memory Operation

Since a memory stores binary data, as per need data is saved into the memory and data are copied from the memory. The write operation saves data into a specified address in the memory and the read operation copies data out of a specified address in the memory. The addressing operation, which is part of both the write and read operation, selects the specified memory address.

11.2.2.1 | The Read Operation

- The code is placed on address bus.
- Address decoder decodes the address and specific memory location is selected.
- Send the read signal.
- Word from the selected memory location is copied to memory data buffer register.

Consider to perform read operation of memory depicted in Figure 11.7.

- 101 is placed on address bus ($A_2A_1A_0$).
- Address decoder decodes the address and specifies the memory location 5 (i.e. Register 5).
- Send high read signal (*RD*) to memory.
- Data of memory location 5 (Register 5) is (0010 0010). This data is copied to output buffer.

11.2.2.2 | The Write Operation

- The code is placed on address bus.
- Address decoder decodes the address and specific memory location is selected.
- Sends the write signal.
- Word from memory data buffer register is copied to the selected memory location.

Consider to perform write operation of memory depicted in Figure 11.7.

- 011 is placed on address bus ($A_2A_1A_0$).
- Address decoder decodes the address and specifies the memory location 3 (i.e. Register 3).
- Send low write signal ($\overline{WR}$) to memory.
- Suppose data in input buffer is (11111111). Data of memory location 3 (i.e. Register 3) is modified from (0000 0010) to (11111111).

11.2.3 | Capacity

The capacity of memory is the total number of data units, those can be stored. Unit gives 1-bit datum. The unit of capacity is bits. The capacity of memory organized in array of two-dimensional or three-dimensional. Organization of memory is represented as $2^n \times m$bits, where n is address lines give address and m is the data lines give the size of word stored at 2^n addresses.

Capacity of memory is 16 bits organized as 4×4 bits ($2^2\times4$ bits) as shown in Figure 11.4(a). It has 4-addresses and size of word stored at each address is of 4 bits or 1 nibble. Such memory has 2-address lines and 4-data lines.

Capacity of memory is 16 bits organized as 8×2 bits ($2^3\times2$ bits) as shown in Figure 11.4(b). It has 8-addresses and size of word stored at each address is of 2 bits. Such memory has 3-address lines and 2-data lines.

Capacity of memory is 16 bits organized as 16×1 bits ($2^4\times1$ bits) as shown in Figure 11.4(c). It has 16-addresses and size of word stored at each address is of 1-bit. Such memory has 4-address lines and 1-data line.

Capacity of memory is 64 bits organized as $(8 \times 8) \times 1$ bits $((2^3 \times 2^3) \times 1$ bits) as shown in Figure 11.5(a). To arrange 64 (8×8) addresses, 8-rows and 8-columns are followed. Size of word stored at each address is of 1-bit. It has 6 address lines and 1 data line. 3-address lines represent number of rows and 3-address lines represent number of columns.

Capacity of memory is 64 bits organized as 8×8 bits or 8×1 Byte as shown in Figure 11.5(b). It has 8-addresses and size of word stored at each address is 8 bits or 1 byte. Such memory has 3-address lines and 8-data line.

Capacity of memory is 256 bits organized as $(8 \times 8) \times 4$ bits or $(8 \times 8) \times 1$ nibble $((2^3 \times 2^3) \times 4$ bits) as shown in Figure 11.6. To arrange 64 addresses, 8-rows and 8-columns are followed. Size of word stored at each address is or 4 bits (or 1 nibble). It has 64-address lines and 4-data lines. 3-address lines represent number of rows and 3-address lines represent number of columns.

Two-dimensional memory: Memory capacity is given by $2^n \times m$ bits, where n is number of address lines and m is data lines. If m is equal to 8, then memory capacity becomes $2^n \times 8$ bits or 2^n bytes or 2^n B. If capacity is given in bits then size of word must be given.

Three-dimensional memory: Memory capacity is given by $(2^{n_1} \times 2^{n_2}) \times m$ bits, where n_1 is number of row address lines, n_2 number of column address lines and m is number of data lines. If m is equal to 8, then memory capacity becomes $(2^{n_1} \times 2^{n_2}) \times 8$ bits or $(2^{n_1} \times 2^{n_2})$ bytes or $(2^{n_1} \times 2^{n_2})$B. If capacity is given in bits, then size of word must be given.

International System of Units (SI), 1 kilo = 1000 (10^3) which approximates 1024 $(2^{10}) \approx 1.024 \times 10^3$. Kilo represents binary multiples 2^{10}. In 1998, the *International Electrotechnical Commission* (IEC) proposed standards for binary prefixes requiring the use of *kilobyte* (kB) to denote 10^3 bytes and kibibyte (kiB) to denote 1024 (2^{10}) bytes. Similarly, *megabyte* (MB) denotes 10^6 bytes and *mebibyte* (MiB) to denote $2^{20}(1024 \times 1024)$ bytes. *Gigabyte* (GB) denotes 10^9 bytes and *Gibibyte* (GiB) to denote 2^{30} (1024^3) bytes. *Terabyte* (TB) denotes 10^9 bytes and *Tebibyte* (TiB) to denote 2^{40} (1024^4) bytes. Table 11.1 explains the use of notation in this book.

TABLE 11.1 | Memory specifications

Name	Symbol	Address Lines	Data Lines	Memory Organization	Memory Capacity	Memory Locations
1 kilobytes	1kB	10	8	2^{10} bytes	$2^{10} \times 8$ bits	2^{10}
1 Megabytes	1MB	20	8	2^{20} bytes	$2^{20} \times 8$ bits	2^{20}
1 Gigabytes	1GB	30	8	2^{30} bytes	$2^{30} \times 8$ bits	2^{30}
1 Terabyte	1TB	40	8	2^{40} bytes	$2^{40} \times 8$ bits	2^{40}
1 Petabyte	1PB	50	8	2^{50} bytes	$2^{50} \times 8$ bits	2^{50}
1 Exabyte	1EB	60	8	2^{60} bytes	$2^{60} \times 8$ bits	2^{60}
1 Zettabyte	1ZB	70	8	2^{70} bytes	$2^{70} \times 8$ bits	2^{70}
1 Yottabyte	1YB	80	8	2^{80} bytes	$2^{80} \times 8$ bits	2^{80}

EXAMPLE 11.1 What is capacity of memory having 32 address lines and 8 data lines?

SOLUTION Number of address lines = 32

'B' is used for byte.
'b' is used for bit.

Number of data lines = 8

Capacity of memory = $2^{32} \times 8$ bits

$= 2^2 \times 2^{30}$ B ($\because$ 1G = 2^{30} and 1 Byte = 8 bits)

= 4 GB or 32 Gb **Ans.**

EXAMPLE 11.2 What is capacity of memory having 16 address lines of row decoder, 8 address lines for column decoder and 8 data lines?

SOLUTION Number of address lines of row decoder = 16

Number of address lines of column decoder = 8

Number of data lines = 8

Capacity of memory = $(2^{16} \times 2^8) \times 8$ bits

$= 2^{24}$ B (1 Byte = 8 bits)

$= 2^4 \times 2^{20}$ B ($\because$ 1 MB = 2^{20} bytes)

= 16 MB or 128 Mb **Ans.**

EXAMPLE 11.3 How many bytes can be stored in 16 Tb memory? How many address lines and data lines this memory has?

SOLUTION Size of word = 1 byte = 8 bits

Capacity of memory = 16 Tb (b means bit)

$= 2T \times 8$ bits (i)

$= 2^1 \times 2^{40} \times 8$ bits ($\because$ 1T = 2^{40})

$= 2^{41} \times 8$ bits (ii)

From Eq. (i): 2T bytes can be stored in the memory having capacity 16Tb. **Ans.**

From Eq. (ii): 41 address lines and 8 data lines **Ans.**

EXAMPLE 11.4 The memory has storage 48 TB. How many address and data lines this memory has?

SOLUTION Storage of memory = 48TB (B means byte)

$= 48\ T \times 8$ bits

$= 6 \times 2^3 \times 2^{40} \times 8$ bits ($\because$ 1T = 2^{40})

$= 2^{3+3+40} \times 8$ bits (To decode 6 lines, 3×2^3 decoder is required so $6 \rightarrow 2^3$)

$= 2^{46} \times 8$ bits

46 address lines and 8 data lines **Ans.**

EXAMPLE 11.5 The memory has storage 32 GB. How many address and data lines this memory has? Row decoder has 20 lines. How many lines are required for column decoder?

SOLUTION Storage of memory = 32 GB

$$= 32\text{ G} \times 8 \text{ bits}$$

$$= 2^5 \times 2^{30} \times 8 \text{ bits} \qquad (\because 1\text{G} = 2^{30})$$

$$= 2^{5+30} \times 8 \text{ bits}$$

$$= 2^{35} \times 8 \text{ bits}$$

35 address lines and 8 data lines **Ans.**
Row address decoder lines = 20

So, $$2^{35} = 2^{20} \times 2^{15}$$

So, column address decoder lines is 15 **Ans.**

11.3 | CLASSIFICATION OF MEMORY DEVICES

The memory devices may be classified based on various characteristics that are listed below:

- Design technology
- Access mode of memory location
- Physical characteristics
- Operational principle

11.3.1 | Design Technology

Memories may be classified based on the components used for the design. Memories based on two components are bipolar transistors and unipolar or MOS transistors.

Static random-access memory (SRAM), ROM and PROM can be designed using bipolar devices (*transistor–transistor logic* (TTL), *Emitter-coupled logic* (ECL) etc.).

Dynamic random-access memory (DRAM), EPROM, EEPROM and flash memory use unipolar devices (*metal oxide, semiconductor field-effect transistor* (MOSFET).

11.3.2 | Access of Memory Location

Basic operation on memory is performed to read the information or store the information. Firstly, memory is located and then information is read from or write into operation are performed on that located memory location given by address. Mode of access deals with the manner in which a memory address is accessed for read or writes operation. In case of ROM, only read operation is possible. Data can be accessed by performing

- Sequential access, and
- Random access

Sequential memories perform to access the memory address in sequence starting from first address. The access time is different for different memory locations to access addresses sequentially. Counter or shift register is used to maintain the sequence to access next memory address. Shift register and *charge-coupled devices* (CCD) are sequential memories. Magnetic tapes being secondary memories are sequential memories. For example, to access the 3rd memory address, more time is required than to access the 2nd memory address. Since the access is always start accessing from 1st location and accesses 2nd memory location, 3rd memory location and so on.

RAM accesses the memory address randomly. In RAM, any memory address can be accessed at any time. Decoder is required to decode the memory location represented by address. RAM, ROM and *content-addressable memory* (CAM) are random access memories. In random access memories, any memory address requires equal time for accessing data.

11.3.3 | Physical Characteristics

Memories can be classified as per their physical characteristics such as:

- Erasable or non-erasable, and
- Volatile or non-volatile.

11.3.3.1 | Erasable or Non-Erasable Memories

Erasable memory is a memory in which the information stored can be erased and new information can be stored. On the other hand, the information stored in the non-erasable memory cannot be erased for example ROM is non-erasable memory. The stored information in the memories can be erased one-by-one or simultaneously all. The erase operation is performed either by electrical signal or using UV rays.

One-by-one Location Erasable Memories In this type, the desired memory locations can be erased one by one location (address) and then the new information is stored. EAROM, RAM and CAM are the examples of one-by-one location erasable memory. Before entering the new information in EAROM, the contents of the memory address (location) must be erased. However, in RAM and CAM, erasing operation is performed automatically when new information is to be entered.

All Locations Simultaneously Erasable Memories These are the memories in which, contents of all the locations of a memory are erased simultaneously. As an example, when we expose an EPROM is exposed to the UV radiation, the contents of all the locations get erased simultaneously.

The erase operation is performed in the following ways:

Electrically Erasable Programmable Read-Only Memory (PROM) Stored information can be erased using electrical signal. Such ROM is known as EEPROM or EAPROM. The ROM can be programmed while it is in the circuit board.

Ultraviolet Erasable Programmable Read-Only Memory PROM Stored information can be erased by exposing the chip to UV radiation. In EPROM, information is entered using electrical impulses and stored information is erased using UV ray. The process is done on separate socket of PROM. In this EPROM, there is a quartz lid or window on the package. By exposing EPROM to the UV light through the quartz window, the stored information is erased. The exposure period is 10 to 15 minutes. The UV ray will erase all the cells simultaneously at a time and all the locations in erased EPROM will store 1.

11.3.3.2 | Volatile or Non-Volatile Memories

On switching off the power, the information stored in a memory is lost. Such memory is known as volatile memory. In other words, in a volatile memory, the information once stored remains available until the power is not turned off. Switch off the power and then switch on the power, the already stored information will not be available. Hence, volatile memories are known as temporary memories. RAM is a volatile memory.

On switching off the power, the information stored in a memory is not lost. Such memory is known as non-volatile memory. In other words, in a non-volatile memory, the information once stored remains available until changed deliberately by user. Switching off the power and then switch on the power, the already stored information is available in the same form. Hence, non-volatile memories are known as permanent memories. All types of ROMs are non-volatile memories.

11.3.4 | Operational Principle

Memories may be classified according to their principle of operation. The basic operation is to access the address of memory and then the information can be accesses or modified or compared. The address may be accessed sequentially or randomly. The time required to find information stored in memory can be reduced if stored information is identified for access by the content of data itself rather than by an address. Such memory is known as CAM. Sometime there is a need to read the data in the same order in which it was stored. Such memory device is known as first-in-first-out (FIFO) memory. In this type of memory, the information which is stored first is read out first. The storage device used for this purpose is a SRAM array with two separate ports that allows data to be written into and read from its array at independent data rates. Flow chart giving classification of memory is shown in Figure 11.10.

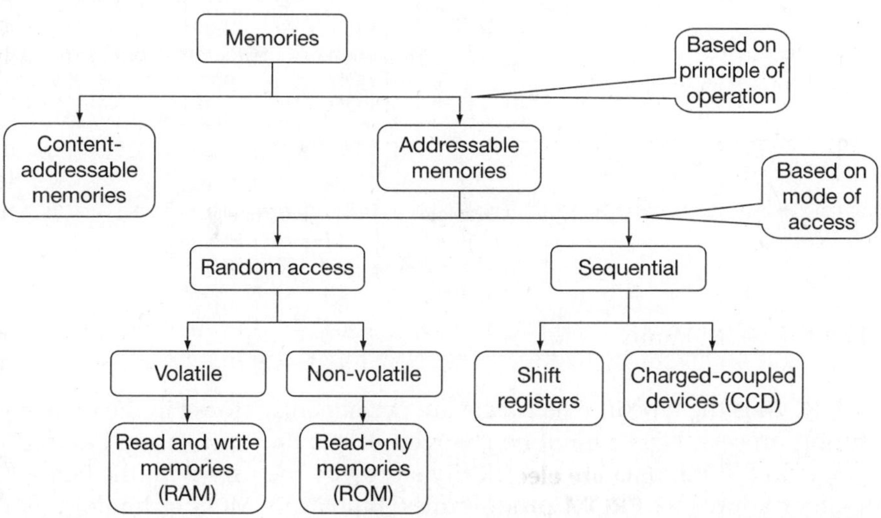

FIGURE 11.10 | Classification of memories

The most commonly used memories are as follows:

- Read-only memories (ROM)
- Read-and-Write memories (RWM or RAM)
- Flash memories
- Content addressable memories (CAM)
- First-in First-out (FIFO) memories

All the above mentioned memories will be discussed later on in this chapter.

11.4 | READ-ONLY MEMORY

A ROM stores information or data permanently or semi-permanently that can be read only. No write operation is possible. So, data cannot be changed. However, there is facility to change the data with specialized equipment known as PROM programmer. ROM stores data that can be used repeatedly like tables, conversion tables, instructions for initialization, etc. ROM retains stored data when power is off. So ROM is non-volatile memories. Figure 11.11 shows the classification of ROM.

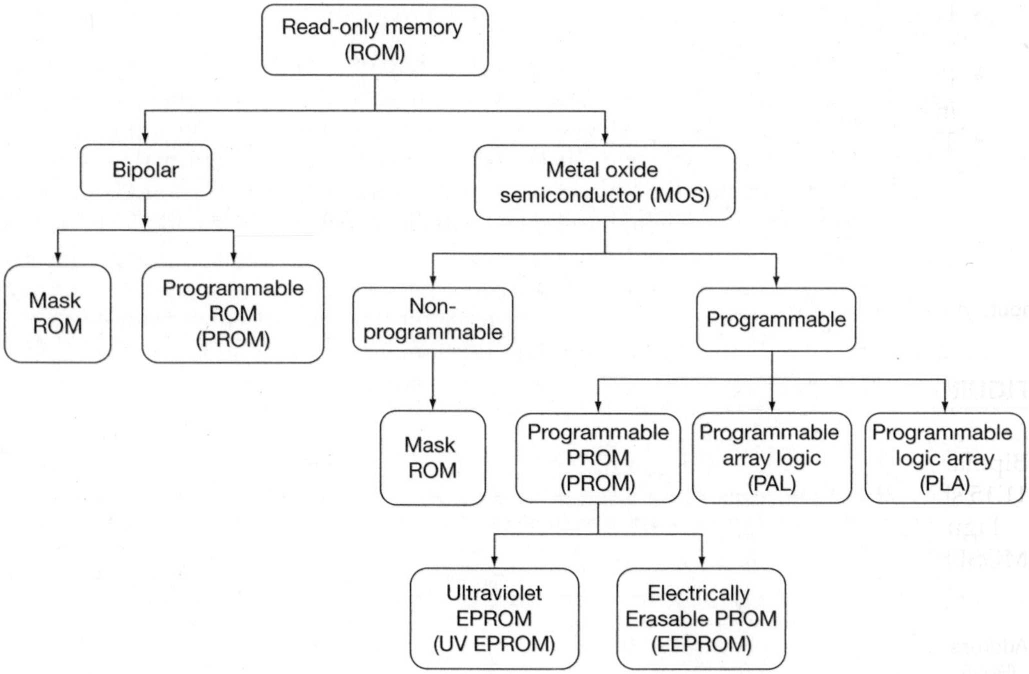

FIGURE 11.11 | ROM family

The mask ROM is the type in which data are permanently stored in the memory during manufacturing process. Data cannot be changed. It has low cost so can be used for high volume usage. In PROM, data are electrically stored by the user with the help of specialized equipment known as PROM programmer. Bipolar or MOS technology is used for both ROM and PROM. Erasable PROM is a MOS device. The UV-EPROM is electrically programmable by the user but the stored data can be erased by exposing to UV light over a period of several minutes (15–30 minutes). The data is erased simultaneously. Electrically erasable PROM (EEPROM or E^2PROM) can be erased in a few minutes by applying voltage of proper amplitude and polarity. Data available at any one memory location can be erased. PAL and PLA comes under the category of PLD.

Basically, ROM are built with an array of AND gates (AND-array) or decoder and an array of OR gates (OR-array). Figure 11.12 explains the design of ROM.

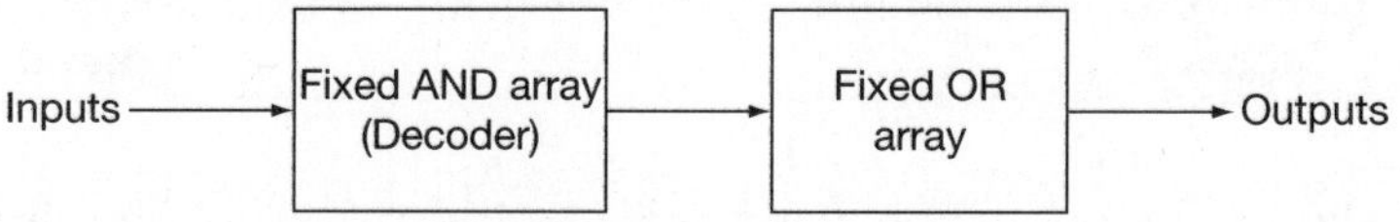

FIGURE 11.12 | Read-only memory (ROM)

Figure 11.13 explains the arrangement of AND gate array, OR gate array and fixed links.

- The product terms constitute a group of 2^n AND gates each of n inputs.
- Fixed links, $2 \times (n \times 2^n)$ are inserted between all n inputs and their complement values to each of the AND gates.
- Fixed links are $(2^n \times m)$ also provided between the outputs of the AND gates and the inputs of the OR gates.
- The sum terms constitute a group of m OR gates each of m outputs.

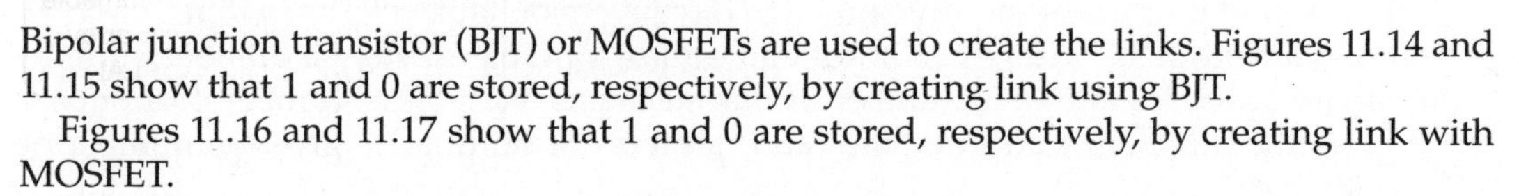

FIGURE 11.13 | Read-only memory (ROM)

Bipolar junction transistor (BJT) or MOSFETs are used to create the links. Figures 11.14 and 11.15 show that 1 and 0 are stored, respectively, by creating link using BJT.

Figures 11.16 and 11.17 show that 1 and 0 are stored, respectively, by creating link with MOSFET.

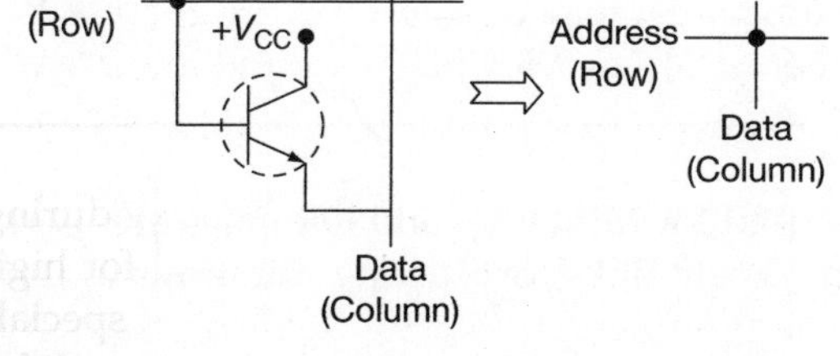

FIGURE 11.14 | BJT for fixed link stores 1

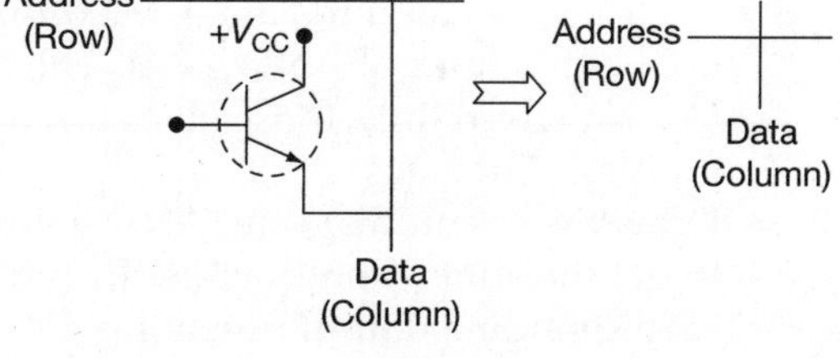

FIGURE 11.15 | BJT for fixed link stores 0

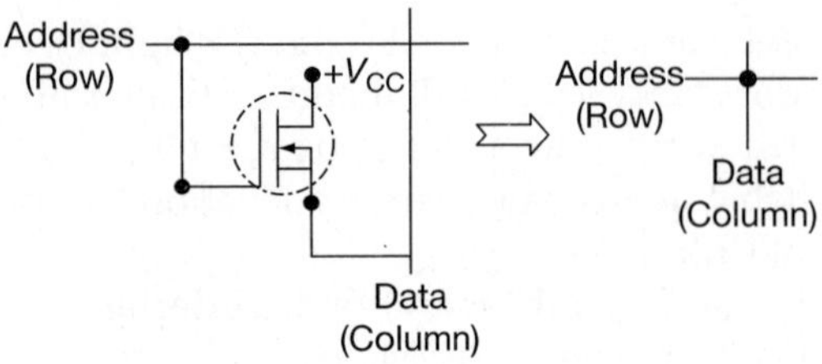

FIGURE 11.16 | MOSFET for fixed link stores 1

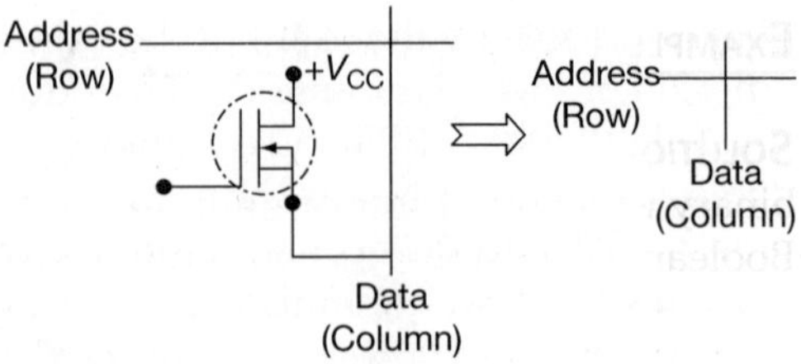

FIGURE 11.17 | MOSFET for fixed link stores 0

Figure 11.18 shows the conventional and array logic symbols for a multiple input AND and a multiple input OR gate.

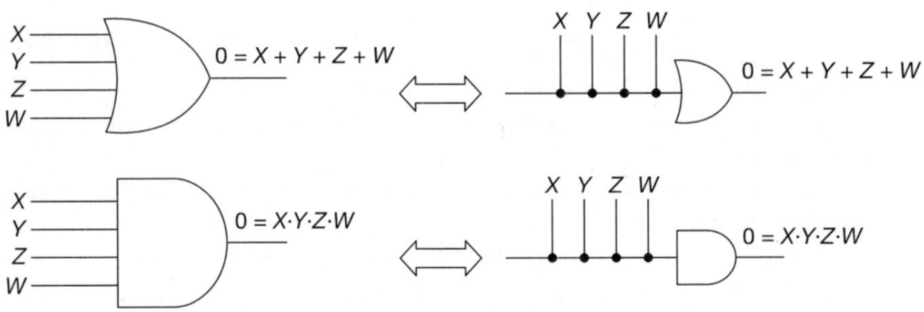

FIGURE 11.18 | Conventional and array logic symbol

11.4.1 | Design Procedure of ROM

Design procedure of ROM is given below.

- Inputs and outputs of the logic function or conversion table, etc. are given.
- Obtain truth table with given input and output variables. Let n be number of input binary variables and m is number of output binary variables. Number of inputs, n gives the number of address lines and numbers of outputs, m give the number of data lines.
- Find expression in standard sum of product form.
- Number of AND gates required is 2^n. Number of OR gates required is m. Number of NOT gates required are n. Maximum number of created fixed links are $2 \times (n \times 2^n)$ before the input of AND gates. The number of BJT or MOSFET required are $2^n \times m$ to create fixed links those give stored 0 or 1 value between output of AND gates and inputs of OR gates.
- Draw the circuit diagram. AND gates are used for Minterms and OR gates are used to get the sum-of-product (SOP) function. Create links for existing minterms in the corresponding output function.

Example 11.6 explains the design procedure.

EXAMPLE 11.6 Implement half-adder using ROM

SOLUTION There are two-input binary variables; augend, a_1 and addend, a_0 and 2-output binary variables as sum, d_0 and carry, d_1. Truth table is constructed and is given in Table 11.2. Boolean expressions for two-outputs in standard minterm form are given below:

$$d_0(a_1, a_0) = \sum m(1,2) \quad (11.1)$$

$$d_1(a_1, a_0) = \sum m(3) \quad (11.2)$$

Logic expression of sum, d_0 for half-adder is given below.

$$d_0(a_1, a_0) = \bar{a}_1 a_0 + a_1 \bar{a}_0 \quad (11.3)$$

Logic expression of sum, d_0 for half-adder is given below.

$$d_1(a_1, a_0) = a_1 a_0 \quad (11.4)$$

TABLE 11.2 | Truth table for half-adder

Minterm	Inputs		Outputs	
	Augend a_1	Addend a_0	Sum d_0	Carry d_1
m_0	0	0	0	0
m_1	0	1	1	0
m_2	1	0	1	0
m_3	1	1	0	1

For ROM design, there is need of two inputs and two outputs.

Number of inputs or address lines, $n = 2$.

Number of outputs or data lines, $m = 2$.

Number of AND gates required = $2^n = 2^2 = 4$.

Number of fixed links before AND gate = $2 \times (n \times 2^n) = 2 \times (2 \times 2^2) = 16$.

Number of OR gates required = $m = 2$

Number of fixed links after AND gate array and before OR gate array = $2^n \times m = 2^2 \times 2 = 8$.

Complete diagram is given in Figure 11.19(a). AND gates and OR gates are having fixed contacts or links. Figure 11.19(b) gives the diagram using decoder. Block diagram of ROM is given in Figure 11.19(c). Two number of tristate buffers are used to read data. Active low signal, $\overline{OE}$ acts as read signal.

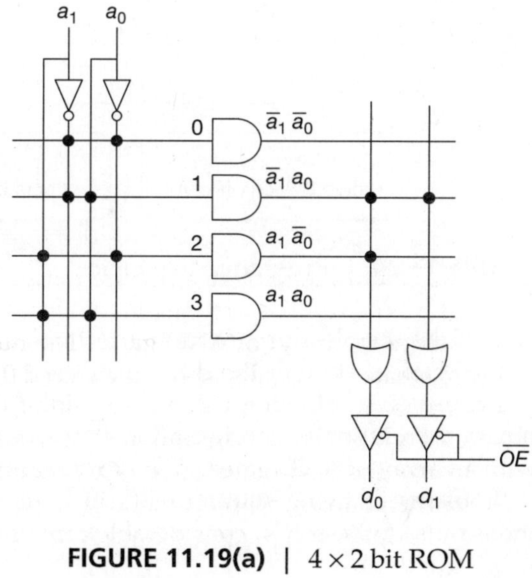

FIGURE 11.19(a) | 4 × 2 bit ROM

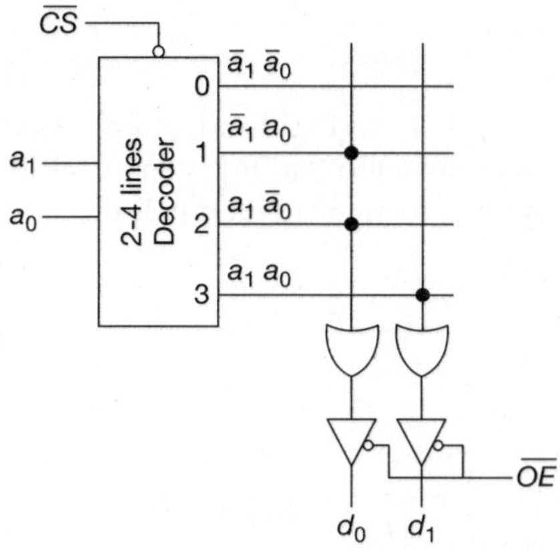

FIGURE 11.19(b) | 4 × 2 bit ROM

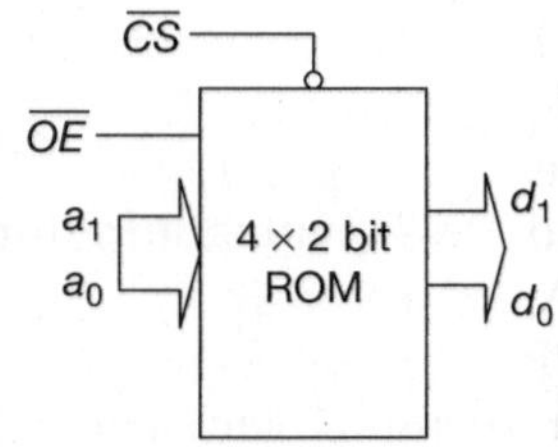

FIGURE 11.19(c) | 4 × 2 bit ROM

11.5 | PROGRAMMABLE LOGIC DEVICE (PLD)

An IC that contains large numbers of gates, flip-flops, etc. that can be configured by the user to perform different functions is called a PLD. The internal logic gates and/or connections of PLDs can be changed/configured by a programming process. Figure 11.20 shows the classification of programmable logic.

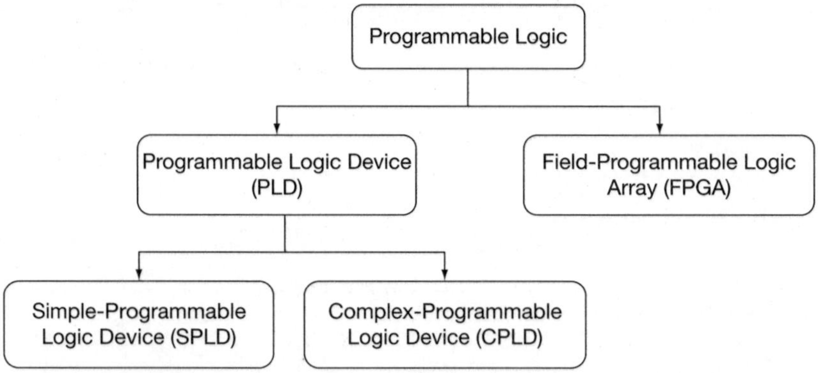

FIGURE 11.20 | Programmable logic

One of the simplest programming technologies uses fuses to store information as 0 or 1. In the original state of the device, all the fuses are intact (1 is stored). Programming the device involves blowing those fuses along the paths that must be removed in order to obtain the particular configuration of the desired logic function. PLDs are typically built with an array of AND gates (AND-array) and an array of OR gates (OR-array).

Problems of using standard ICs in logic design are the requirement of hundreds or thousands of these ICs, considerable amount of circuit board space, a great deal of time and cost in inserting, soldering and testing. So, a significant inventory of ICs is maintained in the store. Advantages of using PLD are less board space, faster read operation, lower power requirements (i.e. smaller power supplies), less costly assembly processes, higher reliability (fewer ICs and circuit connections means easier troubleshooting), and availability of design software.

There are three fundamental types of standard PLDs: PROM, PAL, and PLA. Another type of PLD, which is discussed later, is the *complex-programmable logic device* (CPLD), e.g. *field-programmable gate array* (FPGA). The three fundamental types of PLD differ in the placement of programmable connections (links) in the AND-OR arrays. Figures 11.21, 11.22 and 11.23 show the locations of the programmable connections for the PROM, PAL and PLA, respectively.

The PROM has a fixed AND array (constructed as a decoder) and programmable connections for the output OR gates array (Figure 11.21). The PROM implements Boolean functions in SOP form.

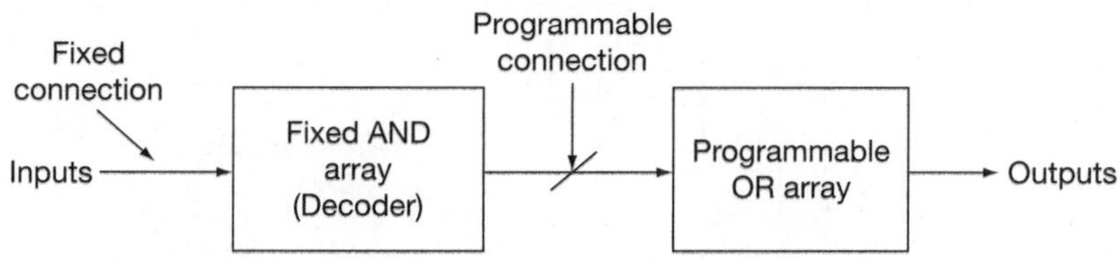

FIGURE 11.21 | Programmable read-only memory (PROM)

The **PAL** device has a programmable AND-gate array and fixed connections for the OR-gate array (Figure 11.22).

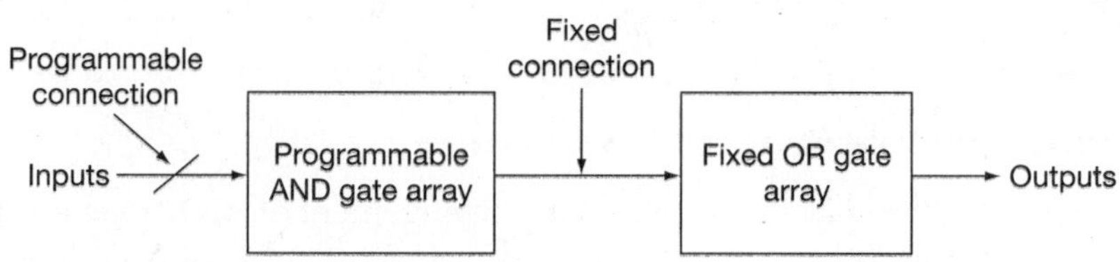

FIGURE 11.22 | Programmable array logic (PAL)

The **PLA** has programmable connections for both AND- and OR-gate arrays (Figure 11.23). So it is flexible type of PLD.

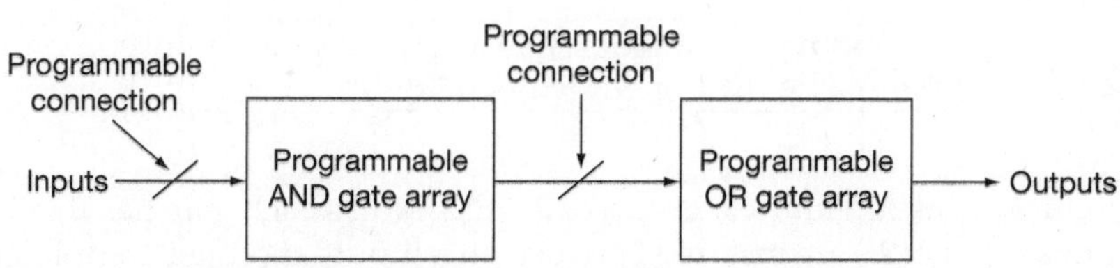

FIGURE 11.23 | Programmable logic array (PLA)

BJTs or MOSFETs are used to create the links. Figures 11.24 and11.25 show that 1 and 0 are stored, respectively, by creating link with BJT.

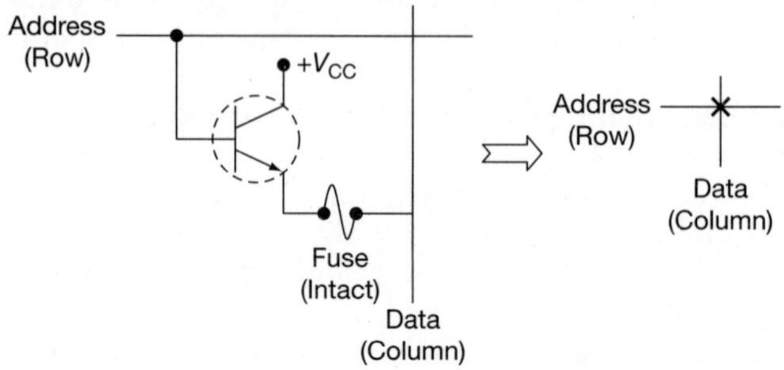

FIGURE 11.24 | Flexible link stores 1

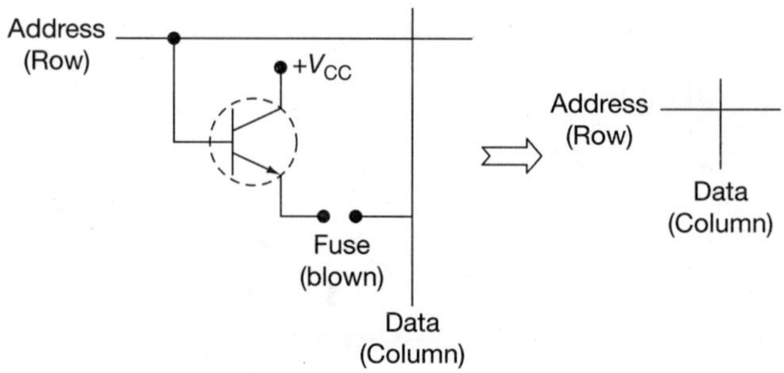

FIGURE 11.25 | Flexible link stores 0

11.5.1 | Programmable Read-Only Memory

PROM is shown in Figure 11.26 that explains the arrangement of AND-gate array, OR-gate array and fixed or fusible links.

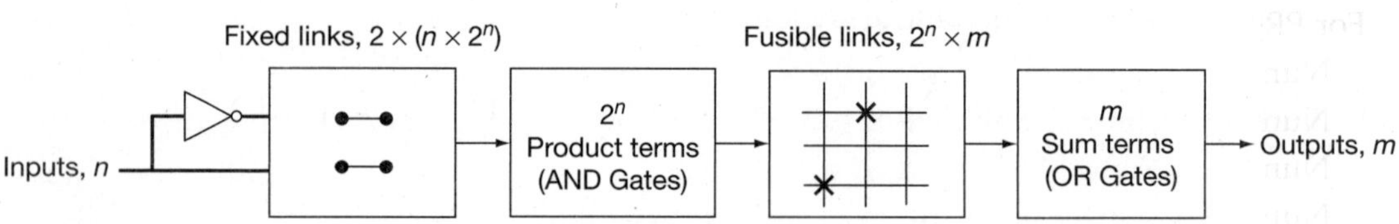

FIGURE 11.26 | Programmable read-only memory (ROM)

- The product terms constitute a group of 2^n AND-gates each gate has n inputs.
- Fixed links, $2 \times (n \times 2^n)$ are inserted between all n inputs and their complement values to each of the AND gates.

- Fusible links are ($2^n \times m$) also provided between the outputs of the AND gates and the inputs of the OR gates.
- The sum terms constitute a group of m OR gates each of m outputs.

11.5.2 | Design Procedure of PROM

Design procedure of PROM is given below.

- Inputs and outputs of the logic function or conversion table etc., are given.
- Obtain truth table with given input and output variables. Let n be number of input binary variables and m is number of output binary variables. Number of inputs, n gives the number of address lines and numbers of outputs, m give the number of data lines.
- Find expression in standard SOP form.
- Number of AND gates required is 2^n. Number of OR gates required are m. Number of NOT gates required is n. Number of $2 \times (n \times 2^n)$ the number of fixed links before the input of AND gates. Number of $2^n \times m$ is the number of BJT or MOSFET are required to create fusible links giving stored 0 or 1 value between outputs of AND gates and inputs of OR gates.
- Draw the circuit diagram. AND gates are used for Minterms and OR gates are used to get the SOP function. Create links for existing minterms in the corresponding output function.

Example 11.7 explains the design procedure.

EXAMPLE 11.7 Implement half-adder using PROM

SOLUTION There are two-input binary variables; augend, a_1 and addend, a_0 and 2-output binary variables as sum, d_0 and carry, d_1. Truth table is constructed and is given in Table 11.2. Logic expression of sum, d_0 for half-adder is given in Eq. (11.3). Logic expression for carry, d_0 for half-adder is given in Eq. (11.4).

For PROM design there is need of two inputs and two outputs.

Number of inputs or address lines, $n = 2$.

Number of outputs or data lines, $m = 2$.

Number of AND gates required $= 2^n = 2^2 = 4$.

Number of fixed links as inputs of AND gate $= 2 \times (n \times 2^n) = 2 \times 2 \times 2^2 = 16$.

Number of OR gates required $= m = 2$.

Number of flexible links between outputs of AND gates and inputs of OR gates $= 2^n \times m = 2^2 \times 2 = 8$.

Complete diagram is given in Figure 11.27(a). AND gates array has fixed links whereas and OR gate array has fusible links. Figure 11.27(b) gives the diagram using decoder.

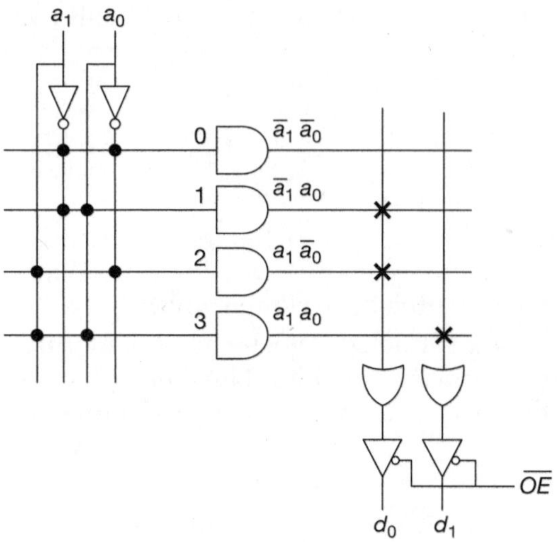

FIGURE 11.27(a) | 4 × 2 bit PROM

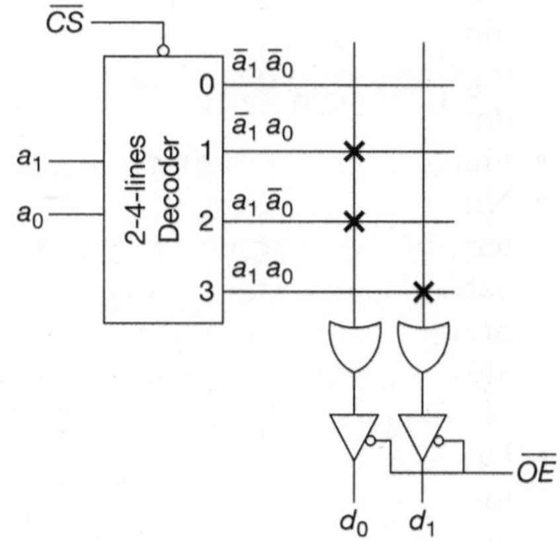

FIGURE 11.27(b) | 4 × 2 bit PROM

11.5.3 | Programmable Array Logic

PAL is shown in Figure 11.28 that explains the arrangement of AND gate array, OR gate array and fixed or fusible links.

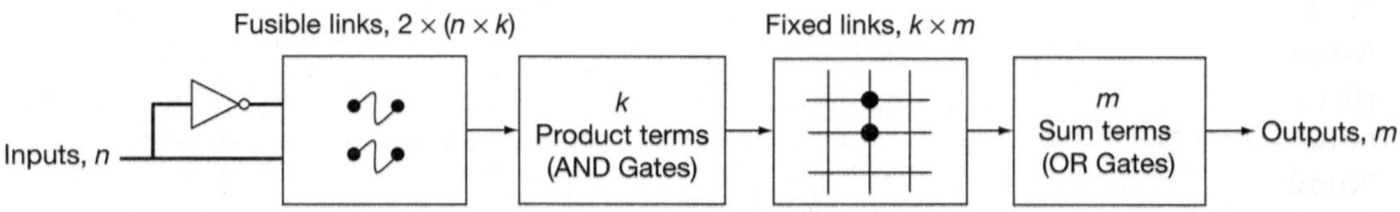

FIGURE 11.28 | Programmable array logic (PAL)

- The product terms constitute a group of k $(l \times m)$ AND gates each of n inputs, where l is the maximum number of product terms in the output function and m is the number of outputs.
- Fusible links, $2 \times (n \times k)$ are inserted between all n inputs and their complement values to each of the AND gates.
- Fixed links are $(l \times m)$ also provided between the outputs of the AND gates and the inputs of the OR gates.
- The sum terms constitute a group of m OR gates each of m outputs.

Design procedure of PAL is given below.

11.5.4 | Design Procedure of PAL

- Given inputs and outputs of the logic function or conversion table, etc.
- Obtain truth table with given input and output variables. Let n be number of input binary variables and m is number of output binary variables. Number of inputs, n gives the number of address lines and numbers of outputs, m give the number of data lines.
- Find expression in simplified sum of product form using K-map or tabular method.
- Number of AND gates, k required are $m \times l$, where l is maximum number of product terms in any output function. Number of OR gates required is m. Number of NOT gates required is n. Number of $2 \times (k \times m)$ the number of fusible links before the input of AND gates. Number of $k \times m$ is the number of BJT or MOSFET are required to create fixed links giving stored 0 or 1 value between outputs of AND gates and inputs of OR gates.
- Draw the circuit diagram. AND gates are used for Minterms and OR gates are used to get the SOP function. Create links for existing minterms in the corresponding output function.

Example 11.8 explains the design procedure.

EXAMPLE 11.8 Implement half-adder using PAL

SOLUTION There are two-input binary variables; augend, a_1 and addend, a_0 and 2-output binary variables as sum, d_0 and carry, d_1. Truth table is constructed and is given in Table 11.2. Outputs are given in Eqs. (11.5) and (11.6).

Number of inputs, $n = 2$.

Number of outputs, $m = 2$.

Maximum product terms in any of the output functions, $l = 2$.

(Because, first function, d_0 has two product terms and second output function, d_1 has one product term.)

Number of AND-gates required $k = m \times l = 2 \times 2 = 4$.

Number of fusible links before AND gate $= 2 \times (n \times k) = 2 \times (2 \times 4) = 16$.

Number of OR gates required $= m = 2$.

Number of fixed links between output of AND gate array and input of OR gate array = $l \times m = 2 \times 2 = 4$. Complete diagram is given in Figures 11.29(a) and 11.29(b) alternative way to draw the circuit diagram. AND gates have flexible links and OR gates have fixed links.

Simple input/output PAL AND-OR array is shown in Figure 11.30. It can have 8-inputs and 1-output. Six minterms are considered in the logic function.

Programmable input/output PAL allows the logic designer to feedback output function as input variable to form a new function. Figure 11.31 illustrates a programmable input/output arrangement, where the output is given back into the AND array for the use as an input. Registered outputs PAL has flip-flops that stores output. Several flip-flops form data register. Tristate buffers control the data register. Flip-flop output is made available to form another product term. Figure 11.32 shows the registered PAL. Some available PALs

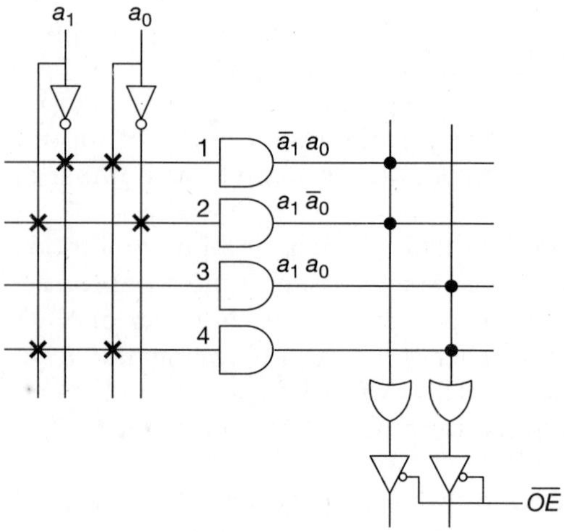

FIGURE 11.29(a) | Two-input and two-output PAL

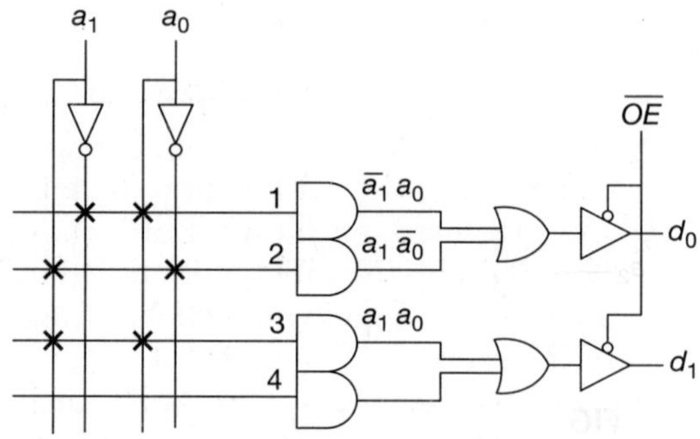

FIGURE 11.29(b) | Two-input and two-output PAL

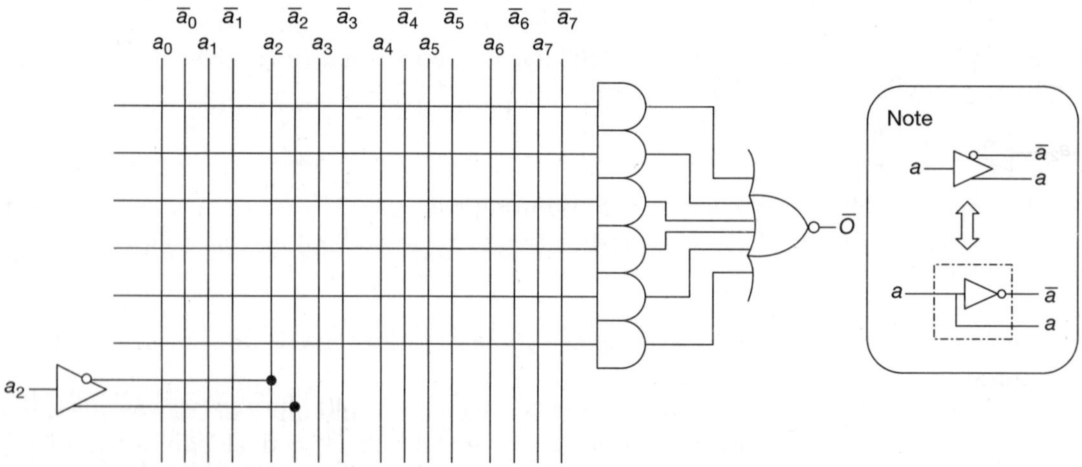

FIGURE 11.30 | Standard input/output programmable array logic (PAL)

are listed in Table 11.3. Prefix PAL identifies device type followed by number that gives number of inputs as given in Figure 11.33. The last number gives the number of outputs. For example, PAL 10H8 is a PAL with 10 input pins and 8 output pins. The *R* in the number PAL16R8 indicates register arrangement. The *X* in PAL 20X8 indicates an AND-OR-XOR logic arrangement.

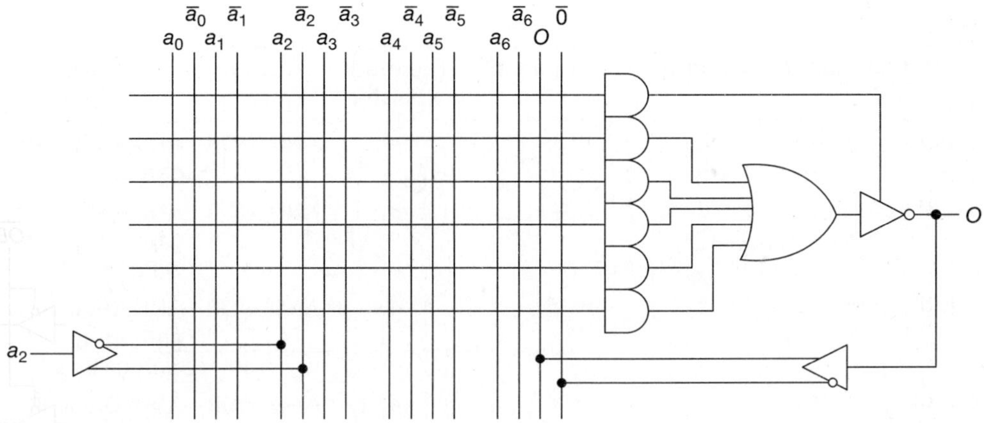

FIGURE 11.31 | Programmable input/output programmable array logic (PAL)

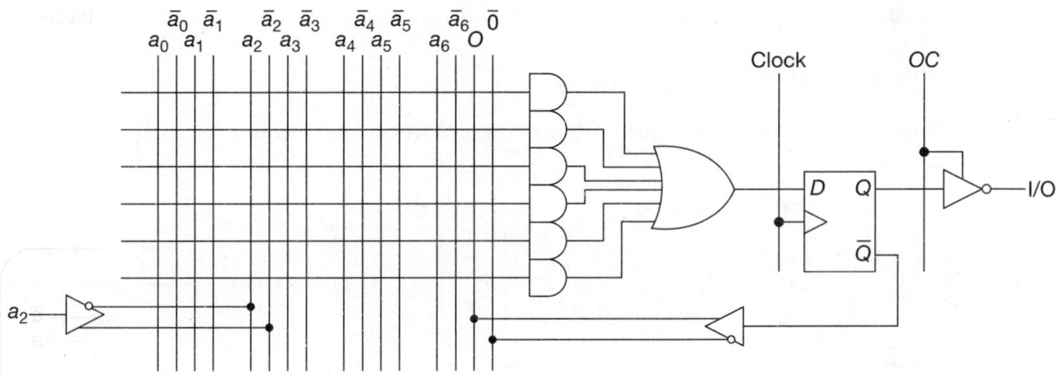

FIGURE 11.32 | Registered output programmable array logic (PAL)

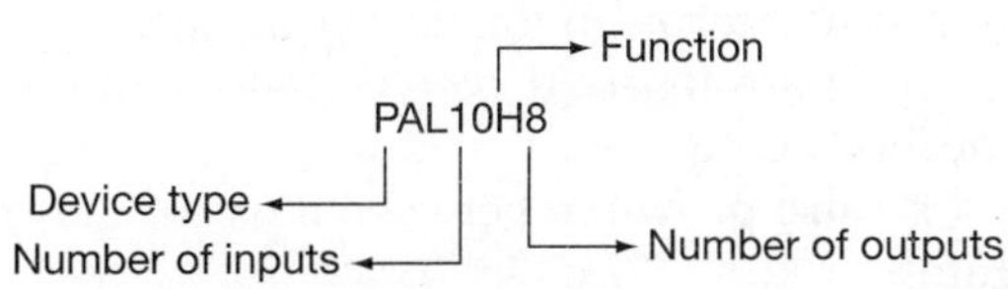

FIGURE 11.33 | Specification of PAL

TABLE 11.3 | List of programmable array logic

Specification	Inputs	Outputs	Programmable I/O	Feedback Register	Output Polarity	Functions
PAL10H8	10	8	—	—	AND–OR	AND–OR gate array
PAL10C8	10	8	—	—	Both	AND–OR gate array
PAL12L6	12	6	—	—	AND–NOR	AND–OR Invert gate array

(Continued)

TABLE 11.3 | (Continued)

Specification	Inputs	Outputs	Programmable I/O	Feedback Register	Output Polarity	Functions
PAL16L8	16	2	6	—	AND–NOR	AND–OR Invert gate array
PAL16R8	16	8	—	8	AND–NOR	AND–OR Invert gate array w/Register's
PAL16R4	16	4	4	4	AND–NOR	AND–OR Invert gate array w/Register's
PAL16X4	16	4	4	4	AND–NOR	AND–OR–XOR Invert gate array w/Register's

11.5.5 | Programmable Logic Array

PLA is shown in Figure 11.34 explains the arrangement of AND gate array, OR gate array and fusible links.

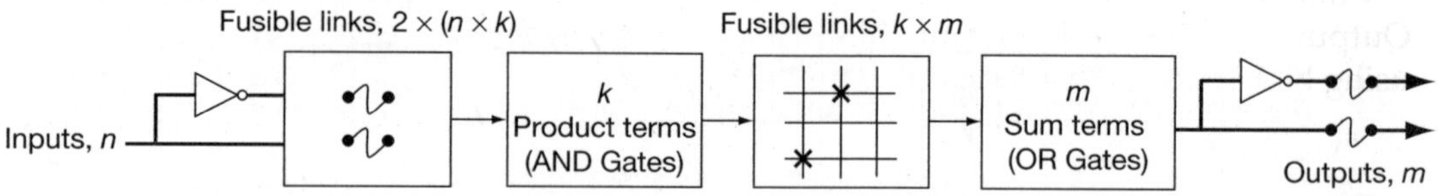

FIGURE 11.34 | Programmable logic array (PLA)

- The product terms constitute a group of $k \leq 2^n$ AND gates each of n inputs, where k are the number of product terms involved in outputs.
- Fusible links, $2 \times (n \times k)$ are inserted between all n inputs and their complement values to each of the AND gates.
- Fusible links are $(k \times m)$ also provided between the outputs of the AND gates and the inputs of the OR gates.
- The sum terms constitute a group of m OR gates each of m outputs.

11.5.6 | Design Procedure of PLA

Design procedure of PLA is given below.

- Inputs and outputs of the logic function or conversion table, etc. is given.
- Obtain truth table with given input and output variables. Let n be number of input binary variables and m is number of output binary variables. Number of inputs, n gives the number of address lines and numbers of outputs, m give the number of data lines.
- Find expression in simplified SOP form using K-map or tabular method.

- Number of AND gates, k required are number of product terms in all output functions. Number of OR gates required are m. Number of NOT gates required is n. Number of $2 \times (k \times m)$ the number of fusible links before the input of AND gates. Number of $k \times m$ is the number of BJT or MOSFETs are required to create fusible links giving stored 0 or 1 value between outputs of AND gates and inputs of OR gates.
- Draw the circuit diagram. AND gates are used for Minterms and OR gates are used to get the SOP function. Create links for existing minterms in the corresponding output function.

Example 11.9 explains the design procedure.

EXAMPLE 11.9 Implement half-adder using PLA.

SOLUTION There are two-input binary variables; augend, a_1 and addend, a_0 and 2-output binary variables as sum, d_0 and carry, d_1. Truth table is constructed and is given in Table 11.2. Outputs are given in Eqs. (11.3) and (11.4).

Number of inputs, $n = 2$.
Number of outputs, $m = 2$.
First function, d_0 has two product terms and second output function, d_1 has one product term.
Number of AND gates required $= 2 + 1 = 3$
Number of OR gates required $= m = 2$.
Complete diagram is given in Figure 11.35. AND and OR gates have flexible links. Output function can be obtained in primed (inverted) or unprimed (non-inverted) form using NOT gate or EX-OR gates.

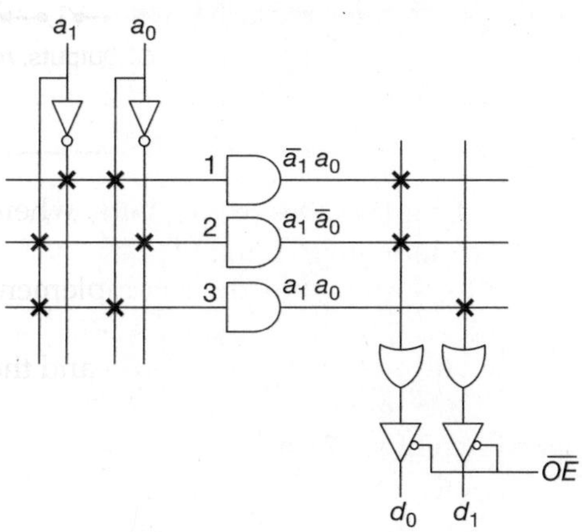

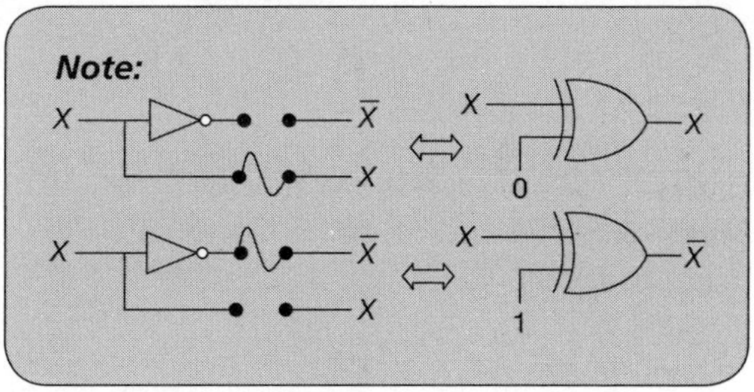

FIGURE 11.35 | Two inputs and two output PLA

EXAMPLE 11.10 Implement half-subtractor using ROM

SOLUTION There are two-input binary variables: minuend, a_1 and subtrahend, a_0 and two-output binary variables as difference, d_0 and borrow, d_1. Truth table is constructed and is given in Table 11.4.

TABLE 11.4 | Truth table for half subtractor

Minterm	Inputs		Outputs	
	Minuend a_1	Subtrahend a_0	Sum d_0	Carry d_1
m_0	0	0	0	0
m_1	0	1	1	1
m_2	1	0	1	0
m_3	1	1	0	0

Boolean equations for two-outputs in standard minterm form are given below:

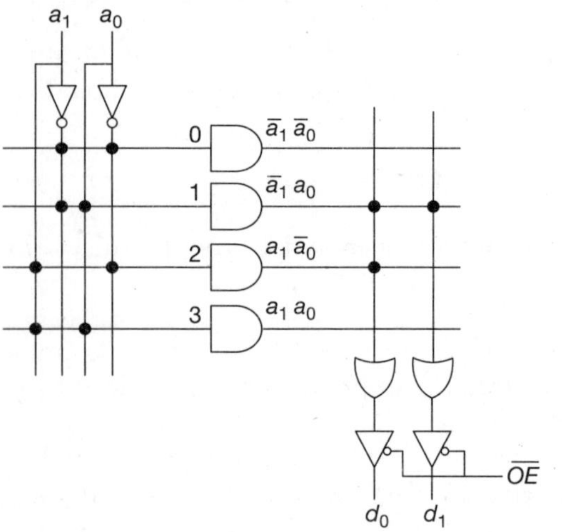

FIGURE 11.36(a) | 4 × 2 bit ROM

$$d_0(a_1, a_0) = \sum m(1,2) \quad (11.5)$$

$$d_1(a_1, a_0) = \sum m(1) \quad (11.6)$$

$$d_0(a_1, a_0) = \bar{a}_1 a_0 + a_1 \bar{a}_0 \quad (11.7)$$

$$d_1(a_1, a_0) = \bar{a}_1 a_0 \quad (11.8)$$

For ROM design there is need of two inputs and two outputs.

Number of inputs, $n = 2$.

Number of outputs, $m = 2$.

Number of AND gates required $= 2^n = 2^2 = 4$.

Number of fixed links as inputs of AND gate $= 2 \times (n \times 2^n) = 2 \times 2 \times 2^2 = 16$.

Number of OR gates required $= m = 2$.

Number of fixed links between outputs of AND gates and inputs of OR gates $= 2^n \times m = 2^2 \times 2 = 8$.

Complete diagram is given in Figure 11.36(a). AND gates and OR gates are having fixed contacts. Figure 11.36(b) gives the diagram using decoder. Block diagram of ROM is given in Figure 11.36(c).

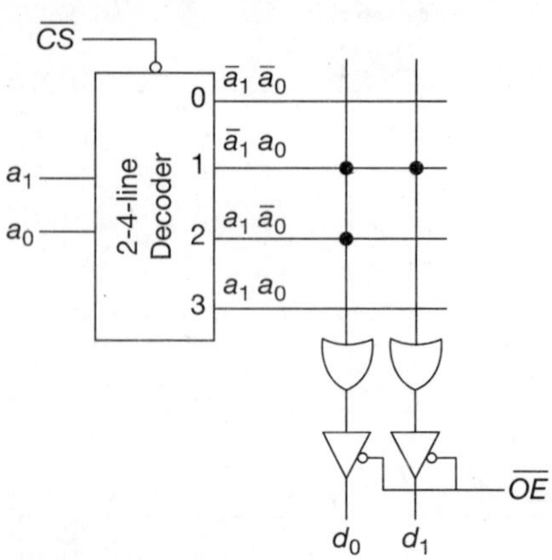

FIGURE 11.36(b) | 4 × 2 bit ROM

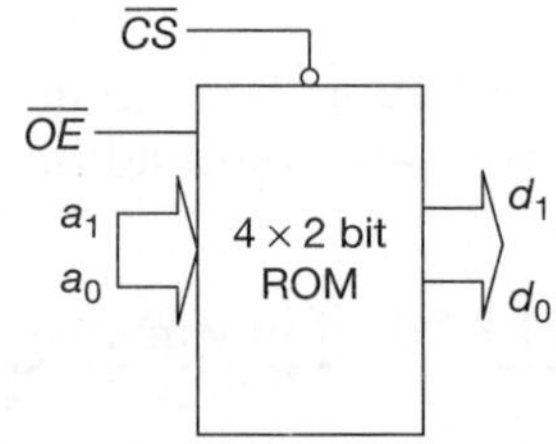

FIGURE 11.36(c) | 4 × 2 bit ROM

EXAMPLE 11.11 Implement half-subtractor using PAL.

SOLUTION There are two-input binary variables; minuend, a_1 and subtrahend, a_0 and 2-output binary variables as difference, d_0 and borrow, d_1. Truth table is constructed and is given in Table 11.4. Outputs are given in Eqs. (11.7) and (11.8).

Number of inputs, $n = 2$.

Number of outputs, $m = 2$.

Maximum product term in the output function, $l = 2$. (Because, first function, d_0 has two product terms and second output function, d_1 has one product term.)

Number of AND gates required $k = m \times l = 2 \times 2 = 4$.

Number of fusible links before AND gate $= 2 \times (n \times k) = 2 \times (2 \times 4) = 16$.

Number of OR gates required $= m = 2$

Number of fixed links between output of AND gate array and input of OR gate array $= l \times m = 2 \times 2 = 4$.

Complete diagram is given in Figure 11.37(a). AND gates have flexible links and OR gates have fixed links. Figure 11.37(b) gives the diagram using decoder. PAL can be designed by feeding inputs considered as state variable where output is used as input for another function. Here, output d_1 can be used in the function d_0. Figure 11.37(c) gives the diagram of PAL using feedback.

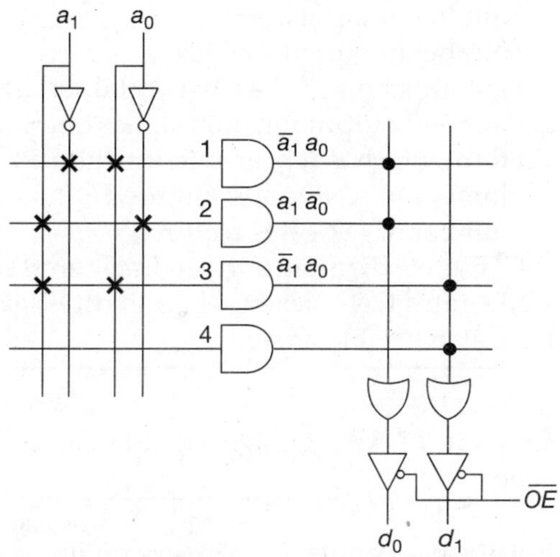

FIGURE 11.37(a) | Two-input and two-output PAL

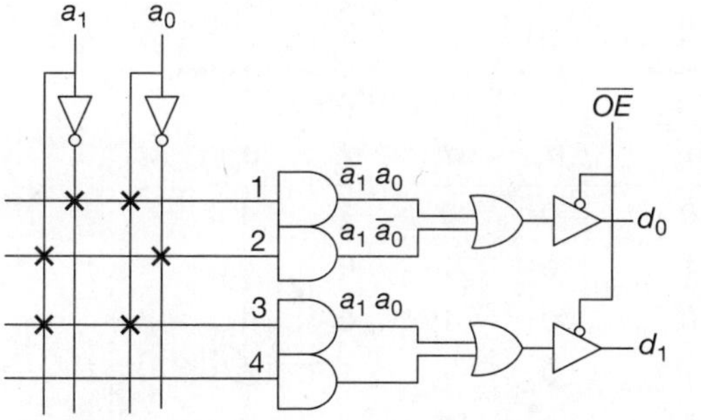

FIGURE 11.37(b) | Two-input and two-output PAL

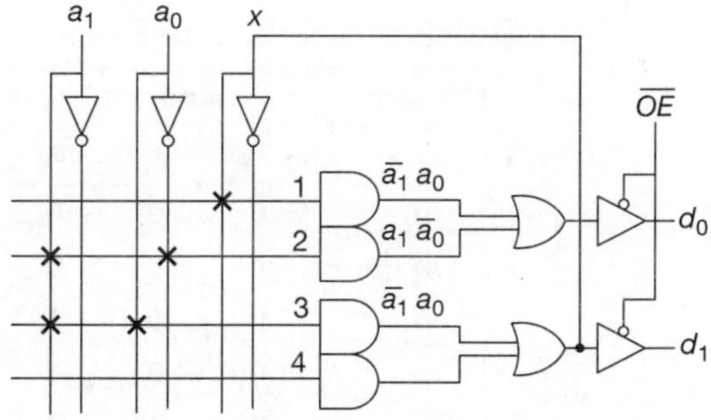

FIGURE 11.37(c) | Two-input and two-output PAL

EXAMPLE 11.12 Implement half-subtractor using PLA.

SOLUTION There are two-input binary variables; minuend, a_1 and subtrahend, a_0 and 2-output binary variables as difference, d_0 and borrow, d_1. Truth table is constructed and is given in Table 11.4. Outputs are given in Eqs. (11.7) and (11.8).

Number of inputs, $n = 2$.
Number of outputs, $m = 2$.
First function, d_0 has two product terms and second output function, d_1 has one product term which is repeated in function d_0.
Number of AND gates required = 2.
Number of OR gates required = m = 2.
Complete diagram is given in Figure 11.38. AND gates have flexible links and OR gates have also flexible links.

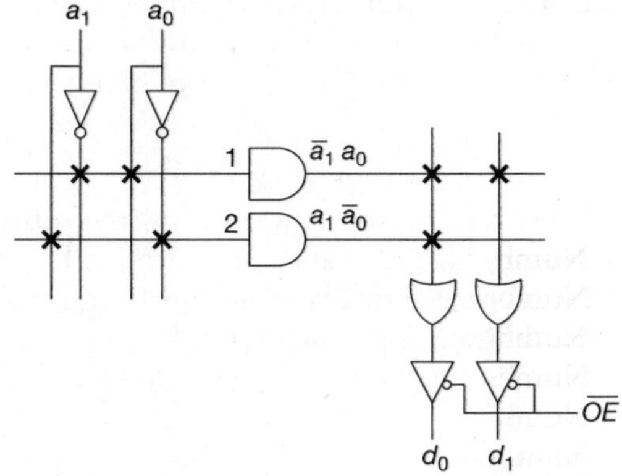

FIGURE 11.38 | Two-input and two-output PLA

EXAMPLE 11.13 Design a ROM that coverts binary-coded decimal (BCD) code to Excess-3 code.

SOLUTION **Inputs:** Four binary digits $a_3a_2a_1a_0$ represent gray number. a_3 gives most significant bit (MSB) and a_0 gives least significant bit (LSB). It is used as address lines of ROM.

Outputs: Four binary digits $d_3d_2d_1d_0$ represent binary number. d_3 gives MSB and d_0 is considered as LSB. It is data contents stored at a particular address. Truth table: truth table is given in Table 11.5. BCD code is used for numeric digits. Four-bit binary values form more values can be considered as don't care conditions.

TABLE 11.5 | BCD to Excess-3 code conversion

Minterm	Address (Input)				Data (Output)			
	a_3	a_2	a_1	a_0	d_3	d_2	d_1	d_0
m_0	0	0	0	0	0	0	1	1
m_1	0	0	0	1	0	1	0	0
m_2	0	0	1	0	0	1	0	1
m_3	0	0	1	1	0	1	1	0
m_4	0	1	0	0	0	1	1	1
m_5	0	1	0	1	1	0	0	0
m_6	0	1	1	0	1	0	0	1
m_7	0	1	1	1	1	0	1	0
m_8	1	0	0	0	1	0	1	1
m_9	1	0	0	1	1	1	0	0

Boolean equations in standard minterms are given below for output functions

$$d_0(a_3,a_2,a_1,a_0)=\sum m(0,2,4,6,8) \tag{11.9}$$

$$d_1(a_3,a_2,a_1,a_0)=\sum m(0,3,4,7,8) \tag{11.10}$$

$$d_2(a_3,a_2,a_1,a_0)=\sum m(1,2,3,4,9) \tag{11.11}$$

$$d_3(a_3,a_2,a_1,a_0)=\sum m(5,6,7,8,9) \tag{11.12}$$

Number of inputs, $n=4$.
Number of outputs, $m=4$.
Number of AND gates required = 10 ($< 2^4 = 16$).
Number of fixed links as inputs of AND gate = $2\times4\times10=80$.
Number of fixed links between outputs of AND gates and inputs of OR gates = $10\times4=40$.
Number of OR gates required = $m=4$.
Complete diagram is given in Figure 11.39. AND gates and OR gates have fixed links.

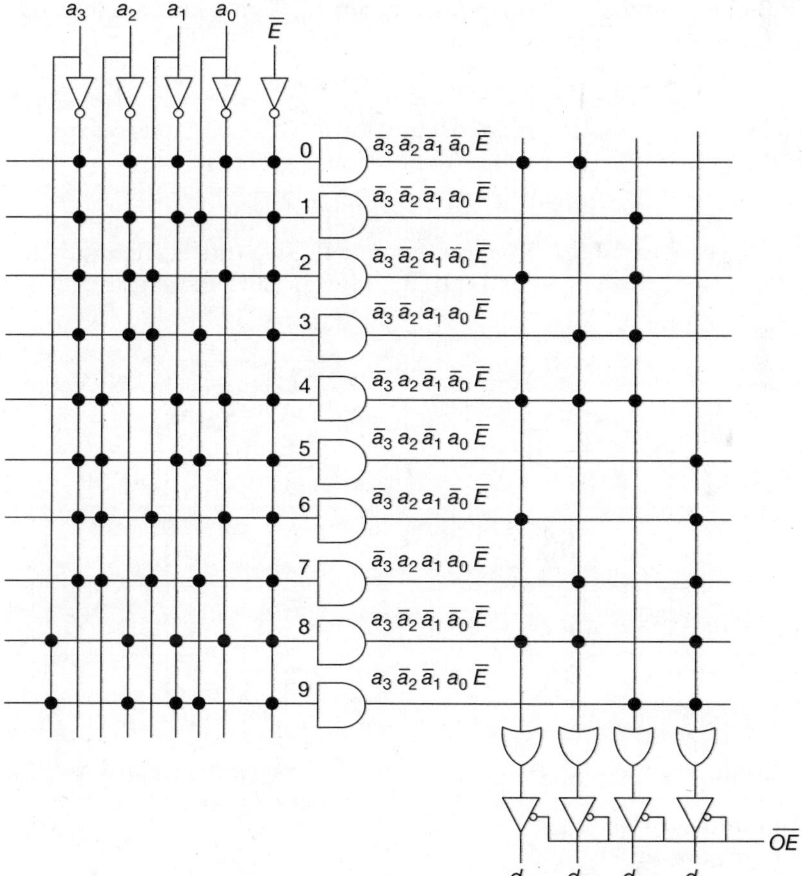

FIGURE 11.39 | 10 × 4 bit ROM

EXAMPLE 11.14 Design a PAL circuit that converts BCD code to Excess-3 code.

SOLUTION Boolean equations given by Eqs. (11.9), (11.10), (11.11) and (11.12) can be simplified using K-map. Six don't care minterms are considered those do not give valid BCD digits. The standard form of outputs for simplification is given below.

$$d_0(a_3,a_2,a_1,a_0)=\sum m(0,2,4,6,8)+\sum d(10,11,12,13,14,15) \tag{11.13}$$

$$d_1(a_3,a_2,a_1,a_0)=\sum m(0,3,4,7,8)+\sum d(10,11,12,13,14,15) \tag{11.14}$$

$$d_2(a_3,a_2,a_1,a_0)=\sum m(1,2,3,4,9)+\sum d(10,11,12,13,14,15) \tag{11.15}$$

$$d_3(a_3,a_2,a_1,a_0)=\sum m(5,6,7,8,9)+\sum d(10,11,12,13,14,15) \tag{11.16}$$

K-maps are used to simplify the logic functions. Figure 11.40(a) gives 4-variable K-map for function y_0. One octet is formed; {0, 4, 8, 12, 2, 6, 10, 14}. The logic expression is given below.

$$d_0(a_3,a_2,a_1,a_0)=\bar{a}_0 \tag{11.17}$$

Figure 11.40(b) gives 4-variable K-map for function y_1. Two quads are formed; {0, 4, 8, 12}, {3, 7, 11,15}. The logic expression is given below.

$$d_1(a_3,a_2,a_1,a_0)=\bar{a}_1\bar{a}_0+a_1a_0 \tag{11.18}$$

Figure 11.40(c) gives 4-variable K-map for function y_2. One pair and two quads are formed; {4, 12}, {1, 3, 9, 11},{2, 3, 10, 11}. The logic expression is given below.

$$d_2(a_3,a_2,a_1,a_0)=\bar{a}_2a_0+\bar{a}_2a_1+a_2\bar{a}_1\bar{a}_0 \tag{11.19}$$

Figure 11.40(d) gives 4-variable K-map for function y_3. Two quads and one octet are formed; {5, 7, 13, 15}, {6, 7, 14, 15} and {8, 9, 10, 11, 12, 13, 14, 15}. The logic expression is given below.

$$d_3(a_3,a_2,a_1,a_0)=a_3+a_2a_0+a_2a_1 \tag{11.20}$$

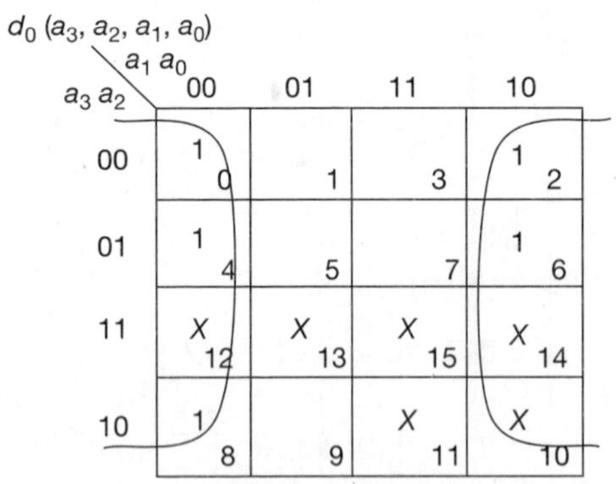

FIGURE 11.40(a) | K-map of d_0 bit

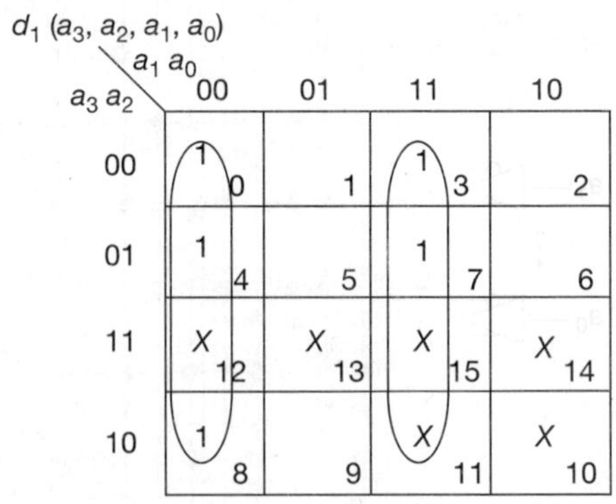

FIGURE 11.40(b) | K-map of d_1 bit

Number of inputs, $n = 4$.
Number of outputs, $m = 4$.
First output function, d_0 has one product terms. Second output function, d_1 has two product terms. Third output function, d_2 has three product terms and fourth output function, d_3 has three product terms.

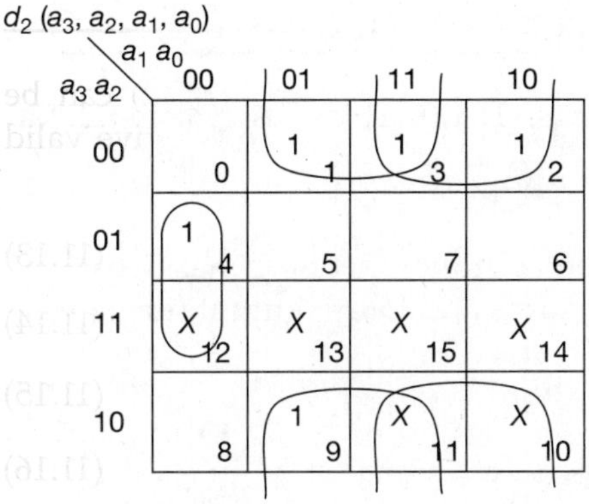

FIGURE 11.40(c) | K-map of d_2 bit

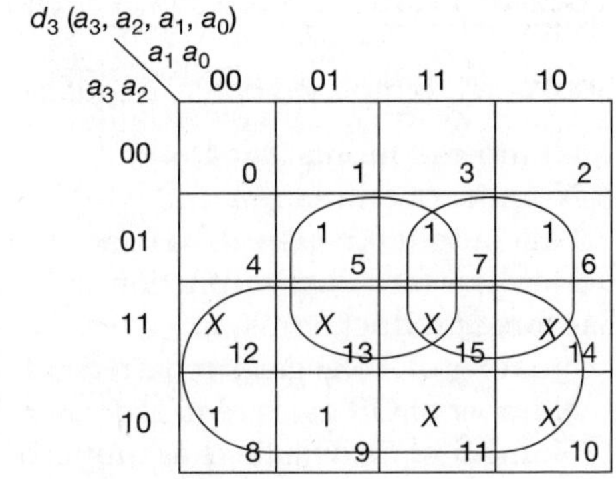

FIGURE 11.40(d) | K-map of d_3 bit

Maximum number of product term in the output functions, $l = 3$.
Number of AND gates required $= m \times l = 4 \times 3 = 12$.
Number of OR gates required $= m = 4$.
Number of flexible links as inputs of AND gate $= 2 \times 4 \times 12 = 96$.
Number of fixed links between outputs of AND gates and inputs of OR gates $= 4 \times 3 = 12$.
Complete diagram is given in Figure 11.41. AND gates have flexible links and OR gates are fixed links.

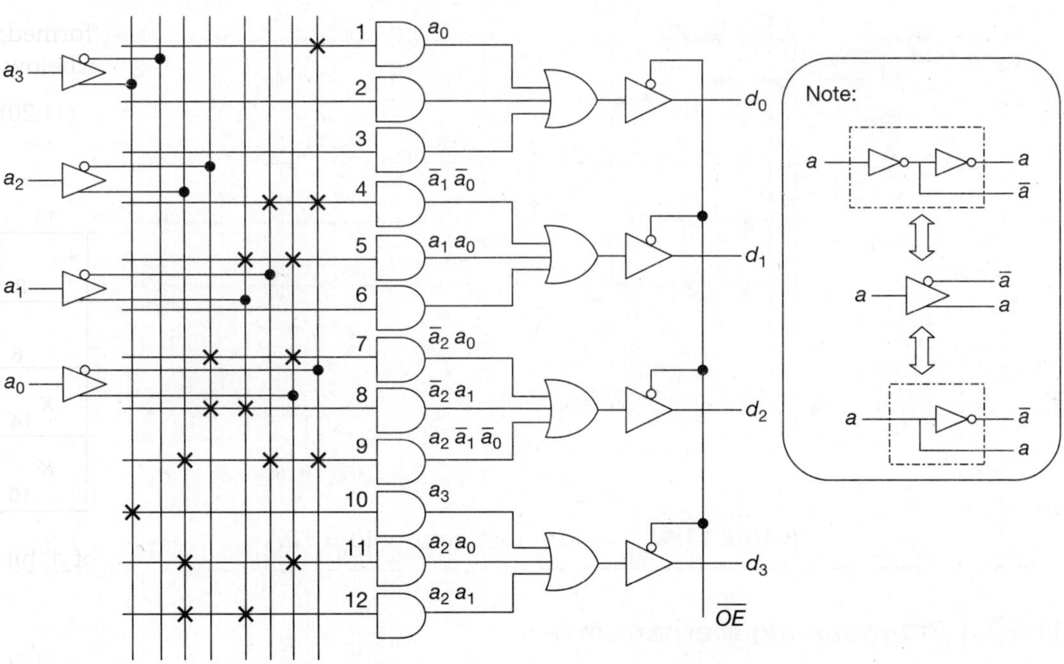

FIGURE 11.41 | Four-input, four-output PAL

Example 11.15 Design a PLA circuit that converts BCD code to Excess-3 code.

Solution Simplified Boolean equations are given by Eqs. (11.17), (11.18), (11.19) and (11.20).

Number of inputs, $n = 4$.

Number of outputs, $m = 4$.

First output function, d_0 has one product terms. Second output function, d_1 has two product terms. Third output function, d_2 has three product terms and fourth output function, d_3 has three product terms.

Number of AND gates required $= 1+2+3+3=9$.

Number of OR gates required $= m = 2$.

Number of flexible links as inputs of AND gate $= 2\times4\times9=72$.

Number of fixed links between outputs of AND gates and inputs of OR gates $= 4\times9 = 36$.

Complete diagram is given in Figure 11.42. AND gates have flexible links and OR gates are also flexible links.

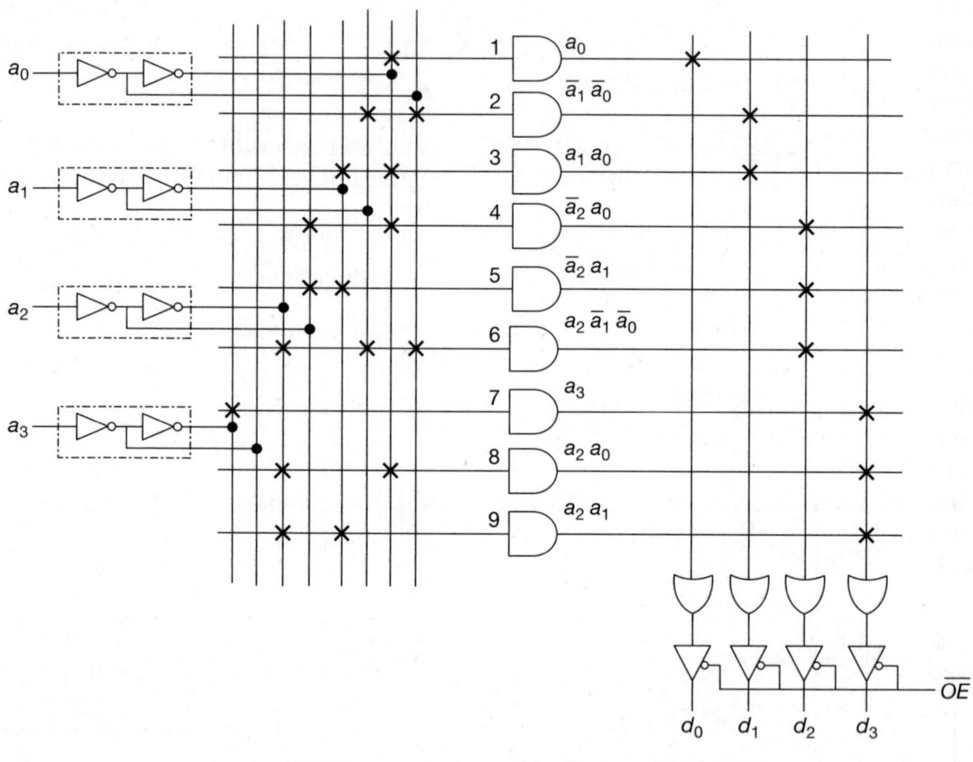

FIGURE 11.42 | Four-input, four-output PLA

11.5.7 | Programming Mechanisms

ICs are fabricated through photo masking, etching, diffusion, etc. A layer of aluminium is deposited on the entire surface of the silicon wafer. The portion of aluminium is etching away to create the desired interconnecting pattern. In the manufacture of mask-programmed read-only memories, the desired row-to-column contacts can be, retained or blown open (etched away), in the final aluminium etching process. The most commonly used programming mechanisms for the bipolar memory devices are: metal link, silicon link and shorted *p-n* junctions or avalanche-induced migration (AIM).

Metal link are made of nichrome material. Fuse is added in the emitter lead of every transistor. During fabrication of memory devices, all the fuse links are left intact and are blown open after packaging. Nichrome is deposited as a very thin layer, which is blown open the connection between the row and column lines. Blown open of fuse is done by flowing through a large current of 20–50 mA for a time, 5 and 200 μs. The programming rate is 5 ms per bit. Power dissipation rate is about 170 μW per bit while operating voltage is 5 V. The nichrome fuse elements are to be left intact for storing 1's and blown open for storing 0's.

Silicon links are made by narrow, notched strips of polycrystalline silicon. It is deposited as a layer (3000 Å) during the manufacturing. Programming of these fuses requires melting of links by passing a sufficient amount of current. The fuse is blown open with a pulse train of successively wider pulses. A current of 20–30 mA is drawn to blow open the fuse. Temperature, 1400°C approximately is attained during the blowing process. The silicon gets oxidized and forms an insulating material.

Shorted p-n junctions or AIM consists of two *p-n*-junctions (diodes) arranged back-to-back as shown in Figure 11.43. One of the *p-n*-junction (diode) is reverse biased and heavy flow of electrons is achieved in reverse order (avalanched) resulting in heat, and voltage migrates aluminium ions and shorts the junction.

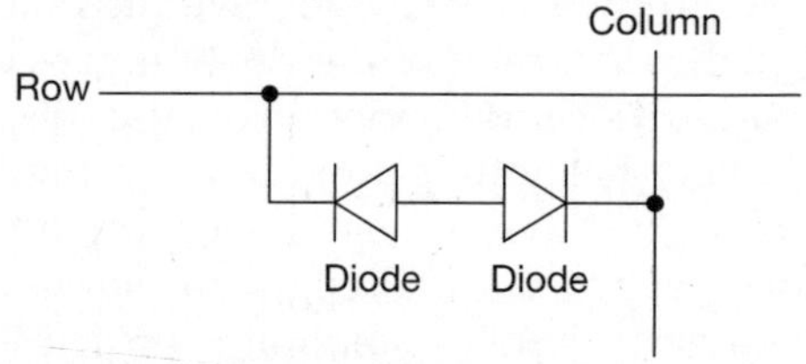

FIGURE 11.43 | Shorted p-n-junction

Floating gate avalanche injection MOS (FAMOS): Using AIM technology, PROM is programmed only once in its life time. It can be used for MOS where resistance and current is required to fuse the links. The cross section of a conventional MOS transistor and a floating gate transistor are shown in Figures 11.44(a) and 11.44(b), respectively. The symbols of a conventional MOS transistor and a floating gate transistor are shown in Figures 11.44(c) and 11.44(d), respectively. The upper gate in Figure 11.44(b) is the control gate and the lower gate, completely isolated within the gate oxide, is the floating gate. C_{FG} and C_{FS} are the capacitances between the floating gate and

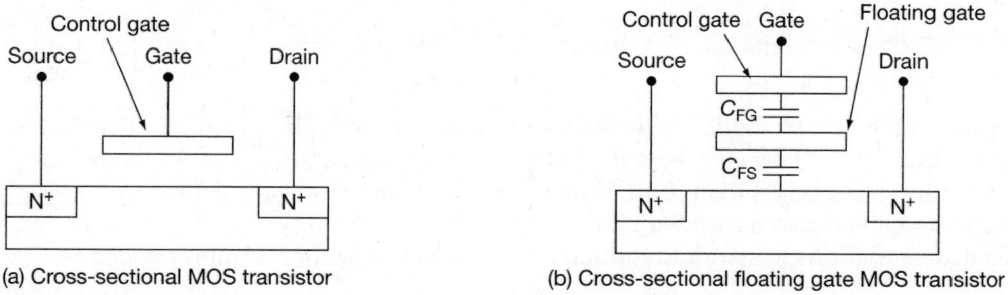

FIGURE 11.44 | (Continued)

(c) Symbol of MOS transistor

(d) Symbol of floating gate MOS transistor

FIGURE 11.44 | MOS and floating gate MOS transistor

the control gate and substrate, respectively. V_G and V_F are the voltages of the control gate and the floating gate, respectively. Q_F is the charge in the floating gate. As electrons have a negative charge, a negative sign is considered.

Floating gate MOS transistors are widely used in digital world to design EPROM and EEPROMSs. To program a floating gate transistor in PROM, a high voltage (16 V for instance) is applied between the drain and source. This causes electron to avalanche through the source towards the drain. As the SiO_2 layer is very thin, electrons travels through this SiO_2 layer to the floating gate, give the floating gate electrode an electrical charge. Hence the threshold voltage of the floating gate transistor is increased. The presence or absence of charge is sensed by measuring the conductance between source and the drain.

There is another programming method which is based on tunnelling effect, and can induce positive or negative charge on the floating gate. When the floating gate transistor is exposed in UV light for some time, the charge on the floating gate disappears. Because of good insulation properties of SiO_2 (silicon oxide), the charge on the floating gate leaks away very slowly, and the threshold voltage is kept high for a long time in room temperature.

Another PROM uses the gate dielectric, such as alumina (Al_2O_3) and silicon nitride for charge storage and provides a reprogramming feature. The device is referred as MAOS. For p-channel device, a positive gate voltage of 50 V amplitude is required for 10–20 μs for programming. Erasing requires a voltage of opposite polarity on the gate.

EXAMPLE 11.16 Design a ROM that compares 2-bit numbers.

SOLUTION

Inputs: Four binary digits $a_3a_2a_1a_0$ represent two 2-bit numbers (a_3a_2) and (a_1a_0). It is used as address lines of ROM.

Outputs: Three-binary digits $d_2d_1d_0$ represent binary number. Binary digit, d_3 sets to 1, if (a_3a_2) is equal to (a_1a_0). Binary digit, d_1 sets to 1, if (a_3a_2) is greater than (a_1a_0). Binary digit, d_0 is sets to 1, if (a_3a_2) is less than (a_1a_0). Data contents are stored at a particular address. Truth table: truth table is given in Table 11.6.

Boolean equations in standard minterms are given below for output functions

$$d_0(a_3,a_2,a_1,a_0)=\sum m(1,2,3,6,7,11) \tag{11.21}$$

$$d_1(a_3,a_2,a_1,a_0)=\sum m(4,8,9,12,13,14) \tag{11.22}$$

$$d_2(a_3,a_2,a_1,a_0)=\sum m(0,5,10,15) \tag{11.23}$$

TABLE 11.6 | Comparison of two-bit binary numbers

Minterm	Address (Input)				Data (Output)		
	a_3	a_2	a_1	a_0	d_2 (=)	d_1 (>)	d_0 (<)
m_0	0	0	0	0	1	0	0
m_1	0	0	0	1	0	0	1
m_2	0	0	1	0	0	0	1
m_3	0	0	1	1	0	0	1
m_4	0	1	0	0	0	1	0
m_5	0	1	0	1	1	0	0
m_6	0	1	1	0	0	0	1
m_7	0	1	1	1	0	0	1
m_8	1	0	0	0	0	1	0
m_9	1	0	0	1	0	1	0
m_{10}	1	0	1	0	1	0	0
m_{11}	1	0	1	1	0	0	1
m_{12}	1	1	0	0	0	1	0
m_{13}	1	1	0	1	0	1	0
m_{14}	1	1	1	0	0	1	0
m_{15}	1	1	1	1	1	0	0

Alternative I:

Number of inputs, $n = 4$.
Number of outputs, $m = 3$.
Number of AND gates required = $2^4 = 16$.
Number of fixed links between outputs of AND gates and inputs of OR gates = $16 \times 4 = 64$.
Number of OR gates required = $m = 3$.
Figure 11.45 illustrates the circuit diagram of 16×3 ROM using decoder 4-to-16 lines.

Alternative II:

Number of inputs, $n = 4$.
Number of outputs, $m = 3$.
Number of fixed links between outputs of AND gates and inputs of OR gates = $16 \times 4 = 64$.
Number of OR gates required = $m = 3$.
Figure 11.46 shows the circuit diagram of 16×3 ROM using row and column decoders 2-to-4 lines.

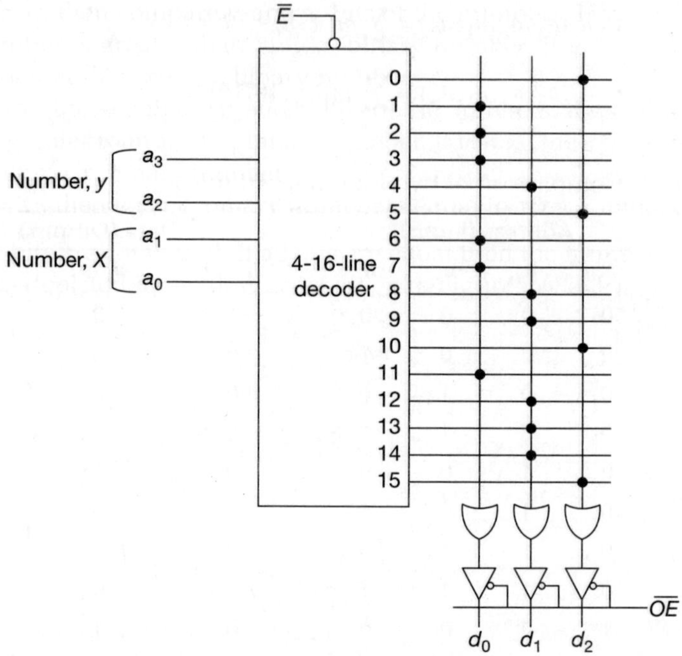

FIGURE 11.45 | 16 × 3 bit ROM

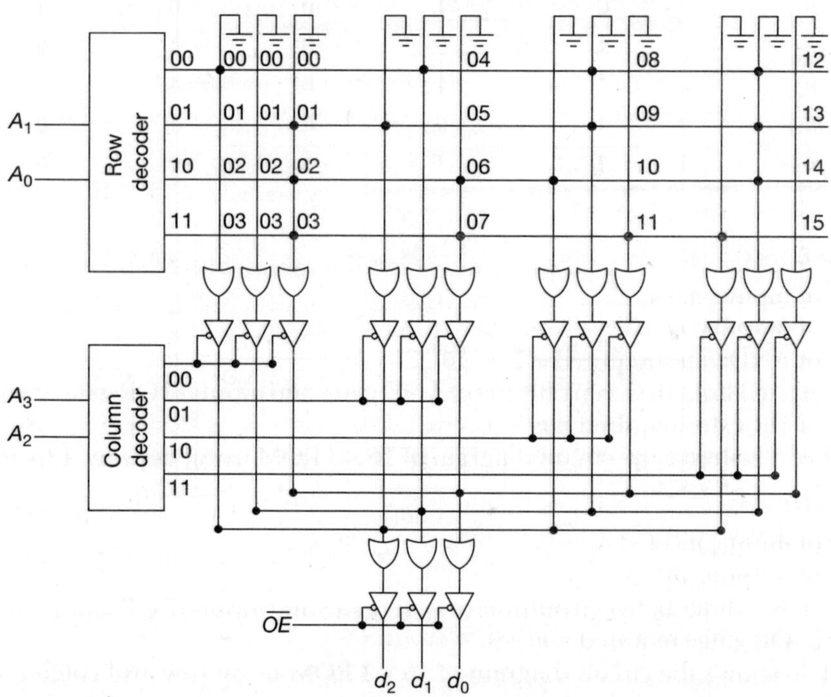

FIGURE 11.46 | 16 × 4 ROM using row and column decoding

11.5.8 | Complex-Programmable Logic Device

A CPLD contains a bunch of PLD blocks whose inputs and outputs are connected together by a global interconnection matrix as shown in Figure 11.47. A CPLD has two levels of programmability: each PLD block can be programmed, and then the interconnections between the PLDs can be programmed.

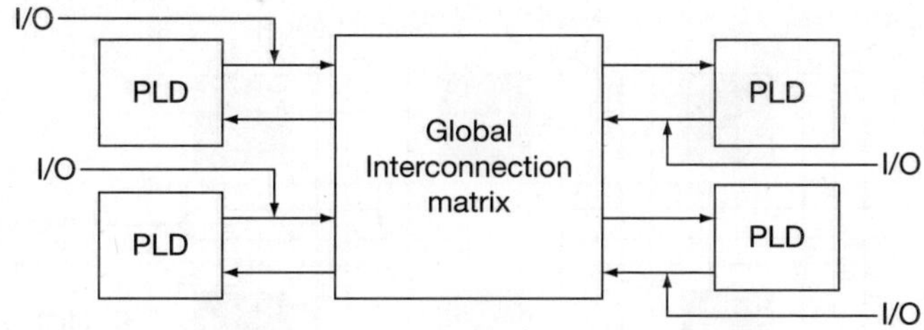

FIGURE 11.47 | Complex-programmable logic device (CPLD)

The expansion of PLD using multiple simple-programmable logic device (SPLD) chips has the disadvantages such as personal computer (PC) board area requirement increases with the number of chips. Connecting wires will result in adverse capacitive effects. Power requirement increases with the number of chips. The system cost increases.

Another method of increasing the input/output and product terms can be designing of PLDs using the architecture SPLDs. This approach was discarded by the designers because of the problems associated with this approach like increase in capacitive effects, increase in leakage currents, and decrease in speed and cost effectiveness. CPLD was evolved. A CPLD is a collection of individual PLDs on a single chip and programmable interconnection structure. By using programming methods, the resource available in various PLDs can be shared in different ways to design complex logic function.

11.5.9 | Field-Programmable Gate Array

FPGA is a high capacity PLD. The gate array of FPGA has the ability to be programmed for a function by the user instead of the manufacturer of device. Moreover, series of columns and rows of gates can be configured (programmed) by the end user. FPGA consists of three configurable (programmable) logic modules (LMs): *configurable logic blocks* (CLBs), input and output blocks and switching matrix for interconnections. The CLB consists of a combinational logic array, data multiplexer (MUX) and flip-flops. The combinational array function is performed by *look-up table* (LUT). The FPGA consists of three main structures:

- Programmable logic structure,
- Programmable routing structure, and
- Programmable input/output (I/O).

11.5.9.1 | Programmable Logic Structure

The programmable logic structure of FPGA consists of a two-dimensional array of CLBs. Each CLB can be configured (programmed) to implement any_Boolean function of its input variables. Typically, CLB has four to six input variables. Functions of larger number of variables are implemented using more than one CLB. In addition, each CLB typically contains one or two flip-flops to allow implementation of sequential logic. Large designs are partitioned and mapped to a number of CLBs with each CLB configured (programmed) to perform a particular function. These CLBs are then connected together to fully implement the target design. Connecting the CLBs is done using the FPGA-programmable routing structure. Figure 11.48 depicts the programmable structure of FPGA.

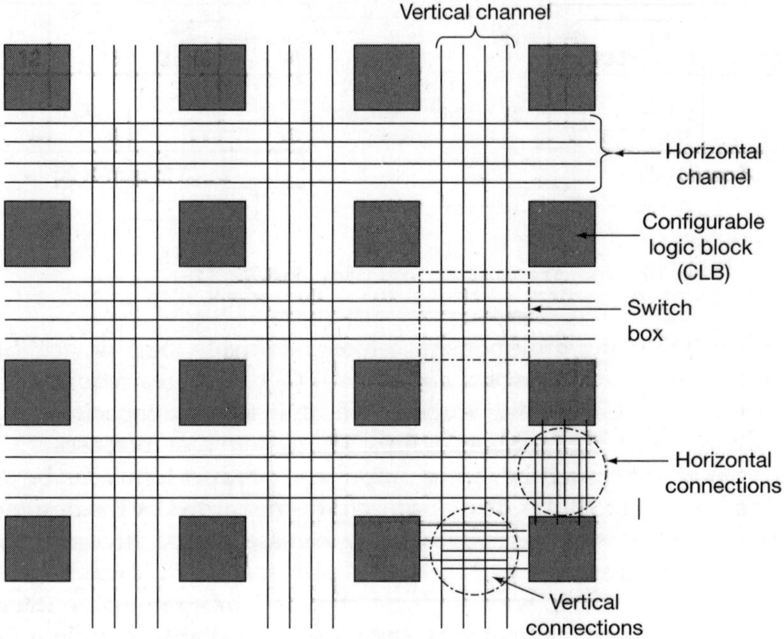

FIGURE 11.48 | Programmable logic structure of field-programmable logic array (FPGA)

11.5.9.2 | Programmable Routing Structure

To allow for flexible interconnection of CLB, FPGA has three programmable routing resources:

Vertical and horizontal routing channels which consist of different length wires that can be connected together if needed. These channels run vertically and horizontally between columns and rows of CLBs as shown in the Figure 11.49.

Connection boxes, which are set of programmable links, can connect input and output pins of the CLBs to wires of the vertical or the horizontal routing channels.

Switch boxes are located at the intersection of the vertical and horizontal channels. These are a set of programmable links that can connect wire segments in the horizontal and vertical channels.

11.5.9.3 | Programmable Input/Output

These are mainly buffers that can be configured either as input buffers, output buffers or input/output buffers as shown in Figure 11.50. These allow the pins of the FPGA chip to function either as input pins, output pins or input/output pins.

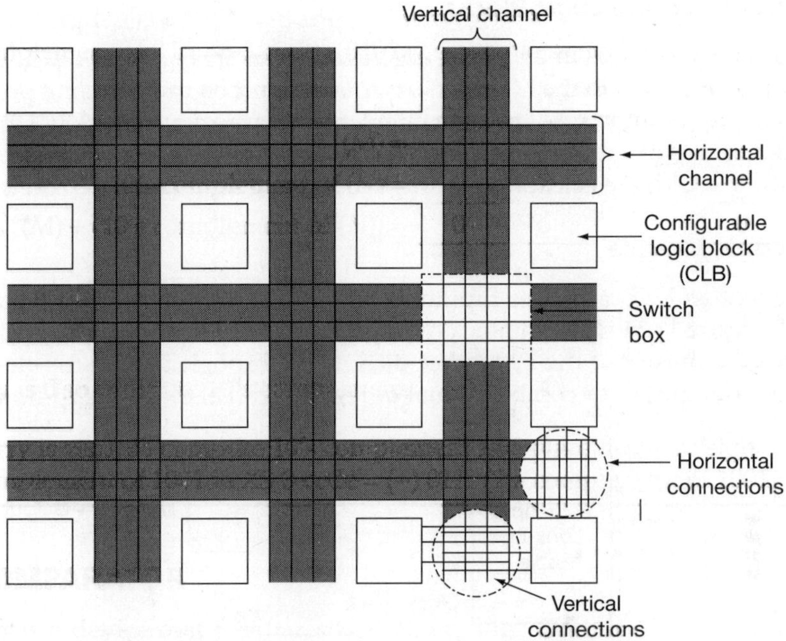

FIGURE 11.49 | Programmable routing structure of field-programmable logic array (FPGA)

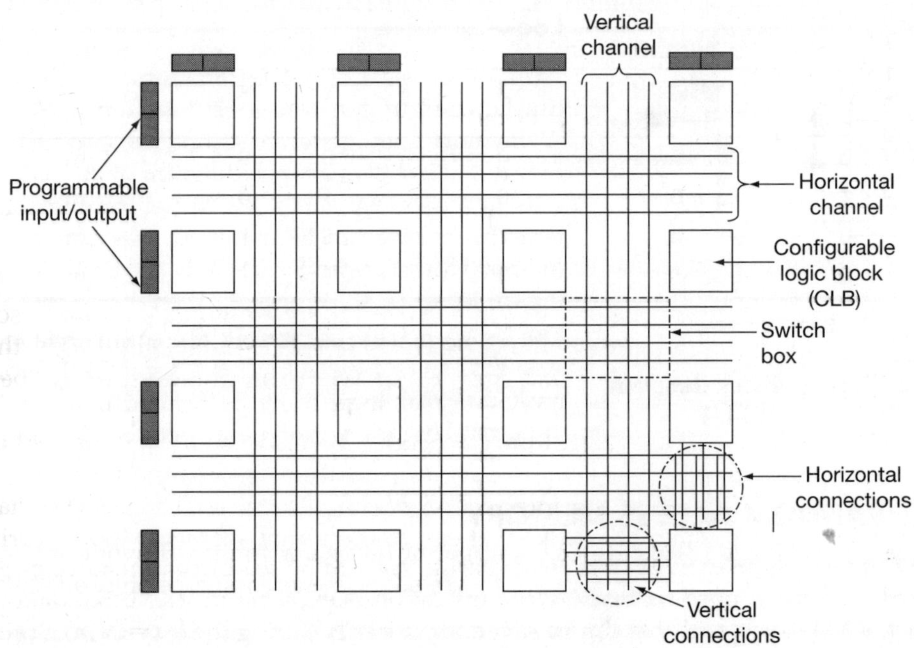

FIGURE 11.50 | Programmable input/output structure of field-programmable logic array (FPGA)

11.5.9.4 | Configurable Logic Blocks

There are a number of CLBs in an FPGA organized as an array of rows and columns. The logic blocks are connected to the I/O blocks through common row/column programmable interconnects. The common row/column interconnects are known as global interconnects. A logic block consists of a number of LMs. The LMs are the basic logic elements in a FPGA. The LMs with in a CLB are connected through local programmable interconnects.

11.5.9.5 | Logic Module

A LM consists of an LUT, a D-type flip-flop and a MUX. Most of the FPGAs are based on 4-input LUT. Figure 11.51 shows a block diagram of a LM with 4-input LUT. Output of the LUT becomes the output of the LM either directly or through D-type flip-flop. Thus, the output can be configured for combinational or registered (i.e. through flip-flop).

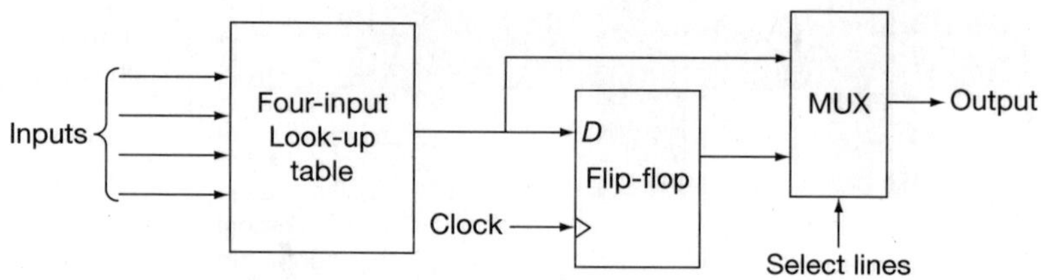

FIGURE 11.51 | Logic module: block diagram

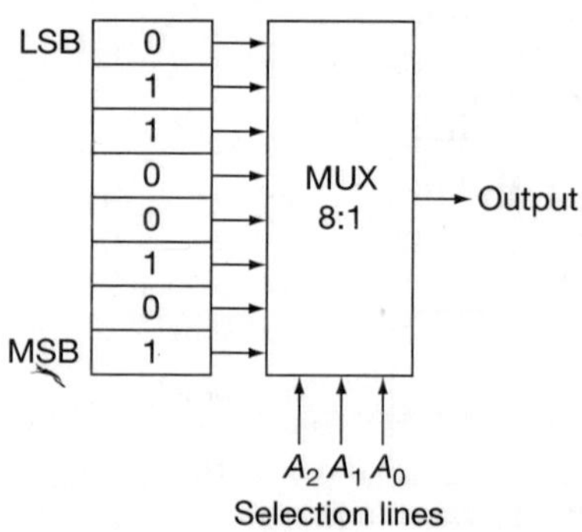

FIGURE 11.52 | Block diagram of LUT

Look-Up Table

An LUT consists of a programmable memory and it can be used to generate logic function in SOP form. A memory can generate canonical product terms. Figure 11.52 shows a block diagram of an LUT. It consists of a memory and a MUX. Let the memory contents be as given in Figure 11.52. Since, it is an 8-bit memory therefore an 8:1 MUX is required. If the 3-bit logical input is A_2, A_1 and A_0, then $Y = \sum m(1,2,5,7)$. The look-up table in most of the commercially available FPGAs is 4-input circuit. Larger LUT allows for more complex logic to be performed per logic block, thus reducing the wiring delay between blocks.

11.6 | RANDOM ACCESS MEMORY

A RAM stores information or data temporarily that can be read from memory as well as modified or written onto memory. Write operation is possible in RAM. So, data can be changed. RAM stores data that can be saved temporarily during the execution of program. RAM cannot retain stored data when power is off. So, RAM is volatile memory. However, there is facility to make the memory as volatile known as NV-RAM (non-volatile RAM). Figure 11.53 shows the classification of RAM.

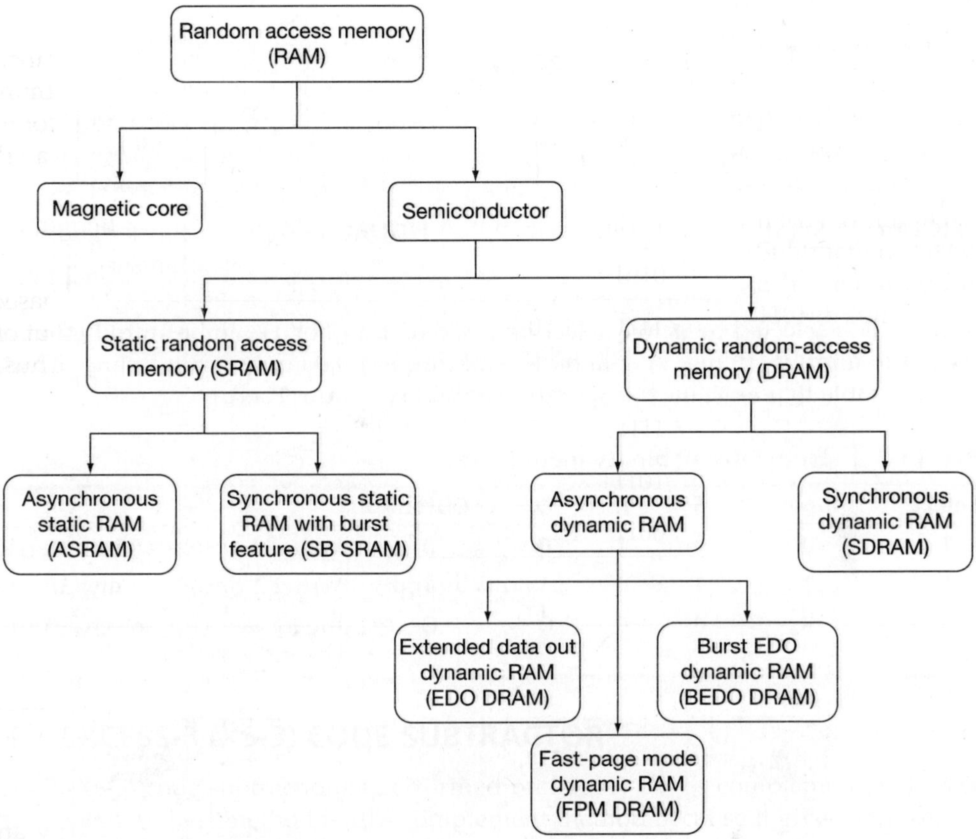

FIGURE 11.53 | RAM family

Semiconductor or magnetic core technology is used for RAM. Semiconductor technology uses BJT or MOSFET. RAM is classified in two major categories known as SRAM and DRAM. SRAM uses latches. Data is stored till dc power is applied. DRAM uses capacitors to store the information. Data cannot be stored for long period without recharging the capacitors due to leakage of charge stored in capacitor being datum. The process is known as refreshing the capacitors. Data can be read faster from SRAM than from DRAM. As compared to SRAM, DRAM can store more data for the given physical size and cost. The construction of DRAM cell is simple. More cells can be filled or arranged into a given chip area.

SRAM can be asynchronous or synchronous with burst feature. Different types of DRAM are fast-page mode DRAM (FPM DRAM), the extended data-out DRAM (EDO DRAM), the burst extended data-out DRAM (DEDO DRAM) and synchronous DRAM (SDRAM).

11.6.1 | Static Random Access Memory

SRAM consists of latch as memory cell. Static memory cell remembers 0 or 1 as long as the dc power is applied.

The stored data bit is lost as and when the dc (direct current) power is removed. Figure 11.54 shows a basic latch memory cell called as binary memory cell. The binary

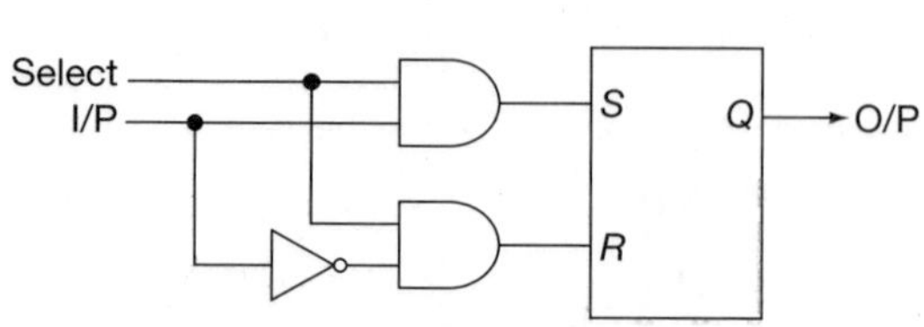

FIGURE 11.54(a) | Binary cell

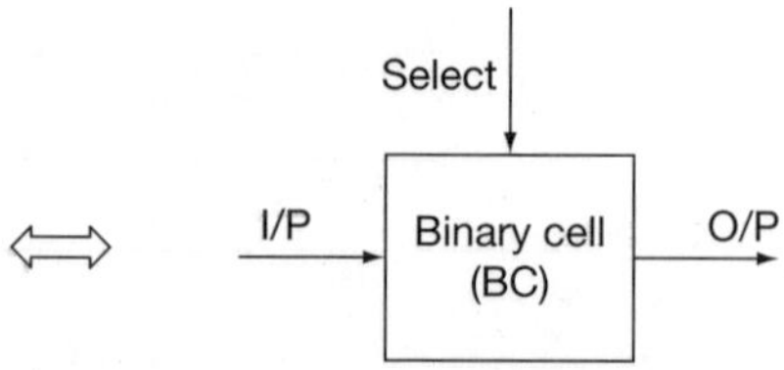

FIGURE 11.54(b) | Block diagram Binary cell

memory cell is selected by active select line. A data bit (0 or 1) can be stored into cell by feeding into input (I/P) line. A data bit is read through data output (O/P) line. Table 11.7 gives truth table that explains the operation of binary memory cell.

TABLE 11.7 | Truth table of binary memory cell

Select	Input	*S*	*R*	*Q*	Output	
1	0	0	1	0	0	Writes 0 or stores 0 into SR latch
1	1	1	0	1	1	Writes 1 or stores 1 into SR latch
0	X	0	0	Q	Q	Binary cell is not selected, no change in state

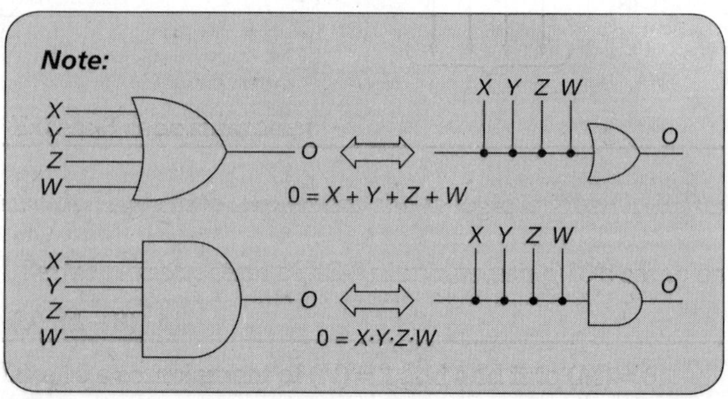

The memory cells in a SRAM are organized in rows and columns, as shown in Figure 11.55 for the case of a 4 × 4 array. All the cells in a row are selected by the same row select line. Each set of Data-in and data-out lines go to each cell in a given column and are connected to a single data line that serves as both an input and output (Data I/O) through the data input and data output buffers. A decoder of 2-to-4 lines is used to provide row select lines. Decoder enable signal can be used as chip select line of memory. Tristate buffers are used to control the flow of input/output data.

Figure 11.56 gives the simplified structure of RAM. It has MAR that is a decoder. MAR gives the number of row or memory locations from where data is available. Input of MAR is known as address of memory. MBR holds the data. The data is written into memory and data is input by MBR. The data are read from memory and is available in MBR. Read and write, R/$\overline{W}$ signal controls the direction of the data to be written into or read from.

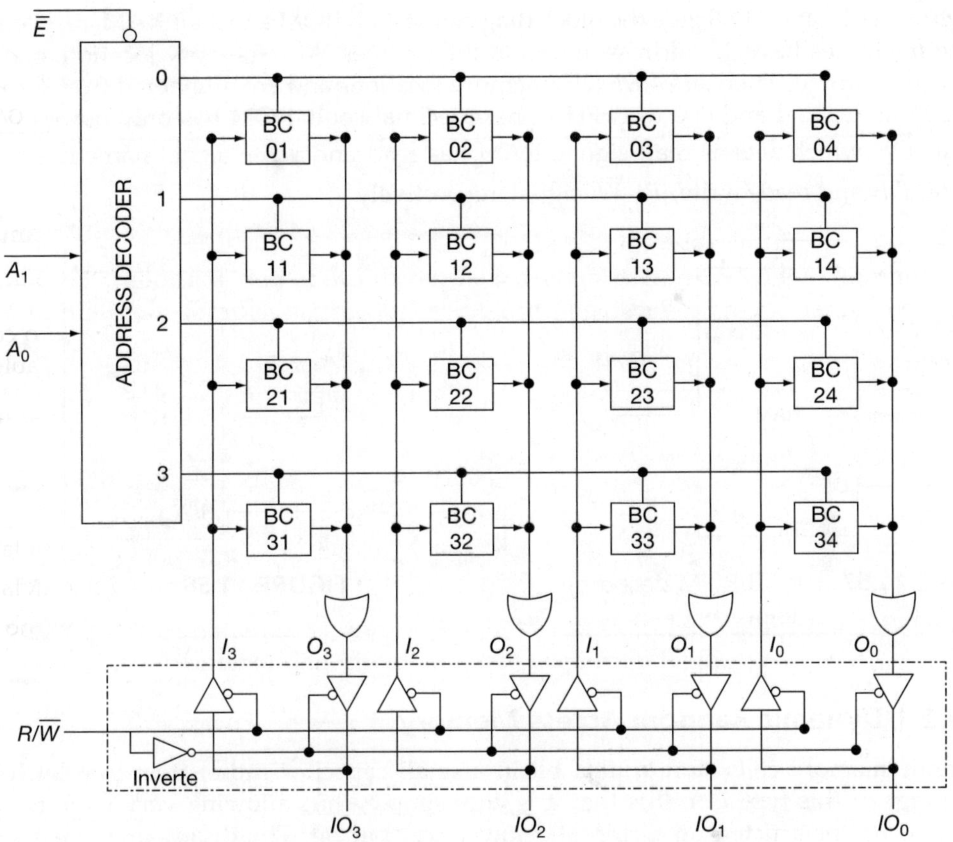

FIGURE 11.55 | $2^3 \times 4$ bits RAM memory

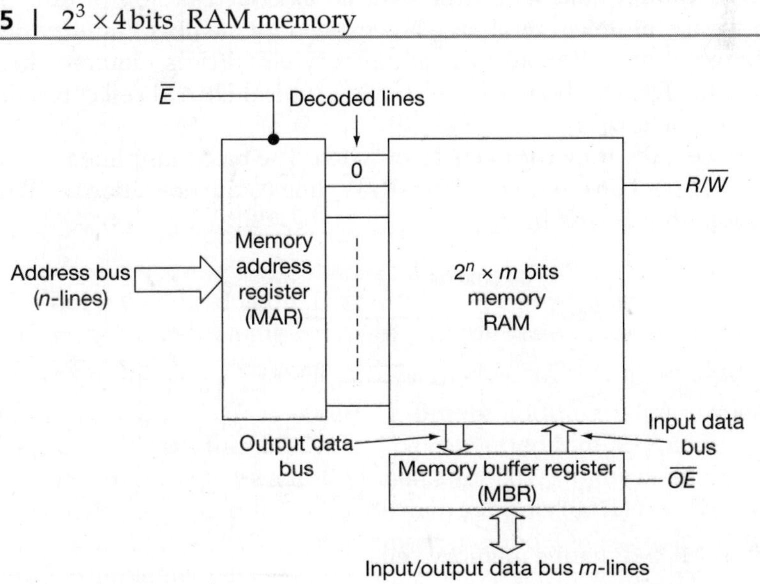

FIGURE 11.56 | RAM memory organization

Figures 11.57 and 11.58 give the block diagram of 4 kB ROM and 4 kB RAM, respectively. These memories have 12 address lines and 8 data lines. At a memory location, one byte data can be stored. There are 4096 (2^{12}) memory locations and are numbered from 0 to 4095 (2^{12} – 1) in decimal and 000 to FFFH in hexadecimal code. ROM has only output enable signal, $\overline{OE}$ which acts as read signal. RAM has read and write signal named as output enable, $\overline{OE}$ and read/write, $R/\overline{W}$ signal, respectively.

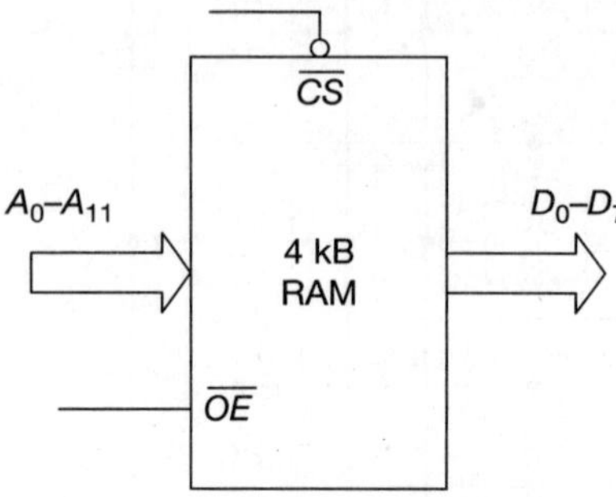

FIGURE 11.57 | 4 kB ROM decoding logic

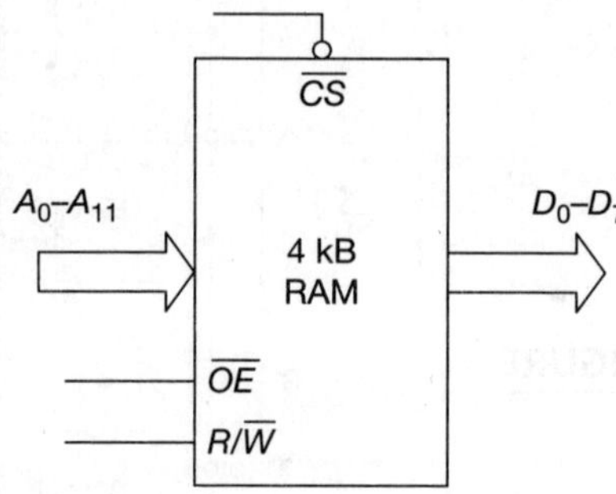

FIGURE 11.58 | 4 kB RAM decoding logic

11.6.2 | Dynamic Random Access Memory

Dynamic memory cells store a data bit in a small capacitor rather than in a latch. The advantage of this type of cell is that it is very simple, thus allowing very large memory arrays to be constructed on a chip at a lower cost per bit. The disadvantage is that the storage capacitor cannot hold its charge over an extended period of time and capacitor loses the stored data bit unless its charge is refreshed by memory circuitry periodically. To refresh the charge of capacitor, additional memory circuitry is required that complicates the operation of the DRAM. Figure 11.59 shows a typical DRAM cell consisting of a single MOSFET and a capacitor.

In this type of cell, the transistor acts as a switch. The basic simplified operation is illustrated in Figures 11.60–11.64. A Low on the (R/$\overline{W}$) line (write operation) enables the tristate buffer and disables the output buffer.

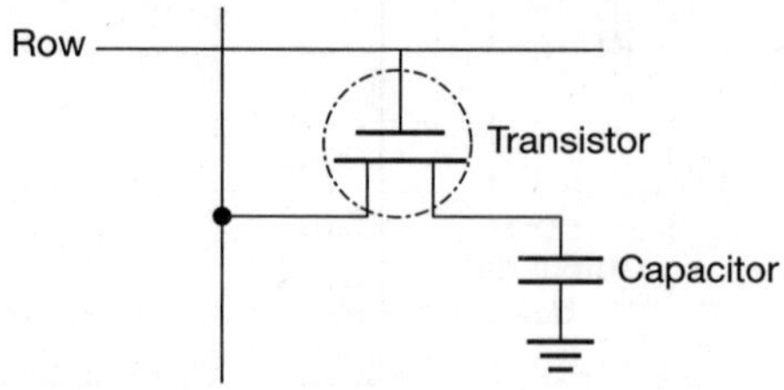

FIGURE 11.59 | MOS dynamic memory cell

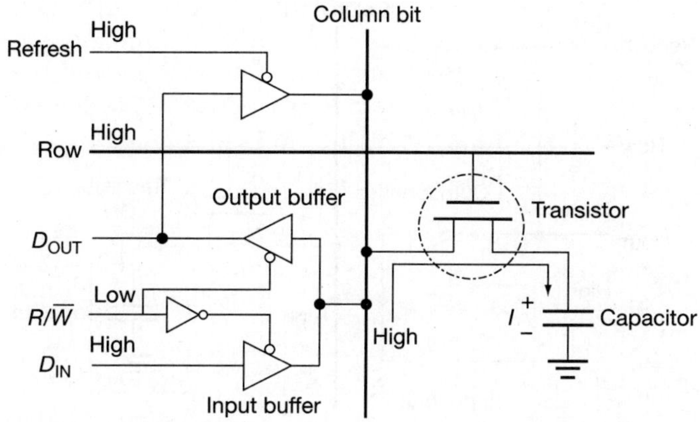

FIGURE 11.60 | Writing 1 into dynamic memory cell

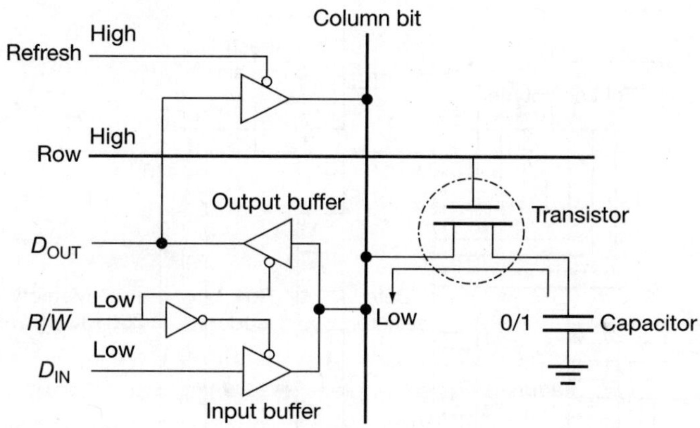

FIGURE 11.61 | Writing '0' into dynamic memory cell

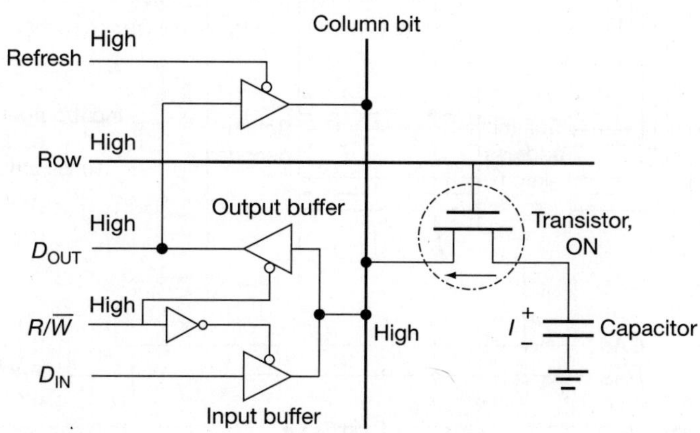

FIGURE 11.62 | Reading 1 into dynamic memory cell

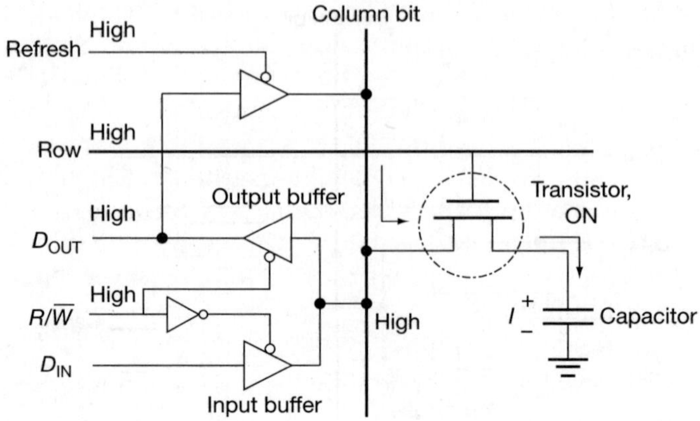

FIGURE 11.63 | Refreshing stored 1 in dynamic memory cell

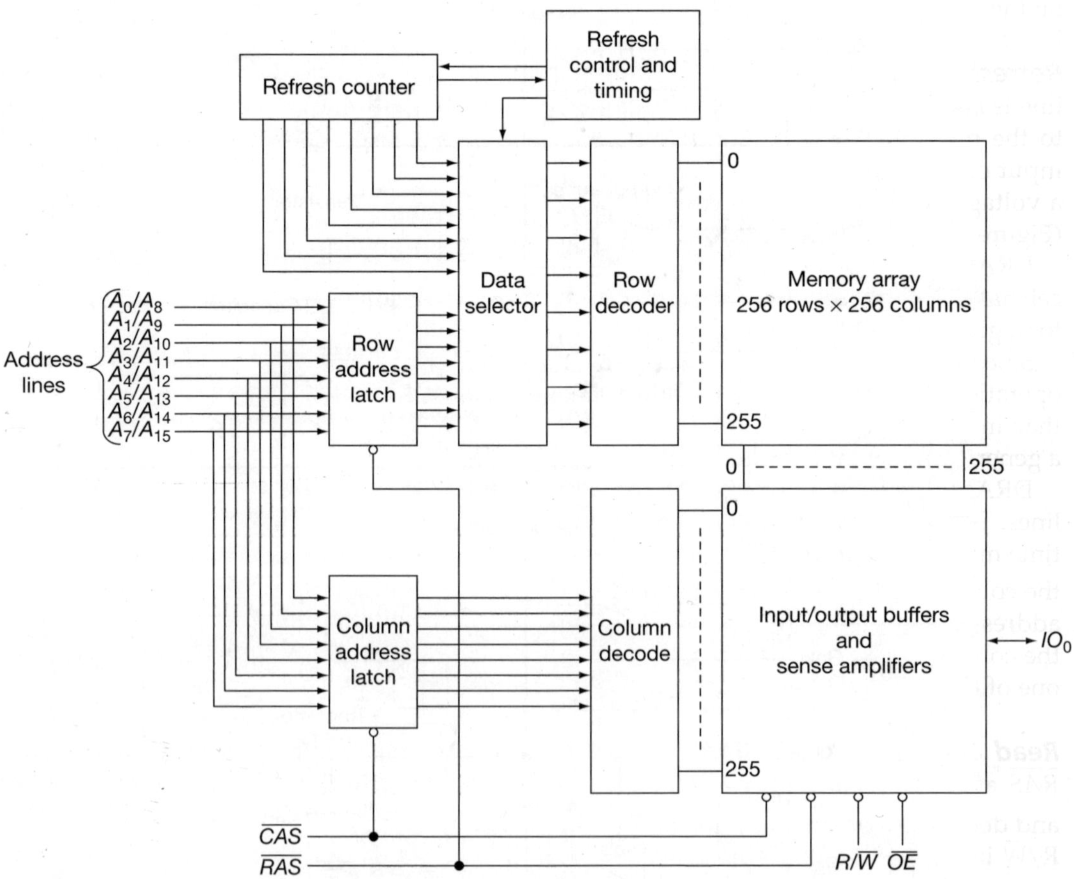

FIGURE 11.64 | Block diagram of 64 k × 1 DRAM

Writing 1 To write 1 into the cell, the D_{in} line is set to *high*. The transistor is turned on by a *high* signal on the row line. The transistor acts as a closed switch connecting the capacitor to the column bit line. This connection allows the capacitor to charge to a positive voltage (Figure 11.60).

Writing 0 To write 0 into the cell, the D_{in} line is set to *low*. The transistor is turned on by a *high* signal on the row line. The transistor acts as a closed switch connecting the capacitor to the column bit line which is at *low*. This connection allows the capacitor to remain uncharged (Figure 11.61). If the capacitor has already been charged to '1' then it is discharged through column bit line.

For the *low* row line, the transistor is switched off. The capacitor is disconnected from the column bit line. The stored charge on capacitor either 0 or 1 is trapped.

Reading 1 To read from the cell, the $R/\overline{W}$ line is set to *high*. It enables the output buffer and disables input buffer. When the row line is taken as *high*, the transistor turns on and connects the capacitor to the bit line and thus to the output buffer, so the data bit appears on the data-output line, D_{out} (Figure 11.62).

Refreshing memory For refreshing the memory cell, the $R/\overline{W}$ line is *high*, the row line is also *high*, and refresh line is *high*. The transistor turns on, connecting the capacitor to the bit line. The output buffer is enabled, and the stored data bit is applied to the input of refresh buffer, which is enabled by the high on the refresh input. This produces a voltage on the bit line corresponding to the stored bit, thus replenishing the capacitor (Figure 11.63).

DRAM memory cell consists of one transistor and a capacitor whereas SRAM memory cell has latch. This allows much greater density in DRAM and results in greater bit capacity for a given chip area, although much slower access time.

Since, stored charge in a capacitor leaks off, the DRAM cell requires a frequent refresh operation to preserve the stored data bit. This requirement results in more complex circuitry than in a SRAM. Several features common to most DRAMs are shown in Figure 11.64 using a generic 64k × 1 bit DRAM.

DRAM uses a technique called address multiplexing to reduce the number of address lines. The blocks that represent the refresh logic are also shown. The eight address lines are time multiplexed at the beginning of a memory cycle by the row address select ($\overline{RAS}$) and the column address select ($\overline{CAS}$) into two separate 8-bit address fields. First, the 8-bit row address is latched into the row address latch. Next, the 8-bit, column address is latched into the column address latch. The row address and the column address are decoded to select one of the 65,536 addresses (2^{16} = 65,536) in the memory array.

Read and Write Operations At the beginning of each read or write memory cycle. $\overline{RAS}$ and $\overline{CAS}$ go active (low) to multiplex the row and column addresses into the latches and decoders. For a read cycle, the $R/\overline{W}$ input is high (Figure 11.65). For a write cycle, the $R/\overline{W}$ input line is *low* (Figure 11.66).

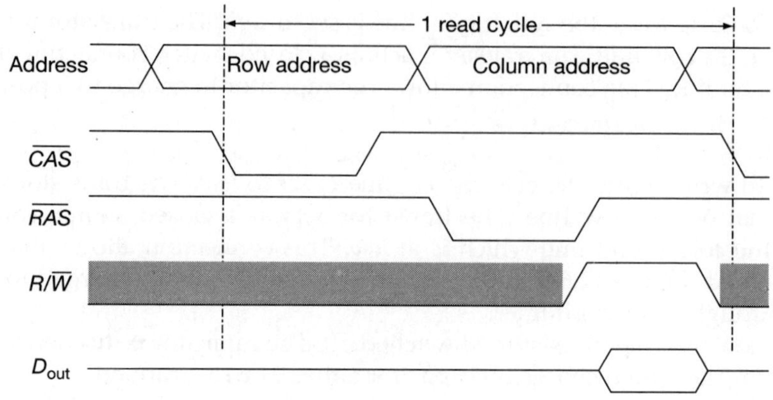

FIGURE 11.65 | Read cycle of DRAM

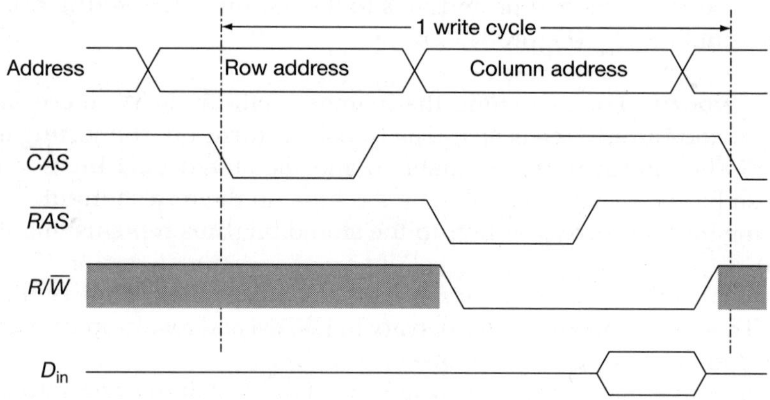

FIGURE 11.66 | Write cycle of DRAM

Fast-Page Mode (FPM) In the normal read or write cycle described previously, the row address for a particular memory location is first loaded by an active-low $\overline{RAS}$ line and then the column address for that location is loaded by an active-low $\overline{CAS}$ line. The next location is selected by another $\overline{RAS}$ followed by a $\overline{CAS}$ line, and so on.

A 'page' is a section of memory available at a single row address and consists of all the columns in a row. FPM allows fast successive read or write operations at each column address in a selected row. A row address is first loaded by $\overline{RAS}$ line going *low* and remains *low* while $\overline{CAS}$ line is toggled between *high* and *low*. A single row address is selected and remains selected while $\overline{RAS}$ line is active. Each successive $\overline{CAS}$ line selects another column in the selected row. So, after a FPM cycle, all of the addresses in the selected row have been read from or written into, depending on R/$\overline{W}$ line. For example, a FPM cycle for the DRAM in Figure 11.64 requires $\overline{CAS}$ to go active 256 times for each row selected by $\overline{RAS}$.

FPM operation for read is illustrated by the timing diagram in Figure 11.65. When $\overline{CAS}$ line goes to its non-asserted state (*High*), it disables the data outputs. Therefore, the transition of $\overline{CAS}$ line to *high* must occur only after valid data are latched by the external system.

Refresh cycle: DRAMs are based on capacitor charge storage for each bit in the memory array. This charge leaks off with time and temperature, so each bit must be periodically refreshed (recharged) to maintain the correct bit state. Typically, a DRAM must be refreshed every 8 ms to 16 ms, although for some devices the refresh period can exceed 100 ms.

A read operation automatically refreshes all the addresses in the selected row. However, in typical applications, it is difficult to predict how often there will be a read cycle. To prevent data loss, frequent occurrence of read cycle is not sufficient. Therefore, special refresh cycles must be implemented in DRAM, systems.

Burst refresh and distributed refresh are the two basic refresh modes for refresh operations.

In burst refresh, all rows in the memory array are refreshed consecutively each refresh period. For a memory with a refresh period of 8 ms, a burst refresh of all rows occurs once every 8 ms. The normal read and write operations are suspended during a burst refresh cycle.

In distributed refresh, each row is refreshed at intervals interspersed between normal read or write cycles. For example, the memory in Figure 11.64 has 256 rows. As an example, for an 8 ms refresh period, each row must be refreshed every 8 ms/256 = 31.25 μs when distributed refresh is used.

The two types of refresh operations are $\overline{RAS}$ only refresh and $\overline{CAS}$ before $\overline{RAS}$ refresh. $\overline{RAS}$-only refresh consists of a $\overline{RAS}$ line transition to the low (active) state, which latches, the address of the row to be refreshed while $\overline{CAS}$ line remains high (inactive) throughout the cycle. An external counter is used to provide the row addresses for this type of operations.

The $\overline{CAS}$ before $\overline{RAS}$ refresh is initiated by $\overline{CAS}$ line going Low before $\overline{RAS}$ goes low. This sequence activates an internal refresh counter that generates the row address to be refreshed. This address is switched by the data selector into the row decoder

11.6.3 | Types of DRAM

The main types of DRAM are the FPM DRAM, the extended data-out (EDO) DRAM, the Burst Extended Data-out (BEDO) DRAM, and the synchronous (S) DRAM.

FPM DRAM This type of DRAM has been the most common and has been the type used in computers until the development of the EDO DRAM. A page in memory is all of the column addresses contained within one row address.

The idea of the FPM DRAM is based on the probability that the next several memory addresses to be accessed are in the same row (on the same page). Fortunately, this happens to save a large percentage of the time. FPM saves time over pure random accessing because in FPM, the row address is specified only once to access several successive column addresses. For pure random accessing, a row address is specified for each column address.

Recall that in a FPM read operation, the $\overline{CAS}$ signal has to wait until the valid data from a given address are accepted (latched) by the external system (CPU) before it can go to its non-asserted state. When $\overline{CAS}$ goes to its non-asserted state, the data outputs are disabled. This means that the next column address cannot occur until the data from the current column address are transferred to the CPU. This limits the rate at which the columns within a page can be addressed.

EDO DRAM The EDO DRAM sometimes called hyper page mode DRAM, is similar to the FPM DRAM. The key difference is that the $\overline{CAS}$ signal in the EDO DRAM does not disable the output data when it goes to its non-asserted state because the valid data from the current address can be held until CAS is asserted again. This means that the next column address can be accessed before the external system accepts the current valid data. The idea is to speed up the access time.

BEDO DRAM The *burst extended data-out DRAM* (BEDO DRAM) is an EDO DRAM with address burst capability. Recall from the discussion of the synchronous burst SRAM that the address burst feature allows up to four addresses to be internally generated from a single external address, which saves some access time. This same concept applies to the BEDO DRAM.

SDRAM Faster DRAMs are needed to keep up with the ever-increasing speed of microprocessors. The *synchronous DRAM* (SDRAM) is one way to accomplish this. Like the synchronous SRAM discussed earlier, the operation of the SDRAM is synchronized with the system clock, which also runs the microprocessor in a computer system. The same basic ideas described in relation to the synchronous burst SRAM, also apply to the SDRAM.

This synchronized operation makes the SDRAM totally different from the other asynchronous DRAM types. With asynchronous memories, the microprocessor must wait for the DRAM to complete its internal operations. However, with synchronous operation, the DRAM latches addresses, data, and control information from the processor is under control of system clock. This allows the processor to handle other tasks while the memory read or write operations are in progress, rather than having to wait for the memory to do its thing as is the case in asynchronous systems.

11.7 | FIRST-IN FIRST-OUT MEMORY

A FIFO memory is a storage device that reads out data from its memory array (SRAM) in the same order in which the data were written into the memory. The first word written into the memory location is the first word that is read out from the memory location. Since, the first word written into memory is read out firstly, so memory is referred to as the FIFO memory. The memory is said to be full when the data is stored at all memory locations. The memory is said to be empty when all memory locations are not having data. No more data can be written into FIFO memory, when memory is full. Similarly, no word can be read from memory, when the memory is empty or having no stored data. The reading and writing operations are internally organized and the appropriate locations for writing into and reading out are pointed to by write pointer and read pointer, respectively. There is no need of address to access the memory for read or write operations.

The write and read operations on memory are performed at independent data rates. This type of memory is used to interface slow input/output (I/O) devices to the fast operating devices. The FIFO memory is used as a data-rate buffer.

A FIFO memory is basically a SRAM array with two ports and control circuitry. One port is for input port for writing into data and the other port is output port for reading data from the memory. Each port has separate access, data, and control signals for accessing a common SRAM array. The SRAM with two ports is known as a dual-port SRAM.

The dual-port SRAMs can be asynchronous dual-port SRAM, synchronous dual-port SRAM and sequential access SRAM.

In asynchronous dual-port SRAM, the operation of both the ports is asynchronous. In synchronous dual-port SRAM, the operation of both ports is synchronized with two different clocks. In a sequential access two-port SRAM, one of the ports is having random (asynchronous) accesses and the other port is following sequential access. A sequential accesses two-port SRAM, is used to interface the asynchronous and synchronous components of a digital system. Normally, the controlling processor is connected to the asynchronous port and the slave processor on the synchronous side. This type of FIFO memory is used for peripheral controllers, networking equipment such as bridges, routers, etc.

FIFO memory is formed by an arrangement of shift registers. One important difference between a conventional shift register and a FIFO register is illustrated in Figure 11.67. In a conventional register, a data bit moves through the register only as new data bits are entered; in a FIFO register, a data bit immediately goes through the register to the right-most bit location that is empty.

Figure 11.67 is a block diagram of a FIFO serial memory. This particular memory has four serial 64-bit data registers and a 64-bit control register (marker register). When data are entered by a shift-in pulse, they move automatically under control of the marker register to the empty location closest to the output. Data cannot advance into occupied positions. However, when a data bit is shifted out by a shift-out pulse, the data bits remaining in the registers automatically move to the next position toward the output. In an asynchronous FIFO, data are shifted out independent of data entry with the use of two separate clocks.

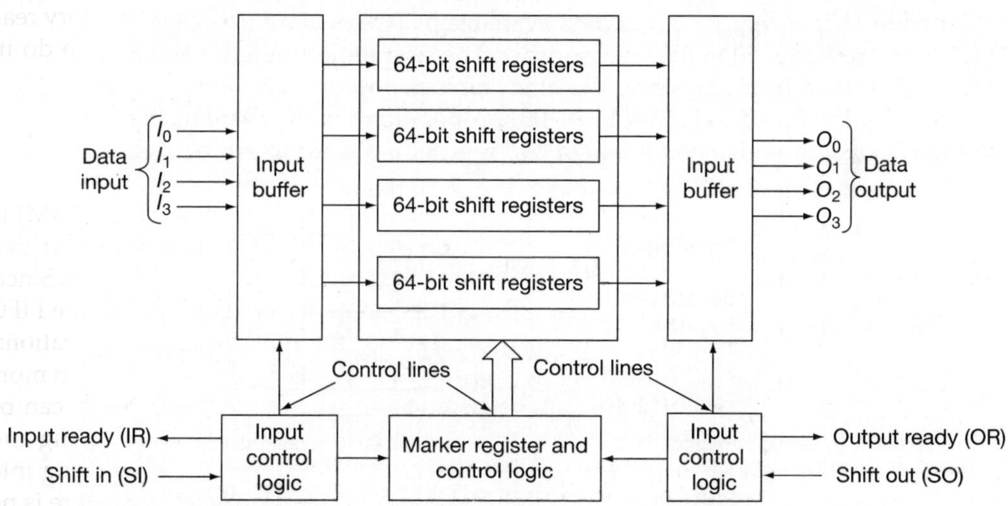

FIGURE 11.67 | Block diagram of a FIFO serial

11.8 | LAST-IN FIRST-OUT MEMORY

The last-in-first out (LIFO) memory is found in applications involving microprocessors and other computing systems. It allows data to be stored and then recalled in reverse order; that is, the last data byte to be stored is the first data byte to be retrieved.

Register Stacks A LIFO memory is commonly referred to as a push-down stack. It is implemented with a group of registers. Group of four 4-bit registers stack is shown in Figure 11.68. A stack can consists of any number of registers, but the register at the top is called the top-of-stack (TOS). First register is considered as TOS. Write into and read from register stack is shown in Figures 11.68(a) and (b), respectively.

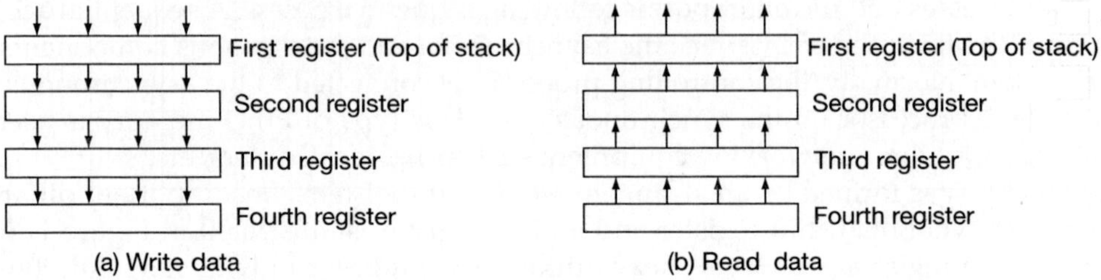

FIGURE 11.68 | Register stack

To demonstrate the principle operation, a byte of data is loaded in parallel onto the TOS. Each successive nibble pushes the previous one down into the next register. This process is illustrated in Figure 11.69. Notice that the new data byte is always loaded into the top register and the previously stored bytes are pushed deeper into the stack. The name push down stack comes from this characteristic. 0001 is pushed into TOS, i.e. register 1. When 0010 data is pushed that is stored into TOS and previous data, i.e. 0001 is pushed into register 2. Similarly, when another data 0011 is pushed into TOS then data available in register 2 is pushed to deeper register 3 and data available in TOS pushed to register 2.

Data bytes are retrieved in the reverse order. The last nibble entered is always at the TOS, so when it is pulled from the stack, the other nibbles pop up into the next lower register. This process is illustrated in Figure 11.70. 0011 is pop up then the available data in 3rd register is shifted to 2nd register and data of 2nd register is shifted to 1st register, that is, TOS.

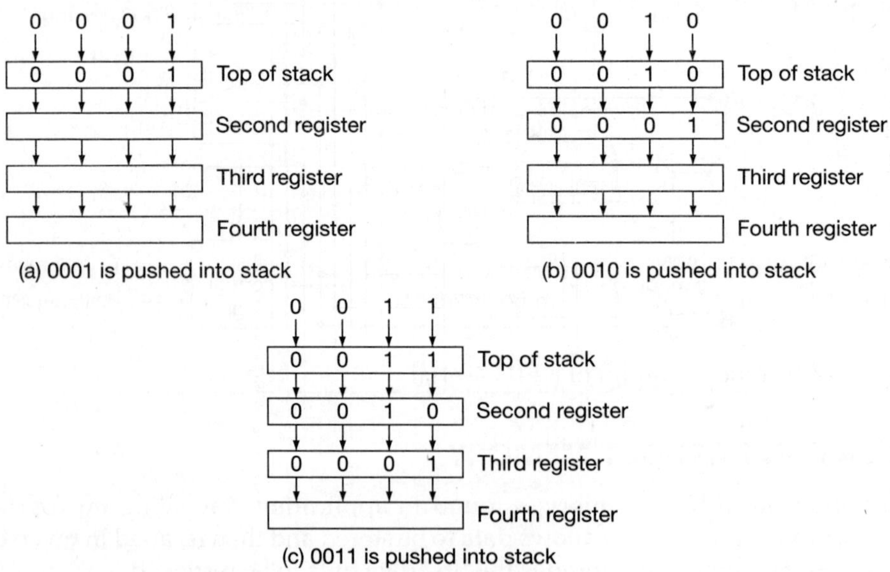

FIGURE 11.69 | Push into register stack operation

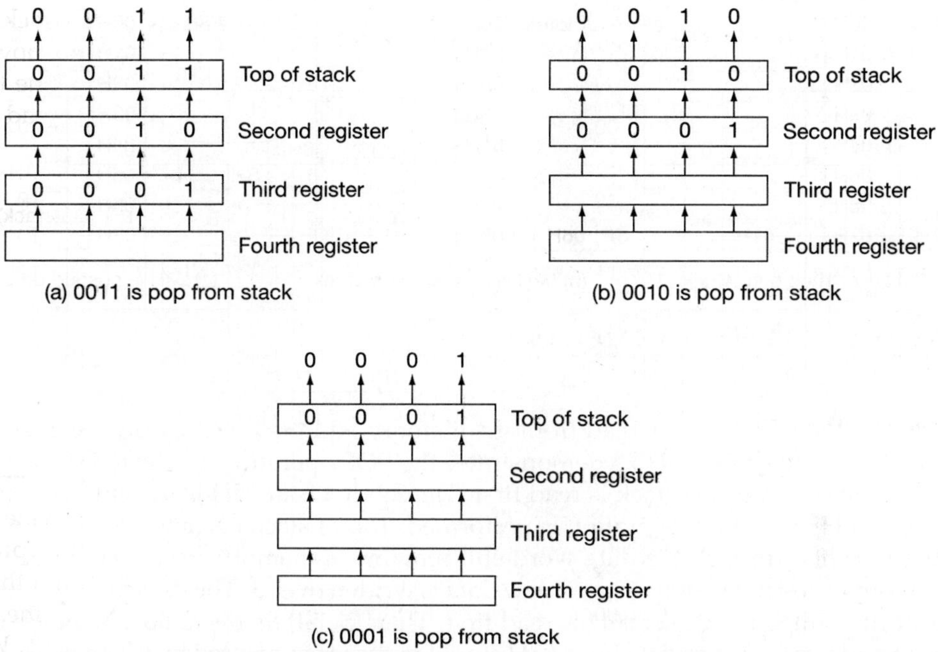

FIGURE 11.70 | Pop from register stack operation

RAM Stack Another approach to LIFO memory is the allocation of a section of RAM as the stack rather than the use of a dedicated set of registers. In a register stack, the data moves up or down from one location to the next whereas TOS remains fix. In a RAM stack, the data itself does not move but the TOS moves under control of a register called the stack pointer (SP).

PUSH Operation The SP is decremented by one and data is stored at the location given by SP. The SP is initially set at address 08H, which is the top of the stack as shown in Figure 11.71. The SP is then decremented (decreased) by one to 07H. This moves the top of the stack to a lower memory address. Top of stack is not stationary as in the fixed register stack but moves downward to lower addresses in the RAM as data words are stored. For 16-bit data SP is decremented by 2. Figure 11.72(a) shows that one data word that is 01H is then pushed onto the stack at the memory address 07. 07H is the contents of SP. To push onto stack another byte, i.e. 02H, the SP is decremented by 1. SP contents become 06H. 02 H is stored at location 06. Similarly 22H is pushed onto stack and SP becomes 05H. 05H is the top of the stack.

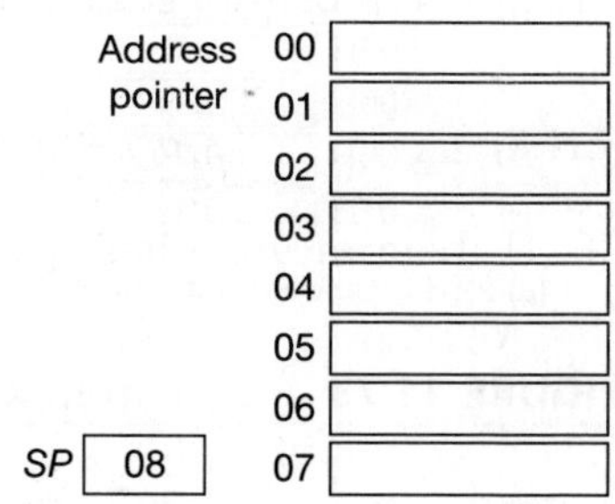

FIGURE 11.71 | RAM stack

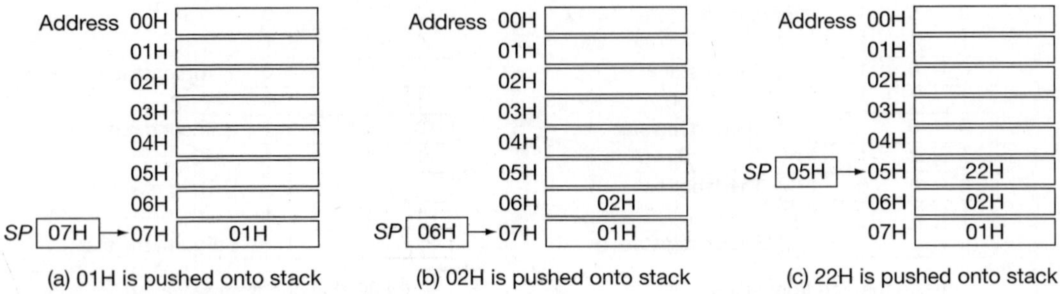

FIGURE 11.72 | PUSH onto RAM stack

POP Operation The data is read from the memory address given by SP and then SP is incremented by one. Figure 11.73 demonstrates the POP operation for the RAM stack. The last data byte stored in the stack is read first. The SP that is at 05H is incremented by one to address 06H and a POP operation is performed. Top of stack becomes 06. RAM is non-destructive when read, so the data word still remains in memory after a POP operation. A data word is destroyed only when a new data is written over it. The data byte, 02H stored in the stack pointed by SP, i.e. 06H is read first. Then the SP that is at 06H is incremented by one to address 07H. The data byte, 01H stored in the stack pointed by SP, i.e. 07H is read first. The SP that is at 07H is incremented by one to address 08H. The movement of SP is observed by number of push and pop operations.

Figure 11.74 illustrates a LIFO stack memory having 32 bytes of storage, can be cascaded for LIFO operation, dedicated port address and fast select feature. It has 3-state TTL outputs. Up/down count counts from 0 to 31 addresses. Address 31 gives empty signal and 0 address represents full. Full output, $\overline{F_{out}}$, full input, $\overline{F_{in}}$ empty output, $\overline{E_{out}}$ and empty input, $\overline{E_{in}}$ are used for cascading purpose.

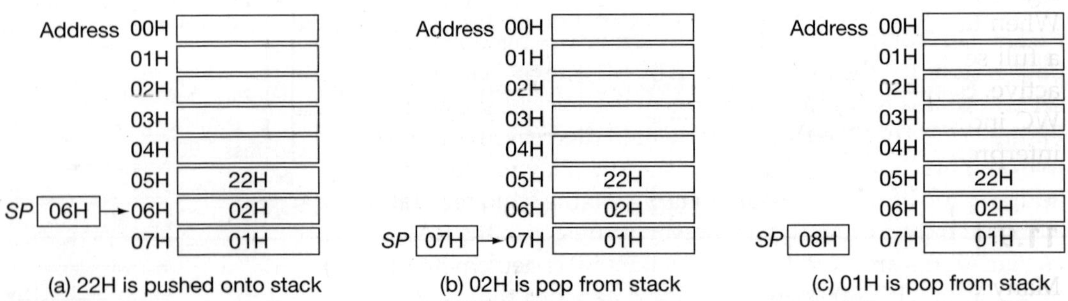

FIGURE 11.73 | POP from RAM stack

When full output $\overline{F_{out}}$ is active low. It means stack is full and no further write (push) operation can be performed until at least one read (pop) operation is executed. On power up, $\overline{F_{out}}$ is high and remains in this inactive state until the last memory location is written to. When empty output, $\overline{E_{out}}$ is active (Low). It means stack is empty and no further read (pop) operation can be performed until at least one write (push) operation is executed.

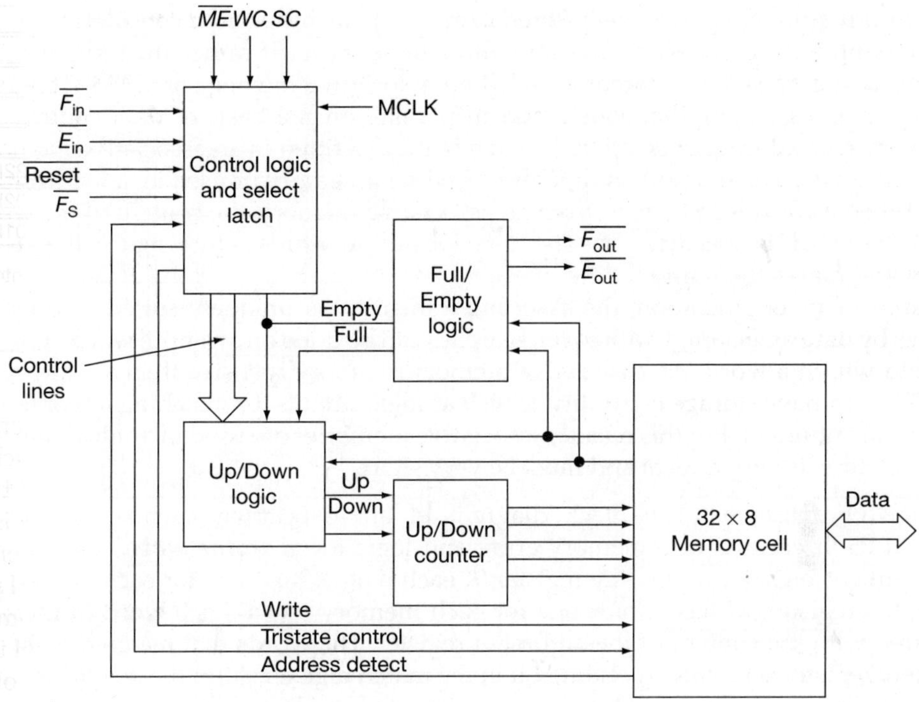

FIGURE 11.74 | 32 × 8 bits LIFO stack memory

On power up, $\overline{F_{out}}$ is high and remains in this inactive state until the last memory location is written to. Full input, $\overline{F_{in}}$ signal indicates that next lower chip is full. Empty input, $\overline{E_{in}}$ signal indicates that next lower chip is empty. Reset operation resets stack to empty status. When fast select, $\overline{FS}$ signal is low, allows read and write operations without performing a full select cycle. On setting master enable, $\overline{ME}$ as low signal, controller bank becomes active. System synchronization can be done with master clock, MCLK. Write command, WC indicates that data bus carries input data. Select command, SC indicates that bus is interpreted as address.

11.9 | ASSOCIATIVE MEMORY OR CONTENT ADDRESS MEMORY

Many data processing applications require the search of items in a table which is stored in memory. An assembler program searches the symbol address table in order to extract the symbol's binary equivalent. The established way to search a table is to store all items where these can be addressed in sequence. The search procedure is a strategy for selecting a sequence of addresses, reading the content of memory at each address, and comparing the information read with the item being searched until a match occurs. The number of accesses to memory depends on the location of the item and the efficiency of the search algorithm.

The time required to find an item stored in memory can be reduced considerably if stored data is identified for access by the content of the data itself rather than by an address. Memory unit accessed by content is called an associative memory or CAM. This type of memory is accessed simultaneously and in parallel on the basis of data content rather than by specific address or location. When a word is written in an associative memory, no address is given. The memory is capable of finding an empty unused location to store the word. When a word is to be read from an associative memory, the content of the word, or part of the word, is specified. The memory locates all words which match the specified content and marks them for reading.

Because of its organization, the associative memory is uniquely suited to do parallel searches by data association. Moreover, searches can be done on an entire word or on a specific field within a word. An associative memory is more expensive than a RAM because each cell must have storage capability as well as logic circuits for matching its content with an external argument. For this reason, associative memories are used in applications where the search time is very critical and must be very short.

Hardware Organization: The block diagram of an associative memory is shown in Figure 11.75. It consists of a memory array and logic for m words with n bits per word. The argument register, X and key register, K each have n bits, one for each bit of a word. The match register, M has m bits, one for each memory word. Each word in memory is compared with the content of the argument register. The words that match the bits of the argument register set a corresponding bit in the match register. After the matching process, set bits in the match register indicate that their corresponding words have been matched. Reading is accomplished by a sequential access to memory for those words whose corresponding bits in the match register have been set.

The key register provides a mask for choosing a particular field or key in the argument word. The entire argument is compared with each memory word if the key register contains all 1's. Otherwise, only those bits in the argument that have 1's in their corresponding position of the key register are compared. Thus, the key provides a mask or identifying piece of information which specifies how the reference memory is made. To illustrate with a numerical example, suppose that the argument register X and the key register K have the bit configuration shown below. Only the three left most bits of X are compared with memory words because K has 1's in these positions.

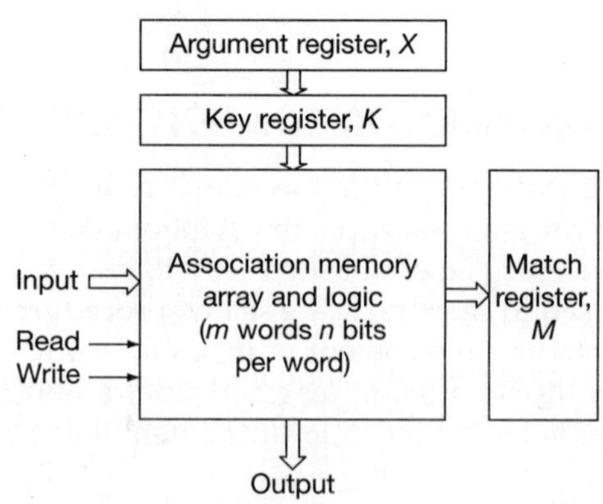

FIGURE 11.75 | Block diagram of associate memory

X:	101 11100	
K:	111 00000	
Word 1:	100 11100	No match
Word 2:	111 00110	Match

Word 2 matches the unmasked argument field because the three left-most bits of the argument and the word are equal.

11.9.1 | Match Logic

The internal organization of a typical cell C_{ij} is shown in Figure 11.76. It consists of a flip-flop a storage element, A_{ij}, and the circuits for reading, writing, and matching the cell. The input bit is transferred into the storage cell during a write operation. The bit stored is read out during a read operation. The match logic compares the content of the storage cell with the corresponding unmasked bit of the argument and provides an output for the decision logic that sets the bit in M_i. Block diagram of the associative memory cell is given in Figure 11.77.

The Boolean expression for setting the match bit for i^{th} word, M_j is given below:

$$M_i = \prod_{j=1}^{n}\left[\left(X_j A_{ij} + \bar{X}_j \bar{A}_{ij}\right) + \bar{K}_j\right] \tag{11.24}$$

where A_{ij} is stored bit in flip-flop.

X_j is the bit of argument register.

K_j is the key register.

and

$$\left(X_j A_{ij} + \bar{X}_j \bar{A}_{ij}\right) + \bar{K}_j = \begin{bmatrix} \left(X_j A_{ij} + \bar{X}_j \bar{A}_{ij}\right) & ;K_j = 1 \\ 1 & ;K_j = 0 \end{bmatrix}$$

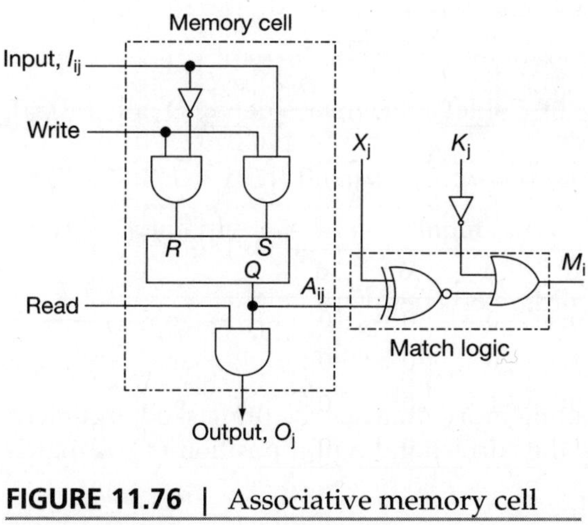

FIGURE 11.76 | Associative memory cell

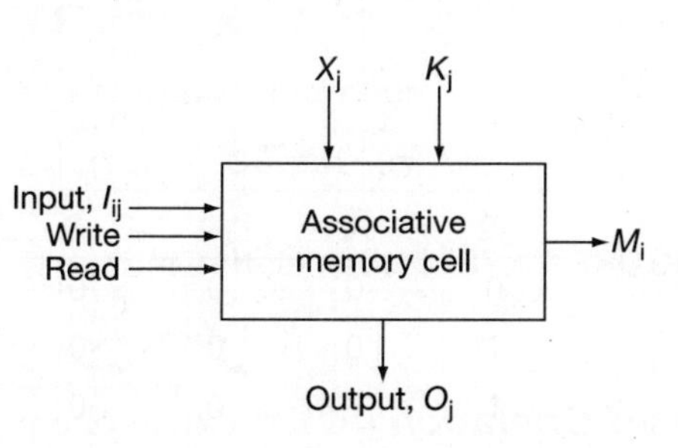

FIGURE 11.77 | Block diagram of associative memory cell

The memory array and external registers in an associative memory are shown in Figure 11.78. The cells in the array are marked by the letter C_{ij}. Cell C_{ij} is the cell for *j*th bit of *i*th word. A bit X_j, in the argument register is compared with all the bits in column *j* of the array provided $K_j = 1$. This is done for all columns using AND gate. If a match occurs between all the unmasked bits of the argument and the bits in *i*th word, the corresponding bit M_i, in the match register is set to 1. For unmasked bits of the argument and the word do not match, M_i, is cleared to 0.

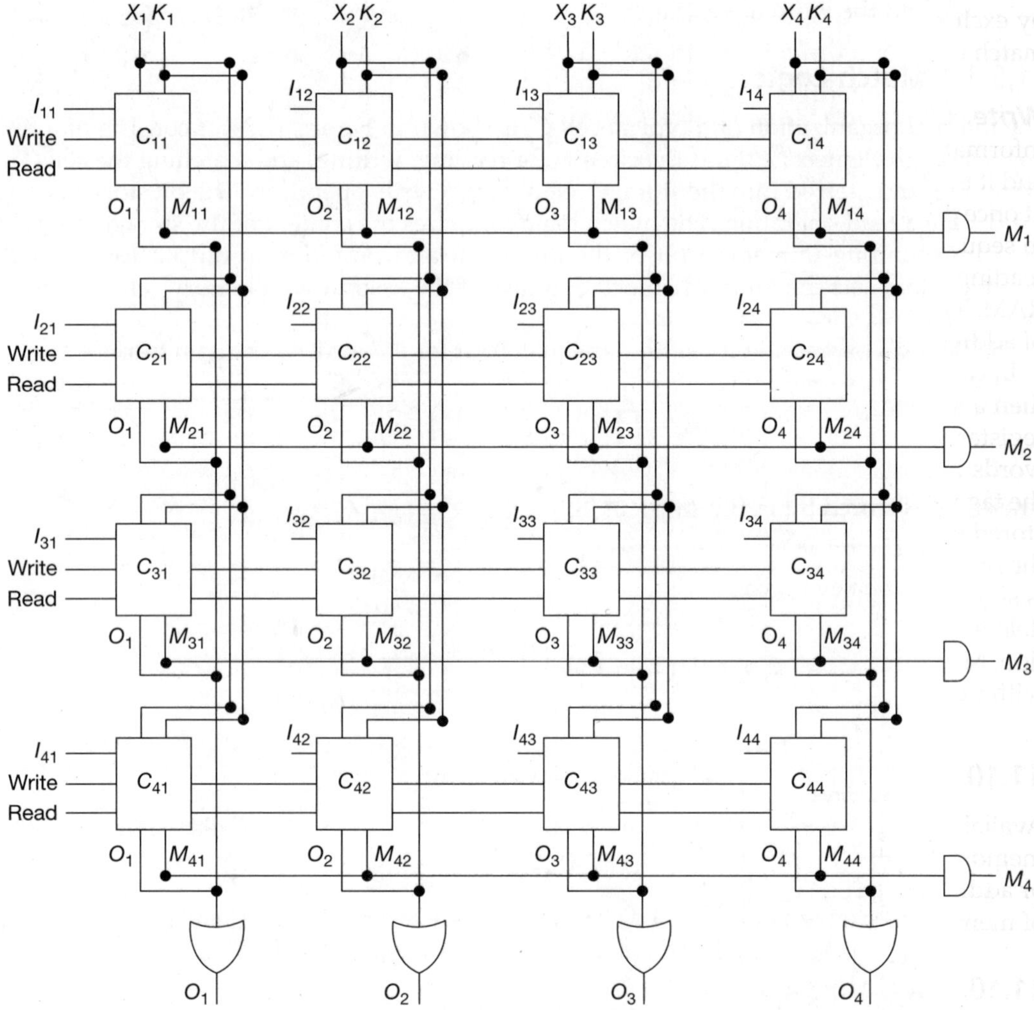

FIGURE 11.78 | Associative memory of 4 words, 4 bits per word

Read Operation If more than one words in memory matches the unmasked argument field, all the matched words will have 1's in the corresponding bit position of the match register. It becomes necessary to scan the bits of the match register one at a time. The matched words are read in sequence by applying a read signal to each word line whose corresponding match bit, M_i is a 1.

In most applications, the associative memory stores a table with no two identical items under a given key. In such a case, only one word may match the unmasked argument field. By connecting output M_i, directly to the line that enables read operation in the same word position, instead of the M register. So, the content of the matched word are presented automatically at the output lines and no special read command signal is required. Furthermore, by excluding words having zero content, an all zero output is indicated. It means that no match is occurred and the searched item is not available in memory.

Write Operation An associative memory also has a write capability to store the information to be searched. Writing in an associative memory takes place in different forms and it is application dependent. In case, the entire memory is loaded with new information at once prior to a search operation then the writing can be done by addressing each location in sequence. The device acts as a RAM for writing and a content addressable memory for reading purpose. The advantage here is that the address for input can be decoded as in a RAM. Thus, instead of having m address lines, one for each word in memory, the number of address lines can be reduced by the decoder to d lines, where $m = 2^d$.

In case unwanted words is to be deleted and new words is to be inserted one at a time, then a special register is required to distinguish between active and inactive words. This register sometimes is known as a tag register. Tag register has as many bits as there are words in the memory. Every active word stored in memory has the corresponding bit in the tag register as 1. By clearing the tag bit to 0, a word is deleted from memory. Words are stored in memory by scanning the tag register until the first 0 bit is encountered. This gives the first available inactive word and a position for writing a new word. After the new word is stored in memory it is made active by setting its tag bit to 1. An unwanted word when deleted from memory can be cleared to all 0's if this value is used to specify an empty location. Moreover, the words that have a tag bit of 0 must be masked (together with the K_j bits) with the argument word so that only active words are compared.

11.10 | MEMORY EXPANSION

Available memory can be expanded to increase the word size (number of bits at each memory location represented by address) or word capacity (number of memory locations or addresses) or both. Memory expansion can be achieved by adding appropriate number of memory IC's to address, data and control.

11.10.1 | Word Size Expansion

To increase the word size of a memory, the number of bits in the data bus is increased. For example, an 8-bit word length can be achieved by using two memories with 4-bit words as shown in Figure 11.79. Two 256 × 4 bits ROM memories are expanded to give 256 × 8 bits.

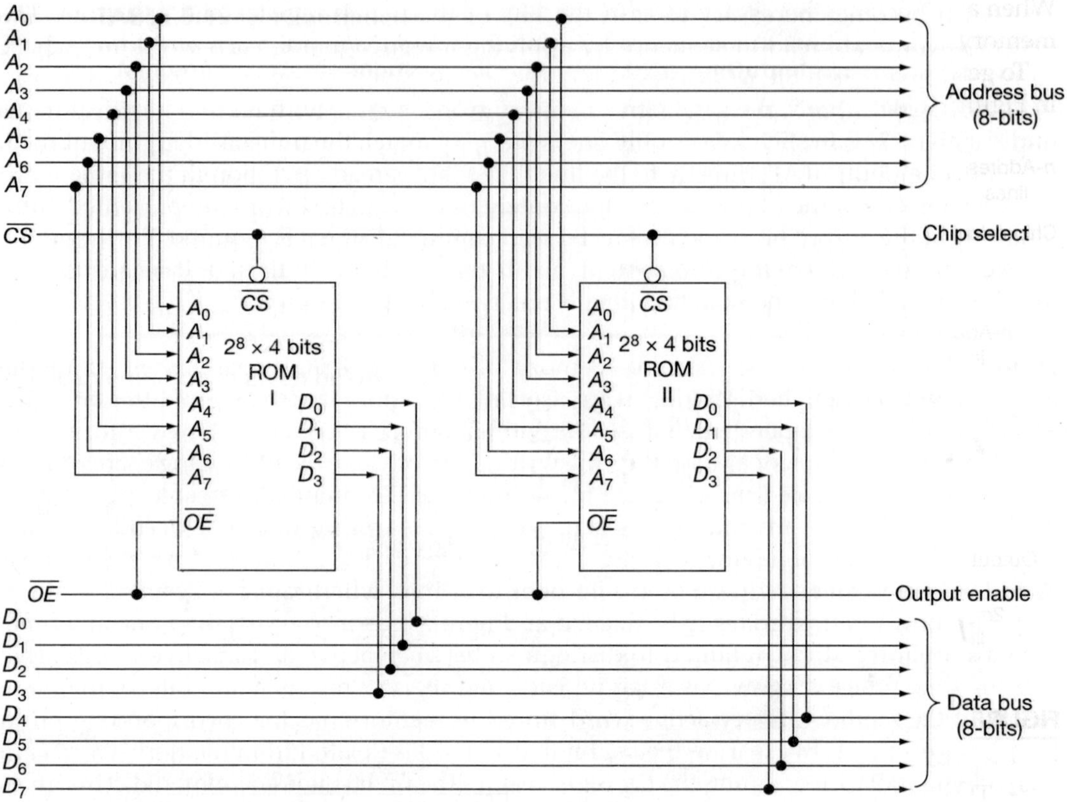

FIGURE 11.79 | Word length expansion (256 × 8 bits ROM from 256× 4 bits ROM)

- The address bus is commonly connected to both ROM memories so that the combination still has the same number of addresses, i.e. $2^8 = 256$ as each individual memory. Address bus is unidirectional.
- Four-bit data buses ($D_0 - D_3$) from the two ROM memories are combined to form an 8-bit data bus ($D_0 - D_7$). On ROM only read operation is performed. Hence, data bus is unidirectional. Table 11.8 explains the data provided by the ROM chip.

TABLE 11.8 | Eight-bit data bus connections

Data Lines	Memory Chip
$D_0 - D_3$	ROM-I
$D_4 - D_7$	ROM-II

- Same chip select, $\overline{CS}$ signal is given to both the ROM memories. Arrangement is given by single row and two columns.
- Same output enable, $\overline{OE}$ signal is given to both the ROM memories.

When an address is selected, eight bits are made available on the data bus-four from each memory.

To generalize, two $2^n \times m$ bits ROM memories are expanded to give $2^n \times 2m$ bits as shown in Figure 11.80.

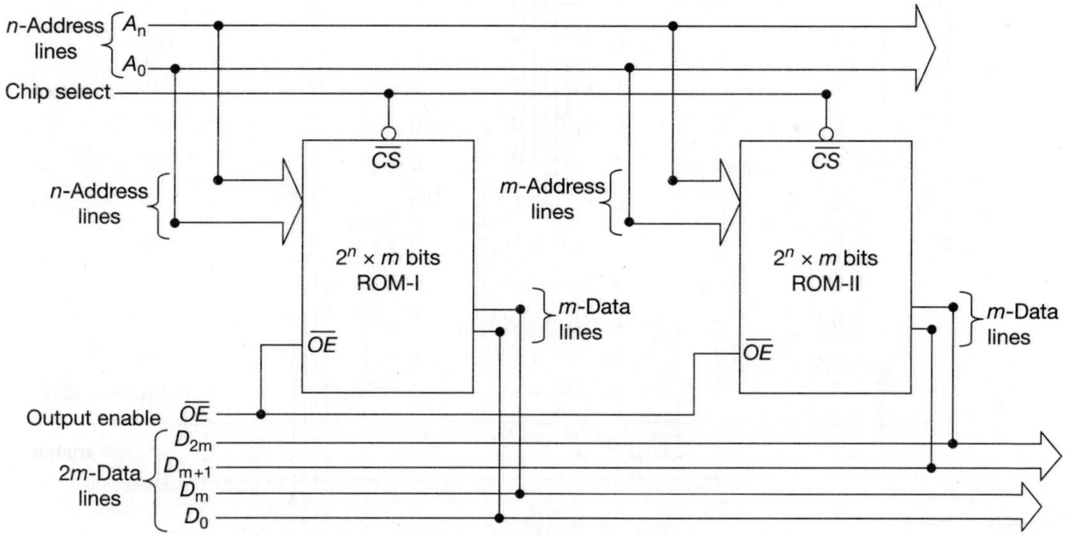

FIGURE 11.80 | Word length expansion ($2^n \times 2m$ bits ROM from $2^n \times m$ bits ROM)

- The address bus is commonly connected to both ROM memories so that the combination still has the same number of addresses, i.e. 2^n as each individual memory. Address bus is unidirectional.
- m-bit data buses ($D_0 - D_m$) from the two ROM memories are combined to form $2m$-bit data bus ($D_0 - D_{2m}$). On ROM only read operation is performed. Hence, data bus is unidirectional. Table 11.9 explains the data provided by the ROM chip.

TABLE 11.9 | Eight-bit data bus connections

Data Lines	Memory Chip
$D_0 - D_m$	ROM-I
$D_{m+1} - D_{2m}$	ROM-II

- Same chip select, $\overline{CS}$ signal is given to both the ROM memories. Arrangement is given by single row and two columns.
- Same output enable, $\overline{OE}$ signal is given to both the ROM memories.

When an address is selected, $2m$ bits are made available on the data bus of m-bits from each memory.

Two 256×4 bits RAM memories are expanded to give 256×8 bits as shown in Figure 11.81.

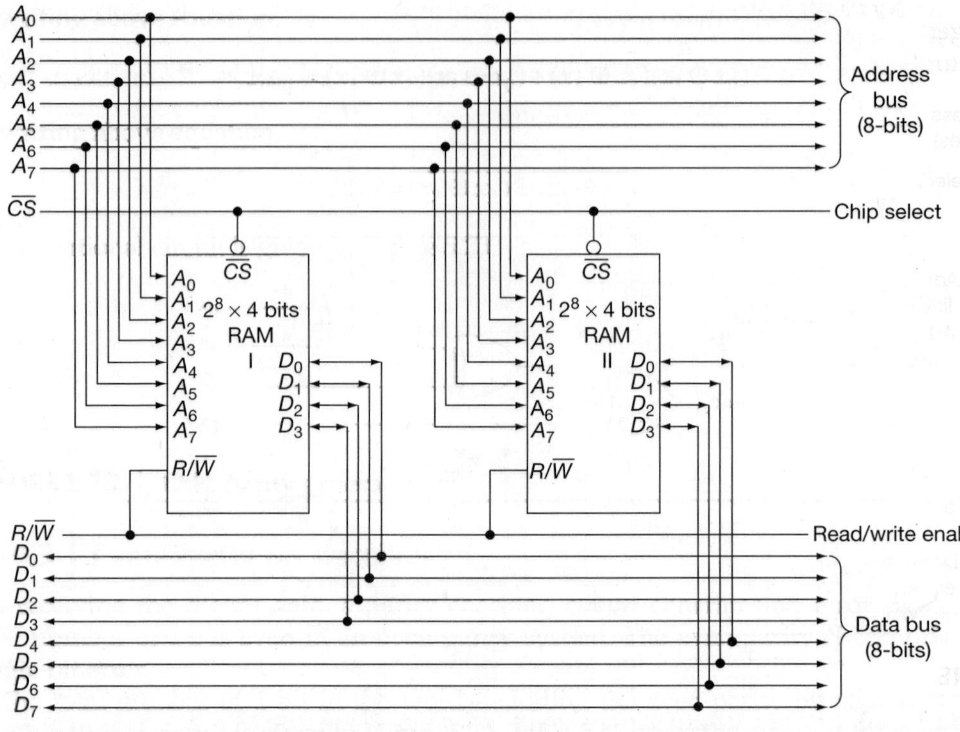

FIGURE 11.81 | Word length expansion (256 × 8-bit RAM from 256 × 4 bits ROM)

- The address bus is commonly connected to both RAM memories so that the combination still has the same number of addresses, i.e. $2^8 = 256$ as each individual memory. Address bus is unidirectional.
- Four-bit data buses ($D_0 - D_3$) from the two RAM memories are combined to form an 8-bit data bus ($D_0 - D_7$). On RAM, read and write operations are performed. Hence, data bus is bidirectional. Table 11.10 explains the data provided by the RAM chip.

TABLE 11.10 | Eight-bit data bus connections

Data Lines	Memory Chip
$D_0 - D_3$	RAM-I
$D_4 - D_7$	RAM-II

- Same chip select, $\overline{CS}$ signal is given to both the RAM memories. Arrangement is given by single row and two columns.
- Same read and write, $R/\overline{W}$ signal is given to both the RAM memories.

When an address is selected, eight bits are made available on the data bus-four from each memory.

To generalize, two $2^n \times m$ bits RAM memories are expanded to give $2^n \times 2m$ bits as shown in Figure 11.82.

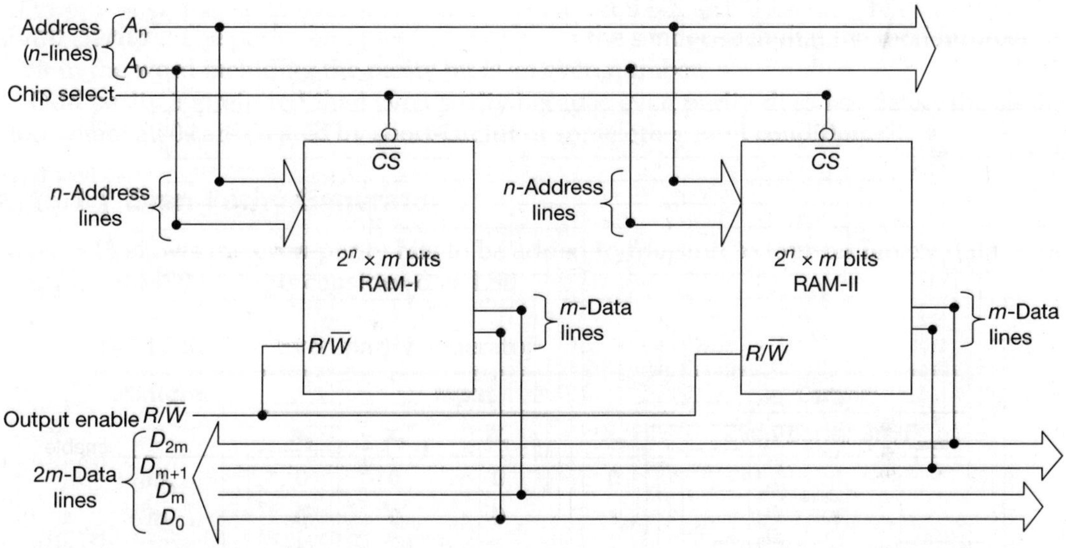

FIGURE 11.82 | Word length expansion ($2^n \times 2\,m$ bits RAM from $2^n \times m$ bits RAM)

- The address bus is commonly connected to both RAM memories so that the combination still has the same number of addresses, i.e. 2^n as each individual memory. Address bus is unidirectional.
- m-bit data buses ($D_0 - D_m$) from the two RAM memories are combined to form $2m$-bit data bus ($D_0 - D_{2m}$). On RAM read and write operations are performed. Hence, data bus is bidirectional. Table 11.11 explains the data provided by the RAM chip.

TABLE 11.11 | Eight-bit data bus connections

Data Lines	Memory Chip
$D_0 - D_m$	RAM-I
$D_{m+1} - D_{2m}$	RAM-II

- Same chip select, $\overline{CS}$ signal is given to both the RAM memories. Arrangement is given by single row and two columns.
- Same read and write, $R/\overline{W}$ signal is given to both the RAM memories.

When an address is selected, $2m$ bits are made available on the data bus of m-bits from each memory.

11.10.2 | Word Capacity Expansion

When memories are expanded to increase the word capacity, the number of addresses (memory locations) is increased. For example, 512 words (addresses) storage can be achieved by using two memories with 256 words (addresses) as shown in Figure 11.83. Two 256 × 8 bits (or 256 Byte or 256 B) ROM memories are expanded to give 512 × 8 bits (or 512 Bytes or 512 B) memory. The 256 B memory has 8 address lines ($2^8 = 256$) whereas 512 B memory requires 9-address lines ($2^9 = 512$).

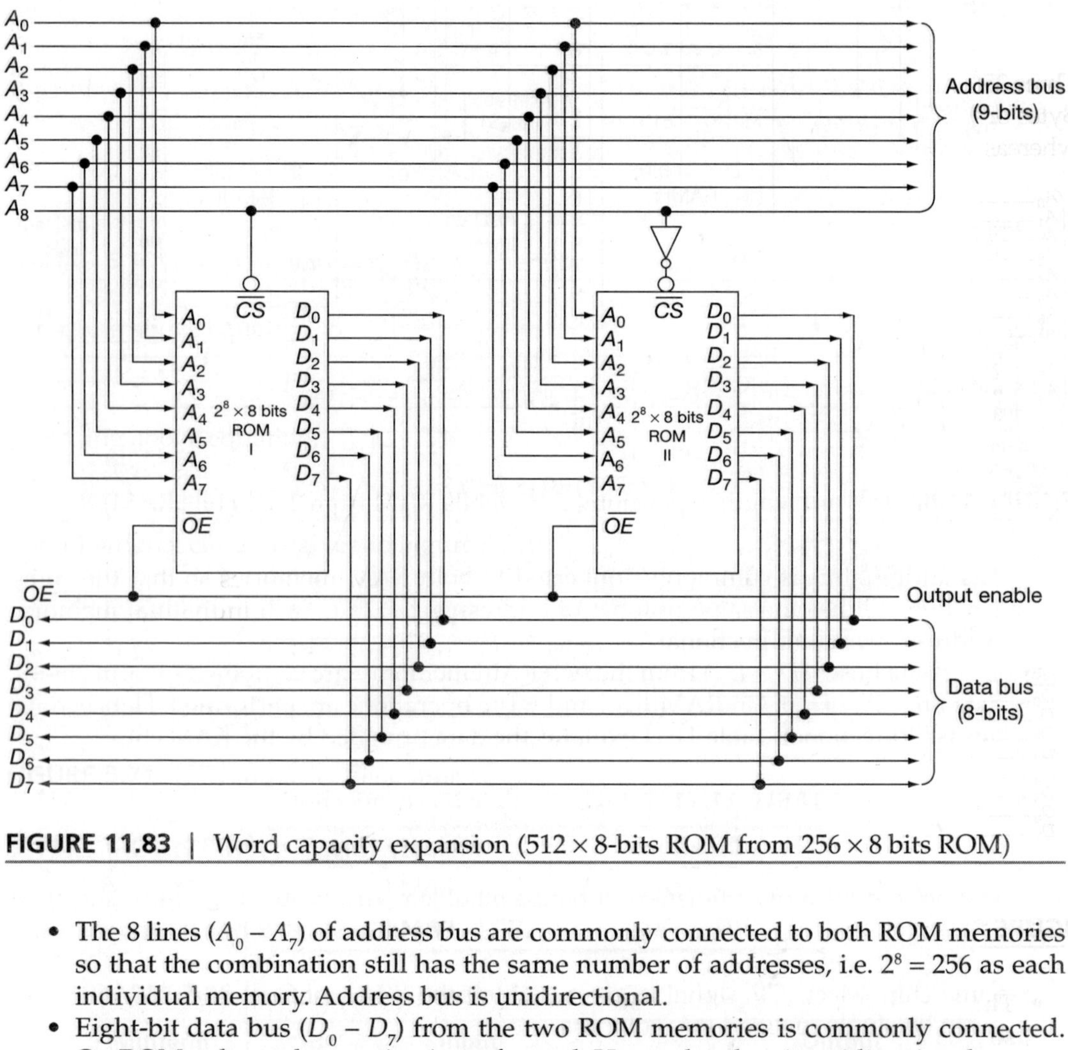

FIGURE 11.83 | Word capacity expansion (512 × 8-bits ROM from 256 × 8 bits ROM)

- The 8 lines ($A_0 - A_7$) of address bus are commonly connected to both ROM memories so that the combination still has the same number of addresses, i.e. $2^8 = 256$ as each individual memory. Address bus is unidirectional.
- Eight-bit data bus ($D_0 - D_7$) from the two ROM memories is commonly connected. On ROM only read operation is performed. Hence data bus is unidirectional.
- Most significant address line, i.e. 9th address line (A_8) act as control line to provide chip select, $\overline{CS}$ signal to the ROM memories. Arrangement is given by two rows and one column. One-to-two line decoder is used where input is address line A_8 and output 0 and 1 is used to select the ROM memory. Table 11.12 explains the range of address provided by the ROM chip.
- Same output enable, $\overline{OE}$ signal is given to both the ROM memories.

TABLE 11.12 | Chip select signal given by address lines

Signal on Address Line, A_8	Selected Memory Chip	Address Range	
		Decimal	Hex
0	ROM-I	0-255	000-0FFH
1	ROM-II	256-511	100-1FFH

Three 256 × 8 bits (or 256 Byte or 256 B) RAM memories are expanded to give 768 × 8 bits (or 768 Bytes or 768 B) memory as shown in Figure 11.84. 256 B memory has 8 address lines ($2^8 = 256$) whereas 768 B memory requires 10-address lines (because $2^9 = 512 < 768$ and 2^{10} is $1024 > 768$).

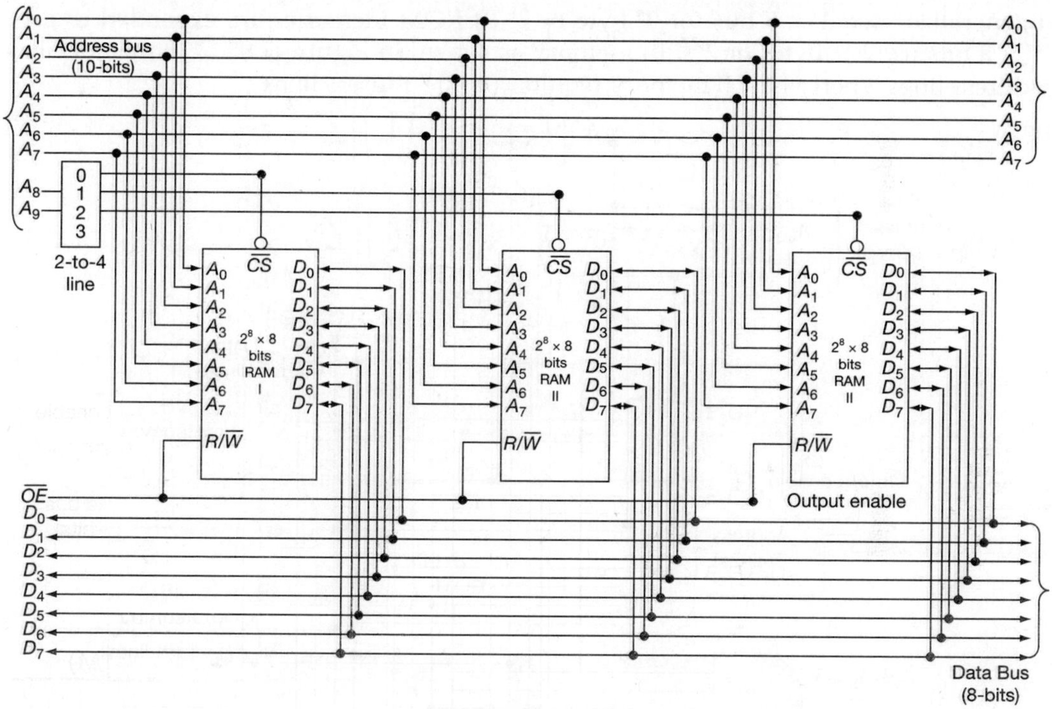

FIGURE 11.84 | Word capacity expansion (768 × 8 bits ROM from 256 × 8 bits ROM)

- The 8 lines ($A_0 - A_7$) of address bus are commonly connected to both RAM memories so that the combination still has the same number of addresses, i.e. $2^8 = 256$ as each individual memory. Address bus is unidirectional.
- Eight-bit data bus ($D_0 - D_7$) from the two RAM memories is commonly connected. On RAM, read and write operations are performed. Hence, data bus is bidirectional.
- Most significant address lines those are 9th and 10th address line (A_8 and A_9) act as control lines to provide chip select, $\overline{CS}$ signal to the three RAM memories. Arrangement is given by three rows and one column. Two-to-four line decoder is used where input is by address lines A_8 and A_9 and outputs 0, 1 and 2 are used to select the RAM memory. Table 11.13 explains the range of address provided by the RAM chip.
- Same read and write, $R/\overline{W}$ signal is given to three RAM memories.

TABLE 11.13 | Chip select signal given by address lines

Signal on Address Lines		Selected Memory Chip	Address Range	
A_9	A_8		Decimal	Hex
0	0	RAM-I	0-255	000-0FFH
0	1	RAM-II	256-511	100-1FFH
1	0	RAM-III	512-767	200-2FFH
1	1	Not used	–	–

To generalize, two $2^n \times 8$ bits (or 2^n byte or 2^n B) ROM memories are expanded to give $2^{n+1} \times 8$ bits (or 2^{n+1} Bytes or 2^{n+1} B) memory as shown in Figure 11.85. 2^n B memory has n-address lines whereas 2^{n+1} B memory requires $(n + 1)$-address lines

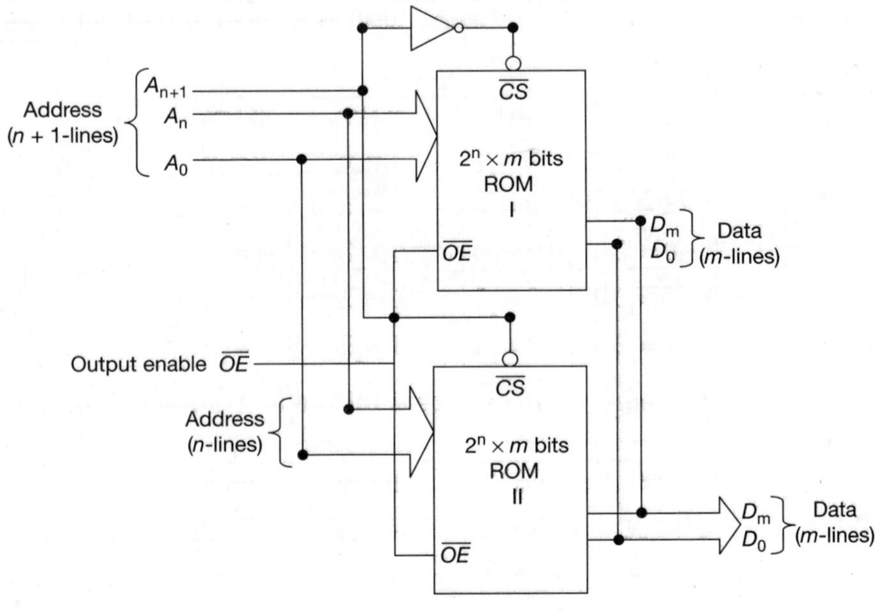

FIGURE 11.85 | Word capacity expansion ($2^{n+1} \times m$ bits ROM from $2^n \times m$ bits ROM)

- The n-lines ($A_0 - A_{n-1}$) of address bus are commonly connected to both ROM memories so that the combination still has the same number of addresses, i.e. 2^n as each individual memory. Address bus is unidirectional.
- Eight-bit data bus ($D_0 - D_7$) from the two ROM memories is commonly connected. On ROM only read operation is performed. Hence data bus is unidirectional.
- Most significant address line, i.e. $(n + 1)$th address line (A_n) act as control line to provide chip select, $\overline{CS}$ signal to the ROM memories. Arrangement is given by two rows and one column. One-to-two line decoder is used where input is address line A_n and output 0 and 1 is used to select the ROM memory. Table 11.14 explains the range of address provided by the ROM chip.
- Same output enable, $\overline{OE}$ signal is given to both the ROM memories.

TABLE 11.14 | Chip select signal given by address lines

Signal on Address Line, A_8	Selected Memory Chip	Address Range
		Decimal
0	ROM-I	$0 - (2^n - 1)$
1	ROM-II	$2^n - (2^{n+1} - 1)$

11.10.3 | Word Size and Capacity Expansion

To increase the word size of a memory, the number of bits in the data bus is increased. When memories are expanded to increase the word capacity, the number of addresses (memory locations) is increased. To generalize, two $2^n \times m$ bits RAM memories are expanded to give $2^{n+1} \times 2m$ bits memory as shown in Figure 11.86. $2^n \times m$ bits memory has n-address lines whereas $2^{n+1} \times 2m$ bits memory requires $(n + 1)$-address lines.

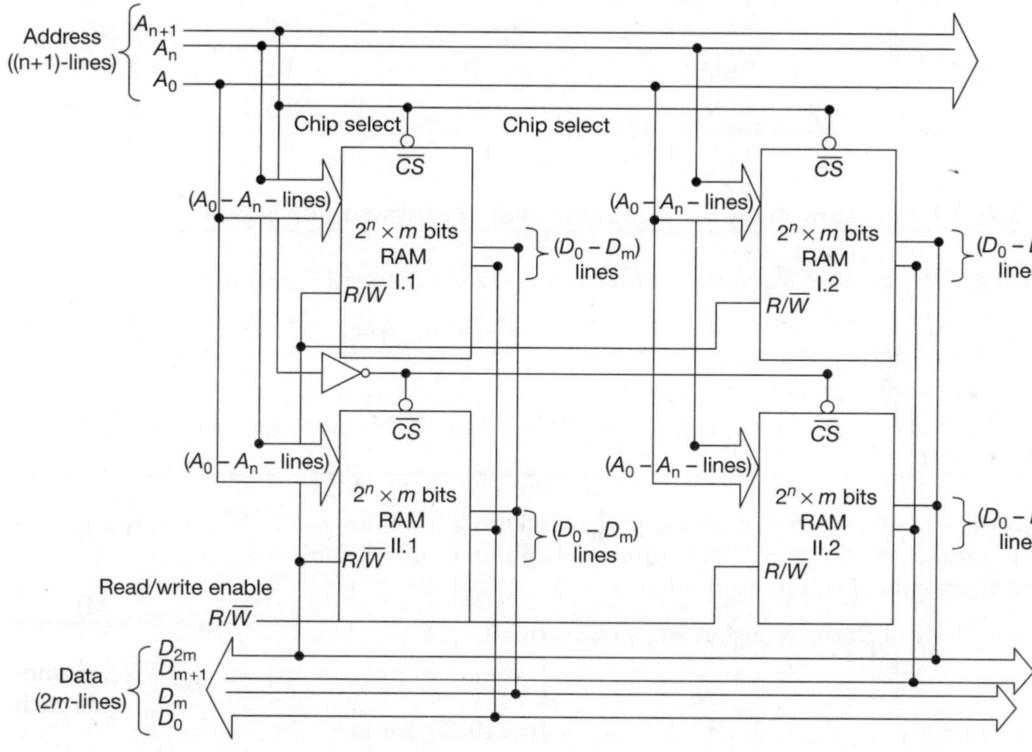

FIGURE 11.86 | Word expansion and word capacity expansion ($2^{n+1} \times 2m$ bits RAM from $2^n \times m$ bits RAM)

- The n-lines ($A_0 - A_7$) of address bus is commonly connected to four RAM memories so that the combination still has the same number of addresses, i.e. 2^n as each individual memory. Address bus is unidirectional.

- m-bit data buses (D_0 – D_m) from the two RAM memories are combined to form an $2m$-bit data bus (D_0 – D_{2m}). On RAM read and write operations are performed. Hence, data bus is bidirectional. Same chip select, $\overline{CS}$ signal is given to both the RAM memories. Arrangement is given by single row and two columns.
- Most significant address line, i.e. 9th address line (A_8) act as control line to provide chip select, $\overline{CS}$ signal to the RAM memories. Arrangement is given by two rows and two columns. One-to-two line decoder is used where input is address line A_8 and output 0 and 1 is used to select the RAM memory. Table 11.15 explains the range of address provided by the RAM chip along with the data.
- Same read and write, $R/\overline{W}$ signal is given to three RAM memories.

TABLE 11.15 | Chip select signal given by address lines

Signal on Address Line, A_8	Selected Memory Chip	Data Lines	Address Range	
			Decimal	Hex
0	RAM-I.1	$D_0 - D_m$	0-255	000-0FFH
0	RAM-I.2	$D_{m+1} - D_{2m}$	0-255	000-0FFH
1	RAM-II.1	$D_0 - D_m$	256-511	100-1FFH
1	RAM-II.2	$D_{m+1} - D_{2m}$	256-511	100-1FFH

EXAMPLE 11.17 How many flip-flops are required to design 4 kB SRAM?

SOLUTION Number of flip-flops $= 4 \times 2^{10} \times 8$ (1 k $= 2^{10}$ and 1 B $=$ 8 bits)

$= 4 \times 1024 \times 8 = 32{,}768$ **Ans.**

EXAMPLE 11.18 Compute the number of 1024×1 bit ROM chips required to design 16 kB ROM memory.

SOLUTION Let A_D is number of addresses of memory to be designed. D_D is the size of word of memory to be designed. A_U is number of addresses of available memory. D_U is the size of word of available memory.

Total capacity of ROM to design $(A_D \times D_D) = 16$ kB

$= 16 \times 2^{10} \times 8$ bits (1 k $= 2^{10}$ and 1 B $=$ 8 bits)

$= 16 \times 1024 \times 8$ bits

Number of addresses of memory to be designed, $A_D = 16 \times 1024$
Size of data of memory to be designed, $D_D = 8$ bits
Capacity of available (unit) ROM $(A_U \times D_U) = 1024 \times 1$ bits
Number of addresses of available memory, $A_D = 16 \times 1024$
Size of data of available memory, $D_D = 8$ bits

$$\text{Number of chips} = \frac{\text{Total capacity of memory to be designed}}{\text{Capacity of available memory}}$$

$$\text{Number of chips} = \frac{A_D}{A_U} \times \frac{D_D}{D_U}$$

$$\text{Number of chips} = \left(\frac{16 \times 1024}{1024}\right) \times \left(\frac{8 \text{ bits}}{1 \text{ bit}}\right) = 16 \times 8 = 128 \quad \textbf{Ans.}$$

> ***Note:***
> Chips are arranged in 16 rows and 8 columns. Number of columns represents the expansion of data size from 1 bit to 8 bits. Number of rows gives the expansion of capacity from 1 k to 16 k.

EXAMPLE 11.19 How many chips of 1 G × 4 bit DRAM chips are required to design 1 GB DRAM memory?

SOLUTION Let A_D is number of addresses of memory to be designed. D_D is the size of word of memory to be designed. A_U is number of addresses of available memory. D_U is the size of word of available memory.

Total capacity of DRAM to be designed = 1 GB

$$= 1 \times 2^{30} \times 8 \text{ bits} \qquad (1\text{ G} = 2^{30} \text{ and } 1\text{ B} = 8 \text{ bits})$$

Number of addresses of memory to be designed, $A_D = 1 \times 2^{30}$

Size of data of memory to be designed, $D_D = 8$ bits

Capacity of available (unit) DRAM = 1 G × 4 bits

$$= 1 \times 2^{30} \times 4 \text{ bits} \qquad (1\text{ G} = 2^{30})$$

Number of addresses of available memory, $A_D = 1 \times 2^{30}$

Size of data of available memory, $D_D = 4$ bits

$$\text{Number of chips} = \frac{\text{Total capacity of memory to be designed}}{\text{Capacity of available memory}}$$

$$\text{Number of chips} = \frac{A_D}{A_U} \times \frac{D_D}{D_U}$$

$$\text{Number of chips} = \left(\frac{1 \times 2^{30}}{1 \times 2^{30}}\right) \times \left(\frac{8 \text{ bits}}{4 \text{ bit}}\right) = 1 \times 2 = 2 \quad \textbf{Ans.}$$

> ***Note:***
> Chips are arranged in 1 row and 2 columns. (Number of columns represents the expansion of data size from 4 bits to 8 bits.)

EXAMPLE 11.20 What is the size of decoder required to expand 1 M × 4 bit DRAM chips to 4GB DRAM memory?

SOLUTION Let A_D be number of addresses of memory to be designed. D_D is the size of word of memory to be designed. A_U is number of addresses of available memory. D_U is the size of word of available memory.

Total capacity of DRAM to be designed = 4 GB.

$$= 4 \times 2^{30} \times 8 \text{ bits} \qquad (1\text{ G} = 2^{30} \text{ and } 1\text{ B} = 8 \text{ bits})$$

Number of addresses of memory to be designed, $A_D = 1 \times 2^{30}$
Size of data of memory to be designed, $D_D = 8$ bits
Capacity of available (unit) DRAM = 1 M × 4 bits

$$= 1 \times 2^{20} \times 4 \text{ bits} \qquad (1\text{ M} = 2^{20})$$

Number of addresses of available memory, $A_D = 1 \times 2^{30}$
Size of data of available memory, $D_D = 4$ bits

$$\text{Number of chips} = \frac{\text{Total capacity of memory to be designed}}{\text{Capacity of available memory}}$$

$$\text{Number of chips} = \frac{A_D}{A_U} \times \frac{D_D}{D_U}$$

$$\text{Number of chips} = \left(\frac{4\times 2^{30}}{1\times 2^{20}}\right) \times \left(\frac{8\text{ bits}}{4\text{ bit}}\right) = 4096 \times 2 = 8192 \quad \textbf{Ans.}$$

Size of decoder is n to 2^n lines
So, $2^n = 4096$ (number of rows)
Taking $\log_{10}$ both sides

$$\log 2^n = \log 4096$$

$$n \log 2 = \log 4096$$

$$n = \frac{\log 4096}{\log 2} = 12$$

Size of decoder is 12 to 4096 lines **Ans.**

> ***Note:***
> Chips are arranged in 4096 rows and 2 columns. Number of columns represents the expansion of data size from 4 bits to 8 bits. Number of rows gives the expansion of capacity from 1 M to 4 G

EXAMPLE 11.21 What is the size of decoder required to expand 1 kB SRAM chips to 64 kbits SRAM of 16-bit word size memory?

SOLUTION Let A_D be number of addresses of memory to be designed. D_D is the size of word of memory to be designed. A_U is number of addresses of available memory. D_U is the size of word of available memory.

Total capacity of SRAM to be designed = 64 kbits of 16-bit word size

$$= 64 \times 2^{10} \times 16 \text{ bits} \qquad (1\text{ k} = 2^{10})$$

Number of addresses of memory to be designed, $A_D = 64 \times 2^{10}$

Size of data of memory to be designed, $D_D = 16$ bits

Capacity of available (unit) SRAM = 1 kB

$$= 1 \times 2^{10} \times 8 \text{ bits} \qquad (1\text{ k} = 2^{10} \text{ and } 1\text{ B} = 8 \text{ bits})$$

Number of addresses of available memory, $A_D = 1 \times 2^{10}$

Size of data of available memory, $D_D = 8$ bits

$$\text{Number of chips} = \frac{\text{Total capacity of memory to be designed}}{\text{Capacity of available memory}}$$

$$\text{Number of chips} = \frac{A_D}{A_U} \times \frac{D_D}{D_U}$$

$$\text{Number of chips} = \left(\frac{64 \times 2^{10}}{1 \times 2^{10}}\right) \times \left(\frac{16 \text{ bits}}{8 \text{ bit}}\right) = 64 \times 2 = 128 \quad \textbf{Ans.}$$

Size of decoder is n to 2^n lines

So $2^n = 64$ (number of rows)

Taking $\log_{10}$ both sides

$$\log 2^n = \log 64$$

$$n \log 2 = \log 64$$

$$n = \frac{\log 64}{\log 2} = 6 \quad \textbf{Ans.}$$

> ***Note:***
> Chips are arranged in 64 rows and 2 columns. Number of columns represents the expansion of data size from 8 bits to 16-bit word. Number of rows gives the expansion of capacity from 1 k to 64 k.

Size of decoder is 12 to 4096 lines **Ans.**

EXAMPLE 11.22 1kB SRAM chips are expanded to 16 kbits SRAM of 16-bit word size memory. What is range of address from where data can be read if 4th line of decoder is selecting the chip?

SOLUTION Let A_D be number of addresses of memory to be designed. D_D is the size of word of memory to be designed. A_U is number of addresses of available memory. D_U is the size of word of available memory.

Total capacity of SRAM to be designed = 16 kbits of 16-bit word size

$$= 16 \times 2^{10} \times 16 \text{ bits} \qquad (1\text{ k} = 2^{10})$$

Number of addresses of memory to be designed, $A_D = 16 \times 2^{10}$

Size of data of memory to be designed, $D_D = 16$ bits

Capacity of available (unit) SRAM = 1 kB

$$= 1 \times 2^{10} \times 8 \text{ bits} \qquad (1\text{ k} = 2^{10} \text{ and } 1\text{ B} = 8 \text{ bits})$$

Number of addresses of available memory, $A_D = 1 \times 2^{10}$

Size of data of available memory, $D_D = 8$ bits

$$\text{Number of chips} = \frac{\text{Total capacity of memory to be designed}}{\text{Capacity of available memory}}$$

$$\text{Number of chips} = \frac{A_D}{A_U} \times \frac{D_D}{D_U}$$

$$\text{Number of chips} = \left(\frac{16 \times 2^{10}}{1 \times 2^{10}}\right) \times \left(\frac{16 \text{ bits}}{8 \text{ bit}}\right) = 16 \times 2 = 128 \quad \textbf{Ans.}$$

Size of decoder is n to 2^n lines gives the line number from 0 to $2^n - 1$.

$$2^n = 64 \text{ or } 2^n = 2^4 \text{ (number of rows), so } n = 4$$

Size of decoder is 4 to 16 lines that give the number from 0 to 15 ($2^4 - 1$).

Fourth line decoder has 3 numbers.

Address lines of expanded memory are computed as

$$2^n = 16 \times 2^{10}$$

$$n = \frac{\log\left(16 \times 2^{10}\right)}{\log 2} = 14$$

Address lines of available memory are computed as

$$2^n = 1 \times 2^{10}$$

$$n = \frac{\log\left(1 \times 2^{10}\right)}{\log 2} = 10$$

> ***Note:***
> Chips are arranged in 64 rows and 2 columns. Number of columns represents the expansion of data size from 8 bits to 16-bit word. Number of rows gives the expansion of capacity from 1 k to 64 k.

Table 11.16 gives address ranges for all four memory chips of 1 kB size. Address range 0000-03FFH belongs to memory chip 1. Memory chip 1 is selected by 0th line of decoder. The data can be read out or written into memory chip 1 of given address lies within the address range 0000-03FFH.

TABLE 11.16 | Address range

Memory Selected by Decoder Line	Range of Address	Address Lines		Hex Value
		Decoded Address Lines $A_{13} - A_{10}$	Common Address Lines of Available Memory $A_9 - A_0$	
0	Minimum	00 00	00 0000 0000	0000H
	Maximum	00 00	11 1111 1111	03FFH
1	Minimum	00 01	00 0000 0000	0400H
	Maximum	00 01	11 1111 1111	07FFH
2	Minimum	00 10	00 0000 0000	0800H
	Maximum	00 10	11 1111 1111	0BFFH
3	Minimum	00 11	00 0000 0000	0C00H
	Maximum	00 11	11 1111 1111	0FFFH

Address is 0C00 to 0FFFH **Ans.**

EXAMPLE 11.23 Implement Full-adder using ROM

SOLUTION There are 3-input binary variables; augend, a_2 and addend, a_1 and carry-in, a_0, and 2-output binary variables as sum, d_0 and carry-out, d_1. Truth table is constructed and is given in Table 11.17.

TABLE 11.17 | Truth table for full-adder

Minterm	Inputs			Outputs	
	Augend A_2	Addend A_1	Carry-in A_0	Carry-out D_1	Sum D_0
m_0	0	0	0	0	0
m_1	0	0	1	0	1
m_2	0	1	0	0	1
m_3	0	1	1	1	0
m_4	1	0	0	0	1
m_5	1	0	1	1	0
m_6	1	1	0	1	0
m_7	1	1	1	1	1

Boolean equations for two-outputs in standard minterm form are given below:

$$d_0(a_2, a_1, a_0) = \sum m(1,2,4,7) \tag{11.25}$$

$$d_1(a_2, a_1, a_0) = \sum m(3,5,6,7) \tag{11.26}$$

Logic expression of sum, D_0 for full-adder is given below.

$$d_0(a_2, a_1, a_0) = \bar{a}_2\bar{a}_1 a_0 + \bar{a}_2 a_1 \bar{a}_0 + a_2\bar{a}_1\bar{a}_0 + a_2 a_1 a_0 \tag{11.27}$$

Logic expression of sum, d_1 for full-adder is given below.

$$d_1(a_2, a_1, a_0) = \bar{a}_2 a_1 a_0 + a_2\bar{a}_1 a_0 + a_2 a_1\bar{a}_0 + a_2 a_1 a_0 \tag{11.28}$$

Number of inputs, $n = 3$.
Number of outputs, $m = 2$.
Number of AND gates required $= 2^n = 2^3 = 8$.
Number of OR gates required $= m = 2$.
Complete diagram is given in Figure 11.87(a). AND gates and OR gates are having fixed contacts in ROM. Figure 11.87(b) gives the diagram using decoder.

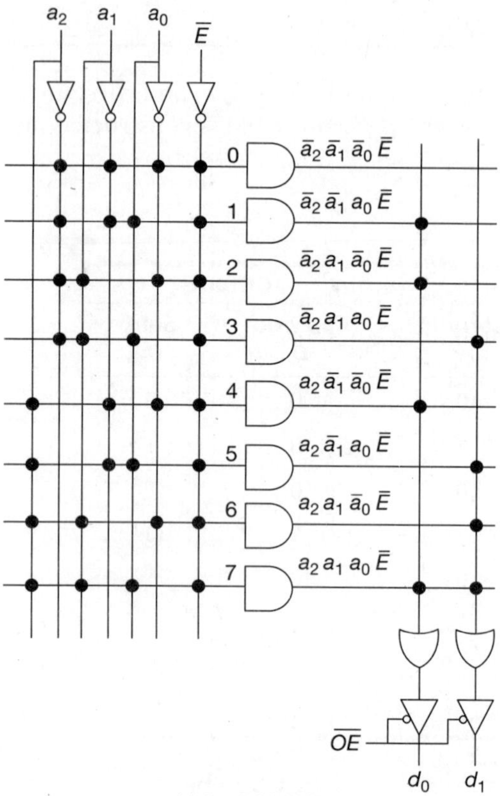

(a) 8 × 2 bit ROM

FIGURE 11.87(a) | 8 × 2 bit ROM

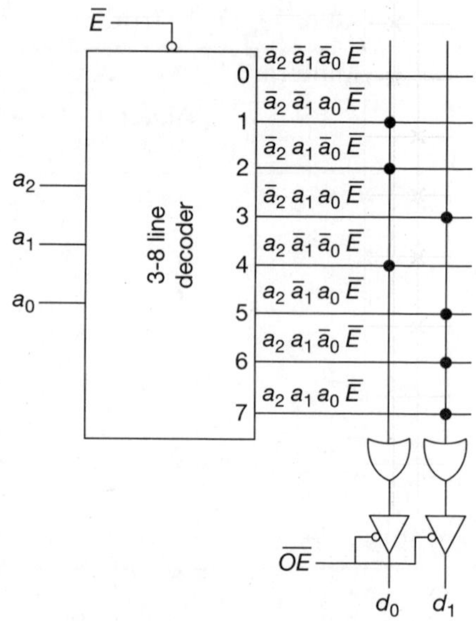

(b) 4 × 2 bit ROM

FIGURE 11.87(b) | 4 × 2 bit ROM

EXAMPLE 11.24 Implement full-adder using PAL.

SOLUTION There are three input binary variables: augend, a_2 and addend, a_1 and carry-in, a_0, and 2-output binary variables as sum, d_0 and carry-out, d_1. Truth table is constructed and is given in Table 11.17.

Outputs are given in Eqs. (11.27) and (11.28). Using K-map, simplified logic expression for carry-out, d_1 given by Eq. (11.28) of full-adder is simplified and is given below.

$$d_1(a_2, a_1, a_0) = a_1a_0 + a_2a_0 + a_2a_1 \tag{11.29}$$

Number of inputs, $n = 3$.
Number of outputs, $m = 2$.
Maximum product terms in the output function, $l = 4$.
(First function, d_0 has four product terms and second output function, d_1 has three product terms.)
Number of AND gates required = $m \times l = 2 \times 4 = 8$.

Number of OR gates required = $m = 2$.

Complete diagram is given in Figure 11.88(a). AND gates have flexible links and OR gates are fixed links. Figure 11.88(b) gives the diagram using decoder.

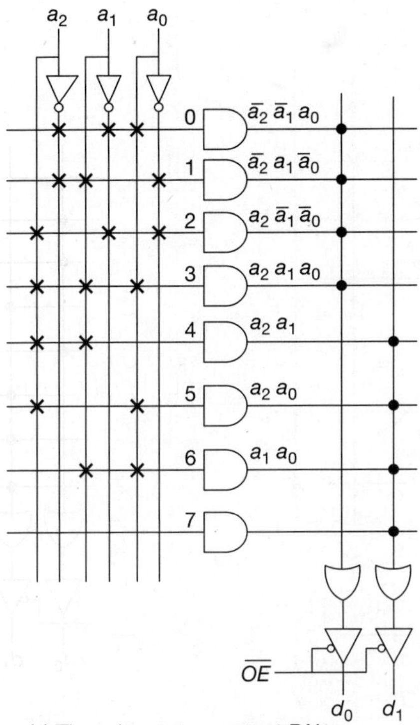

(a) Three-input, two-output PAL

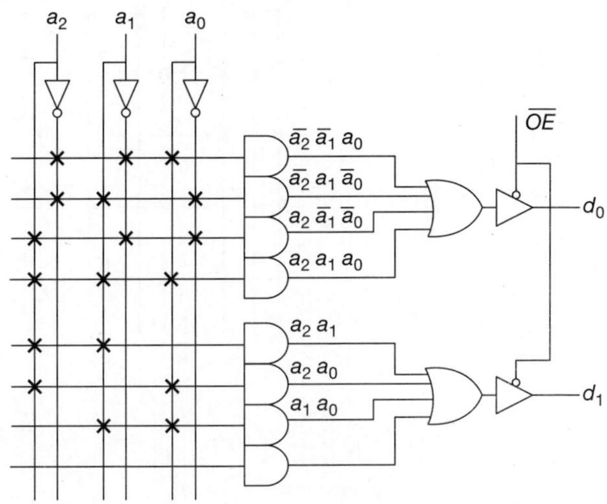

(b) Three-input, two-output PAL

FIGURE 11.88(a) | Three-input, two-output PAL

FIGURE 11.88(b) | Three-input, two-output PAL

EXAMPLE 11.25 Implement full-adder using PLA.

SOLUTION There are three input binary variables; augend, a_2 and addend, a_1 and carry-in, a_0, and two output binary variables as sum, d_0 and carry-out, d_1. Truth table is constructed and is given in Table 11.17. Outputs are given in Eqs. (11.27) and (11.28). Using K-map, simplified logic expression for carry-out, d_1 given by Eq. (11.28) of full-adder is simplified and is given by Eq. (11.29).

Number of inputs, $n = 2$.

Number of outputs, $m = 2$.

First function, d_0 has four product terms and second output function, d_1 has three product terms. Total number of different terms are 7(4 + 3)

Number of AND gates required = $4 + 3 = 7$.

Number of OR gates required = $m = 2$.

Complete diagram is given in Figure 11.89. AND gates have flexible links and OR gates are also flexible links.

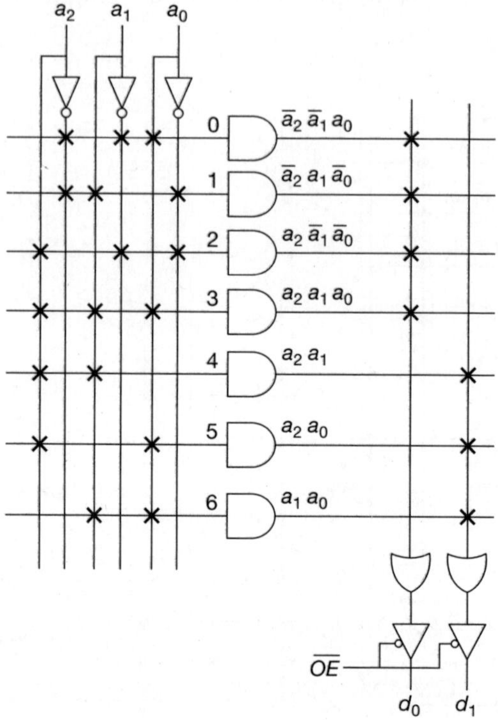

FIGURE 11.89 | Three-input, two-output PLA

EXAMPLE 11.26 Realize the following logical expression using ROM.

$$d_0(a_1, a_0) = \sum m(1,2,3) \text{ and } d_1(a_1, a_0) = \sum m(0,2,3)$$

SOLUTION The following function is given. It has two input variables and two output variables

$$d_0(a_1, a_0) = \sum m(1,2,3) = \bar{a}_1 a_0 + a_1\bar{a}_0 + a_1 a_0 \tag{11.30}$$

$$d_1(a_1, a_0) = \sum m(0,2,3) = \bar{a}_1\bar{a}_0 + a_1\bar{a}_0 + a_1 a_0 \tag{11.31}$$

Number of inputs, $n = 2$.
Number of outputs, $m = 2$.
Number of AND gates required $= 2^2 = 4$.
Number of OR gates required $= m = 2$.

Complete diagram is using AND and OR array is given in Figure 11.90(a). AND gates and OR gates have fixed links. Figure 11.90 (b) uses decoder to realize the function.

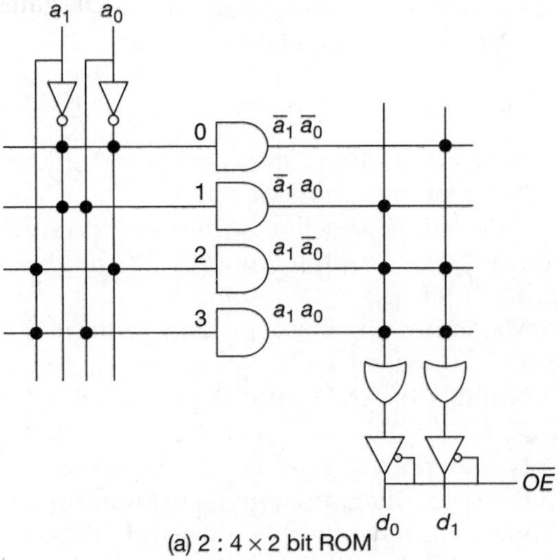

FIGURE 11.90(a) | 4 × 2bit ROM

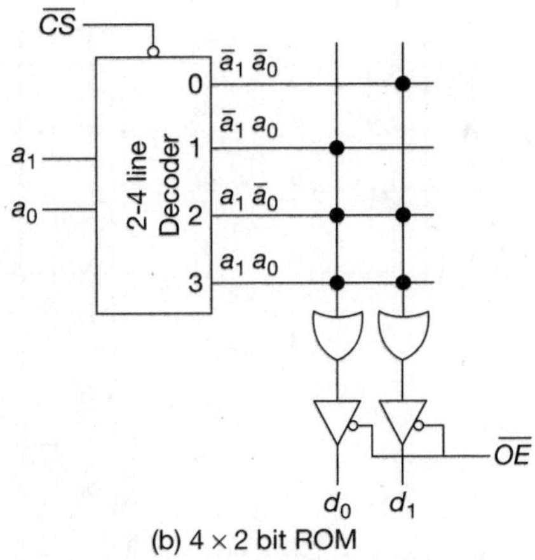

FIGURE 11.90(b) | 4 × 2bit ROM

EXAMPLE 11.27 Realize the following logical expression using PAL.

$$d_0(a_1, a_0) = \sum m(1,2,3) \text{ and } d_1(a_1, a_0) = \sum m(0,2,3)$$

SOLUTION The following function is given. It has two input variables and two output variables.

$$d_0(a_1, a_0) = \sum m(1,2,3) = \bar{a}_1 a_0 + a_1 \bar{a}_0 + a_1 a_0 \tag{11.32}$$

$$d_1(a_1, a_0) = \sum m(0,2,3) = \bar{a}_1 \bar{a}_0 + a_1 \bar{a}_0 + a_1 a_0 \tag{11.33}$$

K-map is used to simplify the expressions given in Figures 11.91(a) and 11.91(b) for function d_0 and d_1 respectively. Simplified logic expression of sum, d_0 of Eq. (11.32) is given below.

$$d_0(a_1, a_0) = a_0 + a_1 \tag{11.34}$$

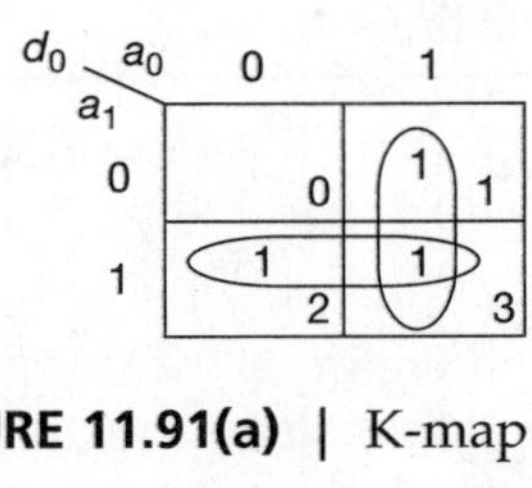

FIGURE 11.91(a) | K-map for d_0

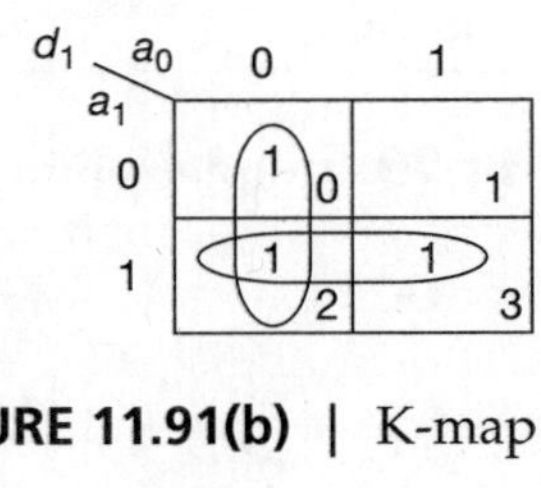

FIGURE 11.91(b) | K-map for d_1

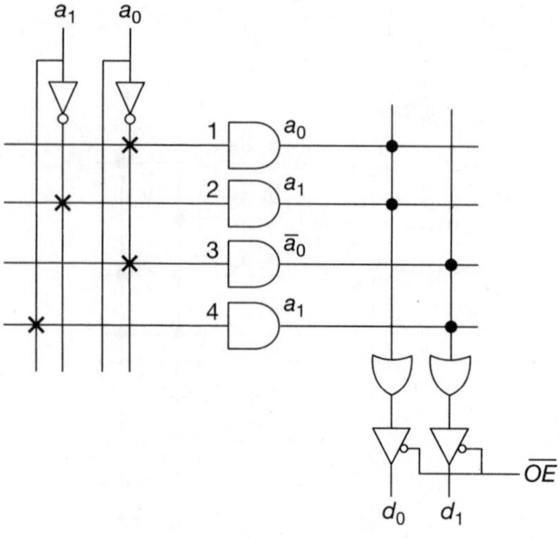

FIGURE 11.92 | Two-input, two-output PAL

Simplified logic expression of sum, $\bar{d}_1$ for Eq. (11.33) is given below.

$$d_1(a_1, a_0) = \bar{a}_0 + a_1 \tag{11.35}$$

Number of inputs, $n = 2$.

Number of outputs, $m = 2$.

First output function, d_0 has two product terms. Second output function, d_1 has two product terms.

Maximum number of product terms in the output functions, $l = 2$.

Number of AND gates required $= m \times l = 2 \times 2 = 4$.

Number of OR gates required $= m = 2$.

Complete diagram is given in Figure 11.92. AND gates have flexible links and OR gates are fixed links. Each function has two fixed OR-links.

EXAMPLE 11.28 Realize the following logical expression using PLA.

$$d_0(a_1, a_0) = \sum m(1,2,3) \text{ and } d_1(a_1, a_0) = \sum m(0,2,3)$$

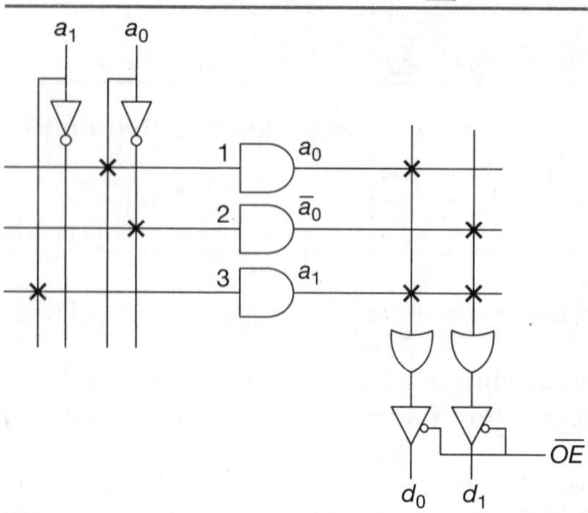

FIGURE 11.93 | Two-input, two-output PLA

SOLUTION The functions are simplified using K-map and are given in Eqs. (11.34) and (11.35) for d_0 and d_1, respectively. It has two input variables and two output variables.

Number of inputs, $n = 2$.

Number of outputs, $m = 2$.

First output function, d_0 has two product terms. Second output function, d_1 has two product terms.

Number of AND gates required $= 2 + 1 = 3$ (Sum of non-repeating terms).

Number of OR gates required $= m = 2$.

Complete diagram is given in Figure 11.93. AND gates have flexible links and OR gates are also flexible links.

EXAMPLE 11.29 Implement the following logics with ROM.

$$d_0(a_2, a_1, a_0) = \sum m(1,3,5,6)$$
$$d_1(a_2, a_1, a_0) = \sum m(2,3,4,5)$$
$$d_2(a_2, a_1, a_0) = \sum m(4,5,6,7)$$

SOLUTION There are three inputs represented by a_2, a_1 and a_0, where a_0 is LSB. There are three outputs represented by d_2, d_1 and d_0, where d_0 is LSB.

Standard minterms are

$$d_0(a_2,a_1,a_0) = \sum m(1,3,5,6) = \bar{a}_2\bar{a}_1 a_0 + \bar{a}_2 a_1 a_0 + a_2\bar{a}_1 a_0 + a_2 a_1\bar{a}_0 \quad (11.36)$$

$$d_1(a_2,a_1,a_0) = \sum m(2,3,4,5) = \bar{a}_2 a_1\bar{a}_0 + \bar{a}_2 a_1 a_0 + a_2\bar{a}_1\bar{a}_0 + a_2\bar{a}_1 a_0 \quad (11.37)$$

$$d_2(a_2,a_1,a_0) = \sum m(4,5,6,7) = a_2\bar{a}_1\bar{a}_0 + a_2\bar{a}_1 a_0 + a_2 a_1\bar{a}_0 + a_2 a_1 a_0 \quad (11.38)$$

Number of inputs, $n = 3$.
Number of outputs, $m = 3$.
Number of AND gates required $= 2^3 = 8$.
Number of OR gates required $= m = 3$.
Complete diagram is using AND and OR array is given in Figure 11.94. AND gates and OR gates have fixed links.

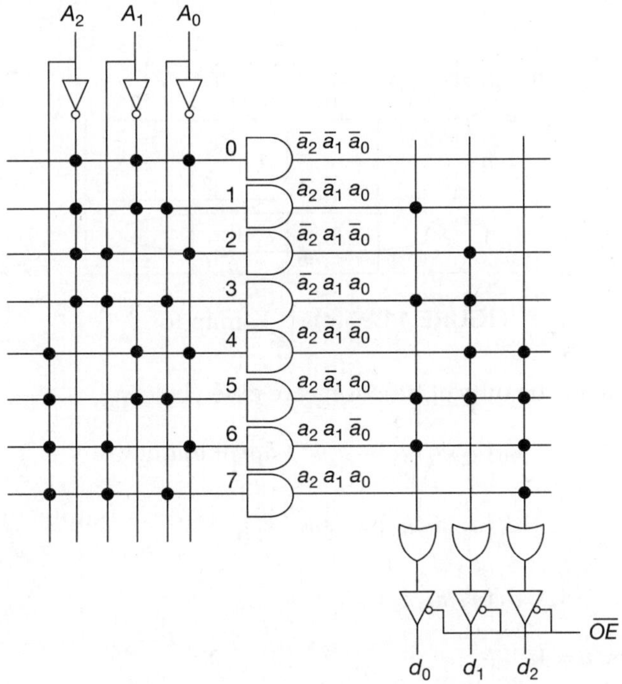

FIGURE 11.94 | 8 × 3 bit ROM

EXAMPLE 11.30 Implement the following logics with PAL.

$$d_0(a_2,a_1,a_0) = \sum m(1,3,5,6)$$

$$d_1(a_2,a_1,a_0) = \sum m(2,3,4,5)$$

$$d_2(a_2,a_1,a_0) = \sum m(4,5,6,7)$$

SOLUTION There are three inputs represented by a_2, a_1 and a_0, where a_0 is LSB. There are three outputs represented by d_2, d_1 and d_0, where d_0 is LSB. Standard minterms are given in Eq. (11.36), Eq. (11.37) and Eq. (11.38) for d_0, d_1 and d_2, respectively. K-map is used to solve the equation. Figure 11.95(a) gives the k-map set up for function d_0. Figure 11.95(b) gives the k-map set up for function d_1. Figure 11.95(c) gives the k-map set up for function d_2.

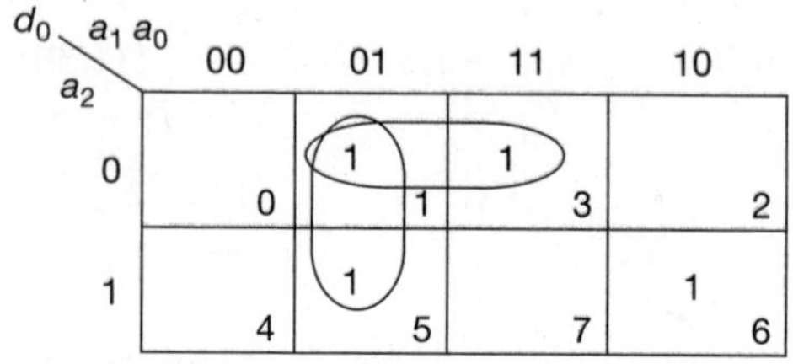

FIGURE 11.95(a) | K-map for d_0

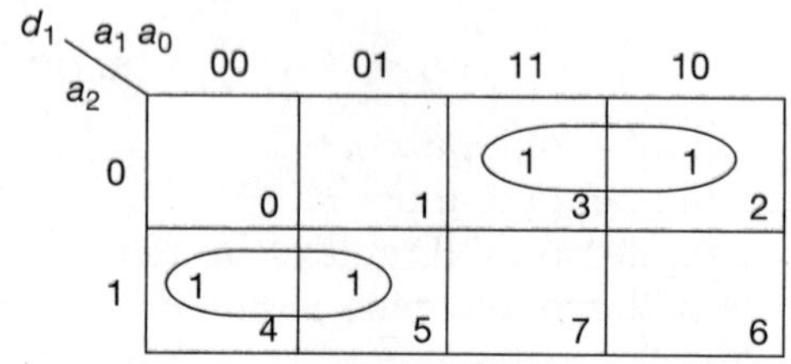

FIGURE 11.95(b) | K-map for d_1

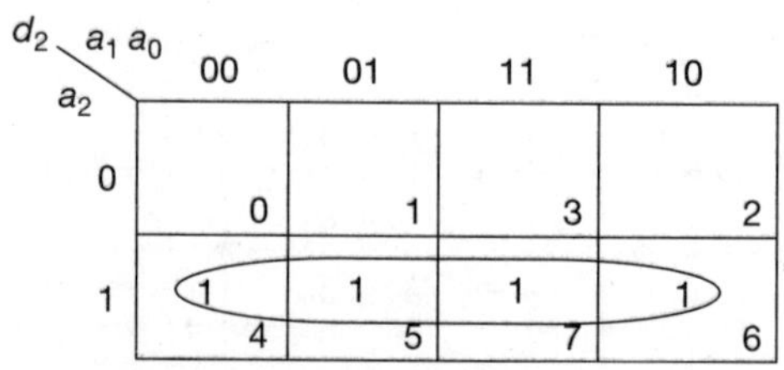

FIGURE 11.95(c) | K-map for d_2

Simplified equations for the output functions are given below.

$$d_0(a_2, a_1, a_0) = \bar{a}_2 a_0 + \bar{a}_1 a_0 + a_2 a_1 \bar{a}_0 \tag{11.39}$$

$$d_1(a_2, a_1, a_0) = \bar{a}_2 a_1 + a_2 \bar{a}_1 \tag{11.40}$$

$$d_2(a_2, a_1, a_0) = a_2 \tag{11.41}$$

Number of inputs, $n = 4$.
Number of outputs, $m = 4$.
First output function, d_0 has three product terms. Second output function, d_1 has two product terms. Third output function, d_2 has one product term. Maximum numbers of product terms are in output function d_1, i.e. 3.
Maximum number of product term in the output functions, $l = 3$.
Number of AND gates required $= m \times l = 3 \times 3 = 9$.
Number of OR gates required $= m = 3$.
Complete diagram is given in Figure 11.96. AND gates have flexible links and OR gates are fixed links. Each function has two fixed OR-links. Each function has four fixed OR-links.

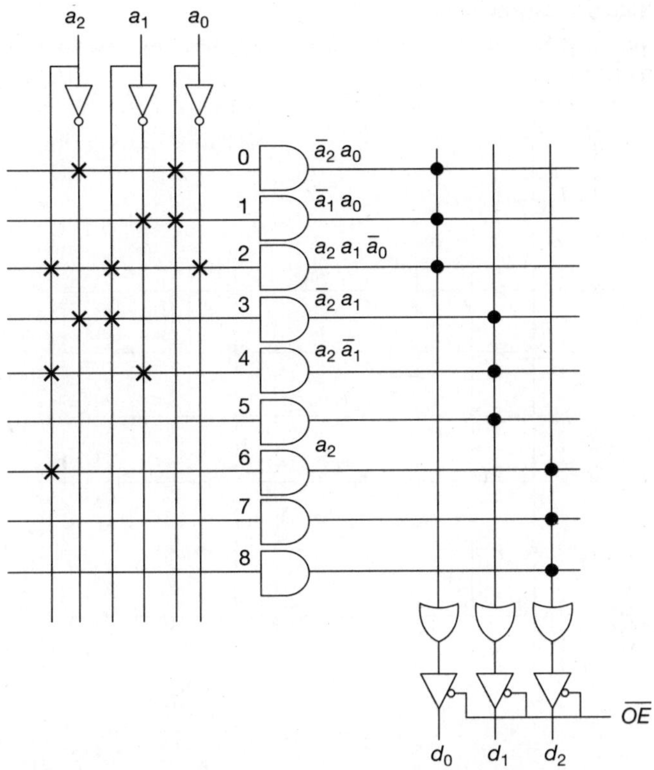

FIGURE 11.96 | Three-input, three-output PAL

EXAMPLE 11.31 Implement the following logics with PLA.

$$d_0(a_2, a_1, a_0) = \sum m(1,3,5,6)$$
$$d_1(a_2, a_1, a_0) = \sum m(2,3,4,5)$$
$$d_2(a_2, a_1, a_0) = \sum m(4,5,6,7)$$

SOLUTION There are three inputs represented by a_2, a_1 and a_0, where a_0 is LSB. There are three outputs represented by d_2, d_1 and d_0, where d_0 is LSB. Standard minterms are given in Eq. (11.36), Eq. (11.37) and Eq. (11.38) for d_0, d_1 and d_2, respectively. K-map is used to solve the equation. Figure 11.95(a) gives the k-map set up for function d_0. Figure 11.95(b) gives the k-map set up for function d_1. Figure 11.95(c) gives the k-map set up for function d_2. Simplified expressions are given in Eq. (11.39) for d_0, Eq. (11.40) for d_1 and Eq. (11.41) for d_2.

Number of inputs, $n = 3$.
Number of outputs, $m = 3$.
First output function, d_0 has three product terms. Second output function, d_1 has two product terms. Third output function, d_2 has one product term.
Number of AND gates required = 3 + 2 + 1 = 6 (Non-repeating terms).

Number of OR gates required = $m = 3$.

Complete diagram is given in Figure 11.97. AND gates have flexible links and OR gates are also flexible links.

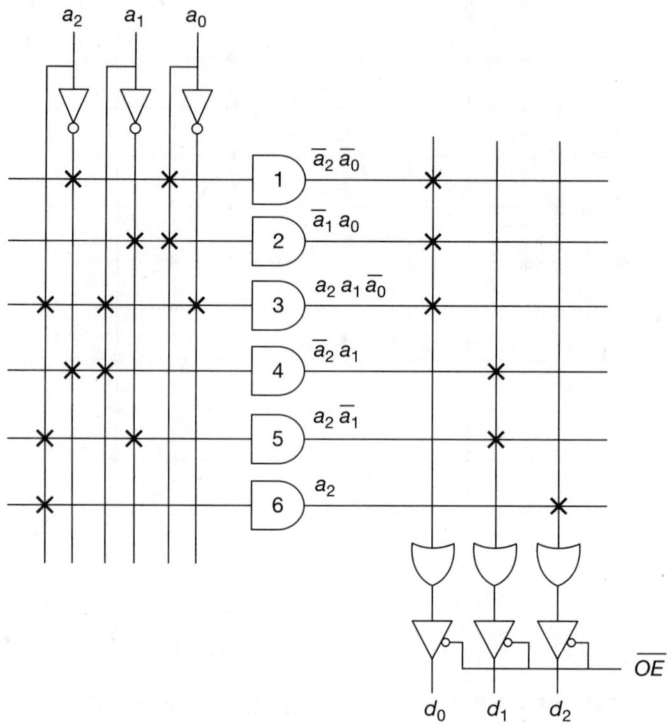

FIGURE 11.97 | Three-input, three-output PLA

SUMMARY

- Access time is the time required for reading or writing a memory location.
- Address is the binary code of a memory location.
- Address bus is a parallel array of conductors used for accessing a memory location.
- Address decoder is a n-line-to-2^n lines decoder used to select a specific memory chip or memory location.
- Antifuse is a programmable element invented by Actel Corporation named as PLICE (programmable low impedance circuit element).
- Asynchronous memory is a memory device in which read and write are unclocked.
- BEDO DRAM is a burst extended data output dynamic random-access memory.
- Bidirectional bus is a bus (group of electrical lines) capable of transmitting data in both the directions.
- Bidirectional FIFO memory is a FIFO memory capable of buffering data between two systems for read-and-write operation.

- BIOS Basic input output system is a set of programs in ROM that interfaces the input/output devices in a digital system.
- Burst is a memory feature which allows read from or write at up to four locations using a single address.
- Cache memory is a high-speed memory (normally SRAM) that stores the most recently used instructions or data from the slower main memory.
- CAM (Content-addressable memory) is a special purpose RAM device which can be accessed by its contents. It is also known as 'associative memory'.
- CCD (Charge-coupled device) is serial-access semiconductor memory device with very high bit packing density.
- Chip is a piece of silicon or other semiconductor material on which an IC is fabricated.
- Chip enable is an input control signal which when activated enables the chip.
- Chip select is an input control signal that selects the chip.
- CLB (Configurable logic block) is a logic block in an FPGA consisting of LMs and local programmable interconnect that is used to connect LMs within the CLB to generate required logic functions.
- Configuration memory is a memory in an FPGA meant to store bit pattern for configuring the FPGA.
- Control bus is a bus used for handling control signals.
- CPLD (Complex-programmable logic device) is a PLD containing a large number of equivalent gates.
- Cycle time is the minimum time between successive read or write cycles in a memory.
- Data bus is a bus used for carrying data.
- Data-rate buffer is a buffer that allows two digital systems with different data rates to communicate.
- DDR refers to Double data rate.
- DRAM refers to dynamic random-access memory.
- Dynamic memory is a memory in which data needs to be refreshed periodically. Data is stored on MOS capacitors.
- EAROM (Electrically alterable read-only memory) is a read-only memory in which the stored words can be erased electrically. Also known as EE^2PROM (Electrically erasable and programmable read-only memory).
- EDO DRAM refers to extended data output dynamic random access memory.
- EPROM refers to Erasable programmable read-only memory. It is a ROM which can be erased by exposure to UV light, and then reloaded with new information.
- First-in, First-out (FIFO) memory is a static random-access memory array with two ports in which words data are read out in the same order in which these were written in.
- Flash memory is a specific type of EEPROM, a non-volatile memory which allows in-circuit writing.
- FPGA (Field-programmable gate array) is a PLD containing very large number of equivalent gates.
- FPM DRAM refers to fast-page mode dynamic random-access memory.
- Fusible link A link of nichrome, or some other materials, used in PROMs and other programmable devices which can be either burnt or kept intact while storing 0's and l's in these devices.

- GAL (Generic array logic) is a type of configurable PAL.
- ISP (in-system programming) is a technique for programming large PLDs such as CPLDs and FPGAs in which the device is programmed on the circuit board itself rather than in a programming unit.
- Logic element (or logic module) is a basic unit of logic in an FPGA that usually contains an LUT, a D-FF, and a MUX.
- LUT (Look-up table) is a type of memory that is programmable for creating the desired logic function in SOP term.
- Macrocell is the basic logic function provided in large PLDs which are used to design logic circuits by configuring them.
- Non-volatile memory is a memory that does not lose its contents when power is turned off such as ROMs.
- Non-volatile RAM is a flash memory is also referred to as a non-volatile random-access memory.
- Programming of memory is to store the desired data in a programmable memory which is of read-only type.
- PAL (Programmable array logic) is a programmable AND array that provides product terms that feed fixed OR gates.
- PLA (Programmable logic array) is a programmable AND array whose product term outputs feed a programmable OR array.
- PLD (Programmable logic device) is an IC which is programmed (configured) for designing digital circuits.
- RAM (Random-access memory) is a read-and-write semiconductor memory in which any memory location can be accessed for reading or writing at random.
- ROM (Read-only memory) is a semiconductor memory in which data can only be read.
- SDRAM refers to synchronous DRAM.
- Serial access memory is a memory in which the access time of a stored bit or word depends on its location in the memory.
- Volatile memory loses its contents when power is turned off.

MULTIPLE CHOICE QUESTIONS

11.1 A 32 × 10 ROM contains a decoder of size
(a) 5 to 32 lines
(b) 32 to 32 lines
(c) 32 to 5 lines
(d) 10 to 32 lines

11.2 A 16 × 5 ROM stores
(a) 4 words of I6 bits each
(b) 16 words of 5 bits each
(c) 16 words of 4 bits each
(d) 5 words of 16 bits each

11.3 A combinational PLD with a fixed AND array and a programmable OR array is called a
(a) PLD (b) PROM
(c) PAL (d) PLA

11.4 A combinational PLD with a programmable AND array and a fixed OR array is called a
(a) PLD (b) PROM
(c) PAL (d) PLA

11.5 A combinational PLD with a programmable AND array and a programmable OR array is called a
(a) PLD (b) PROM
(c) PAL (d) PLA

11.6 The address bus with a ROM of size 1024 × 8 bits is

(a) 8 bits (b) 10 bits
(c) 12 bits (d) 16 bits

11.7 The decimal equivalent of the highest possible address for an 8-bit address bus is
(a) 8 (b) 128
(c) 255 (d) 256

11.8 The data bus width of a ROM of size 2048 × 8 bits is
(a) 8 (b) 10
(c) 12 (d) 16

11.9 A ROM has a 16-bit address bus. The number of locations in this memory is
(a) 16 (b) 32
(c) 1024 (d) 65536

11.10 It is desired to have a 64 × 8 ROM. The ROMs available are of 16 × 4 size. The number of ROMs required will be
(a) 8 (b) 6
(c) 4 (d) 2

11.11 Four ROM chips of 16 × 4 size have their address buses connected together. This system will be of size
(a) 64 × 4 (b) 16 × 16
(c) 32 × 8 (d) 256 × 1

11.12 In a read-only memory information can be stored
(a) at the time of fabrication
(b) by the user only once during its life time
(c) by the user a number of times
(d) in any of the above ways depending upon the type of memory

11.13 The process of entering information in EPROM is commonly known as
(a) writing
(b) programming
(c) storing
(d) none of the above

11.14 An EPROM is fabricated using
(a) TTL bipolar technology
(b) MOS technology
(c) ECL technology
(d) I^2 L technology

11.15 The capacity of a PLA is specified in terms of the
(a) number of inputs only
(b) number of outputs only
(c) number of inputs and outputs only
(d) number of inputs, product terms and outputs

Answers

11.1 (a) 11.2 (b) 11.3 (b) 11.4 (c) 11.5 (d) 11.6 (b) 11.7 (c)
11.8 (a) 11.9 (d) 11.10 (a) 11.11 (b) 11.12 (b) 11.13 (b) 11.14 (b)
11.15 (d)

QUESTIONS

11.1 Give the comparison between PROM, PLA and PAL.
11.2 What are the advantages and disadvantages of using a PROM as a PLD?
11.3 How is the capacity of a PLA specified?
11.4 Are PLAs and FPLAs volatile or non-volatile?
11.5 Are erasable and programmable PLAs available?
11.6 What is a ROM?

11.7 What are the technologies used for the fabrication of ROMs?

11.8 A memory has 16-bit address bus. How many locations are there in this?

11.9 What is a PROM?

11.10 In an EPROM chip, is it possible to erase the contents of only some of the locations? If not, why?

11.11 What is main memory?

11.12 What is peripheral memory?

11.13 What is volatile memory? What is non-volatile memory?

11.14 What is random access memory?

11.15 Distinguish between dynamic and static random access memory.

11.16 What is FIFO memory? What is LIFO memory?

11.17 What is CCD memory? What is its principle of operation and its advantage?

11.18 What is CD? What is its principle of operation and its advantage?

11.19 What is memory expansion? Why is it required?

11.20 Define the following terms: memory cell, address, access time, memory word, and byte.

PROBLEMS

11.1 Design a BCD-to-XS-3 code converter using a (a) PROM, (b) PLA and (c) PAL.

11.2 Design an XS-3-to-BCD code converter using (a) PROM, (b) PLA and (c) PAL.

11.3 Implement the following Boolean functions using PLA:

$$f_1(X,Y,Z) = \sum m(0,1,3,5),\ f_2(X,Y,Z) = \sum m(0,3,5,7)$$

11.4 Implement the following Boolean functions using PAL:

$$f_1(W,X,Y,Z) = \sum m(0,2,5,7,8,10,12,13)$$

$$f_2(W,X,Y,Z) = \sum m(0,2,6,8,9,14,15)$$

11.5 Tabulate the PLA programming table for the Boolean functions listed below:

$$f_1(a,b,c) = \sum m(0,1,2,4,6)$$

$$f_2(a,b,c) = \sum m(0,2,6,7)$$

$$f_3(a,b,c) = \sum m(3,6)$$

12

Analog-to-Digital Conversion

CHAPTER OBJECTIVES

The main goal of this chapter is to impart knowledge about the digital to analog and analog to digital converters. Readers will be able to discuss the following aspects in this chapter:

- Variable resistor networks
- Binary ladders
- Digital-to-analog converters (DAC)
- Digital-to-analog converter (DAC)—accuracy and resolution
- Analog-to-digital converter (ADC)
- Analog-to-digital converter (ADC)—Simultaneous conversion
- Analog-to-digital (A/D) converter(ADC)—Counter method
- Continuous analog to digital converter (ADC)
- Analog-to-digital converter (ADC)—Successive-approximation technique
- Dual-slope analog to digital converter (ADC)
- Analog-to-digital converter (ADC)—Accuracy and resolution

12.1 | INTRODUCTION

Real life systems produce analog signals that carry information. An analog signal is defined over a continuous period of time and its amplitude is assumed to have continuous range of values. It is difficult to store, compare, calculate, and manipulate analog information with good accuracy using purely analog technology. Digital signal processing technology has a number of advantages over analog signal processing. So, digital systems are preferred over analog systems in most of the applications, But mostly the real data is available in analog form, like, voice signal, output of the sensors etc. So, it becomes necessary to convert analog signals into digital signals. An analog-to-digital converter (ADC) is used for this purpose. After processing the digital data by a digital system, it is transmitted through an analog channel or measured by analog equipment. The output of a digital system is converted into an analog signal. Digital-to-analog conversion (DAC) involves conversion of digital information into equivalent analog information. As an example, the output of a digital system is changed to analog form for the purpose of driving a pen recorder. An analog signal is required for the servomotors to drive the cursor arms of a plotter. In this respect, a DAC is considered a decoding device.

The process of changing an analog signal to an equivalent digital signal is accomplished by the use of an ADC. For example, an ADC is used to change the analog output signals from transducers (measuring temperature, pressure, vibration, etc.) into equivalent digital signals. These signals are in suitable form for entry into a digital system. Analog-to-digital converter is referred to as an encoding device since it is used to encode signals that act as input to a digital system.

Digital-to-analog conversion is a straightforward process and is considerably easier than ADC. In fact a DAC is usually an integral part of any ADC. The following section discusses DAC process.

12.2 | VARIABLE RESISTOR NETWORKS

A digital signal is converted into an equivalent analog signal, by changing the '*n*' digital voltage levels into one equivalent analog voltage. A resistive network is designed that changes each digital level into an equivalent binary weighted voltage or current.

Let 2-bit binary signal shown in Figure 12.1 whose analog equivalent is required. Table 12.1 gives four possible combinations from 00 to 11. There is need to convert four possible digital signals into equivalent analog voltage. The output analog signal is in the range of 0 to 1 V as given in Table 12.1. Voltage signal ranging from 0 to 1 V is scaled to 0 to 3 V by scaling factor of $3\left(2^2-1\right)$ V.

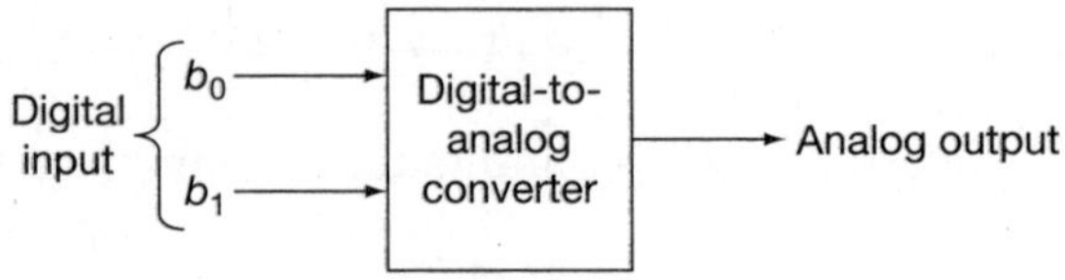

FIGURE 12.1 | Block diagram of 2-bit DAC

TABLE 12.1 | Analog output equivalent to digital input of 2-bits

Digital Input		Analog Output	
$b_1 \times 2^1$	$b_0 \times 2^0$	$\frac{1}{\left(2^2-1\right)}\left(b_1 2^1 + b_0 2^0\right) V_m$	$\frac{1}{\left(2^2-1\right)}\left(b_1 2^1 + b_0 2^0\right) \times 3\text{ V}$
0	0	0/3 V	0 V
0	1	1/3 V	1 V
1	0	2/3 V	2 V
1	1	3/3 V	3 V

The mathematical relation is given as

$$V_A = \frac{1}{\left(2^2-1\right)}\left(b_1 2^1 + b_0 2^0\right) V_m V \tag{12.1}$$

where b_0 and b_1 are binary digits.

Similarly, 3-bit binary signal shown in Figure 12.2 whose analog equivalent is required. Table 12.2 gives eight possible combinations from 000 to 111.

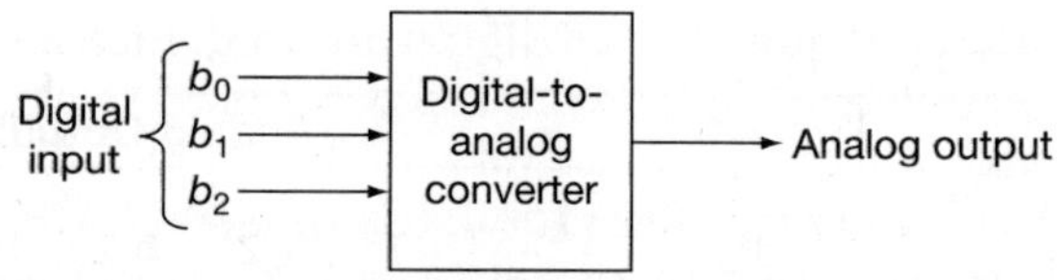

FIGURE 12.2 | Block diagram of 3-bit DAC

TABLE 12.2 | Analog output equivalent to digital input of three bits

Digital Input			Analog Output	
$b_2 \times 2^2$	$b_1 \times 2^1$	$b_0 \times 2^0$	$\frac{1}{(2^3-1)}(b_2 2^2 + b_1 2^1 + b_0 2^0)V_m$	$\frac{1}{(2^3-1)}(b_2 2^2 + b_1 2^1 + b_0 2^0)(2^3-1)\ V$
0	0	0	0/7 V	0 V
0	0	1	1/7 V	1 V
0	1	0	2/7 V	2 V
0	1	1	3/7 V	3 V
1	0	0	4/7 V	4 V
1	0	1	5/7 V	5 V
1	1	0	6/7 V	6 V
1	1	1	7/7 V	7 V

There is need to convert eight possible digital signals into equivalent analog voltage. The output analog signal is in the range of 0 to 1 V as given in Table 12.2. Voltage signal ranging from 0 to 1 V is scaled to 0 to 7 V by scaling factor of $7\left(2^3-1\right)$V.

The mathematical relation is given as

$$V_A = \frac{1}{\left(2^3-1\right)}\left(b_2 2^2 + b_1 2^1 + b_1 2^0\right)V_m V \tag{12.2}$$

where b_0, b_1 and b_2 are binary digits.

Similarly, 4-bit binary signal shown in Figure 12.3 whose analog equivalent is required. Table 12.3 gives 16 possible combinations from 0000 to 1111. There is need to convert sixteen possible digital signals into equivalent analog voltage. The output analog signal is in the range of 0 to 1 V as given in Table 12.3.

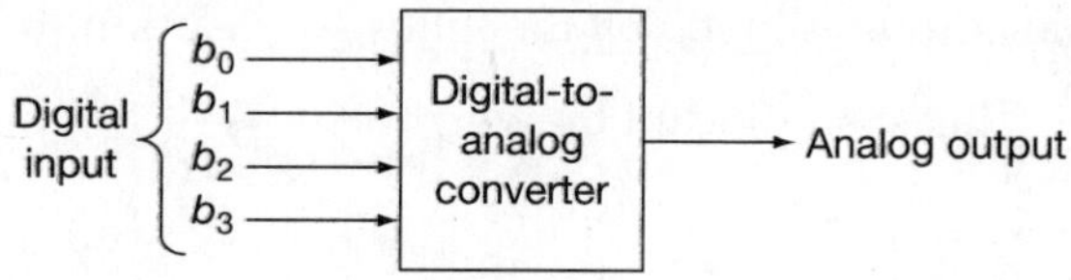

FIGURE 12.3 | Block diagram of 4-bit DAC

TABLE 12.3 | Analog output equivalent to digital input of three bits

Digital Input				Analog Output	
$b_3 \times 2^3$	$b_2 \times 2^2$	$b_1 \times 2^1$	$b_0 \times 2^0$	$V_A = \frac{1}{(2^4-1)}(b_3 2^3 + b_2 2^2 + b_1 2^1 + b_1 2^0)V_m$	$V = V_A \times 15\text{V}$
0	0	0	0	0/15 V	0 V
0	0	0	1	1/15 V	1 V
0	0	1	0	2/15 V	2 V
0	0	1	1	3/15 V	3 V
0	1	0	0	4/15 V	4 V
0	1	0	1	5/15 V	5 V
0	1	1	0	6/15 V	6 V
0	1	1	1	7/15 V	7 V
1	0	0	0	8/15 V	8 V
1	0	0	1	9/15 V	9 V
1	0	1	0	10/15 V	10 V
1	0	1	1	11/15 V	11 V
1	1	0	0	12/15 V	12 V
1	1	0	1	13/15 V	13 V
1	1	1	0	14/15 V	14 V
1	1	1	1	15/15 V	15 V

The mathematical relation is given as

$$V_A = \frac{1}{(2^4-1)}\left(b_3 2^3 + b_2 2^2 + b_1 2^1 + b_0 2^0\right)V_m V \tag{12.3}$$

where b_0, b_1, b_2 and b_3 are binary digits.

The generalized mathematical relation to convert digital information to analog information is given below:

$$V_A = \frac{1}{(2^n-1)}\left(\sum_{i=1}^{n} b_{i-1} 2^{i-1}\right) \times V_m V \tag{12.4}$$

where b_0, b_1, …, b_{n-1} are the binary digits having either 0 or 1 value of n-bit DAC.

Weight to least significant bit is given by the following relation.

$$\text{Least significant bit weight} = \frac{1}{(2^n-1)} \tag{12.5}$$

12.3 | RESISTIVE DIVIDER

The resistive divider circuit is designed to convert the digital input into an equivalent analog output voltage meeting the following requirements:

- The b_0 bit having weightage 2^0 is to be changed to +1 V. The b_1 bit having weightage 2^1 is changed to +2 V, b_2 bit having weightage 2^2 is changed to +4 V and b_{n-1} bit having weightage 2^{n-1} is changed to $+2^{n-1}$ V.
- These 2^{n-1} voltages representing the digital bits are summed together to form the analog output voltage.

Millman's Theorem: Millman's theorem states that the voltage appearing at any node in a resistive network is equal to the sum of currents entering the node (the node voltage is zero) divided by the sum of the conductances connected to the node.

Mathematically Millman's theorem is stated as

$$V_A = \frac{\sum_{i=1}^{n} \frac{V_{i-1}}{R_{i-1}}}{\sum_{i=1}^{n} \frac{1}{R_{i-1}}} \tag{12.6}$$

Let $V_{i-1} = b_{i-1}V,\ R_{i-1} = \frac{R_0}{2^{i-1}}$ $(i = 1, 2, \ldots, n)$ and b_{i-1} is digital input either 0 or 1.

Substitute the assumption in Eq. (12.6) to get

$$V_A = \frac{\sum_{i=1}^{n} \frac{b_{i-1}2^{i-1}}{R_0}V}{\sum_{i=1}^{n} \frac{2^{i-1}}{R_0}} \quad \text{or} \quad V_A = \frac{\sum_{i=1}^{n} b_{i-1}2^{i-1}V}{\sum_{i=1}^{n} 2^{i-1}}$$

$$\text{or } V_A = \frac{\sum_{i=1}^{n} b_{i-1}2^{i-1}V}{2^n - 1} \quad \left(\because 2^n - 1 = \sum_{i=1}^{n-1} 2^{i-1}\right)$$

$$V_A = \frac{1}{(2^n - 1)}\left(\sum_{i=1}^{n} b_{i-1}2^{i-1}\right)V \tag{12.7}$$

A resistive divider can be built to change a digital voltage into an equivalent analog voltage. The following criteria are applied to resistive divider:

- There is one input resistor for each digital bit. Beginning with the least significant bit (LSB), each following resistor value is one-half the size of the previous resistor.
- The full-scale output voltage is equal to the positive voltage of the digital input signal. The divider works with input voltages of 0 and $-V$ also.
- The LSB has a weight of $1/(2^n - 1)$, where n is the number of input bits.
- The change in output voltage due to a change in the LSB is equal to $V/(2^n - 1)$, where V is the digital input voltage level.
- The output voltage V_A can be obtained for any digital input signal by using Eq. (12.7), which is obtained using Millman's theorem.

A resistive divider which performs above-mentioned functions for two bits is shown in Figure 12.4. Resistors R_0 and R_1 form the divider network. Resistance R_L represents the load to which the divider is connected and is considered to be large enough that it does not load the divider network. Presume digital input is 10. Binary bit 0 represents 0 V and binary bit 1 represents 1 volt. Equivalent circuit is shown in Figure 12.5. Resistance R_L being large is neglected.

Apply Millman's theorem to the circuit of Figure 12.5 to get where $b_0 = 0$ and $b_1 = 1$

$$V_A = \frac{1}{\left(2^2-1\right)}\left(\sum_{i=1}^{2} b_{i-1}2^{i-1}V\right) = \frac{1}{3}\left(0\times 2^0 V + 1\times 2^1 V\right) = \frac{2}{3}\text{ V (volts)}$$

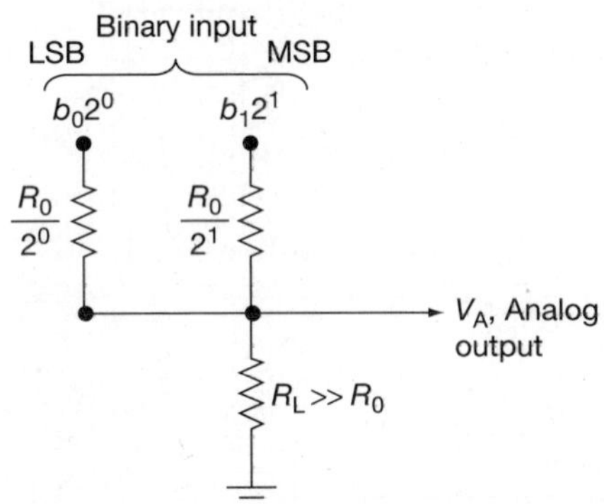

FIGURE 12.4 | Resistive divider

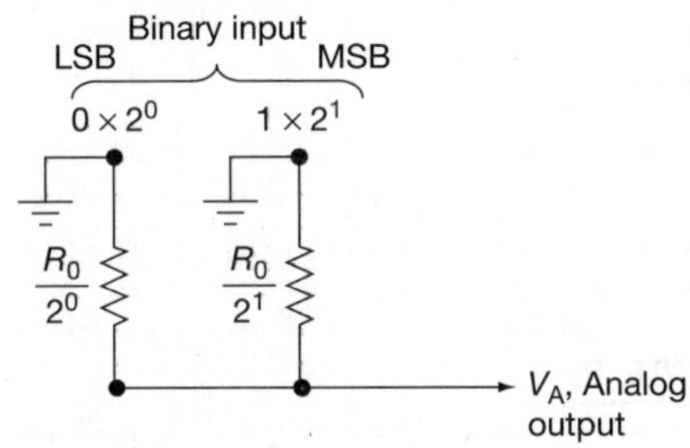

FIGURE 12.5 | Resistive divider

A resistive divider that performs above-mentioned functions for three bits is shown in Figure 12.6. Resistors R_0, R_1 and R_2 form the resistive divider network. Resistance R_L represents the load to which the divider is connected and is considered to be large enough that it does not load the divider network. Equivalent circuit is shown in Figure 12.7.

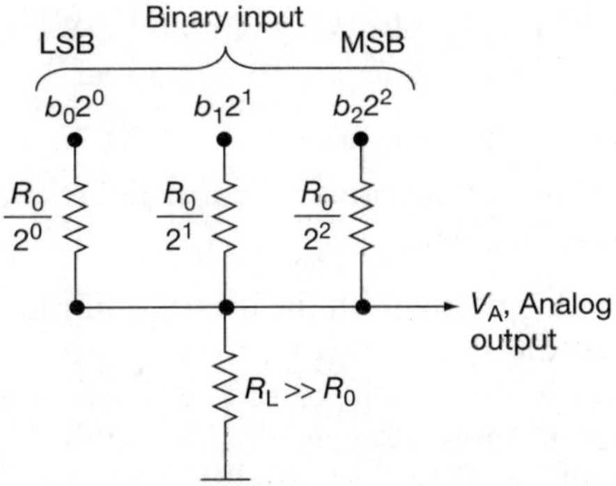

FIGURE 12.6 | Resistive divider

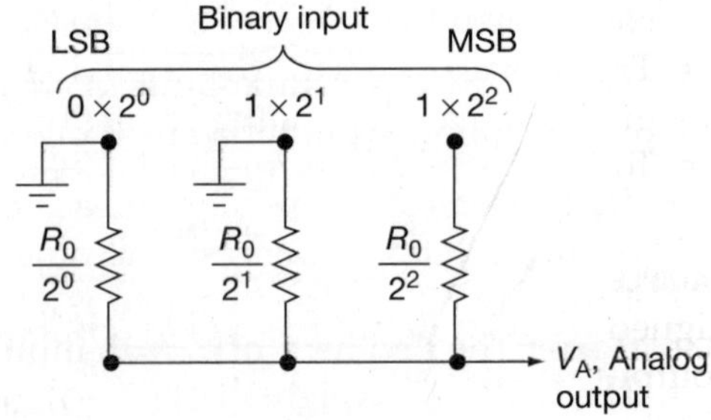

FIGURE 12.7 | Resistive divider

Given $b_0 = 0$, $b_1 = 1$ and $b_2 = 1$ in Figure 12.7. Apply Millman's theorem to the circuit of Figure 12.7 following Eq. (12.7) to get

$$V_A = \frac{1}{(2^3-1)}\left(\sum_{i=1}^{3} b_{i-1}2^{i-1}\ V\right) = \frac{1}{7}\left(0\times 2^0\ V + 1\times 2^1 V + 1\times 2^2 V\right) = \frac{6}{7}\text{V(volts)}$$

A resistive divider that performs above-mentioned functions for four bits is shown in Figure 12.8. Resistors R_0, R_1, R_2 and R_3 form the resistive divider network. Resistance R_L represents the load to which the divider is connected and is considered to be large enough that it does not load the divider network. Equivalent circuit is shown in Figure 12.9.

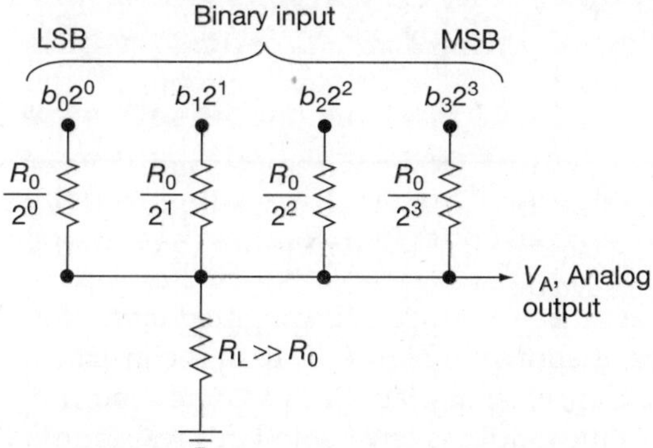

FIGURE 12.8 | Resistive divider

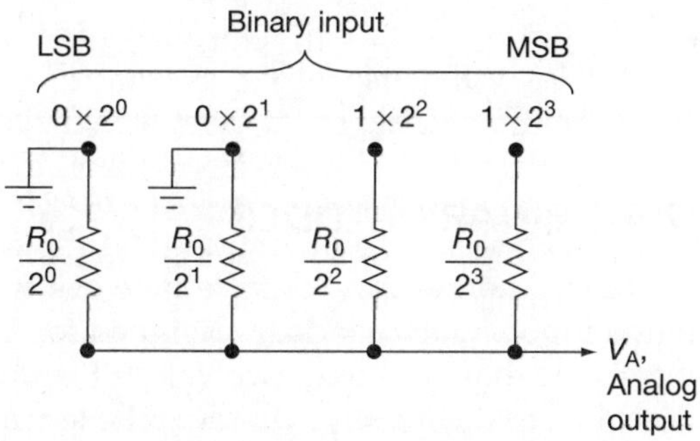

FIGURE 12.9 | Resistive divider

Given $b_0 = 0$, $b_1 = 0$, $b_2 = 1$ and $b_3 = 1$ in Figure 12.9. Apply Millman's theorem to the circuit of Figure 12.9 following Eq. (12.7) to get

$$V_A = \frac{1}{(2^4-1)}\left(\sum_{i=1}^{4} b_{i-1}2^{i-1}V\right) = \frac{1}{15}\left(0\times 2^0 V + 0\times 2^1 V + 1\times 2^2 V + 1\times 2^3 V\right) = \frac{12}{15}\text{V(volts)}$$

Drawback of Resistive Divider:

The resistive divider has the following drawbacks.

- Each resistor in the network has a different value. Since these dividers are usually constructed by using precision resistors that adds cost.
- The resistor used for the most significant bit (MSB) is required to handle a much greater current than used for the LSB resistor.
- To overcome above drawbacks, a ladder has been developed.

EXAMPLE 12.1 Given a 6-bit resistive divider. Determine the following: (a) the weight assigned to the LSB; (b) the weight assigned to the second and third LSB; (c) the change in output voltage due to a change in the LSB, the second LSB, and the third LSB; (d) the output voltage for a digital input of 110101. Assume 0 = 0 V and 1 = +10 V.

SOLUTION

(a) The LSB weight is $1/(2^6 - 1) = 1/63$
(b) The second LSB weight is $2/63$, and the third LSB weight is $4/63$.

(c) The LSB causes a change in the output voltage of 10/63 V.
The second LSB causes an output voltage change of 20/63 V.
The third LSB causes an output voltage change of 40/63 V

(d) The output voltage for a digital input of 110101 is

$$V_A = \frac{1}{(2^6 - 1)}\left(\sum_{i=1}^{6} b_{i-1} 2^{i-1} \mathrm{V}\right)$$

$$= \frac{1}{15}\left(1\times 2^0\,\mathrm{V} + 0\times 2^1\,\mathrm{V} + 1\times 2^2\,\mathrm{V} + 0\times 2^3\,\mathrm{V} + 1\times 2^4\,\mathrm{V} + 1\times 2^5\,\mathrm{V}\right)$$

$$= \frac{53}{63}\times 10\ (\text{volts}) = 8.4\ \mathrm{V}$$

12.4 | BINARY LADDER

The binary ladder is a resistive network whose output voltage is a weighted sum of the digital inputs. Such a ladder, designed for 4 bits, is shown in Figure 12.10. It is constructed of resistors that have only two values *R* and 2*R*. Hence, binary ladder overcomes one of the objections to the resistive divider. The left end of the ladder is terminated in a resistance of 2*R*. Let the right end of the ladder (the output) be open-circuited.

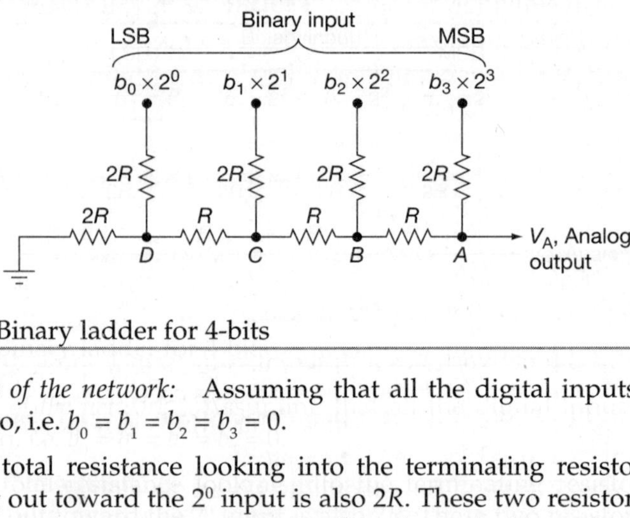

FIGURE 12.10 | Binary ladder for 4-bits

Resistive properties of the network: Assuming that all the digital inputs are at ground or binary bits are zero, i.e. $b_0 = b_1 = b_2 = b_3 = 0$.

At node D: The total resistance looking into the terminating resistor is 2*R*. The total resistance looking out toward the 2^0 input is also 2*R*. These two resistors are connected in parallel to give an equivalent resistor of value *R* as shown in Figure 12.11.

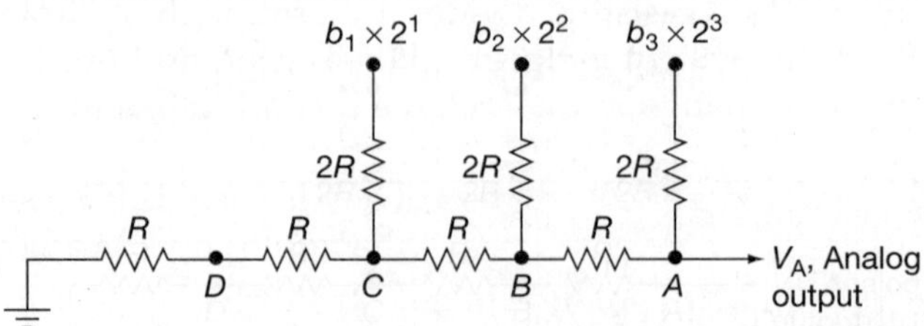

FIGURE 12.11 | Binary ladder

At node C: The total resistance looking into the branch toward node *D* is 2*R*. The total resistance looking out toward the 2^1 input is 2*R*. These resistors are connected in parallel to give an equivalent resistor of value *R* as shown in Figure 12.12.

At node B: The total resistance looking into the branch toward node *C* is 2*R*. The total resistance looking out toward the 2^2 input is 2*R*. These resistors are connected in parallel to give an equivalent resistor of value *R* as shown in Figure 12.13.

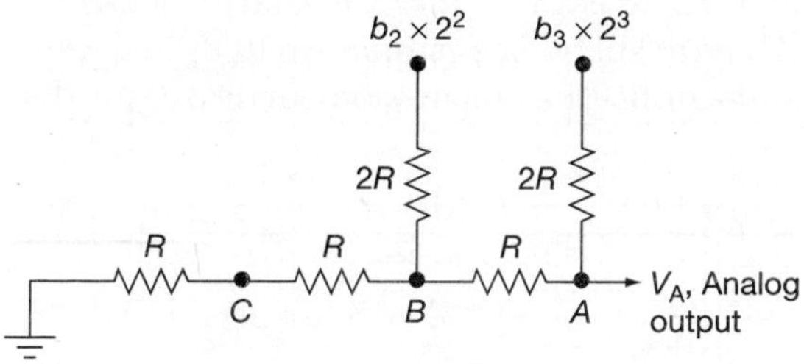

FIGURE 12.12 | Binary ladder

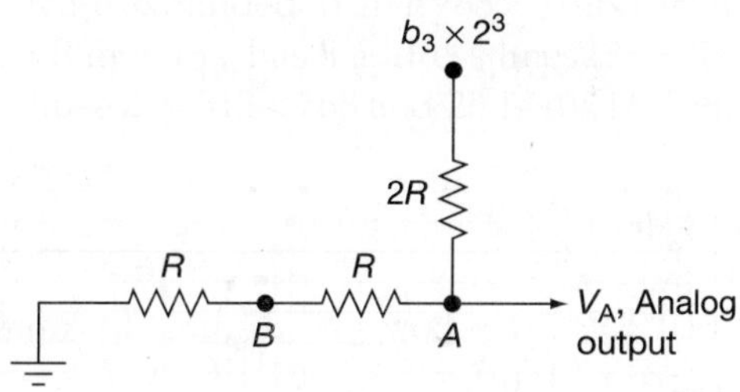

FIGURE 12.13 | Binary ladder

It can be observed that the total resistance looking from any node back toward the terminating resistor or out toward the digital input is 2*R*. It is true regardless of whether the digital inputs are at ground or +V because the internal impedance of an ideal voltage source is 0 Ω because digital inputs are ideal voltage sources.

Figure 12.14 shows the binary ladder circuit that converts digital-to-analog signal. Connecting an operational amplifier with a feedback resistor R_f as shown in Figure 12.14. Amplifier acts as an inverting current-to-voltage amplifier. The output voltage V_A is equal to the negative, of the input current/multiplied by *R*. The input impedance to this amplifier is essentially 0 Ω, when it is connected to an *R*–2*R* ladder. The connecting point is virtually at ground potential. In this configuration, the *R*–2*R* ladder produces a current output, which is a binary-weighted sum of the input digital levels.

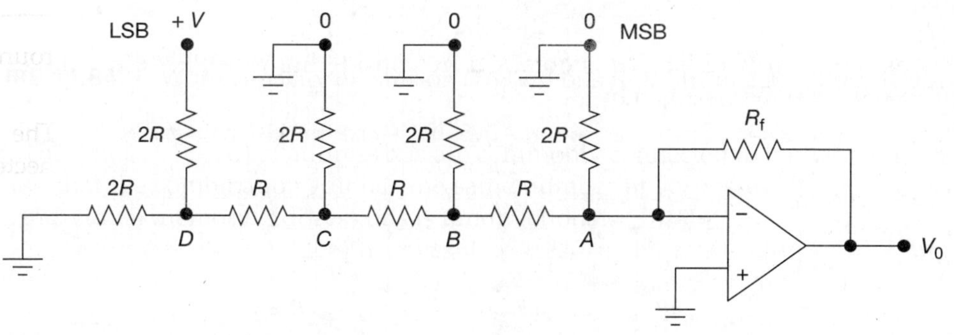

FIGURE 12.14 | Binary ladder

Voltage divider

$$V_B = \frac{2R}{2R+2R} \times V_A = \frac{V_A}{2}$$

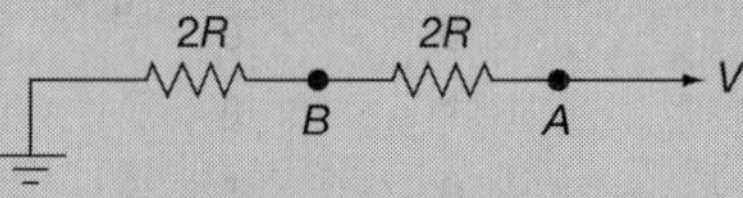

FIGURE | Divider circuit

Parallel circuit

$$\frac{1}{R_{eq}} = \frac{1}{2R} + \frac{1}{2R}$$

$$R_{eq} = \frac{2R \times 2R}{2R+2R} = R$$

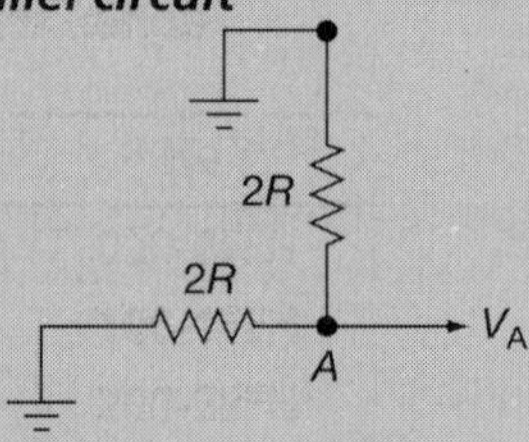

FIGURE | Parallel circuit

EXAMPLE 12.2 Determine the output voltage of 4-bit binary ladder for digital input 0001.

SOLUTION To convert digital to analog voltage, 4-bit binary ladder is shown in Figure 12.15(a).

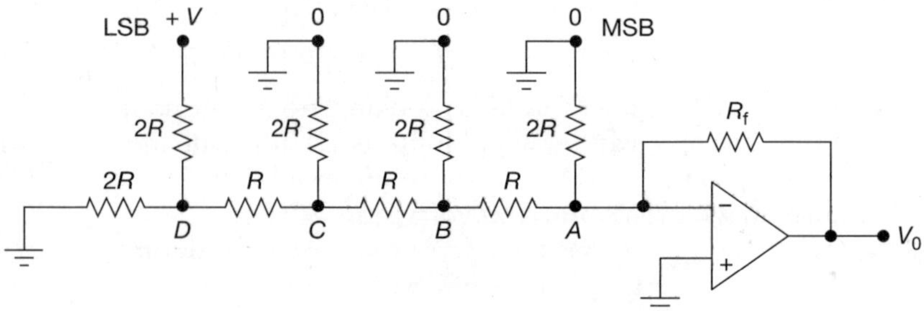

FIGURE 12.15(a) | Binary ladder

At node D: Apply Thevenin theorem looking into node *D* as shown in Figure 12.15(b). Thevenin voltage at *D* is obtained by considering the circuit as voltage divider.

$$V_D = \frac{2R}{2R+2R} \times V = \frac{V}{2}$$

Thevenin equivalent resistance is obtained by considering resistances in parallel and its equivalent resistance is given by

$$R_{eq} = \frac{2R \times 2R}{2R+2R} = R$$

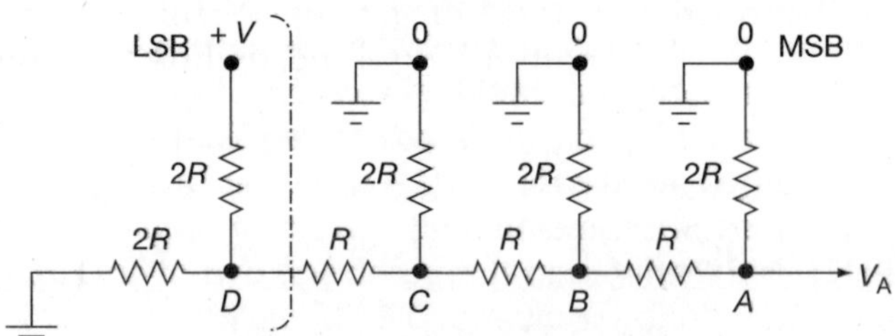

FIGURE 12.15(b) | Binary ladder

The circuit shown in Figure 12.15(b) is redrawn as Figure 12.15(c).

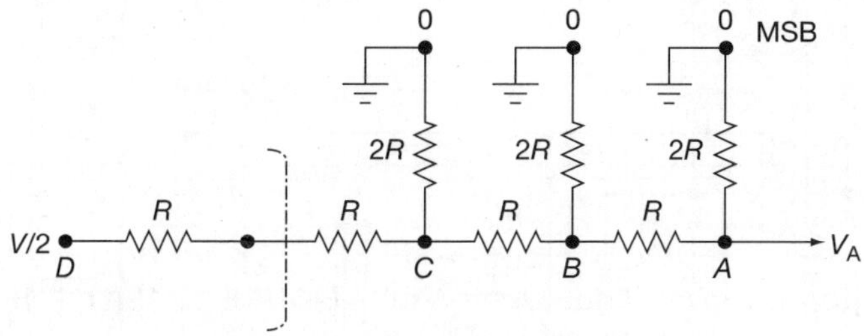

FIGURE 12.15(c) | Binary ladder

At node C: Apply Thevenin theorem looking into node *C* as shown in Figure 12.15(d). Thevenin voltage at *C* is obtained by considering the circuit as voltage divider.

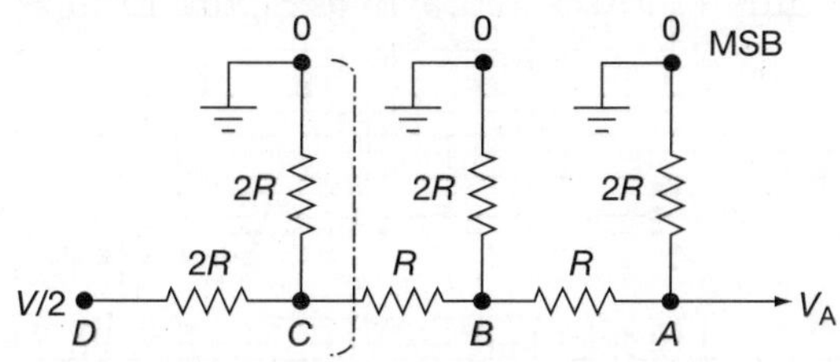

FIGURE 12.15(d) | Binary ladder

$$V_C = \frac{2R}{2R+2R} \times \frac{V}{2} = \frac{V}{4}$$

Thevenin equivalent resistance is obtained by considering resistances in parallel and its equivalent resistance is given by

$$R_{eq} = \frac{2R \times 2R}{2R+2R} = R$$

The circuit shown in Figure 12.15(d) is redrawn as Figure 12.15(e).

At node B: Apply Thevenin theorem looking into node *B* as shown in Figure 12.15(e). Thevenin voltage at *B* is obtained by considering the circuit as voltage divider.

$$V_B = \frac{2R}{2R+2R} \times \frac{V}{4} = \frac{V}{8}$$

Thevenin equivalent resistance is obtained by considering resistances in parallel and its equivalent resistance is given by

$$R_{eq} = \frac{2R \times 2R}{2R+2R} = R$$

The circuit shown in Figure 12.15(e) is redrawn as Figure 12.15(f).

At node A: Apply Thevenin theorem looking into node *B* as shown in Figure 12.15(f). Thevenin voltage at *A* is obtained by considering the circuit as voltage divider.

$$V_A = \frac{2R}{2R+2R} \times \frac{V}{8} = \frac{V}{16}$$

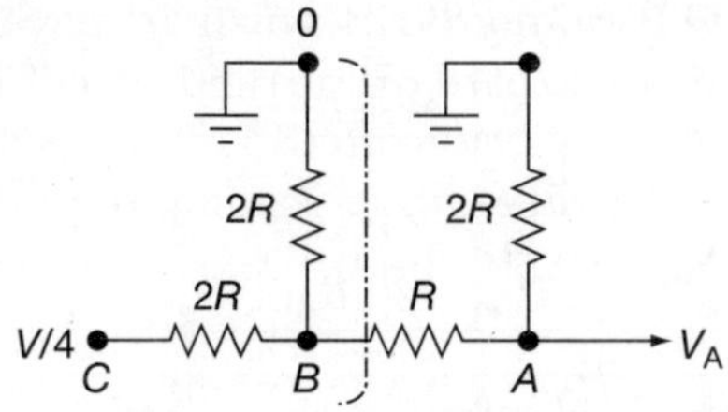

FIGURE 12.15(e) | Binary ladder

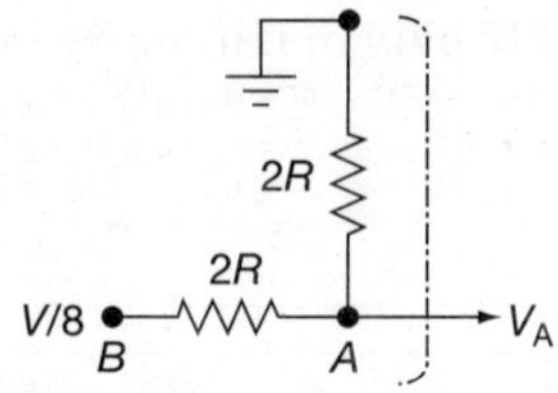

FIGURE 12.15(f) | Binary ladder

Thevenin equivalent resistance is obtained by considering resistances in parallel and its equivalent resistance is given by

$$R_{eq} = \frac{2R \times 2R}{2R + 2R} = R$$

The circuit shown in Figure 12.15(f) is redrawn as Figure 12.15(g).

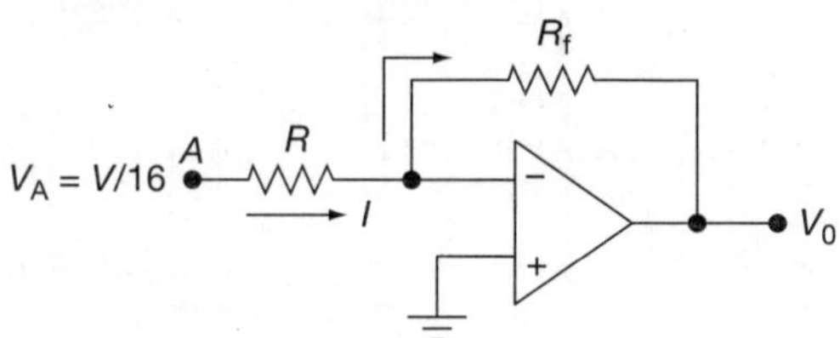

FIGURE 12.15(g) | Binary ladder

Output voltage is obtained as

$$V_0 = \left(-\frac{R_f}{R}\right)\frac{V}{16}V$$

EXAMPLE 12.3 Determine the output voltage of 4-bit binary ladder for digital input 0010.

SOLUTION To convert digital to analog voltage, 4-bit binary ladder is shown in Figure 12.16(a).

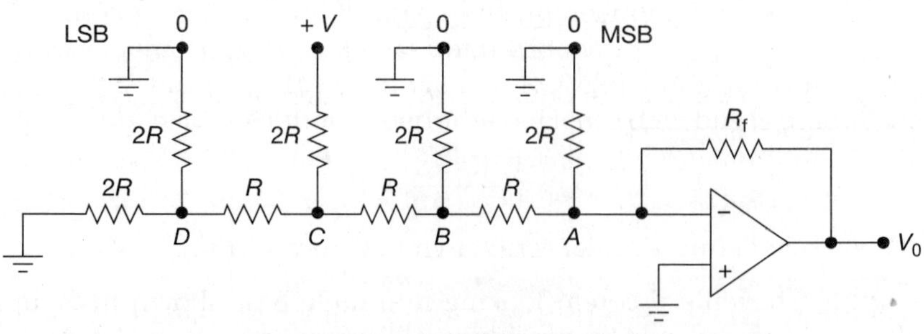

FIGURE 12.16(a) | Binary ladder

At node D: Looking into node D as shown in Figure 12.16(b). The circuit acts as parallel circuit and equivalent resistance is given by

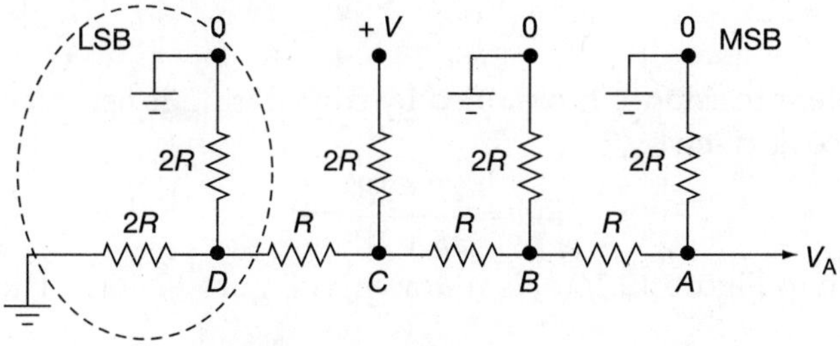

FIGURE 12.16(b) | Binary ladder

$$R_{eq} = \frac{2R}{2} = R$$

The circuit shown in Figure 12.16(b) is redrawn as Figure 12.16(c).

At node C: Apply Thevenin theorem looking into node *C* as shown in Figure 12.16(c). Thevenin voltage at *C* is obtained by considering the circuit as voltage divider.

$$V_C = \frac{2R}{2R+2R} \times V = \frac{V}{2}$$

Thevenin equivalent resistance is obtained by considering resistances in parallel and its equivalent resistance is given by

$$R_{eq} = \frac{2R \times 2R}{2R+2R} = R$$

The circuit shown in Figure 12.16(c) is redrawn as Figure 12.16(d).

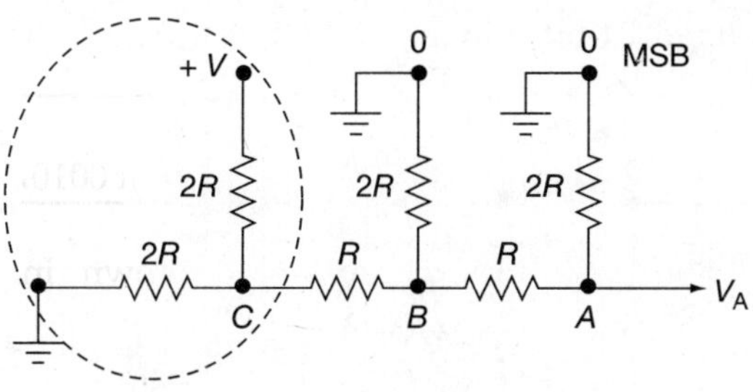

FIGURE 12.16(c) | Binary ladder

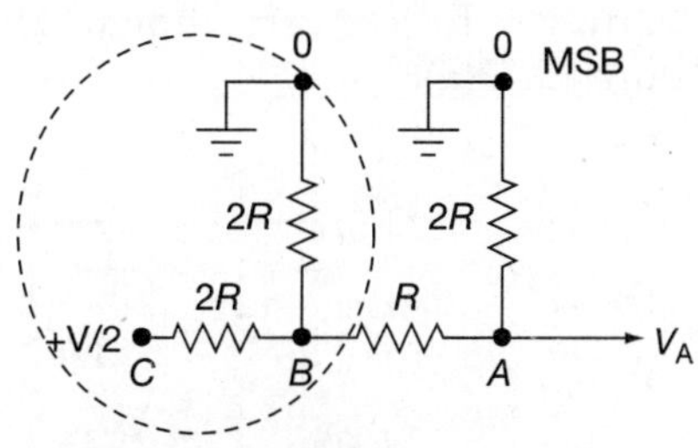

FIGURE 12.16(d) | Binary ladder

At node B: Apply Thevenin theorem looking into node *B* as shown in Figure 12.16(d). Thevenin voltage at *B* is obtained by considering the circuit as voltage divider.

$$V_B = \frac{2R}{2R+2R} \times \frac{V}{2} = \frac{V}{4}$$

Thevenin equivalent resistance is obtained by considering resistances in parallel and its equivalent resistance is given by

$$R_{eq} = \frac{2R \times 2R}{2R+2R} = R$$

The circuit shown in Figure 12.16(c) is redrawn as Figure 12.16(d).

At node A: Apply Thevenin theorem looking into node *A* as shown in Figure 12.16(e). Thevenin voltage at *A* is obtained by considering the circuit as voltage divider.

$$V_A = \frac{2R}{2R+2R} \times \frac{V}{4} = \frac{V}{8}$$

Thevenin equivalent resistance is obtained by considering resistances in parallel and its equivalent resistance is given by

$$R_{eq} = \frac{2R \times 2R}{2R+2R} = R$$

The circuit shown in Figure 12.16(e) is redrawn as Figure 12.16(f). The output voltage is given by

$$V_0 = \left(-\frac{R_f}{R}\right)\frac{V}{8}$$

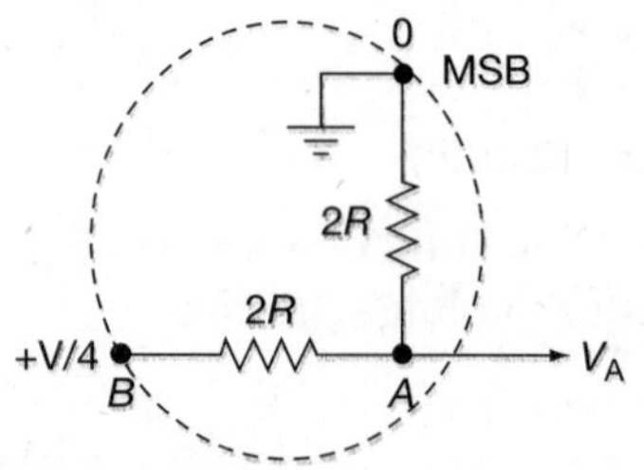

FIGURE 12.16(e) | Binary ladder

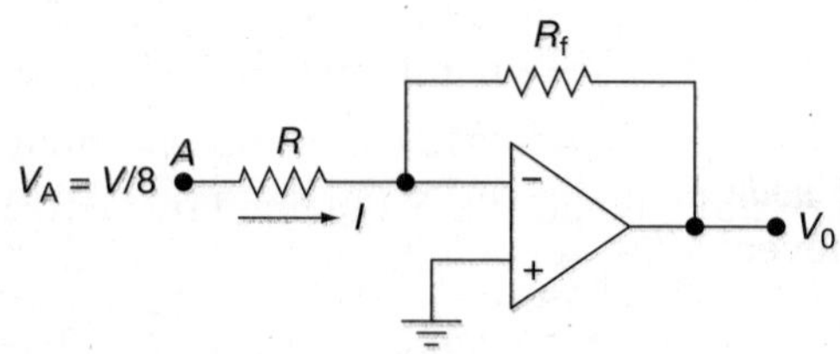

FIGURE 12.16(f) | Binary ladder

EXAMPLE 12.4 Determine the output voltage of 4-bit binary ladder for digital input 0100.

SOLUTION To convert digital to analog voltage, 4-bit binary ladder is shown in Figure 12.17(a).

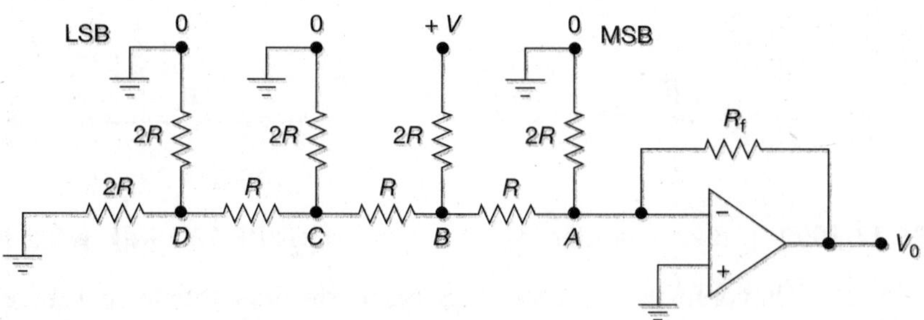

FIGURE 12.17(a) | Binary ladder

At node D: Looking into node *D* as shown in Figure 12.17(b). The circuit act as parallel circuit and equivalent resistance is given by

$$R_{eq} = \frac{2R}{2} = R$$

The circuit shown in Figure 12.17(b) is redrawn as Figure 12.17(c).

At node C: Looking into node C as shown in Figure 12.17(c). The circuit acts as parallel circuit and equivalent resistance is given by

$$R_{\text{eq}} = \frac{2R}{2} = R$$

The circuit shown in Figure 12.17(c) is redrawn as Figure 12.17(d).

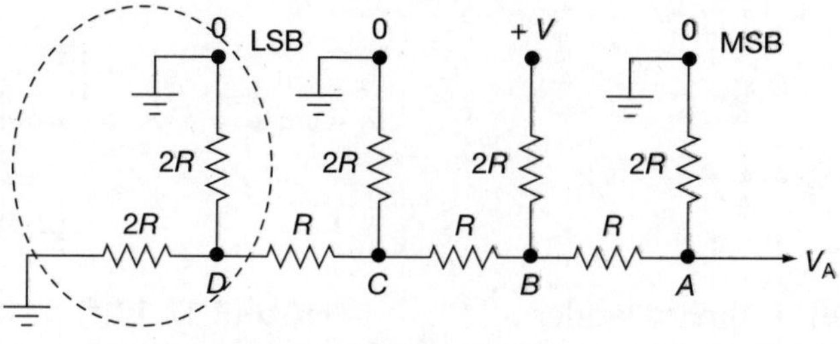

FIGURE 12.17(b) | Binary ladder

At node B: Apply Thevenin theorem looking into node B as shown in Figure 12.17(d). Thevenin voltage at B is obtained by considering the circuit as voltage divider.

$$V_{\text{B}} = \frac{2R}{2R+2R} \times V = \frac{V}{2}$$

Thevenin equivalent resistance is obtained by considering resistances in parallel and its equivalent resistance is given by

$$R_{\text{eq}} = \frac{2R \times 2R}{2R+2R} = R$$

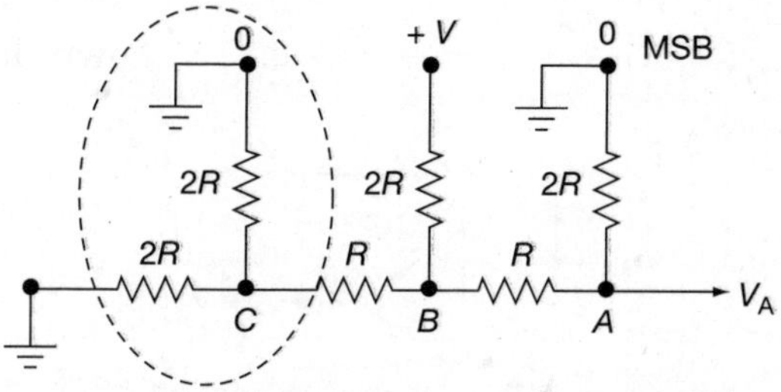

FIGURE 12.17(c) | Binary ladder

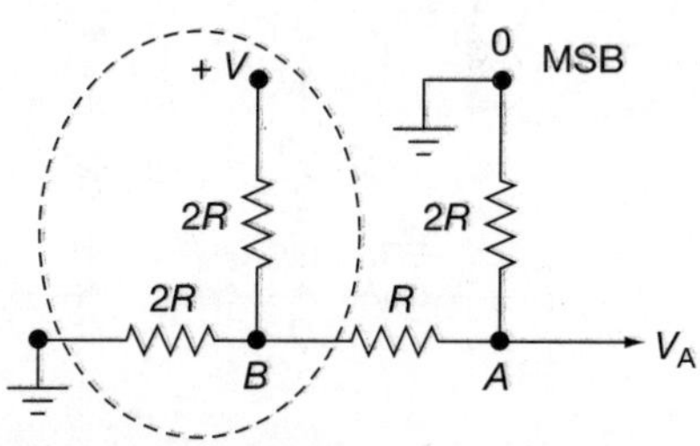

FIGURE 12.17(d) | Binary ladder

The circuit shown in Figure 12.17(d) is redrawn as Figure 12.17(e).

At node A: Apply Thevenin theorem looking into node A as shown in Figure 12.17(e). Thevenin voltage at A is obtained by considering the circuit as voltage divider.

$$V_{\text{A}} = \frac{2R}{2R+2R} \times \frac{V}{2} = \frac{V}{4}$$

Thevenin equivalent resistance is obtained by considering resistances in parallel and its equivalent resistance is given by

$$R_{\text{eq}} = \frac{2R \times 2R}{2R+2R} = R$$

The circuit shown in Figure 12.17(e) is redrawn as Figure 12.17(f). The output voltage is given by

$$V_0 = \left(-\frac{R_f}{R}\right)\frac{V}{4}$$

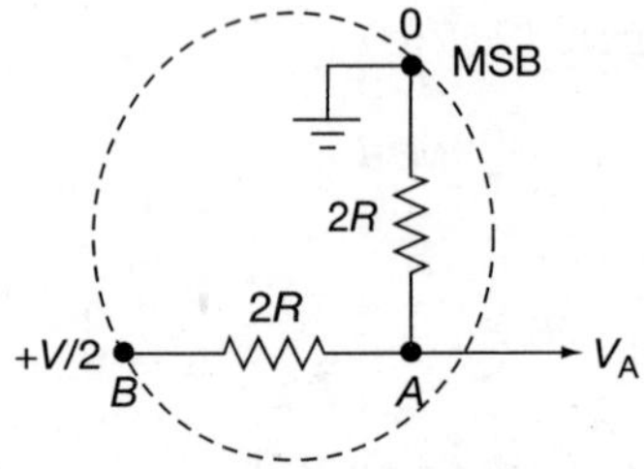

FIGURE 12.17(e) | Binary ladder

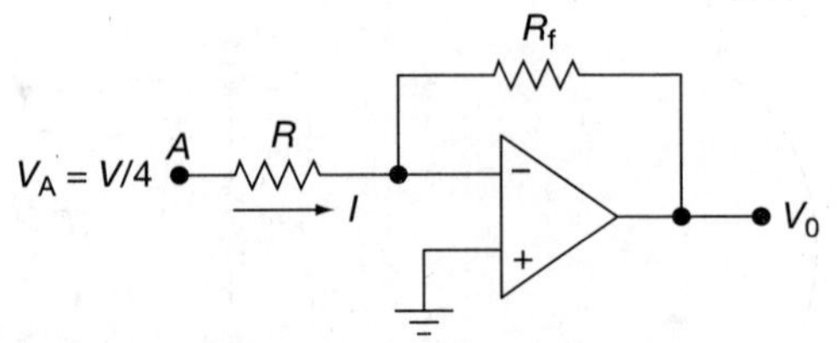

FIGURE 12.17(f) | Binary ladder

EXAMPLE 12.5 Determine the output voltage of 4-bit binary ladder for digital input 1000.

SOLUTION To convert digital to analog voltage, 4-bit binary ladder is shown in Figure 12.18(a).

At node D: Looking into node *D* as shown in Figure 12.18(b). The circuit acts as a parallel circuit and equivalent resistance is given by

$$R_{eq} = \frac{2R}{2} = R$$

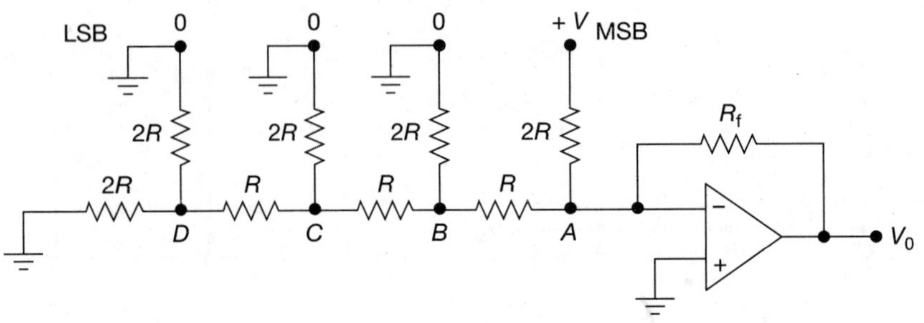

FIGURE 12.18(a) | Binary ladder

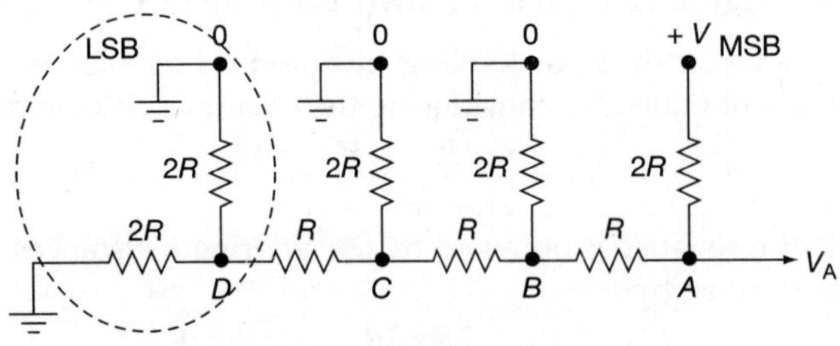

FIGURE 12.18(b) | Binary ladder

The circuit shown in Figure 12.18(b) is redrawn as Figure 12.18(c).

At node C: Looking into node *C* as shown in Figure 12.18(c). The circuit acts as parallel circuit and its equivalent resistance is given by

$$R_{eq} = \frac{2R}{2} = R$$

The circuit shown in Figure 12.18(c) is redrawn as Figure 12.18(d).

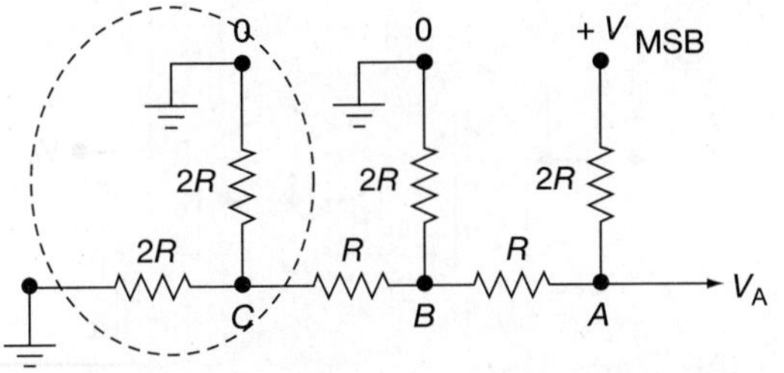

FIGURE 12.18(c) | Binary ladder

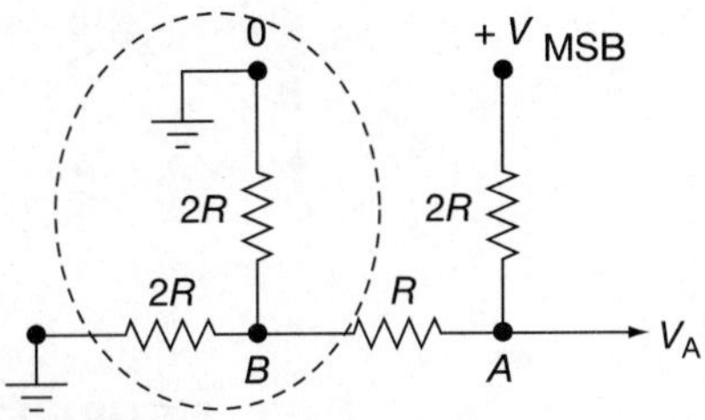

FIGURE 12.18(d) | Binary ladder

At node B: Looking into node *C* as shown in Figure 12.18(d). The circuit acts as parallel circuit and its equivalent resistance is given by

$$R_{eq} = \frac{2R}{2} = R$$

The circuit shown in Figure 12.18(d) is redrawn as Figure 12.18(e).

At node A: Apply Thevenin theorem looking into node *A* as shown in Figure 12.18(e). Thevenin voltage at *A* is obtained by considering the circuit as voltage divider.

$$V_A = \frac{2R}{2R+2R} \times V = \frac{V}{2}$$

Thevenin equivalent resistance is obtained by considering resistances in parallel and its equivalent resistance is given by

$$R_{eq} = \frac{2R \times 2R}{2R+2R} = R$$

The circuit shown in Figure 12.18(e) is redrawn as Figure 12.18(f). The output voltage is given by

$$V_0 = \left(-\frac{R_f}{R}\right)\frac{V}{2}$$

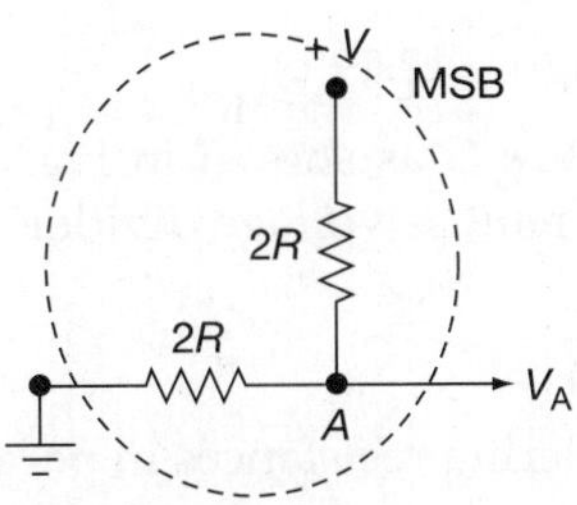

FIGURE 12.18(e) | Binary ladder

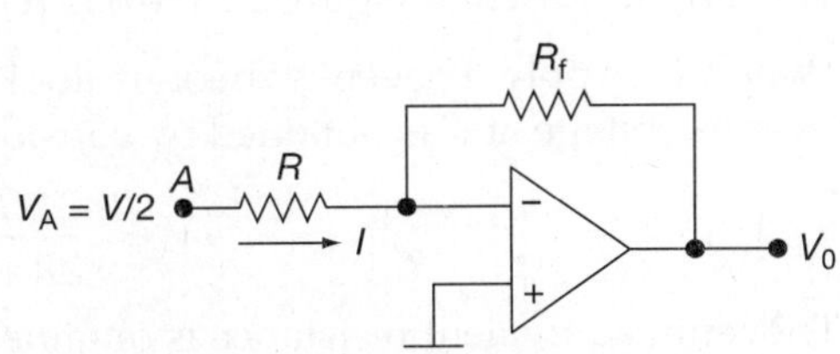

FIGURE 12.18(f) | Binary ladder

EXAMPLE 12.6 Determine the output voltage of 4-bit binary ladder for digital input 1001.

SOLUTION To convert digital to analog voltage, 4-bit binary ladder is shown in Figure 12.19(a).

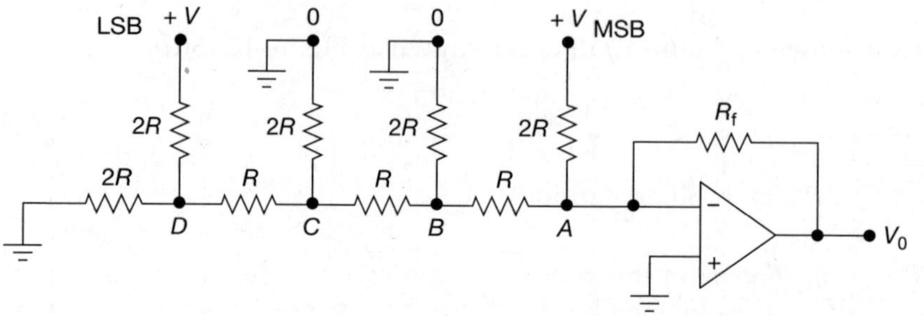

FIGURE 12.19(a) | Binary ladder

At node D: Apply Thevenin theorem looking into node *D* as shown in Figure 12.19(a). Thevenin voltage at *D* is obtained by considering the circuit as voltage divider.

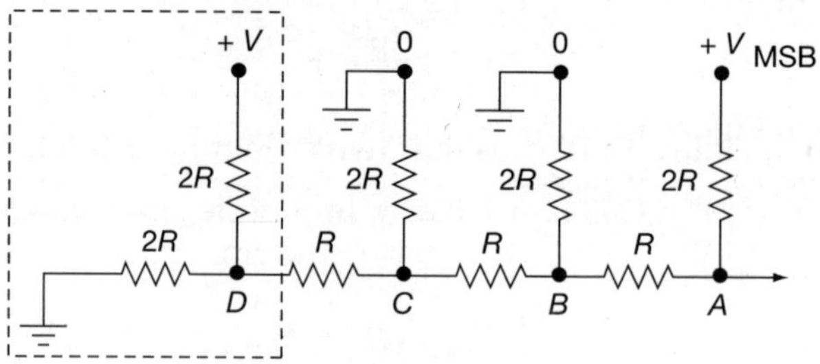

FIGURE 12.19(b) | Binary ladder

$$V_D = \frac{2R}{2R+2R} \times V = \frac{V}{2}$$

Thevenin equivalent resistance is obtained by considering resistances in parallel and its equivalent resistance is given by

$$R_{eq} = \frac{2R \times 2R}{2R+2R} = R$$

The circuit shown in Figure 12.19(b) is redrawn as Figure 12.19(c).

At node C: Apply Thevenin theorem looking into node C as shown in Figure 12.19(c). Thevenin voltage at C is obtained by considering the circuit as voltage divider.

$$V_C = \frac{2R}{2R+2R} \times \frac{V}{2} = \frac{V}{4}$$

Thevenin equivalent resistance is obtained by considering resistances in parallel and its equivalent resistance is given by

$$R_{eq} = \frac{2R \times 2R}{2R+2R} = R$$

The circuit shown in Figure 12.19(c) is redrawn as Figure 12.19(d).

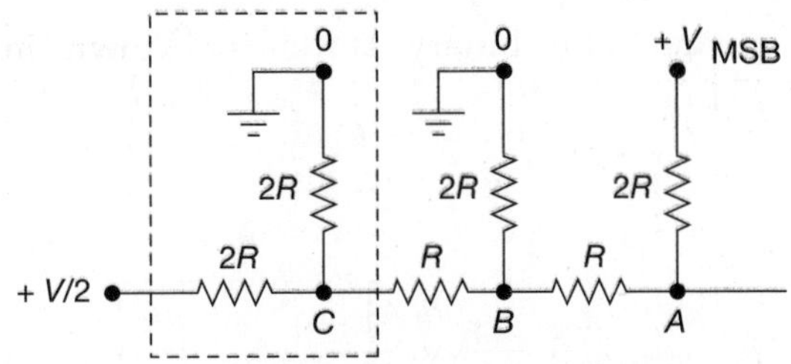

FIGURE 12.19(c) | Binary ladder

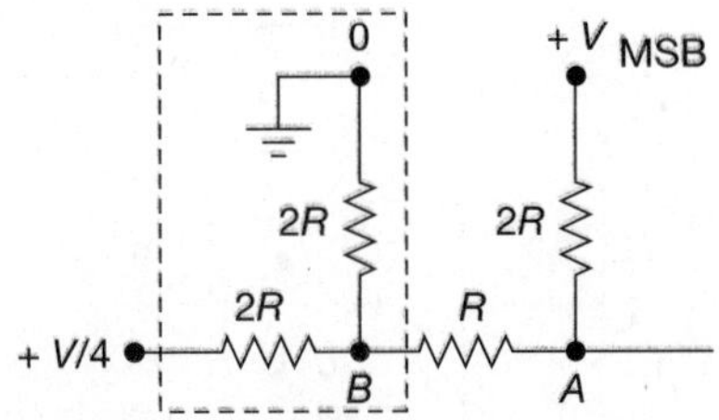

FIGURE 12.19(d) | Binary ladder

At node B: Apply Thevenin theorem looking into node B as shown in Figure 12.19(d). Thevenin voltage at B is obtained by considering the circuit as voltage divider.

$$V_B = \frac{2R}{2R+2R} \times \frac{V}{4} = \frac{V}{8}$$

Thevenin equivalent resistance is obtained by considering parallel circuit and is given by

$$R_{eq} = \frac{2R \times 2R}{2R+2R} = R$$

The circuit shown in Figure 12.19(d) is redrawn as Figure 12.19(e).

At node A: Apply Thevenin theorem equivalent resistance is obtained by considering resistances in parallel and its equivalent is given by

$$R_{eq} = \frac{2R \times 2R}{2R+2R} = R$$

The circuit shown in Figure 12.19(d) is redrawn as Figure 12.19(e).
Looking into node A, as shown in Figure 12.19(e), Thevenin voltage at node A is given by

$$\frac{V - V_A}{2R} = \frac{V_A - \frac{V}{8}}{2R}$$

$$\Rightarrow V - V_A = V_A - \frac{V}{8}$$

$$\Rightarrow 2V_A = V + \frac{V}{8}$$

$$\Rightarrow V_A = \frac{9}{16}V$$

> **Alternative method to calculate V_B:**
> Current, $I = (V - V/8)/4R = 7V/(32R)$
> $V_B - V/8 = 2R \times 7V/(32R)$
> $V_B = V/8 + 7V/16 = 9V/16$

Thevenin equivalent resistance is obtained by considering resistances in parallel and its equivalent resistance is given by

$$R_{eq} = \frac{2R \times 2R}{2R+2R} = R$$

The circuit shown in Figure 12.19(e) is redrawn as Figure 12.19(f). The output voltage is given by

$$V_0 = \left(-\frac{R_f}{R}\right)\frac{9V}{16}$$

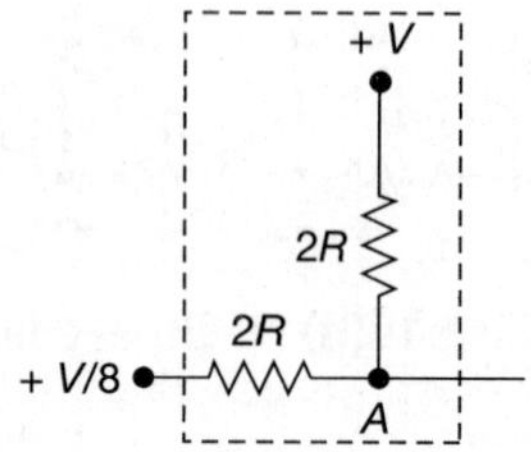

FIGURE 12.19(e) | Binary ladder

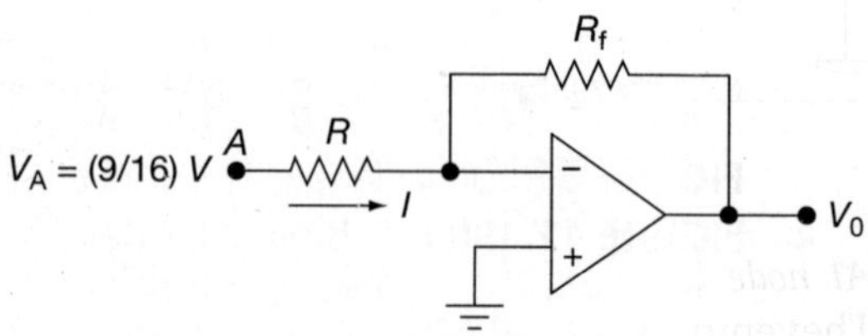

FIGURE 12.19(f) | Binary ladder

EXAMPLE 12.7 Determine the output voltage of 4-bit binary ladder for digital input 1111.

SOLUTION To convert digital to analog voltage, 4-bit binary ladder is shown in Figure 12.20(a).

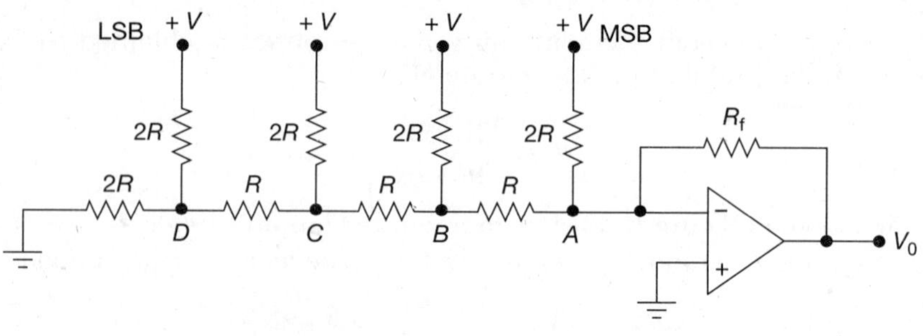

FIGURE 12.20(a) | Binary ladder

At node D: Looking into node *D* as shown in Figure 12.20(b), Thevenin voltage at *D* is obtained by considering the circuit as voltage divider.

$$V_D = \frac{2R}{2R+2R} \times V = \frac{V}{2}$$

Thevenin equivalent resistance is obtained by considering resistances in parallel and its equivalent resistance is given by

$$R_{eq} = \frac{2R \times 2R}{2R+2R} = R$$

The circuit shown in Figure 12.20(b) is redrawn as Figure 12.20(c).

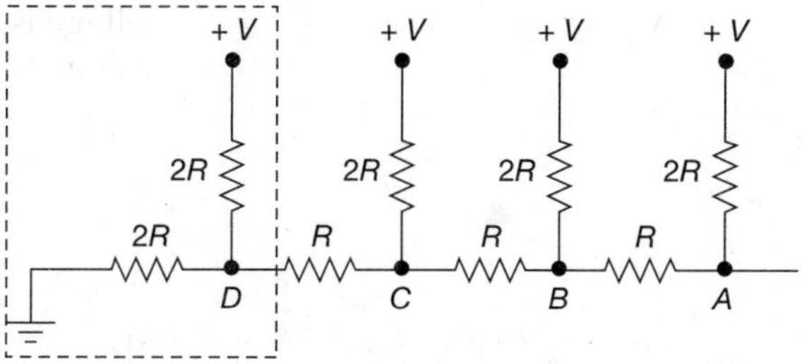

FIGURE 12.20(b) | Binary ladder

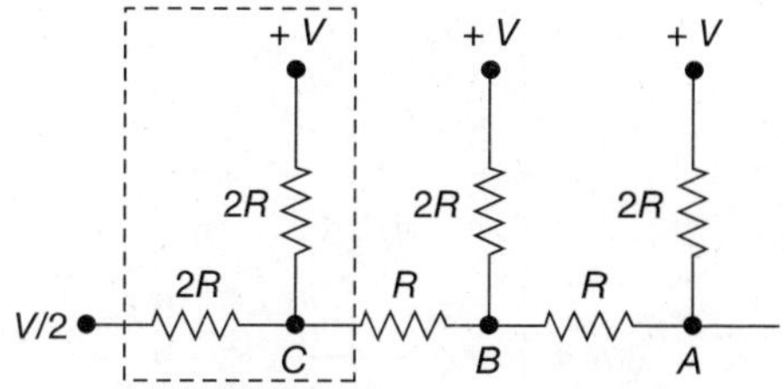

FIGURE 12.20(c) | Binary ladder

At node C: Apply Thevenin theorem looking into node *C* as shown in Figure 12.20(c). Thevenin voltage at *C* is given by nodal analysis

$$\frac{V-V_C}{2R}=\frac{V_C-\frac{V}{2}}{2R}$$

$$\text{or } V-V_C=V_C-\frac{V}{2}$$

$$\text{or } 2V_C=V+\frac{V}{2}$$

$$V_C=\frac{3}{4}V$$

> **Alternative method to calculate V_B:**
>
> Current, $I=(V-V/2)/4R=V/(8R)$
>
> $V_B-V/2=2R\times V/(8R)$
>
> $V_B=V/2+V/4=3V/4$

Thevenin equivalent resistance is obtained by considering resistances in parallel and its equivalent resistance is given by

$$R_{eq}=\frac{2R\times 2R}{2R+2R}=R$$

The circuit shown in Figure 12.20(c) is redrawn as Figure 12.20(d).

At node B: Apply Thevenin looking into node *B* as shown in Figure 12.20(d). Thevenin voltage at *B* is given by

$$\frac{V-V_B}{2R}=\frac{V_B-\frac{3V}{4}}{2R}$$

$$\text{or } V-V_B=V_B-\frac{3V}{4}$$

$$\text{or } 2V_B=V+\frac{3V}{4}$$

$$V_B=\frac{7}{8}V$$

> **Alternative method to calculate V_B:**
>
> Current, $I=(V-3V/4)/4R=V/(16R)$
>
> $V_B-3V/4=2R\times V/(16R)$
>
> $V_B=3V/4+V/8=7V/8$

Thevenin equivalent resistance is obtained by considering resistances in parallel and its equivalent resistance is given by

$$R_{eq}=\frac{2R\times 2R}{2R+2R}=R$$

The circuit shown in Figure 12.20(d) is redrawn as Figure 12.20(e).

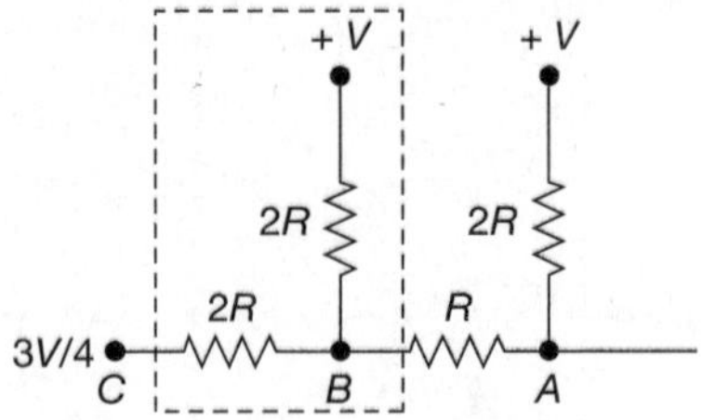

FIGURE 12.20(d) | Binary ladder

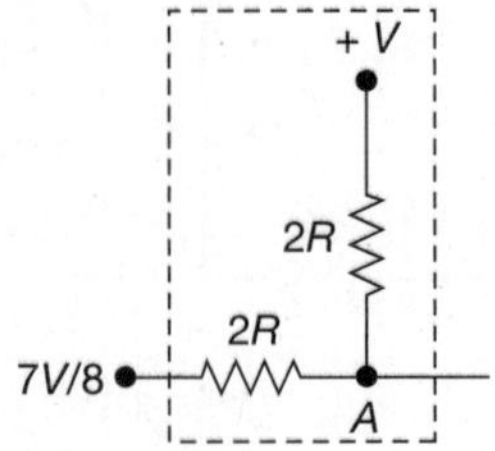

FIGURE 12.20(e) | Binary ladder

At node A: Apply Thevenin looking into node *A* as shown in Figure 12.20(e). Thevenin voltage at *A* is given by

$$\frac{V-V_A}{2R}=\frac{V_A-\dfrac{7V}{8}}{2R} \text{ or } V-V_A=V_A-\frac{7V}{8} \text{ or } 2V_A=V+\frac{7V}{8}$$

$$V_A=\frac{15}{16}V$$

Thevenin equivalent resistance is obtained by considering resistances in parallel and its equivalent resistance is given by

$$R_{eq}=\frac{2R\times 2R}{2R+2R}=R$$

The circuit shown in Figure 12.20(e) is redrawn as Figure 12.20(f).

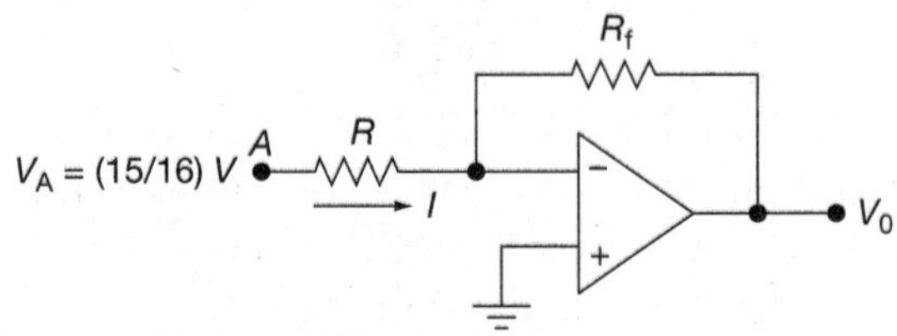

FIGURE 12.20(f) | Binary ladder

The output voltage is given by

$$V_0=\left(-\frac{R_f}{R}\right)\frac{15V}{16}$$

12.4.1 | Analog Output of Binary Ladder Network

The analog voltage for 4-bit binary ladder network is

$$V_A=\frac{V}{2}+\frac{V}{4}+\frac{V}{8}+\frac{V}{16}=\frac{15}{16}V$$

where

$$\text{The value of MSB} = \frac{V}{2}$$

$$\text{The value of 2nd MSB} = \frac{V}{4}$$

$$\text{The value of 3rd MSB} = \frac{V}{8}$$

$$\text{The value of LSB} = \frac{V}{16}$$

$$V_A = \frac{V}{2^1} + \frac{V}{2^2} + \frac{V}{2^3} + \frac{V}{2^4} = \frac{15}{2^4}V$$

$$V_A = \left(\frac{2^3 + 2^2 + 2^1 + 2^0}{2^4}\right)V$$

$$V_A = \frac{1}{2^4}\left(2^3 + 2^2 + 2^1 + 2^0\right)V$$

$$V_A = \frac{1}{2^4}\left(b_3 2^3 + b_2 2^2 + b_1 2^1 + b_0 2^0\right)V$$

where b_0, b_1, b_2 and b_3 are digital bits either be 0 or 1. b_0 is LSB and b_3 is MSB.

$$V_A = \frac{1}{\left(2^4\right)}\left(\sum_{i=1}^{4} b_{i-1} 2^{i-1} V\right)$$

The above equation can be extended for n-bit binary ladder

$$V_A = \frac{1}{\left(2^n\right)}\left(\sum_{i=1}^{n} b_{i-1} 2^{i-1} V\right) \tag{12.8}$$

EXAMPLE 12.8 What are the output voltages caused by each bit in a 6-bit ladder if the input levels are 0 = 0 V and 1 = +10 V?

SOLUTION

$$\text{The value of MSB} = \frac{V}{2} = \frac{10}{2} = 5\text{ V}$$

$$\text{The value of 2nd MSB} = \frac{V}{4} = \frac{10}{4} = 2.5\text{ V}$$

$$\text{The value of 3rd MSB} = \frac{V}{8} = \frac{10}{8} = 1.25\text{ V}$$

$$\text{The value of 4th MSB} = \frac{V}{16} = \frac{10}{16} = 0.625\text{ V}$$

$$\text{The value of 5th MSB} = \frac{V}{32} = \frac{10}{32} = 0.3125 \text{ V}$$

$$\text{The value of LSB} = \frac{V}{64} = \frac{10}{64} = 0.15625 \text{ V}$$

EXAMPLE 12.9 Find the output voltage from a 5-bit ladder that has a digital input of 11001. Assume that 0 = 0 V and 1 = +10 V.

SOLUTION Given binary input is $b_4b_3b_2b_1b_0 = 11001$

$$V_A = \frac{1}{(2^5)}\left(\sum_{i=1}^{5} b_{i-1}2^{i-1}V\right)$$

$$V_A = \frac{1}{(2^5)}\left(b_0 2^0 + b_1 2^1 + b_2 2^2 + b_3 2^3 + b_4 2^4\right) \text{V}$$

$$V_A = \frac{1}{(2^5)}\left(1\times2^0 + 0\times2^1 + 0\times2^2 + 1\times2^3 + 1\times2^4\right)\times 10$$

$$V_A = \frac{1}{32}(1+8+16)\times 10 = \frac{25\times10}{32} = 7.8125 \text{ V}$$

EXAMPLE 12.10 Find the full scale output voltage of a 6-bit ladder. Assume that 0 = 0 V and 1 = +10 V.

SOLUTION Full scale is presented as $b_5b_4b_3b_2b_1b_0 = 111111$

$$V_A = \frac{1}{(2^6)}\left(\sum_{i=1}^{6} b_{i-1}2^{i-1}\text{V}\right)$$

$$V_A = \frac{1}{(2^6)}\left(b_0 2^0 + b_1 2^1 + b_2 2^2 + b_3 2^3 + b_4 2^4 + b_5 2^5\right) \text{V}$$

$$V_A = \frac{1}{(2^6)}\left(1\times2^0 + 1\times2^1 + 1\times2^2 + 1\times2^3 + 1\times2^4 + 1\times2^5\right)\times 10$$

$$V_A = \frac{1}{64}(1+2+4+8+16+32)\times 10 = \frac{63\times10}{64} = 9.84375\ V$$

12.5 | DIGITAL-TO-ANALOG CONVERTER

DAC uses either binary ladder network or resistive divider network as a basic unit. The complete block diagram is given in Figure 12.21.

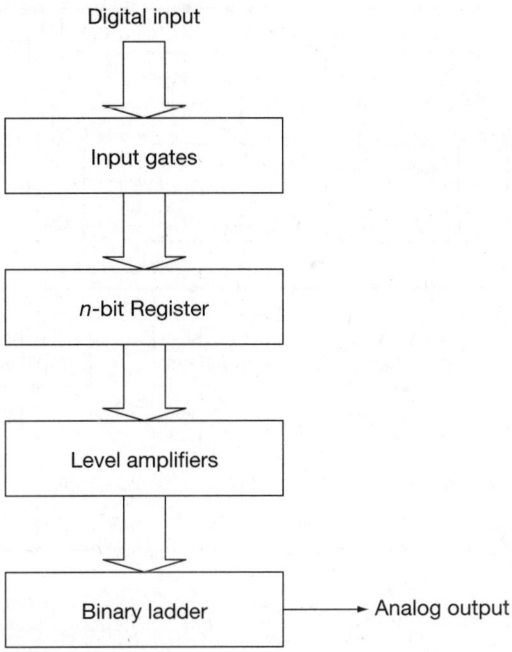

FIGURE 12.21 | Digital to analog converter

Gates: The inputs of the register consisting of the flip-flops are to be set with the proper information from the digital system.

Registers: Register is required to store the digital information. This register is formed by use of RS flip-flops, with one flip-flop per bit.

Level Amplifiers: The level amplifiers are required to ensure that the digital signals presented to the network are all of the same level and are constant.

Resistive divider: Resistive divider is used to convert digital information to analog information.

The schematic of a 4-bit DAC is shown in Figure 12.22. D-flip-flop is used to store the information which is read with the help of control line 'read in' clock pulse. The number of flip-flops defines the size of register. One flip-flop stores 1-bit datum. The output of flip-flop is logic 1 or logic 0, but the values of logic 1 and logic 0 may not be same. To provide the same level voltage for logic 1's and logic 0's, the level amplifiers are used. The level amplifier has two inputs; one is the reference voltage from an external voltage source and the other input is the output of the flip-flop. The amplifier works when the input from a flip-flop is high, the output of the amplifier is V_{ref} and when the input from the flip-flop is low, the output is 0 V.

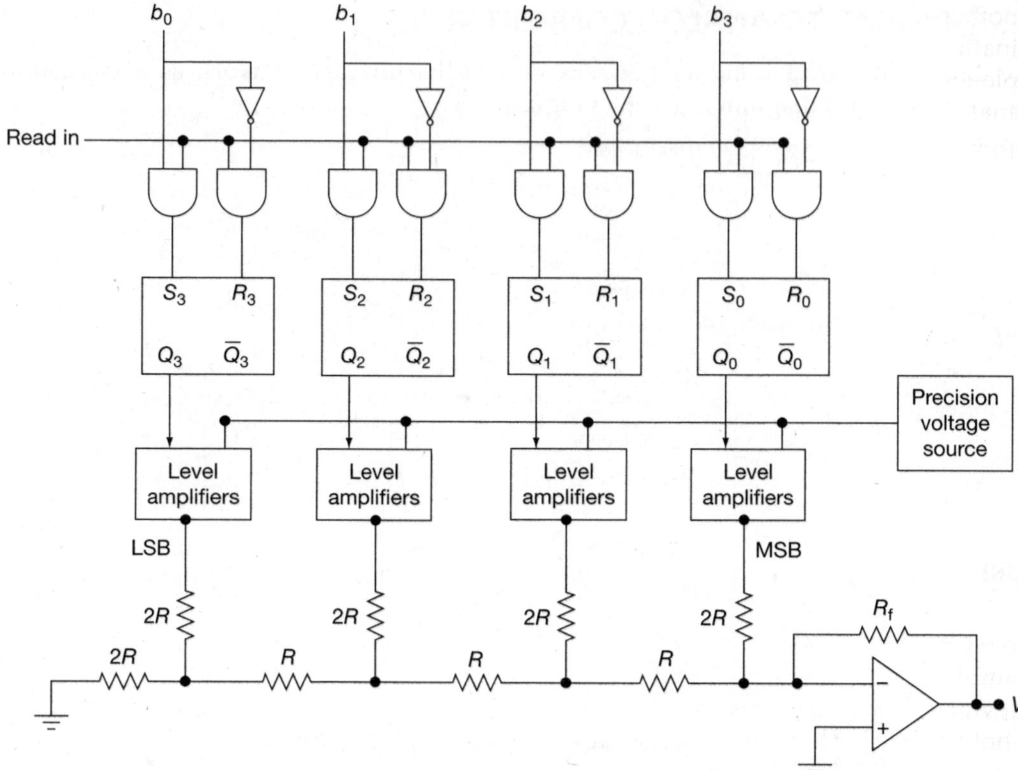

FIGURE 12.22 | Four-bit digital-to-analog converter

12.5.1 | Multiple Signals

To handle many signals, one DAC is used for each signal. Diagram to handle many signals using one DAC for each signal is shown in Figure 12.23. It has the advantage that each signal to be decoded is held in its register and the analog output voltage is held fixed. The digital input lines are connected in parallel to each converter. The proper converter is selected for decoding by the select lines.

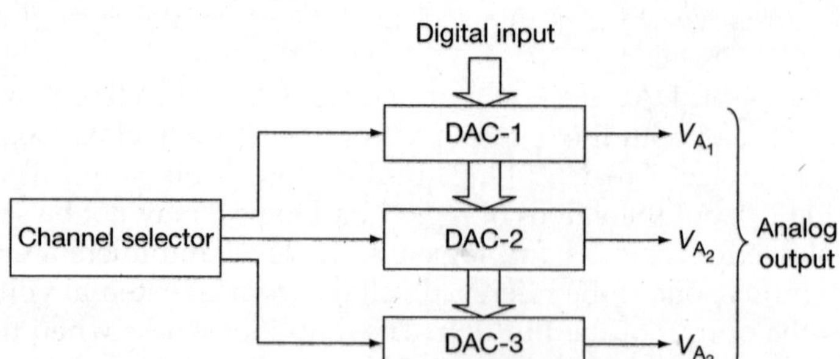

FIGURE 12.23 | Channel selection method to decode number of signals

Another method involves the use of only one DAC and switching its output to desired destination. This is called multiplexing. The diagram using one DAC, multiplexer and sample-and-hold circuit is shown in Figure 12.24. The disadvantage of this method is that the analog output signal must be held between sampling periods. The outputs are to be equipped with sample-and-hold amplifiers.

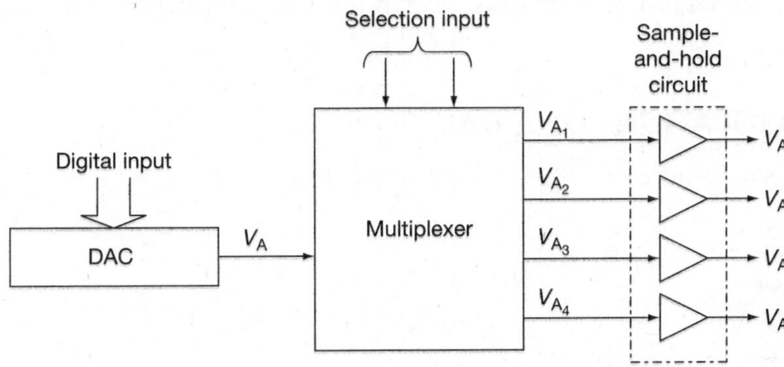

FIGURE 12.24 | Multiplexer method to decode number of signals

An operational amplifier connected as in Figure 12.25 is a unity-gain non-inverting voltage amplifier. The output voltage of unity-gain non-inverting voltage amplifier is equal to input voltage. Two such operational amplifiers are used with a capacitor to form a sample-and-hold amplifier. As and when the switch is closed, the capacitor charges to the DAC output voltage. As and when the switch is opened, the capacitor holds the voltage level until the next sampling time. The operational amplifier provides a large input impedance so that the capacitor is discharged appreciably and at the same time gain is offered to drive external circuits.

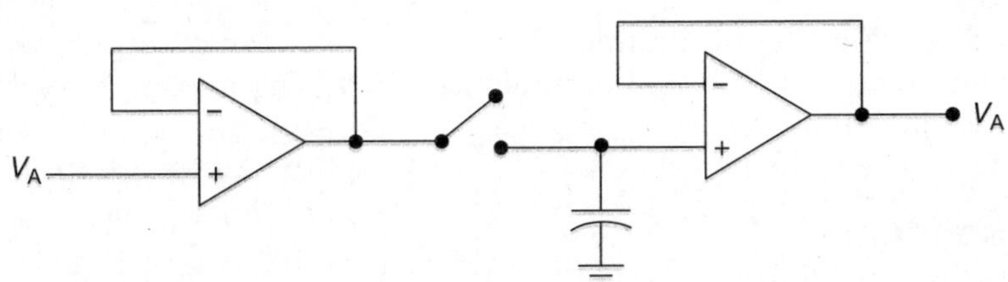

FIGURE 12.25 | Sample-and-hold circuit

When the DAC is used in conjunction with a multiplexer, the maximum rate at which the converter can operate is considered. Each time data is shifted into the register. The transients appear at the output of the converter, because each flip-flop has different rise and fall times. So, a settling time is allowed between the time, data is shifted into the register as well as the time the analog voltage is read out. This settling time is the main factor in determining the maximum rate of multiplexing the output. The worst case is considered when all bits change (e.g. from 1000 to 0111).

Naturally, the capacitors on the sample-and-hold amplifiers are not capable of holding a voltage indefinitely. Therefore, the sampling rate must be sufficient to avoid the decay of voltages appreciably between samples. The sampling rate is a function of the capacitors and the frequency of the analog signal which is expected at the output of the converter. For the sinusoidal signal, it is necessary to sample at only twice the signal frequency. For instance, if the signal is a 2.5-kHz sine wave, it must be sampled at a rate greater than or equal to 5 kHz.

12.6 | SPECIFICATIONS OF A DAC

The general specifications of a DAC are enlisted below:

- Accuracy
- Resolution
- Linearity
- Settling time
- Temperature sensitivity

12.6.1 | Accuracy

It is a measure to determine the closeness of the actual voltage to the theoretical output value. Absolute accuracy defines the maximum deviation of the output from the ideal value. The accuracy of a DAC depends upon the accuracy of the precision resistor used in the resistor divider or ladder network and the precision of the reference voltage. It is specified as a percentage of full-scale or maximum output voltage.

Suppose the theoretical output voltage is 10 V for a full-scale digital input and accuracy is ±5%. It means the output voltage of DAC for the same digital input lies between 9.5 V and 10.5 V.

The accuracy specifies the maximum error that can occur in a output voltage. Let the full-scale output voltage is 10 V and accuracy is ±0.1%. The maximum error comes out 10 mV (i.e. 0001 × 10 V).

12.6.2 | Resolution

It defines the smallest possible change in the output analog voltage due to the change in digital input. The resolution is equal to the weight of LSB. It is also known as the step size. It is a function of the number of bits in the digital input. The resolution of an n-bit DAC using divider is $V/2^n$ and using ladder network, it is $V/2^n$ in volts.

Let the full-scale voltage is 16 V, In a 4-bit DAC using a ladder. The resolution is $16/2^4 = 1$ V. It means that the output voltage changes in a step of 1 V. To produce 5.3 V using 4-bit DAC, the actual output voltage required is 5 V. Similarly, to produce 9.6 V using 4-bit DAC, the actual output voltage required is 10 V.

Resolution is expressed in percentage. The percentage resolution is given by,

$$\%\ \text{resolution} = \frac{\text{Step size}}{\text{Full scale}} \times 100 \tag{12.9}$$

The percentage resolution can be calculated as,

$$\% \text{ resolution} = \frac{1}{\text{Number of steps}} \times 100 \tag{12.10}$$

For an n-bit input, percentage resolution is

$$\% \text{ resolution} = \frac{1}{2^{n-1}} \times 100 \tag{12.11}$$

To get a good resolution, the number of input bits (n) of DAC should be maximum.

12.6.3 | Linearity

In a DAC, the relation between digital input and analog output is supposed to be linear. An equal increment in the numerical significance of the digital input should result in equal increments in the analog output voltage. Due to the errors in resistor values, the input–output relationship is not linear. The linearity error for a digital input is the difference between the expected output voltage and the voltage obtained at the output of DAC. It is termed as monotonicity checking that the output voltage increases regularly as the input digital signal increases. This can be accomplished by using a counter as the digital input signal and observing the analog output on an oscilloscope. For proper monotonicity, a perfect staircase waveform is observed. The steps on the staircase waveform should be equally spaced. Missing steps, steps of different amplitude or steps in a downward fashion indicate malfunctions. The monotonicity test does not check the system for accuracy.

12.6.3.1 | Offset Voltage

For the binary input is zero, the output of DAC should be zero. In practice, there is a very small output voltage under this situation called the offset voltage. There is need to correct the offset error otherwise DAC gives the output after adding it.

12.6.4 | Settling Time

When digital inputs of a DAC change the analog output does not change instantly. Due to the active and passive elements of the circuit, an oscillation occurs at the output. The time required to settle the output within ±½ LSB of the final value, after the change in the digital inputs, is called as the settling time. It limits the frequency of digital input.

12.6.5 | Temperature Sensitivity

The components used in a circuit of DAC such as resistors, reference voltage source, and operational-amplifier are sensitive to temperature. Due to the change in temperature the characteristics of an operational amplifier, the values of resistor and reference voltage are changed.

Therefore, the analog output voltage for any fixed digital input varies with temperature. The change in analog output values with the temperature is known as the temperature sensitivity. It is specified in terms of ±ppm/°C.

EXAMPLE 12.11 What is the exact range of the converter? What is the error of a 4-bit DAC for a range of 0–10 V? Assume that 0 = 0 V and 1 = +10 V.

SOLUTION Full scale is presented as $b_3b_2b_1b_0 = 1111$

$$V_A = \frac{1}{(2^4)}\left(\sum_{i=1}^{4} b_{i-1}2^{i-1}\ V\right)$$

$$V_A = \frac{1}{(2^4)}\left(b_0 2^0 + b_1 2^1 + b_2 2^2 + b_3 2^3\right) V$$

$$V_A = \frac{1}{(2^4)}\left(1\times 2^0 + 1\times 2^1 + 1\times 2^2 + 1\times 2^3\right)\times 10$$

$$V_A = \frac{1}{16}(1+2+4+8)\times 10 = \frac{15\times 10}{16} = 9.375\text{ V } \textbf{Ans.}$$

$$\text{Error} = 10 - 9.375 = 0.625$$

$$\%\ \text{Error} = \frac{0.625}{10}\times 100 = 6.25\% \quad \textbf{Ans.}$$

EXAMPLE 12.12 What is the exact range of the DAC and the error of a 5-bit DAC for a range of 0–10 V? Assume that 0 = 0 V and 1 = +10 V.

SOLUTION Full scale is presented as $b_4b_3b_2b_1b_0 = 11111$

$$V_A = \frac{1}{(2^5)}\left(\sum_{i=1}^{5} b_{i-1}2^{i-1}\ V\right)$$

$$V_A = \frac{1}{(2^5)}\left(b_0 2^0 + b_1 2^1 + b_2 2^2 + b_3 2^3 + b_4 2^4\right) V$$

$$V_A = \frac{1}{(2^5)}\left(1\times 2^0 + 1\times 2^1 + 1\times 2^2 + 1\times 2^3 + 1\times 2^4\right)\times 10$$

$$V_A = \frac{1}{32}(1+2+4+8+16)\times 10 = \frac{31\times 10}{32} = 9.6875\text{ V} \quad \textbf{Ans.}$$

$$Error = 10 - 9.6875 = 0.3125\text{ V}$$

$$\%\ Error = \frac{0.3125}{10}\times 100 = 3.125\% \quad \textbf{Ans.}$$

EXAMPLE 12.13 A 9-bit DAC with full scale of 5.12 V is designed to have ±LSB/2 accuracy. If DAC is calibrated at 25°C and the operating temperature ranges from 0°C to 50°C, what is the maximum net temperature coefficient of DAC?

Solution

$$\text{Range of LSB} = \frac{1}{2^n} \times \text{Range of converter}$$

$$\text{Range of LSB} = \frac{1}{2^9} \times 5.12\ V = 0.01\ \text{V}$$

$$\text{Accuracy} = \pm\frac{\text{LSB}}{2} = \pm\frac{0.01\,\text{V}}{2} = \pm\,0.005\,\text{V} = \pm\,5\text{mV}$$

$$\text{Maximim net temperature coefficient} = \pm\frac{0.005\ V}{25°\text{C}} = \pm 0.0002\ \text{V}/°\text{C} = \pm 200\ \mu\text{V}/°\text{C}$$

Example 12.14 What is the percentage resolution of a 12-bit DAC?

Solution

$$\%\ \text{Resolution} = \frac{1}{2^n - 1} \times 100$$

$$\%\ \text{Resolution} = \frac{1}{2^{12} - 1} \times 100 = 0.0244\%$$

Example 12.15 A DAC has 10 V full-scale output voltage and an accuracy ±0.02%. What is the maximum error for any output voltage?

Solution

Accuracy is defined as percentage of full-scale value.

$$\text{Maximum error in output voltage} = \frac{\pm 0.02}{100} \times 10\ \text{V} = \pm 0.002\ \text{V} = \pm\,2\ \text{mV}$$

Example 12.16 Full-scale output voltage of 8-bit DAC is 12 V. What is its resolution?

Solution

Full scale is presented as $b_7b_6b_5b_4b_3b_2b_1b_0 = 11111111$

$$\text{Full-scale value} = 2^n - 1 = 2^8 - 1 = 255$$

$$V_0 = K\left(2^n - 1\right)$$

$$12 = K\left(2^8 - 1\right)$$

$$K = 12/255 = 0.04706\ \text{V} = 47.06\ \text{mV}$$

$$\text{Resolution} = K = 47.06\ \text{mV}$$

EXAMPLE 12.17 How many bits are required at the input of a DAC if it is necessary to resolve voltages to 5 mV and the ladder has +10 V full scale?

SOLUTION

Resolution = K

$$V_0 = K\left(2^n - 1\right)$$

$$10\ V = 5\ \text{mV}\left(2^n - 1\right)$$

$$2^n = \frac{10}{0.005} - 1 = 2001$$

Taking log on both sides

$$\log_{10} 2^n = \log_{10} 2001$$

$$n(\log_{10} 2) = (\log_{10} 2001)$$

$$n = \frac{(\log_{10} 2001)}{(\log_{10} 2)} = 10.9 \approx 11 \text{ bits} \quad \textbf{Ans.}$$

12.7 | ANALOG-TO-DIGITALCONVERTER

The process of converting an analog voltage into an equivalent digital signal is known as ADC. So, ADC does the inverse function of a DAC. In a DAC, the possible number of digital inputs is fixed to 2^n for n-bit DAC. However, in case of an ADC, the input analog voltage can have any value in a range of 0–5 V, and the digital output can have only 2^n discrete values in the range of 0 to (2^n – 1) for an n-bit ADC. The ADC process includes sampling of input analog signal. Then, each samples converted into its binary equivalent. The block diagram of an ADC is shown in Figure 12.26.

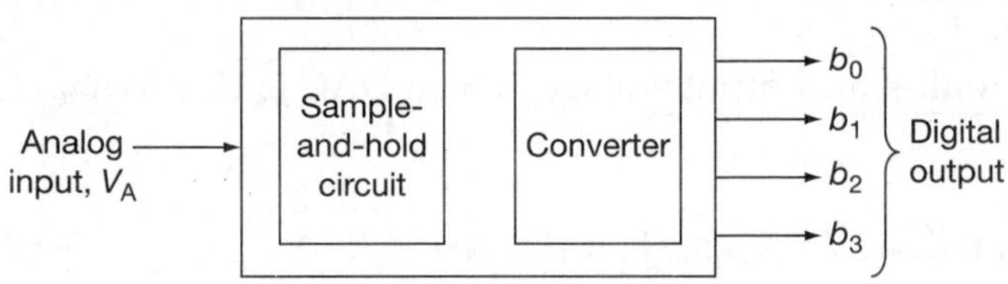

FIGURE 12.26 | Block diagram of ADC

The sample-and-hold circuit captures the sample after a fixed delay. The fixed delay is $T = 1/f$, where f is a sampling frequency. As per the sampling theorem, the sampling frequency should be greater than or equal to twice the band limited frequency of the signal. The result of the sampling process is a series of sampling instants and the amplitude of the signal at that instant of time, i.e. output of the sample circuit is a discrete time signal of different amplitudes. A discrete signal is one that is defined at discrete points of time only.

The amplitude of a discrete signal is a continuous range of values. A discrete signal is the signal that is discrete in time but continuous in amplitude. The discrete signal is converted to binary with the help of a converter. The discrete time is present for a small instant. The converter cannot convert it into binary instantaneously. So, there is a need to hold the sample at least for the duration of conversion time. The sample-and-hold circuit captures the sample and holds for a small time.

12.7.1 | Quantization and Encoding

In an ADC, the input analog voltage has any value in a given range, and it is a function of time. The digital output has only 2^n discrete values for an n-bit ADC. Therefore, the whole range of analog voltage is required to be represented suitable in 2^n intervals. This process is known as quantization. Each interval is then assigned a unique n-bit binary code, which is referred to as encoding. During the process of conversion, the amplitude of a discrete signal is converted to its nearest value of quantization level and then, it is encoded. The process of converting the amplitude of a discrete signal to its nearest value of quantization level is referred as quantization. The process of quantization introduces error. This error is referred to as the quantization error. The maximum quantization error for any analog voltage is $\pm V_m/2^{n+1}$, where V_m is the maximum value of analog input and n is the number bits of ADC. The quantization error can be reduced by increasing the quantization level.

Different methods have been developed to convert an analog signal equivalent digital signal. Commonly used ADCs are:

- Flash/simultaneous ADC
- Counter type ADC
- Continuous type ADC
- Successive-approximation ADC
- Dual-slope ADC

12.8 | SIMULTANEOUS/FLASH ADC

The simultaneous method of ADC is based on the use of a number of comparator circuits. The analog signal to be digitized serves as one of the inputs to each comparator.

Two-bit Simultaneous Converter: A 2-bit binary data have four possible combinations. Four discrete intervals can be achieved by three comparators. The reference voltages used are $+V/4$, $+V/2$ and $+3V/4$. The system is able to accept an analog input voltage between 0 and $+V$. The input voltage ranges are given in Table 12.4.

TABLE 12.4 | Two-bit binary output for input voltage ranges

Input Voltage	Comparator Output			Binary Output	
	C_3	C_2	C_1	b_1	b_0
$0 < V_A \leq +V/4$	Low	Low	Low	0	0
$+V/4 < V_A \leq +V/2$	Low	Low	High	0	1
$+V/2 < V_A \leq +3V/4$	Low	High	High	1	0
$+3V/4 < V_A \leq +V$	High	High	High	1	1

In case all the comparators are off, the analog input signal lies between 0 and $+V/4$. If C_1 is HIGH (it means comparator C_1 is on) and C_2 and C_3 are LOW, the input lies between $+V/4$ and $+V/2$ V. If C_1 and C_2 are high while C_3 is LOW, the input lies between $+V/2$ and $+3V/4$. Finally, if all comparator outputs are HIGH, the input signal lies between $+3V/4$ and $+V$.

Encoding Logic: The line b_1 represents b_1 bit having weightage 2^1 is easiest to determine since it must be high (the b_1 flip-flop must be set) whenever C_2 is HIGH. The logic to set bit b_1 HIGH is given below

$$b_1 = C_2$$

The line b_0 represents b_0 bit having weightage 2^0 must be HIGH whenever C_1 is HIGH and C_2 is LOW or whenever C_3 is HIGH. The logic to set bit, b_0 HIGH is given below.

$$b_0 = C_3 + \bar{C}_2 C_1$$

The three comparator outputs are fed into a coding network to have 2 bits which are equivalent to the input analog voltage. The bits of the coding network are then entered into a flip-flop register for storage. The complete block diagram for 2-bit simultaneous ADC is shown in Figure 12.27.

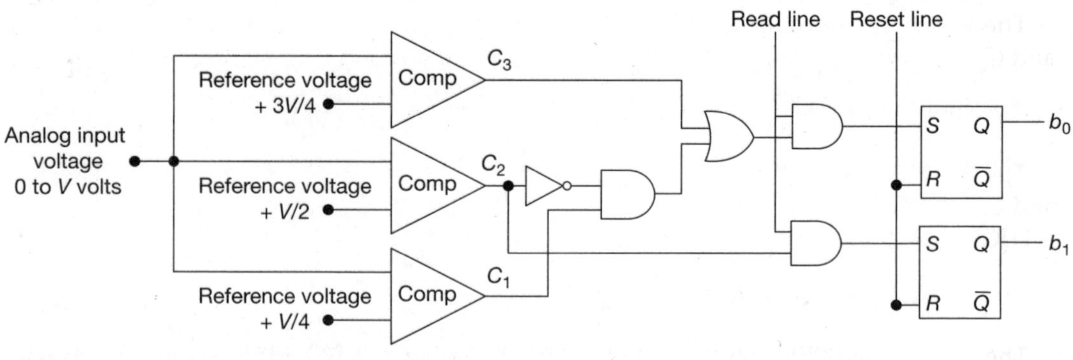

FIGURE 12.27 | Simultaneous 2-bit A/D converter

Three-bit Simultaneous Converter: A 3-bit binary data has eight possible combinations. Eight discrete intervals can be achieved by seven comparators. The reference voltages used are $+V/8$, $+V/4$, $+3V/8$, $+V/2$, $+5V/8$, $+3V/4$ and $+7V/8$. The system is able to accept an analog input voltage between 0 and $+V$. The input voltage ranges are given in Table 12.5.

In case all the comparators are OFF, the analog input signal lies between 0 and $+V/8$. If C_1 is high (it means comparator C_1 is ON) and C_2, C_3, C_4, C_5, C_6 and C_7 are LOW, the input lies between $+V/8$ and $+V/4$V. If C_1 and C_2 are HIGH while C_3, C_4, C_5, C_6 and C_7 are LOW, the input lies between $+V/4$ and $+3V/8$. Finally, if all comparator outputs are HIGH, the input signal must be between $+7V/8$ and $+V$.

TABLE 12.5 | Three-bit binary output for input voltage ranges

Input Voltage	Comparator Output							Binary Output		
	C_7	C_6	C_5	C_4	C_3	C_2	C_1	b_2	b_1	b_0
$0 < V_A \le +V/8$	Low	Low	Low	Low	Low	Low	Low	0	0	0
$+V/8 < V_A \le +V/4$	Low	Low	Low	Low	Low	Low	High	0	0	1
$+V/4 < V_A \le +3V/8$	Low	Low	Low	Low	Low	High	High	0	1	0
$+3V/8 < V_A \le +V/2$	Low	Low	Low	Low	High	High	High	0	1	1
$+V/2 < V_A \le +5V/8$	Low	Low	Low	High	High	High	High	1	0	0
$+5V/8 < V_A \le +3V/4$	Low	Low	High	High	High	High	High	1	0	1
$+3V/4 < V_A \le +7V/8$	Low	High	High	High	High	High	High	1	1	0
$+7V/8 < V_A \le +V$	High	High	High	High	High	High	High	1	1	1

Encoding logic: The line b_2 represents b_2 bit having weightage 2^2 is easiest to determine since it must be high (the b_2 flip-flop is to be set) whenever C_4 is HIGH. The logic to set bit b_2 HIGH is given below

$$b_2 = C_4$$

The b_1 line representing b_1 bit having weightage 2^1 must be HIGH whenever C_2 is HIGH and C_4 is LOW or whenever C_6 is HIGH. The logic to set bit b_1 HIGH is given below.

$$b_1 = C_6 + \bar{C}_4 C_2$$

The line b_0 represents b_0 bit having weightage 2^0 line is to be high whenever C_1 is HIGH and C_2 is LOW or C_3 is HIGH and C_4 is LOW or C_5 is HIGH and C_6 is LOW or whenever C_7 is HIGH. The logic to set bit b_0 HIGH is given below

$$b_0 = C_7 + \bar{C}_6 C_5 + + \bar{C}_4 C_3 + + \bar{C}_2 C_1$$

The seven comparator outputs are fed into a coding network to get 3-bit digital output which is equivalent to the analog input voltage. The bits of the coding network are then entered into a flip-flop register for storage. The complete block diagram for 3-bit simultaneous ADC converter is shown in Figure 12.28.

In general, it can be concluded that $2^n - 1$ number of comparators are required to convert to n-bits digital signal. Some of the comparators have inverters at their outputs since both C and $\bar{C}$ are needed for the encoding matrix.

The construction of a simultaneous ADC is simple and easy. However, as the number of bits in the desired digital number increases, the number of comparators increases following relation ($2^n - 1$), and the number of comparison becomes unmanageable and the cost and size of the circuit increases. Even though this method is simple and has fast conversion rates. Because it is very fast, hence this type of converter is frequently called a flash converter. Its conversion time is 100 ns.

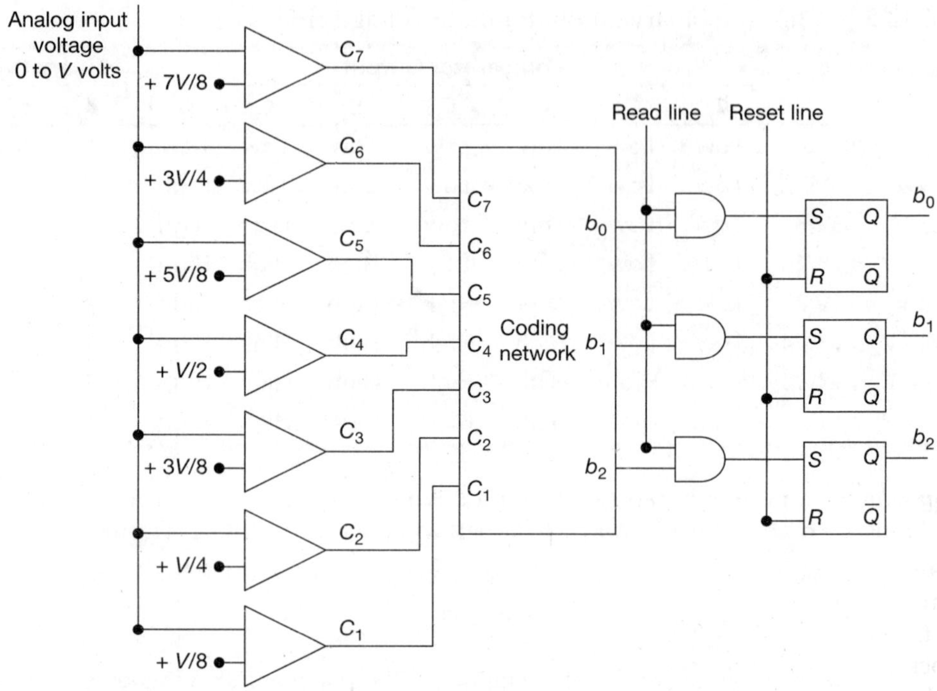

FIGURE 12.28 | Simultaneous 3-bit A/D converter

12.9 | COUNTER TYPE ADC

A higher-resolution ADC uses only one comparator if a variable reference voltage is available. This reference voltage is then applied to the comparator. As and when it becomes equal to the input analog voltage, the conversion is completed.

To construct counter type ADC, a simple MOD-2^n binary counter is used for n-bits binary output. The digital output is taken from binary counter. The output of binary counter is connected to a binary ladder to get analog output and forms a simple DAC. If a clock is now applied to the input of the counter, the output of the binary ladder is the familiar staircase waveform. This waveform is exactly the reference voltage signal, V_R for the comparator. With a minimum of gating and control circuitry, the simple DAC is changed into the desired ADC.

Figure 12.29 shows the block diagram for a counter-type ADC. First, the counter is reset to all 0's. As and when a convert signal appears on the START line, the gate opens and clock pulses are allowed to pass through to the input of the counter. The counter advances binary count sequence and the staircase waveform is generated at the output, V_R of the ladder. This waveform is applied to one side of the comparator and the analog input voltage, V_A is applied to the other side. As and when the reference voltage, V_R becomes equal or exceeds the input analog voltage, V_A the gate is closed. The counter stops and conversion is completed. The number stored in the counter is the digital equivalent of the analog input voltage.

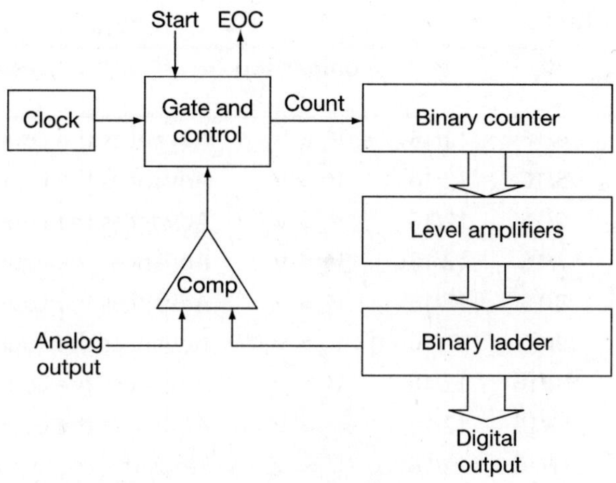

FIGURE 12.29 | Continuous analog to digital converter

This is a closed-loop control system. An error signal is generated at the output of the comparator by taking the difference between the analog input signal and the feedback signal that is a staircase reference voltage. The error is detected by the control circuit, and the clock is issued to advance the counter. The advancement of counter reduces the error signal by increasing the feedback voltage. When the error is reduced to zero, the feedback voltage becomes equal to the analog input signal. The control circuitry stops the clock from advancing the counter, and the system comes to rest.

The counter-type ADC is good method for digitizing a signal to a high resolution. This method is simpler than the simultaneous method for high resolution, but the conversion time required is longer. Since the counter always begins at zero and counts through its normal binary sequence, 2^n counts are necessary before the completion of conversion. The average conversion time is of an n-bit ADC is

$$T_C = \frac{2^n}{2} \times \frac{1}{f_{\text{clock}}} = 2^{n-1} \times \frac{1}{f_{\text{clock}}}$$

where f_{clock} is the clock frequency.

In this conversion process, the counter is reset for each discrete sample undertaken for conversion. Therefore, it takes a long conversion time.

Let an analog input be 0.75 V or 12/16 V and the ADC is a 4-bit counter ADC. Table 12.6 describes the count sequence and action taken by counter type ADC.

TABLE 12.6 | Conversion sequence in counter method of ADC

Clock	Count of Counter	V_R	V_A	Comparison	Description
1	0000	0/16	12/16	$V_A > V_R$	Advances the count of binary counter
2	0001	1/16	12/16	$V_A > V_R$	Advances the count of binary counter
3	0010	2/16	12/16	$V_A > V_R$	Advances the count of binary counter
4	0011	3/16	12/16	$V_A > V_R$	Advances the count of binary counter

(Continued)

TABLE 12.6 | (Continued)

Clock	Count of Counter	V_R	V_A	Comparison	Description
5	0100	4/16	12/16	$V_A > V_R$	Advances the count of binary counter
6	0101	5/16	12/16	$V_A > V_R$	Advances the count of binary counter
7	0110	6/16	12/16	$V_A > V_R$	Advances the count of binary counter
8	0111	7/16	12/16	$V_A > V_R$	Advances the count of binary counter
9	1000	8/16	12/16	$V_A > V_R$	Advances the count of binary counter
10	1001	9/16	12/16	$V_A > V_R$	Advances the count of binary counter
11	1010	10/16	12/16	$V_A > V_R$	Advances the count of binary counter
12	1011	11/16	12/16	$V_A > V_R$	Advances the count of binary counter
13	1100	12/16	12/16	$V_A = V_R$	Stops the count.

12.10 | CONTINUOUS ADC

The counter is reset to zero each time for every conversion in counter type ADC. To speed up the conversion of the signal, the resetting of counter can be avoided. The counter begins at the value of the last converted point. In continuous ADC, the counter is capable of counting either up or down. The control circuit controls the direction of up/down counter based on the result of comparator. Figure 12.30 shows ADC using an up–down counter. This method is known as continuous conversion and thus the converter is called a continuous-type ADC.

The output of the binary ladder is fed into a comparator which has two outputs. When the analog voltage is more positive than the ladder output, the up output of the comparator is set to high. When the analog voltage is more negative than the ladder output, the down output is set to high. If the up output of the comparator is high, the gate and control circuit controls the count-up line of the counter and the counter advances one count. If the down output of the comparator is high, the gate and control circuit controls the count-down line of the counter and the counter decrements one count.

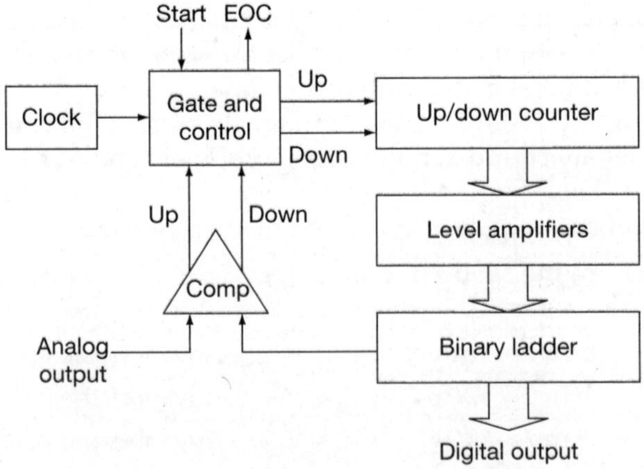

FIGURE 12.30 | Continuous analog to digital converter

Oscillations between up/down count are avoided by adjusting the comparator such that the up output does not set to high unless the binary ladder voltage is more than 1/2 LSB below the analog voltage. Similarly, the down output does not set to high unless the binary ladder voltage is more than 1/2 LSB above the analog voltage. This is called centring on the LSB and provides a digital output within 1/2 LSB.

The continuous converter has a very fast conversion time once it is locked on the signal but loses this advantage when multiplexing inputs are fed.

12.11 | SUCCESIVE APPROXIMATION ADC

The successive-approximation converter is a high resolution ADC that uses only one comparator with a variable reference voltage. The variable reference voltage is obtained by a sequence or ring counter and a DAC. The successive-approximation converter is useful when multiplexing of input is required. The basic principle of this ADC is that the unknown analog input voltage is approximated against an n-bit digital value by trying to change one bit at time, beginning with the MSB. The block diagram for this type of converter is shown in Figure 12.31.

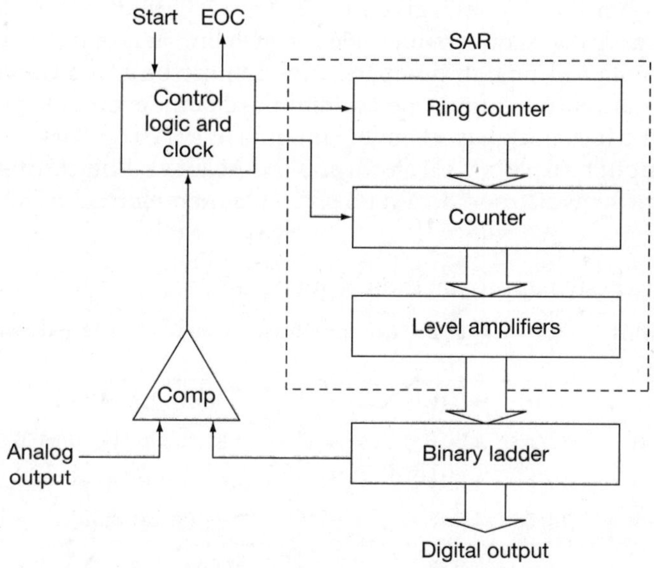

FIGURE 12.31 | Successive approximation ADC

The converter operates by successively dividing the voltage ranges in half. The counter is first reset to all 0s. Then the MSB bit is set. Depending on the output of the comparator, the MSB bit remains set or reset the MSB flip-flop of corresponding bit. In continuation, the second MSB bit is set in, and a comparison is done to determine whether to reset the second MSB flip-flop. The process is repeated down to the LSB bit. The desired number is available in the counter. Since the conversion is based on the operation of one flip-flop at a time, beginning with the MSB, a ring counter is used for flip-flop selection.

The successive-approximation method is the process of approximating the analog voltage by trying one bit at a time beginning with the MSB bit. The operation is shown in diagram form in Figure 12.32.

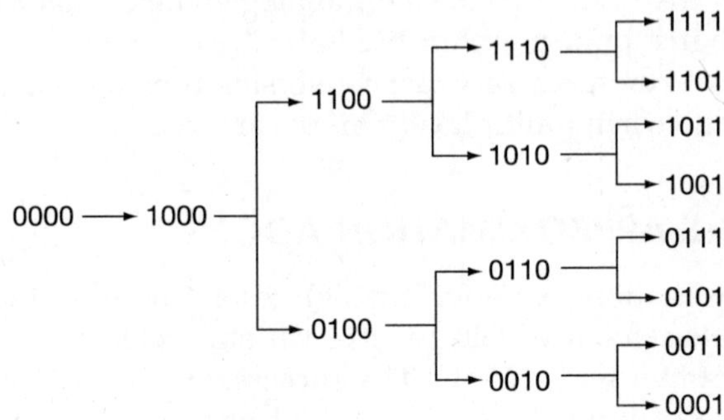

FIGURE 12.32 | Successive-approximation-based 4-bit comparison

It is observed from the diagram given in Figure 12.32 that each conversion takes the same time and requires one conversion cycle for each bit. Hence, the total conversion time is equal to the number of bits, n, times the time required for one conversion cycle. One conversion cycle normally requires one cycle of the clock. As an example, 8-bit converter operating with a 1-MHz clock has a conversion time of $8 \times 10^{-6} = 8$ μs.

Let an analog input be 0.6875 V (11/16 V) and the ADC is a 4-bit counter ADC. Table 12.7 describes the count sequence and action taken by Counter method ADC.

TABLE 12.7 | Successive-approximation sequence in counter method of ADC

Clock	Count of Counter	V_R	V_A	Comparison	Description
1	1000	8/16	11/16	$V_A > V_R$	Set 4th bit (MSB)
2	1100	12/16	11/16	$V_A < V_R$	Set 3rd bit (2nd MSB)
4	1010	10/16	11/16	$V_A > V_R$	Set 2nd bit and reset 3rd bit (2nd MSB)
5	1011	11/16	11/16	$V_A = V_R$	Set 1st bit (LSB)

12.12 | DUAL-SLOPE ADC

The block diagram for a basic dual-slope ADC is given in Figure 12.33. It has four main blocks: (i) a switch driver, (ii) an integrator, (iii) a comparator and (iv) binary counter.

The integrator forms the desired ramp. Basically, two different ramps are formed. One ramp is formed when input is switched to the unknown input voltage, V_A. Second ramp is formed when input is switched to a known reference voltage, V_R. This ramp generator is constructed using an operational amplifier connected as an integrator as shown in Figure 12.33.

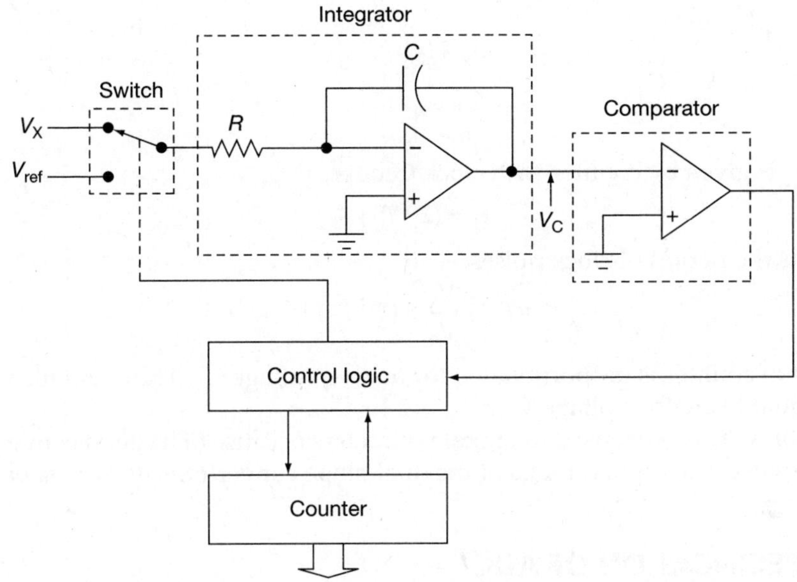

FIGURE 12.33 | Block diagram of dual slope A/D converter

Let the clock is running, and the input voltage V_A is positive. Binary counter is set to zero. The ramp is reset to 0 V. The input is switched to the unknown input voltage V_A. Since V_A is positive, the integrator output V_C is a negative ramp. The comparator output V_0, is positive and the clock is allowed to the counter through control logic. The ramp is formed for fixed time, t_1. Fixed time is detected by count. The voltage V_C at the end of fixed time t_1 is given below.

$$V_C = -\frac{1}{C}\int idt = -\frac{V_A}{RC}t_1 \tag{12.12}$$

When counter reaches the fixed count at time t_1, logic and control unit clears the counter and switch the integrator input to the negative reference voltage V_R. The integrator starts generating a ramp beginning at $-V_C$ and increases steadily upward until it reaches 0 V. During this ramp generation, count is counting. Ramp generation ends when voltage V_C becomes 0 V. Recorded count in the binary counter is the digital output.

The equation for positive ramp is:

$$V_C = \frac{V_R}{RC}\times t_2 \tag{12.13}$$

Here, slope is fixed and time t_2 is variable. Comparing the magnitude of integrator output

$$V_C = \frac{V_A}{RC}\times t_1 = \frac{V_R}{RC}\times t_2$$

Or

$$V_X \times t_1 = V_R \times t_2$$

$$t_2 = \frac{V_A}{V_R} \times t_1 \tag{12.14}$$

The time t_1 is given by counter with clock frequency, f_C.

$$t_1 = (2^N T_C) \tag{12.15}$$

where T_C is the period of clock pulses.

$$nT_C = \frac{V_X}{V_R} \times (2^N T_C) \tag{12.16}$$

So, output of counter is proportional to the analog voltage V_A. The count recorded in the counter is equal to analog voltage V_A if $V_R = 2^N$.

This type of ADC is often used in digital voltmeters because of its good conversion accuracy and low cost. The disadvantage of the dual-slope converter is its slow speed.

12.13 | SPECIFICATION OF ADC

The following specifications are usually specified by the manufactures of ADC.

- Range of input voltage
- Input impedance
- Accuracy
- Conversion time
- Resolution
- Differential linearity

Range of input voltage: The input signal given to the ADC is an analog signal. The amplitude of a signal is the function of time. The range of the input voltage is defined by the maximum and minimum amplitude of an analog signal, which can be applied to the ADC for reliable operation.

Input Impedance: ADC is used to interface the analog circuit with a digital system. To avoid the overloading problem, the output impedance of analog circuit should be matched with the input impedance of an ADC and hence it is required to specify the input impedance of an ADC.

Accuracy: It measures the difference of the actual output voltage to the theoretical voltage. The accuracy depends on the quantization level that is the size of ADC. The accuracy of higher bits is more as compared to lower bits.

Conversion time: It is the time taken to convert the analog data into its digital equivalent. Conversion time is in milliseconds. Ideally, it should be as minimum as possible.

Resolution: It is defined as the voltage input change is necessary for a 1-bit change in output. Resolution in terms of voltage is the full-scale input voltage divided by total number of levels. The resolution of an n-bit ADC is given as,

$$\text{Resolution} = \frac{V}{(2^n - 1)} \tag{12.17}$$

12.14 | DAC AND ADC ICs

Some of the available ICs of DAC and ADC are listed below (Table 12.8).

TABLE 12.8 | ICs related to DAC and ADC

IC Number	Description
ADAC 80	12-bit DAC
A'D7522	10-bit DAC
AD 558	8-bit DAC
ADC O8O9	8-bit ADC
ADADC 80	12-bit ADC
ADC-7109	12-bit binary ADC

EXAMPLE 12.18 An 8-bit DAC has a step size of 50 mV. Determine the full-scale output voltage and percentage resolution.

SOLUTION

Full-scale output of n-bit DAC is $= (2^n - 1)\times \text{step size}$

Full-scale output of 8-bit DAC is $= (2^8 - 1)\times 50\text{ mV} = 12.75\text{ V}$

$$\%\text{ resolution} = \frac{1}{(2^n - 1)}\times 100$$

$$\%\text{ resolution} = \frac{1}{255}\times 100 = 0.39215\% \quad \textbf{Ans.}$$

EXAMPLE 12.19 Determine the conversion time of a 12-bit ADC of the counter type for an input clock frequency of 1 MHz.

SOLUTION

The counter-type ADC has a variable conversion time that is maximum when the input analog voltage is just below the full-scale analog input voltage.

An average conversion time is equal to half the maximum conversion time.

The maximum conversion time equals the time taken by $2^{12} - 1 = 4095$ cycles of clock input.

The clock time period $= \dfrac{1}{1\times 10^6} = 1\ \mu\text{s}$.

Therefore, the maximum conversion time $= 4095 \times (1 \times 10^{-6}) = 4095\ \mu\text{s} = 4.095$ ms.

The average conversion time $= (4.095/2) = 2.047$ ms.

EXAMPLE 12.20 Determine the maximum conversion time that an ADC can have, if it is used to convert signals in the range of 1 kHz to 50 kHz.

SOLUTION

An average conversion time is calculated as half the maximum conversion time. Or an average conversion time is calculated at twice the operating clock frequency.

$$\text{Conversion time} = \frac{1}{2 \times \text{highest frequency}}$$

$$\text{Conversion time} = \frac{1}{2 \times 50 \times 10^3} = 10^{-5} = 10 \times 10^{-6} = 10\ \mu\text{s}$$

EXAMPLE 12.21 An ADC has conversion time 250 μs. What is the highest frequency that its analog input should be allowed to contain?

SOLUTION

An average conversion time is calculated at twice the operating clock frequency.

$$2 \times \text{highest frequency} = \frac{1}{\text{conversion time}}$$

$$\text{Highest frequency} = \frac{1}{2 \times 250 \times 10^{-6}} = \frac{10^6}{500} = 2 \times 10^{-3} = 2\,\text{kHz}$$

EXAMPLE 12.22 Compare the maximum conversion time of 10-bit digital ramp ADC and a 10-bit successive-approximation ADC. Both ADC use 1 MHz clock frequency.

SOLUTION

Clock frequency, $f_c = 1$ MHz

Digital ramp (counter type) ADC:

The maximum conversion time of n-bit ADC = $(2^n - 1) \times f_c$

The maximum conversion time of digital ramp (counter) ADC = $(2^{10} - 1) \times (1 \times 10^6)$ = 1023 μs. **Ans.**

Successive-approximation ADC:

The maximum conversion time of n-bit ADC = $n \times f_c$

The maximum conversion time of n-bit ADC = $10 \times (1 \times 10^6)$ = 8 μs. **Ans.**

EXAMPLE 12.23 How many comparators are needed in a 4-bit flash type ADC?

SOLUTION

Number of bits, $n = 4$
Number of comparators = $2^n - 1$
Number of comparators = $2^4 - 1 = 15$ **Ans.**

EXAMPLE 12.24 Find the successive-approximation ADC output for a 4-bit converter to a 4.67 V input if the reference is 15 V.

SOLUTION Table 12.9 shows the conversion procedure.

TABLE 12.9 | Successive-approximation sequence in counter method of ADC

Clock	Count of Counter	V_R	V_A	Comparison	Description
1	1000	8	4.75 V	$V_A < V_R$	Set 4th bit (MSB)
2	0100	4	4.75 V	$V_A > V_R$	Set 3rd bit (2nd MSB) and reset 4th bit (MSB)
4	0110	6	4.75 V	$V_A < V_R$	Set 2nd bit (3rd MSB)
5	0101	5	4.75 V	$V_A > V_R$	Set 1st bit (LSB) and reset 3rd MSB

EXAMPLE 12.25 What is the conversion time of a 12-bit successive-approximation ADC if its input clock is 4 MHz?

SOLUTION Given $n = 12$ bit

The conversion time of n-bit successive-approximation ADC = nT_C

$$T_C = \frac{1}{4\,\text{MHz}} = \frac{1}{4 \times 10^6} = 0.25 \times 10^{-6}\ \text{s}$$

Conversion time = $nT_C = 12 \times 0.25 \times 10^{-6} = 3 \times 10^{-6} = 2\ \mu\text{s}$

EXAMPLE 12.26 A dual-slope ADC uses a 16-bit counter and a 5 MHz clock rate. The maximum input voltage is +15 V. The maximum integrator output voltage is –8 V when the counter has cycled through $2n$ counts. The capacitor used in the integrator is 0.1 μF. Find the value of resistor of the integrator.

SOLUTION Size of counter = 16 bits.

Operating clock = 5 MHz

$$\text{Time period, } t_1 = \frac{2^{16}}{5\,\text{MHz}} = \frac{2^{16}}{5 \times 10^6} = 13107.2 \times 10^{-6}\ \text{s}$$

Output voltage of integrator, $V_C = \dfrac{V_A}{RC} \times t_1$

$$-8\,\text{V} = \frac{15\,\text{V}}{R \times \left(0.1 \times 10^{-6}\right)} \times 13107.2 \times 10^{-6}$$

$$R = \frac{15}{8 \times 0.1} \times 13107.2 = 245760\ \Omega = 245.76\ \text{k}\Omega \quad \textbf{Ans.}$$

SUMMARY

- **Analog-to-digital conversion (ADC):** The process of converting an analog input voltage to a number of equivalent digital output levels.
- **ADC types are commonly used:** (a) parallel comparator ADC, (b) Successive-approximation ADC and (c) Dual-slope ADC.
- **Binary equivalent weight:** The value assigned to each bit in a digital number expressed as a fraction of the total. The values are assigned in binary fashion according to the sequence 1, 2, 4, 8, …, 2^n, where n is the total number of bits.
- Bit value in digital information is a function of its position.
    - The value of MSB $= \dfrac{V}{2}$
    - The value of 2nd MSB $= \dfrac{V}{4}$
    - The value of 3rd MSB $= \dfrac{V}{8}$
    - The value of LSB $= \dfrac{V}{2^n}$
- Characteristics of a DAC which are generally specified by the designers are (a) accuracy, (b) resolution (c) linearity, (d) settling time and (e) temperature sensitivity.
- **DAC circuits are of two types:** (a) resistor divider DAC and (b) $R/2R$ ladder or binary ladder network DAC.
- **Digital-to-analog conversion (DAC):** The process of converting a number of digital input signals to one equivalent analog output voltage.
- **Differential linearity:** A measure of the variation in size of the input voltage to an ADC which causes the converter to change from one state to the next.
- Equation can be extended for n-bit binary ladder

$$V_A = \frac{1}{\left(2^n\right)}\left(\sum_{i=1}^{n} b_{i-1} 2^{i-1} V\right)$$

- **Millman's theorem:** It states that the voltage at any node in a resistive network is equal to the sum of the currents entering the node divided by the sum of the conductances connected to the node, all determined by assuming that the voltage at the node is zero.

- **Monotonicity:** A consistent increase in output in response to a consistent increase in input (voltage or current).
- **Quantization error:** The error inherent in any digital system due to the size of the LSB.

MULTIPLE CHOICE QUESTIONS

12.1 Among the following, the slowest ADC is
(a) counting type
(b) flash type
(c) integrating type
(d) successive-approximation type

12.2 The number of comparisons carried out in a 4-bit flash-type ADC is
(a) 16 (b) 15
(c) 14 (d) 4

12.3 A 10-bit converter is used to digitize an analog signal in the 0 to 5 V range. The maximum peak-to-peak ripple voltage that can be allowed in the dc supply voltage is
(a) nearly 5 mV
(b) nearly 25 mV
(c) nearly 50 mV
(d) nearly 100 mV

12.4 The number of clocks required for successive approximation with n-output bits is
(a) $2n + 1$
(b) 2^n
(c) $2^n + 1$
(d) $n + 2$

12.5 An analog voltage in the range of 0 to 8 V is divided in eight equal intervals for conversion to 3-bit digital output. The maximum quantization error is
(a) 2 V (b) 1 V
(c) 0.5 V (d) 0 V

12.6 A 12-bit ADC is operating with a 1-μs clock period and the total conversion time is seen to be 14 μs. The ADC must be of
(a) Dual-slope type
(b) Counting type
(c) Parallel-comparator type
(d) Successive-approximation type

12.7 In a 4-bit weighted resistor DAC, the resistor value corresponding is 32 kΩ. The resistor to LSB is value corresponding to MSB will be
(a) 4 kΩ
(b) 8 kΩ
(c) 16 kΩ
(d) 32 kΩ

12.8 A DAC uses a ladder of 10 V full-scale output. The number of bits required of its input for a resolution of 5 mV will be
(a) 7 (b) 8
(c) 11 (d) 16

12.9 Dual-slope integration type analog-to-digital converters type
(a) higher speeds compared to all other type ADC.
(b) very good accuracy without having stability components
(c) good rejection of power supply hum.
(d) none of the above

12.10 An n-bit ADC using V_R as reference voltage has a resolution of
(a) $V_R/(2^n - 1)$
(b) $V_R/(2^n + 1)$
(c) $V_R/(2^{n-1})$
(d) $V_R/(2^n)$

Answers

12.1 (a)	12.2 (b)	12.3 (a)	12.4 (d)	12.5 (c)	12.6 (d)	12.7 (a)
12.8 (c)	12.9 (c)	12.10 (a)				

QUESTIONS

12.1 Explain the basic concepts of an ADC.

12.2 Draw and explain the circuit of (a) weighted resistor DAC, and (b) *R*–2*R* ladder DAC.

12.3 Define the following terms of DAC: (a) Accuracy, (b) Resolution (c) Linearity, (d) Settling time and (e) Temperature sensitivity.

12.4 Draw and explain the circuit of (a) 4-bit parallel comparator ADC (b) Successive-approximation ADC and (c) Dual-slope ADC.

PROBLEMS

12.1 Calculate the values of LSB, MSB, and full-scale output for a 9-bit DAC for 0–12 V range.

12.2 Find the output voltage for a 4-bit ladder, having the digital inputs: (a) 1010 (b) 0110 (c) 0101. Assume logic 0 = 0 V and logic 1 = 12 V.

12.3 What is the resolution of a 12-bit DAC which uses a binary ladder? If the full-scale output is +10 V, what is the resolution in volts?

12.4 How many bits are required in a binary ladder to achieve a resolution of 1 mV if full scale is +5 V?

12.5 What is the conversion time of a 12-bit successive-approximation-type ADC using a 1-MHz clock?

13

Logic Description using VHDL

CHAPTER OBJECTIVES

The main goal of this chapter is to impart knowledge about logic description using very high-speed integrated circuit (VHSIC), very high-speed integrated circuit hardware descriptive language (VHDL). Readers will be able to discuss the following aspects in this chapter:

- Description languages and programming languages
- HDL Format and syntax
- Intermediate signals
- Representing data in VHDL
- Decision control structures in VHDL
- Truth table using VHDL
- Logical operations on bit arrays
- Expanding the bit capacity of a circuit
- VHDL Adders/Subtractor
- VHDL Magnitude comparators
- VHDL Code converters
- Decoders using VHDL
- VHDL 7 Segment decoder/Driver
- VHDL Multiplexers and demultiplexers
- Encoders using VHDL
- Sequential circuits using VHDL
- Edge-triggered devices
- VHDL circuits with multiple components
- Basic counters using VHDL
- Wiring VHDL modules together
- VHDL registers
- VHDL Ring counters

13.1 | INTRODUCTION

'Very high-speed integrated circuit' is abbreviated as 'VHSIC'. The 'VHSIC hardware descriptive language' is abbreviated as 'VHDL'. So, VHDL is a general purpose hardware description language which is used to describe the structure and behaviour of a digital system. In the mid-1980's, the U.S. Department of Defense and the IEEE sponsored the development of this hardware description language with the aim to develop VHSIC. VHDL is used to design a variety of digital systems, ranging in complexity from a few gates to very large number gates. It is the program to document the designs of VHSIC in a concise way. VHDL becomes one of the primary high-level hardware description languages for designing and implementing digital circuits (synthesis) with the development of complex programmable logic devices in digital systems. VHDL allows the digital system to be designed and debugged at a higher level before the implementation to the gate and flip-flop level. This is similar to the use of a high-level language like C to write programs and with the compiler to convert into machine language. Today, VHDL and Verilog are two popular hardware description languages. Another hardware descriptive language (HDL) language is advanced Boolean equation language (ABEL) which was specifically designed for programmable logic devices (PLDs). ABEL is less powerful than the other two languages and is less popular in industry. VHDL established to IEEE standard IEEE-1176 in 1987. An updated standard, IEEE-1164 was adopted in 1993. In 1996, IEEE 1076.3 became a VHDL synthesis standard.

Although these languages look similar as conventional programming languages like BASIC, C, FORTRAN, JAVA etc., but there are some differences. In both cases, a language uses a device to program. Further, computers are complex digital systems that are made up of logic circuits. Computers operate by set of instructions or commands in a sequential order. A hardware description language indented to describe the hardware configuration of a circuit. Programming languages represent a sequence of instructions intended to be carried out by a computer to accomplish some task. A hardware description language is inherently ***parallel,*** i.e. commands, which correspond to logic gates, are executed (computed) in parallel or at a time, as soon as a new input arrives. A HDL program mimics the behaviour of a physical, usually digital, system. It also allows incorporation of timing specifications (gate delays) as well as to describe a system as an interconnection of different components.

VHDL supports both top-down and bottom-up design methodologies for a digital system. With the help of commercially developed programs like synthesis tools, the designs using VHDL creates logic circuit structures from their descriptions. Thus, using VHDL, it is possible to design, simulate, and synthesize anything from a simple combinational circuit to a complete microprocessor system on a chip. Due to highly structured and hierarchical nature of VHDL, it plays an important role in helping the designers to use complex devices like CPLDs and FPGAs very effectively for implementing any digital system.

A digital system is represented at different levels of abstraction as shown in Figure 13.1. This keeps the description and design of complex systems manageable.

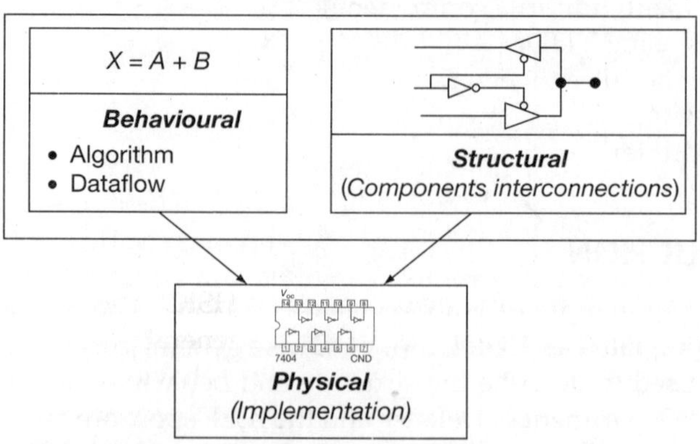

FIGURE 13.1 | Levels of abstraction: behavioural, structural and physical

The highest level of abstraction is the **behavioural** level. A behavioural description specifies the relationship between the input and output signals. This can be a Boolean expression or a more abstract description such as the register transfer or algorithmic level.

The **structural** level describes a system as a collection of gates and components that are interconnected to perform a desired function. A structural description is comparable to a schematic of interconnected logic gates. It is usually closer to the physical realization of a system.

VHDL allows one to describe a digital system at the structural or the behavioural level. The behavioural level can be further divided into two kinds of styles: **Data flow** and **Algorithmic**. The dataflow representation describes the movement of the data through the system. This is typically done in terms of data flow between registers (Register transfer level). The data flow model makes use of concurrent statements that are executed in parallel as soon as data arrives at the input. On the other hand, sequential statements are executed in the sequence as these are specified. VHDL allows both concurrent and sequential signal assignments that will determine the manner in which these are executed.

13.2 | HDL FORMAT AND SYNTAX

A language has unique properties and a proper syntax. A language designed to be interpreted by a computer follows strict rules of syntax. The basic format of any hardware circuit description involves two important elements:

1. The definition of input(s) and output(s).
2. The definition of operations those give the outputs in response to the inputs.

A name is given to the circuit. Symbols (names) are assigned to the inputs and are called ports. Assigned names are defined according to the nature of the port. It may be a single bit from a toggle switch or it may be a four-bit number entered from a keypad. The text based language also conveys the nature of these inputs and outputs. The **mode** of a port defines whether it is input, output, or both. The **type** refers to the number of bits and their grouping. A single bit input has only two possible values: 0 and 1. A four-bit binary number input has any one of 16 different values (0000_2–1111_2). The type determines the range of possible values. The definition of the operation of circuit in a text-based language is contained in a set of statements and is followed by the circuit input/output (I/O) definition.

A digital system in VHDL consists of a design **entity** that contains other entities. Entities are considered components of the top-level entity. Each entity is modelled by an *entity declaration* and an *architecture body*. The entity declaration interfaces to the input and output signals of outside world. The architecture body contains the description of the entity and is composed of interconnected entities, processes and components and is all operating concurrently. The entity is schematically shown in Figure 13.2. There may be many such entities connected together to perform the desired function.

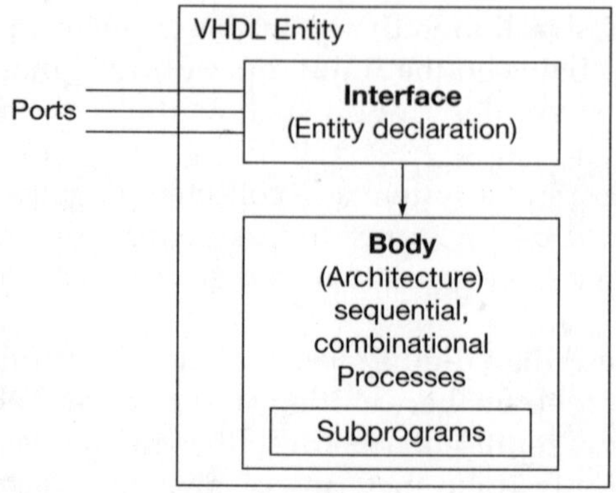

FIGURE 13.2 | VHDL entity

VHDL uses reserved **keywords** that cannot be used as signal names or identifiers. Keywords and user-defined identifiers are **case-sensitive.** Lower-case and upper-case letters are to be considered different. Lines with comments start with two adjacent hyphens (--) and will be ignored by the compiler. VHDL ignores line breaks and extra spaces. VHDL is **a strongly typed** language. Always the type of every object is declared that can have a value, such as signals, constants and variables.

13.2.1 | Identifiers

Several identifiers are used to name the signals, variables, entities, components etc. Generally, identifiers are chosen which are easy to remember. VHDL specifies a few rules to follow to define identifier(s):

The basic rules that apply for defining identifies are;

- Identifier can contain upper-case and lower-case letters, the digits 0 to 9, and special character underscore symbol.
- First and last character of identifier cannot be underscore symbol. Under score (_) must be bracketed by alphabets and digits. Consecutive underscore characters cannot be used.
- Every identifier is a single characters or a string of any desired length.
- Identifiers are case-insensitive.

Examples of legal identifiers:
DELAY, gate, a4, ab12c, Ab5, mux_2, my_circuit_dig
Examples of illegal identifiers:
_ DELAY, ab__12, Ab5_, mux_2_

13.2.2 | Keywords (Reserved Words)

Certain identifiers are used by the system as keywords or reserve words for special use. These keywords or reserve words cannot be used as identifiers for signals or objects to define. Some reserved words are given below

ENTITY, IN, OUT, OR, AND, PORT, MAP, END, etc.

Keywords are often printed in boldface, as is done in this text. A list of keywords is given below:

ABS	*ACCESS*	*AFTER*	*ALIAS*	*ALL*
AND	*ARCHITECTURE*	*ARRAY*	*ASSERT*	*ATTRIBUTE*
BEGIN	*BLOCK*	*BODY*	*BUFFER*	*BUS*
CASE	*COMPONENT*	*CONFIGURATION*	*CONSTANT*	
DISCONNECT	*DOWNTO*	*ELSE*	*ELSIF*	*END*
ENTITY	*EXIT*	*FILE*	*FOR*	*FUNCTION*
GENERATE	*GENERIC*	*GROUP*	*GUARDED*	*IF*
IMPURE	*IN*	*INERTIAL*	*INOUT*	*IS*
LABEL	*LIBRARY*	*LINKAGE*	*LITERAL*	*LOOP*
MAP	*MOD*	*NAND*	*NEW*	*NEXT*
NOR	*NOT*	*NULL*	*OF*	*ON*
OPEN	*OR*	*OTHERS*	*OUT*	*PACKAGE*
PORT	*POSTPONED*	*PROCEDURE*	*PROCESS*	*PURE*
RANGE	*RECORD*	*REGISTER*	*REJECT*	*REM*
REPORT	*RETURN*	*ROL*	*ROR*	*SELECT*
SEVERITY	*SIGNAL*	*SHARED*	*SLA*	*SLL*
SRA	*SRL*	*SUBTYPE*	*THEN*	*TO*
TRANSPORT	*TYPE*	*UNAFFECTED*	*UNITS*	*UNTIL*
USE	*VARIABLE*	*WAIT*	*WHEN*	*WHILE*
WITH	*XNOR*	*XOR*		

13.2.3 | Numbers

The decimal system is the default to represent number. VHDL allows integer literals and real literals. Integer literals consist of whole numbers without a decimal point. Real literals include a decimal point. Exponential notation uses the letter 'E' or 'e'. For integer literals the exponent must always be positive. Examples are:

Integer literals: 12   10   256E3   12e + 6
Real literals: 1.2   256.24   3.14E-2

The number –12 is a combination of a negation operator and an integer literal. To express a number in a base different from the base '10', the following convention is used.

base#number#.

The decimal number 18 is represented in base 2, 8 and 16 and is given below:

Base 2: 2#10010#
Base 8: 8#22#
Base 16: 16#12#

The decimal number 29 is represented in base 2, 8 and 16 and is given below:

Base 2: 2#11101#
Base 8: 8#35#
Base 16: 16#1D#

For easier readability of large numbers easier, underscore is inserted in the numbers. The underscore is not used at the beginning or the end of number. Examples are given below:

Base 2: 2#1001_1101_1100_0010#
Base 10: 215_123

13.2.4 | Characters, Strings and Bit Strings

In a VHDL code, characters literal is put in a single quotation mark, as shown in the examples given below:

'a', 'B', ','

A string of characters are placed in double quotation marks as shown in the following examples:

'This is a string',
'To use a double quotation mark inside a string, use two double quotation marks'
'This is a "String".'

Any printing character can be included inside a string. A bit-string represents a sequence of bit values. In order to indicate that this is a bit string, 'b' or 'B' is placed in front of the string: B"1001" or b"1001". The octal and hexadecimal base strings use 'o' or 'O' and 'x' or 'X', respectively. Some examples are illustrated below:

Binary: B"1100_1001", b"1001011"
Hexadecimal: X"C9", x"4b"
Octal: O"311", o"113"

In the hexadecimal system, each digit represents exactly 4 bits. The number b"1001011" is not the same as X"4b" since the former has only 7 bits while the latter represents a sequence 8 bits. Similarly, O"113" (represents 9 bits) does not have the same sequence as X"4b" (represents 8 bits).

13.2.5 | Entity Declaration

To write a VHDL program, there is need to declare an entity. An entity declaration means declaration of all the input and output signals, and their types in a digital circuit. The basic ports of a digital system and their flow directions are given as:

- IN: Input port. The value of the signal is read.
- OUT: Output port. The value is assigned to the signal.
- INOUT: Bidirectional port. The value is read as well as assigned.
- BUFFER: Output port with read capability (Not a bidirectional port).

The entity declaration defines the NAME of the entity and lists the input and output ports. The general form is shown in Figure 13.3.

```
ENTITY name_of_entity IS [ GENERIC generic_declarations);]
     PORT (signal_names: MODE type;
           signal_names: MODE type;
                  :
           signal_names: MODE type);
END [name_of_entity] ;
```

FIGURE 13.3 | Syntax of entity

An entity always starts with the keyword **ENTITY**, followed by its name and the keyword **IS**. Next are the port declarations using the keyword **PORT**. An entity declaration always ends with the keyword **END**, optionally [] followed by the name of the entity. **ENTITY, IS, PORT** and **END** are the keywords

- The NAME_OF_ENTITY is a user-selected identifier
- signal_names consists of a comma separated list of one or more user-selected identifiers that specify external interface signals.
- **MODE:** is one of the reserved words to indicate the signal direction:
  - **IN** – It indicates input port that the signal is an input.
  - **OUT** – It indicates output port that the signal is an output of the entity whose value is only read by other entities those use it.
  - **BUFFER** – It indicates that the signal is an output of the entity whose value can be read inside the entity's architecture.
  - **INOUT** – It indicates that the signal can be an input or an output.
- **type:** It gives the type of signal. It can be a built-in or user-defined signal type. Examples of types are bit, bit_vector, Boolean, character, std_logic, and std_ulogic.
  - *bit* – This type of signal can have the value 0 and 1
  - *bit_vector* – This type of signal is a vector of bit values, e.g. bit_vector (0 to 7)
  - *std_logic, std_ulogic, std_logic_vector, std_ulogic_vector*: can have 9 values to indicate the value and strength of a signal. Std_ulogic and std_logic are preferred over the bit or bit_vector types.
  - *Boolean* – This type of signal has the value TRUE and FALSE
  - *integer* – This type of signal has a range of integer values
  - *real* – This type of signal has a range of real values
  - *character* – This type of signal is any printing character
  - *time* – This type of signal indicates time. ns (nanosecond) is built in, so no need to specify. others units of time are fs (femto second), ps (pico second), μs (micro second), ms (milli second), s (second), min (minute) and hr (hour)
- **Generic:** Generic declarations are optional and determine the local constants used for timing and sizing (e.g. bus widths) the entity. A generic can have a default value. The syntax for a generic follows,

  generic (

  constant_name: type [:=value];
  constant_name: type [:=value];
  constant_name: type [:=value]);

13.2.6 | Architecture Body

The architecture body specifies the circuit operation and its implementation. An entity or circuit is specified in a variety of ways, such as behavioural, structural (interconnected components), or a combination of the both. An entity can have more architectures but architecture cannot be defined without an entity. The syntax for architecture is given in Figure 13.4.

```
ARCHITECTURE architecture_name OF name_of_entity IS
         [architecture Declarations]
         BEGIN
                  Architecture Statements
         END architecture_name;
```

FIGURE 13.4 | Syntax of architecture

Here **ARCHITECTURE, IS, OF, BEGIN,** and **END** are the keywords. Architecture body contains one or more declarations:

- components declarations
- signal declarations
- constant declarations
- function declarations
- procedure declarations
- type declarations

The architecture body between **BEGIN** and **END** can have one of the following modelling styles:

1. Structural modelling,
2. Data flow modelling, or
3. Behavioural modelling

13.3 | BOOLEAN DESCRIPTION USING VHDL

A circuit given in Figure 13.5 is considered an entity. VHDL code for the circuit given in Figure 13.5 is given in Figure 13.6. For easier reading, a consistent style is followed. Keywords are represented in upper-case letters. Variables are named in lowercase.

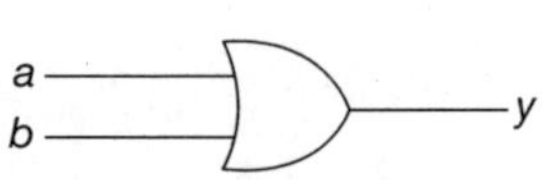

FIGURE 13.5 | Two input AND gate

```
ENTITY or_gate IS
PORT (a, b : IN BIT;
      c    : OUT BIT;)
END or_gate;
ARCHITECTURE circuit OF or_gate IS
BEGIN
c<= a OR b;
END circuit;
```

FIGURE 13.6 | VHDL code for AND gate

The **ENTITY** is a keyword. The name of the entity is *or_gate*. PORT is a keyword. There are two input variables *a* and *b* and one output variable *y*. More than one names used for inputs are separated by commas, ending with a colon (:) and a description of the mode (IN) and type of input (BIT;). In VHDL, the **BIT** description tells the compiler that each variable in the list is a single bit and can have either 0 or 1 value. END or_gate; terminates the ENTITY. Semicolon (;) is used to terminate the line.

The **ARCHITECTURE** declaration is used to describe the operation of the circuit. **ARCHITECTURE** is the key word and 'circuit' is the name for this architectural description. Every **ENTTTY** has at least one architecture associated with it. The words **OF** and **IS** are keywords in this declaration. The body of the architecture description is enclosed between the **BEGIN** and **END** keywords. **END** is followed by the name that has been assigned to this architecture. Within the body there is the description of the operation of block. In this circuit, the operation is given by a very simple Boolean algebra equation. The output, *y* is assigned (<=) the logic level produced by *a* OR *b*. This is referred to as a **concurrent assignment statement**, which means that all such statements between BEGIN and END are evaluated constantly and concurrently. The order in which these statements are listed makes no difference.

13.4 | INTERMEDIATE SIGNALS

In many designs, there is a need to define signal points or nodes 'inside' the circuit block. These points or nodes in the circuit are neither inputs nor outputs for the block but are useful as a reference point. It is a signal that needs to be connected to many other places within the block. In an HDL, these points or nodes are referred to as **buried nodes** or **local signals**. Figure 13.7 shows a very simple circuit that uses an intermediate signal named 'z'. In the HDL, these nodes (signals) are described in the operation of the block. The VHDL code is shown in Figure 13.8.

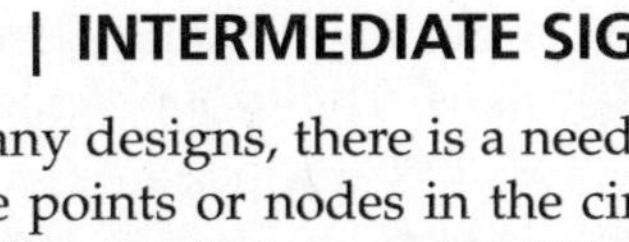

FIGURE 13.7 | Two input AND gate

```
ENTITY Fig13_7 .IS
PORT (a, b, c : IN BIT;
      y       : OUT BIT);
END Fig13_7;
ARCHITECTURE circuit OF Fig13_7 IS
SIGNAL  z :BIT;--local signal
BEGIN
    z<=a OR b;
    y<=z AND C;
END circuit;
```

FIGURE 13.8 | VHDL described in Fig. 13.7

The comments in VHDL follow two dashes (--).Typing two successive dashes allow the designer to write information from that point to the end of the line. The keyword SIGNAL defines intermediate signal z. The keyword BIT designates the type of the signal.

13.5 | REPRESENTING DATA IN VHDL

Numeric data is represented in various ways. The use of hexadecimal number system is a convenient way to communicate bit patterns. However, the use of the decimal number system for numeric data is prevalent in real life. On the other side, computers and digital systems operate only on binary information in VHDL. These various number formats must be understood by the computer which number system is being used. Table 13.1 summarizes the methods of specifying values in binary, hex, and decimal for VHDL.

TABLE 13.1 | Number system in VHDL

Number System	VHDL	Binary Equivalent	Decimal Equivalent
Binary	B"101"	101	5
Hexadecimal	X"101"	1 0000 0001	257
Decimal	101	1100101	101

13.5.1 | SIGNAL

Signal objects are used to connect entities together to form models. Signals are a means to communicate dynamic data between entities. Signals can be declared in entity declaration sections, architecture declarations, and package declarations. The syntax for signal declaration is,

SIGNAL signal_name: signal_type [:= initial value];

The keyword **SIGNAL** is followed by one or more signal names of the same signal type. Signal type is specified after a colon. Finally, a signal can have an initial value which is optional.

13.5.2 | VARIABLE

Variables are used for local storage in process statements and subprograms. All assignments to the variables occur immediately. The syntax for variable declaration is given as

VARIABLE: variable_name: variable_type [: = initial value];

The keyword **VARIABLE** is followed by one or more variable names of the same variable where type is specified after a colon, and an optional initial value can be specified.

13.5.3 | CONSTANT

Constant objects are the names assigned to specific values of a type. If a model requires a fixed value in a number of instances, a constant value is used. The syntax for constant declaration is,

CONSTANT: constant_name: type_name [:= value];

A constant declared in the architecture can be used by any statement inside the architecture, including a process statement. A constant declared in a process is used only in a process.

Variables, signals, and constants can have any of the predefined VHDL types or a user-defined type. Some of the predefined types are given Table 13.2.

TABLE 13.2 | Predefined types

BIT	: 0 or 1
BOOLEAN	: FALSE OR TRUE
INTEGER	: An integer in the range $-(2^{31}-1)$ to $+(2^{31}-1)$
REAL	: Floating point in the range of −1.0E-38 to + 1E + 38
CHARACTER	: Any legal VHDL character including upper- and lower-case characters and special characters
TIME	: An integer with units fs, ps, ns, us, ms, sec
STD_LOGIC	: '0', '1', 'U', 'X', 'Z', 'W', 'L', 'H', '–'

Standard logic values for signals, variables, etc. are given Table 13.3.

TABLE 13.3 | Standard logic (STD_LOGIC) value

'1'	Logic 1 (like BIT type)
'0'	Logic 0 (like BIT type)
'U'	Uninitialized
'X'	Unknown
'Z'	High impedance
'W'	Weak unknown
'L'	Weak '0'
'H'	Weak '1'
'–'	Don't care

13.5.4 | Bit Arrays/Bit Vectors

In an HDL, the type of the signal and its range of acceptable value are defined. Suppose data is to be sent through 8-bit port. It can refer to the individual bits of this port as P1 bit 0 for the least significant bit, on up to P1 bit 7 from the most significant bit.

8-bit data	1	1	0	0	1	1	0	1
8-bit array	P1.7	P1.6	P1.5	P1.4	P1.3	P1.2	P1.1	P1.0
8-bit vector by VHDL	P1[7]	P1[6]	P1[5]	P1[4]	P1[3]	P1[2]	P1[1]	P1[0]

The terms bit array and, bit vector are often used to describe this type of data structure. It simply means that the overall data structure (eight-bit port) has a name (P1). Each individual element (bit) has a unique index number (0–7) to describe the position of bit (numeric weight) in the overall structure. The third bit from the right is designated as P1[2], and it is connected to another signal bit by using an assignment operator.

The syntax for VHDL uses a name for the bit vector followed by the mode (:IN), the type (BIT_VECTOR), and the range of index designations, which are enclosed in parentheses. This declaration is included in the ENTITY section. For example, to declare an eight-bit input port called pl, it is written as given below.

PORT (p1: **IN BIT-VECTOR** (7 DOWNTO 0);
PORT (p2: **IN BIT-VECTOR** (0 TO 7);

EXAMPLE 13.1 Write VHDL code to assign 8-bit data from port p2 to 8-bit local node *t*.

SOLUTION Figure 13.9 shows the VHDL code

```
ENTITY exam13_1 .IS
    PORT (p2    : IN BIT-VECTOR (7 DOWNTO 0);)
END exam13_1;

ARCHITECTURE circuit OF exam13_1 IS
SIGNAL t :BIT-VECTOR (7 DOWNTO 0);
BEGIN
   t<=p2;
END circuit;
```

FIGURE 13.9 | VHDL code for Example 13.1

VHDL follows particular definitions of each type of the data. The type 'bit-vector' describes an array of individual bits. This is interpreted differently than an eight-bit binary number, called a scalar quantity, which has the type integer. VHDL does not assign an integer value to a BIT-VECTOR type signal directly. Data is represented by any of the types shown in Table 13.4. Data assignments and other operations are done between objects of the same type. An integer value is not compatible to BIT_VECTOR declared outputs. Individual BIT and STD_LOGIC data objects like signals, variables, inputs, and outputs are designated by single quotes. Values assigned to BIT_VECTOR and STD_LOGIC_VECTOR type are strings of valid bit values enclosed in double quotes.

TABLE 13.4 | VHDL data types

Data Type	Syntax	Possible Values
BIT	y: OUT BIT;	'0' '1'
STD_LOGIC	driver: STD_LOGIC	'0' '1' 'z' 'X' '–'
BIT_VECTOR	bcd_data: BIT_VECTOR (3 DOWNTO 0);	"0000" "0101" "1001"
STD_LOGIC-VECTOR	dbus: STD_LOIGC _VECTOR (3 DOWNTO0);	"0z1X"
INTEGER	SIGNAL z: INTEGER RANGE –32 TO 31;	–32 ... –2 –1 0 1 2 ... 32

13.5.5 | User-Defined Types

The types of signal, variables, etc. are defined by using the type declaration, which names the type and specifies its value range. The syntax is

TYPE identifier **IS** type_definition;

A few examples of type definitions are given below:

Integer:

TYPE small_int **IS RANGE** 0 **TO** 1024;
TYPE my_word_length **IS RANGE** 31 **DOWNTO** 0;

Floating Point

TYPE cmos_level **IS RANGE** 0.0 **TO** 3.3;
TYPE pmos_level **IS RANGE** –5.0 **to** 0.0;
TYPE probability **IS RANGE** 0.0 **to** 1.0;

Some object declarations that use the above types are given below,

VARIABLE bus_width: small_int:=24;
SIGNAL data_bus: my_word_length;
VARIABLE var1: cmos_level **range** 0.0 **to** 2.5;

An array type is declared as follows:

TYPE *array_name* **IS ARRAY** *indexing scheme* **OF** *element_type;*
TYPE my_word **IS ARRAY** (15 **downto** 0) **OF** std_logic;
TYPE your_word **IS ARRAY** (0 **to** 15) **OF** std_logic;
TYPE var **IS ARRAY** (0 to 7) **OF** integer;
TYPE std_logic_1d **IS ARRAY** (std_ulogic) **OF** std_logic;

Declaration objects of data types defined above are given below

SIGNAL mem_addr: my_word;
SIGNAL data_word: your_word:= B"1101100101010110";
CONSTANT setting: var:= (2,4,6,8,10,12,14,16);

13.6 | LIBRARIES

Libraries are the collection of VHDL code giving logic functions. These libraries offer convenient functions, called macro-functions, like many of the standard TTL devices. Logic function requires standardized data. VHDL offers some standardized data types also. Rather than writing a new description of a familiar TTL device, macro-function out of the library can be used. Of course, there is need to get signals into and out of these macro-functions, and the types of the signals in the code are required to match the types in the functions.

Libraries are collections of predefined quantities and procedures that are used by the compiler to interpret the code. The concept of library is very powerful in VHDL. A library is introduced into the VHDL using a statement of the form

LIBRARY library_name;

For example,' ieee' is a standard library command in VHDL. It is introduced by the command

LIBRARY ieee;

After introducing a library, any definition from the library can be used in writing a VHDL code. Library contains the fundamental definitions of type and basic physical units. The information contained in subgroups is called the standard. A particularly useful package is called IEEE1164, which is based on VHDL language definition. The IEEE library and IEEE 1164 package is initiated in VHDL as

LIBRARY ieee;
USE ieee.std_loqic_1164.**all**;
USE ieee.std_logic_arith.**all**;
USE ieee.std_logic_unsigned.**all**;

In addition, the synopsis library has the attributes package:

LIBRARY SYNOPSYS;
USE SYNOPSYS.attributes.**all**;

The syntax to declare a package is as follows:

```
-- Package declaration
PACKAGE name_of_package IS
      package declarations;
END PACKAGE name_of_package;
-- Package body declarations
PACKAGE BODY name_of_package IS
      package body declarations;
END PACKAGE BODY name_of_package;
```

13.7 | VHDL OPERATORS

Two operands are combined with operator to form expression. Some expression use unary operator because there is need of one operand. Name of expression is based on the type of operator used. VHDL supports different classes of operators that operate on signals, variables and constants (Table 13.5). The different classes of operators are summarized below.

- Binary logical operators
- Relational operators
- Shift operators
- Addition operators
- Unary sign operators
- Multiplying operators
- Miscellaneous operators

TABLE 13.5 | VHDL operators

Class	Operators					
1. Logical operators	AND	OR	NAND	NOR	XOR	XNOR
2. Relational operators	=	/=	<	<=	>	>=
3. Shift operators	SLL	SRL	SLA	SRA	ROL	ROR
4. Addition operators	+	=	&			
5. Unary operators	+	–				
6. Multiplying operators	*	/	MOD	REM		
7. Miscellaneous operators	**	ABS	NOT			

The order of precedence is the highest for the operators of class 7, followed by class 6 with the lowest precedence for class 1. Unless parentheses are used, the operators with the highest precedence are applied first. Operators of the same class have the same precedence and are applied from left to right in an expression.

EXAMPLE 13.2 Consider the following

std_ulogic_vectors, x (='010'), y(='10'), and z ('10101').
Evaluate expression **NOT** x & y **XOR** z **ROL** 1

SOLUTION

Expression is equivalent to
((**NOT** x) & y) **XOR** (z **ROL** 1) = ((101) & 10) **xor** (01011) = (10110) **xor** (01011) = 11101.
The XOR is executed on a bit-per-bit basis.

13.7.1 | Logic Operators

The logic operators (AND, OR, NAND, NOR, XOR and XNOR) are defined for the 'BIT', 'BOOLEAN', 'STD_LOGIC' and 'STD_ULOGIC' types and their vectors. These operators are used to define Boolean logic expression or to perform bit-per-bit operations on arrays of bits. Logic expressions give a result of the same type as the operand (Bit or Boolean). These operators can be applied to signals, variables and constants. Logical operators are summarized in Table 13.6.

TABLE 13.6 | Binary logical operators

Keyword	Operand Types	Result Type	Example	Meaning
AND	BIT or Boolean	BIT or Boolean	*a AND b*	$a \cdot b$
OR	BIT or Boolean	BIT or Boolean	*a OR b*	$a + b$
NAND	BIT or Boolean	BIT or Boolean	*a NAND b*	$\overline{a \cdot b}$
NOR	BIT or Boolean	BIT or Boolean	*a NOR b*	$\overline{a + b}$
XOR	BIT or Boolean	BIT or Boolean	*a XOR b*	$a \oplus b$
XNOR	BIT or Boolean	BIT or Boolean	*a XNOR b*	$a \odot b$

The NAND and NOR operators are not associative. Parentheses are used in a sequence of NAND or NOR operators to prevent a syntax error:

X **NAND** Y **NAND** Z gives a syntax error and is written as (X **NAND** Y) **NAND** Z to avoid error.

13.7.2 | Relational Operators

The relational operators test the relative values of two scalar types and result a Boolean output of 'TRUE' or 'FALSE'. Relational operators are summarized in Table 13.7.

The symbol of the operator '<=' (smaller or equal to) is the same one as the assignment operator used to assign a value to a signal or variable. In the following examples the first '<=' symbol is the assignment operator. Some examples of relational operations are:

```
VARIABLE sts          : BOOLEAN;
CONSTANT a            : INTEGER:=24;
CONSTANT b_count      : INTEGER:=32;
CONSTANT c            : INTEGER:=14;
```

sts <= (a < b_count); -- assigns the value "TRUE" to sts
sts <= ((a >= b_count) **or** (a> c)); -- results in "TRUE"
sts <= (STD_LOGIC ('1', '0', '1') < STD_LOGIC('0', '1','1')); --makes sts "FALSE"

TABLE 13.7 | Relational operators

Operator	Operand Types	Result Type	Example	Meaning
=	Any type	Boolean	*a* = *b*	*a* is equal to *b*
/=	Any type	Boolean	*a* / = *b*	*a* is not equal to *b*
<	Scalar or discrete array types	Boolean	*a* < *b*	*a* is less than *b*
<=	Scalar or discrete array types	Boolean	*a* < = *b*	*a* is less than or equal to *b*
>	Scalar or discrete array types	Boolean	*a* > *b*	*a* is greater than *b*
>=	Scalar or discrete array types	Boolean	*a* > = *b*	*a* is greater than or equal to

For discrete array types, the comparison is done on an element-per-element basis, starting from the left towards the right.

13.7.3 | Shift Operators

Shift operators perform a bit-wise shift or rotate operation on a one-dimensional array elements of the type BIT (or STD_LOGIC) or Boolean. Shift operators are summarized in Table 13.8.

TABLE 13.8 | Shift operators

Operator	Description	Operand Type	Result Type
`SLL`	Shift left logical (fill right vacated bits with the 0)	Left: Any one-dimensional array type with elements of type BIT or Boolean; Right: integer	Same as left type
`SRL`	Shift right logical. Fill left vacated bits with 0.	Left: Any one-dimensional array type with elements of type BIT or Boolean; Right: integer	Same as left type
`SLA`	Shift left arithmetic. Fill right vacated bits with rightmost bit.	Left: Any one-dimensional array type with elements of type BIT or Boolean; Right: integer	Same as left type
`SRA`	Shift right arithmetic. Fill left vacated bits with leftmost bit.	Left: Any one-dimensional array type with elements of type BIT or Boolean; Right: integer	Same as left type
`ROL`	Rotate left (circular)	Left: Any one-dimensional array type with elements of type BIT or Boolean; Right: integer	Same as left type
`ROR`	Rotate right (circular)	Left: Any one-dimensional array type with elements of type BIT or Boolean; Right: integer	Same as left type

The operand is on the left of the operator and the number (integer) of shifts is on the right side of the operator. As an example,

VARIABLE num1:BIT_VECTOR:= "10010110";
num1 **SRL** 2;

The result in the number is "00100101".

When a negative integer is given, the opposite action occurs. Shift to the left performs a shift to the right. As an example

num1 **SRL** –2 is equivalent to num1 **SLL** 2 and gives the result "01011000".

Other examples of shift operations are for the bit_vector A = "101001" are given in Table 13.9.

TABLE 13.9 | Shift operation

VARIABLE A: BIT_VECTOR:="101001";		
Operation	**Intermediate Result**	**Result**
A **SLL** 2	"010010"	"100100"
A **SRL** 2	"010100"	"001010"
A **SLA** 2	"010011"	"100111"
A **SRA** 2	"110100"	"111010"
A **ROL** 2	"010011"	"100110"
A **ROR** 2	"110100"	"011010"

13.7.4 | Addition Operators

The addition operators are used to perform arithmetic operations, i.e. addition and subtraction, on operands of any numeric type. The concatenation (&) operator is used to concatenate (combine) two vectors together to make a longer one. In order to use these operators, one has to specify the package (ieee.std_logic_unsigned.all or std_logic_arith package) in addition to the ieee.std_logic_1164 package.

Addition operators are summarized in Table 13.10.

TABLE 13.10 | Addition operators

Operator	Description	Left Operand Type	Right Operand Type	Result Type
+	Addition	Numeric type	Numeric type	Numeric type
–	Subtraction	Numeric type	Numeric type	Numeric type
&	Concatenation	Array or element type	Array or element type	Same array type

An example of concatenation is the grouping of signals into a single bus [4].

SIGNAL mybus:std_logic_vector (15 **downto** 0);
SIGNAL status:std_logic_vector (2 **downto** 0);
SIGNAL rw, cs1, cs2:std_logic;
SIGNAL mdata:std_logic_vector (0 **to** 9);
mybus <= status & rw & cs1& sc2 & mdata;

Other examples are

MYARRAY (15 **downto** 0) <= "1111_1111" & mdata (2 **to** 9);
NEWWORD <= "VHDL" & "93";

13.7.5 | Unary Operators

The unary operators '+' and '–' are used to specify the sign of a numeric type. Unary operators are summarized in Table 13.11.

TABLE 13.11 | Unary operators

Operator	Description	Operand Type	Result Type
+	Identity	Any numeric type	Same type
–	Negation	Any numeric type	Same type

13.7.6 | Multiplying Operators

The multiplying operators are used to perform mathematical functions on numeric types (integer or floating point).

Multiplying operators are summarized in Table 13.12.

TABLE 13.12 | Multiplying operator

Operator	Description	Left Operand Type	Right Operand Type	Result Type
*	Multiplication	Integer	Integer	Integer
		Floating point	Floating point	Floating point
		Integer	Integer	Integer
		Real type	Real type	Real type
/	Division	Integer	Integer	Integer
		Floating point	Floating point	Floating point
MOD	Modulus	Integer type		Integer type
REM	Remainder	Integer type		Integer type

The remainder (REM) and modulus (MOD) are evaluated as follows:

A **REM** $B = A - (A/B) \times B$ (in which A/B in an integer)
A **MOD** $B = A - B \times N$ (in which N is an integer)

The result of the **REM** operator has the sign of its first operand while the result of the **MOD** operators has the sign of its second operand. Some examples of these operators are given below.

11 **REM** 4 results in 3, i.e. $(11 - (11/4) \times 4)$
(–11) **REM** 4 results in –3, i.e. $(-11 - (-11/4) \times 4)$
9 **MOD** 4 results in 1, i.e. $(9 - 4 \times 2 = 1)$
7 **MOD** (–4) results in –1 $(7 - 4 \times 2 = -1)$.

13.7.7 | Miscellaneous Operators

The absolute and exponentiation operators are applied to numeric type operands. The logical negation (not) results in the inverse polarity but the operand is of same type. Miscellaneous operators are summarized in Table 13.13.

TABLE 13.13 | Miscellaneous operators

Operator	Description	Left Operand Type	Right Operand Type	Result Type
**	Exponentiation	Integer type	Integer type	Integer type
		Floating point	Integer type	Floating point
ABS	Absolute value	Any numeric type		Any numeric type
NOT	Logical negation	BIT or Boolean type		BIT or Boolean type

13.8 | STRUCTURAL MODELLING

An architecture that uses components is called as structural design or structural modelling. Structure defines the precise interconnection structure of signals and entities within that entity. A pure structural description is equivalent to a schematic diagram of the circuit. Structural model is described in Figure 13.10 to obtain the output from the Boolean logic $W = Z(\bar{X}+\bar{Y})$. The circuit diagram is shown in Figure 13.11 along with intermediate points.

```
LIBRARY ieee;
USE ieee.std.logic-1164.ALL;
ENTITY circuit13_11 IS
   PORT (x,y,z:IN BIT;
              W:OUT BIT
        );
END circuit13_11
ARCHITECTURE figure13_10 OF circuit13_11 IS
   -- Declarations
   COMPONENT AND3
       PORT(in1, in2: IN std_logic;
                out1: OUT std_logic
           );
   END COMPONENT;
   COMPONENT OR2
       PORT(in1, in2: IN std_logic;
                out1: OUT std_logic
           );
   END COMPONENT;
   COMPONENT NOT1
       PORT(in1: IN std_logic;
           out1: OUT std_logic
           );
   END COMPONENT;
   -- declaration of signals used to interconnect gates
    SIGNAL x1, y1, z1: std_logic;
BEGIN
   -- Component instantiations statements
   U0: NOT1 PORT MAP(x, x1);
   U1: NOT1 PORT MAP(y, y1);
   U2: OR2  PORT MAP(x1, y1, z1);
   U3: AND3 PORT MAP(z, z1, w);
END figure13_11;
```

FIGURE 13.10 | Structural model of circuit diagram of Figure 13.11

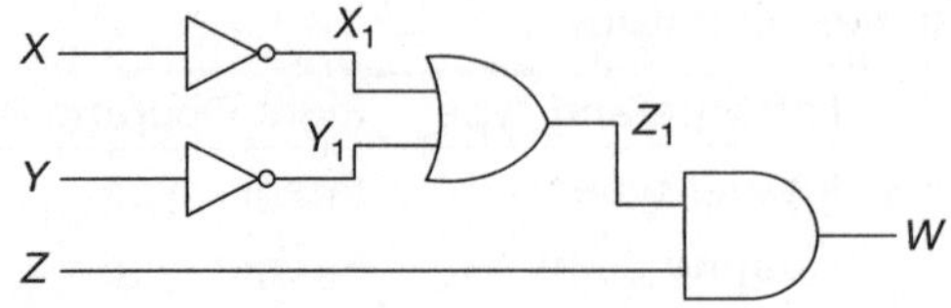

FIGURE 13.11 | Circuit diagram

The name given to the architecture body is Figure 13.11 and the entity declaration is circuit 13.11 which specifies the interface port of this architecture body. The architecture body has two parts:

- Declarative part, and
- Statement part.

Declarative part comes before BEGIN and statement part comes after BEGIN.

13.8.1 | Declarative Part

The component declarations appear in the declarative part. These parts are NOT1, OR2 and AND2. These declarations specify the interface of these components. Either these components are predefined components in the library or define these components separately and add to the library. The declarative part also contains some signal declarations (x_1, y_1 and z_1) which are used only within the architecture for interconnections. Since these signal declarations do not appear in the entity, the MODE is not specified for these signals.

13.8.2 | Statement Part

After the header is the *declarative* part that gives the **COMPONENTS** (gates) that are going to be used in the description of the circuits. In the example, two-input AND gate, two-input OR gate and two inverters. These gates have to be defined first, i.e. these gates need an entity declaration and architecture body. These can be stored in one of the *packages* one refers to in the header of the file. The declarations for the components give the inputs (e.g. *in1*, *in2*) and the output (e.g. *out1*). One has to define internal **SIGNAL** names. In the example these signals are called x_1, y_1 and z_1. The type of the signal is required to be declared.

The declared components and signals in the declarative part of the architecture are instantiated in the statement part of the architecture using component instantiation statements. U0 to U3 are the component labels for the six components which are instantiated. Statements in the body are referred to as component instantiation statements.

A component instantiation statement is a concurrent statement, and hence, the order of this statement is not important. The first statement in which component U0 is instantiated the signal x is connected to the *in1* input ports of the NOT component, while the output port *out1* of this component is connected to the internal signal x_1 which is further connected to OR gate. The syntax of component instantiation statement is given below:

label: *component-name* **PORT MAP** (*signal1*, *signal2*, ..., *signaln*);

e.g. `U2: OR2 PORT MAP (x1, y1, z1);`

An alternative way to use explicit association between the ports is shown as below.

label: *component-name* **PORT MAP** (*port1 => signal1, port2 => signal2 ,... portn => signaln*);

e.g. `U2: OR2 PORT MAP (in1=>x1, in2=>y1, out1=>z1);`

The component name is the name of a previously defined entity that is to be used by declaring **IN** the declarative part of the architecture. One instance of the named entity is created for each component statement that invokes its name and each instance must be named by a unique component label. The **PORT MAP** keyword introduces a list that associates the ports of the named entity with the signals in the current architecture.

EXAMPLE 13.3 Write VHDL code that defines the structure of full-adder.

SOLUTION Diagram of full-adder is given in Figure 13.12. Boolean expression of full-adder is given by

$$sum = augend \oplus addend \oplus previous\ carry$$

$$carryout = (augend)(addend) + (augend \oplus addend)\ Previouscarry$$

There is need of two 2-input XOR gates, two 2-input AND gate and one 2-input OR gate. VHDL code is given in Figure 13.13.

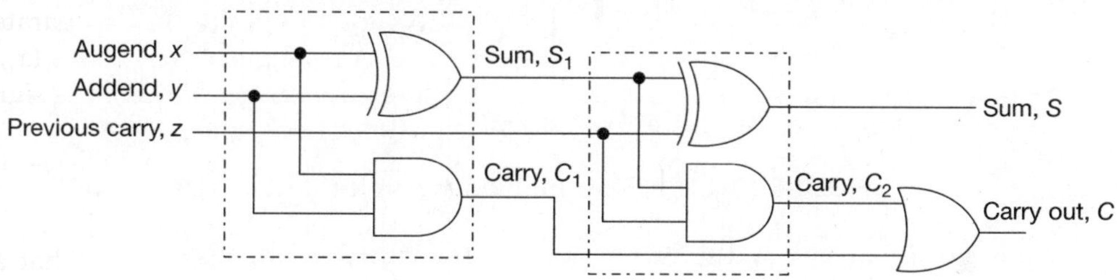

FIGURE 13.12 | Full-adder from half-adders and OR gate

```
-- Structure of a full-adder
LIBRARY  ieee;
USE  ieee.std_logic_1164.ALL;
     -- definition of a full-adder
ENTITY full-adder IS
     PORT (augend, addend, p_carry  : IN BIT;
              sum, carryout            : OUT BIT);
END full-adder;
ARCHITECTURE fa_structure of full-adder is
   COMPONENT XOR1
        PORT (in1, in2  : IN BIT;
                 out1     : OUT BIT);
   END COMPONENT;
   COMPONENT AND2
        PORT (in1, in2  : IN BIT;
                 out1     : OUT BIT);
   END COMPONENT;
   COMPONENT OR3
        PORT (in1, in2  : IN BIT;
                 out1     : OUT BIT);
   END COMPONENT;
```

(Continued)

```
    SIGNAL c1, c2, s1 :BIT
BEGIN
    FA0: XOR1 PORT MAP (augend addend, s1);
    FA1: XOR1 PORT MAP (s1, p_carry, sum);
    FA2: AND2 PORT MAP (augend, addend, c1);
    FA3: AND2 PORT MAP (s1, p_carry, c2);
    FA4: OR3  PORT MAP (c1, c2, carryout,);
END fa_structure;
```

FIGURE 13.13 | VHDL code for structure of full-adder of Example 13.2.

EXAMPLE 13.4 Write a VHDL code for a 4-bit binary adder using structural modelling having full-adder as a component.

SOLUTION Figure 13.14 shows a 4-bit binary adder using full-adder as a component.

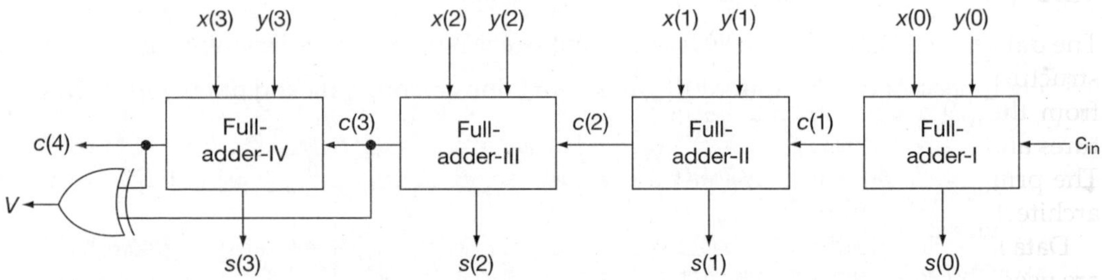

FIGURE 13.14 | Four-bit binary parallel adder

The same input names x and y for the ports of the entity can be used in VHDL since these input names refer to different levels. The internal signals $c(4{:}0)$ are defined to indicate the nets that connect the output carry to the input carry of the next full-adder. For the first input, the input signal c_{in} is used. For the last carry $c(4)$ as an internal signal is defined since the last carry is needed as the input to the XOR gate. VHDL does not allow the use of outputs as internal signals, so the output signal c_{out} cannot be used as input signal for XOR gate. The internal carry $c(4)$ is defined and assigns $c(4)$ to the output carry signal c_{out}.

VHDL code is given in Figure 13.15.

```
-- 4-bit adder
LIBRARY  ieee;
USE  ieee.std_logic_1164.ALL;
ENTITY fourbitadd IS
     PORT (x, y: IN STD_LOGIC_VECTOR(3 DOWNTO0);
            cin : IN STD_LOGIC;
            sum : OUT STD_LOGIC_VECTOR (3 DOWNTO0);
            cout, V: OUT STD_LOGIC);
END fourbitadd;
ARCHITECTURE fouradder_structure OF fourbitadd IS
SIGNAL c: std_logic_vector (4 DOWNTO 0);
COMPONENT full-adder
     PORT(x, y, cin: IN STD_LOGIC;
       sum, carry: OUT STD_LOGIC);
END COMPONENT;
```

(Continued)

```
    COMPONENT XOR2
    PORT(x, y : IN STD_LOGIC;
      z   : OUT STD_LOGIC);
END COMPONENT;

BEGIN
    FA0: full-adder PORT MAP(x(0), y(0), Cin, sum(0), c(1));
    FA1: full-adder PORT MAP(x(1), y(1), c(1), sum(1), c(2));
    FA2: full-adder PORT MAP(x(2), y(2), c(2), sum(2), c(3));
    FA3: full-adder PORT MAP(x(3), y(3), c(3), sum(3), c(4));
    FA4: XOR2       PORT MAP(c(3), c(4), V);
    Cout <= c(4);
END fouradder_structure;
```

FIGURE 13.15 | VHDL code for structure of 4-bit parallel adder

13.9 | DATA FLOW MODELING

The data flow modelling is expressed using concurrent signal assignment statements. The structure of the entity is not explicitly specified in this type of modelling, but it is inferred from the equations. VHDL distinguishes between behavioural and data flow architectures and however data flow architecture is treated together with behavioural description. The primary difference is that, behavioural description uses processes and the data flow architecture does not use processes.

Data flow descriptions are used in cases where the simple design equations of the circuit are used. Let the full-adder circuit discussed in example 13.2 has the data flow equations as

$$\text{sum} = \text{augend} \oplus \text{addend} \oplus \text{previous carry}$$

$$\text{carryout} = (\text{augend})(\text{addend}) + (\text{augend} \oplus \text{addend})\,\text{Previouscarry}$$

These equations can be represented using data flow modelling having concurrent statements as

$$\text{sum} = (\text{augend XOR addend})\,\text{XOR p\_carry}$$

$$\text{carryout} = (\text{augend AND addend})\,\text{OR}\,((\text{augend XOR addend})\,\text{AND p\_carry})$$

Figure 13.16 gives the VHDL code using data flow modelling for full-adder.

```
-- Example of a bit adder
LIBRARY ieee;
USE ieee.std_logic_1164.all;
    -- definition of a full-adder
ENTITY full-adder IS
     PORT (augend, addend, p_carry: IN BIT;
                 sum, carryout    : OUT BIT);
     END full-adder;
ARCHITECTURE fa_dataflow of full-adder is
BEGIN
  sum <= (augend XOR addend) XOR p_carry;
  carryout <= (augend AND addend) OR (p_carry AND (augend XOR addend));
END fa_dataflow;
```

FIGURE 13.16 | VHDL code for data flow of full-adder

It is also possible to use the following conditional signal assignment statements in data flow modelling, since these statements are concurrent.

13.9.1 | WHEN-ELSE Statement

It is conditional signal assignment statement. In a when-else statement, a Boolean expression combines the individual Boolean terms using built-in Boolean operators of VHDL, such as AND, OR and NOT. Boolean terms are typically Boolean variables or the results of comparisons using relational operators:

= (equal to), | = (not equal to), > (greater than), >= (greater than or equal), < (less than), and < = (less than or equal).

The syntax for the conditional signal assignment is given below:

```
Target_signal <= expression1 WHEN condition ELSE expression2;
```

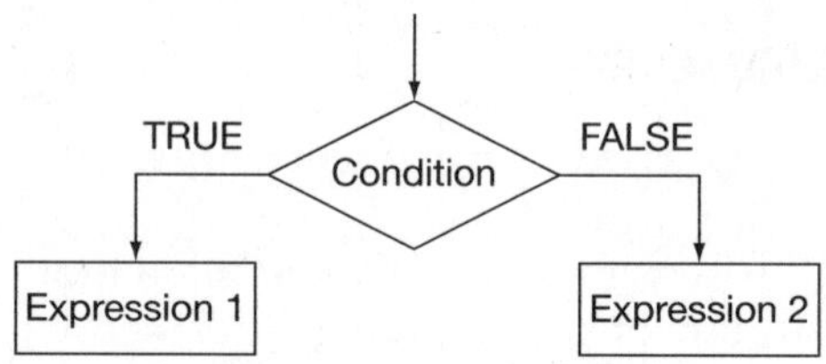

FIGURE 13.17 | When else statement

Condition is logical or relational resulting in Boolean value. The target signal receives the expression1 when condition results in TRUE value. If condition results FALSE value then expression2 is assigned to target signal. Syntax of another form of WHEN-ELSE statement is given below.

```
Target_signal <= expression1 WHEN condition1 ELSE
                 expression2 WHEN condition2 ELSE
                 expression3;
```

The target signal receives the value of the *expression1* whose Boolean or relational *condition1* is TRUE (Figure 13.17). If *condition1* is FALSE then *condition2* is evaluated. TRUE value of *condition2* assigned *expression2* value to target value. FALSE value of *condition2* assigned *expression3* value to target value. So, if no condition is found to be TRUE, the target signal receives the value of the final expression. If more than one condition is TRUE, the value of the corresponding condition that is TRUE will be assigned (Figure 13.18).

A few examples of when-else statements are given below'

1. X <= '1 ' WHEN b <= c ELSE '0 ';
2. Y <= j WHEN X = '0' ELSE
   k WHEN X = '1' ELSE
   m WHEN OTHERS;
3. yout <= i0 WHEN (s0 = '0' AND s1 = '0') ELSE
   i1 WHEN (s0 = '1' AND s1 = '1') ELSE
   '0' WHEN OTHERS;

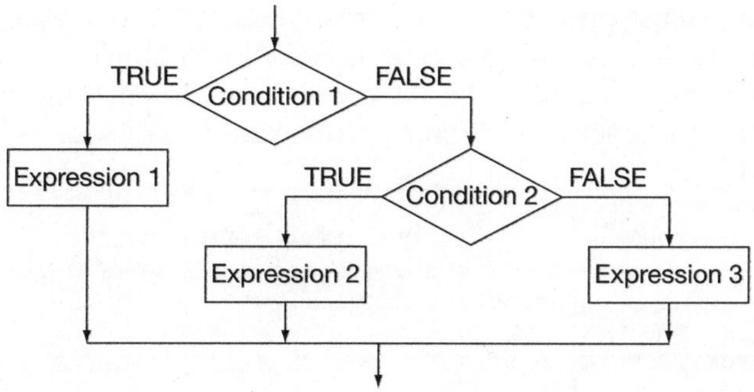

FIGURE 13.18 | When else statement

EXAMPLE 13.5 Write VHDL code using data flow modelling of a 4-to-1 multiplexer using conditional signal assignment.

SOLUTION Block diagram of 4-to-1 MUX is shown in Figure 13.19. VHDL code is given in Figure 13.20 using conditional signal assignment WHEN-ELSE. There are possible selections based on the values of S_1 and S_0.

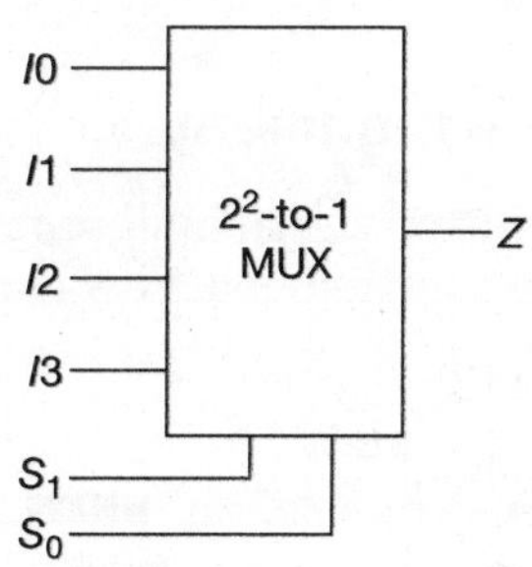

FIGURE 13.19 | Block diagram of 2^2-to-1 MUX

```
ENTITY mux_4_1_dflow IS
    PORT (S1, S0, I0, I1, I2, I3 : IN std_logic;
                Z                : OUT std_logic);
    end mux_4_1_dlow;
ARCHITECTURE dflow_mux41 OF mux_4_1_dlow IS
BEGIN
    Z <= I0 WHEN S1='0' AND S0='0' ELSE
         I1 WHEN S1='0' AND S0='1' ELSE
         I2 WHEN S1='1' AND S0='0' ELSE
         I3;
END dflow_mux41;
```

FIGURE 13.20 | VHDL code of 2^2-to-1 MUX

Name of entity is *mux_4_1_dflow*. Inputs are I0, I1, I2 and I3 of std_logic type. Output is Z and of type std_logic. Architecture name is *dlow_mux41*.

The conditional signal assignment has been re-evaluated as soon as any of the signals in the conditions or expression is changed. The WHEN-ELSE construct is useful to express logic function in the form of a truth table. An example of the same multiplexer as above is given below in a more compact form as shown in Figure 13.21.

```
ENTITY MUX_4_1_ftable is
     PORT (I0, I1, I2, I3: IN std_logic;
            sel: IN std_logic_vector (1 DOWNTO 0);
            Z  : OUT std_logic);
     END MUX_4_1_ funcTab;
ARCHITECTURE dflow1_MUX41 OF MUX_4_1_ ftable IS
BEGIN
   Z <= I0 WHEN SEL = "00" ELSE
        I1 WHEN SEL = "01" ELSE
        I2 WHEN SEL = "10" ELSE
        I3;
END dflow1_MUX41;
```

FIGURE 13.21 | VHDL code of 2^2-to-1 MUX

Two-bit select input variable is defined instead of S_1 and S_0 and is used as condition to check.

13.9.2 | WITH-SELECT Signal Assignments

The WITH-SELECT statement is selected signal assignment statement and is similar to the conditional WHEN-ELSE assignment statement. The syntax of WITH-SELECT statement is given below.

```
WITH choice_expression SELECT
     target_name <= expression WHEN choices,
     target_name <= expression WHEN choices,
                :
     target_name <= expression WHEN choices;
```

The target is a signal that receives the value of an expression whose choice includes the value of the choice_expression. The expression selected is the first with a matching choice. The choice is a fixed or static expression (e.g. 5) or a range expression (e.g. 4 **to** 9). The following rules must be followed for the choices:

- No two choices can overlap
- All possible values of choice_expression must be covered by the set of choices, unless an **OTHERS** choice is present.

EXAMPLE 13.6 Write VHDL code using data flow modelling of a 4-to-1 multiplexer using conditional signal assignment WITH-SELECT.

SOLUTION Block diagram of 4-to-1 MUX is shown in Figure 13.19. VHDL code is given in Figure 13.22 using conditional signal assignment WITH-SELECT. An example of a 4-to-1 multiplexer is given below.

```
ENTITY mux_4_1_dflow1 IS
    PORT (I0, I1, I2, I3: IN std_logic;
        SEL: IN std_logic_vector (1 DOWNTO 0);
        Z  : OUT std_logic);
    END mux_4_1_dflow1;
ARCHITECTURE dflow2_mux41 OF mux_4_1_dflow1 IS
BEGIN
   WITH SEL SELECT
      Z <= I0 WHEN "00",
           I1 WHEN "01",
           I2 WHEN "10",
           I3 WHEN "11";
END dflow1_mux41;
```

FIGURE 13.22 | VHDL code of 2^2-to-1 MUX

When SEL is '00' then I0 is assigned to Z. When SEL is '01' then I1 is assigned to Z and so on. The selected signal assignment is useful to express a function as a truth table.

13.10 | BEHAVIOURAL MODELLING

The behavioural style of modelling specifies the behaviour of an entity as a set of statements that are executed sequentially in the specified order. A process statement is a key element in the behavioural modelling that uses sequential statements to describe the behaviour of a system over time. In other words, a process statement contains a set of sequential statements which specify the functionality of the model but not the exact structure of it. Process statement is a concurrent statement which executes in parallel with other concurrent statements and offer processes. The syntax for a process statement is given below:

```
[process_label:] PROCESS [(sensitivity_list)] [IS]
[process_declarations]
BEGIN
    list of sequential statements such as:
            signal assignments
            variable assignments
            case statement
            exit statement
            if statement
            loop statement
            next statement
            null statement
            procedure call
            wait statement
END PROCESS [process_label];
```

The bracket followed by the keyword **PROCESS** is referred to as the sensitivity list of the process. Whenever one of the signals in the sensitivity list changes, the sequential statements inside the process body are executed in sequence. The process statement also has a declarative part, before the keyword BEGIN along with the statement part in between the keywords **BEGIN** and **END** process. A process is declared within architecture and is a *concurrent* statement. However, the statements inside a process are executed *sequentially*. A process reads and writes signals and values of the interface input and output ports to communicate with the rest of the architecture. The declarative part may contain one or more of the following declarations:

- type declarations;
- variable declarations;
- constant declarations;
- function definitions; or
- procedure definitions.

Variables and constants used inside a process are defined in the *process_declarations* part before the keyword **BEGIN**. The keyword **BEGIN** signals the start of the computational part of the process. The statements are *sequentially* executed, similarly as a conventional software program. It should be noted that variable assignments inside a process are executed immediately and denoted by the ':=' operator. This is in contrast to signal assignments denoted by '<=' and which changes occur after a delay. As a result, changes made to variables are available immediately to all subsequent statements within the same process. Full-adder is composed of two half-adders as shown in Figure 13.23, modelled with two processes P1 and P2.

```
LIBRARY ieee;
USE ieee.std_logic_1164.all;
ENTITY full-adder IS
      PORT (augend, addend, cin : IN std_logic;
            sum, cout : OUT std_logic);
END full-adder;

ARCHITECTURE behave_fa OF full-adder IS
SIGNAL int1, int2, int3: std_logic;
BEGIN
-- Process P1 that defines the first half-adder
P1: PROCESS (augend, addend)
      BEGIN
            int1<= augend XOR addend;
            int2<= augend AND addend;
      END PROCESS;
-- Process P2 that defines the second half-adder and
the OR -- gate
P2: PROCESS (int1, int2, cin)
      BEGIN
            sum <= int1 XOR cin;
            int3 <= int1 AND cin;
            cout <= int2 OR int3;
      END PROCESS;
END behave_fa;
```

FIGURE 13.23 | VHDL code of 2^2-to-1 MUX

13.11 | SEQUENTIAL STATEMENTS FOR BEHAVIOURAL MODELLING

The sequential statements used in behavioural modelling are: **IF, CASE, LOOP, EXIT, NEXT** and **WAIT**. These sequential statements are used inside a process for signal assignments. The process executes whenever there is an event on signals. The process executes all the statements inside the process sequentially, independent of whether there is a change on any signal in its right-hand side expression. The execution of concurrent signal assignment statements gets executed only when an event occurs on a signal in the right-hand side expression.

13.11.1 | IF Statements

The **IF**-statement executes a sequence of statements. The sequence of statements depends on one or more conditions. The result of condition must be Boolean. The syntax is given below:

```
IF condition THEN
        sequential statements
    [ELSIF condition THEN
        sequential statements ]
    [ELSE
        sequential statements ]
END IF;
```

The conditions are generally Boolean or relational expressions. The conditions are tested for a TRUE or FALSE condition. For example (Figure 13.24),

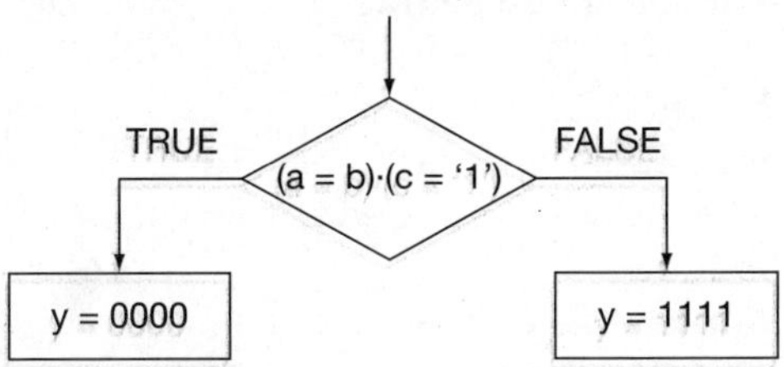

FIGURE 13.24 | When else statement

```
IF((a = b) AND (c = '1')) THEN
      y<="0000";
ELSE
      y<="1111";
END IF;
```

The **IF** statement is performed by checking each condition in the order until a 'TRUE' or 'FALSE' is found. The nesting of IF-statements is allowed. The IF-statement can be used to describe combinational circuits. The following example illustrates 4-to-1 multiplexer with inputs I0, I1, I2 and I3, and select signals S0 and S1. This statement is inside a process construct as given in Figure 13.25.

```
ENTITY MUX_4_1c IS
    PORT (S1, S0, I0, I1, I2, I3: IN std_logic;
         Z: OUT std_logic);
    END MUX_4_1a;
ARCHITECTURE behave_MUX41c OF MUX_4_1c IS
BEGIN
    P1: PROCESS (S1, S0, I0, I1, I2, I3)
    BEGIN
        IF (( NOT S1 AND NOT S0 )= '1') THEN
            Z <= I0;
        ELSIF (( NOT S1 AND S0) = '1') THEN
           Z<=I1;
        ELSIF ((S1 AND NOT S0) = '1') THEN
           Z <=I2;
        ELSE
           Z<=I3;
       END IF;
    END PROCESS P1;
 END behave_MUX41c;
```

FIGURE 13.25 | VHDL code of 2^2-to-1 MUX

13.11.2 | CASE Statement

The case statement executes one of several sequences of statements, based on the value of a single expression. The syntax is given below:

```
CASE expression is
      WHEN choices =>
            sequential statements
      WHEN choices =>
            sequential statements
            -- branches are allowed
      [ WHEN OTHERS => sequential statements ]
END CASE;
```

The expression evaluates to an integer or an enumerated type of a one-dimensional array, such as a bit_vector. The case statement evaluates the expression and compares the value to each of the choices. The WHEN-clause corresponding to the matching choice executes its statement(s). The following rules are followed:

- no two choices can overlap (i.e. each choice can be covered only once)
- if the "**WHEN OTHERS**" choice is not present, all possible values of the expression must be covered by the set of choices.

The following example illustrates 4-to-1 multiplexer with inputs I_0, I_1, I_2 and I_3, and select signals S_0 and S_1. This statement is inside a process construct as given in Figure 13.26.

```
ENTITY mux_4_1d IS
   PORT ( SEL: IN std_logic_vector(2 DOWNTO 1);
          I0, I1, I2, I3: IN std_logic;
          Z: OUT std_logic);
   END mux_4_1;
 ARCHITECTURE behave_mux41d OF mux_4_1d IS
    BEGIN
       pr_mux: PROCESS (SEL, I0, I1, I1, I3)
       BEGIN
          CASE SEL IS
             WHEN "00" => Z <= I0;
             WHEN "01" => Z <= I1;
             WHEN "10" => Z <= I2;
             WHEN "11" => Z <= I3;
             WHEN OTHERS => Z <= 'X';
          END CASE;
       END PROCESS pr_mux;
 END behave_mux41d;
```

FIGURE 13.26 | VHDL code of 2^2-to-1 MUX

The "WHEN OTHERS" covers the cases when SEL="0X", "0Z", "XZ", "UX", etc. It should be noted that these combinational circuits can be expressed in other ways, using concurrent statements such as the "WITH – SELECT" construct. Since the case statement is a sequential statement, one can have nested case statements.

13.11.3 | LOOP Statements

A loop statement is used to repeatedly execute a sequence of sequential statements. The syntax for a loop is as follows:

```
[loop_label:] iteration_scheme LOOP
     sequential statements
     [NEXT [label] [WHEN condition];
     [EXIT [label] [WHEN condition];
END LOOP [loop_label];
```

Labels are optional but are useful when writing nested loops. The next and exit statement are sequential statements that can only be used inside a loop.

- The **NEXT** statement terminates the rest of the current loop iteration and execution will proceed to the next loop *iteration*.
- The **EXIT** statement skips the rest of the statements, terminating the loop entirely, and continues with the next statement after the exited loop.

There are three types of iteration schemes:

- BASIC **LOOP**
- **WHILE-LOOP**
- **FOR-LOOP**

Example of basic loop is illustrated below. This loop uses IF-statement to EXIT (terminate) from loop. Figure 13.27(a) explains its execution.

```
L1: LOOP
      i<=i+1
      IF (i > 100) THEN
              EXIT L1;
      END IF;
END LOOP L1;
```

This loop uses WHEN statement to EXIT (terminate) from loop. Figure 13.27(b) explains its execution.

```
L1: LOOP
      i<=i+1;
      EXIT when i=100;
END LOOP L1;
```

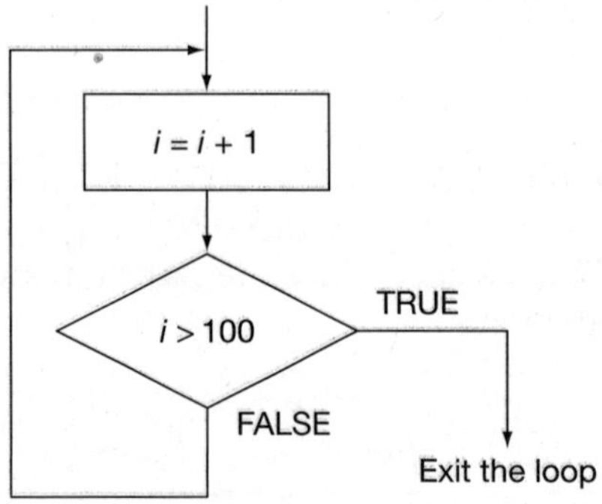

FIGURE 13.27(a) | LOOP with IF-statement

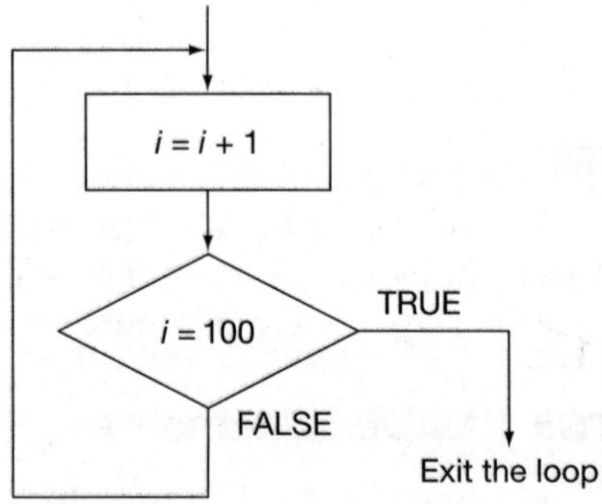

FIGURE 13.27(b) | LOOP with IF-statement

13.11.4 | WHILE-LOOP Statement

The WHILE-LOOP evaluates a Boolean condition. When the condition is TRUE, the loop repeats, otherwise the loop is skipped and the execution halts. The syntax for the WHILE-LOOP is given below:

```
[loop_label:] WHILE condition LOOP
        sequential statements
        [NEXT [label] [WHEN condition];
        [EXIT [label] [WHEN condition];
END LOOP [ loop_label ];
```

For example,

```
WHILE (j<8) LOOP
        x(j)<='1';
        j<=j+1;
END 1OOP,'
```

Figure 13.27(c) gives the execution of the WHILE-LOOP. The condition of the loop is tested before each iteration increment, including the first iteration. If it is false, the loop is terminated.

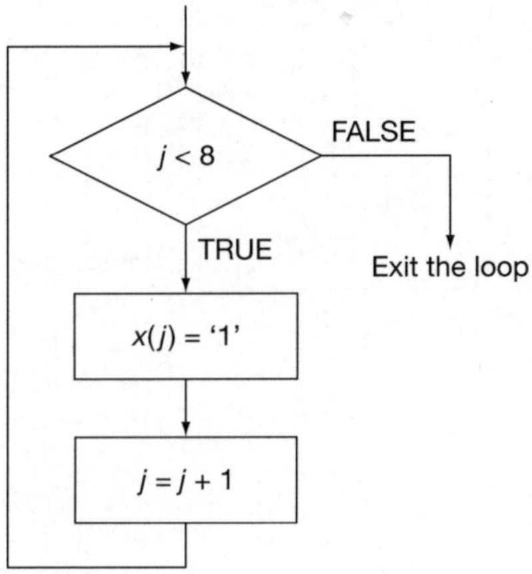

FIGURE 13.27(c) | WHILE LOOP

13.11.5 | FOR-LOOP Statement

The FOR-LOOP uses an integer iteration scheme that is based on known (fixed) number of iterations. The syntax is given below.

```
[loop_label:] FOR identifier IN range LOOP
    sequential statements
    [NEXT [label] [WHEN condition];
    [EXIT [label] [WHEN condition];
END LOOP[loop_label];
```

The *identifier (index)* is automatically declared by the loop itself, so there is no need to declare it separately. The value of the identifier is read inside the loop and is not available outside its loop. The value of the index cannot be assigned or changed. The WHILE-LOOP condition involves variables that can be modified inside the loop. The *range* must be a computable integer range in one of the following forms, in which integer_expression must evaluate to an integer:

- *integer_expression* **TO** *integer_expression*
- *integer_expression* **DOWNTO** *integer_expression*

For example

```
FOR i IN 1 TO 7 LOOP
    y(i)<= '0';
END LOOP;
```

Figure 13.28 gives the execution of the FOR-LOOP of the illustrated example.

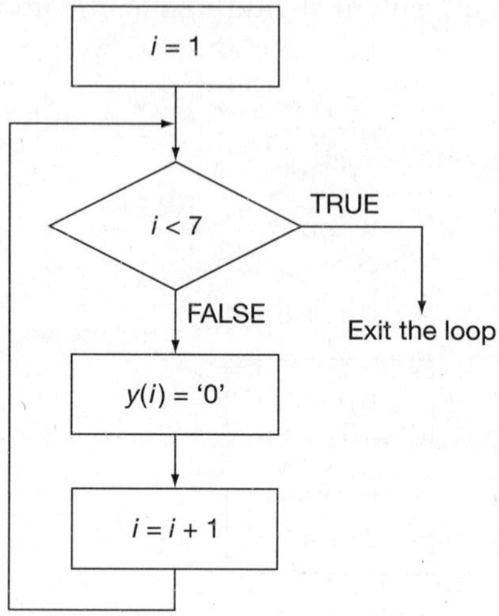

FIGURE 13.28 | FOR LOOP

13.11.6 | NEXT and EXIT Statement

The **NEXT** statement skips execution to the next iteration of a loop statement and proceeds with the next iteration. The syntax is given below:

```
NEXT [label] [WHEN condition];
```

WHEN keyword is optional and executes the next statement when its condition is evaluated to the Boolean value as TRUE.

The **EXIT** statement skips the rest of the statements, terminating the loop entirely, and continues with the next statement after the exited loop. The syntax is given below:

```
EXIT [label] [WHEN condition];
```

WHEN keyword is optional and executes the next statement when its condition evaluates to the Boolean value TRUE. The difference between the next and exit statement, is that the exit statement terminates the loop.

13.11.7 | WAIT STATEMENT

The WAIT statement halts a process until an event occurs. There are several forms of the wait statement,

```
WAIT UNTIL condition;
WAIT FOR time expression;
WAIT ON signal;
WAIT;
```

The condition in the "**wait until**" statement must be TRUE for the process to resume. A few examples follow.

```
WAIT UNTIL CLK='1';
```

The process waits until a positive-going clock edge occurs in above WAIT statement. The process waits until a negative-going clock edge arrives in the WAIT statement given below.

```
WAIT UNTIL CLK='0';
WAIT UNTIL CLK'event and CLK='1';
WAIT UNTIL NOT CLK'stable AND CLK='1';
```

The above are equivalent to the positive-edge or 0-1 transitions. The hardware implementation for these three statements is identical.

A process that contains a wait statement cannot have a sensitivity list.

13.11.8 | NULL Statement

The null statement states that no action takes place. The syntax is given below:

```
NULL;
```

It is useful in a case statement where all choices are to be covered, even if some of them can be ignored. As an example, consider a control signal CNTL in the range 0 to 31. When the value of 'cntl' is 3 or 15, the signals A and B will be XOR-ed, otherwise nothing will occur. Figure 13.29 gives the VHDL code where null statement is used.

```
ENTITY ex_wait IS
   PORT ( cntl: IN INTEGER RANGE 0 TO 31;
         A, B: IN std_logic_vector(7 DOWNTO 0);
         Z: OUT std_logic_vector(7 DOWNTO 0));
   END ex_wait;
ARCHITECTURE a_wait OF ex_wait IS
BEGIN
   P_wait: PROCESS (cntl)
   BEGIN
         Z< =A;
         CASE cntl IS
            WHEN 3 | 15 => Z <= A XOR B;
            WHEN OTHERS => NULL;
         END CASE;
      END PROCESS P_wait;
      END a_wait
```

FIGURE 13.29 | VHDL code using null statement

13.12 | TRUTH TABLE USING VHDL

A truth table is used to express the operation of a circuit block. It relates the output of the circuit to every possible combination of its inputs. A Boolean expression is derived from the truth table and is simplified using K-maps or Boolean algebra. Finally, the circuit is implemented from the final Boolean equation(s). The truth table using HDL can be entered.

13.12.1 | Truth Tables Using VHDL: Selected Signal Assignment

A truth table has three binary inputs and defines output as HIGH if any two or more inputs are HIGH.

The code in Figure 13.30 uses VHDL to implement a truth table given in Table 13.14 using a selected signal assignment WITH-SELECT statement, to describe its operation. It allows the designer to specify the operation of the circuit, filling out a truth table. Three bits are concatenated (combined) and stored in signal *in_bits* which is three bits. Signal *in-bits* is used to select the condition and assigns the corresponding value.

```
LIBRARY ieee;
USE ieee.std_logic_1164.all;
ENTITY table13_14 IS
        PORT (a, b, c : IN BIT; --declare individual input bits
                 y      : OUT BIT);
END table13_14;
ARCHITECTURE truthtable OF table13_14 IS
       SIGNAL in_bits: BIT_VECTOR (2 DOWNTO 0);
       BEGIN
              in_bits <= a &b &c;  --concatenate input bits into bit_vector
              WITH in_bits SELECT
              Y <='0' WHEN "000", --Truth Table
                  '0' WHEN "001",
                  '0' WHEN "010",
                  '1' WHEN "011",
                  '0' WHEN "100",
                  '1' WHEN "101",
                  '1' WHEN "110",
                  '1' WHEN "111";
 END truth table;
```

FIGURE 13.30 | VHDL design file for decoder.

TABLE 13.14 | Truth table

Input			Output
a	*b*	*c*	*Y*
0	0	0	0
0	0	1	0
0	1	0	0
0	1	1	1
1	0	0	0
1	0	1	1
1	1	0	1
1	1	1	1

EXAMPLE 13.7 Write a VHDL code using data flow modelling f or a 3:8 decoder having active high outputs and an enable line.

SOLUTION The entity for a 3:8 lines decoder is shown in Figure 13.31. Truth table of 3:8 lines decoder is given in Table 13.15. There are four binary inputs and eight binary output lines.

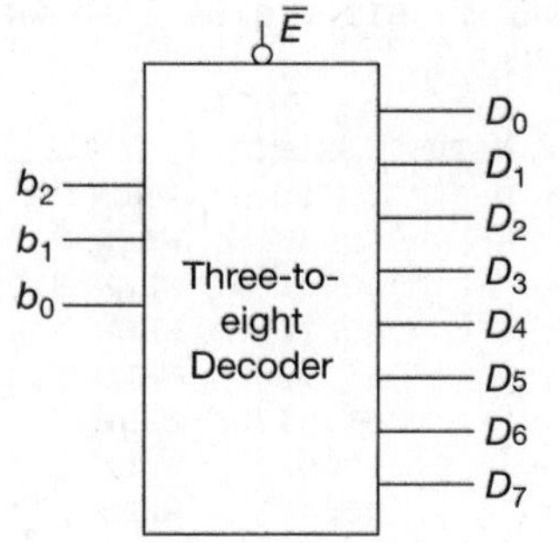

FIGURE 13.31 | Block diagram of 3-to-8 line decoder with enable

TABLE 13.15 | Three-to-eight line decoder with enable

Input				Output							
$\bar{E}$	b_2	b_1	b_0	D_0	D_1	D_2	D_3	D_4	D_5	D_6	D_7
0	0	0	0	1	0	0	0	0	0	0	0
0	0	0	1	0	1	0	0	0	0	0	0
0	0	1	0	0	0	1	0	0	0	0	0
0	0	1	1	0	0	0	1	0	0	0	0
0	1	0	0	0	0	0	0	1	0	0	0
0	1	0	1	0	0	0	0	0	1	0	0
0	1	1	0	0	0	0	0	0	0	1	0
0	1	1	1	0	0	0	0	0	0	0	1
1	X	X	X	0	0	0	0	0	0	0	0

ALTERNATIVE-I: Figure 13.32 gives the VHDL code of 3:8 lines decoder using WITH-SELECT statement. Three-bit input is represented by 3-bit vector and 8-bit output is taken as 8-bit vector. When 3-bit input vector is concatenated with enable, *en* signal it becomes 4-bit vector and act as signal that is declared in the architecture, to assign the selected value.

ALTERNATIVE-II: Figure 13.33 gives the VHDL code of 3:8 lines decoder using WHEN-ELSE statement. Single-bit three inputs and single-bit eight outputs are considered. When three single bit inputs are concatenated with enable, *en* signal it becomes 4-bits input and act as signal, declared in the architecture.

```
LIBRARY ieee;
USE ieee.std_logic_1164.all;
ENTITY decoder3_8 IS
      PORT (en:IN BIT;
            b: IN BIT_VECTOR (2 DOWNTO 0);
            D: OUT BIT_VECTOR (7 DOWNTO 0);
END decoder3_8;
ARCHITECTURE truthtable OF decoder3_8 IS
     SIGNAL T: BIT_VECTOR (3 DOWNTO 0);
     BEGIN
          T <= en & b;
          WITH T SELECT
          D <='00000001' WHEN "1000",
              '00000010' WHEN "1001",
              '00000100' WHEN "1010",
              '00001000' WHEN "1011",
              '00010000' WHEN "1100",
              '00100000' WHEN "1101",
              '01000000' WHEN "1110",
              '10000000' WHEN "1111";
END truthtable;
```

FIGURE 13.32 | VHDL design file for decoder

```
LIBRARY ieee;
USE ieee.std_logic_1164.all;
ENTITY decoder3_8 IS
      PORT (en, b0, b1, b2: IN BIT;
            D0, D1, D2, D3, D4, D5, D6, D7: OUT BIT);
END decoder3_8;
ARCHITECTURE truthtable OF decoder3_8 IS
     SIGNAL T: BIT_VECTOR (3 DOWNTO 0);
     BEGIN
          T <= en & b2 & b1 & b0;
          D0 <='1' WHEN "1000" ELSE '0';
          D1 <='1' WHEN "1001" ELSE '0';
          D2 <='1' WHEN "1010" ELSE '0';
          D3 <='1' WHEN "1011" ELSE '0';
          D4 <='1' WHEN "1100" ELSE '0';
          D5 <='1' WHEN "1101" ELSE '0';
          D6 <='1' WHEN "1110" ELSE '0';
          D7 <='1' WHEN "1111" ELSE '0';
END truthtable;
```

FIGURE 13.33 | HDL design file for decoder with WHEN-ELSE statement

ALTERNATIVE-III: Figure 13.34 gives the VHDL code of 3:8 lines decoder using IF-THEN-ELSE statement. Three-bit input is represented by 3-bit vector and 8-bit output is taken as 8-bit vector. When 3-bit input vector is concatenated with enable, *en* signal it becomes 4-bit vector.

```
LIBRARY ieee;
USE ieee.std_logic_1164.all;
ENTITY decoder3_8 IS
      PORT (en:IN BIT;
            b: IN BIT_VECTOR (2 DOWNTO 0);
            D: OUT BIT_VECTOR (7 DOWNTO 0);
END decoder3_8;
ARCHITECTURE truthtable OF decoder3_8 IS
     BEGIN
         PROCESS (en,b)
           BEGIN
             IF (en ='1')THEN
               IF     b="000" THEN D <='00000001';
               ELSIF  b="001" THEN D <='00000010';
               ELSIF  b="010" THEN D <='00000100';
               ELSIF  b="011" THEN D <='00001000';
               ELSIF  b="100" THEN D <='00010000';
               ELSIF  b="101" THEN D <='00100000';
               ELSIF  b="110" THEN D <='01000000';
               ELSIF  b="111" THEN D <='10000000';
               ENDIF;
             ELSE D <='10000000';
             ENDIF
           END PROCESS
 END truth table;
```

FIGURE 13.34 | VHDL design file for decoder using ELSE-IF statement

ALTERNATIVE-IV: Figure 13.35 gives the VHDL code of 3:8 lines decoder using CASE-WHEN statement. Three bit input is represented by 3-bit vector and 8-bit output is taken as 8-bit vector. When 3-bit input vector is concatenated with enable, en signal it becomes 4-bit vector.

```
LIBRARY ieee;
USE ieee.std_logic_1164.all;
ENTITY decoder3_8 IS
      PORT (en:IN BIT;
            b: IN BIT_VECTOR (2 DOWNTO 0);
            D: OUT BIT_VECTOR (7 DOWNTO 0);
END decoder3_8;
ARCHITECTURE truthtable OF decoder3_8 IS
     BEGIN
         PROCESS (en,b)
           BEGIN
             IF (en ='1')THEN
               CASE b IS
                  WHEN "000" => D <='00000001';
                  WHEN "001" => D <='00000010';
                  WHEN "010" =>D <='00000100';
                  WHEN "011" => D <='00001000';
                  WHEN "100" => D <='00010000';
                  WHEN "101" => D <='00100000';
                  WHEN "110" => D <='01000000';
                  WHEN "111" => D <='10000000';
               END CASE;
             ELSE
                D <='00000000';
             ENDIF
           END PROCESS
END truth table;
```

FIGURE 13.35 | VHDL design file for decoder using CASE statement

13.13 | LOGICAL OPERATIONS ON BIT ARRAYS

Each bit in these data types is designated by an element number. In this example of a bit array named d, the bits are referred to as d_3, d_2, d_1, d_0. Bits can also be grouped into sets. For example, three most significant bits of *d* are as a set by using the expression d (3 DOWNTO 1) in VHDL. Once a value is assigned to the array and the desired set of bits is identified, logical operations can be performed on the entire set of bits. As long as the sets are of the same size (the same number of bits), logical expression can be performed on two sets. This allows one equation to describe the logical operation performed on each bit in a set.

```
LIBRARY ieee;
USE ieee.std_logic_1164.all;
ENTITY bitwise_and IS
  PORT ( d, g : IN BIT_VECTOR (3 DOWNTO 0);
             y : OUTBIT_VECTOR (3 DOWNTO0);
END bitwise_and;
ARCHITECTURE bwa OF bitwise_and IS
BEGIN
   y<= d AND g;
END
```

FIGURE 13.36 | VHDL code for bitwise AND operation of 4-bit data

Figure 13.36 gives the VHDL code for bitwise AND of two 4-bit binary numbers. Assume d_3, d_2, d_1, d_0 has the value 1011 and g_3, g_2, g_1, g_0 has the value 1100. Let us define $d = [d_3, d_2, d_1, d_0]$. and $g = [g_3, g_2, g_1, g_0]$. Let us also define $y = [y_3, y_2, y_1, y_0]$, where *Y* is related to *d* and *g* as follows:

$$y = d \cdot g$$

The value of *y* after this operation comes out:

$$(y_3 y_2 y_1 y_0) = (d_3 d_2 d_1 d_0).(g_3 g_2 g_1 g_0)$$
$$(y_3 y_2 y_1 y_0) = (1011).(1100)$$

Thus, *y* is a set of four bits with value 1000.

13.14 | VHDL SUBTRACTOR

A parallel subtractor circuit is designed that subtracts bit arrays and the logic equation for a single-bit full-subtractor is extended to 4-bit operation. Figure 13.37 shows the basic block diagram with signals labelled to create a four-bit subtractor.

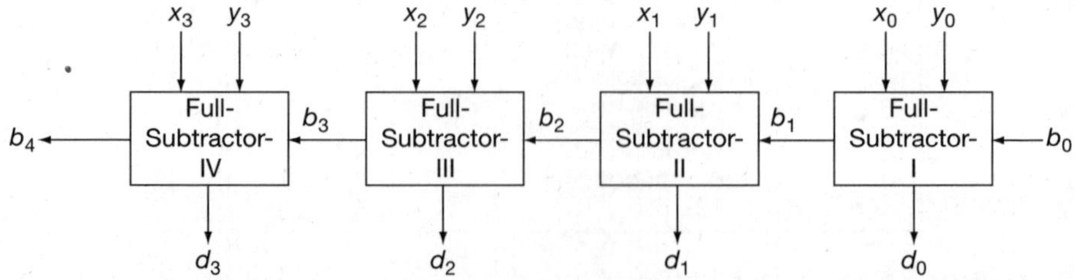

FIGURE 13.37 | Four-bit binary parallel subtractor

Each bit of the minuend [X], subtrahend [Y], borrow [B], and difference [D] are bit array variables and have index numbers associated with them. The equation for the difference and borrow using register notation is

$$\text{Difference} = (\text{Minuend} \oplus \text{Subtrahend}) \oplus \text{PreviousBorrow}$$

$$d(i) = (x(i) \oplus y(i)) \oplus \text{b}(\text{i});\ (i = 0, 1, 2, 3)$$

$$\text{Borrow} = \left(\overline{\text{Minuend}}\right).(\text{Subtrahend}) + \left(\overline{\text{Minuend}} + \text{Subtrahend}\right)\text{Prev.Borrow}$$

$$b(i+1) = \overline{x(\text{i})}.y(i) + \left(\overline{x(\text{i})} + y(i)\right)b(i)\ \ (i = 0, 1, 2, 3)$$

A strategy for labelling the borrow bits is developed so that these operands can be used in an array. Each borrow bit serve as an input to its corresponding subtractor stage, as shown in Figure 13.37. So, b0 is an input to the bit 0 stage, b_1 is the borrow input to the bit 1 stage, and so on. The bits of the borrow array are considered as signals or wires that connect the subtractor. The set of data that represents the borrow outputs would be b_4 through b_1, and the set of data that represents the carry inputs would be b_3 through b_0. In the VHDL code of 4-bit subtractor is given in Figure 13.38. The name is simply used with no bit designations. In order to choose a specific set of elements from the array (i.e. a subset of the array), the name is followed by the range of element numbers in parentheses. The borrow equation in VHDL syntax is:

b(4 DOWNTO 1) <= ((NOT x) AND b) OR ((NOT x) AND b(3 DOWNTO 0)) OR (y AND b (3 DOWNTO 0));

Only four bits of the borrow array *b* are being-assigned, even though the array is five bits long. In this way, the borrow out of each single-bit subtractor is assigned as the borrow in of the next stage

```
LIBRARY ieee;
USE ieee.std_logic_1164.all;
ENTITY subtractor IS
PORT( bin   :IN BIT; -- initial borrow
     x : IN BIT_VECTOR(3 DOWNTO 0); -- minuend vector
     y : IN BIT_VECTOR(3 DOWNTO 0); -- subtrahend vector
     d : OUT BIT_VECTOR(3 DOWNTO 0); -- difference vector
     bout : OUT BIT); -- borrow out
END subtractor;
ARCHITECTURE suba OF subtractor IS
SIGNAL xnot : BIT_VECTOR (n DOWNTO 0);
SIGNAL b: BIT_VECTOR (4 DOWNTO 0); --borrow requires 5 bit array
BEGIN
    xnot<=NOT x;
    b(0) <=  bin;   --Read the borrow into bit array
    d <= (x XOR y)XOR b (3 DOWNTO 0); --Generate the difference bits
    b(4 DOWNTO 1) <= (xnot AND y)
                    OR (xnot AND b(3 DOWNTO 0))
                    OR (y AND b(3 DOWNTO 0));
    bout <= b(4); -- output the borrow of the MSB.
END suba;
```

FIGURE 13.38 | VHDL code for bitwise 4-bit subtractor

13.15 | EXPANDING THE BIT CAPACITY OF A CIRCUIT

The capacity of a circuit can be expanded by cascading the stages. With either of these methods, all the inputs, outputs, and interconnections between blocks are to be specified. In the case of this subtractor circuit, it is easier to start with the HDL file for a four-bit subtractor and simply increase the size of each operand variable in the equation. To obtain an eight-bit adder/subtractor, *x*, *y* and *s* (sum or difference) are expanded to eight bits. The code remains almost identical to the four-bit subtractor given in Figure 13.38. The indices of each signal, and each bit array specification in the equation has to be redefined.

The **constants** are symbolically represented that are used throughout the code. Constants are simply fixed numbers represented by a name (symbol). A symbol is defined that is make up a name, at the top of the source code that is assigned the value for the total number of bits and then this symbol (name) is used throughout the code. It is much easier to modify the circuit. The capacity of adder/subtractor circuit is expanded such as this one reduces the speed of the circuit because of carry propagation.

Constants are included in a **PACKAGE**, as shown in Figure 13.39. Packages are also used to contain component definitions and other information that are available to all entities in the design file. The keyword USE tells the compiler to use the definitions in this package throughout this design file. Inside the package, the keyword CONSTANT is followed by the symbolic name, its type, and the value it is to be assigned using the:= operator. The compiler allows doing some simple math calculations to establish a value of one constant based on another. It is referred to $c(7)$ as $c(n)$ and $c(8)$ as $c(n + 1)$. The size of this adder/subtractor can be expanded by simply changing the value of number-of-bits to the desired number of bits and then recompiling.

```
PACKAGE const IS
  CONSTANT nobits: INTEGER: =8;--set total number of bits
  CONSTANT n : INTEGER: = nobits-1;--MSB index number
END const;
USE work.const.all;
ENTITY add_subIS
PORT (add     : IN BIT;  --add control
      sub     : INBIT;  --subtract control and LSB carry in
      y       : INBIT_VECTOR (n DOWNTO0);
      xin     : INBIT_VECTOR (n DOWNTO0);
      s       : OUTBIT_VECTOR (n DOWNTO0);
      carryout : OUTBIT);
END add_sub;
ARCHITECTURE adsu OF add_sub IS
SIGNAL c : BIT_VECTOR (n+1 DOWNTO 0);--define intermediate
carries
SIGNAL b : BIT_VECTOR (n DOWNTO 0); --define intermediate
operand
SIGNAL xnot : BIT_VECTOR (n DOWNTO 0);
SIGNAL mode : BIT_VECTOR (1 DOWNTO 0);
BEGIN
    xnot <= NOT xin;
    mode <= add & sub;
    WITH mode SELECT
    x <= xin  WHEN "10", --add
         xnot WHEN "01", --sub
         "00" WHEN OTHERS ;
    c(0) <= sub; --read the carry_in to bit array
    s <= (x XOR y) XOR c(n DOWNTO 0); --generate the sum bits
    c (n+1 DOWNTO 1) <= (x AND y) OR
                        (x AND c (n DOWNTO 0)) OR
                        (y AND c (n DOWNTO 0));
     carryout<=c (n+1);--output the carry of the MSB.
END adsu;
```

FIGURE 13.39 | VHDL code for bitwise 4-bit subtractor

The adder select the bin inputs if add = 1 or the 1's complement (xnot) if sub = 1. This is an excellent application of the VHDL selected signal assignment. When *add* is 1, *xin* is channelled to a *x*. When *sub* is 1, *xnot* is channelled to *x*.

13.15.1 | VHDL GENERATE Statement

VHDL **GENERATE** statement can also be used to expand the bit capacity of the circuit. It is another way to tell the compiler to replicate several components that are cascaded together. The adder circuits have been discussed that are cascaded chains of single-bit full-adder modules. The VHDL code for a single-bit full-adder module is shown in Figure 13.40.

Multiple instances of this module are connected to each other to form an *n*-bit adder circuit. The bit-0 adder has inputs that have an index of 0 (e.g. x_0, y_0, c_0, s_0). The carry-out of bit 0 adder is labelled as c_1, and it becomes the carry input for the bit 1 adder module. Each time another component is instantiated for the next bit of the multi-bit adder, the index number of all connections goes up by | (x_1, y_1, c_1, s_1). The GENERATE statement repeats an instantiation of a component n times, increasing the index number by 1 for each instantiation up to *n*. The GENERATE keyword is used in an **iterative loop (FOR loop)**, which means that a set of descriptive actions (PORT MAP) are repeated a certain number of times. The variable *i* represents an index number that starts at 0 (for the first iteration) and ends at *n* (the last iteration). The advantages of this method are code compactness and the ease with which the number of bits can be expanded. The code in Figure 13.41 shows the use of a single-bit adder (Figure 13.40) as a component to generate an eight-bit adder circuit. The file for the single-bit adder (add1b.vhd in Figure 13.40) should be saved in the same folder as the design file that uses it to generate multiple instances of the adder (adde4bit.vhd).

13.16 | MAGNITUDE COMPARATOR

Magnitude comparator device compares the magnitude of two binary numbers and indicates the relationship between the two, i.e. greater than, less than, equal to. Control inputs are provided for the purpose of cascading these chips. These chips are interconnected so that the chip comparing the lower-order bits has its outputs connected to the control inputs of the next higher-order chip. When the highest-order stage detects that its data inputs have equal magnitude, it goes to the next lower stage for comparison and use these control inputs to make the final decision. To compare bigger values using HDL, adjust the size of the comparator input ports, instead of cascading several four-bit comparators. Consequently, there is no need for cascading input controls in the HDL version. IF/ELSE is used to construct because each IF clause evaluates a relationship

```
ENTITY add1b IS
PORT (cin, x, y : INBIT;
      s, cout   : OUTBIT);
END add1b;
ARCHITECTURE a1b OF add1b IS
BEGIN
   s <= (x XOR x) XOR cin;
   cout <= (x AND y) OR (x AND cin) OR (y AND cin);
END a;
```

FIGURE 13.40 | Single-bit full-adder in VHDL

```
PACKAGE const IS
    CONSTANT nobits : INTEGER :=8;
    CONSTANT n : INTEGER : =nobits - 1;
END const;
USE work.const.all;
ENTITY adder4bit IS
PORT (carryin   : IN bit;
     xin        : IN BIT_VECTOR (n DOWNTO0);
     yin        : IN BIT_VECTOR (n DOWNTO0);
     sout       : OUT BIT_VECTOR (n DOWNTO0);
     carryout   : OUT BIT);
END adder4bit;
ARCHITECTURE add4b OF adder4bit IS
COMPONENT add1b
     PORT ( cin, x, y : IN BIT;
           s, cout    : OUT BIT);
END COMPONENT;
SIGNAL c: BIT_VECTOR (n+1 DOWNTO 0);
BEGIN
   c(0) <= carryin;
   repeat : FOR i IN 0 TO n GENERATE
       --cascade them
   casc: add1b PORT MAP (cin=> c(i), x=> xin(i), y=> yin(i),
                         s=> sout (i), cout=> c(i+1));
   END GENERATE;
   carryout <= c(n+1);
END add4b;
```

FIGURE 13.41 | Use of the VHDL GENERATE statement

between two values. The two inputs being compared should definitely be declared as numerical values. The three comparator outputs are declared as individual bits in order to label each bit's purpose clearly.

The VHDL code in Figure 13.42 follows the algorithm that uses IF/ELSE constructs. The data values are declared as four-bit integers. The IF/ELSE constructs are used inside a PROCESS. The PROCESS is evaluated whenever any of the inputs change state. Consequently, each input is listed in the sensitivity list within the parentheses.

```
LIBRARY ieee;
USE ieee.std_logic_1164.all;
ENTITY comparator IS
PORT ( a, b : IN INTEGER RANGE 0 TO 15;
     agtb, altb, aeqb : OUT BIT);
END comparator;

ARCHITECTURE vhdl_comp OF comparatpor IS
BEGIN
    PROCESS (a,b)
    BEGIN
        IF    a < b THEN altb <= '1'; agtb <= '0'; aeqb <= '0';
        ELSIF a > b THEN altb <= '0'; agtb <= '1'; aeqb <= '0';
        ELSE  altb <= '0'; agtb <= '0'; aeqb <= '1';
        END IF;
    END PROCESS;
END vhdl_comp;
```

FIGURE 13.42 | Magnitude comparator in VHDL

13.17 | VHDL BCD-TO-BINARY CODE CONVERTERS

Let number 250 be represented as

$$250 = 2\times100 + 5\times10 + 0\times1$$

Decimal number 250 is represented in binary-coded decimal (BCD) as 0010 0101 0000. Each digit is evaluated in binary as

$$001001010000BCD = (0010\times01100100 + 0101\times1010 + 0000\times0001)B$$

$$001001010000BCD = (11001000 + 00110010 + 00000000\times0001)B$$

$$001001010000BCD = (11111010)B$$

On generalizing

$$\text{Binary number} = d10\times10 + d1$$

where d_{10} is digit having weightage ten. d_1 is digit having weightage unit.

The VHDL solution in Figure 13.43 is very simple due to the powerful math operations available in the language. The inputs and outputs must be declared as integers because it intends to perform arithmetic operations on them. The range is specified based on the largest valid BCD number using two digits. The ten's digit is multiplied by 10, and the ones digit is added to form the binary equivalent of the BCD input.

```
ENTITY BCD_binary IS
PORT (d1, d10: IN INTEGER RANGE 0 TO 9;
        binary : IN INTEGER RANGE 0 TO 99);
END BCD_binary;
ARCHITECTURE vhdl_code OF BCD_binary IS
SIGNAL times10 : INTEGER RANGE 0 TO 90;
BEGIN
    times10 <= d10 * 10;
    binary <= times10 + d1;
END vhdl_code;
```

FIGURE 13.43 | BCD-to-binary code converter in VHDL

13.18 | VHDL SEVEN-SEGMENT DECODER/DRIVER

A BCD-to-7-segment decoder/driver is a 7447. The HDL code produces a device that meets the same criteria as the 7447. The $\overline{BI}$ (blanking input) is the overriding control that turns all segments off regardless of other input levels. The-LT (lamp test) input is used to test all the segments on the display by lighting them up. The $\overline{RBO}$ (ripple blanking output) is designed to go LOW when $\overline{RBI}$ (ripple blanking input) is LOW and the BCD input value is 0. Typically, in multiple-digit display applications, each $\overline{RBO}$ pin is connected to the $\overline{RBI}$ pin of the next digit to the right. This setup creates the feature of blanking all leading zeros in a display value without blanking zeros in the middle of a number.

Seven-segment display consists of seven light emitting diodes (LEDs). These are used to display numeric digits. Segments are illuminated to display a particular digit. Segments are named a through g. Illuminated segment is treated as 1. Figure 13.44 gives the arrangement depicting illuminated segments of 0 to 9 digits. The aim is to design decoder that converts BCD code to seven segment code, as shown in Figure 13.44.

The VHDL code for this circuit is shown in Figure 13.45. This illustration demonstrates the use of a VARIABLE as opposed to a SIGNAL. The BCD input is declared as an INTEGER.

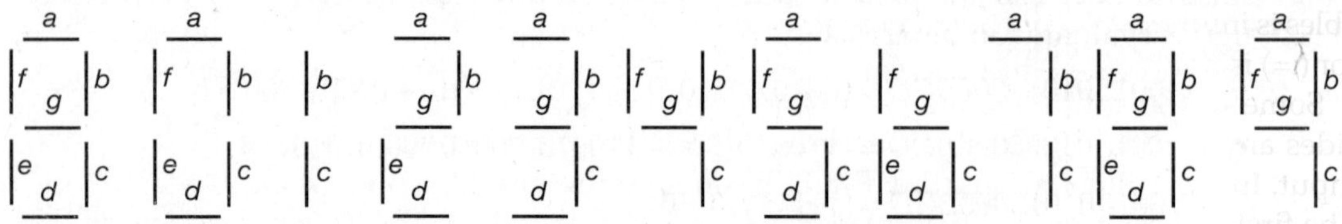

FIGURE 13.44 | Seven-segment display to display numeric digits

```
ENTITY sevenS_decoder IS
PORT (bcd : IN INTEGER RANGE 0 TO 9;
      lt, bi, rbi : IN BIT;
      a, b, c, d, e, f, g, rbo : OUT BIT);
END sevenS_decoder;
ARCHITECTURE vhdl_7s OF sevenS_decoder IS
BEGIN
PROCESS (bcd, lt, bi, rbi)
VARIABLE segments : BIT_VECTOR (0 TO 6);
  BEGIN
     IF bi = '0' THEN
        segments := "1111111"; rbo <= '0'; --blank all
     ELSIF lt = '0' THEN
        segments := "0000000"; rbo <= '1'; --test segments
     ELSIF (rbi = '0' AND bcd = 0) THEN
        segments := "1111111"; rbo <= '0'; --blank leading 0's
     ELSE
        rbo <= '1';
        CASE bcd IS --display 7 segment Common Anode pattern
           WHEN 0 => segments: = "0000001";
           WHEN 1 => segments: = "1001111";
           WHEN 2 => segments: = "0010010";
           WHEN 3 => segments: = "0000110";
           WHEN 4 => segments: = "1001100";
           WHEN 5 => segments: = "0100100";
           WHEN 6 => segments: = "1100001";
           WHEN 7 => segments: = "0001111";
           WHEN 8 => segments: = "0000000";
           WHEN 9 => segments: = "0001100";
           WHEN OTHERS => segments: = "1111111";
        END CASE;
     END IF;
       a <= segments (0);     --assign bits of array to output pins
       b <= segments (1);
       c <= segments (2);
       d <= segments (3);
       e <= segments (4);
       f <= segments (5);
       g <= segments (6);
  END PROCESS;
END vhdl_7s;
```

FIGURE 13.45 | VHDL 7-segment BCD display decoder

This refers to it by its numeric value in decimal rather than being limited to bit pattern references. A PROCESS is employed here in order to use the IF/ELSE constructs to establish the precedence of one input over the other. The sensitivity list contains all the inputs. The code within the PROCESS describes the behavioural operation of the circuit that is necessary whenever any of the inputs in the sensitivity list changes state. Assignment operator for variables is important here. The statement segments: = "1111111". The variable assignment operator (:=) is used for variables in place of the <= operator that was used for signal assignments.

Some of the controls have precedence over other controls. The $\overline{LT}$ (lamp test) overrides any regular digit display, and the $\overline{BI}$ (blinking input) overrides even the lamp test input. In these illustrations, the IF/ELSE control structure is used to establish precedence. The first condition that is evaluated as true will determine the resulting output regardless of the other input levels. Subsequent ELSE statements have no effect. The code tests first $\overline{BI}$, then $\overline{LT}$, then $\overline{RBI}$, and finally determines the correct segment pattern.

13.19 | VHDL ENCODER

A priority encoder uses tristate outputs in VHDL, and describes priority. Figure 13.46 shows the input/output definitions for this encoder circuit. The input switches are defined as bit vectors with indices from 9 to 0. The *d* output is defined as an IEEE standard bit array (std_logic_vector type). This definition is necessary to use of high-impedance states (tristate) on the outputs and assigned. This code uses the **conditional signal assignment statement WHEN-ELSE**. It assigns the value listed to the right of assignment operator to the variable *d* on the left, assuming the condition following WHEN is true. If this clause is not true, the clauses following ELSE are evaluated one at a time until one that is true is found. The value preceding WHEN is assigned to *d*. The conditional signal assignment statement follows sequential evaluation. The precedence of these statements is established

```
LIBRARY ieee;
USE ieee.std_logic_1164.ALL;

ENTITY p_encoder IS
PORT (
        sw : IN BIT_VECTOR (9 DOWNTO 0);--standard logic not needed
        oe : IN BIT;   --standard logic not needed
        d  : OUT STD_LOGIC_VECTOR (3 DOWNTO 0); --std logic for hi-z
      );
END p_encoder;

ARCHITECTURE encode OF p_encoder IS
   BEGIN
      d <= "zzzz" WHEN ((oe = '0') OR (sw = "1111111111")) ELSE
           "1001" WHEN sw (9) = '0') ELSE
           "1000" WHEN sw (8) = '0') ELSE
           "0111" WHEN sw (7) = '0') ELSE
           "0110" WHEN sw (6) = '0') ELSE
           "0101" WHEN sw (5) = '0') ELSE
           "0100" WHEN sw (4) = '0') ELSE
           "0011" WHEN sw (3) = '0') ELSE
           "0010" WHEN sw (2) = '0') ELSE
           "0001" WHEN sw (1) = '0') ELSE
           "0000" WHEN sw (0) = '0')
END encode;
```

FIGURE 13.46 | VHDL priority encoder using conditional signal assignment

by the order in which these statements are listed. The first condition being tested is the enabling of the tristate outputs. When the value "ZZZZ" is assigned to the output, each output is in the high-impedance state. If the outputs are to be disabled (high Z), then none of the other encoding matters. The highest priority input is tested, which is bit 9 of the sw input array. If it is active (LOW), then a value of 9 is output regardless of whether other inputs are being activated at the same time.

13.20 | VHDL MUX AND DEMUX

Figure 13.47 shows the code that creates a four-channel MUX with four bits per channel.

```
ENTITY mux_4_1 IS
PORT (ch0, ch1, ch2, ch3 : IN BIT_VECTOR (3 DOWNTO 0);
          s : IN INTEGER RANGE 0 TO 3;
          dout : OUT BIT_VECTOR (3 DOWNTO 0););
END mux_4_1;
ARCHITECTURE selector OF mux_4_1 IS
   BEGIN
      WITH s SELECT
         dout <= ch0 WHEN 0, --switch channel 0 to output
         ch1 WHEN 1, --switch channel 1 to output
         ch2 WHEN 2, --switch channel 2 to output
         ch3 WHEN 3,    -switch channel 3 to output
   END selector;
```

FIGURE 13.47 | Four-bit four-channel MUX in VHDL

The inputs are declared as bit arrays on line 3. The inputs are declared just as integers ranging from 0 to 15. Whichever way the inputs are declared, the outputs must be of the same type. The select input (s) is declared as a decimal integer from 0 to 3 (equivalent to binary 00 to 11).

The demultiplexer code works in a similar way but has only one input channel and four output channels. In Figure 13.48, the inputs and outputs are declared as usual. The select input(s) is of integer type. The operation of a DEMUX is described using several conditional signal assignment statement WHEN-ELSE. The code for this DEMUX ensures that the outputs are all HIGH when these keys are not selected. This is accomplished with the

```
ENTITY demux IS
PORT (s : IN INTEGER RANGE 0 TO 3;
       din : IN BIT_VECTOR (3 DOWNTO 0);
       ch0, ch1, ch2, ch3 : OUT BIT_VECTOR (3 DOWNTO 0));
END demux;

ARCHITECTURE selector OF demux IS
   BEGIN
      ch0 <= din WHEN s = 0 ELSE "1111";
      ch1 <= din WHEN s = 1 ELSE "1111";
      ch2 <= din WHEN s = 2 ELSE "1111";
      ch3 <= din WHEN s = 3 ELSE "1111";
   END selector;
```

FIGURE 13.48 | Four-bit four-channel DEMUX in VHDL

ELSE clause of each conditional signal assignment. If the ELSE clause is not used, the output values would default automatically to all LOW. For example, channel 2 is connected to the data inputs whenever the select inputs are set to 2. If s is set to any other value, then channel 2 will be forced- to have all bits HIGH.

13.21 | SEQUENTIAL CIRCUITS USING VHDL

The logic circuits use latch and clocked flip-flop circuits that sequence through various states in response to a clock edge. These latching and sequential circuits can also be implemented using PLDs and are described using HDL.

Describing circuits that have outputs that are fed back to the input with Boolean equations or HDL, involves using the output variables in the conditional portion of the description. When writing equations that use feedback requires a special designation for the output. In these cases, the port bit is not only an output but also an output with feedback. Consider SR latch using NAND gates (Figure 13.49). The characteristic table is given in Table 13.16.

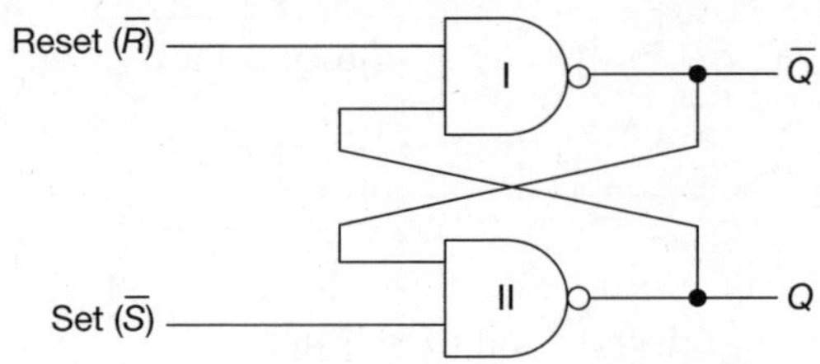

FIGURE 13.49 | SR latch

TABLE 13.16 | Characteristics table of SR flip-flop with NAND gates

Inputs		Outputs		Remarks
$\overline{S}$	$\overline{R}$	$Q(t+1)$	$\overline{Q}(t+1)$	
0	0	Q(t)	$\overline{Q}(t)$	Previous state 0
0	1	0	1	Reset
1	0	1	0	Set
1	1	0	0	(Indeterminate)*

* Complement rule is not followed.

The following statement describes the conditions that make the output HIGH on an SR latch:

IF SET is active, **THEN** Q is **HIGH.**

The following statement describes the conditions that make the output LOW on an SR latch:

IF RESET is active, **THEN** Q is **LOW.**

In case neither input is activated, then the output remains the same and is expressed as $Q = Q$. This expression provides the output state to be combined with input conditions for deciding next to happen. In case both inputs are activated that is the invalid input combination. The latch should not respond to both inputs. If the SET is active, the output is forced to HIGH irrespective of the input on RESET. The ELSIF clause is considered only when SET is not active. The use of the feedback term ($Q = Q$) affects the operation (holding action) only when neither input is active. Figure 13.50 explains the working

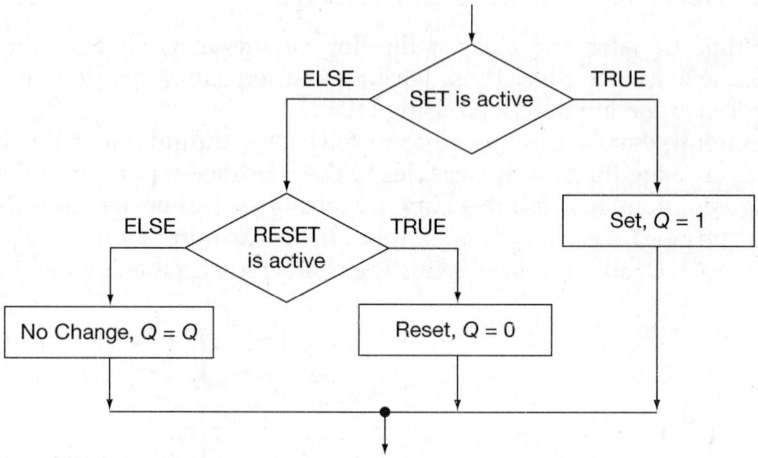

FIGURE 13.50 | Behavioural description of SR latch

When sequential circuits are designed, the output value is fed back to the inputs. It is possible to create an unstable system. A change in the output state is fed back to the inputs, which change the output state again. Output is again fed back to the inputs, which changes the output back again. Obviously, such oscillation is undesirable and so it is ensured that the circuit is stable under all conditions. Describe an active-LOW input SR latch with inputs

```
LIBRARY ieee;
USE ieee.std_logic_1164.all;
ENTITY SRflipflop IS
      PORT (sbar, rbar: IN BIT;
              Q : BUFFER BIT);
END SRflipflop;
ARCHITECTURE behavior OF SRflipflop IS
BEGIN
    PROCESS  (sbar, rbar)
        BEGIN
             IF    sbar = '0' THEN q <= '1'; --set or illegal command
             ELSIF rbar = '0' THEN q <= '0'; --reset
             ELSE  q <=q;                    --hold
             END IF;
        END PROCESS
END behaviour
```

FIGURE 13.51 | A NAND latch using VHDL

named *sbar*, *rbar* and one output named *Q*. The invalid input combination should produce *Q* = 1. Figure 13.51 shows a possible VHDL solution. The following is considered:

i. *Q* is defined as a BUFFER rather than an OUTPUT. This allows it to be fed back in the circuit.
ii. A PROCESS describes action as per the values in the sensitivity list (SBAR, RBAR) change state.
iii. The clause after IF determines which output state occurs when both inputs are active (invalid State). In this code the SET command rules.

13.21.1 | THE D Latch

The transparent D latch can also be easily implemented with VHDL. The VHDL module also shown below is a behavioural description of the D latch function. The characteristic table is given in Table 13.17. The VHDL code is given in Figure 13.52.

TABLE 13.17 | Characteristics of gated or clocked D flip-flop/latch

Inputs		Outputs	Remarks
D (Data)	*E* (Enable)	*Q*(*t* + 1)	
0	1	0	Save 0-state
1	1	1	Save 1-state
X	0	Q(t)	No change

```
LIBRARY ieee;
USE ieee.std_logic_1164.ALL;
ENTITY Dlatch IS
     PORT (enable, din : IN BIT;
           q           : OUT BIT;
END Dlatch;
ARCHITECTURE latch_D OF Dlatch IS
BEGIN
        PROCESS (enable, din)
        BEGIN
          IF enable = '1'THEN
             Q <= din;
          END IF;
        END PROCESS;
END latch_D
```

FIGURE 13.52 | A D-latch using VHDL

13.22 | EDGE-TRIGGERED DEVICES

An edge is a transition from HIGH, or LOW, or vice versa, and is often referred to as an **event.**

The JK flip-flop is a standard building block of clocked (sequential) logic circuits known as logic **primitive.** It has five inputs and one output, as shown in Figure 13.53 and characteristic table is given in Table 13.18. The actual operation of the primitive circuit is defined in a library of components that is available to the HDL compiler as it generates a circuit from our description. VHDL allows the designer to describe the clocked logic circuits operation explicitly in the code. The VHDL code for a JK flip-flop is described in Figure 13.54. VHDL is a very flexible language and it can be used to define the operation of clocked devices explicitly in the code, without relying on logic primitives. The key to edge-triggered sequential circuits in VHDL is the process. Process keyword is followed by a sensitivity list in parentheses.

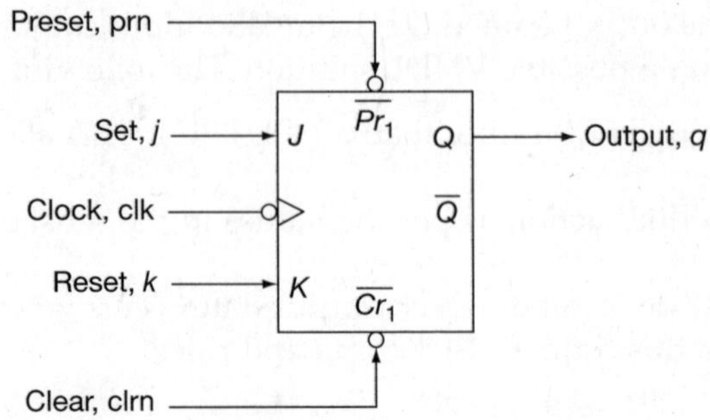

FIGURE 13.53 | JK flip-flop

TABLE 13.18 | Characteristics of J-K flip-flop

Inputs			Outputs	Remarks
J (Set)	*K* (Reset)	CP (Negative Edge)	*Q*(*t* + 1)	
0	0	↓	*Q*(*t*)	Previous state
0	1	↓	0	Set
1	0	↓	1	Clear (Reset)
1	1	↓	$\bar{Q}(t)$	Toggle state
X	X	Otherwise	Q(t)	No change

```
-- JK Flip-Flop Circuit
LIBRARY ieee;
USE ieee.std_logic_1164.ALL;
ENTITY JKflipflop IS
      PORT (clk, j, k, prn, clrn : IN BIT;
            q  :OUT BIT);
END JKflipflop;

ARCHITECTURE flipflop_JK OF JKflipflop IS
SIGNAL qstate    :BIT;
BEGIN
    PROCESS(clk, prn,clrn)
    BEGIN
        -- asynchronous preset
        IF prn = '0' THEN qstate <= '1';
        -- asynchronous clear
        ELSIF clrn = '0' THEN qstate  <= '0';
        --on negative gate triggering clock edge
        ELSIF clk = '0' AND clk' EVENT THEN
```

(Continued)

```
            IF j = '1' AND k = '1' THEN qstate <= NOT qsate;
            ELSIF j = '1' AND k = '0' THEN qstate <= '1';
            ELSIF j = '0' AND k = '1' THEN qstate <='0';
            END IF;
          END IF;
        END PROCESS;
            q <= qstate;  --update output pin
END flipflop_JK ;
```

FIGURE 13.54 | JK flip-flop using VHDL

The term 'clk' **EVENT** evaluates as TRUE only if there has been a transition on clk. Because clk = '0' so it is TRUE only to a falling edge transition on the clock. The next three conditions are evaluated only following a falling edge on clk and serve to update the flip-flop's state. As soon as the PROCESS ends, the flip-flop's state is transferred to the output port *q*.

13.22.1 | D-Flip flop

The *D* flip-flop is clocked (sequential) logic circuits. It has four inputs and one output, as shown in Figure 13.55. The VHDL code is given in Figure 13.56.

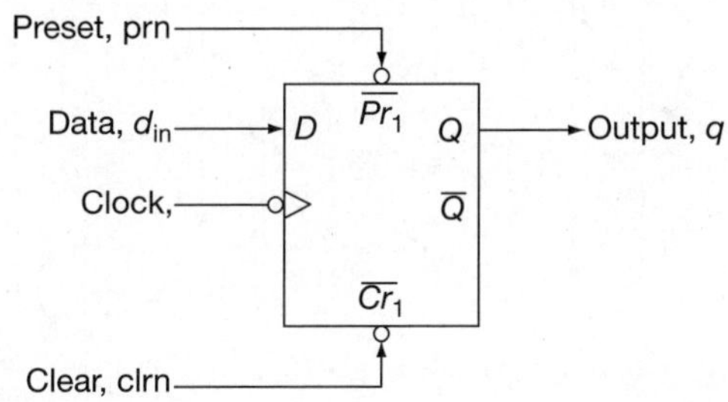

FIGURE 13.55 | D flip-flop

```
LIBRARY ieee;
USE ieee.std_logic_1164.ALL;
ENTITY Dff IS
     PORT (clk, prn, clrn, din : IN BIT; q : OUT BIT);
END Dff;
ARCHITECTURE flipflop_D OF Dff IS
BEGIN
        PROCESS (clk, prn, clrn)
        BEGIN
           IF (clrn='0' AND prn='1'THEN
              q <= '0';
           ELSIF (clrn='1' AND prn='0'THEN
              q <= '1';
           ELSIF (clk 'EVENT AND clk='0' THEN
              q <= din;
           END IF;
        END PROCESS;
END flipflop_D
```

FIGURE 13.56 | D flip-flop using VHDL

13.23 | VHDL CIRCUIT WITH MULTIPLE COMPONENTS

A simple binary up counter is shown in Figure 13.57. The logic symbols are negative edge-triggered. These flip-flops also do not have asynchronous inputs *prn* or *clrn*. The counter circuit is described using HDL by interconnecting three instances of the same JK flip-flop component.

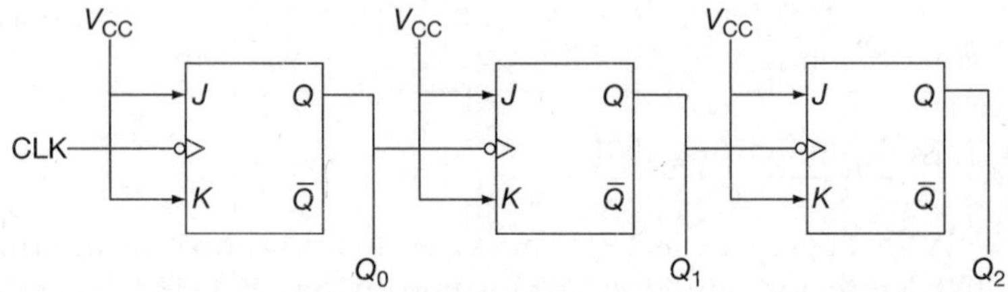

FIGURE 13.57 | MOD 8 ripple counter using JK flip-flop

Figure 13.58 describes the VHDL code for a negative-edge-triggered JK flip-flop with preset and clear controls. The counter in Figure 13.57 is negative edge-triggered and does

```
LIBRARY ieee;
USE ieee.std_logic_1164.ALL;
ENTITY r_counter IS
     PORT (clock : IN BIT; qout : BUFFER BIT_VECTOR (2 DOWNTO 0);
END r-counter;
ARCHITECTURE counter_r OF r_counter IS
   SIGNAL high : BIT;
      COMPONENT neg_jk
      PORT (clk, j, k  : IN BIT; q : OUT BIT);
   END COMPONENT;
BEGIN
   high <= '1';  --connect to Vcc
ff0 : neg_jk PORT MAP (j => high, k => high, clk => clock, q => qout (0));
ff1 : neg_jk PORT MAP (j => high, k => high, clk => qout (0) => qout (1));
ff2 : neg_jk PORT MAP (j => high, k => high, clk => qout (1) => qout (2));
END counter_r;
ENTITY neg_jk IS
     PORT (clk, j , k : IN BIT; q : OUT BIT);
END neg_jk;
ARCHITECTURE simple OF neg_jk IS
     SIGNAL qstate : BIT;
BEGIN
    PROCESS (clk)
    BEGIN
        IF (clk'EVENT AND clk = '0' ) THEN
          IF j = '1'  AND k = '1' THEN qstate <= NOT qstate;
          ELSIF  j = '1' AND k = '0' THEN qstate <= '1';
          ELSIF  j = '0' AND k = '1' THEN qstate <= '0';
          END IF;
        END IF;
    END PROCESS;
    q <= qstate; --connect flip-flop state to output
END simple;
```

FIGURE 13.58 | MOD 8 ripple counter using JK flip-flop

not require asynchronous preset or clear. The VHDL code for one of these flip-flops is described, which represent three instances of the same flip-flop, and interconnect the ports to create the counter.

The VHDL description is given in Figure 13.58. This module of VHDL code is describing the operation of a single JK flip-flop component. The name of the component is *neg_jk* has inputs *clk, j,* and *k* and output *q*. A signal name *qstate* is used to hold the state of the flip-flop and connect it to the *q* output. The PROCESS has only *clk* in its sensitivity list, so it only responds to changes in the *clk* (Positive gate triggering (PGT), negative gate triggering (NGT)).

13.24 | BASIC COUNTERS USING VHDL

The flip-flops have some unique characteristics. The output is normally according to the condition of the synchronous control inputs when the active edge of the clock occurs. It means there is a logic state on the *Q* output before the clock edge (PRESENT state) and potentially a different state on the *Q* output after the clock edge (NEXT state). A flip-flop "remembers", or holds its state between clocks, regardless of changes in the synchronous control inputs (e.g. J and K).

Counter circuits using HDL rely on this basic understanding of a circuit going through a sequence of states in response to the event of a clock edge. Ripple counters are easy to analyse. The problem with ripple counters is the combination of time delay and spurious temporary states that occur when the counter changes state. The methods used to describe counter circuits using HDL primarily use synchronous techniques, where all flip-flops update simultaneously in response to the same clock event. All the bits in a count sequence go from their PRESENT state to their prescribed NEXT state simultaneously, thereby preventing any intermediate, spurious states.

13.24.1 | State Transition Description Methods

Another method to describe counter uses tables. This method assigns values to objects like ports, signals, and variables. With sequential counter circuits, the equivalent of the truth table is the PRESENT state/NEXT state table.

13.24.2 | State Descriptions in VHDL

A simple counter circuit MOD-5 counter is implemented as shown in Figure 13.59 in VHDL. A counter using a control structure similar to a PRESENT state/NEXT state table is followed. Two key tasks are accomplished in VHDL: (i) detecting the desired clock edge, and (ii) assigning the proper NEXT state to the counter. A PROCESS is used to respond to a transition of an input signal. A CASE construct evaluates an expression and, for any valid input value. Assign a corresponding value to another signal. The inputs and outputs are defined in the ENTITY declaration, as in the past.

When VHDL is used to describe a counter, the state of the counter between clock pulses, i.e. the action of a flip-flop is stored. This is done in one of two ways: (i) using SIGNALs, or (ii) using VARIABLEs. SIGNALs are used that operate concurrently. A SIGNAL in VHDL holds the last value that was assigned to it like a flip-flop. Consequently a SIGNAL is used

```
LIBRARY ieee;
USE ieee.std_logic_1164.ALL;
ENTITY s_counter IS
   PORT (Clock : IN BIT; q : OUT BIT_VECTOR (2 DOWNTO 0) );
END s_counter;

ARCHITECTURE sc OF s_counter IS
BEGIN
    PROCESS (clock)   --respond to clk input
    VARIABLE count: BIT_VECTOR (2 DOWNTO 0); --create a 3-bit register
    BEGIN
        IF (clock = '1' AND clock' EVENT) THEN --rising edge trigger
          CASE count IS
              WHEN "000" => count: = "001";
              WHEN "001" => count: = "010";
              WHEN "010" => count: = "011";
              WHEN "011" => count: = "100";
              WHEN "100" => count: = "000";
              WHEN OTHERS => count: = "000";
          END CASE;
       END IF;
       q <= count; --assign register to output pins
     END PROCESS;
 END sc;
```

FIGURE 13.59 | VHDL MOD-5 counter

as the data object representing value. This SIGNAL can then be used to connect the counter value to any other elements in the architecture description.

A **VARIABLE** is used instead of a SIGNAL as the data object that stores the counter value. VARIABLEs are not exactly like SIGNALs because these variables are not used to connect various parts of the design. Variables are considered to be local data objects because these variables are recognized only within the PROCESS in which these are declared. In Figure 13.59, the variable named Count is declared within the PROCESS before BEGIN. Its type is the same as the output port *q*.

13.24.3 | Behavioural Description

The **behavioural level of abstraction** is a way to describe a circuit by describing its behaviour. A counter circuit operation is described without using flip-flops or logic gates. When the counter input changes from LOW to HIGH, the number on the output counts up by 1. This level of description deals more with cause-and-effect relationships than with the path of data flow or wiring details. In VHDL, the first important step in this description method is to declare properly the counter output port, as shown in Figure 13.60. The data type of the output port must match the type of the counter variable, and it must be a type that allows arithmetic operations.

BIT_VECTORS are treated as a string of bits, not as a binary numeric quantity. In order to recognize the signal as a numeric quantity, the data object is typed as an INTEGER. The compiler looks at the RANGE 0 TO 4 and knows that the counter needs three bits. The first statement after BEGIN in the PROCESS responds to the rising edge of the clock as in the previous examples. It then uses behavioural description methods to define the counter's response to the clock edge. If the counter has not reached its maximum, then it is incremented. Otherwise, it should recycle the counter to zero. The last statement connects the counter value to the output pins of the device.

```
LIBRARY ieee;
USE ieee.std_logic_1164.ALL;
ENTITY counter IS
     PORT (clock : IN BIT; q : OUT INTEGER RANGE 0 TO 4);
END counter;

ARCHITECTURE counter_a OF counter IS
BEGIN
     PROCESS (clock)
     VARIABLE count: INTEGER RANGE 0 TO 4;
          BEGIN
             IF (clock = '1' AND clock' EVENT) THEN
                If count < 4 THEN
                  count: = count + 1; --increment value
                ELSE
                  count : = 0; --recycle to zero
                END IF;
             END IF;
        q <= count; --transfer register contents to outputs
        END PROCESS;
END counter_a;
```

FIGURE 13.60 | Behavioural description of a MOD-5 counter in VHDL

13.25 | FULL-FEATURED COUNTERS IN VHDL

The standard IC counters have many other features including count enable, up/down counting, parallel loading (preset to any count), and clearing.

In addition, these counters have been designed to easily cascade synchronously to create larger counters. MOD-16 binary counter is to change states on the rising edge of the clock input when the counter is enabled with a HIGH level. A direction control input makes the counter count up when it is LOW or count down when it is HIGH. The counter has an active–HIGH, asynchronous clear to reset the counter immediately when the control input is activated. The counter can be synchronously loaded with a number of data input pins when the load control is HIGH.

The priority of the input control functions, from highest to lowest, will be clearing, loading and counting. The counter includes an active-HIGH output that detects the terminal state of the counter when the count functions are enabled. The code in Figure 13.61 implements all the features of a four-bit counter, and expandable in size. The PROCESS statement is the key to all clocked circuits described in VHDL. It also determines whether the circuit responds synchronously or asynchronously to its inputs. This circuit responds immediately to transitions on the clock, clear, and down inputs. The variable count is defined as an INTEGER so it is incremented and decremented. Variables are declared within the PROCESS and are used within the PROCESS only.

```
LIBRARY ieee;
USE ieee.std_logic_1164.ALL;
ENTITY counter_U IS
   PORT (clock, clear, load, cntenbl, down : IN BIT;
          din      : IN INTEGER RANGE 0 TO 15
          q        : OUT INTEGER RANGE 0 TO 15);
          term_ct : OUT BIT
END counter_U;

ARCHITECTURE counter OF counter_U IS
   BEGIN
      PROCESS (clock, clear, down)
      VARIABLE count: INTEGER RANGE 0 TO 15;
          BEGIN
             IF clear = '1' THEN count: =0; --asynchronous clear
             ELSIF (clock = '1' AND clock 'EVENT') THEN
                IF load = '1' THEN count: =din; --parallel load
                ELSIF cntenabl = '1' THEN
                   IF down = 0 THEN count : = count+1; --increment
                   ELSE count : = count-1; --decrement
                   END IF;
                END IF;
             END IF;
             IF (((count = 0) AND (down = '1')) OR
             ((count = 15) AND (down = '0'))) AND cntenabl = '1'
             THEN term_ct <= '1';
             ELSE  term_ct <= '0';
             END IF;
                     q <= count;--transfer register contents to outputs
      END PROCESS;
  END counter;
```

FIGURE 13.61 | Full-featured counter in VHDL

The clear input is given precedence by evaluating it with the first IF statement. In order to make the load function operate synchronously, it is evaluated after detecting the clock edge. The clock edge is detected, and the circuit checks immediately to see if load is active. If load is active, the count is loaded from din, regardless of whether or not the counter is enabled. Consequently, the conditional decision (IF) evaluates the load input; only if it is inactive does it evaluate to see if the counter is enabled. If the counter is enabled, the count is incremented or decremented.

The next issue is detecting the terminal count. One of the key concepts of using HDLs is that it is generally very easy to expand the size of a logic module. The necessary changes to this VHDL design to increase the binary counter modulus to 256. Only four modifications to Figure 13.61 are required to make this change in counter modulus:

Line #	Modification
5	RANGE 0 TO 255
6	RANGE 0 TO 255
12	RANGE 0 TO 255
23	(count = 255)

13.26 | WIRING VHDL MODULES TOGETHER

Counter circuits can be connected to other digital modules to create larger systems. Designing large digital systems becomes much easier if the system is subdivided into smaller, more manageable modules that are then interconnected. This is the essence of the concept of **hierarchical design.**

13.26.1 | Decoding the VHDL MOD-5 Counter

Block diagram of decoding the mod-5 counter is given in Figure 13.62.

Figure 13.63 gives the VHDL code for the 3-line-to-5-lines decoder module. Figure 13.64 gives the VHDL code of MOD-5 up synchronous counter module. These two modules

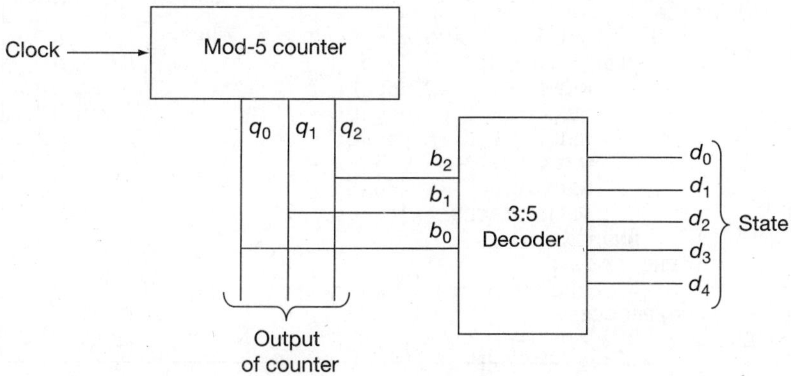

FIGURE 13.62 | Block diagram of Mod-5 counter and decoder circuit

```
LIBRARY ieee;
USE ieee.std_logic_1164.ALL;
ENTITY decode5 IS
    PORT (c, b, a: IN BIT; state : OUT BIT_VECTOR (0 TO 4));
END decode5;

ARCHITECTURE aa OF decode5 IS
SIGNAL input : BIT_VECTOR (2 DOWNTO 0);
BEGIN
        input <= (c & b & a);    --combine inputs into bit vector
        PROCESS (c, b, a)
        BEGIN
                CASE input IS
                   WHEN "000" => state: <= "10000";
                   WHEN "001" => state: <= "01000";
                   WHEN "010" => state: <= "00100";
                   WHEN "011" => state: <= "00010";
                   WHEN "100" => state: <= "00001";
                   WHEN OTHERS => state: <= "00000";
             END CASE;
    END PROCESS;
END aa;
```

FIGURE 13.63 | VHDL decoder module

are used as components to combine the circuit. The VHDL code is given in Figure 13.65. Three bit count output, q acts as input to decoder to generate a 5-bit control bits names as state.

```
LIBRARY ieee;
USE ieee.std_logic_1164.ALL;
ENTITY mod5 IS
   PORT (Clock : IN BIT; q : OUT BIT_VECTOR (2 DOWNTO 0));
END s_counter;
ARCHITECTURE sc OF mod5 IS
BEGIN
    PROCESS (clock)  --respond to clk input
    VARIABLE count: BIT_VECTOR (2 DOWNTO 0);
    BEGIN
        IF (clock = '1' AND clock' EVENT) THEN
           CASE count IS
               WHEN "000" => count: = "001";
               WHEN "001" => count: = "010";
               WHEN "010" => count: = "011";
               WHEN "011" => count: = "100";
               WHEN "100" => count: = "000";
               WHEN OTHERS => count: = "000";
           END CASE;
        END IF;
        q <= count; --assign register to output pins
     END PROCESS;
  END sc;
```

FIGURE 13.64 | VHDL MOD-5 counter module

```
LIBRARY ieee;
USE ieee.std_logic_1164.ALL;
ENTITY mod5decoded1 IS
       PORT (clk : IN BIT;
             q   : BUFFER BIT_VECTOR (2 DOWNTO 0);
             cntr_state  : OUT BIT_VECTOR (0 TO 4));
END mod5decoded1;
ARCHITECTURE toplevel OF mod5decoded1 IS
COMPONENT mod5
        PORT (clock : IN BIT;
              q      : OUT BIT_VECTOR (2 DOWNTO 0));
END COMPONENT;
COMPPONENT decode5
       PORT ( c, b, a : IN BIT;
             state    : OUT BIT_VECTOR (0 TO 4));
END COMPONENT;
BEGIN
counter: mod5 PORT MAP (clock => clk, q => q1);
decoder: decode5 PORT MAP
     (c => q(2) , b => q(1) , a => q(0) , state =>
cntr_state);
END toplevel;
```

FIGURE 13.65 | Higher-level VHDL file to connect mod5 and decode5 together

13.27 | REGISTERS

HDL techniques use bit arrays to describe a register's data and to transfer that data in a parallel or serial format. The diagrams shown in Figure 13.66 shows four flip-flops performing transfer operations of parallel load, shift right, shift left, and hold data. For all of these diagrams, the bits are transferred synchronously, which means that these bits move simultaneously on a single clock edge.

In Figure 13.66(a), the parallel data is loaded parallel into the register and is presented to the *D* inputs, and on the next clock pulse, it is transferred to the *q* outputs. Shifting data right means that each bit is transferred to the bit location to its immediate right, while a new bit is transferred into the left end and the last bit on the right end is lost. This situation is depicted in Figure 13.66(b). The data set that is required in the NEXT state is made up of the new serial input and three of the four bits in the PRESENT state array. This data simply needs to move over and overwrite the four data bits of the register. The same operation occurs in Figure 13.66(c), but it is moving data to the left. The key to shifting the contents of the register to the right or left is to group the appropriate three PRESENT state data bits in correct order with the serial input bit so that these four bits can be loaded in parallel into the register. Concatenation (grouping together in a specific sequence) of the desired set of data bits is used to describe the necessary data movement for serial shifting in either direction. The last possibility is called the hold data mode and is shown in Figure 13.66(d). The *Q* outputs are tied back to the *D* inputs for each flip-flop so that the old data is reloaded on each clock.

13.27.1 | VHDL SISO Register

A four-bit serial in/serial out (SISO) register in VHDL is listed in Figure 13.67. A register is created with the declaration of the variable *q* on line 8 and the serial output is obtained from the register's last bit or *q*(0). If the shift control is HIGH, serial_in will be shifted into the register and the other bits will move to the right. Concatenating serial_in and register bits *q*(3), *q*(2), and *q*(1) together in that order creates the proper shift-right data input bit pattern. If the shift control is LOW VHDL will assume that the variable stays the same and will therefore hold the current data.

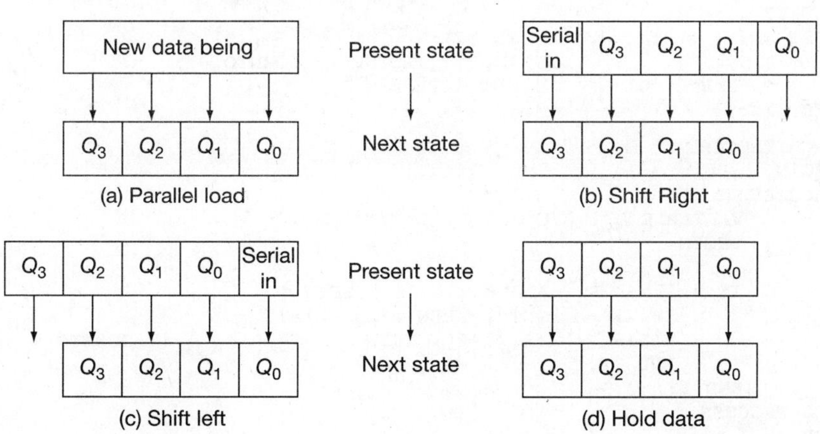

FIGURE 13.66 | Data transfer with shift register

```
LIBRARY ieee;
USE ieee.std_logic_1164.ALL;
ENTITY siso IS
    PORT ( clk, shift, load : IN BIT;
           data             : IN BIT_VECTOR (3 DOWNTO 0);
           serial_out       : OUT BIT );
END siso;
ARCHITECTURE vhdl_siso OF siso IS
BEGIN
    PROCESS (clk)
    VARIABLE q : BIT_VECTOR (3 DOWNTO 0);
          BEGIN
          serial_out <= q (0); --output last register bit
          IF (clk' EVENT AND clk = '1') THEN
               IF (shift = '1') THEN
                     --concatenate for shift
                     q: = (serial_in & q (3 DOWNTO 1));
               END IF;
          END IF;
    END PROCESS;
END vhdl_siso;
```

FIGURE 13.67 | SERIAL-in/serial-out register using VHDL

13.27.2 | VHDL PISO Register

A four-bit parallel-in/serial-out (PISO) register in VHDL is listed in Figure 13.68. The register is created with the variable declaration for *q*, and the serial output from *q*(0). The register has separate parallel load and serial shift controls. If load is HIGH, the external input data is synchronously loaded. Load has priority and is LOW to serial-shift the register's contents on each PGT of *clk* when shift is HIGH. The pattern for shifting data right is created by concatenation. A constant LOW will be the serial data input for a shift operation. If neither load nor shift is HIGH, the register will hold the current data value by VHDLs implied operation.

```
LIBRARY ieee;
USE ieee.std_logic_1164.ALL;
ENTITY piso IS
PORT (clk, shift, load : IN BIT;
       data            : IN BIT_VECTOR (3 DOWNTO 0);
       serial_out      : OUT BIT);
END piso;

ARCHITECTURE vhdl_piso OF piso IS
BEGIN
PROCESS (clk)
        VARIABLE q : BIT_VECTOR (3 DOWNTO 0);
        BEGIN
        serial_out <= q (0);
        IF (clk' EVENT AND clk = '1') THEN
             IF (load = '1') THEN q: = data;
             ELSIF (shift = '1') THEN q: = ('0' & q (3 DOWNTO 1));
             END IF;
        END IF;
END PROCESS;
END vhdl_piso;
```

FIGURE 13.68 | Parallel-in/serial-out register using VHDL

Universal shift register is given in Figure 13.69. That performs serial left, serial right, parallel load and hold data.

```
LIBRARY ieee;
USE ieee.std_logic_1164.ALL;
ENTITY univRegister IS
PORT (clock : IN BIT;
     din : IN BIT_VECTOR (3 DOWNTO 0); --parallel data in
     ser_out : IN BIT;--serial data in L or R
     q : OUT BIT_VECTOR (3 DOWNTO 0)); --0=hold 1=right 2=left 3=load
END univRegister;
ARCHITECTURE uR OF univRegister IS
BEGIN
    PROCESS (clock) --respond to clock
    VARIABLE ff : BIT_VECTOR (3 DOWNTO 0);
    BEGIN
        IF (clock' EVENT AND clock = '1') THEN
        CASE mode IS --hold data
            WHEN 0 => ff := ff;
            WHEN 1 => ff (2 DOWNTO 0) := ff(3 DOWNTO 1);
                      ff (3) := ser_in;
            WHEN 2 => ff (3 DOWNTO 1) := ff (2 DOWNTO 0);
                      ff (0) := ser_in;
            WHEN 3 => ff := din; --parallel load
        END CASE;
     END IF;
    q <= ff; --update outputs
    END PROCESS
END uR;
```

FIGURE 13.69 | VHDL universal shift register

13.28 | VHDL RING COUNTERS

The shift registers make the register shift one position on each clock. The main feature of this code is the method of completing the 'ring' by driving the *ser_in* line of the shift register. In order to make this counter self-start without using asynchronous inputs, the *ser_in* act as control line of the shift register. It is implemented using an IF/ELSE construct. The VHDL code is shown in Figure 13.70.

```
LIBRARY ieee;
USE ieee.std_logic_1164.ALL;
ENTITY ringcounter IS
PORT (clk : IN BIT;
         q : OUT BIT_VECTOR (3 DOWNTO 0));
END ringcounter;
ARCHITECTURE vhdl_ring OF ringcounter IS
SIGNAL ser_in: BIT;
BEGIN
PROCESS (clk)
    VARIABLE ff : BIT_VECTOR (3 DOWNTO 0);
    BEGIN
      IF (ff (3 DOWNTO 1) = "000")   THEN  ser_in <= '1';
      ELSE ser_in <= '0';
      END IF ;
      IF (clk' EVENT  AND clk = '1') THEN
         ff (3 DOWNTO 0) := (ser_in & ff (3  DOWNTO 1));
      END IF ;
    q <= ff;
END PROCESS ;
END vhdl;
```

FIGURE 13.70 | VHDL four-bit ring counter

SUMMARY

- AHDL (Altera hardware description language) is a hardware descriptive language (HDL) developed by Altera corporation of USA for programming their PLDs.
- ARCHITECTURE is a key word in VHSIC HDL and defines the operation of a circuit declared using an ENTITY.
- Architecture body: A body associated with an entity declaration to describe the internal organization or operation of a design entity. It is used to describe the behaviour, data flow, or structure of a design entity.
- Behavioural modelling is a method of describing a digital circuit/system which gives the behaviour of the circuit/system to its inputs.
- Binding is the process of associating a design entity and, optionally, an architecture with an instance of a component. It can be specified in an explicit or default binding indication.
- Buried Node is point identified in a circuit that is not accessible from outside this circuit. Also known as a local signal.
- Character literal is a character of the literal type formed by enclosing one of the graphic characters (including a space and non-breaking space characters) between two apostrophe (') characters.
- Character type is an enumeration type with at least one character literal among its enumeration literals.
- Clocked flip-flops are available as logic primitives.
- Component is VHDL keyword used to provide information about a component declared in a library.
- Component instantiation statement is a VHDL concurrent statement that instantiates a component in the architecture body.
- Concatenation is a term used to describe the linking of two or more data objects into an ordered set.
- Concurrent assignment statement is a statement in VHDL used to describe a circuit that works concurrently with other concurrent statements.
- Conditional signal assignment is a VHDL concurrent construct that evaluates a series of conditions sequentially to determine the first true condition evaluated and assigns its value to a signal.
- Configuration is a construct that defines how component instances in a given block are bound to design entities in order to describe the way design entities are put together to form a complete design.
- Constant is an object whose value does not change.
- Counters can be described in many different ways using HDL, including structural wiring descriptions, PRESENT state/NEXT state tables and behavioural descriptions.
- Custom code can be written in HDL to describe each of the common logic functions presented in this chapter.
- Declaration is a construct that defines a declared entity and associates an identifier with it.
- Declarative part is a syntactic component of certain declarations or statements such as entity declarations and architecture bodies.
- Declarative region is a semantic component of certain declarations or statements.

- Design entity is an entity declaration together with an associated architecture body. Different design entities share the same entity declaration.
- Design library is a host-dependent storage facility for intermediate form representations analysed design units.
- Design unit is a construct that can be independently analysed and stored in a design library. A design unit can be an entity declaration, an architecture body, a configuration declaration, a package declaration, or a package body declaration.
- Digital systems can be subdivided into smaller modules or blocks that can be interconnected as a hierarchical design.
- Entity declaration is a definition of the interface between a given design entity and the environment in which it is used. It may also specific declarations and statements that are part of the design entity. A given entity declaration may be shared by many design entities, each of which has a different architecture.
- Enumeration literal is a literal of an enumeration type. It may be either an identifier or a character literal.
- Enumeration type is a type whose values are defined by listing (enumerating) them. The values of the types are represented by enumeration literals.
- EVENT is a change in the current value of a signal, which occurs when the signal is updated with its effective value.
- Fundamental building blocks called logic primitives are available in the Altera library to help implement larger systems.
- Generic is an interface constant declared in the block header of a block statement, a component declaration or an entity declaration.
- Hardware description languages have become an important method of describing digital circuits.
- HDL can be used to describe interconnected components in a manner much like a graphic schematic capture tool.
- HDL code contains comments that document its vital characteristics so a person reading it later can understand-what it does.
- HDL circuit description contains a definition of the inputs and outputs, followed by a section that describes the circuit's operation. In addition to inputs and outputs, intermediate connections that are buried within the circuit can be defined. These intermediate connections are called nodes or signals.
- HDL macro functions are available for many MSI standard parts.
- Hierarchical design is a method of designing a system/circuit by breaking it into its constituent modules each of which can be broken further into simpler constituent modules.
- Identifier defines an alias for some other name.
- Instance is a subcomponent of a design entity whose prototype is a component declaration, design entity, or configuration declaration. A component instantiation statement whose instantiated unit denotes a component that creates an instance of the corresponding component.
- Library parts in the HDL circuits can be interconnected using either graphic or structural HDL techniques.
- Libraries of parameterized modules (LPMs) offer a flexible, easily modified or expanded solution for many types of digital circuits.

- Logical control structures such as IE, ELSE, and CASE are used to describe the operation of a logic - circuit, making the code and the problem's solution much more straightforward.
- Logical operations can be performed on all the bits in a set using Boolean equations.
- Mode is the direction of signal flow through a port.
- Object gives various ways of representing data in the HDL's code.
- PACKAGE is a VHDL keyword used to define a set of global elements that are available to other modules.
- Programmable logic device (PLD) offers an alternative to the use of MSI circuits to implement digital systems. Boolean equations can be used to describe the operation of these circuits, but HDLs also offer high-level language constructs.
- PORT is a VHDL keyword used to list the connections between components.
- Practicing good software engineering techniques, specifically the use of symbols to represent constants, allows for easy code modification and expansion of the bit capacity of circuits such as full-adders.
- Prefabricated functions are available in the Altera libraries.
- Priority and precedence can be established in VHDL using conditional signal assignments or using a PROCESS containing IF/ELSE or CASE decisions.
- PROCESS is a VHDL keyword used to define a set of sequential statements that describes the functionality of a portion of an entity.
- Programmable logic devices can be programmed to operate as latching circuits and sequential circuits.
- Sensitivity list is a list of signals used to invoke the sequence of statements in a PROCESS.
- Sequential statements are the statements that execute in sequence and in the order in which these appear. These are used for algorithmic descriptions.
- Shift register counters such as Johnson and ring counters can be implemented easily in HDL. Decoding and self-starting features are easy to write into the description.
- Structural modelling is a method of describing a digital circuit system using various physical components of the circuit.
- The Altera system allows convenient hierarchical design techniques using any form of hardware description.
- The ELSE clause in the conditional signal assignment statement of VHDL can be used to define the default state of an output.
- The type of data objects must be specified so that the HDL compiler knows the range of numbers to be represented.
- Tristate outputs can be created in HDL. VHDL assigns Z (high impedance) as a valid state for STD_LOGIC outputs.
- Truth tables can be entered directly into the source file using the features of HDL.
- VHDL code can be written to describe clocked logic explicitly without using logic primitives.
- VHDL allows HDL files to be used as components in larger systems. Prefabricated components are available in the Altera library.

MULTIPLE CHOICE QUESTIONS

13.1 In an entity statement BUFFER is an example of
- (a) signal type
- (b) signal PORT
- (c) signal mode
- (d) concurrent statement

13.2 Which of the following is an invalid name?
- (a) DECODER8
- (b) _WHAT_4
- (c) INVALID
- (d) all are valid

13.3 The RANGE keyword is associated with which signal type
- (a) BIT-VECTOR
- (b) STD_LOGIC
- (c) INTEGER
- (d) BUFFER

13.4 In STD_LOGIC, forcing unknown is represented by what value?
- (a) 'Z' (b) 'U'
- (c) 'W' (d) 'X'

13.5 7 MOD (−4) results in
- (a) −1 (b) +1
- (c) −3 (d) +3

13.6 The statement Y<='1' WHEN (D=0 AND ng='0') ELSE '0'
- (a) ARCHITECTURE
- (b) ENTITY
- (c) PORT MAP
- (d) Selected signal assignment

13.7 A ______ is a construct containing statements that are executed in a simulation if a signal in the sensitivity list changes
- (a) PROCESS
- (b) ENTITY
- (c) PORT MAP
- (d) CASE Statement

13.8 A VHDL design technique that connects pre-packaged components using internal or buried
- (a) Structural design
- (b) Data flow design
- (c) Behavioural design
- (d) Block design

13.9 Which of the following is true about parameters?
- (a) The default size of a parameter in most synthesizers is the size of an integer, 32 bits
- (b) Parameters enable Verilog code to be compatible with VHDL.
- (c) Parameters cannot accept a default value
- (d) All of the above

13.10 Which of following is not reserve word?
- (a) COMPONENTS
- (b) POSTPONED
- (c) UNAFFECTED
- (d) none of the above

Answers

13.1 (b)	13.2 (b)	13.3 (c)	13.4 (d)	13.5 (a)	13.6 (d)	13.7 (a)
13.8 (a)	13.9 (a)	13.10 (d)				

QUESTIONS

13.1 Explain in brief the three modelling styles in VHDL.

13.2 With suitable examples, explain the conditions are signal assignments in VHDL.

13.3 Explain in brief the following sequential statements in VHDL
(a) if-then-else, (b) Exit, (c) case-when, (d) Next (e) for-loop (f) Wait

13.4 What are the various classes and types of data objects in VHDL?

13.5 List the various VHDL operators.

13.6 What is meant by configuration in VHDL?

13.7 Explain the syntax of process statement in VHDL and its importance.

13.8 What is component instantiation in VHDL? Explain with a suitable example.

13.9 What is GENERATE keyword in VHDL? Use GENERATE with a suitable example.

13.10 How tables are implemented using VHDL?

PROBLEMS

13.1 Write VHDL codes for BCD-7 segment decoder combinational circuit.

13.2 Write VHDL codes for 3-bit binary-to-gray code converter using 3:8 decoder as a component.

13.3 Write VHDL codes for 4-bit BCD adder using 4-bit binary adder as a component.

13.4 Write VHDL codes for 8:1 line multiplexer.

13.5 Write VHDL codes for 16:1 multiplexer using 8:1 multiplexer as a component.

13.6 Write VHDL codes for the master-slave JK flip-flop using NAND gates as components.

13.7 Write VHDL codes for master-slave JK flip-flop in behavioural modelling.

13.8 Write VHDL codes for 3-bit up-down counter structural and behavioural).

13.9 Write VHDL codes for 3-bit shift register (structural).

13.10 Write entity declarations for the gates: (i) 2-input NAND gate and (ii) 3-input NOR gate.

13.11 Write entity declaration for a 4: I multiplexer circuit.

13.12 Write VHDL code for the full-adder using data flow architecture.

13.14 Write VHDL code for the truth table of full-subtractor.

13.15 Write VHDL code for the following circuits.

14

Digital Logic Families

CHAPTER OBJECTIVES

The main goal of this chapter is to impart knowledge about the digital logic families. Readers will be able to discuss the following aspects in this chapter:

- Digital IC specifications
- Logic families
- Transistor-transistor logic (TTL) Family: TTL Parameters
- Open-collector gates; Three-state TTL gates; Schottky TTL Gates
- TTL subfamilies
- External drive for TTL loads; TTL Driving external loads
- Integrated injection logic (IIL or I^2L)
- Emitter-coupled logic (ECL)
- Metal oxide semiconductor (MOS) family
- Complementary metal oxide semiconductor (CMOS) Family
- CMOS Characteristics
- TTL to CMOS Interface
- CMOS to TTL Interface

14.1 | INTRODUCTION

The complex digital functions are realized using basic electronic components such as diodes, transistors, *metal-oxide-semiconductor field-effect transistor* (MOSFETs), registers, and capacitors as a logic family. Digital integrated circuits (ICs) are used in modern digital systems. General purpose digital computer has interconnections among digital components and modules. Mostly, the digital circuits are constructed on a single chip, which is called as ICs. ICs contain a large number of interconnected digital circuits within a single small chip or package. The digital IC technology has advanced from *small-scale integration* (SSI), through medium-scale integration (MSI), *large-scale integration* (LSI), *very large-scale integration* (VLSI), to *ultra large-scale integration* (ULSI). *Giant-scale integration* (GSI) is the recent thrust of technology, in which millions of gate equivalent circuits are integrated on single chip. ICs are fabricated using various technologies such as transistor-transistor logic (TTL), emitter-coupled logic (ECL), and integrated injection Logic (IIL or I^2L) which use bipolar transistors, whereas the *metal-oxide-semiconductor* (MOS) and *complementary metal-oxide-semiconductor* (CMOS) technologies use unipolar MOSFETs. Popularity of ICs is due to their small size, low cost, less power consumption, high noise margin, high reliability and high speed. ICs have certain limitations. ICs are unable to handle large voltages or currents. It is not possible to implement precision resistors, inductors, transformer, and large-scale capacitors on chips. The operations requiring high power level devices cannot be integrated. Such operations are accomplished by discrete components.

14.2 | LOGIC FAMILIES

The set of compatible ICs with the same logic levels and same supply voltages have been fabricated to perform the various logic functions known as logic family. Based on the fabrication technology, logic families are classified into two types:

- Bipolar logic family
- Unipolar logic family

14.2.1 | Bipolar Logic Family

Transistors and diodes are bipolar devices, in which the current flows because of both electrons and holes being charge carriers. On the basis of operations of transistors in ICs, bipolar logic families are further classified as:

- Saturated bipolar logic families
- Unsaturated bipolar logic families

In saturated bipolar logic families, transistors operate in saturation region. The speed of saturated bipolar logic family is low. Saturated bipolar logic families are listed below:

- Resistor-transistor logic (RTL)
- Direct-coupled transistor logic (DCTL)
- Diode-transistor logic (DTL)
- High-threshold logic (HTL)
- Transistor-transistor logic (TTL)
- Integrated injection logic (IIL or I^2L)

In non-saturated bipolar logic families, transistors operate in active region. The speed of non-saturated bipolar logic families is high as compared to saturated logic families. Unsaturated bipolar logic families are listed below:

- Schottky transistor-transistor logic
- Emitter-coupled logic (ECL)

The transistor is one of the key elements used in logic families. One of the important applications of the transistor is the switch.

14.2.2 | Unipolar Logic Family

In unipolar logic families, unipolar devices are the key elements. MOSFET is a unipolar device, in which the current flows because of only one type of charge carrier that is electrons or holes. The unipolar families include *p-channel metal-oxide semiconductor field-effect transistor* (PMOS), *n-channel metal-oxide semiconductor field-effect transistor* NMOS and CMOS.

RTL, DCTL, DTL, and HTL families are obsolete. The logic families TTL, ECL, IIL, MOS, and CMOS are currently in use. The basic function of any type of gate is always the same regardless of the circuit technology used. The TTL and CMOS are suitable for SSI and MSI. The MOS and CMOS are particularly suitable for LSI. The IIL is mainly suitable for VLSI and ULSI. The ECL is mainly used in superfast computers. The logic families currently in use are compared in Table 14.1 in terms of the commonly used specification parameters.

TABLE 14.1 | Comparison of logic families

Logic Families	Propagation Delay Time (ns)	Power Dissipation Per Gate (mW)	Noise Margin (V)	Fan-in	Fan-out	Cost
TTL	9	10	0.4	8	10	Low
ECL	1	50	0.25	5	10	High
MOS	50	0.1	1.5		10	Low
CMOS	<50	0.01	5	10	50	Low
IIL	1	0.1	.35	5	8	Very low

14.2.3 | Requirement of a Logic Family

The main features of a logic family are given below:

High fan out: High fan-out is advantageous as it reduces the need of additional drivers to drive more gates.

High noise immunity: The ability of circuits to operate reliably in noisy environments is important in many applications. Therefore, it should be highly immune to noise.

Low-power dissipation and high speed: To achieve low-power dissipation and high speed of operation, the digital circuit must have a minimum number of gates between input and output.

Small Size: The size of IC should be small as much as possible.

Flexibilities available: It must be considered while selecting a logic family for a particular application for the purpose of interfacing with other logic family devices. It should have good interfacing.

14.3 | DIGITAL IC SPECIFICATIONS

Nomenclature and terminology of digital ICs are fairly standardized. There are various logic families like TTL, ECL, IIL, MOS, CMOS logic families. The selection of a logic family for a particular application is based on its characteristics. Real time applications demand high-speed logic families. It becomes necessary to study the characteristics of digital ICs. The nomenclature and terminology terms are defined below.

- Threshold voltage
- Propagation delay
- Power dissipation

- Speed power product
- Current and voltage
- Fan-out
- Fan-in
- Noise immunity
- Operating temperature

14.3.1 | Threshold Voltage

The threshold voltage is defined as the voltage at the input of a gate that causes a change in the state of the output from one logic level to the other logic level.

14.3.2 | Propagation Delay

To propagate a pulse from input to output of a gate takes a certain amount of time, this interval of time is known as the *propagation delay* of the gate.

It is the average transition delay time, t_{pd}, expressed by

$$t_{pd} = \frac{t_{PLH} + t_{PHL}}{2} \tag{14.1}$$

where t_{PLH} is the signal delay time when the output goes from a logic 0 to a logic 1 state (Figure 14.1).
t_{PHL} is the signal delay time when the output goes from a logic 1 to a logic 0 state (Figure 14.1).

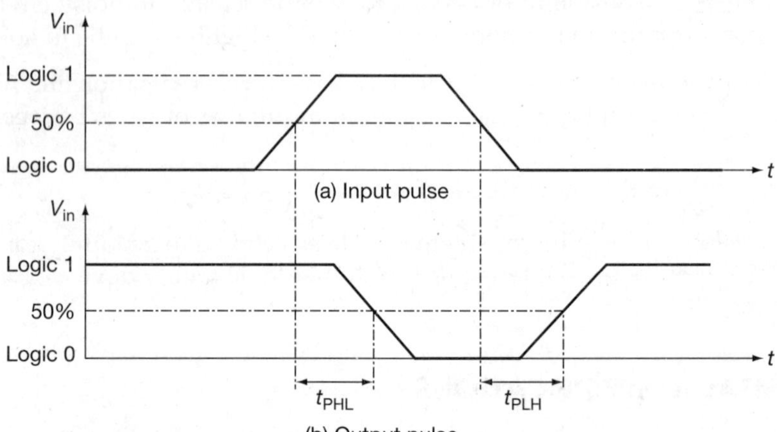

FIGURE 14.1 | Propagation delay in an inverter

14.3.3 | Power Dissipation

Every logic gate draws some current from the supply for its operation. The current drawn in HIGH state is different from that drawn in LOW state. The power dissipation, P_D, of a logic gate is the power required by the gate to operate with 50 per cent duty cycle at a given frequency and is expressed in milliwatts (mW). This means that logic 1 and logic 0 periods of the output are equal. The power dissipation of a gate is given by

$$P_D = V_{CC} \times (I_{CC}/n) \tag{14.2}$$

where V_{CC} is the gate supply voltage,
I_{CC} is the average current drawn from the supply by the entire IC and,
n is the number of gates in the IC.

Average current drawn by the IC is the average current of all the gates in the IC when these are in HGH as well as LOW state

$$I_{CC} = \frac{I_{CCH} + I_{CCL}}{2} \tag{14.3}$$

where I_{CCH} is the current drawn by the IC when all the gates in the IC are in HIGH state.
I_{CCL} is the current drawn by the IC when all the gates in the IC are in LOW state.

The total power consumed by an IC is equal to the product of the power dissipated by each gate and the number of gates in that IC.

14.3.4 | Speed Power Product

To reduce the power requirements, the power dissipation of logic families is usually set to minimum. On reducing the power dissipation of an IC, the propagation delay is reduced. Hence, the speed of power product, also known as Figure of Merit, is a parameter considered for comparison instead of using the propagation delay and power dissipation. The speed of power product is a product of propagation delay and power dissipation. Let propagation delay is 10 ns and power dissipation is 5 mW. So, the speed power product or figure of merit is 50 Pico-Joules (10 ns × 5 mW = 50 pJ).

14.3.5 | Voltage and Current Parameters

The definitions of voltage and current levels corresponding to the logic 0 and logic 1 states are as follows:

V_{IH} *(HIGH-level input voltage).* It is the minimum voltage level required at the input of a gate so that input to be accepted as a logic 1. Any voltage below this level is not accepted as logic 1 by the logic circuit.

V_{OH} *(HIGH-level output voltage).* It is the minimum voltage level required at the output of a gate so that output to be accepted as a logic 1. Any voltage below this level is not accepted as logic 1 output.

V_{IL} *(LOW-level input voltage).* It is the maximum voltage level that can be accepted as logic 0 at the input of the gate. Any voltage above this level is not accepted as a logic 0 input by the logic gate.

V_{OL} *(LOW-level output voltage).* It is the maximum voltage level that can be accepted as logic 0 at the output of the gate. Any voltage above this level is not accepted as a logic 0 output.

I_{IH} *(HIGH-level input current).* The minimum input current that flows into gate as input when a specified HIGH level voltage is applied to that input.

I_{IL} *(LOW-level input current).* The maximum input current that flows into gate as input when a specified LOW level voltage is applied to that input.

I_{OH} *(HIGH-level output current).* The minimum output current that flows from gate as output in logic 1 state under specified load conditions.

I_{OL} *(LOW-level output current).* The maximum output current that flows from gate as output in a logic 0 state under specified load conditions.

Figures 14.2 and 14.3 illustrate the currents and voltages in the HIGH and LOW states, respectively.

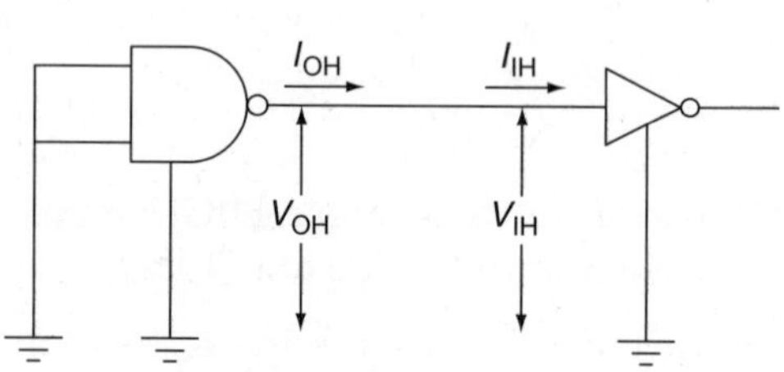

FIGURE 14.2 | Current and voltage in High state

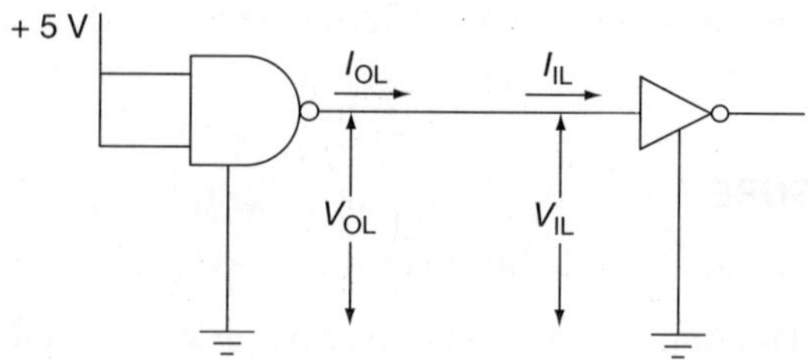

FIGURE 14.3 | Current and voltage in Low state

14.3.6 | Fan-out

The *fan-out* of a logic gate is defined as the maximum number of standard loads that the output of the gate can drive without impairing its normal operation. Fan-out is also called the *loading factor.* A standard load is usually specified as the amount of current needed by an input of another gate of the same IC family. If a gate is made to drive more than this number of gate inputs, the performance of the gate is not guaranteed. The gate may malfunction.

HIGH state fan-out is the fan-out of the gate when its output is logic 1. LOW state fan-output is the fan-out of the gate when its output is logic 0. The smaller of these two numbers is taken as the actual fan-out. The fan-out of a gate affects the propagation delay time as well as saturation. The driving gate sinks current when it is in LOW state and sources current when it is in HIGH state. High state fan-out is given by

$$\text{HIGH state fan-out} = \frac{I_{OH}}{I_{IH}} \tag{14.4}$$

where I_{OH} is the maximum current that the driver gate can source when it is in a 1 state.
I_{IH} is the current drawn by each driven gate from the driver gate.

Similarly, low state fan-out is given by

$$\text{LOW state fan-out} = \frac{I_{OL}}{I_{IL}} \tag{14.5}$$

where I_{OL} is the maximum current that the driver gate can sink when its output is a logic 0.
I_{IL} is the current drawn from each driven gate by the driver gate.

Figures 14.4 and 14.5 depict the current sourcing and current sinking actions, respectively, for the TTL 7400 NAND gate.

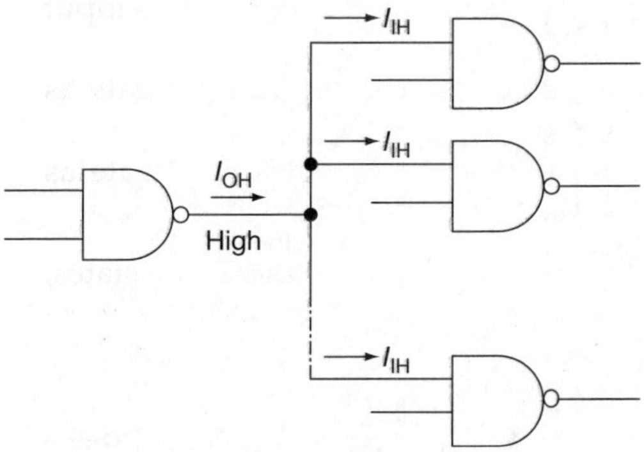

FIGURE 14.4 | Current sourcing in High state

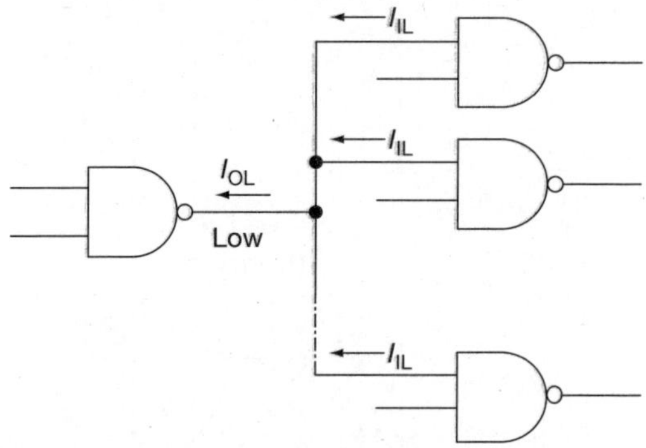

FIGURE 14.5 | Current sinking in Low state

The fan-out of a logic family can be calculated as

$$\text{Fan-out} = \text{minimum of}\left\{\frac{I_{OH}}{I_{IH}}, \frac{I_{OL}}{I_{IL}}\right\} \tag{14.6}$$

14.3.7 | Fan-in

The *fan-in* of a logic gate is defined as the number of inputs that the gate is designed to handle. For a two-input gate, fan-in is two; and for a four-input gate, fan-in is four.

14.3.8 | Noise Immunity

When the digital circuits operate in noisy environment the gates malfunction if the noise is beyond certain limits. The *noise immunity* of a logic circuit refers to the circuit's ability to tolerate noise voltages at its inputs.

A quantitative measure of noise immunity is called *noise margin.* Unwanted spurious signals are called *noise.* Noise may be ac noise or dc noise. A drift in the voltage levels of signals is called dc noise. The ac noise is a random pulse caused by other switching signals. Noise margin is expressed in volts. Noise margin represents the maximum noise signal that can be added to the input signal of a digital circuit without causing an undesirable change in the circuit output. This is an important criterion for the selection of a logic family for certain applications where environment noise may be high.

Figure 14.6 shows the range of output voltages that can occur in a logic circuit. Voltages greater than V_{OH} are considered as a logic I and voltages lower than V_{OL} are considered as a logic 0. Voltages in the disallowed range should not appear at a logic circuit output under normal conditions. Figure 14.6 shows the input voltage requirements of a logic circuit. The logic circuit will respond to any input greater than V_{IH} as a logic 1 and to any input lower than V_{IL} as a logic 0. Voltages in the indeterminate range produces an unpredictable response and is avoided to use. Noise margin is HIGH state noise margin or LOW state noise margin

$$\text{HIGH state noise margin } (NM_H) \text{ is, } V_{NH} = V_{OH} - V_{IH}$$

$$\text{LOW state noise margin } (NM_L) \text{ is, } V_{NL} = V_{IL} - V_{OL}$$

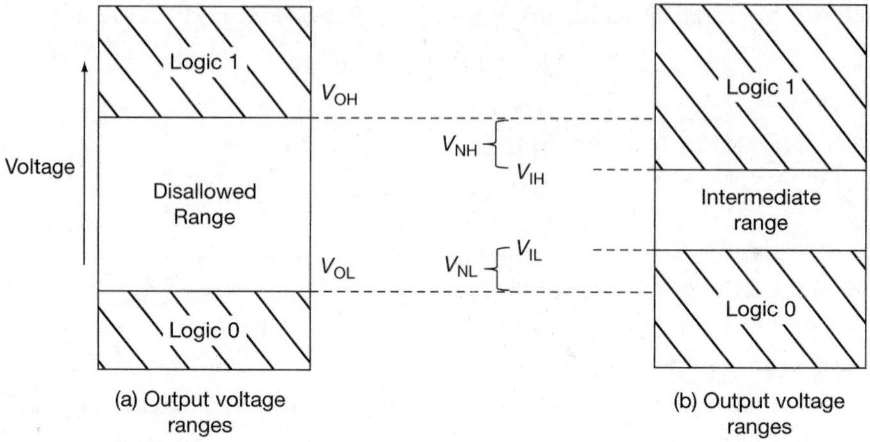

FIGURE 14.6 | DC noise margins

High state noise margin is the difference between the lowest possible high output and the minimum input voltage required for a HIGH. Low state noise margin is the difference between the largest possible LOW output and the, maximum input voltage for a LOW.

The noise leads to an undesired output. Consider an example of a TTL AND gate. The TTL gate has $V_{OH} = 2.4$ V, $V_{OL} = 0.4$ V, $V_{IH} = 2$ V and $V_{IL} = 0.8$ V. The noise introduced in the signal (V_{NH} or V_{NL}) is shown in Figure 14.7.

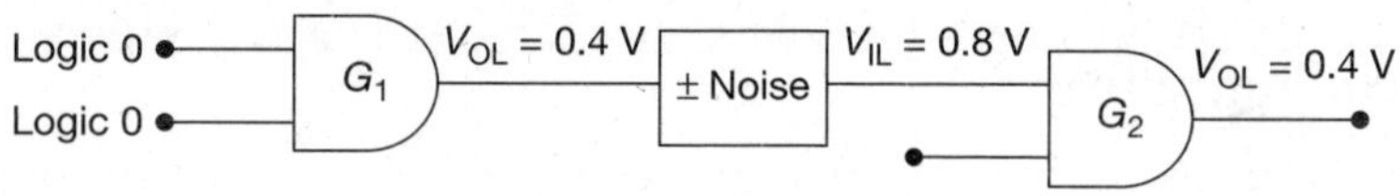

FIGURE 14.7 | Noise

Let the inputs of gate, G_1 cause output as logic 0 (Figure 14.7). This output acts as input for gate, G_2. Due to noise, actual input given to gate, G_2 is

$$V_{NL} = V_{IL} - V_{OL}$$

V_{IL} in above equation acts as input to gate, G_2. Maximum LOW-level noise level is

$$V_{NL} = 0.8 \text{ V} - 0.4 \text{ V} = 0.4 \text{ V}$$

Let the inputs of gate, G_1 cause output as logic 1 (Figure 14.8). This output acts as input for gate, G_2. Due to noise, actual input given to gate, G_2 is

$$V_{NH} = V_{OH} - V_{IH}$$

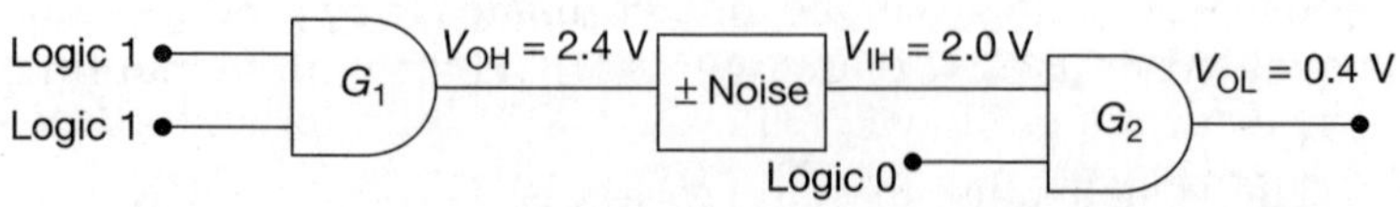

FIGURE 14.8 | Noise

V_{IL} in above equation that acts as input to gate G_2. Minimum high level noise level is

$$V_{NH} = 2.4\text{ V} - 2.0\text{ V} = 0.4\text{ V}$$

14.3.8.1 | Operating Temperatures

The IC gates and other circuits are temperature sensitive being semiconductor devices. However, these devices are designed to operate satisfactorily over a specified range of temperatures. The range specified for commercial applications is 0°C to 70°C, for industrial applications it is 0°C to 85°C, and for military applications it is –55°C to 125°C.

14.4 | TRANSISTOR-TRANSISTOR LOGIC

The TTL family depends on transistors alone to perform basic logic operations. It is a popular logic family. It uses bipolar transistor for digital IC family. Transistors operating in saturated mode are used in TTL. It is the fastest of the saturated logic families. The basic TTL logic circuit is the NAND gate. TTL logic family has good speed, low manufacturing cost, wide range of circuits, and the availability in SSI and MSI. Limitations TTL logic family are less V_{CC} tolerance, relatively high power consumption, moderate packing density, generation of noise spikes and susceptibility to power transients.

The TTL logic family consists of several subfamilies or series such as: Standard TTL, high-speed TTL, low-power TTL, Schottky TTL, low-power Schottky TTL, advanced Schottky TTL, advanced low-power Schottky TTL and F (fast) TTL.

The differences between the various TTL subfamilies are in their electrical characteristics such as propagation delay time, power dissipation, switching speed, fan-out, fan-in, noise margin as given in Table 14.1. Moreover, standard TTL considers logic 0 for a voltage signal in the range of (0–0.8 V) and logic 1 for a voltage signal in the range of (2–5 V). Voltage signals in the range of (0.8–2 V) are considered indeterminate or are not accepted. If a terminal is left open in TTL, it is equivalent to connecting it to HIGH, i.e. +5 V. But such a practice is not recommended, because the floating TTL is susceptible to pick-up noise signals that adversely affect the operation of the device.

14.4.1 | The Bipolar Junction Transistor

The bipolar junction transistor (BJT) is the active switching element used in all TTL circuits. Figure 14.9(a) shows the symbol for an *npn* BJT with its three terminals: base, emitter and collector. A BJT has two junctions, the base–emitter junction and the base–collector junction.

The switch is activated by applying a voltage between base and emitter. Working of BJT as switch is explained below:

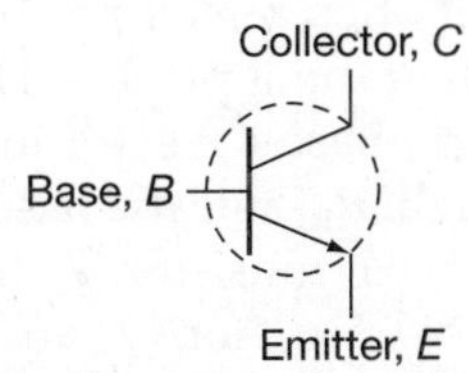

FIGURE 14.9(a) | *npn* BJT

- The voltage between base and emitter is zero or less than 0.7 V (<0.7 V). The switch is open, and no current is allowed to flow between collector and emitter. The transistor acts OFF (Figure 14.9(b)).

- A voltage is applied between base and emitter more than 0.7 V (≥0.7). The switch closes and a current is allowed to flow between collector and emitter. The transistor turns ON (Figure 14.9(c)). The voltage between emitter and collector, across a closed switch, is zero.

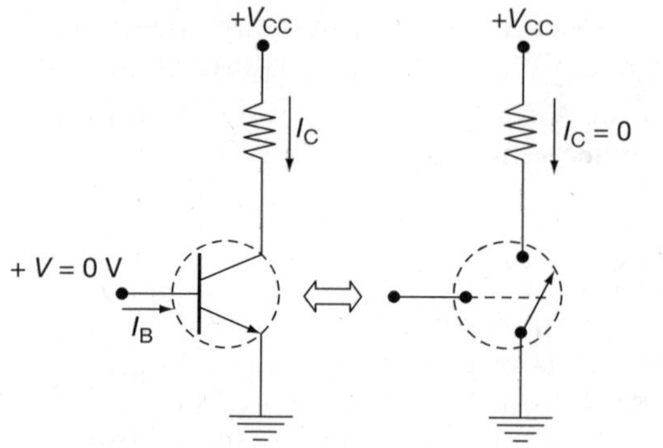

FIGURE 14.9(b) | *npn* BJT as open switch

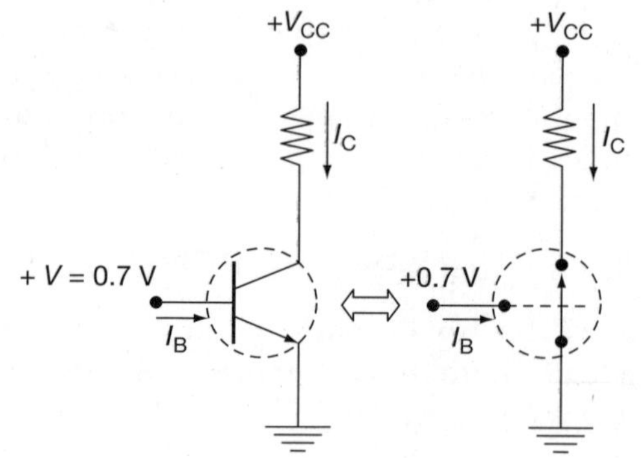

FIGURE 14.9(c) | *npn* BJT as closed switch

For the *npn* BJT, the base is more positive than the emitter. This *base-emitter* voltage is applied across a forward-biased *pn* junction, a diode, and is thus limited to about 0.7 V_{dc}. In case, voltage exceeds 0.7 V (dc), the BJT may be destroyed. The current through the *npn* transistor flows from collector to emitter as shown in Figure 14.9(c). For the *pnp* BJT, the base is more negative than the emitter. The current flows from emitter to collector when *pnp* BJT acts as closed switch.

14.4.2 | TTL Inverter

The logic function of an inverter or any type of gate is always the same, regardless of the type of circuit technology that is used. Figure 14.10 shows a standard TTL circuit for an inverter. Transistor, Q_1 is the input *coupling* transistor, and, D_1 is the input clamp diode. Transistor, Q_2 is called a *phase splitter*, and the combination of Q_3 and Q_4 forms the output circuit often referred to as a *totem-pole* arrangement.

When the input is a HIGH, the base-emitter junction of transistor, Q_1 is reverse-biased and the base-collector junction is forward-biased. This condition permits current to flow through R_{B_1} and the base-collector junction of transistor, Q_1 into the base of transistor, Q_2. It derives transistor, Q_2 into saturation. As a result, transistor, Q_3 is turned ON due to ON state of transistor, Q_2, and its collector voltage, which is the output, is near to ground potential. Therefore a LOW output is produced for a HIGH input. At the same time, the collector of transistor, Q_2 is at a sufficiently LOW voltage level to keep transistor, Q_4 OFF.

When the input is LOW, the base-emitter junction of transistor, Q_1 is forward-biased, and the base-collector junction is reverse-biased. The current flows through resistor, R_{B_1} and the base-emitter junction of transistor, Q_1 to the LOW input. A LOW provides a path to ground for the current. No current flows into the base of transistor, Q_2, so it is OFF. The collector of transistor, Q_2 is HIGH, thus turning transistor, Q_4 ON. A saturated transistor, Q_4 provides a low-resistance path from V_{CC} to the output. Therefore, a HIGH on the output is produced for a LOW on the input. At the same time, the emitter of transistor, Q_2 is at ground potential, keeping transistor, Q_3 OFF.

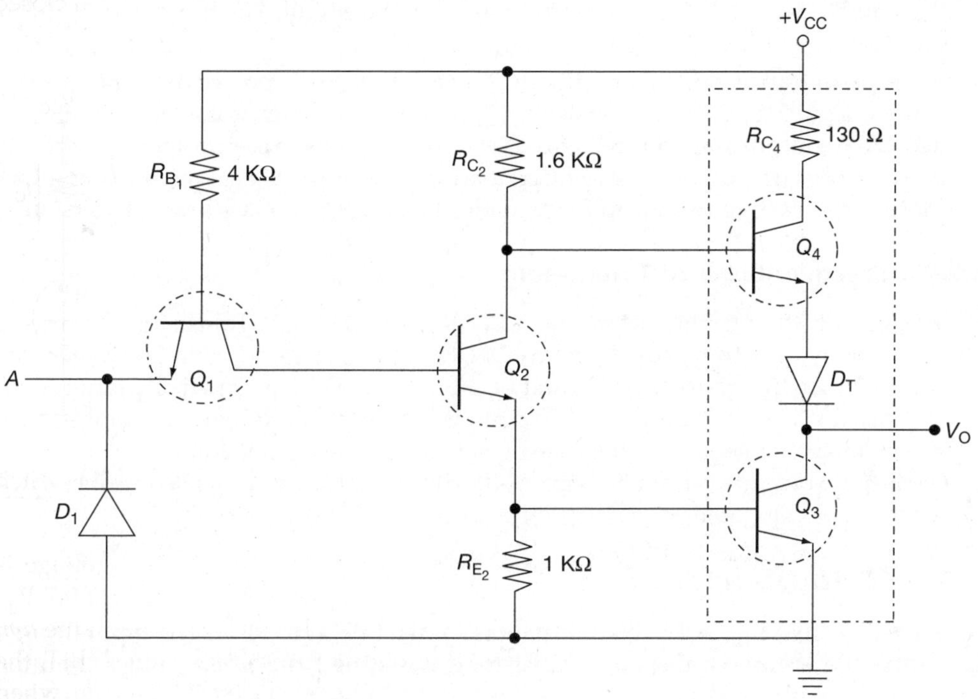

FIGURE 14.10 | TTL inverter gates

Diode D_1 in the TTL circuit prevents negative spikes of voltage on the input from damaging Q_1. Diode D_T ensures that transistor; Q_4 turns OFF when transistor, Q_2 is on HIGH input. In this condition, the collector voltage of transistor, Q_2 is equal to the base-to-emitter voltage, V_{BE} of transistor, Q_3 plus the collector-to-emitter voltage, V_{CE}, of transistor, Q_2. Diode D_T provides an additional V_{BE} equivalent drop in series with the base–emitter junction of transistor, Q_4 to ensure its turn-OFF state when transistor, Q_2 is ON.

The operation of the TTL inverter for the two input states is illustrated in Table 14.2. In the circuit, the base of transistor, Q_1 is 2.1 V above ground, so transistors, Q_2 and Q_3 are ON. The base of transistor, Q_1 is about 0.7 V above ground-not enough to turn transistors, Q_2 and Q_3 ON.

TABLE 14.2 | Operation of TTL inverter

Inputs	Transistors				Output
A	Q_1	Q_2	Q_3	Q_4	V_o
	Emitter junction, A				
Logic 0	Forward bias (ON)	OFF	OFF	ON	Logic 1
Logic 1	Reverse bias (OFF)	ON	ON	OFF	Logic 0

14.4.2.1 | Advantages of Totem-pole

- The TTL inverter can work with transistor, Q_3 and resistor, R_{C4} connected directly to the collector of transistor, Q_4, with transistor, Q_3, in the circuit. The transistor, Q_4, ensures that there is no flow of current through resistor, R_{C_4} in the output LOW state. So, the inclusion of transistor, Q_4, and diode, D_T keeps the circuit power dissipation low.
- In the output HIGH state, transistor, Q_4, acts as an emitter follower with its associated LOW output impedance. This LOW output impedance provides a small time constant for charging any capacitive load on the output. This action is commonly referred to as active *pull-up* and it provides fast rise time waveforms at TTL output.

14.4.2.2 | Disadvantages of Totem-pole

- During transition of the output from LOW to HIGH logic, Transistor, Q_4 turns OFF more slowly than transistor, Q_3 turns ON. So, there is a period of a few nanoseconds during which the transistors, Q_3, and Q_4 are conducting and, relatively large currents are drawn from the supply. So, TTL circuits suffer from internally generated current transients or current spikes because of the totem-pole connection.
- Totem-pole outputs cannot be wire ANDed, that is, the outputs of a number of gates cannot be tied together to obtain AND operation of those outputs.

14.4.3 | TTL NAND Gate

The circuit of the two-input TTL NAND gate is shown in Figure 14.11. The input transistor, Q_1 is a multiple emitter transistor. Transistor Q_2 is called the *phase splitter*. Emitter of

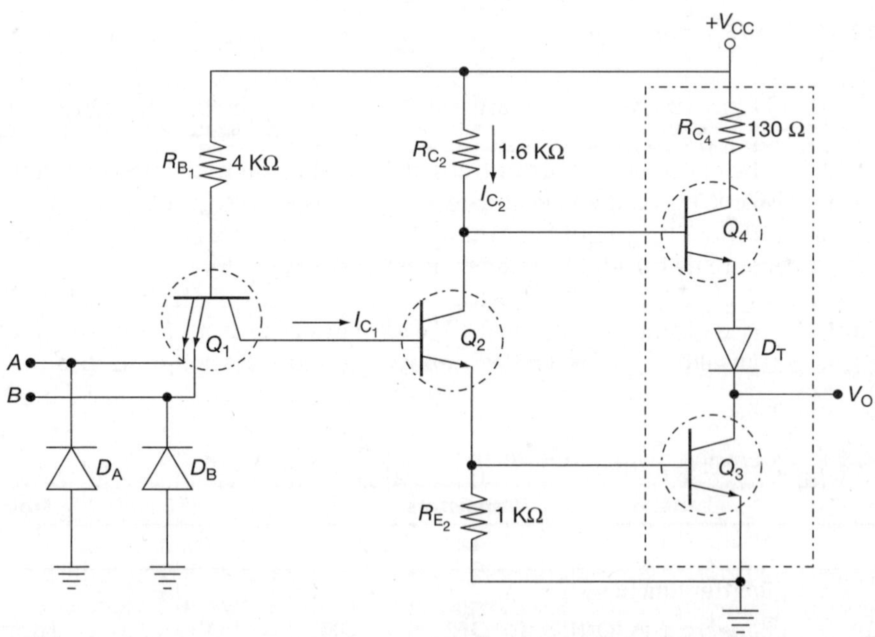

FIGURE 14.11 | TTL NAND gate

transistor, Q_3 is connected to collector of transistor, Q_4 through diode D_T. Transistors Q_3 and Q_4 form a *totem pole* arrangement. Diodes, D_A and D_B protect transistor, Q_1 from being damaged by the negative spikes of voltages at the inputs. When negative spikes appear at the input terminals, the diodes conduct and bypass the spikes to ground. Diode D_T ensures that transistors, Q_3 and Q_4 do not conduct simultaneously. Transistor, Q_3 acts as an emitter follower.

14.4.3.1 | Operation

Table 14.3 explains the operation of TTL NAND gate with totem-pole having two inputs. The circuit is redrawn when TTL multiple transistor Q_1 (Figure 14.12(a)) is replaced by equivalent diode circuit shown in Figure 14.12(b).

- A LOW voltage at either emitter E_1 or emitter E_2 forward-biases the corresponding diode D_1 or D_2 and reverse-biases diode D_3 which is a base-collector junction of transistor Q_1. There is no flow of current from base to collector of transistor Q_1.
- A LOW on both emitters of transistor Q_1 does the same action.
- A HIGH on both emitters reverse-biases both input diodes D_1 and D_2 and forward-biases D_3. The current flows from base to collector of transistor Q_1.

TABLE 14.3 | Operation of TTL NAND gate

Inputs		Transistors					Output
A	B	Q_1 Emitter junction, A	Q_1 Emitter junction, B	Q_2	Q_3	Q_4	V_O
Logic 0	Logic 0	Forward bias (ON)	Forward bias (ON)	OFF	OFF	ON	Logic 1
Logic 0	Logic 1	Forward bias (ON)	Reverse bias (OFF)	OFF	OFF	ON	Logic 1
Logic 1	Logic 0	Reverse bias (OFF)	Forward bias (ON)	OFF	OFF	ON	Logic 1
Logic 1	Logic 1	Reverse bias (OFF)	Reverse bias (OFF)	ON	ON	OFF	Logic 0

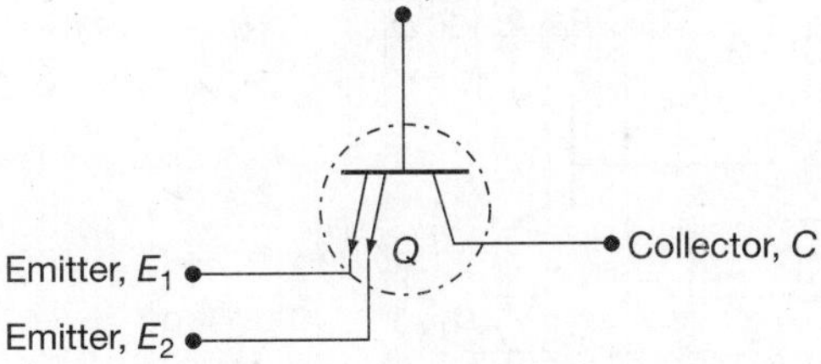

FIGURE 14.12(a) | TTL multiple emitters

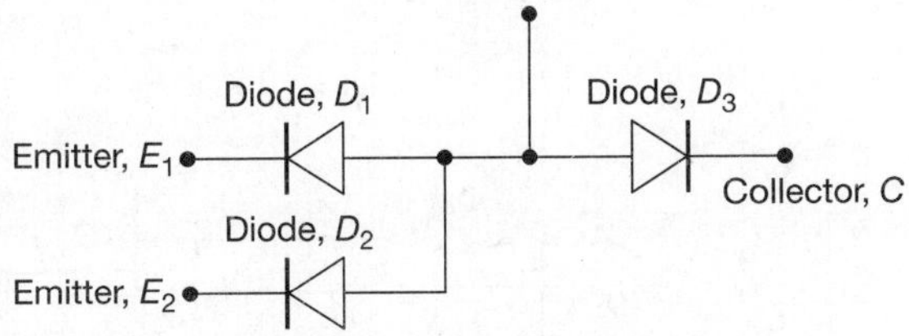

FIGURE 14.12(b) | Diode equivalent of TTL multiple emitters

Inputs A and B are HIGH: Both the inputs A and B are HIGH (+5 V), both the base–emitter junctions of Q_1 are *reverse-biased.* No current flows to the emitters of transistor of Q_1. The collector–base junction of Q_1 is *forward-biased.* So, a current, I_{C_1} flows through R_{B_1} to the base of transistor Q_2, and transistor Q_2 turns ON. Current from emitter of transistor Q_2

flows into the base of transistor Q_3. So transistor Q_3 is turned ON. The collector current I_{C_2} of transistor Q_2 flows through resistor, R_{C_2} and, produces a voltage drop across it thereby reducing the voltage at the collector of transistor, Q_2. So, transistor Q_4 acts OFF. Since transistor Q_3 is ON, V_o is at its LOW level (V_{CE}, saturation). So, the output is logic 0.

Inputs A and B are LOW: Both inputs A and B are LOW (0V). The corresponding base–emitter junctions are forward-biased and the collector–base junction of transistor Q_1 is reverse-biased. So, the current flows to ground through the emitters of transistor Q_1. Therefore, the base of transistor Q_1 is at 0.7 V, which cannot forward bias the base–emitter junction of transistor Q_2. So, transistor Q_2 cuts OFF. With transistor Q_2 OFF, transistor Q_3 does not get the required base drive. So, transistor Q_3 cuts OFF. Transistor Q_4 gets enough base drive because transistor Q_2 is OFF. Since, no current flows into the collector of transistor Q_2, all the current flows into the base of transistor Q_4. Therefore, transistor Q_4 turns ON. The output voltage, $V_0 = V_{CC} - V_{C2} - V_{BE4} - V_D$, which is a HIGH level or Logic 1.

Inputs A is HIGH and B is LOW: High input A and LOW input B do the same action as described when both inputs A and B are LOW. The output voltage V_0 is a Logic HIGH level.

Inputs A is LOW and B is HIGH: LOW input A and HIGH input B do the same action as described when both inputs A and B are LOW. The output voltage V_0 is a Logic HIGH level.

When transistor Q_4 is OFF, no current flows through it, but the stray and output capacitances between the output terminal, i.e. the collector of Q_4 and ground get charged to this voltage ranging from 3.4 V to 3.8 V.

14.4.4 | TTL NOR Gate

The circuit of the two-input TTL NOR gate is shown in Figure 14.13. Two input transistors Q_A and Q_B are emitter transistors. Transistor Q_1 and Q_2 are called the *phase splitters*. Emitter of transistor Q_3 is connected to collector of transistor Q_4 through diode D_T. Transistors Q_3

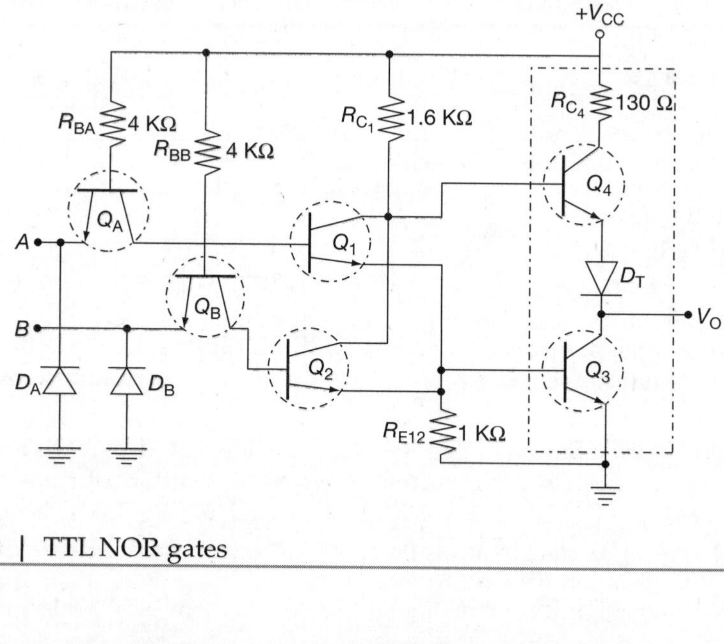

FIGURE 14.13 | TTL NOR gates

and Q_4 form a *totem pole* arrangement. Diodes D_A and D_B protect transistor Q_1 from being damaged by the negative spikes of voltages at the inputs. When negative spikes appear at the input terminals, the diodes conduct and bypass the spikes to ground. Diode D_T ensures that transistors Q_3 and Q_4 do not conduct simultaneously. Transistor Q_3 acts as an emitter follower.

14.4.4.1 | Operation

Table 14.4 explains the operation of TTL NOR gate with totem-pole. The working is explained below.

Inputs A and B are LOW: Both inputs *A* and *B* are LOW (0 V). The corresponding base–emitter junctions are *forward-biased* and the collector–base junctions of transistors Q_A and Q_B are *reverse-biased*. So, the current flows to ground through the emitters of transistors Q_A and Q_B. Therefore, the base of transistors Q_A and Q_B is at 0.7 V, which cannot forward bias the base–emitter junctions of transistors Q_1 and Q_2. So, transistors Q_1 and Q_2 are OFF. With transistors Q_1 and Q_2 OFF, transistor Q_3 does not get the required base drive. So, transistor Q_3 keeps OFF. Transistor Q_4 gets enough base drive because transistors Q_1 and Q_2 are OFF, i.e. since no current flows into the collectors of transistors Q_1 and Q_2, all the current flows into the base of transistor Q_4. Therefore, transistor Q_4 is ON. The output voltage, $V_0 = V_{CC} - V_{C1} - V_{BE4} - V_D$, which is a Logic 1 (HIGH).

TABLE 14.4 | Operation of TTL NOR gate

Inputs		Transistors						Output
A	*B*	Q_A	Q_B	Q_1	Q_2	Q_3	Q_4	V_0
		Emitter junction, A	Emitter junction, B					
Logic 0	Logic 0	Forward bias (ON)	Forward bias (ON)	OFF	OFF	OFF	ON	Logic 1
Logic 0	Logic 1	Forward bias (ON)	Reverse bias (OFF)	OFF	ON	ON	OFF	Logic 0
Logic 1	Logic 0	Reverse bias (OFF)	Forward bias (ON)	ON	OFF	ON	OFF	Logic 0
Logic 1	Logic 1	Reverse bias (OFF)	Reverse bias (OFF)	ON	ON	ON	OFF	Logic 0

Inputs A and B are HIGH: Both the inputs A and B are HIGH (+5 V), so, both the base–emitter junctions of transistor Q_A and Q_B are *reverse-biased*. No current flows to the emitters of Q_A and Q_B transistors. The collector–base junctions of Q_A and Q_B transistors are *forward-biased*. So, currents, I_{CA} and I_{CB} flow through resistors R_{BA} and R_{BB} to the base of transistors Q_A and Q_B and transistors Q_1 and Q_2 turn ON. The currents flow from emitters of transistors Q_1 and Q_2 into the base of transistor Q_3. So, transistor Q_3 is turned ON. The collector current I_{C_1} of transistors Q_1 and Q_2 flows through R_{C_1} and, produces a drop across it thereby reducing the voltage at the collector of transistors Q_1 and Q_2. So, transistor Q_4 acts OFF. Since transistor Q_3 is ON, V_0 is at its LOW level (V_{CE}, saturation). So, the output is Logic 0 (LOW).

Inputs A is HIGH and B is LOW: When input A is HIGH (+5 V) and input B is LOW (0 V), the base–emitter junction of transistor Q_A is *reverse-biased* and the base–emitter junction of transistor Q_B is *forward-biased*. No current flows to the emitter of Q_A transistors. The collector–base junction of Q_A transistor is *forward-biased*. So, current, I_{CA} flows through resistor R_{BA} to the base of Q_A and transistors Q_1 turns ON. On the other hand, the collector–base junction of transistor Q_B is *reverse biased*. So, the current flows to ground through the emitter of transistor Q_B. Therefore, the base of transistor Q_B is at 0.7 V, which cannot forward bias the base–emitter junctions of transistors Q_2. So, transistor Q_2 acts OFF.

Currents from emitter of transistor Q_1 flows into the base of transistor Q_3. So transistor Q_3 is turned ON. The collector current I_{C_1} of transistor Q_1 flows through resistor R_{C_1} and, produces a drop across it thereby reducing the voltage at the collector of transistors Q_1. So, transistor Q_4 acts OFF. Since transistor Q_3 is ON, V_0 is at its LOW level (V_{CE}, saturation). So, the output is Logic 0 (LOW).

Inputs A is LOW and B is HIGH: When input B is HIGH (+5 V) and input A is LOW (0 V), the base-emitter junction of transistor Q_B is *reverse-biased* and the base–emitter junction of transistor Q_A is *forward-biased*. No current flows to the emitter of Q_B transistors. The collector–base junction of Q_B transistor is *forward-biased*. So, current, I_{CB} flows through resistor R_{BB} to the base of Q_B and transistors Q_2 turns ON. On the other hand, the collector–base junction of transistor Q_A is *reverse-biased*. So, the current flows to ground through the emitter of transistor Q_A. Therefore, the base of transistor Q_A is at 0.7 V, which cannot forward bias the base–emitter junctions of transistor Q_1. So, transistor Q_1 acts OFF.

Currents from emitter of transistor Q_2 flows into the base of transistor Q_3. So transistor Q_3 is turned ON. The collector current I_{C_1} of transistor Q_2 flows through resistor R_{C_1} and, produces a drop across it, thereby reducing the voltage at the collector of transistors Q_2. So, transistor Q_4 acts OFF. Since transistor Q_3 is ON, V_0 is at its LOW level (V_{CE} (saturation)). So, the output is Logic 0 (LOW).

14.5 | TTL PARAMETERS

The low and high states for the input and output voltages for TTL logic family for LOW and HIGH states are given below:

$$V_{IL} = 0.8\text{ V},\ V_{IH} = 2\text{ V},\ V_{OL} = 0.4\text{ V and } V_{OH} = 2.4\text{ V}$$

These are the worst-case values. On the input side, a voltage has to be less than 0.8 V to qualify as a LOW-state input, and it must be more than 2 V to be considered a HIGH state input. On the output side, the voltage has to be less than 0.4 V to be a low-state output and more than 2.4 V to be a HIGH-state output.

14.5.1 | Current Sinking

A TTL circuit acts as a *current sink* in LOW state, as, it receives current from the input of the gate by which it is driving (Figure 14.14). Transistor Q_3 is the current-sinking transistor or the pull-down transistor, because it brings the output voltage down to its LOW state.

14.5.2 | Current Sourcing

A TTL circuit acts as a *current source* in the HIGH state, as, it supplies current to the gate by which it is driving (Figure 14.15). Transistor Q_3 is the current-sourcing transistor or the pull-up transistor, because it pulls up the output voltage to its HIGH state.

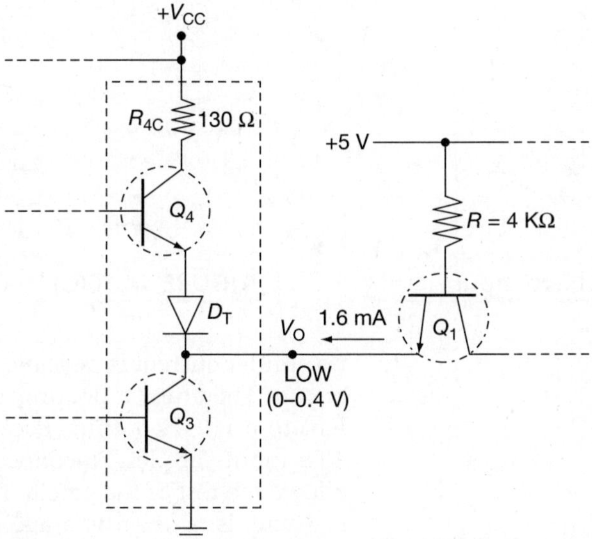

FIGURE 14.14 | Sinking current

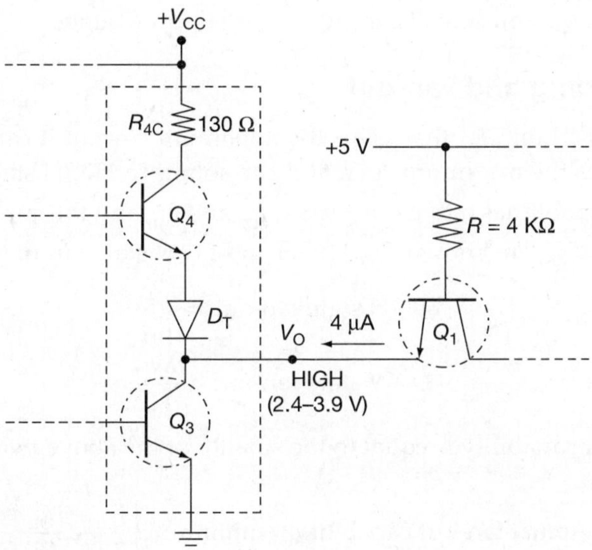

FIGURE 14.15 | Sourcing current

14.5.3 | Floating Inputs

When a TTL input is HIGH (ideally +5 V), the emitter current is approximately zero (Figure 14.16(a)). When a TTL input is floating (unconnected as shown in Figure 14.16(b)),

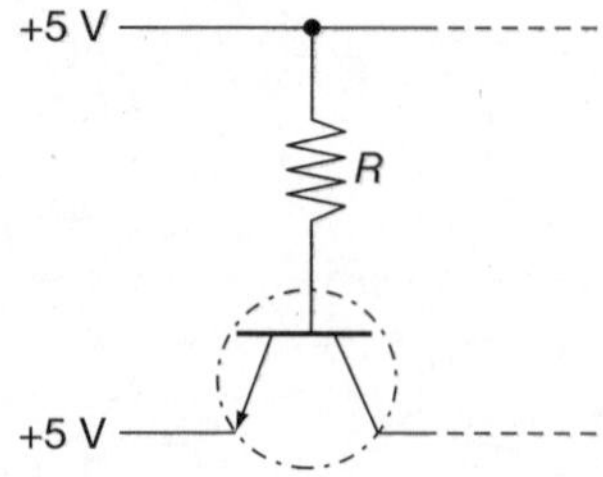

FIGURE 14.16(a) | High input

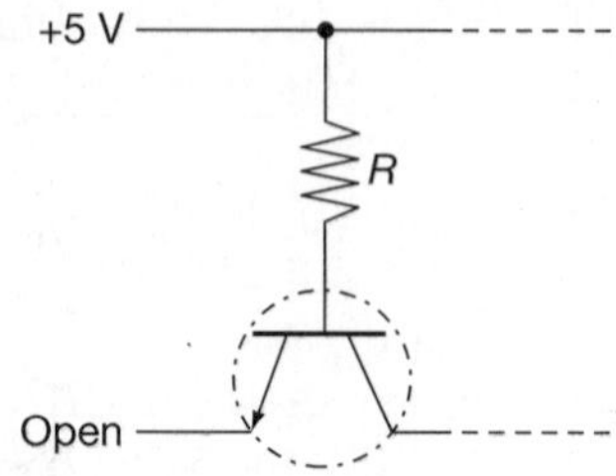

FIGURE 14.16(b) | Open as high input

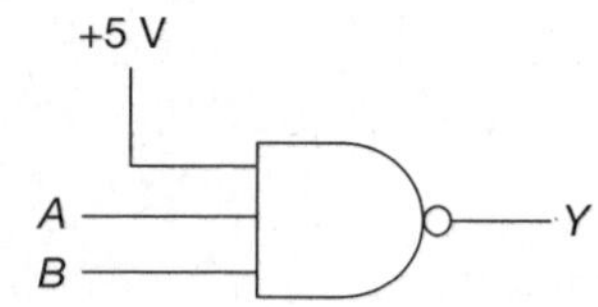

FIGURE 14.16(c) | Direct connection to supply

no emitter current is possible because of the open circuit. Therefore, a floating TTL input is equivalent to a HIGH output. Because of this, unused TTL inputs are left unconnected: an open input allows the rest of the gate to function properly.

There is a disadvantage to floating inputs. When an input is left open, it acts as a small antenna. Therefore, it picks up stray electromagnetic noise voltages. In some environments, the noise pickup is large enough to cause erratic operation of logic circuits. For this reason, it is preferred to connect unused TTL inputs to the supply voltage.

14.5.4 | TTL Loading and Fan-out

The TTL output has a limit, I_{OL} that gives the maximum current it can sink in LOW state and a limit, I_{OH}, gives the maximum current it can source in HIGH state. To determine the fan-out, the drive capabilities of the output, i.e. I_{OL} and I_{OH} and the current requirements of each input, i.e. I_{IL} and I_{IH} are known. So, HIGH and LOW state fan-outs are given by:

$$\text{HIGH state fan-out} = \frac{I_{OH}}{I_{IH}}$$

$$\text{LOW state fan-out} = \frac{I_{OL}}{I_{IL}}$$

The actual fan-out capability is equal to the smaller of the above two fan-out values and is given by

$$\text{Actual fan-out capability} = \text{minimum}\left\{\frac{I_{OH}}{I_{IH}}, \frac{I_{OL}}{I_{IL}}\right\}$$

A standard TTL output in LOW state connected to drive several standard TTL inputs. Transistor Q_4 is ON and is acting as a current sink for an amount of current I_{OL}, that is, the sum of the currents from each input. In its ON state, collector–emitter resistance of transistor Q_4 is very small but not zero. So, the current will produce a voltage drop V_{OL}. This voltage drop must not exceed the V_{OL} (max) limit of the IC. This limits the maximum value of I_{OL} and, thus, the number of loads that can be driven, that is, LOW state fan-out.

Suppose, transistor Q_3 can sink up to 16 mA (Figure 14.14) before its output reaches $V_{OL} = 0.4$ V.

Suppose I_{IL} = 1.6 mA. This means that transistor Q_4 can sink the current from up to 16/1.6 = 10 loads. If it is connected to more than 10 loads, V_{OL} increases above 0.4 V, and so, the noise margin is reduced and V_{OL} goes to the indeterminate state.

When the TTL output is in a HIGH state, Q_3 is acting as an emitter follower, that is, sourcing a total current I_{OH}, which is the sum of the I_{IH} currents of the different TTL inputs as shown in Figure 14.15.

If too many loads are being driven, the current I_{OH} will become large enough to cause a large voltage drop across R_2, to bring $V_{0H}\left(=V_{CC}-V_{C4}-V_{BE3}-V_{D}\right)$ below V_{OH}(min). This is undesirable because it reduces the HIGH state noise margin and could even cause V_{0H} to go into the indeterminate range.

Suppose Q_3 can source 0.4 mA of current before V_{OH} falls below V_{OH} (min), and each load receives I_{IH} = 40 μA. This means Q_3 can source upto 0.4 mA/40 μA = 10 loads.

14.5.5 | Unit Load

Unit load means the current drawn or sourced back by similar gates. For 7400,

One unit load is 40 μA in HIGH state that is known as I_{IH}

One unit load is 1.6 mA in LOW state that is known as I_{IL}

If the output of 7400 IC is rated at I_{OH} = 800 μA and I_{OL} = 48 mA, then HIGH state and LOW state fan-outs are given by:

$$\text{HIGH state fan-out} = \frac{I_{OH}}{I_{IH}} = \frac{800\ \mu\text{A}}{40\ \mu\text{A}} = 20 \text{ unit loads}$$

$$\text{LOW state fan-out} = \frac{I_{OL}}{I_{IL}} = \frac{48\text{ mA}}{1.6\text{ mA}} = 30 \text{ unit loads}$$

Therefore, the actual fan-out is equal to the smaller of the above two fan-out values and is given by

$$\text{Actual fan-out capability} = \text{minimum}\left\{\frac{I_{OH}}{I_{IH}}, \frac{I_{OL}}{I_{IL}}\right\} = \text{minimum}\{20, 30\} = 20 \text{ units}$$

14.6 | OPEN-COLLECTOR GATES

The open-collector inverter circuit is shown in Figure 14.17. In open-collector TTL inverter, the output is at the collector of transistor Q_3. No transistor, diode, and resistors are connected to it, therefore, the name *open collector.* In order to get the proper HIGH and LOW logic levels out of the circuit, an external pull-up resistor is connected to V_{CC} from the collector of Q_3 as shown in Figure 14.18.

When transistor Q_3 is OFF, the output is pulled to V_{CC} through the external resistor R. When transistor Q_3 is ON, the output is connected to near ground through the saturated transistor. The value of R must be so chosen that when one gate output goes LOW while the others are HIGH, the sink current through the LOW output does not exceed the I_{OL} maximum limit. Since the output is pulled to logic HIGH level through a resistor, it is called the *passive pull-up.* Table 14.5 gives the operation of open collector TTL inverter.

The open-collector arrangement is much slower than the totem-pole arrangement, because the time constant with which the load capacitance charges in this case is considerably larger. In the case of totem-pole, it is active pull-up, i.e. pull-up is through transistor Q_3.

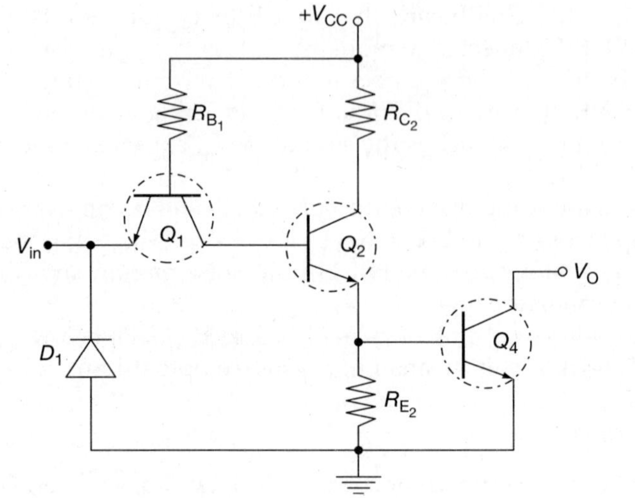

FIGURE 14.17 | Open collector inverter

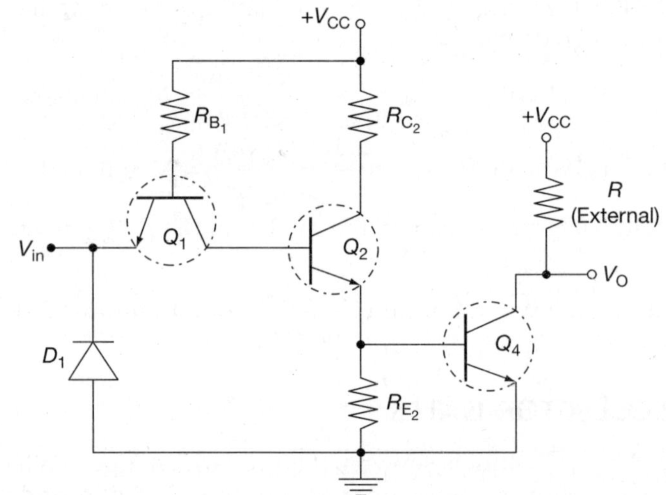

FIGURE 14.18 | Open-collector inverter with the external resistor

TABLE 14.5 | Operation of open-collector TTL inverter

Inputs	Transistors			Output
V_{in}	Q_1	Q_2	Q_3	V_o
	Emitter junction, A			
Logic 0	Forward bias (ON)	OFF	OFF	Open/Logic 1
Logic 1	Reverse bias (OFF)	ON	ON	Grounded/Logic 0

The resistance in saturation of transistor Q_4 is very small; so, the charging time constant is low and the output rises fast. The speed can be increased only by selecting a smaller resistance. Because of this reason, the open-collector circuits should not be used in applications where switching speed is a principal consideration.

14.6.1 | Wired AND Operation

The outputs of a number of NAND gates needs to be ANDed. This can be achieved by using additional two NAND gates (4 and 5) as shown in Figure 14.19(a). The same logic operation can be performed by simply tying the outputs of NAND gates 1, 2, and 3 as shown in Figure 14.19(b). This is called *wired AND operation*, because the AND operation is obtained by simply connecting the output wires together. When any of the gate outputs go to a LOW state, the common output point also goes LOW as a result of its shorting to ground through the ON transistor. The common output will be HIGH only when all the gate outputs are in a HIGH state. This arrangement has the advantage of needing fewer gates compared to the conventional arrangement.

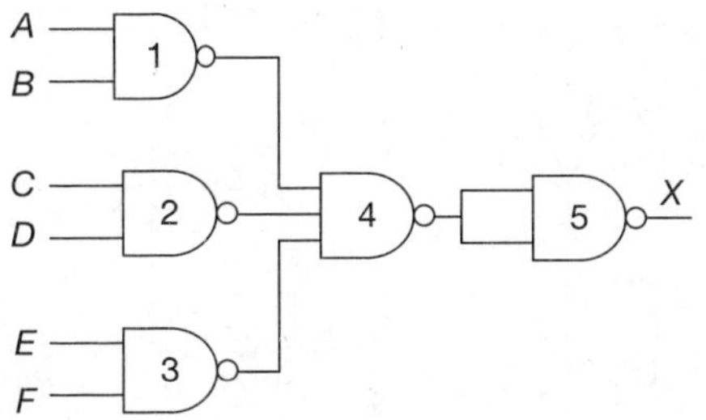

FIGURE 14.19(a) | Conventional AND operation

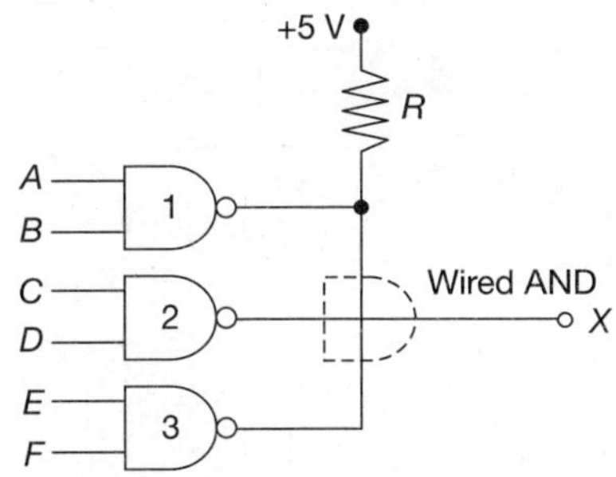

FIGURE 14.19(b) | Wired AND operation

The totem-pole outputs of gates cannot be wired ANDed, because when one output is HIGH and the other is LOW. These connections are wired ANDed, a large current flows from supply to ground through transistor Q_4 of the HIGH-state gate and transistor Q_3 of the LOW-state gate. This is because transistor Q_3 of the LOW-state gate acts as a very low resistance load on transistor Q_4 of the HIGH-state gate (see Figure 14.11). This large current can easily damage any of these transistors. The situation is even worse when more than two TTL outputs are tied together. The open-collector gates, on the other hand, may be wired ANDed without any problem.

Because of the absence of pull-up transistors, the wired-AND connections significantly reduce switching speeds. However, they are useful in reducing the chip count of a system when speed is not a consideration.

14.6.2 | Three-state TTL

Another type of TTL configuration is the tri-state configuration. It utilizes the advantage of high speed of operation of the totem-pole configuration and wire ANDing of the open-collector configuration. It is called the *tri-state* TTL, because it allows three possible output states: HIGH, LOW, and high-impedance. In the high-impedance state, both the transistors in the totem-pole arrangement are turned OFF, so that the output terminal is high-impedance to ground or V_{CC}. In fact, the output is an open or floating terminal, that is, neither a LOW nor a HIGH. In practice, the output terminal is not an exact open circuit, but has a resistance of several MΩ or more relative to ground and V_{CC}. The circuit of a tri-state inverter is shown in Figure 14.20. The tri-state operation is obtained by modifying the basic totem-pole circuit. The circuit has two inputs – *A* is the normal logic input and enable, *E* is an enable input that produces the high-impedance.

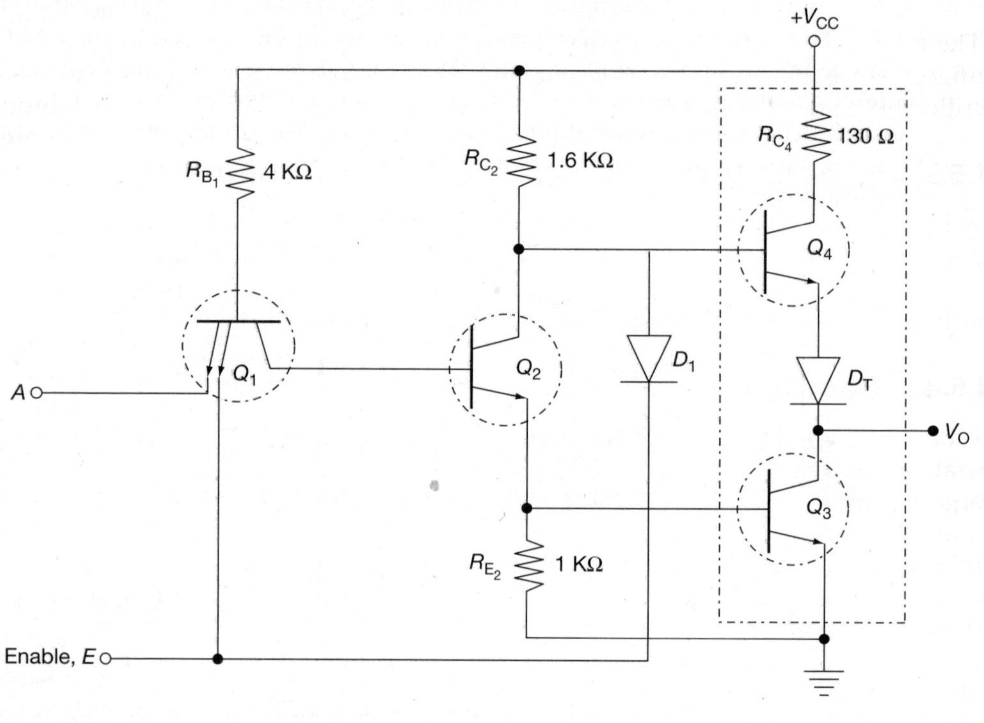

FIGURE 14.20 | Tri-state TTL inverter

The enabled state: With enable, E = 1, the circuit operates as a normal inverter because the high voltage at E has no effect on Q_1 or Q_2. In this enabled condition, the output is simply the inversion of logic input A.

The disabled state (High-impedance): When enable, E = 0, the circuit goes into its high-impedance state regardless of the state of logic input *A*. The LOW at E forward biases the emitter base junction of transistor, Q_1 and shunts the current in resistor, R_{B_1} away from transistor, Q_2, so that, transistor, Q_2 turns OFF, which turns transistor, Q_3 OFF. The LOW at *E* also forward biases diode D_1 to shunt current away from the base of transistor, Q_4, and therefore, transistor, Q_4 also turns OFF. With both totem-pole transistors in the non-conducting state, the output terminal is essentially an open circuit. Table 14.6 summarizes the operation of tri-state TTL inverter.

TABLE 14.6 | Operation of tri-state TTL inverter

Inputs		Transistors				Diode	Output
Enable, E	A	Q_1	Q_2	Q_3	Q_4	D_1	V_o
		Emitter junction, A					
Logic 1	Logic 0	Forward bias (ON)	OFF	OFF	ON	OFF	Logic 1
Logic 1	Logic 1	Reverse bias (OFF)	ON	ON	OFF	OFF	Logic 0
Logic 0	X	Forward bias (ON)	OFF	OFF	OFF	ON	Open state

There are many ICs that are designed with tri-state outputs. The advantage of the tri-state configuration is that the outputs of the tri-state ICs can be connected together without sacrificing the switching speed.

14.6.3 | Buffer/Drivers

Any logic circuit that is called a *buffer*, a *driver*, or a *buffer/driver* is designed to have a greater output current and/or voltage capability than that of an ordinary logic circuit. Buffer/driver ICs are available with totem-pole outputs, open-collector outputs, or tri-state outputs. Tri-state buffers also invert the signal as it goes through. They are called *inverting* tri-state buffers.

14.6.4 | Schottky TTL

All the transistors in the circuits of standard TTL, low-power TTL, and high-speed TTL operate in saturation or cut-off region. When the transistor is in saturation it stores the charge and the operation causes a storage-time delay during the transistor transition from ON to OFF; and this limits the circuit's switching speed.

In Schottky TTL families, Schottky transistors are used instead of normal transistors. The Schottky transistor is operated in active region or cut-off region it never goes into saturation and the storage time delay is negligible. The Schottky transistor is obtained by using a Schottky barrier diode (SBD) between the base and collector terminals of the transistor as shown in Figure 14.21. The Schottky diode has a forward biased voltage of 0.25 V. The diode is connected between the base and the collector terminals of the transistor. The collector junction of the transistor cannot get forward biased and the transistor never goes in saturation. The transistor operates in cut-off or active region. The symbol of Schottky transistor is shown in Figure 14.21.

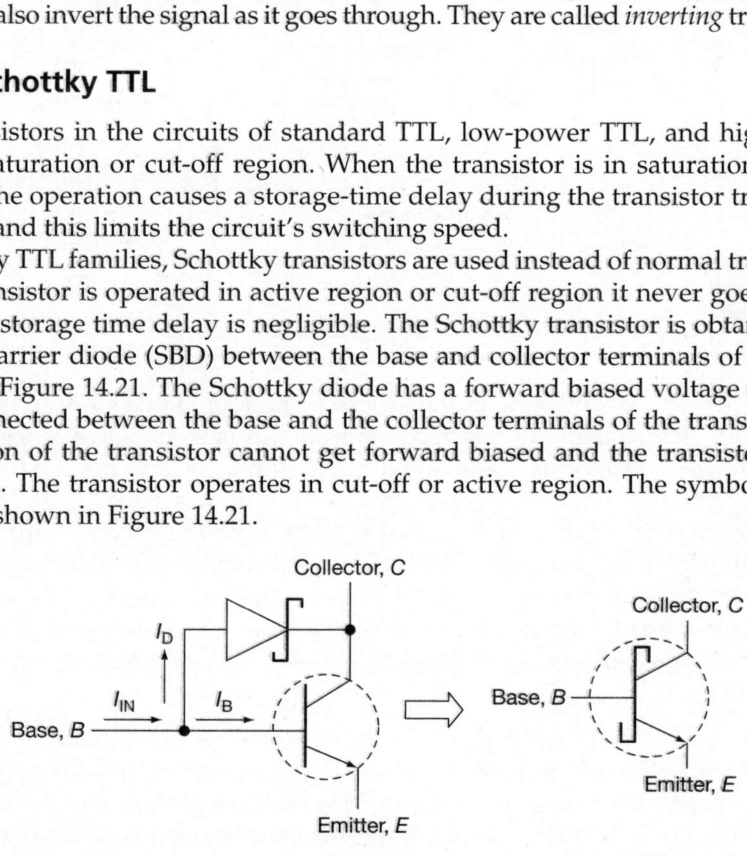

FIGURE 14.21 | Symbol of Schottky transistor

The Schottky clamped transistors are used to construct 2-input Schottky-TTL NAND gate shown in Figure 14.22. Diodes D_1 and D_2 are the clamping diodes to protect the negative input voltage. Transistors Q_5 and Q_6 are included. Q_4 and Q_5 form Darlington pair to provide a shorter output rise time when switching from ON to OFF.

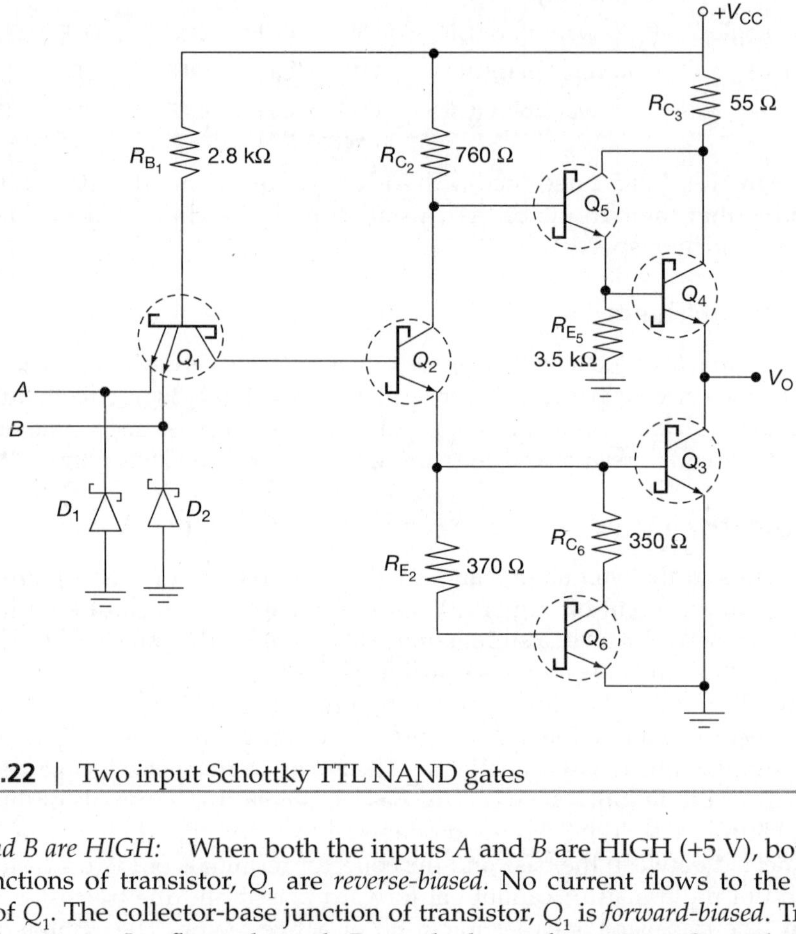

FIGURE 14.22 | Two input Schottky TTL NAND gates

Inputs A and B are HIGH: When both the inputs *A* and *B* are HIGH (+5 V), both the base-emitter junctions of transistor, Q_1 are *reverse-biased.* No current flows to the emitters of transistor of Q_1. The collector-base junction of transistor, Q_1 is *forward-biased*. Transistor Q_1 acts OFF. A current, I_{C_1} flows through R_{B_1} to the base of transistor Q_2, and transistor Q_2 turns ON. Current from emitter of transistor Q_2 flows into the base of transistors Q_3 and Q_6. So transistors Q_3 and Q_6 are turned ON. The collector current I_{C_2} of transistor Q_2 flows through resistor R_{C_2} and, produces a drop across it thereby reducing the voltage at the collector of transistor Q_2. So, transistor Q_5 acts OFF. Since Q_5 is OFF so transistor, Q_4 goes OFF. Since transistor Q_3 is ON, V_0 is at its LOW level (V_{CE}, saturation). So, the output is Logic 0 (LOW).

Inputs A and B are LOW: When A and B are LOW (0 V), the corresponding base–emitter junctions are forward-biased and the collector–base junction of transistor Q_1 is reverse-biased. So, the current flows to ground through the emitters of transistor Q_1. Therefore, the base of transistor Q_1 is at 0.7 V, which cannot forward bias the base–emitter junction of transistor Q_2. So, transistor Q_2 is OFF. With transistor Q_2 OFF, transistors Q_3 and Q_6 do not get the required base drive. So, transistors Q_3 and Q_6 act as OFF. Transistor Q_5 gets enough base drive because transistor Q_2 is OFF, i.e. since no current flows into the collector of transistor, Q_2. The current flows into the base of transistor Q_5. Therefore, transistor Q_5 is ON and transistor Q_4 turns ON. The output voltage, V_O is a Logic HIGH level.

Inputs A is HIGH and B is LOW: High input A and LOW input B do the same action as described when both inputs *A* and *B* are LOW. The output voltage V_0 is a Logic HIGH level.

Inputs A is LOW and B is HIGH: LOW input A and HIGH input B do the same action as described when both inputs *A* and *B* are LOW. The output voltage V_0 is a Logic HIGH level. The operation of Schottky TTL NAND gate summarizes Table 14.7.

TABLE 14.7 | Operation of Schottky TTL NAND gate

Inputs		Schottky Transistors							Output
A	*B*	Q_1	Q_1	Q_2	Q_6	Q_3	Q_5	Q_4	V_O
		Emitter junction, A	Emitter junction, B						
Logic 0	Logic 0	Forward bias (ON)	Forward bias (ON)	OFF	OFF	OFF	ON	ON	Logic 1
Logic 0	Logic 1	Forward bias (ON)	Reverse bias (OFF)	OFF	OFF	OFF	ON	ON	Logic 1
Logic 1	Logic 0	Reverse bias (OFF)	Forward bias (ON)	OFF	OFF	OFF	ON	ON	Logic 1
Logic 1	Logic 1	Reverse bias (OFF)	Reverse bias (OFF)	ON	ON	ON	OFF	OFF	Logic 0

14.7 | TTL SUBFAMILIES

The TTL logic family is recognized by 54 series/74 series that are popular and commonly used series of digital ICs. These series have the limitation of speed and power dissipation. These limitations are overcome to a certain limit in the improved TTL series. The improved TTL series are given as follows:

- 74L series (low-power TTL)
- 74H series (high-speed TTL)
- 74S series (Schottky TTL)
- 74LS series (low-power Schottky TTL)
- 74AS series (advanced Schottky TTL)
- 74ALS series (advanced low-power Schottky TTL)
- 74F series (fast TTL)

14.7.1 | Standard TTL, 74 Series

The standard TTL ICs follow 74 series. This series offer a combination of speed and power dissipation suited for many applications. The 54 series is the counterpart of the 74 series. The 54 series are functionally equivalent to the 74 series, but can be operated over wider temperature and voltage ranges, as required by military specifications. In these days, the standard TTL is rarely used in new systems.

14.7.2 | Low-power TTL, 74L Series

The low-power TTL ICs are designated by 74L series. The 74L series and 74 have the same basic circuit except all resistance that have increased values like $R_1 = 40$ kΩ, $R_2 = 20$ kΩ, $R_3 =$ 12 kΩ, and $R_4 = 500$ Ω. The larger resistors reduce the current. Therefore, the power requirement is reduced, but the speed is slower. The power consumption of low-power TTL is about 1/10 of that of standard TTL, whereas the standard TTL is three times or faster than the low-power TTL. The low-power version is now not available in 7400 series. Nowadays, low-power Schottky TTL and CMOS versions of the 7400 series are widely in use.

14.7.3 | High-speed TTL, 74H Series

The high-speed TTL ICs are designated by 74H series. The 74H series and the standard 74 series have the same basic circuit. The smaller resistance values ($R_1 = 2.8$ kΩ, $R_2 = 760$ Ω, $R_3 = 470$ Ω, and $R_4 = 58$ Ω) are used in 74H series. The emitter follower transistor Q_4 is replaced by a Darlington pair and emitter to base joining of Darlington pair ($Q_5 - Q_4$) is connected to ground through a resistance of 4 kΩ (Figure 14.22). The switching speed of the 74H series is approximately two times more than that of the standard TTL, as also the power consumption. Newer Schottky versions are superior in both speed and power consumption.

14.7.4 | Schottky TTL, 74S Series

The standard TTL, low-power TTL, and high-speed TTL series operate using saturated switching. When a transistor is saturated, excess charge carriers are stored in the base region and these are removed before the transistor is turned OFF. Owing to storage time delay, the speed is reduced. The Schottky TTL 74S series reduces this storage time delay by not allowing the transistor to go into full saturation. This is accomplished by using a SBD between the base and the collector of each transistor. Virtually, all modern TTL devices incorporate this so-called Schottky clamp. The SBD has a forward voltage of only 0.25 V. The circuits in the 74S series also use smaller resistance values to improve the speed of operation. The speed of the 74S series is twice that of the 74H series. Schottky TTL has more than three times the switching speed of standard TTL, at the expense of approximately doubling the power consumption.

14.7.5 | Low-power Schottky TTL, 74LS Series

The low-power Schottky TTL logic family is designated by 74LS. The 74LS series is a low-powered, slower speed version of the 74S series. It uses the Schottky clamped transistor, but with larger resistance values than those in the 74S series. The larger resistance values reduce the circuit power requirement but at the expense of reduction in speed. The switching speed of low-power Schottky TTL is about the same as that of the standard TTL, but the power consumption is about 1/5 of the standard TTL. The 74LS NAND gate does not use the multiple emitter input transistor. Instead it uses diodes. This series is replacing the 74 series.

14.7.6 | Advanced Schottky TTL, 74AS Series

The advanced Schottky TTL logic family is designated by 74AS. Owing to recent innovations in IC design, 74AS series has been developed. It provides considerable improvement in speed over the 74S series and at a much lower power requirement. It is the fastest TTL series and its speed–power product is significantly lower than that of the 74S series. Its fan-out is larger than that of the 74S series because of its lower input current requirement. It is twice as fast and consumes less than half as much power as the 74S series.

14.7.7 | Advanced Low-power Schottky TTL, 74ALS Series

The advanced low-power Schottky TTL logic family is designated by 74ALS. This series is a low-power version of the advanced Schottky TTL. This series offers an improvement over the 74LS series in both speed and power dissipation. The 74ALS has the lowest speed–power product of all the TTL series, and it is very close to having the lowest gate power dissipation.

14.7.8 | Fast TTL, 74F Series

The fast TTL logic family is designated by 74F. This series is the newest and fastest TTL series. Devices in this series have the letter 'F' inserted in their designations. The ALS and AS technologies are the recent enhancements in Schottky TTL circuitry. Among other refinements, the advanced Schottky devices are fabricated with an improved doping technique and Schottky-clamped transistors provide improved isolation. These enhancements reduce capacitance and, thus, improve switching times.

Also, it is a more complex circuit design which uses additional active devices to speed up switching, reduce power consumption, and increase fan-out. The inverter IC in different TTL subfamilies is 7404, 74L04, 74H04, 74LS04, 74AS04, 74ALS04, and 74F04. The typical characteristics of TTL subfamilies are shown in Table 14.8.

TABLE 14.8 | The typical characteristics of TTL subfamilies

Performance Rating	74	74L	74H	74S	74LS	74AS	74ALS	74F
Propagation delay (ns)	9	33	6	3	9.5	1.7	4	3
Power dissipation (mW)	10	1	23	20	2	8	1.2	6
Speed power product (pJ)	90	33	138	60	19	13.6	4.8	18
Max. clock rate (MHz)	35	3	50	125	45	200	70	100
Fan-out (Same series)	10	20	10	20	20	40	20	33
Noise margin (V)	0.4	0.4	0.4	0.7	0.7	0.5	0.5	0.5

14.8 | EXTERNAL DRIVE FOR TTL LOADS

To drive a TTL load with an external source, TTL input requirements for voltage and current are to be satisfied. For standard TTL in the low state, an input voltage between 0 and 0.8 V with a current of approximately 1.6 mA is the requirement. In the high state, the voltage has to be from 2 to 5 V with a current of approximately 40 μA. A brief look at some of the ways to drive a TTL load is discussed below.

14.8.1 | Switch Drive

Figure 14.23 shows the preferred method for driving a TTL input from a switch. With the open switch, the input is pulled up to +5 V. In the worst case, only 40 μA of input current exists. Therefore, the voltage appearing at the input pin is slightly less than the supply voltage because of the small voltage drop across the pull-up resistor:

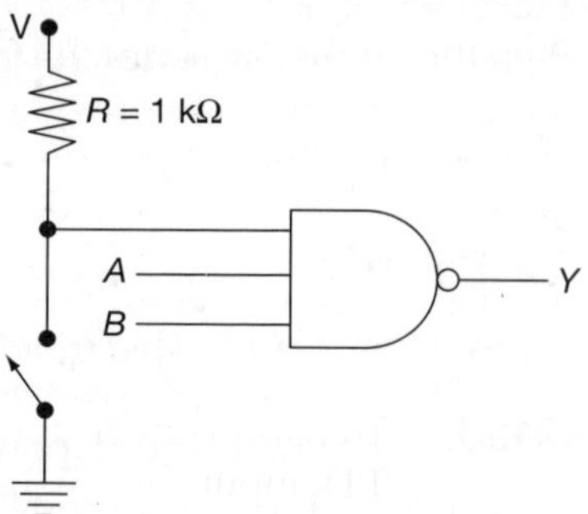

FIGURE 14.23 | Switch drive for TTL input

$$V_i = 5\text{ V} - (40\ \mu\text{A})(1\text{ k}\Omega) = 4.96\text{ V}$$

This is more than the minimum requirement of 2 V. It means that the noise immunity is good.

When the switch is closed, the input is pulled down to ground. In the worst case, the input current is 1.6 mA. This sink current creates no problem because it flows through the closed switch to ground. The noise immunity is good because the input voltage is 0 V that is below the maximum allowable value of 0.8 V.

14.8.2 | Size of Pull-Up Resistance

A pull-up resistance of 1 kΩ is nominal. In the Figure 14.23, the current drain with a closed switch is 5 mA (5V/1kΩ). The smaller the pull-up resistance, the larger the current drain. At some point, too much current drain becomes a problem for the power supply. Large resistance is used to keep the current drain to tolerable levels.

A large pull-up resistance causes speed problems. The worst case occurs when the switch is opened. Let the input capacitance is 10 picofarads (pF) in Figure 14.23. The time constant is 10 ns (RC = 1 kΩ × 10 pF). Larger the pull-up resistance, larger is the time constant. A larger time constant causes a slower switching speed because the input capacitance has to charge through the pull-up resistance. Pull-up resistances between 1 and 10 kΩ are typical. It results in current drains and time constants that are acceptable in most applications.

14.8.3 | Transistor Drive

A transistor switch is used to control the state of the TTL input as shown in Figure 14.24(a). When V_i is low, the transistor is OFF. The transistor acts equivalent to an open switch. Then the TTL input is pulled up to +5 V through a resistance of 1 kΩ. When V_i is high, the transistor is ON. The transistor acts equivalent to a closed switch. In this case, it easily sinks the 1.6 mA of input current.

The transistor inverts the input signal V_i. An inverter is inserted as shown in Figure 14.24(b). Now, the double inversion produces an in-phase control signal at the TTL.

14.8.4 | Operational Amplifier Drive

Sometimes, the output of an operational amplifier (OP-AMP) is used to control a TTL input. Operational amplifier splits supply voltages of +15 V and −15 V. Figure 14.25 shows

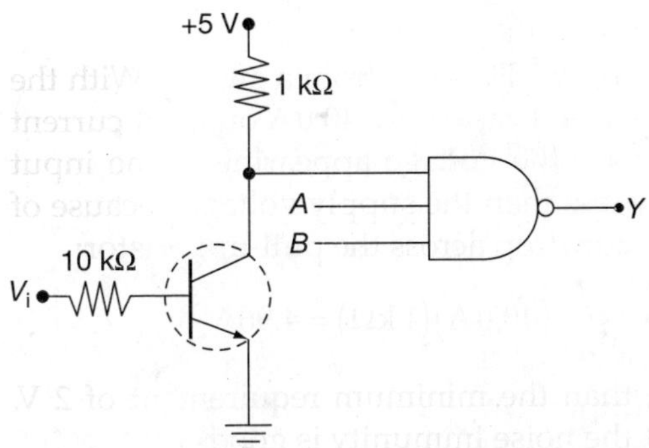

FIGURE 14.24(a) | Transistor drive for TTL input

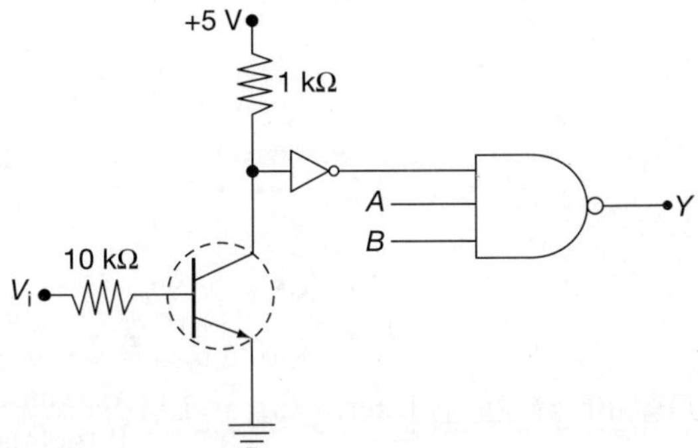

FIGURE 14.24(b) | Transistor-invert for TTL input

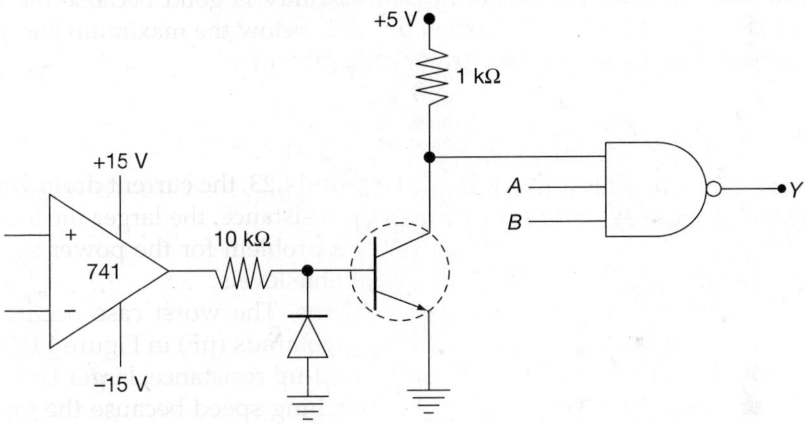

FIGURE 14.25 | Operational amplifier and transistor drive for TTL input

the output of a 741 operational amplifier that controls a TTL input. The output of the operational amplifier swings between +15 V and –15 V. The positive swing closes the transistor switch, producing a TTL input of approximately 0 V. The negative swing drives the transistor into cut-off, producing a TTL input of +5 V. Diode in the base circuit protects the base against excessive reverse voltage. The diode clamps the base voltage at approximately –0.7 V on the negative swing.

14.8.5 | Comparator Drive

Figure 14.26 shows the connection of a comparator LM339 to a TTL load. Because of the open-collector output, the output pin of the comparator is connected to a supply voltage of +5 V through a pull-up resistance of 1 kΩ. When V_i is positive, the sink transistor goes OFF and the TTL input is pulled HIGH. When V_i is negative, the sink transistor goes on and the TTL output is pulled LOW.

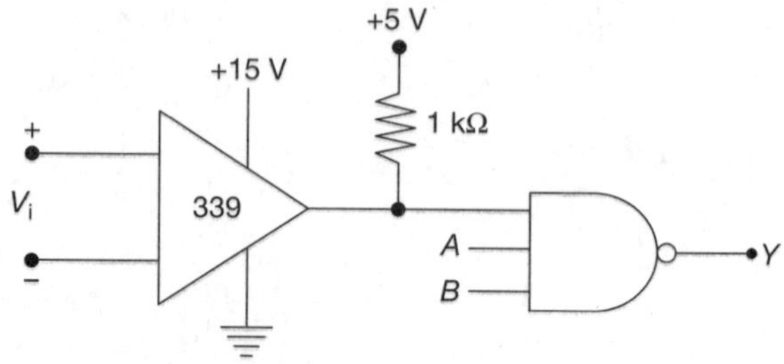

FIGURE 14.26 | Interfacing an LM339 to a TTL input

14.9 | TTL DRIVING EXTERNAL LOADS

Because standard TTL can sink up to 16 mA, TTL driver can be used to control an external load such as a relay, an LED, etc. Figure 14.27(a) illustrates the idea. When the TTL output is HIGH, there is no load current. But when the TTL output is LOW, the lower end of R_L is ideally grounded. This sets up a load current of approximately

$$I_L = \frac{5\text{ V}}{R_L}$$

Since standard TTL can sink a maximum of 16 mA, the load resistance is limited to a minimum value of about $R_L = 312\ \Omega$ (5 V/16 mA).

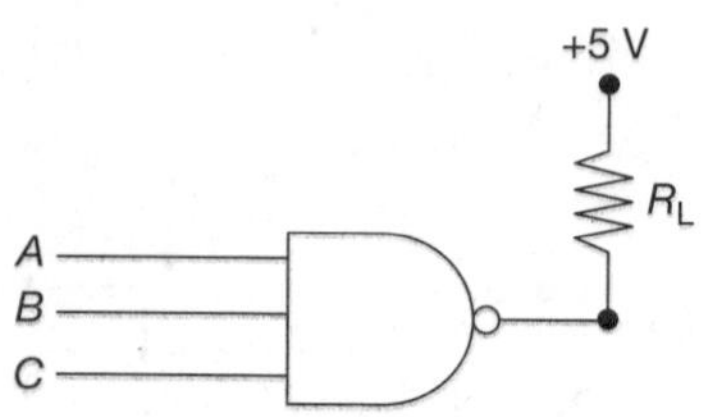

FIGURE 14.27(a) | TTL output drives load resistor

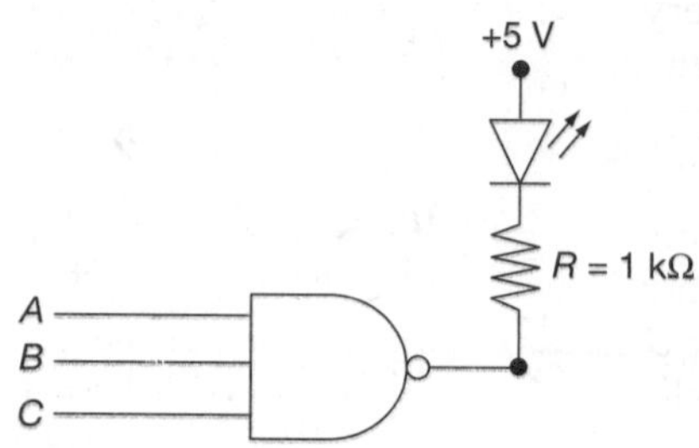

FIGURE 14.27(b) | TTL output drives load resistor

14.9.1 | Supply Voltage Different from +5 V

A supply voltage different from +5 V can be used by using an open-collector TTL device. For instance, Figure 14.28(a) shows an open-collector gate driving a load resistor that is returned to +15 V. Since an open-collector device can sink a maximum of 16 mA, the minimum load resistance in Figure 14.28(a) is slightly less than 1 kΩ.

More than 16 mA of load current can be tackled by using an external transistor, as shown in Figure 14.28(b). When the open-collector device has a low output, the external transistor goes off and the load current is zero. When the device has a high output, the external transistor goes on and the load current is maximum.

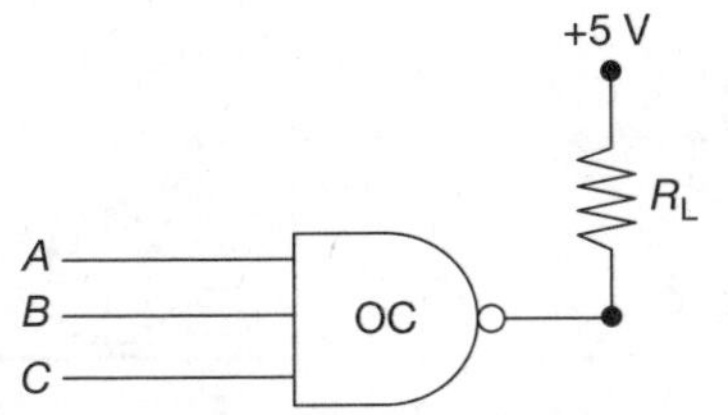

FIGURE 14.28(a) | Open collector device for higher voltage

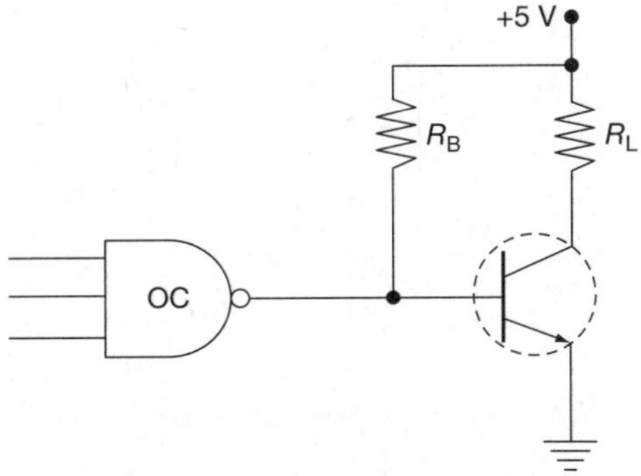

FIGURE 14.28(b) | Transistor increases current drive

14.10 | INTEGRATED INJECTION LOGIC

Latest among logic families is integrated injection logic (IIL or I²L) or current injection logic (CIL) family, which is used in LSI and VLSI circuits. It is not suitable for discrete gate ICs. The I²L logic gates are constructed using bipolar transistors only. The absence of resistors makes it possible to integrate a large number of gates on a single package. Complete microprocessors can be obtained on a single chip. The I²L circuits are easily fabricated and are economical. Their power consumption is also low. The speed power product is constant and very small of the order of 4 pJ, comparable to advanced low-power Schottky TTL. The I²L logic family has propagation delay, t_{pd} = 1 ns, power dissipation, P_D = 1 mW, noise margin, NM = 0.35 V, fan-out = 8, and the relative cost is very low.

In I²L logic family, no transients are produced like in TTL and MOS since the currents are constant. It can easily be integrated on the same chip with bipolar analog circuits such as operational amplifiers. By programming the injector currents, the propagation delay and power dissipation can be varied over a wide range. The disadvantage is that, it requires extra step in its manufacturing process than those used in MOS.

14.10.1 | IIL OR I²L Inverter

Since discrete gates are not available in I²L, the operation of an I²L inverter is explained by considering the inverter of Figure 14.29(a) that behaves like inverter. The *pnp* transistor Q_1 serves as a constant current source that 'injects' current into node X. The direction in which the current flows after entering node X depends on the input level.

A LOW input is a current sink. When the input is LOW, the injected current flows into the input, thus, diverting current from the base of transistor *npn*-transistor Q_2. The *npn*-transistor, Q_2 is, therefore, OFF and the output is HIGH. If the input is HIGH, the injected current flows into the base of *npn*-transistor Q_2 turning it ON and making the output LOW as shown in Figure 14.29(b). Table 14.9 summarizes the operation of IIL inverter.

Figure 14.30 shows an actual I²L inverter. The output transistor has two collectors (sometimes three), making it equivalent to two transistors with parallel bases and emitters. Thus, it produces two equal outputs. Instead of a collector resistor, the outputs are connected directly to the inputs of other I²L gates.

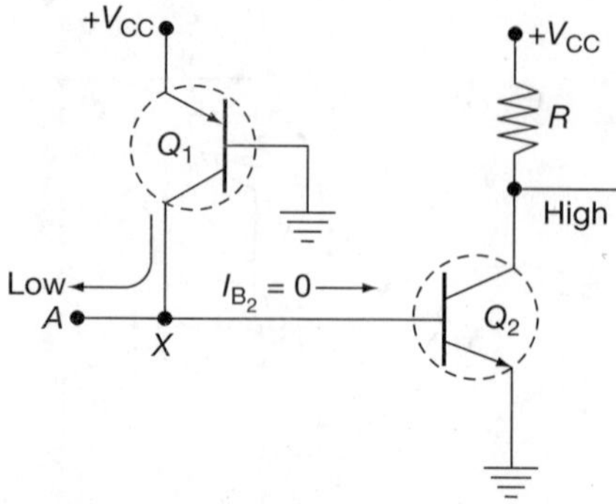

FIGURE 14.29(a) | I²L inverter for Low input

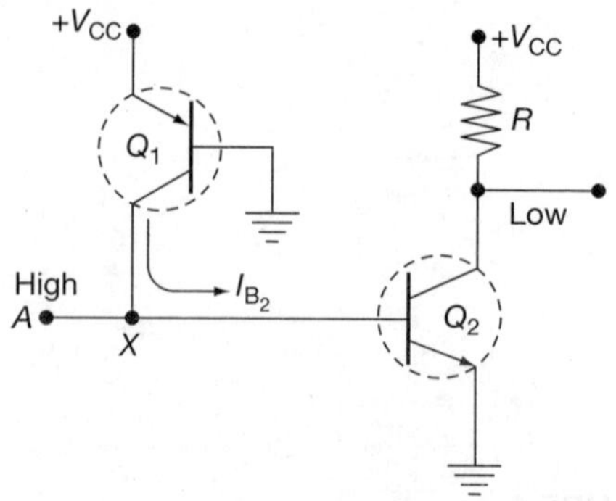

FIGURE 14.29(b) | I²L inverter for High input

TABLE 14.9 | Operation of IIL inverter

Input	*pnp*-Transistor	Base Current	*npn*-Transistor	Output
A	Q_1	I_{B_2}	Q_2	*Y*
Logic 0	ON	$I_{B_2} = 0$	OFF	Logic 1
Logic 1	ON	I_{B_2}	ON	Logic 0

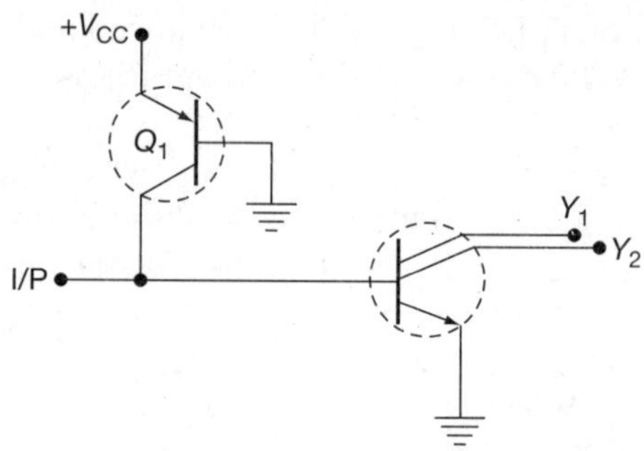

FIGURE 14.30 | I²L inverter

14.10.2 | IIL OR I²L NAND Gate

The I²L NAND gate is shown in Figure 14.31. The circuit uses an inverter with inputs connected directly together at the input of inverter. In case, either input A, or input B, or both the inputs A and B are LOW (current sinks), the injected current flows into the inputs and *npn*-transistor, Q_2 remains OFF (HIGH). If both the inputs are HIGH, the injected current turns on npn-transistor, Q_2 making the output LOW.

Thus, NAND operation is performed. The transistor *pnp*-transistor Q_1 is called a current injector transistor, because when its emitter is connected to an external power source, it can supply current to the base of *npn*-transistor, Q_2. Table 14.10 summarizes the operation of IIL NAND gate.

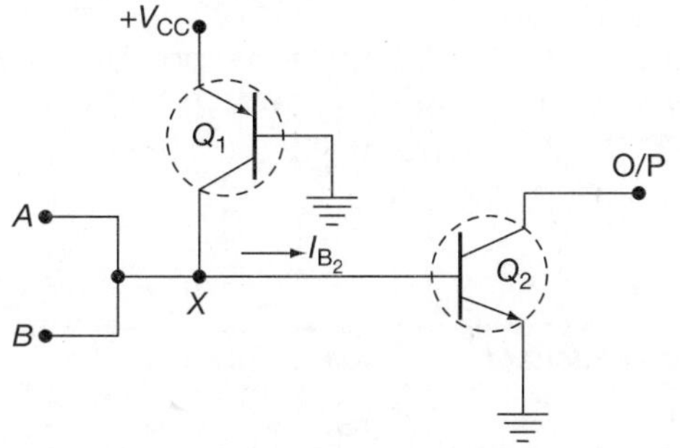

FIGURE 14.31 | I²L NAND gate

TABLE 14.10 | Operation of IIL NAND gate

Input		*pnp*-Transistor	Base Current	*npn*-Transistor	Output
A	B	Q_1	I_{B_2}	Q_2	Y
Logic 0	Logic 0	ON	$I_{B_2} = 0$	OFF	Logic 1
Logic 0	Logic 1	ON	$I_{B_2} = 0$	OFF	Logic 1
Logic 1	Logic 0	ON	$I_{B_2} = 0$	OFF	Logic 1
Logic 1	Logic 1	ON	I_{B_2}	ON	Logic 0

14.10.3 | IIL OR I²L NOR Gate

The I^2L NOR gate is shown in Figure 14.32. The circuit uses two inverters with their outputs connected together. Inverter with input A, consists of *pnp*-transistor Q_{A_1} and

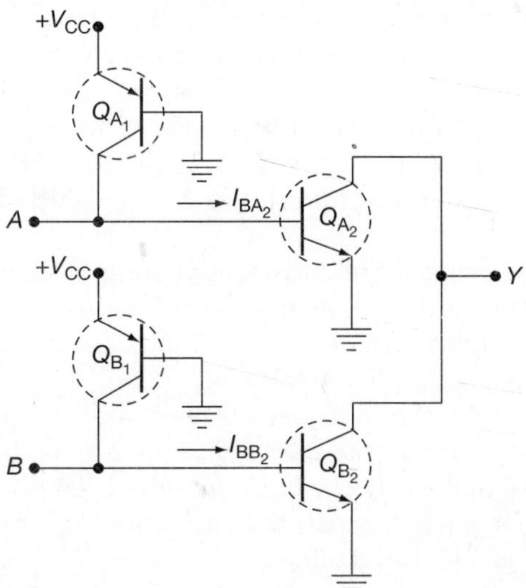

FIGURE 14.32 | I²L NOR gate

npn-transistor Q_{A_2}. Inverter with input B, consists of *pnp*-transistor Q_{B_1} and *npn*-transistor Q_{B_2}. If either or both the inputs are HIGH, the corresponding output transistor is ON and the output is a current sink. So, the output is LOW. If both the inputs are LOW, both the output transistors are OFF. So, the output is HIGH. This is a NOR operation. Table 14.11 summarizes the operation of IIL NOR gate.

TABLE 14.11 | Operation of IIL NOR gate

Input		*pnp*-Transistor		Base Currents		*npn*-Transistor		Output
A	*B*	Q_{A_1}	Q_{B_1}	I_{BA_2}	I_{BB_2}	Q_{A_2}	Q_{B_2}	*Y*
Logic 0	Logic 0	ON	ON	$I_{BA_2} = 0$	$I_{BB_2} = 0$	OFF	OFF	Logic 1
Logic 0	Logic 1	ON	ON	$I_{BA_2} = 0$	I_{BB_2}	OFF	ON	Logic 0
Logic 1	Logic 0	ON	ON	I_{BA_2}	$I_{BB_2} = 0$	ON	OFF	Logic 0
Logic 1	Logic 1	ON	ON	I_{BA_2}	I_{BB_2}	ON	ON	Logic 0

14.11 | EMITTER-COUPLED LOGIC

Emitter-coupled logic (ECL) is another bipolar logic family. ECL prevents transistor saturation, thereby increasing overall switching speed. The ECL is formed when BJTs are coupled at their emitters. It is also called current-mode logic or current-steering logic. Its fast speed of operation is due to following reasons:

- It is a non-saturated logic, because the transistors are not allowed to go into saturation. It eliminates storage time delays and increases speed of operation.
- To charge and discharge stray capacitances quickly, currents are kept high, and the output impedance is so low.
- It has limited voltage swing.

The transistors are avoided saturation when the input changes from LOW to HIGH. To achieve this, logic levels are selected very close to each other. On having logic levels close to each other, it is difficult to achieve good noise immunity. As the transistors are not-saturated, the power consumption is increased. The advantage of ECL is that the current drawn from the supply is steady. ECL gates do not experience large switching transients. As compared to other families, the ECL family has considerably greater power consumption.

The ECL operates on the principle of current switching. A fixed bias current is switched from one transistor's collector to another transistor. The fixed bias current is less than I_C saturated. Due to the current mode operation, ECL logic form is also called as *current mode logic* (CML). It is also known as *current-steering logic* (CSL), because current is steered from one device to another. The ECL family is used in very high-frequency applications where its speed is superior but is not so popular and in wide use as the TTL and MOS families. It has some drawbacks like high cost, low noise margin, high power dissipation and problem of cooling. Moreover, it is not compatible with other logic families due to its negative supply voltage and logic levels. Interfacing of ECL with TTL and MOS circuits becomes difficult.

Despite some drawbacks, the ECL is used in superfast computers and high-speed special purpose applications. The ECL gates can be wired ORed, no noise spikes are generated, and complementary outputs are also available. The important characteristics of ECL gates are:

- Transistors never saturate. So, speed is high with propagation delay, t_{pd} is 360 ps.
- Noise margin is less than near to 250 mV. So, ECL is unreliable for the use in heavy industrial environment.
- Eliminates the use of inverters because ECL circuits produce the output and its complement.
- Fan-out is large because the output impedance is low. Typically, it is almost 25.
- Power dissipation per gate is large, P_D is 40 mW.
- The total current that flows in ECL is more or less constant. So, no noise spikes are generated internally.

14.11.1 | Basic ECL Circuit

The basic circuit for emitter-coupled logic is a differential amplifier configuration as shown in Figure 14.33. The V_{EE} supply produces a fixed current I_E, which remains around 3 mA during normal operation. This current is allowed to flow through either transistor Q_1 or transistor Q_2, depending on the voltage level at V_{IN}. In other words, this current switches between collector of Q_1 transistor and collector of Q_2 transistor as V_{IN} switches between its two logic levels of −1.7 V (logical 0 for ECL) and −0.8 V (logical 1 for ECL). Table 14.12

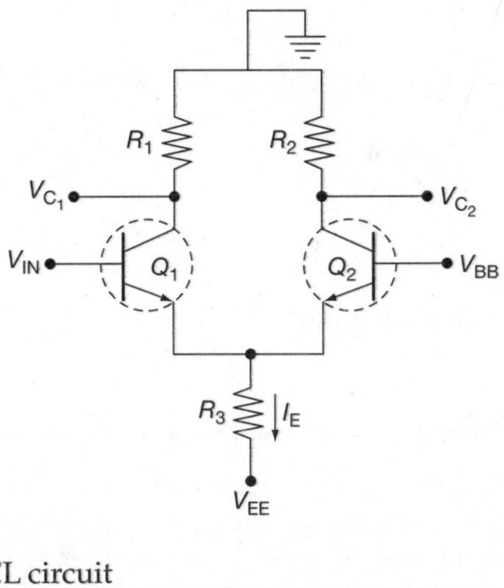

FIGURE 14.33 | Basic ECL circuit

TABLE 14.12 | Operating states of ECL

V_{IN}		Outputs		Remarks
Voltage Level	**Binary Logic**	V_{C_1}	V_{C_1}	
−1.7 V	Logic 0	0 V	−0.9 V	Q_2 conducts
−0.8 V	Logic 1	−0.9 V	0 V	Q_1 conducts

shows the resulting output voltages for these two conditions at V_{IN}. Two important points are noted: (i) V_{C_1} and V_{C_2} are the complements of each other, and (ii) the output voltage levels are not the same as the input logic levels. It is taken care of by connecting V_{C_1} and V_{C_2} to emitter-follower stages (Q_4 and Q_5), as shown in Figure 14.34.

The emitter followers perform two functions:

- Emitter followers subtract approximately 0.8 V from V_{C_1} and V_{C_2} to shift the output levels to the correct ECL logic levels; and
- Emitter followers provide very low output impedance (typically 7 Ω), which provides for large fan-out and fast charging of load capacitance.

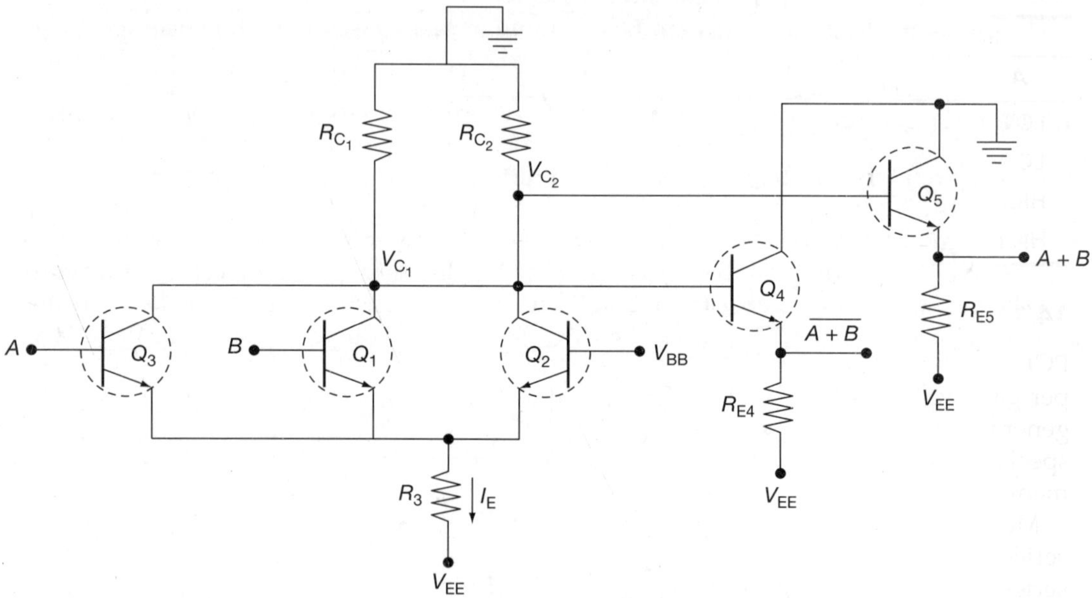

FIGURE 14.34 | ECL NOR and OR gates

14.11.2 | ECL OR/NOR Gate

A two-input ECL OR/NOR gate is shown in Figure 14.34. It has two outputs which are complements of each other. Transistors Q_2 and Q_1 form a differential amplifier. Transistors Q_1 and Q_3 are in parallel. Transistors Q_4 and Q_5 are emitter followers whose emitter voltages are the same as the base voltages (less than 0.8 V base to emitter drops). Inputs are applied to transistors Q_1 and Q_3, and transistor Q_2 is supplied with constant –1.3 V.

When inputs *A* and *B* are LOW (–1.7 V), transistor Q_2 is more forward-biased than transistors Q_3 and Q_1 and transistor Q_2 is ON. Transistors Q_3 and Q_1 are OFF. The value of resistor R_{C_2} is such that current flowing through transistor Q_2 puts the collector at about –0.9 V. Therefore, the emitter of transistor Q_5 is at, –0.9 V – 0.8 V = –1.7 V, and the OR output is LOW. The base current of transistor, Q_4 passing through resistor, R_{C_1} is very small. The value of resistor R_{C_1} is such that this current puts the collectors of transistors Q_3 and Q_1 at about –0.1 V. So, the emitter of transistor Q_4 is at, –0.1 V – 0.8 V = –0.9 V, that is, the NOR output is HIGH.

When input A is HIGH, or input B is HIGH, or both inputs A and B are HIGH, the corresponding transistors are ON, because these transistors are more forward-biased than transistor Q_2 and transistor Q_2 is OFF. So, the collectors of transistor Q_3 and transistor Q_1 are at −0.9 V, which makes the NOR output = −0.9 V − 0.8 V = −1.7 V, that is a logic 0. Only the small base current of transistor Q_5 flows through resistor R_{C_2}. So, the collector of transistor, Q_2 is approximately at −0.1 V, and therefore, the OR output is, −0.1 V − 0.8 V = −0.9 V, that is a logic 1. So, the circuit shown in Figure 14.34 works as an OR/NOR gate.

Table 14.13 summarizes the operation of ECL OR/NOR gate.

TABLE 14.13 | Operation of ECL OR/NOR gate

Inputs		Transistors			Transistors		Output	
A	B	Q_3	Q_1	Q_2	Q_4	Q_5	$A+B$	$\overline{A+B}$
LOW	LOW	OFF	OFF	ON	ON	OFF	LOW	HIGH
LOW	HIGH	OFF	ON	OFF	OFF	ON	HIGH	LOW
HIGH	LOW	ON	OFF	OFF	OFF	ON	HIGH	LOW
HIGH	HIGH	ON	ON	OFF	OFF	ON	HIGH	LOW

14.11.3 | ECL Subfamilies

ECL subfamilies differ in characteristics such as propagation delay, power dissipation per gate, and speed–power product. The ECL subfamilies do not include wide range of general purpose logic gates as is included by TTL and CMOS families. ECL has complex special purpose circuits used in high-speed digital data transmission, arithmetic units, and memories.

Motorola is the first marketed MECL-I series, followed by MECL-II series. Both these series are now obsolete. The MECL-III series carrying MC1600 numbers, the MECL10K series carrying MC10000 numbers, and the recent MECL10KH series with MCI0H000 numbers are in use presently. The MECLI0KH series has propagation delay, t_{pd} of 1 ns and power dissipation, P_D of 25 mW, giving a speed–power product of 25 pJ (least of all ECL series).

14.11.4 | Wired OR Connections

The ECL gates are available with open-emitter outputs, that is, with resistors in the output emitter followers omitted. The open-emitter outputs can be connected together directly, and the common emitter output terminal may be connected through an external resistor to a negative supply voltage (−5.2 V) as shown in Figure 14.35 to perform a wired OR operation. The transistors labelled Q_1 and Q_2 are the output transistors of gates 1 and 2.

When the bases of both the transistors are at −0.9V, both the transistors conduct and make the common emitter voltage to be, −0.9 V − 0.8 V = −1.7 V. When both the bases are at −0.1 V, again both the transistors conduct and make the output voltage to be, −0.1 V − 0.8 V = −0.9 V. When only one base is at −0.1 V and the other at −0.9 V, the output transistor with −0.1 V base voltage conducts and makes the common emitter voltage −0.9 V, preventing the second transistor from conducting. Hence, the circuit provides OR operation.

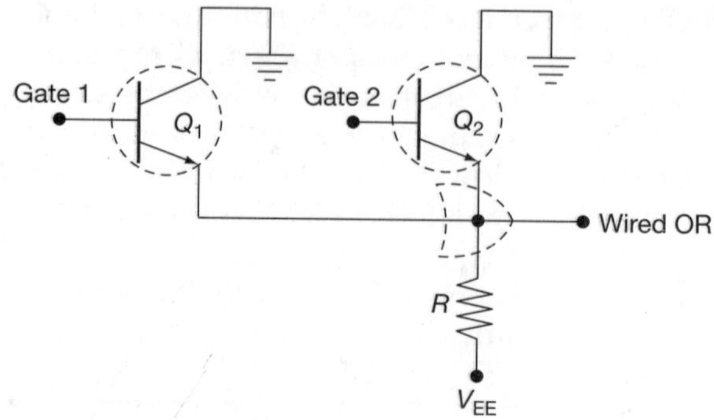

FIGURE 14.35 | Wired OR operation of ECL gates

14.11.5 | Interfacing ECL Gates

Since ECL logic levels are so different from those of TTL and CMOS circuits, interfacing ECL gates with other logic families require special level shifter circuits called *level translators.* Level translators are available in various ECL series to facilitate interfacing of ECL with other families.

14.12 | MOS LOGIC

In comparison to bipolar logic families, the MOS families are simpler and inexpensive to fabricate, require much less power, have a better noise margin, a greater supply voltage range, a higher fan-out, and require much less chip area. The MOS families use *metal oxide semiconductor field effect transistors* (MOSFETs) as active switching element. But these are slower in operating speed and are susceptible to static charge damage. For MOS logic, propagation delay t_{pd} is 50 ns, noise margin NM is 1.5 V (for +5 V supply), power dissipation P_D is 0.1 mW, and fan-out is 50 for frequencies greater than 100 Hz and it is virtually unlimited for dc or low frequencies. The propagation delay associated with MOS gates is large (50 ns) because of their high output resistance (100 kΩ) and capacitive loading presented by the driven gates.

The MOS logic occupies very small space, because it requires only one basic element-an *n-channel* metal oxide semiconductor (NMOS) or a PMOS transistor. It does not require other elements like resistors and diodes, which occupy large space. Because of its ease of fabrication and lower power dissipation per gate, P_D, it is ideally suited for LSI, VLSI, and ULSI for dedicated applications such as large memories, calculator chips, large microprocessors, etc. The operating speed of MOS is slower than that of TTL, so, these are hardly used in SSI and MSI applications. The greater packing density of MOS ICs results in higher reliability because of the reduction in the number of external connections.

Because of the very high impedance present at a MOSFET's input, the MOS logic families are more susceptible to static charge damage. The CMOS family is less susceptible to static charge damage.

MOSFETs are of two types-*depletion type* and *enhancement type.* The MOS digital ICs use enhancement MOSFETs exclusively. The MOSFETs are of *n-channel* MOSFET (NMOS) type or PMOS type. MOSFET circuitry is constructed using NMOS devices, because these operate at about three times the speed of their PMOS counterparts. NMOS has twice the packing density of PMOS. NMOS and PMOS have greater packing density than that of CMOS, and are more economical than CMOS. The CMOS family has the greatest complexity and the lowest packing density of all the MOS families, but it possesses the important advantages of higher speed and much lower power dissipation. The CMOS can be operated at high voltage resulting in improved noise margin.

14.12.1 | Symbols and Switching Action of MOS

Figure 14.36(a) shows the circuit symbol of *n-channel* MOSFET (NMOS). The three terminals of MOSFET are ***gate, drain*** and ***source***. The arrow in the symbols of MOSFETs indicates *n channel*. In the channel, the broken line between the source and the drain indicates that normally there is no conducting channel between these electrodes. The separation between the gate and the other terminals indicates the existence of very high resistance (10^4 MΩ or 10^{10} Ω) between the gate and the channel. The switch in a MOSFET is between the drain and source terminals. The gate to source voltage V_{GS} controls the switch. In an *n-channel* MOSFET, switch closes and current flows from drain to source when V_{GS} is positive, and switch opens when V_{GS} is negative or zero with respect to the source.

When the gate voltage of an *n-channel* MOSFET is more positive than the source, the NMOS is ON (saturation). Figure 14.36(b) shows its equivalent as a closed switch when it is ON and offers resistance of 1 kΩ. When the gate to source voltage is zero, NMOS is OFF (cut-off). Figure 14.36(c) shows its equivalent as an open switch when it is OFF.

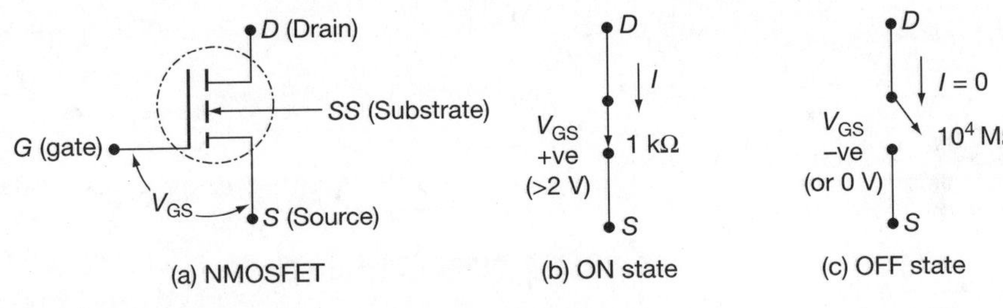

FIGURE 14.36 | Symbol and operation of NMOSFET

Figure 14.37(a) shows the circuit symbol of *p-channel* MOSFET (PMOS). The three terminals of MOSFET are ***gate, drain*** and ***source***. The arrow in the symbols of MOSFETs indicates *p channel*. In the channel, the broken line between the source and the drain indicates that normally there is no conducting channel between these electrodes. The separation between the gate and the other terminals indicates the existence of very high resistance (10^4 MΩ) between the gate and the channel. The switch in a MOSFET is between the source and drain terminals. The gate to source voltage V_{GS} controls the switch. In an *p-channel* MOSFET, switch closes and current flows from source to drain when V_{GS} is negative, and switch opens when V_{GS} is positive or zero with respect to the source.

When the gate voltage of a *p-channel* MOSFET is more negative than the source, the PMOSFET is ON (saturation). Figure 14.37(b) shows its equivalent as a closed switch when it is ON. When the gate to source voltage is zero, PMOSFET is OFF (cut-off). Figure 14.37(c) shows its equivalent as an open switch when it is OFF. MOSFET is also called MOS transistor.

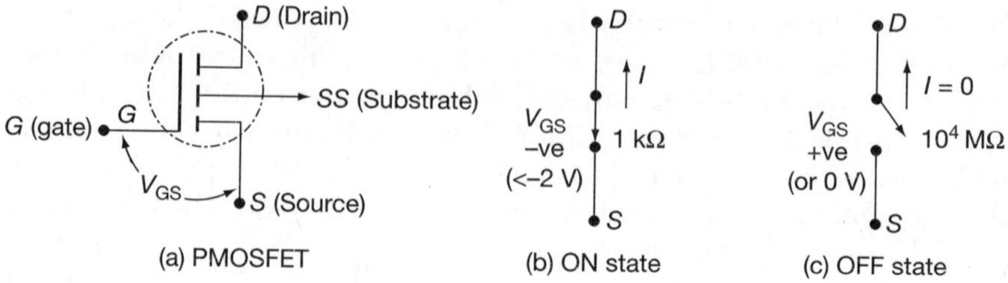

FIGURE 14.37 | Symbol and operation of PMOSFET

14.12.2 | Resistor

A MOS transistor is utilized as an active resistor as shown in Figure 14.38. The value of the resistance presented by a resistor connected NMOS device depends on the current through it. The gate is permanently connected to +5 V, and so, it is always in the ON state and the transistor acts as a resistor of value R_{ON} The load resistor is designed to have a narrower channel, so, its R_{ON} is much greater than the R_{ON} of the switching transistor. Typically, resistance value of R_{ON} is 100 kΩ.

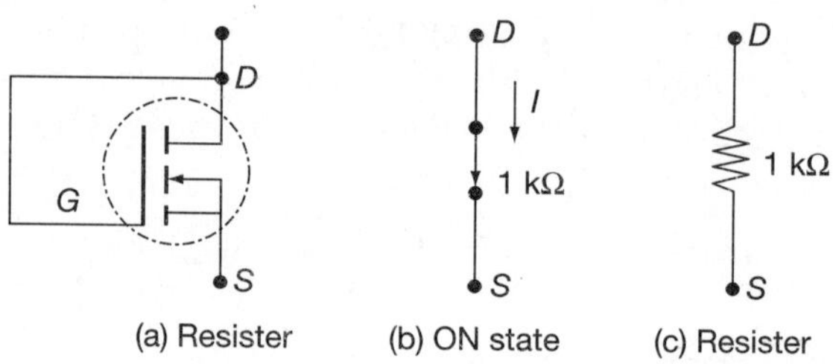

FIGURE 14.38 | NMOSFET as resistor

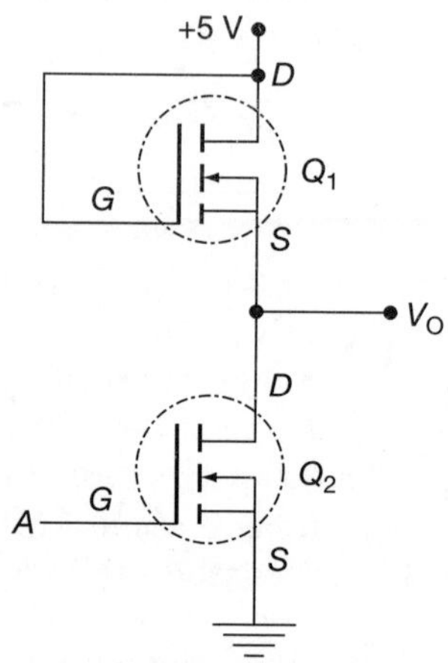

FIGURE 14.39 | NMOSFET as Inverter

14.12.3 | NMOS Inverter

The basic n-channel metal oxide semiconductor (NMOS) transistor is used as inverter is shown in Figure 14.39 that contains two n-channel MOSFETs. Transistor Q_1 is called the load MOSFET or load resistor and transistor Q_2 the switching MOSFET. Transistor Q_2 switches from ON to OFF in response to A (+5 V). These two MOSFETs are considered as resistors and the circuit as a potential divider.

- When A is LOW (0 V), Transistor Q_2 is OFF (Figure 14.40). So, its resistance $R_2 = 10^{10}\ \Omega$ when transistor Q_2 is OFF but transistor Q_1 is ON and acts as register with $R_1 = 100\ \text{k}\Omega$. Therefore, output voltage is

$$V_o = \frac{V_{DD}R_2}{R_1 + R_2} \approx \frac{5\times 10^{10}}{100\times 10^3 + 10^{10}} \approx 5\ \text{V}$$

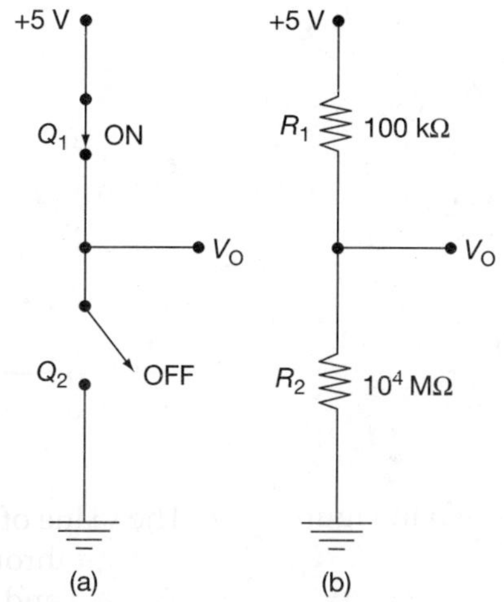

FIGURE 14.40 | Equivalent circuits when $A = 0$

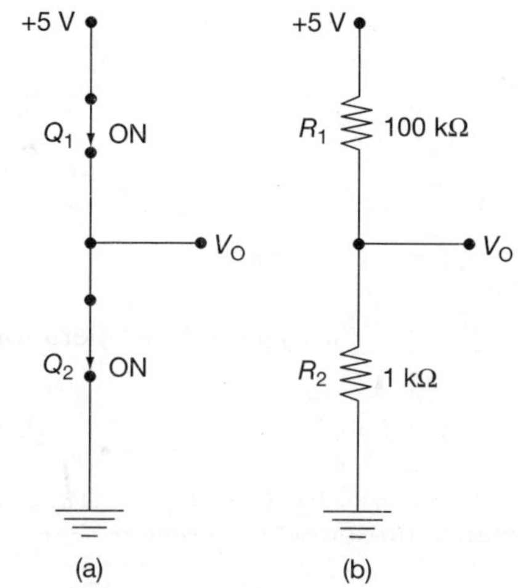

FIGURE 14.41 | Equivalent circuits when $A = 1$

- When A is HIGH (5V). Q_2 is ON (Figure 14.41). So, it offers resistance $R_2 = 1\,\text{k}\Omega$, when transistor Q_2 is ON and $R_1 = 100\,\text{k}\Omega$ when transistor Q_1 is ON and it acts as register. Therefore, output voltage is

$$V_o = \frac{V_{DD}R_2}{R_1 + R_2} \approx \frac{5\times10^3}{100\times10^3 + 1\times10^3} \approx 0\text{ V}$$

This shows that the above circuit acts as an inverter. The operation and the truth table are given in Tables 14.14(a) and 14.14(b), respectively.

TABLE 14.14(a) | Operation of NMOS inverter

Input, A	Transistor, Q_2	Output, V_o
LOW (0 V)	OFF	+5 V (HIGH)
HIGH (5 V)	ON	0 V (LOW)

TABLE 14.14(b) | Truth table

A	V_o
0	1
1	0

14.12.4 | NMOS NAND Gate

Figure 14.42 shows an *n-channel* metal oxide semiconductor (NMOS) two-input NAND gate. Transistor Q_1 is acting as a load resistor and transistors Q_2 and Q_3 act as switches controlled by input levels A and B, respectively. Equivalent circuits of NMOS 2-input NAND gate for different possible combinations of inputs in terms of resistance values of transistors in ON and OFF positions are shown in Figure 14.43.

Table 14.15 summarizes the operation of 2-input NMOS NAND gate.

- When both A and B are LOW (0 V), both transistors Q_2 and Q_3 act as OFF. The equivalent circuit shown in Figure 14.43(a) results with X as HIGH (logic 1) that is +5 V.

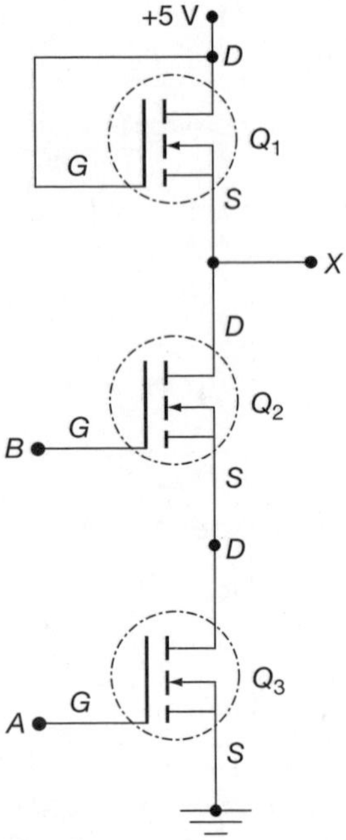

FIGURE 14.42 | NAND gate using NMOSFET

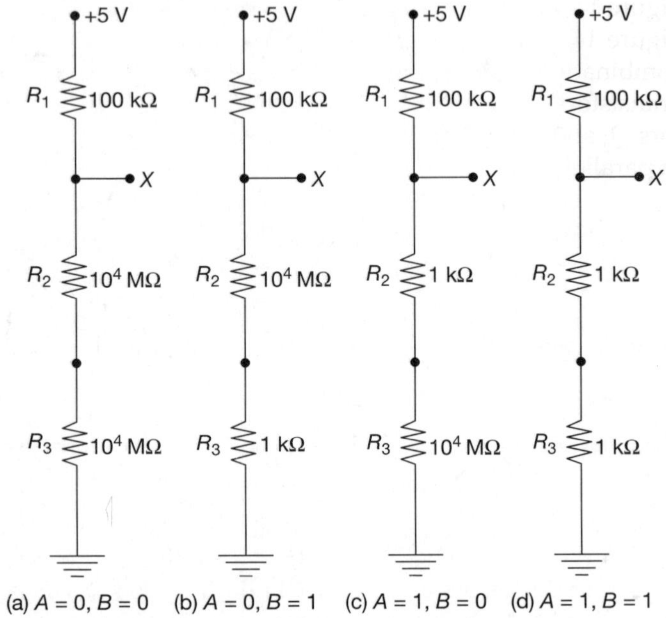

FIGURE 14.43 | Equivalent circuit for various inputs

TABLE 14.15 | Operation of 2-input NMOS NAND gate

Inputs		Transistors		Output
A	*B*	Q_2	Q_3	*X*
LOW	LOW	OFF	OFF	+5 V (HIGH)
LOW	HIGH	OFF	ON	+5 V (HIGH)
HIGH	LOW	ON	OFF	+5 V (HIGH)
HIGH	HIGH	ON	ON	0 V (LOW)

- When *A* is LOW (0 V) and *B* is HIGH (+5 V), transistor Q_2 cuts OFF and transistor Q_3 turns ON. The equivalent circuit shown in Figure 14.43(b) results with *X* as HIGH (logic 1) that is +5 V.
- When A is HIGH (+5 V) and B is LOW (0 V), transistor Q_2 turns ON and transistor Q_3 turns OFF. The equivalent circuit shown in Figure 14.43c results with X as HIGH (logic 1) that is +5 V.
- When A is HIGH (+5 V) and B is HIGH (+5 V), both the transistors Q_2 and Q_3 turn ON. So, the equivalent circuit shown in Figure 14.43d results with X as LOW (logic 0) that is 0 V.

Thus, the above circuit works as a positive logic two-input NAND gate.

14.12.5 | NMOS NOR Gate

Figure 14.44 shows an *n-channel* metal oxide semiconductor (NMOS) two-input NOR gate. Figure 14.45 shows equivalent circuits of NMOS 2-input NOR gate for different possible combinations of inputs in terms of resistance values of transistors in ON and OFF positions. Transistors Q_1 is the resistor-connected NMOS transistor that serves as a load and transistors Q_2 and Q_3 are the switching transistors controlled by the inputs *A* and *B*, respectively in parallel.

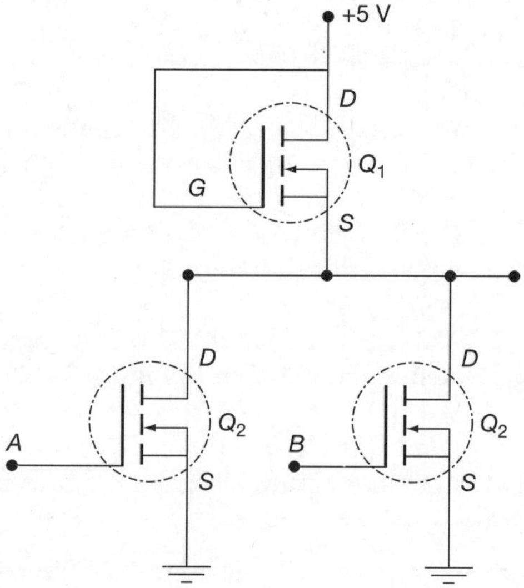

FIGURE 14.44 | NOR gate using NMOSFET

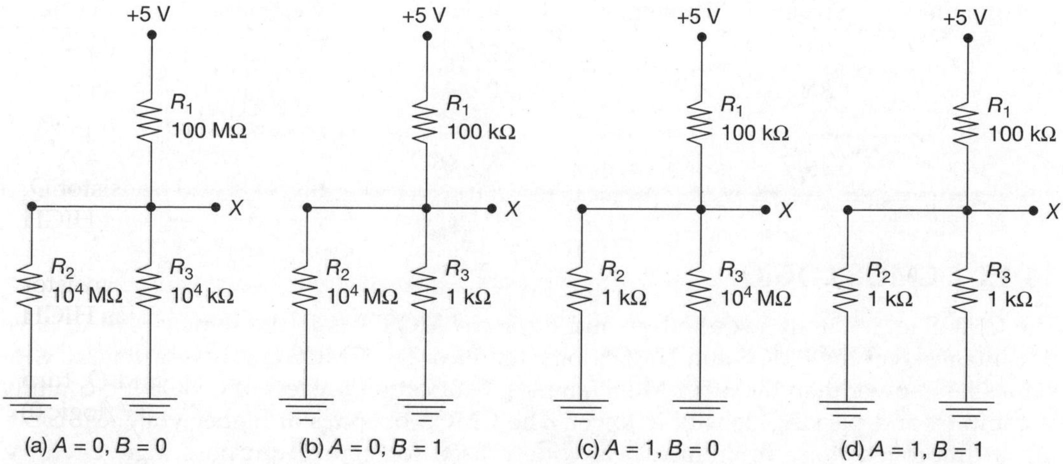

FIGURE 14.45 | Equivalent circuit for various inputs

The operation of 2-input NMOS NOR gate is shown in Table 14.16.

TABLE 14.16 | Operation of 2-input NMOS NOR gate

Inputs		Transistors		Output
A	*B*	Q_2	Q_3	*X*
LOW	LOW	OFF	OFF	+5 V (HIGH)
LOW	HIGH	OFF	ON	0 V (LOW)
HIGH	LOW	ON	OFF	0 V (LOW)
HIGH	HIGH	ON	ON	0 V (LOW)

- When A is LOW (0 V) and B is LOW (0 V), Transistors Q_2 and Q_3 are OFF. So, the equivalent circuit shown in Figure 14.45(a) results with *X* as HIGH (logic 1) that is +5 V.
- When A is LOW (0 V) and B is HIGH (+5 V), Transistors Q_2 is OFF and transistors Q_3 is ON. So, the equivalent circuit shown in Figure 14.45(b) results with *X* as LOW (logic 0) that is 0 V.
- When A is HIGH (+5 V) and B is LOW (0 V), Transistors Q_2 is ON and transistors Q_3 is OFF. So, the equivalent circuit shown in Figure 14.45(c) results with *X* as LOW (logic 0) that is 0 V.
- When *A* is HIGH (+5 V) and *B* is HIGH (+5 V), Transistors Q_2 is ON and transistors Q_3 turns ON. So, the equivalent circuit shown in Figure 14.45(d) results with *X* as LOW (logic 0) that is 0 V.

Thus, the above circuit works as a positive logic 2-input NOR gate.

Characteristics of NMOS are summarized in Table 14.17.

TABLE 14.17 | Characteristics of NMOS

Parameter	Value	Parameter	Value	Parameter	Value
V_{IH}	2.0 V	I_{OH}	–400 μA	t_{PHL}	60 ns
V_{IL}	1.8 V	I_{OL}	2 mA	t_{PLH}	45 ns
V_{OH}	2.4 V	Fan-out	30	Power dissipation	0.1 mW
V_{OL}	0.45 V	Noise margin	1.5 V		

14.13 | CMOS LOGIC

The CMOS logic family uses both *p*- and *n-channel* MOSFETs in the same circuit to have advantages over the PMOS and NMOS logic families. The CMOS family is faster and consumes less power than the other MOS families. Fabrication process of CMOS logic family is complex and packing density is lower. The CMOS operates at higher voltages resulting in improved noise immunity. It is widely used for general purpose logic circuitry. The CMOS technology has been used to construct SSI, MSI and VLSI design for various applications ranging from general-purpose logic to microprocessors. Extremely small power consumption leads to the application in watches and calculators. But, CMOS cannot compete with MOS in applications requiring the use in LSI. The CMOS has very high input resistance. Thus, it draws almost zero current from the driving gate, and therefore, its fan-out is very high. Its output resistance is small (1 kΩ) compared to that of NMOS (100 kΩ). So, it is faster than NMOS. The speed of CMOS decreases with increase in load. In CMOS, there is always a very high resistance between the V_{DD} terminal and ground, because of the MOSFET in the current path. Hence, its power consumption is very low. The noise margin of CMOS is the same in both the LOW and HIGH states and it is 30 per cent of V_{DD}, indicating that noise margin increases with an increase in power supply voltage. So in noisy environments, CMOS with large V_{DD} is preferred. However, an increase in V_{DD} results in the corresponding increase in P_D. The CMOS loses some of its advantages at high frequencies.

In MSI, the CMOS is also competitive with TTL. The CMOS fabrication process is simpler than that of the TTL and it has greater packing density, thereby permitting more circuitry in a given area and reducing the cost per function. The CMOS uses only a fraction of the power needed even for low-power TTL and is, thus, ideally suited for applications requiring battery power or battery backup power. The CMOS is, however, generally slower than TTL.

14.13.1 | CMOS Inverter

Figure 14.46 shows a CMOS inverter. Figure 14.47 shows the equivalent circuits of CMOS inverter for different inputs. It consists of an NMOS transistor Q_1 and a PMOS transistor Q_2. The input is connected to the gates of both the devices and the output is at the drain of both the devices. The positive supply voltage is connected to the source of the PMOS transistor Q_2, and the source of transistor Q_1 is grounded.

- When A is LOW (0 V). Gate to source voltage V_{GS_2} of transistor Q_2 is −5 V, and gate to source voltage V_{GS_1} of transistor Q_1 is 0 V. So, transistor Q_2 acts as ON and transistor Q_1 acts as OFF. Therefore, the switching circuit shown in Figure 14.47(a) results with V_O as logic HIGH that is 5 V.

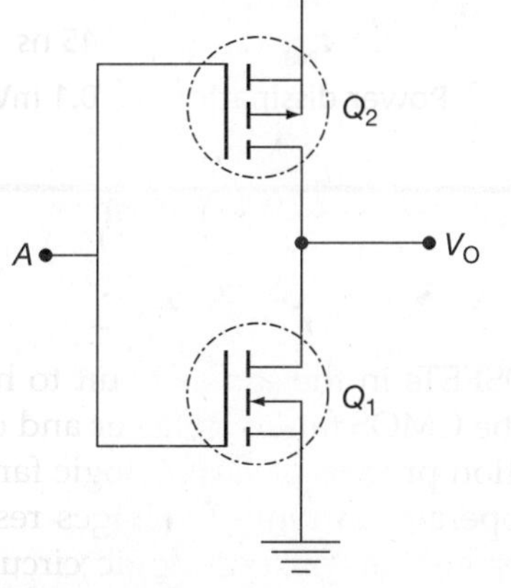

FIGURE 14.46 | CMOS as inverter

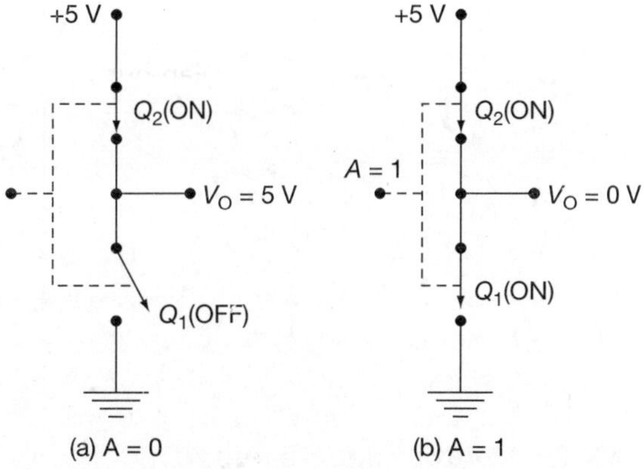

FIGURE 14.47 | Equivalent circuit

- When A is HIGH (+5 V), Gate to source voltage V_{GS_2} of transistor Q_2 is 0 V, and gate to source voltage V_{GS_1} of transistor Q_1 is +5 V. So, transistor Q_2 acts as OFF and transistor Q_1 acts as ON. Therefore, the switching circuit shown in Figure 14.47(b) results with V_O as logic LOW that is 0 V.

Thus, the above circuit acts as an inverter.

Operation of CMOS inverter is given in Table 14.18(a) and truth table of inverter is given in Table 14.18(b).

TABLE 14.18(a) | Operation of CMOS inverter

Input, *A*	***p-channel* MOSFET, Q_1**	***n-Channel* MOSFET, Q_2**	**Output, V_O**
LOW (0 V)	ON	OFF	+5 V (HIGH)
HIGH (5 V)	OFF	ON	0 V (LOW)

TABLE 14.18(b) | Truth table

A	**V_O**
0	1
1	0

14.13.2 | CMOS NAND Gate

Figure 14.48 shows a CMOS two-input NAND gate. Figure 14.49 gives equivalent circuits of 2-input CMOS NAND gate for various input combinations. Here, *p-channel* MOSFETs Q_1 and Q_2 are connected in parallel and *n-channel* MOSFETs Q_3 and Q_4 are connected in series. During the operation of CMOS NAND gate, voltage levels for biasing of *p-channel* and *n-channel* MOSFETs gate and source are given in Table 14.19.

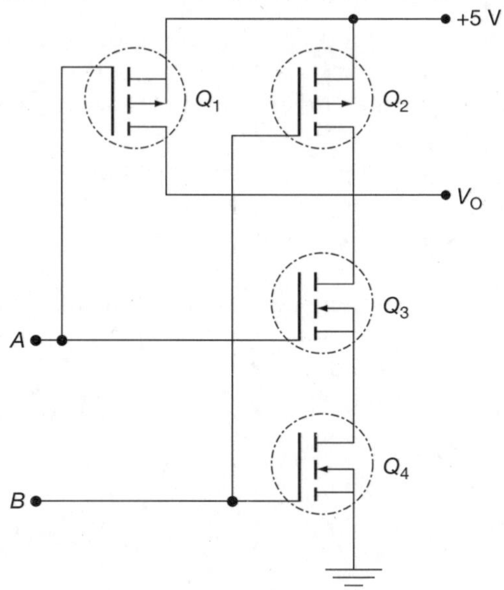

FIGURE 14.48 | CMOS as NAND

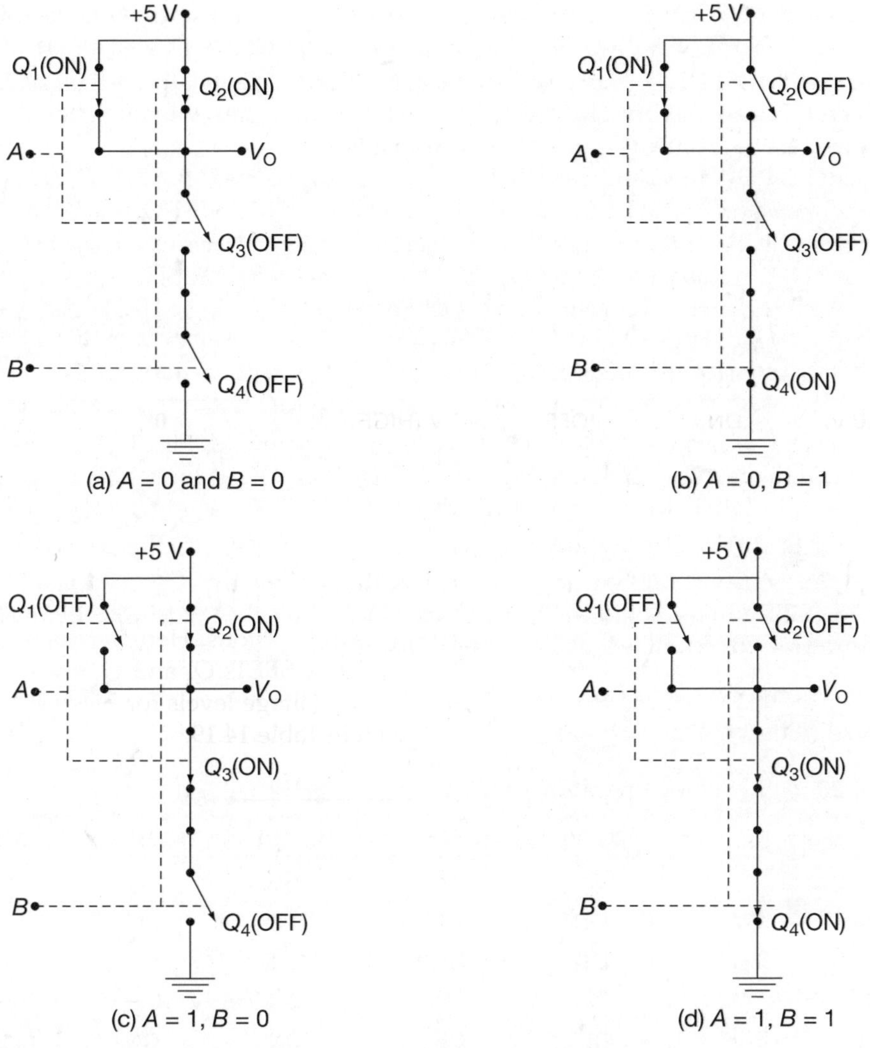

FIGURE 14.49 | Equivalent switching circuit of CMOS NAND gate

TABLE 14.19 | Voltage level during operation of CMOS NAND gate

Inputs		*p-channel* MOSFET		*2-channel* MOSFET		Output
A	*B*	Q_1	Q_2	Q_1	Q_2	*X*
0 V	0 V	–5 V	–5 V	0 V	0 V	+5 V
0 V	+5 V	–5 V	0 V	0 V	+5 V	+5 V
+5 V	0 V	0 V	–5 V	+5 V	0 V	+5 V
+5 V	+5 V	0 V	0 V	+5 V	+5 V	0 V

- When A is LOW (0 V) and B is also LOW (0V). V_{GS_1} and V_{GS_2} of *p-channel* MOSFETs Q_1 and Q_2 are at –5 V. V_{GS_3} and V_{GS_4} of *p-channel* MOSFETs Q_3 and Q_4 are at 0 V. So, *p-channel* MOSFET Q_1 acts ON, *n-channel* MOSFET Q_3 acts OFF, *p-channel* MOSFET Q_2 acts ON and *n-channel* MOSFET Q_4 acts OFF. Thus, the switching circuit shown in Figure 14.49(a) results V_o as logic HIGH that is +5 V.
- When A is LOW (0 V) and B is HIGH (5 V). V_{GS_1} of *p-channel* MOSFET Q_1 is at –5 V and V_{GS_2} of *p-channel* MOSFET Q_2 is at 0 V. V_{GS_3} of *n-channel* MOSFET Q_3 is at 0 V and V_{GS_4} of *n-channel* MOSFET Q_4 are at +5 V. So, *p-channel* MOSFET Q_1 acts ON, *n-channel* MOSFET Q_3 acts OFF, *p-channel* MOSFET Q_2 acts OFF and *n-channel* MOSFET Q_4 acts ON. Thus, the switching circuit shown in Figure 14.49(b) results V_o as logic HIGH that is +5 V.
- When A is HIGH (5 V) and B is LOW (0 V). V_{GS_1} of *p-channel* MOSFET Q_1 is at 0 V and V_{GS_2} of *p-channel* MOSFET Q_2 is at –5 V. V_{GS_3} of *n-channel* MOSFET Q_3 is at +5 V and V_{GS_4} of NMOSFET Q_4 are at 0 V. So, *p-channel* MOSFET Q_1 acts OFF, NMOSFET Q_3 acts ON, *p-channel* MOSFET Q_2 acts ON and *n-channel* MOSFET Q_4 acts OFF. Thus, the switching circuit shown in Figure 14.49(c) results V_o as logic HIGH that is +5 V.
- When A is HIGH (+5 V) and B is also HIGH (+5 V). V_{GS_1} and V_{GS_2} of *p-channel* MOSFETs Q_1 and Q_2 are at 0 V. V_{GS_3} and V_{GS_4} of *n-channel* MOSFETs Q_3 and Q_4 are at +5 V. So, *p-channel* MOSFET Q_1 acts OFF, NMOSFET Q_3 acts ON, *p-channel* MOSFET Q_2 acts OFF and *n-channel* MOSFET Q_4 acts ON. Thus, the switching circuit shown in Figure 14.49(d) results V_o as logic LOW that is 0V.

Switching actions of *p-channel* and *n-channel* MOSFETs used for 2-input CMOS NAND gate are given in Table 14.20.

TABLE 14.20 | Switching operation of CMOS NAND gate

Inputs		*p-channel* MOSFET		*2-channel* MOSFET		Output
A	*B*	Q_1	Q_2	Q_1	Q_2	*X*
LOW	LOW	ON	ON	OFF	OFF	HIGH
LOW	HIGH	ON	OFF	OFF	ON	HIGH
HIGH	LOW	OFF	ON	ON	OFF	HIGH
HIGH	HIGH	OFF	OFF	ON	ON	LOW

14.13.3 | CMOS NOR Gate

Figure 14.50 shows a CMOS two-input NOR gate. The equivalent circuits for various input combinations are shown in Figure 14.51. Here, *p-channel* MOSFETs Q_1 and Q_2 are connected in series and *n-channel* MOSFETs Q_3 and Q_4 are connected in parallel. During the operation of CMOS NOR gate, voltage levels for biasing of *p-channel* and *n-channel* MOSFETs gate and source are given in Table 14.21.

- When A is LOW (0 V) and B is also LOW (0 V). V_{GS_1} and V_{GS_2} of *p-channel* MOSFETs Q_1 and Q_2 are at –5 V. V_{GS_3} and V_{GS_4} of *p-channel* MOSFETs Q_3 and Q_4 are at 0 V. So, *p-channel* MOSFET Q_1 acts ON, *n-channel* MOSFET Q_3 acts OFF, *p-channel* MOSFET Q_2 acts ON and *n-channel* MOSFET Q_4 acts OFF. Thus, the switching circuit shown in Figure 14.51(a) results V_o as logic HIGH that is +5 V.

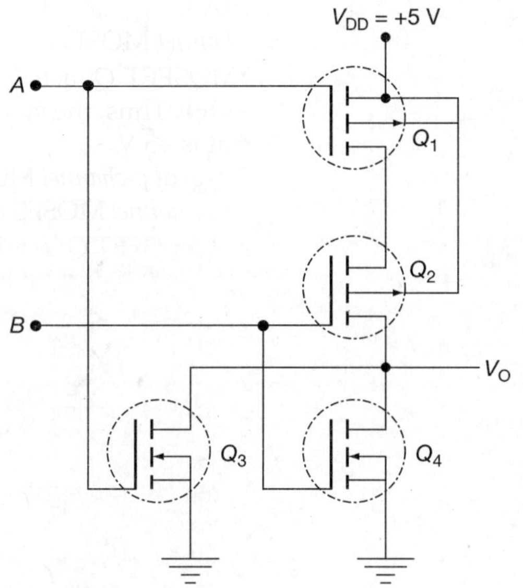

FIGURE 14.50 | CMOS as NOR

- When A is LOW (0 V) and B is HIGH (5 V). V_{GS_1} of *p-channel* MOSFET Q_1 is at −5 V and V_{GS_2} of *p-channel* MOSFET Q_2 is at 0 V. V_{GS_3} of *n-channel* MOSFET Q_3 is at 0 V and V_{GS_4} of *n-channel* MOSFET Q_4 are at +5 V. So, *p-channel* MOSFET Q_1 acts ON, *n-channel* MOSFET Q_3 acts OFF, *p-channel* MOSFET Q_2 acts OFF and *n-channel* MOSFET Q_4 acts ON. Thus, the switching circuit shown in Figure 14.51(b) results V_O as logic HIGH that is +5 V.

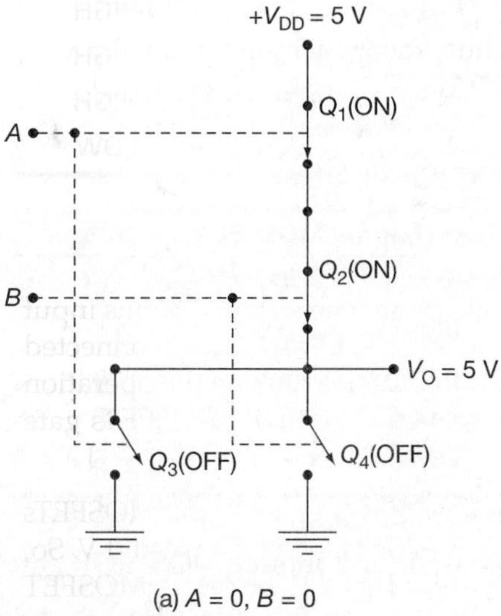

FIGURE 14.51(a) | A = 0, B = 0

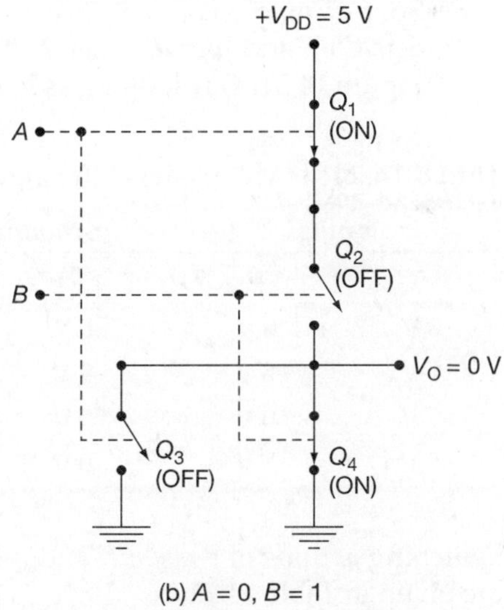

FIGURE 14.51(b) | A = 0, B = 1

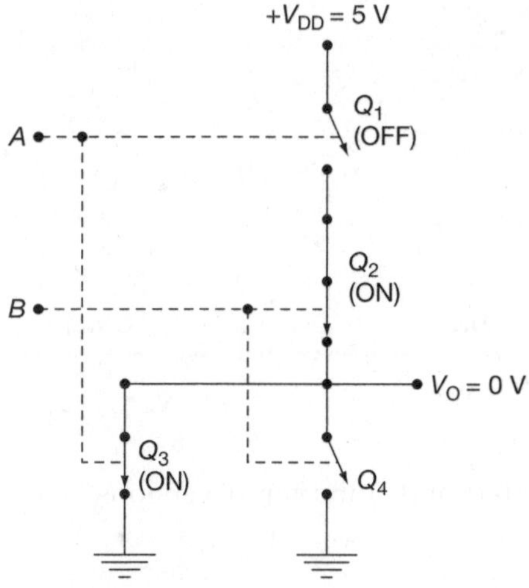

(c) $A = 1, B = 0$

FIGURE 14.51(c) | A = 1, B = 0

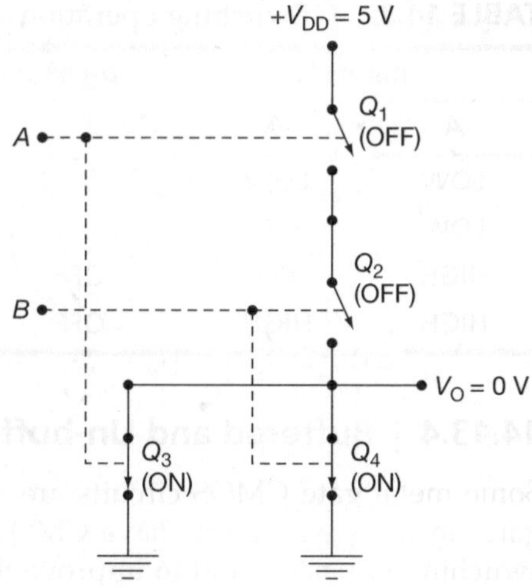

(d) $A = 1, B = 1$

FIGURE 14.51(d) | A = 1, B = 1

- When A is HIGH (5 V) and A is LOW (0 V). V_{GS_1} of *p-channel* MOSFET Q_1 is at 0 V and V_{GS_2} of *p-channel* MOSFET Q_2 is at –5 V. V_{GS_3} of *n-channel* MOSFET Q_3 is at +5 V and V_{GS_4} of NMOSFET Q_4 are at 0 V. So, *p-channel* MOSFET Q_1 acts OFF, NMOSFET Q_3 acts ON, *p-channel* MOSFET Q_2 acts ON and *n-channel* MOSFET Q_4 acts OFF. Thus, the switching circuit shown in Figure 14.51(c) results V_O as logic HIGH that is +5 V.
- When A is HIGH (+5 V) and B is also HIGH (+5 V). V_{GS_1} and V_{GS_2} of *p-channel* MOSFETs Q_1 and Q_2 are at 0 V. V_{GS_3} and V_{GS_4} of *n-channel* MOSFETs Q_3 and Q_4 are at +5 V. So, *p-channel* MOSFET Q_1 acts OFF, NMOSFET Q_3 acts ON, *p-channel* MOSFET Q_2 acts OFF and *n-channel* MOSFET Q_4 acts ON. Thus, the switching circuit shown in Figure 14.51(d) results V_O as logic LOW that is 0 V.

TABLE 14.21 | Voltage level during operation of CMOS NAND gate

Inputs		*p-channel* MOSFET		*Two-channel* MOSFET		Output
A	*B*	Q_1	Q_2	Q_1	Q_2	*X*
0 V	0 V	–5 V	–5 V	0 V	0 V	+5 V
0 V	+5 V	–5 V	0 V	0 V	+5 V	0 V
+5 V	0 V	0 V	–5 V	+5 V	0 V	0 V
+5 V	+5 V	0 V	0 V	+5 V	+5 V	0 V

Switching actions of *p-channel* and *n-channel* MOSFETs used for 2-input CMOS NOR gate are given in Table 14.22.

TABLE 14.22 | Switching operation of CMOS NAND gate

Inputs		*p-channel* MOSFET		*Two-channel* MOSFET		Output
A	*B*	Q_1	Q_2	Q_1	Q_2	*X*
LOW	LOW	ON	ON	OFF	OFF	HIGH
LOW	HIGH	ON	OFF	OFF	ON	LOW
HIGH	LOW	OFF	ON	ON	OFF	LOW
HIGH	HIGH	OFF	OFF	ON	ON	LOW

14.13.4 | Buffered and Un-buffered Gates

Some metal gate CMOS circuits are available in buffered and unbuffered versions. The gates in buffered circuits have CMOS inverters in series with their outputs to suppress switching transients and to improve the sharpness of the voltage transition at the output. The gates discussed above are unbuffered gates.

14.13.5 | Transmission Gate

A transmission gate is simply a digitally controlled CMOS switch. When the switch is open (OFF), the impedance between its terminals is very large. It is used to implement special logic functions. Since the CMOS gate can transmit signals in both directions, it is called a *bilateral transmission gate*. It is also called a bilateral switch. It is useful for digital and analog applications. The TTL and ECL gates are essentially unidirectional.

Figure 14.52 shows the schematic diagram and logic symbols of a CMOS transmission gate. The *n-channel* MOS and *p-channel* MOS transistors are connected in parallel. So, both polarities of input voltages can be switched. The control, C signal is connected to the *n-channel* MOSFET and its inverse is connected to the *p-channel* MOSFET. When the control, C is HIGH, the gate of *p-channel* MOSFET Q_1 is LOW and the gate of *n-channel* MOSFET Q_2 is HIGH. If the input (data) is LOW, V_{GS_1} is 0 V and V_{GS_2} is positive. So, *p-channel* MOSFET Q_1 is OFF and *n-channel* MOSFET is ON. In case the input is HIGH, V_{GS_1} is negative and V_{GS_2} is 0 V. So, Q_1 is ON and Q_2 is OFF. Thus, there is always one conducting path from input to output when the control, C is HIGH.

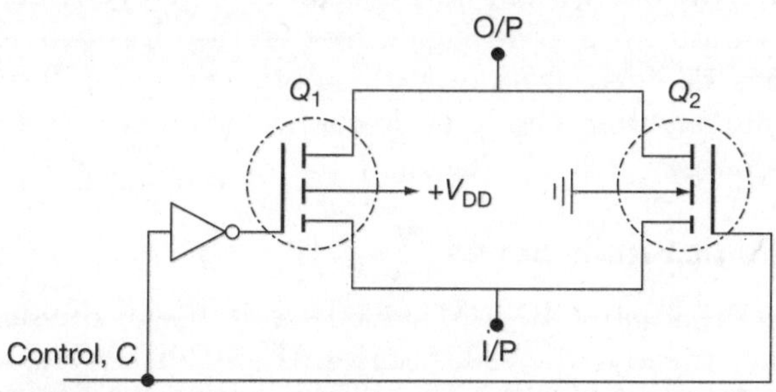

FIGURE 14.52 | CMOS transmission gate

On the other hand, when the control, C is LOW, the gate of *p*-channel MOSFET Q_1 is HIGH and the gate of *n*-channel MOSFET Q_2 is LOW. In case the input (data) is LOW, V_{GS_1} is positive and V_{GS_2} is 0 V. Therefore, Q_1 is OFF and Q_2 is also OFF. If the input (data) is HIGH, V_{GS_1} is 0 V and V_{GS_2} is negative. So, again Q_1 is OFF and Q_2 is also OFF. Thus, there is no conducting path from input to output when the control, C is LOW. Table 14.23 explains the operation of CMOS transmission gate.

TABLE 14.23 | CMOS transmission gate

Inputs		*p-channel* MOSFET		*n-channel* MOSFET		Action
C	I/P	V_{G_1}	Q_1	V_{G_2}	Q_2	
LOW	LOW	+ve	OFF	0 V	OFF	No transmission
LOW	HIGH	0 V	OFF	−ve	OFF	No transmission
HIGH	LOW	0 V	OFF	+ve	ON	Transmission
HIGH	HIGH	−ve	ON	0 V	OFF	Transmission

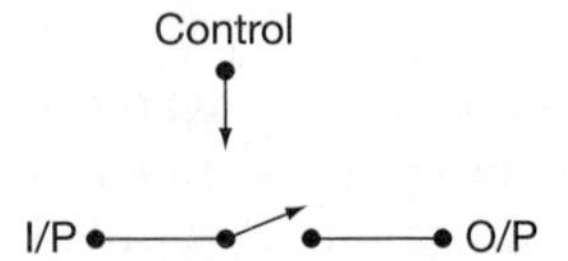

FIGURE 14.53 | Switch

So, it can be concluded that when the control, C is HIGH, the circuit acts as a closed switch and allows the transmission of the signal from input to output (Figure 14.53). When the control, C is LOW, the circuit acts as an open switch and blocks the transmission of the signal from input to output. The control, C acts as an active-HIGH enabling signal. Active-LOW enabling is possible, if the control, C is connected to the gate of *p*-channel MOS and control, $\bar{C}$ to the gate of n-channel MOS.

Since the input and output terminals are interchangeable, the circuit can also transmit signals in the opposite direction. So, it acts as a bilateral switch.

14.13.6 | Open Drain and High Impedance Outputs

The CMOS logic gates are available with open-drain outputs similar to their TTL counter parts with open collector outputs. In these devices, the output stage consists only of an *n-channel* MOSFET whose drain is unconnected, since the upper MOSFET has been eliminated. An external pull-up resistor is needed to produce a HIGH state voltage level. Like open-collector outputs, the open-drain outputs can be wired ANDed. Figure 14.54 shows two inverters at the output of a CMOS gate that are used to provide the buffering. As shown in Figure 14.55, the open-drain output is obtained by omitting the *p-channel* MOS transistor in the output inverter. Diode D_1 is connected internally to provide protection from electrostatic discharge.

14.13.6.1 | CMOS Logic Family Series

CMOS logic family is designated by 4000/14000 series. It was the first CMOS series. The original 4000 series is changed to the 4000A series. The 4000B series is an improved version of the 4000A. The 4000B series has a higher output current capability than that of the 4000A series. The 4000A and 4000B series are still widely used despite the emergence of CMOS series.

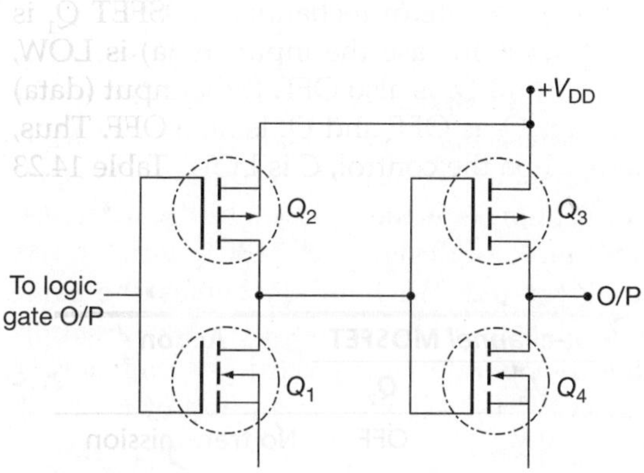

FIGURE 14.54 | Output buffering

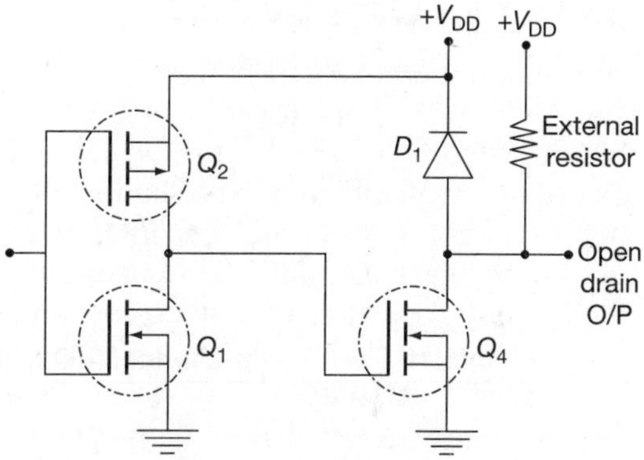

FIGURE 14.55 | Open drain CMOS inverter

14.13.6.2 | 74C Series

74C series is compatible pin for pin and function for function with TTL devices having the same number. Many, but not all, functions that are available in TTL are also available in this CMOS series. This makes it possible to replace some TTL circuits by an equivalent CMOS design. The performance characteristics of the 74C series are about the same as those of the 4000 series.

14.13.6.3 | 74HC Series (High-speed Series)

74HC series is an improved version of the 74C series. High speed (10 times that of the 74C series) and higher output current capabilities are its main features. The speed of the devices in this series is compatible with that of the 74LS TTL series.

14.13.6.4 | 74HCT Series

74HCT series is a high-speed CMOS series and it is designed to be voltage compatible with TTL devices. In other words, it can be directly driven by a TTL output.

14.13.6.5 | 74AC/ACT (Advanced CMOS)

This series is often referred to as ACL for advanced CMOS logic. The series is functionally equivalent to the various TTL series. However, it is not pin compatible with TTL because the pin placements on 74AC or 74ACT chips have been chosen to improve noise immunity 74AC devices are not electrically compatible with TTL; 74ACT devices can be connected directly to TTL. ACL offers advantages over the HC series in noise immunity, propagation delay, and maximum clock speed.

14.13.6.6 | 74LHC/AHCT (Advanced High-speed CMOS)

Advanced high-speed CMOS designated as 74LHC/AHCT series of CMOS devices offers a natural migration path from the HC series to faster, lower-power, low-drive applications. The devices in this series are three times faster. These devices offer similar noise immunity to HC without the overshoot/undershoot problems.

14.13.6.7 | BiCMOS 5-V Logic

The combination of the best features of bipolar and CMOS logic is called BiCMOS logic. The low-power characteristics of CMOS and the high-speed characteristics of bipolar circuits are combined to create an extremely low-power, high-speed logic family. BiCMOS ICs are not available in most SSI and MSI functions, but are limited to functions that are used in microprocessor and bus interfacing applications such as latches, buffers, drivers, and transceivers. The BiCMOS bus-interface technology designated 74BCT series offers 75 per cent reduction in power consumption over the 74F family whereas the same speed and drive characteristics are maintained. Devices in this series are pin-compatible with industry standard TTL devices. These devices operate on standard 5-V logic levels. The 74ABT (advanced BiCMOS technology) series is the second generation of BiCMOS bus-interface devices.

14.14 | CHARACTERISTICS OF CMOS

Electrical characteristic of CMOS are summarized in Table 14.24.

TABLE 14.24 | Characteristics of CMOS

Parameter	Value	Parameter	Value
V_{IH}	3.5 V	t_{PHL}	60 ns
V_{IL}	1.5 V	t_{PLH}	45 ns
V_{OH}	4.5 V	Fan-out	30
V_{OL}	0.5 V	Noise margin	1.5 V
I_{IH}	1 μA	Power dissipation	10 mW
I_{IL}	−1 mA		
I_{OH}	−100 μA		
I_{OL}	360 μA		

Supply voltage: The 4000 and 74C series can operate with V_{DD} values ranging from 3 to 15 V. The 74HC and 74HCT series can operate with V_{DD} values ranging from 2 to 6 V.

Voltage levels: When a CMOS output drives only a CMOS input and CMOS gate has an extremely high input resistance, the current drawn is almost zero and, therefore, the output voltage levels will be very close to zero for LOW state and V_{DD} for HIGH state, i.e. $V_{OL}(\text{max}) = 0$ V, $V_{OH}(\text{min}) = V_{DD}$. Usually, the input voltage levels are expressed as percentage of V_{DD} values, for example, $V_{IL}(\text{max}) = 30$ per cent of V_{DD}, $V_{IH}(\text{min}) = 70$ per cent of V_{DD}.

Power dissipation: When a CMOS circuit is in a static state, its power dissipation per gate is extremely small, but it increases with increase in operating frequency and supply voltage level. For dc, CMOS power dissipation is only 2.5 nW per gate when $V_{DD} = 5$ V, and it increases to 10 nW per gate when $V_{DD} = 10$ volts. With a V_{DD} of 10 V at a frequency of 100 kpps (kilo pulse per seconds), power dissipation is 0.1 mW/gate, and at 1 MHz, $P_D = 1$ mW.

Noise margins: Since for a CMOS gate, $V_{OL}(\text{max}) = 0$ V, $V_{OH}(\text{min}) = V_{DD}$ and $V_{IL}(\text{max})$ is 30 per cent of V_{DD} and $V_{IH}(\text{min})$ is 70 per cent of V_{DD}, the low level and high level noise margins will be the same (30 per cent of V_{DD}) and increase with an increase in the value of V_{DD}. Of course, the higher values of V_{DD} result in higher power dissipations.

$$V_{NH} = V_{OH}(\text{min}) - V_{IH}(\text{min}) = V_{DD} - 70 \text{ per cent of } V_{DD} = 30 \text{ per cent of } V_{DD}$$

$$V_{NL} = V_{IL}(\text{max}) - V_{OL}(\text{max}) = 30 \text{ per cent of } V_{DD} - 0\text{ V} = 30 \text{ per cent of } V_{DD}$$

Fan-out: The CMOS fan-out depends on the permissible maximum propagation delay. For low frequencies (:S; I MHz), the fan-out is 50, and for high frequencies it will be less.

Switching speed: The speed of the CMOS gate increases with increase in V_{DD}. The 4000 series has $t_{pd} = 50$ ns at $V_{DD} = 5$ V and $t_{pd} = 25$ ns at $V_{DD} = 10$ V. The increase in V_{DD} results in increase in power dissipation too.

Unused inputs: The CMOS inputs should never be left disconnected. All CMOS inputs have to be tied either to a fixed voltage level (0 V or V_{DD}) or to another input.

Static charge susceptibility: The high input resistance of CMOS inputs makes CMOS gates prone to static charge build up, that can produce voltages large enough to break down the dielectric insulation between the MOSFET gate and the channel. Most of the newer CMOS devices are protected against static charge damage by the inclusion of protective Zener diodes on each input.

The CMOS and TTL families are compared and the comparison is given in Table 14.25. Typically fan-out of CMOS and TTL is 50 and 10, respectively.

TABLE 14.25 | Comparison of CMOS and TTL

Parameter	CMOS		TTL			
	Silicon Gate CMOS	Metal Gate CMOS	74	74L	74AS	74ALS
V_{IH}	3.5 V	3.5 V	2.0 V	2.0 V	2.0 V	2.0 V
V_{IL}	1.0 V	1.5 V	0.8 V	0.8 V	0.8 V	0.8 V
V_{OH}	4.9 V	4.95 V	2.4 V	2.7 V	2.7 V	2.7 V
V_{OL}	0.1 V	0.05 V	0.4 V	0.5 V	0.5 V	0.4 V
V_{NH}	1.4 V	1.45 V	0.4 V	0.7 V	0.7 V	0.7 V
V_{NL}	0.9 V	1.45 V	0.4 V	0.3 V	0.3 V	0.4 V
t_{PD}	8 ns	105 ns	10 ns	10 ns	1.5 ns	4 ns
PD (per gate)	0.17 mW	0.1 mW	10 mW	2 mW	8.5 mW	1 mW
$t_{pd} \times PD$	1.4 pJ	10.5 pJ	100 pJ	20 pJ	12.8 pJ	4 pJ

14.15 | DYNAMIC MOS LOGIC

When power consumption and physical size are the prime design considerations as in digital watches and calculators, dynamic MOS logic meets these requirements. Each transistor used in a dynamic MOS circuit is identical to the other, and each can be fabricated in a very small amount of space on a chip. Consequently, large and very large-scale integrations are possible.

In dynamic MOS logic, power consumption is minimized by relying on the inherent capacitance of the MOS transistors to store logic levels, i.e. to remain charged or discharged-and by using clock signals to turn on transistors for very brief intervals of time only. The clock signals turn transistors on to allow the capacitance to recharge or discharge at periodic intervals. Since a transistor is OFF during most of any given time interval, the average power consumption is quite small.

The NMOS transmission gate shown in Figure 14.56 is a fundamental component of dynamic logic circuits. Because the NMOSFET is completely symmetrical, the drain and source terminals are in distinguishable, i.e. current can flow in either direction. In dynamic logic applications there is a shunt capacitance at each of these terminals identified in the Figure 14.56 as C_1, and C_2.

- When the gate terminal G is LOW, the NMOS transistor cuts OFF irrespective of the potentials at drain and source, i.e. irrespective of charges on C_1, and C_2, because the gate-to-source voltage may be either 0 or negative. Once the transistor is OFF, it acts as an open switch as shown in Figure 14.57 and the charges on capacitors remain unaltered. That is, no transfer of charge takes place and therefore, no signal transmission takes place.

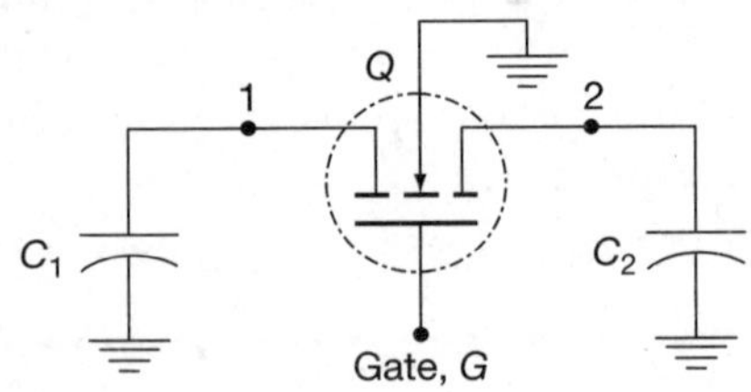

FIGURE 14.56 | NMOS transmission gate

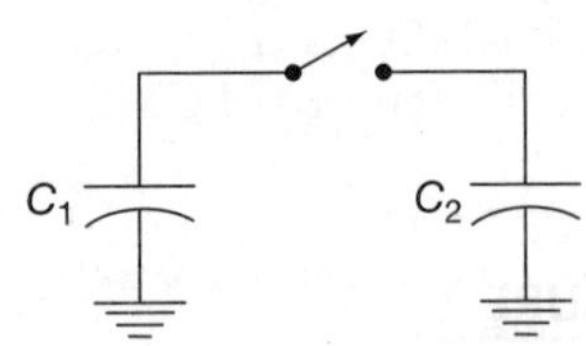

FIGURE 14.57 | Gate is low

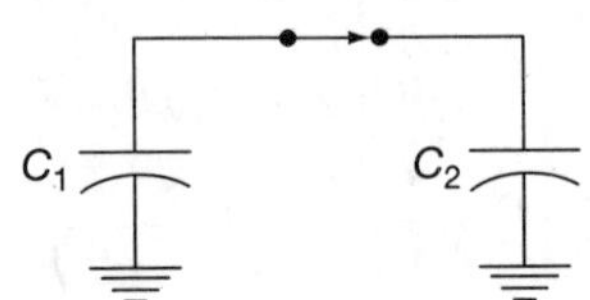

FIGURE 14.58 | Gate is high

- When G is HIGH, the transistor is ON and acts as a closed switch as shown in Figure 14.58. If capacitors C_1 and C_2 are charged to the same level, no transfer of charge takes place. But if one capacitor is charged and the other is discharged, transfer of charge takes place from one charged capacitor to the other discharged capacitor, i.e. the input is transmitted to the output.

Table 14.26 summarizes the operation to transfer the charge from charged capacitor C_1 to C_2 or vice versa.

14.15.1 | Dynamic MOS Inverter

Dynamic inverter is shown in Figure 14.59. The capacitance shown by dotted lines represents the inherent device (inter-electrode) capacitance. The V_{G_1} and V_{G_2} are the control signals that are used to control switching operation of *n-channel* MOSFET Q_1 and Q_2. These are two-phase non-overlapping clock, because V_{G_1} and V_{G_2} are not HIGH simultaneously at

TABLE 14.26 | Operation of NMOS transmission gate

Gate, V_G	NMOSFET, Q	Capacitor, C_1	Capacitor, C_2	Remarks
LOW	OFF	Charged	Charged	No transfer of charge
LOW	OFF	Discharged	Charged	No transfer of charge
LOW	OFF	Charged	Discharged	No transfer of charge
LOW	OFF	Discharged	Discharged	No transfer of charge
HIGH	ON	Charged	Charged	No transfer of charge
HIGH	ON	Discharged	Charged	Transfer charge from C_2 to C_1
HIGH	ON	Charged	Discharged	Transfer charge from C_1 to C_2

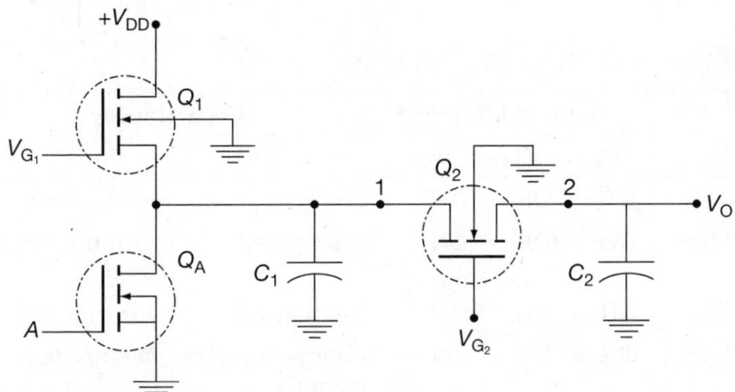

FIGURE 14.59 | Dynamic MOS inverter

the same time. An *n-channel* MOSFET Q_1 act as active load only when clock V_{G_1} is HIGH, at other time when V_{G_1} is LOW, transistor Q_1 turns OFF and does not allow any current to pass through it. AN *n-channel* MOSFET Q_2 acts as transmission gate. It allows transferring charge only when clock V_{G_2} is HIGH. For other time when V_{G_2} is LOW, no transfer of charge takes place as transistor Q_2 turns OFF.

- As and when A is LOW, *n-channel* MOSFET Q_A is OFF. When V_{G_1} goes HIGH, Q_1 conducts and C_1 is charged, but when V_{G_1} goes LOW, C_1 remains charged because there is no path for C_1 to discharge. As and when V_{G_2} goes HIGH, this charge on capacitor C_1 is transferred to capacitor C_2. So, output voltage V_O goes HIGH. Thus, a LOW at input, A results in a HIGH at the output.
- As and when A is HIGH, *n-channel* MOSFET Q_A is ON. When V_{G_1} goes HIGH, Q_1 conducts and C_1 cannot be charged and remains discharged. As and when V_{G_2} goes HIGH, *n-channel* MOSFET Q_2, is ON this charge on C_2 is transferred to capacitor, C_1 and capacitor, C_2 is discharged. So, output voltage V_O goes LOW. Thus, a HIGH at input, A results in a LOW at the output.

Equivalent switching circuit of dynamic MOS inverter is shown in Figure 14.60. Table 14.27 summarizes the operation of dynamic MOS inverter.

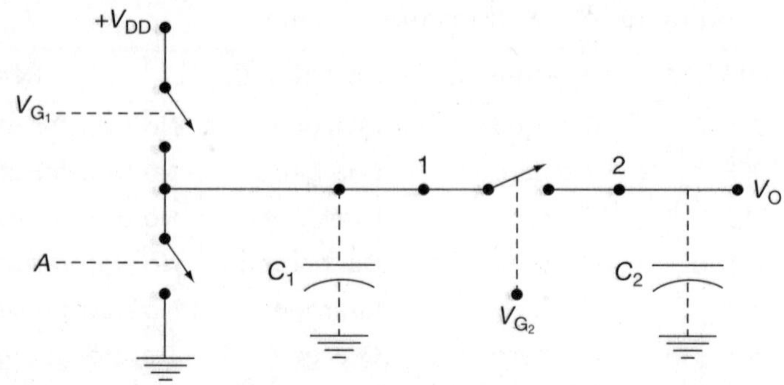

FIGURE 14.60 | Equivalent switching circuit of dynamic MOS inverter

TABLE 14.27 | Operation of dynamic MOS inverter

Inputs			n-channel MOSFET			Capacitances		Output
A	V_{G_1}	V_{G_2}	Q_A	Q_1	Q_2	C_1	C_2	V_o
LOW	HIGH	LOW	OFF	ON	OFF	Charged	No transfer	—
LOW	HIGH	HIGH	OFF	ON	ON	Discharged	Charge transfer from C_1	HIGH
HIGH	HIGH	LOW	ON	ON	OFF	Discharged	No transfer	—
HIGH	HIGH	HIGH	ON	ON	ON	Charge transfer from C_2	Discharged	LOW
LOW	LOW	LOW	OFF	OFF	OFF	No change	No change	—

The output of a dynamic logic gate is 'valid' only when V_{G_2} is HIGH. Thus, gates are sampled at the frequency of V_{G_2}. A sampled output becomes the input to other gates, whose responses become available only at the next sampling time. The disadvantage of dynamic logic is the complexity added by the clocking requirements. The capacitors need to be recharged periodically so that the charge on the capacitors does not decay very much. This process of recharging is called refreshing. The minimum clock frequency is, therefore, determined by the amount of time taken by the capacitance to decay significantly. A typical period is 1 ms, giving a minimum clock frequency of 1 kHz.

14.15.2 | Dynamic MOS NAND Gate

A dynamic two-input MOS NAND gate is shown in Figure 14.61 and its equivalent switching circuit is shown in Figure 14.62. The only difference between this and the static n-channel MOSFET NAND gate is that, the load MOSFET is clocked by V_{G_1}. A transmission gate is added at the output and the outputs are clocked through the transmission gate by V_{G_2}.

- When V_{G_1} goes HIGH, C_1 is charged. According to the NAND logic of inputs A and B and when V_{G_2} goes HIGH, this charge is transferred from capacitance C_1 to C_2. So, the output V_O follows the NAND logic.

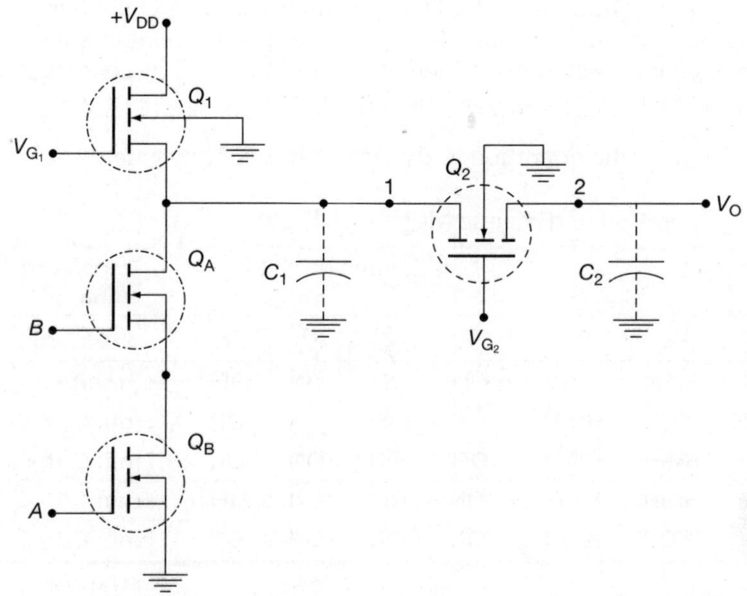

FIGURE 14.61 | Dynamic MOS NAND gate

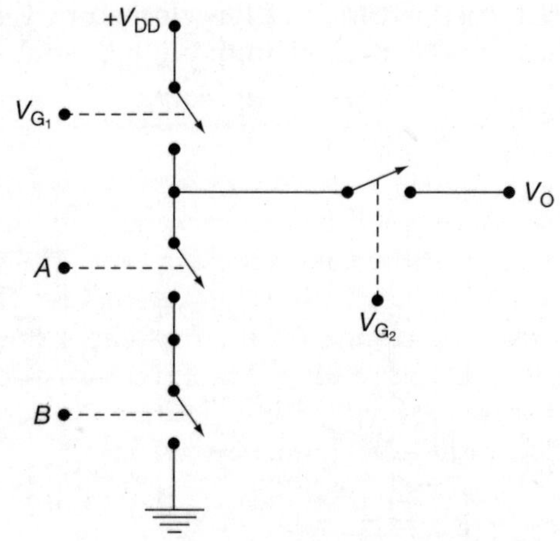

FIGURE 14.62 | Equivalent switching circuit

- When either A is LOW, or B is LOW, or both A and B are LOW, the corresponding *n-channel* MOSFETs (Q_A and Q_B) will be OFF and no current passes through them. Thus, C_1, is charged when V_{G_1} goes HIGH, and this charge on C_1 is transferred to C_2, when V_{G_2} goes HIGH. Therefore, the output goes HIGH and C_1 remains discharged after V_{G_2} goes LOW.

- Only when both A and B are HIGH. Q_A, and Q_B will turn ON when V_{G_1} goes HIGH and, therefore, no current flows through C_1, and it does not charge and remains in the discharged condition only. When V_{G_2} goes HIGH, C_2 discharges into C_1, and the output goes LOW. Hence, this circuit works as a two-input NAND gate.

Table 14.28 summarizes the operation of dynamic MOS NAND gate.

TABLE 14.28 | Operation of dynamic MOS NAND gate

Inputs				*n*-channel MOSFET				Transfer of Charge	Output
A	B	V_{G_1}	V_{G_2}	Q_A	Q_B	Q_1	Q_2	C_2	V_O
LOW	LOW	HIGH	LOW	OFF	OFF	ON	OFF	C_1 charged	—
LOW	LOW	HIGH	HIGH	OFF	OFF	ON	ON	From C_1 to C_2	HIGH
LOW	HIGH	HIGH	HIGH	OFF	ON	ON	ON	From C_1 to C_2	HIGH
HIGH	LOW	HIGH	LOW	ON	OFF	ON	OFF	From C_1 to C_2	HIGH
HIGH	HIGH	HIGH	HIGH	ON	ON	ON	ON	From C_2 to C_1	LOW

14.15.3 | Dynamic MOS NOR Gate

A dynamic two-input MOS NOR gate is shown in Figure 14.63 and its equivalent switching circuit is shown in Figure 14.64. The only difference between this and the static n-channel MOSFET NAND gate is that, the load MOSFET is clocked by V_{G_1}. A transmission gate is added at the output and the outputs are clocked through the transmission gate by V_{G_2}.

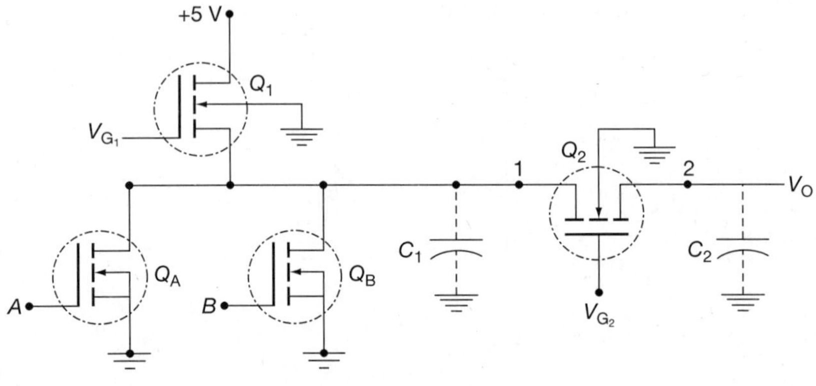

FIGURE 14.63 | Dynamic MOS NOR gate

- When V_{G_1} goes HIGH, C_1 is charged. According to the NOR logic of inputs A and B and when V_{G_2} goes HIGH, this charge is transferred from capacitance C_1 to C_2. So, the output V_O follows the NOR logic.
- Only when both A and B are LOW. Q_A, and Q_B will turn OFF when V_{G_1} goes HIGH and, Thus, C_1 is charged. When V_{G_2} goes HIGH, the charge of capacitance C_1 transfers to capacitance C_2 charges and the output goes HIGH.

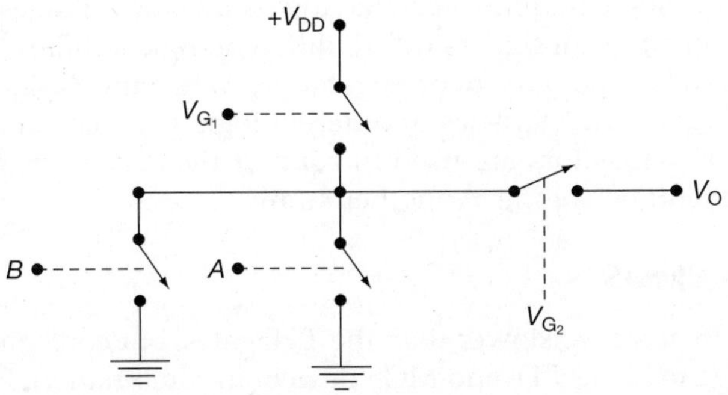

FIGURE 14.64 | Equivalent switching circuit of dynamic MOS NOR gate

- When either A is HIGH or B is HIGH, or both A and B are HIGH, the corresponding *n-channel* MOSFETs (Q_A and Q_B) will be ON and no current passes through them. Thus, capacitance C_1 is charged when V_{G_1} goes HIGH, and capacitance C_1 loses charge. Charge of capacitance C_2 is transferred to capacitance C_1, when V_{G_2} goes HIGH. Therefore, the output goes LOW. Hence, this circuit works as a two-input NOR gate.

Table 14.29 summarizes the operation of dynamic MOS NAND gate.

TABLE 14.29 | Operation of dynamic MOS NOR gate

Inputs				n-channel MOSFET				Transfer of Charge	Output
A	B	V_{G_1}	V_{G_2}	Q_A	Q_B	Q_1	Q_2	C_2	V_O
LOW	LOW	HIGH	LOW	OFF	OFF	ON	OFF	C_1 charged	V_{G_2}
LOW	LOW	HIGH	HIGH	OFF	OFF	ON	ON	From C_1 to C_2	HIGH
LOW	HIGH	HIGH	HIGH	OFF	ON	ON	ON	From C_2 to C_1	LOW
HIGH	LOW	HIGH	LOW	ON	OFF	ON	OFF	From C_2 to C_1	LOW
HIGH	HIGH	HIGH	HIGH	ON	ON	ON	ON	From C_2 to C_1	LOW

14.16 | INTERFACING

Interfacing means connecting two different systems or devices, having different electrical characteristics. In such a case, direct connection is not possible. The circuit used to connect driver and load circuits is called as the interface circuit. The interface circuit takes the input from the driver and converts it, so that it is compatible with the requirements of the load.

Following are the important factors to be considered:

- The driver output must satisfy the voltage and current requirements of the load circuit.
- The driver and load circuit may require different power supplies. In such cases, the output of both circuits must swing between their specified voltage ranges.

There are a number of logic families, each having its own advantages. In designing more complex digital systems, the designers utilize different logic families for different parts of the system in order to take advantage of each family. When the designed parts are assembled, since the electrical characteristics of different logic families vary widely, interfacing circuits or logic level translators are used to connect the driver circuit belonging to one family to the load circuit belonging to another family.

14.16.1 | TTL to CMOS

The MOS and CMOS gates are slower than the TTL gates, but consume less space. Hence, there is an advantage in using TTL and MOS devices in combination. From Table 14.30 it is clear that the TTL has no problem to drive the CMOS, as the input current requirement of CMOS is very low as compared to the output current capabilities of TTL.

TABLE 14.30 | Comparison of current levels of CMOS and TTL family

Parameter	CMOS		TTL		
	4000B	74HCT	74	74L	74AS
I_{IH} (min)	1 μA	1 μA	40 μA	20 μA	200 μA
I_{IL} (max)	1 μA	1 μA	1.6 mA	0.4 mA	2 mA
I_{OH} (min)	0.4 mA	4 mA	0.4 mA	0.4 mA	2 mA
I_{OL} (max)	0.4 mA	4 mA	16 mA	8 mA	20 mA

The input current values of CMOS are low as compared to the output current capabilities of any TTL series. Thus, TTL has no problem in meeting the CMOS input current requirements. So, a level translator is used to raise the level of the output voltage of the TTL gate to an acceptable level for CMOS. The arrangement is shown in Figure 14.65, where the TTL output is connected to a +5 V source with a pull-up resistor. The presence of the pull-up resistor will cause the TTL output to rise to approximately +5 V in the HIGH state, thereby providing an adequate CMOS input.

In case TTL has to drive a high voltage CMOS, the pull-up resistor cannot be used to raise the level of the TTL output to the level of the CMOS input, since the TTL is sensitive to voltage levels. In such a case, an open collector buffer can be used to interface TTL to a high voltage CMOS as shown in Figure 14.66.

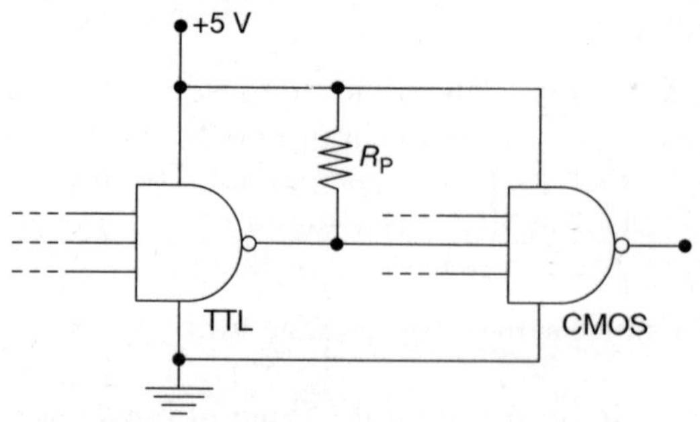

FIGURE 14.65 | TTL to CMOS interfacing

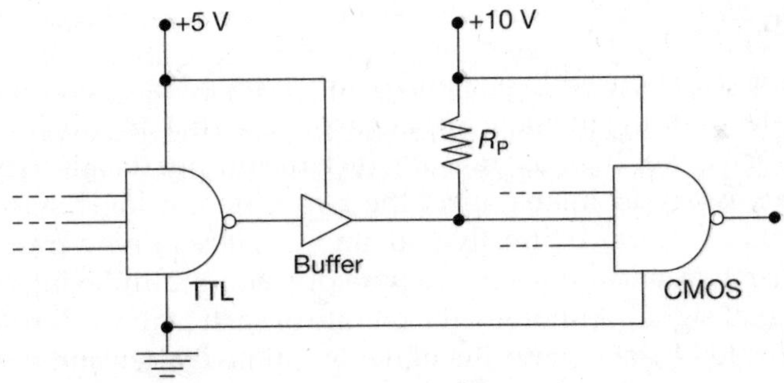

FIGURE 14.66 | TTL to CMOS interfacing

14.16.2 | CMOS to TTL

The CMOS output can supply enough voltage and current to satisfy the TTL input requirements in the HIGH state (Table 14.31). Hence, no special consideration is required for the HIGH state.

TABLE 14.31 | CMOS output and TTL input requirements

CMOS (4000 B)		TTL	
Parameter	**Value**	**Parameter**	**Value**
V_{OH} (min)	4.95 V	V_{IH} (min)	2.0 V
V_{OL} (max)	0.05 V	V_{IL} (max)	0.8 V
I_{OH} (min)	0.4 mA	I_{IH} (min)	40 μA
I_{OL} (max)	0.4 mA	I_{IL} (max)	1.6 mA

But the TTL input current requirements at LOW state cannot be met directly. Therefore, an interface circuit with a LOW input current requirement and a sufficiently high output current rating is required. A CMOS buffer serves this purpose. The arrangement is shown in Figure 14.67.

When a high voltage CMOS has to drive a TTL gate, a voltage level translator that converts the high voltage input to a +5 V output is used between CMOS and TTL as shown in Figure 14.68.

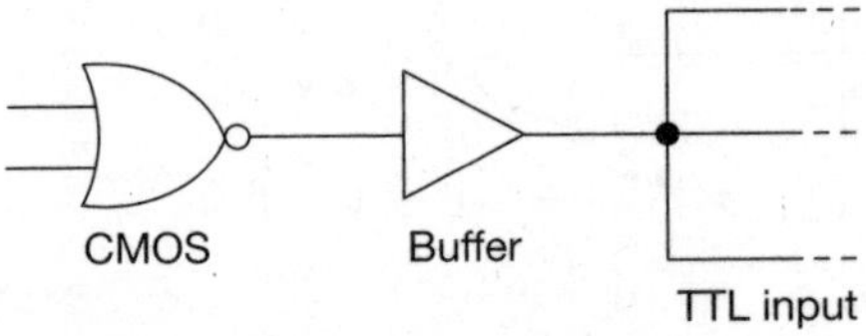

FIGURE 14.67 | CMOS to TTL interfacing

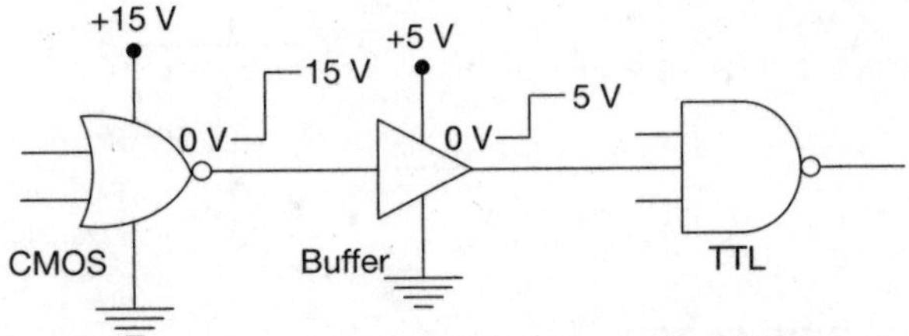

FIGURE 14.68 | CMOS to TTL interfacing

14.16.3 | TTL to ECL

The TTL is the most widely used logic family, but its speed of operation is not very high. The ECL is the fastest family. In some applications, the rate at which input data is to be handled may be much lower than the rate at which the output data is to be handled. Therefore, it becomes necessary to interconnect the two different logic systems, such as TTL and ECL. A TTL cannot interface directly with an ECL; it requires a translator as shown in Figure 14.69. One such application is in the time division multiplexing of n digital signals to form a single digital signal. Although, the bit rate of each of the n signals may be handled using TTL, the bit rate of the composite signal is n times faster and may require ECL to process it.

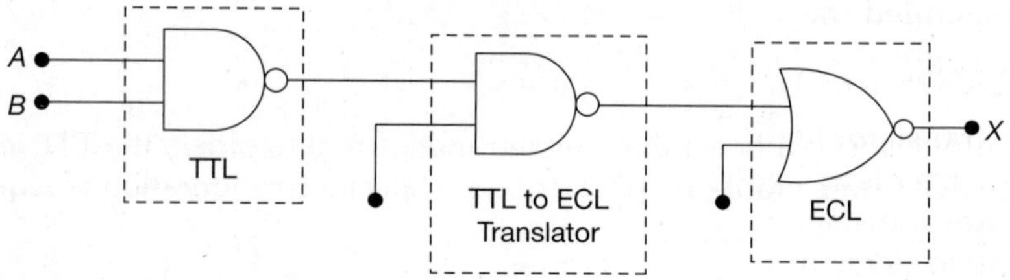

FIGURE 14.69 | TTL driving ECL

It is observed that

$$V_{IH}(\text{Translator}) < V_{OH}\ (\text{TTL}),\ V_{IL}(\text{Translator>})\ V_{OL}\ (\text{TTL})$$

and

$$V_{IH}\ (\text{ECL}) < V_{OH}\ (\text{Translator}),\ V_{IL}(\text{ECL}) > V_{OL}\ (\text{Translator})$$

14.16.4 | ECL to TTL

Sometimes, the input data is at a faster rate, but the output data is at a slower rate like in demultiplexers. An ECL to TTL logic translator will be of use in such cases. It shows that the input logic levels of a translator are compatible with the output logic levels of ECL and the output logic levels of a translator are compatible with the input logic levels of a TTL. Figure 14.70 shows the ECL gate driving a TTL gate.

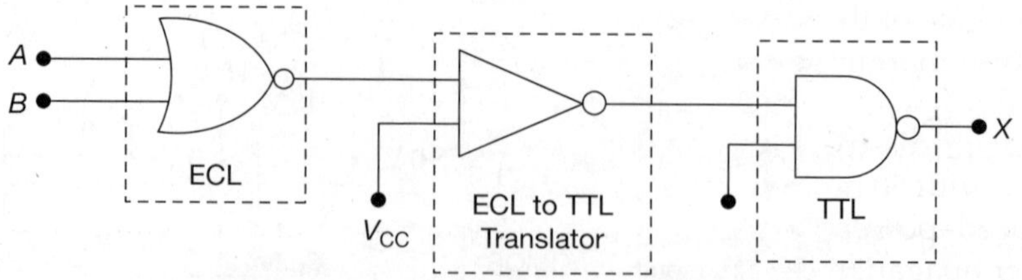

FIGURE 14.70 | ECL driving TTL

SUMMARY

- **Integrated circuit:** Most of the digital circuits are constructed on a single chip, which are referred to as integrated circuits (ICs).
- **Logic family:** The set of compatible ICs with the same logic levels and same supply voltages, fabricated to perform various logic functions, is known as a logic family.
- **Unipolar logic family:** The logic family having unipolar devices like MOSFETs as its key element is referred to as unipolar logic family.
- **Bipolar logic family:** The logic family with transistors and diodes as its key element is referred to as bipolar logic family.
- **Resistor-transistor logic (RTL):** The logic family that consists of resistors and transistors is called as resistor-transistor logic.
- **Direct-coupled transistor logic (DCTL):** The logic family in which the input signal is directly given to the base of the transistor is known as the direct-coupled transistor logic.
- **Diode Transistor logic (DTL):** The logic family with diodes and transistors is called as diode-transistor logic.
- **Transistor-transistor logic (TTL):** The saturated logic family consisting transistor as its basic element, which operates either in cut-off or saturation region, is called transistor-transistor logic.
- **Emitter-coupled logic (ECL):** The transistors of emitter-coupled logic are operated in cut-off or active region, it never goes in saturation and therefore the storage time is eliminated.
- **Integrated injection logic (I^2L):** It uses only transistors for the construction of gate and hence, it becomes possible to integrate a large number of gates in a single package.
- **MOSFET Logic:** MOSFETs are very popular for logic circuit because of their low-power dissipation and high density of fabrication.
- **CMOS Logic:** CMOS family uses n-channel and *p*-channel MOSFETs. In CMOS, *p*-channel and *n*-channel MOS devices are fabricated on the same chip, which makes its fabrication complicated but it reduces the packaging density. It has small power consumption.
- Propagation delay is the average transition delay time for a pulse to propagate from the input to the output of a switching circuit.
- The fan-in of a logic gate is defined as the number of inputs the gate can handle.
- The fan-out of a logic gate specifies the number of standard loads the output of a gate can drive without impairing its operation.
- A standard load is usually defined as the amount of current needed by an input of another gate of the same logic family.
- The noise margin represents the maximum noise signal that can be tolerated by a gate.
- The power dissipation of a logic gate is the supply power required by the gate to operate with 50 per cent duty cycle at a specified frequency.
- The speed–power product is called the *figure of merit* of an IC family. It is the product of the propagation delay and the power dissipation per gate.
- The RTL, DTL, DCTL, and HTL logic families are now obsolete.
- The TTL is the most widely used logic family. There are eight TTL subfamilies.

- The TTL gates may be totem-pole type or open-collector type or tri-state type.
- The totem-pole outputs cannot be wired ANDed whereas open-collector outputs can be.
- A TTL circuit acts as a current sink in the LOW state and as a current source in the HIGH state.
- The TTL is the fastest of all saturated logic families. It is used in SSI and MSI ICs.
- Low-power TTL uses large resistors to reduce the power requirements, which consequently reduce its speed of operation.
- High-speed TTL uses smaller resistors and a Darlington pair in place of Q_3, to increase its speed.
- The speed of Schottky TTL is higher than that of standard TTL, because in Schottky TTL the transistors are not allowed to go into saturation by connecting a Schottky barrier diode (SBD) between the base and the collector of each transistor.
- The totem-pole configuration has the advantages of high-speed and low-power dissipation but the disadvantages are those of generation of current spikes and the inability to be wired ANDed.
- In TTL gates, Q_3 is called the current-sourcing transistor or pull-up transistor because it pulls up the output voltage to its logic HIGH level and Q_4 is called the current sinking transistor or pull-down transistor because it pulls down the output voltage to its logic LOW level.
- A tri-state gate has three states-HIGH, LOW, and HIGH impedance state. It utilizes the advantages of high speed of operation of the totem-pole configuration and wire ANDing of the open-collector configuration.
- The resistor R connected between V_{OUT} and $+ V_{CC}$ of an open-collector gate which pulls the V_{OUT} to V_{CC} level when Q_4 is OFF is called the pull-up resistor.
- The IIL has the highest packing density and is mainly suitable for VLSI and ULSI.
- The ECL is the fastest of all logic families and is mainly used in supercomputers. It is a non-saturated logic. Its logic levels are negative and complementary outputs are available as well.
- ECL gates, which are available in open-emitter configuration, can be wired ORed.
- The MOS logic is the simplest to fabricate.
- A MOS transistor can be connected as a resistor.
- Most modem MOSFET circuitry is constructed using NMOS devices because they are three times faster and two times denser than their PMOS counterparts.
- The MOS is ideally suited for LSI/VLSI/ULSI and for dedicated applications such as large memories, large microprocessor chips, etc.
- The MOS digital ICs use enhancement MOSFETs exclusively.
- The CMOS uses both n-channel and p-channel MOSFETs in the same circuit.
- The CMOS requires the least power, whereas the ECL requires the maximum power.
- The CMOS is ideally suited for battery-powered circuits.
- The CMOS has the greatest complexity and the lowest packing density of the MOS families, but it possesses the important advantages of higher speed and much less power dissipation.
- The CMOS can be operated at high voltage resulting in improved noise margin.
- The CMOS inputs should never be left disconnected. All CMOS inputs should be tied either to a fixed voltage level (0 V or V_{DD}) or to another input.

- The speed of the CMOS gate increases with increase in V_{DD}.
- The noise margin of CMOS gates is 30 per cent of V_{DD}. Therefore, they are very much preferred in noisy environments.
- At higher frequencies, the CMOS loses some of its advantages over the other logic families.
- The fan-out of CMOS gates depends on the maximum permissible value of the propagation delay. The propagation delay of a CMOS gate increases with increase in load.
- Each CMOS load increases the propagation delay of the driving circuit by 3 ns, because each CMOS input typically presents a 5 pF load to ground.
- The CMOS has large fan-out because of its low output resistance.
- The CMOS is faster than NMOS because of its smaller output resistance than that of the MOS.
- The CMOS logic gates are available with open-drain outputs, similar to their TIL counterparts with open-collector outputs.
- The open-drain CMOS gates can be wired ANDed.
- A transmission gate is simply a digitally-controlled CMOS switch. It is a bilateral device.
- Both NMOS and PMOS are more economical than CMOS because of their greater packing density than that of CMOS.
- When power consumption and physical size are the prime design considerations, as in digital watches and calculators, the dynamic MOS logic is preferred.
- Interfacing means connecting the outputs of one circuit or system to the inputs of another system with different electrical characteristics.
- Level shifters are specially designed ICs which are used to make devices from different logic families compatible with each other.

MULTIPLE CHOICE QUESTIONS

14.1 Positive logic in a logic circuit is one in which ......
   (a) logic 0 and 1 are represented by negative and positive voltages, respectively
   (b) logic 0 voltage level is higher than logic 1 voltage level
   (c) logic 0 voltage level is lower than logic 1 voltage level
   (d) logic 0 and 1 are represented by 0 and positive voltages, respectively

14.2 In standard TTL gates, the totem-pole output stage is primarily used to
   (a) increase the noise margin of the gate
   (b) decrease the output switching delay
   (c) facilitate a wired-OR logic connection
   (d) increase the output impedance of the circuit

14.3 The switching speed of ECL is very high, because ......
   (a) the transistors are switched between cut-off and saturation regions
   (b) the transistors are switched between active and saturation regions
   (c) the transistors are switched between cut-off and active regions
   (d) the transistors may operate in any of the three regions

14.4 In standard TTL, the totem-pole stage refers to ......
(a) the multi-emitter input stage
(b) the phase-splitter
(c) the output buffer
(d) the open collector output stage

14.5 The figure of merit of a logic family is given by ......
(a) gain-bandwidth product
(b) product of propagation delay time and power dissipation
(c) product of fan-out and propagation delay time
(d) product of noise margin and power dissipation

14.6 Which of the following uses the least power?
(a) TTL
(b) ECL
(c) CMOS
(d) all use same power

14.7 Which of the following logics is the fastest?
(a) TTL (b) ECL
(c) CMOS (d) LSI

14.8 In Schottky-clamped TTL, the purpose of Schottky diode is
(a) to increase propagation delay
(b) to achieve efficient non-saturated switching
(c) to improve noise margin
(d) to decrease dissipation

14.9 The logic family with both logic levels negative is
(a) TTL
(b) ECL
(c) CMOS
(d) MOS

14.10 The logic family preferred in super-fast computers is
(a) TTL (b) ECL
(c) IIL (d) MOS

Answers

14.1 (d) 14.2 (b) 14.3 (c) 14.4 (c) 14.5 (b) 14.6 (c) 14.7 (b)
14.8 (b) 14.9 (b) 14.10 (b)

QUESTIONS

14.1 Explain the following terms with reference to a gate:
(a) propagation delay
(b) fan-in, (c) fan-out
(d) noise margin, and
(e) speed–power product

14.2 Describe what happens to each of the following CMOS characteristics as V_{DD} is increased?
(a) noise margin
(b) power dissipation and
(c) switching speed

14.3 In which application is CMOS ideally suited?

14.4 Which special CMOS circuit has no TTL or ECL counterpart?

14.5 What do you mean by a transmission gate?

14.6 Explain the operation of a CMOS bilateral switch. Is there any TTL bilateral switch?

14.7 Compare the following technologies:
(a) Bipolar and CMOS
(b) TTL and MOS, and
(c) MOS and CMOS

14.8 Compare TTL, ECL, IIL, MOS and CMOS with respect to fan-in, fan-out, noise margin, t_{pd} and P_D.

14.9 What do you mean by dynamic MOS logic? When is it preferred?

14.10 What do you mean by interfacing?

14.11 With the help of neat circuit diagrams, explain the working of:
(a) A two input TTL NAND gate
(b) A two input ECL OR/NOR gate and
(c) IIL NAND and NOR gates

14.12 With the help of a neat circuit diagram, explain the working of:
(a) A MOS inverter
(b) A two-input MOS NAND gate and
(c) A two-input MOS NOR gate

14.13 With the help of a neat circuit diagram, explain the working of:
(a) A CMOS inverter
(b) A two-input CMOS NAND gate and
(c) A two-input CMOS NOR gate

15

Clocks and Timing Circuits

CHAPTER OBJECTIVES

The main goal of this chapter is to impart knowledge about clocks and timing circuits. Readers will be able to discuss the following aspects in this chapter:

- Clock waveforms
- Logic gates in timing circuits
- Multivibrators: astable, monostable and bistable
- Schmitt trigger
- 555 Timer and its use for astable and monostable multivibrators
- Monostables with input logic

15.1 | INTRODUCTION

Most digital systems require some kind of a timing waveform, for example, a source of trigger pulses is required for all clocked sequential systems. In digital systems, a rectangular waveform is most desirable like analog systems where sinusoidal signals are used. The generators of rectangular waveforms are referred to as multivibrators. Multivibrators are classified into three types:

- Astable (or free-running) multivibrator,
- Monostable multivibrator (or one-shot), and
- Bistable multivibrator (or FLIP-FLOP).

15.1.1 | Astable Multivibrator

An astable multivibrator is an oscillator which generates rectangular waveform. It goes quasi-stable states and does not require any triggering. So, it is called as a free-running multivibrator. This is used as a source of clock pulses in sequential circuits. The astable multivibrator generates a continuous flow of pulses as depicted in Figure 15.1. Astable multivibrator can be used for the following applications.

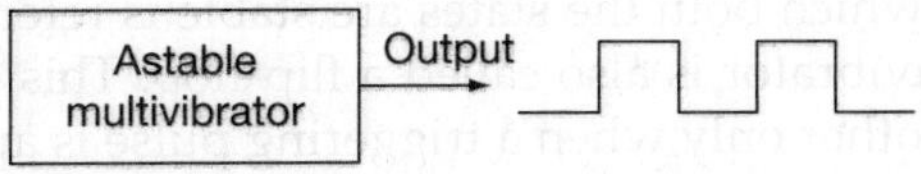

FIGURE 15.1 | Astable multivibrators

- Rectangular-wave oscillator
- Square-wave oscillator
- Ramp generator
- Voltage-controlled oscillator (VCO)
- Schmitt trigger
- Frequency shift keying (FSK) generator

15.1.2 | Monostable Multivibrator

A monostable multivibrator has one stable state, that is under steady-state condition its output is fixed. It is in either the LOW or the HIGH state. When the circuit is triggered with an externally applied pulse, it goes into the other state. LOW state goes to the HIGH state and HIGH state goes to LOW state. The circuit remains in this state for a time duration depending upon the values of the elements used in the circuit. This state of the circuit is referred to as quasi-stable state since it recovers back to the stable state without any external triggering pulse. The width of the trigger pulse is very small and the width of the output pulse depends upon the time duration for which the circuit remains in the quasi-stable state. This circuit is also referred to as a one-shot since one trigger pulse produces only one pulse but of a different pulse width. A monostable multivibrator is also called a one-shot multivibrator. When the one-shot is triggered, as shown in Figure 15.2, the multivibrator generates a single short pulse.

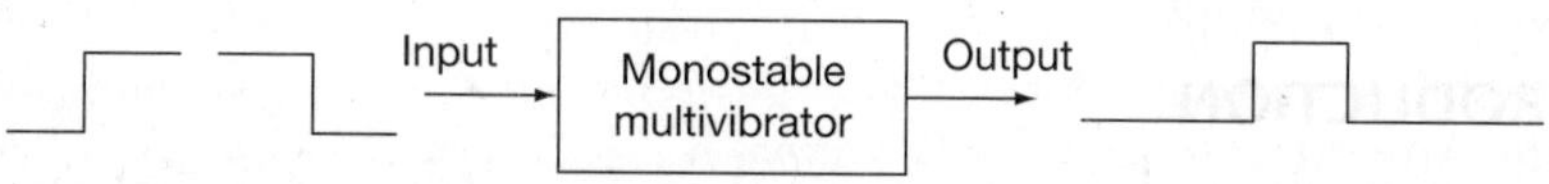

FIGURE 15.2 | Monostable multivibrators

Monostable vibrator circuit is useful because it generates a relatively long pulse (tens of ms) from a narrow pulse, therefore, it is also known as pulse stretcher. For instance, a microprocessor or microcontroller sends a signal to an output device to print data by transmitting a pulse. The electromechanical output device is slower than the microprocessor or microcontroller and requires the signal pulse for a longer duration. This is achieved by the interface circuitry consisting of a monostable multivibrator. A monostable multivibrator can be used in the following applications

- Pulse width modulator
- Missing pulse detector
- Re-triggerable monostable multivibrator.

15.1.3 | Bistable Multivibrator

A multivibrator circuit in which both the states are stable is referred to as a bistable multivibrator. A bistable multivibrator is also called a flip-flop. This circuit makes transitions from one stable state to another only when a triggering pulse is applied. The bistable multivibrator gives one out of two stable states that is set (HIGH) or reset (LOW). The basic of a bistable multivibrator is depicted in Figure 15.3, where the input pulse triggers a change in output from LOW to HIGH.

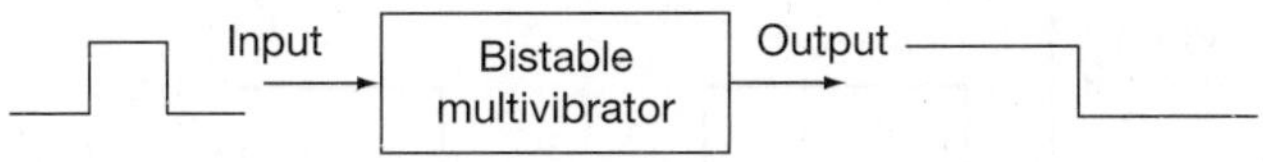

FIGURE 15.3 | Bistable multivibrators

Multivibrators were designed using discrete devices like vacuum triodes, bipolar junction transistors (BJTs), field-effect transistors (FETs), etc. These methods are obsolete in these days because of the availability of various integrated circuits (ICs). The ICs used are:

- Logic gates,
- Operational amplifiers (OP-AMPs),
- Monostable multivibrators and
- Timers

15.2 | LOGIC GATES IN TIMING CIRCUITS

To generate the pulse required in digital systems, logic gates can be used. These circuits are easy to design and analyse but their applications are rather restricted due to the lack of precision.

15.2.1 | Astable (Free-running) Multivibrator

A simple astable multivibrator also called as free-running multivibrator is shown in Figure 15.4. Astable multivibrator circuit is designed by directly feeding back output of a last NOT gate. An odd number of NOT gates are connected in series.

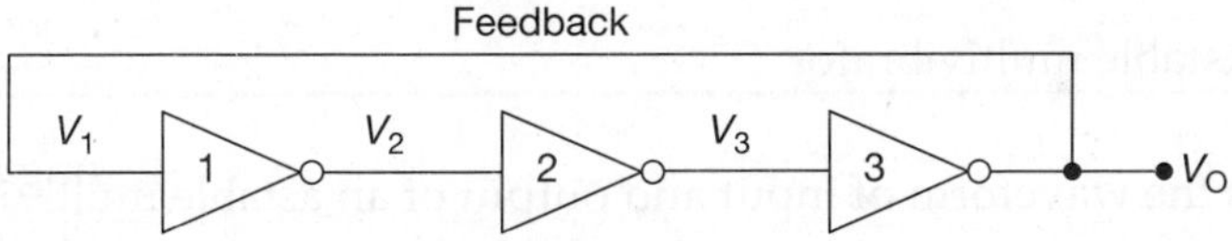

FIGURE 15.4 | Astable multivibrator using inverters

This circuit works on the principle of RC phase shift oscillator. In this circuit, the phase shift is provided by the propagation delay time of the inverters (NOT gates).

Let the input V_I to the inverter 1 makes a transition from logic 0 to logic 1 at t_0. Now after propagation delay of t_p its output changes from logic 1 to logic 0. This makes the output of the inverter 2 to change from logic 0 to logic 1 after a propagation delay of t_p, and after another time delay of t_p, propagation delay of the inverter 3. The output voltage $V_O = V_I$, changes from logic 1 to logic 0. This process will go on indefinitely. The waveform is drawn as shown Figure 15.5. Since the propagation delay of each inverter is t_p, hence the input signal takes $3t_p$, to propagate from input to output.

It is possible to have some control over the frequency of the square wave by using timing elements-a resistance and a capacitance as shown in Figure 15.6. The frequency of the astable multivibrator clock is given by

$$f = \frac{1}{T} = \frac{1}{2.2RC} \tag{15.1}$$

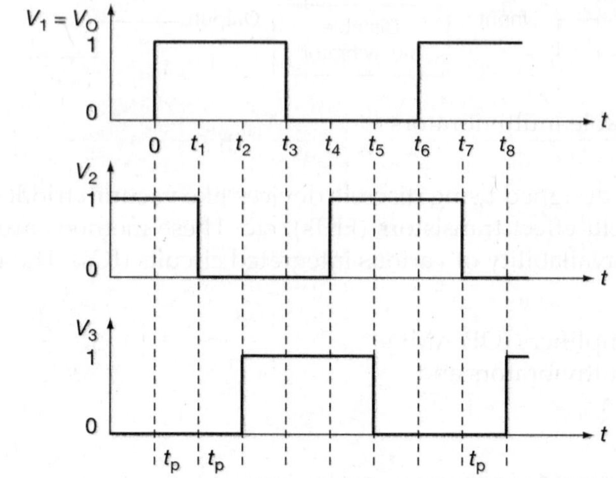

FIGURE 15.5 | Waveforms of V_1, V_2 and V_3 of an astable multivibrator using inverters

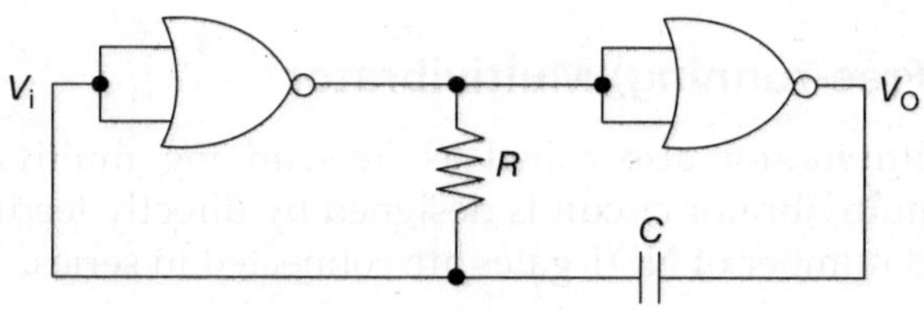

FIGURE 15.6 | Astable multivibrator

Figure 15.7 shows the waveform of input and output of an astable multivibrator shown in Figure 15.6.

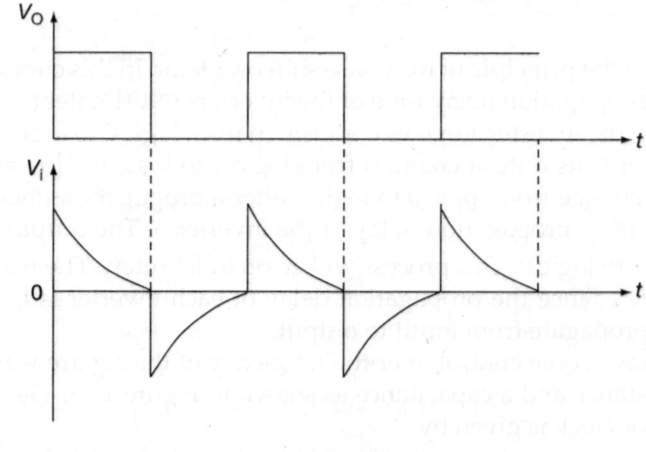

FIGURE 15.7 | Waveform of an astable multivibrator

15.2.2 | Monostable Multivibrator

A monostable multivibrator or one logic short circuit using gates is shown in Figure 15.8. To consider its operation, let trigger input is at HIGH (logic 1). As and when both input of the NAND gate are HIGH (logic 1), it results its output as LOW that is considered logic 0. Therefore, the capacitor remains uncharged and the voltage across capacitor, V_C remains zero. Under steady-state condition, the voltage V_R across the resistance R is zero. The inverter changes LOW (logic 0) to HIGH (logic 1) and makes the output V_O in HIGH (logic 1). HIGH (Logic 1) state is the stable state of this circuit.

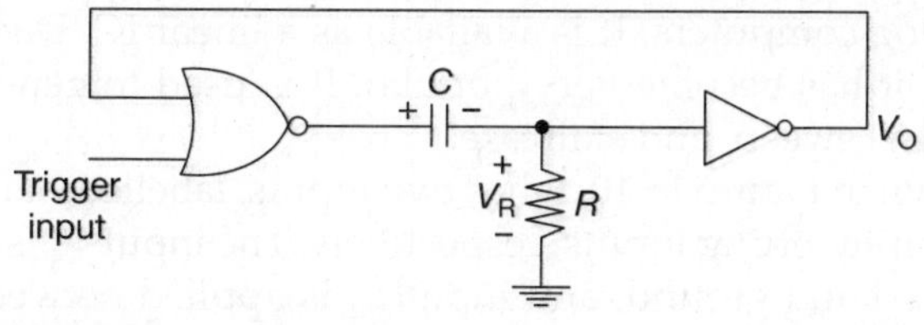

FIGURE 15.8 | Monostable multivibrator

When the trigger input goes from HIGH (logic 1) to LOW (logic 0), as a result the output of NAND gate becomes HIGH (Logic 1). The capacitor starts charging with the time constant, $\tau = RC$ and V_C increases exponentially. The voltage V_R decreases and goes to LOW (logic 0) and the output of the NOT gate goes to HIGH.

When both the inputs of the NAND gate become HIGH, its output again becomes LOW. The capacitor C starts discharging making V_R LOW and the circuit comes to the stable state. The waveforms of trigger, V_C, V_R and output V_O, are given in Figure 15.9.

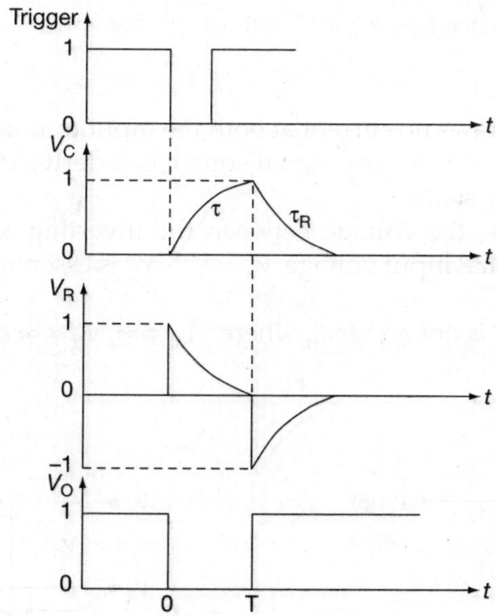

FIGURE 15.9 | Waveforms of a monostable multivibrator

It is not possible to obtain the exact value of the pulse duration, T because of the uncertain values of the input/output voltages corresponding to 0/1 level and the propagation delay time of the gates. However, the circuit is useful because of its simplicity and low cost.

15.3 | OPERATIONAL AMPLIFIER

An operational amplifier, popularly known as OP-AMP, is a very high gain d.c. amplifier. Originally, it was designed using vacuum tubes to perform mathematical operations such as addition, multiplication by a constant, differentiation, integration, etc. It was a basic building block of analog computers. It is available as a linear IC. Due to its low cost, versatility and reliability, it has become very popular. It is used to generate waveforms like square, triangular, pulse, sweep, and staircase.

An OP-AMP is shown in Figure 15.10. It has two inputs, labelled – and +, and are referred to as inverting, and non-inverting inputs, respectively. The input V_1 is applied between the inverting input terminal and ground, and input V_2 is applied between the non-inverting input terminal and ground. If $V_1 = 0$, output V_O is 180° out of phase with input signal V_2. When $V_2 = 0$, the output V_O will be in phase with the input signal applied at V_1, The ground terminal is not there in this circuit but is the ground terminal of the supplies $+V$ and $-V$. The output V_O is obtained between the output terminal and ground. The input terminals are called differential inputs and the output is single-ended.

OP-AMP is said to be ideal (Figure 15.11), if it has the following characteristics:

- Open loop voltage gain, $A_{OL} \to \infty$
- Input impedance, $R_i \to \infty$
- Output impedance, $R_O \to 0$
- Band width, BW $= \infty$
- Zero offset, i.e. $V_O = 0$ when $V_1 = V_2 = 0$.

It can be observed that

- An ideal OP-AMP draws no current at both the input terminal that $i_1 = i_2 = 0$. Because of infinite input impedance, any signal source can derive it and there is no loading on proceeding driver stage.
- Since gain is infinite, the voltage between the inverting and non-inverting terminals, that is, differential input voltage $V_d = V_1 - V_2$ is essentially zero for finite output voltage, V_O.
- A physical amplifier is not an ideal, where $A_{ol} \neq \infty$, $R_i \neq \infty$ and $R_O \neq 0$.

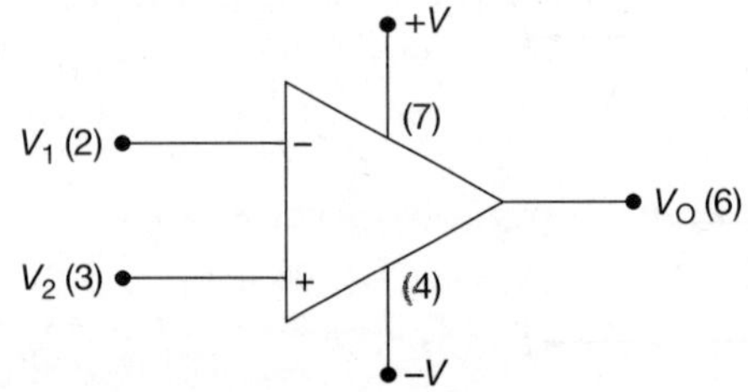

FIGURE 15.10 | Operational amplifier

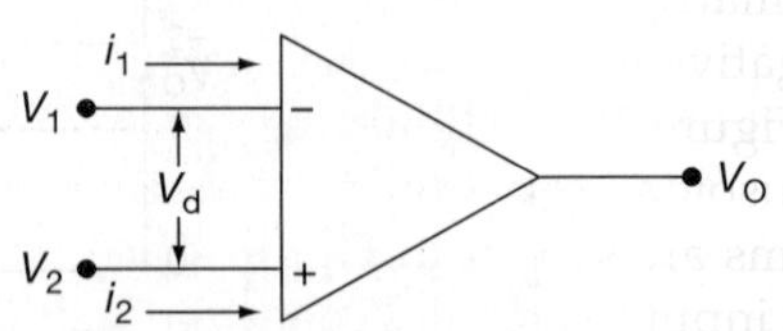

FIGURE 15.11 | Ideal operational amplifier

The output voltage depends on the difference voltage and is given by

$$V_O = A_{OL}(V_2 - V_1) \tag{15.2}$$

where A_{OL} is the voltage gain of the amplifier, which has a large negative value (ideally, $A_{OL} \to \infty$).

The OP-AMP can be used as an analog comparator to compare two analog signals. The analog signals to be compared are applied at the two inputs and the polarity of the output voltage indicates the comparison. The magnitude of the output voltage is V_{sat}. This is the basic building block required for non-sinusoidal waveform generators.

Figure 15.12 shows the OP-AMP where inverted input terminal connected to ground and sinusoidal signal of V_m peak at the non-inverting input. The input and the output waveforms are shown in Figure 15.13. As and when the input crosses 0 V while the sinusoidal wave is changing from positive to negative the output changes from $+V_{sat}$ to $-V_{sat}$. As and when the input crosses 0 V while the sinusoidal wave is changing from negative to positive, then the output changes from $-V_{sat}$ to $+V_{sat}$.

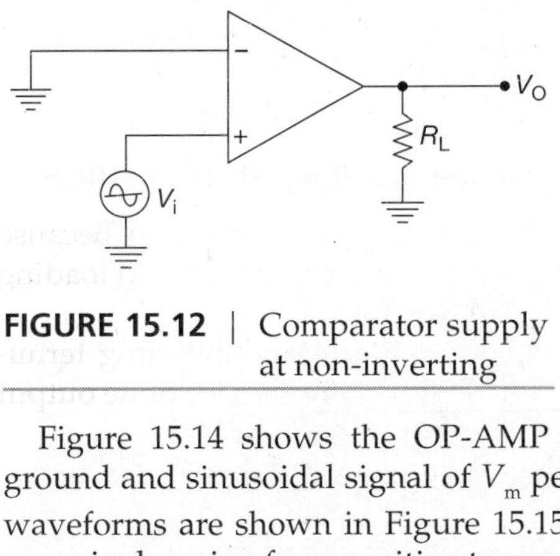

FIGURE 15.12 | Comparator supply at non-inverting

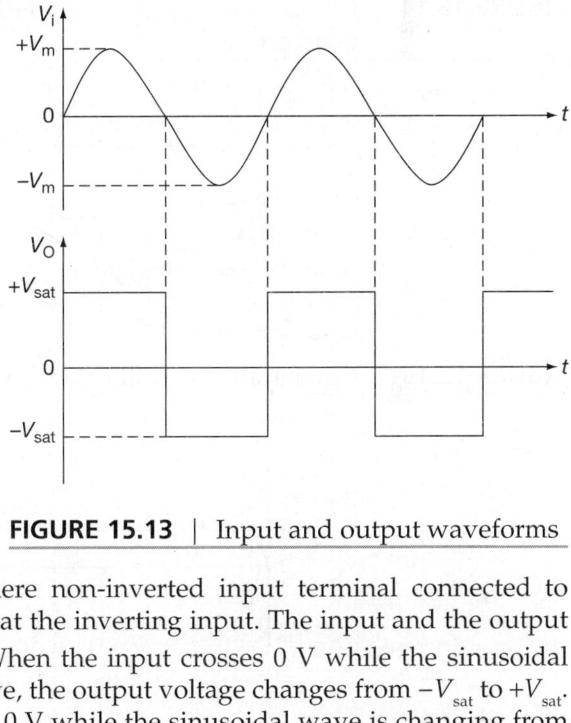

FIGURE 15.13 | Input and output waveforms

Figure 15.14 shows the OP-AMP where non-inverted input terminal connected to ground and sinusoidal signal of V_m peak at the inverting input. The input and the output waveforms are shown in Figure 15.15. When the input crosses 0 V while the sinusoidal wave is changing from positive to negative, the output voltage changes from $-V_{sat}$ to $+V_{sat}$. Similarly, when the input passes through 0 V while the sinusoidal wave is changing from negative to positive, the output changes from $+V_{sat}$ to $-V_{sat}$.

Figure 15.16 shows the OP-AMP where inverted input terminal connected to $+V_R$ and sinusoidal signal of V_m peak at the non-inverting input. The input and the output waveforms are shown in Figure 15.17. In this case, a reference, V_R is maintained at the inverting input terminal. Whenever the voltage V_i is greater than the V_R, the output is $+V_{sat}$ and whenever it is less than V_R, the output is $-V_{sat}$.

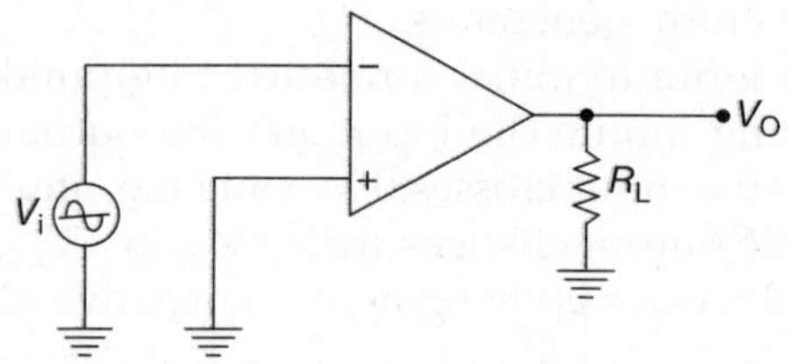

FIGURE 15.14 | Comparator supply at inverting

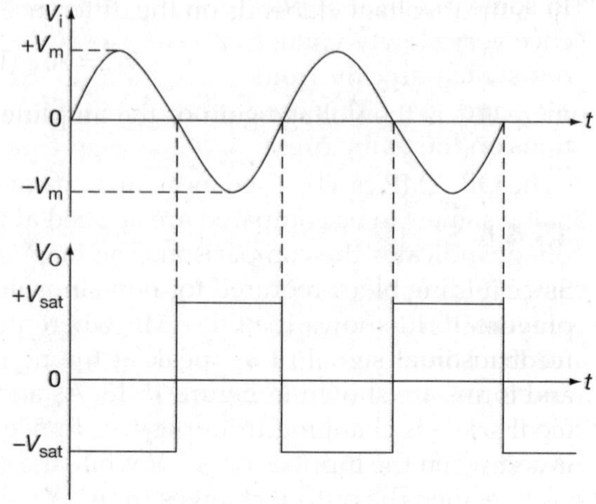

FIGURE 15.15 | Input and output waveforms

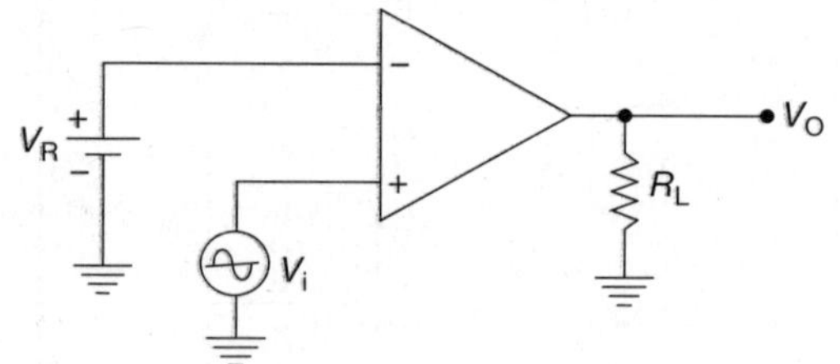

FIGURE 15.16 | Comparator with sinusoidal supply at non-inverting, $+V$ at inverting

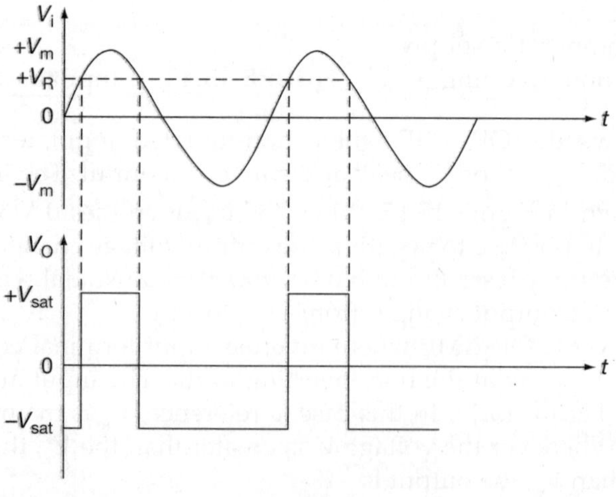

FIGURE 15.17 | Input and output waveform

In some practical applications, the input voltage to the comparator approaches the reference very slowly voltage and/or oscillates around V_R. In such a situation, V_O either does not switch quickly from one saturation voltage to the other or would oscillate between $+V_{sat}$ and $-V_{sat}$. This oscillation occurs due to ringing caused from the fast voltage transitions or due to the presence of noise on wires leading to the OP-AMP's input terminals.

15.4 | SCHMITT TRIGGER (REGENERATIVE COMPARATOR)

Since it is not be possible to completely eliminate the noise voltage, therefore, the circuit is prevented from sensing these false changes at the input. This is achieved by rising positive feedback in the circuit and the resulting circuit is referred to as a regenerative comparator and is also known as Schmitt trigger. Figure 15.18 shows a Schmitt trigger circuit. Positive feedback is applied by taking a fraction of the output voltage and applying it to the non-inverting input. Fraction of output voltage is

$$V_f = \frac{R_2}{R_1 + R_2} V_O \tag{15.3}$$

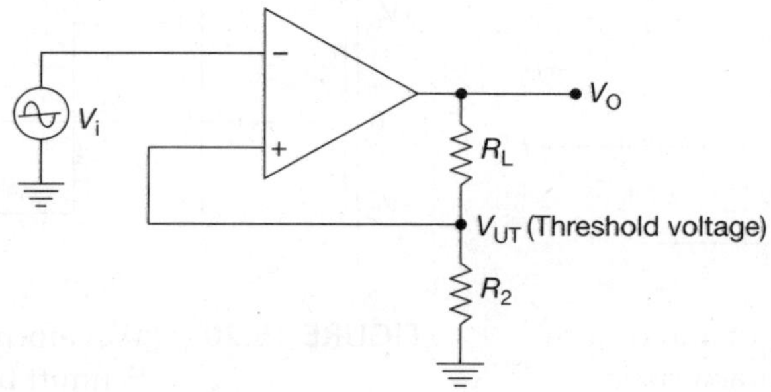

FIGURE 15.18 | Inverting Schmitt trigger

Let $V_O = +V_{sat}$ and $V_i < V_f$. If V_i is increased, then V_O remains constant at $+V_{sat}$, and V_f. V_{sat} is constant until

$$V_i = V_{UT} = \frac{R_2}{R_1 + R_2} V_{sat} \tag{15.4}$$

At this voltage V_{UT}, known as upper-threshold voltage or upper triggering voltage, the output switches to $+V_{sat}$ and remains at this value as long as $V_i > V_{UT}$. This input–output characteristic is shown in Figure 15.19. The voltage at the non-inverting input terminal for $V_i > V_{UT}$, is

$$V_{UT} = \frac{R_2}{R_1 + R_2} V_{sat} \tag{15.5}$$

If V_O is decreased, then the output remains at $-V_{sat}$ until V_i equals V_{LT}, where

$$V_{LT} = \frac{R_2}{R_1 + R_2} (-V_{sat}) \tag{15.6}$$

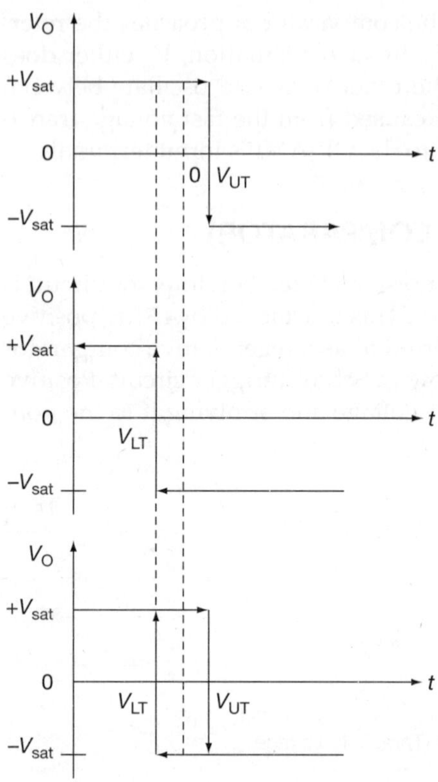

FIGURE 15.19 | Input and output characteristic

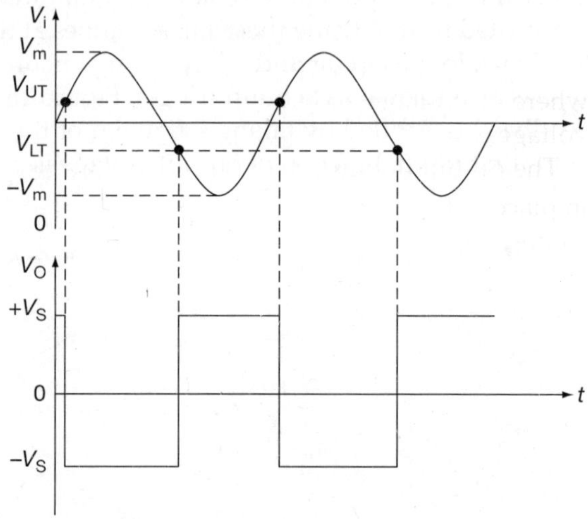

FIGURE 15.20 | Waveform for V_i and V_O of Schmitt trigger

It is known as lower-threshold voltage or lower triggering voltage V_{LT}. At this input voltage, the output switches to V_{sat} and remains at this value as long as $V_i < V_{LT}$. This input–output characteristic is shown in Figure 15.19. The complete input–output characteristic is shown in Figure 15.19, where the portions without arrows may be traversed in either direction, but the segments with arrows can only be obtained if V_i, varies as indicated by the arrows. This characteristic exhibits the hysteresis action in this circuit. The difference in voltage of V_{UT} and V_{LT} is called the hysteresis voltage, V_H.

In case input is sinusoidal having V_m peak. The waveform of input and output is shown in Figure 15.20.

15.4.1 | Limiting Output Voltage

For an OP-AMP, the output voltage levels are limited to $+V_{sat}$ and $-V_{sat}$, and are dependent on supply voltages. These output voltages may not be compatible with voltage levels required by a particular load. For example, TTL requires input voltages that are approximately +5 V and 0 V.

To limit the output voltages, which are independent of the power supply voltages, two back-to-back Zener diodes are used at the output as shown in Figure 15.21. The resistance R is used to limit the current in the Zener diodes. For this circuit, the upper and lower threshold voltages are given by Eqs. (15.7) and (15.8)

$$V_{UT} = \frac{R_2}{R_1 + R_2}(V_{Z_1} + V_D) + \frac{R_2}{R_1 + R_2}V_R \tag{15.7}$$

and

$$V_{LT} = -\frac{R_2}{R_1 + R_2}(V_{Z_2} + V_D) + \frac{R_2}{R_1 + R_2}V_R \tag{15.8}$$

where V_D voltage across a forward-biased diode (–0.7 V) and V_{Z_1} and V_{Z_2} are the Zener voltages.

The output voltages will be ($V_{Z_1} + V_D$) and $-(V_{Z_2} + V_D)$. If the input signal V_i is applied in place of the reference voltage V_R and $V_{R'}$ is applied at the inverting input terminal, then a non-inverting comparator is obtained.

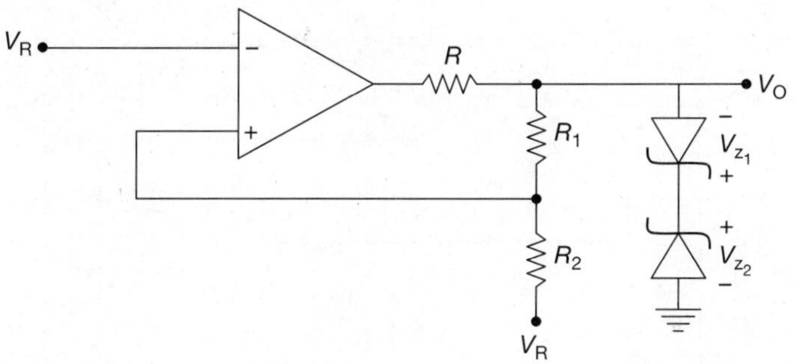

FIGURE 15.21 | A comparator circuit with output voltage limiting

EXAMPLE 15.1 In the Schmitt trigger, $R_2 = 100\ \Omega$, $R_1 = 50\ \text{k}\Omega$, $V_i = 1\ \text{V}_1$ (peak-to-peak) sine wave and saturation voltage = ±14 V. Determine threshold voltages V_{VT} and V_{LT}.

SOLUTION

$$V_{VT} = \left(\frac{R_2}{R_1 + R_2}\right)(+V_{sat})$$

$$V_{VT} = \left(\frac{100}{50 \times 10^3 + 100}\right)(14) = 0.027944\ \text{V} = 27.944\ \text{mV} \quad \textbf{Ans.}$$

$$V_{LT} = \left(\frac{R_2}{R_1 + R_2}\right)(-V_{sat})$$

$$V_{LT} = \left(\frac{100}{50 \times 10^3 + 100}\right)(-14) = -0.027944\ \text{V} = -27.944\ \text{mV} \quad \textbf{Ans.}$$

15.5 | ASTABLE MULTIVIBRATOR USING OP-AMP

An astable or free-running multivibrator is a square-wave generator. The circuit of Figure 15.21 is a comparator circuit in which an RC low-pass circuit (combination of R and C) is used to integrate the output voltage V_o and the voltage across the capacitor C, is applied to the inverting input terminal in place of the external signal.

A simple op-amp square-wave generator is as shown in Figure 15.22. The principle of generation of square-wave output is to force an op-amp to operate in the saturation region. Fraction $\beta = \left(\dfrac{R_2}{R_1 + R_2}\right)$ of the output voltage is fed back to the non-inverting (+) input terminal. Thus, the reference voltage V_{ref} is βV_o and may take values as $+\beta V_{sat}$ or $-\beta V_{sat}$. The output is also feedback to the inverting (–) input terminal after integrating by means of a low pass RC combination. Whenever input of the inverting (–) input terminal just exceeds V_{ref}, switching takes place resulting in a square-wave output. In an astable multivibrator, both the states are quasi-stable (unstable). Waveforms of output and capacitor voltage for an astable multivibrator are shown in Figure 15.23.

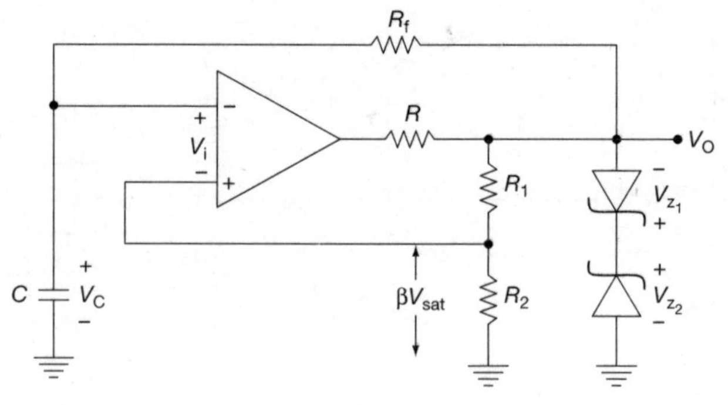

FIGURE 15.22 | Astable multivibrator

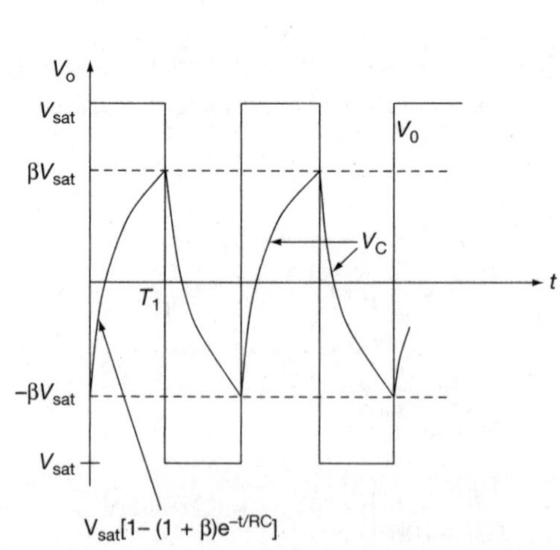

FIGURE 15.23 | Waveform of output and capacitor voltage for an astable multivibrator

Working: Consider an instant of time when the output is at V_{sat}. The capacitor now starts charging towards $+V_{sat}$ through resistance R. The voltage at non-inverting (+) input terminal is held at $+\beta V_{sat}$ by R_1 and R_2 combination. This condition continues as the charge on C raises, until it has just exceeded $+\beta V_{sat}$, the reference voltage. When the voltage at the inverting (–) input terminal becomes just greater than this reference voltage, the output is driven to $-V_{sat}$. At this instant, the voltage on the capacitor is $+\beta V_{sat}$. It begins to discharge through R that is charged toward $-V_{sat}$.

When the output voltage switches to $-V_{sat}$, the capacitor charges more and more negatively until its voltage just exceeds $-\beta V_{sat}$. The output switches back to $+V_{sat}$. The cycle repeats itself as shown.

The frequency is determined by the time it takes the capacitor to charge from $(-\beta V_{sat})$ to $(+\beta V_{sat})$ and vice versa.

The voltage across the capacitor as a function of time is given by,

$$V_c(t) = V_f + (V_i - V_f)e^{-t/RC} \tag{15.9}$$

where the final value, $V_f = +V_{sat}$ and initial value, $V_i = -\beta V_{sat}$

Therefore,

$$V_c(t) = (+V_{sat}) + (-\beta V_{sat} - V_{sat})e^{-t/RC}$$

$$V_c(t) = V_{sat} - V_{sat}(1+\beta)e^{-t/RC} \tag{15.10}$$

At $t = T_1$, voltage across the capacitor reaches βV_{sat} and switching takes place. Therefore, Eq. (15.10) becomes

$$\beta V_{sat} = V_{sat} - V_{sat}(1+\beta)e^{-T_1/RC}$$

On simplifying

$$e^{-T_1/RC} = \frac{(1-\beta)}{(1+\beta)}$$

Taking ln on both sides, we get

$$T_1 = -RC\ln\left(\frac{1-\beta}{1+\beta}\right)$$

Further

$$T_1 = RC\ln\left(\frac{1-\beta}{1+\beta}\right) \tag{15.11}$$

This gives only one half of the period. Total time period is

$$T = 2T_1$$

$$T = 2RC\ln\left(\frac{1-\beta}{1+\beta}\right) \tag{15.12}$$

The output waveform is symmetrical. The output swings from $+V_{sat}$ to $-V_{sat}$, so, V_o peak to peak $= 2\,V_{sat}$.

15.6 | MONOSTABLE MULTIVIBRATOR USING OP-AMP

Monostable multivibrator has one stable state and the other is quasi stable (unstable) state. The circuit shown in Figure 15.24 is useful for generating single output pulse of adjusting time duration. The width of the output pulse depends only on external components connected to the OP-AMP. A diode D_1 clamps the capacitor voltage to 0.7 V when the output is at $+V_{sat}$. A negative-going pulse signal of magnitude V_1 passes through the differentiator R_4C_4 and diode D_2 produces a negative-going triggering impulse and is applied to the non-inverting (+) input terminal.

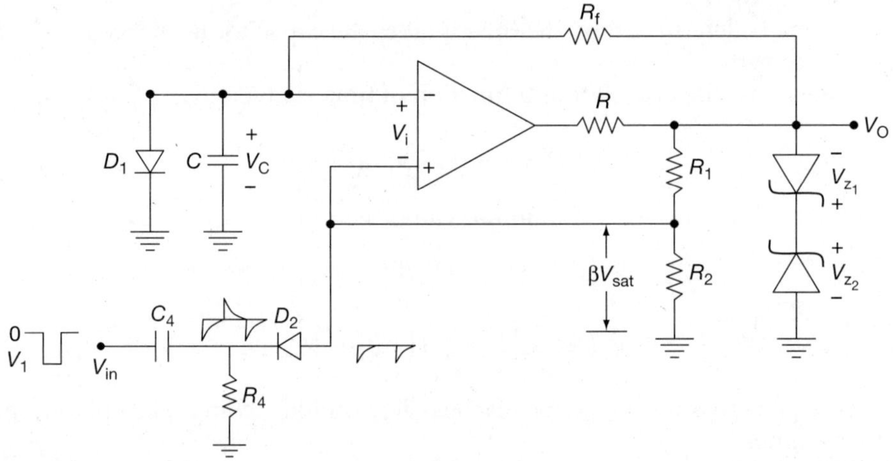

FIGURE 15.24 | Mono multivibrator using OP-AMP

To analyse the circuit, it is assumed that in the stable state, the output V_o is at $+V_{sat}$. The diode D_1 conducts and V_c the voltage across the capacitor C gets clamp at +0.7 V. The voltage at the non-inverting (+) input terminal through R_1R_2 is $+\beta V_{sat}$. Now if signal at this terminal is less than 0.7 V, i.e. $([\beta V_{sat} + (-V_1)] < 0.7\text{ V})$, the output switches from $+V_{sat}$ to $-V_{sat}$. The diode now gets reverse biased and the capacitor starts charging exponentially to $-V_{sat}$ through the resistance R. The voltage at the non-inverting (+) input terminal is now $-\beta V_{sat}$. When the capacitor voltage V_c becomes just slightly more negative than $-\beta V_{sat}$, the output of the OP-AMP switches back to $+V_{sat}$. The capacitor C now starts charging to $+V_{sat}$ through R until V_c is 0.7 V as capacitor gets clamped to the voltage. Waveforms of output and capacitor voltages for monostable multivibrator are shown in Figure 15.25. The pulse width T of mono-stable multi-vibrator is calculated as follows:

$$V_o = V_f + (V_i - V_f)e^{-t/RC} \tag{15.13}$$

Here, final value $V_f = -V_{sat}$ and initial value, $V_i = V_D$, (diode forward voltage)

$$V_o = -V_{sat} + (V_D + V_{sat})e^{-t/RC} \tag{15.14}$$

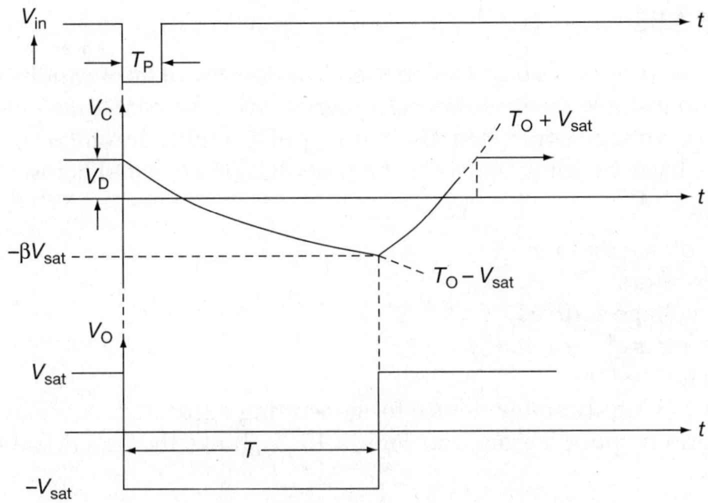

FIGURE 15.25 | Waveforms of capacitor voltage and output

At $t = T$, $V_c = -\beta V_{sat}$, Eq. (15.14) becomes

$$-\beta V_{sat} = -V_{sat} + (V_D + V_{sat})\, e^{-t/RC}$$

On simplifying above equation,

$$(1-\beta)V_{sat} = V_{sat}\left(\frac{V_D}{V_{sat}} + 1\right) e^{-t/RC}$$

$$e^{-t/RC} = \left(\frac{1-\beta}{1+\dfrac{V_D}{V_{sat}}}\right)$$

Taking ln both sides,

$$T = -RC \ln\left(\frac{1-\beta}{1+\dfrac{V_D}{V_{sat}}}\right)$$

The above equation is rewritten as

$$T = RC \ln\left(\frac{1+\dfrac{V_D}{V_{sat}}}{1-\beta}\right) \tag{15.15}$$

where $\beta = \dfrac{R_2}{R_1 + R_2}$

15.7 | TIMER 555

This is a monolithic timer IC which can be used in a number of applications such as astable multivibrator, monostable multivibrators, square-wave and triangular-wave generators, ramp generators, voltage-controlled oscillator, pulse width modulator, missing pulse detector, etc. The basic building blocks of the timer IC 555 are listed below:

- RS flip-flop
- Resistive voltage divider
- Two comparators
- Reference voltage source
- Output buffer and
- Discharge transistor.

The 555 timer is a highly stable device for generating accurate time delay or oscillation. Figure 15.26 shows its pin diagram and Figure 15.27 shows the functional block diagram of the 555 timer.

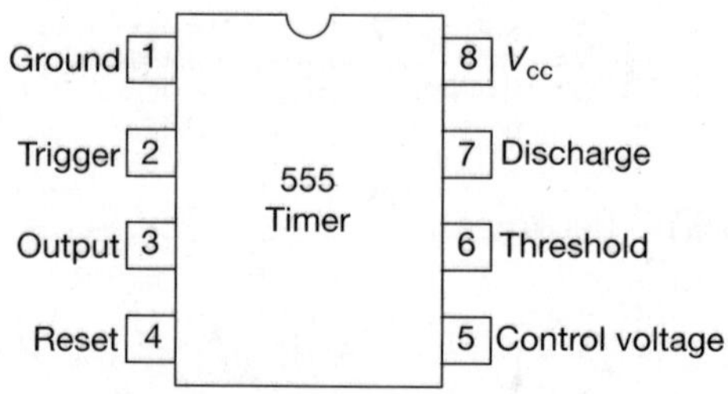

FIGURE 15.26 | Pin diagram of 555 timer

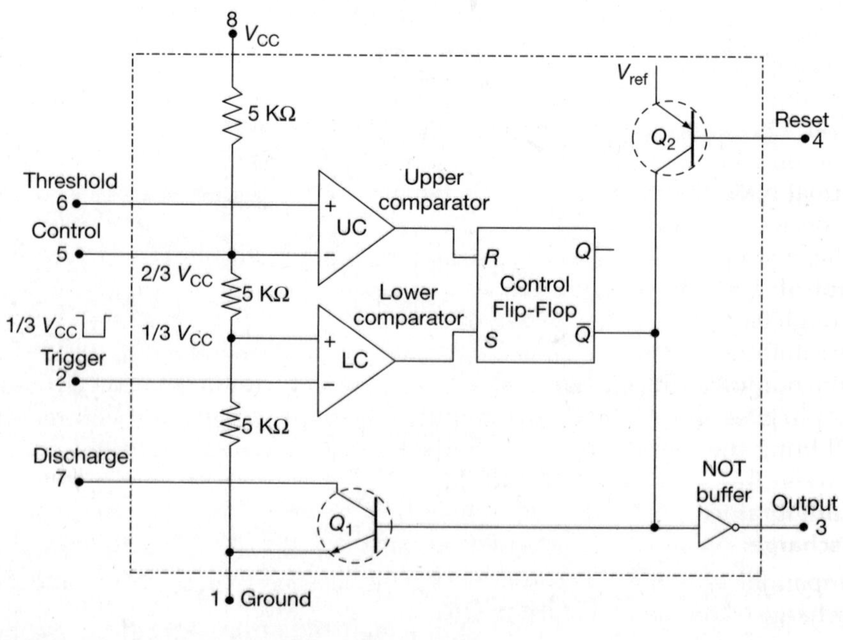

FIGURE 15.27 | Functional diagram of 555 timer

Three 5 kΩ internal resistors act as voltage divider, providing bias voltage of $\left(\frac{2}{3}\right)V_{cc}$ to the upper comparator (UC) and $\left(\frac{1}{3}\right)V_{cc}$ to lower comparator (LC), where V_{cc} is the supply voltage. Since these two voltages fix the necessary comparator threshold voltage, these voltages also aid in the timing interval. In applications, where no such modulation is intended it is recommended that a capacitor (0.01μF) be connected between control voltage terminal (Pin-5) and ground to by-pass noise or ripple from the supply.

Working: In the stand by (stable) state, the output $\bar{Q}$ of the control SR flip-flop is HIGH. This makes the output LOW because of an inverter. A negative going trigger pulse is applied to pin 2 and should have its dc level greater than the threshold level of the lower comparator (i.e. $V_{cc}/3$). At the negative going edge of trigger, as the trigger passes through ($V_{cc}/3$) the output of the lower comparator goes HIGH and sets the flip-flop ($Q = 1, \bar{Q} = 0$). During the positive excursion, when the threshold voltage at pin 6 passes through $\left(\frac{2}{3}\right)V_{cc}$, the output of the upper comparator goes HIGH and resets the flip-flop ($Q = 0, \bar{Q} = 1$). The reset input (Pin 4) provides a mechanism to reset flip-flop in a manner which overrides the effect of any instruction coming to flip-flop from lower comparator. When this reset is not used, it is returned to V_{cc}. The transistor Q_2 serves as a buffer to isolate the reset input from the flip-flop and transistor Q_1. The transistor Q_2 is driven by an internal reference voltage V_{ref} obtained from supply voltage V_{cc}. The operation and function of the 555 pins are:

Pin 1: System ground: This is the reference or ground point of the IC and it is to be connected to the ground terminal of the dc voltage source used to supply power to the timer IC.

Pin 2: Trigger: Input to lower comparator which is used to set flip-flop. When the voltage at this pin crosses above to below $1/3V_{CC}$ the comparator switches to a high, setting the flip-flop. Transistor Q_1 turns ON.

Pin 3: Output: The output of the 555 is driven by an inverting buffer capable of sinking or sourcing 200 mA.

Pin 4: Reset: On connected this pin to a low voltage then output of the timer is forced to become low irrespective of any value. The timer is then said to be reset. For the normal operation of the timer, this pin should be connected to $+V_{CC}$. If reset pin is connected to a low voltage, the discharge transistor Q_1 is turned ON to bring the voltage at pin number 7 to zero. Also, the output voltage is forced to 0.

Pin 5: Control: Used to override the $2/3V_{CC}$ level, if required. Usually, it is grounded through 0.01 pF capacitor to bypass noise.

Pin 6: Threshold: Input to the upper comparator which is used to reset the flip-flop. This is the non-inverting (+) terminal of the upper comparator. As soon as the voltage at this pin goes above $2/3V_{CC}$, the output of the upper comparator becomes high. This will bring the output voltage to a low level. An external capacitor C is connected between this pin and ground in the applications such as astable and monostable multivibrators.

Pin 7: Discharge: As soon as, threshold becomes equal to $2/3V_{CC}$, the output of upper comparator goes high. This will set the SR flip-flop to make $Q = 1$ and $\bar{Q} = 0$. The discharge transistor Q_1 is turned ON.

Pin 8: Supply voltage (V_{CC}): Supply voltage can range from 4.5 to 18 V.

Some of the important features of IC 555 are listed as follows:

- Timing from microseconds through hours.
- Duty cycle of the output is adjustable.
- Sourcing and sinking capacity of the output is 200 mA.
- Output and supply voltages are TTL compatible.
- Very good temperature stability, better than 0.005% per °C.
- Output can be normally ON or normally OFF.
- It is available in 8 pin micro small outline package (MSOP).
- Supply voltage: single polarity 18 V maximum.
- Power dissipation: 600 to 1100 mW.
- Operating temperature range: 0°C to 70°C.

Some of applications of 555 timer are listed below:

- Oscillator
- Pulse generator
- Ramp and square-wave generator
- Mono-shot multi-vibrator
- Burglar alarm
- Traffic light control
- Voltage monitor, etc.

15.8 | MONOSTABLE MULTIVIBRATOR USING TIMER

The monostable circuit has only one permanently stable state and one quasi-stable state. For a monostable circuit, we require an external triggering signal to induce a transition of its output from the stable state to the quasi stable state.

The circuit remains in the quasi-stable state for some time and eventually returns to its stable state on its own. The circuit configuration for a monostable multivibrator using IC 555 has been shown in Figure 15.28. Figure 15.29 shows the simplified diagram of monostable multivibrator. This is called as monostable multivibrator because the output has only one stable state. R_o and C are the externally connected components. The pin 6 (threshold) and pin 7 discharge are connected to each other. The trigger pulse is applied to the pin 2 (trigger input) of the timer IC. Whenever voltage at pin 2 goes below $1/3V_{CC}$, the output goes HIGH. This circuit does not show the internal diagram of IC 555, but it shows only the external components to be connected. Pin number 4 is the reset pin and it is made inactive by connecting it to V_{CC}. A capacitor of 0.01 μF is connected from pin number 5 (control voltage) to ground in order to filter out any noise coupled there.

15.8.1 | Operation of the Monostable Multivibrator

In the standby state, flip-flop holds transistor Q_1 ON, thus clamping the external timing capacitor C to ground. The output remains at ground potential, i.e. LOW. As the trigger passes through $V_{cc}/3$, the flip-flop is set, i.e. $\overline{Q} = 0$. This makes the transistor Q_1 OFF and the short circuit across the timing capacitor C is released. As $\bar{Q}$ is LOW, output goes HIGH (= V_{cc}). The timing cycle now begins. Since C is unclamped voltage across it rises

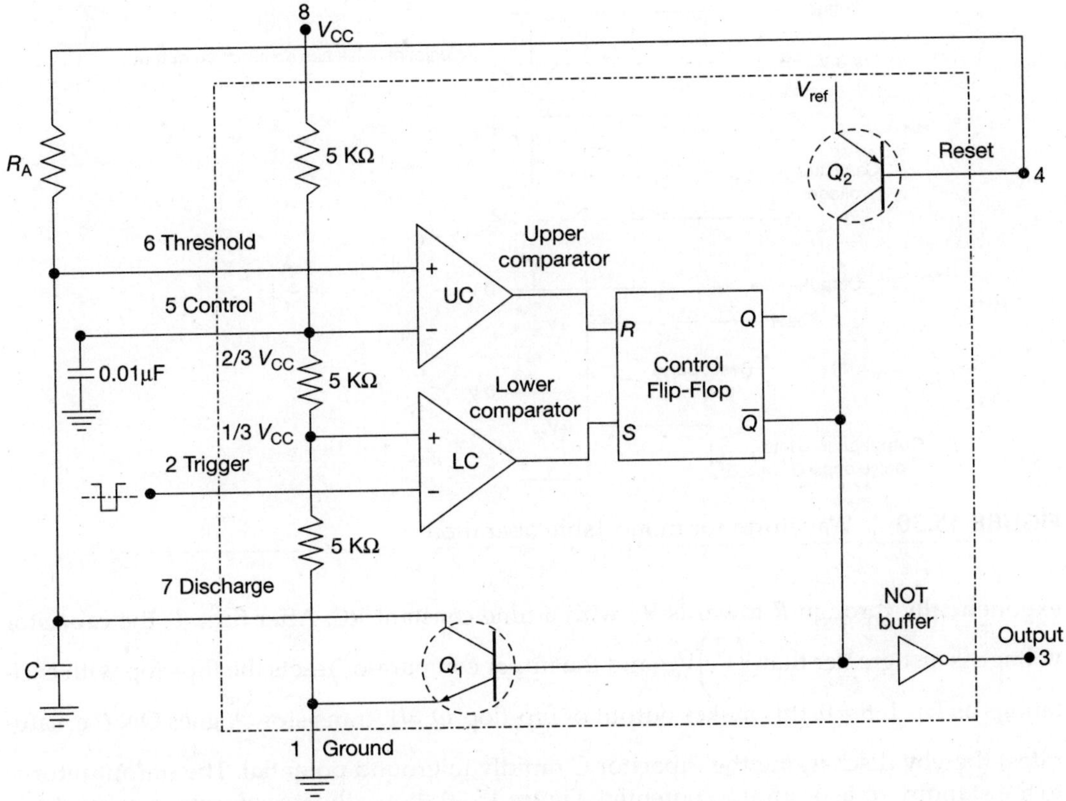

FIGURE 15.28 | Monostable multivibrator using 555 Timer

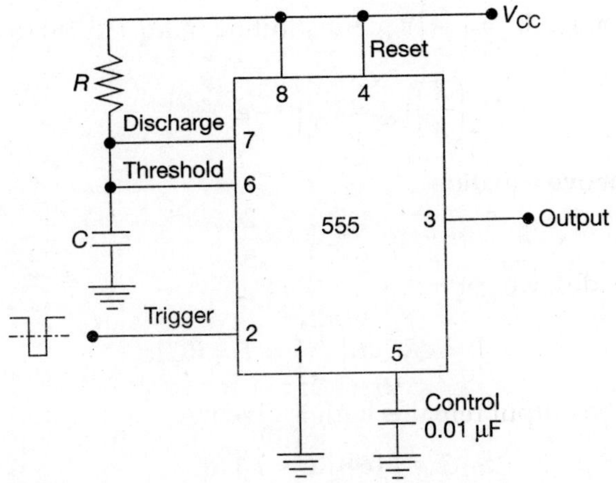

FIGURE 15.29 | Monostable multivibrator using 555 Timer

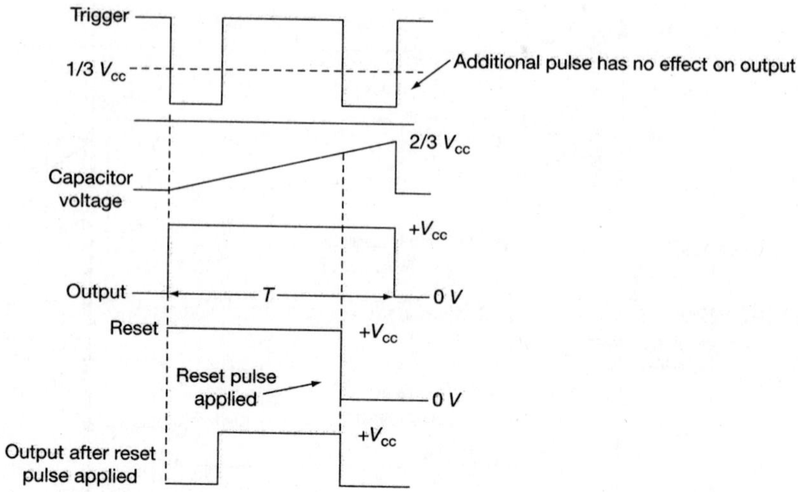

FIGURE 15.30 | Waveform for monostable operation

exponentially through R towards V_{cc} with a time constant RC. After time T, the capacitor voltage is just greater than $\left(\frac{2}{3}\right)V_{cc}$ and the upper comparator resets the flip-flop with excitations as $R = 1$, $S = 0$, this makes output of flip-flop $\bar{Q} = 1$, transistor Q_1 goes ON (i.e. saturates) thereby discharging the capacitor C rapidly to ground potential. The output returns to the standby state or ground potential. Figure 15.30 shows the waveforms of monostable multivibrator relating its operation. The voltage across the capacitor is given by

$$V_c = V_{cc}\left(1 - e^{-t/RC}\right) \tag{15.16}$$

At $t = T$, capacitor voltage, $V_c = \left(\frac{2}{3}\right)V_{cc}$. Substituting in Eq. (15.16) to get

$$\left(\frac{2}{3}\right)V_{cc} = V_{cc}\left(1 - e^{-t/RC}\right)$$

On simplifying the above equation,

$$e^{-t/RC} = 1 - \frac{2}{3}$$

Taking $\log_e$ on both sides, we get

$$T = -RC \ln\left(\frac{1}{3}\right) \Rightarrow T = RC \ln 3$$

Time during which the output remains high is given by

$$T = 1.098\,RC \approx 1.1\,RC \tag{15.17}$$

Figure 15.31 shows variation in capacitance, C to generate time delay, T in monostable multivibrator for a known resistance value, R. Table 15.1 gives the values of capacitance for the given value of resistance to produce a delay by monostable multivibrator.

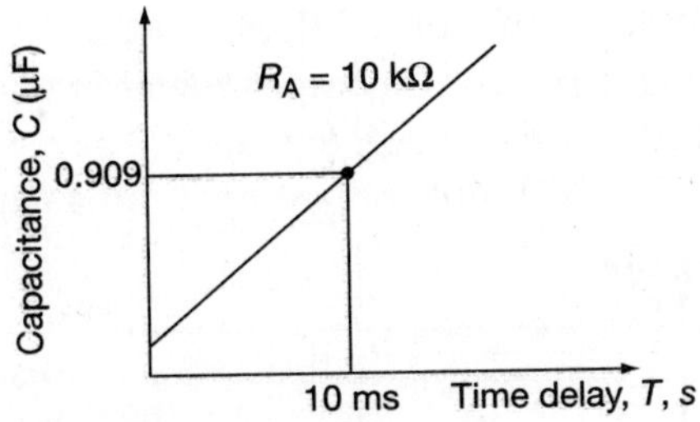

FIGURE 15.31 | Time delay versus capacitance for given resistance of monostable multivibrator

TABLE 15.1 | Capacitance for given resistance and delay of monostable multivibrator

Resistance, R_A	Time Delay, T	Capacitance, C
10 kΩ	100 μs	0.00909 μF
	1 ms	0.0909 μF
	10 ms	0.909 μF
	100 ms	9.09 μF
100 kΩ	1 ms	9.09 pF
	10 ms	90.9 μF
	100 ms	0.909 μF
	1 s	9.09 μF
	10 s	90.9 μF
1 MΩ	10 ms	0.00909 μF
	100 ms	0.0909 μF
	1 s	0.909 μF
	10 s	9.09 μF

EXAMPLE 15.2 In the mono-stable 555 timer IC, $R = 100$ kΩ and the time delay $T = 100$ ms. Calculate the value of C.

SOLUTION

$$T = 1.1RC$$

$$C = \left(\frac{T}{1.1R}\right) = \frac{100 \times 10^{-3}}{1.1 \times 100 \times 10^{3}} = 0.909\ \mu\text{F}$$

15.9 | ASTABLE MULTIVIBRATOR USING TIMER

Circuit diagram of an astable multivibrator is shown in Figure 15.32. Figure 15.33 shows the simplified diagram of an astable multivibrator. Comparing with mono-stable operation, the timing resistor is now split into two sections R_A and R_B. Pin 7 of discharging transistor Q_1 is connected to the junction of R_A and R_B. When the power supply V_{cc} is connected, the external timing capacitor C charges towards V_{CC} with a time constant $(R_A + R_B)C$. During this time, output (Pin 3) is high (equals V_{CC}) as $R = 0$, $S = 1$ (i.e. FF is set), and this combination makes $\bar{Q} = 0$, which unclamped the timing capacitor C.

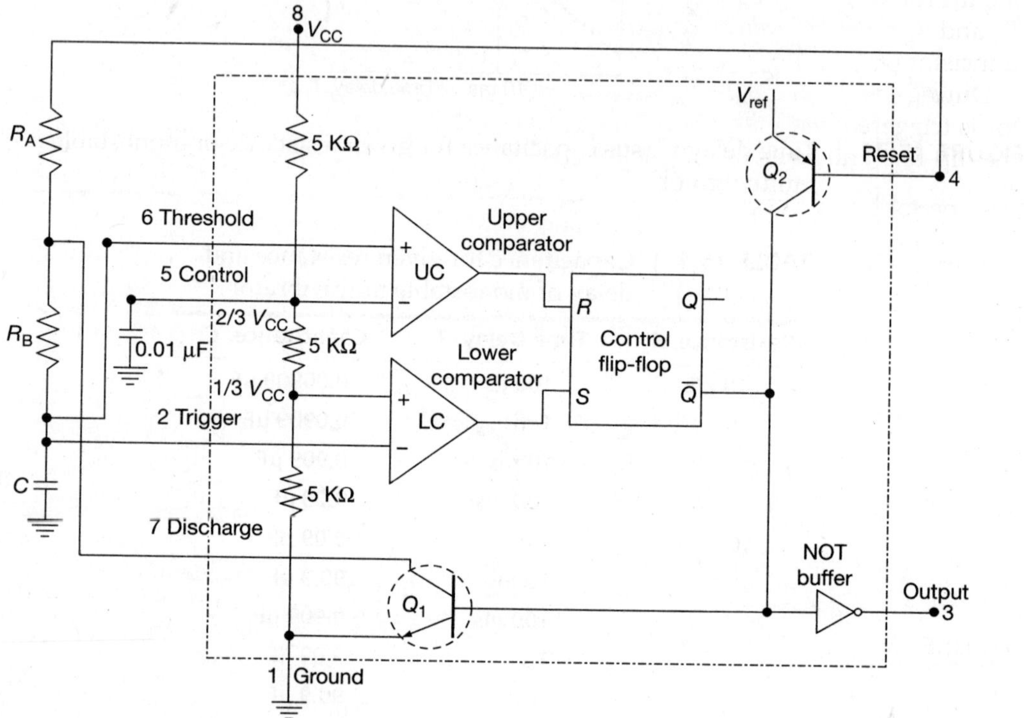

FIGURE 15.32 | Astable multivibrator using 555 Timer

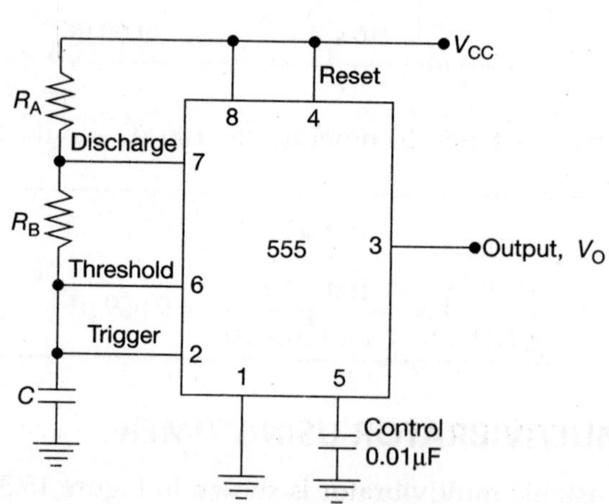

FIGURE 15.33 | Astable multivibrator using 555 Timer

Working: When the capacitor voltage equals $\left(\frac{2}{3}\right)V_{cc}$, the upper comparator triggers the control flip-flop so that $\overline{Q} = 1$. This in turn, makes transistor Q_1 ON and capacitor C starts discharging towards ground through R_B and transistor Q_1 with a time constant R_BC, neglecting the forward resistance of Q_1. Current also flows into transistor Q_1 through R_A. Resistors R_A and R_B must be large enough to limit this current and prevent damage to the discharge transistor Q_1.

During the discharge of the timing capacitor C, as it reaches $V_{cc}/3$, the lower comparator is triggered and at this stage excitations of SR flip-flop are $S = 1$ and $R = 0$, which sets the flip-flop and turns $\overline{Q} = 0$, unclamps the external timing capacitor C. The capacitor C is thus periodically charged and discharged between $\left(\frac{2}{3}\right)V_{cc}$ and $\left(\frac{1}{3}\right)V_{cc}$, respectively. The output and capacitance voltage waveforms are shown in Figure 15.34.

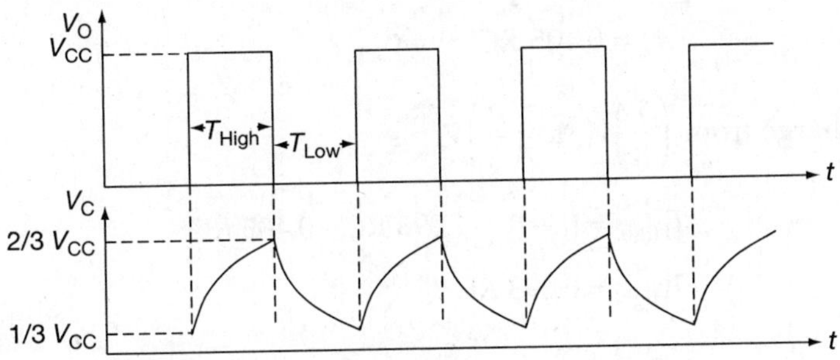

FIGURE 15.34 | Output and capacitor voltage waveforms

The length of time that the output remains HIGH is the time for capacitor to charge from $\left(\frac{1}{3}\right)V_{cc}$ and $\left(\frac{2}{3}\right)V_{cc}$. The capacitor voltage for a low-pass RC circuit subjected to a step input of V_{cc} is:

$$V_{cc} = V_{cc}\left(1 - e^{-t/RC}\right) \tag{15.18}$$

At time t_1, capacitor charges from 0 to $\left(\frac{2}{3}\right)V_{cc}$. Equation (15.18) becomes

$$\left(\frac{2}{3}\right)V_{cc} = V_{cc}\left(1 - e^{-t_1/RC}\right)$$

On simplifying the above equation, we get

$$e^{-T/RC} = 1 - \frac{2}{3}$$

Taking $\log_e$ (i.e. ln) on both sides, we get

$$T = -RC\ln\left(\frac{1}{3}\right) \Rightarrow T = RC\ln 3$$

$$t_1 = 1.098\ RC \tag{15.19}$$

And at time t_2, capacitor charges from 0 to $\left(\frac{1}{3}\right)V_{cc}$

$$\left(\frac{1}{3}\right)V_{cc} = V_{cc}\left(1 - e^{-t_2/RC}\right) \tag{15.20}$$

On simplifying the above equation, we get

$$e^{-t_2/RC} = 1 - \frac{1}{3}$$

Taking $\log_e$ (ln) on both sides, we get

$$t_2 = -RC\ln\left(\frac{2}{3}\right) \Rightarrow t_2 = RC\ln 1.5$$

$$t_2 = 0.405\ RC \tag{15.21}$$

As the time to charge from $\left(\frac{1}{3}\right)V_{cc}$ to $\left(\frac{2}{3}\right)V_{cc}$ is

$$T_{High} = t_1 - t_2 = 1.098\ RC - 0.405\ RC$$

$$T_{High} = 0.693\ RC \tag{15.22}$$

For the circuit

$$T_{High} = 0.693(R_A + R_B)C \tag{15.23}$$

The output is low while the capacitor discharges from $\left(\frac{2}{3}\right)V_{cc}$ to $\left(\frac{1}{3}\right)V_{cc}$ and the voltage across the capacitor is given by:

$$\left(\frac{1}{3}\right)V_{cc} = \left(\frac{2}{3}\right)V_{cc}\,e^{-t/RC} \tag{15.24}$$

On simplifying the above equation, we get

$$e^{-t/RC} = \frac{1}{2}$$

Taking $\log_e$ (ln) on both sides, we get

$$t = -RC\ln\left(\frac{1}{2}\right) \Rightarrow t = RC\ln 2$$

$$t = 0.693RC \tag{15.25}$$

So, for the given circuit, we have

$$T_{Low} = 0.693R_BC \tag{15.26}$$

Here, R_A and R_B are in charge path, but only R_B is in the discharge path.

Total time is given by

$$T = T_{HIGH} + T_{LOW}$$

$$T = 0.693(R_A + 2R_B)C \tag{15.27}$$

So,

$$f = \frac{1}{T} = \frac{1.45}{(R_A + 2R_B)C} \tag{15.28}$$

So, frequency is independent of the supply. Figure 15.35 shows variation in capacitance, C with respect to free-running frequency of an astable multivibrator, *f* for a known resistance value, *R*. The graph is drawn on log–log scale. Table 15.2 gives the values of capacitance for the given value of resistance to produce a frequency by astable multivibrator.

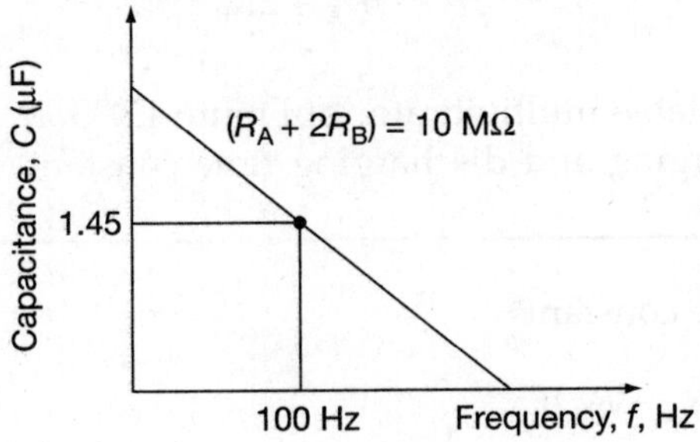

FIGURE 15.35 | Free-running frequency versus capacitance for given resistance of astable multivibrator

TABLE 15.2 | Capacitance for given resistance and free-running frequency of an astable multivibrator

Resistance, $(R_A + 2R_B)$	Frequency, f	Capacitance, C
10 MΩ	0.1 Hz	1.45 μF
	1 Hz	0.145 μF
	10 Hz	0.0145 μF
	100 Hz	0.00145 μF
1 MΩ	0.1 Hz	14.5 μF
	1 Hz	1.45 μF
	10 Hz	0.145 μF
	100 Hz	0.0145 μF
	1 kHz	0.00145 μF
100 kΩ	1 Hz	14.5 μF
	10 Hz	1.45 μF
	100 Hz	0.145 μF
	1 kHz	0.0145 μF
	10 kHz	0.00145 μF

15.9.1 | Duty Cycle

Duty cycle term is associated with astable multivibrator. Duty cycle is defined as the ratio of ON time the output remains high to the total period ($T = T_{ON} + T_{OFF}$).

$$\%\ \text{Duty} = \frac{T_{ON}}{T} \times 100 \tag{15.29}$$

When the transistor Q_1 is ON, the output goes Low.

$$\%\ \text{Duty} = \frac{T_{High}}{T} \times 100 \tag{15.30}$$

$$\%\ \text{Duty cycle} = \frac{R_A + R_B}{R_A + 2R_B} \times 100 \tag{15.31}$$

EXAMPLE 15.3 In the astable multivibrator of Figure 15.33. $R_A = 2.2\ \text{k}\Omega$, $R_B = 3.3\ \text{k}\Omega$ and $C = 0.1\ \mu\text{F}$. Calculate charging and discharging time constant, total time, frequency and duty cycle.

SOLUTION Charging time constant

$$T_{High} = 0.693(R_A + R_B)C$$

$$T_{High} = 0.693\left(2.2\times10^3\,\Omega + 3.3\times10^3\,\Omega\right)\times\left(0.1\times10^{-6}\ \text{F}\right) = 0.38115\ \text{ms}$$

Discharging time constant

$$T_{Low} = 0.693 R_B C$$

$$T_{Low} = 0.693\left(3.3\times10^3\,\Omega\right)\times\left(0.1\times10^{-6}\,\text{F}\right) = 0.22869\ \text{ms}$$

Total time, $T = T_A + T_B = (0.38115\ \text{ms} + 0.22869\ \text{ms}) = 0.60984\ \text{ms}$ **Ans.**

Frequency, $f = \frac{1}{T} = \frac{1}{0.60984\ \text{ms}} = 1.63977\ \text{kHz}$ **Ans.**

Duty cycle $= \frac{R_A + R_B}{R_A + 2R_B} \times 100 = \frac{2.2\ \text{k}\Omega + 3.3\ \text{k}\Omega}{2.2\ \text{k}\Omega + 2\times3.3\ \text{k}\Omega} \times 100 = 62.5\%$ **Ans.**

EXAMPLE 15.4 Determine the free-running frequency if the duty cycle is 40% and ON period is 0.38 ms.

SOLUTION Duty cycle $= \frac{T_{ON}}{T} \times 100 = T_{ON} \times f \times 100$

$$f = \frac{\text{Duty cycle}}{T_{ON} \times 100} = \frac{40}{(0.38\times10^{-3})\times100} = 1.05\times10^3\ \text{Hz} = 1.05\ \text{kHz}\quad \textbf{Ans.}$$

EXAMPLE 15.5 Design an astable multivibrator using 555 timer for a free-running frequency of 1.07 kHz and a duty cycle of 75%.

SOLUTION Design and find the values of R_A and R_B.

Free-running frequency, $f = \dfrac{1}{T}$

Total time, $$T = \frac{1}{f} = \frac{1}{1.07\text{ kHz}} = \frac{1}{1.07 \times 10^3} = 0.93458\text{ ms}$$

$$\text{Duty cycle} = \frac{T_{ON}}{T} \times 100$$

$$T_{ON} = \frac{\text{Duty cycle} \times T}{100} = \frac{75 \times 0.93458\text{ ms}}{100} = 0.700935\text{ ms}$$

$$T_{High} = T_{ON} = 0.700935\text{ ms}$$

$$T_{Low} = T - T_{ON} = 0.93458\text{ ms} - 0.700935\text{ ms} = 0.23365\text{ ms}$$

Let $$C = 0.1\ \mu\text{F}$$

$$T_{Low} = 0.693 R_B C$$

$$R_B = \frac{T_{Low}}{0.693C} = \frac{0.23365\text{ ms}}{0.693 \times 0.1\ \mu\text{F}} = 3.37157\text{ k}\Omega \quad \textbf{Ans.}$$

$$T_{High} = 0.693(R_A + R_B)C$$

$$R_A = \frac{T_{High}}{0.693C} - R_B = \frac{0.700935\text{ ms}}{0.693 \times 0.1\ \mu\text{F}} - 3.37157\text{ k}\Omega = 6.74293\text{ k}\Omega \quad \textbf{Ans.}$$

EXAMPLE 15.6 In the monostable multivibrator using 555 timer, R is 180 kΩ and time period is 120 ms. Calculate the capacitor value.

SOLUTION Time period is

$$T = 1.1RC$$

$$C = \frac{T}{1.1R} = \frac{120\text{ ms}}{1.1 \times 189\text{ k}\Omega} = \frac{120 \times 10^{-3}}{1.1 \times 180 \times 10^3} = 0.606 \times 10^{-6} = 0.606\ \mu\text{F} \quad \textbf{Ans.}$$

SUMMARY

- **Analog comparator:** An analog circuit which compares two analog signals and produces an output whose polarity indicates the result of comparison.
- **Astable multivibrator:** A multivibrator circuit having both the states as quasi-stable states. It is a square-wave oscillator. It is also known as free-running multivibrator.

- **Bistable multivibrator:** A multivibrator circuit that has both the states as stable. It goes from one stable state to another when triggered. It is same as the FLIP-FLOP.
- **Free-running multivibrator:** A multivibrator circuit having both the states as quasi-stable states. It is a square-wave oscillator. It is also known as the astable multivibrator.
- **Hysteresis:** The difference between the upper and lower triggering voltages in a Schmitt trigger circuit.
- **Jitter:** When the successive pulse durations of a monostable multivibrator are not same, it is said to jitter.
- **Monostable multivibrator:** An electronic circuit that produces an output pulse for a fixed time period in response to a trigger pulse and then returns to its quiescent state. The width of the output pulse produced depends upon the timing elements (*R* and *C*) used in the circuit. It is also known as one-shot circuit.
- **One-shot:** An electronic circuit that produces an output pulse for a fixed time period in response to a trigger pulse and then returns to its quiescent state. The width of the output pulse produced depends upon the timing elements (*R* and *C*) used in the circuit. It is also known as one-shot circuit.
- **Operational amplifier (OP-AMP):** A high-gain differential input amplifier.
- **Pulse stretcher:** An electronic circuit that produces an output pulse for a fixed time period in response to a trigger pulse and then returns to its quiescent state. The width of the output pulse produced depends upon the timing elements (*R* and *C*) used in the circuit. It is also known as one-shot circuit.
- **Quasi-stable state:** The state which is not a stable state. When a circuit is forced to go to a quasi-stable state, it comes back to its stable state after a time depending upon the elements of the circuit.
- **Recovery time:** The time required for a timing circuit to come back to steady state after having been triggered.
- **Regenerative comparator:** An analog comparator circuit with different upper and lower triggering voltages. It exhibits hysteresis effect. It produces a single, sharp transition from a slowly changing input.
- **Re-triggerable monostable multivibrator:** A monostable multivibrator that is capable of getting triggered even when it is in its non-quiescent (or quasi-stable) state.
- **Schmitt trigger:** An analog comparator circuit with different upper and lower triggering voltages. It exhibits hysteresis effect. It produces a single, sharp transition from a slowly changing input. It is the same as regenerative comparator.
- **Timer 555:** Timer 555 is a monolithic timer IC which can be used in a number of applications such as an astable multivibrator, monostable multivibrators, square-wave and triangular-wave generators, ramp generators, voltage-controlled oscillator, pulse width modulator, missing pulse detector and many more.
- A monostable multivibrator can be used for the applications of (i) Frequency divider, (ii) Missing pulse detector, (iii) pulse width modulator and (iv) Re-triggerable monostable multivibrator.
- Astable multivibrator can be used for the applications of (i) Rectangular-wave oscillator, (ii) Square-wave oscillator, (iii) Ramp generator, (iv) Voltage-controlled oscillator (VCO) and (v) Schmitt trigger.

MULTIPLE CHOICE QUESTIONS

15.1 Which of the following can work as square-wave generator?
(a) Astable multivibrator
(b) Monostable multivibrator
(c) Bistable multivibrator
(d) Schmitt trigger

15.2 Which of the following is a free-running multivibrator?
(a) Astable multivibrator
(b) Monostable multivibrator
(c) Bistable multivibrator
(d) Schmitt trigger

15.3 Which of the following is also known as one-shot?
(a) Schmitt trigger
(b) 555 timer
(c) Astable multivibrator
(d) Monostable multivibrator

15.4 The frequency of an astable multi-vibrator is given by
(a) 1.449*RC*
(b) 1.38*RC*
(c) 0.725*RC*
(d) 0.69*RC*

15.5 Which of the following can work as memory elements in digital circuits?
(a) Astable multivibrator
(b) Monostable multivibrator
(c) Bistable multivibrator
(d) 555 Timer

15.6 Pin 6 of 555 timer is
(a) Discharge (b) Threshold
(c) Output (d) Reset

15.7 The basic operating modes of the 555 timer are
(a) Astable and monostable mode
(b) Astable and bistable mode
(c) Bistable and monostable mode
(d) Astable, monostable and bistable mode.

15.8 1 μs pulse can be converted into 1 ms pulse by using
(a) Astable multivibrator
(b) Monostable multivibrator
(c) Bistable multivibrator
(d) 555 Timer

15.9 Which of the following can be used for generation of a rectangular waveforms form a sine wave or any other waveform?
(a) Astable multivibrator
(b) Monostable multivibrator
(c) Schmitt Trigger
(d) 555 Timer

15.10 A device one-shot has
(a) two stable states
(b) one stable state
(c) either one or two stable states
(d) no stable state

Answers

15.1 (a) 15.2 (a) 15.3 (d) 15.4 (c) 15.5 (c) 15.6 (b) 15.7 (a)
15.8 (a) 15.9 (d) 15.10 (b)

QUESTIONS

15.1 Draw and explain the functional diagram of a 555 timer.

15.2 Draw the circuit diagram of a Schmitt trigger and explain its operation.

15.3 What is multivibrator? What are different types of multivibrators? What are the uses of multivibrators?

15.4 What is an astable multivibrator? Draw the circuit diagram of an astable multivibrator and explain it operation.

15.5 What are gated astable multivibrators? Draw its waveform.

15.6 What is monostable multivibrator? Draw the circuit diagram of monostable multivibrator and explain it operation.

15.8 Discuss the application of 555 timer as monostable multivibrator.

15.9 Discuss astable multivibrator making use of logic gates with the help of circuit diagram and waveforms.

15.10 Discuss the application of 555 timer as an astable multivibrator.

15.11 What is timer 555? Give its pin configuration and its special features.

PROBLEMS

15.1 Design an astable multivibrator using 555 timer for a free-running frequency of 1.04 kHz and a duty cycle of 40.3% if $C = 0.01$ μF.

15.2 Find the free-running frequency if the duty cycle is 62.5% and the OFF period is 0.23 ms.

15.3 Calculate the duty cycle of the output of an astable multivibrator if frequency is 1.05 kHz and T_{Low} is 0.23 ms.

15.4 In the astable multivibrator register connected between pin 8 and 7 is $R_A = 30$ kΩ and register connected between pin 7 and pin 6 is $R_B = 70$ kΩ and capacitor connected to pin 5 is 0.01 μF. Calculate charging and discharging time constant, total time, frequency and duty cycle.

15.5 Design monostable multivibrator using 555 timer, if the time duration of the output pulse100 ms.

15.6 Design a 100 kHz, 60% duty cycle square-wave generator using 555 timer.

Bibliography

Bartee Thomas C. (1985), *Digital Computer Fundamentals*, 6th Edition (New Delhi, India: McGraw-Hill).

Bhasker, J. (1999), *VHDL Primer* (Singapore: Addison Wesley Longman).

Bignell, James and Donovan, Robert (2011) *Digital Electronics: Logic and Systems*, 2nd Reprint (New Delhi: Cengage Learning) .

Blakeslee, T. R. (1979), *Digital Design with Standard MSI and LSI* (New York: John Wiley).

Brown, Stephen and Vranesic, Zvonko (2009), *Fundamentals of Digital Logic with VHDL Design*, 3rd edition (New Delhi: McGraw-Hill Higher Education).

Coughlin, R. F. and Driscoll, F. F. (1977), *Operational Amplifiers and Linear Integrated Circuits* (Englewood Cliffs, NJ: Prentice-Hall).

Fletcher, William I. (2009), *An Engineering Approach to Digital Design* (New Delhi: Prentice-Hall).

Floyd Thomas L. (2011), *Digital Fundamentals*, 10th edition (New Delhi: Pearson Education).

Garrett, Lane S. (1970a), *Integrated-Circuit Digital Logic Families I - Requirements and Features of Logic Families, RTL, DTL, and HTL Devices*, IEEE Spectrum, vol. 7, no. 10, pp. 46–62.

Garrett, Lane S. (1970b), *Integrated-Circuit Digital Logic Families II - TTL Devices*, IEEE Spectrum, vol. 7, no. 11, pp. 63–72.

Garrett, Lane S. (1970c), *Integrated-Circuit Digital Logic Families III - ECL and MOS Devices*, IEEE Spectrum, vol. 7, no. 12, p. 30–42.

Green, D. C. (2011), *Digital Electronics*, 5th edition (New Delhi: Pearson Education).

Hill, F. J. and Peterson, G. R. (1974), *Introduction to Switching Theory and Logical Design* (New York: John Wiley).

IEEE Standard VHDL Language Reference Manual, New York, June 1994.

Jain, R. P. and Anand, M. M. S. (1983), *Digital Electronics Practice Using Integrated Circuits* (New Delhi: Tata McGraw-Hill).

Jain, R. P. (2010), *Modern Digital Electronics*, 4th edition (New Delhi: Tata McGraw-Hill).

Leach, Donald P., Malvino, Albert Paul and Saha, Goutam (2006), *Digital Principals and Applications*, 6th edition (New Delhi: Tata McGraw-Hill).

Karim Mohammad A. and Chen, Xinghao (2008), *Digital Design: Basic Concepts and Principles* (Boca Raton: CRC Press).

Kohavi, Zvi (1978), *Switching and Finite Automata Theory* (New Delhi: Tata McGraw-Hill).

Kumar, A. Anand (2012), *Fundamental of Digital Circuits*, 2nd edition (New Delhi: Prentice-Hall).

Malvino, Albert Paul (1995), *Electronics Principles*, 3rd edition (New Delhi: Tata McGraw-Hill).

Mano, M. Morris (2006), *Digital Logic and Computer Design* (India: Pearson Prentice-Hall).

Marcovitz, Alan B. (2002), *Introduction to Logic Design* (New Delhi: Tata McGraw-Hill).

Millman, J. and Halkias, C. C. (1972), *Integrated Electronics* (New York: McGraw-Hill).

Millman, J. and Taub, H. (1965), *Pulse, Digital, and Switching Waveforms* (New York: McGraw-Hill).

Millman, J. and Grabel, A. (1987), *Microelectronics*, 2nd edition (New York: McGraw-Hill).

Navabi, Zainalabedin (1998), *VHDL Analysis and Modeling of Digital Systems* (New York: McGraw-Hill).

Pellerin, David and Holley, Michael (1991), *Practical Design Using Programmable Logic* (New Jersey: Prentice-Hall Inc.).

Perry, Douglas (2001), *VHDL*, 3rd edition (New Delhi: Tata McGraw-Hill).

Roth, Charles H. (2006), *Fundamentals of Logic Design*, 4th edition (New Delhi: Jaico Publishing House).

Semiconductor Industry Association, National Technology Roadmap for Semiconductors, http://www.semichips.org/

Taub, H. and Schilling, D. (1977), *Digital Integrated Electronics* (New York: McGraw-Hill).

Texas Instruments, Inc. (1976), *The TTL Data Book for Design Engineers*.

Texas Instruments Staff (1971), *Designing with TTL Integrated Circuits* (New York: McGraw-Hill).

Tocci Ronald J., Widmer, Neal S. and Moss, Gregory L. (2009), *Digital Systems Principles and Applications*, 10th edition (New Delhi: Pearson Education).

Tokheim Roger L. (2004), *Digital Electronics: Principles and Applications*, 6th edition (New Delhi: Tata McGraw-Hill).

Unger, Stephen H. (1989), *The Essence of Logic Circuits* (New Jersey: Prentice-Hall).

Wakerly, John F. (2011), *Digital Design Principles and Practices*, 4th edition (New Delhi: Pearson Education).

Yarbrough, John M. (2011), *Digital Logic Applications and Design*, 8th Reprint (New Delhi: Cengage Learning).

WEBSITES

http ://en.wikipedia. org

http://focus.ti.com

www.actel.com

www.alliancememory.com

www.altera.com

www.amd.com

www.atmel.com

www.cypress.com

www2.electronicproducts.com

www.epn-online.com

www.fairchild.com

www.fufurlec.com

www.idt.com

www.issi.com

www.jedec.org

www.latticesemi.com

www.micron.com

www.necel.com

www.philips.com

www.vantis.com

Index

C

F

G

H

I

J

K

N

O

P

Q

R

S